Giving Biology a Personal Perspective

Connections:
Where Biology Impacts Society

Biology and Staying Healthy

Bird and Swine Flu

The influenza virus has been one of the most lethal viruses in human history. Flu viruses are animal RNA viruses containing 11 genes. An individual flu virus resembles a sphere studded with spikes. The spikes are of two kinds, H and N, and are called H3N2.

How New Flu Strains Arise

Worldwide epidemics of the flu in the last century have been caused by shifts in flu virus H-N combinations as mutation creates new versions of H and N. The "killer flu" of 1918, H1N1, thought to have passed directly from birds to humans, killed between 40 and 100 million people worldwide. The Asian flu of 1957, H2N2, killed over 100,000 Americans, and the Hong Kong flu of 1968, H3N2, killed 70,000 Americans.

It is no accident that new strains of flu usually originate in the Far East. The most common hosts of influenza virus are ducks, chickens, and pigs, which in Asia often live in close proximity to each other and to humans. Pigs are subject to infection by both bird and human strains of the virus, and individual animals are often simultaneously infected with multiple strains. This creates conditions favoring genetic recombination between strains, as illustrated above, sometimes putting together novel combinations of H and N spikes unrecognizable by human immune defenses specific for the old configuration. The Hong Kong flu, for example, arose from recombination between H3N8 from ducks and H2N2 from humans. The new strain of influenza, in this case H3N2, then passed back to humans, creating an epidemic because the human population had never experienced that H-N combination before.

The new strain of flu undergoes explosive population growth in order to produce the deadly epidemic. This accounts for the low mortality rate...

Conditions f...

Not every new st...
Three conditions...
novel combinatio...
has no significan...
be able to repli...
enza viruses in...
in human cells;...
between humans...
exhaled from inf...

Why did so many die? Because so much of the world's population was infected.

Bird Flu

A potentially deadly new strain of flu virus emerged in Hong Kong in 1997, H5N1. Like the 1918 pandemic strain, H5N1 passes to humans directly from infected birds, usually chickens or ducks, and for this reason has been dubbed "bird flu." Bird flu has gotten an unusual amount of public attention, because it satisfies the first two conditions of a pandemic. Bird flu is a novel combination of H and N spikes for which humans have little immunity, and the resulting strain is particularly deadly. Of 256 individuals infected by handling birds by the end of 2006, 151 died of the infection, a mortality of 59% (recall, the mortality of the 1918 strain was only 2%). Fortunately, the third condition for a pandemic is not yet met: The H5N1 strain of flu virus...

Recombination within humans
A person infected with a flu virus can become infected with another type of flu virus by direct contact with birds. The two viruses can undergo genetic recombination to produce a third type of virus, which can spread from human to human.

Recombination within pigs
Pigs can contract flu viruses from both birds and humans. The flu viruses can undergo genetic recombination in the pig, to produce a new kind of flu virus, which can spread from pigs to humans.

Applications (Apps):
Biology in Your World

Heme group   Beta (β) chains   Oxygen (O₂)

25.4 How Respiration Works: Gas Exchange

CONCEPT PREVIEW: Oxygen molecules move through the circulatory system carried by the protein hemoglobin within red blood cells. Most CO₂ is transported in the blood plasma as bicarbonate.

When oxygen has diffused from the air into the moist cells lining the inner surface of the lung, its journey has just begun. Passing from these cells into the bloodstream, the oxygen travels throughout the body in the circulatory system, described in chapter 24. It has been estimated that it would take a molecule of oxygen three years to diffuse from your lung to your toe if it moved only by diffusion, unassisted by a circulatory system.

O_2 Transport

Oxygen (O_2) moves within the circulatory system carried piggyback on the protein **hemoglobin**. Hemoglobin molecules contain iron, which binds oxygen, as shown in figure 25.5. Hemoglobin molecules act like little oxygen sponges, soaking up oxygen within red blood cells and causing more to diffuse in from the blood plasma. The oxygen binds in a reversible way, which is necessary so that the oxygen can unload when it reaches the tissues of the body. Hemoglobin is manufactured within red blood cells and never leaves these cells, which circulate in the bloodstream like ships bearing cargo.

At the high O_2 levels that occur in the lung (panel 1 in *Essential Biological Process 25B*), most hemoglobin molecules carry a full load of oxygen atoms. In tissue, the presence of carbon dioxide (CO_2) causes the hemoglobin molecule to assume a different shape, one that gives up its oxygen more easily (as in panel 3). The effect of CO_2 on oxygen unloading is important, because CO_2 is produced by the tissues through cell metabolism. For this reason, the blood unloads oxygen more readily within tissues undergoing metabolism.

CO_2 Transport

At the same time the red blood cells are unloading oxygen in panel 3, they are also absorbing CO_2 from the tissue. About 8% of the CO_2 in blood is simply dissolved in plasma. Another 20% is bound to hemoglobin; however, this CO_2 does not bind to the heme group but rather to another site on the hemoglobin molecule, and so it does not compete with oxygen binding. The remaining 72% of the CO_2 diffuses into the red blood cells. To keep CO_2 from diffusing out of the red blood cells back to the plasma where CO_2 levels are low, the enzyme *carbonic anhydrase* combines CO_2 molecules with water molecules (panel 4) to form **carbonic acid** (H_2CO_3) in the cell. This acid dissociates out of the red blood cell into the plasma. This transporter protein exchanges one chloride ion (Cl^-) for a bicarbonate, a process called the *chloride shift*. This reaction keeps the levels of CO_2 in the blood plasma low, facilitating the diffusion of more CO_2 into it from the surrounding tissue. The facilitation is critical to CO_2 removal, because the difference in CO_2 concentration between blood and tissue is not large (only 5%). The formation of carbonic acid and bicarbonate also important in maintaining the acid-base balance of the blood because the molecules act as a buffering system.

The blood plasma carries bicarbonate ions back to the lungs. The lower CO_2 concentration in the air inside the lungs causes the carbonic anhydrase reaction to proceed in...

As discussed on page 40 a buffer is an acid-base pair of molecules that adjusts the levels of H⁺ in a solution, thereby controlling pH levels. The reversible reaction between carbonic acid and bicarbonate regulates the amount of H⁺ in the blood, thereby adjusting pH levels in the body.

BIOLOGY & YOU

Hiccups. A hiccup (pronounced "HICK-up") is a spasmodic contraction of the diaphragm that repeats several times a minute. The abrupt rush of air into your lungs causes your epiglottis to close, making the "hic" noise. Hiccups are caused by irritation of the phrenic and vagus nerves, which activates reflexive motor pathways to the diaphragm muscles. Hiccups often occur after drinking carbonated soda or alcohol, but can be initiated by any of a host of other reasons, like eating too fast or taking a cold drink while eating a hot meal. What do you do when you get the hiccups? In most cases they can be stopped simply by forgetting about them—the basis of the common home remedy of "scaring them away" with a surprise or fright. Increasing respired CO₂ by breathing into a paper bag also often works, for the interesting reason that hiccups may be an evolutionary remnant of earlier amphibian respiration. Air gulping in frogs is inhibited by CO₂ just as your hiccuping is. Frogs and other amphibians don't have a diaphragm, and instead gulp air via a simple motor reflex much like your hiccuping reflex. In humans the hiccuping motor pathway forms early in fetal development, well before the motor pathways that drive normal breathing. Premature infants born before their lung motor pathways are fully functional spend 2.5% of their time hiccuping, gulping air just like amphibians.

Implications:
Questioning Yourself

...is not to say that the characteristics may be... The organisms... survive more often,... population, and...

...ation of biological... places—because..., the organisms... will tend to vary in different places. As we will discuss later in this chapter, section 14.9, there are five evolutionary forces that can affect biological diversity, although natural selection is the only evolutionary force that produces *adaptive* changes.

Darwin Drafts His Argument

Darwin drafted the overall argument for evolution by natural selection in a preliminary manuscript in 1842. After showing the manuscript to a few of his closest scientific friends, however, Darwin put it in a drawer for 16 years and turned to other research.

Wallace Has the Same Idea

The stimulus that finally brought Darwin's theory into print was... he received in 1858. A young English naturalist named Alfred Russel Wallace (1823–1913) sent the essay to Darwin from Malaysia... forth the theory of evolution by means of natural selection... lace had developed independently of Darwin. Like Darwin, Wallace had been greatly influenced by Malthus's 1798 book. After reading Wallace's essay, Darwin arranged for a joint presentation of their ideas... London. Darwin then completed his own book, expanding the... script that he had written so long ago, and submitted it for...

Publication of Darwin's Theory

Darwin's book appeared in November 1859 and caused an immediate sensation. Although people had long... resembled apes in many characteristics,... be a direct evolutionary relationship... did not actually discuss this idea in his... from the principles he outlined. In a sub... *Man*, Darwin presented the argument... that humans and living apes have common... deeply disturbed with the suggestion that... from the same ancestor as apes, and Darwin... him to become a victim of the satirists of his day—the ca... 14.8 is a vivid example. Darwin's arguments for the theory... by natural selection were so compelling, however, that h... almost completely accepted within the intellectual comm... Britain after the 1860s.

Concept Check

1. How long after his original voyage on HMS *Beagle* did Darwin do his work and publish *On the Origin of Species?*
2. What did Darwin see on the Galápagos Islands that hinted at evolution?
3. How did Malthus influence both Darwin and Wallace?

Figure 14.8   Darwin greets his monkey ancestor.

In his time, Darwin was often portrayed unsympathetically, as in this drawing from an 1874 publication.

IMPLICATION In 2008 the Spanish parliament approved resolutions granting to gorillas, chimpanzees, and orangutans statutory rights currently only applicable to humans. This was the first time a country has taken such action. The resolutions were based on the Great Ape Project, a framework designed by scientists and philosophers to provide humans' closest relatives with the right to life, liberty, and protection from torture. Zoos could still legally hold apes, but living conditions must be "optimal." Using apes in performances will be illegal. The law will also ban using apes in potentially useful research, if it might hurt the ape in any way. Do you think this last condition of the law is appropriate? Explain.

Experiments:
Thinking Like a Scientist

Inquiry & Analysis

Why Do Human Cells Age?

Human cells appear to have built-in life spans. In 1961 cell biologist Leonard Hayflick reported the startling result that skin cells growing in tissue culture, such as those growing in culture flasks in the photo below, will divide only a certain number of times. After about 50 population doublings cell division stops (a doubling is a round of cell division producing two daughter cells for each dividing cell, for example... This led to the hypothesis that a run of some 16 TTAGGGs was where the DNA replicating enzyme, called polymerase, first sat down on the DNA (16 TTAGGGs being the size of the enzyme's "footprint"), and...

TTAGGG TTAGGG TTAGGG TTAGGG TTAGGG........

because of being its docking spot, the polymerase was unable to copy that bit. Thus a 100-base portion of the telomere was lost by a chromosome during each doubling as DNA replicated. Eventually, after some 50 doubling cycles, each with a round of DNA replication, the telomere would be used up and there would be no place for the DNA replication enzyme to sit. The cell line would then enter senescence, no longer able to proliferate.

This hypothesis was tested in 1998. Using genetic engineering, researchers transferred into newly established human cell cultures a gene that leads to expression of an enzyme called *telomerase* that all cells possess but no body cells uses. This enzyme adds TTAGGG sequences back to the end of telomeres, in effect rebuilding the lost portions of the telomere. Laboratory cultures of cell lines with (telomerase plus) and without (normal) this gene were then monitored for many generations. The graph above displays the results.

Effect of Telomerase on Cell Culture Growth

Population doublings
30   60   90   120   150

— Normal
— Telomerase plus

Relative growth rate

0   3   6   9   12   15   18   21   24   27   30   33
Months

Analysis

1. **Applying Concepts** Comparing continuous processes, how do normal skin cells (blue line) differ in their growth history from telomerase plus cells with the telomerase gene (red line)?
2. **Interpreting Data** After how many doublings do the normal cells cease to divide? the telomerase plus cells?
3. **Making Inferences** After 9 population doublings, would the rate of cell division be different between the two cultures? after 15? Why?
4. **Drawing Conclusions** How does the addition of the telomerase gene affect the senescence (death by old age) of skin cells growing in culture? Does this result confirm the telomerase hypothesis this experiment had set out to test?

Brief Contents

Essentials of
The Living World

THIRD EDITION

George B. Johnson
Washington University

 **Higher Education**

Boston   Burr Ridge, IL   Dubuque, IA   New York   San Francisco   St. Louis
Bangkok   Bogotá   Caracas   Kuala Lumpur   Lisbon   London   Madrid   Mexico City
Milan   Montreal   New Delhi   Santiago   Seoul   Singapore   Sydney   Taipei   Toronto

Contents

v

Preface

No one who teaches biology today can fail to appreciate how important a subject it has become for our modern world. From global warming to stem cell iniatives to teaching intelligent design in classrooms, biology permiates the news, and in large measure will define students' futures. As a teacher, I have stood in front of classrooms for over 30 years and attempted to explain biology to puzzled and sometimes uninterested students, an experience that has been both fun and frustrating: Fun because biology is a joy to teach, rich in ideas and interesting concepts, and increasingly key to many important public issues; frustrating because in every biology class there are always some students who will not pay attention, who not only miss out on the fun but also fail to acquire a tool that will be essential to their futures.

This text, *Essentials of The Living World*, is my attempt to address this problem. It is short enough to use in one semester, without a lot of technical details to intimidate wary students. I have tried to write it in an informal, friendly way, to engage as well as to teach. The focus of the book is on the biology each student ought to know to live as an informed citizen in the 21st century. I have at every stage addressed ideas and concepts, rather than detailed information, trying to teach *how* things work and *why* things happen the way they do rather than merely naming parts or giving definitions.

Focusing on the Essential Concepts

More than most subjects, biology is at its core a set of ideas, and if students can master these basic ideas, the rest comes easy. Unfortunately, while most of today's students are very interested in biology, they are put off by the terminology. When you don't know what the words mean, it's easy to slip into thinking that the content is difficult, when actually the ideas are simple, easy to grasp, and fun to consider. It's the terms that get in the way, that stand as a wall between students and science. With this text I have tried to turn those walls into windows, so that readers can peer in and join the fun.

Analogies have been my tool. In writing *Essentials of The Living World* I have searched for simple analogies that relate the matter at hand to things we all know. As science, analogies are not exact, but I do not count myself compromised. Analogies trade precision for clarity. If I do my job right, the key idea is not compromised by the analogy I use to explain it, but rather revealed.

There is no way to avoid the fact that some of the important ideas of biology are complex. No student encountering photosynthesis for the first time gets it all on the first pass. To aid in learning the more difficult material, I have given special attention to key concepts and processes like photosynthesis and osmosis that form the core of biology. The essential processes of biology are not optional learning. A student must come to understand every one of them if he or she is to master biology as a science. A student's learning goal should not be simply to memorize a list of terms, but rather to be able to visualize and understand what's going on. With this goal in mind, I have prepared nearly two dozen "this is how it works" *Essential Biological Process* illustrations explaining the important concepts and processes that students encounter in introductory biology. Each of these *Essential Biological Process* illustrations walks the student through a complex process, one step at a time, so that the central idea is not lost in the details.

Essential Biological Process 4C

Facilitated Diffusion

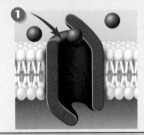

Particular molecules can bind to special protein carriers in the plasma membrane.

The protein carrier helps (facilitates) the diffusion process and does not require energy.

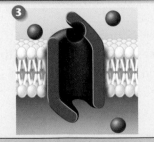

The molecule is released on the far side of the membrane. Protein carriers transport only certain molecules across the membrane but will take them in either direction down their concentration gradients.

An example of an *Essential Biological Process* illustration

Teaching Biology as an Evolutionary Journey

This text, and its companion concepts text *The Living World*, were the first texts to combine evolution and diversity into one continuous narrative. Traditionally, students had been exposed to weeks of evolution before being dragged through a detailed tour of the animal phyla, the two areas presented as if unrelated to each other. I chose instead to combine these two areas, presenting biological diversity as an evolutionary journey. This has proven a very powerful way to teach evolution's role in biology, and today you would be hard pressed to find a text that does not organize the material in this way.

Evolution not only organizes biology, it explains it. It is not enough to say that a frog is an amphibian, transitional between fish and reptiles. This correctly organizes frogs on the evolutionary spectrum, but fails to explain *why* frogs are the way they are, with a tadpole life stage and wet skin. Only when the student is taught that amphibians evolved as highly successful land animals, often as big as ponys and armour plated, can students get the point: Of 37 families of amphibians, all but the three that lived in water (frogs, salamanders, and caecilians) were driven extinct with the advent of reptiles. A frog has evolved to invade water, not escape it. It is in this way that evolution explains biology, and that is how I have tried to use evolution in this text, to explain.

A red-eyed tree frog

Linking Essentials Concepts to Everyday Life

One of the principal roles of nonmajor biology courses is to create educated citizens. In writing *Essentials of The Living World* I have endeavored to relate what the student is learning to the biology each student ought to know to live as an informed citizen in the 21st century. Students also engage much more actively in the course when they can see how what they are studying relates to their own everyday lives.

Throughout the text, *Essentials of The Living World* presents full-page **connections,** readings written by the author that make connections between a chapter's contents and the everyday world: *Biology and Staying Healthy* discusses health issues that impact each student; *Today's Biology* examines advances in biology that importantly affect society; *A Closer Look* examines interesting points in more detail; and *Author's Corner* takes a more personal view (the author's) of how science relates to our everyday lives.

It is impossible to thumb through the pages of this new edition without seeing the second way *Essentials of The Living World* links what a student is learning to the world that the student knows. In the margins of many pages are short **apps**—application dialogues: *In the News* relates a page's content to today's news; *Evolution* points to evolutionary connections; and *Biology & You* explains how the page's content is linked to something in a student's everyday life. These short essays do not attempt to teach the content of the page, but rather to relate it to something the student knows or cares about. There are several apps in every chapter, some of them surprising, all of them interesting.

A third way this new edition links what the student is learning to everyday life is the **Implication questions** found below key illustrations in each chapter. These questions are not directed at assessing the student's understanding of the illustration, but rather push the student to think about the implications of the illustration to their own lives. Many are open-ended, probing a student's own opinions on a subject. All of them link the illustration to the student's everyday life.

BIOLOGY & YOU

Vegans. We humans are omnivores, meaning we can eat a broad range of plant and animal tissues—but not all of us choose to do so. Some people don't like spinach and love steak, while others called vegetarians choose not to eat meat. Some become vegetarians because they judge it a more healthy diet—plants are low in saturated fats linked to heart disease. Others make the choice for ethical reasons, sensitive to the animal rights issues associated with livestock agriculture. Still others simply don't like meat. The most extreme form of vegetarian diet is the "vegan" diet. Vegans avoid all animal proteins. They don't eat red meat, poultry, fish, eggs, or milk. Instead, they obtain all protein and nutrients from grains, vegetables, fruits, legumes, nuts, and seeds. The vegan diet, mirroring that of our early human ancestors, is very challenging, because no single fruit, vegetable, or grain contains all the essential amino acids which humans require in their diet. Vegetal foods must be eaten in particular combinations to provide this necessary balance. Beans and rice together provide a balanced diet, but neither food does so when eaten alone. For calcium, which is usually obtained from milk, vegans must eat green leafy vegetables like broccoli or spinach. In practice it is not difficult to achieve this balance if a vegan eats a variety of plants.

An example of an "apps," an application dialogue feature

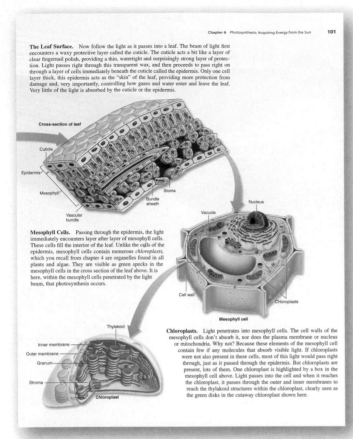

Integration of art into the text

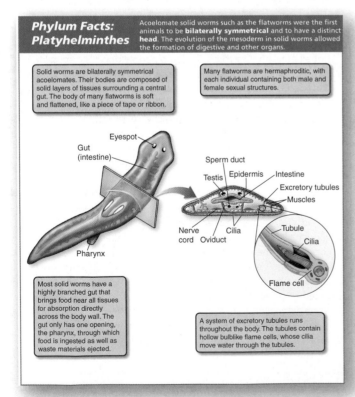

An example of a Phylum Facts

Using Visuals to Teach Concepts

Art has always been a core component of this text, as today's students are visual learners. To help students learn, *Essentials of The Living World* has a clean and simple art style that focuses on concepts and minimizes detail. In this edition I have sought to amplify the power of illustrations to teach concepts by linking the interior content of illustrations directly to the text that describes that part of the illustration. I have set about doing this in three ways: **1. Bubble numbers.** In complex diagrams where there is a lot going on, I have placed numbers (set off in colored balls) at key positions, and the same "bubble numbers" at those locations in the text where that element of the illustration is being described. This makes it much easier for a student to use the illustration as it was intended, to walk through the process and see how the parts are related. **2. Integration of art into text.** In some places like the introduction to photosynthesis (treated on pages 100 to 103) many different processes are covered, each with its own illustration. In these instances, bouncing back and forth between illustration and text makes it difficult for a student to gain or retain perspective, and so I have chosen in these instances to integrate the illustrations directly into the text, providing a single narrative. **3. Phylum facts.** The biggest problem students encounter in studying animal diversity is the mass of detail filling every page. To ease the student's task of sorting through all this, I have constructed "Phylum Facts" illustrations that highlight the key points.

In setting out to improve this edition of *Essentials of The Living World*, my focus has been on improving it as a learning tool. I have made many changes, some obvious, others more subtle, all of them aimed at making it easier for students using this text to understand and appreciate what they learn in lecture, and so do better in the course.

New to This Edition

Making the Text More Accessible. The most obvious change in this edition is that it no longer presents two columns of text crowded side-by-side onto each page. A single wider column of text, with illustrations in the margin, allows a cleaner, clearer organization of material, making it easier for a student to follow an argument or extended discussion.

Linking to Related Topics. On many pages I have placed "link" arrows that point a student to an earlier place in the text where an important related concept or finding has been discussed, or to a place later in the text where a topic will play an important role. In a course like introductory biology a student is barraged with terms and ideas, and linking them together is an important aspect of mastering them. Link arrows provide the student help in this key task.

Learning How to Learn. In over 30 years of teaching I have seen students do well and others do poorly, and one of the best predictors of who would do well has been how well a student is prepared to learn. Entering a large freshman course, does a student know how to take notes? Does a student know how to use these notes effectively with the textbook? Can a student read a graph? In this edition I decided to tackle this problem head-on, and have added a "chapter 0" at the beginning of the text to help students with these very basic but essential learning tools.

Inquiry & Analysis. One of the most useful things a student can take away from his or her biology class is the ability to judge scientific claims that they encounter as citizens, long after college is over.

As a way of teaching that important skill, I have greatly expanded the *Inquiry & Analysis* features I introduced in the previous edition. Most chapters now end with a full-page presentation of an actual scientific investigation that requires the student to analyze the data and reach conclusions. Few pages in this text provide more bang for the buck in learning that lasts.

Updating the Content.   Biology as a science has advanced rapidly in the years since the last revision of *Essentials of The Living World*. Four examples serve to make the point clearly:

RNA Interference (page 210)   This discovery, so important that it won a Nobel Prize in the shortest length of time ever, has totally altered our view of how genes are regulated, and is revolutionizing medicine.

Ethanol and Biofuels (pages 449 to 450)   A topic very much in the news, a careful explanation is needed for a student to understand the issue.

Curing Cancer (pages 138 to 139)   The families of many students are affected by cancer, and one quarter of all students will someday experience it.

Transforming Adult Tissue Cells into Embryonic-like Stem Cells (page 233)   Using embryonic stem cells to cure disease is a very controversial topic, indeed a political hot potato. This advance of using adult tissue cells may provide a less controversial alternative.

A Closer Look at Content Changes.   Many chapters of this revision of *Essentials of The Living World* have been updated to reflect these advances, and to improve the text as a learning tool:

Part 1

• Moved "How Scientists Analyze and Present Experimental Results" to chapter 0, where it is integrated into the discussion of how to study biology.

Part 2

• Moved the content of chapter 2 "Evolution and Ecology" into the appropriate evolution and ecology chapters. The placement of this chapter in the second edition was an attempt to give students an early introduction to the topics of evolution and ecology, providing students with a "macro" view to begin their study, but was seen as confusing when placed out of context.

• Added a new "Biology and Staying Healthy" feature in chapter 3, "Anabolic Steroids in Sports."

• The *Essential Biological Process* art features provide a clear and uncluttered illustration of a process that is essential to the study of biology. These art pieces are called out so they are easily recognized.

• A new visual overview of photosynthesis at the beginning of chapter 6 walks the student through the general process of photosynthesis.

• The discussions of photosynthesis and cellular respiration have been streamlined, focusing on the general concepts and processes.

Part 3

• Added a new "Biology and Staying Healthy" feature in chapter 11, "Protecting Your Genes," which talks about the cancer risks of smoking and sun tanning.

• Separated the discussion of "regulating gene expression" in chapter 12 into regulating mechanisms of prokaryotes and regulating mechanisms of eukaryotes. This new organization is clearer, helping students understand the differences between the processes in prokaryotes and in eukaryotes.

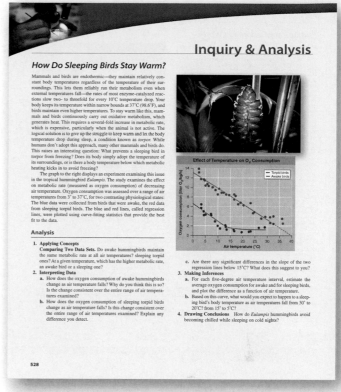

An example of an Inquiry & Analysis feature

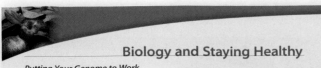
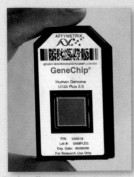
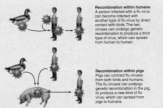

- Added more information on RNA-level control in chapter 12, as this is an area of active research and medical interest. Also added a new "Biology and Staying Healthy" feature on "Silencing Genes to Treat Disease," showing medical applications of RNA interference.
- Added two new "Biology and Staying Healthy" features in chapter 13, "Putting Your Genome to Work" and "DNA and the Innocence Project."
- Expanded information on adult stem cell therapies in chapter 13, including the possibility of using adult cells with embryonic stem cell-like properties.

Part 4

- The discussion of evolution in chapter 14 now begins with the information on Darwin. This information has been moved to chapter 14 from chapter 2 in the previous edition. The story of Darwin provides the students with a historical perspective before delving into the details of evolutionary processes.
- "Evolution of Microbial Life," chapter 16, has been streamlined to help students focus on the general groups, and not overload them with excessive detail that is not needed at this level.
- A new "Biology and Staying Healthy: Bird and Swine Flu" feature explains how flu viruses can become quite deadly.
- "Evolution of Plants," chapter 17, has been enhanced with the addition of general phylogenetic trees of plant evolution at the beginnings of each major group so that the student knows where each plant group fits into the evolution of the kingdom.
- "Evolution of Animals," chapter 18, has been streamlined to help students focus on the general groups, and general phylogenetic trees have also been added to the beginnings of each major animal group.
- Much of the detailed features of the animal groups in chapter 18 have been incorporated into visual presentations called "Phylum Facts." These features comprise an illustrated catalog of the major animal groups. Each major group has a representative Phylum Fact that provides a succinct overview of the features of that group.
- A new section in chapter 18 entitled "The Animal Family Tree" covers some of the new molecular analyses of the relationships among animals.
- The section "Overview of Vertebrate Evolution" has been enhanced with the addition of art and photos that will help the student visualize the process of vertebrate evolution.

Part 5

- Chapter 19 "Populations and Communities" has been moved ahead of the ecosystem chapter, as it makes more sense to introduce populations and communities, which inhabit ecosystems, before covering ecosystems in detail.
- The material on populations and communities has been streamlined to focus the students' attention on key examples that illustrate the concepts.
- In chapter 22, the discussions of global warming and the ozone hole have been separated, so that students don't confuse the two.
- The discussion of alternative energy sources in chapter 22 has been expanded, with a lengthier discussion of biofuels.

Part 6 and Part 7

- The chapters on animal biology and plant biology have been streamlined throughout, allowing the students to focus on the key concepts underlying how animal and plant bodies function.

Acknowledgments

I have for several years enjoyed and profited from collaborations with Jonathan Losos of Harvard University, a friend of long standing and a delight to work with. My coauthor on the previous edition, Jonathan had to relinquish that role on this edition. A coauthor on *Biology*, classroom teacher, and active researcher, this text has proven one load too many. I miss the fun of working with him.

Every author knows that he or she labors on the shoulders of many others; the text you see is the result of hard work by an army of "behind-the-scenes" editors, spelling and grammar checkers, photo researchers, and artists that perform their magic on our manuscript; and an even larger army of production managers and staff that then transform this manuscript into a bound book. I cannot thank them all. Michael Hackett, Rose Koos, and Kris Tibbetts were my editorial team, with whom I worked every day. Publisher Janice Roerig-Blong solved the many management problems her author inadvertently created in his excess of enthusiasm, and provided valuable advice and support. Marty Lange, the editor-in-chief, oversaw all of this with humor and consistent support. Sheila Frank spearheaded our production team, which for several editions now has made a habit of working miracles with a tight schedule. The photo program was carried out by Lori Hancock, who as always did a super job. The art program, conceived several editions ago by William Ober, M.D., and Claire Garrison, R.N., continues to develop with the clarity and excitment they had envisioned. Laurie Janssen did a great job with the design—she really seems to love frogs. This edition was produced by Electronic Publishing Services Inc.

My long-time, off-site developmental editors and right arms Liz Sievers and Megan Berdelman have again played an invaluable role in overseeing every detail of a complex revision. Their intelligence and perseverance continue to play a major role in the quality of this book.

The marketing of this new edition was planned and supervised by Tamara Maury, a battle-wise general not afraid to fight in the trenches alongside the many able sales reps that present our book to instructors.

Last but not least, I would like to extend a special thanks to Michael Lange, vice president–new product launches at McGraw-Hill, for his continued strong support of this project.

George Johnson

Reviewers

I have authored other texts, and all of my writing efforts have taught me the great value of reviewers in improving my texts. Scientific colleagues from around the country have provided numerous suggestions on how to improve the content of the third edition of *Essentials,* and many instructors and students using the second edition have suggested ways to clarify explanations, improve presentations, and expand on important topics. The instructors listed below provided detailed comments. I have tried to listen carefully to all of you. Everyone of you has my thanks!

Sylvester Allred
   Northern Arizona University
Lena Ballard
   Rock Valley College

Dennis Bell
   University of Louisiana–Monroe
Linda L. Bergen-Losee
   Saint Leo University

Lisa L. Boggs
   Southwestern Oklahoma State University
Cheryl Boice
   Lake City Community College

Nancy Bowers
Park University
Carol A. Britson
University of Mississippi
Steven G. Brumbaugh
Green River Community College
Lisa Bryant
Arkansas State University
Matthew Burnham
Jones County Junior College
Chantae M. Calhoun
Lawson State Community College
Jocelyn Cash
Central Piedmont Community College
Bane W. Cheek
Polk Community College
Denise L. Chung
Long Island University–Brooklyn Campus
Craig W. Clifford
Northeastern State University
Yvonne E. Cole
Florissant Valley Community College
George R. Davis
Minnesota State University–Moorhead
Chris Davison
Long Beach City College
Buffany DeBoer
Wayne State College
Kristiann Dougherty
Valencia Community College
Michael J. Dougherty
Hampden–Sydney College
William E. Dunscombe
Union County College
Bruce Edinger
West Liberty State College
Marirose T. Ethington
Genesee Community College
Tracy M. Felton
Union County College
Tullio Ferretti
Hinds Community College
Victor Fet
Marshall University
Teresa G. Fischer
Indian River College
Melanie Florence
Dixie State College
Brandon L. Foster
Wake Technical Community College
Debra L. Foster
Park University
Nancy A. Freeman
El Camino Community College
Dennis W. Fulbright
Michigan State University
Michael D. Gottlieb
LaGuardia Community College
Tammy Greene
Arkansas State University–Beebe

Carla Guthridge
Cameron University
Sue Habeck
Tacoma Community College
Richard Hanke
Rose State College
Nixie Hnetkovsky
Frontier Community College
Michael E. S. Hudspeth
Northern Illinois University
Amy G. Hurst
Rose State College
Jeremiah N. Jarrett
Central Connecticut State University
Adeline Jasinski
Bristol Community College
Angela Jones
California State University–Long Beach
Robyn Jordan
University of Louisiana–Monroe
Judy Kaufman
Monroe Community College
Ronald Keiper
Valencia Community College
Amy Kennedy
Central Carolina Community College
Amine Kidane
Columbus State Community College
Karl Kleiner
York College of Pennsylvania
Roger C. Klockziem
Martin Luther College
Mary Lehman
Longwood University
Susan Lewandowski
Westmoreland County Community College
Suzanne Long
Monroe Community College
Eric Lovely
Arkansas Tech University
Richard Maloof
County College of Morris
Mark Manteuffel
St. Louis Community College
Roy J. Marler
Cascade College
Kamau W. Mbuthia
Bowling Green State University
Melissa Meador
Arkansas State University–Beebe
Linda Meeks
Lake Michigan College
Judith Megaw
Indian River State College
Eric R. Myers
South Suburban College
Steven Mark Norris
California State University–Channel Islands
Igor V. Oksov
Union County College

Theodore J. O'Tanyi
Widener University
Joe Petti
Albertus Magnus College
Mary Phillips
Tulsa Community College
Crystal Pietrowicz
Southern Maine Community College
Karen Plucinski
Missouri Southern State University
Wendy M. Rappazzo
Harford Community College
Pamela Riddell
Macomb Community College
Carlton Rockett
Bowling Green State University
Robert J. Schodorf
Lake Michigan College
Roy D. Schodtler
Lake Land College
Roger Seeber, Jr.
West Liberty State College
William A. Shear
Hampden–Sydney College
Greg Sievert
Emporia State University
William Simcik
Lone Star College–Tomball
Beatrice Sirakaya
Pennsylvania State University
Erika Stephens
Arkansas State University–Beebe
Judith L. Stewart
College of Southern Nevada
Irina Stroup
Shasta College
George E. Veomett
University of Nebraska–Lincoln
Adil M. Wadia
The University of Akron Wayne College
Timothy S. Wakefield
John Brown University
Suzanne Wakim
Butte–Glenn Community College District
Jamie Welling
South Suburban College
Jennifer Wiatrowski
Pasco–Hernando Community College
Daniece Williams
Hinds Community College–Rankin Campus
David Williams
Valencia Community College
Harry E. Womack
Salisbury University
Calvin Young
Fullerton College
Kevin V. Young
Utah State University–Brigham City

Dedicated to providing high-quality and effective supplements for instructors and students, the following supplements were developed for *Essentials of The Living World*.

For Instructors

Connect Biology

 McGraw-Hill Connect Biology is a web-based assignment and assessment platform that gives students the means to better connect with their coursework, with their instructors, and with the important concepts that they will need to know for success now and in the future. With Connect Biology, instructors can deliver assignments, quizzes, and tests easily online. Students can practice important skills at their own pace and on their own schedule.

Companion Web Site
www.mhhe.com/esstlw3

The companion Web site contains the following resources for instructors:

Presentation Tools   Everything you need for outstanding presentations in one place! This easy-to-use table of assets includes:
- Animation PowerPoints—numerous full-color animations illustrating important processes are also provided. Harness the visual impact of concepts in motion by importing these files into classroom presentations or online course materials.
- Lecture PowerPoints—with animations fully embedded
- Labeled and unlabeled JPEG images—full-color digital files of all illustrations that can be readily incorporated into presentations, exams, or custom-made classroom materials.
- Tables—tables from the text are available in electronic format.

Presentation Center   In addition to the images from your book, this online digital library contains photos, artwork, animations, and other media from an array of McGraw-Hill textbooks that can be used to create customized lectures, visually enhanced tests and quizzes, compelling course Web sites, or attractive printed support materials. All assets are copyrighted by McGraw-Hill Higher Education, but can be used by instructors for classroom purposes.

Instructor's Manual   The instructor's manual contains chapter outlines, lecture enrichment ideas, and critical thinking questions.

Computerized Test Bank   A comprehensive bank of test questions is provided within a computerized test bank powered by McGraw-Hill's flexible electronic testing program EZ Test Online.

EZ Test Online allows you to create paper and online tests or quizzes in this easy-to-use program! Imagine being able to create and access your test or quiz anywhere, at any time, without installing the testing software. Now, with EZ Test Online, instructors can select questions from multiple McGraw-Hill test banks or author their own, and then either print the test for paper distribution or give it online.

Test Creation
- Author/edit questions online using the 14 different question-type templates.
- Create question pools to offer multiple versions online—great for practice.
- Export your tests for use in WebCT, Blackboard, PageOut, and Apple's iQuiz.
- Sharing tests with colleagues, adjuncts, TAs is easy.

Online Test Management
- Set availability dates and time limits for your quiz or test.
- Assign points by question or question type with dropdown menu.
- Provide immediate feedback to students or delay feedback until all finish the test.
- Create practice tests online to enable student mastery.
- Your roster can be uploaded to enable student self-registration.

Online Scoring and Reporting
- Automated scoring for most of EZ Test's numerous question types.
- Allows manual scoring for essay and other open-response questions.
- Manual rescoring and feedback are also available.
- EZ Test's grade book is designed to easily export to your grade book.
- View basic statistical reports.

Support and Help
- Flash tutorials for getting started on the support site.
- Support Web site: **www.mhhe.com/eztest**
- Product specialist available at 1-800-331-5094.
- Online training: **http://auth.mhhe.com/mpss/workshops**

Virtual Labs
A complete set of more than 20 online introductory biology laboratory exercises are available on this text's companion Web site. Each exercise is self-contained with all instructions and assessment for students. For use either to stand-alone or as a supplement to a traditional lab course, this interactive learning tool will help students understand key biological concepts and processes.

McGraw-Hill: Biology Digitized Video Clips

McGraw-Hill is pleased to offer an outstanding presentation tool to text-adopting instructors—digitized biology video clips on DVD! Licensed from some of the highest-quality science video producers in the world, these brief segments range from about five seconds to just under three minutes in length and cover all areas of general biology from cells to ecosystems. Engaging and informative, McGraw-Hill's digitized videos will help capture students' interest while illustrating key biological concepts and processes such as mitosis, how cilia and flagella work, and how some plants have evolved into carnivores.

Student Response System

Wireless technology brings interactivity into the classroom or lecture hall. Instructors and students receive immediate feedback through wireless response pads that are easy to use and engage students. This system can be used by instructors to take attendance, administer quizzes and tests, create a lecture with intermittent questions, manage lectures and student comprehension through the use of the grade book, and integrate interactivity into their PowerPoint presentations.

For Students

Companion Web Site
www.mhhe.com/esstlw3

The Johnson: *Essentials of The Living World* companion Web site is an electronic study system that offers students a digital portal of knowledge. Students can readily access a variety of digital learning objects that include:

- Chapter-level quizzing
- Bio Tutorial animations with quizzing
- Vocabulary flashcards
- Virtual labs

Biology Prep, also available on the companion Web site, helps students prepare for their upcoming coursework in biology. This Web site enables students to perform self-assessments, conduct self-study sessions with tutorials, and perform a postassessment of their knowledge in the following areas:

- Introductory Biology Skills
- Basic Math Review I and II
- Chemistry
- Metric System
- Lab Reports and Referencing

Electronic Book

If you or your students are ready for an alternative version of the traditional textbook, McGraw-Hill has partnered with CourseSmart and VitalSource to bring you innovative and inexpensive electronic textbooks. Students can save up to 50% off the cost of a print book, reduce their impact on the environment, and gain access to powerful Web tools for learning including full-text search, notes and highlighting, and e-mail tools for sharing notes between classmates. eBooks from McGraw-Hill are smart, interactive, searchable, and portable.

To review comp copies or to purchase an eBook, go to either www.CourseSmart.com or www.VitalSource.com.

How to Study Science

This workbook offers students helpful suggestions for meeting the considerable challenges of a science course. It gives practical advice on such topics as how to take notes, how to get the most out of laboratories, and how to overcome science anxiety.

Photo Atlas for General Biology

This atlas was developed to support our numerous general biology titles. It can be used as a supplement for a general biology lecture or laboratory course.

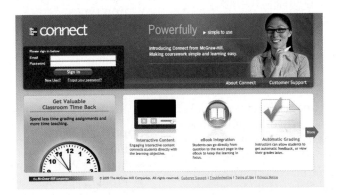

Chapter 0

Studying Biology

Learning

0.1 How to Study

CONCEPT PREVIEW: Studying biology successfully is an active process. To do well, you should attend lectures, do assigned readings before lecture, take complete class notes, rewrite those notes soon after class, and study for exams in short, focused sessions.

Taking Notes

Listening to lectures and reading the text are only the first steps in learning enough to do well in a biology course. The key to mastering the mountain of information and concepts you are about to encounter is to take careful notes. Studying from poor-quality notes that are sparse, disorganized, and barely intelligible is not a productive way to approach preparing for an exam.

There are three simple ways to improve the quality of your notes:

1. **Take many notes.** Always attempt to take the most complete notes possible during class. If you miss class, take notes yourself from a tape of the lecture, if at all possible. It is the process of taking notes that promotes learning. Using someone else's notes is but a poor substitute. When someone else takes the notes, that person tends to do most of the learning as well.
2. **Take paraphrased notes.** Develop a legible style of abbreviated note taking. Obviously, there are some things that cannot be easily paraphrased (referred to in a simpler way), but using abbreviations and paraphrasing will permit more comprehensive notes. Attempting to write complete organized sentences in note taking is frustrating and too time consuming—people just talk too fast!
3. **Revise your notes.** As soon as possible after lecture, you should decipher and revise your notes. Nothing else in the learning process is more important, because this is where most of your learning will take place. By revising your notes, you meld the information together, and put it into a context that is understandable to you. As you revise your notes, organize the material into major blocks of information with simple "heads" to identify each block. Add ideas from your reading of the text, and note links to material in other lectures. Clarify terms and concepts that might be confusing with short notes and definitions. Thinking through the ideas of the lecture in this organized way will crystallize them for you, which is the key step in learning. Also, simply rewriting your notes to make them legible, neat, and tidy can be a tremendous improvement that will further enhance your ease of learning.

Remembering and Forgetting

Learning is the process of placing information in your memory. Just as in your computer, there are two sorts of memory. The first, *short-term memory*, is analogous to the RAM (random access memory) of a computer, holding information for only a short period of time. Just as in your computer, this memory is constantly being "written over" as new information comes in. The second kind of memory, *long-term memory*, consists of information that you have stored in your memory banks for future retrieval, like storing files on your computer's hard drive. In its simplest context, learning is the process of transferring information to your hard drive.

BIOLOGY & YOU

Improving Memory. There is an active market on the internet for commercial products that claim to improve your memory. Many of these products involve repetitive games or other gimmicks; few have any lasting impact on memory. Psychologists have carried out considerable research on this subject, and have found that the best way to improve memory seems to be to increase the supply of oxygen to the brain. How do you do this? These researchers recommend aerobic exercise. Walking for three hours each week significantly increases brain oxygen levels, as does swimming or cycling. One study found that chewing gum while studying will supply the brain with enough oxygen to improve memorizing items simply because of the muscle movement.

Forgetting is the loss of information stored in memory. Most of what we forget when taking exams is the natural consequence of short-term memories not being effectively transferred to long-term memory. Forgetting occurs very rapidly, dropping to below 50% retention within one hour after learning and leveling off at about 20% retention after 24 hours.

There are many things you can do to slow down the forgetting process. Here are two important ones:

1. **Recopy your notes as soon as possible after lecture.** Remember, there is about a 50% memory loss in the first hour. You should use your textbook as well when recopying your notes.
2. **Establish a purpose for reading.** When you sit down to study your textbook, have a definite goal to learn a particular concept. Each chapter begins with a preview of its key concepts—let them be your guides. Do not try and learn the entire contents of a chapter in one session; break it up into small pieces that are "easily digested."

Learning

Learning may be viewed as the efficient transfer of information from your short-term memory to your long-term memory. This transfer is referred to as *rehearsal* by learning strategists. As its name implies, rehearsal always involves some form of repetition. There are four general means of rehearsal in the jargon of education called "critical thinking skills" (figure 0.1).

Repeating.   The most obvious form of rehearsal is repetition. To learn facts, the sequence of events in a process, or the names of a group of things, you write them down, say them aloud, and mentally repeat them over and over, until you have "memorized" them. This often is a first step on the road to learning. Many students mistake this as the only step. It is not, as it involves only rote memory, not understanding. If all you do in this course is memorize facts, you will not succeed.

Organizing.   It is important to organize the information you are attempting to learn, because the process of sorting and ordering increases retention. For example, if you place a sequence of events in order, like the stages of mitosis, the entire sequence can be recalled if you can remember what gets the sequence started.

Linking.   Biology has a natural hierarchy of information, with terms and concepts nested within other terms and concepts. You will learn facts and concepts more easily if you attempt to connect them with something you already know, linking them to some information that is already stored in your memory. Throughout this textbook, you will see arrows, like the one in figure 0.2, indicating such links. Use them to check back over concepts and processes you have already learned. You will be surprised how much doing this will help you learn the new material.

Connecting.   You will learn biology much more effectively if you relate what you are learning to the world about you. The many challenges of living in today's world are often related to the information presented in this course, and understanding these relationships will help you learn. In each chapter of this textbook you will encounter several Apps (Application dialogs) in the outer margins (there is a "BIOLOGY & YOU" App on the facing page) that allow you to briefly explore a "real-world" topic related to what you are learning. Read them. You may not be tested on these Apps, but reading them will provide you with another "hook" to help you learn the material on which you will be tested.

Figure 0.1   Learning requires work.

Learning is something you do, not something that happens to you.

IMPLICATION   If you are honest with yourself, how many of the four rehearsal techniques (critical thinking skills) do you use when you take a science course like this one? Do you think they are as important in non-science classes like English or History? Why?

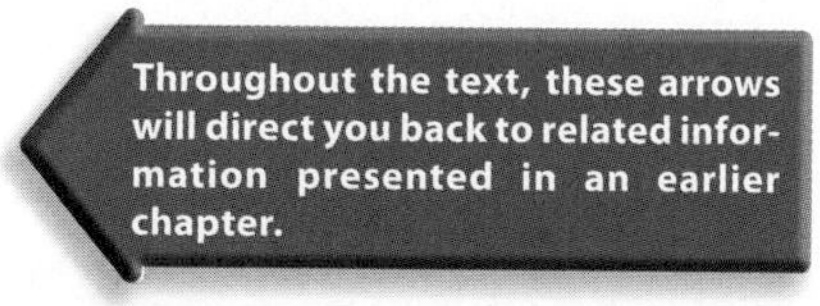

Throughout the text, these arrows will direct you back to related information presented in an earlier chapter.

Figure 0.2   Linking concepts.

These linking arrows, found throughout the text, will help you to form connections between seemingly discrete topics covered earlier in the text.

Figure 0.3 Critical learning occurs in the classroom.

Learning occurs in at least four distinct stages: attending class; doing assigned textbook readings before lecture; listening and taking notes during lecture; and recopying notes shortly after lecture. If you are diligent in these steps, then studying lecture notes and text assignments before exams is much more effective. Skipping any of these stages makes successful learning far less likely.

Studying to Learn

If I have heard it once, I have heard it a thousand times, "Gee, Professor Johnson, I studied for 20 hours straight and I still got a D." By now, you should be getting the idea that just throwing time at the material does not necessarily ensure a favorable outcome. Many students treat studying for biology like penance: If you do it, you will be rewarded for having done so. Not always.

The length of time spent studying and the spacing between study or reading sessions directly affects how much you learn. If you had 10 hours to spend studying, you would be better off if you broke it up into 10 one-hour sessions than to spend it all in one or two sessions. There are two good reasons for this:

First, we know from formal cognition research (as well as from our everyday life experiences) that we remember "beginnings" and "endings" but tend to forget "middles." Thus, the learning process can benefit from many "beginnings" and "endings."

Second, unless you are unusual, after 30 minutes or an hour your ability to concentrate is diminished. Concentration is a critical component of studying to learn. Many short, topic-focused study sessions maximize your ability to concentrate effectively. For most of us, effective concentration also means a comfortable, quiet environment with no outside distractions like loud music or conversations.

Learning Is an Active Process

It is important to realize that learning biology is not something you can do passively. Many students think that simply possessing a lecture video or a set of class notes will get them through. In and of themselves, videos and notes are no more important than the Nautilus machine an athlete works out on. It is not the machine *per se*, but what happens when you use it effectively that is of importance.

Common sense will have a great deal to do with your success in learning biology, as it does in most of life's endeavors. Your success in this biology course will depend on simple, obvious things (figure 0.3):

- *Attend class.* Go to all the lectures and be on time.
- *Read the assigned readings before lecture.* If you have done so, you will hear things in lecture that will be familiar to you, a recognition that is a vital form of learning reinforcement. Later you can go back to the text to check details.
- *Take comprehensive notes.* Recognizing and writing down lecture points is another form of recognition and reinforcement. Later, studying for an exam, you will have already forgotten lecture material you did not record, and so even if you study hard you will miss exam questions on this material.
- *Revise your notes soon after lecture.* Actively interacting with your class notes while you still hold much of the lecture in short-term memory provides perhaps the most powerful form of reinforcement, and will be a key to your success.

As you proceed through this textbook, you will encounter a blizzard of terms and concepts. Biology is a field rich with ideas and the technical jargon needed to describe them. What you discover reading this textbook is intended to support the lectures that provide the core of your biology course. Integrating what you learn here with what you learn in lecture will provide you with the strongest possible tool for successfully mastering the basics of biology. The rest is just hard work.

Pulling an All-Nighter

At some point in the next months you will face that scary rite, the first exam in this course. As a university professor I get to give the exams rather than take them, but I can remember with crystal clarity when the shoe was on the other foot. I didn't like exams a bit as a student—what student does? But in my case I was often practically paralyzed with fear. What scared me about exams was the possibility of unanticipated questions. No matter how much I learned, there was always something I didn't know, some direction from which my teacher could lob a question I had no chance of answering.

I lived and died by the all-nighter. Black coffee was my closest friend in final exam week, and sleep seemed a luxury I couldn't afford. My parents urged me to sleep more, but I was trying to cram enough in to meet any possible question, and couldn't waste time sleeping.

Now, driven by time (often kicking and screaming), I find I did it all wrong. In work published over the last few years, researchers at Harvard Medical School have demonstrated that our memory of newly-learned information improves only after sleeping at least six hours. If I wanted to do well on final exams, I could not have chosen a poorer way to prepare. The gods must look after the ignorant, as I usually passed.

Learning is, in its most basic sense, a matter of forming memories. The Harvard researchers' experiments showed that a person trying to learn something does not improve his or her knowledge until after they have had more than six hours of sleep (preferably eight). It seems the brain needs time to file new information and skills away in the proper slots so they can be retrieved later. Without enough sleep to do all this filing, new information does not get properly encoded into the brain's memory circuits.

To sort out the role of sleep in learning, the Harvard Medical School researchers used Harvard undergrads as guinea pigs. The undergraduates were trained to look for particular visual targets on a computer screen, and to push a button as soon as they were sure they had seen one. At first, responses were relatively sluggish—it typically took 400 milliseconds for a target to reach a student's conscious awareness. With an hour's training, however, many students were hitting the button correctly in 75 milliseconds.

How well had they learned? When retested from 3 to 12 hours later on the same day, there was no further improvement past a student's best time in the training session. If the researchers let a student get a little sleep, but less than six hours, then retested the next day, the student still showed no improvement in performing the target identification.

For students who slept more than six hours, the story was very different. Sleep greatly improved performance. Students who achieved 75 milliseconds in the training session would reliably perform the target identification in 62 milliseconds after a good night's sleep! After several nights of ample sleep, they often got even more proficient.

Why six or eight hours, and not four or five? The sort of sleeping you do at the beginning of a night's sleep and the sort you do at the end are different, and both, it appears, are required for efficient learning.

The first two hours of sleeping are spent in deep sleep, what psychiatrists call slow wave sleep. During this time, certain brain

chemicals become used up, which allows information that has been gathered during the day to flow out of the memory center of the brain, the hippocampus, and into the cortex, the outer covering of the brain where long-term memories are stored. Like moving information in a computer from active memory to the hard drive, this process preserves experience for future reference. Without it, long-term learning cannot occur.

Over the next hours, the cortex sorts through the information it has received, distributing it to various locations and networks. Particular connections between nerve cells become strengthened as memories are preserved, a process that is thought to require the time-consuming manufacturing of new proteins.

If you halt this process before it is complete, the day's memories do not get fully "transcribed," and you don't remember all that you would have, had you allowed the process to continue to completion. A few hours are just not enough time to get the job done. Four hours, the Harvard researchers estimate, is a minimum requirement.

The last two hours of a night's uninterrupted sleep are spent in rapid-eye-movement (rem) sleep. This is when dreams occur. The brain shuts down the connection to the hippocampus and runs through the data it has stored over the previous hours. This process is also important to learning, as it reinforces and strengthens the many connections between nerve cells that make up the new memory. Like a child repeating a refrain to memorize it, the brain goes over what it has learned, till practice makes perfect.

That's why my college system of getting by on three or four hours of sleep during exam week and crashing for 12 hours on weekends didn't work. After a few days, all of the facts I had memorized during one of my "all-nighters" faded away. Of course they did. I had never given them a chance to integrate properly into my memory circuits.

As I look back, I now see that how well I did on my exams probably had far less to do with how hard I studied than with how much I slept. It doesn't seem fair that after all these years, with my own kids now in college pulling all-nighters, I have to admit that my parents were right all along.

0.2 Using Your Textbook

CONCEPT PREVIEW: Your text is a tool to reinforce and clarify what you learn in lecture. Your use of it will only be effective if coordinated with your development of recopied lecture notes.

A Textbook Is a Tool

A student enrolled in an introductory biology course as you are, almost never learns everything from the textbook. Your text is a tool to explain and amplify what you learn in lecture. No textbook is a substitute for attending lectures, taking notes, and studying them. Success in your biology course is like a stool with three legs: lectures, class notes, and text reading—all three are necessary. Used together, they will take you a long way towards success in the course.

When to Use Your Text. While you can glance at your text at any time to refresh your memory or answer a question that pops into your mind, your use of your text as a learning tool should focus on providing support for the other two "legs" of course success, lectures and class notes.

Do the Assigned Reading. Many instructors assign reading from the text, reading that is supposed to be done before lecture. The timing here is very important: If you already have a general idea of what is being discussed in lecture, it is much easier to follow the discussion and take better notes.

Link the Text to Your Lecture Notes. Few lectures cover exactly what is in the text, and much of what is in the text may not be covered in lecture. That said, much of what you will hear in lecture is covered in your text. This coverage provides you with a powerful tool to reinforce ideas and information you encounter in lecture. Text illustrations and detailed explanations can pound home an idea quickly grasped in lecture, and answer any questions that might occur to you as you sort through the logic of an argument. Thus it is absolutely essential that you follow along with your text as you recopy your lecture notes, keying your notes to the textbook as you go. Annotating your notes in this way will make them far better learning tools as you study for exams later.

Review for Exams. It goes without saying that you should review your recopied lecture notes to prepare for an exam. But that is not enough. What is often missed in gearing up for an exam is the need to also review that part of the text that covers the same material. Reading the chapter again, one last time, helps place your lecture notes in perspective, so that it will be easier to remember key points when a topic explodes at you off the page of your exam.

How to Use Your Text. The single most important way to use your text is to read it. As your biology course proceeds and you move through the text, read each assigned chapter all the way through at one sitting. This will give you valuable perspective. Then, guided by your lecture notes, go back through the chapter one topic at a time, and focus on learning that one topic as you recopy your notes. Pay attention to the "linking" arrows in the text, as using them will reinforce what you are learning. As discussed earlier, building a bridge between text and lecture notes is a very powerful way to learn. Remember, your notes don't take the exam, and neither does the textbook; you do, and the learning that occurs as you integrate text pages and lecture notes in your mind will go a long way toward your taking it well.

Learning Tools at Your Disposal

A textbook is more than just words. What do you see when you flip through the pages of this text? Pictures, lots of them. And questions, scattered through each chapter and clustered at chapter's end. The pictures and quiz questions you will encounter within each chapter can be an important part of your learning experience.

Let the Illustrations Teach You.   All introductory biology texts are rich with colorful photographs and diagrams. They are not there to decorate, but to aid your comprehension of ideas and concepts. When the text refers you to a specific figure, look at it—the visual link will help you remember the idea much better than restricting yourself to cold words on a page.

Three sorts of illustrations offer particularly strong reinforcement:

Essential Biological Process Illustrations.   While you will be asked to learn many technical terms in this course, learning the names of things is not your key goal. Your goal is to master a small set of concepts. There are several essential biological processes that explain how organisms work the way they do. When you have understood these processes, much of the heavy lifting in learning biology is done. Every time you encounter one of these essential biological processes in the text, you will be provided with an illustration to help you better understand. These *Essential Biological Process* illustrations break the process down into easily-understood stages, so that you can grasp how the overall process works without being lost in a forest of details (figure 0.4*a*).

Bubble Links. Illustrations teach best when they are simple. Unfortunately, some of the structures and processes being illustrated just aren't simple. Every time you encounter a complex diagram in the text, it will be "predigested" for you, the individual components of the diagram are each identified with a number in a colored circle, or bubble. This same number is also placed in the text narrative right where that component is discussed. These bubble links allow the text to step you through the illustration, explaining what is going on at each stage—the illustration is a feast you devour one bite at a time.

Phylum Facts. Not all of what you will learn are concepts. Sometimes you will need to soak up a lot of information, painting a picture with facts. Nowhere is this more true than when you study animal diversity. In chapter 18 you will encounter a train of animal phyla (a phylum is a major category of organisms) with which you must become familiar. In such a sea of information, what should you learn? Every time you encounter a phylum in chapter 18, you will be provided with a *Phylum Facts* illustration that selects the key bits of information about the body and lifestyle of that kind of animal (figure 0.4*b*). If you learned and understood only the items highlighted there, you would have mastered much of what you need to know.

Check What You Know.   As you move through a chapter, addressing first one topic and then another, it will be important that you monitor your progress—not only what you have read, but how well you have understood it.

Concept Checks. At the end of each segment of a chapter you will encounter Concept Checks, a few questions you can use to assess how you are doing. They are not comprehensive, but rather representative. If you miss any of them, you should go back and have another go at that section.

End-of Chapter Questions. When you complete a chapter, you can gauge how well you have learned the material by answering the "Self-Test" and "Visual Understanding" questions at the end of the chapter. If you get a question wrong, that indicates where you need to go to finish the job. If you don't get any wrong, you have probably done your job well. You can further test your knowledge with the "Challenge Questions."

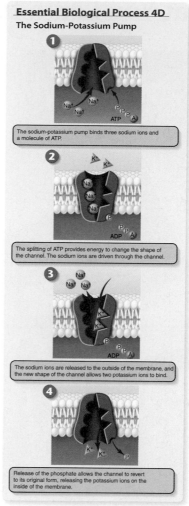

(a)

Figure 0.4   Visual learning tools.

(a) An example of an *Essential Biological Process* illustration.
(b) An example of a Phylum Facts illustration.

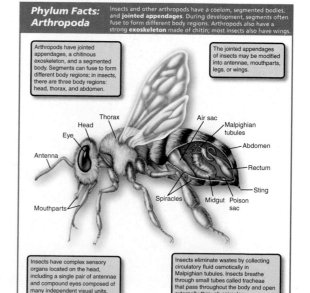

(b)

Putting What You Learn to Work

Biofuels

Bird flu

Tanning and skin cancer

Embryonic stem cells

Darwin versus intelligent design

Steroids in sports

Figure 0.5 Biology in the news.

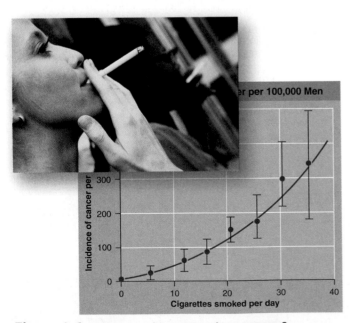

Figure 0.6 Does smoking cause lung cancer?

0.3 Science Is a Way of Thinking

CONCEPT PREVIEW: Scientists investigate by gathering data and analyzing it to form possible explanations they can test.

In the study of biology you will encounter a great deal of information, a forest of new terms and definitions, and a lot of descriptions of how things work—body processes, evolutionary relationships, interactions within ecosystems, and many others. In all of this, you will be asked to accept that what you are being taught is "true," that it accurately reflects reality. In fact, what it accurately reflects is what we know about reality. Of the things that you learn in this course, some will be altered by future scientists as they learn more. As we will discuss in chapter 1, our knowledge of science is always incomplete, the picture of reality we construct always a rough draft.

One of the most important things you can learn in a biology course is how these adjustments are made. Long after this class is completed, you will be making decisions that involve biology, and they will be better, more informed decisions if you have acquired the skill to evaluate scientific claims for yourself. Because it is printed in the newspaper or cited on a web site doesn't make a scientific claim valid. Figure 0.5 illustrates the sorts of biology you encounter today in the news—and this is just a small sample. They are, all of them, important issues that will affect your own life. How do you reach informed opinions about them?

You do it by asking the question, "How do we know this?" Science is a way of thinking that demands to see the evidence, that challenges the validity of every claim. If you can learn to do this, to apply this skill in the future to personal decisions about biology as it impacts your life, you will have taken from this course a valuable lesson.

How Do We Know What We Know?

A useful way to learn how scientists think, how they constantly check and question what they know, is to look at real cases. What follows are four instances where biologists have come to a conclusion. These conclusions will be taught in this textbook, reflecting the world about us as best as science can determine. All four of these cases will be treated at length in later chapters—here they serve only to introduce the process of scientific questioning.

Does Cigarette Smoking Cause Lung Cancer? According to the American Cancer Society, 564,830 Americans died of cancer in 2006. Fully one in four of the students using this textbook can be expected to die of it. Twenty-nine percent of these cancer victims, almost a third, die of lung cancer.

As you might imagine, something that kills so many of us has been the subject of much research. The first step biologists took was to ask a simple question: "Who gets lung cancer?" The answer came back loud and clear: Fully 87% of lung cancer deaths are cigarette smokers. Delving into this more closely, researchers looked to see if the incidence of lung cancer (that is, how many people contract it per 100,000 people) can be predicted by how many cigarettes a person smokes each day. As you can see in the graph in figure 0.6 (which is shown again on page 590), it can. The more cigarettes smoked, the higher the occurrence of lung cancer. Based on this study, and lots of others like it, examined in more detail on pages 194 and 498, biologists concluded that smoking cigarettes causes lung cancer.

Does Carbon Dioxide Cause Global Warming?   Our world is getting warmer—a lot warmer. The great ice caps that cover Antarctica and Greenland are melting, and sea levels are rising. Looking for the cause, atmospheric scientists soon began to suspect what might at first seem an unlikely culprit: carbon dioxide (CO_2), a gas that is a minor component (0.03%) of the air we breathe. As you will learn in this course, burning coal and other fossil fuels releases CO_2 into the atmosphere. Problems arise because CO_2 traps heat. As the modern world industrializes, more and more CO_2 is released. Does this lead to a hotter earth? To find out, researchers looked to see if the rise in global temperature reflected a rise in the atmosphere's CO_2. As you can see in the graph in figure 0.7 (which is shown again on page 436), it does. After these and other careful studies we will explore in detail on pages 436 and 452, scientists concluded that rising CO_2 levels are indeed the cause of global warming.

Does Obesity Lead to Type 2 Diabetes?   The United States is in the midst of an obesity epidemic. Over the last 16 years, the percentage of Americans who are obese has almost tripled, from 12% to over 34%. Coincidentally (or is it a coincidence?), the number of Americans suffering from type 2 diabetes (a disorder in which the body loses its ability to regulate glucose levels in the blood, often leading to blindness and amputation of limbs) has more than tripled over the same 16 year period, from 7 million to more than 23 million (that's one in every fourteen Americans!).

What is going on here? When researchers compared obesity levels with type 2 diabetes levels, they found a marked correlation, clearly visible in the graph in figure 0.8 (which is shown again on page 585). Investigating more closely, the researchers found that an estimated 80% of people who develop type 2 diabetes are obese. Detailed investigations described on page 585 have now confirmed the relationship that these early studies hinted at: Overeating triggers changes in the body that lead to type 2 diabetes.

What Causes the Ozone Hole?   Twenty five years ago, atmospheric scientists first reported a loss of ozone (O_3 gas) high in the atmosphere over Antarctica. This was scary, because ozone in the upper atmosphere absorbs ultraviolet radiation from the sun, protecting the earth's surface from these harmful rays. Trying to understand the reason for this "ozone hole," researchers considered many possibilities, and one of them seemed a good candidate: chlorofluorocarbons, or CFCs. CFCs are supposedly inert chemicals that are widely used as heat exchangers in air conditioners. However, further studies, detailed on pages 20 and 21, indicated that CFCs are not inert after all—in the intense cold temperatures high over Antarctica, they cause O_3 to be converted to O_2. Scientists concluded that CFCs were indeed causing the ozone hole over Antarctica. This finding led to international treaties beginning in 1990 blocking the further manufacture of CFC chemicals. As you can see in the graph in figure 0.9 (which is shown again on page 28), the size of the ozone hole soon stopped expanding in size.

Looking at the Evidence

One thing these four cases have in common is that in each, scientists reached their conclusion not by applying established rules but rather by looking in detail at what was going on and then testing possible explanations. In short, they gathered data and analyzed it. If you are going to think independently about scientific issues in the future, then you will need to learn how to analyze data and understand what it is telling you. In each case above, the data is presented in the form of a graph. Said simply, you will need to learn to read a graph.

Figure 0.7   **Does carbon dioxide cause global warming?**

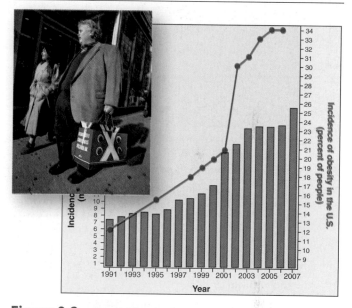

Figure 0.8   **Does obesity lead to type 2 diabetes?**

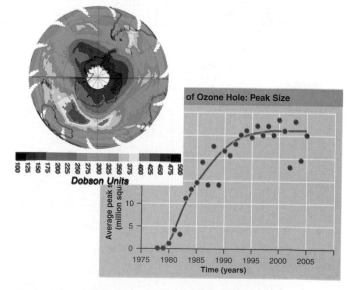

Figure 0.9   **What causes the ozone hole?**

0.4 How to Read a Graph

CONCEPT PREVIEW: Scientists often present data in standardized graphs, which portray how a dependent variable changes when an independent variable is changed.

Variables and Graphs

In the previous section, you encountered four graphs illustrating what happened to variables like global temperature, when other variables like atmospheric carbon dioxide change. A **variable,** as its name implies, is something that can change. Variables are the tools of science, and you will encounter many different kinds as you proceed through this text. Many of the variables biologists study are examined in graphs like you saw on the previous pages. A **graph** shows what happens to one variable when another one changes.

There are two types of variables. The first kind, an **independent variable,** is one that a researcher deliberately changes—for example, the concentration of a chemical in a solution, or the number of cigarettes smoked per day. The second kind, a **dependent variable,** is what happens in response to the changes in the independent variable—for example, the intensity of a solution's color, or the incidence of lung cancer. Importantly, the change in a dependent variable that is measured in an experiment is not predetermined by the investigator.

In science, all graphs are presented in a consistent way. The independent variable is always presented and labeled across the bottom, called the *x axis.* The dependent variable is always presented and labeled along the side (usually the left side), called the *y axis* (figure 0.10).

Some research involves examining correlations between sets of variables, rather than the deliberate manipulation of a variable. For example, a researcher who measures both diabetes and obesity levels (as described in the previous section) is actually comparing two dependent variables. While such a comparison can reveal correlations and so suggest potential relationships, *correlation does not prove causation*. What is happening to one variable may actually have nothing to do with what happens to the other variable. Only by manipulating a variable (making it an independent variable) can you test for causality. Just because people that are obese tend to also have diabetes does not establish that obesity *causes* diabetes. Other experiments are needed to determine causation.

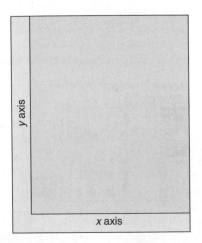

Figure 0.10   The two axes of a graph.

The independent variable is almost always presented along the x axis, and the dependent variable is usually shown along the y axis.

Using the Appropriate Scale and Units

A key aspect of presenting data in a graph is the selection of proper scale. Data presented in a table can utilize many scales, from seconds to centuries, with no problems. A graph, however, typically has a single scale on the *x* axis and a single scale on the *y* axis, which might consist of molecular units (for example, nanometers, microliters, micrograms) or macroscopic units (for example, feet, inches, liters, days, milligrams). In each instance, a scale must be chosen that fits what is being measured. Changes in centimeters would not be obvious in a graph scaled in kilometers. Also, if a variable changes a great deal over the course of the experiment, it is often useful to use an expanding scale. A **log** or **logarithmic scale** is a series of numbers plotted as powers of 10 (1, 10, 100, 1,000,...) rather than in the linear progression seen on most graphs (2,000, 4,000, 6,000...). Consider the two graphs in figure 0.11, where the *y* axis is plotted on a linear scale on the left and on a log scale on the right. You can see that the log scale more clearly displays changes in the dependent variable (the *y* axis) for the upper

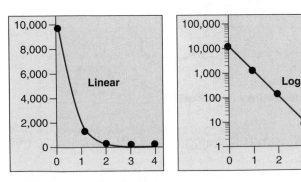

Figure 0.11   Linear and log scale: two ways of presenting the same data.

values of the independent variable (the *x* axis, values 2, 3, and 4). Notice that the interval *between* each *y* axis number is not linear either—the interval between each number is itself subdivided on a log scale. Thus, 50 (the fourth tick mark between 10 and 100) is plotted much closer to 100 than to 10.

Individual graphs use different units of measurement, each chosen to best display the experimental data. By international convention, scientific data are presented in **metric units,** a system of units expressed as powers of 10. For example, weight is expressed in units called *grams*. Ten grams make up a decagram, and 1,000 grams is a kilogram. Smaller weights are expressed as a portion of a gram—for example, a centigram is a hundredth of a gram, and a milligram is a thousandth of a gram. The units of measurement employed in a graph are by convention indicated in parentheses next to the independent variable label on the *x* axis and the dependent variable label on the *y* axis.

Drawing a Line

Most of the graphs that you will find in this text are **line graphs,** which are graphs composed of data points and one or more lines. Line graphs are typically used to present *continuous data*—that is, data that are discrete samples of a continuous process. An example might be data measuring how quickly the ozone hole develops over Antarctica in August and September of each year. You could in principle measure the area of the ozone hole every day, but to make the project more manageable in time and resources, you might actually take a measurement only once a week. Measurements reveal that the ozone hole increases in area rapidly for about six weeks before shrinking, yielding six data points during its expansion. These six data points are like individual frames from a movie—frozen moments in time. The six data points might indicate a very consistent pattern, or they might not.

Consider the hypothetical data in the graphs of figure 0.12. The data points on the left graph are changing in a very consistent way, with little variation from what a straight line (drawn in red) would predict. The graph in the middle shows more experimental variation, but a straight line still does a good job of revealing the overall pattern of how the data are changing. Such a straight "best-fit line" is called a **regression line** and is calculated by estimating the distance of each point to possible lines, adding the values, and selecting the line with the lowest sum. The data points in the graph on the right are randomly distributed and show no overall pattern, indicating that there is no relationship between the dependent and the independent variables.

Other Graphical Presentations of Data

Sometimes the independent variable for a data set is not continuous but rather represents discrete sets of data. A line graph, with its assumption of continuity, cannot accurately represent the variation occurring in discrete sets of data, where the data sets are being compared with one another. In these cases, the preferred presentation is that of a **histogram,** a kind of bar graph. For example, if you were surveying the heights of pine trees in a park, you might group their heights (the independent variable) into discrete "categories" such as 0 to 5 meters tall, 5 to 10 meters, and so on. These categories are placed on the *x* axis. You would then count the number of trees in each category and present that dependent variable on the *y* axis, as shown in figure 0.13.

Some data represent proportions of a whole data set, for example the different types of trees in the park as a percentage of all the trees. This type of data is often presented in a **pie chart** (figure 0.14).

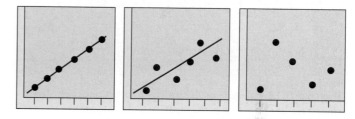

Figure 0.12   Line graphs: hypothetical growth in size of the ozone hole.

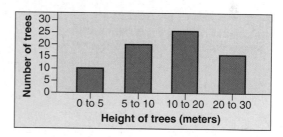

Figure 0.13   Histogram: the frequency of tall trees.

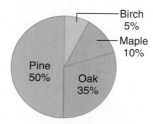

Figure 0.14   Pie chart: the composition of a forest.

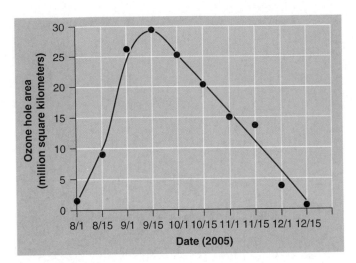

Figure 0.15 **Changes in the size of the ozone hole over one year.**

This graph shows how the size of the ozone hole changes over time, first expanding in size and then getting smaller.

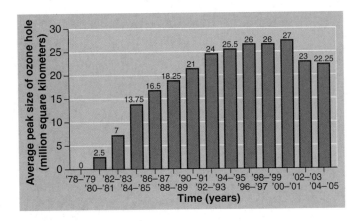

Figure 0.16 **The peak sizes of the ozone hole.**

This histogram shows how the size of the ozone hole increased in size (determined by its peak size) for 20 years before it then started to decrease.

Putting Your Graph-Reading Skills to Work: Inquiry & Analysis

The sorts of graphs you have encountered here in this brief introduction are all used frequently by scientists in analyzing and presenting their experimental results, and you will encounter them often as you proceed through this text.

Learning to read a graph and understanding what it does and does not tell you is one of the most important things you can take away from a biology course. To help you develop this skill, most of the chapters of this text end with a full-page *Inquiry & Analysis* feature. Each of these end-of-chapter features describes a real scientific investigation. You will be introduced to a question, and then given an hypothesis posed by a researcher (a hypothesis is a kind of explanation) to answer that question. The feature will then tell you how the researcher set about evaluating his or her hypothesis with an experiment and will present a graph of the data the researcher obtained. You are then challenged to analyze the data and reach a conclusion about the validity of the hypothesis.

As an example, consider research on the ozone hole, which will be the subject of the *Inquiry & Analysis* feature at the end of chapter 1. What sort of graph might you expect to see? A *line graph* can be used to present data on how the size of the ozone hole changes over the course of one year. Because the dependent variable is the size of the ozone hole measured continuously over a single season, a smooth curve accurately portrays what is actually going on. In this case, the regression line is not a straight line, but rather a curve (figure 0.15).

Can line graphs and histograms be used to present the same data? In some cases, yes. The mode of its presentation does not alter the data; it only serves to emphasize the point being investigated. The *histogram* in figure 0.16 presents data on how the peak size of the ozone hole has changed in two-year intervals over the past 26 years. However, this same data could also have been presented as a line graph—as in the *Inquiry & Analysis* of chapter 1 (also shown in figure 0.9).

Presented with a graph of the data obtained in the investigation of the *Inquiry & Analysis*, it will be your job to analyze it. Every analysis of a graph involves four distinct steps, some more complex than others but all essential to the process.

Applying Concepts. Your first task is to make sure you understand the nature of the variables, and the scale at which they are being presented in the graph. As a self-test, it is always a good idea to ask yourself to identify the dependent variable.

Interpreting Data. Look at the graph. What is changing? How much? How quickly? Is the change continuous? Progressive? What in fact has happened?

Making Inferences. Looking at what has happened, can you logically infer that the independent variable has caused the change you see in the dependent variable?

Drawing Conclusions. Does the inference you were able to make support the hypothesis that the experiment set out to test?

This process of inquiry and analysis is the nuts and bolts of science, and by mastering it, you will go a long way toward learning how a scientist thinks.

Chapter **1**

The Science of Biology

Biology and the Living World

What are these? Throughout this text you will encounter blue and green linking arrows. These "Flash-back" and "Flash-forward" concept links will direct you to related material in earlier chapters or in later chapters. Using them will help you master the interconnected nature of concepts in biology.

1.1 The Diversity of Life

CONCEPT PREVIEW: The living world is very diverse, but all organisms share key properties.

In its broadest sense, biology is the study of living things—the science of life. The living world is teeming with a breathtaking variety of creatures—whales, algae, mushrooms, bacteria, pine trees—all of which can be categorized into six groups, or **kingdoms,** of organisms (figure 1.1). Organisms that are placed into a kingdom possess similar characteristics with all other organisms in that same kingdom and are very different from organisms in the other kingdoms.

Biologists study the diversity of life in many different ways. They live with gorillas, collect fossils, and listen to whales. They isolate bacteria, grow mushrooms, and examine the structure of fruit flies. They read the messages encoded in the long molecules of heredity. In the midst of all this diversity, it is easy to lose sight of the key lesson of biology, which is that all living things have much in common.

Figure 1.1   The six kingdoms of life.

Biologists assign all living things to six major categories called *kingdoms.* Each kingdom is profoundly different from the others.

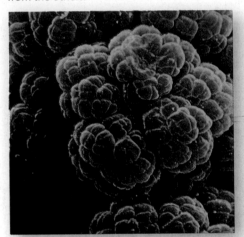

Archaea.   This kingdom of prokaryotes (simple cells that do not have nuclei) includes this methanogen, which manufactures methane.

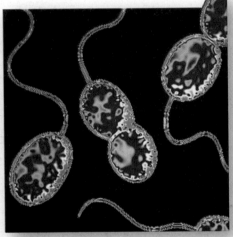

Bacteria.   This group is the second of the two prokaryotic kingdoms. Shown here are purple sulfur bacteria, which are able to convert light energy into chemical energy.

Protista.   Most of the unicellular eukaryotes (those whose cells contain a nucleus) are grouped into this kingdom, and so are the multicellular algae pictured here.

Fungi.   This kingdom contains nonphotosynthetic organisms, mostly multicellular, that digest their food externally, such as these mushrooms.

Plantae.   This kingdom contains photosynthetic multicellular organisms that are primarily terrestrial, such as the flowering plant pictured here.

Animalia.   Organisms in this kingdom are nonphotosynthetic multicellular organisms that digest their food internally, such as this ram.

1.2 Properties of Life

CONCEPT PREVIEW: All living things possess cells that carry out metabolism, maintain stable internal conditions, reproduce, and use DNA to transmit hereditary information to offspring.

Biology is the study of life—but what does it mean to be alive? What are the properties that define a living organism? This is not as simple a question as it seems because some of the most obvious properties of living organisms are also properties of many nonliving things—for example, *complexity* (a computer is complex), *movement* (clouds move in the sky), and *response to stimulation* (a soap bubble pops if you touch it). To appreciate why these three properties, so common among living things, do not help us define life, imagine a mushroom standing next to a television: The television seems more complex than the mushroom, the picture on the television screen is moving while the mushroom just stands there, and the television responds to a remote control device while the mushroom continues to just stand there—yet it is the mushroom that is alive. All living things share five basic properties:

1. **Cellular organization.** All living things are composed of one (figure 1.2) or more cells. A cell is a tiny compartment with a thin covering called a *membrane*. Some cells have simple interiors, while others are complexly organized, but all are able to grow and reproduce. A human body contains about 10–100 trillion cells (depending on how big you are)—that is a lot, a string 100 trillion centimeters long could wrap around the world 1,600 times!

2. **Metabolism.** All living things use energy. Moving, growing, thinking—everything you do requires energy. Where does all this energy come from? It is captured from sunlight by plants, algae, and certain bacteria through photosynthesis. To get the energy that powers our lives, we extract it from plants or from animals that eat plants or that eat plant-eating animals (figure 1.3). The transfer of energy from one form to another in cells is an example of *metabolism*. All organisms transfer energy from one place to another within cells using a special energy-carrying molecule called ATP.

3. **Homeostasis.** All living things maintain stable internal conditions so that their complex processes can be better coordinated. While the environment often varies a lot, organisms act to keep their interior conditions relatively constant, a process called *homeostasis*. For example, your body acts to maintain an internal temperature of approximately 37°C (98.6°F), regardless of how hot or cold the weather might be.

4. **Growth and reproduction.** All living things grow and reproduce. Bacteria increase in size and simply split in two, as often as every 15 minutes. More complex organisms grow by increasing the number of cells and reproduce sexually by producing gametes that combine giving rise to offspring (some, like the bristlecone pine of California, have reproduced after 4,600 years).

5. **Heredity.** All organisms possess a genetic system that is based on a long molecule called *DNA* (*deoxyribonucleic acid*). The information that determines what an individual organism will be like is contained in a code that is dictated by the order of the subunits making up the DNA molecule, just as the order of letters on this page determines the sense of what you are reading. Each set of instructions within the DNA is called a *gene*. DNA is faithfully copied from one generation to the next, and so any change in a gene is preserved and passed on to future generations. The transmission of characteristics from parent to offspring is a process called *heredity*.

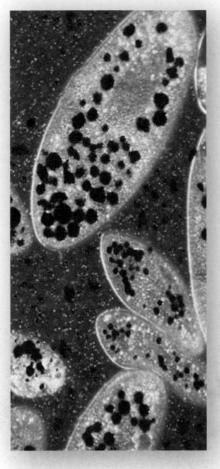

Figure 1.2
Cellular organization.

These paramecia are complex single-celled protists that have just ingested several yeast cells. Like these paramecia, many organisms consist of just a single cell, while others are composed of trillions of cells.

IMPLICATION In your body, bacterial cells outnumber the body's own cells by a 10 to 1 ratio. What do you imagine these bacteria are doing?

Figure 1.3 **Metabolism.**

This kingfisher obtains the energy it needs to move, grow, and carry out its body processes by eating fish that eat algae. The bird metabolizes this food using chemical processes that occur within cells.

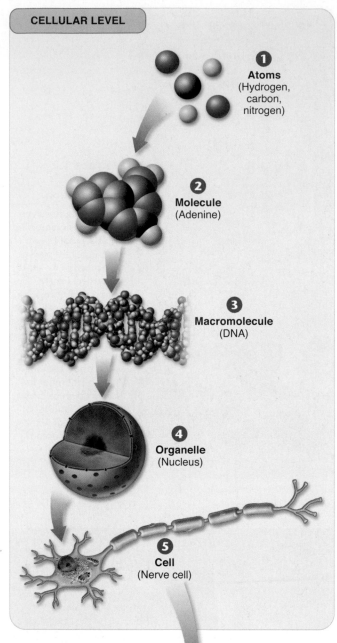

CELLULAR LEVEL

1 **Atoms**
(Hydrogen, carbon, nitrogen)

2 **Molecule**
(Adenine)

3 **Macromolecule**
(DNA)

4 **Organelle**
(Nucleus)

5 **Cell**
(Nerve cell)

1.3 The Organization of Life

CONCEPT PREVIEW: Cells, multicellular organisms, and ecological systems each are organized in a hierarchy of increasing complexity. Life's hierarchical organization is responsible for the emergent properties that characterize the living world.

A Hierarchy of Increasing Complexity

A key factor in organizing living things is the degree of complexity. We will examine the complexity of life at three levels: cellular, organismal, and populational.

Cellular Level. Following down the first section of figure 1.4, you can see that structures within cells get more and more complex—that there is a *hierarchy* of increasing complexity within cells.

> This text follows this hierarchy of increasing complexity. The cellular level is discussed in chapters 2 through 13; the organismal level is discussed in chapters 23 through 33; and the population level is discussed in chapters 14 through 22.

1 **Atoms.** The fundamental elements of matter are atoms.

2 **Molecules.** Atoms are joined together into complex clusters called molecules.

3 **Macromolecules.** Large complex molecules are called macromolecules, such as DNA that stores hereditary information.

4 **Organelles.** Complex biological molecules are assembled into tiny compartments within cells called organelles, such as the nucleus within which the cell's DNA is stored.

5 **Cells.** Organelles and other elements are assembled into membrane-bounded units we call cells. Cells are the smallest level of organization that can be considered alive.

Figure 1.4 **Levels of organization.**

A traditional and very useful way to sort through the many ways in which the organisms of the living world interact is to organize them in terms of levels of organization, proceeding from the very small and simple to the very large and complex. Here we examine organization within the cellular, organismal, and populational levels.

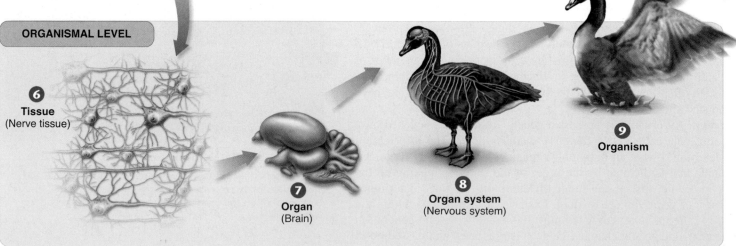

ORGANISMAL LEVEL

6 **Tissue**
(Nerve tissue)

7 **Organ**
(Brain)

8 **Organ system**
(Nervous system)

9 **Organism**

Organismal Level. At the organismal level, in the second section of figure 1.4, cells are organized into four levels of complexity.

6 **Tissues.** The most basic level is that of tissues, which are groups of similar cells that act as a functional unit. Nerve tissue is one kind of tissue, composed of cells called neurons that carry electrical signals.

7 **Organs.** Tissues, in turn, are grouped into organs, which are body structures composed of several different tissues that form a structural and functional unit. Your brain is an organ composed of nerve cells and a variety of connective tissues that form protective coverings and distribute blood.

8 **Organ systems.** At the third level of organization, organs are grouped into organ systems. The nervous system, for example, consists of sensory organs, the brain and spinal cord, neurons that convey signals throughout the body, and supporting cells.

9 **Organism.** Organ systems function together to form an organism.

Populational Level. Organisms are further organized into several hierarchical levels within the living world, as you can see to the right.

10 **Population.** The most basic of these is the population, which is a group of organisms of the same species living in the same place. A flock of geese living together on a pond is a population.

11 **Species.** All the populations of a particular kind of organism together form a species, its members similar in appearance and able to interbreed. All Canada geese, whether found in Canada, Minnesota, or Missouri, are basically the same, members of the species *Branta canadensis*. Sandhill cranes are a different species.

12 **Community.** At a higher level of biological organization, a community consists of all the populations of different species living together in one place. Geese, for example, may share their pond with ducks, fish, grasses, and many kinds of insects. All interact in a single pond community.

13 **Ecosystem.** At the highest tier of biological organization, a biological community and the soil and water within which it lives together constitute an ecological system, or ecosystem.

Emergent Properties

At each higher level in the living hierarchy, novel properties emerge, properties that were not present at the simpler level of organization. These **emergent properties** result from the way in which components interact, and often cannot be guessed just by looking at the parts themselves. Humans have the same array of cell types as a giraffe, for example, yet, examining a collection of its individual cells gives little clue of what your body is like.

The emergent properties of life are not magical or supernatural. They are the natural consequence of the hierarchy or structural organization which is the hallmark of life. Water, which makes up 50–75% of your body's weight, and ice are both made of H_2O molecules, but one is liquid and the other solid because the H_2O molecules in ice are more organized.

Functional properties emerge from more complex organization. Metabolism is an emergent property of life. The chemical reactions within a cell arise from interactions between molecules that are orchestrated by the orderly environment of the cell's interior. Consciousness is an emergent property of the brain that results from the interactions of many neurons in different parts of the brain.

POPULATIONAL LEVEL

10 Population

11 Species

12 Community

13 Ecosystem

E V O L U T I O N

Computer Games Evolve.   In 2008 the designer of the best-selling video game of all time (THE SIMS) unveiled his new creation, SPORE. In this computer game, players experience life unfolding across billions of years, its course guided by evolutionary biology. The game starts with single-celled bacteria living in the ocean, and follows their evolution into intelligent multicelled creatures that populate the land. Based on the mathematics of game theory, SPORE performs a digital version of natural selection, and challenges the player to survive by adapting to a changing world.

1.4 Biological Themes

CONCEPT PREVIEW: The five general themes of biology are (1) evolution, (2) the flow of energy, (3) cooperation, (4) structure determines function, and (5) homeostasis.

Just as a house is organized into thematic areas such as bedroom, kitchen, and bathroom, so the living world is organized by major *themes,* such as how energy flows within the living world from one part to another. As you study biology in this text, five general themes will emerge repeatedly, themes that serve to both unify and explain biology as a science.

Evolution

Evolution is genetic change in a species over time. Charles Darwin was an English naturalist who, in 1859, proposed the idea that this change is a result of a process called **natural selection.** Simply stated, those organisms whose characteristics make them better able to survive the challenges of their environment live to reproduce, passing their favorable characteristics on to their offspring. Darwin was thoroughly familiar with variation in domesticated animals (in addition to many nondomesticated organisms). He knew that varieties of pigeons could be selected by breeders to exhibit exaggerated characteristics, a process called *artificial selection.* You can see some of these extreme-looking pigeons pictured here. We now know that the characteristics selected are passed on through generations because DNA is transmitted from parent to offspring. Darwin visualized how selection in nature could be similar to that which had produced the different varieties of pigeons. Thus, the many forms of life we see about us on earth today, and the way we ourselves are constructed and function, reflect a long history of natural selection.

The Flow of Energy

All organisms require energy to carry out the activities of living—to build bodies and do work and think thoughts. All of the energy used by most organisms comes from the sun and is passed in one direction through ecosystems. The simplest way to understand the flow of energy through the living world is to look at who uses it. The first stage of energy's journey is its capture by green plants, algae, and some bacteria by the process of photosynthesis. This process uses energy from the sun to synthesize sugars that photosynthetic organisms like plants store in their bodies. Plants then serve as a source of life-driving energy for animals that eat them. Other animals, like the eagle shown here, may then eat the plant eaters. At each stage, some energy is used for the processes of living, some is transferred, and much is lost primarily as heat. The flow of energy is a key factor in shaping ecosystems, affecting how many and what kinds of organisms live in a community.

Cooperation

The ants cooperating in the photo on the facing page protect the plant on which they live from predators and shading by other plants, while this plant returns the favor by providing the ants with nutrients (the yellow structures at the tips of the leaves). This type of cooperation between different kinds of organisms has played a critical role in the evolution of life on earth. For example, organisms of two different species that live in direct contact, like the ants and the plant on which they live, form a type of relationship called **symbiosis.** Many types of cells possess organelles that are the descendants of symbiotic bacteria, and symbiotic fungi helped plants first invade land from the sea. The coevolution of flowering plants and insects—where changes in flowers influenced insect evolution and in turn, changes in insects influenced flower evolution—has been responsible for much of life's great diversity.

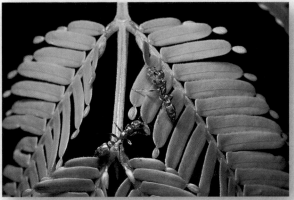

Structure Determines Function

One of the most obvious lessons of biology is that biological structures are very well suited to their functions. You will see this at every level of organization: Enzymes, which are macromolecules that cells use to carry out chemical reactions, are precisely structured to match the shapes of the chemicals the enzymes must manipulate. Within the many kinds of organisms in the living world, body structures seem carefully designed to carry out their functions—the long tongue with which the moth to the right sucks nectar from deep inside a flower is one example. The superb fit of structure to function in the living world is no accident. Life has existed on earth for over 2 billion years, a long time for evolution to favor changes that better suit organisms to meet the challenges of living. It should come as no surprise to you that after all this honing and adjustment, biological structures carry out their functions well.

Homeostasis

The high degree of specialization we see among complex organisms is only possible because these organisms act to maintain a relatively stable internal environment, a process introduced earlier called homeostasis. Without this constancy, many of the complex interactions that need to take place within organisms would be impossible, just as a city cannot function without rules to maintain order. Maintaining homeostasis in a body as complex as yours requires a great deal of signaling back-and-forth between cells. For example, homeostasis often involves water balance to maintain proper blood chemistry. All complex organisms need water—some, like this hippo, luxuriate in it. Others, like the kangaroo rat that lives in arid conditions where water is scarce, obtain water from food and never actually drink. But, they both need to maintain water balance in their bodies.

As already stated, you will encounter these biological themes repeatedly in this text. But just as a budding architect must learn more than the parts of buildings, so your study of biology should teach you more than a list of themes, concepts, and parts of organisms. Biology is a dynamic science that will affect your life in many ways, and that lesson is one of the most important you will learn.

Concept Check

1. What five basic life properties does a mushroom have that a television lacks?
2. Name one emergent property of life.
3. Identify a function in your body where maintaining homeostasis is important.

The Scientific Process

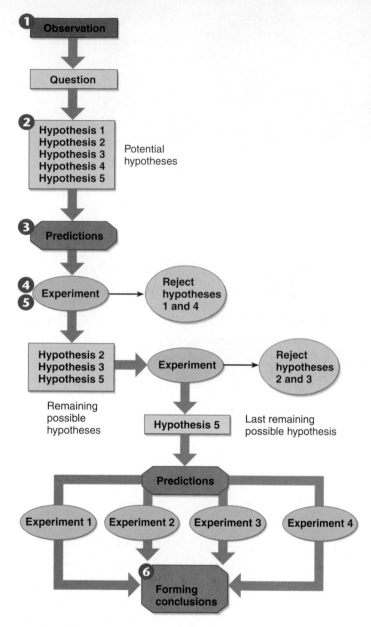

Figure 1.5   The scientific process.

This diagram illustrates the stages of a scientific investigation. First, observations are made that raise a particular question. Then a number of potential explanations (hypotheses) are suggested to answer the question. Next, predictions are made based on the hypotheses, and several rounds of experiments (including control experiments) are carried out in an attempt to eliminate one or more of the hypotheses. Finally, any hypothesis that is not eliminated is retained. Further predictions can be made based on the accepted hypothesis and tested with experiments. If a hypothesis is validated by numerous experiments and stands the test of time, it may eventually become a theory.

1.5 Stages of a Scientific Investigation

CONCEPT PREVIEW: Science progresses by systematically eliminating hypotheses that are not consistent with observation.

How Science Is Done

Scientists establish general principles as a way to explain the world around us, but how do scientists determine which general principles are true from among the many that might be? They do this by systematically testing alternative proposals. If these proposals prove inconsistent with experimental observations, they are rejected as untrue. After making careful observations concerning a particular area of science, scientists construct a hypothesis, which is a suggested explanation that accounts for those observations, an "educated guess." A hypothesis is a proposition that might be true. Those hypotheses that have not yet been disproved are retained. They are useful because they fit the known facts, but they are always subject to future rejection in the light of new information.

We call the test of a hypothesis an experiment. To evaluate alternative hypotheses, you would conduct an experiment designed to eliminate one or more of the hypotheses. Note that this does not prove that any of the other hypotheses are true; it merely demonstrates that one of them is not. A successful experiment is one in which one or more of the alternative hypotheses is demonstrated to be inconsistent with the results and is thus rejected.

In this text, you will encounter a great deal of information, often accompanied by explanations. These explanations are hypotheses that have withstood the test of experiment. Many will continue to do so; others will be revised as new observations are made. Biology, like all science, is in a constant state of change, with new ideas appearing and replacing old ones.

The Scientific Process

Science is a particular way of investigating the world, of forming general rules about why things happen by observing particular situations. A scientist is an observer, someone who looks at the world to understand how it works.

Scientific investigations can be said to have six stages as illustrated in figure 1.5: ❶ observing what is going on; ❷ forming a set of hypotheses; ❸ making predictions; ❹ testing them and ❺ carrying out controls, until one or more of the hypotheses have been eliminated; and ❻ forming conclusions based on the remaining hypothesis. To better understand how a scientist progresses through these stages, let's examine an actual scientific investigation, the discovery and analysis of the ozone hole.

1. **Observation.** The key to any successful scientific investigation is careful *observation*. Scientists had studied the skies over the Antarctic for many years, noting a thousand details about temperature, light, and levels of chemicals. You can see an example of these types of observations in figure 1.6, which is a recording of the levels of ozone in the atmosphere (the purple areas represent the lowest levels of ozone). Had these scientists not kept careful records of what they observed, they might not have noticed that ozone levels were dropping.

2. **Hypothesis.** When the unexpected drop in ozone was reported, scientists made a guess why—perhaps something was destroying

the ozone; maybe the culprit was chlorofluorocarbons (CFCs), an industrial chemical used as a coolant in air conditioners, propellants in aerosols, and foaming agents in making Styrofoam. Of course, this was not a guess in the true sense; scientists had some working knowledge of CFC and what it might be doing in the upper atmosphere. We call such a guess a **hypothesis.** What the scientists guessed was that chlorine from the breakdown of CFCs was reacting chemically with ozone over the Antarctic, converting ozone (O_3) into oxygen gas (O_2), removing the ozone shield from our earth's atmosphere. Often, scientists will form *alternative hypotheses* if they have more than one guess about what they observe. In this case, there were several other hypotheses advanced to explain the ozone hole. One suggestion explained it as a normal consequence of the spinning of the earth, the ozone spinning away from the polar regions much as water spins away from the center as a clothes washer moves through its spin cycle. Another hypothesis was that the ozone hole was simply due to sunspots, and would soon disappear.

3. **Predictions.** If the CFC hypothesis is correct, then several consequences can reasonably be expected. We call these expected consequences *predictions*. A prediction is what you expect to happen if a hypothesis is true. The CFC hypothesis predicts that if CFCs are responsible for producing the ozone hole, then it should be possible to detect CFCs in the upper Antarctic atmosphere as well as the chlorine released from CFCs that attack the ozone.

4. **Testing.** Scientists set out to test the CFC hypothesis by attempting to verify some of its predictions. We call the test of a hypothesis an *experiment.* To test the hypothesis, atmospheric samples were collected from the stratosphere over 6 miles up by a high-altitude balloon. Analysis of the samples revealed the presence of CFCs, as predicted. Were the CFCs interacting with the ozone? The samples contained free chlorine and fluorine, confirming the breakdown of CFC molecules. The results of the experiment thus support the hypothesis.

5. **Controls.** Events in the upper atmosphere can be influenced by many factors. We call each factor that might influence a process a **variable.** To evaluate alternative hypotheses about one variable, all the other variables must be kept constant so that we do not get misled or confused by these other influences. This is done by carrying out two experiments in parallel: In the first experimental test, we alter one variable in a known way to test a particular hypothesis; in the second, called a *control experiment,* we do *not* alter that variable. In all other respects, the two experiments are the same. To further test the CFC hypothesis, scientists carried out control laboratory experiments in which they detected no drop in ozone levels in the absence of CFCs. The result of the control was consistent with the predictions of the hypothesis.

6. **Conclusion.** A hypothesis that has been tested and not rejected is tentatively accepted. The hypothesis that CFCs released into the atmosphere are destroying the earth's protective ozone shield is now supported by a great deal of experimental evidence and is widely accepted. A collection of related hypotheses that have been tested many times and not rejected is called a **theory.** A theory indicates a higher degree of certainty; however, in science, nothing is "certain." For example, the theory of the ozone shield—that ozone in the upper atmosphere shields the earth's surface from harmful UV rays by absorbing them—is supported by a wealth of observation and experimentation and is widely accepted.

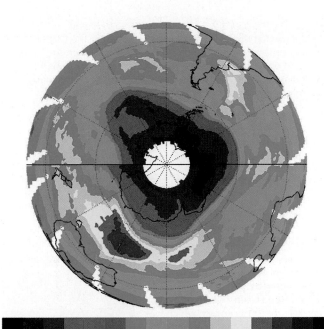

Dobson Units

Figure 1.6   The ozone hole.

The swirling colors represent different concentrations of ozone over the South Pole as viewed from a satellite on September 15, 2001. As you can easily see, there is an "ozone hole" (the *purple* areas) over Antarctica covering an area about the size of the United States. (The color *white* indicates areas where no data were available.)

IMPLICATION   If the ozone hole indicated by the purple area only extends over Antarctica (and sometimes the tip of South America), why should people living in North America worry about it?

BIOLOGY & YOU

Ozone Therapy.   Ozone is claimed by some to have remarkable healing properties and holistic health benefits. For example, it is asserted that ozone inhibits the growth of lung and breast cancer cells. Like other claims of health benefits, this claim is widely discounted by the medical community. Most states prohibit the medical use of ozone and consider the marketing of ozone generators as dangerous (ozone has toxic effects on the lungs) and providing no medical benefit.

The theory that the Earth was flat—the idea that the Earth was flat was generally accepted until about 330 B.C. when Aristotle provided observational evidence for a spherical Earth using constellations. About 100 years later Eratosthenes provided quantitative evidence by estimating the circumference of the Earth.

The theory that the sun circles Earth—this geocentric view of the universe was accepted as fact until Copernicus first proposed a sun-centered solar system in the 16th century, which was supported by quantitative evidence in the early 1600s by Galileo.

The theory of creationism—this explanation of the origin and diversity of life on Earth following the account presented in the Bible was widely accepted, although other explanations that proposed that life evolved on Earth were also put forth. The theory of creationism was questioned and then dispelled in 1859, when Charles Darwin proposed an explanation for evolution, a process called natural selection, which is the widely accepted scientific explanation of the origin and diversity of life on Earth.

The theory of acquired traits—Jean-Baptiste Lamarck, a predecessor of Darwin, proposed an explanation of evolution. He stated that traits were acquired during one's life and were then passed on to offspring. Darwin later proposed the explanation of natural selection as the driving force of evolution.

The hypothesis of cold nuclear fusion—the experimental results that led to this hypothesis, that nuclear reactions could be performed at near room temperature and pressure, were presented in 1989. The results could have revolutionized the energy industry, but the results could not be duplicated and widespread support for the hypothesis never developed.

Figure 1.7   Rejected theories.

To illustrate how experimentation changes scientific thought, consider these scientific theories that were once accepted as true, but have since been rejected.

IMPLICATION   The theory of global warming was controversial when first proposed, but has gained widespread support in recent years. How has the acceptance of this theory affected your own life, or has it?

1.6 Theory and Certainty

CONCEPT PREVIEW: A scientist does not follow a fixed method to form hypotheses but relies also on judgment and intuition.

A theory is a unifying explanation for a broad range of observations. Thus we speak of the theory of gravity, the theory of evolution, and the theory of the atom. Theories are the solid ground of science, that of which we are the most certain. However, there is no absolute truth in science, only varying degrees of uncertainty. A scientist's acceptance of a theory is always provisional, because the possibility always remains that future evidence will cause a theory to be revised; some "once-accepted" theories are discussed in figure 1.7.

The word "theory" is thus used very differently by scientists than by the general public. To a scientist, a theory represents that of which he or she is most certain; to the general public, the word theory implies a *lack* of knowledge or a guess. How often have you heard someone say, "It's only a theory!"? As you can imagine, confusion often results. In this text the word theory will always be used in its scientific sense, in reference to a generally accepted scientific principle.

The Scientific "Method"

It was once fashionable to claim that scientific progress is the result of applying a series of steps called the *scientific method;* that is, a series of logical "either/or" predictions tested by experiments to reject one alternative. The assumption was that trial-and-error testing would inevitably lead one through the maze of uncertainty that always slows scientific progress. If this were indeed true, a computer would make a good scientist—but science is not done this way! If you ask successful scientists how they do their work, you will discover that without exception they design their experiments with a pretty fair idea of how they will come out. Environmental scientists understood the chemistry of chlorine and ozone when they formulated the CFC hypothesis, and they could imagine how the chlorine in CFCs would attack ozone molecules. A hypothesis that a successful scientist tests is not just any hypothesis. Rather, it is a "hunch" or educated guess in which the scientist integrates all that he or she knows, in an attempt to get a sense of what *might* be true. It is because insight and imagination play such a large role that some scientists are so much better at science than others—just as Beethoven and Mozart stand out among composers.

The Limitations of Science

Scientific study is limited to organisms and processes that we are able to observe and measure. Supernatural and religious hypotheses are beyond the realm of scientific analysis because they cannot be scientifically studied, analyzed, or explained. Supernatural hypotheses can be used to explain any result, and cannot be disproven by experiment or observation. Scientists in their work are limited to objective interpretations of observable phenomena.

Concept Check

1. What laboratory control experiment did scientists carry out to test the CFC hypothesis?
2. Why don't computers, which are good at trial-and-error testing, make good scientists?
3. What is the difference between a hypothesis and a theory?

Where Are All My Socks Going?

All my life, for as far back as I can remember, I have been losing socks. Not pairs of socks, mind you, but single socks. I first became aware of this peculiar phenomenon when as a young man I went away to college. When Thanksgiving rolled around that first year, I brought an enormous duffle bag of laundry home. My mother, instead of braining me, dumped the lot into the washer and dryer, and so discovered what I had not noticed—that few of my socks matched anymore.

That was over forty years ago, but it might as well have been yesterday. All my life, I have continued to lose socks. This last Christmas I threw out a sock drawer full of socks that didn't match, and took advantage of sales to buy a dozen pairs of brand-new ones. Last week, when I did a body count, three of the new pairs had lost a sock!

Enough. I set out to solve the mystery of the missing socks. How? The way Sherlock Holmes would have, scientifically. Holmes worked by eliminating those possibilities that he found not to be true. A scientist calls possibilities "hypotheses" and, like Sherlock, rejects those that do not fit the facts. Sherlock tells us that when only one possibility remains unrejected, then—however unlikely—it must be true.

Hypothesis 1: It's the socks. I have four pairs of socks bought as Christmas gifts but forgotten until recently. Deep in my sock drawer, they have remained undisturbed for five months. If socks disappear because of some intrinsic property (say the manufacturer has somehow designed them to disappear to generate new sales), then I could expect at least one of these undisturbed ones to have left the scene by now. However, when I looked, all four pairs were complete. Undisturbed socks don't disappear. Thus I reject the hypothesis that the problem is caused by the socks themselves.

Hypothesis 2: Transformation, a fanciful suggestion by science fiction writer Avram Davidson in his 1958 story "Or All the Seas with Oysters" that I cannot get out of the quirky corner of my mind. I discard the socks I have worn each evening in a laundry basket in my closet. Over many years, I have noticed a tendency for socks I have placed in the closet to disappear. Over that same long period, as my socks are disappearing, there is something in my closet that seems to multiply—COAT HANGERS! Socks are larval coat hangers! To test this outlandish hypothesis, I had only to move the laundry basket out of the closet. Several months later, I was still losing socks, so this hypothesis is rejected.

Hypothesis 3: Static cling. The missing single socks may have been hiding within the sleeves of sweat shirts or jackets, inside trouser legs, or curled up within seldom-worn garments. Rubbing around in the dryer, socks can garner quite a bit of static electricity, easily enough to cause them to cling to other garments. Socks adhering to the outside of a shirt or pant leg are soon dislodged, but ones that find themselves within a sleeve, leg, or fold may simply stay there, not "lost" so much as misplaced. However, after a diligent search, I did not run across any previously lost socks hiding in the sleeves of my winter garments or other seldom-worn items, so I reject this hypothesis.

Hypothesis 4: I lose my socks going to or from the laundry. Perhaps in handling the socks from laundry basket to the washer/dryer and back to my sock drawer, a sock is occasionally lost. To test this hypothesis, I have pawed through the laundry coming into the

©2006 Eric Lewis from cartoonbank.com. All rights reserved.

washer. No single socks. Perhaps the socks are lost after doing the laundry, during folding or transport from laundry to sock drawer. If so, there should be no single socks coming out of the dryer. But there are! The singletons are first detected among the dry laundry, before folding. Thus I eliminate the hypothesis that the problem arises from mishandling the laundry. It seems the problem is in the laundry room.

Hypothesis 5: I lose them during washing. Perhaps the washing machine is somehow "eating" my socks. I looked in the washing machine to see if a sock could get trapped inside, or chewed up by the machine, but I can see no possibility. The clothes slosh around in a closed metal container with water passing in and out through little holes no wider than a pencil. No sock could slip through such a hole. There is a thin gap between the rotating cylinder and the top of the washer through which an errant sock might escape, but my socks are too bulky for this route. So I eliminate the hypothesis that the washing machine is the culprit.

Hypothesis 6: I lose them during drying. Perhaps somewhere in the drying process socks are being lost. I stuck my head in our clothes dryer to see if I could see any socks, and I couldn't. However, as I look, I can see a place a sock could go—behind the drying wheel! A clothes dryer is basically a great big turning cylinder with dry air blowing through the middle. The edges of the turning cylinder don't push hard against the side of the machine. Just maybe, every once in a while, a sock might get pulled through, sucked into the back of the machine.

To test this hypothesis, I should take the back of the dryer off and look inside to see if it is stuffed with my missing socks. My wife, knowing my mechanical abilities, is not in favor of this test. Thus, until our dryer dies and I can take it apart, I shall not be able to reject hypothesis 6. Lacking any other likely hypothesis, I take Sherlock Holmes' advice and tentatively conclude that the dryer is the culprit.

Core Ideas of Biology

Figure 1.8   **Life in a drop of pond water.**

All organisms are composed of cells. Some organisms, including these protists, are single celled, while others, such as plants, animals, and fungi, consist of many cells.

1.7 Four Theories Unify Biology as a Science

CONCEPT PREVIEW: The theories uniting biology state that cellular organisms store hereditary information in DNA. Sometimes DNA alterations occur, which when preserved result in evolutionary change. Today's biological diversity is the product of a long evolutionary journey.

The Cell Theory: Organization of Life

All organisms are composed of cells, life's basic units. Cells were discovered by Robert Hooke in England in 1665. Hooke was using one of the first microscopes, one that magnified 30 times. Looking through a thin slice of cork, he observed many tiny chambers, which reminded him of monks' cells in a monastery. Not long after that, the Dutch scientist Anton van Leeuwenhoek used microscopes capable of magnifying 300 times, and discovered an amazing world of single-celled life in a drop of pond water like you see in figure 1.8. He called the bacterial and protist cells he saw "wee animalcules." However, it took almost two centuries before biologists fully understood their significance. In 1839, the German biologists Matthias Schleiden and Theodor Schwann, summarizing a large number of observations by themselves and others, concluded that all living organisms consist of cells. Their conclusion forms the basis of what has come to be known as the *cell theory*. Later, biologists added the idea that all cells come from other cells. The cell theory, one of the basic ideas in biology, is the foundation for understanding the reproduction and growth of all organisms. You will learn much more about cells in chapter 4.

The Gene Theory: Molecular Basis of Inheritance

Even the simplest cell is incredibly complex, more intricate than a computer. The information that specifies what a cell is like—its detailed plan—is encoded in a long cablelike molecule called **DNA (deoxyribonucleic acid)**. Researchers James Watson and Francis Crick discovered in 1953 that each DNA molecule is formed from two long chains of building blocks, called nucleotides, wound around each other. You can see in figure 1.9 that the two chains face

> How DNA functions in the cell is directly related to its structure and is discussed in detail in chapter 3, pages 52 and 53 and again in chapter 11, page 186.

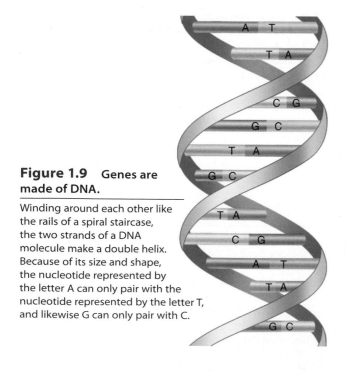

Figure 1.9   **Genes are made of DNA.**

Winding around each other like the rails of a spiral staircase, the two strands of a DNA molecule make a double helix. Because of its size and shape, the nucleotide represented by the letter A can only pair with the nucleotide represented by the letter T, and likewise G can only pair with C.

each other, like two lines of people holding hands. The chains contain information in the same way this sentence does—as a sequence of letters. There are four different nucleotides in DNA (symbolized as A, T, C, and G in the figure), and the sequence in which they occur encodes the information. Specific sequences of several hundred to many thousand nucleotides make up a *gene*, a discrete unit of hereditary information. A gene might encode a particular protein, or a different kind of unique molecule called RNA, or a gene might act to regulate other genes. All organisms on earth encode their genes in strands of DNA. This prevalence of DNA led to the development of the *gene theory*. Illustrated in figure 1.10, the gene theory states that the proteins and RNA molecules encoded by an organism's genes determine what it will be like. How genes function is the subject of chapter 12. In chapter 13 you will explore how detailed knowledge of genes is revolutionizing biology and having an impact on the lives of all of us.

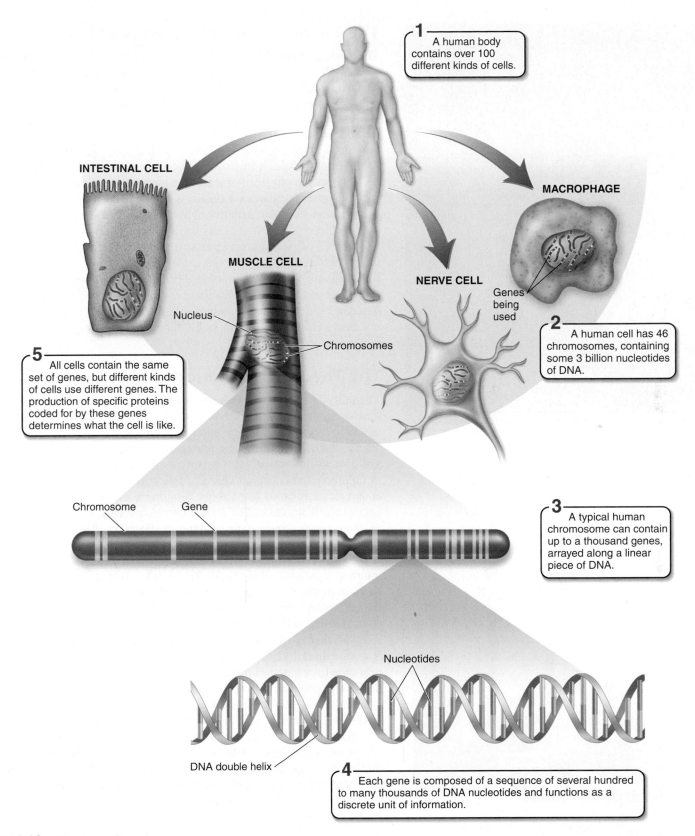

1 A human body contains over 100 different kinds of cells.

INTESTINAL CELL

MACROPHAGE

MUSCLE CELL

NERVE CELL

Genes being used

2 A human cell has 46 chromosomes, containing some 3 billion nucleotides of DNA.

Nucleus

Chromosomes

5 All cells contain the same set of genes, but different kinds of cells use different genes. The production of specific proteins coded for by these genes determines what the cell is like.

Chromosome   Gene

3 A typical human chromosome can contain up to a thousand genes, arrayed along a linear piece of DNA.

Nucleotides

DNA double helix

4 Each gene is composed of a sequence of several hundred to many thousands of DNA nucleotides and functions as a discrete unit of information.

Figure 1.10   The gene theory.

The gene theory states that what an organism is like is determined in large measure by its genes. Here you see how the many kinds of cells in the body of each of us are determined by which genes are used in making each particular kind of cell.

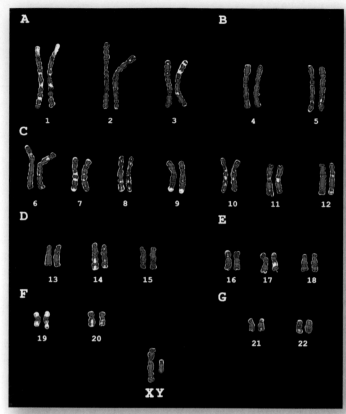

Figure 1.11   **Human chromosomes.**

The chromosomal theory of inheritance states that genes are located on chromosomes. This human karyotype (an ordering of chromosomes) shows banding patterns on chromosomes that represent clusters of genes.

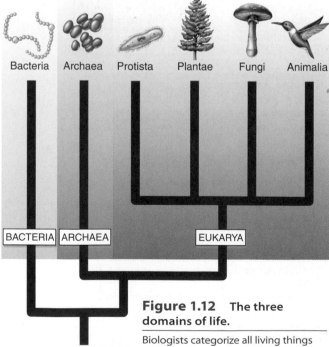

Figure 1.12   **The three domains of life.**

Biologists categorize all living things into three overarching groups called domains: Bacteria, Archaea, and Eukarya. Domain Bacteria contains the kingdom Bacteria, and domain Archaea contains the kingdom Archaea. Domain Eukarya is composed of four more kingdoms: Protista, Plantae, Fungi, and Animalia.

The Theory of Heredity: Unity of Life

The storage of hereditary information in genes composed of DNA is common to all living things. The *theory of heredity* first advanced by Gregor Mendel in 1865 states that the genes of an organism are inherited as discrete units. A triumph of experimental science developed long before genes and DNA were understood, Mendel's theory of heredity is the subject of chapter 10. Soon after Mendel's theory gave rise to the field of genetics, other biologists proposed what has come to be called the *chromosomal theory of inheritance,* which in its simplest form states that the genes of Mendel's theory are physically located on chromosomes, and that it is because chromosomes are parceled out in a regular manner during reproduction that Mendel's regular patterns of inheritance are seen. In modern terms, the two theories state that genes are a component of a cell's chromosomes (like the 23 pairs of human chromosomes you see in figure 1.11), and that the regular duplication of these chromosomes during sexual reproduction is responsible for the pattern of inheritance we call Mendelian segregation. Sometimes a character is conserved essentially unchanged in a long line of descent, reflecting a fundamental role in the biology of the organism, one not easily changed once adopted. Other characters might be modified due to changes in DNA.

The Theory of Evolution: Diversity of Life

The unity of life, which we see in the retention of certain key characteristics among many related life-forms, contrasts with the incredible diversity of living things that have evolved to fill the varied environments of earth. These diverse organisms are sorted by biologists into six kingdoms, as you learned in section 1.1. Organisms placed in the same kingdom have in common some general characteristics. In recent years, biologists have added a classification level above kingdoms, based on fundamental differences in cell structure. The six kingdoms are each now assigned into one of three great groups called *domains:* Bacteria, Archaea, and Eukarya. Bacteria (the yellow zone in figure 1.12) and Archaea (the pink zone) each consist of one kingdom of prokaryotes (single-celled organisms with little internal structure). Four more kingdoms composed of organisms with more complexly-organized cells are placed within the domain Eukarya, the eukaryotes (the purple zone in figure 1.12). The kingdoms of life are discussed in detail in chapter 15.

The **theory of evolution,** advanced by Charles Darwin in 1859, attributes the diversity of the living world to natural selection. Those organisms best able to respond to the challenges of living will leave more offspring, he argued, and thus their traits become more common in the population. It is because the world offers diverse opportunities that it contains so many different life-forms.

Today scientists can decipher all the genes (the genome) of an organism. One of the great triumphs of science in the century and a half since Darwin is the detailed understanding of how Darwin's theory of evolution is related to the gene theory—of how changes in life's diversity result from changes in individual genes (figure 1.13).

Concept Check

1. Why was it not possible to develop the cell theory until the end of the 17th century?
2. How is the gene theory related to the theory of heredity?
3. The theory of heredity is closely related to the chromosomal theory of inheritance. Explain the relationship between these two theories.

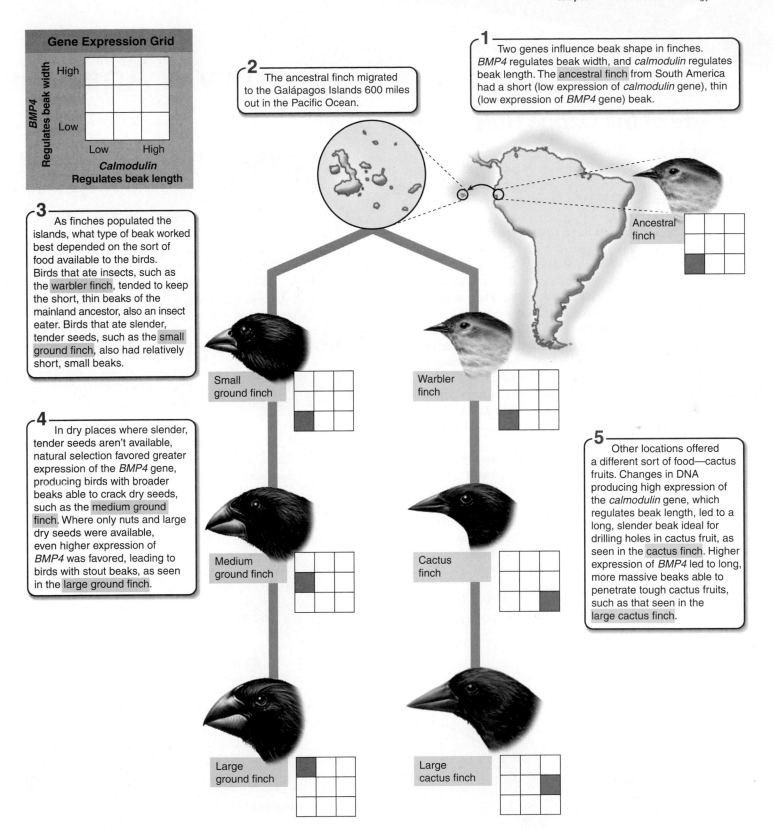

Gene Expression Grid

BMP4 Regulates beak width — High / Low
Calmodulin Regulates beak length — Low / High

1 Two genes influence beak shape in finches. *BMP4* regulates beak width, and *calmodulin* regulates beak length. The ancestral finch from South America had a short (low expression of *calmodulin* gene), thin (low expression of *BMP4* gene) beak.

2 The ancestral finch migrated to the Galápagos Islands 600 miles out in the Pacific Ocean.

3 As finches populated the islands, what type of beak worked best depended on the sort of food available to the birds. Birds that ate insects, such as the warbler finch, tended to keep the short, thin beaks of the mainland ancestor, also an insect eater. Birds that ate slender, tender seeds, such as the small ground finch, also had relatively short, small beaks.

4 In dry places where slender, tender seeds aren't available, natural selection favored greater expression of the *BMP4* gene, producing birds with broader beaks able to crack dry seeds, such as the medium ground finch. Where only nuts and large dry seeds were available, even higher expression of *BMP4* was favored, leading to birds with stout beaks, as seen in the large ground finch.

5 Other locations offered a different sort of food—cactus fruits. Changes in DNA producing high expression of the *calmodulin* gene, which regulates beak length, led to a long, slender beak ideal for drilling holes in cactus fruit, as seen in the cactus finch. Higher expression of *BMP4* led to long, more massive beaks able to penetrate tough cactus fruits, such as that seen in the large cactus finch.

Ancestral finch

Small ground finch

Warbler finch

Medium ground finch

Cactus finch

Large ground finch

Large cactus finch

Figure 1.13   The theory of evolution.

Darwin's theory of evolution proposes that many forms of a gene may exist among members of a population, and that those members with a form better suited to their particular habitat will tend to reproduce more successfully, and so their traits become more common in the population, a process Darwin dubbed "natural selection." Here you see how this process is thought to have worked on two pivotal genes that helped generate the diversity of finches on the Galápagos Islands, visited by Darwin in 1831 on his round-the-world voyage on HMS *Beagle*. This example of evolution was a key to Darwin's thinking. The way in which modern knowledge of genes has expanded our understanding of evolution among Darwin's finches is examined in greater detail in chapter 14.

Is the Size of the Ozone Hole Increasing?

In 1985 Joseph Farman, a British atmospheric scientist working in Antarctica, made an unexpected discovery. Analyzing the Antarctic sky, he found far less ozone (O_3, a form of oxygen gas) than should be there—a 30% drop from a reading recorded five years earlier in the Antarctic!

At first it was argued that this thinning of the ozone (soon dubbed the "ozone hole") was an as-yet-unexplained weather phenomenon. Evidence soon mounted, however, implicating synthetic chemicals as the culprit. Detailed analysis of chemicals in the Antarctic atmosphere revealed a surprisingly high concentration of chlorine, a chemical known to destroy ozone. The source of the chlorine was a class of chemicals called **chlorofluorocarbons (CFCs).** CFCs have been manufactured in large amounts since they were invented in the 1920s, largely for use as coolants in air conditioners, propellants in aerosols, and foaming agents in making Styrofoam. CFCs were widely regarded as harmless because they are chemically unreactive under normal conditions. But in the atmosphere over Antarctica, CFCs break down and produce chlorine, which acts as a catalyst, attacking and destroying ozone, as shown in the figure below.

The thinning of the ozone layer in the upper atmosphere 25 to 40 kilometers above the surface of the earth is a serious matter. The ozone layer protects life from the harmful ultraviolet (UV) rays of the sun that bombard the earth continuously. Like invisible sunglasses, the ozone layer filters out these dangerous rays. So when ozone is destroyed, the UV rays are able to pass through to the earth. When UV rays damage the DNA in skin cells, it can lead to skin cancer. It is estimated that every 1% drop in the atmospheric ozone concentration leads to a 6% increase in skin cancers.

As scientific observations have become widely known, governments have rushed to correct the situation. By 1990, worldwide agreements to phase out production of CFCs by the end of the century had been signed. Production of CFCs declined by 86% in the following 10 years. The world currently produces less than 200,000 tons of CFCs annually, down from 1986 levels of 1.1 million tons.

While ozone depletion is still producing major ozone holes over the Antarctic, researchers' models predict that the situation should gradually improve as less CFCs are produced, with the period of maximum ozone depletion peaking now and beginning to fall in the next few years.

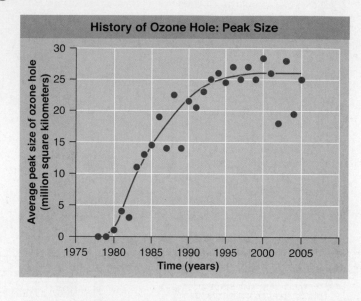

Are they right? The graph above presents measurements of the average peak size of the ozone hole centered over Antarctica, data collected by atmospheric scientists since 1976.

Analysis

1. **Making Inferences** Did the size of the ozone hole increase from 1980 to 1990? from 1995 to 2005?
2. **Drawing Conclusions** Does the graph support the conclusion that the size of the ozone hole over Antarctica is presently increasing? Explain. What might account for the difference you observe between 1980–90 and 1995–2005?

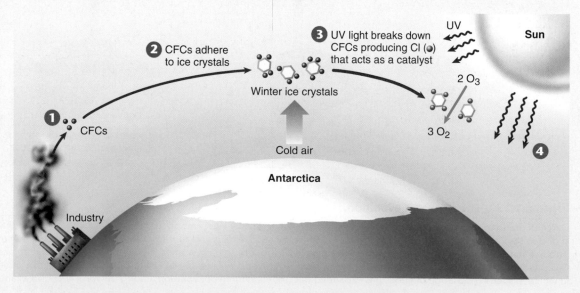

Concept Summary

Biology and the Living World

1.1   The Diversity of Life

- Biology is the study of life. All living organisms share common characteristics, but they are also diverse and are categorized into six groups called kingdoms. The six kingdoms are Bacteria, Archaea, Protista, Fungi, Plantae, and Animalia (**figure 1.1**).

1.2   Properties of Life

- All living organisms share five basic properties: cellular organization, metabolism, homeostasis, growth and reproduction, and heredity. Cellular organization indicates that all living organisms are composed of cells. Metabolism means that all living organisms, like the kingfisher shown here from **figure 1.3,** use energy. Homeostasis is the process  whereby all living organisms maintain stable internal conditions. The properties of growth and reproduction indicate that all living organisms grow in size and reproduce. The property of heredity describes how all living organisms possess genetic information in DNA that determines how each organism looks and functions, and this information is passed on to future generations.

1.3   The Organization of Life

- Living organisms exhibit increasing levels of complexity within their cells (cellular level), within their bodies (organismal level), and within ecosystems (population level) (**figure 1.4**).
- Novel properties that appear in each level of the hierarchy of life are called emergent properties. These properties are the natural consequences of ever more complex structural organization.

1.4   Biological Themes

- Five themes emerge from the study of biology: evolution, the flow of energy, cooperation, structure determines function, and homeostasis. These themes are used to examine the similarities and differences among organisms.

The Scientific Process

1.5   Stages of a Scientific Investigation

- Scientists use observations to formulate hypotheses. Hypotheses are possible explanations that are used to form predictions. These predictions are tested experimentally. Some hypotheses are rejected based on experimental results, while others are tentatively accepted.
- Scientific investigations often use a series of stages, called the scientific process, to study a scientific question. These stages are observations, forming hypotheses, making predictions, testing, establishing controls, and drawing conclusions (**figure 1.5**).
- The discovery of the hole in the ozone required careful observations of data collected from the atmosphere. Based on these observations (**figure 1.6**), scientists proposed an explanation of what caused a decrease in the levels of ozone over the Antarctic. This explanation is called a hypothesis. They then formed predictions and tested the hypothesis against controls. The hypothesis that CFCs released into the atmosphere were causing the breakdown of ozone to oxygen gas was supported by the data. Further experimentation allowed scientists to form the conclusion that CFCs were responsible for the loss of ozone over the Antarctic.

1.6   Theory and Certainty

- Hypotheses that hold up to testing over time are combined into statements called theories. Theories carry a higher degree of certainty, although no theory in science is absolute.
- The process of science was once viewed as a series of "either/or" predictions that were tested experimentally. This process, referred to as the "scientific method," did not take into account the importance of insight and imagination that are necessary to good scientific investigations.
- Science can only study what can be tested experimentally. A hypothesis can only be established through science if it can be tested and potentially disproven.

Core Ideas of Biology

1.7   Four Theories Unify Biology as a Science

- There are four unifying theories in biology: cell theory, gene theory, the theory of heredity, and the theory of evolution.
- The cell theory states that all living organisms are composed of cells, which grow and reproduce to form other cells (**figure 1.8**).
- The gene theory states that long molecules of DNA carry instructions for producing cellular components. These instructions are encoded in the nucleotide sequences in the strands of DNA, like this section of DNA from **figure 1.9.** The nucleotides are organized into discrete units called genes, and the genes determine how an organism looks and functions (**figure 1.10**).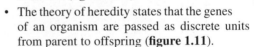
- The theory of heredity states that the genes of an organism are passed as discrete units from parent to offspring (**figure 1.11**).
- Organisms are organized into kingdoms based on similar characteristics. The organisms within a kingdom show similarities but exhibit differences from those of other kingdoms. The kingdoms are further organized into three major groups called domains based on their cellular characteristics. The three domains are Bacteria, Archaea, and Eukarya (**figure 1.12**).
- The theory of evolution states that modifications in genes that are passed from parent to offspring result in changes in future generations. These changes lead to greater diversity among organisms over time, ultimately leading to the formation of new groups of organisms (**figure 1.13**).

Self-Test

1. Biologists categorize all living things based on related characteristics into large groups, called
   - **a.** kingdoms.
   - **b.** species.
   - **c.** populations.
   - **d.** ecosystems.

2. Living things can be distinguished from nonliving things because they have
   - **a.** complexity.
   - **b.** movement.
   - **c.** cellular organization.
   - **d.** response to a stimulus.

3. Living things are organized. Choose the answer that illustrates this organization and that is arranged from smallest to largest.
   - **a.** cell, atom, molecule, tissue, organelle, organ, organ system, organism, population, species, community, ecosystem
   - **b.** atom, molecule, organelle, cell, tissue, organ, organ system, organism, population, species, community, ecosystem
   - **c.** atom, molecule, organelle, cell, tissue, organ, organ system, organism, community, population, species, ecosystem
   - **d.** atom, molecule, cell wall, cell, organ, organelle, organism, species, population, community, ecosystem

4. At each level in the hierarchy of living things, properties occur that were not present at the simpler levels. These properties are referred to as
   - **a.** novelistic properties.
   - **b.** complex properties.
   - **c.** incremental properties.
   - **d.** emergent properties.

5. The five general biological themes include
   - **a.** evolution, energy flow, competition, structure determines function, and homeostasis.
   - **b.** evolution, energy flow, cooperation, structure determines function, and homeostasis.
   - **c.** evolution, growth, competition, structure determines function, and homeostasis.
   - **d.** evolution, growth, cooperation, structure determines function, and homeostasis.

6. When trying to figure out explanations for observations, you usually construct a series of possible hypotheses. Then you make predictions of what will happen if each hypothesis is true, and
   - **a.** test each hypothesis, using appropriate controls, to determine which hypothesis is true.
   - **b.** test each hypothesis, using appropriate controls, to rule out as many as possible.
   - **c.** use logic to determine which hypothesis is most likely true.
   - **d.** reject those that seem unlikely.

7. Which of the following statements is correct regarding a hypothesis?
   - **a.** After sufficient testing, you can conclude that it is true.
   - **b.** If it explains the observations, it doesn't need to be tested.
   - **c.** After sufficient testing, you can accept it as probable, being aware that it may be revised or rejected in the future.
   - **d.** You never have any degree of certainty that it is true; there are too many variables.

8. Cell theory states that
   - **a.** all organisms have cell walls and all cell walls come from other cells.
   - **b.** all cellular organisms undergo sexual reproduction.
   - **c.** all living organisms use cells for energy, either their own or they ingest cells of other organisms.
   - **d.** all living organisms consist of cells, and all cells come from other cells.

9. The gene theory states that all the information that specifies what a cell is and what it does
   - **a.** is different for each cell type in the organism.
   - **b.** is passed down, unchanged, from parents to offspring.
   - **c.** is contained in a long molecule called DNA.
   - **d.** All of the above.

10. The theory of evolution is based on the hypothesis put forth by
    - **a.** Mendel.
    - **b.** Watson and Crick.
    - **c.** Darwin.
    - **d.** Schleiden and Schwann.

Visual Understanding

1. For over two centuries global temperatures have been warming, and over this same period of time the number of pirate ship attacks has steadily decreased. Does this graph support the conclusion that the number of pirate ship attacks has decreased because of warmer temperatures? [Hint: Read section 0.4 in chapter 0 to help answer the question.]

2. **Figure 1.6** Scientists identified the hole in the ozone layer through observations, by analyzing data collected over time. What other step in the scientific process involves the collection of data?

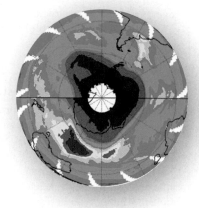

Challenge Questions

1. You are the biologist in a group of scientists who have traveled to a distant star system and landed on a planet. You see an astounding array of shapes and forms. You have three days to take samples of living things before returning to earth. How do you decide what is alive?

2. St. John's wort is an herb that has been used for hundreds of years as a remedy for mild depression. How might a modern-day scientist research its effectiveness?

Chapter **2**

The Chemistry of Life

Some Simple Chemistry

Figure 2.1   Replacing electrolytes.

During extreme exercise, athletes will often consume drinks that contain "electrolytes," chemicals such as calcium, potassium, and sodium that play an important role in muscle contraction. Electrolytes can also be depleted in other types of dehydration.

Hydrogen
Nucleus contains
1 proton

1 electron in orbit
around nucleus

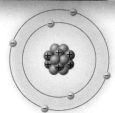

Carbon
Nucleus contains
6 protons
6 neutrons

6 electrons in orbit
around nucleus

Proton ⊕   Neutron ●   Electron ⊖
(Positive charge)   (No charge)   (Negative charge)

Figure 2.2   Basic structure of atoms.

All atoms have a nucleus consisting of protons and neutrons, except hydrogen, the smallest atom, which has only one proton and no neutrons in its nucleus. Carbon, for example, has six protons and six neutrons in its nucleus. Electrons spin around the nucleus in orbitals a far distance away from the nucleus. The electrons determine how atoms react with each other.

TABLE 2.1	Elements Common in Living Organisms		
Element	**Symbol**	**Atomic Number**	**Mass Number**
Hydrogen	H	1	1.008
Carbon	C	6	12.011
Nitrogen	N	7	14.007
Oxygen	O	8	15.999
Sodium	Na	11	22.989
Phosphorus	P	15	30.974
Sulfur	S	16	32.064
Chlorine	Cl	17	35.453
Potassium	K	19	39.098
Calcium	Ca	20	40.080
Iron	Fe	26	55.847

2.1 Atoms

CONCEPT PREVIEW: Atoms, the smallest particles into which a substance can be divided, are composed of electrons orbiting a nucleus that contains protons and neutrons. Electrons determine the chemical behavior of atoms.

Biology is the science of life, and all life, in fact even all nonlife, is made of substances. Chemistry is the study of the properties of these substances. So, while it may seem tedious or unrelated to examine chemistry in a biology text, it is essential. Organisms are chemical machines (figure 2.1), and to understand them we must learn a little chemistry.

Any substance in the universe that has mass and occupies space is defined as matter. All matter is composed of extremely small particles called **atoms.** An atom is the smallest particle into which a substance can be divided and still retain its chemical properties.

Every atom has the same basic structure you see in figure 2.2. At the center of every atom is a small, very dense nucleus formed of two types of subatomic particles, **protons** (illustrated by purple balls) and **neutrons** (the pink balls in the figure). Whizzing around the nucleus is an orbiting cloud of a third kind of subatomic particle, the **electron** (depicted by yellow balls on concentric rings). Neutrons have no electrical charge, whereas protons have a positive charge and electrons have a negative one. In each atom, there is an orbiting electron for every proton in the nucleus. The electron's negative charge balances the proton's positive charge so that the atom is electrically neutral.

An atom is typically described by the number of protons in its nucleus or by the overall mass of the atom. The terms *mass* and *weight* are often used interchangeably, but they have slightly different meanings. Mass refers to the amount of a substance, whereas weight refers to the force gravity exerts on a substance. Hence, an object has the same mass whether it is on the earth or the moon, but its weight will be greater on the earth, because the earth's gravitational force is greater than the moon's. For example, an astronaut weighing 180 pounds on earth will weigh about 30 pounds on the moon. He didn't lose any significant mass during his flight to the moon, there is just less gravitational pull on his mass.

The number of protons in the nucleus of an atom is called the **atomic number.** For example, the atomic number of carbon is 6 because it has six protons. Atoms with the same atomic number (that is, the same number of protons) have the same chemical properties and are said to belong to the same **element.** Formally speaking, an element is any substance that cannot be broken down into any other substance by ordinary chemical means.

Neutrons are similar to protons in mass, and the number of protons and neutrons in the nucleus of an atom is called the **mass number.** A carbon atom that has six protons and six neutrons has a mass number of 12. An electron's contribution to the overall mass of an atom is negligible. The atomic numbers and mass numbers of some of the most common elements in living organisms are shown in table 2.1.

Electrons Determine What Atoms Are Like

Electrons have very little mass (only about 1/1,840 the mass of a proton). Of all the mass contributing to your weight, the portion that is contributed by electrons is less than the mass of your eyelashes. And yet electrons determine the chemical behavior of atoms because they are the parts of atoms that come close enough to each other in nature to interact. Almost all the volume of an atom is empty space. Protons and neutrons lie at the core of this space, while orbiting electrons are very far from the nucleus. If the nucleus of an atom were the size of an apple, the orbit of the nearest electron would be more than a mile out!

Electrons Carry Energy

Because electrons are negatively charged, they are attracted to the positively charged nucleus, but they also repel the negative charges of each other. It takes work to keep them in orbit, just as it takes work to hold an apple in your hand when gravity is pulling the apple down toward the ground. The apple in your hand is said to possess **energy,** the ability to do work, because of its position—if you were to release it, the apple would fall. Similarly, electrons have energy of position, called *potential energy*. It takes work to oppose the attraction of the nucleus, so moving the electron farther out away from the nucleus, as shown by the set of arrows on the right side of figure 2.3, requires an input of energy and results in an electron with greater potential energy.

Moving an electron in toward the nucleus has the opposite effect (the set of arrows on the left side); energy is released, and the electron has less potential energy. Consider again an apple held in your hand. If you carry the apple up to a second-story window, it has a greater potential energy when

> Potential energy is discussed in more detail on page 88 and examples of how the potential energy from electrons is used in biological systems are described in chapters 6 and 7.

you drop it, compared to six inches from the ground. Cells use the potential energy of atoms to drive chemical reactions.

While the energy levels of an atom are often visualized as well-defined circular orbits around a central nucleus as was shown in figure 2.2, such a simple picture is not accurate. These energy levels, called *electron shells,* often consist of complex three-dimensional shapes, and the exact location of an individual electron at any given time is impossible to specify. However, some locations are more probable than others, and it is often possible to say where an electron is *most likely* to be located. The volume of space around a nucleus where an electron is most likely to be found is called the *orbital* of that electron.

Each electron shell has a specific number of orbitals, and each orbital can hold up to two electrons. The first shell in any atom contains one orbital. Helium, shown in figure 2.4*a*, has one electron shell with one orbital that corresponds to the lowest energy level. The orbital contains two electrons, shown above and below the nucleus. In atoms with more than one electron shell, the second shell contains four orbitals and holds up to eight electrons. Nitrogen, shown in figure 2.4*b*, has two electron shells; the first one is completely filled with two electrons, but three of the four orbitals in the second electron shell are not filled because nitrogen's second shell contains only five electrons (openings in orbitals are indicated with dotted circles). In atoms with more than two electron shells, subsequent shells also contain up to four orbitals with a maximum of eight electrons. Atoms with unfilled electron orbitals tend to be more reactive because they lose, gain, or share electrons in order to fill their outermost electron shell. An atom with a completely filled outermost shell is more stable. Losing, gaining, or sharing electrons is the basis for chemical reactions in which chemical bonds form between atoms.

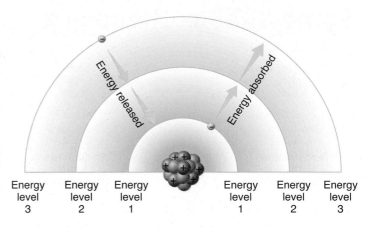

Energy level 3 Energy level 2 Energy level 1 Energy level 1 Energy level 2 Energy level 3

Figure 2.3   The electrons of atoms possess potential energy.

Electrons that circulate rapidly around the nucleus contain energy, and depending on their distance from the nucleus, they may contain more or less energy. Energy level 1 is the lowest potential energy level because it is closest to the nucleus. When an electron absorbs energy, it moves from level 1 to the next higher energy level (level 2). When an electron loses energy, it falls to a lower energy level closer to the nucleus.

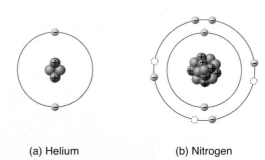

(a) Helium   (b) Nitrogen

Figure 2.4   Electrons in electron shells.

(a) An atom of helium has two protons, two neutrons, and two electrons. The electrons fill the one orbital in its one electron shell, the lowest energy level. (b) An atom of nitrogen has seven protons, seven neutrons, and seven electrons. Two electrons fill the orbital in the innermost electron shell, and five electrons occupy orbitals in the second electron shell (the second energy level). The orbitals in the second electron shell can hold up to eight electrons; therefore there are three vacancies in the outer electron shell of a nitrogen atom.

IMPLICATION   Helium gas is often used to inflate party balloons, and if you release these balloons they rise up into the air. If you fill the party balloons with nitrogen gas, they don't rise. Why this difference in party balloon behavior?

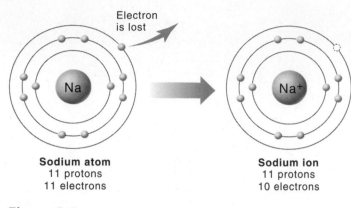

Electron
is lost

Sodium atom
11 protons
11 electrons

Sodium ion
11 protons
10 electrons

Figure 2.5   Making a sodium ion.

An electrically neutral sodium atom has 11 protons and 11 electrons. Sodium ions bear a positive charge when they ionize and lose one electron. Sodium ions have 11 protons and only 10 electrons.

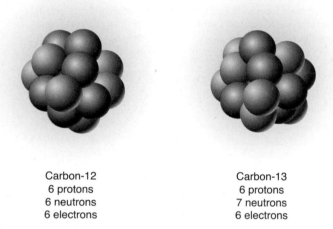

Carbon-12
6 protons
6 neutrons
6 electrons

Carbon-13
6 protons
7 neutrons
6 electrons

Figure 2.6   Isotopes of the element carbon.

The three most abundant isotopes of carbon are carbon-12, carbon-13, and carbon-14. The yellow "clouds" in the diagrams represent the orbiting electrons, whose numbers are the same for all three isotopes. Protons are shown in purple, and neutrons are shown in pink.

Figure 2.7   Using a radioactive tracer to identify cancer.

In certain medical imaging procedures, the patient is injected intravenously with a radioactive tracer that is absorbed in greater amounts by cancer cells. The tracer emits radioactivity that is detected using PET and PET/CT equipment. A cancerous area in the neck is seen as bright yellow glowing areas.

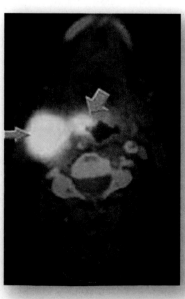

2.2 Ions and Isotopes

CONCEPT PREVIEW: Ions, which are electrically-charged, result when an atom gains or loses electrons. Isotopes of an element differ in the number of neutrons they contain, but all have the same chemical properties.

Ions

Sometimes an atom may gain or lose an electron from its outer shell (we will look at why this happens in the next section). Atoms in which the number of electrons does not equal the number of protons because they have gained or lost one or more electrons are called **ions.** All ions are electrically charged. For example, an atom of sodium (on the left in figure 2.5) becomes a positively charged ion (Na^+), called a *cation*, when it loses an electron (on the right); one proton in the nucleus is left with an unbalanced charge (11 positively charged protons and only 10 negatively charged electrons). Negatively charged ions, called *anions*, form when an atom gains one or more electrons from another atom.

Isotopes

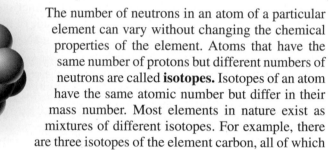

Carbon-14
6 protons
8 neutrons
6 electrons

The number of neutrons in an atom of a particular element can vary without changing the chemical properties of the element. Atoms that have the same number of protons but different numbers of neutrons are called **isotopes.** Isotopes of an atom have the same atomic number but differ in their mass number. Most elements in nature exist as mixtures of different isotopes. For example, there are three isotopes of the element carbon, all of which possess six protons (the purple balls in figure 2.6). The most common isotope of carbon (99% of all carbon) has six neutrons (the pink balls). Because its mass number is 12 (six protons plus six neutrons), it is referred to as carbon-12 (on the left). The isotope carbon-14 (on the right) is rare (1 in 1 trillion atoms of carbon) and unstable, such that its nucleus tends to break up into particles with lower atomic numbers, a process called **radioactive decay.** Radioactive isotopes are used in dating fossils, as discussed in the *Inquiry & Analysis* at the end of the chapter, and in medicine.

Medical Uses of Radioactive Isotopes

Radioactive isotopes are used in many medical procedures. Short-lived isotopes, those that decay fairly rapidly and produce harmless products, are commonly used as tracers in the body. A *tracer* is a radioactive substance that is taken up and used by the body. Emissions from the radioactive isotope tracer are detected using special laboratory equipment, and can reveal key diagnostic information about the functioning of the body. For example, PET and PET/CT (positron emission tomography/computerized tomography) imaging procedures can be used to identify a cancerous area in the body. First, a radioactive tracer is injected into the body. This tracer is taken up by all cells, but it is taken up in larger amounts in cells with higher metabolic activities, such as cancer cells. Images are then taken of the body, and areas emitting greater amounts of the tracer can be seen. For example, in the image in figure 2.7, the radioactive-emitting cancer site appears as yellow glowing areas. There are many other uses of radioactive isotopes in medicine both in detection and treatment of disorders.

2.3 Molecules

CONCEPT PREVIEW: Molecules are atoms linked together by chemical bonds. Ionic bonds, covalent bonds, and hydrogen bonds are the three principal types of bonds; van der Waals forces are weaker interactions.

A **molecule** is a group of atoms held together by energy. The energy acts as "glue," ensuring that the various atoms stick to one another. The energy or force holding two atoms together is called a **chemical bond.** There are three principal kinds of chemical bonds: ionic bonds, where the force is generated by the attraction of oppositely charged ions; covalent bonds, where the force results from the sharing of electrons; and hydrogen bonds, where the force is generated by the attraction of opposite partial electrical charges. Another type of chemical attraction called van der Waals forces will be discussed later, but keep in mind that this type of interaction is not considered a chemical bond.

Ionic Bonds

Chemical bonds called **ionic bonds** form when atoms are attracted to each other by opposite electrical charges. Just as the positive pole of a magnet is attracted to the negative pole of another, so an atom can form a strong link with another atom if they have opposite electrical charges. Because an atom with an electrical charge is an ion, these bonds are called ionic bonds.

Everyday table salt is built of ionic bonds. The sodium and chlorine atoms that make up table salt are ions. The sodium you see in the yellow panels of figure 2.8a gives up the sole electron in its outermost shell (leaving a filled outer shell) and chlorine, in the light green panels, gains an electron to complete its outermost shell. Recall from section 2.1 that an atom is more stable when its outermost electron shell is filled. As a result of this electron hopping, sodium atoms in table salt are positive sodium ions and chlorine atoms are negative chloride ions. Because each ion is electrically attracted to all surrounding ions of opposite charge, they form an elaborate matrix of ionic bonds between alternating sodium and chloride ions—a crystal (figure 2.8b). That is why table salt is composed of tiny crystals and is not a powder.

The two key properties of ionic bonds that make them form crystals are that they are strong (although not as strong as covalent bonds) and that they are *not* directional. An ion is attracted to the electrical field contributed by all nearby ions of opposite charge. Ionic bonds do not play an important part in most biological molecules because of this lack of directionality. They require the more specific associations made possible by directional bonds.

Covalent Bonds

Strong chemical bonds called **covalent bonds** form between two atoms when they share electrons (figure 2.9). Most of the atoms in your body are linked to other atoms by covalent bonds. Why do atoms in molecules share electrons? Remember, all atoms seek to fill up their outermost shell of orbiting electrons, which in all atoms (except tiny hydrogen and helium) takes eight electrons. For example, an atom with six outer shell electrons seeks to share them with an atom that has two outer shell electrons or with two atoms that have single outer shell electrons. The carbon atom has four electrons in its outermost shell, and so carbon can form as many as four covalent bonds in its attempt to fully populate its outermost shell of electrons. Because there are many ways four covalent bonds can form, carbon atoms participate in many different kinds of molecules. The strength of covalent bonds increases as more electrons are

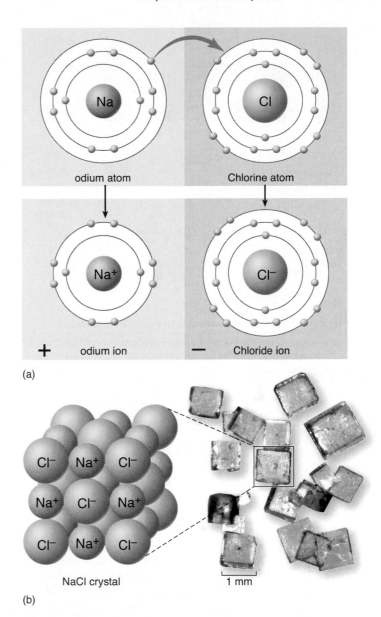

odium atom   Chlorine atom

Na⁺   Cl⁻

+   odium ion   **—**   Chloride ion

(a)

Cl⁻ Na⁺ Cl⁻

Na⁺ Cl⁻ Na⁺

Cl⁻ Na⁺ Cl⁻

NaCl crystal   1 mm

(b)

Figure 2.8   The formation of ionic bonds in table salt.

(a) When a sodium atom donates an electron to a chlorine atom, the sodium atom, lacking that electron, becomes a positively charged sodium ion. The chlorine atom, having gained an extra electron, becomes a negatively charged chloride ion. (b) Sodium chloride forms a highly regular lattice of alternating sodium ions and chloride ions. You are familiar with these crystals as everyday table salt.

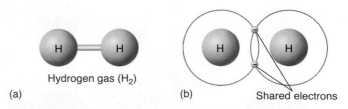

H — H

Hydrogen gas (H_2)

(a)   (b)   Shared electrons

Figure 2.9   Covalent bonds.

Covalent bonds involve the sharing of electrons, indicated by the blur bar in (a). Hydrogen gas consists of two atoms of hydrogen, each having one electron. The covalent bond forms when the two atoms share the two electrons, as shown in (b).

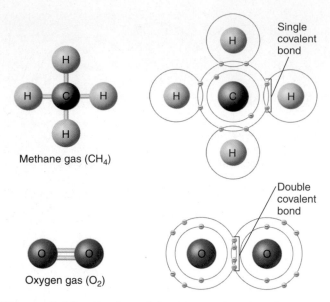

Figure 2.10   Single and double covalent bonds.

Single covalent bonds involve the sharing of one pair of electrons, as in methane (CH_4). In double covalent bonds, two pairs of electrons are shared, as in oxygen gas.

Figure 2.11   Water molecules.

(a) Each water molecule is composed of one oxygen atom (red) and two hydrogen atoms (blue). Because electrons are more attracted to oxygen atoms than hydrogen atoms, the shared electrons spend more time near the oxygen atom, making the water molecule polar. (b) Each oxygen has a partial negative charge (δ^-), and each hydrogen has a partial positive charge (δ^+). Hydrogen bonds (dashed lines) form between the positive end of one water molecule and the negative end of another water molecule.

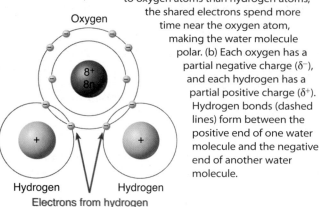

(a) Electron shells in a water molecule

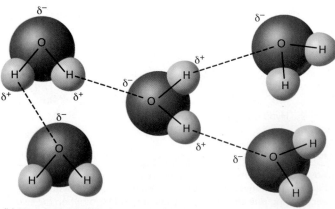

(b) Hydrogen bonding in water molecules

shared. If only one pair of electrons is shared between two atoms (as in the carbon-hydrogen bonds in the methane molecule in figure 2.10), the covalent bond is called a *single covalent bond*. If two pairs of electrons are shared (as in oxygen gas in figure 2.10), it is a stronger bond, called a *double covalent bond;* and if three pairs of electrons are shared it is a very strong *triple covalent bond.*

The atoms in nitrogen gas (N_2) are held together by a very strong triple covalent bond. As described on page 394, only a few kinds of bacteria are able to break this bond and make the nitrogen available to other organisms.

The two key properties of covalent bonds that make them ideal for their molecule-building role in living systems are that (1) they are strong, involving the sharing of lots of energy; and (2) they are very directional—allowing bonds to form between two specific atoms, rather than creating a generalized attraction of one atom for its neighbors. For example, compare the methane molecule in figure 2.10 with the NaCl crystal in figure 2.8.

Polar and Nonpolar Covalent Bonds. When a covalent bond forms between two atoms, one nucleus may be much better at attracting the shared electrons than the other. In water, for example, the shared electrons are much more strongly attracted to the oxygen atom than to the hydrogen atoms. When this happens, shared electrons spend more time in the vicinity of the oxygen atom, making it somewhat negative in charge; they spend less time in the vicinity of the hydrogens, and these become somewhat positive in charge (figure 2.11). The charges are not full electrical charges, as in ions, but rather tiny *partial charges*. What you end up with is a sort of molecular magnet, with positive and negative ends, or "poles." Molecules like this are said to be **polar molecules.**

Hydrogen Bonds

Polar molecules like water are attracted to one another through a special type of weak chemical bond called a **hydrogen bond** (figure 2.11*b*). Hydrogen bonds occur when the positive end of one polar molecule is attracted to the negative end of another, like two magnets drawn to each other. Two key properties of hydrogen bonds cause them to play an important role in the molecules found in organisms. First, hydrogen bonds are highly directional. Second, they are weak and so are not effective over long distances like more powerful covalent and ionic bonds. Too weak to actually form stable molecules, they act more like Velcro, forming a tight bond by the additive effects of many weak interactions.

van der Waals Forces

A kind of weak chemical attraction (not a bond) is a nondirectional attractive force called **van der Waals forces.** These chemical forces come into play only when two atoms are very close to one another. The attraction is very weak, and disappears if the atoms move even a little apart. It becomes significant when numerous atoms in one molecule simultaneously come close to numerous atoms of another molecule—that is, when their shapes match precisely.

Concept Check

1. The three common isotopes of carbon all have the same chemical behavior, meaning that they form the same types of chemical bonds. Explain why they are able to do this?
2. Of the attractive forces discussed here, which is not a chemical bond?
3. Both ionic and covalent bonds are strong. Why are molecules in living organisms usually made of covalent but not ionic bonds?

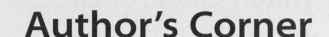

How Tropical Lizards Climb Vertical Walls

Science is most fun when it tickles your imagination. This is particularly true when you see something you know just can't be true. A few years ago, my wife, Barbara, and I were on a tropical vacation, and I was lying on the bed when a little lizard walked up the wall beside me and across the ceiling, stopping right over my head, and looked down at me.

This was no special effect, no trick with mirrors. I was seeing it with my own eyes, real as day, right there above me. The lizard, a green gecko about the size of a toothbrush, stood upside down on the ceiling and seemed to laugh at me for several minutes before trotting over to the far wall and down.

How did my gecko friend perform this gripping feat? Investigators have puzzled over the adhesive properties of geckos for decades. What force prevented gravity from dropping the gecko on my nose?

The most reasonable hypothesis seemed suction—salamanders' feet form suction cups that let them climb walls, so maybe geckos' do too. The way to test this is to see if the feet adhere in a vacuum, with no air to create suction. Salamander feet don't adhere, but gecko feet do. It's not suction.

How about friction? Cockroaches climb using tiny hooks that grapple onto irregularities in the surface, much as rock climbers use crampons. Geckos, however, happily run up walls of smooth polished glass that no cockroach can climb. It's not friction.

Electrostatic attraction? Clothes in a dryer stick together because of electrical charges created by their rubbing together. You can stop this by adding a "static remover" that is itself heavily ionized. But a gecko's feet still adhere in ionized air. It's not electrostatic attraction.

Could it be glue? Many insects use adhesive secretions from glands in their feet to aid climbing. But there are no gland cells in gecko feet, no secreted chemical. It's not glue.

There was one tantalizing clue, however, the kind that experimenters love. Gecko feet seem to get stickier on surfaces with highly ordered molecules. This suggests that geckos are tapping directly into the molecular structure of the surfaces they walk on!

Tracking down this clue, Robert Full of the University of California, Berkeley, and his research team took a closer look at gecko feet. Geckos have rows of tiny hairs on the bottoms of their feet, like the bristles of a toothbrush. There are about half a million of these hairs on each foot, pointed toward the heel.

When you look at these hairs under a microscope, the end of each hair is divided into between 400 and 1,000 fine projections, the projections sticking out from the tip like tiny stiff brushes.

When a gecko takes a step, it drives the sole of its foot onto the surface and pushes it backward. This shoves the forest of tips directly against the surface. The atoms of each gecko tip become closely engaged with the atoms of the surface, and *that* is the force that defies gravity. When two atoms approach each other very closely—closer than the diameter of an atom—a subtle nuclear attraction called "van der Waals forces" comes into play. These forces are individually very weak, but when lots of atoms add their little bits, the sum can add up to quite a lot.

Full and his team used microelectrical mechanical sensors, originally designed to be used with atomic force microscopes that map a surface by "feeling" it with a mechanical probe. Using the sensors, they were able to measure the force exerted by a single hair removed from a gecko's foot. It was 200 micronewtons, a tiny force but stupendous for a single hair. Enough to hold up an ant. A million hairs could support a small child. My little gecko, ceiling-walking with 2 million of them, could have carried an 80-pound backpack—talk about being overengineered!

If they stick that well, how do geckos ever become unstuck? For a gecko's foot to stick, each hair projection must butt up squarely against the surface, so the hair's individual atoms can come into play. Tipped past a critical angle—30 degrees—the attractive forces between hair and surface atoms weaken to nothing. The trick is to tip the foot hairs until the projections let go. Geckos release their feet by curling up each toe and peeling it off, sort of like undoing Velcro.

Now I can laugh with my little gecko friend, should I see it again, for I know its secret.

Water: Cradle of Life

TABLE 2.2	The Properties of Water
Property	**Explanation**
Heat storage	Hydrogen bonds require considerable heat before they break, minimizing temperature changes.
Ice formation	Water molecules in an ice crystal are spaced relatively far apart because of hydrogen bonding.
High heat of vaporization	Many hydrogen bonds must be broken for water to evaporate, which requires considerable heat.
Cohesion	Hydrogen bonds hold molecules of water together.
High polarity	Water molecules are attracted to ions and other polar compounds.

Iceberg in ocean

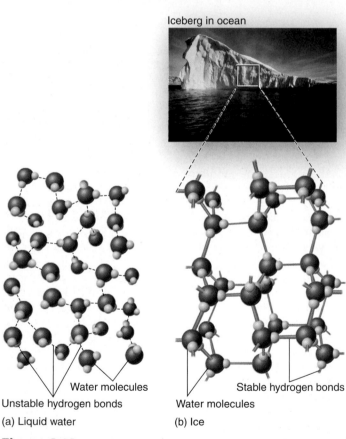

(a) Liquid water   Water molecules   Unstable hydrogen bonds

(b) Ice   Water molecules   Stable hydrogen bonds

Figure 2.12   Ice formation.

When water (a) cools below 0°C, it forms a regular crystal structure (b) that floats. The individual water molecules are spaced apart and held in position by hydrogen bonds.

IMPLICATION: Will the melting of the polar ice cap (ice floating in the ocean) due to global warming raise sea level? [Hint: If you leave a glass of ice water on the table, does the level of water in the glass change as the ice melts?]

2.4 Hydrogen Bonds Give Water Unique Properties

CONCEPT PREVIEW: Water molecules form a network of hydrogen bonds in liquid and dissolve other polar molecules. Many of the key properties of water arise because it takes considerable energy to break liquid water's many hydrogen bonds.

Three-fourths of the earth's surface is covered by liquid water. About two-thirds of your body is water, and you cannot exist long without it. All other organisms also require water. It is no accident that tropical rain forests are bursting with life, whereas dry deserts seem almost lifeless except after it rains. The chemistry of life, then, is water chemistry.

Water has a simple atomic structure, an oxygen atom linked to two hydrogen atoms by single covalent bonds (see figure 2.11). The chemical formula for water is thus H_2O. It is because the oxygen atom attracts the shared electrons more strongly than the hydrogen atoms that water is a *polar molecule* and so can form *hydrogen bonds*. Water's ability to form hydrogen bonds is responsible for much of the organization of living chemistry, from the structure of membranes that encase every cell to how large molecules, called proteins, fold.

The weak hydrogen bonds that form between a hydrogen atom of one water molecule and the oxygen atom of another produce a lattice of hydrogen bonds within liquid water. Each of these bonds is individually very weak and short-lived—a single bond lasts only 1/100,000,000,000 of a second. However, like the grains of sand on a beach, the cumulative effect of large numbers of these bonds is enormous and is responsible for many of the important physical properties of water (table 2.2).

Heat Storage

The temperature of any substance is a measure of how rapidly its individual molecules are moving. Because of the many hydrogen bonds that water molecules form with one another, a large input of thermal energy is required to disrupt the organization of liquid water and raise its temperature. Because of this, water heats up more slowly than almost any other compound and holds its temperature longer. That is a major reason why your body, which is mostly water, is able to maintain a relatively constant internal temperature.

Ice Formation

If the temperature is low enough, very few hydrogen bonds between water molecules will break. Instead, the lattice of these bonds assumes a crystal-like structure, forming a solid we call ice. Interestingly, ice is less dense than water—that is why icebergs and ice cubes float. Why is ice less dense? This is best understood by comparing the molecular structures of water and ice that you see in figure 2.12. At temperatures above freezing (0°C or 32°F), water molecules in figure 2.12*a* move around each other with hydrogen bonds breaking and forming. As temperatures drop, the movement of water molecules decreases, allowing hydrogen bonds to stabilize, holding individual molecules farther apart, as in figure 2.12*b*, making the ice structure less dense.

High Heat of Vaporization

If the temperature is high enough, many hydrogen bonds between water molecules will break, with the result that the liquid is changed into vapor (a gas). A considerable amount of heat energy is required to do this—every gram of water that evaporates from your skin removes 2,452 joules of heat from your body, which is equal to the energy released by lowering the temperature of 586 grams of water 1°C (which is quite a lot of heat). That is why sweating cools you off; as the sweat evaporates (vaporizes) it takes energy with it, in the form of heat, cooling the body.

Cohesion

Because water molecules are very polar, they are attracted to other polar molecules—hydrogen bonds bind polar molecules to each other. When the other polar molecule is another water molecule, the attraction is called **cohesion.** The surface tension of water is created by cohesion. Surface tension is the force that causes water to bead, like on the spider web in figure 2.13, or supports the weight of the water strider. When the other polar molecule is a different substance, the attraction is called **adhesion.** Capillary action—such as water moving up into a paper towel—is created by adhesion. Water clings to any substance, such as paper fibers, with which it can form hydrogen bonds. Adhesion is why things get "wet" when they are dipped in water and why waxy substances do not—they are composed of nonpolar molecules that don't form hydrogen bonds with water molecules.

> Cohesion and adhesion are properties of water that are necessary for the movement of water in plants, from the roots to the leaves as described on page 627.

High Polarity

Water molecules in solution always tend to form the maximum number of hydrogen bonds possible. Polar molecules form hydrogen bonds and are attracted to water molecules. Polar molecules are called **hydrophilic** (from the Greek *hydros,* water, and *philic,* loving). Water molecules gather closely around any molecule that exhibits an electrical charge, whether a full charge (ion) or partial charge (polar molecule). When a salt crystal dissolves in water as you see happening in figure 2.14, what really happens is that individual ions break off from the crystal and become surrounded by water molecules. The partial positive charge of blue hydrogen atoms of water are attracted to the negative charge of the chloride ions and the partial negative charge of the red oxygen atoms are attracted to the positive charge of the sodium ions. Water molecules orient around each ion like a swarm of bees attracted to honey, and this shell of water molecules, called a *hydration shell,* prevents the ions from reassociating with the crystal. Similar shells of water form around all polar molecules, and polar molecules that dissolve in water in this way are said to be **soluble** in water.

Nonpolar molecules like oil do not form hydrogen bonds and are not water-soluble. When nonpolar molecules are placed in water, the water molecules shy away, instead forming hydrogen bonds with other water molecules. The nonpolar molecules are forced into association with one another, crowded together to minimize their disruption of the hydrogen bonding of water. It seems almost as if the nonpolar compounds shrink from contact with water, and for this reason they are called **hydrophobic** (from the Greek *hydros,* water, and *phobos,* fearing). Many biological structures are shaped by such hydrophobic forces, as will be discussed in chapter 3.

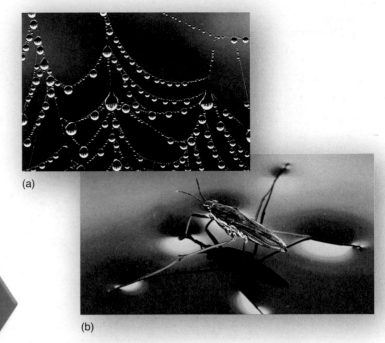

(a)

(b)

Figure 2.13   Cohesion.

(a) Cohesion allows water molecules to stick together and form droplets. (b) Surface tension is a property derived from cohesion—that is, water has a "strong" surface due to the force of its hydrogen bonds. Some insects, such as this water strider, literally walk on water.

IMPLICATION   If you were to devise very large footpads made of lightweight film, would you be able to walk on water like this water strider? Why would you have to wear footpads and not just bare feet?

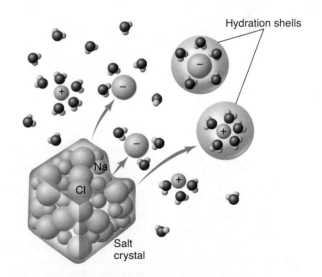

Hydration shells

Na

Cl

Salt crystal

Figure 2.14   How salt dissolves in water.

Salt is soluble in water because the partial charges on water molecules are attracted to the charged sodium and chloride ions. The water molecules surround the ions, forming what are called hydration shells. When all of the ions have been separated from the crystal, the salt is said to be dissolved.

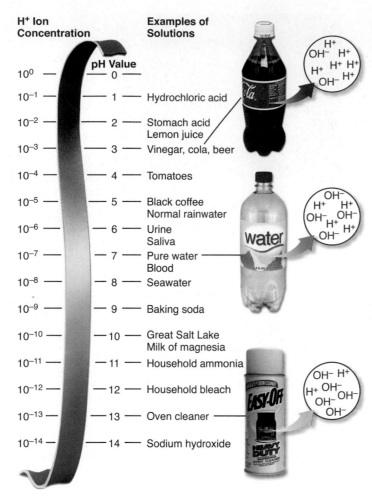

H⁺ Ion Concentration | Examples of Solutions

H^+ Ion Concentration	pH Value	
10^0	0	
10^{-1}	1	Hydrochloric acid
10^{-2}	2	Stomach acid / Lemon juice
10^{-3}	3	Vinegar, cola, beer
10^{-4}	4	Tomatoes
10^{-5}	5	Black coffee / Normal rainwater
10^{-6}	6	Urine / Saliva
10^{-7}	7	Pure water / Blood
10^{-8}	8	Seawater
10^{-9}	9	Baking soda
10^{-10}	10	Great Salt Lake / Milk of magnesia
10^{-11}	11	Household ammonia
10^{-12}	12	Household bleach
10^{-13}	13	Oven cleaner
10^{-14}	14	Sodium hydroxide

Figure 2.15 **The pH scale.**

A fluid is assigned a value according to the number of hydrogen ions present in a liter of that fluid. The scale is logarithmic, so that a change of only 1 means a 10-fold change in the concentration of hydrogen ions; thus lemon juice with a pH of 2 is 100 times more acidic than tomatoes with a pH of 4, and seawater is 10 times more basic than pure water.

BIOLOGY & YOU

Heartburn. Your stomach contains large amounts of hydrochloric acid, used to digest food. Sometimes this acid backs up from the stomach into the esophagus (the food pipe) stretching up from the stomach to the throat. This escape of acid from the stomach, called acid reflux, causes a painful burning sensation as the acid attacks the inner lining of the esophagus. Because the esophagus lies just behind the heart, the burning sensation is informally referred to as "heartburn." Nearly one-third of the adult population of the U.S. experiences acid reflux to some degree at least once a month. For minor heartburn you might take an antacid, such as Tums, which is a base that counteracts stomach acidity.

2.5 Water Ionizes

CONCEPT PREVIEW: A tiny fraction of water molecules spontaneously ionize at any moment, forming H^+ and OH^-. The pH of a solution is a measure of its H^+ concentration. Low pH values indicate high H^+ concentrations (acids), and high pH values indicate low H^+ concentrations (bases).

The covalent bonds within a water molecule sometimes break spontaneously. When it happens, a proton (hydrogen atom nuclei) dissociates from the molecule as a positively charged ion, *hydrogen ion* (H^+). The rest of the dissociated water molecule, which has retained the shared electron from the covalent bond, is a negatively charged *hydroxide ion* (OH^-).

$$H_2O \longleftrightarrow OH^- + H^+$$

water / hydroxide ion / hydrogen ion

pH

A convenient way to express the hydrogen ion concentration of a solution is to use the **pH scale** (figure 2.15). This scale ranges from 0 (highest hydrogen ion concentration) to 14 (lowest hydrogen ion concentration). Pure water has a pH of 7. Each pH unit represents a 10-fold change in hydrogen ion concentration. This means that a solution with a pH of 4 has *10 times* the H^+ concentration of one with a pH of 5, and *100 times* the H^+ concentration of one with a pH of 6.

Acids. Any substance that dissociates in water to increase the concentration of H^+ is called an **acid.** Acidic solutions have pH values below 7. The stronger an acid, the more H^+ and so the lower its pH. For example, hydrochloric acid (HCl), which is abundant in your stomach, ionizes completely in water, giving the solution a pH of 1.

Bases. A substance that combines with H^+ when dissolved in water is called a **base.** By combining with H^+, a base lowers the H^+ concentration in the solution. Basic (or alkaline) solutions, therefore, have pH values above 7. Very strong bases, such as sodium hydroxide (NaOH), have pH values of 12 or more.

Buffers

The pH inside almost all living cells, and in the fluid surrounding cells in multicellular organisms, is fairly close to 7. The many proteins that govern metabolism are all extremely sensitive to pH, and slight alterations in pH can cause the molecules to take on different shapes that disrupt their activities. For this reason, it is important that a cell maintain a constant pH level. The pH of your blood, for example, is 7.4, and you would survive only a few minutes if it were to fall to 7.0 or rise to 7.8.

What keeps an organism's pH constant? Cells contain chemical substances called **buffers** that minimize changes in concentrations of H^+ and OH^- by taking up or releasing hydrogen ions into solution as the hydrogen ion concentration of the solution changes. As the "Today's Biology" feature on the facing page explains, when acid in rain or snow exceeds the buffering capacity of a tree or other organism, death may result.

Concept Check

1. What is the difference between cohesion and adhesion?
2. What property of water makes it advantageous for an athlete to sweat?
3. When you drink a coke you are consuming an acid. Why doesn't your body's pH go down as a result?

Today's Biology

Acid Rain

As you study biology, you will learn that hydrogen ions play many roles in the chemistry of life. When conditions become overly acidic—too many hydrogen ions—serious damage to organisms often results. One important example of this is acid precipitation, more informally called **acid rain.** Acid precipitation is just what it sounds like, the presence of acid in rain or snow. Where does the acid come from? Tall smokestacks from coal-burning power plants send smoke high into the atmosphere through these stacks, each of which is over 65 meters tall. The smoke the stacks belch out contains high concentrations of sulfur dioxide (SO_2), because the coal that the plants burn is rich in sulfur. The sulfur-rich smoke is dispersed and diluted by winds and air currents. Since the 1950s, such tall stacks have become popular in the United States and Europe—there are now over 800 of them in the United States alone.

In the 1970s, 20 years after the stacks were introduced, ecologists began to report evidence that the tall stacks were not eliminating the problems associated with the sulfur, just exporting the ill effects elsewhere. The lakes and forests of the Northeast suffered drastic drops in biodiversity, forests dying and lakes becoming devoid of life. It turned out that the SO_2 introduced into the upper atmosphere by high smokestacks combines with water vapor to produce sulfuric acid (H_2SO_4). When this water later falls back to earth as rain or snow, it carries the sulfuric acid with it. When schoolchildren measured the pH of natural rainwater as part of a nationwide project in 1989, rain and snow in the Northeast often had a pH as low as 2 or 3—more acidic than vinegar.

After accumulating in soils for over 50 years, the effects of acid rain are now only too evident. The impact of acid rain on forests first became apparent in the Northeast. Some 15% of the lakes in New England have become chronically acidic and are dying biologically as their pH levels fall to below 5.0. Many of the forests of the northeastern United States and Canada have also been seriously damaged. The trees in this photo show the ill effects of acid precipitation. In the last decades, acid added to forest soils has caused the loss from these soils of over half the essential plant nutrients, calcium and magnesium. Researchers blame excess acids for dissolving Ca^{++} and Mg^{++} ions into drainage waters much faster than weathering rocks can replenish them. Without them, trees stop growing and die.

Now, some 30 years later, acid rain effects are becoming apparent in the Southeast as well. Researchers suggest the reason for the delay is that southern soils are generally thicker than northern ones and thus able to sponge up far more acid. But now that southern forest soils are becoming saturated, they too are beginning to die. In a third of the southeastern streams studied, fish are declining or already gone.

The solution is straightforward: capture and remove the emissions instead of releasing them into the atmosphere. Progressively tougher pollution laws over the past three decades have reduced U.S. emissions of sulfur dioxide by about 40% from its 1973 peak of 28.8

metric tons a year. Despite this significant progress, much remains to be done. Unless levels are cut further, researchers predict forests may not recover for centuries.

An informed public will be essential. While textbook treatments have in the past tended to minimize the impact of this issue on students ("the vast majority of North American forests are not suffering substantially from acid precipitation"), it is important that we face the issue squarely and support continued efforts to address this serious problem.

Using Radioactive Decay to Date the Iceman

In the fall of 1991, sticking out of the melting snow on the crest of a high pass near the mountainous border between Italy and Austria, two Austrian hikers found a corpse. Right away it was clear the body was very old, frozen in an icy trench where he had sought shelter long ago and only now released as the ice melted. In the years since this startling find, scientists have learned a great deal about the dead man, who they named Ötzi. They know his age, his health, the clothing he wore, what he ate, and that he died from an arrow that ripped through his back. Its tip is still embedded in the back of his left shoulder. From the distribution of chemicals in his teeth and bones, we know he lived his life within 60 kilometers of where he died.

When did this Iceman die? Scientists answered this key question by measuring the degree of decay of the short-lived carbon isotope ^{14}C in Ötzi's body. While most carbon atoms are the stable isotope ^{12}C, a tiny proportion are the unstable radioactive isotope ^{14}C, created by the bombardment of nitrogen-14 (^{14}N) atoms with cosmic rays. This proportion of ^{14}C is captured by plants in photosynthesis and is present in the carbon molecules of the animal's body that eats the plant. After the plant or animal dies, it no longer accumulates any more carbon, and the ^{14}C present at the time of death decays over time back to ^{14}N. Over time the ratio of ^{14}C to ^{12}C decreases. It takes 5,730 years for half of the ^{14}C present to decay, a length of time called the **half-life** of the ^{14}C isotope. Because the half-life is a constant that never changes, the extent of radioactive decay allows you to date a sample. Thus a sample that had one quarter of its original proportion of ^{14}C remaining would be approximately 11,460 years old (two half-lives).

The graph to the right displays the radioactive decay curve of the carbon isotope ^{14}C. Scientists know it takes 5,730 years for half of the ^{14}C present in a sample to decay to nitrogen-14 (^{14}N). When Ötzi's carbon isotopes were analyzed, researchers determined that the ratio of ^{14}C to ^{12}C (a **ratio** is the size of one variable relative to another), also written as the fraction $^{14}C/^{12}C$, in Ötzi's body was 0.435 of the fraction found in tissues of a person who has recently died.

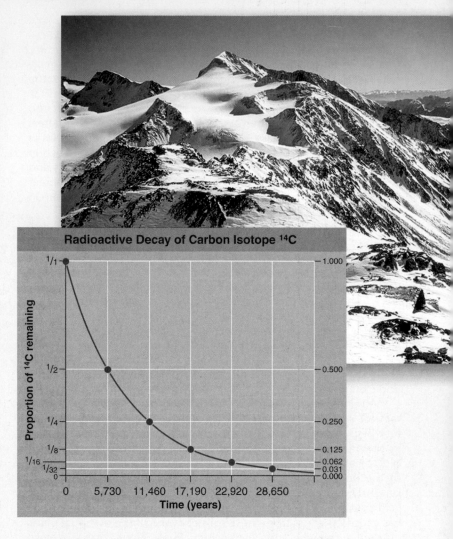

Radioactive Decay of Carbon Isotope ^{14}C

Proportion of ^{14}C remaining — Time (years)

Analysis

1. **Applying Concepts**   What proportion (a **proportion** is the size of a variable relative to the whole) of the ^{14}C present in Ötzi's body when he died is still there today? When he died, it would have been 1.0.

2. **Interpreting Data**   Plot this proportion on the ^{14}C radioactive decay curve above. How many half-lives does this point represent?

3. **Making Inferences**   If Ötzi were indeed a recent corpse, made to look old by the harsh weather conditions found on the high mountain pass, what would you expect the ratio of ^{14}C to ^{12}C to be, relative to that in your own body?

4. **Drawing Conclusions**   When did Ötzi the Iceman die?

Concept Summary

Some Simple Chemistry

2.1 Atoms

- An atom is the smallest particle that retains the chemical properties of its substance. Atoms contain a core nucleus of protons and neutrons; electrons spin around the nucleus (**figure 2.2**). The number of electrons equals the number of protons in an atom.

- The number of protons in an atom is called its atomic number. The mass that is contributed by the protons and neutrons is called the atom's mass number. All atoms that have the same atomic number are said to be the same element.

- Protons are positively charged particles and neutron particles carry no charge. Electrons are negatively charged particles that orbit around the nucleus at different energy levels. Electrons determine the chemical behavior of an atom because they are the subatomic particles that interact with other atoms.

- It takes energy to hold the electrons in their orbits; this energy of position is called potential energy. The amount of potential energy of an electron is based on its distance from the nucleus (**figure 2.3**).

- Most electron shells hold up to eight electrons and atoms will undergo chemical reactions in order to fill the outermost electron shell, either by gaining, losing, or sharing electrons (**figure 2.4**).

2.2 Ions and Isotopes

- Ions are atoms that have either gained one or more electrons (negative ions called anions) or lost one or more electrons (positive ions called cations) (**figure 2.5**).

- Isotopes are atoms that have the same number of protons but differing numbers of neutrons (**figure 2.6**). Isotopes tend to be unstable and break up into other elements through a process called radioactive decay. Some isotopes have applications in medicine (**figure 2.7**).

2.3 Molecules

- Molecules form when atoms are held together with energy. The force holding atoms together is called a chemical bond. There are three main types of chemical bonds.

- Ionic bonds form when ions of opposite charge are attracted to each other. Table salt is formed by ionic bonds between positive sodium ions and negative chloride ions (**figure 2.8**).

- Covalent bonds form when two atoms share electrons, attempting to fill empty electron orbitals (**figure 2.9**). Covalent bonds are stronger when more electrons are shared. The sharing of two pairs of electrons in a double covalent bond, shown here from **figure 2.10,** is stronger then a single covalent bond, and a triple bond is even stronger.

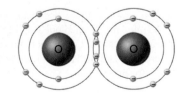

- Hydrogen bonds form between polar molecules. The atoms in a polar molecule are held together by covalent bonds in which the shared electrons are unevenly distributed around their nuclei, giving the molecule a slightly positive end and a slightly negative end. Hydrogen bonds form when the positive end of one molecule is attracted to the negative end of another (**figure 2.11**).

- Weak chemical attractions, called van der Waals forces, can hold two atoms together temporarily when they come into contact. However, van der Waals forces are not considered chemical bonds.

Water: Cradle of Life

2.4 Hydrogen Bonds Give Water Unique Properties

- Water molecules are polar molecules that form hydrogen bonds with each other and with other polar molecules. Many of the physical properties of water are attributed to hydrogen bonding.

- Water molecules held together through hydrogen bonding are more difficult to separate, and as a result a significant amount of heat energy is needed to pull the molecules apart. For this reason, water heats up slowly and holds it temperature longer.

- The hydrogen bonds that hold water molecules together become more stable at lower temperatures and as a result they lock water molecules into place in solid crystal structures called ice, such as that shown here from **figure 2.12.** Individual water molecules maintain a set distance from neighboring molecules due to the stable formation of the hydrogen bonds. In liquid water, hydrogen bonds break and reform, allowing individual water molecules more movement, bringing them into closer proximity to each other. This results in a substance that is more dense than ice.

- In order for water to vaporize into a gas, a significant input of heat energy is needed to break the hydrogen bonds. This high heat of vaporization is a property of water used by our bodies in regulating body temperature.

- Because water molecules are polar molecules, they will form hydrogen bonds with other polar molecules. If the other polar molecules are water molecules, the process is called cohesion (**figure 2.13**). If the other polar molecules are some other substance, the process is called adhesion.

- When water molecules form hydrogen bonds with other polar molecules, water molecules will tend to surround other polar molecules, forming a barrier around them called a hydration shell. Polar molecules are said to be hydrophilic and are water-soluble (**figure 2.14**). Nonpolar molecules do not form hydrogen bonds and will cluster together when placed in water. They are said to be hydrophobic and are water-insoluble.

2.5 Water Ionizes

- Water molecules dissociate forming negatively charged hydroxide ions (OH^-) and positively charged hydrogen ions (H^+). This property of water is significant because the concentration of H^+ in a solution determines its pH and effects its chemical properties.

- A solution with a higher H^+ concentration is an acid and has a pH below 7. A solution with a lower H^+ concentration is a base and has a pH above 7 (**figure 2.15**). A buffer is a chemical substance that minimizes changes in pH by taking up excess H^+ in acidic solution, or releasing H^+ in basic solutions.

Self-Test

1. The smallest particle into which a substance can be divided and still retain all of its chemical properties is
   a. matter.
   c. a molecule.
   b. an atom.
   d. mass.

2. An atom that has gained or lost one or more electrons is
   a. an isotope.
   c. an ion.
   b. a neutron.
   d. radioactive.

3. Atoms are held together by a force called a bond. The three types of bonds are
   a. positive, negative, and neutral.
   b. hydrophobic, hydrophilic, and van der Waals interactions.
   c. magnetic, electric, and radioactive.
   d. ionic, covalent, and hydrogen.

4. Carbon has four electrons in its outer electron shell, therefore
   a. it has a completely filled outer electron shell.
   b. it can form four single covalent bonds.
   c. it does not react with any other atom.
   d. it has a positive charge.

5. The partial separation of charge in the water molecule
   a. results from the electrons' greater attraction to the oxygen atom.
   b. means the molecule has a positive end and a negative end.
   c. indicates that the water molecule is a polar molecule.
   d. All of these are correct.

6. Water has some very unusual properties. These properties occur because of the
   a. hydrogen bonds between the individual water molecules.
   b. covalent bonds between the individual water molecules.
   c. hydrogen bonds within each individual water molecule.
   d. ionic bonds between the individual water molecules.

7. Which of the following properties are somehow related to the need for significant heat energy to break hydrogen bonds?
   a. cohesion and adhesion
   b. hydrophobic and hydrophilic
   c. heat storage and heat of vaporization
   d. ice formation and high polarity

8. The attraction of water molecules to other water molecules is called
   a. cohesion.
   c. solubility.
   b. capillary action.
   d. adhesion.

9. Water sometimes ionizes, a single molecule breaking apart into a hydrogen ion and a hydroxide ion. Other materials may dissociate in water, resulting in either (1) an increase of hydrogen ions or (2) decrease of hydrogen ions in the solution. We call the results
   a. (1) acids and (2) bases.
   b. (1) bases and (2) acids.
   c. (1) neutral solutions and (2) neutronic solutions.
   d. (1) hydrogen solutions and (2) hydroxide solutions.

10. Which of the following is *not* true about buffers?
    a. A buffer takes up H^+ from the solution.
    b. A buffer keeps the pH relatively constant.
    c. A buffer stops water from ionizing.
    d. A buffer releases H^+ into the solution.

Visual Understanding

1. **Figure 2.11a** This figure shows an oxygen atom forming covalent bonds with two hydrogen atoms. A carbon atom, like oxygen, has two electrons in its innermost shell, but only four electrons in its outermost shell. Using this water molecule and figure 2.10 as guides, draw a diagram showing how carbon forms covalent bonds with two oxygen atoms in a carbon dioxide (CO_2) molecule.

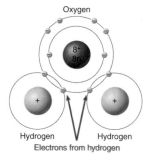

Oxygen

8+
8n

Hydrogen   Hydrogen
Electrons from hydrogen

2. **Figure 2.13** How can these things—drops of water clinging to a web and insects walking on water—occur?

Challenge Questions

1. What results if an atom gains or loses electrons? What results if an atom gains or loses neutrons? What results if an atom gains or loses protons?

2. You are on a 10-day backpacking trip with a small group of friends. Yvonne has washed out a set of water bottles with bleach. Before she can rinse them, Carlos, hot and thirsty, picks one up and drinks it, drops it, and begins to choke. What is the problem, and what can you do for him?

Molecules of Life

Forming Macromolecules

Nutrition Facts

Serving Size 2 tbsp (33g)
(makes 3.5 cups popped)
Servings Per Bag about 3
Servings Per Box about 9

Amount Per Serving	2 tbsp (33g) Unpopped	Per 1 cup Popped
Calories	170	35
Calories from Fat	90	20
	% Daily Value**	
Total Fat 10g*	15%	3%
Saturated Fat 2g	10%	0%
Trans Fat 3.5g		
Cholesterol 0mg	0%	0%
Sodium 440mg	18%	4%
Total Carbohydrate 9g	6%	1%
Dietary Fiber 3g	12%	4%
Sugars 0g		
Protein 3g		
Iron	6%	0%

Figure 3.1 What's in a nutritional label?

Fats, cholesterol, carbohydrates, and proteins are just some of the molecules found in popcorn and in other foods and are discussed in this chapter.

Group	Structural Formula	Ball-and-Stick Model	Found In
Hydroxyl	—OH		Carbohydrates
Carbonyl	C=O		Lipids
Carboxyl	—C(=O)OH		Proteins
Amino	—N(H)H		Proteins
Phosphate	—O—P(O⁻)(O)—O⁻		DNA, ATP

Figure 3.2 Five principal functional groups.

These functional groups can be transferred from one molecule to another and are common in organic molecules.

IMPLICATION: Which chemical elements appear to be common in organic molecules, based on these functional groups?

3.1 Polymers Are Built of Monomers

CONCEPT PREVIEW: Macromolecules are formed by linking subunits together into long chains. The chemical reaction involves removing a water molecule as each link is formed. Macromolecules are broken down into their subunits by adding the water molecules back.

The bodies of organisms contain thousands of different kinds of molecules and atoms. Organisms obtain many of these molecules from their surroundings and from what they consume. You might be familiar with some of the substances listed on nutritional labels, such as the one shown in figure 3.1. But what do the words on these labels mean? Some of them are names of minerals, such as calcium and iron (discussed in chapters 23, 25, 32). Others are vitamins, which are discussed in chapter 26. Still others are the subject of this chapter: large molecules that are found in our food and that make up the bodies of organisms, such as proteins, carbohydrates (including sugars), and lipids (including, fats, trans fats, saturated fats, and cholesterol). These molecules, called *organic molecules,* are formed by living organisms and consist of a carbon-based core with special groups attached. These groups of atoms have special chemical properties and are referred to as *functional groups.* Functional groups, like those listed in figure 3.2, tend to act as units during chemical reactions and confer specific chemical properties on the molecules that possess them.

The bodies of organisms contain thousands of different kinds of organic molecules, but much of the body is made of just four kinds: *proteins, nucleic acids, carbohydrates,* and *lipids.* Called **macromolecules** because they can be very large, these four are the building materials of cells, the "bricks and mortar" that make up the body of a cell and the machinery that runs within it.

Macromolecules are assembled by sticking smaller bits, called **monomers,** together much as a train is built by linking rail-cars together. A molecule built up of long chains of similar subunits is called a **polymer.**

Making (and Breaking) Macromolecules

The four different kinds of macromolecules (proteins, nucleic acids, carbohydrates, and lipids) are built from different monomers, as shown in figure 3.3, but all have their subunits put together in the same way. A covalent bond is formed between two subunits in which a hydroxyl group (OH) is removed from one subunit and a hydrogen (H) is removed from the other. This process, illustrated in figure 3.4a, is called *dehydration synthesis* because, in effect, the removal of the OH and H groups (highlighted by the blue oval) constitutes removal of a molecule of water—the word *dehydration* means "taking away water." This process requires the help of a special class of proteins called **enzymes** to facilitate the positioning of the molecules so that the correct chemical bonds are stressed and broken. The process of tearing down a molecule

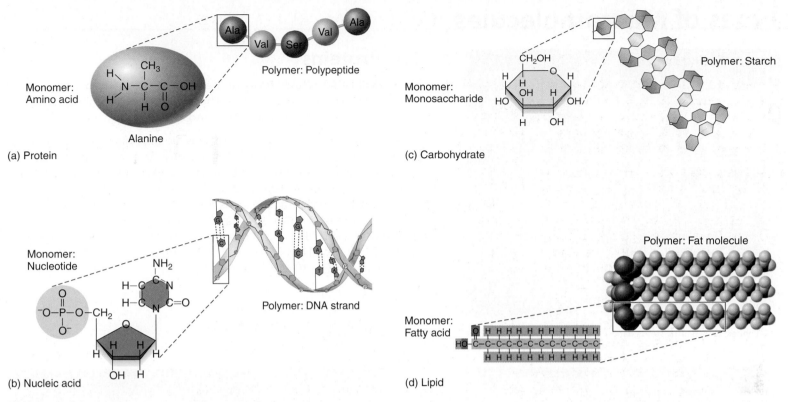

Figure 3.3   Polymers are built from monomers.

Each macromolecule polymer is built from different monomers. (a) A protein polymer, called a polypeptide, is built from amino acid monomers. (b) A nucleic acid polymer, such as a strand of DNA, is built from nucleotide monomers. (c) A carbohydrate polymer, such as a starch molecule, is built from monosaccharide monomers. (d) A lipid polymer, such as a fat molecule, is built from fatty acids.

such as the protein or fat contained in the food you eat is essentially the reverse of dehydration synthesis: instead of removing a water molecule, one is added. When a water molecule comes in, as shown in figure 3.4b, a hydrogen becomes attached to one subunit and a hydroxyl to another, and the covalent bond is broken. The breaking up of a polymer in this way is called **hydrolysis.**

> Recall from page 40 that water molecules will undergo a process of ionization, where the molecule dissociates into H⁺ and OH⁻ ions. Enzymes can help facilitate this process thereby facilitating dehydration and hydrolysis reactions in the cell.

Concept Check

1. What type of atom makes up the core in all organic molecules?
2. Which functional group is not present in either proteins or carbohydrates?
3. When subunits link together to form a polymer, what molecule is produced as a by-product?

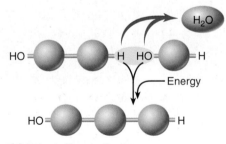

(a) Dehydration synthesis

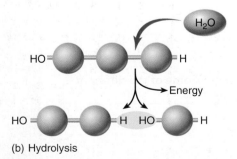

(b) Hydrolysis

Figure 3.4   Dehydration and hydrolysis.

(a) Biological molecules are formed by linking subunits with a covalent bond in a dehydration synthesis, during which a water molecule is released. (b) Breaking such a bond requires the addition of a water molecule, a reaction called hydrolysis.

Types of Macromolecules

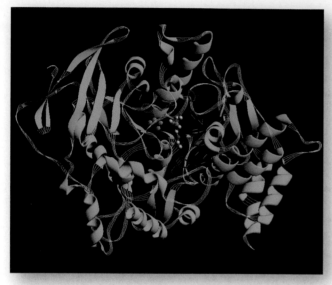

(a) Enzymes: Globular proteins called enzymes play a key role in many chemical reactions. This is a computer model of an enzyme.

3.2 Proteins

CONCEPT PREVIEW: Proteins are made up of chains of amino acids that fold into complex shapes. The sequence of its amino acids determines a protein's structure and its function.

Complex macromolecules called **proteins** are important biological macro-molecules within the bodies of all organisms. One of the most important types of proteins are *enzymes,* which have the key role in cells of helping to carry out particular chemical reactions. Other proteins play structural roles. Cartilage, bones, and tendons all contain a structural protein called collagen. Keratin, another structural protein, forms hair, the horns of a rhinoceros, and feathers. Still other proteins act as chemical messengers within the brain and throughout the body. Figure 3.5 presents an overview of the wide-ranging functions of proteins.

(b) Structural proteins (keratin): Keratin forms hair, nails, feathers, and components of horns.

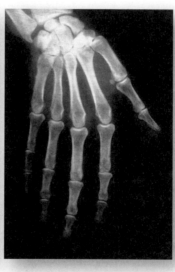

(c) Structural proteins (collagen): Collagen is present in bones, tendons, and cartilage.

(d) Contractile proteins: Proteins called actin and myosin are present in muscles.

Figure 3.5   Some of the different types of proteins.

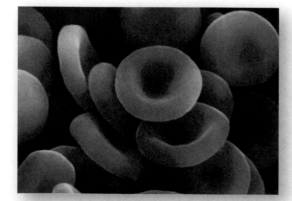

(e) Transport proteins: Red blood cells contain the protein hemoglobin, which transports oxygen in the body.

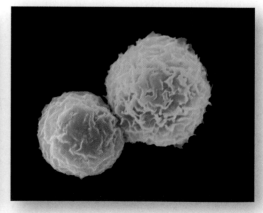

(f) Defensive proteins: White blood cells destroy foreign cells in the body and make antibody proteins that attack invaders.

Amino Acids

Despite their diverse functions, all proteins have the same basic structure: a long polymer chain made of subunits called amino acids. Amino acids are small molecules with a simple basic structure: a central carbon atom attached to an amino group ($-NH_2$), a carboxyl group ($-COOH$), a hydrogen atom (H), and a functional group, designated "R" (figure 3.6).

There are 20 common amino acids that differ from one another by the identity of their functional R group. The R group of an amino acid largely determines its chemical properties. Some amino acid R groups are polar, interacting with water; some are non-polar, shying away from water; and others have special chemical groups that are important in forming links between protein chains or in forming kinks in their shapes.

Amino acid

Figure 3.6 Basic structure of an amino acid.

Linking Amino Acids

An individual protein is made by linking specific amino acids together in a particular order, just as a word is made by putting letters of the alphabet together in a particular order. The covalent bond linking two amino acids together is called a **peptide bond.** You can see in figure 3.7 that a water molecule is released as the peptide bond forms. Long chains of amino acids linked by peptide bonds are called **polypeptides.** Functional polypeptides are more commonly called proteins. The keratin in human hair is a kind of protein.

> A peptide bond, like all covalent bonds discussed on page 35, involves the sharing of electrons between atoms. The carbon atom from the carboxyl group of one amino acid shares electrons with the nitrogen atom from the amino group of another amino acid.

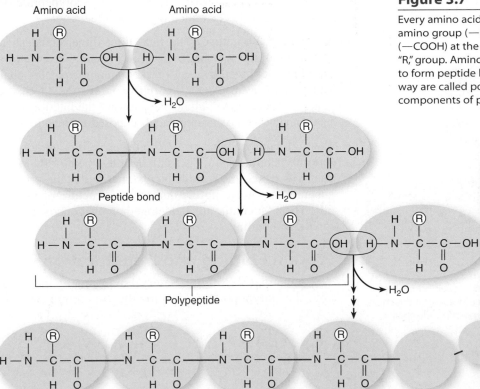

Figure 3.7 The formation of a peptide bond.

Every amino acid has the same basic structure, with an amino group ($-NH_2$) at one end and a carboxyl group ($-COOH$) at the other. The only variable is the functional, or "R," group. Amino acids are linked by dehydration synthesis to form peptide bonds. Chains of amino acids linked in this way are called polypeptides and are the basic structural components of proteins, such as keratin in human hair.

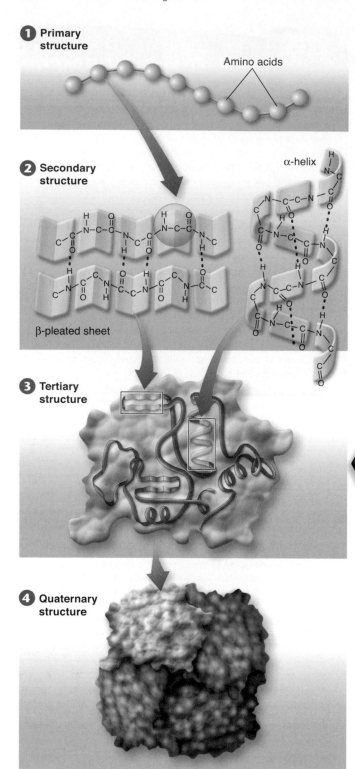

❶ Primary structure

Amino acids

❷ Secondary structure

α-helix

β-pleated sheet

❸ Tertiary structure

❹ Quaternary structure

Figure 3.8 Levels of protein structure.

The *primary structure* of a protein is its sequence of amino acids. Twisting or pleating of the chain of amino acids, called *secondary structure,* is due to the formation of localized hydrogen bonds (the *red* dotted lines) within the chain. More complex folding of the chain is referred to as *tertiary structure.* Two or more polypeptide chains associated together form a *quaternary structure.*

Protein Structure

Some proteins form long, thin fibers, whereas others are globular, with the long strands of polypeptides coiled up and folded back on themselves or intertwined with other polypeptides. The shape of a protein is very important because it determines the protein's function. If we picture a polypeptide as a long strand similar to a strand of yarn, a protein might be the sweater knitted from it. Importantly, the sequence of amino acids in a polypeptide determines the protein's structure. There are four general levels of protein structure: primary, secondary, tertiary, and quaternary (figure 3.8).

❶ Primary Structure. The sequence of amino acids of a polypeptide chain is termed the polypeptide's **primary structure.** The amino acids are linked together by peptide bonds, forming long chains like the "beaded strand" at the top of figure 3.8. The primary structure of a protein, the sequence of its amino acids, determines all other levels of protein structure.

❷ Secondary Structure. Because some of the amino acids are nonpolar and others are polar, a polypeptide chain folds up in solution as the nonpolar regions are forced together. To understand this, recall the polar properties of water. Water is a polar molecule that is attracted to and forms hydrogen bonds with other polar molecules but repels nonpolar molecules. This polar attraction and repulsion will push nonpolar amino acid functional groups away from the watery environment, leaving the polar amino acid functional groups to interact with water molecules and each other. Hydrogen bonds forming between different parts of the chain then stabilize the folding of the polypeptide. As you can see in ❷, these stabilizing hydrogen bonds do not involve the R groups themselves, but rather the polypeptide backbone. This initial folding is called the **secondary structure** of a protein. Hydrogen bonding within this secondary structure can fold the polypeptide into coils, called α-helices, and sheets, called β-pleated sheets.

> Hydrogen bonding, as discussed on page 36, occurs between polar molecules. The polar portions of the amino acids that are in close proximity are held in place with hydrogen bonds.

❸ Tertiary Structure. The final three-dimensional shape, or **tertiary structure,** of the protein, folded and twisted in the case of a globular molecule, is determined by exactly where in a polypeptide chain the nonpolar amino acids occur. Again, the repulsion of the nonpolar amino acids by water will force these amino acids toward the interior of the globular protein, leaving the polar amino acids exposed to the outside.

❹ Quaternary Structure. When a protein is composed of more than one polypeptide chain, the spatial arrangement of the several component chains is called the **quaternary structure** of the protein, like the four subunits that make up the quaternary structure of the protein in figure 3.8. These subunits are held together by noncovalent forces, such as hydrogen bonding.

Protein Folding and Denaturation

The polar nature of the watery environment in the cell influences how the polypeptide folds into the functional protein. The protein in figure 3.9 is folded in such a way that allows it to carry out its function. If the polar nature of the protein's environment changes by either increasing temperature or lowering pH, both of which alter hydrogen bonding, the protein may unfold, as in the lower right of the figure. When this happens the protein is said to be *denatured*. When proteins are denatured, they usually lose their ability to function properly. When the polar nature of the solvent is reestablished, some proteins may spontaneously refold, but most don't. Cooking an egg is an example of denaturing proteins that do not refold. The egg proteins denature as temperature increases, but do not refold as the egg cools down—they are permanently denatured. Protein denaturation is also the rationale behind traditional methods of preserving food. Prior to the ready availability of refrigerators and freezers, a practical way to keep microorganisms from growing in food was to keep the food in a solution containing a high concentration of vinegar, a treatment called pickling. The low pH of the vinegar denatures proteins in microorganisms and so keeps them from growing on the food.

> You will discover on page 92 how enzyme function is affected by temperature and pH. Most enzymes have optimal ranges of temperature and pH. Conditions above or below their ranges can interfere with enzyme function.

Protein Structure Determines Function

The three dimensional shape of a protein determines its function. For example, many structural proteins assume long cable-like shapes that let them play architectural roles within cells. The spiderweb-like structures you see in figure 3.10*a* are such structural proteins, tagged so that they are visible through a microscope. The cables that these proteins form within the cell help maintain the cell's shape, and also function as rails through the interior used to rapidly transport materials from one area to another.

Globular proteins are polypeptides that fold and twist into complex three-dimensional shapes, and for them to function properly they need to fold correctly. *Enzymes* are globular proteins that help particular chemical reactions to occur in the cell. When the polypeptide folds correctly the enzyme surface has a groove or depression that precisely fits a particular molecule. For example, the red molecule binding to the groove on the surface of the enzyme in figure 3.10*b* is a sugar. Once within the groove, the molecule is induced to undergo a chemical reaction, such as the formation or breaking of one of its covalent bonds. By bringing two atoms close together, an enzyme can make it easier for them to share electrons and form covalent bonds. Other enzymes function by positioning a molecule so that there is stress on a particular bond so that it breaks. To better understand this, consider a high-heel shoe or boot. It is structured so that your foot fits into it, but (as anyone who has worn one knows) the foot is stressed by the shape of the shoe. An enzyme chemically stresses a molecule lying within a groove on its surface.

Because the primary structure of a protein (its sequence of amino acids) determines how the protein folds into its functional shape, a change in the identity of even one amino acid can have profound effects on a protein's ability to function properly.

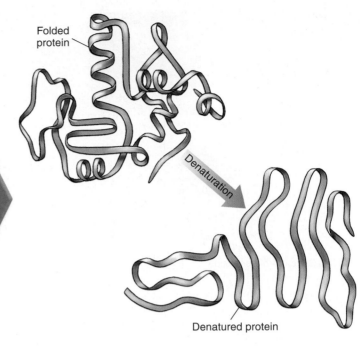

Figure 3.9 **Protein denaturation.**

Changes in a protein's environment, such as variations in temperature or pH, can cause a protein to unfold and lose its shape in a process called denaturation. In this denatured state, proteins are biologically inactive.

IMPLICATION   A tiger can eat raw meat with relish, but most of us humans prefer our steak cooked. Is this just a matter of taste, or does cooking meat before eating it have any benefit?

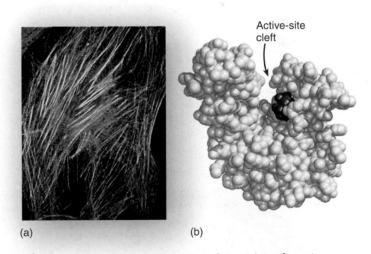

(a)   (b)

Figure 3.10 **Protein structure determines function.**

(a) Fluorescently-labeled structural proteins within a cell.
(b) Enzymes are globular proteins that aid chemical reactions in the cell. This enzyme (*blue*) has a deep groove that binds a specific chemical (*red*) at a site on the enzyme called the active site.

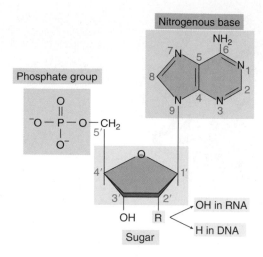

Phosphate group

Nitrogenous base

Sugar

OH in RNA

H in DNA

(a) Structure of nucleotide

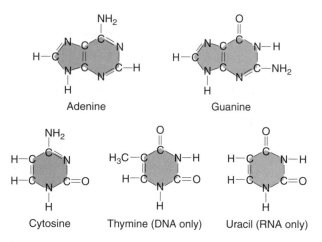

Adenine

Guanine

Cytosine

Thymine (DNA only)

Uracil (RNA only)

(b) Nitrogenous bases

Figure 3.11   The structure of a nucleotide.

(a) Nucleotides are composed of three parts: a five-carbon sugar, a phosphate group, and an organic nitrogenous base. The nitrogenous base can be one of five, shown in (b).

3.3 Nucleic Acids

CONCEPT PREVIEW: Nucleic acids like DNA are composed of long chains of nucleotides. The sequence of nucleotides in a DNA molecule specifies the amino acid sequence of proteins.

Very long polymers called nucleic acids serve as the genetic information storage devices of cells, just as DVDs or hard drives store the information that computers use. Nucleic acids are long polymers of repeating subunits called **nucleotides.** Each nucleotide is a complex organic molecule composed of three parts shown in figure 3.11a: a five-carbon sugar (in blue), a phosphate group (in yellow, PO_4), and an organic nitrogen-containing base (in orange). In the formation of a nucleic acid, the individual sugars are linked through dehydration reactions with the phosphate groups forming very long polynucleotide chains. The nitrogenous bases extend out from the backbone of the chain.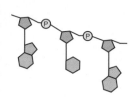

How does the long, chainlike structure of a nucleic acid permit it to store the information necessary to specify what an organism is like? If nucleic acids were simply a monotonous repeating polymer, it could not encode the message of life. Imagine trying to write a story using only the letter *E* and no spaces or punctuation. All you could ever say is "EEEEEEE. . . ." You need more than one letter to communicate—the English alphabet uses 26 letters. Nucleic acids can encode information because they contain more than one kind of nucleotide. There are five different nucleotides found in nucleic acids: two larger ones that contain the nitrogenous bases adenine and guanine (shown in the top row of figure 3.11b), and three smaller ones (in the bottom row) that contain the nitrogenous bases cytosine, thymine, and uracil. Nucleic acids encode information by varying the identity of the nucleotide at each position in the polymer.

DNA and RNA

Nucleic acids come in two varieties, **deoxyribonucleic acid (DNA)** and **ribonucleic acid (RNA),** both are polymers of nucleotides but they have different functions in the cell and they differ in their structures. RNA is similar to DNA, but with two major chemical differences. First, RNA molecules contain the sugar ribose in which four of the five carbons bond to a hydroxyl group (—OH). In DNA, one of the hydroxyl groups is replaced with a hydrogen atom (this is the carbon labeled 2′ in figure 3.11a). Second, DNA contains the thymine nucleotide, RNA molecules do not; they

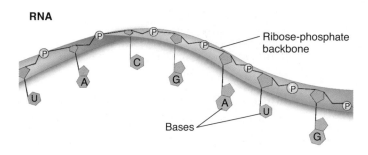

Figure 3.12   How DNA structure differs from RNA.

RNA is a single-strand of nucleotides (*left*), while DNA (*right*) contains two polynucleotide strands wrapped around each other.

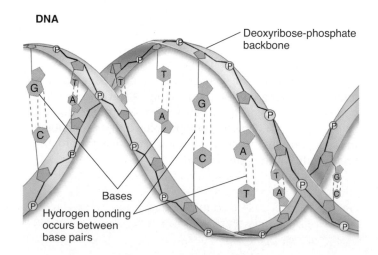

contain the uracil nucleotide instead. Structurally, RNA is also different. RNA is a long, single strand of nucleotides (on the left in figure 3.12) while DNA consists of *two* polynucleotide chains wound around each other in a double helix, like strands of a pearl necklace twisted together (shown on the right in figure 3.12). They also have different roles to play in the cell. DNA stores the genetic information that determines what a cell is like. The sequence of nucleotides in DNA determines the order of amino acids in the primary structure of the protein. However, the DNA doesn't convey this information directly, rather RNA carries this information from the DNA to the protein-making machinery in the cell.

The Double Helix

How does DNA form a double helix? When scientists looked carefully at the structure of DNA, they found that the bases of each chain point inward toward the other (like the DNA strands shown in figure 3.13). The bases of the two chains are linked in the middle of the molecule by hydrogen bonds (the dotted lines between the two strands), like two columns of people holding hands across. The key to understanding the double helix structure of DNA is revealed by looking at the nucleotide bases: *only two base pairs are possible.* Because the distance between the two strands is consistent, this suggests that two big bases cannot pair together—the combination is simply too bulky to fit; similarly, two little ones cannot pair, as they would pinch the helix inward too much. To form a double helix, it is necessary to pair a big base with a little one. *In every DNA double helix, adenine (A) pairs with thymine (T) and guanine (G) pairs with cytosine (C).* The reason A doesn't pair with C and G doesn't pair with T is that these base pairs cannot form proper hydrogen bonds—the electron-sharing atoms are not aligned with each other.

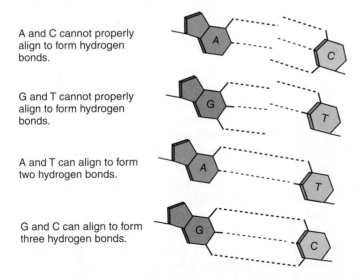

A and C cannot properly align to form hydrogen bonds.

G and T cannot properly align to form hydrogen bonds.

A and T can align to form two hydrogen bonds.

G and C can align to form three hydrogen bonds.

The simple A–T, G–C base pairs within the DNA double helix allow the cell to copy the information in a very simple way. It just unzips the helix and adds the nucleotides with complementary bases to each strand! That is the great advantage of a double helix—it actually contains two copies of the information, one the mirror image of the other. If the sequence of one chain is ATTGCAT, the sequence of its partner in the double helix *must* be TAACGTA. The fidelity with which hereditary information is passed from one generation to the next is a direct result of this simple double-entry bookkeeping, which makes accurate copying of the genetic message possible.

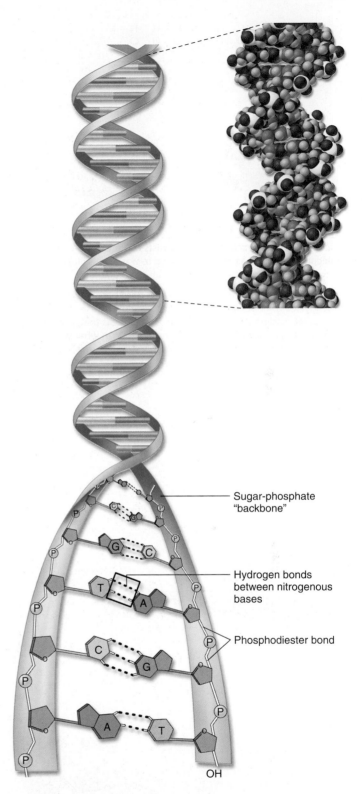

Sugar-phosphate "backbone"

Hydrogen bonds between nitrogenous bases

Phosphodiester bond

OH

Figure 3.13   The DNA double helix.

The DNA molecule is composed of two polynucleotide chains twisted together to form a double helix. The two chains of the double helix are joined by hydrogen bonds between the A–T and G–C base pairs. The section of DNA on the upper right is a space-filling model of DNA, where atoms are indicated by colored balls.

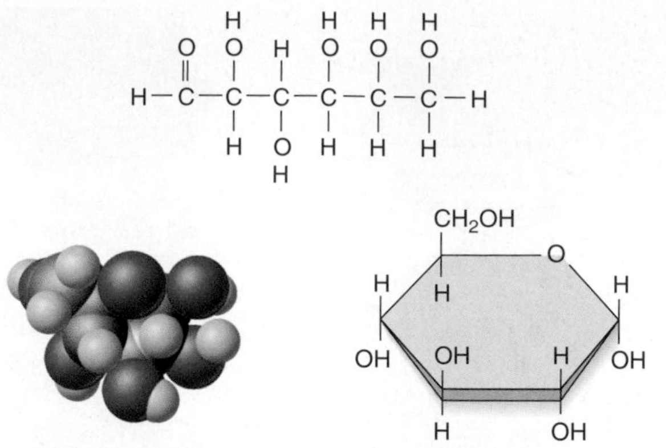

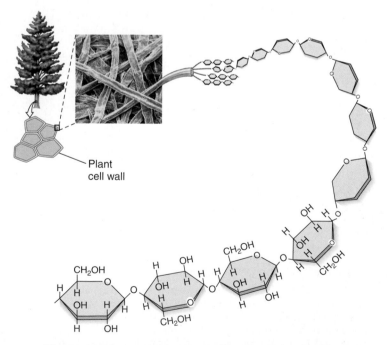

Figure 3.14 **The structure of glucose.**

Glucose is a monosaccharide and consists of a linear six-carbon molecule that forms a ring when placed in water. This illustration shows three ways glucose can be pictured in a diagram.

IMPLICATION Many vitamin and fruit drinks are marketed as being healthier because they are "all natural," with "no refined sugars." Yet reading the nutrition facts label reveals that a single 12oz bottle typically contains 29 grams of sugar. In what way do you think "natural" sugar is different from "refined" sugar?

Figure 3.15 **A polysaccharide: cellulose.**

The polysaccharide cellulose is found in the cell walls of plant cells and is composed of glucose subunits.

3.4 Carbohydrates

CONCEPT PREVIEW: Carbohydrates are molecules made of C, H, and O atoms. As sugars they store energy in C—H bonds, and as long chains, they can provide structural support in some types of organisms.

Polymers called **carbohydrates** make up the structural framework of certain cells and play a critical role in energy storage. A carbohydrate is any molecule that contains carbon, hydrogen, and oxygen in the ratio 1:2:1. Some carbohydrates are simple, small monomers or dimers and are called simple carbohydrates. Others are long polymers and are called complex carbohydrates. Because they contain many carbon-hydrogen (C—H) bonds, carbohydrates are well-suited for energy storage. Such C—H bonds are the ones most often broken by organisms to obtain energy. Table 3.1 on the facing page shows some examples of carbohydrates.

Simple Carbohydrates

The simplest carbohydrates are the *simple sugars* or **monosaccharides** (from the Greek *monos*, single, and *saccharon*, sweet). These molecules consist of one subunit. For example, glucose, the sugar that carries energy to the cells of your body, is made of six carbons and has the chemical formula $C_6H_{12}O_6$. A molecule of glucose is pictured in several ways in figure 3.14. The long chain of carbon atoms at the top of the figure is its formal chemical structure. When placed in water, the chain folds into the ring structure shown on the lower right. The individual atoms are depicted in the "3-D" space-filling model you see in the lower left. Another type of simple carbohydrate is a **disaccharide,** which forms when two monosaccharides link together through a dehydration reaction. Table sugar is a disaccharide, sucrose, made by linking two six-carbon sugars together, a glucose and a fructose (see table 3.1).

Complex Carbohydrates

Organisms store their metabolic energy by converting sugars, which are water-soluble, into insoluble forms that can be deposited in specific storage areas in the body. This trick is achieved by linking the sugars together into long polymer chains called **polysaccharides.** Plants and animals store energy in polysaccharides formed from glucose. The glucose polysaccharide that plants use to store energy is called *starch*—that is why potatoes are referred to as "starchy" food. In animals, energy is stored in *glycogen*, a highly insoluble macromolecule formed of glucose polysaccharides that are very long and highly branched.

Plants and animals also use glucose chains as building materials, linking the subunits together in different orientations not recognized by most enzymes. These structural polysaccharides are chitin in animals and *cellulose* in plants. The cellulose deposited in the cell walls of the plant cells, like the cellulose strand shown in figure 3.15, cannot be digested by humans and makes up the fiber in our diets. Microbes in the digestive tracts of cows and horses, however, have the cellulose-digesting enzymes we humans lack. The microbial enzyme breaks the bonds holding the glucose molecules together so that they can be used by the animal cells for energy. These animals can thrive on a diet of grass, but you can't. Termites have these enzymes too, which is why a termite can eat wood while you would starve on a diet of tree limbs or lumber.

As discussed on page 307, fungi possess enzymes that can break down cellulose, which is why fungi often grow on dead trees. Some bacteria are also able to break down cellulose and live in the digestive system of certain animals as described on page 507.

TABLE 3.1 | Carbohydrates and Their Functions

Carbohydrate	Example	Description

Transport Disaccharides

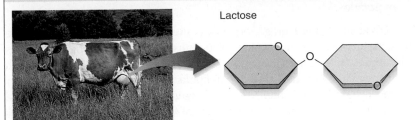

Lactose

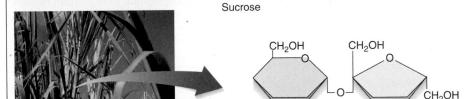

Sucrose

Glucose is transported within some organisms as a disaccharide. In this form, it is less readily metabolized because the normal glucose-utilizing enzymes of the organism cannot break the bond linking the two monosaccharide subunits. One type of disaccharide is called lactose. Many mammals supply energy to their young in the form of lactose, which is found in milk. Another transport disaccharide is sucrose. Many plants transport glucose throughout the plant in the form of sucrose, which is harvested from sugarcane to make granulated sugar.

Storage Polysaccharides

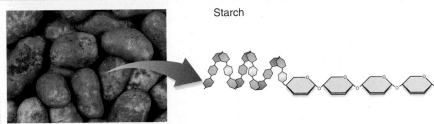

Starch

Organisms store energy in long chains of glucose molecules called polysaccharides. The chains tend to coil up in water, making them insoluble and ideal for storage. The storage polysaccharides found in plants are called starches, which can be branched or unbranched. Starch is found in potatoes and in grains, such as corn and wheat.

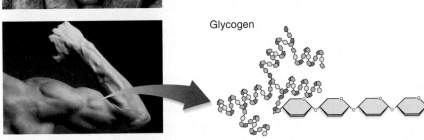

Glycogen

In animals, glucose is stored as glycogen. Glycogen is similar to starch in that it consists of long chains of glucose that coil up in water and are insoluble. But glycogen chains are much longer and highly branched. Glycogen can be stored in muscles and the liver.

Structural Polysaccharides

Cellulose

Cellulose is a structural polysaccharide found in the cell walls of plants; its glucose subunits are joined in a way that cannot be broken down readily. Cleavage of the links between the glucose subunits in cellulose requires an enzyme most organisms lack. Some animals, such as cows, are able to digest cellulose by means of bacteria and protists they harbor in their digestive tract, which provide the necessary enzymes.

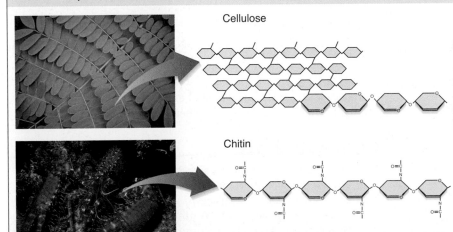

Chitin

Chitin is a type of structural polysaccharide found in the external skeletons of many invertebrates, including insects and crustaceans, and in the cell walls of fungi. Chitin is a modified form of cellulose with a nitrogen group added to the glucose units. When cross-linked by proteins, it forms a tough, resistant surface material.

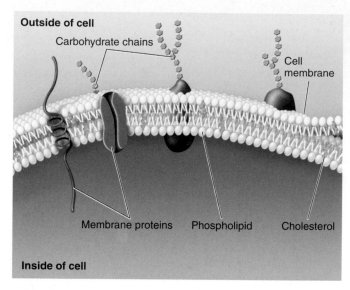

Figure 3.16   **Lipids are a key component of biological membranes.**

Lipids are one of the most common molecules in the human body, because the membranes of all cells are composed of phospholipids. Membranes also contain cholesterol, another type of lipid.

3.5 Lipids

CONCEPT PREVIEW: Lipids are not water-soluble. A type of lipid called a fat contains chains of fatty acid subunits that store energy. Other lipids include phospholipids, steroids, rubber, and pigment molecules.

For long-term energy storage, organisms usually convert glucose into fats, another kind of storage molecule that contains more energy-rich C—H bonds than carbohydrates. Fats and all other biological molecules that are not soluble in water are called **lipids.** Lipids are nonpolar; in water, fat molecules cluster together because they cannot form hydrogen bonds with water molecules. This is why oil forms into a layer on top of water when the two substances are mixed. A special type of lipid called a phospholipid is important because it forms boundary layers in cells called membranes (figure 3.16).

Fats

Fat molecules are lipids composed of two kinds of subunits: fatty acids and glycerol. A *fatty acid* is a long chain of carbon and hydrogen atoms. Glycerol contains three carbons and forms the backbone to which three fatty acids are attached through dehydration reactions.

The chemical composition of the fatty acids that make up a fat molecule can affect its physical properties. Fats whose fatty acid chains are composed of the maximum number of hydrogen atoms (figure 3.17*b*) are said to be *saturated.* Saturated fats are solid at room temperature. Animal fats are often saturated and occur as hard fats. On the other hand, fats composed of fatty acids with double bonds between one or more pairs of carbon atoms (figure 3.17*c*) contain fewer than the maximum number of hydrogen atoms and are called *unsaturated.* Unsaturated fats are liquid at room temperature. Many plant fats are unsaturated and occur in oils. Unsaturated fats in food products may be artificially *hydrogenated* (industrial addition of hydrogens), extending the shelf life of products like peanut butter. In some cases, the hydrogenation creates *trans fats,* a type of unsaturated fat linked to heart disease.

> Covalent bonds, as discussed on page 35, form when atoms share electrons. The single covalent bonds in saturated fats result when the carbons share one pair of electrons. The double covalent bonds in unsaturated fats result when two pairs of electrons are shared between carbon atoms.

Other Types of Lipids

Other types of lipids include phospholipids and cholesterol, which play key roles in the membranes that encase all cells of your body. Cholesterol is a type of lipid called a *steroid.* The male and female sex hormones testosterone and estradiol are also steroids. Rubber, waxes, and light-absorbing pigments are other important biological lipids.

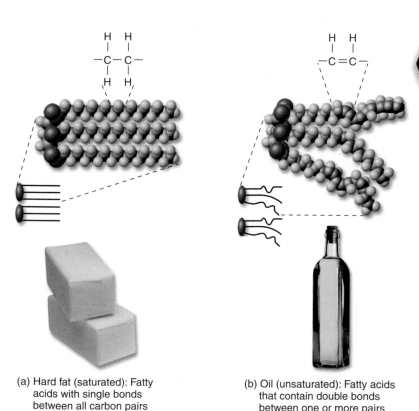

(a) Hard fat (saturated): Fatty acids with single bonds between all carbon pairs

(b) Oil (unsaturated): Fatty acids that contain double bonds between one or more pairs of carbon atoms

Figure 3.17   **Saturated and unsaturated fats.**

Fat molecules each contain a three-carbon glycerol to which is attached three fatty acid tails. (a) Most animal fats are "saturated" (every carbon atom carries the maximum load of hydrogens). Their fatty acid chains fit closely together and form immobile arrays called hard fats. (b) Most plant fats are unsaturated, which prevents close association between chains and so results in oils.

Concept Check

1. Would you expect to find a polar or nonpolar amino acid in the interior of a globular protein? Why?
2. If one chain of a DNA strand has the sequence ACCTGGAAT, what is the sequence of the other strand?
3. Why can you digest starch but not cellulose?
4. Which fat is liquid at room temperature, a saturated or unsaturated fat?

Anabolic Steroids in Sports

Among the most notorious of lipids in recent years has been the class of synthetic hormones known as anabolic steroids. Since the 1950s some athletes have been taking these chemicals to build muscle and so boost athletic performance. Both because of the intrinsic unfairness of this and because of health risks, the use of anabolic steroids has been banned in sports for decades. Controversy over their use in professional baseball has recently returned anabolic steroids to the nation's front pages.

Anabolic steroids were developed in the 1930s to treat hypogonadism, a condition in which the male testes do not produce sufficient amounts of the hormone testosterone for normal growth and sexual development. Scientists soon discovered that by slightly altering the chemical structure of testosterone, they could produce synthetic versions that facilitated the growth of skeletal muscle in laboratory animals. The word "anabolic" means growing or building. Further tweaking reduced the added impact of these new chemicals on sexual development. More than 100 different anabolic steroids have been developed, most of which have to be injected to be effective. All require a prescription to be used legally in the United States, and all are banned in professional, college, and high school sports.

Another way to increase the body's level of testosterone is to use a chemical which is not itself anabolic but one that the body converts to testosterone. One such chemical is 4-androstenedione, more commonly called "andro." It was first developed in the 1970s by East German scientists to try to enhance their athletes' Olympic performances. Because andro does not have the same side effects as anabolic steroids, it was legally available until 2004. It was used by Mark McGwire, but it is now banned in all sports, and possession of andro is a federal crime.

Anabolic steroids work by signalling muscle cells to make more protein. They bind to special "androgenic receptor" proteins within the cells of muscle tissue. Like jabbing these proteins with a poker, the binding prods the receptors into action, causing them to activate genes on the cell's chromosomes that produce muscle tissue proteins, triggering an increase in protein synthesis. At the same time, the anabolic steroid molecules bind to so-called "cortisol receptor" proteins in the cell, preventing these receptors from doing their job

Homerun slugger Barry Bonds was involved in a steroid controversy in 2006.

of causing protein breakdown, the muscle cell's way of suppressing inflammation and promoting the use of proteins for fuel during exercise. By increasing protein production and inhibiting the breakdown of proteins in muscle cells after workouts, anabolic steroids significantly increase the mass of an athlete's muscle tissue.

If the only effect of anabolic steroids on your body was to enhance your athletic performance by increasing your muscle mass, using them would still be wrong, for one very simple and important reason: fairness. To gain advantage in competition by concealed use of anabolic steroids—"doping"—is simply cheating. That is why these drugs are banned in sports.

The use of anabolic steroids by athletes and others is not only wrong, but also illegal, because increased muscle mass is not the only effect of using these chemicals. Among adolescents, anabolic steroids can also lead to premature termination of the adolescent growth spurt, so that for the rest of their lives, users remain shorter than they would have been without the drugs. Adolescents and adults are also affected by steroids in the following ways. Anabolic steroids can lead to potentially fatal liver cysts and liver cancer (the liver is the organ of the body that attempts to detoxify the blood), cholesterol changes and hypertension (both of which can promote heart attack and stroke), and acne. Other signs of steroid use in men include reduced size of testicles, balding, and development of breasts. In women, signs include the growth of facial hair, lowering of the voice, and cessation of menstruation.

In the fall of 2003, athletic organizations learned that some athletes were using a new performance-enhancing anabolic steroid undetectable by standard antidoping tests, tetrahydrogestrinone (THG). The use of THG was only discovered because an anonymous coach sent a spent syringe to U.S. antidoping officials. THG's chemical structure is similar to gestrinone, a drug used to treat a form of pelvic inflammation, and can be made from it by simply adding four hydrogen atoms, an easy chemical task. THG tends to break down when prepared for analysis by standard means, which explains why antidoping tests had failed to detect it. New urine tests for THG that were developed in 2004 have been used to catch several well-known sports figures, including British athlete Dwain Chambers and baseball slugger Rafael Palmeiro.

How Does pH Affect a Protein's Function?

The red blood cells you see to the lower right carry oxygen to all parts of your body. These cells are red because they are chock full of a large iron-rich protein called *hemoglobin*. The iron atoms in each hemoglobin molecule provide a place for oxygen gas molecules to stick to the protein. When oxygen levels are highest (in the lungs), oxygen atoms bind to hemoglobin tightly, and a large percent of the hemoglobin molecules in a cell possess bound oxygen atoms. When oxygen levels are lower (in the tissues of the body), hemoglobin doesn't bind oxygen atoms as tightly, and as a consequence hemoglobin releases its oxygen to the tissues. What causes this difference between lungs and tissues in how hemoglobin loads and unloads oxygen? Oxygen concentration is not the only factor that might be responsible. A protein's function can be affected by pH, and blood pH, for example, also differs between lungs and body tissues (**pH** is a measure of how many H^+ ions a solution contains). Tissues are slightly more acidic (that is, they have more H^+ ions and a lower pH). Their metabolic activities release CO_2 into the blood, which quickly becomes converted to carbonic acid and lowers the pH.

The graph to the right displays so-called "oxygen loading curves" that reveal the effectiveness with which hemoglobin binds oxygen. The more effective the binding, the less oxygen required before hemoglobin becomes fully loaded, and the further to the left a loading curve is shifted. To assess the impact of pH on this process, O_2 loading curves were carried out at three different blood pH values. In the graph, oxygen levels in the blood are presented on the *x* axis, and for each data point the corresponding % hemoglobin saturation (a **%**, or **percent,** is the numerator [top part] of a fraction whose denominator [bottom part] is 100—in this case, a measure of the fraction of the hemoglobin that is bound to oxygen) is presented on the *y* axis. The oxygen-loading curve was repeated at pH values of 7.6, 7.4, and 7.2, corresponding to the blood pH that might be expected in resting, exercising, and very active muscle tissue, respectively.

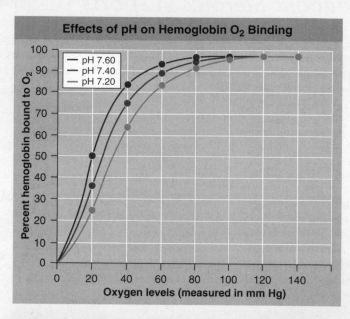

Effects of pH on Hemoglobin O_2 Binding

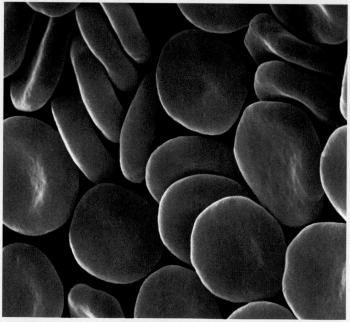

0.8 µm

Analysis

1. **Applying Concepts**   Which of the three pH values represents the highest concentration of hydrogen ions? (The **concentration** of a substance is the amount of that substance present in a given volume.) Is this value more acidic or more basic than the other two?

2. **Interpreting Data**   What is the percent hemoglobin bound to O_2 for each of the three pH concentrations at saturation (where the lines flatten out)? at an oxygen level of 20 mm Hg? at 40 mm Hg? at 60 mm Hg?

3. **Making Inferences**   At an oxygen level of 40 mm Hg, would hemoglobin bind oxygen more tightly at a pH of 7.8 or 7.0?

4. **Drawing Conclusions**   How does pH affect the release of oxygen from hemoglobin?

Concept Summary

Forming Macromolecules

3.1 Polymers Are Built of Monomers

- Living organisms produce organic macromolecules, which are large carbon-based molecules. The chemical properties of these molecules are due to unique functional groups that are attached to the carbon core (**figure 3.2**).

- Macromolecules are formed by the linking together of subunits, called monomers, to form larger polymers. Amino acids are the subunits that link together to form polypeptides. Nucleotide monomers link together to form nucleic acids. Monosaccharide monomers link together to form carbohydrates. Fatty acids are the monomers that link together to form a type of lipid called fats (**figure 3.3**).

- Polymers are formed by dehydration reactions, like the reaction shown here from **figure 3.4a.** Dehydration reactions produce covalent bonds that link monomers together. The reaction is called a dehydration reaction because a water molecule is a product of the reaction.

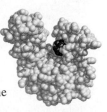

- The breakdown of macromolecule involves hydrolysis reactions, where a water molecule is broken down into H^+ and OH^-. These ions disrupt the covalent bonds that hold two monomers together, causing the bond to break (**figure 3.4b**).

Types of Macromolecules

3.2 Proteins

- Proteins are macromolecules that carry out many functions in the cell (**figure 3.5**). They are produced by the linking together of amino acid subunits that form a chain called a polypeptide.

- There are 20 different amino acids found in proteins. All amino acids have the same basic core structure (**figure 3.6**). They differ in the type of functional group attached to the core. The functional groups are referred to as R groups. Some functional groups are polar, some are nonpolar, and still others give the amino acid unique chemical properties.

- Amino acids are linked together with covalent bonds referred to as peptide bonds (**figure 3.7**).

- The sequence of amino acids within the polypeptide is the primary structure of the protein (**figure 3.8**). The chain of amino acids can twist into a secondary structure, where hydrogen bonding holds portions of the polypeptide in a coiled shape, called an α-helix, or in sheets called β-pleated sheets. Further bending and folding of the polypeptide results in its tertiary structure. When two or more polypeptides are present in a protein, the interaction of these polypeptide subunits is its quaternary structure.

- Changes in environmental conditions that disrupt hydrogen bonding can cause a protein to unfold, a process called denaturation (**figure 3.9**). Globular proteins, like the enzyme shown here from **figure 3.10,** cannot function if they are denatured. Some denatured proteins can refold back into their functional shapes. The structure of the protein determines its function.

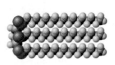

3.3 Nucleic Acids

- Nucleic acids, such as DNA and RNA, are long chains of nucleotides. Nucleotides contain three parts: a 5-carbon sugar, a phosphate group, and a nitrogenous base (**figure 3.11**). DNA and RNA function in information storage and retrieval in the cell, carrying the information needed to build proteins. The information is stored as different sequences of nucleotides that determine the order of amino acids in proteins.

- DNA and RNA differ chemically in that the sugar found in DNA is deoxyribose and in RNA it is ribose. The nitrogenous bases in DNA are cytosine, adenine, guanine, and thymine, and the same are found in RNA except for thymine, which is substituted in RNA with uracil (**figure 3.11b**).

- DNA and RNA also differ structurally. DNA contains two strands of nucleotides wound around each other, called a double helix. RNA, as shown here from **figure 3.12,** is a single strand of nucleotides.

- The two nucleotide strands of the DNA double helix are held together through hydrogen bonding between nitrogenous bases: adenine (A) pairs with thymine (T), and cytosine (C) pairs with guanine (G) (**figure 3.13**).

3.4 Carbohydrates

- Carbohydrates, also referred to as sugars, are macromolecules that serve two primary functions in the cell: structural framework and energy storage.

- Carbohydrates that consist of only one or two monomers are called simple carbohydrates, such as the glucose monosaccharide shown here from **figure 3.14,** and disaccharides. Carbohydrates that consist of long chains of monomers are called complex carbohydrates or polysaccharides (**figure 3.15**).

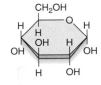

- Polysaccharides such as starch and glycogen provide a means of storing energy in the cell. They are broken down in the cells when energy is needed. Carbohydrates such as cellulose and chitin provide structural integrity (**table 3.1**) and are not broken down by animals because they lack the enzyme necessary. Certain microbes in the guts of some animals are able to breakdown cellulose.

3.5 Lipids

- Lipids are large nonpolar molecules that are insoluble in water. Lipids called phospholipids are components of biological membranes. Fats that function in long-term energy storage include saturated fats, as shown here from **figure 3.17,** and unsaturated fats. Saturated fats are solid at room temperature and are found in animals, while unsaturated fats are liquid (oils) at room temperature and are found in plants.

- Other lipids include the steroids (including the sex steroids and cholesterol), rubber, and pigments.

Self-Test

1. The four kinds of organic macromolecules are
   a. hydroxyls, carboxyls, aminos, and phosphates.
   b. proteins, carbohydrates, lipids, and nucleic acids.
   c. DNA, RNA, simple sugars, and amino acids.
   d. carbon, hydrogen, oxygen, and nitrogen.

2. Organic molecules are made up of monomers. Which of the following is *not* considered a monomer of organic molecules?
   a. amino acids   c. polypeptides
   b. simple sugars   d. nucleotides

3. Your body is filled with many types of proteins. Each type has a distinctive sequence of amino acids that determines both its unique _____ and its specialized _____.
   a. number, weight   c. structure, function
   b. length, mass   d. charge, pH

4. A peptide bond forms
   a. by the removal of a water molecule.
   b. by a dehydration reaction.
   c. between two amino acids.
   d. All of the above.

5. Nucleic acids
   a. are the energy source for our bodies.
   b. act on other molecules, breaking them apart or building new ones to help us function.
   c. are only found in a few, specialized locations within the body.
   d. are information storage devices found in body cells.

6. The two strands of a DNA molecule are held together through hydrogen bonds between nucleotide bases. Which of the following best describes this base pairing in DNA?
   a. Adenine forms hydrogen bonds with thymine.
   b. Adenine forms hydrogen bonds with cytosine.
   c. Cytosine forms hydrogen bonds with thymine.
   d. Guanine forms hydrogen bonds with adenine.

7. Carbohydrates are used for
   a. structure and energy.   c. fat storage and hair.
   b. information storage.   d. hormones and enzymes.

8. Which of the following carbohydrates is *not* found in plants?
   a. glycogen   c. starch
   b. cellulose   d. All are found in plants.

9. A characteristic common to fat molecules is
   a. that they contain long chains of C—H bonds.
   b. that they are insoluble in water.
   c. that they have a glycerol backbone.
   d. All of these are characteristics of fat molecules.

10. Lipids are used for
    a. motion and defense.
    b. information storage.
    c. energy storage and for some hormones.
    d. enzymes and for some hormones.

Visual Understanding

1. **Figure 3.7 and table 3.1** The molecule below, on the left, is a peptide made from monomers of amino acids. The molecule below, on the right, is a disaccharide made from monomers of simple sugars. Both molecules were synthesized using a common chemical reaction. What is the chemical reaction that formed these molecules and what is the common by-product of both these reactions?

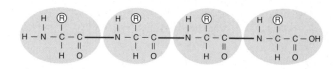
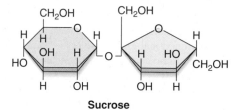

Sucrose

2. **Figure 3.17** The following are two lipid molecules. The lipid on the left is a saturated fat and the one on the right is an unsaturated fat. What is the difference in the chemical structure of their fatty acid tails and how does this affect their physical properties?

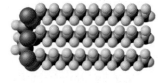

Challenge Questions

1. How many molecules of water are used up in the breakdown of a polypeptide that is 15 amino acids in length?

2. The enzyme present in the cells of your body can break the bonds between the glucose monomers in starch but it cannot break the bonds between the glucose monomers in cellulose. Enzymes are very specific in what molecules they bind to. After examining the chemical structures of starch and cellulose in table 3.1, explain why the same enzyme that breaks down starch cannot break down cellulose.

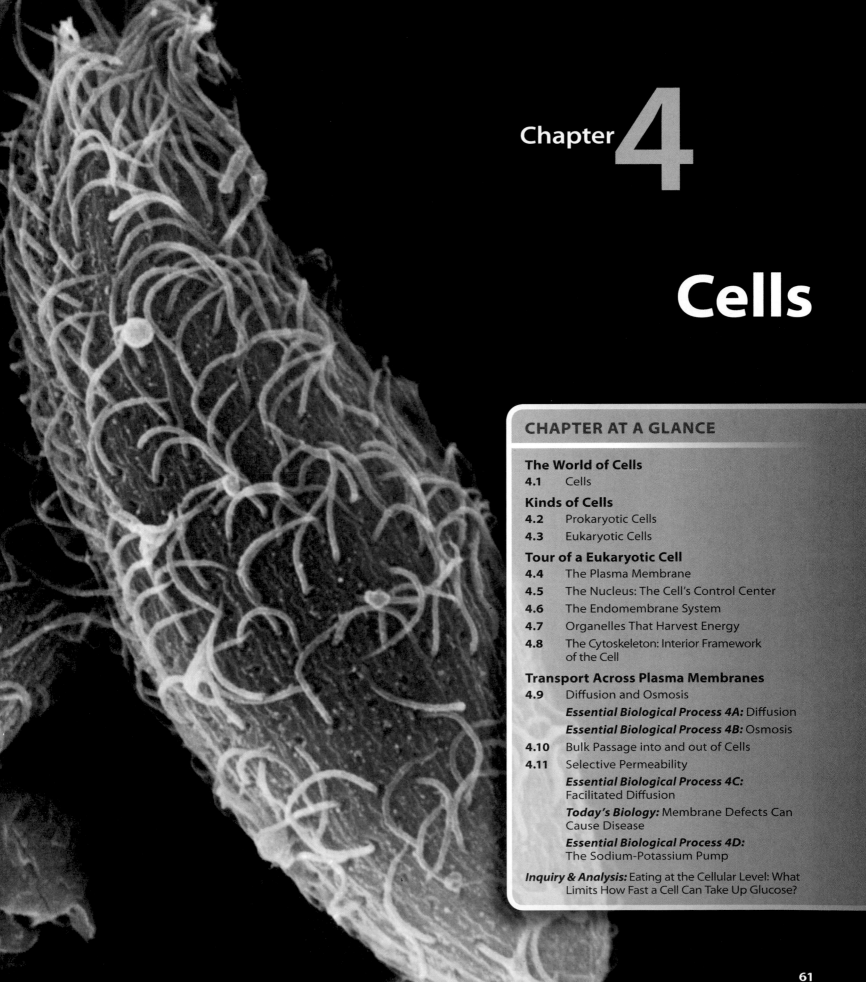

Chapter 4

Cells

The World of Cells

4.1 Cells

CONCEPT PREVIEW: All living things are composed of one or more cells. Most cells and their components are so small they can only be viewed using microscopes.

Hold your finger up and look at it closely. What do you see? Skin. It looks solid and smooth, creased with lines and flexible to the touch. But if you were able to remove a bit and examine it under a microscope, it would look very different. Figure 4.1 takes you on a journey into your fingertip. The crammed bodies you see in panels ❸ and ❹ are skin cells, laid out like a tiled floor. As your journey continues, you travel inside one of the cells and see organelles, structures in the cell that perform specific functions. Proceeding even further inward, you encounter the molecules of which the structures are made, and finally the atoms shown in panels ❽ and ❾. While some organisms are composed of a single cell, your body is composed of many cells. A human body has as many cells as there are stars in a galaxy, between 10 and 100 trillion, depending on your size. In this chapter we look more closely at cells and learn something of their internal structure and how they communicate with their environment.

Figure 4.1   The size of cells and their contents.

This diagram shows the size of human skin cells, organelles, and molecules. In general, the diameter of a human skin cell is a little less than 20 micrometers (μm), a mitochondrion is 2 μm, a ribosome is 20 nanometers (nm), a protein molecule is 2 nm, and an atom is 0.2 nm.

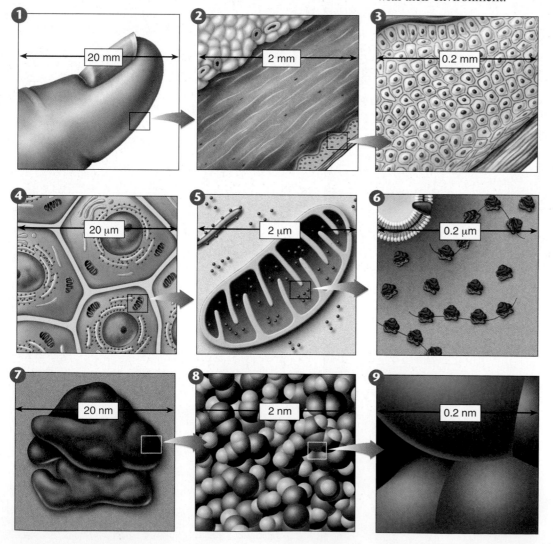

The Cell Theory

Cells are small, so small that no one observed them until microscopes were invented in the mid-seventeenth century. Robert Hooke first described cells in 1665, when he used a microscope he had built to examine a thin slice of nonliving plant tissue called cork. Hooke observed a honeycomb of tiny, empty (because the cells were dead) compartments. He called the compartments in the cork *cellulae* (Latin, small rooms), and the term has come down to us as **cells.** For another century and a half, however, biologists failed to recognize the importance of cells. In 1838, botanist Matthias Schleiden made a careful study of plant tissues and developed the first statement of the cell theory. He stated that all plants "are aggregates of fully individualized, independent, separate beings, namely the cells themselves." In 1839, Theodor Schwann reported that all animal tissues also consist of individual cells.

The idea that all organisms are composed of cells is called the cell theory. In its modern form, the cell theory includes three principles:

1. All organisms are composed of one or more cells, within which the processes of life occur.

2. Cells are the smallest living things. Nothing smaller than a cell is considered alive.
3. Cells arise only by division of a previously existing cell. Although life likely evolved spontaneously in the environment of the early earth, biologists have concluded that no additional cells are originating spontaneously at present. Rather, life on earth represents a continuous line of descent from those early cells.

Most Cells Are Very Small

Cells are not all the same size. Individual marine alga cells, for example, can be up to 5 centimeters long—as long as your little finger. In contrast, the cells of your body are typically from 5 to 20 micrometers (μm) in diameter, too small to see with the naked eye. It would take anywhere from 100 to 400 human cells to span the diameter of the head of a pin. The cells of bacteria are even smaller than your cells, only a few micrometers thick (figure 4.2).

Why Aren't Cells Larger?

Why are most cells so tiny? Most cells are small because larger cells do not function as efficiently. In the center of every cell is a command center that must issue orders to all parts of the cell, directing the synthesis of certain enzymes, the entry of ions and molecules from the exterior, and the assembly of new cell parts. These orders must pass from the core to all parts of the cell, and it takes them a very long time to reach the periphery of a large cell. For this reason, an organism made up of relatively small cells has an advantage over one composed of larger cells.

Another reason cells are not larger is the advantage of having a greater surface area. A cell's surface provides the interior's only opportunity to interact with the environment, as its surface provides the only way for substances to pass into and out of the cell. As cells grow larger, their interior volume increases much more than their surface area, and as a result there is far less surface available to service each unit of volume. In the same way, before airplanes and trains were invented the size of cities was limited because the surrounding countryside could not support all the people living in a big city—only so many roads could be built into the city, only so many farms were close enough to the city to use them.

Visualizing Cells

How many cells are big enough to see with the unaided eye? Not many (see figure 4.3). Most are less than 50 micrometers in diameter, far smaller than the period at the end of this sentence.

The Resolution Problem. How do we study cells if they are too small to see? The key is to understand why we can't see them. The reason we can't see such small objects is the limited resolution of the human eye. *Resolution* is defined as the minimum distance two points can be apart and still be distinguished as two separated points. The limit of resolution of the human eye is about 100 micrometers. This limit occurs because when two objects are closer together than about 100 micrometers, the light reflected from each strikes the same "detector" cell at the rear of the eye. Only when the objects are farther apart than 100 micrometers will the light from each strike different cells, allowing your eye to resolve them as two objects rather than one.

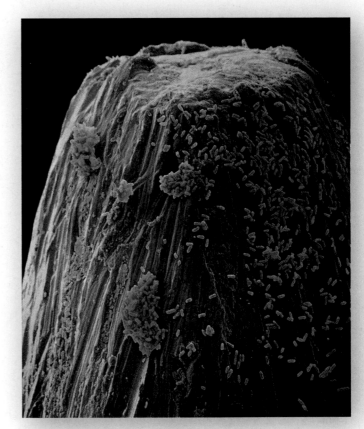

Figure 4.2   **Bacteria on the point of a pin (175×).**

IN THE NEWS

High Resolution TV. A new generation of "high definition" televisions is providing remarkable pictures of very high resolution. Computer and television engineers define resolution very simply as the number of individual points of color (or "pixels") you can see. A traditional 15-inch television displays 640 dots on each of 480 vertical lines, so-called 480 resolution. Until recently, DVD discs provided the best resolution for a TV picture. DVDs use red-laser light beams to read and project 720 vertical lines. This was as many dots per line as you could get without two dots being read by adjacent red-laser beams overlapping. With the invention of the much narrower blue laser, it became possible to squeeze dots much more closely together on a true high-resolution HD TV screen—1,080 vertical lines, yielding over 2 million dots of color. This produces an image of vivid clarity difficult to distinguish from what your eye would deliver.

Light microscope

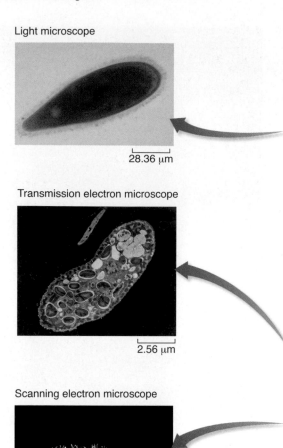

28.36 μm

Transmission electron microscope

2.56 μm

Scanning electron microscope

6.76 μm

Microscopes. One way to increase resolution is to increase magnification, so that small objects appear larger. Modern *light microscopes* use two magnifying lenses (and a variety of correcting lenses) to achieve very high magnification and clarity (figure 4.3). The first lens focuses the image of the object on the second lens, which magnifies it again and focuses it on the receptor cells that line the inside of back of the eye. Microscopes that magnify in stages using several lenses are called **compound microscopes.** They can resolve structures that are separated by as little as 200 nanometers (nm).

Increasing Resolution. Light microscopes, even compound ones, are not powerful enough to resolve many structures within cells. For example, a membrane is only 5 nanometers thick. Why not just add another magnifying stage to the microscope and so increase its resolving power? Because when two objects are closer than a few hundred nanometers, the light beams reflecting from the two images start to overlap. The only way two light beams can get closer together and still be resolved is if their wavelengths are shorter.

One way to avoid overlap is by using a beam of electrons rather than a beam of light. Electrons have a much shorter wavelength, and a microscope employing electron beams has 1,000 times the resolving power of a light microscope. A **transmission electron microscope (TEM),** so called because the electrons used to visualize the specimens are transmitted through the material, is capable of resolving objects only 0.2 nanometers apart—just twice the diameter of a hydrogen atom!

A second kind of electron microscope, the **scanning electron microscope (SEM),** beams the electrons onto the surface of the specimen. The electrons reflect back, are amplified, and the image created is transmitted to a screen and photographed, producing an often striking three-dimensional picture.

Concept Check

1. All of the cells in your body developed from a single fertilized egg cell. Which principle of the cell theory relates to this fact?
2. Why is it advantageous for your body to be made up of small cells?
3. Why can your eye not see two objects as different if closer than 100 micrometers?

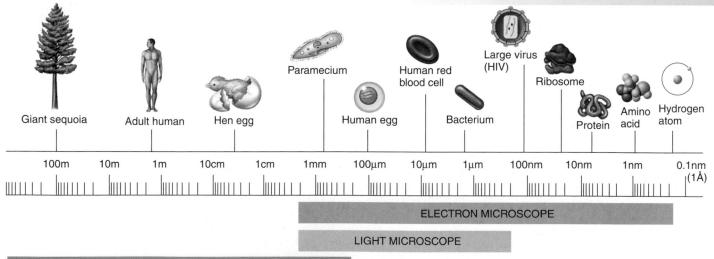

Figure 4.3 A scale of visibility.

Most cells are microscopic in size, although many vertebrate eggs can be seen with the unaided eye. Prokaryotic cells are generally 1 to 2 micrometers (μm) across.

Kinds of Cells

4.2 Prokaryotic Cells

CONCEPT PREVIEW: Prokaryotic cells lack a nucleus and do not have an extensive system of interior membranes. They are encased in a cell wall and may contain external structures such as a capsule, flagellum, or pili.

> Recall from the discussion of the diversity of life on page 26, that all organisms are classified based on their characteristics into three domains and further divided into six kingdoms. Two of the three domains contain prokaryotic organisms: Bacteria and Archaea.

There are two major kinds of cells: prokaryotes and eukaryotes. **Prokaryotes** have a relatively uniform interior that is not subdivided by internal membranes into separate compartments. They do not, for example, have special membrane-bounded compartments, called *organelles*, or a *nucleus* (a membrane-bounded compartment that holds hereditary information). All *bacteria* and *archaea* are prokaryotes; all other organisms are eukaryotes.

Prokaryotes are the simplest cellular organisms. Over 5,000 species are recognized, but doubtless many times that number actually exist and have not

> The plasma membrane, as described on page 57, is composed of phospholipids that are arranged to form two layers, where the polar heads are oriented to the outside and the nonpolar tails extend in toward the interior of the membrane.

yet been described. Although these species are diverse in form, their organization is fundamentally similar: They are single-celled organisms, with small cells typically about 2 micrometers thick; the cells are enclosed by a plasma membrane but have no distinct interior compartments. Outside of almost all bacteria is a *cell wall*, a framework of carbohydrates cross-linked into a rigid structure. In some bacteria another layer called the *capsule* encloses the cell wall. Archaea are an extremely diverse group that inhabit diverse environments. Bacteria are abundant and play critical roles in many biological processes. Bacteria assume many shapes, like the sausage or spiral shapes shown in figure 4.4a,b. They can also adhere in masses or chains, as shown in figure 4.4c, but in these cases the individual cells remain functionally separate from one another.

If you were able to peer into a prokaryotic cell, you would be struck by its simple organization. The entire interior of the cell is one unit, with little or no internal support structure (the rigid wall, the purple layer surrounding the cell in figure 4.5, supports the cell's shape) and no internal compartments bounded by membranes. Scattered throughout the cytoplasm of prokaryotic cells are small structures called *ribosomes*, the small spherical structures you see inside the cell in figure 4.5. Ribosomes are the sites where proteins are made, but they are not considered organelles because they lack a membrane boundary. The DNA is found in a region of the cytoplasm called the *nucleoid region*. Although the DNA is localized in this region of the cytoplasm, it is not considered a nucleus because, as you can see in figure 4.5, the nucleoid region and its associated DNA are not enclosed within an internal membrane.

Some prokaryotes use a **flagellum** (plural, **flagella**) to move. Flagella are long, threadlike structures, made of protein fibers that project from the surface of a cell. They are used in locomotion and feeding. There may be none, one, or more per cell depending on the species. Bacteria can swim at speeds up to 20 cell diameters per second, rotating their flagella like screws.

Some prokaryotic cells contain **pili** (singular, **pilus**), which are short flagella (only several micrometers long, and about 7.5 to 10 nanometers thick). Pili help the prokaryotic cell attach to appropriate substrates and aid in the exchange of genetic information between cells.

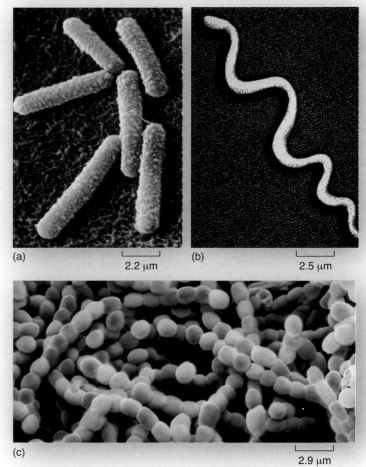

(a)   ⊢ 2.2 μm ⊣   (b)   ⊢ 2.5 μm ⊣

(c)   ⊢ 2.9 μm ⊣

Figure 4.4 **Bacterial cells have different shapes.**

(a) *Bacillus* is a rod-shaped bacterium. (b) *Treponema* is a coil-shaped bacterium; rotation of internal filaments produces a cork-screw movement. (c) *Streptomyces* is a more or less spherical bacterium in which the individuals adhere in chains.

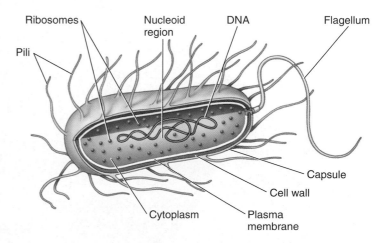

Figure 4.5 **Organization of a prokaryotic cell.**

Prokaryotic cells lack internal compartments. Not all prokaryotic cells have a flagellum or a capsule like the one illustrated here, but all have a nucleoid region, ribosomes, a plasma membrane, cytoplasm, and a cell wall.

4.3 Eukaryotic Cells

CONCEPT PREVIEW: Eukaryotic cells have a system of interior membranes and membrane-bounded organelles that subdivide the interior into functional compartments.

Eukaryotic cells are much larger and profoundly different from prokaryotic cells, with a complex interior organization. Figures 4.6 and 4.7 present cross-sectional diagrams of idealized animal and plant cells. As you can see, the interior of a eukaryotic cell is much more complex than the prokaryotic cell you encountered in figure 4.5. The **plasma membrane** ❶ encases a semi-fluid matrix called the **cytoplasm** ❷, which contains within it the nucleus and various cell structures called organelles. An **organelle** is a specialized structure within which particular cell processes occur. Each organelle, such as a **mitochondrion** ❸, has a specific function in the eukaryotic cell. The organelles are anchored at specific locations in the cytoplasm by an interior scaffold of protein fibers, the **cytoskeleton** ❹.

One of the organelles is very visible when these cells are examined with a microscope, filling the center of the cell like the pit of a peach, the **nucleus** ❺ (plural, *nuclei*), from the Latin word for "kernel." Inside the nucleus, the DNA is wound tightly around proteins and packaged into compact units called chromosomes. It is the nucleus that gives **eukaryotes** their name, from the Greek words *eu*, true, and *karyon*, nut; by way of contrast, the earlier-evolving bacteria and archaea are called prokaryotes ("before the nut").

If you examine the organelles in figures 4.6 and 4.7, you can see that most of them form separate compartments within the cytoplasm, bounded by

Figure 4.6 **Structure of an animal cell.**

In this generalized diagram of an animal cell, the plasma membrane encases the cell, which contains the cytoskeleton and various cell organelles and interior structures suspended in a semifluid matrix called the cytoplasm. Some kinds of animal cells possess fingerlike projections called microvilli. Other types of eukaryotic cells, for example many protist cells, may possess flagella, which aid in movement, or cilia, which can have many different functions.

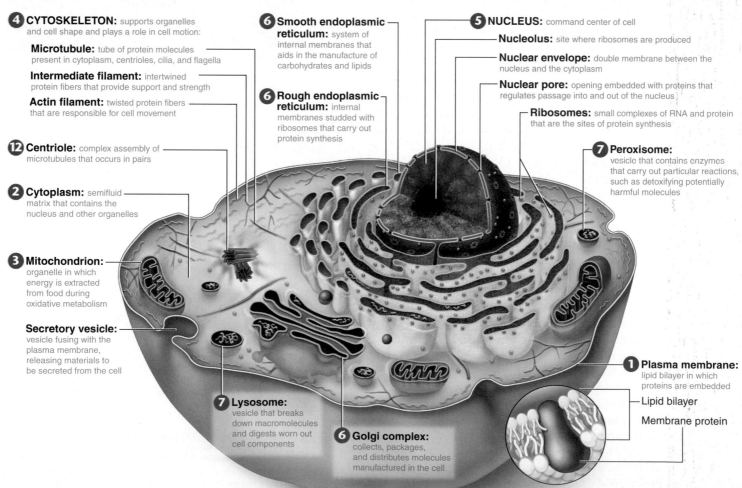

❹ **CYTOSKELETON:** supports organelles and cell shape and plays a role in cell motion:

Microtubule: tube of protein molecules present in cytoplasm, centrioles, cilia, and flagella

Intermediate filament: intertwined protein fibers that provide support and strength

Actin filament: twisted protein fibers that are responsible for cell movement

⓬ **Centriole:** complex assembly of microtubules that occurs in pairs

❷ **Cytoplasm:** semifluid matrix that contains the nucleus and other organelles

❸ **Mitochondrion:** organelle in which energy is extracted from food during oxidative metabolism

Secretory vesicle: vesicle fusing with the plasma membrane, releasing materials to be secreted from the cell

❻ **Smooth endoplasmic reticulum:** system of internal membranes that aids in the manufacture of carbohydrates and lipids

❻ **Rough endoplasmic reticulum:** internal membranes studded with ribosomes that carry out protein synthesis

❼ **Lysosome:** vesicle that breaks down macromolecules and digests worn out cell components

❻ **Golgi complex:** collects, packages, and distributes molecules manufactured in the cell

❺ **NUCLEUS:** command center of cell

Nucleolus: site where ribosomes are produced

Nuclear envelope: double membrane between the nucleus and the cytoplasm

Nuclear pore: opening embedded with proteins that regulates passage into and out of the nucleus

Ribosomes: small complexes of RNA and protein that are the sites of protein synthesis

❼ **Peroxisome:** vesicle that contains enzymes that carry out particular reactions, such as detoxifying potentially harmful molecules

❶ **Plasma membrane:** lipid bilayer in which proteins are embedded

Lipid bilayer

Membrane protein

their own membranes. *The hallmark of the eukaryotic cell is this compartmentalization.* This internal compartmentalization is achieved by an extensive **endomembrane system** ❻ that weaves through the cell interior.

Vesicles ❼ (small membrane-bounded sacs that store and transport materials) form closed-off compartments in the cell, allowing different processes to proceed simultaneously without interfering with one another, just as rooms do in a house. For example, organelles called *lysosomes* are recycling centers that have acidic interiors in which old organelles are broken down and their component molecules recycled. This acid would be very destructive if released into the cytoplasm.

Comparing figure 4.6 with figure 4.7, you will see the same set of organelles, with a few interesting exceptions. For example, the cells of plants, fungi, and many protists have strong exterior **cell walls** ❽ composed of cellulose or chitin fibers, while the cells of animals lack cell walls. All plants and many kinds of protists have **chloroplasts** ❾, within which photosynthesis occurs. No animal or fungal cells contain chloroplasts. Plant cells also contain a large **central vacuole** ❿ that stores water, and **plasmodesmata** ⓫, which are openings in the cell wall that create cytoplasmic connections between cells. **Centrioles** ⓬, which will be described later, are present in animal cells but absent in plant and fungal cells.

Concept Check

1. Why can't you see a ribosome with a light microscope?
2. Are there membrane-bounded organelles within a bacterial cell?
3. What important differences are there between plant and animal cells?

Figure 4.7   Structure of a plant cell.

Most mature plant cells contain large central vacuoles that occupy a major portion of the internal volume of the cell and organelles called chloroplasts, within which photosynthesis takes place. The cells of plants, fungi, and some protists have cell walls, although the composition of the walls varies among the groups. Plant cells have cytoplasmic connections through openings in the cell wall called plasmodesmata. Flagella occur in sperm of a few plant species, but are otherwise absent in plant and fungal cells. Centrioles are also absent in plant and fungal cells.

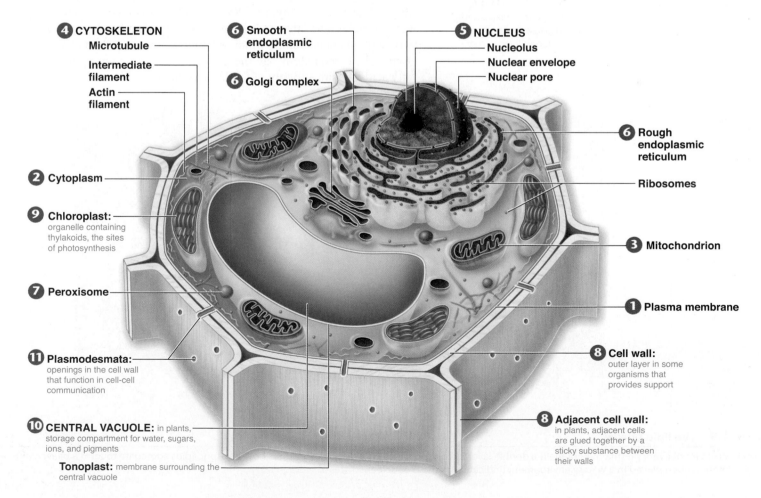

❹ **CYTOSKELETON**
Microtubule
Intermediate filament
Actin filament

❻ **Smooth endoplasmic reticulum**
❻ **Golgi complex**

❺ **NUCLEUS**
Nucleolus
Nuclear envelope
Nuclear pore

❻ **Rough endoplasmic reticulum**

Ribosomes

❷ Cytoplasm

❾ **Chloroplast:** organelle containing thylakoids, the sites of photosynthesis

❸ **Mitochondrion**

❼ **Peroxisome**

❶ **Plasma membrane**

⓫ **Plasmodesmata:** openings in the cell wall that function in cell-cell communication

❽ **Cell wall:** outer layer in some organisms that provides support

❿ **CENTRAL VACUOLE:** in plants, storage compartment for water, sugars, ions, and pigments

❽ **Adjacent cell wall:** in plants, adjacent cells are glued together by a sticky substance between their walls

Tonoplast: membrane surrounding the central vacuole

Tour of a Eukaryotic Cell

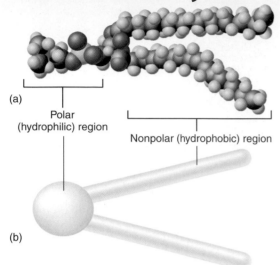

(a)

Polar (hydrophilic) region

Nonpolar (hydrophobic) region

(b)

Figure 4.8 **Phospholipid structure.**

One end of a phospholipid molecule is polar and the other is nonpolar. (a) The molecular structure is shown with colored spheres representing individual atoms (*gray* for carbon, *light blue* for hydrogen, *red* for oxygen, *yellow* for phosphorus, and *blue* for nitrogen). (b) The phospholipid is often depicted diagrammatically as a ball with two tails.

IMPLICATION Sometimes pregnancies end tragically in the second trimester when the mother's immune system unexpectedly attacks fetal membrane phospholipids. Would you expect a specific fetal organ to be primarily affected, or a more generalized organ failure? Explain.

4.4 The Plasma Membrane

CONCEPT PREVIEW: All cells are encased within a delicate lipid bilayer sheet, the plasma membrane, within which are embedded a variety of proteins that act as markers or channels through the membrane.

Encasing all living cells, both prokaryotes and eukaryotes, is a delicate sheet of molecules called the **plasma membrane.** It would take more than 10,000 of these molecular sheets, which are about 5 nanometers thick, piled on top of one another to equal the thickness of this sheet of paper. However, the sheets are not simple in structure, like a soap bubble's skin. Rather, they are made up of a diverse collection of proteins floating within a lipid framework like small boats bobbing on the surface of a pond. Regardless of the kind of cell they enclose, all plasma membranes have the same basic structure of proteins embedded in a sheet of lipids, called the **fluid mosaic model.**

> Recall from the discussion of lipids on pages 56 and 57 that a phospholipid is a modified fat molecule where an end carbon in the glycerol backbone forms a chemical bond with a phosphate functional group instead of a third fatty acid tail.

The lipid layer that forms the foundation of a plasma membrane is composed of modified fat molecules called **phospholipids.** A phospholipid molecule can be thought of as a polar head with two nonpolar tails attached to it (figure 4.8). The head of a phospholipid molecule has a phosphate chemical group linked to it—the yellow sphere in figure 4.8a—making it extremely polar (and thus water-soluble). The other end of the phospholipid molecule is composed of two long fatty acid chains, which are long chains of carbon and hydrogen atoms. The carbon atoms are the gray spheres you see in figure 4.8a. The fatty acid tails are strongly nonpolar and thus water-insoluble.

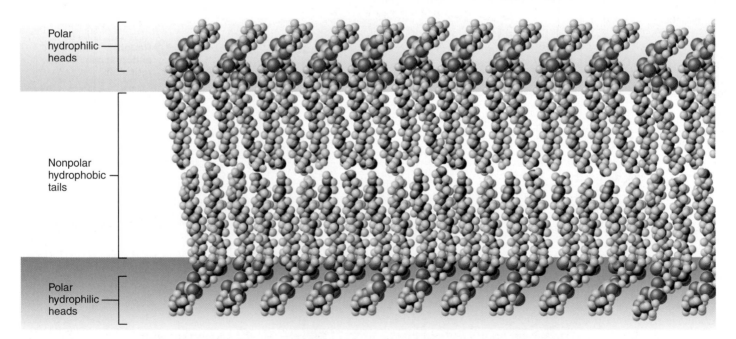

Polar hydrophilic heads

Nonpolar hydrophobic tails

Polar hydrophilic heads

Figure 4.9 **The lipid bilayer.**

The basic structure of every plasma membrane is a double layer of phospholipids. This diagram illustrates how phospholipids aggregate to form a bilayer with a nonpolar interior when placed in a watery environment (indicated by the blue screened areas).

Imagine what happens when a collection of phospholipid molecules is placed in water. A structure called a **lipid bilayer** forms spontaneously. How does this happen? The long nonpolar tails of the phospholipid molecules are pushed away by the water molecules that surround them, shouldered aside as the water molecules seek partners that can form hydrogen bonds. After much shoving and jostling, every phospholipid molecule ends up with its polar head facing water and its nonpolar tail facing away from water. The phospholipid molecules form a *double* layer, called a bilayer. The shaded blue areas in figure 4.9 represent watery environments inside and outside the plasma membrane that push the nonpolar tails to the interior of the bilayer. Because there are two layers with the tails facing each other, no tails are ever in contact with water. Because the interior of a lipid bilayer is completely nonpolar, it repels any water-soluble molecules that attempt to pass through it, just as a layer of oil stops the passage of a drop of water (that's why ducks do not get wet—oily feathers).

Cholesterol, another nonpolar lipid molecule, resides in the interior portion of the bilayer. Cholesterol is a multiringed molecule that affects the fluid nature of the membrane. Although cholesterol is important in maintaining the integrity of the plasma membrane, it can accumulate in blood vessels, forming plaques that lead to cardiovascular disease.

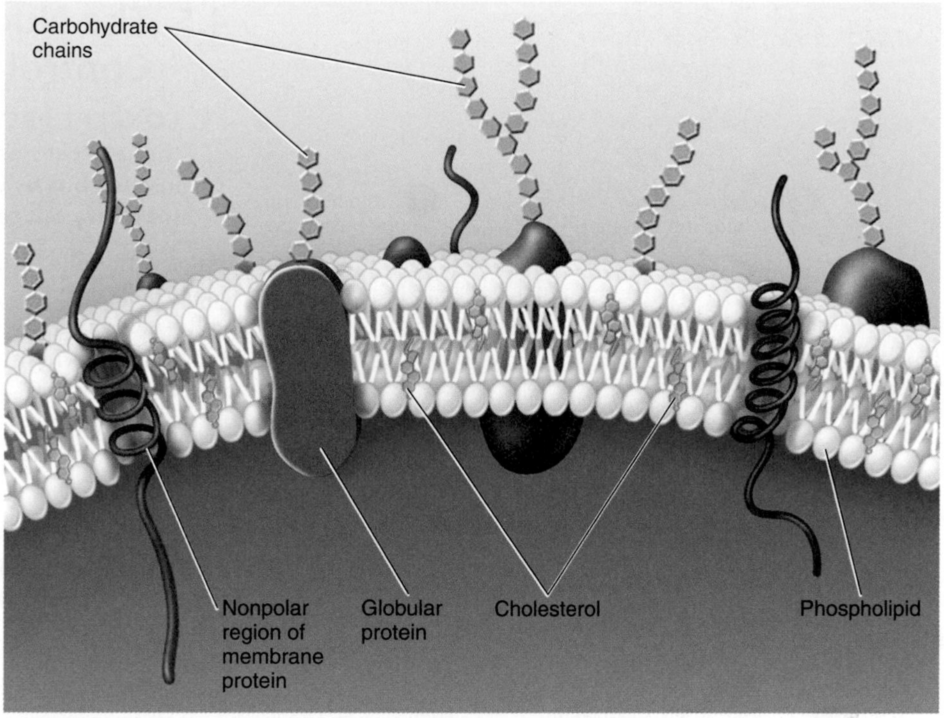

Proteins Within the Membrane

The second major component of every biological membrane is a collection of membrane proteins that float within the lipid bilayer. As you can see in figure 4.10, some proteins (the purple structures) pass through the lipid bilayer, providing channels through which molecules and ions pass. While some membrane proteins are fixed into position, others move about freely.

Many membrane proteins project up from the surface of the plasma membrane like buoys, often with carbohydrate chains (the orange chains in figure 4.10) or lipids attached to their tips like flags. These *cell surface proteins* act as markers to identify particular types of cells, or as beacons to bind specific hormones or proteins to the cell.

> Cell surface markers called MHC proteins discussed on page 537 identify a person's cells as "self" so the immune system won't attack them. Transmembrane proteins are present in nerve cells and function as ion channels, as discussed on page 554.

Protein channels that extend all the way across the bilayer provide passageways for ions and polar molecules so they can pass into and out of the cell. How do these *transmembrane proteins* manage to span the membrane, rather than just floating on the surface in the way that a drop of water floats on oil? The part of the protein that actually traverses the lipid bilayer is a specially constructed spiral helix of nonpolar amino acids—the red coiled areas of the transmembrane protein you see in figure 4.11. Water responds to these nonpolar amino acids much as it does to nonpolar lipid chains, and as a result the helical spiral is held within the lipid interior of the bilayer, anchored there by the strong tendency of water to avoid contact with these nonpolar amino acids.

Carbohydrate chains

Nonpolar region of membrane protein

Globular protein

Cholesterol

Phospholipid

Figure 4.10 **Proteins are embedded within the lipid bilayer.**

A variety of proteins protrude through the lipid bilayer. Membrane proteins function as channels, receptors, and cell surface markers. Carbohydrate chains are often bound to these proteins and to phospholipids in the membrane. These chains serve as distinctive identification tags, unique to particular types of cells.

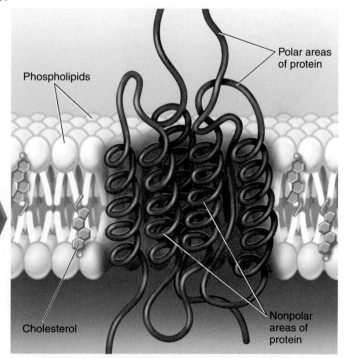

Polar areas of protein

Phospholipids

Cholesterol

Nonpolar areas of protein

Figure 4.11 **Nonpolar regions lock proteins into membranes.**

A spiral helix of nonpolar amino acids (*red*) extends across the nonpolar lipid interior, while polar (*purple*) portions of the protein protrude out from the bilayer. The protein cannot move in or out because such a movement would drag polar segments of the protein into the nonpolar interior of the membrane.

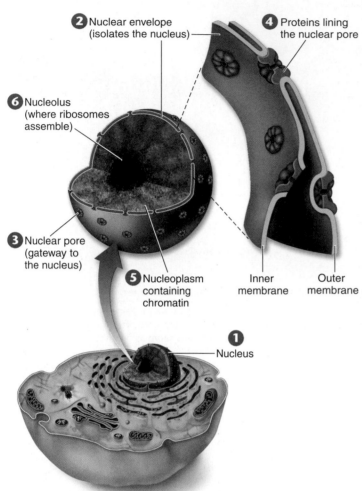

② Nuclear envelope (isolates the nucleus)

④ Proteins lining the nuclear pore

⑥ Nucleolus (where ribosomes assemble)

③ Nuclear pore (gateway to the nucleus)

⑤ Nucleoplasm containing chromatin

Inner membrane

Outer membrane

① Nucleus

Figure 4.12 The nucleus.

The nucleus is composed of a double membrane, called a nuclear envelope, enclosing a fluid-filled interior containing the chromosomes. In cross section, the individual nuclear pores are seen to extend through the two membrane layers of the envelope. A pore is lined with protein, which acts to control access through the pore.

IMPLICATION A widely held view about the origin of the nucleus explains that the nucleus evolved when an ancient archaea invaded a bacterial cell. Does the fact that the nucleus is bounded by two membranes support this theory?

4.5 The Nucleus: The Cell's Control Center

CONCEPT PREVIEW: The nucleus is the command center of the cell. It stores the cell's hereditary information and issues instructions by way of RNA to guide the production of the proteins that carry out cell activities.

Eukaryotic cells have many structural components and organelles in common (table 4.1). If you were to journey far into the interior of one of your cells, you would eventually reach the center of the cell. There you would find, cradled within a network of fine filaments like a ball in a basket, the **nucleus ①** (figure 4.12). The nucleus is the command and control center of the cell, directing all of its activities. It is also the genetic library where the hereditary information is stored.

Nuclear Membrane

The surface of the nucleus is bounded by a special kind of membrane called the **nuclear envelope ②**. The nuclear envelope is actually *two* membranes, one outside the other, like a sweater over a shirt. The nuclear envelope acts as a barrier between the nucleus and the cytoplasm, but substances need to pass through the envelope. The exchange of materials occurs through openings scattered over the surface of this envelope. Called **nuclear pores ③**, these openings form when the two membrane layers of the nuclear envelope pinch together. A nuclear pore is not an empty opening, however; rather, it has many proteins embedded within it that permit proteins and RNA to pass into and out of the nucleus ④.

Chromosomes

In both prokaryotes and eukaryotes, all hereditary information specifying cell structure and function is encoded in DNA. However, unlike prokaryotic DNA that forms an enclosed circle, the DNA of eukaryotes is divided into several segments and associated with protein, forming **chromosomes.** The proteins in the chromosome permit the DNA to wind tightly and condense during cell division. Under a light microscope, these condensed chromosomes are readily seen in dividing cells as densely staining rods. After cell division, eukaryotic chromosomes uncoil and fully extend into threadlike strands called **chromatin ⑤** that can no longer be distinguished individually with a light microscope within the nucleoplasm. Once uncoiled, the chromatin is available for protein synthesis. The process begins when RNA copies of genes are made from the DNA in the nucleus. The RNA molecules leave the nucleus through the nuclear pores and enter the cytoplasm where proteins are synthesized. These proteins, carrying out many different functions, determine what the cell is like.

Ribosomes

To make its many proteins, the cell employs a special structure called a **ribosome,** a kind of platform on which the proteins are built. Ribosomes read the RNA copy of a gene and use that information to direct the construction of a protein. Ribosomes are made up of a special form of RNA called *ribosomal RNA*, or *rRNA*, that is bound up within a complex of several dozen different proteins. Ribosome subunits are assembled in a region within the nucleus called the **nucleolus ⑥**.

Protein synthesis is described in detail in chapter 12. The ribosome is the physical structure that holds the RNA in place while it is "read," being translated into the protein encoded by the RNA molecule.

TABLE 4.1 | Eukaryotic Cell Structures and Their Functions

Structure		Description	Function
Structural Elements			
Cell wall		Outer layer of cellulose or chitin; absent in animal cells	Protection; support
Cytoskeleton		Network of protein filaments	Structural support; cell movement
Flagella and cilia		Cellular extensions with 9 + 2 arrangement of pairs of microtubules	Motility or moving fluids over surfaces
Plasma Membrane and Endomembrane System			
Plasma membrane		Lipid bilayer in which proteins are embedded	Regulates what passes into and out of cell; cell-to-cell recognition
Endoplasmic reticulum		Network of internal membranes	Forms compartments and vesicles; participates in protein and lipid synthesis
Nucleus		Structure (usually spherical) that contains chromosomes; surrounded by double membrane	Control center of cell; directs protein synthesis and cell reproduction
Golgi complex		Stacks of flattened vesicles	Packages proteins for export from the cell; forms secretory vesicles
Lysosomes		Vesicles derived from Golgi complex that contain hydrolytic digestive enzymes	Digest worn-out organelles and cell debris; play role in cell death
Energy-Producing Organelles			
Mitochondria		Bacteria-like elements with double membrane	Sites of oxidative metabolism; provides ATP for cellular energy
Chloroplasts		Bacteria-like organelles found in plants and algae; complex inner membrane consists of stacked vesicles	Sites of photosynthesis
Elements of Gene Expression			
Chromosomes		Long threads of DNA that form a complex with protein	Contain hereditary information
Nucleolus		Site of genes for rRNA synthesis	Assembles ribosomes
Ribosomes		Small, complex assemblies of protein and RNA, often bound to endoplasmic reticulum	Sites of protein synthesis

4.6 The Endomembrane System

CONCEPT PREVIEW: An extensive system of interior membranes organizes the interior of the cell into functional compartments that manufacture and deliver proteins and carry out a variety of specialized chemical processes.

Surrounding the nucleus within the interior of the eukaryotic cell is a tightly packed mass of membranes. They fill the cell, dividing it into compartments, channeling the transport of molecules through the interior of the cell and providing the surfaces on which enzymes act. The system of internal compartments created by these membranes in eukaryotic cells constitutes the most fundamental distinction between the cells of eukaryotes and prokaryotes.

Figure 4.13 The endoplasmic reticulum (ER).

The endoplasmic reticulum provides the cell with an extensive system of internal membranes for the synthesis and transport of materials. Ribosomes are associated with only one side of the rough ER; the other side is the boundary of a separate compartment within the cell into which the ribosomes extrude newly made proteins destined for secretion. Smooth endoplasmic reticulum has few to no bound ribosomes.

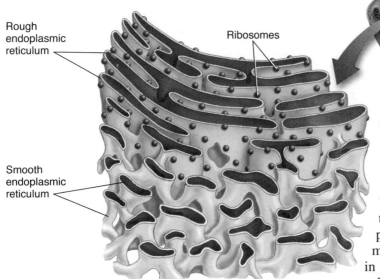

Rough endoplasmic reticulum

Ribosomes

Smooth endoplasmic reticulum

Endoplasmic Reticulum: The Transportation System

The extensive system of internal membranes is called the **endoplasmic reticulum,** often abbreviated **ER.** The term *endoplasmic* means "within the cytoplasm," and the term *reticulum* is a Latin word meaning "little net." Looking at the sheets of membrane weaving through the interior of the cell in figure 4.13, you can see how the ER got its name. The ER creates a series of channels and interconnections, and it also isolates some spaces as membrane-enclosed sacs called **vesicles.**

The surface of the ER is the place where the cell makes proteins intended for export (such as enzymes secreted from the cell surface). The surface of those regions of the ER devoted to the synthesis of such transported proteins is heavily studded with ribosomes and appears pebbly, like the surface of sandpaper, when seen through an electron microscope. For this reason, these regions are called **rough ER.** Regions in which ER-bound ribosomes are relatively scarce are correspondingly called **smooth ER.** The surface of the smooth ER is embedded with enzymes that aid in the manufacture of carbohydrates and lipids.

Figure 4.14 Golgi complex.

This vesicle-forming system, called the Golgi complex after its discoverer, is an integral part of the cell's internal membrane system. The Golgi complex processes and packages materials for transport to another region within the cell and/or for export from the cell. It receives material for processing in transport vesicles on one side and sends the material packaged in secretory vesicles off the other side. (a) Diagram of a Golgi complex. (b) Micrograph of a Golgi complex showing vesicles.

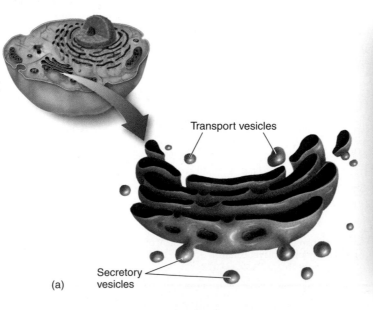

Transport vesicles

Secretory vesicles

(a)

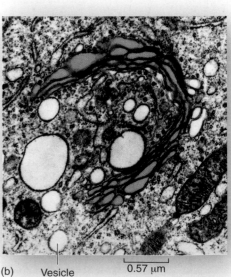

(b) Vesicle

0.57 μm

The Golgi Complex: The Delivery System

As new molecules are made on the surface of the ER, they are passed from the ER to flattened stacks of membranes called **Golgi bodies.** These structures, which in figure 4.14 look like pancakes stacked one on top of the other, are named for Camillo Golgi, the nineteenth-century Italian physician who first called attention to them. The number of Golgi bodies a cell contains ranges from 1 or a few in protists, to 20 or more in animal cells and several hundred in certain plant cells. Golgi bodies function in the collection, packaging, and distribution of molecules manufactured in the cell. Scattered through the cytoplasm, Golgi bodies are collectively referred to as the **Golgi complex.**

The rough ER, smooth ER, and Golgi work together as an endomembrane transport system in the cell. Figure 4.15 walks you through the path that molecules take from the ER through the Golgi and out to their final destinations. Proteins and lipids that are manufactured on the ER membranes are transported through the channels of the ER and are packaged into transport vesicles that bud off from the ER ❶. The vesicles fuse with the membrane of the Golgi bodies, dumping their contents into the Golgi ❷. Within the Golgi bodies the molecules may take one of many paths, indicated by the branching arrows in figure 4.15. Many of these molecules become tagged with carbohydrates. The molecules collect at the ends of the membranous folds of the Golgi bodies; these folds are given the special name *cisternae* (Latin, collecting vessels). Vesicles that pinch off from the cisternae carry the molecules to the different compartments of the cell ❸ and ❹, or to the inner surface of the plasma membrane, where molecules to be secreted are released to the outside ❺.

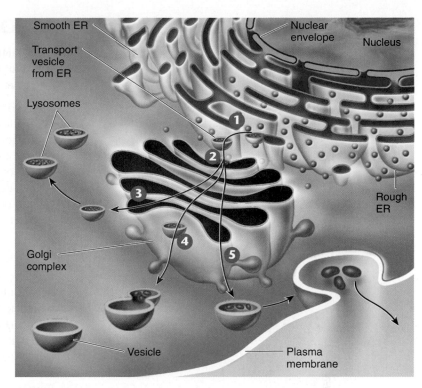

Figure 4.15 **How the endomembrane system works.**

A highly efficient highway system within the cell, the endomembrane system transports material from the ER to the Golgi and from there to other destinations.

Lysosomes: Recycling Centers

Other organelles called **lysosomes** arise from the Golgi complex (the light orange vesicle budding at ❸) and contain a concentrated mix of the powerful enzymes manufactured in the rough ER that break down macromolecules. Lysosomes are the recycling centers of the cell, breaking down failing organelles and other structures within cells, digesting worn-out cell components, and recycling the proteins and other materials of the old parts.

Vacuoles: Storage Compartments

The interiors of plant and many protist cells contain membrane-bounded storage compartments called **vacuoles.** The center of the plant cell shown in figure 4.16, as in all plant cells, contains a large, apparently empty space, called the *central vacuole*. This vacuole is not really empty; it contains large amounts of water and other materials, such as sugars, ions, and pigments. The central vacuole functions as a storage center for these important substances.

In protists like *Paramecium,* cells contain a *contractile vacuole* near the cell surface that accumulates excess water. This vacuole has a small pore that opens to the outside of the cell. By rhythmic contractions, it pumps accumulated water out through the pore.

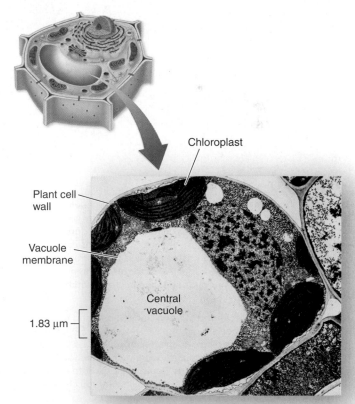

Figure 4.16 **A plant central vacuole.**

A plant's central vacuole stores dissolved substances and can increase in size to increase the surface area of a plant cell.

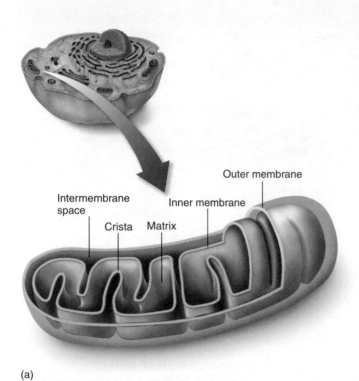

Figure 4.17   Mitochondria.

The mitochondria of a cell are sausage-shaped organelles within which oxidative metabolism takes place, and energy is extracted from food using oxygen. (a) A mitochondrion has a double membrane. The inner membrane is shaped into folds called cristae. The space within the cristae is called the matrix. The cristae greatly increase the surface area for oxidative metabolism. (b) Micrograph of two mitochondria, one in cross section, the other cut lengthwise.

IMPLICATION   Mitochondrial medicine is a new and rapidly developing medical speciality of diseases resulting from failure of the mitochondria. These diseases affect almost all organs, for the simple reason that almost all organs depend on mitochondria for the energy they use. Which organs would you expect to exhibit the most serious mitochondrial diseases? Explain why you chose the organs you did.

4.7 Organelles That Harvest Energy

CONCEPT PREVIEW: Eukaryotic cells contain complex organelles that harvest energy. They have their own DNA and are thought to have arisen by endosymbiosis from ancient bacteria.

Eukaryotic cells contain several kinds of complex, energy harvesting organelles that contain their own DNA and appear to have been derived from ancient bacteria. These bacteria were taken up by ancestral eukaryotic cells in the distant past and remained living inside the host cell, forming a relationship called symbiosis. These organelles include mitochondria (which occur in the cells of all but a very few eukaryotes) and chloroplasts (which occur only in algae and plant cells—they do not occur in animal or fungal cells).

Mitochondria: Powerhouses of the Cell

Eukaryotic organisms extract energy from organic molecules ("food") in a complex series of chemical reactions called **oxidative metabolism,** which takes place only in their mitochondria. **Mitochondria** (singular, **mitochondrion**) are sausage-shaped organelles about the size of a bacterial cell. Mitochondria are bounded by two membranes. The outer membrane, shown partially cut away in figure 4.17, is smooth and apparently derives from the plasma membrane of the host cell that first took up the bacterium long ago. The inner membrane, apparently the plasma membrane of the bacterium that gave rise to the mitochondrion, is bent into numerous folds called **cristae** (singular, **crista**) that resemble the folded plasma membranes in various groups of bacteria. The cutaway view of the figure shows how the cristae partition the mitochondrion into two compartments, an inner **matrix** and an outer compartment, called the **intermembrane space.** As you will learn in chapter 7, this architecture is critical to successfully carrying out oxidative metabolism.

During the 1.5 billion years in which mitochondria have existed in eukaryotic cells, most of their genes have been transferred to the chromosomes of the host cells. But mitochondria still have some of their original genes, contained in a circular, closed, naked molecule of DNA (called *mitochondrial DNA,* or *mtDNA*) that closely resembles the circular DNA molecule of a bacterium. On this mtDNA are several genes that produce some of the proteins essential for oxidative metabolism. In both mitochondria and bacteria, the circular DNA molecule is replicated during the process of division. When a mitochondrion divides, it copies its DNA located in the matrix and splits into two by simple fission, dividing much as bacteria do.

Chloroplasts: Energy-Capturing Centers

All photosynthesis in plants and algae takes place within another bacteria-like organelle, the **chloroplast** (figure 4.18). There is strong evidence that chloroplasts, like mitochondria, were derived from bacteria by symbiosis. A chloroplast is bounded, like a mitochondrion, by two membranes, the inner derived from the original bacterium and the outer resembling the host cell's ER. Chloroplasts are larger than mitochondria, and have a more complex organization. Inside the chloroplast, another series of membranes are fused to form stacks of closed vesicles called **thylakoids,** the green disklike structures visible in the interior of the chloroplast in

The light-dependent reactions of photosynthesis are discussed on pages 104 to 107. Proteins embedded in the thylakoid membrane use energy from the sun to make two molecules, ATP and NADPH. These molecules are used to make sugar molecules.

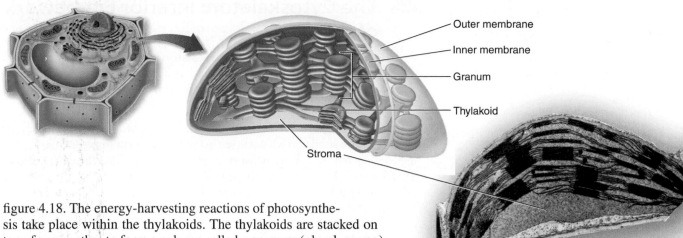

figure 4.18. The energy-harvesting reactions of photosynthesis take place within the thylakoids. The thylakoids are stacked on top of one another to form a column called a **granum** (plural, **grana**). The interior of a chloroplast is bathed with a semiliquid substance called the **stroma.**

Like mitochondria, chloroplasts have a circular DNA molecule. This DNA contains many of the genes coding for the proteins necessary to carry out photosynthesis. Plant cells can contain from one to several hundred chloroplasts, depending on the species. Neither mitochondria nor chloroplasts can be grown in a cell-free culture; they are totally dependent on the cells within which they occur.

Figure 4.18   A chloroplast.

Bacteria-like organelles called chloroplasts are the sites of photosynthesis in photosynthetic eukaryotes. Like mitochondria, they have a complex system of internal membranes on which chemical reactions take place. The internal membranes of a chloroplast are fused to form stacks of closed vesicles called thylakoids. Photosynthesis occurs within these thylakoids. Thylakoids are stacked one on top of the other in columns called grana. The interior of the chloroplast is bathed in a semiliquid substance called the stroma.

Endosymbiosis

Symbiosis is a close, integrated relationship between organisms of different species that live together. The theory of **endosymbiosis** proposes that some of today's eukaryotic organelles evolved by a symbiosis in which one cell of a prokaryotic species was engulfed by and lived inside the cell of another species of prokaryote that was a precursor to eukaryotes. Figure 4.19 shows how this is thought to have occurred. Many cells take up food or other substances through endocytosis, a process whereby the plasma membrane of a cell wraps around the substance, enclosing it within a vesicle inside the cell. Oftentimes, the contents in the vesicle are broken down with digestive enzymes. According to the endosymbiont theory this did not occur; instead, the engulfed prokaryotes provided their hosts with certain advantages associated with their special metabolic abilities. Two key eukaryotic organelles just described are believed to be the descendants of these endosymbiotic prokaryotes: mitochondria, which are thought to have originated as bacteria capable of carrying out oxidative metabolism; and chloroplasts, which apparently arose from photosynthetic bacteria.

The endosymbiont theory is supported by a wealth of evidence. Both mitochondria and chloroplasts are surrounded by two membranes; the inner membrane probably evolved from the plasma membrane of the engulfed bacterium, while the outer membrane is probably derived from the plasma membrane or endoplasmic reticulum of the host cell. Mitochondria are about the same size as most bacteria, and the cristae formed by their inner membranes resemble the folded membranes in various groups of bacteria. Mitochondrial ribosomes are also similar to bacterial ribosomes in size and structure. Both mitochondria and chloroplasts contain circular molecules of DNA similar to those in bacteria. Finally, mitochondria divide by simple fission, splitting in two just as bacterial cells do, and they apparently replicate and partition their DNA in much the same way as bacteria.

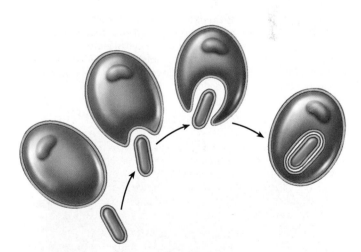

Figure 4.19   Endosymbiosis.

This figure shows how a double membrane may have been created during the symbiotic origin of mitochondria or chloroplasts.

IMPLICATION   When first proposed (by Professor Lynn Margulis of Amherst College), the endosymbiosis theory was once widely ridiculed as absurd speculation, until molecular evidence was found to support it. Her more recent Gaia hypothesis that our planet is a super-organism is being similarly discounted. How do you feel about such an idea?

4.8 The Cytoskeleton: Interior Framework of the Cell

CONCEPT PREVIEW: The cytoskeleton is a latticework of protein fibers that determines a cell's shape and anchors organelles to particular locations within the cytoplasm. Cells can move with the aid of cytoskeletal proteins.

If you could shrink down and enter into the interior of a eukaryotic cell, your view would be similar to what you see in figure 4.20, your immediate surroundings dominated by a dense network of protein fibers called the **cytoskeleton.** This framework supports the shape of the cell. It also anchors organelles like the mitochondria in this figure to fixed locations within the cell interior. This fiber network cannot be seen with a light microscope because the individual fibers are single chains of protein, much too fine for microscopes to resolve. To "see" the cytoskeleton, scientists attach fluorescent antibodies to the protein fibers and then photograph them under fluorescent light.

The protein fibers of the cytoskeleton are a dynamic system, constantly being formed and disassembled. There are three different kinds of protein fibers shown as enlargements in the figure: thick ropes of intertwined protein called **intermediate filaments,** hollow tubes called **microtubules** made of the protein tubulin, and long, slender **microfilaments** made of the protein **actin.** The intermediate filaments, microtubules, and microfilaments are anchored to membrane proteins embedded within the plasma membrane.

The cytoskeleton plays a major role in determining the shape of animal cells, which lack rigid cell walls. Because filaments can form and disassemble readily, the shape of an animal cell can change rapidly. If you examine the surface of an animal cell with a microscope, you will often find it alive with motion, projections shooting out from the surface and then retracting, only to shoot out elsewhere moments later.

The cytoskeleton is not only responsible for the cell's shape, but it also provides a scaffold both for ribosomes to carry out protein synthesis and for enzymes to be localized within defined areas of the cytoplasm. By anchoring particular enzymes near one another, the cytoskeleton participates with organelles in organizing the cell's activities.

Figure 4.20   The protein fibers of the cytoskeleton.

Organelles such as mitochondria are anchored to fixed locations in the cytoplasm by the cytoskeleton, a network of protein fibers. *Intermediate filaments* are composed of overlapping proteins that form a ropelike structure, providing tremendous mechanical strength to the cell. *Microtubules* are composed of tubulin protein subunits arranged side by side to form a tube. Microtubules are comparatively stiff cytoskeletal elements that function in intracellular transport and stabilization of cell structure. *Actin microfilaments* are made of two strands of the protein actin twisted together and usually occur in bundles. Actin microfilaments are responsible for cell movement.

IMPLICATION   Athletic muscle injuries can result from overstretching before hard running or other activities that extend muscles. Damage appears to occur at the cytoskeleton level, with temporary disorganization of the intermediate filaments. Do you think such injuries are more likely to occur among overenthusiastic joggers or experienced athletes? Explain.

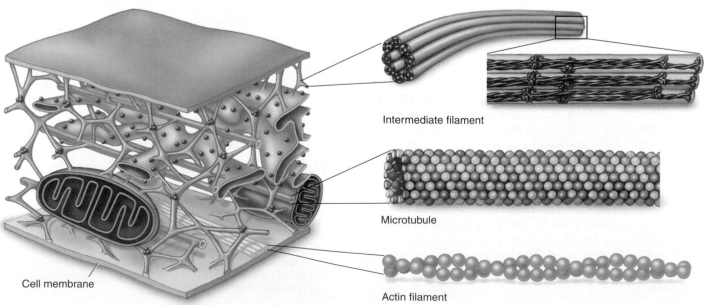

Cell membrane

Intermediate filament

Microtubule

Actin filament

Centrioles

Complex structures called **centrioles** assemble microtubules from tubulin subunits in the cells of animals and most protists. Centrioles occur in pairs within the cytoplasm, usually located at right angles to one another as you can see in figure 4.21. They are usually found near the nuclear envelope and are among the most structurally complex microtubular assemblies of the cell.

> Centrioles are involved in assembling a network of microtubules that attach to and separate the chromosomes during cell division. This network of microtubules, called the spindle, is discussed in more detail in chapters 8 and 9.

Figure 4.21   Centrioles.

Centrioles anchor and assemble microtubules. Centrioles usually occur in pairs and are composed of nine triplets of microtubules.

Microtubule triplet

Cell Movement

Like a tendon that anchors a muscle to a bone, intermediate filaments act as intracellular tendons, preventing excessive stretching of cells. Actin microfilaments play a major role in determining the shape of cells. Essentially, all cell motion is tied to the movement of actin microfilaments, microtubules, or both. Because actin microfilaments can form and dissolve so readily, they enable some cells to change shape quickly and move from place to place.

Some Cells Crawl. It is the arrangement of actin microfilaments within the cell cytoplasm that allows cells to "crawl," literally! Crawling is a significant cellular phenomenon, essential to inflammation, clotting, wound healing, and the spread of cancer. White blood cells in particular exhibit this ability. Produced in the bone marrow, these cells are released into the circulatory system and eventually crawl out of capillaries and into the tissues to destroy potential pathogens. The crawling mechanism is an exquisite example of cellular coordination.

Swimming with Flagella and Cilia. Some eukaryotic cells contain **flagella** (singular, **flagellum**), fine, long, threadlike organelles protruding from the cell surface. Figure 4.22*a* shows how a flagellum arises from a microtubular structure called a **basal body,** with groups of microtubules arranged in rows of three, shown in the cross-sectional view. Some of these microtubules extend up into the flagellum, which consists of a circle of nine microtubule pairs surrounding two central microtubules. This **9 + 2 arrangement** is a fundamental feature of eukaryotes and apparently evolved early in their history. In humans, we find a single long flagellum on each sperm cell that propels the cell in a swimming motion. If flagella are numerous and organized in dense rows, they are called **cilia.** Cilia do not differ from flagella in their structure, but cilia are usually shorter. The paramecium in figure 4.22*b* is covered with cilia, giving it a furry appearance. In humans, dense mats of cilia project from cells that line our breathing tube, the trachea, to move mucus and dust particles out of the respiratory tract into the throat (where we can expel these contaminants by spitting or swallowing). Eukaryotic flagella serve a similar function as the bacterial flagella discussed in section 4.2, but are very different structurally.

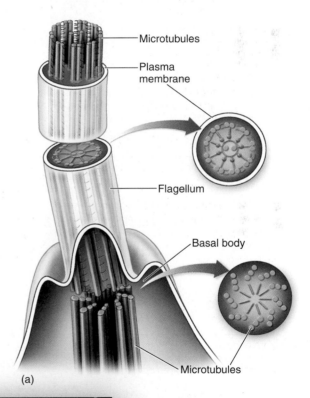

Microtubules

Plasma membrane

Flagellum

Basal body

Microtubules

(a)

(b)

Figure 4.22   Flagella and cilia.

(a) A eukaryotic flagellum springs directly from a basal body and is composed of a ring of nine pairs of microtubules with two microtubules in its core. (b) The surface of this paramecium is covered with a dense forest of cilia.

Concept Check

1. How are proteins able to insert into the plasma membrane?
2. How is smooth ER different from rough ER?
3. What is the evidence that mitochondria evolved from ancient bacteria?
4. Which element of the cytoskeleton is active in the crawling of white blood cells?

Transport Across Plasma Membranes

Essential Biological Process 4A

Diffusion

1

Lump of sugar

A lump of sugar is dropped into a beaker of water.

2

Sugar molecule

Sugar molecules begin to break off from the lump.

3

More and more sugar molecules move away and randomly bounce around.

4

Eventually, all of the sugar molecules become evenly distributed throughout the water.

4.9 Diffusion and Osmosis

CONCEPT PREVIEW: Random movements of molecules cause them to move to areas of lower concentration, a process called diffusion. Water associated with polar solutes is not free to diffuse, and there is a net movement of water across a membrane toward the side with less "free" water, a process called osmosis.

For cells to survive, nutrients, water, and other materials must pass into the cell, and waste materials must be eliminated. All of this moving back and forth across the cell's plasma membrane occurs in one of three ways: (1) water and other substances diffuse through the membrane, (2) food particles and sometimes liquids are engulfed by the membrane folding around them, or (3) proteins in the membrane act as doors that admit certain molecules only. First we will examine diffusion.

Diffusion

How a molecule moves—just where it goes—is totally random, like shaking marbles in a cup, so if two kinds of molecules are added together, they soon mix. The random motion of molecules always tends to produce uniform mixtures because a substance moves from regions where its concentration is high to regions where its concentration is lower (that is, *down* the **concentration gradient**). How does a molecule "know" in what direction to move? It doesn't—molecules don't "know" anything. A molecule is equally likely to move in any direction and is constantly changing course in random ways. There are simply more molecules able to move from where they are common than from where they are scarce. This mixing process is called **diffusion.** Diffusion (*Essential Biological Process 4A*) is the net movement of molecules down a concentration gradient toward regions of lower concentration (that is, where there are relatively fewer of them) as a result of random motion. For example, a lump of sugar dropped into a beaker of water will break apart into individual sugar molecules that will move about randomly. However, they will tend to travel away from the area of high concentration (the sugar cube) to an area of lower concentration (the rest of the beaker). Eventually, the substance will achieve a state of *equilibrium,* where there is no net movement toward any particular direction (as shown in panel 4). The individual molecules of the substance are still in motion, but there is no net change in direction.

Osmosis

Diffusion allows molecules like oxygen, carbon dioxide, and nonpolar lipids to cross the plasma membrane. Ions and polar molecules, by contrast, cannot cross the very nonpolar environment found in the lipid core of the membrane bilayer. However, the movement of water molecules, which are very polar, is not blocked—water diffuses freely across the plasma membrane. How is this possible? Water molecules pass through small channels, called **aquaporins,** that traverse the membrane. These water channels are very selective, even blocking the passage of protons (hydrogen ions), which are smaller than water molecules. This selectivity is due to a cluster of positively charged amino acids that line the pore and repel protons, which are also positively charged.

As in diffusion, water passes into and out of a cell down its concentration gradient, a process called **osmosis** (*Essential Biological Process 4B*). However, the movement of water into and out of a cell is dependent upon the concentration of other substances in solution. To understand how water moves into and out of a cell, let's focus on the water molecules already present inside a cell. What are they doing? Many of them are interacting with the sugars, proteins, and other polar molecules inside. Remember, water is very polar itself and readily interacts with other polar molecules. These social water molecules are not randomly moving about as they were outside; instead, they remain clustered around the polar molecules inside the cell. As a result, while water molecules keep coming into the cell by random motion, they don't randomly come out again. The simple experiment shown in *Essential Biological Process 4B* illustrates what happens. Think of the right side of the beaker as the inside of a cell, and the left side is a watery environment. Like the plasma membrane, a semipermeable membrane that allows some substances to pass but not all substances separates the two areas. When the polar molecule urea is present in the cell, water molecules cluster around each urea molecule and are no longer able to pass through the membrane to the "outside." In effect, the polar solute has reduced the number of free water molecules. Because the "outside" of the cell (on the left) has more unbound water molecules, water moves by diffusion into the cell (to the right).

The concentration of *all* molecules dissolved in a solution (the **solutes**) is called the osmotic concentration of the solution. If the osmotic concentrations of two solutions are equal, the solutions are **isotonic** (Greek *iso,* the same). If two solutions have unequal osmotic concentrations, the solution with the higher solute concentration is said to be **hypertonic** (Greek *hyper,* more than), and the solution with the lower one is **hypotonic** (Greek *hypo,* less than).

Movement of water into a cell by osmosis creates pressure, called osmotic pressure, which can cause a cell to swell and burst (figure 4.23). Most animal cells cannot withstand osmotic pressure unless their plasma membranes are braced to resist the swelling. If placed in pure water, they soon burst like overinflated balloons. That is why the cells of so many kinds of organisms have cell walls to stiffen their exteriors. In fact, this osmotic pressure, called turgor pressure in plants, is important for plant cells to maintain their shape. Without adequate water inside the cells, the plants wilt. In animals, the fluids bathing the cells have as many polar molecules dissolved in them as the cells do, making them isotonic, so the problem doesn't arise.

Hypertonic Solution	Isotonic Solution	Hypotonic Solution

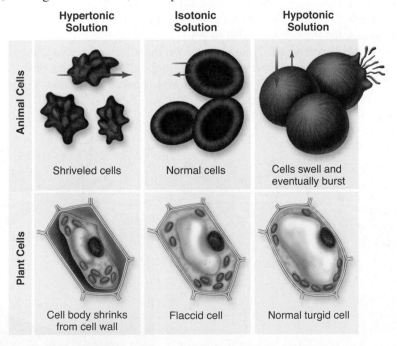

Animal Cells		
Shriveled cells	Normal cells	Cells swell and eventually burst
Plant Cells		
Cell body shrinks from cell wall	Flaccid cell	Normal turgid cell

Figure 4.23   **Osmotic pressure in animal and plant cells.**

Essential Biological Process 4B
Osmosis

1

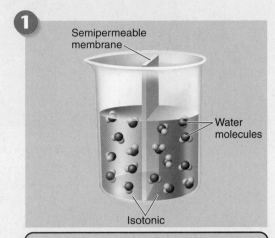

Semipermeable membrane

Water molecules

Isotonic

Diffusion causes water molecules to distribute themselves equally on both sides of a semipermeable membrane.

2

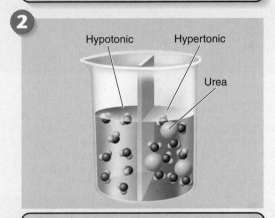

Hypotonic   Hypertonic

Urea

Addition of solute molecules that cannot cross the membrane reduces the number of free water molecules on that side, as they bind to the solute.

3

Diffusion then causes free water molecules to move from the side where their concentration is higher to the solute side, where their concentration is lower.

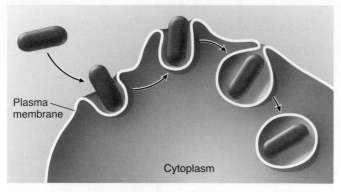

Plasma membrane

Cytoplasm

(a) Phagocytosis

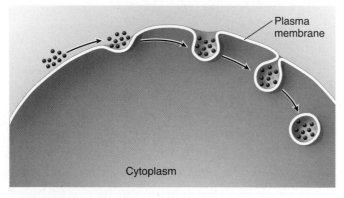

Plasma membrane

Cytoplasm

(b) Pinocytosis

Figure 4.24   Endocytosis.

Endocytosis is the process of engulfing material by folding the plasma membrane around it, forming a vesicle. (a) When the material is an organism or some other relatively large fragment of organic matter, the process is called phagocytosis. (b) When the material is a liquid, the process is called pinocytosis.

4.10 Bulk Passage into and out of Cells

CONCEPT PREVIEW: The plasma membrane can engulf materials by endocytosis, folding the membrane around the material to encase it within a vesicle. Exocytosis is essentially this process in reverse, using vesicles to expel substances.

Endocytosis

The cells of many eukaryotes take in food and liquids by extending their plasma membranes outward toward food particles. The membrane engulfs the particle and forms a vesicle—a membrane-bounded sac—around it. This process is called **endocytosis** (figure 4.24).

If the material the cell takes in is particulate (made up of discrete particles), such as an organism, like the red bacterium in figure 4.24*a*, or some other fragment of organic matter, the process is called **phagocytosis** (Greek *phagein,* to eat, and *cytos,* cell). If the material the cell takes in is liquid or substances dissolved in a liquid, like the small particles in figure 4.24*b*, it is called **pinocytosis** (Greek *pinein,* to drink). Pinocytosis is common among animal cells. Mammalian egg cells, for example, "nurse" from surrounding cells; the nearby cells secrete nutrients that the maturing egg cell takes up by pinocytosis. Virtually all eukaryotic cells constantly carry out these kinds of endocytosis, trapping particles and extracellular fluid in vesicles and ingesting them. Endocytosis rates vary from one cell type to another. They can be surprisingly high: Some types of white blood cells ingest 25% of their cell volume each hour!

Exocytosis

The reverse of endocytosis is **exocytosis,** the discharge of material from vesicles at the cell surface. The vesicle in figure 4.25 contains a substance to be discharged, or released, from the cell. The purple particles remain suspended in the vesicle as it fuses with the plasma membrane. The membrane that forms the vesicle is made of phospholipids, and as it comes in contact with the plasma membrane, the phospholipids of both membranes interact, forming a pore through which the contents leave the vesicle to the outside. In plant cells, exocytosis is an important means of exporting the materials needed to construct the cell wall that lies outside the plasma membrane. Among protists, the discharge of a contractile vacuole is a form of exocytosis. In animal cells, exocytosis provides a mechanism for secreting many hormones, neurotransmitters, digestive enzymes, and other substances.

Figure 4.25   Exocytosis.

Exocytosis is the discharge of material from vesicles at the cell surface. Proteins and other molecules are secreted from cells in small pockets called secretory vesicles, whose membranes fuse with the plasma membrane, thereby allowing the secretory vesicles to release their contents to the cell surface. In the photomicrograph, you can see exocytosis taking place somewhat explosively.

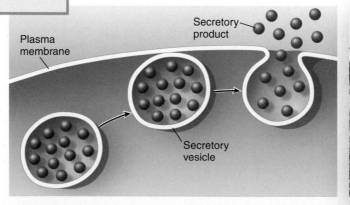

Plasma membrane

Secretory product

Secretory vesicle

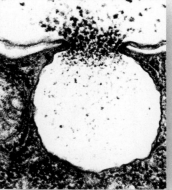

4.11 Selective Permeability

CONCEPT PREVIEW: Cells are selectively permeable, admitting only certain molecules. Facilitated diffusion is selective transport across a membrane in the direction of lower concentration. Active transport is energy-driven transport across a membrane toward a region of higher concentration.

From the point of view of efficiency, the problem with endocytosis is that it is expensive to carry out—the cell must make and move a lot of membrane. Also, endocytosis is not picky—in pinocytosis particularly, engulfing liquid does not allow the cell to choose which molecules come in. Cells solve this problem by using proteins in the plasma membrane as channels to pass molecules into and out of the cell. Because each kind of channel allows passage of only a certain kind of molecule, the cell can control what enters and leaves, an ability called **selective permeability.**

Selective Diffusion

Some channels act like open doors. As long as a molecule fits the channel, it is free to pass through in either direction. Diffusion tends to equalize the concentration of such molecules on both sides of the membrane, with the molecules moving toward the side where they are scarcest. This mechanism of transport is called **selective diffusion.** One class of selectively open channels consists of ion channels, which are pores that span the membrane. Ions that fit the pore can diffuse through it in either direction. Such ion channels play an essential role in signaling by the nervous system.

Facilitated Diffusion

Most diffusion occurs through use of a special carrier protein. This protein binds only certain kinds of molecules, such as a particular sugar, amino acid, or ion. The molecule physically binds to the carrier on one side of the membrane and is released to the other side. The direction of the molecule's net movement depends on its concentration gradient across the membrane. If the concentration is greater outside the cell, the molecule is more likely to bind to the carrier on the extracellular side of the membrane, as shown in panel 1 of the *Essential Biological Process 4C*, and be released on the cytoplasmic side, as in panel 3. If the concentration of the molecule is greater inside the cell, the net movement will be from inside to outside. Thus the net movement always occurs from high concentration to low, just as it does in simple diffusion, but the process is facilitated by the carriers. For this reason, this mechanism of transport is given a special name, **facilitated diffusion.**

A characteristic feature of transport by carrier proteins is that its rate can be saturated. If the concentration of a substance is progressively increased, the rate of transport of the substance increases up to a certain point and then levels off. There are a limited number of carrier proteins in the membrane, and when the concentration of the transported substance is raised high enough, all the carriers will be in use. The transport system is then said to be "saturated." When an investigator wishes to know if a particular substance is being transported across a membrane by a carrier system, or is diffusing across, he or she conducts experiments to see if the transport system can be saturated. If it can be saturated, it is carrier-mediated; if it cannot be saturated, it is not.

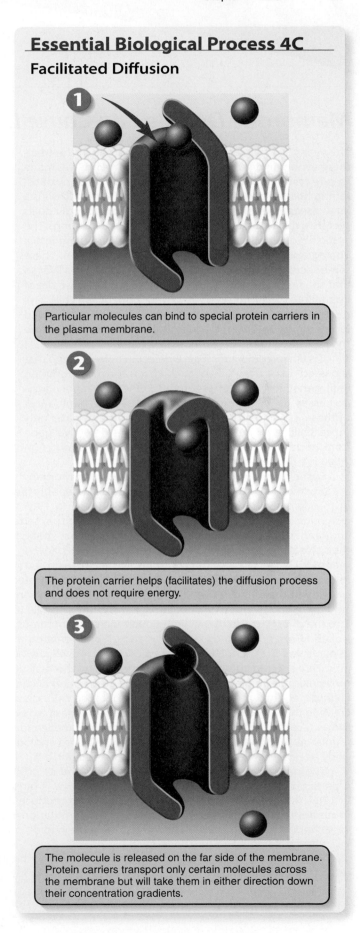

Essential Biological Process 4C

Facilitated Diffusion

1 Particular molecules can bind to special protein carriers in the plasma membrane.

2 The protein carrier helps (facilitates) the diffusion process and does not require energy.

3 The molecule is released on the far side of the membrane. Protein carriers transport only certain molecules across the membrane but will take them in either direction down their concentration gradients.

Membrane Defects Can Cause Disease

The year 1993 marked an important milestone in the treatment of human disease. That year the first attempt was made to cure **cystic fibrosis** (**CF**), a deadly genetic disorder, by transferring healthy genes into sick individuals. Cystic fibrosis is a fatal disease in which the body cells of affected individuals secrete a thick mucus that clogs the airways of the lungs. The cystic fibrosis patient in the photograph is breathing into a Vitalograph, a device that measures lung function. These same secretions block the ducts of the pancreas and liver so that the few patients who do not die of lung disease die of liver failure. Cystic fibrosis is usually thought of as a children's disease because until recently few affected individuals lived long enough to become adults. Even today half die before their mid-twenties. There is no known cure.

Cystic fibrosis results from a defect in a single gene that is passed down from parent to child. It is the most common fatal genetic disease of Caucasians. One in 20 individuals possesses at least one copy of the defective gene. Most of these individuals are not afflicted with the disease; only those children who inherit a copy of the defective gene from each parent succumb to cystic fibrosis—about 1 in 2,500 infants.

Cystic fibrosis has proven difficult to study. Many organs are affected, and until recently it was impossible to identify the nature of the defective gene responsible for the disease. In 1985 the first clear clue was obtained. An investigator, Paul Quinton, seized on a commonly observed characteristic of cystic fibrosis patients, that their sweat is abnormally salty, and performed the following experiment. He isolated a sweat duct from a small piece of skin and placed it in a solution of salt (NaCl) that was three times as concentrated as the NaCl inside the duct. He then monitored the movement of ions. Diffusion tends to drive both the sodium (Na^+) and the chloride (Cl^-) ions into the duct because of the higher outer ion concentrations. In skin isolated from normal individuals, Na^+ and Cl^- both entered the duct, as expected. In skin isolated from cystic fibrosis individuals, however, only Na^+ entered the duct—no Cl^- entered. For the first time, the molecular nature of cystic fibrosis became clear. Water accompanies chloride, and was not entering the ducts because chloride was not, creating thick mucus. Cystic fibrosis is a defect in a plasma membrane protein called CFTR (*cystic fibrosis transmembrane conductance regulator*) that normally regulates passage of Cl^- into and out of the body's cells.

The defective *cf* gene was isolated in 1987, and its position on a particular human chromosome (chromosome 7) was pinpointed in 1989. Interestingly, many cystic fibrosis patients produce a CFTR protein with a normal amino acid sequence. The *cf* mutation in these cases appears to interfere with how the CFTR protein folds, preventing it from folding into a functional shape.

Soon after the *cf* gene was isolated, experiments were begun to see if it would be possible to cure cystic fibrosis by gene therapy—that is, by transferring healthy *cf* genes into the cells with defective ones. In 1990 a working *cf* gene was successfully transferred into human lung cells growing in tissue culture, using adenovirus, a cold virus, to carry the gene into the cells. The CFTR-defective cells were "cured," becoming able to transport chloride ions across their plasma membranes. Then in 1991 a team of researchers successfully transferred a normal human *cf* gene into the lung cells of a living animal—a rat. The *cf* gene was first inserted into the adenovirus genome because adenovirus is a cold virus and easily infects lung cells. The treated virus was then inhaled by the rat. Carried piggyback, the *cf* gene entered the rat lung cells and began producing the normal human CFTR protein within these cells!

These results were very encouraging, and at first the future for all cystic fibrosis patients seemed bright. Clinical tests using adenovirus to introduce healthy *cf* genes into cystic fibrosis patients were begun with much fanfare in 1993.

They were not successful. As described in detail in chapter 13, there were insurmountable problems with the adenovirus being used to transport the *cf* gene into cystic fibrosis patients. The difficult and frustrating challenge that cystic fibrosis researchers had faced was not over. Research into clinical problems is often a time-consuming and frustrating enterprise, never more so than in this case. Recently, new ways of introducing the healthy *cf* gene have been tried with better results. The long, slow journey toward a cure has taught us not to leap to the assumption that a cure is now at hand, but the steady persistence of researchers has taken us a long way, and again the future for cystic fibrosis patients seems bright.

Active Transport

Other channels through the plasma membrane are closed doors. These channels open only when energy is provided. They are designed to enable the cell to maintain high or low concentrations of certain molecules, much more or less than exists outside the cell. Like motor-driven turnstiles, the channels operate to move a certain substance *up* its concentration gradient. The operation of these one-way, energy-requiring channels results in **active transport,** the movement of molecules across a membrane to a region of higher concentration by the expenditure of energy.

You might think that the plasma membrane possesses all sorts of active transport channels for the transport of sugars, amino acids, and other molecules, but in fact, most of the active transport in cells is carried out by one kind of channel, the sodium-potassium pump.

The Sodium-Potassium Pump. The most important active transport channel is the *sodium-potassium (Na⁺-K⁺) pump,* which expends metabolic energy to actively pump sodium ions (Na^+) out of cells, and potassium ions (K^+) into cells (*Essential Biological Process 4D*). More than one-third of all the energy expended by your body's cells is spent driving Na^+-K^+ pump channels. This energy is derived from *adenosine triphosphate (ATP),* a molecule we will learn more about in chapter 5. The transportation of two different ions in opposite directions happens because energy causes a change in the shape of the membrane protein carrier. The panels to the right walk you through one cycle of the pump. Each channel can move over 300 sodium ions per second when working full tilt. As a result of all this pumping, there are far fewer sodium ions in the cell. This concentration gradient, paid for by the expenditure of considerable metabolic energy in the form of ATP molecules, is exploited by your cells in many ways. Two of the most important are (1) the conduction of signals along nerve cells (discussed in detail in chapter 29) and (2) the pulling of valuable molecules such as sugars and amino acids into the cell *against* their concentration gradient!

Cells of the nervous system, called neurons, have an electrical charge difference across their plasma membranes that allows them to function properly, as discussed on page 553. This charge difference, called the resting membrane potential, is maintained through the actions of Na^+/K^+ pumps.

We will focus for a moment on this second process. The plasma membranes of many cells are studded with facilitated diffusion channels, which offer a path for sodium ions that have been pumped out by the Na^+-K^+ pump to diffuse back in. There is a catch, however; these channels require that the sodium ions have a partner in order to pass through—like a dancing party where only couples are admitted through the door—which is why these are called coupled channels. Coupled channels won't let sodium ions across unless another molecule tags along, crossing hand in hand with the sodium ion. In some cases the partner molecule is a sugar, in others an amino acid or other molecule. Because the concentration gradient for sodium is so large, many sodium ions are trying to get back in, this diffusion pressure drags in the partner molecules as well, even if they are already in high concentration within the cell. In this way, sugars and other actively transported molecules enter the cell—via special coupled channels.

Concept Check

1. How does water diffuse across the plasma membrane?
2. What happens when a red blood cell is placed in a: hyper-, hypo-, and isotonic solution?
3. Distinguish between selective diffusion and facilitated diffusion.

Essential Biological Process 4D

The Sodium-Potassium Pump

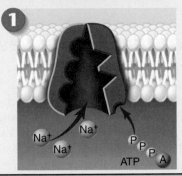

The sodium-potassium pump binds three sodium ions and a molecule of ATP.

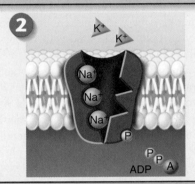

The splitting of ATP provides energy to change the shape of the channel. The sodium ions are driven through the channel.

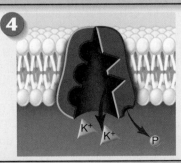

The sodium ions are released to the outside of the membrane, and the new shape of the channel allows two potassium ions to bind.

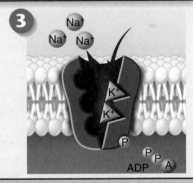

Release of the phosphate allows the channel to revert to its original form, releasing the potassium ions on the inside of the membrane.

Eating at the Cellular Level: What Limits How Fast a Cell Can Take Up Glucose?

All animal cells need food molecules to power their lives. The food molecules you eat (proteins, fats, and carbohydrates) are broken into simpler bits that can be taken up and used as fuel by your cells. But many of these food molecules are polar (partially charged), and the lipid bilayer that surrounds all cells is a nonpolar barrier across which polar molecules cannot readily diffuse. The blue line on the graph to the right is the calculated curve for the rate of glucose uptake if it enters a cell solely by simple diffusion across a lipid bilayer membrane. As you can see, the rate is very low, not enough to keep a red blood cell alive. And yet your body's red blood cells do not starve. So how do these polar molecules get in?

When scientists learned that the membranes of human cells contain many proteins inserted into them, an answer suggested itself: Perhaps some of these proteins form channels which let polar food molecules like glucose pass in across the membrane. When they looked, they found that glucose-associated channels account for some 2% of the total protein in a red blood cell's plasma membrane! As you can see in the diagram, a channel is composed of a long rope-like protein consisting of alternating segments of polar and nonpolar amino acids, looping back and forth to create a passage spanning the membrane. In the glucose channels, 12 such polar-nonpolar segments form a polar "door," like the hole in a donut, through which glucose molecules could diffuse.

But do they actually do so? Said more formally, is the hypothesis that cells take up glucose using these "glucose channels" valid? To evaluate this hypothesis, researchers tested one of its key predictions: If glucose molecules enter a cell through special protein channels, then there must be an upper limit to how many glucose molecules can enter at any given time. When all the channels are occupied transporting glucose molecules, there is no other way for glucose molecules to enter the cell, no matter how much glucose is available. In technical terms, the rate of a red blood cell's glucose uptake would be expected to "saturate" at some maximum value, determined by the number of channels.

The red line on the graph above presents an experiment designed to investigate this question. Isolated blood cells were placed in a culture fluid containing glucose. Cells were removed and the amount of glucose they contain measured. The researchers plot the amount of glucose versus time and determine the rate of glucose entry into the cell (the **rate** of a molecular process is simply the speed at which it occurs, in this case, measured as the number of molecules diffused per unit volume [ml] per unit time [hour]). The rate is then plotted

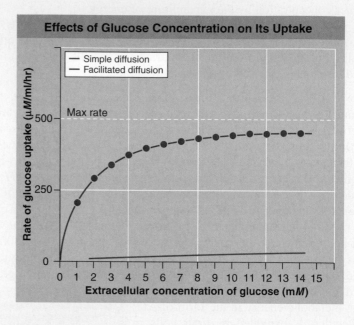

Effects of Glucose Concentration on Its Uptake

as a single red dot on the graph above. Repeating the experiment using increasing amounts of glucose in the culture fluid (measured as mM [mM is a **measure of concentration** which means millimolar, or one-thousandth of a mole per liter]), the researchers obtained the red curve seen above, in which the rate of glucose uptake is plotted against the extracellular concentration of glucose.

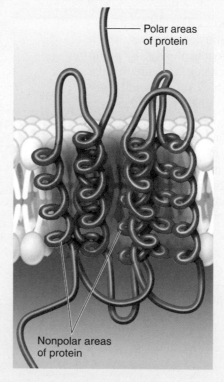

Polar areas of protein

Nonpolar areas of protein

Analysis

1. **Applying Concepts** Is the rate of glucose uptake affected by the extracellular glucose concentration? How?

2. **Interpreting Data** What is the difference in the uptake rate between 1 mM and 4 mM? between 10 mM and 14 mM?

3. **Making Inferences** If the external glucose concentration was increased to 30 mM, what would you predict would be the rate of glucose uptake? to 100 mM? Why?

4. **Drawing Conclusions** Why do you think the rate of glucose uptake by red blood cells in this experiment never exceeds 500 μM/ml/hr? Cells in other tissues of the human body take up glucose more rapidly. How would you predict the plasma membranes of these cells to be different from those of red blood cells?

Concept Summary

The World of Cells

4.1 Cells

- Cells are the smallest living structure. They consist of cytoplasm enclosed in a plasma membrane. Organisms may be composed of a single cell or multiple cells (**figure 4.1**).

- A smaller cell can function more efficiently, from transporting materials throughout the cell to transporting materials across the plasma membrane. A smaller cell has a larger surface-to-volume ratio, which increases the area through which materials may pass. Because cells are so small, a microscope is needed to view and study them (**figure 4.3**).

- There are many different types of microscopes, each providing a slightly different view. Light microscopes magnify and enhance resolution. Transmission and scanning electron microscopes increase resolution.

Kinds of Cells

4.2 Prokaryotic Cells

- Prokaryotic cells are simple unicellular organisms that lack nuclei or other internal organelles and are usually encased in a rigid cell wall. They vary in shape (**figure 4.4**), and some contain external structures, as shown here from **figure 4.5.**

4.3 Eukaryotic Cells

- Eukaryotic cells are larger and more structurally complex compared with prokaryotic cells. They contain nuclei and have internal membrane systems (**figures 4.6** and **4.7**).

Tour of a Eukaryotic Cell

4.4 The Plasma Membrane

- The plasma membrane that encloses all cells consists of a double layer of lipids, called the lipid bilayer, in which proteins are embedded. The structure of the plasma membrane is called the fluid mosaic model.

- The lipid bilayer is made up of special lipid molecules called phospholipids (**figure 4.8**), which have a polar (water-soluble) end and a nonpolar (water-insoluble) end. The bilayer forms because the nonpolar ends move away from the watery surroundings, forming the two layers (**figure 4.9**). Membrane proteins are either attached to the surface of the cell or are embedded within the membrane, as shown here from **figure 4.10.** Transmembrane proteins are held in place by the interaction of nonpolar sections of amino acids with the interior lipid portion of the bilayer (**figure 4.11**).

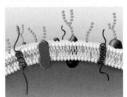

4.5 The Nucleus: The Cell's Control Center

- The nucleus is the command and control center of the cell. It contains the cell's DNA, which encodes the hereditary information that runs the cell (**figure 4.12**).

4.6 The Endomembrane System

- The endomembrane system is a collection of interior membranes that organize and divide the cell's interior into functional areas. The endoplasmic reticulum (**figure 4.13**) is a transport system that modifies and moves proteins and other molecules produced in the ER to the Golgi complex. The Golgi complex is a delivery system that carries molecules to the surface of the cell where they are released to the outside (**figures 4.14** and **4.15**).

- Lysosomes and vacuoles are other compartments in the cell. Lysosomes contain enzymes that digest worn out organelles. Vacuoles are storage compartments (**figure 4.16**).

4.7 Organelles That Harvest Energy

- The mitochondrion, shown here from **figure 4.17,** is called the powerhouse of the cell because it is the site of oxidative metabolism, an energy-extracting process. Chloroplasts are the site of photosynthesis and are present in plant and algal cells (**figure 4.18**). Mitochondria and chloroplasts are cell-like organelles that appear to be ancient bacteria that formed endosymbiotic relationships with early eukaryotic cells (**figure 4.19**).

4.8 The Cytoskeleton: Interior Framework of the Cell

- The interior of the cell contains a network of protein fibers, called the cytoskeleton, that supports the shape of the cell and anchors organelles in place (**figure 4.20**). Centrioles are paired structures that assemble microtubules in the cell (**figure 4.21**).

- Cells are dynamic structures. Cilia and flagella propel the cell through its environment (**figure 4.22**).

Transport Across Plasma Membranes

4.9 Diffusion and Osmosis

- Materials pass into and out of the cell passively through diffusion and osmosis. Diffusion is the movement of molecules from an area of high concentration to an area of low concentration, down their concentration gradients (***Essential Biological Process 4A***). Osmosis is the movement of water into and out of the cell, driven by differing concentrations of solute. Water molecules move to areas of higher solute concentrations (***Essential Biological Process 4B*** and **figure 4.23**).

4.10 Bulk Passage into and out of Cells

- Larger structures or larger quantities of material move into and out of the cell through endocytosis and exocytosis, respectively (**figures 4.24** and **4.25**). Endocytosis can involve bringing in larger particles, which is called phagocytosis, or bringing in liquids, which is called pinocytosis.

4.11 Selective Permeability

- Selective transport of materials across the membrane is accomplished by facilitated diffusion and active transport.

- Facilitated diffusion is driven by the concentration gradient, transporting substances down their concentration gradient, but substances must bind to a membrane transporter, called a carrier, in order to pass across the membrane (***Essential Biological Process 4C***).

- Active transport involves the input of energy to transport substances against (or up) their concentration gradients. Examples include the sodium-potassium pump (***Essential Biological Process 4D***), which pumps sodium ions out of the cell and potassium ions into the cell. Coupled channels are powered by the large sodium concentration gradient created by the actions of the sodium-potassium pump.

Self-Test

1. Cell theory includes the principle that
   a. cells are the smallest living things. Nothing smaller than a cell is considered alive.
   b. all cells are surrounded by cell walls that protect them.
   c. all organisms are made up of many cells arranged in specialized, functional groups.
   d. all cells contain membrane-bounded structures called organelles.

2. Organisms that have cells with a relatively uniform cytoplasm and no nucleus are called _____, and organisms whose cells have organelles and a nucleus are called _____.
   a. cellulose, nuclear
   b. eukaryotes, prokaryotes
   c. flagellated, streptococcal
   d. prokaryotes, eukaryotes

3. The plasma membrane is
   a. a carbohydrate layer that surrounds groups of cells to protect them.
   b. a double lipid layer with proteins inserted in it, which surrounds every cell individually.
   c. a thin sheet of structural proteins that encloses cytoplasm.
   d. composed of proteins that form a protective barrier.

4. Within the nucleus of a cell you can find
   a. a nucleolus.
   b. a cytoskeleton.
   c. mitochondria.
   d. All of these.

5. The endomembrane system within a cell includes the
   a. cytoskeleton and the ribosomes.
   b. prokaryotes and the eukaryotes.
   c. endoplasmic reticulum and the Golgi complex.
   d. mitochondria and the chloroplasts.

6. It was once thought that only the nucleus of each cell contained DNA. We now know that DNA is also found in the
   a. cytoskeleton and the ribosomes.
   b. prokaryotes and the eukaryotes.
   c. endoplasmic reticulum and the Golgi bodies.
   d. mitochondria and the chloroplasts.

7. Which of the following structures is not a component of the cytoskeleton?
   a. microtubules
   b. cristae
   c. intermediate filaments
   d. actin

8. If you put a drop of food coloring into a glass of water, the drop of color will
   a. fall to the bottom of the glass and sit there unless you stir the water; this is because of hydrogen bonding.
   b. float on the top of the water, like oil, unless you stir the water; this is because of surface tension.
   c. instantly disperse throughout the water; this is because of osmosis.
   d. slowly disperse throughout the water; this is because of diffusion.

9. When large molecules, such as food particles, need to get into a cell, they cannot easily pass through the plasma membrane, and so they move across the membrane through the processes of
   a. diffusion and osmosis.
   b. endocytosis and phagocytosis.
   c. exocytosis and pinocytosis.
   d. facilitated diffusion and active transport.

10. Active transport of specific molecules involves
    a. facilitated diffusion.
    b. endocytosis and pinocytosis.
    c. energy and specialized pumps or channels.
    d. permeability and a concentration gradient.

Visual Understanding

1. **Figure 4.3** The first microscope was used in about 1590. Electron microscopes came into common use about 70 years ago. Just over 100 years ago most physicians did not wash up between patients, even when someone had just died, or was very sick. Explain why early physicians didn't think it important to wash their hands.

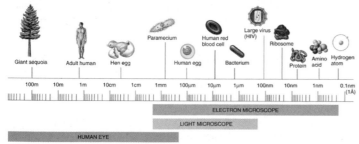

2. ***Essential Biological Process 4C*** In the lungs, there are steep concentration gradients for oxygen and carbon dioxide molecules such that large numbers of these molecules move across the plasma membrane of the cells that line the lungs. These molecules pass through the plasma membranes by simple diffusion. This process is fast and efficient. Would this process be just as efficient if the oxygen and carbon dioxide molecules passed through the membranes by facilitated diffusion? Why or why not?

Challenge Questions

1. You are using a computer program to design a new single-celled organism. Discuss why a smaller cell will be more efficient in transporting materials than a larger cell.

2. Antibiotics are medicines that target bacterial infections in vertebrates. How can an antibiotic kill all the bacterial cells and not harm vertebrate cells? Hint: what part of the bacterial cell must antibiotics be targeting?

3. Compare the cellular organelles and other structures to the parts of a city—for example, the nucleus is city hall and the DNA is all the city's laws and instructions.

4. A ribosome contains two subunits. The ribosome subunits are assembled within the nucleus, but ribosomes act in the cytoplasm. How do you imagine the subunits get out of the nucleus and into the cytoplasm?

Chapter 5

Energy and Life

Cells and Energy

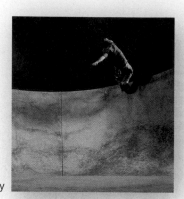

(a) Potential energy

(b) Potential energy

(c) Kinetic energy

Figure 5.1   Potential and kinetic energy.

Objects that have the capacity to move but are not moving have potential energy, while objects that are in motion have kinetic energy. (a) The kinetic energy of this skateboarder becomes potential energy at the peak of this slope. (b) The energy required to move the ball up the hill is stored as potential energy. (c) This stored energy is released as kinetic energy as the ball rolls down the hill.

5.1 The Flow of Energy in Living Things

CONCEPT PREVIEW: Energy is the capacity to do work, either actively (kinetic energy) or stored for later use (potential energy). Chemical reactions occur when the covalent bonds linking atoms together are formed or broken.

We are about to begin our discussion of energy and cellular chemistry. Although these subjects may seem difficult at first, remember that all life is driven by energy. The concepts and processes discussed in the next three chapters are key to life. We are chemical machines, powered by chemical energy, and for the same reason that a successful race car driver must learn how the engine of a car works, we must look at cell chemistry. Indeed, if we are to understand ourselves, we must "look under the hood" at the chemical machinery of our cells and see how it operates.

As described in chapter 2, **energy** is defined as the ability to do work. It can be considered to exist in two states: kinetic energy and potential energy. **Kinetic energy** is the energy of motion. Objects that are not in the process of moving but have the capacity to do so are said to possess **potential energy,** or stored energy (figure 5.1*a*). A boulder perched on a hilltop (figure 5.1*b*) has potential energy; after the man pushes the boulder and it begins to roll downhill (figure 5.1*c*), some of the boulder's potential energy is converted into kinetic energy. All of the work carried out by living organisms also involves the transformation of potential energy to kinetic energy.

Energy exists in many forms: mechanical energy, heat, sound, electric current, light, or radiation. Because it can exist in so many forms, there are many ways to measure energy. The most convenient is in terms of heat, because all other forms of energy can be converted to heat. Thus the study of energy is called *thermodynamics,* meaning "heat changes."

Energy flows into the biological world from the sun, which shines a constant beam of light on the earth. It is estimated that the sun provides the earth with more than 13×10^{23} calories per year, or 40 million billion calories per second! Plants, algae, and certain kinds of bacteria capture a fraction of this energy through photosynthesis. In photosynthesis, energy garnered from sunlight is used to combine small molecules (water and carbon dioxide) into more complex molecules (sugars). These complex sugar molecules have potential energy due to the arrangement of their atoms. This potential energy, in the form of chemical energy, will eventually be used by the cell to do its work. Recall from chapter 2 that an atom consists of a central nucleus surrounded by one or more orbiting electrons, and a covalent bond forms when two atomic nuclei share electrons. Breaking such a bond requires energy to pull the nuclei apart. Indeed, the strength of a covalent bond is measured by the amount of energy required to break it. For example, it takes 98.8 kcal to break 1 mole (6.023×10^{23}) of carbon–hydrogen (C—H) bonds.

> You will see in chapter 6 how plants use the sun's energy to build carbohydrates, storing this energy as potential energy. Chapter 7 explains how this potential energy is used by the plants and other organisms to fuel the functions of living.

All the chemical activities within cells can be viewed as a series of chemical reactions between molecules. A **chemical reaction** is the making or breaking of chemical bonds—gluing atoms together to form new molecules, or tearing molecules apart and sometimes sticking the pieces onto other molecules.

5.2 The Laws of Thermodynamics

CONCEPT PREVIEW: The first law of thermodynamics states that energy cannot be created or destroyed; it can only undergo conversion from one form to another. The second law states that disorder (entropy) in the universe tends to increase.

Running, thinking, singing, reading these words—all activities of living organisms involve changes in energy. A set of universal laws we call the laws of thermodynamics govern these and all other energy changes in the universe.

The First Law of Thermodynamics

The first of these universal laws, the **first law of thermodynamics,** concerns the amount of energy in the universe. It states that energy can change from one state to another (from potential to kinetic, for example) but it can never be destroyed, nor can new energy be made. The total amount of energy in the universe remains constant.

The lion eating a giraffe in figure 5.2 is in the process of acquiring energy. The lion isn't creating new energy, rather it is merely transferring some of the potential energy stored in the giraffe's tissues to its own body. Within any living organism, this chemical potential energy can be shifted to other molecules and stored in chemical bonds, or it can be converted into kinetic energy, or into other forms of energy. During each conversion, some of the energy dissipates into the environment as heat energy, a measure of the random motions of molecules (and, hence, a measure of one form of kinetic energy). Energy continuously flows through the biological world in one direction, with new energy from the sun constantly entering the system to replace the energy dissipated as heat.

The Second Law of Thermodynamics

The **second law of thermodynamics** concerns this transformation of potential energy into heat, or random molecular motion. It states that the disorder in a closed system like the universe is continuously increasing. Put simply, disorder is more likely than order. For example, it is much more likely that a column of bricks will tumble over than that a pile of bricks will arrange themselves spontaneously to form a column. In general, energy transformations proceed spontaneously to convert matter from a more ordered, less stable form, to a less ordered, more stable form. Without an input of energy from the teenager (or a parent), the ordered room in figure 5.3 falls into disorder. When the input of energy is localized, one area can become far more organized than its disordered surroundings, like cleaning one room of a messy house. It is in just this way that a cell uses energy to keep more organized than its surroundings.

Entropy is a measure of the degree of disorder of a system, so the second law of thermodynamics can also be stated simply as "entropy increases." When the universe formed 10 to 20 billion years ago, it held all the potential energy it will ever have. It has become progressively more disordered ever since, with every energy exchange increasing the entropy of the universe.

Concept Check

1. Is the energy in a "high energy" food bar kinetic or potential? Explain.
2. When Tiger Woods strikes a golf ball, is what happens to the golf ball a chemical reaction?
3. When an organism dies, what happens to its entropy?

Figure 5.2   Giraffe for dinner.

EVOLUTION

Creationists and the Law.   One objection often made by creationists to the theory of evolution is that it violates the second law of thermodynamics. If disorder is continually increasing in the universe, making disorder more likely than order, then they assert that a more ordered system (cells) could not arise spontaneously without direction. Evolutionary biologists respond to this criticism by pointing out that this creationist objection misstates the second law of thermodynamics by omitting a key provision: *"in a closed system."* The living world is by no means a closed system, as energy is continually being supplied to it by the sun. It is this energy, captured in photosynthesis, that powers the organized system we call life. In no sense does this in any way violate the second law.

Disorder happens "spontaneously"

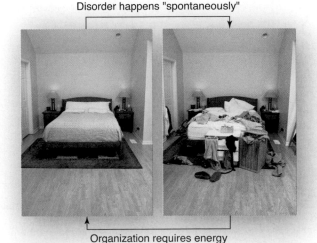

Organization requires energy

Figure 5.3   Entropy in action.

As time elapses, a teenager's room becomes more disorganized. It takes energy to clean it up.

Cell Chemistry

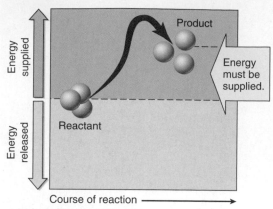

(a) Endergonic reaction

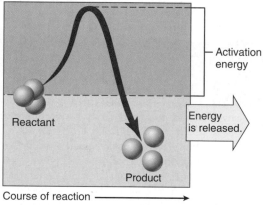

(b) Exergonic reaction

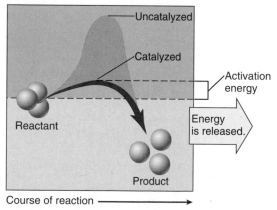

(c) Catalyzed reaction

Figure 5.4   **Chemical reactions and catalysis.**

(a) The products of endergonic reactions contain more energy than the reactants. (b) The products of exergonic reactions contain less energy than the reactants, but exergonic reactions do not necessarily proceed rapidly because it takes energy to get them going. The "hill" in this energy diagram represents energy that must be supplied to destabilize existing chemical bonds. (c) Catalyzed reactions occur faster because the amount of activation energy required to initiate the reaction—the height of the energy hill that must be overcome—is lowered.

5.3 Chemical Reactions

CONCEPT PREVIEW: The products of endergonic reactions contain more energy than the reactants, while the products of exergonic reactions contain less energy than the reactants. Catalysis can lower the activation energy needed to initiate a reaction.

In a chemical reaction, the original molecules before the chemical reaction occurs are called **reactants,** or sometimes **substrates,** whereas the molecules that result after the reaction has taken place are called the **products** of the reaction. Not all chemical reactions are equally likely to occur. Just as a boulder is more likely to roll downhill than uphill, so a reaction is more likely to occur if it releases energy than if it needs to have energy supplied. Consider how the chemical reaction proceeds in figure 5.4a. Like rolling a boulder uphill, energy needs to be supplied. This is because the product of the reaction contains more energy than the reactant. This type of chemical reaction, called **endergonic,** does not occur spontaneously. By contrast, an **exergonic** reaction, shown in figure 5.4b, tends to occur spontaneously because the product has less energy than the reactant, like a boulder that has rolled downhill.

> Covalent bonds form when two or more atoms share electrons, as discussed on page 35. Many important chemical reactions involve the formation or breaking of these bonds.

Activation Energy

If all chemical reactions that release energy tend to occur spontaneously, it is fair to ask, "Why haven't all exergonic reactions occurred already?" Why doesn't all the gasoline in all the automobiles in the world just burn up right now? It doesn't because the burning of gasoline, and almost all other chemical reactions, requires an input of energy to get it started—it is first necessary to break existing chemical bonds in the reactants, and this takes energy. The extra energy required to destabilize existing chemical bonds and so initiate a chemical reaction is called **activation energy,** indicated by brackets in figure 5.4b and c. You must first nudge a boulder out of the hole it sits in before it can roll downhill. Activation energy is simply a chemical nudge.

Catalysis

One way to make a reaction more likely to happen is to lower the necessary activation energy. Like digging away the ground below your boulder, lowering activation energy reduces the nudge needed to get things started. The process of lowering the activation energy of a reaction is called **catalysis.** Catalysis cannot make an endergonic reaction occur spontaneously—you cannot avoid the need to supply energy—but it can make a reaction, endergonic or exergonic, proceed much faster. Compare the activation energy levels (the red arched arrows) in the second and third panels to the left: The catalyzed reaction in figure 5.4c has a lower barrier to overcome.

Concept Check

1. If exergonic reactions tend to occur spontaneously, why haven't they all done so? What stops the world's gasoline from burning?
2. Can catalysis make endergonic reactions occur spontaneously?
3. Why is the speed of a chemical reaction affected by the amount of activation energy required to initiate it?

Enzymes

5.4 How Enzymes Work

CONCEPT PREVIEW: Enzymes catalyze chemical reactions within cells. Sometimes enzymes are organized into biochemical pathways. Enzymes are sensitive to temperature and pH, because both of these variables influence enzyme shape.

Macromolecules called **enzymes** are the catalysts used by cells to touch off particular chemical reactions. By controlling which enzymes are present, and when they are active, cells are able to control what happens within themselves, just as a conductor controls the music an orchestra produces by dictating which instruments play when.

An enzyme works by binding to a specific molecule in such a way as to make a particular reaction more likely (*Essential Biological Process 5A*). The key to this activity is the shape of the enzyme. An enzyme is specific for a particular reactant, or substrate, because the enzyme surface provides a mold that very closely fits the shape of the desired reactant. For example, the blue-colored lysozyme enzyme in figure 5.5 is contoured to fit a specific sugar molecule (the yellow reactant). Other molecules that fit less perfectly simply don't adhere to the enzyme's surface. The site on the enzyme surface where the reactant fits is called the **active site.** The site on the reactant that binds to an enzyme is called its **binding site.** In figure 5.5b, the edges of the lysozyme hug the sugar molecule, leading to an "induced fit" between the enzyme and its reactant, like a hand wrapping around a baseball.

An enzyme lowers the activation energy of a particular reaction. In the case of lysozyme, an enzyme found in human tears, the enzyme has an antibacterial function, encouraging the breaking of a particular chemical bond in molecules that make up the cell wall of

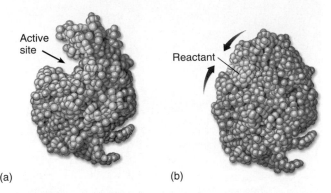

(a)   (b)

Figure 5.5 **An enzyme's shape determines its activity.**

(a) A groove runs through the lysozyme enzyme that fits the shape of the reactant (in this case, a chain of sugars). (b) When such a chain of sugars slides into the groove, it induces the protein to change its shape slightly and embrace the substrate more intimately. This induced fit causes a chemical bond between two sugar molecules within the chain to break.

IMPLICATION University of Chicago researchers have found the activity levels of protein kinase C (an enzyme linked to depression) to be much lower in the brains of teenage suicides. What sort of investigations would you propose to look into the link between this enzyme and teen suicide?

Essential Biological Process 5A

How Enzymes Work

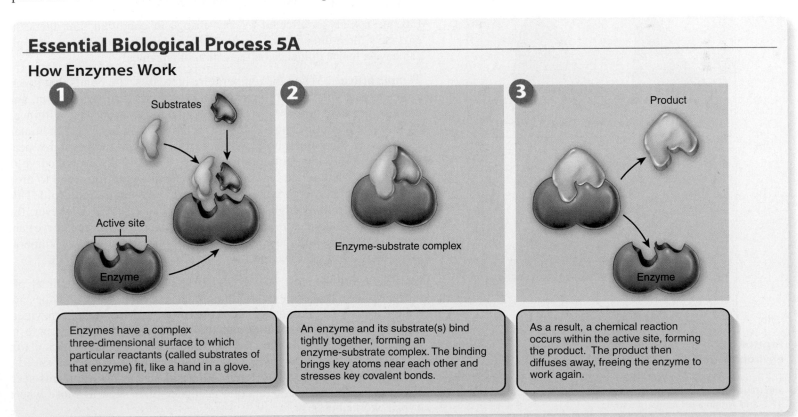

1
Substrates
Active site
Enzyme

Enzymes have a complex three-dimensional surface to which particular reactants (called substrates of that enzyme) fit, like a hand in a glove.

2
Enzyme-substrate complex

An enzyme and its substrate(s) bind tightly together, forming an enzyme-substrate complex. The binding brings key atoms near each other and stresses key covalent bonds.

3
Product
Enzyme

As a result, a chemical reaction occurs within the active site, forming the product. The product then diffuses away, freeing the enzyme to work again.

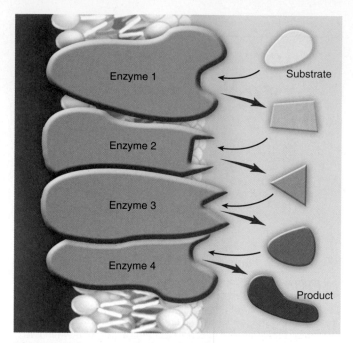

Figure 5.6 A biochemical pathway.

The original substrate is acted on by enzyme 1, changing the substrate to a new form recognized by enzyme 2. Each enzyme in the pathway acts on the product of the previous stage.

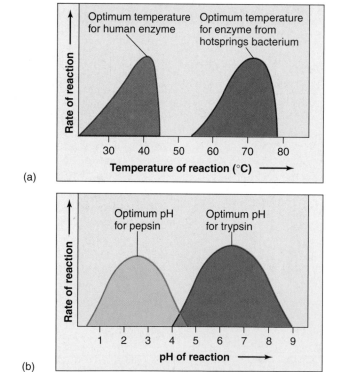

(a)

(b)

Figure 5.7 Enzymes are sensitive to their environment.

The activity of an enzyme is influenced by both (a) temperature and (b) pH. Most human enzymes work best at temperatures near 37°C and within a pH range of 6 to 8.

bacteria. The enzyme weakens the bond by drawing away some of its electrons. Alternatively, an enzyme may encourage the formation of a link between two reactants, like in *Essential Biological Process 5A,* by holding them near each other. Regardless of the type of reaction, the enzyme is not affected by the chemical reaction and is available to be used again.

Biochemical Pathways

Every organism contains thousands of different kinds of enzymes that together catalyze a bewildering variety of reactions. Often several of these reactions occur in a fixed sequence called a **biochemical pathway,** the product of one reaction becoming the substrate for the next. You can see in figure 5.6 how the initial substrate is altered by enzyme 1 so that it now fits into the active site of another enzyme, becoming the substrate for enzyme 2, and so on until the final product is produced. Because these reactions occur in sequence, the enzymes involved are often positioned near each other in the cell. For example, the enzymes involved in this biochemical pathway are all embedded in a membrane near each other. Many biochemical pathways occur in membranes, although enzyme assemblies also occur in organelles and within the cytoplasm. The close proximity of the enzymes allows the reactions of a biochemical pathway to proceed faster. Biochemical pathways are the organizational units of metabolism. We will explore them in chapters 6 and 7.

The Calvin cycle on page 108 is a biochemical pathway in photosynthesis where the product of one reaction is the substrate for the next reaction. Similarly, glycolysis on page 116 and the Krebs cycle on page 118 are biochemical pathways.

Factors Affecting Enzyme Activity

Enzyme activity is affected by any change in condition that alters the enzyme's three-dimensional shape. For this reason, temperature and pH can have a major influence on the action of enzymes.

Temperature. When the temperature increases, the bonds that determine enzyme shape are too weak to hold it in the proper position, and the enzyme denatures. As a result, enzymes function best within an optimum temperature range, which is relatively narrow for most human enzymes. In the human body, enzymes work best at temperatures near the normal body temperature of 37°C, as shown by the brown curve in figure 5.7a. Also notice that the rates of enzyme reactions tend to drop quickly at higher temperatures, when the enzyme begins to unfold. This is why an extremely high fever in humans can be fatal. However, the shapes of the enzymes found in hotsprings bacteria (the red curve) are more stable, allowing the enzymes to function at much higher temperatures. This allows the bacteria to live in water that is near 70°C.

pH. In addition, most enzymes also function within an optimal pH range, because the shape-determining polar interactions of enzymes are quite sensitive to hydrogen ion (H[+]) concentration. Most human enzymes, such as the protein-degrading enzyme trypsin (the dark blue curve in figure 5.7b) work best within the range of pH 6 to 8. Blood has a pH of 7.4. However, some enzymes, such as the digestive enzyme pepsin (the light blue curve) are able to function in very acidic environments such as the stomach, but can't function at higher pHs, such as those where trypsin works best.

Essential Biological Process 5B

Regulating Enzyme Activity

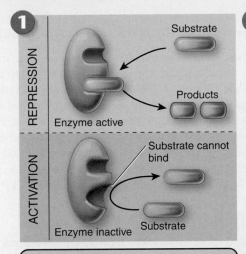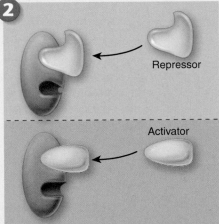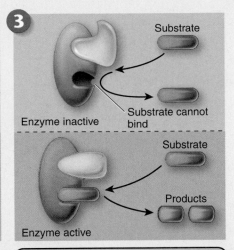

Allosteric enzymes subject to repression are active in the absence of signal molecules, while allosteric enzymes that rely on activation are not active in the absence of signal molecules.

When signal molecules bind allosteric enzymes, they change the shape of the active site. Repressors disrupt the active site, while activators restore it.

Allosteric enzymes subject to repression are not active in the presence of signal molecules, while allosteric enzymes that rely on activation require signal molecules to be active.

5.5 How Cells Regulate Enzymes

CONCEPT PREVIEW: An enzyme's activity can be affected by signal molecules that bind to it, changing its shape.

Because an enzyme must have a precise shape to work correctly, it is possible for the cell to control when an enzyme is active by altering its shape. Many enzymes have shapes that can be altered by the binding of "signal" molecules to their surfaces, making them work better (activation) or worse (inhibition). These are called *allosteric enzymes*. For example, the upper panels of *Essential Biological Process 5B* show an enzyme that is inhibited. The binding of a signal molecule called a **repressor** (the yellow molecule in panel 2) alters the shape of the enzyme's active site such that it cannot bind the substrate (the red molecule). In other cases, the enzyme may not be able to bind the reactants *unless* the signal molecule is bound to the enzyme. The lower set of panels shows a signal molecule serving as an **activator.** The red substrate cannot bind to the enzyme's active site unless the activator (the yellow molecule) is first in place.

Enzymes are often regulated by a mechanism called **feedback inhibition,** where the product of the reaction acts as a repressor. Feedback inhibition can occur in two ways: *competitive* and, much more commonly, *noncompetitive*. The blue molecule in figure 5.8*a* functions as a competitive inhibitor, blocking the active site so that the substrate cannot bind. The yellow molecule in figure 5.8*b* functions as a noncompetitive inhibitor. It binds to an allosteric site, changing the shape of the enzyme such that it is unable to bind to the substrate.

Concept Check

1. What is the difference between an active site and a binding site?
2. Explain how hotsprings bacteria can live in near-boiling water (70° C = 158° F) that would kill a human bather.
3. Can a signal molecule activate one enzyme and repress another?

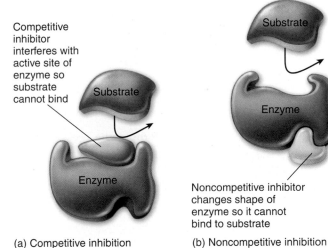

(a) Competitive inhibition   (b) Noncompetitive inhibition

Figure 5.8   How enzymes can be inhibited.

(a) In competitive inhibition, the inhibitor interferes with the active site of the enzyme. (b) In noncompetitive inhibition, the inhibitor binds to the enzyme at a place away from the active site, effecting a conformational change in the enzyme so that it can no longer bind to its substrate.

IMPLICATION   Many antibiotics work by inhibiting enzymes. The antibiotic penicillin inhibits an enzyme bacteria use in making cell walls. Imagine a confused patient mistakenly took two to three times the number of prescribed penicillin pills at one time. Is it likely the patient would be seriously harmed? Explain.

How Cells Use Energy

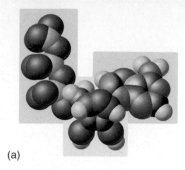

(a)

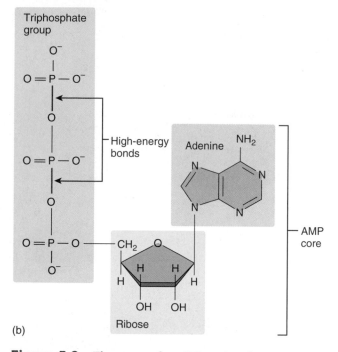

(b)

Figure 5.9 **The parts of an ATP molecule.**

The model (a) and structural diagram (b) both show that ATP consists of three phosphate groups attached to a ribose (five-carbon sugar) molecule. The ribose molecule is also attached to an adenine molecule (also one of the nitrogenous bases of DNA and RNA). When the endmost phosphate group is split off from the ATP molecule, considerable energy is released.

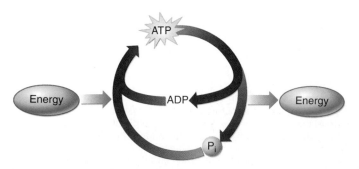

Figure 5.10 **The ATP-ADP cycle.**

5.6 ATP: The Energy Currency of the Cell

CONCEPT PREVIEW: Cells store chemical energy in ATP molecules and then use the breakdown of ATP to drive chemical reactions.

Cells use energy to do all those things that require work, but how does the cell use energy from the sun or the potential energy stored in molecules to power its activities? The sun's radiant energy and the energy stored in molecules are energy sources, but like money that is invested in stocks and bonds or real estate, these energy sources cannot be used directly to run a cell, any more than money invested in stocks can be used to buy a candy bar at the store. To be useful, the energy from the sun or food molecules must first be converted to a source of energy that a cell can use, like someone converting stocks and bonds to ready cash. The "cash" molecule in the body is **adenosine triphosphate (ATP).**

Structure of the ATP Molecule

Each ATP molecule is composed of three parts (figure 5.9): (1) a sugar (colored blue) serves as the backbone to which the other two parts are attached, (2) adenine (colored peach), which is also one of the four nitrogenous bases in DNA and RNA, and (3) a chain of three phosphates (colored yellow) contain high-energy bonds. As you can see in the figure, the phosphates carry negative electrical charges, and so it takes considerable chemical energy to hold the line of three phosphates next to one another at the end of ATP. Like a compressed spring, the phosphates are poised to push apart. It is for this reason that the chemical bonds linking the phosphates are such chemically reactive bonds. When the endmost phosphate is broken off an ATP molecule, a sizable packet of energy is released. The reaction converts ATP to adenosine diphosphate, ADP, and P_i, inorganic phosphate:

$$ATP \longleftrightarrow ADP + P_i + energy$$

Chemical reactions require activation energy, and endergonic reactions require the input of even more energy, and so these reactions in the cell are usually coupled with the breaking of the phosphate bond in ATP, called *coupled reactions.* Because almost all chemical reactions in cells require less energy than is released by this reaction, ATP is able to power many of the cell's activities. Table 5.1 introduces you to some of the key cell activities powered by the breakdown of ATP. ATP is continually recycled from ADP and P_i via the ATP-ADP cycle (figure 5.10).

Cells use two different but complementary processes to convert energy from the sun and food molecules into potential energy stored in the chemical bonds of ATP. Some cells convert energy from the sun into molecules of ATP through the process of **photosynthesis,** the subject of chapter 6. This ATP is then used to manufacture sugar molecules, converting the energy from ATP into potential energy stored in the bonds that hold the atoms in the sugar molecule together. All cells convert the potential energy found in food molecules like sugar into ATP through **cellular respiration,** the subject of chapter 7.

Concept Check

1. Compare and contrast ATP with the nucleotides found in DNA and RNA. [Hint: see figure 3.11.]
2. Why are the chemical bonds linking the phosphates in ATP so reactive?
3. How do the roots of a plant, far from light, get the ATP they need?

TABLE 5.1 | How Cells Use ATP Energy to Power Cellular Work

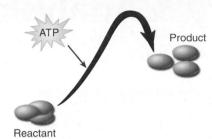

Biosynthesis

Cells use the energy released from the exergonic hydrolysis of ATP to drive endergonic reactions like those of protein synthesis, an approach called energy coupling.

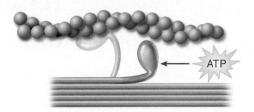

Contraction

In muscle cells, filaments of protein repeatedly slide past each other to achieve contraction of the cell. An input of ATP is required for the filaments to reset and slide again.

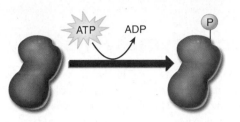

Chemical Activation

Proteins can become activated when a high-energy phosphate from ATP attaches to the protein, activating it. Other types of molecules can also become phosphorylated by transfer of a phosphate from ATP.

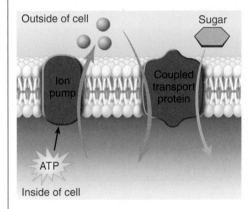

Importing Metabolites

Metabolite molecules such as amino acids and sugars can be transported into cells against their concentration gradients by coupling the intake of the metabolite to the inward movement of an ion moving down its concentration gradient, this ion gradient being established using ATP.

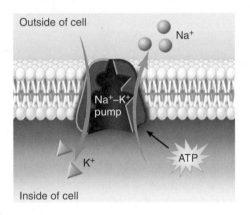

Active Transport: Na⁺–K⁺ Pump

Most animal cells maintain a low internal concentration of Na⁺ relative to their surroundings, and a high internal concentration of K⁺. This is achieved using a protein called the sodium-potassium pump, which actively pumps Na⁺ out of the cell and K⁺ in, using energy from ATP.

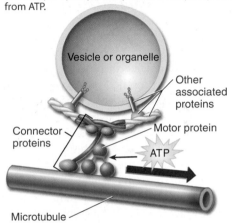

Cytoplasmic Transport

Within a cell's cytoplasm, vesicles or organelles can be dragged along microtubular tracks using molecular motor proteins, which are attached to the vesicle or organelle with connector proteins. The motor proteins use ATP to power their movement.

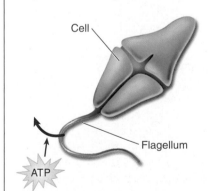

Flagellar Movements

Microtubules within flagella slide past each other to produce flagellar movements. ATP powers the sliding of the microtubules.

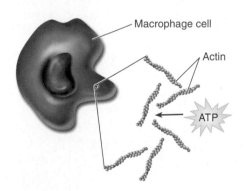

Cell Crawling

Actin filaments in a cell's cytoskeleton continually assemble and disassemble to achieve changes in cell shape and to allow cells to crawl over substrates or engulf materials. The dynamic character of actin is controlled by ATP molecules bound to actin filaments.

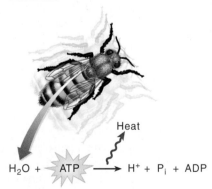

Heat Production

The hydrolysis of the ATP molecule releases heat. Reactions that hydrolyze ATP often take place in mitochondria or in contracting muscle cells and may be coupled to other reactions. The heat generated by these reactions can be used to maintain an organism's temperature.

Do Enzymes Physically Attach to Their Substrates?

When scientists first began to examine the chemical activities of organisms, no one knew that biochemical reactions were catalyzed by enzymes. The first enzyme was discovered in 1833 by French chemist Anselme Payen. He was studying how beer is made from barley: First barley is pressed and gently heated so its starches break down into simple two-sugar units; then yeasts convert these units into ethanol. Payen found that the initial breakdown requires a chemical factor that is not alive, and which does not seem to be used up during the process—a catalyst. He called this first enzyme *diastase* (we call it amylase today).

Did this catalyst operate at a distance, increasing reaction rate all around it, much as raising the temperature of nearby molecules might do? Or did it operate in physical contact, actually attaching to the molecules whose reaction it catalyzed (its "substrate")?

The answer was discovered in 1903 by French chemist Victor Henri. He saw that the hypothesis that an enzyme physically binds to its substrate makes a clear and testable prediction: In a solution of substrate and enzyme, there must be a maximum reaction rate. When all the enzyme molecules are working full tilt, the reaction simply cannot go any faster, no matter how much more substrate you add to the solution. To test this prediction, Henri carried out the experiment whose results you see in the graph, measuring the reaction rate (V) of diastase at different substrate concentrations (S).

Analysis

1. **Making Inferences** As S increases, does V increase? If so, in what manner—steadily, or by smaller and smaller amounts? Is there a maximum reaction rate?
2. **Drawing Conclusions** Does this result provide support for the hypothesis that an enzyme binds physically to its substrate? Explain. If the hypothesis were incorrect, what would you expect the graph to look like?
3. **Further Analysis** If the smaller amounts by which V increases are strictly the result of fewer unoccupied enzymes being available at higher values of S, then the curve in Henri's experiment should show a pure exponential decline in V—mathematically, meaning a reciprocal plot ($1/V$ versus $1/S$) should be a straight line. If some other factor is also at work that reacts differently to substrate concentration, then the reciprocal plot would curve upward or downward. Fill in the reciprocal values in the table to the right, and then plot the values on the lower graph ($1/S$ on the x axis and $1/V$ on the y axis). Is a reciprocal plot of Henri's data a straight line?

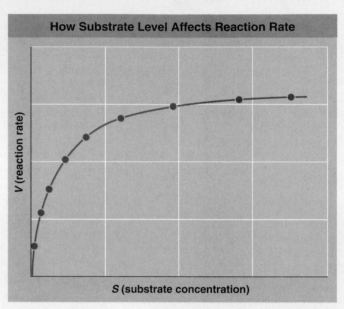

How Substrate Level Affects Reaction Rate

V (reaction rate) vs. *S* (substrate concentration)

Trial	S	1/S	V	1/V
1	5	0.200	7.7	0.130
2	10	___	15.4	___
3	25	___	23.1	___
4	50	___	30.8	___
5	75	___	38.5	___
6	125	___	40.7	___
7	200	___	46.2	___
8	275	___	47.7	___
9	350	___	48.5	___

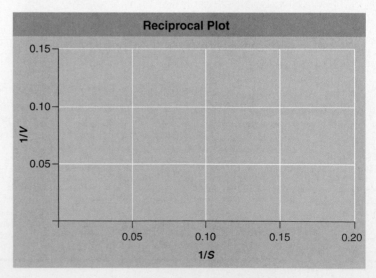

Reciprocal Plot

$1/V$ vs. $1/S$

Concept Summary

Cells and Energy

5.1 The Flow of Energy in Living Things

- Energy is the ability to do work. Energy exists in two states: kinetic energy and potential energy.

- Kinetic energy is the energy of motion. Potential energy is stored energy, which exists in objects that aren't in motion but have the capacity to move, like a ball poised at the top of a hill shown here from **figure 5.1.** Work carried out by living organisms involves the transformation of potential energy into kinetic energy. Energy also exists in different forms in the universe, such as light, electrical, or heat energy.

- Energy flows from the sun to the earth, where it is trapped by photosynthetic organisms and stored in carbohydrates as potential energy. This energy is transferred during chemical reactions.

5.2 The Laws of Thermodynamics

- The laws of thermodynamics describe changes in energy in our universe. The first law of thermodynamics explains that energy can not be created or destroyed, only changed from one state to another. The total amount of energy in the universe remains constant.

- The second law of thermodynamics explains that the conversion of potential energy into random molecular motion is constantly increasing. This conversion of energy progresses from an ordered but less stable form to a disordered but stable form. Entropy, which is a measure of disorder in a system, is constantly increasing such that disorder is more likely than order. Energy must be used to maintain order (**figure 5.3**).

Cell Chemistry

5.3 Chemical Reactions

- Chemical reactions involve the breaking or formation of covalent bonds. The starting molecules are called the reactants, and the molecules produced by the reaction are called the products. Chemical reactions in which the products contain more potential energy than the reactants are called endergonic reactions (**figure 5.4a**). Chemical reactions that release energy are called exergonic reactions (**figure 5.4b**) and are more likely to occur.

- All chemical reactions require an input of energy. The energy required to start a reaction is called activation energy and is indicated by the red arrow in the chemical reaction shown here from **figure 5.4.** A chemical reaction proceeds faster when its activation energy is lowered, and this occurs through a process called catalysis (**figure 5.4c**).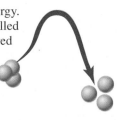

Enzymes

5.4 How Enzymes Work

- Enzymes are molecules that lower the activation energy of chemical reactions in the cell. Enzymes are cellular catalysts.

- An enzyme, like the lysozyme shown here from **figure 5.5,** binds the reactant, or substrate. The substrate binds to the enzyme's active site. The enzyme forms around the reactant and acts to stress covalent bonds or bring atoms into closer proximity (**Essential Biological Process 5A**). The actions of the enzyme increase the likelihood that chemical bonds will break or form. An enzyme lowers the activation energy of the reaction. The enzyme is not affected by the reaction and can be used over and over again.

- Sometimes enzymes work in a series of reactions called a biochemical pathway. The product of one reaction becomes the substrate for the next reaction. The enzymes that are involved are usually located near each other in the cell (**figure 5.6**).

- In the cell, chemical reactions are regulated by controlling which enzymes are active. Other factors, such as temperature and pH, can also affect enzyme function, and so most enzymes have an optimal temperature and pH range (**figure 5.7**).

- Higher temperatures can disrupt the bonds that hold the enzyme in its proper shape, decreasing its ability to catalyze a chemical reaction. The bonds that hold the enzyme's shape are also affected by hydrogen ion concentrations, and so increasing or decreasing the pH can disrupt the enzyme's function.

5.5 How Cells Regulate Enzymes

- An enzyme can be inhibited or activated in the cell as a means of regulation by temporarily altering the enzyme's shape (**Essential Biological Process 5B**). An enzyme can be inhibited when a molecule, called a repressor, binds to the enzyme, altering the shape of the active site so that it cannot bind the substrate. Some enzymes need to be activated, or turned on, in order to bind to their substrate. A molecule called an activator binds to the enzyme, changing the shape of the active site so that it is able to bind the substrate. Enzymes that are controlled in this way are allosteric enzymes.

- A repressor molecule can bind to the active site of the enzyme, blocking it. This is called a competitive inhibition. In noncompetitive inhibition, the repressor binds to a different site on the enzyme, altering the shape of the active site so it cannot bind its substrate (**figure 5.8**).

How Cells Use Energy

5.6 ATP: The Energy Currency of the Cell

- Cells require energy to do work in the form of ATP (**table 5.1**). ATP contains a sugar, an adenine, and a chain of three phosphates, as shown here from **figure 5.9.** The three phosphates are held together with high-energy bonds. When the endmost phosphate bond breaks, considerable energy is released. A cell uses this energy to drive reactions in the cell by coupling the breakdown of ATP with other chemical reactions in the cell.

Self-Test

1. The ability to do work is the definition for
   a. thermodynamics.
   b. radiation.
   c. energy.
   d. entropy.

2. The first law of thermodynamics
   a. says that energy recycles constantly, as organisms use and reuse it.
   b. says that entropy, or disorder, continually increases in a closed system.
   c. is a formula for measuring entropy.
   d. says that energy can change forms, but cannot be created or destroyed.

3. The second law of thermodynamics
   a. says that energy recycles constantly, as organisms use and reuse it.
   b. says that entropy, or disorder, continually increases in a closed system.
   c. is a formula for measuring entropy.
   d. says that energy can change forms, but cannot be made nor destroyed.

4. Chemical reactions that occur spontaneously are called
   a. exergonic and release energy.
   b. exergonic and their products contain more energy.
   c. endergonic and release energy.
   d. endergonic and their products contain more energy.

5. The catalysts that help an organism carry out needed chemical reactions are called
   a. hormones.
   b. enzymes.
   c. reactants.
   d. substrates.

6. Factors that affect the activity of an enzyme molecule include
   a. peptides and energy.
   b. activation energy.
   c. temperature and pH.
   d. entropy and thermodynamics.

7. In order for an enzyme to work properly
   a. it must have a particular shape.
   b. the temperature must be within certain limits.
   c. the pH must be within certain limits.
   d. All of the above.

8. In competitive inhibition
   a. an enzyme molecule has to compete with other enzyme molecules for the necessary substrate.
   b. an enzyme molecule has to compete with other enzyme molecules for the necessary energy.
   c. an inhibitor molecule competes with the substrate for the active site on the enzyme.
   d. two different products compete for the same active site on the enzyme.

9. Which of the following is not a component of ATP?
   a. active site
   b. ribose
   c. adenine
   d. phosphate groups

10. Endergonic reactions can occur in the cell because they are coupled with
    a. the breaking of phosphate bonds in ATP.
    b. uncatalyzed reactions.
    c. activators.
    d. all of the above.

Visual Understanding

1. **Figure 5.7** Examine the graphs shown here. Describe what happens to a human enzyme at a temperature of 50°C. Looking at part (b), what happens to trypsin's ability to function as the surrounding concentration of H⁺ ions increases? How would the enzyme pepsin respond to a change in pH from 4 to 3?

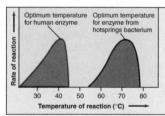
(a)

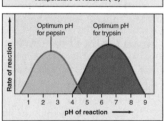
(b)

2. **Table 5.1** ATP forms primarily from the breakdown of glucose. If your blood glucose level drops, what sorts of problems can that cause in your body?

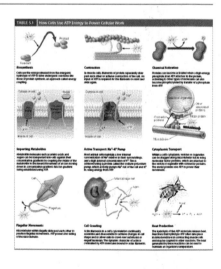

Challenge Questions

1. Photosynthetic organisms, such as plants, algae, and bacteria, capture the sun's energy and use it to build sugar molecules that other organisms can use. The formation of these molecules involves endergonic reactions. Explain what this means and where the sun's energy is stored in these sugar molecules.

2. What sorts of things can keep an enzyme from doing its job?

3. When a baseball thrown by a pitcher encounters the swinging bat of a slugger, what happens to the ball's kinetic energy? What happens to the bat's kinetic energy?

Photosynthesis: Acquiring Energy from the Sun

Photosynthesis

6.1 An Overview of Photosynthesis

CONCEPT PREVIEW: Photosynthesis uses energy from sunlight to power the synthesis of organic molecules from CO_2 in the air. In plants this takes place in specialized compartments within chloroplasts.

Life is powered by sunshine. All of the energy used by almost all living cells comes ultimately from the sun, captured by plants, algae, and some bacteria through the process of **photosynthesis.** Every oxygen atom in the air we breathe was once part of a water molecule, liberated by photosynthesis as you will discover in this chapter. Life as we know it is only possible because our earth is awash in energy streaming inward from the sun. Each day, the radiant energy that reaches the earth is equal to that of about 1 million Hiroshima-sized atomic bombs. About 1% of it is captured by photosynthesis and provides the energy needed to synthesize carbohydrates that drives almost all life on earth. Use the arrows on this page and the next three pages to follow the path of energy from the sun through photosynthesis.

Trees.   Many kinds of organisms carry out photosynthesis, not only the diversity of plants that make our world green, but also bacteria and algae. Photosynthesis is somewhat different in bacteria, but we will focus our attention on photosynthesis in plants, starting with this maple tree crowned with green leaves. Later we will look at the grass growing beneath the maple tree—it turns out that grasses and other related plants sometimes take a different approach to photosynthesis depending on the conditions.

Leaves.   To learn how this maple tree captures energy from sunlight, follow the light. It comes beaming in from the sun, down through earth's atmosphere, bathing the top of the tree in light. What part of the maple tree is actually being struck by this light? Its green leaves. Each branch at the top of the tree ends in a spread of these leaves, each leaf flat and thin like the page of a book. Within these green leaves is where photosynthesis occurs. No photosynthesis occurs within this tree's stem, covered with bark, and none in the roots, buried within the soil—no light reaches these parts of the plant. The tree has a very efficient internal plumbing system that transports the products of photosynthesis to the stem, roots, and other parts of the plant so that they too may benefit from the capture of the sun's energy.

The Leaf Surface.   Now follow the light as it passes into a leaf. The beam of light first encounters a waxy protective layer called the cuticle. The cuticle acts a bit like a layer of clear fingernail polish, providing a thin, watertight and surprisingly strong layer of protection. Light passes right through this transparent wax, and then proceeds to pass right on through a layer of cells immediately beneath the cuticle called the epidermis. Only one cell layer thick, this epidermis acts as the "skin" of the leaf, providing more protection from damage and, very importantly, controlling how gases and water enter and leave the leaf. Very little of the light is absorbed by the cuticle or the epidermis.

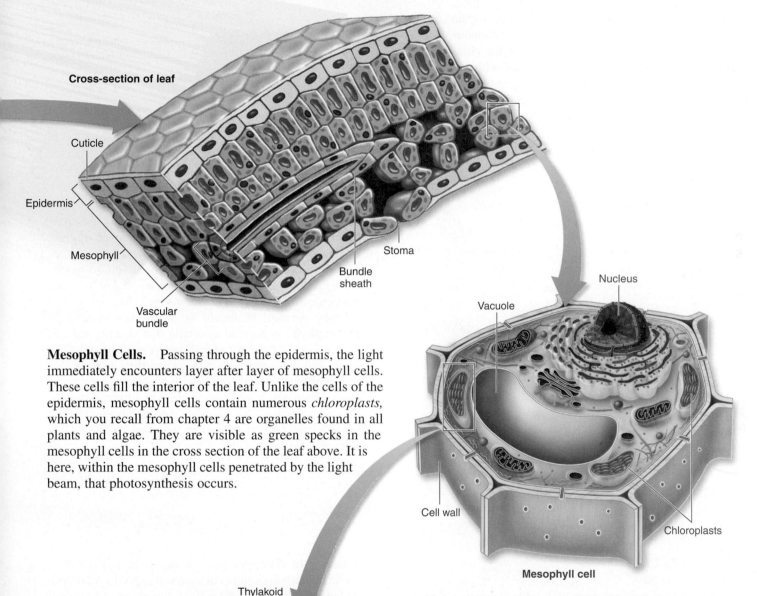

Cross-section of leaf

Cuticle

Epidermis

Mesophyll

Vascular bundle

Bundle sheath

Stoma

Vacuole

Nucleus

Cell wall

Chloroplasts

Mesophyll cell

Mesophyll Cells.   Passing through the epidermis, the light immediately encounters layer after layer of mesophyll cells. These cells fill the interior of the leaf. Unlike the cells of the epidermis, mesophyll cells contain numerous *chloroplasts,* which you recall from chapter 4 are organelles found in all plants and algae. They are visible as green specks in the mesophyll cells in the cross section of the leaf above. It is here, within the mesophyll cells penetrated by the light beam, that photosynthesis occurs.

Thylakoid

Inner membrane

Outer membrane

Granum

Stroma

Chloroplast

Chloroplasts.   Light penetrates into mesophyll cells. The cell walls of the mesophyll cells don't absorb it, nor does the plasma membrane or nucleus or mitochondria. Why not? Because these elements of the mesophyll cell contain few if any molecules that absorb visible light. If chloroplasts were not also present in these cells, most of this light would pass right through, just as it passed through the epidermis. But chloroplasts are present, lots of them. One chloroplast is highlighted by a box in the mesophyll cell above. Light passes into the cell and when it reaches the chloroplast, it passes through the outer and inner membranes to reach the thylakoid structures within the chloroplast, clearly seen as the green disks in the cutaway chloroplast shown here.

Inside the Chloroplast

All the important events of photosynthesis happen inside the chloroplast. The journey of light into the chloroplasts ends when the light beam encounters a series of internal membranes within the chloroplast organized into flattened sacs called *thylakoids*. Often, numerous thylakoids are stacked on top of one another in columns called *grana*. In the drawing below, the grana look not unlike piles of dishes. While each thylakoid is a separate compartment that functions more-or-less independently, the membranes of the individual thylakoids are all connected, part of a single continuous membrane system. Occupying much of the interior of the chloroplast, this thylakoid membrane system is submerged within a semi-liquid substance called *stroma*, which fills the interior of the chloroplast in much the same way that cytoplasm fills the interior of a cell. Suspended within the stroma are many enzymes and other proteins, including the enzymes that act later in photosynthesis to assemble organic molecules from carbon dioxide (CO_2) in reactions that do not require light and which are discussed later.

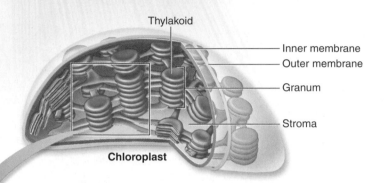

Thylakoid

Inner membrane
Outer membrane

Granum

Stroma

Chloroplast

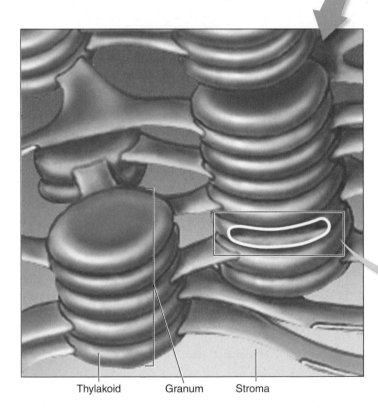

Thylakoid   Granum   Stroma

Penetrating the Thylakoid Surface. The first key event of photosynthesis occurs when a beam of sunlight strikes the surface membrane of a thylakoid. Embedded within this membrane, like icebergs on an ocean, are clusters of light-absorbing pigments. A pigment molecule is a molecule that absorbs light energy. The primary pigment molecule in most photosystems is **chlorophyll,** an organic molecule that absorbs red and blue light, but does not absorb green wavelengths. The green light is instead reflected, giving the thylakoid and the chloroplast which contains it an intense green color. Plants are green because they are rich in green chloroplasts. Except for some alternative pigments also present in thylakoids, no other parts of the plant absorb visible light with such intensity.

Striking the Photosystem. Within each pigment cluster, the chlorophyll molecules are arranged in a network called a *photosystem*. The light-absorbing chlorophyll molecules of a photosystem act together as an antenna to capture photons (units of light energy). A lattice of structural proteins, indicated by the purple structure inserted into the thylakoid membrane in the diagram on the facing page, anchors each of the chlorophyll molecules of a photosystem into a precise position, such that every chlorophyll molecule is touching several others. Wherever light strikes the photosystem, some chlorophyll molecule will be in position to receive it.

Energy Absorption. When sunlight strikes any chlorophyll molecule in the photosystem, the chlorophyll molecule absorbs energy. The energy becomes part of the chlorophyll molecule, boosting some of its electrons to higher energy levels. Possessing these more energetic electrons, the chlorophyll molecule is said to now be "excited." With this key event, the biological world has captured energy from the sun.

Excitation of the Photosystem. The excitation that the absorption of light creates is then passed from the chlorophyll molecule which was hit to another, and then to another, like a hot potato being passed down a line of people. This shuttling of excitation is not a chemical reaction, in which an electron physically passes between atoms. Rather, it is energy that passes from one chlorophyll molecule to its neighbor. A crude analogy to this form of energy transfer is the initial "break" in a game of pool. If the cue ball squarely hits the point of the triangular array of 15 billiard balls, the two balls at the far corners of the triangle fly off, and none of the central balls move at all. The kinetic energy is transferred through the central balls to the most distant ones. In much the same way, the sun's excitation energy moves through the photosystem from one chlorophyll to the next.

Energy Capture. As the energy shuttles from one chlorophyll molecule to another within the photosystem network, it eventually arrives at a key chlorophyll molecule, the only one that is touching a membrane-bound protein. Like shaking a marble in a box with a walnut-sized hole in it, the excitation energy will find its way to this special chlorophyll just as sure as the marble will eventually find its way to and through the hole in the box. The special chlorophyll then transfers an excited (high-energy) electron to the acceptor molecule it is touching.

The Light-Dependent Reactions. Like a baton being passed from one runner to another in a relay race, the electron is then passed from that acceptor protein to a series of other proteins in the membrane that put the energy of the electron to work making ATP and NADPH. In a way you will explore later in this chapter, the energy is used to power the movement of protons across the thylakoid membrane to make ATP and another key molecule, NADPH. So far, photosynthesis has consisted of two stages, indicated by numbers in the diagram to the lower left: ❶ capturing energy from sunlight—accomplished by the photosystem; and ❷ using the energy to make ATP and NADPH. These first two stages of photosynthesis take place only in the presence of light, and together are traditionally called the **light-dependent reactions.** ATP and NADPH are important energy-rich chemicals, and after this, the rest of photosynthesis becomes a chemical process.

The Light-Independent Reactions. The ATP and NADPH molecules generated by the light-dependent reactions are then used to power a series of chemical reactions in the stroma of the chloroplast, each catalyzed by an enzyme present there. Acting together like the many stages of a manufacturing assembly line, these reactions accomplish the synthesis of carbohydrates from CO_2 in the air ❸. This third stage of photosynthesis, the formation of organic molecules like glucose from atmospheric CO_2, is called the **Calvin cycle,** but is also referred to as the **light-independent reactions** because it doesn't require light directly. We will examine the Calvin cycle in detail later in this chapter.

This completes our brief overview of photosynthesis. In the rest of the chapter we will revisit each stage and consider its elements in more detail. For now, the overall process may be summarized by the following simple equation:

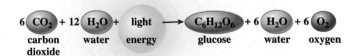

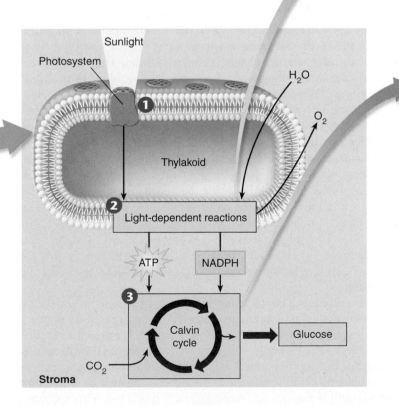

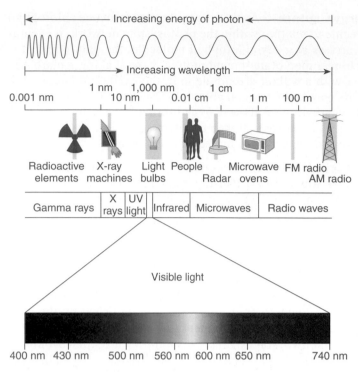

Figure 6.1 Photons of different energy: the electromagnetic spectrum.

Light is composed of packets of energy called photons. Some of the photons in light carry more energy than others. The shorter the wavelength of light, the greater the energy of its photons. Visible light represents only a small part of the electromagnetic spectrum, that with wavelengths between about 400 and 740 nanometers.

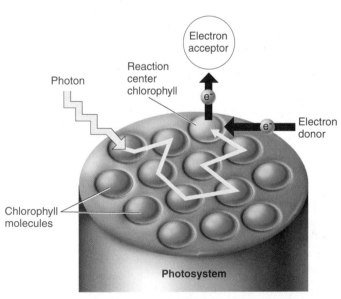

Figure 6.2 How a photosystem works.

When light of the proper wavelength strikes any pigment molecule within a photosystem, the light is absorbed and its excitation energy is then transferred from one molecule to another within the cluster of pigment molecules until it encounters the reaction center, which exports the energy as high-energy electrons to an acceptor molecule.

6.2 How Plants Capture Energy from Sunlight

CONCEPT PREVIEW: Photons from red and blue light are captured by chlorophyll pigments and used to excite electrons in the reaction center. The excited electrons are used to do the chemical work of producing ATP and NADPH.

Where is the energy in light? What is there about sunlight that a plant can use to create chemical bonds? The revolution in physics in the twentieth century taught us that light actually consists of tiny packets of energy called **photons,** which have properties both of particles and of waves. When light shines on your hand, your skin is being bombarded by a stream of these photons.

Sunlight contains photons of many energy levels, only some of which we "see." We call the full range of these photons the **electromagnetic spectrum.** As you can see in figure 6.1, some of the photons in sunlight have shorter wavelengths (toward the left side of the spectrum) and carry a great deal of energy—for example, gamma rays and ultraviolet (UV) light. Others such as radio waves carry very little energy and have longer wavelengths (hundreds to thousands of meters long). Molecules that absorb light energy are called **pigments.** When we speak of visible light, we refer to those wavelengths that the pigment within human eyes, called *retinal,* can absorb—roughly with wavelengths from 400 nanometers (violet) to 740 nanometers (red). Plants are even more picky, absorbing mainly blue and red light and reflecting back what is left of the visible light. Plants are perceived by our eyes as green simply because only the green wavelengths of light are reflected off the plant leaves.

Pigments and Photosystems

Other animals use different pigments for vision and thus "see" a different portion of the electromagnetic spectrum. For example, the pigment in insect eyes absorbs at shorter wavelengths than retinal. That is why bees can see ultraviolet light, which we cannot see, but are blind to red light, which we can see. The main pigment in plants that absorbs light is chlorophyll, present in two versions: chlorophyll *a* and chlorophyll *b*. While chlorophyll absorbs fewer kinds of photons than our eye pigment retinal, it is much more efficient at capturing them.

The light-dependent reactions of photosynthesis occur on membranes. In most photosynthetic bacteria, the proteins involved in the light-dependent reactions are embedded within the plasma membrane. In algae, intracellular membranes contain the proteins that drive the light-dependent reactions. In plants, photosynthesis occurs in specialized organelles called chloroplasts. The chlorophyll molecules and proteins involved in the light-dependent reactions are embedded in the thylakoid membranes inside the chloroplasts. This complex of proteins and pigment molecules makes up the **photosystem.**

Like a magnifying glass focusing light on a precise point, a photosystem channels the excitation energy gathered by any one of its pigment molecules to a specific chlorophyll *a* molecule, which is called the reaction center chlorophyll. For example, in figure 6.2, a chlorophyll molecule on the outer edge of the photosystem is excited by the photon, and this energy passes from one chlorophyll molecule to another, indicated by the yellow zigzag arrow, until it reaches the reaction center molecule. This molecule then passes the energy, in the form of an excited electron, out of the photosystem to drive the synthesis of ATP and organic molecules.

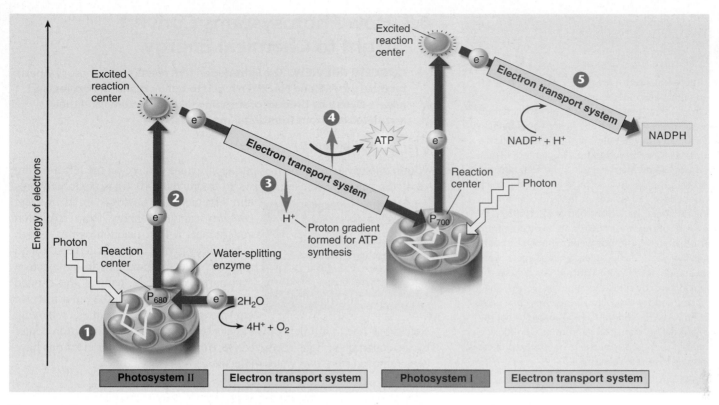

Figure 6.3   Plants use two photosystems.

In stage ❶, a photon excites pigment molecules in photosystem II. In stage ❷, a high-energy electron from photosystem II is transferred to the electron transport system. In stage ❸, the excited electron is used to pump a proton across the membrane. In stage ❹, the concentration gradient of protons is used to produce a molecule of ATP. In stage ❺, the ejected electron then passes to photosystem I, which uses it with a photon of light energy, to drive the formation of NADPH.

IMPLICATION: Photosystem I consists entirely of chlorophyll *a* pigment molecules, while photosystem II is made up of half chlorophyll *a* and half chlorophyll *b* molecules. Both chlorophyll *a* and *b* absorb blue and red light but reflect green light. What color would plant leaves be if they absorbed all wavelengths of visible light?

Using Two Photosystems

Plants and algae use two photosystems, photosystems I and II, indicated by the two purple cylinders in figure 6.3. Photosystem II captures the energy that is used to produce the ATP needed to build sugar molecules. The light energy that it captures is used in stages ❶ and ❷ to transfer the energy of a photon of light to an excited electron; the energy of this electron is then used by the electron transport system ❸ to produce ATP ❹.

Photosystem I powers the production of the hydrogen atoms needed to build sugars and other organic molecules from CO_2 (which has no hydrogen atoms). Photosystem I uses an energized electron, carried by a hydrogen ion (a proton), to form NADPH from $NADP^+$ in stage ❺. NADPH shuttles hydrogens to the Calvin cycle where sugars are made.

The photosystems are not numbered in the order in which they are used. Photosystem II actually acts first in the series, and photosystem I acts second. The confusion arises because the photosystems were named in the order in which they were discovered, and photosystem I was discovered before photosystem II.

Figure 6.4   The photosynthetic electrons are used to produce ATP and NADPH.

The energy of the electron absorbed by photosystem II powers the pumping of protons into the thylakoid space. These protons then pass back out through ATP synthase channels, their movement powering the production of ATP. The energy of the electron absorbed by photosystem I powers the attachment of a proton to $NADP^+$, forming NADPH.

6.3 How Photosystems Convert Light to Chemical Energy

CONCEPT PREVIEW: The light-dependent reactions of photosynthesis produce the ATP and NADPH needed to build organic molecules, and release O_2 as a by-product of stripping hydrogen atoms and their associated electrons from water molecules.

Photosystem II

Within photosystem II, the first purple structure you see on the left in figure 6.4, the reaction center consists of more than 10 transmembrane protein subunits. The photosystem II *antenna complex* captures energy from a photon and funnels it to a reaction center, which responds by giving up an excited electron to a primary electron acceptor in the electron transport system. The path of the excited electron is indicated with the red arrow. An enzyme splits water molecules, removing electrons one at a time to fill the electron hole left in the reaction center by the departure of light-energized electrons. As soon as four electrons have been removed from two water molecules, O_2 is released.

> To better understand how energy from the sun is captured by the plant, look back to figure 2.3 on page 33. An electron in the reaction center absorbs energy from the sun and is moved to a higher energy level. This energized electron is now carrying energy from the sun.

Electron Transport System

The light-energized electrons leaving photosystem II are passed to a series of electron-carrier molecules called the *electron transport system*. These proteins are embedded within the thylakoid membrane. Importantly, one of them is a "proton pump" protein. The energy of the electron is used by this protein to pump a proton from the stroma into the thylakoid space (indicated by the blue arrow through the electron transport system). A nearby protein in the membrane then carries the now energy-depleted electron on to photosystem I.

> The transfer of electrons from one carrier protein to another along the electron transport system is a type of chemical reaction called oxidation-reduction, or redox. Redox reactions involve the transfer of energy along with electrons and are discussed in chapter 7 on page 114.

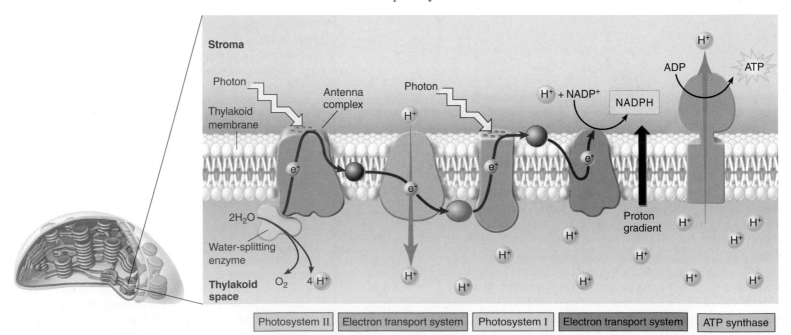

| Photosystem II | Electron transport system | Photosystem I | Electron transport system | ATP synthase |

Making ATP: Chemiosmosis

Before progressing onto photosystem I, let's see what happens with the protons that were pumped into the thylakoid by the electron transport system. The process of ATP formation is called **chemiosmosis,** and involves three elements: photosystem II, the first electron transport system, and a protein channel called ATP synthase. The thylakoid membrane you see in figure 6.4 is impermeable to protons, so protons that have been pumped across the membrane by the first electron transport system build up inside the thylakoid space, creating a very large concentration gradient. These protons now diffuse back out of the thylakoid space, down their concentration gradient, passing through a special channel protein called *ATP synthase*.

> As you may recall from the discussion of diffusion on page 78, molecules in solution diffuse from areas of higher concentration to areas of lower concentration, down their concentration gradient.

The ATP synthase is the membrane protein on the far right side of figure 6.4, protruding like a knob out of the external surface of the thylakoid membrane. As protons pass out of the thylakoid through the ATP synthase channels, a phosphate group is added onto ADP to form ATP in a process called phosphorylation. ATP is released into the stroma (the fluid matrix inside the chloroplast). The stroma contains the enzymes that catalyze the light-independent reactions, where ATP is used to build carbohydrate molecules.

> The ATP molecule, as described on page 94, contains high-energy phosphate bonds. The chemical reaction that forms the outermost high-energy bond from ADP and a phosphate group uses energy from the sun harvested by exciting electrons.

Photosystem I

Now, with ATP formed, let's return to the middle of figure 6.4 where photosystem I accepts an electron from the electron transport system. Energy is fed to photosystem I by an antenna complex of chlorophyll molecules. The electron arriving from the first electron transport system has by no means lost all of its light-excited energy; almost half remains. Thus, the absorption of another photon of light energy by photosystem I boosts the electron leaving its reaction center to a very high energy level.

Making NADPH

Like photosystem II, photosystem I passes electrons to an electron transport system. When two of these electrons reach the end of this electron transport system, they are then donated to a molecule of $NADP^+$ along with a proton (a hydrogen ion) to form NADPH. Because the reaction occurs on the stromal side of the membrane and involves the uptake of a proton in forming NADPH, it contributes further to the proton concentration gradient established during photosynthetic electron transport.

Although photosynthetic bacteria carry out the light-dependent reactions in a slightly different way, photosynthesis in algae, such as those shown in figure 6.5, functions in the same way as in plants.

Products of the Light-Dependent Reactions

The ATP and NADPH produced in the light-dependent reactions end up being passed on to the Calvin cycle in the stroma of the chloroplast. There, ATP is used to power chemical reactions that build carbohydrates. NADPH provides the hydrogens and electrons used in building carbohydrates. In the next section we examine how the Calvin cycle of photosynthesis puts these two molecules to work.

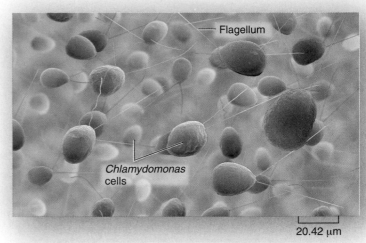

Flagellum

Chlamydomonas cells

20.42 µm

Figure 6.5   Plants aren't the only ones that carry out photosynthesis.

Each of these green balls is a single-celled photosynthetic organism, a green alga called *Chlamydomonas* that is common in pond water.

IMPLICATION   Photosynthetic algae and bacteria in the oceans, called phytoplankton, carry out much of the earth's photosynthesis, removing far more CO_2 from the atmosphere than their cellular respiration adds to it. However, increases in ocean temperatures speed up cellular respiration much faster than photosynthesis. What would you expect the effect of global warming to be on the ocean's influence on atmospheric CO_2? Why is this consequence important?

IN THE NEWS

Using Algae to Fix Global Warming.   Engineers at Ohio University have come up with a novel solution to the problem of what to do with the massive amounts of CO_2 that will be generated by burning coal. They have created a photo bioreactor that passes streams of CO_2 over sheets of a woven material made of living algae. The algae use photosynthesis to remove CO_2 from the air stream! When an algal sheet has grown substantially, it is removed and the algae dried and used as feed for farm animals. Test facilities are already up and running. A full-scale reactor that produces 1.25 million square meters of algae sheets may be operational by 2010.

EVOLUTION

Global Warming and the Calvin Cycle. Photosynthesis removes CO_2 from air by binding it to the compound RuBP in the first step of the Calvin cycle. The enzyme that catalyzes this reaction, with a long chemical name usually shortened to RuBisCO, is the most abundant enzyme in the world, and also one of the least efficient. It is a thousand times slower than most other enzymes. Its inefficiency limits the amount of CO_2 plants can remove from the atmosphere. For years scientists have tried to engineer a speedier variant of the RuBisCO enzyme by altering specific sites on the enzyme and then looking to see if the change improved the enzyme. Nothing they tried has worked—there are just too many possible changes to test. However, scientists at Emory University have found a way through the thicket of possibilities. With a nod to Darwin, they have used a process called "directed evolution." They added the gene for RuBisCO to a bacterium in such a way that the bacterium could not survive unless the enzyme functioned; then they randomly altered the gene. The fastest growing bacteria were those with the most efficient RuBisCO! The best experimental results exhibited a 500% increase in enzyme speed. The next step . . . to get this engineered enzyme into plants.

6.4 Building New Molecules

CONCEPT PREVIEW: In a series of reactions, called the Calvin cycle, CO_2, ATP, and NADPH are used to assemble new organic molecules.

The Calvin Cycle

Stated very simply, photosynthesis is a way of making organic molecules from carbon dioxide (CO_2). The actual assembly of new molecules employs a complex battery of enzymes in what is called the **Calvin cycle,** or **C₃ photosynthesis** (C₃ because the first molecule produced in the process is a three-carbon molecule). The process takes place in three stages, indicated by the three panels in *Essential Biological Process 6A*. Three turns of the cycle are needed to produce one molecule of glyceraldehyde 3-phosphate. In any *one* turn of the cycle, a carbon atom from a carbon dioxide molecule is first added to a five-carbon sugar, producing two three-carbon sugars. This process is called **carbon fixation** because it attaches a carbon atom that was in a gas to an organic molecule. The cycle has to "turn" six times in order to form a new glucose molecule. The cycle is driven by energy from ATP and hydrogen atoms are supplied by NADPH, both produced in the light-dependent reactions.

Concept Check

1. In what cells of a leaf does photosynthesis occur?
2. Which photosystem powers the production of ATP, I or II?
3. What three substances are needed by the Calvin cycle to produce a molecule of glucose?

Essential Biological Process 6A

The Calvin Cycle

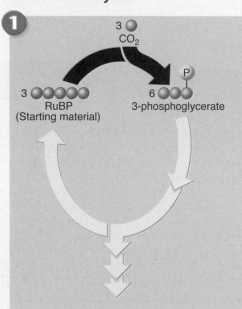

The Calvin cycle begins when a carbon atom from a CO_2 molecule is added to a five-carbon molecule (the starting material). The resulting six-carbon molecule is unstable and immediately splits into three-carbon molecules. (Three "turns" of the cycle are indicated here with three molecules of CO_2 entering the cycle.)

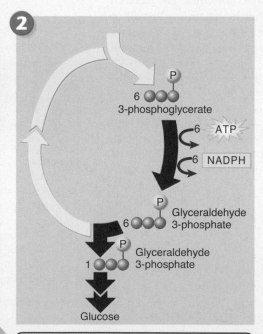

Then, through a series of reactions, energy from ATP and hydrogens from NADPH (the products of the light-dependent reactions) are added to the three-carbon molecules. The now-reduced three-carbon molecules either combine to make glucose or are used to make other molecules.

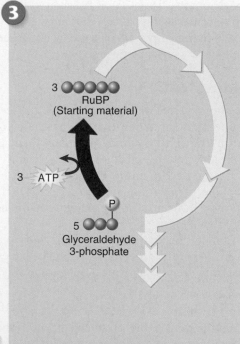

Most of the reduced three-carbon molecules are used to regenerate the five-carbon starting material, thus completing the cycle.

Photorespiration

6.5 Photorespiration: Putting the Brakes on Photosynthesis

CONCEPT PREVIEW: Photorespiration occurs due to a buildup of oxygen within photosynthetic cells. C_4 plants get around photorespiration by synthesizing sugars in bundle-sheath cells, and CAM plants delay the light-independent reactions until night, when stomata are open.

Many plants have trouble carrying out C_3 photosynthesis when the weather is hot. As temperatures increase in hot, arid weather, plants partially close their **stomata** (singular, **stoma**) to conserve water. The stomata are openings in the epidermis through which water vapor and O_2 pass out of the leaf and CO_2 passes in. As a result, when stomata close, CO_2 and O_2 are not able to enter and exit the leaves (figure 6.6). The concentration of CO_2 in the leaves falls, while the concentration of O_2 in the leaves rises. Under these conditions the active site of the enzyme that carries out the first step of the Calvin cycle (called rubisco) binds O_2 instead of CO_2. When this occurs, CO_2 is ultimately released as a by-product of an alternate reaction, effectively short-circuiting the Calvin cycle. This failure of photosynthesis is called **photorespiration.**

> Oxygen and carbon dioxide both bind to rubisco at the same active site. In this way, oxygen is a competitive inhibitor of the enzyme, a topic discussed on page 93.

C_4 Photosynthesis

Some plants are able to adapt to climates with higher temperatures by performing **C_4 photosynthesis.** In this process, plants such as sugarcane, corn, and many grasses are able to fix carbon using different types of cells and chemical reactions within their leaves, thereby avoiding this reduction in photosynthesis due to higher temperatures.

A cross section of a leaf from a C_4 plant is shown in figure 6.7. Examining it, you can see how these plants solve the problem of photorespiration. In the enlargement, you see two cell types: The green cell is a mesophyll cell and the tan cell is a bundle-sheath cell. In the mesophyll cell, CO_2 combines with a three-carbon molecule instead of RuBP (the five-carbon starting molecule in the Calvin cycle). This reaction produces a four-carbon molecule, oxaloacetate (hence the name, C_4 photosynthesis), rather than the three-carbon molecule, phosphoglycerate, shown in panel 1 of *Essential Biological Process 6A*. C_4 plants carry out this process using a different enzyme. The oxaloacetate is then converted to malate, which is transferred to the bundle-sheath cells of the leaf. In the tan bundle-sheath cell, malate is broken down to regenerate CO_2, which enters the Calvin cycle and sugars are synthesized. Why go to all this trouble? Because the bundle-sheath cells are impermeable to CO_2 and so the concentration of CO_2 increases within them, which substantially lowers the rate of photorespiration.

A second strategy to decrease photorespiration is used by many succulent (water-storing) plants such as cacti and pineapples. Called **crassulacean acid metabolism (CAM)** plants after the plant family Crassulaceae in which it was first discovered, these plants open their stomata and fix CO_2 into organic compounds during the night when it's cooler, and then close the stomata during the day.

Concept Check

1. Are C_4 plants more likely to be common in Arizona or Maine? Why?
2. In what phase of photosynthesis does photorespiration occur?
3. In what two types of cells does C_4 photosynthesis occur? Why these?

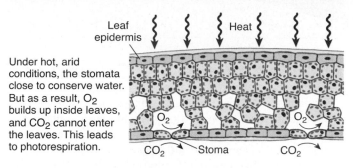

Under hot, arid conditions, the stomata close to conserve water. But as a result, O_2 builds up inside leaves, and CO_2 cannot enter the leaves. This leads to photorespiration.

Figure 6.6   Plant response in hot weather.

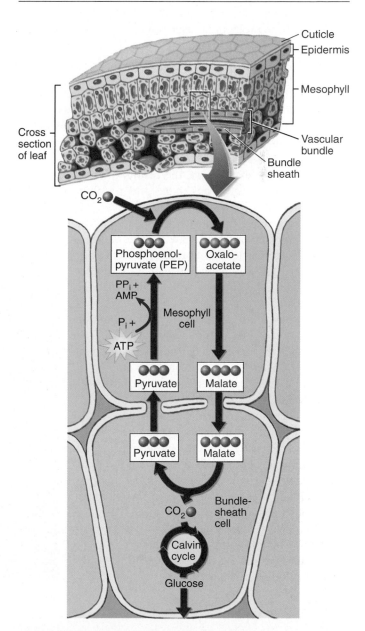

Figure 6.7   Carbon fixation in C_4 plants.

In mesophyll cells, oxaloacetate is converted into malate that is transported into bundle-sheath cells. Once there, malate undergoes a chemical reaction producing carbon dioxide. The carbon dioxide is trapped in the bundle-sheath cell, where it enters the Calvin cycle.

Does Iron Limit the Growth of Ocean Phytoplankton?

Phytoplankton are microscopic organisms that live in the oceans, carrying out much of the earth's photosynthesis. The photo below is of *Chaetoceros,* a phytoplankton. Decades ago, scientists noticed "dead zones" in the ocean where little photosynthesis occurred. Looking more closely, they found that phytoplankton collected from these waters are not able to efficiently fix CO_2 into carbohydrates. In an attempt to understand why not, the scientists hypothesized that lack of iron was the problem (an electron carrier in the electron transport system requires iron to function properly), and they predicted that fertilizing these ocean waters with iron could trigger an explosively rapid growth of phytoplankton.

To test this idea, they carried out a field experiment, seeding large areas of phytoplankton-poor ocean waters with iron crystals to see if this triggered phytoplankton growth. Other similarly phytoplankton-poor areas of ocean were not seeded with iron and served as controls.

In one such experiment, the results of which are presented in the graph to the right, a 72-km² grid of phytoplankton-deficient ocean water was seeded with iron crystals and a tracer substance in three successive treatments, indicated with arrows on the *x* axis of the graph (on days 0, 3, and 7). The multiple seedings were carried out to reduce the effect of the iron crystals dissipating over time. A smaller control grid, 24 km², was seeded with just the tracer substance.

To assess the numbers of phytoplankton organisms carrying out photosynthesis in the ocean water, investigators did not actually count organisms. Instead, they estimated the amount of chlorophyll *a* in water samples as an easier-to-measure index. An **index** is a parameter that accurately reflects the quantity of another less easily measured parameter. In this instance, the level of chlorophyll *a*, easily measured by monitoring the wavelengths of light absorbed by a liquid sample, is a suitable index of phytoplankton, as this pigment is found nowhere else in the ocean other than within phytoplankton.

Chlorophyll *a* measurements were made periodically on both test and control grids for 14 days. The results are plotted on the graph. Red points indicate chlorophyll *a* concentrations in iron-seeded

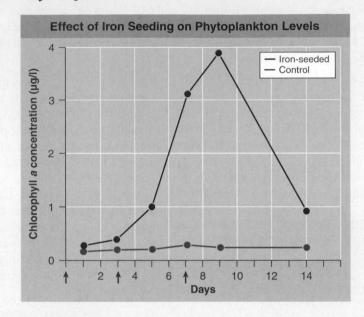

Effect of Iron Seeding on Phytoplankton Levels

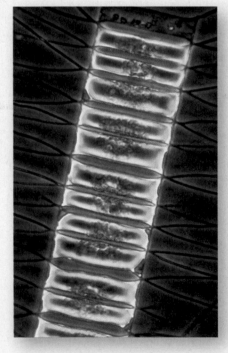

waters; blue points indicate chlorophyll *a* levels in the control grid waters that were not seeded.

Analysis

1. **Applying Concepts** What substance is lacking in the waters sampled in the blue-dot plots ?
2. **Interpreting Data** Comparing the red line to the blue line, about how many times more numerous are phytoplankton in iron-seeded waters on each of the three days of seeding?
3. **Making Inferences**
   a. What general statement can be made regarding the effect of seeding phytoplankton-poor regions of the ocean with iron?
   b. Why did chlorophyll *a* levels drop by day 14?
4. **Drawing Conclusions** Do these results support the claim that lack of iron is limiting the growth of phytoplankton, and thus of photosynthesis, in certain areas of the oceans?

Concept Summary

Photosynthesis

6.1 An Overview of Photosynthesis

- Photosynthesis is a biochemical process whereby, energy from the sun is captured and used to build carbohydrates from CO_2 gas and water.

- Photosynthesis consists of a series of chemical reactions that occurs in two stages (overview figure from **page 103**): the light-dependent reactions that produce ATP and NADPH occur on the thylakoid membranes of chloroplasts in plants, while the light-independent reactions (the Calvin cycle) that synthesize carbohydrates occur in the stroma.

6.2 How Plants Capture Energy from Sunlight

- Sunlight contains packets of energy called photons that contain varying amounts of energy. Photons of shorter wavelengths, such as gamma rays and UV light, have a lot of energy. As the wavelengths increase in size, the amount of energy in the photons decrease (**figure 6.1**). Visible light are wavelengths absorbed by pigments in the human eye (between 400 and 740 nanometers). Pigments are molecules that capture light energy. Plants use the pigment chlorophyll to absorb light energy.

- Plants appear green because of their chlorophyll pigments. Chlorophyll absorbs wavelengths in the far ends of the visual spectrum (the blue and red wavelengths) and reflect the green wavelengths, which is why leaves appear green.

- The light-dependent reactions occur on the thylakoid membranes of chloroplasts in plants. The chlorophyll molecules and other pigments involved in photosynthesis are embedded in a complex of proteins within the membrane called a photosystem.

- The energy from a photon of light is absorbed by a chlorophyll molecule and is transferred between chlorophyll molecules in the photosystem, as shown here from **figure 6.2**. Once the energy is passed to the reaction center it excites an electron which is transferred to the electron transport system.

- The energized electron is used to generate ATP and NADPH. ATP powers the Calvin cycle and NADPH donates hydrogen atoms toward the building of carbohydrate molecules. Plants utilize two photosystems that occur in series (**figure 6.3**). Photosystem II leads to the formation of ATP, and photosystem I leads to the formation of NADPH.

6.3 How Photosystems Convert Light to Chemical Energy

- The excited electron that leaves the reaction center of photosystem II is replenished with an electron from the breakdown of a water molecule. Oxygen gas is a by-product of this reaction.

- The excited electron is passed from one protein to another in the electron transport system, where energy from the electron is used to operate a proton pump that pumps hydrogen ions across the membrane against a concentration gradient (**figure 6.4**).

- The hydrogen ion concentration gradient is used as a source of energy to generate molecules of ATP. This energy is used to drive H^+ back across the membrane through a specialized channel protein called ATP synthase, which catalyzes the formation of ATP, as shown here from **figure 6.4.**

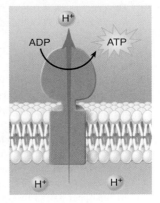

- After the electron passes along the first electron transport system, it is then transferred to a second photosystem, photosystem I, where it gets an energy boost from the capture of another photon of light (**figure 6.4**). This reenergized electron is passed along another electron transport system to an ultimate electron acceptor, $NADP^+$. $NADP^+$ binds electrons and a H^+ to produce NADPH, which is shuttled to the Calvin cycle.

6.4 Building New Molecules

- ATP and NADPH, from the light-dependent reactions, are shuttled to the stroma where they are used in the Calvin cycle.

- The Calvin cycle is carried out by a series of enzymes that use the energy from ATP and electrons and hydrogen ions from NADPH to build molecules of carbohydrates by reducing CO_2 from the air (***Essential Biological Process 6A***).

Photorespiration

6.5 Photorespiration: Putting the Brakes on Photosynthesis

- In hot dry weather, plants will close the stomata in their leaves to conserve water. As a result, the levels of O_2 increase in the leaves, and CO_2 levels drop, as shown here from **figure 6.6.** Under these conditions, the Calvin cycle, also called C_3 photosynthesis, is disrupted. When there is a higher internal concentration of oxygen, O_2 rather than CO_2 enters the Calvin cycle in a process called photorespiration. In this case, the first enzyme in the Calvin cycle, rubisco, binds oxygen instead of carbon dioxide.

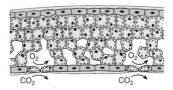

- C_4 plants reduce the effects of photorespiration by modifying the carbon-fixation step, splitting it into two steps that take place in different cells. The C_4 pathway produces malate in mesophyll cells. Malate is then transferred to bundle-sheath cells, where it breaks down to produce carbon dioxide. This CO_2 then enters the Calvin cycle in the bundle-sheath cells (**figure 6.7**).

- In CAM plants, carbon dioxide is processed into organic molecules through the C_4 pathway during the night when stomata are open.

Self-Test

1. The energy that is used by almost all living things on our planet comes from the sun. It is captured by plants, algae, and some bacteria through the process of
   a. thylakoid.
   c. photosynthesis.
   b. chloroplasts.
   d. the Calvin cycle.

2. Plants capture the energy from sunlight
   a. through photorespiration.
   b. with molecules called pigments that absorb photons and use their energy.
   c. with the light-independent reactions.
   d. with the electron transport system.

3. Visible light occupies what part of the electromagnetic spectrum?
   a. the entire spectrum
   b. the upper half of the spectrum (with longer wavelengths)
   c. a small portion in the middle of the spectrum
   d. the lower half of the spectrum (with shorter wavelengths)

4. The colors of light that are absorbed by chlorophyll are
   a. red and blue.
   b. green and yellow.
   c. infrared and ultraviolet.
   d. All colors are equally absorbed.

5. Once a plant has initially captured the energy of a photon,
   a. a series of reactions occurs in thylakoid membranes of the cell.
   b. the energy is transferred through several steps into a molecule of ATP.
   c. a water molecule is broken down, releasing oxygen.
   d. All of the above.

6. Plants use two photosystems to capture energy used to produce ATP and NADPH. The electrons used in these photosystems
   a. recycle through the system, with energy added from the photons.
   b. recycle through the system several times and then are lost due to entropy.
   c. only go through the system once; they are obtained by splitting a water molecule.
   d. only go through the system once; they are obtained from the photon.

7. During photosynthesis, ATP molecules are generated by
   a. the Calvin cycle.
   b. chemiosmosis.
   c. the splitting of a water molecule.
   d. photons of light being absorbed by chlorophyll molecules.

8. NADPH is recycled during photosynthesis. It is produced during the _____ and used in the _____.
   a. electron transport system of photosystem I, Calvin cycle
   b. process of chemiosmosis, Calvin cycle
   c. electron transport system of photosystem II, electron transport system of photosystem I
   d. light-independent reactions, light-dependent reactions

9. The overall purpose of the Calvin cycle is to
   a. generate molecules of ATP.
   b. generate NADPH.
   c. build sugar molecules.
   d. produce oxygen.

10. Many plants cannot carry out the typical C_3 photosynthesis in hot weather, so some plants
    a. use the ATP cycle.
    b. use C_4 photosynthesis or CAM.
    c. shut down photosynthesis completely.
    d. All of these are true for different plants.

Visual Understanding

1. This figure shows the areas of the visible spectrum that are absorbed by two forms of chlorophyll. The green and yellow wavelengths are reflected back and so plants appear green. The red skin of an apple contains different pigments. What areas of the spectrum do you think are absorbed and reflected by these pigments?

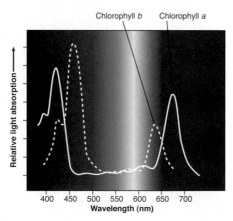

2. **Figure 6.4** Could a plant cell produce ATP through chemiosmosis if the thylakoid membrane was "leaky" with regards to protons? Explain.

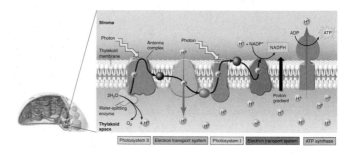

Challenge Questions

1. To reduce six molecules of carbon dioxide to glucose via photosynthesis, how many molecules of NADPH and ATP are required?

2. In theory, a plant kept in total darkness could still manufacture glucose if it were supplied with which molecules?

3. If you were going to design a plant that would survive in the deserts of Arizona and New Mexico, how would you balance its need for CO_2 with its need to avoid water loss in the hot summer temperatures?

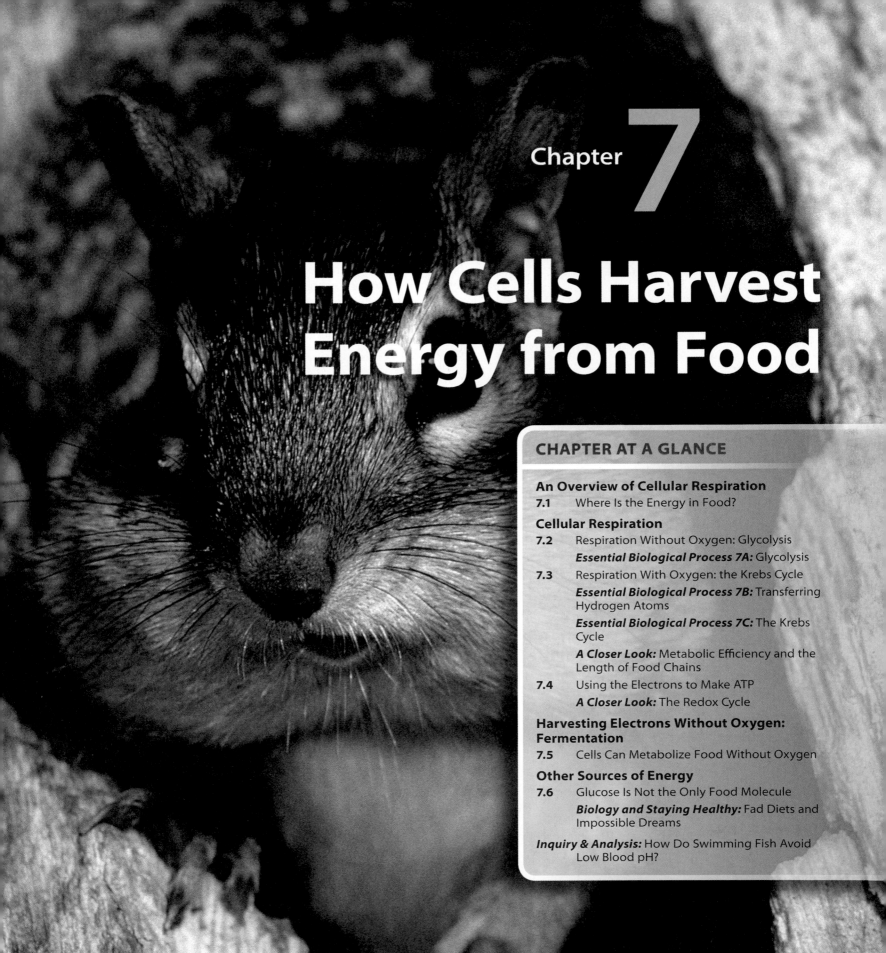

Chapter **7**

How Cells Harvest Energy from Food

An Overview of Cellular Respiration

Figure 7.1 **A human acquiring energy.**

Energy that this teen extracts from the hamburger he is eating will be used to power his singing, fuel his running, and build a bigger teenager.

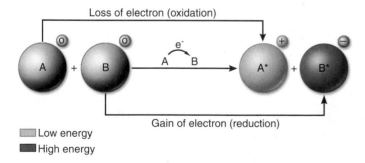

Loss of electron (oxidation)

Gain of electron (reduction)

Low energy
High energy

Figure 7.2 **Redox reactions.**

Oxidation is the loss of an electron; reduction is the gain of an electron. Here the charges of molecules A and B are shown in small circles to the upper right of each molecule. Molecule A loses energy as it loses an electron, while molecule B gains energy as it gains an electron.

7.1 Where Is the Energy in Food?

CONCEPT PREVIEW: Cellular respiration is the dismantling of food molecules to obtain energy. In aerobic respiration, the cell harvests energy from glucose molecules in two stages, glycolysis and oxidation.

In plants and animals, and in fact in almost all organisms, the energy for living is obtained by breaking down the organic molecules originally produced by photosynthetic organisms, such as plants, algae, and certain bacteria. The energy invested in building the organic molecules is retrieved by stripping away the energetic electrons and using them to make ATP, a process called **cellular respiration.** Do not confuse the term cellular respiration with the breathing of oxygen gas that your lungs carry out, which is called simply respiration.

> All cells use ATP to power biological functions. The depth and breadth of some of these functions were illustrated in table 5.1, page 95.

The cells of plants fuel their activities with sugars and other molecules that they produce through photosynthesis which they break down in cellular respiration. Nonphotosynthetic organisms eat plants, extracting energy from plant tissue in cellular respiration. Other animals, like the teenager gobbling up the hamburger in figure 7.1, eat these animals.

Eukaryotes produce the majority of their ATP by harvesting electrons from chemical bonds of the food molecule glucose. The electrons are transferred along an electron transport chain (similar to the electron transport system in photosynthesis), and eventually donated to oxygen gas. Chemically, there is little difference between this process in a cell and the burning of wood in a fireplace. In both instances, the reactants are carbohydrates and oxygen, and the products are carbon dioxide, water, and energy:

$$C_6H_{12}O_6 + 6\ O_2 \longrightarrow 6\ CO_2 + 6\ H_2O + energy\ (heat\ or\ ATP)$$

In a chemical reaction, when an atom or molecule loses an electron, it is said to be *oxidized,* and the process by which this occurs is called **oxidation.** The name reflects the fact that in biological systems, oxygen, which attracts electrons strongly, is the most common electron acceptor. This is certainly the case in cellular respiration, where oxygen is the final electron acceptor. Conversely, when an atom or molecule gains an electron, it is said to be *reduced,* and the process is called **reduction.** Oxidation and reduction always take place together, because every electron that is lost by an atom through oxidation is gained by some other atom through reduction. Therefore, chemical reactions of this sort are called **oxidation-reduction (redox) reactions.** In redox reactions, energy follows the electron, as shown in figure 7.2.

Cellular respiration is carried out in two stages, illustrated in figure 7.3. The first stage uses coupled reactions to make ATP. This stage, *glycolysis,* takes place in the cell's cytoplasm (the blue area in figure 7.3). Importantly, it is anaerobic (that is, it does not require oxygen). This ancient energy-extracting process is thought to have evolved over 2 billion years ago, when there was no oxygen in the earth's atmosphere.

The second stage is aerobic (requires oxygen) and takes place within the mitochondrion (the tan sausage-shaped structure in figure 7.3). The focal point of this stage is the *Krebs cycle,* a cycle of chemical reactions that harvests electrons from C—H chemical bonds and passes the energy-rich electrons to carrier molecules, NADH and FADH$_2$. These molecules deliver the electrons to the electron transport chain, which uses their energy to power the production of ATP. The harvesting of electrons, a form of *oxidation*, is far more powerful than glycolysis at recovering energy from food molecules, and is how the bulk of the energy used by eukaryotic cells is extracted from food molecules.

Concept Check

1. When an atom is oxidized, does it gain or lose an electron?
2. The Krebs cycle harvests electrons from what kind of chemical bond?
3. Which stage of cellular respiration occurs in the cytoplasm?

Figure 7.3   An overview of cellular respiration.

Electrons harvested from C—H chemical bonds are first transferred to NADH and FADH$_2$; these then carry the electrons to the electron transport chain, as indicated by the long red arrow on the left. The energy-depleted electron is finally donated with a proton to oxygen, forming a molecule of water.

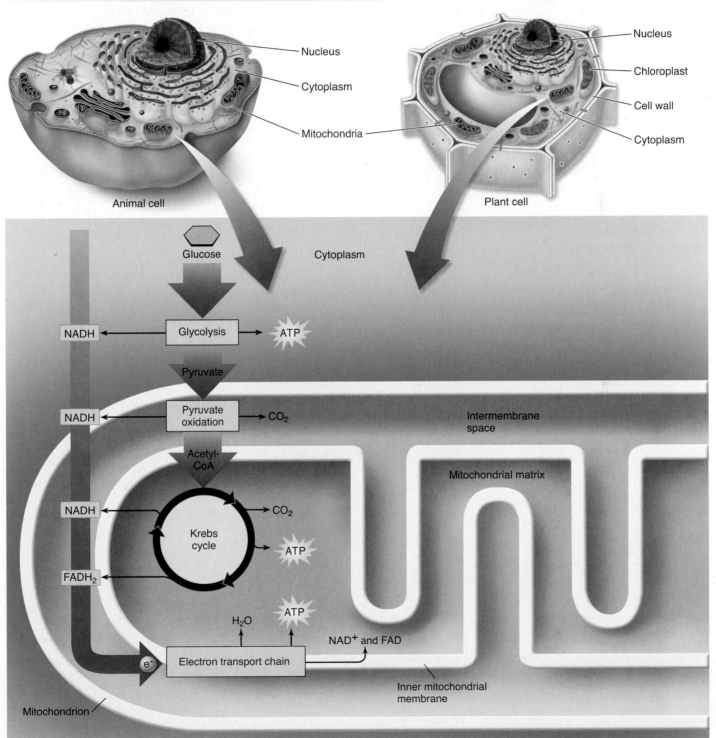

Cellular Respiration

7.2 Respiration Without Oxygen: Glycolysis

CONCEPT PREVIEW: In glycolysis, cells shuffle around chemical bonds in glucose to produce ATP by substrate-level phosphorylation.

The first stage in cellular respiration, called **glycolysis,** is a series of sequential biochemical reactions, a *biochemical pathway.* In 10 enzyme-catalyzed reactions, the six-carbon sugar glucose is cleaved into two three-carbon molecules called pyruvate. *Essential Biological Process 7A* presents a conceptual overview of the process. Where is the energy extracted? In each of two "coupled" reactions (panel 3), the breaking of a chemical bond in an exergonic reaction releases enough energy to drive the formation of an ATP molecule from ADP (an endergonic reaction). This transfer of a high-energy phosphate group from a substrate to ADP is called **substrate-level phosphorylation.** In the absence of oxygen, this is the only way organisms can get energy from food.

Essential Biological Process 7A

Glycolysis

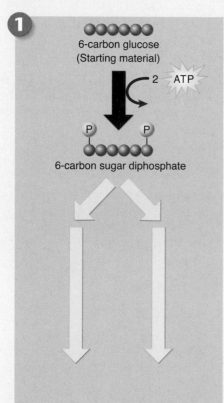

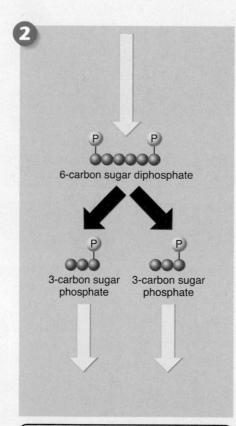

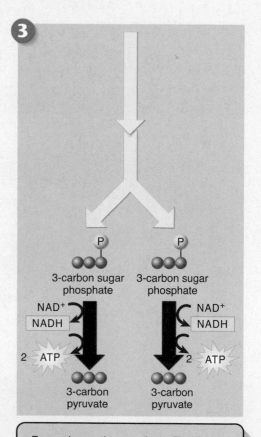

Priming reactions. Glycolysis begins with the addition of energy. Two high-energy phosphates from two molecules of ATP are added to the six-carbon molecule glucose, producing a six-carbon molecule with two phosphates.

Cleavage reactions. Then, the phosphorylated six-carbon molecule is split in two, forming two three-carbon sugar phosphates.

Energy-harvesting reactions. Finally, in a series of reactions, each of the two three-carbon sugar phosphates is converted to pyruvate. In the process, an energy-rich hydrogen is harvested as NADH, and two ATP molecules are formed for each pyruvate.

7.3 Respiration With Oxygen: the Krebs Cycle

CONCEPT PREVIEW: Pyruvate, the end product of glycolysis, is oxidized in the Krebs cycle yielding ATP and many energized electrons.

The first step of oxidative respiration in the mitochondrion is the oxidation of the three-carbon molecule called pyruvate, which is the end product of glycolysis. The cell harvests electrons from pyruvate in two steps: first, by oxidizing pyruvate to form acetyl-CoA, and then by oxidizing acetyl-CoA in the Krebs cycle.

Step One: Producing Acetyl-CoA

Pyruvate is oxidized in a single reaction that cleaves off one of pyruvate's three carbons. This carbon then departs as part of the CO_2 molecule shown coming off the pathway in figure 7.4. Pyruvate dehydrogenase, the complex of enzymes that removes CO_2 from pyruvate, is one of the largest enzymes known. It contains 60 subunits! In the course of the reaction, a hydrogen and electrons are removed from pyruvate and donated to NAD^+ to form NADH. *Essential Biological Process 7B* shows how an enzyme catalyzes this redox reaction, bringing the substrate (pyruvate) into proximity with NAD^+. As with all redox reactions, the oxidation and reduction reactions are coupled—pyruvate is oxidized when an electron and its energy, along with a hydrogen atom, is transferred to NAD^+, reducing it to NADH. Now focus again on figure 7.4. The two-carbon fragment (called an acetyl group) that remains after removing CO_2 from pyruvate is joined to a cofactor called coenzyme A (CoA) by pyruvate dehydrogenase, forming a compound known as **acetyl-CoA.** If the cell has a plentiful supply of ATP, acetyl-CoA is funneled into fat synthesis, with its energetic electrons preserved for later needs. If the cell needs ATP, the fragment is directed instead into ATP production through the Krebs cycle.

> Carbon dioxide, a by-product of cellular respiration, is released from organisms as waste. As you will see on page 496, vertebrates expel this waste as carbon dioxide gas from the lungs.

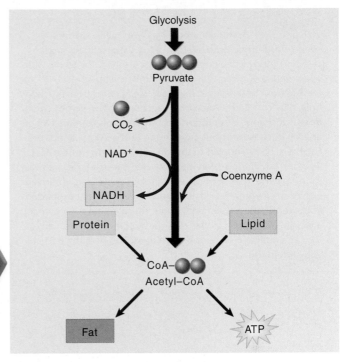

Figure 7.4   Producing acetyl-CoA.

Pyruvate, the three-carbon product of glycolysis, is oxidized to the two-carbon molecule acetyl-CoA, and in the process loses one carbon atom as CO_2 and an electron (donated to NAD^+ to form NADH). Almost all the molecules you use as foodstuffs are converted to acetyl-CoA; the acetyl-CoA is then channeled into fat synthesis or into ATP production, depending on your body's needs.

IMPLICATION   What do you think might determine how much of your acetyl-CoA is channeled into fat?

Essential Biological Process 7B

Transferring Hydrogen Atoms

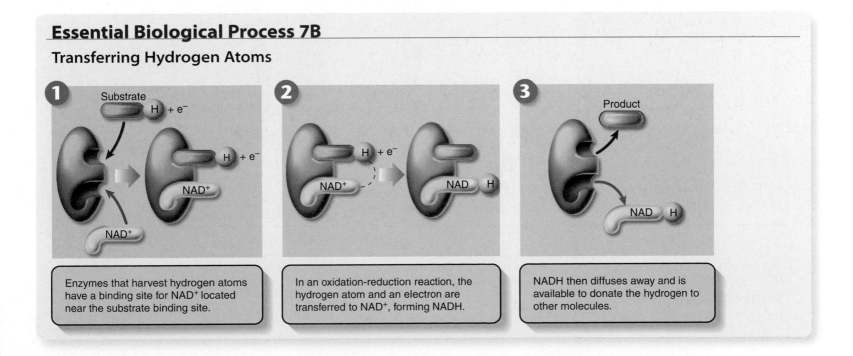

1 Enzymes that harvest hydrogen atoms have a binding site for NAD^+ located near the substrate binding site.

2 In an oxidation-reduction reaction, the hydrogen atom and an electron are transferred to NAD^+, forming NADH.

3 NADH then diffuses away and is available to donate the hydrogen to other molecules.

BIOLOGY & YOU

Burning Carbs. Did you ever wonder why, even though they both exercise a lot, weight lifters are massive and long distance runners thin? It has a lot to do with when and how often they use the Krebs cycle. At rest, 70% of the ATP your body uses to maintain itself comes from fats, and only 30% from carbs. Anaerobic exercise such as weight lifting does not change this proportion, but does miss out on a lot of energy because without oxygen the Krebs cycle cannot run. Fat energy is channeled to the job of increasing muscle mass. By contrast, during high-intensity aerobic exercise such as running almost 100% of the energy your body uses is derived from carbohydrates, and all of this energy is fully expended in powering the movement of your muscle fibers. In essence, a weight lifter puts calories to work building muscles, while a runner simply burns them. If you want to lose weight, it seems you should run.

Step Two: The Krebs Cycle

The next stage in oxidative respiration is called the **Krebs cycle,** named after the man who discovered it. The Krebs cycle (not to be confused with the Calvin cycle in photosynthesis) takes place within the mitochondrion. While a complex process, its nine reactions can be broken down into three stages, as indicated by the overview presented in *Essential Biological Process 7C*:

Stage 1. The cycle starts when the two-carbon acetyl-CoA fragment produced from pyruvate is stuck onto a four-carbon sugar, producing a six-carbon molecule.

Stage 2. Then, in rapid-fire order, two carbons are removed as CO_2, their electrons donated to NAD^+, and a four-carbon molecule is left. A molecule of ATP is also produced.

Stage 3. When it is all over, two carbon atoms have been expelled as CO_2, more energetic electrons are extracted and taken away as NADH or on other carriers such as $FADH_2$ (which serves the same function as NADH), and we are left with the same four-carbon sugar we started with.

The process is a cycle—that is, a circle of reactions. In each turn of the cycle, a new acetyl group replaces the two CO_2 molecules lost, and more electrons are extracted. The Krebs cycle makes two turns for every molecule of glucose broken down during respiration.

In the process of cellular respiration, glucose is entirely consumed. All that is left to mark the passing of the glucose molecule into six CO_2 molecules is its energy, preserved in four ATP molecules (two from glycolysis and two from the Krebs cycle) and electrons carried by 10 NADH and two $FADH_2$ carriers.

Essential Biological Process 7C

The Krebs Cycle

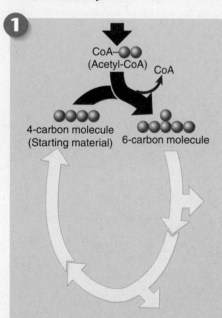

1

The Krebs cycle begins when a two-carbon fragment is transferred from acetyl-CoA to a four-carbon molecule (the starting material).

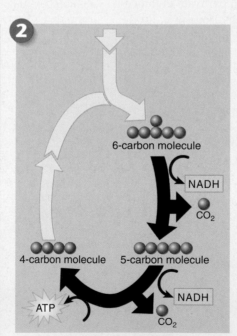

2

Then, the resulting six-carbon molecule is oxidized (a hydrogen removed to form NADH) and decarboxylated (a carbon removed to form CO_2). Next, the five-carbon molecule is oxidized and decarboxylated again, and a coupled reaction generates ATP.

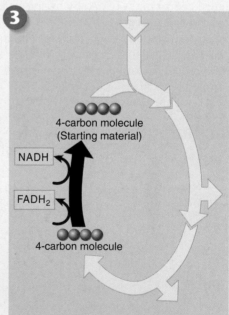

3

Finally, the resulting four-carbon molecule is further oxidized (hydrogens removed to form $FADH_2$ and NADH). This regenerates the four-carbon starting material, completing the cycle.

Metabolic Efficiency and the Length of Food Chains

In the earth's ecosystems, the organisms that carry out photosynthesis are often consumed as food by other organisms. We call these "organism-eaters" *heterotrophs*. Humans are heterotrophs, as no human photosynthesizes.

It is thought that the first heterotrophs were ancient bacteria living in a world where photosynthesis had not yet introduced much oxygen into the oceans or atmosphere. The only mechanism they possessed to harvest chemical energy from their food was glycolysis. Neither oxygen-generating photosynthesis nor the oxidative stage of cellular respiration had evolved yet. It has been estimated that a heterotroph limited to glycolysis, as these ancient bacteria were, captures only 3.5% of the energy in the food it consumes. Hence, if such a heterotroph preserves 3.5% of the energy in the photosynthesizers it consumes, then any other heterotrophs that consume the first heterotroph will capture through glycolysis 3.5% of the energy in it, or 0.12% of the energy available in the original photosynthetic organisms. A very large base of photosynthesizers would thus be needed to support a small number of heterotrophs.

When organisms became able to extract energy from organic molecules by oxidative cellular respiration, which we discuss on the next page, this constraint became far less severe, because the efficiency of oxidative respiration is estimated to be about 32%. This increased efficiency results in the transmission of much more energy from one trophic level to another than does glycolysis. (A *trophic level* is a step in the movement of energy through an ecosystem.) The efficiency of oxidative cellular respiration has made possible the evolution of food chains, in which photosynthesizers are consumed by heterotrophs, which are consumed by other heterotrophs, and so on. You will read more about food chains in chapter 20.

Even with this very efficient oxidative metabolism, approximately two-thirds of the available energy is lost at each trophic level, and that puts a limit on how long a food chain can be. Most food chains, like the East African grassland ecosystem illustrated here, involve only three or rarely four trophic levels. Too much energy is lost at each transfer to allow chains to be much longer than that. For example, it would be impossible for a large human population to subsist by eating lions captured from the grasslands of East Africa; the amount of grass available there would not support enough zebras and other herbivores to maintain the number of lions needed to feed the human population. Thus, the ecological complexity of our world is fixed in a fundamental way by the chemistry of oxidative cellular respiration.

Photosynthesizers. The grass under this yellow fever tree grows actively during the hot, rainy season, capturing the energy of the sun and storing it in molecules of glucose, which are then converted into starch and stored in the grass.

Herbivores. These zebras consume the grass and transfer some of its stored energy into their own bodies.

Carnivores. The lion feeds on zebras and other animals, capturing part of their stored energy and storing it in its own body.

Scavengers. This hyena and the vultures occupy the same stage in the food chain as the lion. They also consume the body of the dead zebra, after it has been abandoned by the lion.

A food chain in the savannas, or open grasslands, of East Africa.

At each of these levels in the food chain, only about a third or less of the energy present is used by the recipient.

Refuse utilizers. These butterflies, mostly *Precis octavia*, are feeding on the material left in the hyena's dung after the food the hyena consumed had passed through its digestive tract.

7.4 Using the Electrons to Make ATP

CONCEPT PREVIEW: The electrons harvested by oxidizing food molecules are used to power proton pumps that chemiosmotically drive the production of ATP.

Mitochondria use chemiosmosis to make ATP in much the same way that chloroplasts do, although a mitochondrion's proton pumps transport protons *out of* an enclosed space (the matrix) while the electron transport system in chloroplasts transport protons *into* an enclosed space (the thylakoid). Mitochondria use energetic electrons extracted from food molecules to power proton pumps that drive protons across the inner mitochondrial membrane. As protons become far more scarce inside than outside, the concentration gradient drives protons back in through special ATP synthase channels. Their passage powers the production of ATP from ADP. The ATP then passes out of the mitochondrion through ATP-transport channels.

Figure 7.5   **The electron transport chain.**

High-energy electrons are transported (*red arrows*) along a chain of electron-carrier molecules. Three of these molecules are protein complexes that use portions of the electrons' energy to pump protons (*blue arrows*) out of the matrix and into the intermembrane space. The electrons are finally donated to oxygen to form water.

IMPLICATION   In harvesting the energy contained within the chemical bonds of a molecule of glucose, a cell gains a net of four ATP molecules from glycolysis and the Krebs cycle. Assuming that each proton pumped out of the mitochondrial matrix reenters via ATP synthase to produce an ATP molecule, how many ATP molecules in total are obtained by metabolizing the glucose molecule?

> A similar electron carrier is used in photosynthesis as described on page 107. In the light-dependent reactions, two electrons and a hydrogen ion are transferred to NADP⁺ producing NADPH, which carries these to the Calvin cycle.

Moving Electrons Through the Electron Transport Chain

The NADH and $FADH_2$ molecules formed during the first stages of aerobic respiration each contain electrons and hydrogens

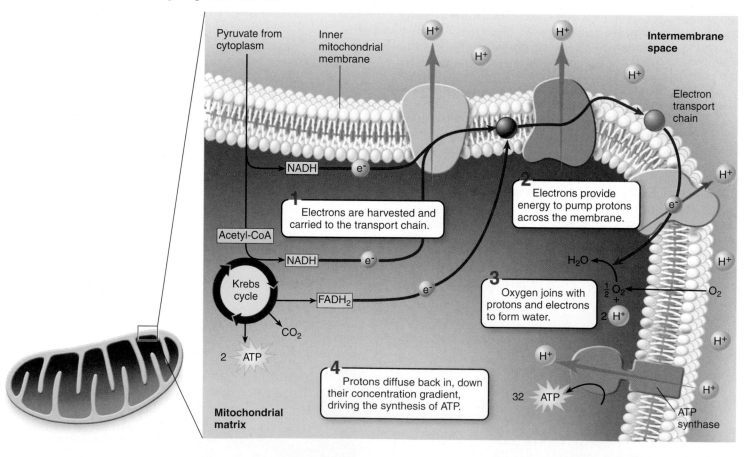

that were gained when NAD$^+$ and FAD were reduced (refer back to figure 7.2). The NADH and FADH$_2$ molecules carry their electrons to the inner mitochondrial membrane (an enlarged area of the membrane is shown in figure 7.5), where they transfer the electrons to a series of membrane-associated molecules collectively called the **electron transport chain.** The electron transport chain works much as does the electron transport system you encountered in studying photosynthesis.

> There are two electron transport systems in photosynthesis, one produces ATP and the other produces NADPH, as described on page 106. In both cases, an energized electron is passed along a series of membrane proteins, and the energy drives proton gradient formation.

A protein complex (the pink structure in figure 7.5) receives the electrons and, using a mobile carrier, passes these electrons to a second protein complex (the purple structure). This protein complex, along with others in the chain, operates as a proton pump, using the energy of the electron to drive a proton out across the membrane into the **intermembrane space.** The arrows indicate the transport of the protons out of the matrix of the mitochondrion into the intermembrane space.

The electron is then carried by another carrier to a third protein complex (the light blue structure). This complex uses electrons such as this one to link oxygen atoms with hydrogen ions to form molecules of water.

Producing ATP: Chemiosmosis

In eukaryotes, aerobic respiration takes place within the mitochondria present in virtually all cells. The internal compartment, or **matrix,** of a mitochondrion contains the enzymes that carry out the reactions of the Krebs cycle. As described earlier, the electrons harvested by oxidative respiration are passed along the electron transport chain, and the energy they release transports protons out of the matrix and into the intermembrane space. Proton pumps in the inner mitochondrial membrane accomplish the transport. The electrons contributed by NADH activate three of these proton pumps, and those contributed by FADH$_2$ activate two, as indicated in figure 7.5. As the proton concentration in the intermembrane space rises above that in the matrix, the concentration gradient induces the protons to reenter the matrix by diffusion through a special proton channel called **ATP synthase.** This channel is embedded in the inner mitochondrial membrane, as shown in figures 7.5 and 7.6. As the protons pass through, the channel synthesizes ATP from ADP and P$_i$ within the matrix. The ATP is then transported by facilitated diffusion out of the mitochondrion into the cell's cytoplasm. This ATP synthesizing process is the same **chemiosmosis** process that you encountered in studying photosynthesis in chapter 6.

> This pumping of protons is an example of active transport, discussed on page 83, where energy drives the transport of a substance against a concentration gradient. In active transport, energy can come from ATP, or as in this case, from energized electrons.

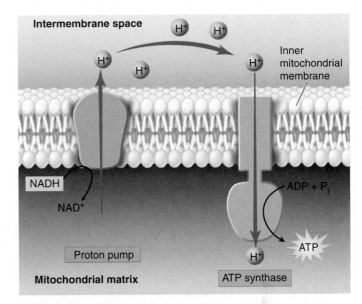

Figure 7.6 Chemiosmosis.

NADH transports high-energy electrons harvested from macromolecules to "proton pumps" that use the energy to pump protons out of the mitochondrial matrix. As a result, the concentration of protons outside the inner mitochondrial membrane rises, inducing protons to diffuse back into the matrix. Many of the protons pass through ATP synthase channels that couple the reentry of protons to the production of ATP.

EVOLUTION

Evolution of ATP Synthase. ATP synthase is an example of "molecular evolution," where two proteins with their own functions have become associated to form a new two-subunit protein with an entirely new function. In this instance, the gene for one subunit of ATP synthase has a very similar DNA sequence to a DNA-binding enzyme called helicase, allowing ATP synthase to bind the adenine of ADP. The gene for the other subunit of ATP synthase is very similar to that of the proton-driven motor that bacteria use to power the rotation of their flagella, allowing ATP synthase to join P$_i$ to ADP by running the motor in reverse.

Concept Check

1. In glycolysis, by what process is ATP produced?
2. What molecule is the starting material for the Krebs cycle?
3. If the electron transport chain uses the energy harvested from C—H bonds to drive protons out of the matrix, how is it that the ATP molecules formed as a consequence are *within* the matrix?

The Redox Cycle

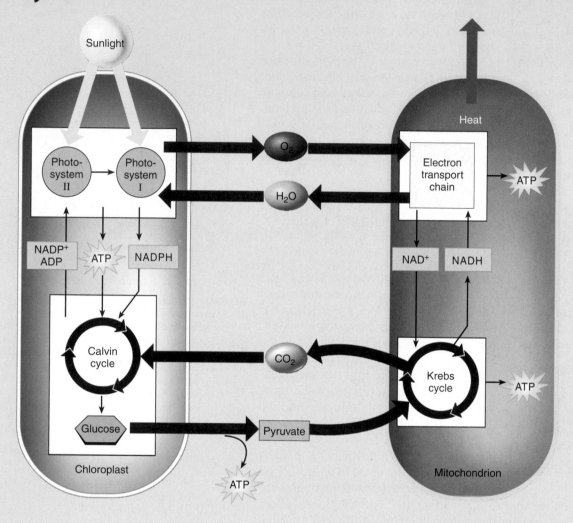

The energy-capturing metabolism of the chloroplasts studied in the previous chapter and the energy-utilizing metabolism of the mitochondria studied in this chapter are intimately related. Photosynthesis carried out by chloroplasts uses the products of cellular respiration as starting substrates and reduces carbon by adding hydrogen atoms. Cellular respiration carried out by mitochondria uses the products of photosynthesis as its starting substrates and oxidizes carbon by removing hydrogen atoms. Together, photosynthesis and cellular respiration form an oxidation-reduction cycle, as diagrammed here. Note that it is electrons that cycle between chloroplasts and mitochondria. Energy passes through the redox cycle, flowing into the cycle from the sun, passing through newly assembled molecules, and eventually flowing out of the cycle as heat.

The evolutionary history of the redox cycle can be seen in its elements. The Calvin cycle of photosynthesis uses part of the glycolytic pathway of cellular respiration, run in reverse, to produce glucose. The principal proteins involved in electron transport in chloroplasts are related to those in mitochondria, and in many cases are actually the same. Biologists believe the glycolysis stage of cellular respiration, which takes place in the cytoplasm, converting glucose to pyruvate and requiring no oxygen, to be the most ancient process in the redox cycle. The chloroplast's oxygen-generating photosynthesis is thought to have evolved next, followed later by the oxidative reactions of the mitochondria's cellular respiration.

Harvesting Electrons Without Oxygen: Fermentation

7.5 Cells Can Metabolize Food Without Oxygen

CONCEPT PREVIEW: In fermentation, which occurs in the absence of oxygen, the electrons that result from the glycolytic breakdown of glucose are donated as hydrogen atoms to an organic molecule, regenerating NAD$^+$ from NADH.

Fermentation

In the absence of oxygen, aerobic metabolism (the Krebs cycle and electron transport chain) cannot occur, and cells must rely exclusively on glycolysis to produce ATP. Under these conditions, the hydrogen atoms and electrons that were involved in the oxidation of NAD$^+$ to NADH in glycolysis are donated to organic molecules instead of the electron transport chain, in a process called **fermentation.** Fermentation recycles NAD$^+$ so that glycolysis can continue.

Ethanol Fermentation.   Bacteria carry out more than a dozen kinds of fermentations, all using some form of organic molecule to accept the hydrogen atom from NADH. By contrast, eukaryotic cells are capable of only a few types of fermentation. In one type, which occurs in single-celled fungi called yeast, the molecule that accepts hydrogen from NADH is pyruvate, the end product of glycolysis itself. Yeast enzymes remove a CO_2 group from pyruvate through decarboxylation, producing a two-carbon molecule called acetaldehyde. The CO_2 released causes bread made with yeast to rise, while bread made without yeast (unleavened bread) does not. The acetaldehyde accepts a hydrogen atom from NADH, producing NAD$^+$ and ethanol (figure 7.7, *upper panel*). This particular type of fermentation is of great interest to humans, because it is the source of the ethanol in wine and beer. Ethanol is a by-product of fermentation that is actually toxic to yeast; as it approaches a concentration of about 12%, it begins to kill the yeast. That is why naturally fermented wine contains only about 12% ethanol.

> Humans have been using yeast for thousands of years to produce bread and alcohol, but yeast has many other commercial uses. The manufacturing potential of yeast will be discussed in more detail on page 309.

Lactic Acid Fermentation.   Most animal cells regenerate NAD$^+$ by a second type of fermentation, using an enzyme called lactate dehydrogenase to transfer a hydrogen atom from NADH back to the pyruvate that is produced by glycolysis. This reaction converts pyruvate into lactic acid and regenerates NAD$^+$ from NADH (figure 7.7, *lower panel*). It therefore closes the metabolic circle, allowing glycolysis to continue as long as glucose is available. Circulating blood removes excess lactate (the ionized form of lactic acid) from muscles. It was once thought that during strenuous exercise, when the removal of lactic acid cannot keep pace with its production, the accumulation induces muscle fatigue. However, scientists now believe that lactic acid is actually used by muscles as another source of fuel.

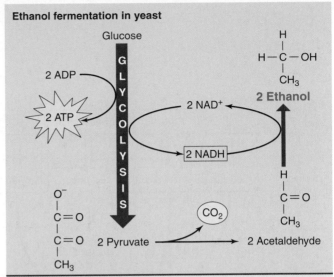

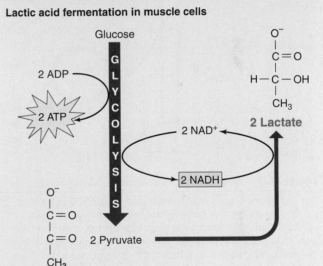

Figure 7.7   **Fermentation.**

Yeasts carry out the conversion of pyruvate to ethanol. Muscle cells convert pyruvate into lactate, which is less toxic than ethanol. In both cases, NAD$^+$ is regenerated to allow glycolysis to continue.

IMPLICATION   Despite our best intentions, once in a while most of us consume a little more alcohol than we should and wake up the next morning with a "hangover"—a pounding headache, nausea, shakiness, and often a very dry mouth. Many of these symptoms are those of dehydration. What sort of hangover prevention does this hypothesis suggest?

Concept Check

1. Why can't the Krebs cycle function in the absence of oxygen?
2. In eukaryotic fermentations, the hydrogen and electron from NADH are donated to what molecules?
3. What would happen to glycolysis if NAD$^+$ wasn't recycled?

Other Sources of Energy

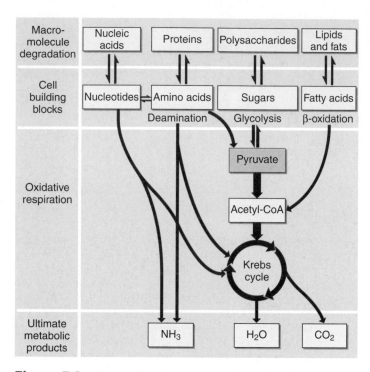

Figure 7.8   How cells obtain energy from foods.

Most organisms extract energy from organic molecules by oxidizing them. The first stage of this process, breaking down macromolecules into their subunits, yields little energy. The second stage, cellular respiration, extracts energy, primarily in the form of high-energy electrons. The subunit of many carbohydrates, glucose, readily enters glycolysis and passes through the biochemical pathways of oxidative respiration. However, the subunits of other macromolecules must be converted into products that can enter the biochemical pathways found in oxidative respiration.

IMPLICATION   Many fad diets assert that weight loss will result from a high protein–low carb diet such as the ones discussed on the facing page. Have you ever tried this sort of diet? Did you lose weight? For how long?

7.6 Glucose Is Not the Only Food Molecule

CONCEPT PREVIEW: Cells also garner energy from proteins and fats, which are broken down into products that feed into cellular respiration.

We have considered in detail the fate of a molecule of glucose, a simple sugar, in cellular respiration. But how much of what you eat is sugar? As a more realistic example of the food you eat, consider the fate of a fast-food hamburger. The hamburger is composed primarily of carbohydrates, fats, and proteins. This diverse collection of complex molecules is broken down by the process of digestion in your stomach and intestines into simpler molecules. Carbohydrates are broken down into simple sugars, fats into fatty acids, and proteins into amino acids. Nucleic acids are also present in the food you eat, but these macromolecules store little energy that the body actually uses.

Cellular Respiration of Protein

Proteins (the second category in figure 7.8) are first broken down into their individual amino acids. A series of *deamination* reactions removes the nitrogen side groups (called amino groups) and converts the rest of the amino acid into a molecule that takes part in the Krebs cycle. For example, alanine is converted into pyruvate, glutamate into α-ketoglutarate, and aspartate into oxaloacetate. The reactions of the Krebs cycle then extract the high-energy electrons from these molecules and put them to work making ATP.

Cellular Respiration of Fat

Lipids and fats (the fourth category in figure 7.8) are first broken down into fatty acids. A fatty acid typically has a long tail of sixteen or more —CH_2 links, and the many C—H bonds in these long tails provide a rich harvest of energy. Enzymes in the matrix of the mitochondrion first remove one two-carbon acetyl group from the end of a fatty acid tail, and then another, and then another, in effect chewing down the length of the tail in two-carbon bites. Eventually the entire fatty acid tail is converted into acetyl groups. Each acetyl group then combines with coenzyme A to form acetyl-CoA, which feeds into the Krebs cycle. This process is known as *β-oxidation*.

> Recall from the discussion of lipids on page 56 that fats are composed of a three-carbon glycerol backbone attached to three fatty acid tails. The breakdown of fats, as described here, occurs with the removal of two carbons at a time from the fatty acid tails.

Concept Check

1. How can your body obtain energy from proteins or fats when they don't go through glycolysis?
2. Proteins and fatty acids have to undergo chemical reactions before they can enter into oxidative respiration. What are the names of these two types of reactions?
3. What part of a hamburger would yield energy in the absence of oxygen?

Biology and Staying Healthy

Fad Diets and Impossible Dreams

Most Americans put on weight in middle age, slowly adding 30 or more pounds. They did not ask for that weight, do not want it, and are constantly looking for a way to get rid of it. It is not a lonely search—it seems like everyone past the flush of youth is trying to lose weight. Many have been seduced by fad diets, investing hope only to harvest frustration. The much discussed Atkins' diet is the fad diet most have tried—*Dr. Atkins' Diet Revolution* is one of the 10 best-selling books in history, prominently displayed in bookstores. The reason this diet doesn't deliver on its promise of pain-free weight loss is well understood by science, but not by the general public. Only hope and hype make it a perpetual best seller.

The secret of the Atkins' diet, stated simply, is to avoid carbohydrates. Atkins' basic proposition is that your body, if it does not detect blood glucose (from metabolizing carbohydrates), will think it is starving and start to burn body fat, even if there is lots of fat already circulating in your bloodstream. You may eat all the fat and protein you want, all the steak and eggs and butter and cheese, and you will still burn fat and lose weight—just don't eat any carbohydrates, any bread or pasta or potatoes or fruit or candy. Despite the title of Atkins' book, this diet is hardly revolutionary. A basic low-carbohydrate diet was first promoted over a century ago in the 1860s by William Banting, an English casket maker, in his best-selling book *Letter on Corpulence*. Books promoting low-carbohydrate diets have continued to be best sellers ever since.

Those who try the Atkins' diet often lose 10 pounds in two to three weeks. In three months it is all back, and then some. So what happened? Where did the pounds go, and why did they come back? The temporary weight loss turns out to have a simple explanation. Carbohydrates act as water sponges in your body, and so forcing your body to become depleted of carbohydrates causes your body to lose water. The 10 pounds lost on this diet was not fat weight but water weight, quickly regained with the first starchy foods eaten.

The Atkins' diet is the sort of diet the American Heart Association tells us to avoid (all those saturated fats and cholesterol), and it is difficult to stay on. If you do hang in there, you will lose weight, simply because you eat less. Other popular diets these days, *The Zone* diet of Dr. Barry Sears and *The South Beach Diet* of Dr. Arthur Agatston, are also low-carbohydrate diets, although not as extreme as the Atkins' diet. Like the Atkins' diet, they work not for the bizarre reasons claimed by their promoters, but simply because they are low-calorie diets.

There are two basic laws that no diet can successfully violate:

1. All calories are equal.
2. (calories in) – (calories out) = fat.

The fundamental fallacy of the Atkins' diet, the Zone diet, the South Beach diet, and indeed of all fad diets, is the idea that somehow carbohydrate calories are different from fat and protein calories. This is scientific foolishness. Every calorie you eat contributes equally to your eventual weight, whether it comes from carbohydrate, fat, or protein.

To the extent these diets work at all, they do so because they obey the second law. By reducing calories in, they reduce fat. If that were all there was to it, we should all go out and buy a diet book. Unfortunately, losing weight isn't that simple, as anyone who has seriously tried already knows. The problem is that your body will not cooperate.

If you try to lose weight by exercising and eating less, your body will attempt to compensate by metabolizing more efficiently. It has a fixed weight, what obesity researchers call a "set point," a weight to which it will keep trying to return. A few years ago, a group of researchers at Rockefeller University in New York, in a landmark study, found that if you lose weight, your metabolism slows down and becomes more efficient, burning fewer calories to do the same work—your body will do everything it can to gain the weight back! Similarly, if you gain weight, your metabolism speeds up. In this way your body uses its own natural weight control system to keep your weight at its set point. No wonder it's so hard to lose weight!

Clearly our bodies don't keep us at one weight all our adult lives. It turns out your body adjusts its fat thermostat—its set point—depending on your age, food intake and amount of physical activity. Adjustments are slow, however, and it seems to be a great deal easier to move the body's set point up than to move it down. Apparently higher levels of fat reduce the body's sensitivity to the leptin hormone that governs how efficiently we burn fat. That is why you can gain weight, despite your set point resisting the gain—your body still issues leptin alarm calls to speed metabolism, but your brain doesn't respond with as much sensitivity as it used to. Thus the fatter you get, the less effective your weight control system becomes.

This doesn't mean that we should give up and learn to love our fat. Rather, now that we are beginning to understand the biology of weight gain, we must accept the hard fact that we cannot beat the requirements of the two diet laws. The real trick is not to give up. Eat less and exercise more, and keep at it. In one year, or two, or three, your body will readjust its set point to reflect the new reality you have imposed by constant struggle. There simply isn't any easy way to lose weight.

How Do Swimming Fish Avoid Low Blood pH?

Animals that live in oxygen-poor environments, like worms living in the oxygen-free mud at the bottom of lakes, are not able to obtain the energy required for muscle movement from the Krebs cycle. Their cells lack the oxygen needed to accept the electrons stripped from food molecules. Instead, these animals rely on glycolysis to obtain ATP, donating the electron to pyruvate, forming lactic acid. While much less efficient than the Krebs cycle, glycolysis does not require oxygen. Even when oxygen is plentiful, the muscles of an active animal may use up oxygen more quickly than it can be supplied by the bloodstream and so be forced to temporarily rely on glycolysis to generate the ATP for continued contraction.

This presents a particular problem for fish. Fish blood is much lower in carbon dioxide than yours is, and as a consequence, the amount of sodium bicarbonate acting as a buffer in fish blood is also quite low. Now imagine you are a trout, and need to suddenly swim very fast to catch a mayfly for dinner. The vigorous swimming will cause your muscles to release large amounts of lactic acid into your poorly buffered blood; this could severely disturb the blood's acid-base balance and so impede contraction of your swimming muscles before the prey is captured.

The graph to the right presents the results of an experiment designed to explore how a trout solves this dilemma. In the experiment, the trout was made to swim vigorously for 15 minutes in a laboratory tank, and then allowed a day's recovery. The lactic acid concentration in its blood was monitored periodically during swimming and recovery phases.

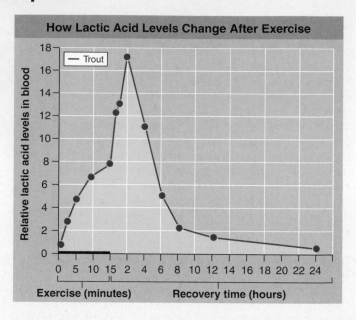

Analysis

1. **Applying Concepts** Lactic acid levels are presented for both swimming and recovery periods. In what time units are the swimming data presented? The recovery data?

2. **Interpreting Data** What is the effect of exercise on the level of lactic acid in the trout's blood? How does the level of lactic acid change after exercise stops?

3. **Making Inferences** About how much of the total lactic acid created by vigorous swimming is released after this exercise stops? [Hint: Notice the x axis scale changes from minutes to hours.]

4. **Drawing Conclusions** Is this result consistent with the hypothesis that fish maintain blood pH levels by delaying the release of lactic acid from muscles? Why might this be beneficial to the fish?

Concept Summary

An Overview of Cellular Respiration

7.1 Where Is the Energy in Food?

- Nonphotosynthetic organisms acquire energy from the breakdown of food, either by eating plants that store the food or by eating animals that have eaten plants. Energy stored in carbohydrate molecules is extracted through the process of cellular respiration and is stored in the cell as ATP.

- Coupled reactions, called oxidation-reduction or redox reactions, involve the transfer of electrons from one atom or molecule to another. The atom or molecule that loses an electron is said to be oxidized and loses energy. The atom or molecule that gains the electron is said to be reduced and gains energy (**figure 7.2**).

- Cellular respiration is carried out in two stages: glycolysis occurring in the cytoplasm and oxidation occurring in the mitochondria (**figure 7.3**).

Cellular Respiration

7.2 Respiration Without Oxygen: Glycolysis

- Glycolysis is an energy-extracting process. It is a series of 10 chemical reactions in which glucose is broken down into two three-carbon pyruvate molecules. The energy is extracted from glucose by two exergonic reactions that are coupled with an endergonic reaction that leads to the formation of ATP. This is called substrate-level phosphorylation and is shown in *Essential Biological Process 7A.*

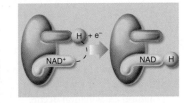

- Electrons extracted from glucose are donated to a carrier molecule, NAD⁺, which becomes NADH. NADH carries electrons and hydrogen atoms to be used in a later stage of oxidative respiration.

7.3 Respiration With Oxygen: The Krebs Cycle

- The two molecules of pyruvate formed in glycolysis are passed into the mitochondrion, where they are converted into two molecules of acetyl-coenzyme A (**figure 7.4**). What the cell does with acetyl-CoA depends on the needs of the cell. If the cell has enough ATP, acetyl-CoA is used in synthesizing fat molecules. If the cell needs energy, acetyl-CoA is directed to the Krebs cycle.

- The formation of NADH is an enzyme catalyzed reaction, as shown here from *Essential Biological Process 7B.* The enzyme brings the substrate and NAD⁺ into close proximity. Through a redox reaction, a hydrogen atom and an electron are transferred to NAD⁺, reducing it to NADH. NADH then carries the electrons and hydrogen to a later step in oxidative respiration.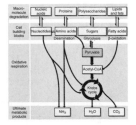

- Acetyl-CoA enters a series of chemical reactions called the Krebs cycle, where one molecule of ATP is produced in a coupled reaction.

- Energy is also harvested in the form of electrons that are transferred to molecules of NAD⁺ and FAD to produce NADH and FADH₂, respectively (*Essential Biological Process 7C*).

- The Krebs cycle makes two turns for every molecule of glucose that is oxidized.

7.4 Using the Electrons to Make ATP

- The molecules of NADH and FADH₂ that were produced during glycolysis and the Krebs cycle carry electrons to the inner mitochondrial membrane. Here they give up electrons to the electron transport chain. The electrons, along with their energy, are passed along the electron transport chain. The energy from the electrons drives proton pumps that pump H⁺ across the inner membrane from the matrix to the intermembrane space, creating an H⁺ concentration gradient (**figure 7.5**).

- When the electrons reach the end of the electron transport chain, they bind with oxygen and hydrogen to form water molecules.

- ATP is produced in the mitochondrion through chemiosmosis. The H⁺ concentration gradient in the intermembrane space drives H⁺ back across the membrane through ATP synthase channels, as shown here from **figure 7.6**. The energy from the movement of H⁺ through the channel is transferred to the chemical bonds in ATP. Thus, the energy stored in the glucose molecule is harvested through glycolysis and the Krebs cycle with the formation of NADH and FADH₂ and ultimately the formation of ATP. The energy carried by NADH and FADH₂ is transferred to the electron transport chain and is stored in ATP.

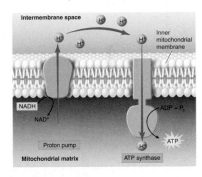

Harvesting Electrons Without Oxygen: Fermentation

7.5 Cells Can Metabolize Food Without Oxygen

- In the absence of oxygen, other molecules can be used as electron acceptors. When the electron acceptor is an organic molecule, the process is called fermentation. Depending on what type of organic molecule accepts the electrons, either ethanol or lactic acid, in the form of lactate is formed (**figure 7.7**).

Other Sources of Energy

7.6 Glucose Is Not the Only Food Molecule

- Food sources other than glucose are also used in oxidative respiration. Macromolecules, such as proteins, lipids, and nucleic acids, are broken down into intermediate products that enter cellular respiration in different reaction steps, as shown here from **figure 7.8**.

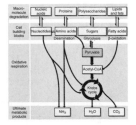

Self-Test

1. In animals, the energy for life is obtained by cellular respiration. This involves
   a. breaking down the organic molecules that were consumed.
   b. capturing photons from plants.
   c. obtaining ATP from plants.
   d. breaking down CO_2 that was produced by plants.

2. During glycolysis, ATP forms by
   a. the breakdown of pyruvate.
   b. chemiosmosis.
   c. substrate-level phosphorylation.
   d. NAD^+.

3. Which of the following processes can occur in the absence of oxygen?
   a. the Krebs cycle   c. chemiosmosis
   b. glycolysis   d. All of the above

4. Every living creature on this planet is capable of carrying out the rather inefficient biochemical process of glycolysis, which
   a. makes glucose, using the energy from ATP.
   b. makes ATP by splitting a molecule of glucose in half and capturing the energy.
   c. phosphorylates ATP to make ADP.
   d. makes glucose, using oxygen and carbon dioxide and water.

5. The electrons generated from the Krebs cycle are transferred to _____ which then carries them to _____.
   a. NAD^+, oxygen
   b. NAD^+, the electron transport chain
   c. NADH, oxygen
   d. NADH, the electron transport chain

6. After glycolysis, the pyruvate molecules go to the
   a. nucleus of the cell and provide energy.
   b. membranes of the cell and are broken down in the presence of CO_2 to make more ATP.
   c. mitochondria of the cell and are broken down in the presence of O_2 to make more ATP.
   d. Golgi bodies and are packaged and stored until needed.

7. The vast majority of the ATP molecules produced within a cell are produced
   a. during pyruvate oxidation.
   b. during glycolysis.
   c. during the Krebs cycle.
   d. during the electron transport chain.

8. NAD^+ is recycled during
   a. glycolysis.   c. the Krebs cycle.
   b. fermentation.   d. the formation of acetyl-CoA.

9. The final electron acceptor in lactic acid fermentation is
   a. pyruvate.   c. lactic acid.
   b. NAD^+.   d. O_2.

10. Cells can extract energy from foodstuffs other than glucose because
   a. proteins, fatty acids, and nucleic acids get converted to glucose and then enter oxidative respiration.
   b. each type of macromolecule has its own oxidative respiration pathway.
   c. each type of macromolecule is broken down into its subunits, which enter the oxidative respiration pathway.
   d. they can all enter the glycolytic pathway.

Visual Understanding

1. Consider the structure of a mitochondrion, as shown here in a cutaway view. If you poke a hole in a mitochondrion, can it still perform oxidative respiration? Explain. Can fragments of a mitochondrion perform oxidative respiration? Explain.

2. **Figure 7.8** Your friend wants to go on a low-carbohydrate diet so that he can lose some of the "baby fat" he's still carrying. He asks your advice; what do you tell him?

Challenge Questions

1. If cellular respiration were the stock market (you're investing ATPs and getting ATP dividends), where would you get the most return on your investment: glycolysis, the Krebs cycle, or the electron transport chain? Explain your answer.

2. How much less ATP would be generated in the cells of a person who consumed a diet of pyruvate instead of glucose (use one molecule of each for your calculation)?

3. Soft drinks are artificially carbonated, which is what causes them to fizz. Beer and sparkling wines are naturally carbonated. How does this natural carbonation occur?

4. Which of the following food molecules would generate the most ATP molecules, assuming that glycolysis, the Krebs cycle, and the electron transport chain were all functioning and that the foods were consumed in equal amounts; carbohydrates, proteins, or fats? Explain your answer.

Chapter **8**

Mitosis

CHAPTER AT A GLANCE

Cell Division

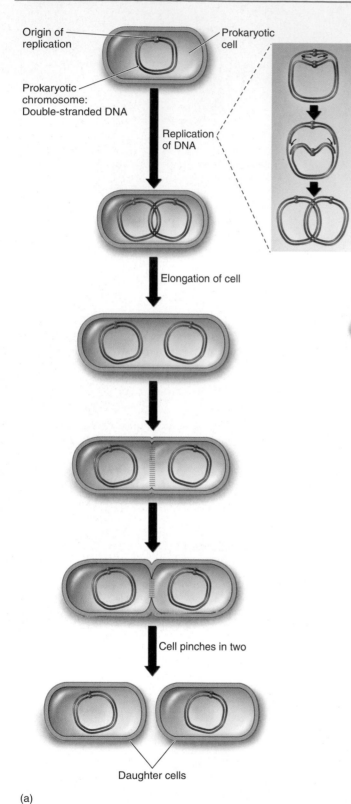

Origin of replication

Prokaryotic cell

Prokaryotic chromosome: Double-stranded DNA

Replication of DNA

Elongation of cell

Cell pinches in two

Daughter cells

(a)

8.1 Prokaryotes Have a Simple Cell Cycle

CONCEPT PREVIEW: Prokaryotes divide by binary fission after the DNA has replicated.

All species reproduce, passing their hereditary information on to their offspring. In this chapter, we begin our consideration of heredity with a look at how cells reproduce. Cell division in prokaryotes takes place in two stages, which together make up a simple cell cycle. First the DNA is copied, and then the cell splits in half by a process called **binary fission.** The cell in figure 8.1 is undergoing binary fission.

In prokaryotes, the hereditary information—that is, the genes that specify the prokaryote—is encoded in a single circle of DNA, called a prokaryotic chromosome. Before the cell itself divides, the DNA circle makes a copy of itself, a process called *replication*. Starting at one point, the origin of replication (indicated at the top of figure 8.1*a*), the double helix of DNA begins to unzip, exposing the two strands. The enlargement on the right of figure 8.1*a* shows how the DNA replicates. The purple strand is from the original DNA and the red strand is the newly formed DNA. The new double helix is formed from each naked strand by placing on each exposed nucleotide its complementary nucleotide (that is, A with T, G with C, as discussed in chapter 3). DNA replication is discussed in more detail in chapter 11. When the unzipping has gone all the way around the circle, the cell possesses two copies of its hereditary information.

> As described on pages 52 and 53, DNA is a type of nucleic acid. The hereditary information is encoded in the DNA molecule through the order of the nucleotides that make up the long double-stranded DNA molecule.

When the DNA has been copied, the cell grows, resulting in elongation. The newly replicated DNA molecules are partitioned toward each end of the cell. This partitioning process involves DNA sequences near the origin of replication, and results in these sequences being attached to the membrane. When the cell reaches an appropriate size, the prokaryotic cell begins to split into two equal halves. New plasma membrane and cell wall are added at a point between where the two DNA copies are partitioned, indicated by the green divider in figure 8.1*a*. As the growing plasma membrane pushes inward, the cell is constricted in two, eventually forming two *daughter cells*. Each contains one prokaryotic chromosome that is genetically identical to the parent cell's and each is a complete living cell in its own right.

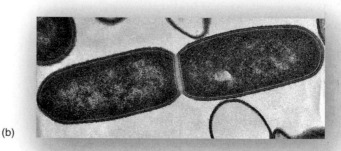

(b)

Figure 8.1   **Cell division in prokaryotes.**

Prokaryotic cells divide by a process of binary fission. (a) Before the cell splits, the circular DNA molecule of a prokaryote initiates replication at a single site, called the origin of replication, moving out in both directions. When the two moving replication points meet on the far side of the molecule its replication is complete. The cell then undergoes binary fission, where the cell divides into two daughter cells. (b) Here, a prokaryotic cell has divided in two and is about to be pinched apart by the growing plasma membrane.

8.2 Eukaryotes Have a Complex Cell Cycle

CONCEPT PREVIEW: Eukaryotic cells divide by separating copies of their chromosomes into daughter cells.

Cell division in eukaryotes is more complex than in prokaryotes, both because eukaryotes contain far more DNA and because the DNA is packaged in chromosomes that reside in the nucleus. A eukaryotic **chromosome** is a single, long DNA molecule wound tightly around proteins that condense into a compact shape. The cells of eukaryotic organisms either undergo mitosis or meiosis to divide up the DNA. **Mitosis** is the mechanism of cell division that occurs in an organism's nonreproductive cells, or *somatic cells*. **Meiosis,** divides the DNA in cells that participate in sexual reproduction, or *germ cells*.

The events that prepare the eukaryotic cell for division and the division process itself constitute a **complex cell cycle.** *Essential Biological Process 8A* walks you through the phases of the cell cycle:

Interphase. Interphase is composed of three phases:

 G_1 phase. This "first gap" phase is the cell's primary growth phase. For most organisms, this phase occupies much of the cell's life span.
 S phase. In this "synthesis" phase, the DNA replicates, producing two copies of each chromosome.
 G_2 phase. Cell division preparation continues in the "second gap" phase with the replication of mitochondria, chromosome condensation, and the synthesis of microtubules.

M phase. In mitosis, a microtubular apparatus binds to the chromosomes and moves them apart.

C phase. In cytokinesis, the cytoplasm divides, creating two daughter cells.

IN THE NEWS

Alzheimer's Disease Linked to Cell Cycle. Alzheimer's disease, common among the elderly, involves the degeneration of brain nerve cells. The dying cells become clogged with masses of protein called amyloid plaques, but why the plaques appear is not known. Recently several research laboratories have reported findings that shed light on this critical point. What they report is that the nerve cell degeneration of Alzheimer's appears to be a disease of inappropriate cell cycle control. Six months or more before the first amyloid deposits appear, proteins associated with the cell cycle, such as proteins called cyclins, are seen within affected nerve cells in the frontal cortex of the brain. What is important to note is that normally adult brain cells don't divide; after development they go into a resting state. Apparently, this initiation of the cell cycle leads to amyloid deposits and cell death because nerve cells that reenter the cell cycle die rather than divide. These findings suggest that therapies targeted toward preventing mitotic changes may have a profound and positive impact on Alzheimer's disease progression.

Essential Biological Process 8A

The Cell Cycle

Interphase. The chromosomes are extended and in use during the G_1, S, and G_2 phases.

Prophase. The chromosomes condense, the nuclear envelope breaks down, and the spindle forms.

Metaphase. The chromosomes line up on the central plane of the cell.

Interphase (G_1, S, G_2 phases)

Cytokinesis (C phase)

Mitosis (M phase)

Cytokinesis. The cytoplasm of the cell is cleaved in half.

Telophase. The chromosomes uncoil, and a new nuclear envelope forms. The spindle fibers disappear.

Anaphase. The centromeres divide, and the chromatids move toward opposite poles.

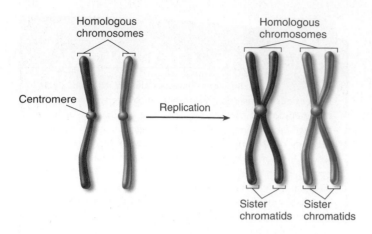

Figure 8.2 **The difference between homologous chromosomes and sister chromatids.**

Homologous chromosomes are a pair of the same chromosome—say, chromosome number 16. Sister chromatids are the two replicas of a single chromosome held together by the centromere after DNA replication. A duplicated chromosome looks somewhat like an X.

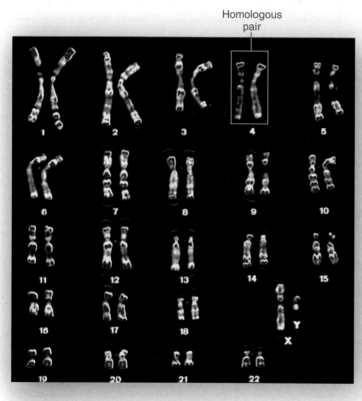

Figure 8.3 **The 46 chromosomes of a human.**

In this presentation, photographs of the individual chromosomes of a human male have been cut out and paired with their homologues, creating an organized display called a karyotype. The chromosomes are in a duplicated state, and the sister chromatids can actually be seen in many of the homologous pairs. The different sizes and shapes of chromosomes allow scientists to pair together the ones that are homologous. For example, chromosome 1 is much larger than chromosome 14, and its centromere is more centrally located on the chromosome.

8.3 Chromosomes

CONCEPT PREVIEW: All eukaryotic cells store their hereditary information in chromosomes. Coiling of the DNA into chromosomes allows it to fit in the nucleus.

Chromosomes were first observed by the German embryologist Walther Fleming in 1882, while he was examining the rapidly dividing cells of salamander larvae. When Fleming looked at the cells through what would now be a rather primitive light microscope, he saw minute threads within their nuclei that appeared to be dividing lengthwise. Fleming called their division *mitosis,* based on the Greek word *mitos,* meaning "thread."

Chromosome Number

Since their initial discovery, chromosomes have been found in the cells of all eukaryotes examined. Their number may vary enormously from one species to another. A few kinds of organisms—such as the Australian ant *Myrmecia* spp.; the plant *Haplopappus gracilis,* a relative of the sunflower that grows in North American deserts; and the fungus *Penicillium*—have only 1 pair of chromosomes, while some ferns have more than 500 pairs. Most eukaryotes have between 10 and 50 chromosomes in their body cells.

Homologous Chromosomes

Chromosomes exist in somatic cells as pairs, called **homologous chromosomes,** or **homologues.** Homologues carry information about the same traits at the same locations on each chromosome but the information can vary between homologues, which will be discussed in chapter 10. Cells that have two of each type of chromosome are called **diploid cells.** One chromosome of each pair is inherited from the mother (colored green in figure 8.2) and the other from the father (colored purple). Before cell division, each homologous chromosome replicates, resulting in two identical copies, called **sister chromatids.** You see in figure 8.2 that the sister chromatids remain joined together after replication at a special linkage site called the **centromere,** the knoblike structure in the middle of each chromosome. Human body cells have a total of 46 chromosomes, which are actually 23 pairs of homologous chromosomes. In their duplicated state, before mitosis, there are still only 23 pairs of chromosomes, but each chromosome has duplicated and consists of two sister chromatids, for a total of 92 chromatids. The duplicated sister chromatids can make it confusing to count the number of chromosomes in an organism, but keep in mind that the number of centromeres doesn't increase with replication, and so you can always determine the number of chromosomes simply by counting the centromeres.

The Human Karyotype

The 46 human chromosomes can be paired as homologues by comparing size, shape, location of centromeres, and so on. This arrangement of chromosomes is called a **karyotype.** An example of a human karyotype is shown in figure 8.3. A chromosome can contain thousands of genes that play important roles in determining how a person's body develops and functions. For this reason, possession of all the chromosomes is essential to survival. Humans missing even one chromosome, a condition called monosomy, do not usually survive embryonic development. Nor does the human embryo develop properly with an extra copy of any one chromosome, a condition called trisomy. For all but a few of the smallest chromosomes, trisomy is fatal; even in those cases, serious problems result. We will revisit this issue of differences in chromosome number in chapter 10.

Chromosome Structure

Chromosomes are composed of **chromatin,** a complex of DNA and protein; most are about 40% DNA and 60% protein. A significant amount of RNA is also associated with chromosomes because chromosomes are the sites of RNA synthesis. The DNA of a chromosome is one very long, double-stranded fiber that extends unbroken through the entire length of the chromosome. A typical human chromosome contains about 140 million (1.4×10^8) nucleotides in its DNA. Furthermore, if the strand of DNA from a single chromosome were laid out in a straight line, it would be about 5 centimeters (2 inches) long. The amount of information in one human chromosome would fill about 2,000 printed books of 1,000 pages each! Fitting such a strand into a nucleus is like cramming a string the length of a football field into a baseball—and that's only 1 of 46 chromosomes! In the cell, however, the DNA is coiled, allowing it to fit into a much smaller space than would otherwise be possible.

Chromosome Coiling

The DNA of eukaryotes is divided into several chromosomes, although the chromosomes you see in figure 8.3 hardly look like long double-stranded molecules of DNA. These chromosomes, duplicated as sister chromatids, are formed into the shape we see here by winding and twisting the long DNA strands into a much more compact form. Winding up DNA presents an interesting challenge. Because the phosphate groups of DNA molecules have negative charges, it is impossible to just tightly wind up DNA because all the negative charges would simply repel one another. As you can see in figure 8.4, the DNA helix wraps around proteins with positive charges called **histones.** The positive charges of the histones counteract the negative charges of the DNA, so that the complex has no net charge. Every 200 nucleotides, the DNA duplex is coiled around a core of eight histone proteins, forming a complex known as a **nucleosome.** The nucleosomes are further coiled into a solenoid. This solenoid is then organized into looped domains. The final organization of the chromosome is not known, but it appears to involve further radial looping into rosettes around a preexisting scaffolding of protein. This complex of DNA and histone proteins, coiled tightly, forms a compact chromosome.

Figure 8.4 **Levels of eukaryotic chromosomal organization.**

Compact, rod-shaped chromosomes are in fact highly wound-up molecules of DNA. The arrangement illustrated here is one of many possibilities.

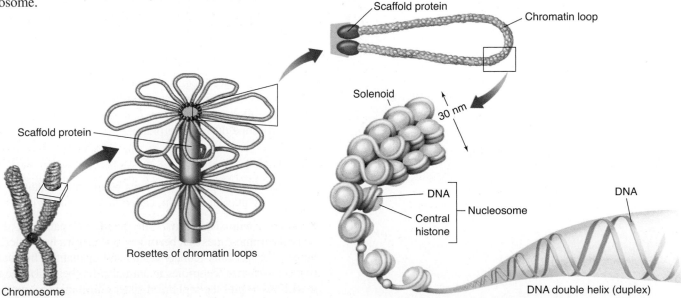

Scaffold protein
Chromatin loop
Solenoid
30 nm
DNA
Nucleosome
Central histone
DNA
DNA double helix (duplex)
Scaffold protein
Rosettes of chromatin loops
Chromosome

Essential Biological Process 8B

Cell Division

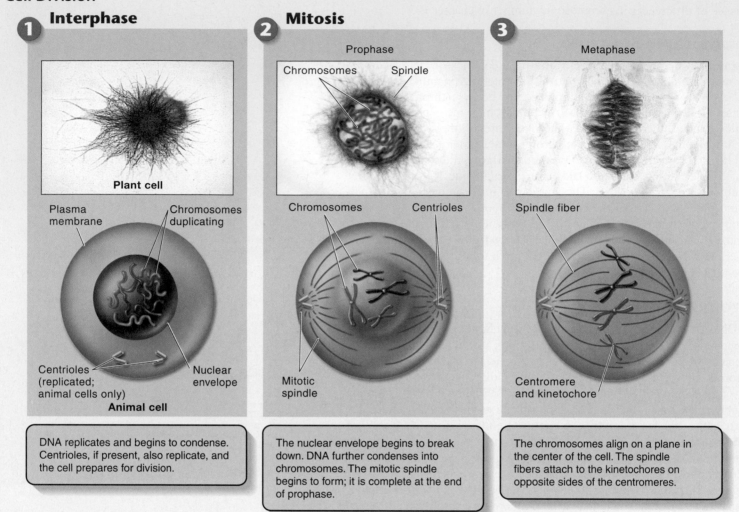

1 **Interphase**

Plant cell

Plasma membrane

Chromosomes duplicating

Centrioles (replicated; animal cells only)

Nuclear envelope

Animal cell

DNA replicates and begins to condense. Centrioles, if present, also replicate, and the cell prepares for division.

2 **Mitosis**

Prophase

Chromosomes   Spindle

Chromosomes   Centrioles

Mitotic spindle

The nuclear envelope begins to break down. DNA further condenses into chromosomes. The mitotic spindle begins to form; it is complete at the end of prophase.

3

Metaphase

Spindle fiber

Centromere and kinetochore

The chromosomes align on a plane in the center of the cell. The spindle fibers attach to the kinetochores on opposite sides of the centromeres.

8.4 Cell Division

CONCEPT PREVIEW: In interphase, the replicated chromosomes begin to condense. In mitosis, these chromosomes are drawn by microtubules to opposite ends of the cell; in cytokinesis, the cell is split into two daughter cells.

Interphase

When cell division begins in interphase (panel 1 of *Essential Biological Process 8B*), chromosomes first replicate, and then begin to wind up tightly, a process called **condensation.** Chromosomes are not usually visible under the microscope during interphase.

Mitosis

Interphase is not a phase of mitosis, but it sets the stage for cell division. It is followed by nuclear division, called *mitosis,* subdivided into four stages:

Prophase: Mitosis Begins. In **prophase** (panel 2), the individual condensed chromosomes first become visible with a light microscope. As the replicated chromosomes condense, the cell dismantles the nuclear envelope and two centrosomes (centrioles in animal cells) begin to assemble the apparatus it will use to pull the replicated sister chromatids to opposite ends ("poles")

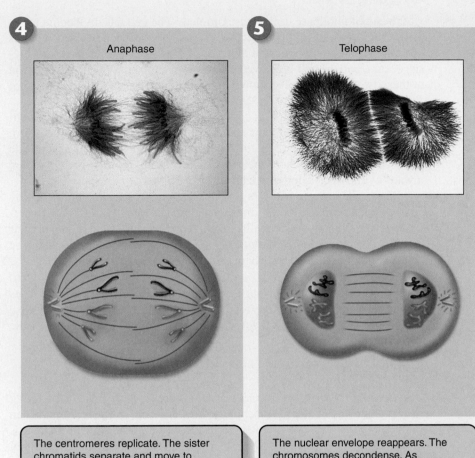

The centromeres replicate. The sister chromatids separate and move to opposite poles.

The nuclear envelope reappears. The chromosomes decondense. As telophase progresses, cytokinesis also occurs.

In cytokinesis two daughter cells form. Each cell is a replicate of the parent cell and is diploid.

of the cell. In the center of an animal cell, the pairs of centrioles separate and move apart toward opposite poles of the cell, forming between them as they move apart a network of protein cables called the **spindle.** Each cable is called a *spindle fiber* and is made of microtubules, which are long, hollow tubes of protein. Plant cells lack centrioles and instead brace the ends of the spindle toward the poles. The spindle fibers attach to the chromosomes, and when the process is complete, one sister chromatid of each pair is attached by microtubules to one pole and the other sister chromatid to the other pole.

Metaphase: Alignment of the Chromosomes. The second phase of mitosis, **metaphase,** begins when the chromosomes, each consisting of a pair of sister chromatids, align in the center of the cell along an imaginary plane that divides the cell in half, referred to as the equatorial plane. Microtubules attached to the centromeres extend back toward the opposite poles of the cell.

Anaphase: Separation of the Chromatids. In **anaphase,** the centromeres split, and the sister chromatids are freed from each other. Cell division is now simply a matter of reeling in the microtubules, dragging the sister chromatids (now referred to as daughter chromosomes) to the poles.

Telophase: Re-formation of the Nuclei. In **telophase,** the mitotic spindle disassembles, and a nuclear envelope forms around each set of chromosomes while they begin to uncoil. The nucleolus also reappears.

EVOLUTION

Doing Mitosis More Efficiently. Fungi carry out mitosis in a strikingly different way than the process described here. Instead of completely disassembling the nuclear envelope before chromosome segregation, fungi and other lower eukaryotes carry out a more primitive form of mitosis in which the nuclear membrane remains intact, and all of mitosis occurs within the nucleus. Why the difference? In fungi and other lower eukaryotes, the bodies from which the spindle microtubules emanate are embedded within the nuclear envelope, with the tubulin molecules used to build the spindle being imported from the cytoplasm through special channels in the nuclear envelope. Higher eukaryotes dispense with this complex arrangement. Their spindle microtubules emanate from centrioles or other structures outside the nucleus, not the nuclear envelope. Removal of the nuclear envelope allows chromosomes within the nucleus to have direct access to microtubules being made in the cytoplasm without passing the microtubule subunits across a membrane.

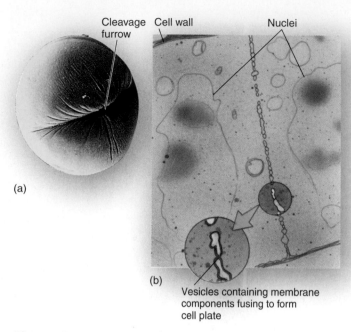

Cleavage furrow   Cell wall   Nuclei

(a)

(b)

Vesicles containing membrane components fusing to form cell plate

Figure 8.5 Cytokinesis.

The division of cytoplasm that occurs after mitosis is called cytokinesis and cleaves the cell into roughly equal halves. (a) In an animal cell, such as this sea urchin egg, a cleavage furrow forms around the dividing cell. (b) In this dividing plant cell, a cell plate is forming between the two newly forming daughter cells.

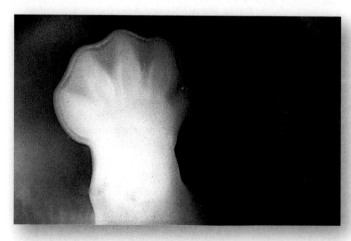

Figure 8.6 Programmed cell death.

In the human embryo, programmed cell death results in the formation of fingers and toes from paddlelike hands and feet.

Cytokinesis

At the end of telophase, mitosis is complete. The cell has divided its replicated chromosomes into two nuclei, which are positioned at opposite ends of the cell. As mitosis ends, the division of the cytoplasm, called **cytokinesis,** occurs, and the cell is cleaved into roughly equal halves. The formation of these two daughter cells, shown in the last panel of *Essential Biological Process 8B,* signals the end of cell division.

In animal cells, which lack cell walls, cytokinesis is achieved by pinching the cell in two with a contracting belt of actin filaments (figure 8.5*a*). As contraction proceeds, a *cleavage furrow* becomes evident around the cell's circumference, where the cytoplasm is being progressively pinched inward by the decreasing diameter of the actin belt.

Plant cells have rigid walls that are far too strong to be deformed by actin filament contraction. A different approach to cytokinesis has therefore evolved in plants. Plant cells assemble membrane components in their interior, at right angles to the mitotic spindle. In figure 8.5*b*, you can see how membrane is deposited between the daughter cells by vesicles that fuse together. This expanding partition, called a *cell plate*, grows outward until it reaches the interior surface of the plasma membrane and fuses with it, at which point it has effectively divided the cell in two. Cellulose is then laid down over the new membranes, forming the cell walls of the two new cells.

Cell Death

Despite the ability to divide, no cell lives forever. The ravages of living slowly tear away at a cell's machinery. To some degree damaged parts can be replaced, but no replacement process is perfect. If food supplies are cut off, animal cells cannot obtain the energy necessary to maintain their lysosome membranes. The cells die, digested from within by their own enzymes.

> Lysosomes are vesicles within the cell described on page 73. They are filled with digestive enzymes that break down cellular debris, and if the lysosome's membrane becomes leaky, the enzymes spill out into the cell, killing it.

During fetal development, many cells are programmed to die. In human embryos, hands and feet appear first as "paddles" (figure 8.6), but the skin cells between bones die as programmed to form the separated toes and fingers. The cells in the tissue between the bones will later die, leaving behind a set of fingers. In ducks, this cell death is not part of the developmental program, which is why ducks have webbed feet.

Human cells appear to be programmed to undergo only so many cell divisions and then die, following a plan written into the genes. In tissue culture, cell lines divide about 50 times, and then the entire population of cells dies off. Even if some of the cells are frozen for years, when they are thawed they simply resume where they left off and die on schedule. Only cancer cells appear to thwart these instructions, dividing endlessly. All other cells in your body contain a hidden clock that keeps time by counting cell divisions, and when the alarm goes off the cells die.

Concept Check

1. What are the three phases of interphase, and what happens in each?
2. At the end of interphase, how many chromatids does a human cell contain?
3. What are the four stages of mitosis, and what happens in each?

Cancer and the Cell Cycle

8.5 What Is Cancer?

CONCEPT PREVIEW: Cancer is unrestrained cell growth and division caused by damage to genes regulating the cell division cycle.

Cancer is a growth disorder of cells. It starts when an apparently normal cell begins to divide in an uncontrolled way. The result is a cluster of cells, called a **tumor,** that constantly expands in size. The cluster of pink lung cells in the photo in figure 8.7 have begun to form a malignant tumor called a *carcinoma.* Malignant tumors are invasive, their cells able to break away from the tumor, enter the bloodstream, and spread to other areas of the body (figure 8.8), forming new tumors at distant sites called **metastases.**

Cancer is perhaps the most devastating and deadly disease. Most of us have had family or friends affected by the disease. In 2007, 1,444,920 American men and women were diagnosed with cancer; in that same year, 559,650 Americans died of cancer. One in every two Americans born in 2009 will be diagnosed with some form of cancer during their lifetime. In the U.S., the three deadliest human cancers are lung cancer, cancer of the colon and rectum, and breast cancer. Lung cancer, responsible for the most cancer deaths, is largely preventable; most cases result from smoking cigarettes. Colorectal cancers appear to be fostered by the high-meat diets so favored in the United States. The cause of breast cancer is still a mystery.

Not surprisingly, researchers are expending a great deal of effort to learn the cause of cancer. Scientists have made considerable progress in the last 30 years using molecular biological techniques, and the rough outlines of understanding are now emerging. We now know that cancer is a gene disorder of somatic tissue, in which damaged genes fail to properly control cell growth and division. The cell division cycle is regulated by a sophisticated group of proteins called growth factors. Cancer results from damage to the genes encoding these proteins. Damage to DNA, such as damage to these genes, is called **mutation.** Cancer can be caused by chemicals that alter DNA like the tars in cigarette smoke, by environmental factors such as UV rays that damage DNA, or in some instances by viruses that circumvent the cell's normal growth and division controls.

> Mutations are errors in DNA and are discussed in detail in chapter 11, pages 192–193. Damage to DNA often involves an incorrect nucleotide being inserted during DNA replication, which changes the information encoded in the DNA.

There are two general classes of growth factor genes that are usually involved in cancer: proto-oncogenes and tumor-suppressor genes. Genes known as **proto-oncogenes** encode proteins that stimulate cell division. Mutations to these genes can cause cells to divide excessively. Mutated proto-oncogenes become cancer-causing genes called **oncogenes.**

The second class of cancer-causing genes are called **tumor-suppressor genes.** Cell division is normally turned off in healthy cells by proteins encoded by tumor-suppressor genes. Mutations to these genes essentially "release the brakes," allowing the cell containing the mutated gene to divide uncontrolled. The cell cycle never stops in a cancerous line of cells.

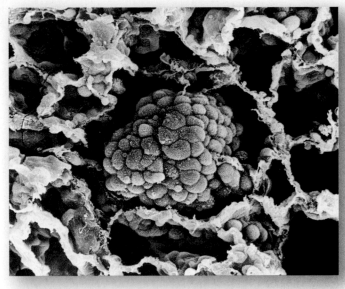

Figure 8.7 Lung cancer cells (300×).

These cells are from a tumor located in the alveolus (air sac) of a human lung.

IMPLICATION 160,390 people died of lung cancer in the United States in 2007, almost all of them cigarette smokers. Fully 7 1/2 % of pack-a-day smokers will die of lung cancer within 30 years of their first cigarette. That's 1 in 13. Do you smoke? Do any of your friends? Can you think of a reason that would justify the risk?

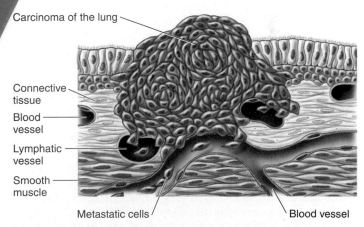

Figure 8.8 Portrait of a tumor.

This ball of cells is a carcinoma (cancer tumor) developing from epithelial cells that line the interior surface of a human lung. As the mass of cells grows, it invades surrounding tissues, eventually penetrating lymphatic and blood vessels, both of which are plentiful within the lung. These vessels carry metastatic cancer cells throughout the body, where they lodge and grow, forming new masses of cancerous tissue.

Concept Check

1. What are the three deadliest human cancers?
2. Cancer results from damage to which two general classes of genes?
3. Name three sorts of things that cause cancer.

Biology and Staying Healthy

Curing Cancer

Half of all Americans will face cancer at some point in their lives. Potential cancer therapies are being developed on many fronts. Some act to prevent the start of cancer within cells. Others act outside cancer cells, preventing tumors from growing and spreading. The figure on the right indicates targeted areas for the development of cancer treatments. The following discussion will examine each of these areas.

Preventing the Start of Cancer

Many promising cancer therapies act within potential cancer cells, focusing on different stages of the cell's "Shall I divide?" decision-making process.

1 Receiving the Signal to Divide. The first step in the decision process is receiving a "divide" signal, usually a small protein called a growth factor released from a neighboring cell. The growth factor, the red ball at #1 in the figure, is received by a protein receptor on the cell surface. Like banging on a door, its arrival signals that it's time to divide. Mutations that increase the number of receptors on the cell surface amplify the division signal and so lead to cancer. Over 20% of breast cancer tumors prove to overproduce a protein called HER2 associated with the receptor for epidermal growth factor (EGF).

Therapies directed at this stage of the decision process utilize the human immune system to attack cancer cells. Special protein molecules called *monoclonal antibodies,* created by genetic engineering, are the therapeutic agents. These monoclonal antibodies are designed to seek out and stick to HER2. Like waving a red flag, the presence of the monoclonal antibody calls down attack by the immune system on the HER2 cell. Because breast cancer cells overproduce HER2, they are killed preferentially. The biotechnology research company Genentech's recently approved monoclonal antibody, called herceptin, has given promising results in clinical tests.

Up to 70% of colon, prostate, lung, and head/neck cancers have excess copies of a related receptor, epidermal growth factor 1 (HER1). The monoclonal antibody C225, directed against HER1, has succeeded in shrinking 22% of advanced, previously incurable colon cancers in early clinical trials. Apparently blocking HER1 interferes with the ability of tumor cells to recover from chemotherapy or radiation.

2 Passing the Signal via a Relay Switch. The second step in the decision process is the passage of the signal into the cell's interior, the

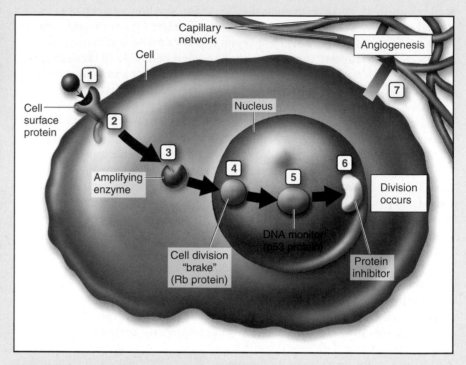

Seven different stages in the cancer process.

(1) On the cell surface, a growth factor's signal to divide is increased. (2) Just inside the cell, a protein relay switch that passes on the divide signal gets stuck in the "ON" position. (3) In the cytoplasm, enzymes that amplify the signal are amplified even more. In the nucleus, (4) a "brake" preventing DNA replication is inoperable, (5) proteins that check for damage in the DNA are inactivated, and (6) other proteins that inhibit the elongation of chromosome tips are destroyed. (7) The new tumor promotes angiogenesis, the formation of new blood vessels that promote growth.

cytoplasm. This is carried out in normal cells by a protein called Ras that acts as a relay switch, #2 in the figure. When growth factor binds to a receptor like EGF, the adjacent Ras protein acts like it has been "goosed," contorting into a new shape. This new shape is chemically active, and initiates a chain of reactions that passes the "divide" signal inward toward the nucleus. Mutated forms of the Ras protein behave like a relay switch stuck in the "ON" position, continually instructing the cell to divide when it should not. Thirty percent of all cancers have a mutant form of Ras. So far, no effective therapies have been developed targeting this step.

3 Amplifying the Signal. The third step in the decision process is the amplification of the signal within the cytoplasm. Just as a TV signal needs to be amplified in order to be received at a distance, so a "divide" signal must be amplified if it is to reach the nucleus at the interior of the cell, a very long journey at a molecular scale. To get a signal all the way into the nucleus, the cell employs a sort of pony express. The "ponies" in this case are enzymes called *tyrosine kinases,* #3 in the figure. These enzymes add phosphate groups to proteins, but only at a particular amino acid, tyrosine. No other enzymes in the cell do this, so the tyrosine kinases form an elite corps of signal carriers not confused by the myriad of other molecular activities going on around them.

Cells use an ingenious trick to amplify the signal as it moves toward the nucleus. Ras, when "ON," activates the initial protein kinase. This protein kinase activates other protein kinases that in their turn activate still others. The trick is that once a protein kinase enzyme is activated, it goes to work like a demon, activating hoards of others every second! And each and every one it activates behaves the same way too, activating still more, in a cascade of ever-widening effect. At each stage of the relay, the signal is amplified a thousandfold.

Mutations stimulating any of the protein kinases can dangerously increase the already amplified signal and lead to cancer. Some 15 of the cell's 32 internal tyrosine kinases have been implicated in cancer. Five percent of all cancers, for example, have a mutant hyperactive form of the protein kinase Src. The trouble begins when a mutation causes one of the tyrosine kinases to become locked into the "ON" position, sort of like a stuck doorbell that keeps ringing and ringing.

To cure the cancer, you have to find a way to shut the bell off. Each of the signal carriers presents a different problem, as you must quiet it without knocking out all the other signal pathways the cell needs. The cancer therapy drug Gleevec, a monoclonal antibody, has just the right shape to fit into a groove on the surface of the tyrosine kinase called "abl." Mutations locking abl "ON" are responsible for chronic myelogenous leukemia, a lethal form of white blood cell cancer. Gleevec totally disables abl. In clinical trials, blood counts revert to normal in more than 90% of cases.

4 **Releasing the Brake.** The fourth step in the decision process is the removal of the "brake" the cell uses to restrain cell division. In healthy cells this brake, a tumor-suppressor protein called Rb, blocks the activity of a protein called E2F, #4 in the figure. When free, E2F enables the cell to copy its DNA. Normal cell division is triggered to begin when Rb is inhibited, unleashing E2F. Mutations that destroy Rb release E2F from its control completely, leading to ceaseless cell division. Forty percent of all cancers have a defective form of Rb.

Therapies directed at this stage of the decision process are only now being attempted. They focus on drugs able to inhibit E2F, which should halt the growth of tumors arising from inactive Rb. Experiments in mice in which the *E2F* genes have been destroyed provide a model system to study such drugs, which are being actively investigated.

5 **Checking That Everything Is Ready.** The fifth step in the decision process is the mechanism used by the cell to ensure that its DNA is undamaged and ready to divide. This job is carried out in healthy cells by the tumor-suppressor protein p53, which inspects the integrity of the DNA, #5 in the figure. When it detects damaged or foreign DNA, p53 stops cell division and activates the cell's DNA repair systems. If the damage doesn't get repaired in a reasonable time, p53 pulls the plug, triggering events that kill the cell. In this way, mutations such as those that cause cancer are either repaired or the cells containing them eliminated. If p53 is itself destroyed by mutation, future damage accumulates unrepaired. Among this damage are mutations that lead to cancer. Fifty percent of all cancers have a disabled p53. Fully 70% to 80% of lung cancers have a mutant inactive p53—the chemical benzo[*a*]pyrene in cigarette smoke is a potent mutagen of p53.

6 **Stepping on the Gas.** Cell division starts with replication of the DNA. In healthy cells, another tumor suppressor "keeps the gas tank nearly empty" for the DNA replication process by inhibiting production of an enzyme called *telomerase.* Without this enzyme, a cell's chromosomes lose material from their tips, called *telomeres.* Every time a chromosome is copied, more tip material is lost. After some 30 divisions, so much is lost that copying is no longer possible. Cells in the tissues of an adult human have typically undergone 25 or more divisions. Cancer can't get very far with only the five remaining cell divisions, so inhibiting telomerase is a very effective natural brake on the cancer process, #6 in the figure. It is thought that almost all cancers involve a mutation that destroys the telomerase inhibitor, releasing this brake and making cancer possible. It should be possible to block cancer by reapplying this inhibition. Cancer therapies that inhibit telomerase are just beginning clinical trials.

Preventing the Spread of Cancer

7 **Stopping Tumor Growth.** Once a cell begins cancerous growth, it forms an expanding tumor. As the tumor grows ever-larger, it requires an increasing supply of food and nutrients, obtained from the body's blood supply. To facilitate this necessary grocery shopping, tumors leak out substances into the surrounding tissues that encourage the formation of small blood vessels, a process called angiogenesis, #7 in the figure. Chemicals that inhibit this process are called *angiogenesis inhibitors.* Two such natural angiogenesis inhibitors, angiostatin and endostatin, caused tumors to regress to microscopic size in mice, but initial human trials were disappointing.

Laboratory drugs are more promising. A monoclonal antibody drug called Avastin, targeted against a blood vessel growth promoting substance called vascular endothelial growth factor (VEGF), destroys the ability of VEGF to carry out its blood-vessel-forming job. Given to hundreds of advanced colon cancer patients as part of a large clinical trial, Avastin improved colon cancer patients' chance of survival by 50% over chemotherapy.

Inquiry & Analysis

Why Do Human Cells Age?

Human cells appear to have built-in life spans. In 1961 cell biologist Leonard Hayflick reported the startling result that skin cells growing in tissue culture, such as those growing in culture flasks in the photo below, will divide only a certain number of times. After about 50 population doublings cell division stops (a **doubling** is a round of cell division producing two daughter cells for each dividing cell, for example going from a population of 30 cells to 60 cells). If a cell sample is taken after 20 doublings and frozen, when thawed it resumes growth for 30 more doublings, and then stops. An explanation of the "Hayflick limit" was suggested in 1986 when researchers first glimpsed an extra length of DNA at the end of chromosomes. Dubbed *telomeres*, these lengths proved to be composed of the simple DNA sequence TTAGGG, repeated nearly a thousand times. Importantly, telomeres were found to be substantially shorter in the cells of older body tissues. This led to the hypothesis that a run of some 16 TTAGGGs was where the DNA replicating enzyme, called polymerase, first sat down on the DNA (16 TTAGGGs being the size of the enzyme's "footprint"), and

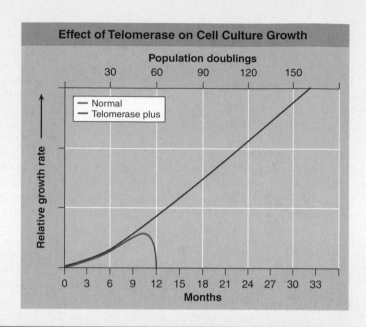

TTAGGG TTAGGG TTAGGG TTAGGG TTAGGG--------

because of being its docking spot, the polymerase was unable to copy that bit. Thus a 100-base portion of the telomere was lost by a chromosome during each doubling as DNA replicated. Eventually, after some 50 doubling cycles, each with a round of DNA replication, the telomere would be used up and there would be no place for the DNA replication enzyme to sit. The cell line would then enter senescence, no longer able to proliferate.

This hypothesis was tested in 1998. Using genetic engineering, researchers transferred into newly established human cell cultures

a gene that leads to expression of an enzyme called *telomerase* that all cells possess but no body cell uses. This enzyme adds TTAGGG sequences back to the end of telomeres, in effect rebuilding the lost portions of the telomere. Laboratory cultures of cell lines with (telomerase plus) and without (normal) this gene were then monitored for many generations. The graph above displays the results.

Analysis

1. **Applying Concepts** Comparing continuous processes, how do normal skin cells (blue line) differ in their growth history from telomerase plus cells with the telomerase gene (red line)?
2. **Interpreting Data** After how many doublings do the normal cells cease to divide? the telomerase plus cells?
3. **Making Inferences** After 9 population doublings, would the rate of cell division be different between the two cultures? after 15? Why?
4. **Drawing Conclusions** How does the addition of the telomerase gene affect the senescence (death by old age) of skin cells growing in culture? Does this result confirm the telomerase hypothesis this experiment had set out to test?

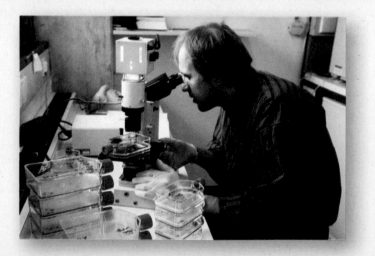

Concept Summary

Cell Division

8.1 Prokaryotes Have a Simple Cell Cycle

- Prokaryotic cells divide in a two-step process: DNA replication and binary fission. The genetic information in a prokaryotic cell is present as a single loop of DNA. The DNA begins replication at a site called the origin of replication. The DNA double strand unzips, and new strands form along the original strands, producing two circular prokaryotic chromosomes that separate to the ends of the cell. New plasma membrane and cell wall are added down the middle of the cell, as shown here from **figure 8.1,** splitting the cell in two. This cell division, called binary fission, produces two daughter cells that are genetically identical to the parent cell.

8.2 Eukaryotes Have a Complex Cell Cycle

- Cell division in eukaryotes is more complex than in prokaryotes because eukaryotic cells contain more DNA and their DNA is packaged into linear chromosomes.

- Eukaryotic cells divide by one of two methods, mitosis or meiosis. Mitosis occurs in nonreproductive cells, called somatic cells. Meiosis occurs in cells that are involved in sexual reproduction, forming germ cells, such as sperm and eggs.

- The complex cell cycle in eukaryotic cells occurs in several phases: interphase, M phase, and C phase. Interphase is the first portion of the cell cycle. Interphase is also broken down into phases. The G_1 phase is the growing phase and takes up the major portion of the cell's life cycle. The S phase is the synthesis phase and is when the DNA is replicated. The G_2 phase involves the final preparations for cell division with the replication of mitochondria, chromosome condensation, and synthesis of microtubules.

- During the M phase the chromosomes are distributed into opposite sides of the cell. During the C phase the cell divides its cytoplasm into two separate daughter cells (***Essential Biological Process 8A***).

8.3 Chromosomes

- Eukaryotic DNA is organized into chromosomes. Chromosomes are found in all eukaryotic cell, but the number of chromosomes varies greatly between species. Most eukaryotes have between 10 and 50 chromosomes.

- Chromosomes exist in cells as pairs called homologous chromosomes. Two chromosomes that carry copies of the same genes, like the two shown here from **figure 8.2,** are homologous chromosomes.

- Before cells divide, the DNA replicates forming two identical copies of each chromosome, called sister chromatids. Sister chromatids stay connected at an area called the centromere. Human somatic cells have 46 chromosomes but just before cell division, their DNA replicates forming 92 sister chromatids.

- Chromosomes are not uniform; they vary in size, shape, and placement of centromeres. These variations allow researchers to match up homologues making an array, called a karyotype, where homologues are positioned next to each other (**figure 8.3**).

- The DNA in a chromosome is one long double-stranded fiber. After the DNA is replicated, it associates with proteins, forming chromatin. Chromatin begins to coil up in a process called condensation. The negatively charged DNA can coil up tightly because it wraps around positively charged histone proteins. There are several levels of chromosomal organization. The DNA wraps around a histone complex forming a nucleosome and then further folds and loops on itself forming a compact chromosome (**figure 8.4**).

8.4 Cell Division

- Interphase begins the cell cycle followed by mitosis, which consists of four phases: prophase, metaphase, anaphase, and telophase (***Essential Biological Process 8B***).

- Prophase signals the beginning of mitosis. The DNA that was replicated during interphase condenses into chromosomes. The sister chromatids stay attached at the centromeres. The nuclear envelope disappears. Centrioles, when present, migrate to opposite sides of the cell, called the poles, and begin forming the spindle. Microtubules that form the spindle extend from the poles and attach to the chromosomes at the centromeres, anchoring sister chromatids to opposite poles.

- Metaphase involves the alignment of sister chromatids along the equatorial plane.

- During anaphase, the centromeres split, freeing the sister chromatids. The microtubules shorten, pulling the sister chromatids apart and toward opposite poles.

- Telophase signals the completion of nuclear division. The microtubule spindle is dismantled, the chromosomes begin to uncoil; nuclear envelopes form.

- Following mitosis, the cell separates into two daughter cells in a process called cytokinesis (**figure 8.5**). Cytokinesis in animal cells involves a pinching in of the cell around its equatorial plane until the cell eventually splits into two cells. Cytokinesis in plant cells involves the assembly of plasma membranes and cell walls between the two poles, eventually, forming two separate cells.

- Many cells are programmed to die, either as part of development or after a set number of cell divisions (usually about 50 divisions). Only cancer cells appear to divide endlessly.

Cancer and the Cell Cycle

8.5 What Is Cancer?

- Cancer is a growth disorder of cells, where there is a loss of control over cell division. Cells begin to divide in an uncontrolled way, forming a mass of cells called a tumor (**figure 8.7**). A tumor in which cells break away from the mass and spread to other tissues is called metastasis (**figure 8.8**). Cancer results when genes that encode proteins that control the cell cycle, such as proto-oncogenes and tumor-suppressor genes, are damaged.

Self-Test

1. Prokaryotes reproduce by
   a. copying DNA then undergoing binary fission.
   b. splitting in half.
   c. undergoing mitosis.
   d. copying DNA then undergoing the M phase.
2. The eukaryotic cell cycle is different from prokaryotic cell division in all the following ways *except*
   a. the amount of DNA present in the cells.
   b. how the DNA is packaged.
   c. in the production of genetically identical daughter cells.
   d. the involvement of microtubules.
3. In eukaryotes, the genetic material is found in chromosomes
   a. and the more complex the organism, the more pairs of chromosomes it has.
   b. and many organisms have only one chromosome.
   c. and most eukaryotes have between 10 and 50 pairs of chromosomes.
   d. and most eukaryotes have between 2 and 10 pairs of chromosomes.
4. Homologous chromosomes
   a. are also referred to as sister chromatids.
   b. are genetically identical.
   c. carry information about the same traits located in the same places on the chromosomes.
   d. are connected to each other at their centromeres.

5. Chromosomes are composed of
   a. DNA.   c. chromatin.
   b. proteins.   d. all of the above.
6. In mitosis, when the duplicated chromosomes line up in the center of the cell, that stage is called
   a. prophase.   c. anaphase.
   b. metaphase.   d. telophase.
7. The division of the cytoplasm in the eukaryotic cell cycle is called
   a. interphase.   c. cytokinesis.
   b. mitosis.   d. binary fission.
8. Which of the following pairings is correct?
   a. interphase/DNA replication
   b. animal cells/cell plate
   c. plant cells/cleavage furrow
   d. cytokinesis/cell death
9. The cell cycle is controlled by
   a. growth factors.   c. tumor-suppressor genes.
   b. proto-oncogenes.   d. all of the above.
10. When cell division becomes unregulated, and a cluster of cells begins to grow without regard for the normal controls, that is called
   a. a mutation.   c. metastases.
   b. cancer.   d. oncogenes.

Visual Understanding

1. **Figure 8.3** This karyotype shows a complete set of human chromosomes of an individual. At what stage of the cell cycle are such photos taken? Explain.

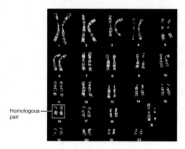

2. **Figure 8.4** During interphase the DNA is not visible through a microscope. Why isn't it visible, and why would you expect this to be the case?

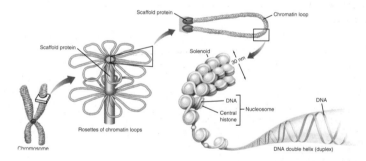

Challenge Questions

1. Why does the DNA in a cell need to change periodically from a long, double-helix chromatin molecule into a tightly wound-up chromosome? What does it do in one configuration that it cannot do in the other?

2. Despite all we know about cancer today, some types of cancers are still increasing in frequency. Lung cancer in women is one of those. What reason(s) might there be for this increasing problem? Can you suggest a solution?

Meiosis

2500×

Meiosis

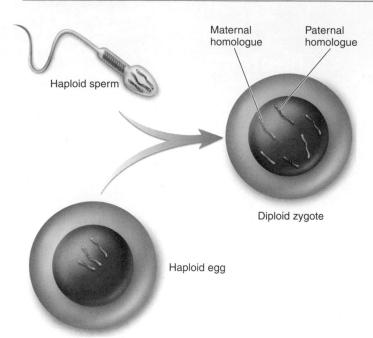

Figure 9.1 **Diploid cells carry chromosomes from two parents.**

A diploid cell contains two versions of each chromosome, a maternal homologue contributed by the haploid egg of the mother, and a paternal homologue contributed by the haploid sperm of the father.

IMPLICATION Many male veterans of the Vietnam War claim that their children born years later have birth defects caused by the herbicide Agent Orange used as a defoliant in the war. What types of cells would the chemical have to have affected in these men to cause the birth defects? How might these cells have become exposed to the chemical?

Figure 9.2 **Sexual and asexual reproduction.**

Reproduction in an organism is not always either sexual or asexual. The strawberry reproduces both asexually (runners) and sexually (flowers).

9.1 Discovery of Meiosis

CONCEPT PREVIEW: Meiosis is a process of cell division in which the number of chromosomes in certain cells is halved during gamete formation.

Only a few years after Walther Fleming's discovery of chromosomes in 1882, Belgian cytologist Pierre-Joseph van Beneden was surprised to find different numbers of chromosomes in different types of cells in the roundworm *Ascaris*. Specifically, he observed that the **gametes** (eggs and sperm) each contained two chromosomes, whereas the somatic (nonreproductive) cells of embryos and mature individuals each contained four. From his observations, van Beneden proposed in 1887 that an egg and a sperm, each containing half the complement of chromosomes found in other cells, fuse to produce a single cell called a **zygote.** The zygote, like all of the somatic cells ultimately derived from it, contains two copies of each chromosome. The fusion of gametes to form a new cell is called **fertilization,** or **syngamy.**

Meiosis

It was clear even to early investigators that gamete formation must involve some mechanism that reduces the number of chromosomes to half the number found in other cells. If it did not, the chromosome number would double with each fertilization, and after only a few generations, the number of chromosomes in each cell would become impossibly large. For example, after one generation the 46 chromosomes present in human cells would increase to 92 (46×2^1) chromosomes, after two generations it would increase to 184, and by 10 generations, the 46 chromosomes present in human cells would increase to over 47,000 (46×2^{10}) chromosomes.

The number of chromosomes does not explode in this way because of a special reduction division that occurs during gamete formation, producing cells with half the normal number of chromosomes. The subsequent fusion of two of these cells ensures a consistent chromosome number from one generation to the next. This reduction division process is known as **meiosis.**

The Sexual Life Cycle

Meiosis and fertilization together constitute a cycle of reproduction. Two sets of chromosomes are present in the somatic cells of adult individuals, making them **diploid** cells (Greek, *di,* two and often indicated by $2n$, where "*n*" is the number of sets of chromosomes), but only one set is present in the gametes, which are thus **haploid** (Greek, *haploos,* one and often indicated by $1n$). Figure 9.1 shows how two haploid cells, a sperm cell containing three chromosomes from the father and an egg cell containing three chromosomes from the mother, fuse to form a diploid zygote with six chromosomes. Reproduction that involves this alternation of meiosis and fertilization is called **sexual reproduction.** Some organisms however, reproduce by mitotic division and don't involve the fusion of gametes. Reproduction in these organisms is referred to as **asexual reproduction.** Some organisms are able to reproduce both asexually and sexually (figure 9.2).

9.2 The Sexual Life Cycle

CONCEPT PREVIEW: In the sexual life cycle, there is an alternation of diploid and haploid phases.

Alternation of Generations

The life cycles of all sexually reproducing organisms follow the same basic pattern of alternation between diploid chromosome numbers and haploid ones (figure 9.3). In unicellular eukaryotic organisms like the protist shown in figure 9.4*a*, individuals are haploid for most of their lives. When they encounter environmental stress, haploid cells sometimes fuse with other haploid cells to form a diploid cell. Later, this cell undergoes meiosis to reform the haploid phase.

In most animals like the frog you see in figure 9.4*b*, fertilization results in the formation of a diploid zygote. This single diploid cell divides by mitosis, eventually gives rise to the adult frog shown in the photo. All animals are diploid for the muticellular stage of their life cycle.

In plants like the fern you see in figure 9.4*c*, haploid and diploid individuals alternate. Certain cells of a diploid individual undergo meiosis making haploid gametes that divide repeatedly by mitosis to form a multicellular haploid individual. Some cells of this haploid individual eventually differentiate into eggs or sperm, which fuse to form a diploid zygote. Dividing by mitosis, the zygote forms a diploid individual.

Germ-Line Tissues

In animals, the cells that will eventually undergo meiosis to produce gametes are set aside from other cells early in the course of development. The cells of the body are called **somatic** cells, from the Latin word for "body," while gamete-forming cells, located in the reproductive organs of males and females, are referred to as **germ-line** cells. Both the somatic cells and the gamete-producing germ-line cells are diploid. Somatic cells undergo mitosis to form genetically identical, diploid daughter cells. The germ-line cells undergo meiosis, producing haploid gametes.

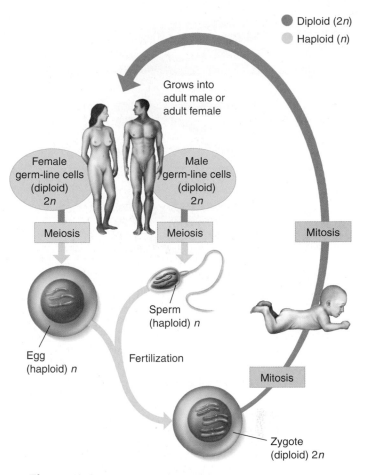

● Diploid (2*n*)
○ Haploid (*n*)

Grows into adult male or adult female

Female germ-line cells (diploid) 2*n* — Meiosis

Male germ-line cells (diploid) 2*n* — Meiosis

Mitosis

Egg (haploid) *n*

Sperm (haploid) *n*

Fertilization

Mitosis

Zygote (diploid) 2*n*

Figure 9.3   The sexual life cycle in animals.

In animals, the completion of meiosis is followed soon by fertilization. Thus, the vast majority of the life cycle is spent in the diploid stage. In this text, *n* stands for haploid and 2*n* stands for diploid. Germ-line cells are set aside early in development and undergo meiosis to form haploid gametes (eggs or sperm). The rest of the body cells are called somatic cells.

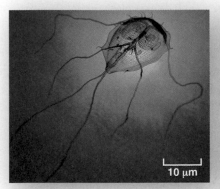

(a) Protist: spends most of its life cycle as a haploid individual

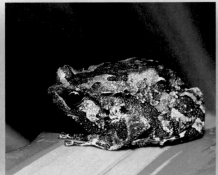

(b) Animal: spends most of its life cycle as a diploid individual

(c) Plant: spends significant portions of its life cycle as haploid and diploid individuals

Figure 9.4   Three types of sexual life cycles.

In sexual reproduction, haploid cells or organisms alternate with diploid cells or organisms.

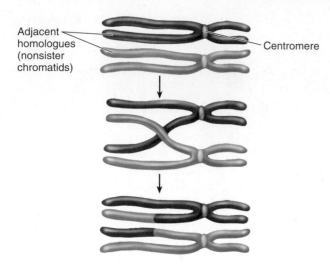

Adjacent homologues (nonsister chromatids)

Centromere

Figure 9.5   Crossing over.

In crossing over, the two homologues of a chromosome exchange portions. During the crossing over process, nonsister chromatids that are next to each other exchange chromosome arms or segments.

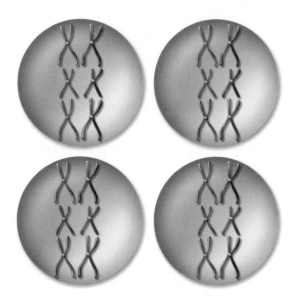

Figure 9.6   Independent assortment.

Independent assortment occurs because the orientation of chromosomes on the metaphase plate is random. Shown here are four possible orientations of chromosomes in a hypothetical cell. Each of the many possible orientations results in gametes with different combinations of parental chromosomes.

IMPLICATION   The number of different combinations of gametes can be calculated using 2^n, where "n" is the number of pairs of chromosomes. A dog has 39 pairs of chromosomes. Considering only the independent assortment of homologous chromosomes to the gametes, how many genetically different gametes are possible?

9.3 The Stages of Meiosis

CONCEPT PREVIEW: During meiosis I, homologous chromosomes move to opposite poles of the cell. At the end of meiosis II, each of the four haploid cells contains one copy of every chromosome in the set, rather than two. Because of crossing over, no two cells are the same.

Now, let's look more closely at the process of meiosis. Just as in mitosis, the chromosomes have replicated before meiosis begins, during a period called interphase. The first of the two divisions of meiosis, called **meiosis I,** serves to separate the homologous chromosomes (or homologues); the second division, **meiosis II,** serves to separate the *sister chromatids.* Thus when meiosis is complete, what started out as one diploid cell ends up as four haploid cells. Because there was one replication of DNA but *two* cell divisions, the process reduces the number of chromosomes by half.

Meiosis I

Meiosis I is traditionally divided into four stages (see the left side of *Essential Biological Process 9A* on page 148):

1. **Prophase I.**   The two versions of each chromosome (the two homologues) pair up and exchange segments.
2. **Metaphase I.**   The chromosomes align on a central plane.
3. **Anaphase I.**   One homologue with its two sister chromatids still attached moves to a pole of the cell, and the other homologue moves to the opposite pole.
4. **Telophase I.**   Individual chromosomes gather together at the two poles.

In **prophase I,** individual chromosomes first become visible, when viewed with a light microscope, as their DNA coils more and more tightly. Because the DNA replicates before the onset of meiosis, each of the threadlike chromosomes actually consists of two sister chromatids associated along their lengths and joined at their centromeres, just as in mitosis. However, now meiosis begins to differ from mitosis. During prophase I, the two homologous chromosomes line up side by side, physically touching one another, as you see in figure 9.5 and in the photo opening this chapter. It is at this point that a process called **crossing over** is initiated, in which the chromosomes actually break in the same place on both nonsister chromatids and sections of chromosomes are swapped between the homologous chromosomes, producing a hybrid chromosome that is part maternal chromosome (the green sections) and part paternal chromosome (the purple sections).

> Crossing over between homologous chromosomes can occur because, as discussed in chapter 8 on page 132, homologues carry information about the same traits at the same locations on the chromosomes, although details of the information may vary between the homologues.

In **metaphase I,** the spindle apparatus forms, but because homologues are held close together by crossovers, spindle fibers can attach to only the outward-facing portion of each centromere. For each pair of homologues, the orientation on the metaphase plate is random. Each orientation of homologues results in gametes with different combinations of parental chromosomes. This process is called **independent assortment** (figure 9.6).

In **anaphase I,** the spindle attachment is complete, and homologues are pulled apart and move toward opposite poles. At the end of anaphase I, each pole has half as many chromosomes as were present in the cell when meiosis began. Remember that the chromosomes replicated and thus contained two sister chromatids before the start of meiosis, but sister chromatids are not counted as separate chromosomes. As in mitosis, count the number of centromeres to determine the number of chromosomes.

A Closer Look

Evolutionary Consequences of Sex

Meiosis is a lot more complicated than mitosis. Why has evolution gone to so much trouble? While our knowledge of how meiosis and sex evolved is sketchy, it is abundantly clear that meiosis and sexual reproduction have an enormous impact on how species continue to evolve today, because of their ability to rapidly generate new genetic combinations. Three mechanisms each make key contributions: independent assortment, crossing over, and random fertilization.

Independent Assortment

The reassortment of genetic material that takes place during meiosis is the principal factor that has made possible the evolution of eukaryotic organisms, in all their bewildering diversity, over the past 1.5 billion years. Sexual reproduction represents an enormous advance in the ability of organisms to generate genetic variability. To understand, recall that most organisms have more than one pair of chromosomes. For example, the organism represented in the figure below has three pairs of chromosomes, each offspring receiving three homologues from each parent, purple from the father and green from the mother. The offspring in turn produces gametes, but the distribution of homologues into the gametes is completely random. A gamete could receive all homologues that are paternal in origin, as on the far left; or it could receive all maternal homologues, as on the far right, or any combination. Independent assortment alone leads to eight possible gamete combinations in this example. In humans, each gamete receives one homologue of each of the 23 chromosomes, but which homologue of a particular chromosome it receives is determined randomly. Each of the 23 pairs of chromosomes migrates independently, so there are 2^{23} (more than 8 million) different possible kinds of gametes that can be produced.

To make this point to his class, one professor offers an "A" course grade to any student who can write down all the possible combinations of heads and tails (an "either/or" choice, like that of a chromosome migrating to one pole or the other) with flipping a coin 23 times (like 23 chromosomes moving independently). No student has ever won an "A;" there are over 8 million possibilities.

Crossing Over

The DNA exchange that occurs when the arms of nonsister chromatids cross over adds even more recombination to the independent assortment of chromosomes that occurs later in meiosis. Thus, the number of possible genetic combinations that can occur among gametes is virtually unlimited.

Random Fertilization

Also, the zygote that forms a new individual is created by the fusion of two gametes, each produced independently, so fertilization squares the number of possible outcomes ($2^{23} \times 2^{23} = 70$ trillion).

Importance of Generating Diversity

Paradoxically, the evolutionary process is both revolutionary and conservative. It is revolutionary in that the pace of evolutionary change is quickened by genetic recombination, much of which results from sexual reproduction. It is conservative in that change is not always favored by selection, which may instead preserve existing combinations of genes. These conservative pressures appear to be greatest in some asexually reproducing organisms that do not move around freely and that live in especially demanding habitats. In vertebrates, on the other hand, the evolutionary premium appears to have been on versatility, and sexual reproduction is the predominant mode of reproduction.

Whatever the forces that led to sexual reproduction, its evolutionary consequences have been profound. No genetic process generates diversity more quickly; and as you will see in chapter 10, genetic diversity is the raw material of evolution, the fuel that drives it and determines its potential directions.

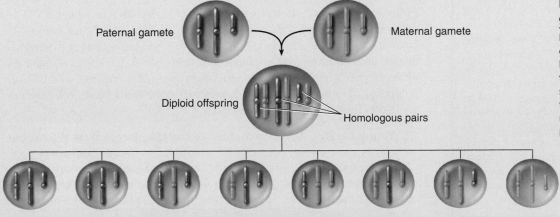

Paternal gamete — Maternal gamete

Diploid offspring

Homologous pairs

Potential gametes

Independent assortment increases genetic variability.

Independent assortment contributes new gene combinations to the next generation because the orientation of chromosomes on the metaphase plate is random. In the cell shown here with three chromosome pairs, there are eight different gametes that can result, each with different combinations of parental chromosomes.

Meiosis

Meiosis I

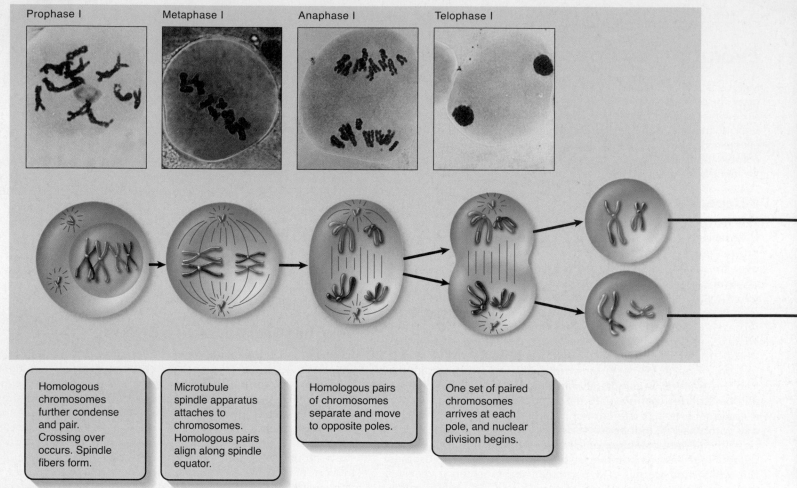

Prophase I	Metaphase I	Anaphase I	Telophase I

Homologous chromosomes further condense and pair. Crossing over occurs. Spindle fibers form.

Microtubule spindle apparatus attaches to chromosomes. Homologous pairs align along spindle equator.

Homologous pairs of chromosomes separate and move to opposite poles.

One set of paired chromosomes arrives at each pole, and nuclear division begins.

IN THE NEWS

Bloom's Syndrome. A rare inherited disorder called Bloom's syndrome has long puzzled investigators. Affected people have severe growth deficiencies, are predisposed to a wide variety of cancers, and are often sterile—a very unhappy state of affairs. What sort of gene disorder could lead to such a broad array of defects? Recent reports implicate events early in meiosis. It turns out that the gene defective in Bloom's syndrome patients encodes a protein dubbed BLM that binds to homologues early in meiotic prophase I. A properly functioning BLM protein is required to release homologues that are interlocked during crossover, in preparation for homologous chromosome separation during anaphase I. In Bloom's syndrome individuals, the BLM protein is defective and cannot remove cross-links that are created during crossover, resulting in a high incidence of gaps and breaks in homologous chromosomes as they separate.

In **telophase I,** the chromosomes gather at their respective poles to form two chromosome clusters. After an interval, meiosis II occurs, in which the sister chromatids are separated as in mitosis.

Meiosis II

Following meiosis I, the cells enter a brief interphase, in which no DNA synthesis occurs, and then the second meiotic division begins. Meiosis II (shown in *Essential Biological Process 9A* on the facing page) is simply a mitotic division involving the products of meiosis I. At the end of telophase I, each pole has a haploid complement of chromosomes, each of which is still composed of two sister chromatids attached at the centromere. Like meiosis I, meiosis II is divided into four stages:

1. **Prophase II.** At the two poles of the cell, the clusters of chromosomes enter a brief prophase II, where a new spindle forms.
2. **Metaphase II.** In metaphase II, spindle fibers bind to both sides of the centromeres and the chromosomes line up along a central plane.
3. **Anaphase II.** The spindle fibers shorten, splitting the centromeres and moving the sister chromatids to opposite poles.
4. **Telophase II.** Finally, the nuclear envelope re-forms around the four sets of daughter chromosomes.

Meiosis II

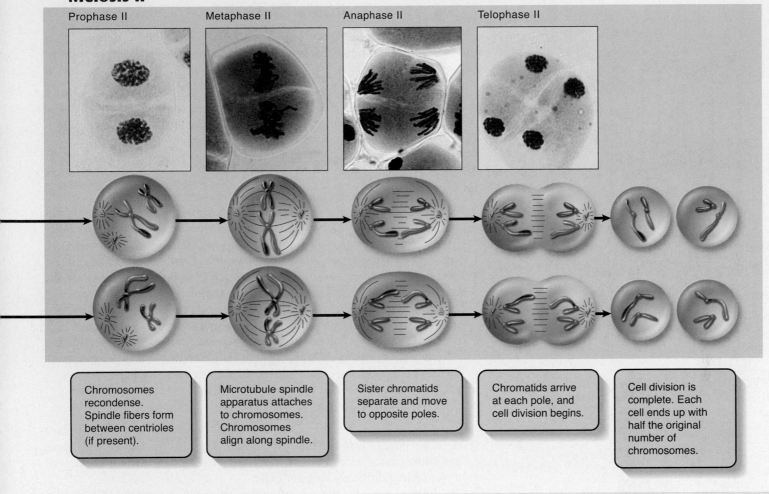

Prophase II	Metaphase II	Anaphase II	Telophase II

| Chromosomes recondense. Spindle fibers form between centrioles (if present). | Microtubule spindle apparatus attaches to chromosomes. Chromosomes align along spindle. | Sister chromatids separate and move to opposite poles. | Chromatids arrive at each pole, and cell division begins. | Cell division is complete. Each cell ends up with half the original number of chromosomes. |

The main outcome of the four stages of meiosis II is to separate the sister chromatids. The final result of this division is four cells containing haploid sets of chromosomes. No two are alike because of the crossing over in prophase I.

If you think about it, the key to meiosis is that the sister chromatids of each chromosome do not separate from each other in the first division, prevented by the crossing over that occurred in prophase I. Imagine two people dancing closely—you can tie a rope to the back of each person's belt, but you cannot tie a second rope to their belt buckles in front because the two dancers are facing each other and are very close. In just the same way, microtubules cannot attach to the inner sides of the centromeres to pull the two sister chromatids apart because crossing over holds the homologous chromosomes together like dancing partners.

> If you compare metaphase of mitosis in panel 3 of *Essential Biological Process 8B* on page 134 with metaphase of meiosis I, you will see how they differ. In meiosis I shown above, the homologous pairs align side by side so the spindle can't attach to the inside of the centromeres.

Concept Check

1. Are human gamete-producing germ line cells haploid? Explain.
2. How is the plant sexual life cycle different from the animal sexual life cycle?
3. Do sister chromatids separate in meiosis I or II?

Comparing Meiosis and Mitosis

9.4 How Meiosis Differs from Mitosis

CONCEPT PREVIEW: In meiosis, homologous chromosomes become intimately associated and do not replicate between the two nuclear divisions.

While there are differences between eukaryotes in the details of meiosis, two consistent features are seen in the meiotic processes of every eukaryote: synapsis and reduction division. Indeed, these two unique features are the key differences that distinguish meiosis from mitosis, which you studied in chapter 8.

Synapsis

The first of these two features happens early during the first nuclear division. Following chromosome replication, homologous chromosomes or homologues *pair all along their lengths,* with sister chromatids being held together by proteins called cohesin. While homologues are thus physically joined, *genetic exchange occurs at one or more points between them.* The process of forming these complexes of homologous chromosomes is called **synapsis,** and the exchange process between paired homologues, as described earlier, is crossing over. Figure 9.7*a* shows how the homologous chromosomes are held together close enough that they are able to physically exchange segments of their DNA. Sister chromatids do not separate from each other in the first nuclear division, so each homologue is still composed of two chromatids joined at the centromere, and still considered one chromosome.

Reduction Division

The second unique feature of meiosis is that *the chromosome homologues do not replicate between the two nuclear divisions,* so that chromosome assortment in the second division separates sister chromatids of each chromosome into different daughter cells.

In most respects, the second meiotic division is identical to a normal mitotic division. However, because of the crossing over that occurred during the first division, the sister chromatids in meiosis II are not identical to each other. Also, there are only half the number of chromosomes in each cell at the beginning of meiosis II because only one of the homologues is present. Figure 9.7*b* shows how reduction division occurs. The diploid cell contains four chromosome (two homologous pairs). After meiosis I, the cells contain just two chromosomes (remember to count the number of *centromeres,* because sister chromatids are not considered separate chromosomes). During meiosis II, the sister chromatid separate, but each gamete still only contains two chromosomes, half as many of the germ-line cell.

Because mitosis and meiosis use similar terminology, it is easy to confuse the two processes. Figure 9.8 compares the two processes side-by-side. Both processes start with a diploid cell, but you can see that early during meiosis I crossing over occurs, and that as a consequence homologous *pairs,* not individual centromeres, line up along the meiosis I metaphase plate. These two differences result in haploid cells in meiosis and diploid cells in mitosis.

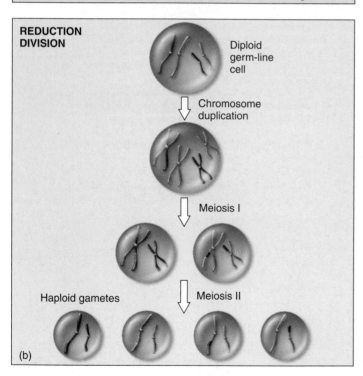

Figure 9.7   Unique features of meiosis.

(a) Synapsis draws homologous chromosomes together, all along their lengths, creating a situation (indicated by the circle) where two homologues can physically exchange portions of arms, a process called crossing over. (b) Reduction division, omitting a chromosome duplication before meiosis II, produces haploid gametes, thus ensuring that the chromosome number remains the same as that of the parents, following fertilization.

Concept Check

1. How many chromatids does each homologue contain after meiosis I?
2. How do two sister chromatids entering meiosis II differ?
3. Compare the arrangement of chromosomes at the metaphase plate of meiosis I with that of mitosis.

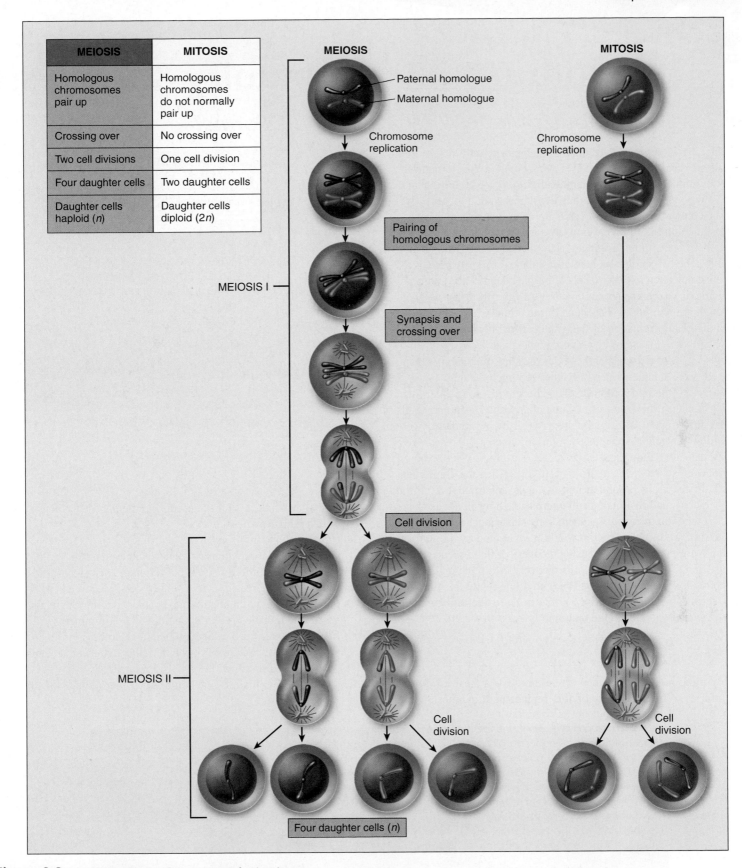

MEIOSIS	MITOSIS
Homologous chromosomes pair up	Homologous chromosomes do not normally pair up
Crossing over	No crossing over
Two cell divisions	One cell division
Four daughter cells	Two daughter cells
Daughter cells haploid (*n*)	Daughter cells diploid (2*n*)

MEIOSIS

MITOSIS

Paternal homologue
Maternal homologue

Chromosome replication

Chromosome replication

MEIOSIS I

Pairing of homologous chromosomes

Synapsis and crossing over

Cell division

MEIOSIS II

Cell division

Cell division

Four daughter cells (*n*)

Figure 9.8   A comparison of meiosis and mitosis.

Meiosis differs from mitosis in several key ways, highlighted by the orange boxes. Meiosis involves two nuclear divisions with no DNA replication between them. It thus produces four daughter cells, each with half the original number of chromosomes. Also, crossing over occurs in prophase I of meiosis. Mitosis involves a single nuclear division after DNA replication. Thus, it produces two daughter cells, each containing the original number of chromosomes, which are genetically identical to those in the parent cell.

Are New Microtubules Made When the Spindle Forms?

During interphase, before the beginning of meiosis, only a few long microtubules extend from the so-called *centrosome* (a zone around the centrioles of animal cells where microtubules are organized) to the cell periphery. Like most microtubules, they are refreshed at a low rate with resynthesis. Late in prophase, however, a dramatic change can be seen—the centrosome appears to divide into two, and a large increase is seen in the number of microtubules radiating from each of the two daughter centrosomes. The two clusters of new microtubules are easily seen as the green fibers connecting to the two sets of purple daughter chromosomes in the micrograph of early prophase below (a **micrograph** is a photo taken through a microscope). This burst of microtubule assembly marks the beginning of the formation of the spindle at the beginning of metaphase. When these clusters of new microtubules first became known to cell biologists, they asked whether these were existing microtubules being repositioned in the spindle, or newly synthesized microtubules.

The graph to the upper right displays the results of an experiment designed to answer this question. Mammalian cells in culture (cells **in culture** are growing in the laboratory on artificial medium) were injected with microtubule subunits (tubulin) to which a fluorescent dye had been attached (a **fluorescent dye** is one that glows when exposed to ultraviolet or short-wavelength visual light). After the fluorescent subunits had become incorporated into the cells' microtubules, all the fluorescence in a small region of a cell was bleached by an intense laser beam, destroying the microtubules there. Any subsequent rebuilding of microtubules in the bleached region would have to employ the fluorescent subunits present in the cell, causing recovery of fluorescence in the bleached region. The graph reports this recovery as a function of time, for interphase and metaphase cells. The dotted line represents the time for 50% recovery of fluorescence ($t_{1/2}$) (that is, $t_{1/2}$ is the time required for half of the microtubules in the region to be resynthesized).

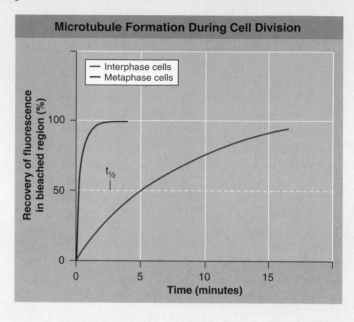

Microtubule Formation During Cell Division

Analysis

1. **Applying Concepts** Are new microtubules synthesized during interphase? What is the $t_{1/2}$ of this replacement synthesis? Are new microtubules synthesized during metaphase? What is the $t_{1/2}$ of this replacement synthesis?

2. **Interpreting Data** Is there a difference in the rate at which microtubules are synthesized during interphase and metaphase? How big is the difference? What might account for it?

3. **Making Inferences**
   a. What general statement can be made regarding the relative rates of microtubule production before and during meiosis?
   b. Is there any difference in the final amount of microtubule synthesis which would occur if this experiment were to be continued for an additional 15 minutes?

4. **Drawing Conclusions** When are the microtubules of the spindle assembled?

Reprinted with permission from *Science* Vol. 311, no. 5759 20 January 2006. Image: Khodjakov. Copyright 2006 AAAS.

10 μm

Concept Summary

Meiosis

9.1 Discovery of Meiosis

- In sexually reproducing organisms, a gamete from the male fuses with a gamete from the female to form a cell called the zygote. This process is called fertilization or syngamy. The number of chromosomes in gametes must be halved to maintain the correct number of chromosomes in offspring (**figure 9.1**). Organisms accomplish this through a cell division process called meiosis.

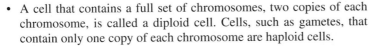

- A cell that contains a full set of chromosomes, two copies of each chromosome, is called a diploid cell. Cells, such as gametes, that contain only one copy of each chromosome are haploid cells.

- Sexual reproduction involves meiosis, but some organisms also undergo asexual reproduction, which is reproducing by mitosis or binary fission.

9.2 The Sexual Life Cycle

- Sexual life cycles alternate between diploid and haploid stages, with variation in the amount of time devoted to each stage. Three types of sexual life cycles exist: In many protists the majority of the life cycle is devoted to the haploid stage; in most animals the majority of the life cycle is devoted to the diploid stage (**figure 9.3**); and in plants the life cycle is split more equally between a haploid stage and a diploid stage.

- Germ-line cells of an organism are diploid but produce haploid gametes through meiosis.

9.3 The Stages of Meiosis

- Meiosis involves two nuclear divisions, meiosis I and meiosis II, each containing a prophase, metaphase, anaphase, and telophase. Like mitosis, the DNA replicates during interphase, before meiosis begins. Because there are two nuclear divisions but only one round of DNA replication, the four daughter cells contain half the number of chromosomes as the parent cell.

- Meiosis I is divided into four stages: prophase I, metaphase I, anaphase I, and telophase I. Prophase I is distinguished by the exchange of genetic material between homologous chromosomes, a process called crossing over. In this process, homologous chromosomes align with each other along their lengths, and sections of non-sister chromatids are physically exchanged, as shown here from **figure 9.5**. This recombines the genetic information contained in the chromosomes.

- During metaphase I, microtubules in the spindle apparatus attach to homologous chromosomes, and chromosome pairs align along the metaphase plate. The alignment of the chromosomes is random: there is shuffling in the arrangement of paternal and maternal chromosomes along the metaphase plate, leading to the independent assortment of chromosomes into the gametes (**figure 9.6**).

- The homologous chromosomes separate during anaphase I, being pulled apart by the spindle apparatus toward their respective poles. This differs from mitosis and later in meiosis II, where sister chromatids separate in anaphase.

- In telophase I, the chromosomes cluster at the poles. This leads to the next phase of meiosis, called meiosis II.

- Meiosis II mirrors mitosis in that it involves the separation of sister chromatids through the phases of prophase II, metaphase II, anaphase II, and telophase II. Meiosis II differs from mitosis in that there is no DNA replication before meiosis II. Because homologous pairs were separated during meiosis I, each daughter cell, shown forming here in telophase II from **Essential Biological Process 9A,** has only one-half the number of chromosomes. Also, the chromosomes in the daughter cells at the end of meiosis II are not genetically identical because of crossing over.

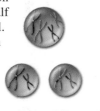

Comparing Meiosis and Mitosis

9.4 How Meiosis Differs from Mitosis

- Two processes that distinguish meiosis from mitosis are crossing over through synapsis and reduction division.

- When homologous chromosomes come together during prophase I, they associate with each other along their lengths, a process called synapsis (**figure 9.7a**). Synapsis does not occur in mitosis. During synapsis, sections of homologous chromosomes are physically exchanged in crossing over. Crossing over results in daughter cells that are not genetically identical to the parent cell or to each other. In contrast, mitosis results in daughter cells that are genetically identical to the parent cell and to each other.

- Meiosis also differs from mitosis in reduction division, where the daughter cells contain half the number of chromosomes as the parent cell. As shown here from **figure 9.7b,** reduction division occurs because meiosis contains two nuclear divisions (the first in meiosis I where the parent cell divides in two and the second in meiosis II, where the cells produced in meiosis I divide producing four daughter cells) but only one round of DNA replication during interphase.

- The primary reasons for the differences in meiosis and mitosis stem from the synapsis of homologous chromosomes in prophase I. Because of synapsis, the arms of homologous chromosomes are close enough to undergo crossing over. Also, the close association of the homologous chromosomes in synapsis blocks the inner centromeres from attaching to the spindle. As a result, sister chromatids do not separate during meiosis I, resulting in reduction division (**figure 9.8**).

Self-Test

1. An egg and a sperm unite to form a new organism. To prevent the new organism from having twice as many chromosomes as its parents,
   a. half of the chromosomes in the new organism quickly disassemble, leaving the correct number.
   b. half of the chromosomes from the egg, and half from the sperm, are ejected from the new cell.
   c. the large egg contains all the chromosomes, the tiny sperm only contributes some DNA.
   d. the egg and sperm have only half the number of chromosomes found in the parents because of meiosis.

2. The diploid number of chromosomes in humans is 46. The haploid number is
   a. 138.   c. 46.
   b. 92.   d. 23.

3. In organisms that have sexual life cycles, there is a time when there are
   a. 1*n* gametes (haploid), followed by 2*n* zygotes (diploid).
   b. 2*n* gametes (haploid), followed by 1*n* zygotes (diploid).
   c. 2*n* gametes (diploid), followed by 1*n* zygotes (haploid).
   d. 1*n* gametes (diploid), followed by 2*n* zygotes (haploid).

4. Crossing over occurs during prophase I and is when
   a. homologous chromosomes exchange sections of chromosomes.
   b. homologous chromosomes cross over to opposite sides of the cell.
   c. sister chromatids exchange genetic information.
   d. the DNA replicates forming two sister chromatids that are attached at the centromere.

5. Which of the following occurs in meiosis I?
   a. All chromosomes duplicate.
   b. Homologous chromosomes randomly orient themselves on the metaphase plate, called independent assortment.
   c. The duplicated sister chromatids separate.
   d. The original cell divides into four diploid cells.

6. Which of the following occurs in meiosis II?
   a. All chromosomes duplicate.
   b. Homologous chromosomes randomly separate, called independent assortment.
   c. The duplicated sister chromatids separate.
   d. Genetically identical daughter cells are produced.

7. During which stage of meiosis does synapsis occur?
   a. prophase I   c. metaphase II
   b. anaphase I   d. interphase

8. Synapsis is the process whereby
   a. homologous pairs of chromosomes separate and migrate toward a pole.
   b. homologous chromosomes exchange chromosomal material.
   c. homologous chromosomes become closely associated.
   d. the daughter cells contain half of the number of chromosomes of the parent cell.

9. Mitosis results in ____, while meiosis results in _____.
   a. cells that are genetically identical to the parent cell/haploid cells
   b. haploid cells/diploid cells
   c. four daughter cells/two daughter cells
   d. cells with half the number of chromosomes as the parent cell/cells which vary in chromosome number

10. Sister chromatids don't separate during meiosis I because
    a. they are held together through synapsis.
    b. of crossing over.
    c. the spindle fibers can only attach to the outward-facing side of the centromeres.
    d. All of the above.

Visual Understanding

1. **Figure 9.5** How is it that, in meiosis, you can end up with four "daughter cells" that are all genetically different from one another?

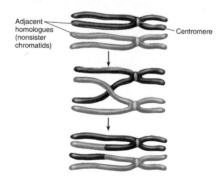

Adjacent homologues (nonsister chromatids)
Centromere

2. **Figure 9.7a** Referring to the homologous chromosomes shown here during prophase I, and knowing that they stay in synapsis during metaphase I, explain why it is that the sister chromatids don't separate as they do in mitosis.

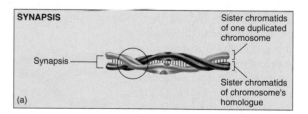

SYNAPSIS

Synapsis

Sister chromatids of one duplicated chromosome

Sister chromatids of chromosome's homologue

(a)

Challenge Questions

1. It would seem that you only need one set of instructions for your body to do all the jobs it needs to carry out. So why aren't organisms simply haploid all their lives?

2. Are the gamete cells of your body haploid or diploid? Why not the alternative?

3. An organism has 56 chromosomes in its diploid stage. Indicate how many chromosomes are present in the following, and explain your reasoning:
   a. somatic cells   c. metaphase II
   b. metaphase I   d. gametes

Chapter 10

Foundations of Genetics

Mendel

Figure 10.1   Families look alike.

These two young women are mother and daughter. It is no accident that they look so much alike, as the daughter shares half her mother's genes.

IMPLICATION   Not all members of families look as much alike as this. In your family, do your brothers and sisters resemble you a lot? See if you can learn in this chapter why all the children in your family might not look alike.

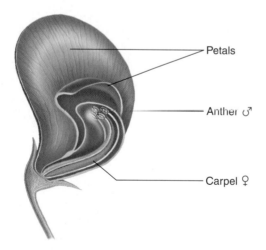

- Petals
- Anther ♂
- Carpel ♀

Figure 10.2   The garden pea flower.

Because it is easy to cultivate and because there are many distinctive varieties, the garden pea, *Pisum sativum,* was a popular choice as an experimental subject in investigations of heredity for as long as a century before Mendel's studies. The flower shown here is partially cut away to show internal structures.

10.1 Mendel and the Garden Pea

CONCEPT PREVIEW: Mendel studied heredity by crossing true-breeding garden peas that differed in easily scored alternative traits and then allowing the offspring to self-fertilize.

When you were born, many things about you resembled your mother or father (figure 10.1). This tendency for traits to be passed from parent to offspring is called **heredity.** *Traits* are the expressions of a character, or a heritable feature. How does heredity happen? Before DNA and chromosomes were discovered, this puzzle was one of the greatest mysteries of science. The key to understanding the puzzle of heredity was found in the garden of an Austrian monastery over a century ago by a monk named Gregor Mendel. Mendel used the scientific process described in chapter 1 as a powerful way of analyzing the problem. Crossing pea plants with one another, Mendel made observations that allowed him to form a simple but powerful hypothesis that accurately predicted patterns of heredity—that is, how many offspring would be like one parent and how many like the other. When Mendel's rules, introduced in chapter 1 as the theory of heredity, became widely known, investigators all over the world set out to discover the physical mechanism responsible for them. They learned that hereditary traits are instructions carefully laid out in the DNA a child receives from each parent. Mendel's solution to the puzzle of heredity was the first step on this journey of understanding and one of the greatest intellectual accomplishments in the history of science.

Early Ideas About Heredity

Mendel was not the first person to try to understand heredity by crossing pea plants (figure 10.2). Over 200 years earlier British farmers had performed similar crosses and obtained results similar to Mendel's. They observed that in crosses between two types—tall and short plants, for example—one type would disappear in one generation, only to reappear in the next. In the 1790s, for example, the British farmer T. A. Knight crossed a variety of the garden pea that had purple flowers with one that had white flowers. All the offspring of the cross had purple flowers. If two of these offspring were crossed, however, some of *their* offspring were purple and some were white. Knight noted that the purple had a "stronger tendency" to appear than white, but he did not count the numbers of each kind of offspring.

Mendel's Experiments

Gregor Mendel was born in 1822 to peasant parents and was educated in a monastery. He became a monk and was sent to the University of Vienna to study science and mathematics. Although he aspired to become a scientist and teacher, he failed his university exams for a teaching certificate and returned to the monastery, where he spent the rest of his life, eventually becoming abbot. Upon his return, Mendel joined an informal neighborhood science club, a

Gregor Mendel

group of farmers and others interested in science. Under the patronage of a local nobleman, each member set out to undertake scientific investigations, which were then discussed at meetings and published in the club's own journal. Mendel undertook to repeat the classic series of crosses with pea plants done by Knight and others, but this time he intended to count the numbers of each kind of offspring in the hope that the numbers would give some hint of what was going on. Quantitative approaches to science—measuring and counting—were just becoming fashionable in Europe.

Mendel's Experimental System: The Garden Pea

Mendel chose to study the garden pea because several of its characteristics made it easy to work with:

1. Many varieties were available. Mendel selected seven pairs of lines that differed in easily distinguished traits (including the white versus purple flowers that Knight had studied 60 years earlier).
2. Mendel knew from the work of Knight and others that he could expect the infrequent version of a character to disappear in one generation and reappear in the next. He knew, in other words, that he would have something to count.
3. Pea plants are small, easy to grow, produce large numbers of offspring, and mature quickly.
4. The reproductive organs of peas are enclosed within their flowers (see figure 10.2). Left alone, the flowers do not open. They simply fertilize themselves with their own pollen (male gametes). To carry out a cross, Mendel had only to pry the petals apart, reach in with a scissors, and snip off the male organs (anthers); he could then dust the female organs (the tip of the carpel) with pollen from another plant to make the cross.

> Flowers are the reproductive structures in a group of plants called angiosperms. Angiosperm reproduction, through the pollination and fertilization of flowers, is discussed in more detail on pages 634 through 637.

Mendel's Experimental Design

Mendel's experimental design was the same as Knight's, only Mendel counted his plants. The crosses were carried out in three steps that are presented in figure 10.3:

1. Mendel began by letting each variety self-fertilize for several generations (step 1). This ensured that each variety was **true-breeding,** meaning that it contained no other varieties of the trait, and so would produce only offspring of the same variety when it self-pollinated. The white flower variety, for example, produced only white flowers and no purple ones in each generation. Mendel called these lines the **P generation** (P for parental).
2. Mendel then conducted his experiment: He crossed two pea varieties exhibiting alternative traits, such as white versus purple flowers in step 2. The offspring that resulted he called the **F_1 generation** (F_1 for "first filial" generation, from the Latin word for "son" or "daughter").
3. Finally, Mendel allowed the plants produced in the crosses of step 2 to self-fertilize, and he counted the numbers of each kind of offspring that resulted in this **F_2** ("second filial") **generation.** As reported by Knight and shown in step 3, the white flower trait reappeared in the F_2 generation, although not as frequently as the purple flower trait.

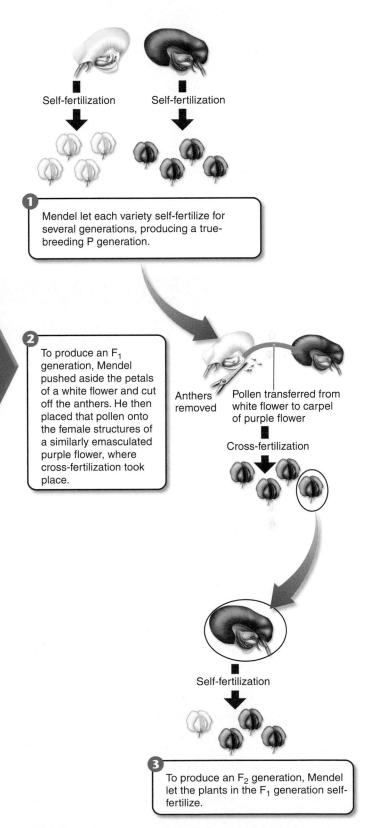

1 Mendel let each variety self-fertilize for several generations, producing a true-breeding P generation.

2 To produce an F_1 generation, Mendel pushed aside the petals of a white flower and cut off the anthers. He then placed that pollen onto the female structures of a similarly emasculated purple flower, where cross-fertilization took place.

Anthers removed

Pollen transferred from white flower to carpel of purple flower

Cross-fertilization

Self-fertilization

3 To produce an F_2 generation, Mendel let the plants in the F_1 generation self-fertilize.

Figure 10.3   How Mendel conducted his experiments.

10.2 What Mendel Observed

CONCEPT PREVIEW: When Mendel crossed two contrasting traits and counted the offspring in the subsequent generations, he observed that all of the offspring in the first generation (F_1) exhibited one (dominant) trait, and none exhibited the other (recessive) trait. In the following generation (F_2), 25% were true-breeding for the dominant trait, 50% were not-true-breeding and appeared dominant, and 25% were true-breeding for the recessive trait.

Mendel experimented with a variety of traits in the garden pea and repeatedly made similar observations. In all, Mendel examined seven pairs of contrasting traits as shown in table 10.1. For each pair of contrasting traits that Mendel crossed he observed the same result, shown in figure 10.3, where a trait disappeared in the F_1 generation only to reappear in the F_2 generation. We will examine in detail Mendel's crosses with flower color.

TABLE 10.1	Seven Characters Mendel Studied in His Experiments				
	Character			F_2 Generation	
	Dominant Form	×	Recessive Form	Dominant: Recessive	Ratio
	Purple flowers	×	White flowers	705:224	3.15:1 (3/4:1/4)
	Yellow seeds	×	Green seeds	6022:2001	3.01:1 (3/4:1/4)
	Round seeds	×	Wrinkled seeds	5474:1850	2.96:1 (3/4:1/4)
	Green pods	×	Yellow pods	428:152	2.82:1 (3/4:1/4)
	Inflated pods	×	Constricted pods	882:299	2.95:1 (3/4:1/4)
	Axial flowers	×	Terminal flowers	651:207	3.14:1 (3/4:1/4)
	Tall plants	×	Dwarf plants	787:277	2.84:1 (3/4:1/4)

The F₁ Generation

In the case of flower color, when Mendel crossed purple and white flowers, all the F₁ generation plants he observed were purple; he did not see the contrasting trait, white flowers. Mendel called the trait expressed in the F₁ plants **dominant** and the trait not expressed **recessive.** In this case, purple flower color was dominant and white flower color recessive. Mendel studied several other characters in addition to flower color, and for every pair of contrasting traits Mendel examined, one proved to be dominant and the other recessive. The dominant and recessive traits for each character he studied are indicated in table 10.1.

Figure 10.4   Round versus wrinkled seeds.

One of the differences among varieties of pea plants that Mendel studied was the shape of the seed. In some varieties the seeds were round, whereas in others they were wrinkled.

The F₂ Generation

After allowing individual F₁ plants to mature and self-fertilize, Mendel collected and planted the seeds from each plant to see what the offspring in the F₂ generation would look like. Mendel found (as Knight had earlier) that some F₂ plants exhibited white flowers, the recessive trait. The recessive trait had disappeared in the F₁ generation, only to reappear in the F₂ generation. It must somehow have been present in the F₁ individuals but unexpressed!

At this stage Mendel instituted his radical change in experimental design. He *counted* the number of each type among the F₂ offspring. He believed the proportions of the F₂ types would provide some clue about the mechanism of heredity. In the cross between the purple-flowered F₁ plants, he counted a total of 929 F₂ individuals (see table 10.1). Of these, 705 (75.9%) had purple flowers and 224 (24.1%) had white flowers. Approximately one-fourth of the F₂ individuals exhibited the recessive form of the trait. Mendel carried out similar experiments with other traits, such as round versus wrinkled seeds (figure 10.4) and obtained the same result: Three-fourths of the F₂ individuals exhibited the dominant form of the character, and one-fourth displayed the recessive form. In other words, the dominant:recessive ratio among the F₂ plants was always approximately 3:1.

A Disguised 1:2:1 Ratio

Mendel let the F₂ plants self-fertilize for another generation and found that the one-fourth that were recessive were true-breeding—future generations showed nothing but the recessive trait. Thus, the white F₂ individuals described previously showed only white flowers in the F₃ generation (as shown on the right in figure 10.5). Among the three-fourths of the plants that had shown the dominant trait in the F₂ generation, only one-third of the individuals were true-breeding in the F₃ generation (as shown on the left). The others showed both traits in the F₃ generation (as shown in the center)—and when Mendel counted their numbers, he found the ratio of dominant to recessive to again be 3:1! From these results Mendel concluded that the 3:1 ratio he had observed in the F₂ generation was in fact a disguised 1:2:1 ratio:

1 true-breeding : 2 not-true-breeding : 1 true-breeding
dominant   dominant   recessive

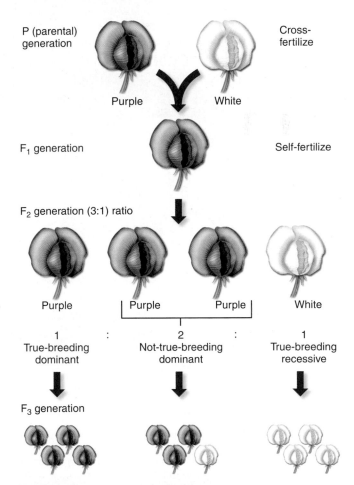

Figure 10.5   The F₂ generation is a disguised 1:2:1 ratio.

By allowing the F₂ generation to self-fertilize, Mendel found from the offspring (F₃) that the ratio of F₂ plants was one true-breeding dominant, two not-true-breeding dominant, and one true-breeding recessive.

Look-alike Meter. Have you ever wondered who you resemble more, your mother or your father? Using state-of-the-art face recognition technology, internet services now allow you to answer this question in a very objective way. You upload a photo of your face, looking straight into the camera, and similar photos of your mom and dad, and the service analyzes them to provide you with your own personal "look-alike meter." Typically your mother's photo is shown on the left end of the meter scale, your father's on the right. If you resemble both parents equally, the meter's needle points to the middle (straight up). If, on the other hand, you look more like one parent than the other, then the needle will swing over toward the side of that parent. The more you resemble her or him, the closer to that side of the meter the needle points. When the baby girl of Tom Cruise and Katie Holmes was analyzed in this way with the look-alike meter shown below, Suri proved to look more like Tom, by 15%. Why do you think it is possible for Suri to look more like her father than her mother? Didn't she get half her genes from each parent?

10.3 Mendel Proposes a Theory

CONCEPT PREVIEW: The genes that an individual has are referred to as its genotype; the outward appearance of the individual is referred to as its phenotype. The phenotype is determined by the alleles inherited from the parents. Analyses using Punnett squares determine all possible genotypes and phenotypes of a particular cross. A test cross determines the genotype of a dominant trait.

To explain his results, Mendel proposed a simple set of hypotheses that would faithfully predict the results he had observed. Now called Mendel's theory of heredity, it has become one of the most famous theories in the history of science. Mendel's theory is composed of five simple hypotheses:

Hypothesis 1: *Parents do not transmit traits directly to their offspring.* Rather, they transmit information about the traits, what Mendel called *merkmal* (the German word for "factor"). These factors act later, in the offspring, to produce the trait. In modern terminology, we call Mendel's factors **genes.**

Hypothesis 2: *Each parent contains two copies of the factor governing each trait.* The two copies may or may not be the same. If the two copies of the factor are the same (both encoding purple or both white flowers, for example) the individual is said to be **homozygous.** If the two copies of the factor are different (one encoding purple, the other white, for example), the individual is said to be **heterozygous.**

Hypothesis 3: *Alternative forms of a factor lead to alternative traits.* Alternative forms of a factor are called **alleles.** Mendel used lowercase letters to represent recessive alleles and uppercase letters to represent dominant ones. In modern terms, we call the appearance of an individual its **phenotype.** Appearance is determined by which alleles of the flower-color gene the plant receives from its parents, and we call those particular alleles the individual's **genotype.** Thus a pea plant might have the phenotype "white flower" and the genotype *pp*.

Hypothesis 4: *The two alleles that an individual possesses do not affect each other,* any more than two letters in a mailbox alter each other's contents. Each allele is passed on unchanged when the individual matures and produces its own gametes (egg and sperm). At the time, Mendel did not know that his factors were carried from parent to offspring on chromosomes.

Hypothesis 5: *The presence of an allele does not ensure that a trait will be expressed in the individual that carries it.* In heterozygous individuals, only the dominant allele achieves expression; the recessive allele is present but unexpressed.

These five hypotheses, taken together, constitute Mendel's model of the hereditary process. Many traits in humans exhibit dominant or recessive inheritance similar to the traits Mendel studied in peas (table 10.2).

TABLE 10.2	Some Dominant and Recessive Traits in Humans
Recessive Traits	**Phenotypes**
Common Baldness	M-shaped hairline receding with age
Albinism	Lack of melanin pigmentation
Alkaptonuria	Inability to metabolize homogenistic acid
Red-green color blindness	Inability to distinguish red and green wavelengths of light
Dominant Traits	**Phenotypes**
Mid-digital hair	Presence of hair on middle segment of finger
Brachydactyly	Short fingers
Phenylthiocarbamide (PTC) sensitivity	Ability to taste PTC as bitter
Camptodactyly	Inability to straighten the little finger
Polydactyly	Extra fingers and toes

Analyzing Mendel's Results

To analyze Mendel's results, it is important to remember that each trait is determined by the inheritance of alleles from the parents, one allele from the mother and the other from the father. These alleles, present on chromosomes, are distributed to gametes during meiosis. Each gamete receives one copy of each chromosome, and therefore one copy of an allele.

> During sexual reproduction, discussed on page 144, haploid gametes, that contain one allele for each trait, combine to form a diploid offspring. The genetic make up of the offspring is thus half maternal and half paternal.

Consider again Mendel's cross of purple-flowered with white-flowered plants. Like Mendel, we will assign the symbol P, written in uppercase, to the dominant allele associated with the production of purple flowers, and the symbol p, written in lowercase, to the recessive allele associated with the production of white flowers.

In this system, the genotype of an individual true-breeding for the recessive white-flowered trait would be designated pp, as both copies of the allele specify the white phenotype. Similarly, the genotype of a true-breeding purple-flowered individual would be designated PP, and a heterozygote would be designated Pp (dominant allele first). Using these conventions, and denoting a cross between two strains with ×, we can symbolize Mendel's original cross as $pp \times PP$.

Punnett Squares

The possible results from a cross between a true-breeding, white-flowered plant (pp) and a true-breeding, purple-flowered plant (PP) can be visualized with a **Punnett square.** In a Punnett square, the possible gametes of one individual are listed along the horizontal side of the square, while the possible gametes of the other individual are listed along the vertical side. The genotypes of potential offspring are represented by the cells within the square. Figure 10.6 walks you through the set-up of a Punnett square crossing two individual plants that are heterozygous for flower color ($Pp \times Pp$). The genotypes of the parents are placed along the top and side and the genotypes of potential offspring appear in the cells.

The frequency that these genotypes occur in the offspring is usually expressed by a **probability.** For example, in a cross between a homozygous white-flowered plant (pp) and a homozygous purple-flowered plant (PP), Pp is the only possible genotype for all individuals in the F_1 generation as shown by the Punnett square on the left of figure 10.7. Because P is dominant to p, all individuals in the F_1 generation have purple flowers. When individuals from the F_1 generation are crossed, as shown by the Punnett square on the right, the probability of obtaining a homozygous dominant (PP) individual in the F_2 is 25% because one-fourth of the possible genotypes are PP. Similarly, the probability of an individual in the F_2 generation being homozygous recessive (pp) is 25%. Because the heterozygous genotype has two possible ways of occurring (Pp and pP, but both written as Pp), it occurs in half of the cells within the square; the probability of obtaining a heterozygous (Pp) individual in the F_2 is 50% (25% + 25%).

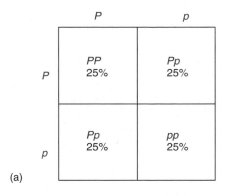

(a)

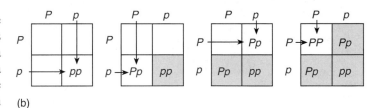

(b)

Figure 10.6   **A Punnett square analysis.**

(a) Each square represents 1/4 or 25% of the offspring from the cross. The squares in (b) show how the square is used to predict the genotypes of all potential offspring.

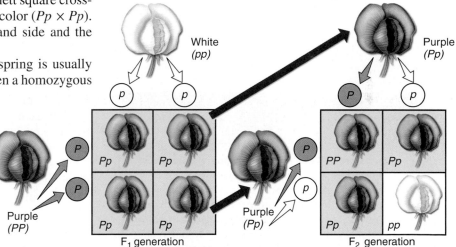

Figure 10.7   **How Mendel analyzed flower color.**

The only possible offspring of the first cross are Pp heterozygotes, purple in color. These individuals are known as the F_1 generation. When two heterozygous F_1 individuals cross, three kinds of offspring are possible: PP homozygotes (purple flowers); Pp heterozygotes (also purple flowers), which may form two ways; and pp homozygotes (white flowers). Among these individuals, known as the F_2 generation, the ratio of dominant phenotype to recessive phenotype is 3:1.

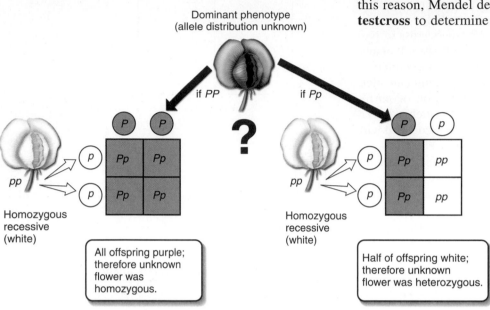

Figure 10.8   How Mendel used the testcross to detect heterozygotes.

To determine whether an individual exhibiting a dominant phenotype, such as purple flowers, was homozygous (*PP*) or heterozygous (*Pp*) for the dominant allele, Mendel devised the testcross. He crossed the individual in question with a known homozygous recessive (*pp*)—in this case, a plant with white flowers.

The Testcross

How did Mendel know which of the purple-flowered individuals in the F$_2$ generation (or the P generation) were homozygous (*PP*) and which were heterozygous (*Pp*)? It is not possible to tell simply by looking at them. For this reason, Mendel devised a simple and powerful procedure called the **testcross** to determine an individual's actual genetic composition. Consider a purple-flowered plant. It is impossible to determine such a plant's genotype simply by looking at its phenotype. To learn its genotype, you must cross it with some other plant. What kind of cross would provide the answer? If you cross it with a homozygous dominant individual, all of the progeny will show the dominant phenotype whether the test plant is homozygous or heterozygous. It is also difficult (but not impossible) to distinguish between the two possible test plant genotypes by crossing with a heterozygous individual. However, if you cross the test plant with a homozygous recessive individual, the two possible test plant genotypes will give totally different results. To see how this works, step through a testcross of a purple-flowered plant with a white-flowered plant. Figure 10.8 shows you the two possible alternatives:

Alternative 1 (on left): unknown plant is homozygous (*PP*). *PP* × *pp:* all offspring have purple flowers (*Pp*) as shown by the four purple squares.

Alternative 2 (on right): unknown plant is heterozygous (*Pp*). *Pp* × *pp:* one-half of offspring have white flowers (*pp*) and one-half have purple flowers (*Pp*) as shown by the two white and two purple squares.

To perform his testcross, Mendel crossed heterozygous F$_1$ individuals back to the parent homozygous for the recessive trait. He predicted that the dominant and recessive traits would appear in a 1:1 ratio, and that is what he observed, as you can see illustrated in alternative 2 above.

For each pair of alleles he investigated, Mendel observed phenotypic F$_2$ ratios of 3:1 (see table 10.1) and testcross ratios very close to 1:1, just as his model predicted.

Testcrosses can also be used to determine the genotype of an individual when two genes are involved. Mendel carried out many two-gene crosses, some of which we will soon discuss. He often used testcrosses to verify the genotypes of particular dominant-appearing F$_2$ individuals. Thus an F$_2$ individual showing both dominant traits (*A_ B_*) might have any of the following genotypes: *AABB*, *AaBB*, *AABb*, or *AaBb*. By crossing dominant-appearing F$_2$ individuals with homozygous recessive individuals (that is, *A_ B_* × *aabb*), Mendel was able to determine if either or both of the traits bred true among the progeny and so determine the genotype of the F$_2$ parent.

AABB	trait A breeds true	trait B breeds true
AaBB		trait B breeds true
AABb	trait A breeds true	
AaBb		

10.4 Mendel's Laws

CONCEPT PREVIEW: Mendel's theories of segregation and independent assortment are well supported and are considered "laws."

Mendel's First Law: Segregation

Mendel's model brilliantly predicts the results of his crosses, accounting in a neat and satisfying way for the ratios he observed. Similar patterns of heredity have since been observed in countless other organisms. Traits exhibiting this pattern of heredity are called *Mendelian traits*. Because of its overwhelming importance, Mendel's theory is often referred to as Mendel's first law, or the **law of segregation.** In modern terms, Mendel's first law states that *the two alleles of a trait separate during the formation of gametes, so that half of the gametes will carry one copy and half will carry the other copy.*

Mendel's Second Law: Independent Assortment

Mendel went on to ask if the inheritance of one factor, such as flower color, influences the inheritance of other factors, such as plant height. To investigate this question, he followed the inheritance of two separate traits, called a **dihybrid** cross (*di* meaning two; a *monohybrid* cross examines one trait). He first established a series of true-breeding lines of peas that differed from one another with respect to two of the seven pairs of characteristics, and then crossed them. Figure 10.9 shows an experiment in which the P generation consists of homozygous individuals with round, yellow seeds (*RRYY* in the figure) that are crossed with individuals that are homozygous for wrinkled, green seeds (*rryy*). This cross produces F$_1$ individuals that have round, yellow seeds and are heterozygous for both of these traits (*RrYy*). The chromosomes are allocated to the gametes during meiosis such that there are four types of gametes for these two traits.

Mendel then allowed the dihybrid individuals to self-fertilize. If the segregation of alleles affecting seed shape and alleles affecting seed color were independent, the probability that a particular pair of seed-shape alleles would occur together with a particular pair of seed-color alleles would simply be a product of the two individual probabilities that each pair would occur separately. For example, the probability of an individual with wrinkled, green seeds appearing in the F$_2$ generation would be equal to the probability of an individual with wrinkled seeds (1 in 4) multiplied by the probability of an individual with green seeds (1 in 4), or 1 in 16.

In his dihybrid crosses, Mendel found that the frequency of phenotypes in the F$_2$ offspring closely matched the 9:3:3:1 ratio predicted by the Punnett square analysis shown in figure 10.9. He concluded that for the pairs of traits he studied, the inheritance of one trait does not influence the inheritance of the other trait, a result often referred to as Mendel's second law, or the **law of independent assortment.** We now know that this result is only valid for genes not located near one another on the same chromosome. Thus in modern terms, Mendel's second law is often stated as follows: *genes located on different chromosomes are inherited independently of one another.*

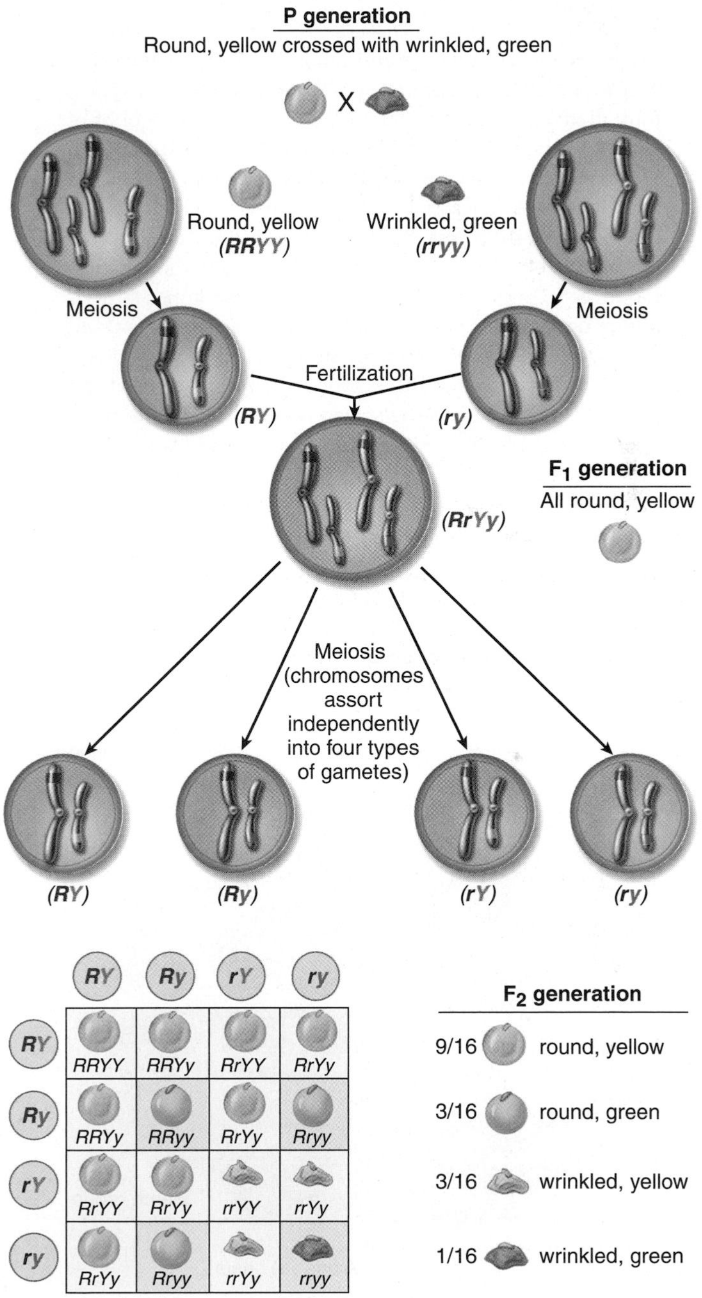

Concept Check

1. In Mendel's crosses, the offspring of the F$_1$ generation self-fertilized to produce the F$_2$ generation. Why didn't Mendel allow the parent plants to self-fertilize like this in making the F$_1$ generation?
2. Mendel's model of inheritance rests on five hypotheses. Name them.
3. How did Mendel know which F$_2$ generation peas were heterozygous?

Figure 10.9 **Analysis of a dihybrid cross.**

This dihybrid cross shows round (*R*) versus wrinkled (*r*) seeds and yellow (*Y*) versus green (*y*) seeds. The ratio of the four possible phenotypes in the F$_2$ generation is predicted to be 9:3:3:1.

From Genotype to Phenotype

Structure of hemoglobin

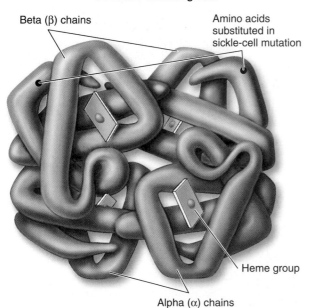

Beta (β) chains

Amino acids substituted in sickle-cell mutation

Heme group

Alpha (α) chains

10.5 How Genes Influence Traits

CONCEPT PREVIEW: Genes determine phenotypes by specifying the amino acid sequences, and thus the functional shapes, of the proteins that carry out cell activities.

It is useful, before considering Mendelian genetics further, to gain a brief understanding of how genes work. With this in mind, we will sketch, in broad strokes, a picture of how a Mendelian trait is influenced by a particular gene. We will use the protein hemoglobin as our example—you can follow along on figure 10.10.

From DNA to Protein

Each body cell of an individual contains the same set of DNA molecules, called the genome of that individual. The human genome contains about 20,000 to 25,000 genes, parcelled out into 23 pairs of chromosomes. You can see on figure 10.10 that the hemoglobin gene (*Hb*) is located on chromosome 11.

Individual genes are "read" from the chromosomal DNA by enzymes that create an RNA strand of the same sequence. After editing out unnecessary bits, this RNA transcript of the hemoglobin gene leaves the nucleus as messenger RNA (mRNA) and is delivered to ribosomes in the cytoplasm. Each ribosome is a tiny protein-assembly plant, and uses the nucleotide sequence of the mRNA to determine the amino acid sequence of a particular polypeptide. In the case of beta-hemoglobin, the mRNA encodes a strand of 146 amino acids.

How Proteins Determine the Phenotype

The beta-hemoglobin amino acid chain, that resembles beads on a string in the figure, spontaneously folds into a complex three-dimensional shape. This beta-hemoglobin associates with another beta chain and two alpha chains to form the active four-subunit hemoglobin protein (shown to the left) that binds oxygen in red blood cells. As a general rule, genes influence the phenotype by specifying the kind of proteins present in the body, which determines in large measure how that body looks and functions.

How Mutation Alters Phenotype

A change in the identity of a single nucleotide within a gene, called a mutation, can have a profound effect if the new version of the protein folds differently, as this may alter or destroy its function. How well hemoglobin performs its oxygen transport duties depends on the precise shape that the protein subunits assume when they fold. A change in the sixth amino acid of beta-hemoglobin from glutamic acid to valine causes the hemoglobin molecules to aggregate into long chains that cause the blood cells to deform into a sickle shape which can no longer carry oxygen efficiently. The resulting sickle-cell disease can be fatal, as discussed further on page 176.

Populations usually contain several versions of a gene, most of them rare. Sometimes one of the rare versions functions better under new conditions. When that happens, natural selection (see chapter 1, page 26) will favor the rare allele, which will then become more common in the population. The sickle-cell version of the beta-hemoglobin gene, rare throughout most of the world, is common in Central Africa because heterozygous individuals are resistant to malaria, a deadly disease common there.

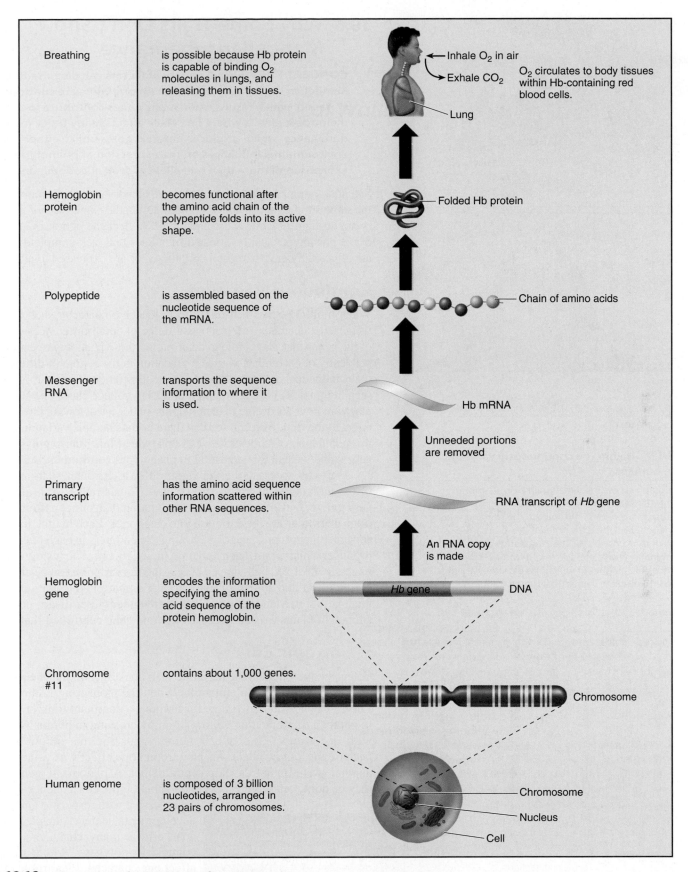

Figure 10.10 **The journey from DNA to phenotype.**

What an organism is like is determined in large measure by its genes. Here you see how one gene of the 20,000 to 25,000 in the human genome plays a key role in allowing oxygen to be carried throughout your body. The many steps on the journey from gene to trait are the subject of chapters 11 and 12.

(a)

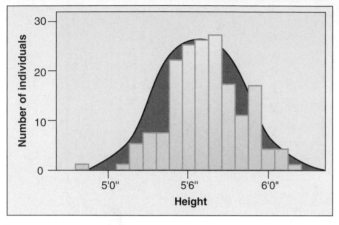

(b)

Figure 10.11   **Height is a continuously varying character in humans.**

(a) This photograph shows the variation in height among students of the 1914 class of the Connecticut Agricultural College. Because many genes contribute to height and tend to segregate independently of each other, there are many possible combinations of those genes. (b) The cumulative contribution of different combinations of alleles for height forms a continuous spectrum of possible heights, in which the extremes are much rarer than the intermediate values. This is quite different from the 3:1 ratio seen in Mendel's F₂ peas.

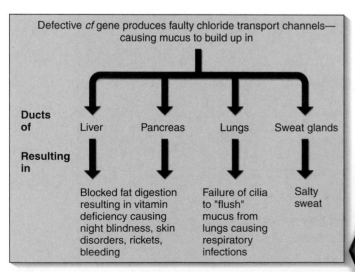

Figure 10.12   **Pleiotropic effects of the cystic fibrosis gene, *cf.***

10.6 Why Some Traits Don't Show Mendelian Inheritance

CONCEPT PREVIEW: A variety of factors can disguise the Mendelian segregation of alleles. Among them are continuous variation, which results when many genes contribute to a trait; pleiotropic effects where one allele affects many traits; incomplete dominance, which produces heterozygotes unlike either parent; environmental influences on the expression of phenotypes; and the expression of more than one allele as seen in codominance.

Scientists attempting to confirm Mendel's theory often had trouble obtaining the same simple ratios he had reported. Often the expression of the genotype is not straightforward. Most phenotypes reflect the action of many genes. Some phenotypes can be affected by alleles that lack complete dominance, are affected by environmental conditions, or are expressed together.

Continuous Variation

When multiple genes act jointly to influence a character such as height or weight, the character often shows a range of small differences. Because all of the genes that play a role in determining these phenotypes segregate independently of each other, we see a gradation in the degree of difference when many individuals are examined. A classic illustration of this sort of variation is seen in figure 10.11, a photograph of a 1914 college class. The students were placed in rows according to their heights, under 5 feet toward the left and over 6 feet to the right. You can see that there is considerable variation in height in this population of students. We call this type of inheritance **polygenic** (many genes) and we call this gradation in phenotypes **continuous variation.**

How can we describe the variation in a character such as the height of the individuals in figure 10.11a? Individuals range from quite short to very tall, with average heights more common than either extreme. What we often do is to group the variation into categories. Each height, in inches, is a separate phenotypic category. Plotting the numbers in each height category produces a histogram, such as that in figure 10.11b. The histogram approximates an idealized bell-shaped curve, and the variation can be characterized by the mean and spread of that curve. Compare this to the inheritance of plant height in Mendel's peas; they were either tall or dwarf, no intermediate height plants existed because only one gene controlled that trait.

Pleiotropic Effects

Often, an individual allele has more than one effect on the phenotype. Such an allele is said to be **pleiotropic.** When the pioneering French geneticist Lucien Cuenot studied yellow fur in mice, a dominant trait, he was unable to obtain a true-breeding yellow strain by crossing individual yellow mice with one another. Individuals homozygous for the yellow allele died, because the yellow allele was pleiotropic: one effect was yellow color, but another was a lethal developmental defect. A pleiotropic gene alteration may be dominant with respect to one phenotypic consequence (yellow fur) and recessive with respect to another (lethal developmental defect). In pleiotropy, one gene affects many characters, in marked contrast to polygeny, where many genes affect one character. Pleiotropic effects are difficult to predict, because the genes that affect a character often perform other functions we may know nothing about.

Cystic fibrosis, discussed on page 82, is caused by a mutation in a single gene that disrupts the functions of a chloride channel in the plasma membrane. This allele has pleiotropic effects because the chloride channel is found in many different types of cells in the body.

Figure 10.13   **Incomplete dominance.**

In a cross between a red-flowered Japanese four o'clock, genotype $C^R C^R$, and a white-flowered one ($C^W C^W$), neither allele is dominant. The heterozygous progeny have pink flowers and the genotype $C^R C^W$. If two of these heterozygotes are crossed, the phenotypes of their progeny occur in a ratio of 1:2:1 (red:pink:white).

IMPLICATION   Palomino is a golden coat color in horses. The color is created by the action of one allele of a gene called "cream" working on a red (chestnut) base coat. Two cream alleles create an almost white horse called "cremello." Is it possible to breed a true-breeding line of palomino horses? Explain.

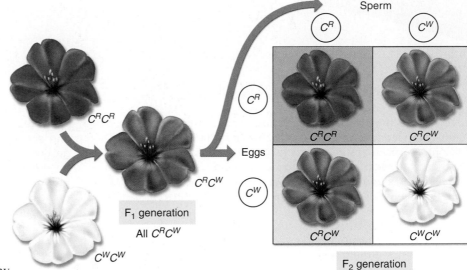

Sperm

C^R C^W

C^R

Eggs

C^W

$C^R C^R$ $C^R C^W$

$C^R C^W$ $C^W C^W$

$C^R C^R$

$C^R C^W$

F$_1$ generation

All $C^R C^W$

$C^W C^W$

F$_2$ generation

1 : 2 : 1
$C^R C^R$: $C^R C^W$: $C^W C^W$

Pleiotropic effects are characteristic of many inherited disorders, such as cystic fibrosis and sickle-cell disease, discussed later in this chapter. In these disorders, multiple symptoms can be traced back to a single gene defect. As shown in figure 10.12, cystic fibrosis patients exhibit overly sticky mucus, salty sweat, liver and pancreas failure, and a battery of other symptoms. All are pleiotropic effects of a single defect, a mutation in a gene that encodes a chloride ion transmembrane channel. In sickle-cell disease, a defect in the oxygen-carrying hemoglobin molecule causes anemia, heart failure, increased susceptibility to pneumonia, kidney failure, enlargement of the spleen, and many other symptoms.

Incomplete Dominance

Not all alternative alleles are fully dominant or fully recessive in heterozygotes. Some pairs of alleles exhibit **incomplete dominance** and produce a heterozygous phenotype that is intermediate between those of the parents. For example, the cross of red- and white-flowered Japanese four o'clocks described in figure 10.13 produced red-, pink-, and white-flowered F$_2$ plants in a 1:2:1 ratio—heterozygotes are intermediate in color. This is different than in Mendel's pea plants that didn't exhibit incomplete dominance; the heterozygotes expressed the dominant phenotype.

Environmental Effects

The degree to which many alleles are expressed depends on the environment. Some alleles are heat-sensitive, for example. Traits influenced by such alleles are more sensitive to temperature or light than are the products of other alleles. The arctic foxes in figure 10.14, for example, make fur pigment only when the weather is warm. Can you see why this trait would be an advantage for the fox? Imagine a fox that didn't possess this trait and was white all year round. It would be very visible to predators in the summer, standing out against its darker surroundings. Similarly, the *ch* allele in Himalayan rabbits and Siamese cats encodes a heat-sensitive version of tyrosinase, one of the enzymes mediating the production of melanin, a dark pigment. The ch version of the enzyme is inactivated at temperatures above about 33°C. At the surface of the main body and head, the temperature is above 33°C and the tyrosinase enzyme is inactive, while it is more active at body extremities such as the tips of the ears and tail, where the temperature is below 33°C. The dark melanin pigment this enzyme produces causes the ears, snout, feet, and tail to be black.

(a)

(b)

Figure 10.14   **Environmental effects on an allele.**

(a) An arctic fox in winter has a coat that is almost white, so it is difficult to see the fox against a snowy background. (b) In summer, the same fox's fur darkens to a reddish brown, so that it resembles the color of the surrounding tundra. Heat-sensitive alleles control this color change.

Does Environment Affect I.Q.?

Nowhere has the influence of environment on the expression of genetic traits led to more controversy than in studies of I.Q. scores. I.Q. is a controversial measure of general intelligence based on a written test that many feel to be biased toward white middle-class America. However well or poorly I.Q. scores measure intelligence, a person's I.Q. score has been believed for some time to be determined largely by his or her genes.

How did science come to that conclusion? Scientists measure the degree to which genes influence a multigene trait by using an off-putting statistical measure called the *variance*. Variance is defined as the square of the standard deviation (a measure of the degree-of-scatter of a group of numbers around their mean value), and has the very desirable property of being additive—that is, the total variance is equal to the sum of the variances of the factors influencing it.

What factors can contribute to the total variance of I.Q. scores? There are three: 1. The first factor is variation at the gene level, some gene combinations leading to higher I.Q. scores than others. 2. The second factor is variation at the environmental level, some environments leading to higher I.Q. scores than others. 3. The third factor is what a statistician calls the covariance, the degree to which environment affects genes.

The degree to which genes influence a trait like I.Q., the *heritability* of I.Q., is given the symbol H and is defined simply as the fraction of the total variance that is genetic.

So how heritable is I.Q.? Geneticists estimate the heritablity of I.Q. by measuring the environmental and genetic contributions to the total variance of I.Q. scores. The environmental contributions to variance in I.Q. can be measured by comparing the I.Q. scores of identical twins reared together with those reared apart (any differences should reflect environmental influences). The genetic contributions can be measured by comparing identical twins reared together (which are 100% genetically identical) with fraternal twins reared together (which are 50% genetically identical). Any differences should reflect genes, as twins share identical prenatal conditions in the womb and are raised in virtually identical environmental circumstances, so when traits are more commonly shared between identical twins than fraternal twins, the difference is likely genetic.

When these sorts of "twin studies" have been done in the past, researchers have uniformly reported that I.Q. is highly heritable, with values of H typically reported as being around 0.7 (a very high value). While it didn't seem significant at the time, almost all the twins available for study over the years have come from middle-class or wealthy families.

The study of I.Q. has proven controversial, because I.Q. scores are often different when social and racial groups are compared. What is one to make of the observation that I.Q. scores of poor children measure lower as a group than do scores of children of middle-class and wealthy families? This difference has led to the controversial suggestion by some that the poor are genetically inferior.

What should we make of such a harsh conclusion? To make a judgment, we need to focus for a moment on the fact that these measures of the heritability of I.Q. have all made a critical assumption, one to which population geneticists, who specialize in these sorts of things, object strongly. The assumption is that environment does not affect gene expression, so that covariance makes no contribution to the total variance in I.Q. scores—that is, that the covariance contribution to H is zero.

Studies have allowed a direct assessment of this assumption. Importantly, it proves to be flat wrong.

In November of 2003, researchers reported an analysis of twin data from a study carried out in the late 1960s. The National Collaborative Prenatal Project, funded by the National Institutes of Health, enrolled nearly 50,000 pregnant women, most of them black and quite poor, in several major U.S. cities. Researchers collected abundant data, and gave the children I.Q. tests seven years later. Although not designed to study twins, this study was so big that many twins were born, 623 births. Seven years later, 320 of these pairs were located and given I.Q. tests. This thus constitutes a huge "twin study," the first ever conducted of I.Q. among the poor.

When the data were analyzed, the results were unlike any ever reported. The heritability of I.Q. was different in different environments! Most notably, the influence of genes on I.Q. was far less in conditions of poverty, where environmental limitations seem to block the expression of genetic potential. Specifically, for families of high socioeconomic status, H = 0.72, much as reported in previous studies, but for families raised in poverty, H = 0.10, a very low value, indicating genes were making little contribution to observed I.Q. scores. The lower a child's socioeconomic status, the less impact genes had on I.Q.

These data say, with crystal clarity, that the genetic contributions to I.Q. don't mean much in an impoverished environment.

How does poverty in early childhood affect the brain? Neuroscientists reported in 2008 that many children growing up in very poor families experience poor nutrition and unhealthy levels of stress hormones, both of which impair their neural development. This affects language development and memory for the rest of their lives.

Clearly, improvements in the growing and learning environments of poor children can be expected to have a major impact on their I.Q. scores. Additionally, these data argue that the controversial differences reported in mean I.Q. scores between racial groups may well reflect no more than poverty, and are no more inevitable.

Codominance

A gene may have more than two alleles in a population, and in fact most genes possess several different alleles. Often in heterozygotes there isn't a dominant allele; instead, the effects of both alleles are expressed. In these cases, the alleles are said to be **codominant.** Codominance is seen in the color patterning of some animals, such as the "roan" pattern exhibited by some varieties of horses and cattle. A roan animal expresses both white and colored hairs because both alleles are being expressed. The gray horse in figure 10.15 is exhibiting the roan pattern. It looks like it has gray hairs, but if you were able to examine its coat closely, you would see both white hairs and black hairs.

A human gene that exhibits more than one dominant allele is the gene that determines ABO blood type. This gene encodes an enzyme that adds sugar molecules to lipids on the surface of red blood cells. These sugars act as recognition markers for cells in the immune system and are called cell surface antigens. The gene that encodes the enzyme, designated I, has three common alleles: I^B, whose product adds galactose; I^A, whose product adds galactosamine; and i, whose product does not add a sugar.

Different combinations of the three I gene alleles occur in different individuals because each person may be homozygous for any allele or heterozygous for any two. An individual heterozygous for the I^A and I^B alleles produces both forms of the enzyme and adds both galactose and galactosamine to the surfaces of red blood cells. Because both alleles are expressed simultaneously in heterozygotes, the I^A and I^B alleles are codominant. Both I^A and I^B are dominant over the i allele because both I^A or I^B alleles lead to sugar addition and the i allele does not. The different combinations of the three alleles produce four different phenotypes (figure 10.16):

1. Type A individuals add only galactosamine. They are either $I^A I^A$ homozygotes or $I^A i$ heterozygotes (the three darkest boxes).
2. Type B individuals add only galactose. They are either $I^B I^B$ homozygotes or $I^B i$ heterozygotes (the three lightest-colored boxes).
3. Type AB individuals add both sugars and are $I^A I^B$ heterozygotes (the two intermediate-colored boxes).
4. Type O individuals add neither sugar and are ii homozygotes (the one white box).

These four different cell surface phenotypes are called the **ABO blood groups.** A person's immune system can distinguish between these four phenotypes. If a type A individual receives a transfusion of type B blood, the recipient's immune system recognizes that the type B blood cells possess a "foreign" antigen (galactose) and attacks the donated blood cells, causing the cells to clump or agglutinate. This also happens if the donated blood is type AB. However, if the donated blood is type O, it contains no galactose or galactosamine antigens on the surfaces of its blood cells, and so elicits no immune response to these antigens. For this reason, the type O individual is often referred to as a "universal donor." Because neither galactose nor galactosamine is foreign to type AB individuals (whose red blood cells have both sugars), those individuals ("universal recipients") may receive any type of blood.

Figure 10.15 Codominance in color patterning.

This roan horse is heterozygous for coat color. The offspring of a cross between a white homozygote and a black homozygote, it expresses both phenotypes. Some of the hairs on its body are white and some are black.

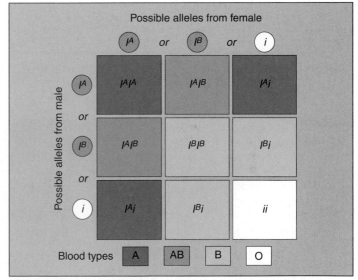

Figure 10.16 Multiple alleles controlling the ABO blood groups.

Three common alleles control the ABO blood groups. The different combinations of the three alleles result in four different blood type phenotypes: type A (either $I^A I^A$ homozygotes or $I^A i$ heterozygotes), type B (either $I^B I^B$ homozygotes or $I^B i$ heterozygotes), type AB ($I^A I^B$ heterozygotes), and type O (ii homozygotes).

Concept Check

1. Explain how a single nucleotide change in the gene for beta-hemoglobin can lead to organ damage and death.
2. Why is human height not a dominant/recessive trait?
3. Is human blood type A dominant over blood type B? Explain.

Chromosomes and Heredity

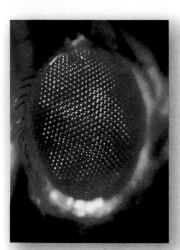

Figure 10.17 Red-eyed (wild type) and white-eyed (mutant) *Drosophila*.

The white-eye defect (shown on the *right*) is hereditary, the result of a mutation in a gene located on the X chromosome. By studying this mutation, Morgan first demonstrated that genes are on chromosomes.

10.7 Chromosomes Are the Vehicles of Mendelian Inheritance

CONCEPT PREVIEW: Mendelian traits segregate because they are determined by genes located on chromosomes that segregate into the gametes during meiosis.

The Chromosomal Theory of Inheritance

A central role for chromosomes in heredity was first suggested in 1900 by the German geneticist Karl Correns, in one of the papers announcing the rediscovery of Mendel's work. Soon observations that similar chromosomes paired with one another during meiosis led to the *chromosomal theory of inheritance*, first formulated by American Walter Sutton in 1902.

Several pieces of evidence supported Sutton's theory. One was that reproduction involves the initial union of only two cells, egg and sperm. If Mendel's model was correct, then these two gametes must make equal hereditary contributions. Sperm, however, contain little cytoplasm, suggesting that the hereditary material must reside within the nuclei of the gametes. Furthermore, while diploid individuals have two copies of each pair of homologous chromosomes, gametes have only one. This observation was consistent with Mendel's model, in which diploid individuals have two copies of each heritable gene and gametes have one. Finally, chromosomes segregate during meiosis, and each pair of homologues orients on the metaphase plate independently of every other pair. Segregation and independent assortment were two characteristics of the genes in Mendel's model.

Problems with the Chromosomal Theory

Investigators soon pointed out one problem with this theory, however. If Mendelian traits are determined by genes located on the chromosomes, and if the independent assortment of Mendelian traits reflects the independent assortment of chromosomes in meiosis, why does the number of traits that assort independently in a given kind of organism often greatly exceed the number of chromosome pairs the organism possesses? This seemed a fatal objection, and it led many early researchers to have serious reservations about Sutton's theory.

Morgan's White-Eyed Fly

The essential correctness of the chromosomal theory of heredity was demonstrated by a single small fly. In 1910 Thomas Hunt Morgan, studying the fruit fly *Drosophila melanogaster,* detected a mutant male fly that differed strikingly from normal fruit flies: Its eyes were white instead of red (figure 10.17).

Morgan immediately set out to determine if this new trait would be inherited in a Mendelian fashion. He first crossed the mutant male with a normal female to see if red or white eyes were dominant. All of the F_1 progeny had red eyes, so Morgan concluded that red eye color was dominant over white. Following the experimental procedure that Mendel had established long ago, Morgan then crossed the red-eyed flies from the F_1 generation with each other. Of the 4,252 F_2 progeny Morgan examined, 782 (18%) had white eyes. Although the ratio of red eyes to white eyes in the F_2 progeny was greater than 3:1, the results of the cross nevertheless provided clear evidence that eye color segregates. However, there was something about the outcome that was strange and totally unpredicted by Mendel's theory—*all of the white-eyed F_2 flies were males!*

How could this result be explained? Perhaps it was impossible for a white-eyed female fly to exist; such individuals might not be viable for some unknown reason. To test this idea, Morgan testcrossed the female F_1 progeny with the original white-eyed male. He obtained white-eyed and red-eyed males and females in a 1:1:1:1 ratio, just as Mendelian theory predicted. Hence, a female could have white eyes. Why, then, were there no white-eyed females among the progeny of the original cross?

Sex Linkage Confirms the Chromosomal Theory

The solution to this puzzle involved sex. In *Drosophila,* the sex of an individual is determined by the number of copies of a particular chromosome, the X chromosome, that an individual possesses. A fly with two X chromosomes is a female, and a fly with only one X chromosome is a male. In males, the single X chromosome pairs in meiosis with a large, dissimilar partner called the Y chromosome. The female thus produces only X gametes, while the male produces both X and Y gametes. When fertilization involves an X sperm, the result is an XX zygote, which develops into a female; when fertilization involves a Y sperm, the result is an XY zygote, which develops into a male.

The solution to Morgan's puzzle is that the gene causing the white-eye trait in *Drosophila* resides only on the X chromosome—it is absent from the Y chromosome. (We now know that the Y chromosome in flies carries almost no functional genes.) A trait determined by a gene on the sex chromosome is said to be **sex-linked.** Knowing the white-eye trait is recessive to the red-eye trait, we can now see that Morgan's result was a natural consequence of the Mendelian assortment of chromosomes (figure 10.18). In this experiment, the F_1 generation all had red eyes, while the F_2 generation contained flies with white eyes—but they were all males. This at-first-surprising result happens because the segregation of the white-eye trait has a one-to-one correspondence with the segregation of the X chromosome. In other words, the white-eye gene is on the X chromosome. In humans, traits such as color-blindness (see chapter 29) and hemophilia (a blood-clotting disease discussed later in this chapter) are sex-linked.

Morgan's experiment presented the first clear evidence that the genes determining Mendelian traits reside on chromosomes, just as Sutton had proposed. Now we can see that the reason Mendelian traits segregate is because chromosomes segregate. When Mendel observed the segregation of alternative traits in pea plants, he was observing a reflection of the meiotic segregation of the chromosomes, which contained the characters he was observing.

If genes are located on chromosomes, you might expect that two genes on the same chromosome would segregate together. However, if the two genes are located far from each other on the chromosome, like genes *A* and *I* in figure 10.19, the likelihood of crossing over occurring between them is very high, leading to independent assortment. Conversely, the closer two genes are to each other on a chromosome, like genes *I* and *T*, the less likely it is that a cross over event will occur between them. Genes that are located quite close to each other almost always segregate together, and so are inherited together. The tendency of close-together genes to segregate together is called **linkage.**

> Crossing over, discussed on page 146, occurs during meiosis when homologous chromosomes physically exchange segments. As a result, an allele located on one chromosome may be transferred to its homologue, causing the alleles to segregate independently.

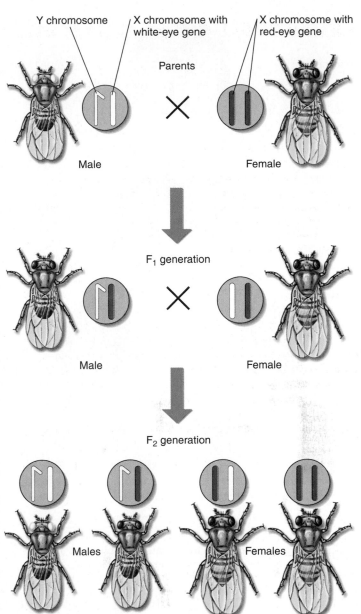

Figure 10.18 Morgan's experiment demonstrating the chromosomal basis of sex linkage.

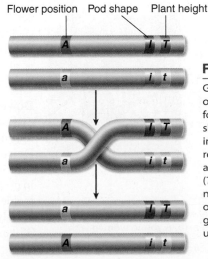

Figure 10.19 Linkage.

Genes that are located farther apart on a chromosome, like the genes for flower position (*A*) and pod shape (*I*) in Mendel's peas, will assort independently because crossing over results in recombination of these alleles. Pod shape (*I*) and plant height (*T*), however, are positioned very near each other, such that crossing over usually would not occur. These genes are said to be linked and do not undergo independent assortment.

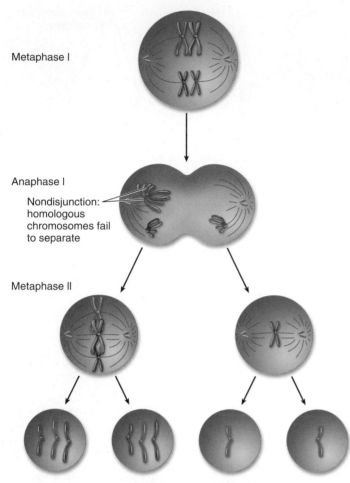

Metaphase I

Anaphase I

Nondisjunction: homologous chromosomes fail to separate

Metaphase II

Results in four gametes: two are $n+1$ and two are $n-1$

Figure 10.20 **Nondisjunction in anaphase I.**

In nondisjunction that occurs during meiosis I, one pair of homologous chromosomes fails to separate in anaphase I, and the gametes that result have one too many or one too few chromosomes. Nondisjunction can also occur in meiosis II, when sister chromatids fail to separate during anaphase II.

(a)   (b)

Figure 10.21 **Down syndrome.**

(a) In this karyotype of a male individual with Down syndrome, the trisomy at position 21 can be clearly seen. (b) A young woman with Down syndrome.

10.8 Human Chromosomes

CONCEPT PREVIEW: Autosome loss is always lethal and an extra autosome is, with few exceptions, lethal too. Additional sex chromosomes have less serious consequences, although they can lead to sterility.

Each human somatic cell normally has 46 chromosomes, which in meiosis form 23 pairs. Homologous chromosomes can be identified, according to size, shape, and appearance. Of the 23 pairs of human chromosomes, 22 are perfectly matched in both males and females and are called **autosomes.** The remaining pair, the **sex chromosomes,** consist of two similar chromosomes in females and two dissimilar chromosomes in males. In humans, females are designated XX and males are XY. The genes present on the Y chromosome determine "maleness" and therefore humans who inherit the Y chromosome develop into males.

Recent evidence suggests that a gene called *SRY* for "*Sex*-determining *Region* of the Y chromosome" may be responsible for the development of "maleness." This is discussed in more detail on page 595.

Nondisjunction

Sometimes during meiosis, homologous chromosomes or sister chromatids that paired up during metaphase remain stuck together instead of separating. The failure of chromosomes to separate correctly during either meiosis I or II is called **nondisjunction.** Nondisjunction leads to **aneuploidy,** an abnormal number of chromosomes. The nondisjunction you see in figure 10.20 occurs because the homologous pair of larger chromosomes failed to separate in anaphase I. The gametes that result from this division have unequal numbers of chromosomes. Under normal meiosis (refer back to *Essential Biological Process 9A*), all gametes would be expected to have two chromosomes, but as you can see two of these gametes have three chromosomes, while the others have just one.

Almost all humans of the same sex have the same karyotype (refer back to figure 8.3) simply because other arrangements don't work well. Humans who have lost even one copy of an autosome (called **monosomics**) do not survive development. In all but a few cases, humans who have gained an extra autosome (called **trisomics**) also do not survive. However, five of the smallest chromosomes—those numbered 13, 15, 18, 21, and 22—can be present in humans as three copies and still allow the individual to survive for a time. The presence of an extra chromosome 13, 15, or 18 causes severe developmental defects, and infants with such a genetic makeup die within a few months. In contrast, individuals who have an extra copy of chromosome 21 or, more rarely, chromosome 22, usually survive to adulthood. In such individuals, the maturation of the skeletal system is delayed, so they generally are short and have poor muscle tone. Their mental development is also affected, and children with trisomy 21 or trisomy 22 are always mentally impaired.

Down Syndrome. The developmental defect produced by trisomy 21, an extra copy of chromosome 21 seen in the karyotype in figure 10.21, was first described in 1866 by J. Langdon Down; for this reason, it is called Down syndrome. About 1 in every 750 children exhibits Down syndrome, and the frequency is similar in all racial groups. It is much more common in children of older mothers. The graph in figure 10.22 shows the increasing incidence in older mothers. In mothers under 30 years old, the incidence is only about 0.6 per 1,000 (or 1 in 1,500 births), while in mothers 30 to 35 years old, the incidence doubles to about 1.3 per 1,000 births (or 1 in

750 births). In mothers over 45, the risk is as high as 63 per 1,000 births (or 1 in 16 births). The reason that older mothers are more prone to Down syndrome babies is that all the eggs that a woman will ever produce are present in her ovaries by the time she is born, and as she gets older they may accumulate damage that can result in nondisjunction.

Nondisjunction Involving the Sex Chromosomes

As noted, 22 of the 23 pairs of human chromosomes are perfectly matched in both males and females and are called autosomes. The remaining pair are the sex chromosomes, X and Y. In humans, as in *Drosophila* (but by no means in all diploid species), females are XX and males XY; any individual with at least one Y chromosome is male. The Y chromosome is highly condensed and bears few functional genes in most organisms. Some of the active genes the Y chromosome does possess are responsible for the features associated with "maleness." Individuals who gain or lose a sex chromosome do not generally experience the severe developmental abnormalities caused by changes in autosomes. Such individuals may reach maturity, but with somewhat abnormal features.

Nondisjunction of the X Chromosome. When X chromosomes fail to separate during meiosis, some of the gametes that are produced possess both X chromosomes and so are XX gametes; the other gametes that result from such an event have no sex chromosome and are designated "O."

Figure 10.23 shows what happens if gametes from X chromosome nondisjunction combine with sperm. If an XX egg combines with an X sperm, the resulting XXX zygote develops into a female who is taller than average but other symptoms can vary greatly. Some are normal in most respects, others may have lower reading and verbal skills, and still others are mentally retarded. If an XX egg combines with a Y sperm, the XXY zygote develops into a sterile male who has many female body characteristics and, in some cases, diminished mental capacity. This condition, called *Klinefelter syndrome,* occurs in about 1 in 500 male births.

If an O egg fuses with a Y sperm, the OY zygote is nonviable and fails to develop further because humans cannot survive when they lack the genes on the X chromosome. If an O egg fuses with an X sperm, the XO zygote develops into a sterile female of short stature, with a webbed neck and immature sex organs that do not undergo changes during puberty. The mental abilities of XO individuals are normal in verbal learning but lower in nonverbal/math-based problem solving. This condition, called *Turner syndrome,* occurs roughly once in every 5,000 female births.

Nondisjunction of the Y Chromosome. The Y chromosome can also fail to separate in meiosis, leading to the formation of YY sperm. When these sperm combine with X eggs, the XYY zygotes develop into fertile males of normal appearance. The frequency of the XYY genotype is about 1 per 1,000 newborn males.

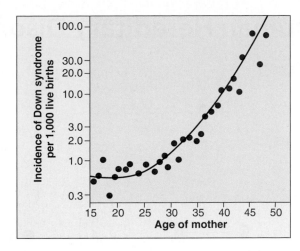

Figure 10.22 **Correlation between maternal age and the incidence of Down syndrome.**

As women age, the chances they will bear a child with Down syndrome increase. After a woman reaches age 35, the frequency of Down syndrome increases rapidly.

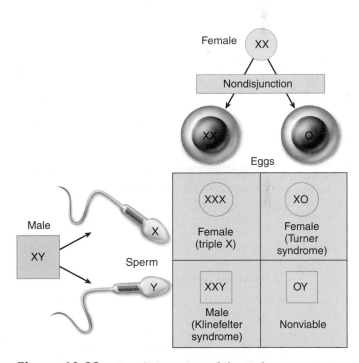

Figure 10.23 **Nondisjunction of the X chromosome.**

Nondisjunction of the X chromosome can produce sex chromosome aneuploidy—that is, abnormalities in the number of sex chromosomes.

Concept Check

1. Mendel reported that flower position and pod shape, traits located on the same chromosome, segregated independently in dihybrid crosses. Explain how this can occur.
2. What is the cause of Down syndrome?
3. Is it possible for a female to have Klinefelter syndrome? Explain.

Human Hereditary Disorders

10.9 The Role of Mutations in Human Heredity

CONCEPT PREVIEW: Many human hereditary disorders reflect the presence of rare (and sometimes not so rare) mutations within human populations.

The proteins encoded by most of your genes must function in a very precise fashion for you to develop properly and for the many complex processes of your body to function correctly. Unfortunately, genes sometimes sustain damage or are copied incorrectly. We call these accidental changes in genes **mutations.** Mutations occur only rarely, because your cells police your genes and attempt to correct any damage they encounter. Still, some mutations get through. Many of them are bad for you in one way or another. It is easy to see why. Mutations hit genes at random—imagine that you randomly changed the number of a part on the design of a jet fighter. Sometimes it won't matter critically—a seat belt becomes a radio, say. But what if a key rivet in the wing becomes a roll of toilet paper? The chance of a random mutation in a gene improving the performance of its protein is about the same as that of a randomly selected part making the jet fly faster.

Most mutations are rare in human populations. Almost all result in recessive alleles, and so they are not eliminated from the population by evolutionary forces—because they are not expressed in most individuals (heterozygotes) in which they occur. Do you see why they occur mostly in heterozygotes? Because mutant alleles are rare, it is unlikely that a person carrying a copy of the mutant allele will marry someone who also carries

> Mutations can occur in somatic cells or in germ-line cells. Both can have dramatic effects, but as discussed on page 193, only mutations in germ-line tissues are passed on to offspring and affect future generations.

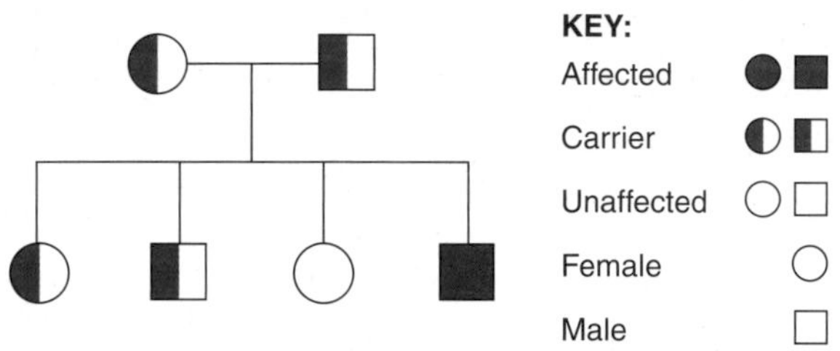

Figure 10.24   A general pedigree.

This pedigree is consistent with the inheritance of a recessive trait.

KEY:

	Female	Male
Affected	●	■
Carrier	◐	◧
Unaffected	○	□

TABLE 10.3	Some Important Genetic Disorders				
Disorder	**Symptom**	**Defect**	**Dominant/ Recessive**	**Frequency Among Human Births**
Cystic fibrosis	Mucus clogs lungs, liver, and pancreas	Failure of chloride ion transport mechanism	Recessive	1/2,500 (Caucasians)
Sickle-cell disease	Poor blood circulation	Abnormal hemoglobin molecules	Recessive	1/625 (African Americans)
Tay-Sachs disease	Deterioration of central nervous system in infancy	Defective enzyme (hexosaminidase A)	Recessive	1/3,500 (Ashkenazi Jews)
Phenylketonuria (PKU)	Brain fails to develop in infancy	Defective enzyme (phenylalanine hydroxylase)	Recessive	1/12,000
Hemophilia	Blood fails to clot	Defective blood-clotting factor VIII	Sex-linked recessive	1/10,000 (Caucasian males)
Huntington's disease	Brain tissue gradually deteriorates in middle age	Production of an inhibitor of brain cell metabolism	Dominant	1/24,000
Muscular dystrophy (Duchenne)	Muscles waste away	Degradation of myelin coating of nerves stimulating muscles	Sex-linked recessive	1/3,700 (males)
Congenital hypothyroidism	Increased birth weight, puffy face, constipation, lethargy	Failure of proper thyroid development	Recessive	1/1,000 (Hispanics) 1/700 (Native Americans)
Hypercholesterolemia	Excessive cholesterol levels in blood, leading to heart disease	Abnormal form of cholesterol cell surface receptor	Dominant	1/500

it. Instead, he or she will typically marry someone homozygous dominant, and so none of their children would be homozygous for the mutant allele. While most mutations are harmful to normal functions and are usually recessive, that is not to say all mutations are undesirable; some mutations can lead to enhanced function. Nor are all mutations recessive; rarely they can occur as dominant alleles.

In some cases, particular mutant alleles have become more common in human populations. In these cases, the harmful effects that they produce are called *genetic disorders.* Some of the most common genetic disorders are listed in table 10.3. To study human heredity, scientists look at the results of crosses that have already been made. They study family trees, or **pedigrees,** to identify which relatives exhibit a trait. Figure 10.24 shows a general pedigree. Females are indicated with circles and males with squares. The lines connect the parents and display the offspring. Solid shapes indicate an individual affected by a disorder; half-filled shapes indicate a carrier (someone who is heterozygous); open shapes indicate an individual not affected. Pedigrees can often indicate whether the gene producing the trait is sex-linked or autosomal and whether the trait's phenotype is dominant or recessive. Frequently, they can infer which individuals are homozygous and which are heterozygous for the allele specifying the trait.

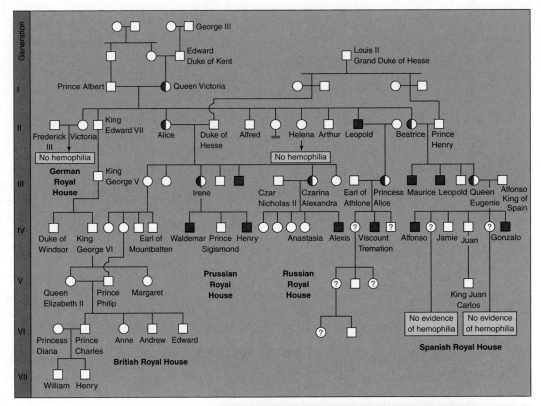

Hemophilia: A Sex-Linked Trait

Blood in a cut clots as a result of the polymerization of protein fibers circulating in the blood. A dozen proteins are involved in this process, and all must function properly for a blood clot to form. A mutation causing any of these proteins to lose their activity leads to a form of **hemophilia,** a hereditary condition in which the blood clots slowly or not at all.

Hemophilias are recessive disorders, expressed only when an individual does not possess any copy of the normal allele and so cannot produce one of the proteins necessary for clotting. Most of the genes that encode the blood-clotting proteins are on autosomes, but two (designated VIII and IX) are on the X chromosome. These two genes are sex-linked (see section 10.7). Any male who inherits a mutant allele will develop hemophilia, because his other sex chromosome is a Y chromosome that lacks any alleles of those genes.

The most famous instance of hemophilia, often called the Royal hemophilia, is a sex-linked form that arose in the royal family of England. This hemophilia was caused by a mutation in gene IX that occurred in one of the parents of Queen Victoria of England (1819–1901). The pedigree in figure 10.25 shows that in the six generations since Queen Victoria, 10 of her male descendants have had hemophilia (the solid squares). The present British royal family has escaped the disorder because Queen Victoria's son, King Edward VII, did not inherit the defective allele, and all the subsequent rulers of England are his descendants. Three of Victoria's nine children did receive the defective allele, however, and they carried it by marriage into many of the other royal families of Europe.

Figure 10.25   The Royal hemophilia pedigree.

Queen Victoria's daughter Alice introduced hemophilia into the Russian and Prussian royal houses, and her daughter Beatrice introduced it into the Spanish royal house. Victoria's son Leopold, himself a victim, also transmitted the disorder in a third line of descent. Half-shaded symbols represent carriers with one normal allele and one defective allele; fully shaded symbols represent affected individuals. Squares represent males; circles represent females. In this photo, Queen Victoria of England is surrounded by some of her descendants in 1894. Standing behind Victoria and wearing feathered boas are two of Victoria's granddaughters, Alice's daughter's: Princess Irene of Prussia *(right)*, and Alexandra *(left)*, who would soon become Czarina of Russia. Both Irene and Alexandra were also carriers of hemophilia.

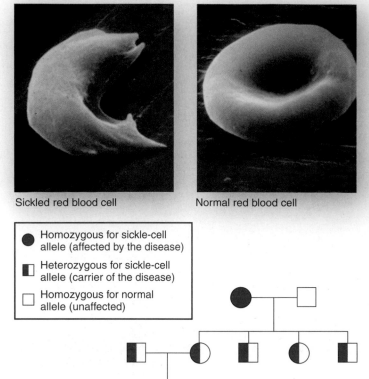

Sickled red blood cell   Normal red blood cell

● Homozygous for sickle-cell
   allele (affected by the disease)

◧ Heterozygous for sickle-cell
   allele (carrier of the disease)

□ Homozygous for normal
   allele (unaffected)

Figure 10.26   **Inheritance of sickle-cell disease.**

Sickle-cell disease is a recessive autosomal disorder. If one parent is homozygous for the recessive trait, all of the offspring will be carriers (heterozygotes) like the F$_1$ generation of Mendel's testcross. A normal red blood cell is shaped like a flattened sphere. In individuals homozygous for the sickle-cell trait, many of the red blood cells have sickle shapes.

IMPLICATION   If the sickle-cell allele (*S*) when heterozygous with the normal allele (*A*) confers resistance to malaria, then the phenotype of homozygous individuals (*AA*) is different from the phenotype of heterozygous individuals (*AS*). *AS* individuals are resistant to malaria while *AA* individuals are not. In light of this, how can the *S* allele be considered recessive to the *A* allele? Discuss.

Sickle-Cell Disease: Recessive Trait

Sickle-cell disease is a recessive hereditary disorder. Its inheritance is shown in the pedigree in figure 10.26, where affected individuals are homozygous, carrying two copies of the mutated gene. Affected individuals have defective molecules of hemoglobin, the protein within red blood cells that carries oxygen. Consequently, these individuals are unable to properly transport oxygen to their tissues.

The hemoglobin in the defective red blood cells differs from that in normal red blood cells in only one of beta-hemoglobin's 146 amino acid subunits. In the defective hemoglobin, the amino acid valine replaces a glutamic acid at a single position in each of the two beta-subunit proteins. Interestingly, the position of the change is far from the active site of hemoglobin where the iron-bearing heme group binds oxygen. Instead, the change occurs on the outer corners of the 4-subunit hemoglobin molecule. Why then is the result so catastrophic? The sickle-cell mutation puts a very nonpolar amino acid (valine) on the surface of the hemoglobin protein, creating a "sticky patch" that sticks to other such patches—nonpolar amino acids tend to associate with one another in polar environments like water. The defective hemoglobin molecules adhere to one another in chains. These hemoglobin chains form stiff, rodlike structures that result in sickle-shaped red blood cells (see figure 10.26). As a result of their stiffness and irregular shape, these cells have difficulty moving through the smallest blood vessels; they tend to accumulate in those vessels and form clots. People who have large proportions of sickle-shaped red blood cells tend to have intermittent illness and a shortened life span.

Individuals heterozygous for the sickle-cell allele are generally indistinguishable from normal persons. However, some of their red blood cells show the sickling characteristic when they are exposed to low levels of oxygen. The allele responsible for sickle-cell disease is particularly common among people of African descent because the sickle-cell allele is more common in Africa. About 9% of African Americans are heterozygous for this allele, and about 0.2% are homozygous and therefore have the disorder. In some groups of people in Africa, up to 45% of all individuals are heterozygous for this allele, and fully 6% are homozygous and express the disorder. What factors determine the high frequency of sickle-cell disease in Africa? It turns out that heterozygosity for the sickle-cell allele increases resistance to malaria, a common and serious disease in Central Africa. Comparing the two maps shown in figure 10.27, you can see that the area of the sickle-cell trait matches well with the incidence of malaria. The interactions of sickle-cell disease and malaria are discussed further in chapter 14.

Figure 10.27   **The sickle-cell allele confers resistance to malaria.**

The distribution of sickle-cell disease closely matches the occurrence of malaria in central Africa. This is not a coincidence. The sickle-cell allele, when heterozygous, confers resistance to malaria, a very serious disease.

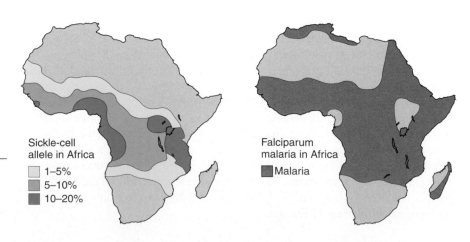

Sickle-cell
allele in Africa
  ░ 1–5%
  ▒ 5–10%
  ▓ 10–20%

Falciparum
malaria in Africa
  ■ Malaria

Tay-Sachs Disease: Recessive Trait

Tay-Sachs disease is an incurable hereditary disorder in which the brain deteriorates. Affected children appear normal at birth and usually do not develop symptoms until about the eighth month, when signs of mental deterioration appear. The children are blind within a year after birth, and they rarely live past five years of age.

The Tay-Sachs allele produces the disease by encoding a nonfunctional form of the enzyme hexosaminidase A. This enzyme breaks down *gangliosides,* a class of lipids occurring within the lysosomes of brain cells. As a result, the lysosomes fill with gangliosides, swell, and eventually burst, releasing oxidative enzymes that kill the cells. There is no known cure for this disorder.

> Lysosomes are a component of the endomembrane system, as discussed on page 73. They are membrane-bounded vesicles that contain powerful enzymes that break down cellular debris. If the lysosome enzymes leak out, they will kill the cell from within.

Tay-Sachs disease is rare in most human populations, occurring in only 1 in 300,000 births in the United States. However, the disease has a high incidence among Jews of Eastern and Central Europe (Ashkenazi) and among American Jews, 90% of whom trace their ancestry to Eastern and Central Europe. In these populations, it is estimated that 1 in 28 individuals is a heterozygous carrier of the disease, and approximately 1 in 3,500 infants has the disease. Because the disease is caused by a recessive allele, most of the people who carry the defective allele do not themselves develop symptoms of the disease because, as shown by the middle bar in figure 10.28, their one normal gene produces enough enzyme activity (50%) to keep the body functioning normally.

Huntington's Disease: Dominant Trait

Not all hereditary disorders are recessive. **Huntington's disease** is a hereditary condition caused by a dominant allele that causes the progressive deterioration of brain cells. Perhaps 1 in 24,000 individuals develops the disorder. Because the allele is dominant, every individual who carries the allele expresses the disorder. Nevertheless, the disorder persists in human populations because its symptoms usually do not develop until the affected individuals are more than 30 years old, and by that time most of those individuals have already had children. Consequently, as illustrated by the pedigree in figure 10.29, the allele is often transmitted before the lethal condition develops.

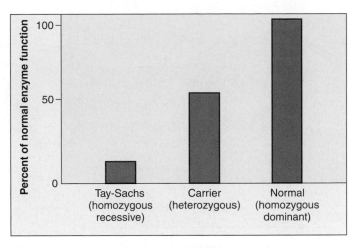

Figure 10.28   Tay-Sachs disease.

Homozygous individuals (*left bar*) typically have less than 10% of the normal level of hexosaminidase A (*right bar*), while heterozygous individuals (*middle bar*) have about 50% of the normal level—enough to prevent deterioration of the central nervous system.

IMPLICATION   Tay-Sachs disease is named for Warren Tay, an eye doctor who first described the cherry-red spot on the retina that is now the marker for the disease, and Bernard Sachs, a neurologist who first described the changes in the brain and the prevalence among Ashkenazi Jews. Can you think of another disease named for its discoverer?

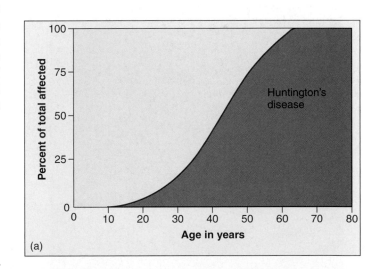

(a)

Figure 10.29   Huntington's disease is a dominant genetic disorder.

(a) Because of the late age of onset of Huntington's disease, the allele causing it persists despite being both dominant and fatal. (b) The pedigree illustrates how a dominant lethal allele can be passed from one generation to the next. Although the mother was affected, we can tell that she was heterozygous because if she were homozygous dominant, all of her children would have been affected. However, by the time she found out that she had the disease, she had probably already given birth to her children. In this way the trait passes on to the next generation even though it is fatal.

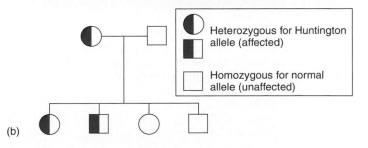

(b)

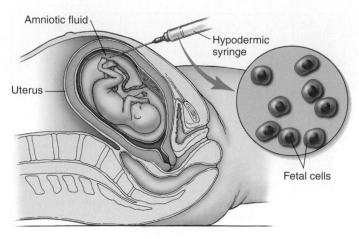

Figure 10.30 Amniocentesis.

A needle is inserted into the amniotic cavity, and a sample of amniotic fluid, containing some free cells derived from the fetus, is withdrawn into a syringe. The fetal cells are then grown in culture and their karyotype and many of their metabolic functions are examined.

IMPLICATION The risk of amniocentesis-related miscarriage due to infection, puncture leakage, or other complications is estimated to be as low as 1 in 600. In contrast, the risk of miscarriage for chorionic villus sampling is approximately 1 in 100, although the sampling may be done up to four weeks earlier. If you have a child, would you carry out genetic screening? If so, which procedure would you use? Why?

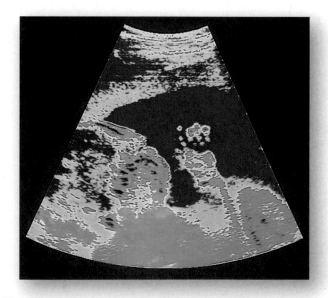

Figure 10.31 An ultrasound view of a fetus.

During the fourth month of pregnancy, when amniocentesis is normally performed, the fetus usually moves about actively. The head of the fetus (visualized in *green*) is to the *left*.

10.10 Genetic Counseling and Therapy

CONCEPT PREVIEW: It has recently become possible to detect genetic defects early in pregnancy, allowing appropriate planning by prospective parents.

The process of identifying parents at risk of producing children with genetic defects and of assessing the genetic state of early embryos is called **genetic counseling.** Genetic counseling can help prospective parents determine their risk of having a child with a genetic disorder, and advise them on medical treatments or options if a genetic disorder is determined to exist in an unborn child.

Genetic Screening

When a pregnancy is diagnosed as being high risk, many women elect to undergo **amniocentesis,** a procedure that permits the prenatal diagnosis of many genetic disorders. Figure 10.30 shows how an amniocentesis is performed. In the fourth month of pregnancy, a sterile hypodermic needle is inserted into the expanded uterus of the mother, and a small sample of the amniotic fluid bathing the fetus is removed. Within the fluid are free-floating cells derived from the fetus; once removed, these cells can be grown in cultures in the laboratory. During amniocentesis, the position of the needle and that of the fetus are usually observed by means of **ultrasound.** The ultrasound image in figure 10.31 clearly reveals the fetus's position in the uterus.

In recent years, physicians have increasingly turned to another invasive procedure for genetic screening called **chorionic villus sampling.** In this procedure, the physician removes cells from the chorion, a membranous part of the placenta that nourishes the fetus. This procedure can be used earlier in pregnancy (by the eighth week) and yields results much more rapidly than does amniocentesis, but can increase the risk of miscarriage.

Genetic counselors look at three things in the cultures of cells obtained from amniocentesis or chorionic villus sampling:

1. **Chromosomal karyotype.** Analysis of the karyotype can reveal aneuploidy (extra or missing chromosomes) and gross alterations.
2. **Enzyme activity.** In many cases, it is possible to test directly for the proper functioning of enzymes involved in genetic disorders. The lack of normal enzymatic activity signals the presence of the disorder.
3. **Genetic markers.** Genetic counselors can look for an association with known genetic markers. For sickle-cell disease, Huntington's disease, and one form of muscular dystrophy (a genetic disorder characterized by weakened muscles), investigators have found other mutations on the same chromosomes that, by chance, occur at about the same place as the mutations that cause those disorders. By testing for the presence of these other mutations, a genetic counselor can identify individuals with a high probability of possessing the disorder-causing mutations.

Concept Check

1. If most recessive mutations disrupt genes and so are bad for you, why aren't they eliminated from the human population by natural selection?
2. If Huntington's disease is caused by a dominant allele and is lethal, why doesn't the disease disappear from the human population?
3. Why not use ultrasound instead of amniocentesis for prenatal diagnosis? Isn't it safer?

Why Woolly Hair Runs in Families

The woman in the photo on the right does not cut her hair. Her hair breaks off naturally as it grows, keeping it from getting long. Other members of her family have the same sort of hair, suggesting it is a hereditary trait. Because of its curly, fuzzy texture, this trait has been given the name "woolly hair."

While the woolly hair trait is rare, it flares up in certain families. The extensive pedigree below (drawn curved so as to fit in the large families produced by the second and subsequent generations) records the incidence of woolly hair in five generations (indicated by the Roman numerals on the left) of a Norwegian family. As is the convention, affected individuals are indicated by solid symbols, with circles indicating females and squares indicating males. The pedigree will provide you with all the information you need to discover how this trait is inherited within human families.

Analysis

1. **Applying Concepts**   In the diagram below, how many individuals are documented? Are all of them related?
2. **Interpreting Data**
   a. Does the woolly hair trait appear in both sexes equally?
   b. Does every woolly hair child have a woolly hair parent?
   c. What percentage of the offspring born to a woolly haired parent are also woolly haired?
3. **Making Inferences**
   a. Is woolly hair sex-linked or autosomal?
   b. Is woolly hair dominant or recessive?
   c. Is the woolly hair trait determined by a single gene, or by several?
4. **Drawing Conclusions**
   a. How many copies of the woolly hair allele are necessary to produce a detectable change in a person's hair?
   b. Are there any woolly hair homozygous individuals in the pedigree? Explain.

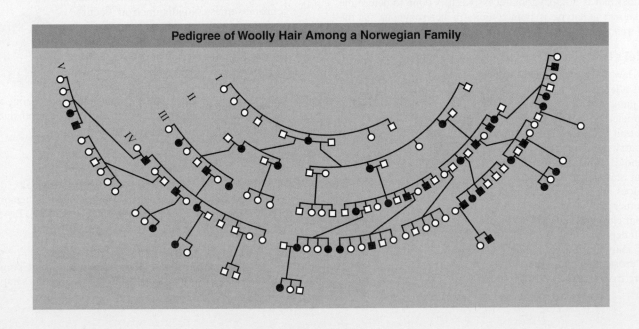

Pedigree of Woolly Hair Among a Norwegian Family

Concept Summary

Mendel

10.1   Mendel and the Garden Pea

- Mendel studied inheritance using the garden pea (**figure 10.2**) and an experimental system that included counting his results.

- Mendel used plants that were true-breeding for a particular characteristic; these plants were the P generation. He then crossed two P generation plants that expressed alternate traits (different forms of a characteristic). Their offspring were called the F_1 generation. He then allowed the F_1 plants to self-fertilize, giving rise to the F_2 generation (**figure 10.3**).

10.2   What Mendel Observed

- In Mendel's experiments, the F_1 generation plants all expressed the same alternative form, called the dominant trait. In the F_2 generation, 3/4 of the offspring expressed the dominant trait and 1/4 expressed the other form, called the recessive trait. Mendel found this 3:1 ratio in the F_2 generation in all the seven traits he studied (**table 10.1**). Mendel then found that this 3:1 ratio was actually a 1:2:1 ratio—1 true-breeding dominant: 2 not-true-breeding dominant: 1 true-breeding recessive (**figure 10.5**).

10.3   Mendel Proposes a Theory

- Mendel's theory of heredity explains that characteristics are passed from parent to offspring as alleles, one allele inherited from each parent. If both of the alleles are the same the individual is homozygous for the trait. If the individual has one dominant and one recessive allele, it is heterozygous for the trait. An individual's alleles are its genotype, and the expression of those alleles is its phenotype.

- A Punnett square can be used to predict the probabilities of inheriting certain genotypes and phenotypes in the offspring of a cross. The alleles are assigned letters: an uppercase letter for the dominant allele and a lowercase letter for the recessive allele (**figures 10.6** and **10.7**).

- A testcross is the mating of an individual of unknown genotype with an individual that is homozygous recessive. It is done to determine if the unknown genotype is homozygous or heterozygous for the dominant trait (**figure 10.8**).

10.4   Mendel's Laws

- Mendel's law of segregation states that alleles are distributed into gametes so that half of the gametes will carry one copy of a trait and the remaining gametes carry the other copy of the trait. Mendel's law of independent assortment states that the inheritance of one trait does not influence the inheritance of other traits. Genes located on different chromosomes are inherited independent of each other, as shown through a dihybrid cross, like the one shown here from **figure 10.9**.

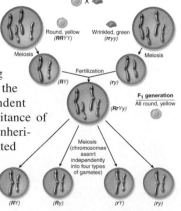

From Genotype to Phenotype

10.5   How Genes Influence Traits

- Genes coded in DNA determine phenotype because DNA encodes the amino acid sequences of proteins, and proteins are the outward expression of genes (**figure 10.10**). Alleles, which are alternative forms of a gene, result from mutations.

10.6   Why Some Traits Don't Show Mendelian Inheritance

- Not all traits follow the inheritance patterns outlined by Mendel. Continuous variation results when more than one gene contributes in a cumulative way to a phenotype, resulting in a continuous array of phenotypes. This pattern of inheritance is called polygenic (**figure 10.11**). Pleiotropic effects result when one gene influences more than one trait (**figure 10.12**). Incomplete dominance results when alternative alleles are not fully dominant or fully recessive such that heterozygous individuals express a phenotype that is inter-mediate between the dominant and recessive phenotypes. An example is the Japanese four o'clock flowers shown here from **figure 10.13**. The expression of some genes is influenced by environmental factors, such as the changing of fur color triggered by heat-sensitive alleles (**figure 10.14**). Codominance occurs when there isn't a dominant allele—two alleles are expressed resulting in pheno-typic expression of both alleles (**figures 10.15** and **10.16**).

Chromosomes and Heredity

10.7   Chromosomes Are the Vehicles of Mendelian Inheritance

- Genes segregate because they are located on chromosomes that seg-regate during meiosis. Morgan demonstrated this using an X-linked gene in fruit flies (**figure 10.18**). However, the farther apart two genes are on a chromosome, the more likely they are to assort independently because of crossing over (**figure 10.19**).

10.8   Human Chromosomes

- Humans have 23 pairs of homologous chromosomes, for a total of 46 chromosomes. They have 22 pairs of autosomes and one pair of sex chromosomes. Nondisjunction occurs when sister chromatids or homologous pairs, as shown here from **figure 10.20**, fail to separate during meiosis, result-ing in gametes with too many or too few chromosomes. Nondisjunction of autosomes is usually fatal, Down syndrome being an exception (**figures 10.21** and **10.22**), but the effects of nondisjunction of sex chromosomes are less severe (**figure 10.23**).

Human Hereditary Disorders

10.9   The Role of Mutations in Human Heredity

- Mutations can lead to genetic disorders such as hemophilia (**figure 10.25**), sickle-cell disease (**figures 10.26** and **10.27**), Tay-Sachs (**figure 10.28**), and Huntington's disease (**figure 10.29**).

10.10   Genetic Counseling and Therapy

- Some genetic disorders can be detected during pregnancy using amniocentesis (**figure 10.30**) and chorionic villus sampling.

Self-Test

1. Gregor Mendel studied the garden pea plants because
   a. pea plants are small, easy to grow, grow quickly, and produce lots of flowers and seeds.
   b. he knew about studies with the garden pea that had been done for hundreds of years, and wanted to continue them, using math— counting and recording differences.
   c. he knew that there were many varieties available with distinctive characteristics.
   d. all of these.

2. Mendel examined seven characteristics, such as flower color. He crossed plants with two different forms of a character (purple flowers and white flowers). In every case the first generation of offspring (F_1) were
   a. all purple flowers.
   b. half purple flowers and half white flowers.
   c. 3/4 purple and 1/4 white flowers.
   d. all white flowers.

3. Following question 2, when Mendel allowed the F_1 generation to self-fertilize, the offspring in the F_2 generation were
   a. all purple flowers.
   b. half purple flowers and half white flowers.
   c. 3/4 purple and 1/4 white flowers.
   d. all white flowers.

4. Mendel then studied his results, and proposed a set of hypotheses to explain them. The basis of these hypotheses is that parents transmit
   a. traits directly to their offspring, and they are expressed.
   b. some factor, or information, about traits to their offspring, and it may or may not be expressed.
   c. some factor, or information, about traits to their offspring, and it will always be expressed.
   d. some factor, or information, about traits to their offspring, and both traits expressed in every generation, perhaps in a "blended" form with information from the other parent.

5. A cross between two individuals results in a ratio of 9:3:3:1 for four possible phenotype combinations. This is an example of a
   a. dihybrid cross.   c. testcross.
   b. monohybrid cross.   d. None of these are correct.

6. Human height shows a continuous variation from the very short to the very tall. Height is most likely controlled by
   a. a single gene.   c. sex-linked genes.
   b. environmental factors.   d. multiple genes.

7. In the human ABO blood grouping, the four basic blood types are type A, type B, type AB, and type O. The enzymes that produce types A and B are
   a. simple dominant and recessive traits.
   b. incomplete dominant traits.
   c. codominant traits.
   d. sex-linked traits.

8. What finding finally determined that genes were carried on chromosomes?
   a. heat sensitivity of certain enzymes that determined coat color
   b. sex-linked eye color in fruit flies
   c. the finding of complete dominance
   d. establishing pedigrees

9. Nondisjunction
   a. occurs when homologous chromosomes or sister chromatids fail to separate during meiosis.
   b. may lead to Down syndrome.
   c. results in aneuploidy.
   d. all of the above.

10. Which of the following analyses can detect aneuploidy?
   a. enzyme activity
   b. chromosomal karyotyping
   c. pedigrees
   d. genetic markers

Visual Understanding

1. **Figure 10.9** Using the four gametes shown from the F_1 generation, how many possible crosses are there? Remember that a gamete type could cross with another of the same type (such as $RY \times RY$). Draw a Punnett square for each cross, and list the ratios of genotypes and phenotypes for the F_2 generation of that cross.

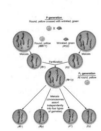

2. **Figure 10.16** Referring to this figure, use Punnett squares to illustrate whether a type A female and a type B male can have a child with type O blood.

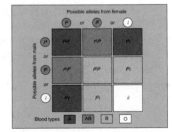

Challenge Questions

1. As Mendel struggled with understanding inheritance and formed his laws, how would the outcome have been different if he had chosen two traits that are carried on the same chromosome, with loci that are in close proximity to each other?

2. Kim and Su-Ling are doing fruit fly crosses in their biology class. Kim wants to test whether the female they have is homozygous or heterozygous for red eyes by mating with a red-eyed male. How should Su-Ling explain the difficulty to him?

3. Your biology class is collecting information on heredity. Michael realizes that he, along with three of his four brothers, are color blind, but his four sisters are not, and neither are his parents nor his grandparents. Can you help Michael understand what happened?

4. Kuzungu is a child orphaned by civil war in her country and raised in a group home. She has sickle-cell disease, and type AB blood. Two couples who believe they are her grandparents ask you, a genetic counselor, to help them determine who her real grandparents are. What do you suggest?

Additional Genetics Problems

These genetics problems will help you see the far-reaching effects of Mendel's experiments. If you need help, the answers appear at www.mhhe.com/tlwessentials3.

1. Silky feathers in chickens is a single-gene recessive trait whose effect is to produce shiny plumage.
   a. If 108 birds were raised from a cross between individuals heterozygous for this gene, how many would be expected to be silky and how many normal?
   b. If you had a normal-feathered bird, what would be the easiest cross to perform in order to determine if the bird is homozygous or heterozygous for the silky allele?

2. Among Hereford cattle there is a dominant allele called *polled;* the individuals that have this allele lack horns. Suppose you acquire a herd consisting entirely of polled cattle, and you carefully determine that no cow in the herd has horns. Some of the calves born that year, however, grow horns. You remove them from the herd and make certain that no horned adult has gotten into your pasture. Despite your efforts, more horned calves are born the next year. What is the reason for the appearance of the horned calves? If your goal is to maintain a herd consisting entirely of polled cattle, what should you do?

3. An inherited trait among humans in Norway causes affected individuals to have very curly hair, not unlike that of a sheep. The trait, called *woolly,* is very evident when it occurs in families (see *Inquiry & Analysis*); no child possesses woolly hair unless at least one parent does. Imagine you are a Norwegian judge, and you have before you a woolly haired man suing his normal-haired wife for divorce because their first child has woolly hair but their second child has normal hair. The husband claims this constitutes evidence of his wife's infidelity. Do you accept his claim? Justify your decision.

4. Brachydactyly is a rare human trait that causes a shortening of the length of the fingers by a third. A review of medical records reveals that the progeny of marriages between a brachydactyl person and a normal person are approximately half brachydactylous. What proportion of offspring in matings between two brachydactylous individuals would be expected to be brachydactylous?

5. Many animals and plants bear recessive alleles for *albinism,* a condition in which homozygous individuals lack certain pigments. An albino plant, for example, lacks chlorophyll and is white, and an albino human lacks melanin. If two normally pigmented persons heterozygous for the same albinism allele marry, what proportion of their children would you expect to be albino?

6. You inherit a racehorse and decide to put him out to stud. In looking over the stud book, however, you discover that the horse's grandfather exhibited a rare disorder that causes brittle bones. The disorder is hereditary and results from homozygosity for a recessive allele. If your horse is heterozygous for the allele, it will not be possible to use him for stud because the genetic defect may be passed on. How would you determine whether your horse carries this allele?

7. Your instructor presents you with a *Drosophila* (fruit fly) with red eyes, as well as a stock of white-eyed flies and another stock of flies homozygous for the red-eye allele. You know that the presence of white eyes in *Drosophila* is caused by homozygosity for a recessive allele. How would you determine whether the single red-eyed fly was heterozygous for the white-eye allele?

8. Hemophilia is a recessive sex-linked human blood disease that leads to failure of blood to clot normally. One form of hemophilia has been traced to the royal family of England, from which it spread throughout the royal families of Europe. For the purposes of this problem, assume that it originated as a mutation either in Prince Albert or in his wife, Queen Victoria.
   a. Prince Albert did not have hemophilia. If the disease is a sex-linked recessive abnormality, how could it have originated in Prince Albert, a male, who would have been expected to exhibit sex-linked recessive traits?
   b. Alexis, the son of Czar Nicholas II of Russia and Empress Alexandra (a granddaughter of Victoria), had hemophilia, but their daughter Anastasia did not. Anastasia died, a victim of the Russian revolution, before she had any children. Can we assume that Anastasia would have been a carrier of the disease? Would your answer be different if the disease had been present in Nicholas II or in Alexandra?

9. A normally pigmented man marries an albino woman. They have three children, one of whom is an albino. What is the genotype of the father?

10. A man works in an atomic energy plant, and he is exposed daily to low-level background radiation. After several years, he has a child who has Duchenne muscular dystrophy, a recessive genetic defect caused by a mutation on the X chromosome. Neither the parents nor the grandparents have the disease. The man sues the plant, claiming that the abnormality in their child is the direct result of radiation-induced mutation of his gametes, and that the company should have protected him from this radiation. Before reaching a decision, the judge hearing the case insists on knowing the sex of the child. Which sex would be more likely to result in an award of damages, and why?

Chapter **11**

DNA:
The Genetic Material

CHAPTER AT A GLANCE

Genes Are Made of DNA

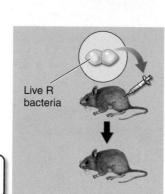

1 S bacteria have a polysaccharide capsule and are pathogenic. When they are injected into mice, the mice die.

2 R bacteria do not have the capsule and do not kill mice.

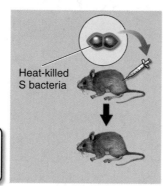

3 Heat-killed bacteria are dead but still have the capsule. They do not kill mice.

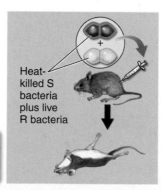

4 A mixture of live R bacteria and heat-killed S bacteria does cause mice to die.

Figure 11.1   How Griffith discovered transformation.

Transformation, the movement of a gene from one organism to another, provided some of the key evidence that DNA is the genetic material. Griffith found that extracts of dead pathogenic strains of the bacterium *Streptococcus pneumoniae* can "transform" live harmless strains into live pathogenic strains.

11.1 The Griffith Experiment

CONCEPT PREVIEW: Hereditary information can pass from dead cells to living ones and transform them.

As we learned in chapters 8, 9, and 10, chromosomes contain genes, which, in turn, contain hereditary information. However, Mendel's work left a key question unanswered: What *is* a gene?

When biologists began to examine chromosomes in their search for genes, they soon learned that chromosomes are made of two kinds of macromolecules, both of which you encountered in chapter 3: **proteins** (long chains of *amino acid subunits* linked together in a string) and **DNA** (deoxyribonucleic acid—long chains of *nucleotide* subunits linked together in a string). It was possible to imagine that either of the two was the stuff that genes are made of—information might be stored in a sequence of different amino acids, or in a sequence of different nucleotides. But which one is the stuff of genes, protein or DNA? This question was answered clearly in a variety of different experiments, all of which shared the same basic design: If you separate the DNA in an individual's chromosomes from the protein, which of the two materials is able to change another individual's genes?

In a key experiment in 1928, British microbiologist Frederick Griffith made a series of unexpected observations while experimenting with pathogenic (disease-causing) bacteria. Figure 11.1 takes you stepwise through his discoveries. When he infected mice with a virulent strain of *Streptococcus pneumoniae* bacteria (then known as *Pneumococcus*), the mice died of blood poisoning **1**. However, when he infected similar mice with a mutant strain of *S. pneumoniae* that lacked the virulent strain's polysaccharide capsule, the mice showed no ill effects **2**. The capsule was apparently necessary for infection. The normal pathogenic form of this bacterium is referred to as the S form because it forms smooth colonies in a culture dish. The mutant form, which lacks an enzyme needed to manufacture the polysaccharide capsule, is called the R form because it forms rough colonies.

To determine whether the polysaccharide capsule itself had a toxic effect, Griffith injected dead bacteria of the virulent S strain into mice, and the mice remained perfectly healthy **3**. Finally, he injected mice with a mixture containing dead S bacteria (the virulent strain) and live, capsuleless R bacteria, each of which by itself did not harm the mice. Unexpectedly, the mice developed disease symptoms and many of them died **4**. The blood of the dead mice was found to contain high levels of live, virulent *Streptococcus* type S bacteria, which had surface proteins characteristic of the live (previously R) strain. Somehow, the information specifying the polysaccharide capsule had passed from the dead, virulent S bacteria to the live, capsuleless R bacteria in the mixture, permanently transforming the capsuleless R bacteria into the virulent S variety. Griffith called this process, **transformation.**

11.2 The Avery and Hershey-Chase Experiments

CONCEPT PREVIEW: Several key experiments demonstrated conclusively that DNA, not protein, is the hereditary material.

The Avery Experiments

The agent responsible for transforming *Streptococcus* went undiscovered until 1944. In a classic series of experiments, Oswald Avery and his coworkers Colin MacLeod and Maclyn McCarty characterized what they referred to as the "transforming principle." Avery and his colleagues prepared the same mixture of dead S *Streptococcus* and live R *Streptococcus* that Griffith had used, but first they removed as much of the protein as they could from their preparation of dead S *Streptococcus*, eventually achieving 99.98% purity. Despite the removal of nearly all protein from the dead S *Streptococcus*, the transforming activity was not reduced. Moreover, the chemical properties of the transforming principle resembled those of DNA, and protein-digesting enzymes did not affect the principle's activity, while a DNA-digesting enzyme destroyed all transforming activity. They concluded that "a nucleic acid of the deoxyribose type is the fundamental unit of the transforming principle of *Pneumococcus* Type III"—in essence, that DNA is the hereditary material.

The Hershey-Chase Experiment

Avery's result was not widely appreciated at first, because most biologists still preferred to think that genes were made of proteins. In 1952, however, a simple experiment carried out by Alfred Hershey and Martha Chase was impossible to ignore. The team studied the genes of viruses that infect bacteria. These viruses attach themselves to the surface of bacterial cells and inject their genes into the interior; once inside, the genes take over the genetic machinery of the cell and conduct the manufacturing of hundreds of new viruses. When mature, the progeny viruses burst out to infect other cells. These bacteria-infecting viruses have a very simple structure: a core of DNA surrounded by a coat of protein.

In this experiment shown in figure 11.2, Hershey and Chase used radioactive isotopes to "label" the DNA and protein of the viruses. Radioactively tagged molecules are indicated in red in the figure. In the preparation on the right, the viruses were grown so that their DNA contained radioactive phosphorus (^{32}P), which is present in DNA but not in proteins; in another preparation on the left side of the figure, the viruses were grown so that their protein coats contained radioactive sulfur (^{35}S), present in proteins but not in DNA. After the labeled viruses were allowed to infect bacteria, Hershey and Chase shook the suspensions forcefully to dislodge attacking viruses from the surface of bacteria. They then used a rapidly spinning centrifuge to isolate the bacteria, and asked a very simple question: What did the viruses inject into the bacterial cells, protein or DNA? They found that the bacterial cells infected by viruses containing the ^{32}P label had labeled tracer in their interiors; cells infected by viruses containing the ^{35}S labeled tracer did not. The conclusion was clear: The genes that viruses use to specify new viruses are made of DNA and not protein.

> Isotopes, as discussed on page 34, are atoms of an element that contain varying numbers of neutrons. Some isotopes are unstable, undergoing radioactive decay, which can be detected when the isotopes are incorporated into DNA or proteins.

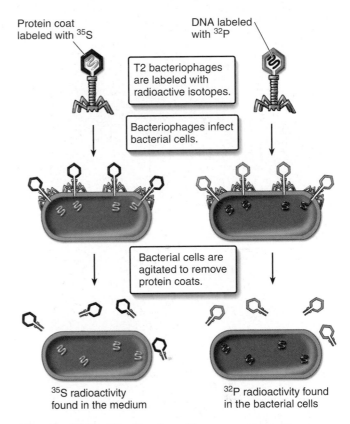

Protein coat labeled with ^{35}S

DNA labeled with ^{32}P

T2 bacteriophages are labeled with radioactive isotopes.

Bacteriophages infect bacterial cells.

Bacterial cells are agitated to remove protein coats.

^{35}S radioactivity found in the medium

^{32}P radioactivity found in the bacterial cells

Figure 11.2   The Hershey-Chase experiment.

The experiment that convinced most biologists that DNA is the genetic material was carried out soon after World War II, when radioactive isotopes were first becoming commonly available to researchers. Hershey and Chase used different radioactive labels to "tag" and track protein and DNA. They found that when bacterial viruses inserted their genes into bacteria to guide the production of new viruses, it was DNA and not protein that was inserted. More specifically, ^{35}S radioactivity did not enter infected bacterial cells and ^{32}P radioactivity did. Clearly the virus DNA, not the virus protein, was responsible for directing the production of new viruses.

IMPLICATION   Household smoke detectors contain a very small amount (3 millionths of a gram) of the radioactive isotope Americium-241, a decay product of plutonium. Am-241 emits a stream of alpha particles which ionize oxygen and nitrogen atoms in the air, creating a small steady electric current. When smoke enters the smoke detector, smoke particles absorb the alpha radiation; this halts ion production, causing the current to fall and setting off the alarm. Why do you think Am-241 is used, rather than some other, possibly less expensive radioisotope?

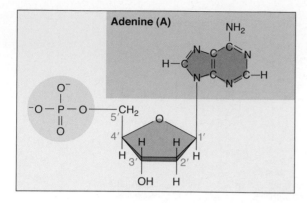

Adenine (A)

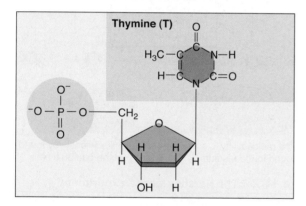

Guanine (G)

Thymine (T)

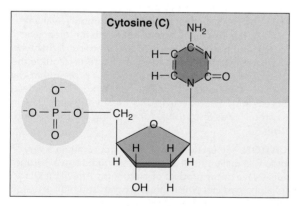

Cytosine (C)

Figure 11.3   The four nucleotide subunits that make up DNA.

The nucleotide subunits of DNA are composed of three parts: a central five-carbon sugar called deoxyribose, a phosphate group, and an organic, nitrogen-containing base.

11.3 Discovering the Structure of DNA

CONCEPT PREVIEW: The DNA molecule has two strands of nucleotides held together by hydrogen bonds between bases. The two strands wind into a double helix.

As it became clear that DNA stored the hereditary information, researchers began to question how a molecule like DNA could carry out the complex function of inheritance. Scientists at that time did not know what the DNA molecule looked like.

We now know that DNA is a long, chainlike molecule made up of subunits called **nucleotides.** As you can see in figure 11.3, each nucleotide has three parts: a central sugar called deoxyribose to which a phosphate (PO_4) group and an organic base are attached. The sugar (lavender pentagon structure) and the phosphate group (yellow-circled structure) are the same in every nucleotide of DNA. However, there are four different kinds of bases: two large ones with double-ring structures, and two small ones with single rings. The large bases, called **purines,** are **A** (adenine) and **G** (guanine). The small bases, called **pyrimidines,** are **C** (cytosine) and **T** (thymine). The carbon atoms that make up the sugar of the deoxyribose are numbered from 1′ to 5′, as shown in the top panel. The phosphate group binds to the 5′ carbon and the organic base binds to the 1′ carbon. You will learn about the significance of this numbering system later.

Early in the analysis of DNA, a key observation was made by Erwin Chargaff. He noted that DNA molecules always had equal amounts of purines and pyrimidines. In fact, with slight variations due to imprecision of measurement, the amount of A always equals the amount of T, and the amount of G always equals the amount of C. This observation (A = T, G = C), known as **Chargaff's rule,** suggested that DNA had a regular structure.

In 1953 the British chemist Rosalind Franklin carried out the first X-ray diffraction experiments on DNA. In her experiments, DNA molecules bombarded with X-ray beams created a pattern on photographic film, as shown in figure 11.4a, that looked like the ripples created by tossing a rock into a smooth lake. Franklin's results suggested that the DNA molecule had the shape of a coiled spring or a corkscrew, a form called a **helix.**

Franklin's work was shared with two researchers at Cambridge University, Francis Crick and James Watson, before it was published. Using Tinkertoy-like models of the bases, Watson and Crick deduced the structure of DNA (figure 11.4b): The DNA molecule is a **double helix,** a structure that resembles a winding staircase (figure 11.4c). The sugar and phosphate groups form the stringers of the staircase, and the bases of the nucleotides form the steps. The significance of Chargaff's rule is now clear, a direct reflection of this structure—every bulky purine on one strand is paired with a slender pyrimidine on the other strand. Specifically, A (blue bases) pairs with T (orange bases), and G (purple bases) pairs with C (pink bases).

> The constant distance between the strands in DNA could also be explained by A pairing with C. The reason why A doesn't pair with C, or G with T, is because of how hydrogen bonds form between the nitrogen bases, described on page 53.

Concept Check

1. If dead S bacteria and live R bacteria are each harmless when injected into mice, why is injecting them both together lethal to the mouse?
2. How could Hershey and Chase conclude that DNA was the hereditary material based on their results?
3. What subunits make up the DNA nucleotides and which nucleotides contain purine bases and which contain pyrimidine bases?

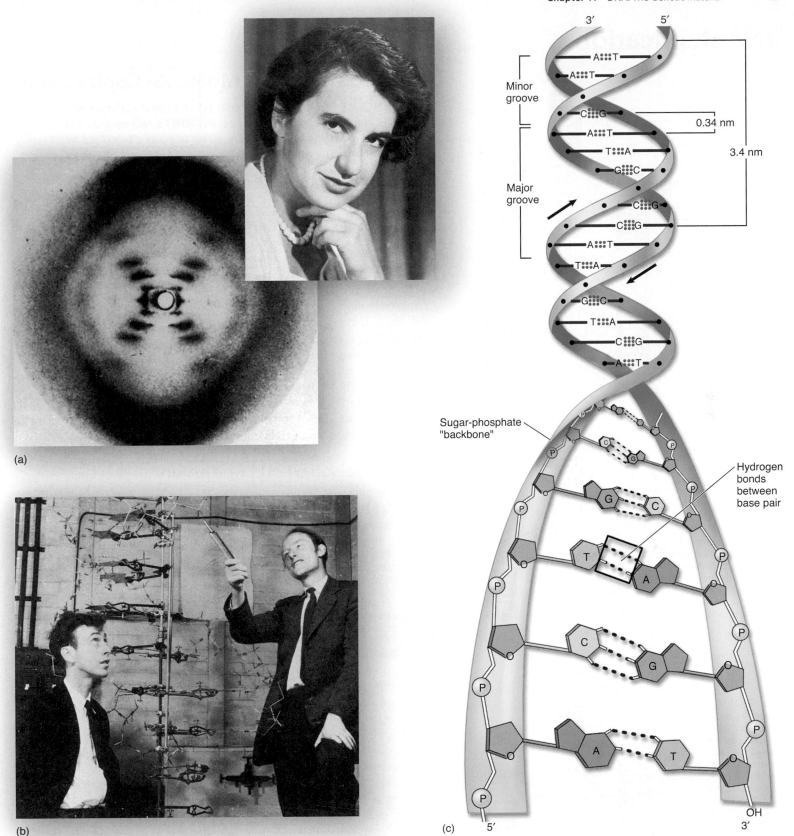

Figure 11.4   The DNA double helix.

(a) This X-ray diffraction photograph was made in 1953 by Rosalind Franklin (inset) in the laboratory of Maurice Wilkins. It suggested to Watson and Crick that the DNA molecule was a helix, like a winding staircase. (b) In 1953 Watson and Crick deduced the structure of DNA. James Watson (seated and peering up at their homemade model of the DNA molecule) was a young American postdoctoral student, and Francis Crick (pointing) was an English scientist. (c) The dimensions of the double helix were suggested by the X-ray diffraction studies. In a DNA duplex molecule, only two base pairs are possible: adenine (A) with thymine (T) and guanine (G) with cytosine (C). A G–C base pair has three hydrogen bonds; an A–T base pair has only two.

DNA Replication

Parent DNA

Replicating DNA

Conservative replication: The two strands of the double helix separate and serve as templates for the assembly of two new strands by base pairing A with T and G with C. After replicating, the original strands rejoin, preserving the parent DNA and forming an entirely new duplex.

Rejoined original DNA

Daughter DNA composed of two new strands

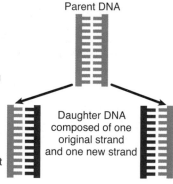

Parent DNA

Semiconservative replication: The double helix need only "unzip" and assemble a new complementary chain along each single strand. The sequence of the original duplex is conserved after one round of replication, but the duplex itself is not. Instead, each strand of the parent duplex becomes part of another duplex.

Daughter DNA composed of one original strand and one new strand

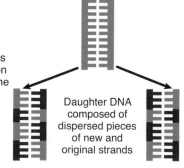

Parent DNA

Dispersive replication: The original DNA serves as a template for the formation of new DNA strands, but the new and old DNA are dispersed among the two daughter strands. Each daughter strand is made up of sections of original strands and new strands.

Daughter DNA composed of dispersed pieces of new and original strands

Figure 11.5   Alternative mechanisms of DNA replication.

There are three possible alternatives as to how the DNA could serve as a template for the assembly of new DNA molecules. In all three, the original strands of DNA are colored blue, and the newly synthesized strands are colored red.

11.4 How the DNA Molecule Copies Itself

CONCEPT PREVIEW: The basis for the great accuracy of DNA replication is complementarity. DNA's two strands are complementary so either one can be used as a template to reconstruct the other.

What holds the two DNA strands together? The "glue" is provided by the weak hydrogen bonds that form between the bases that face each other from the two strands. That is why A pairs with T and not C. The A base can only form hydrogen bonds with the T base. Similarly, G can form hydrogen bonds with C but not T. In the Watson-Crick model of DNA, the two strands of the double helix are said to be *complementary* to each other. One chain of the helix can have any sequence of bases, of A, T, G, and C, but this sequence completely determines that of its partner in the helix. If the sequence of one chain is ATTGCAT, the sequence of its partner in the double helix must be TAACGTA. Each chain in the helix is a complementary mirror image of the other. This complementarity makes it possible for the DNA molecule to copy itself during cell division in a very direct manner; but there are several different ways that this could occur. Figure 11.5 walks you through three possible mechanisms of DNA replication.

> The complementarity of the DNA structure is universal, with all living cells showing the same complementarity in their DNA. As you will see on page 220, this allows the DNA of different organisms to be combined, which is the basis for genetic engineering.

The Meselson-Stahl Experiment

The three alternative hypotheses of DNA replication were tested in 1958 by Matthew Meselson and Franklin Stahl of the California Institute of Technology. These two scientists grew bacteria in a medium containing the heavy isotope of nitrogen, ^{15}N, which became incorporated into the bases of the bacterial DNA (figure 11.6). After several generations, samples were taken from this culture and grown in a medium containing the normal lighter isotope ^{14}N, which became incorporated into the newly replicating DNA. Bacterial samples were taken from the ^{14}N media at 20 minute intervals (❷ through ❹). DNA was extracted from all three samples and from a fourth sample, ❶, that served as a control.

> Isotopes, as discussed on page 34, are atoms that contain varying numbers of neutrons. Isotopes have the same atomic number but different masses. In this experiment, ^{15}N contains one more neutron than ^{14}N and so it is heavier; DNA that contains ^{15}N is heavier.

By dissolving the DNA they had collected in a heavy salt called cesium chloride, and then spinning the solution at very high speeds in an ultracentrifuge, Meselson and Stahl were able to separate DNA strands of different densities. The centrifugal forces caused the cesium ions to migrate toward the bottom of the centrifuge tube, creating a gradient of cesium concentration, and thus a gradation of density. Each DNA strand floats or sinks in the gradient until it reaches the position where its density exactly matches the density of the cesium there. Because ^{15}N strands are denser than ^{14}N strands, they migrate farther down the tube to a denser region of cesium.

The DNA collected immediately after the transfer was all dense, as shown in test tube ❷. However, after the bacteria completed their first round of DNA replication in the ^{14}N medium, the density of their DNA had decreased to a value intermediate between ^{14}N-DNA and ^{15}N-DNA, as shown in test tube ❸. After the second round of replication, two density classes of DNA were observed, one intermediate and one equal to that of ^{14}N-DNA, as shown in test tube ❹.

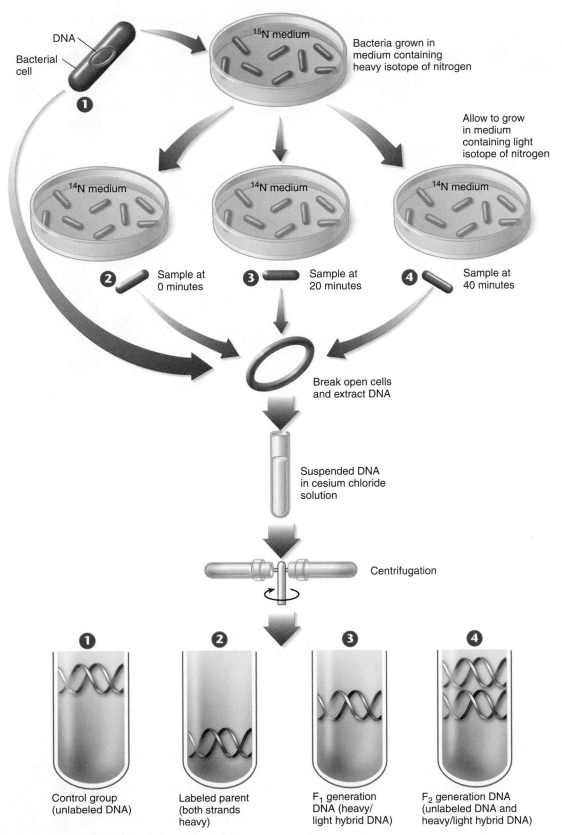

Figure 11.6   The Meselson-Stahl experiment.

Bacterial cells were grown for several generations in a medium containing a heavy isotope of nitrogen (^{15}N) and then were transferred to a new medium containing the normal lighter isotope (^{14}N). At various times thereafter, samples of the bacteria were collected, and their DNA was dissolved in a solution of cesium chloride, which was spun rapidly in a centrifuge. The labeled and unlabeled DNA settled in different areas of the tube because they differed in weight. The DNA with two heavy strands settled down toward the bottom of the tube. The DNA with two light strands settled higher up in the tube. The DNA with one heavy and one light strand settled in between the other two.

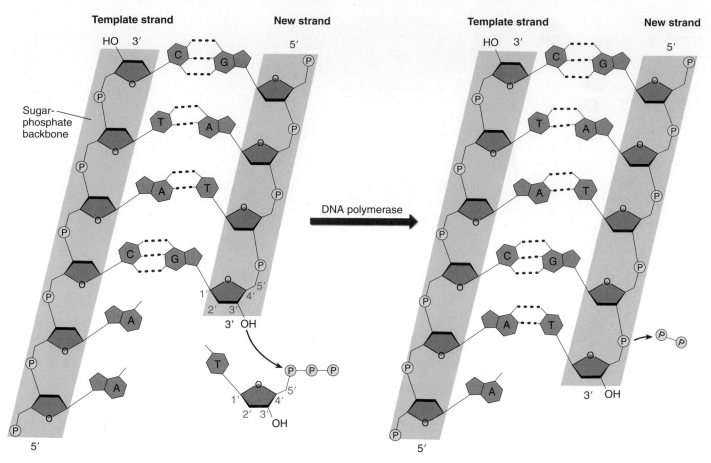

Figure 11.7   **How nucleotides are added in DNA replication.**

Nucleotides are added to the new growing strand of DNA by DNA polymerase. The addition of the nucleotides follows base pairing.

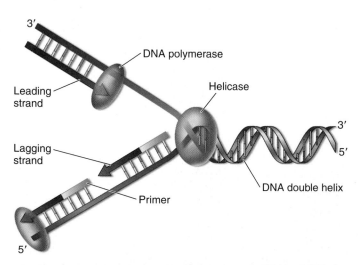

Figure 11.8   **Building the leading and lagging strands.**

DNA polymerase builds the leading strand as a continuous strand moving into the replication fork growing 5′ to 3′, but the lagging strand, also growing 5′ to 3′, is assembled moving away from the replication fork, in segments, each beginning with a primer.

Meselson and Stahl interpreted their results as follows: After the first round of replication, each daughter DNA duplex was a hybrid possessing one of the heavy strands of the parent molecule and one light strand; when this hybrid duplex replicated, it contributed one heavy strand to form another hybrid duplex and one light strand to form a light duplex. Thus, this experiment clearly ruled out conservative and dispersive DNA replication, and confirmed the prediction of the Watson-Crick model that DNA replicates in a semiconservative manner.

How DNA Copies Itself

The copying of DNA before cell division is called **DNA replication** and is carried out by an enzyme called *DNA polymerase*. An enzyme called *helicase* first unwinds the DNA double helix, then DNA polymerase reads along each single strand (the blue strand in figure 11.7) and adds the correct complementary nucleotide (A pairs with T, G with C) at each position as it moves, creating a complementary strand (the pink strand).

However, there are some limitations of the actions of DNA polymerase. First, it can only add to an existing strand; it cannot begin a strand. Another enzyme circumvents this difficulty by beginning the new strand with a section of nucleotides called a *primer*. These are the green segments in figure 11.8. This happens at the place where the parent DNA molecule becomes unzipped, called the *replication fork*. At the replication fork, the polymerase very actively shuttles several

This process of unwinding and separating the two strands of the DNA molecule happens again during protein synthesis. A section of the DNA that contains a gene is opened during transcription, discussed on page 200, and an RNA copy of the DNA is made by a different enzyme.

hundred nucleotides up one strand, building a new strand of DNA called the **leading strand,** adding on to the primer in a continuous fashion. The leading strand is the upper red strand in figure 11.8, the primer is off further to the left out of the frame of the diagram. The new strand is built when the phosphate group of a nucleotide, called the 5′ end, attaches to the sugar end, called the 3′ end, of the nucleotide at the end of the growing strand. The carbon atoms in the sugar component of a nucleotide have a numbering scheme, shown in figure 11.7 (or refer back to figure 11.3). In a nucleotide, the phosphate group is attached to the 5′ carbon atom of the sugar, and an OH group is attached to the 3′ carbon atom. So, each nucleotide has a 5′ end and a 3′ end, and when the nucleotides attach to each other in a long chain, there will be a 5′ phosphate end to the chain on one side and a 3′ OH end on the other side. In the DNA double helix, the two strands of nucleotides pair up in opposite orientations, with one strand running 5′ to 3′ and the other running 3′ to 5′.

The directionality of building the new strand is apparent in figure 11.7, where the phosphate group of the incoming T nucleotide attaches to the OH group of the sugar of the G nucleotide. The new strand assembles in a 5′ to 3′ direction, and new nucleotides can only be added to the 3′ end of an existing strand. This reveals a second limitation: DNA polymerase can only build a strand of DNA in one direction, and so it assembles the other DNA strand, called the **lagging strand,** in segments. Each lagging strand segment begins with a primer (the green segments you saw in figure 11.8), and the DNA polymerase then builds it away from the replication fork until it encounters the previous section.

Eukaryotic chromosomes each contain a single, very long molecule of DNA, one far too long to copy all the way from one end to the other with a single replication fork. Each eukaryotic chromosome is instead copied in sections of about 100,000 nucleotides, each with its own replication origin and fork.

Before the newly formed DNA molecules wind back into the double helix shape, the primers must be removed and the segments of DNA assembled in sections on the lagging strand need to be covalently linked together. The enzyme that performs this sealing function is *DNA ligase.* DNA ligase joins the ends of newly synthesized segments of DNA after the primers have been removed, resulting in one continuous strand of DNA. This process is summarized in *Essential Biological Process 11A*, with the DNA ligase in panel 3 sealing the gaps between the DNA sections of the lagging strand.

The enormous amount of DNA that resides within the cells of your body represents a long series of DNA replications, starting with the DNA of a single cell—the fertilized egg. Living cells have evolved many mechanisms to avoid errors during DNA replication and to preserve the DNA from damage. These mechanisms of DNA repair proofread the strands of each DNA molecule against one another for accuracy and correct any mistakes. But the proofreading is not perfect. If it were, no mistakes such as mutations would occur. Mutation will be discussed in more detail in the next section and in chapter 14, where we consider mutation as the raw material of evolution.

Concept Check

1. Explain complementarity and why it occurs.
2. In Meselson and Stahl's experiment, what would be the expected result if DNA replication had been conservative, as illustrated in figure 11.5?
3. If DNA polymerase cannot begin a strand, explain how DNA replication gets started.

Essential Biological Process 11A

DNA Replication

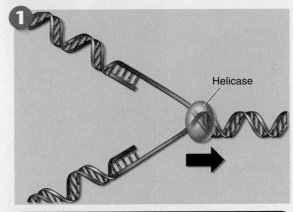

Helicase

Helicase unwinds the DNA double helix for about 1,000 nucleotides.

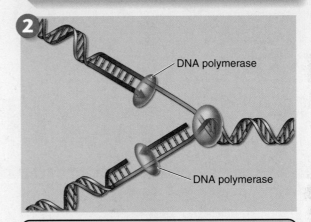

DNA polymerase

DNA polymerase

DNA polymerase assembles a complementary new strand on each old one, building the two strands in opposite directions.

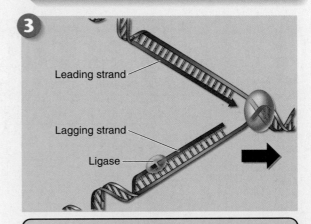

Leading strand

Lagging strand

Ligase

DNA ligase attaches one new strand to the previously replicated segment on the lagging strand, and helicase unwinds another segment.

Altering the Genetic Message

Figure 11.9
Mutation.

Fruit flies normally have one pair of wings, extending from the thorax. This fly is a *bithorax* mutant. Because of a mutation in a gene regulating a critical stage of development, it possesses two thorax segments and thus two sets of wings.

200 μm

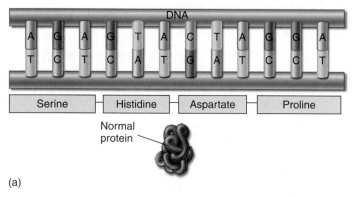

(a)

Serine — Histidine — Aspartate — Proline

Normal protein

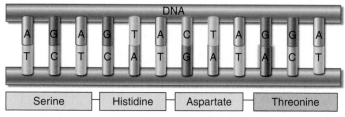

Serine — Histidine — Aspartate — Threonine

Base substitution (red) in DNA: changes proline to threonine in the protein.

Mutated protein

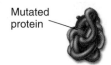

(b)

Figure 11.10   Base substitution mutation.

(a) Often, DNA sequences code for a particular protein. (b) A mutation that substitutes one base for another can result in a change in a single amino acid. This can produce a mutated protein that may not function the same as the normal protein.

11.5 Mutation

CONCEPT PREVIEW: Rare changes in genes, called mutations, can have significant effects on the individual when they occur in somatic tissue, but they are inherited only if they occur in germ-line tissue. Inherited changes provide the raw material for evolution.

A change in the content of the genetic message—the base sequence of one or more genes—is referred to as a **mutation.** As you learned in the previous section, DNA copies itself by forming complementary strands along single strands of DNA when they are separated. The template strand directs the formation of the new strand. However, this replication process is not foolproof. Sometimes errors are made and these are called mutations. Some mutations alter the identity of a particular nucleotide, while others remove or add nucleotides to a gene. The cells of eukaryotes contain an enormous amount of DNA, and the mechanisms that protect and proofread the DNA are not perfect. If they were, no variation would be generated.

Mistakes Happen

In fact, cells do make mistakes during replication, as shown in figure 11.9. Other mutations occur because of DNA alteration by chemicals like those in cigarette smoke, or by radiation like the ultraviolet light from the sun or in tanning beds. However, mutations are rare. Typically, a particular gene is altered in only one of a million gametes. Limited as it might seem, however, this steady trickle of change is the very stuff of evolution. Every difference in the genetic messages that specify different organisms arose as the result of genetic change.

Kinds of Mutation

The message that DNA carries in its genes is the "instructions" of how to make proteins. The sequence of nucleotides in a strand of DNA translates into the sequence of amino acids that makes up a protein. This process was introduced in section 10.5 on page 164. If the core message in the DNA is altered through mutation as

Gene expression occurs in two steps: transcription, discussed on page 200 transfers the nucleotide message in the DNA to a molecule of RNA. In translation, discussed on page 201, the nucleotide message is converted into a string of amino acids that make up the protein.

shown in figure 11.10, where an A nucleotide (in red) is inserted instead of a C nucleotide during DNA replication, then the protein product can also be altered, sometimes to the point where it can no longer function properly. Because mutations can occur randomly in a cell's DNA, most mutations are detrimental, just as making a random change in a computer program usually worsens performance. The consequences of a detrimental mutation may be minor or catastrophic, depending on the function of the altered gene.

Mutations in Germ-Line Tissues.   The effect of a mutation depends critically on the identity of the cell in which the mutation occurs. During the embryonic development of all multicellular organisms, there comes a point when cells destined to form gametes (germ-line cells) are segregated from those that will form the other cells of the body (somatic cells). Only when a mutation occurs within a germ-line cell is it passed to subsequent generations as part of the hereditary endowment of the gametes derived from that cell. Mutations in germ-line tissue are of enormous biological importance because they provide the raw material from which natural selection produces evolutionary change.

Mutations in Somatic Tissues. Change can occur only if there are new, different allele combinations available to replace the old. Mutation produces new alleles, and recombination puts the alleles together in different combinations. In animals, it is the occurrence of these two processes in germ-line tissue that is important to evolution, because mutations in somatic cells (somatic mutations) are not passed from one generation to the next. However, a somatic mutation may have drastic effects on the individual organism in which it occurs, because it is passed on to all of the cells that are descended from the original mutant cell. Thus, if a mutant lung cell divides, all cells derived from it will carry the mutation. Somatic mutations of lung cells are the principal cause of lung cancer.

Altering the Sequence of DNA. One category of mutational changes affects the message itself, producing alterations in the sequence of DNA nucleotides (table 11.1). If alterations involve only one or a few base pairs in the coding sequence, they are called **point mutations.** Sometimes the identity of a nucleotide changes (*base substitution*), while other times one or a few nucleotides are added (*insertion*) or lost (*deletion*). If an insertion or deletion throws the reading of the gene message out of register, a **frame-shift mutation** results. Figure 11.10 shows a base substitution mutation that results in the change of an amino acid, from proline to threonine. This could be a minor change or catastrophic. However, suppose that this had been the deletion of a nucleotide, that the cytosine base nucleotide had been skipped during replication. This would shift the register of the DNA message (imagine removing the "w" from this sentence, yielding "*This oulds hiftt her egistero fth eDN Amessag*") and you can see the problem.

Many point mutations result from damage to the DNA caused by mutagens, usually radiation or chemicals. The latter is of particular importance because modern industrial societies often release many chemical mutagens into the environment.

The Importance of Genetic Change

All evolution begins with alterations in the genetic message that create new alleles or alter the organization of genes on chromosome. Some changes in germ-line tissue produce alterations that enable an organism to leave more offspring, and those changes tend to be preserved as the genetic endowment of future generations. Other changes reduce the ability of an organism to leave offspring. Those changes tend to be lost, as the organisms that carry them contribute fewer members to future generations. Evolution can be viewed as the selection of particular combinations of alleles from a pool of alternatives. The rate of evolution is ultimately limited by the rate at which these alternatives are generated. Genetic change through mutation and recombination provides the raw material for evolution.

Concept Check

1. In order for a mutation to be passed on to offspring, it has to occur in germ-line tissue. What cells are these in the body?
2. What type of mutation would most likely cause major changes in the cell?
3. If somatic mutations are not inherited, why are they important?

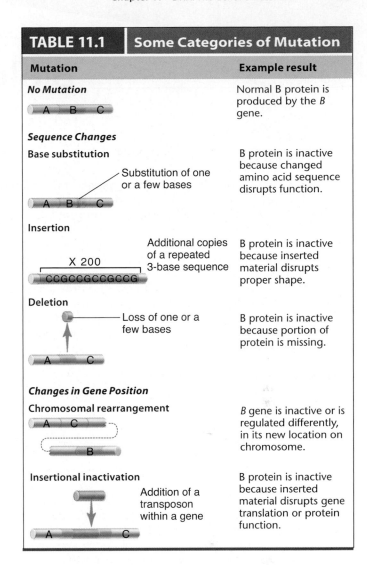

TABLE 11.1 | **Some Categories of Mutation**

Mutation	Example result
No Mutation	Normal B protein is produced by the *B* gene.
Sequence Changes	
Base substitution — Substitution of one or a few bases	B protein is inactive because changed amino acid sequence disrupts function.
Insertion — Additional copies of a repeated 3-base sequence (X 200, CCGCCGCCGCCG)	B protein is inactive because inserted material disrupts proper shape.
Deletion — Loss of one or a few bases	B protein is inactive because portion of protein is missing.
Changes in Gene Position	
Chromosomal rearrangement	*B* gene is inactive or is regulated differently, in its new location on chromosome.
Insertional inactivation — Addition of a transposon within a gene	B protein is inactive because inserted material disrupts gene translation or protein function.

BIOLOGY & YOU

Almost All Inherited Mutations Occur in Males. Molecular analysis of DNA allows us to determine whether a mutation occurred in the mother or the father. The results are dramatic. In studies to determine the parent of origin, a total of 92 new human mutations were analyzed, and ALL were paternal. How can we account for a far higher mutation rate in males than in females? The explanation seems to lie in the far greater number of cell divisions in the male germ line. In females germ cell divisions stop by the time of birth, while in males cell divisions are continuous, so that many divisions have occurred before an adult male's sperm is produced. Because most germ-line mutations are replication errors, the far greater number of replications in males leads to a much bigger opportunity for mistakes. This predicts that human mutation rates should increase with the age of the father. They do. At age 20, the mutation rate for a man is about 8 times the female rate. The difference is some 30-fold by age 30, and as much as 100-fold by age 40.

Protecting Your Genes

This text's discussion of changes in genes—mutations—has largely focused on heredity, how changes in the information encoded in DNA can affect offspring. It is important, however, to realize that inherited mutations occur only in germ-line tissue, in the cells that generate your eggs or sperm. Mutations in the other cells of your body, in so-called somatic tissues, are not inherited. This does not, however, mean that such mutations are not important. In fact, somatic mutations can have a disastrous impact upon your health, because they can lead to cancer. Protecting the DNA of your body's cells from damaging mutation is perhaps the most important thing you can do to prolong your life. Here we will examine two potential threats.

Smoking and Lung Cancer

The association of particular chemicals in cigarette smoke with lung cancer, particularly chemicals that are potent mutagens (see chapters 8 and 25), led researchers early on to suspect that lung cancer might be caused, at least in part, by the action of chemicals on the cells lining the lung.

The hypothesis that chemicals in tobacco cause cancer was first advanced over 200 years ago in 1761 by Dr. John Hill, an English physician. Hill noted unusual tumors of the nose in heavy snuff users and suggested tobacco had produced these cancers. In 1775, a London surgeon, Sir Percivall Pott, made a similar observation, noting that men who had been chimney sweeps exhibited frequent cancer of the scrotum. He suggested that soot and tars might be responsible. These observations led to the hypothesis that lung cancer results from the action of tars and other chemicals in tobacco smoke.

It was over a century before this hypothesis was directly tested. In 1915, Japanese doctor Katsusaburo Yamagiwa applied extracts of tar to the skin of 137 rabbits every two or three days for three months. Then he waited to see what would happen. After a year, cancers appeared at the site of application in seven of the rabbits. Yamagiwa had induced cancer with the tar, the first direct demonstration of chemical carcinogenesis. In the decades that followed, this approach demonstrated that many chemicals can cause cancer.

But do these lab studies apply to people? Do tars in cigarette smoke in fact induce lung cancer in humans? In 1949, the American physician Ernst Winder and the British epidemiologist Richard Doll independently reported that lung cancer showed a strong link to the smoking of cigarettes, which introduces tars into the lungs. Winder interviewed 684 lung cancer patients and 600 normal controls, asking whether each had ever smoked. Cancer rates were 40 times higher in heavy smokers than in nonsmokers. From these studies, it seemed likely as long as 50 years ago that tars and other chemicals in cigarette smoke induce cancer in the lungs of persistent smokers. While this

suggestion was resisted by the tobacco industry, the evidence that has accumulated since these pioneering studies makes a clear case, and there is no longer any real doubt. Chemicals in cigarette smoke cause cancer.

As you will learn in chapter 25 (page 498), tars and other chemicals in cigarette smoke cause lung cancer by mutating DNA, disabling genes that in normal lung cells restrain cell division. Lacking these restraints, the altered lung cells begin to divide ceaselessly, and lung cancer results. Over 160,000 Americans died of lung cancer last year, and almost all of them were cigarette smokers.

If cigarette smoking is so dangerous, why do so many Americans smoke? Fully 23% of American men smoke, and 18% of women. Are they not aware of the danger? Of course they are. But they are not able to quit. Tobacco smoke, you see, also contains another chemical, nicotine, which is highly addictive. The nature of drug addiction is discussed in chapter 29 (page 557). Basically, what happens is that a smoker's brain makes physiological compensations to overcome the effects of nicotine, and once these adjustments are made the brain does not function normally without nicotine. The body's physiological response to nicotine is profound and unavoidable; there is no way to prevent addiction to nicotine with willpower.

Many people attempting to quit smoking use patches containing nicotine to help them, the idea being that providing nicotine removes the craving for cigarettes. This is true, it does—as long as you keep using the patch. Actually, using such patches simply substitutes one (admittedly less dangerous) nicotine source for another. If you are going to quit smoking, there is no way to avoid the necessity of eliminating the drug to which you are addicted, nicotine. There is no easy way out. The only way to quit is to quit.

Clearly, if you do not smoke, you should not start. When asked what three things were most important to improve Americans' health, a prominent physician replied: "Don't smoke. Don't smoke. Don't smoke."

Tanning and Skin Cancer

Almost all cells in the human body undergo cell division, replacing themselves as they wear out. Some adult cells do this very frequently, others rarely if ever. Skin cells divide quite frequently. Exposed to a lot of wear and tear, they divide about every 27 days to replace dead or damaged cells. The skin sloughs off dead cells from the surface and replaces these with new cells from beneath. The average person will lose about 105 pounds of skin by the time he or she turns 70.

While skin can become damaged in many ways, the damage that seems to have the most long-term affect is caused by the sun. The skin contains cells called melanocytes that produce a pigment called melanin when exposed to UV light. Melanin produces a yellow-to-brown color in the skin. The type of melanin and the amount produced is genetically determined. People with darker skin types have more melanocytes and produce a melanin that is dark brown in color. Protected by UV-absorbing melanin, they almost never sunburn. Fair skinned people have fewer melanocytes and produce melanin that is more yellow in color. Unprotected by melanin, these people sunburn easily and rarely tan. When cells on the body's surface are badly damaged by the sun, called a sunburn, the cells slough off. Recall the peeling that you experience if you have ever had a bad sunburn.

Up until the early 20th century, a tan was a condition that people went to great lengths to avoid. A tanned body was a sign of the working class, people who had to work in the sun. The wealthy elite avoided the sun with pale skin being in fashion. All of this changed in the 1920s, when tans became a status symbol, with the wealthy able to travel to warm, sunny destinations, even in the middle of winter. That tan, bronzed glow that people would sit in the sun for hours to achieve was thought to be both healthy and attractive.

During the 1970s, doctors started to see an uptick in the number of cases of melanoma, a deadly form of skin cancer. New cases were increasing about 6% each year. Researchers proposed that UV rays from the sun were the underlying cause of this epidemic of skin cancer and warned people to avoid the sun when possible and protect themselves with sunscreen.

Malignant melanoma is the most deadly of skin cancers, although treatable if caught early. Melanoma is cancer of melanocyte cells. Melanoma lesions usually appear as shades of tan, brown, and black and often begin in or near a mole, and so changes in a mole is a symptom of melanoma. Melanoma is most prevalent in fair-skinned people, but unlike the other forms of skin cancer, it can also affect people with darker complexions.

The public has been slow to respond to warnings about avoiding sun exposure, perhaps because the cosmetic benefits of tanning are immediate while the health hazards are much delayed. The desire to achieve that tanned, bronzed body is as strong as ever.

A good tan requires regular exposure to the sun to maintain it, so indoor tanning salons have become popular. Tanning beds emit concentrated UV rays from two sides, allowing a person to tan in less time and in all weather conditions (sun, rain, snow). The indoor tanning business has grown in the U.S. to a $2-billion-a-year industry with an estimated 28 million Americans tanning annually.

People thought that building up a tan through the use of tanning beds would protect a person's skin from burning and would reduce the time exposed to the UV radiation; both leading to a reduced risk of skin cancer. However, recent research does not support these assumptions. A 2003 study of 106,000 Scandinavian women showed that exposure to UV rays in a tanning bed as little as once a month can increase your risk of melanoma by 55%, especially when the exposure is during early adulthood. Those women who were in their 20s and used sun lamps to tan were at the highest risk, about 150% higher than those who didn't use a tanning bed. As with other studies, fair-skinned women were at the greatest risk. In fact, tanning beds, even for those people who tan more easily, heighten the risk for skin cancer because people use the tanning beds year-round, increasing their cumulative exposure.

It is difficult to avoid the conclusion that to protect your genes you should avoid tanning beds. Like smoking cigarettes, excessive tanning is gambling with your life.

Are Mutations Random or Directed By the Environment?

Once biologists appreciated that Mendelian traits were in fact alternative versions of DNA sequences, which resulted from mutations, a very important question arose and needed to be answered—Are mutations random events that might happen anywhere on a DNA chromosome, or are they directed to some degree by the environment? Do the mutagens in cigarettes, for example, damage DNA at random locations, or do they preferentially seek out and alter specific sites such as those regulating the cell cycle?

This key question was addressed and answered in an elegant and deceptively simple experiment carried out in 1943 by two of the pioneers of molecular genetics, Salvadore Luria and Max Delbruck. They chose to examine a particular mutation that occurs in laboratory strains of the bacterium *E. coli*. These bacterial cells are susceptible to T1 viruses, tiny chemical parasites that infect, multiply within, and kill the bacteria. If 10^5 bacterial cells are exposed to 10^{10} T1 viruses, and the mixture spread on a culture dish, not one cell grows—every single *E. coli* cell is infected and killed. However, if you repeat the experiment using 10^9 bacterial cells, lots of cells survive! When tested, these surviving cells prove to be mutants, resistant to T1 infection. The question is, did the T1 virus cause the mutations, or were the mutations present all along, too rare to be present in a sample of only 10^5 cells but common enough to be present in 10^9 cells?

To answer this question, Luria and Delbruck devised a simple experiment they called a "fluctuation test," illustrated here. Five cell generations are shown for each of four independent bacterial cultures, all tested for resistance in the fifth generation. If the T1 virus causes the mutations (top row), then each culture will have more or less the same number of resistant cells, with only a little fluctuation (that is, variation among the four). If, on the other hand, mutations are spontaneous and so equally likely to occur in any generation, then bacterial cultures in which the T1 resistance mutation occurs in earlier generations will possess far more resistant cells by the fifth generation than cultures in which the mutation occurs in later generations, resulting in wide fluctuation among the four cultures. The table presents the data they obtained for 20 individual cultures.

Number of Bacteria Resistant to T1 Virus			
Culture number	Resistant colonies found	Culture number	Resistant colonies found
1	1	11	107
2	0	12	0
3	3	13	0
4	0	14	0
5	0	15	1
6	5	16	0
7	0	17	0
8	5	18	64
9	0	19	0
10	6	20	35

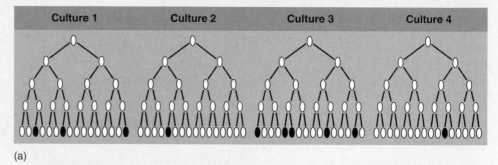

(a)

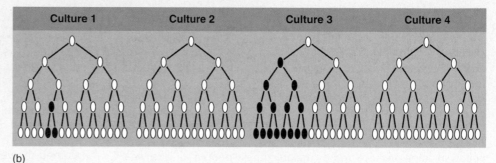

(b)

Analysis

1. **Interpreting Data** What is the mean number of T1-resistant bacteria found in the 20 individual cultures?

2. **Making Inferences**
   a. Comparing the twenty individual cultures, do the cultures exhibit similar numbers of T1-resistant bacterial cells?
   b. Which of the two alternative outcomes illustrated above, (*a*) or (*b*), is more similar to the outcome obtained by Luria and Delbruck in this experiment?

3. **Drawing Conclusions** Are these data consistent with the hypothesis that the mutation for T1 resistance among *E. coli* bacteria is caused by exposure to T1 virus? Explain.

Concept Summary

Genes Are Made of DNA

11.1 The Griffith Experiment

- Using *Streptococcus pneumoniae* bacteria, Griffith showed that information that controls physical characteristics can be passed from one bacterium to another, even from a dead bacterium.

- By injecting mice with different strains of *S. pneumoniae,* Griffith determined that some strains were pathogenic, resulting in the mice's deaths. Bacteria of the pathogenic strains contained polysaccharide capsules (S strain), like the bacteria shown here from **figure 11.1,** while those without the capsules (R strain) were non-lethal. When he mixed dead pathogenic bacteria (S), which usually would not cause death, and live nonpathogenic bacteria (R) and injected them into mice, the mice died. The dead mice contained living S strains.

- Something passed from the dead lethal bacteria to the live nonlethal bacteria, causing them to turn deadly.

11.2 The Avery and Hershey-Chase Experiments

- Avery and colleagues showed that protein was not the source of this transformation. They replicated Griffith's experiment but removed all protein from the preparation. The virulent strain with its protein coat removed was still able to transform nonvirulent bacteria. This experimental result supported the hypothesis that DNA, not protein, was the transforming principle.

- Using bacterial viruses, Hershey and Chase showed that genes were carried on DNA and not proteins. They used two different radioactively tagged preparations, DNA in one, as shown here from **figure 11.2,** and protein in the other. Each preparation was used to infect bacteria. When they screened the two bacterial cultures, they discovered that the infected bacteria contained radioactively tagged DNA.

11.3 Discovering the Structure of DNA

- The structure of DNA was not known. The basic chemical components of DNA were determined to be nucleotides. Each nucleotide has a similar structure: a deoxyribose sugar attached to a phosphate group and one of four organic bases (**figure 11.3**).

- Erwin Chargaff observed that two sets of bases are always present in equal amounts in a molecule of DNA (the amount of A nucleotides equals the amount of T nucleotides, and C nucleotides equals G nucleotides). This observation, called Chargaff's rule, gave some insight into the structure of DNA—that there was some regularity to the structure.

- Using X-ray diffraction, Rosalind Franklin was able to form a "picture" of DNA. The image suggested that the DNA molecule was coiled, a form called a helix.

- Using Chargaff's and Franklin's research, Watson and Crick determined that DNA is a double helix, two strands that are connected by base pairing between the nucleotide bases. An A nucleotide on one strand pairs with T on the other, and similarly G pairs with C (**figure 11.4**).

DNA Replication

11.4 How the DNA Molecule Copies Itself

- The complementarity of DNA (A pairs with T and C with G) implies a method of replication where a single strand of DNA can serve as the template for production of another strand, but there are several ways in which this could occur (**figure 11.5**).

- Meselson and Stahl showed that DNA replicates semiconservatively, using each of the original strands as templates to form new strands. In semiconservative replication, each new strand of DNA consists of a template strand from the parent DNA and a newly synthesized strand that is complementary to the template stand (**figure 11.6**).

- In replication, the DNA molecule first unwinds by the actions of an enzyme called helicase. Each DNA strand is then copied by the actions of an enzyme called DNA polymerase. The two original strands serve as templates to the new DNA strands. DNA polymerase adds nucleotides to the new DNA strands that are complementary to the original single strands. DNA polymerase can only add on to an existing strand, and so the new strand begins after a section of nucleic acids called a primer is added. A different enzyme builds the primer. Nucleotides are added to the growing strand in a 5′ to 3′ direction (**figure 11.7**).

- The point where the DNA separates is called the replication fork. Because nucleotides can only be added onto the 3′ end of the growing DNA strand, DNA copies in a continuous manner on one of the strands, called the leading strand, and in a discontinuous manner on the other strand, called the lagging strand (**figure 11.8**). On the lagging strand, primers are inserted at the replication fork, and nucleotides are added in sections, as shown here on the lower strand from ***Essential Biological Process 11A.*** Before the new DNA strands rewind, the primers are removed and the DNA segments are linked together with another enzyme called DNA ligase.

- Errors can occur during the replication of DNA. The cell has many mechanisms to correct damage to the DNA or mistakes made during replication. This proofreading process compares one strand against its complementary strand and corrects errors, but this system is not foolproof.

Altering the Genetic Message

11.5 Mutation

- A mutation is a change in the nucleotide sequence of the genetic message. Mutations that change one or only a few nucleotides are called point mutations (**figure 11.10**). Some mutations cause only minor changes while others can have dramatic consequences (**table 11.1**).

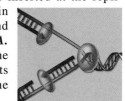

Self-Test

1. In the experiments performed by Frederick Griffith, they found that
   a. hereditary information within a cell cannot be changed.
   b. hereditary information can be added to cells from other cells.
   c. mice infected with live R strains die.
   d. mice infected with heat-killed S strains die.
2. The experiment performed by Alfred Hershey and Martha Chase showed that the molecule viruses use to specify new viruses is
   a. a protein.   c. ATP.
   b. a carbohydrate.   d. DNA.
3. Erwin Chargaff, Rosalind Franklin, Francis Crick, and James Watson all worked on pieces of information relating to the
   a. structure of DNA.   c. inheritance of DNA.
   b. function of DNA.   d. mutations of DNA.
4. The four DNA nucleotides are all different in terms of
   a. their sizes.
   b. the number of hydrogen bonds they can form with their base pair.
   c. the type of nitrogen base.
   d. the type of sugar.
5. Which of the following lists the organic bases found in the purine nucleotide bases?
   a. adenine and cytosine   c. cytosine and thymine
   b. guanine and thymine   d. adenine and guanine
6. If one strand of a DNA molecule has the base sequence ATTGCAT, its complementary strand will have the sequence

   a. ATTGCAT.   c. GCCATGC.
   b. TAACGTA.   d. CGGTACG.
7. Regarding the duplication of DNA, we now know that each double helix
   a. reforms after replicating.
   b. splits down the middle into two single strands, and each one then acts as a template to build its complement.
   c. fragments into small chunks that duplicate and reassemble.
   d. All of these are true for different types of DNA.
8. DNA polymerase can only add nucleotides to an existing chain, so _____ is required.
   a. a primer   c. a lagging strand
   b. helicase   d. a leading strand
9. Genetic messages can be altered in two ways:
   a. through semiconservative replication or conservative replication.
   b. through the chromosome or through the protein.
   c. by mutation or by transformation.
   d. by activation or by repression.
10. Mutations can occur in
    a. germ-line tissues and are passed on to future generations.
    b. somatic tissues and are passed on to future generations.
    c. germ-line tissues but not in somatic tissues.
    d. somatic tissues but not in germ-line tissues.

Visual Understanding

1. **Figure 11.4c** What are some of the possible problems that could occur if the cytosine nucleotide indicated with the red arrow is accidentally replaced with an adenine nucleotide?

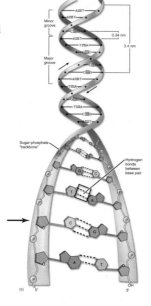

2. **Table 11.1** What types of mutations in the table result in a shift in the reading frame of the DNA? Explain.

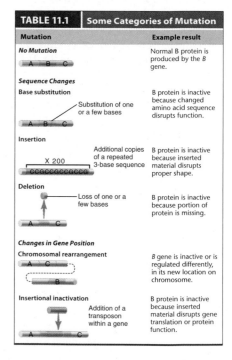

Challenge Questions

1. Based on the experiments and discoveries discussed in chapter 10 and here in chapter 11, defend the statement, attributed to Sir Isaac Newton in 1676 (though some say that Bernard of Chartres said it first, way back in about 1130!) that scientists build new ideas in science by "standing on the shoulders of giants."

2. Mutations can occur in somatic (body) cells or in germ-line cells. What problems or opportunities does each kind of mutated cell face?

Chapter **12**

How Genes Work

From Gene to Protein

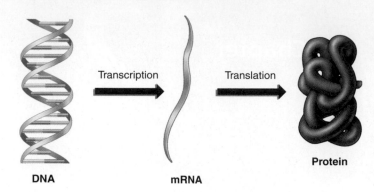

Figure 12.1   The central dogma: DNA to RNA to protein.

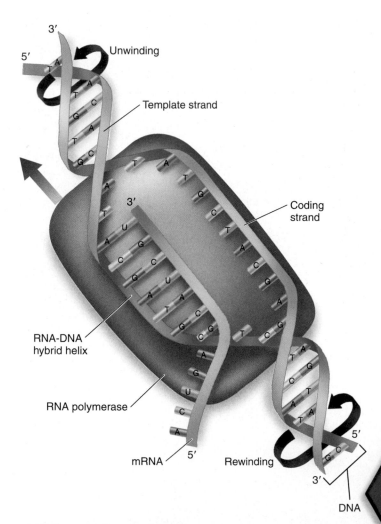

Figure 12.2   Transcription.

One of the strands of DNA functions as a template on which nucleotide building blocks are assembled into mRNA by RNA polymerase as it moves along the DNA strand.

The production of an RNA molecule during transcription is similar to the process of DNA replication, with the RNA chain of nucleotides growing in a 5′ to 3′ direction as is seen in DNA synthesis in figure 11.7 on page 190.

12.1 Transcription

CONCEPT PREVIEW: Transcription is the production of an mRNA copy of a gene by the enzyme RNA polymerase.

The discovery that genes are made of DNA, discussed in chapter 11, left unanswered the question of how the information in DNA is used. How does a string of nucleotides in a spiral molecule determine if you have red hair or green eyes? We now know that the information in DNA is arrayed in little blocks, like entries in a dictionary, and each block is a gene that specifies the sequence of amino acids for a polypeptide. These polypeptides form the proteins that determine what a particular cell will be like.

Just as an architect protects building plans from loss or damage by keeping them safe in a central place and issuing only blueprint copies to on-site workers, so your cells protect their DNA instructions by keeping them safe within a central DNA storage area, the nucleus. The DNA never leaves the nucleus. Instead, "blueprint" copies of particular genes within the DNA instructions are sent out into the cell to direct the assembly of proteins. These working copies of genes are made of ribonucleic acid (RNA) rather than DNA. Recall that RNA, like DNA, is made of nucleotides except that the sugars in RNA have an extra oxygen atom and T is replaced by a similar pyrimidine base called uracil, U (see figure 3.11). The path of information is thus: **DNA $\longrightarrow$ RNA $\longrightarrow$ protein.** This information path is often called the *central dogma,* because it describes the key organization used by your cells to express their genes (figure 12.1).

A cell uses three kinds of RNA in the synthesis of proteins: messenger RNA (mRNA), ribosomal RNA (rRNA), and transfer RNA (tRNA). The use of information in DNA to direct the production of particular proteins is called **gene expression.** Gene expression occurs in two stages: in the first stage, called **transcription,** mRNA molecules are synthesized from genes within the DNA; in the second stage, called **translation,** the mRNA is used to direct the production of polypeptides, the components of proteins.

The Transcription Process

The RNA copy of a gene used in the cell to produce a polypeptide is called **messenger RNA (mRNA)**—it is the messenger that conveys the information from the nucleus to the cytoplasm. The copying process that makes the mRNA is called transcription—just as monks in monasteries used to make copies of manuscripts by faithfully transcribing each letter, so enzymes make mRNA copies of your genes by faithfully complementing each nucleotide.

In cells, the transcriber is a large and very sophisticated protein called **RNA polymerase.** It binds to the DNA double helix at a particular site called a *promoter.* It separates the two strands and then moves along one of the DNA strand like a train engine on a track. Although DNA is double-stranded, the two strands have complementary rather than identical sequences, so RNA polymerase is only able to bind one of the two DNA strands (the one with the promoter-site sequence it recognizes). As RNA polymerase goes along the DNA strand, it pairs each nucleotide with its complementary RNA version (G with C, A with U—recall that RNA uses a nucleotide with uracil, U, in place of thymine), building an mRNA chain in the 5′ to 3′ direction (figure 12.2).

12.2 Translation

CONCEPT PREVIEW: The genetic code dictates how a nucleotide sequence specifies a particular amino acid sequence. A gene is transcribed into mRNA, which is then translated into a polypeptide. The sequence of mRNA codons dictates the corresponding sequence of amino acids in a growing polypeptide chain.

The Genetic Code

The essence of Mendelian genetics is that the information determining hereditary traits, traits passed from parent to child, is encoded information, written within the chromosomes in blocks called genes. To correctly read a gene, a cell must translate the information encoded in DNA into the language of proteins—that is, it must convert the order of the gene's nucleotides into the order of amino acids in a polypeptide, a process called **translation.** The rules that govern this translation are called the **genetic code.**

> Genes and the proteins they encode underlie the expression of phenotypes, as described on page 164. RNAs are used to make proteins encoded in the DNA, and the proteins determine phenotype. A mutation in DNA can result in an error in the mRNA and its protein product.

The mRNA nucleotide sequence is "read" by a ribosome in three nucleotide units called **codons.** Each codon, with the exception of three, codes for a particular amino acid. Biologists worked out which codon corresponds to which amino acid by trial-and-error experiments carried out in test tubes. In these experiments, investigators used artificial mRNAs to direct the synthesis of polypeptides in the tube, and then looked to see the sequence of amino acids in the newly formed polypeptides. An mRNA that was a string of UUUUUU . . . , for example, produced a polypeptide that was a string of phenylalanine (Phe) amino acids, telling investigators that the codon UUU corresponded to the amino acid Phe. Most amino acids are specified by more than one codon. The entire genetic code dictionary is presented in figure 12.3.

> Amino acids, discussed on page 49, are the monomers that make up a protein. They form long chains, held together with peptide bonds. The 20 amino acids are similar in structure except for their functional groups that give each amino acid its unique characteristics.

Figure 12.3 **The genetic code (RNA codons).**

A codon consists of three nucleotides read in sequence. For example, ACU codes for threonine. To determine which amino acid corresponds to a codon you read the chart first down the left side, then across the top, then down the right side. For the codon ACU, find the first letter, A, in the First Letter column; then follow the row over to the second letter, C, in the Second Letter row across the top; and then in that column trace down to the third letter, U, in the Third Letter column. Most amino acids are specified by more than one codon. For example, threonine is specified by four codons, which differ only in the third nucleotide (ACU, ACC, ACA, and ACG).

The Genetic Code

First Letter	Second Letter — U		Second Letter — C		Second Letter — A		Second Letter — G		Third Letter
U	UUU UUC	Phenylalanine	UCU UCC	Serine	UAU UAC	Tyrosine	UGU UGC	Cysteine	U C
	UUA UUG	Leucine	UCA UCG		UAA	Stop	UGA	Stop	A
					UAG	Stop	UGG	Tryptophan	G
C	CUU CUC	Leucine	CCU CCC	Proline	CAU CAC	Histidine	CGU CGC	Arginine	U C
	CUA CUG		CCA CCG		CAA CAG	Glutamine	CGA CGG		A G
A	AUU AUC	Isoleucine	ACU ACC	Threonine	AAU AAC	Asparagine	AGU AGC	Serine	U C
	AUA		ACA		AAA AAG	Lysine	AGA AGG	Arginine	A
	AUG	Methionine; Start	ACG						G
G	GUU GUC	Valine	GCU GCC	Alanine	GAU GAC	Aspartate	GGU GGC	Glycine	U C
	GUA GUG		GCA GCG		GAA GAG	Glutamate	GGA GGG		A G

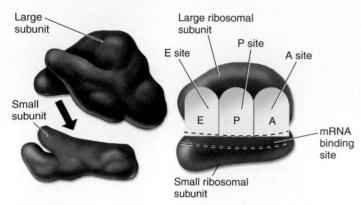

Figure 12.4 **A ribosome is composed of two subunits.**

The smaller subunit fits into a depression on the surface of the larger one. The A, P, and E sites on the ribosome play key roles in protein synthesis.

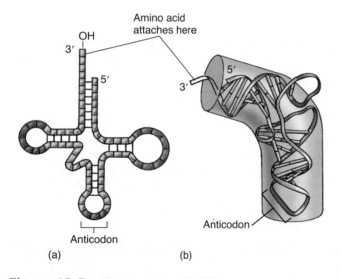

Figure 12.5 **The structure of tRNA.**

tRNA, like mRNA, is a long strand of nucleotides. However, unlike mRNA, hydrogen bonding occurs between its nucleotides, causing the strand to form hairpin loops, as seen in (a). The loops then fold up on each other to create the compact, three-dimensional shape seen in (b). Amino acids attach to the free, single-stranded —OH end of a tRNA molecule. A three-nucleotide sequence called the anticodon in the lower loop of tRNA interacts with a complementary codon on the mRNA.

The genetic code is universal, the same in practically all organisms. GUC codes for valine in bacteria, in fruit flies, in eagles, and in your own cells. The only exception biologists have ever found to this rule is in the way in which cell organelles that contain DNA (mitochondria and chloroplasts) and a few microscopic protists read the "stop" codons. In every other instance, the same genetic code is employed by all living things.

Translating the RNA Message into Proteins

The final result of the transcription process is the production of an mRNA copy of a gene. Like a photocopy, the mRNA can be used without damage or wear and tear on the original. After transcription of a gene is finished, the mRNA passes out of the nucleus into the cytoplasm through pores in the nuclear envelope. There, translation of the genetic message occurs. In translation, organelles called **ribosomes** use the mRNA produced by transcription to direct the synthesis of a polypeptide following the genetic code.

The Protein-Making Factory. Ribosomes are the polypeptide-making factories of the cell. Ribosomes use mRNA, the "blueprint" copies of nuclear genes, to direct the assembly of polypeptides, which are then combined into proteins.

Ribosomes are very complex, containing over 50 different proteins (shown in gold in the image on page 199) and three chains of **ribosomal RNA (rRNA)** of some 3,000 nucleotides (shown in gray). It had been traditionally assumed that the proteins in a ribosome act as enzymes to catalyze the assembly process, with the RNA acting as a scaffold to position the proteins. In the year 2000 powerful atomic-resolution X-ray diffraction studies unexpectedly revealed the many proteins of a ribosome to be scattered over its surface like decorations on a Christmas tree. The role of these proteins seems to be to stabilize the many bends and twists of the RNA chains, the proteins acting like spot-welds between RNA strands. Importantly, there are no proteins on the inside of the ribosome where the chemistry of protein synthesis takes place—just twists of RNA. Thus it is the ribosome's RNA, not its proteins, that catalyzes the synthesis of polypeptides!

Ribosomes are composed of two parts, or subunits, one nested into the other like a fist in the palm of your hand. The "fist" is the smaller of the two subunits, the pink structure in figure 12.4. Its rRNA has a short nucleotide sequence exposed on the surface of the subunit. This exposed sequence is identical to a sequence called the leader region that occurs at the beginning of all genes. Because of this, an mRNA molecule binds to the exposed rRNA of the small subunit like a fly sticking to flypaper.

The Key Role of tRNA. Directly adjacent to the exposed rRNA sequence are three small pockets or dents, designated the A, P, and E sites, in the surface of the ribosome (shown in figure 12.4 and discussed shortly). These sites have just the right shape to bind yet a third kind of RNA molecule, **transfer RNA (tRNA).** It is tRNA molecules that bring amino acids to the ribosome used in making proteins. tRNA molecules are chains about 80 nucleotides long. The string of nucleotides folds back on itself, forming a 3-looped structure shown in figure 12.5a. The looped structure further folds into a compact shape shown in figure 12.5b, with a three-nucleotide sequence at the bottom (the pink loop) and an amino acid attachment site at the top (the 3′ end).

The three-nucleotide sequence on the tRNA, called the **anticodon,** is very important: It is the complementary sequence to 1 of the 64 codons of the genetic code! Special enzymes, called *activating enzymes,* match amino acids in the cytoplasm with their proper tRNAs. The anticodon determines which amino acid will attach to a particular tRNA.

Essential Biological Process 12A

Translation

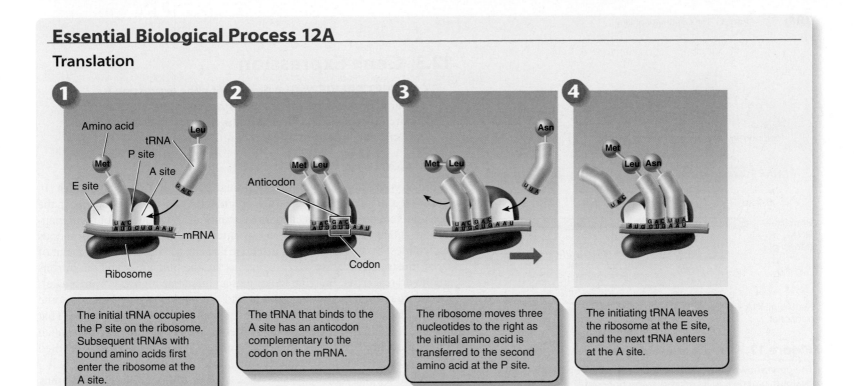

1 The initial tRNA occupies the P site on the ribosome. Subsequent tRNAs with bound amino acids first enter the ribosome at the A site.

2 The tRNA that binds to the A site has an anticodon complementary to the codon on the mRNA.

3 The ribosome moves three nucleotides to the right as the initial amino acid is transferred to the second amino acid at the P site.

4 The initiating tRNA leaves the ribosome at the E site, and the next tRNA enters at the A site.

Because the first dent in the ribosome, called the A site (the attachment site where amino-acid-bearing tRNAs will bind) is directly adjacent to where the mRNA binds to the rRNA, three nucleotides of the mRNA are positioned directly facing the anticodon of the tRNA. Like the address on a letter, the anticodon ensures that an amino acid is delivered to its correct "address" on the mRNA where the ribosome is assembling the polypeptide.

Making the Polypeptide. Once an mRNA molecule has bound to the small ribosomal subunit, the other larger ribosomal subunit binds as well, forming a complete ribosome. The ribosome then begins the process of translation, shown in *Essential Biological Process 12A*. Panel 1 of the figure shows how the mRNA begins to thread through the ribosome like a string passing through the hole in a doughnut. The mRNA passes through in short spurts, three nucleotides at a time, and at each burst of movement a new three-nucleotide codon on the mRNA is positioned opposite the A site in the ribosome, where a tRNA molecule first binds, as shown in panel 2.

As each new tRNA brings in an amino acid to each new codon presented at the A site, the old tRNA paired with the previous codon is passed over to the P site where peptide bonds form between the incoming amino acid and the growing polypeptide chain. The tRNA in the P site eventually shifts to the E site (the exit site), as shown in panel 3, and the amino acid it carried is attached to the end of a growing amino acid chain. The tRNA is then released in panel 4. So as the ribosome proceeds down the mRNA, one tRNA after another is selected to match the sequence of mRNA codons. In figure 12.6 you can see the ribosome traveling along the length of the mRNA, the tRNAs bringing the amino acids into the ribosome and the growing polypeptide chain extending out from the ribosome. Translation continues until a "stop" codon is encountered, which signals the end of the polypeptide. The ribosome complex falls apart, and the newly made polypeptide is released into the cell.

> A peptide bond is the name given to the chemical bond that forms between two amino acids in a polypeptide, as discussed on page 49. Two amino acids are linked together through a dehydration reaction where a water molecule is a by-product of the reaction as described on page 47.

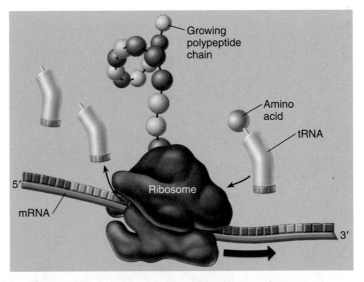

Figure 12.6   Ribosomes guide the translation process.

tRNA binds to an amino acid as determined by the anticodon sequence. Ribosomes bind the loaded tRNAs to their complementary sequences on the strand of mRNA. tRNA adds its amino acid to the growing polypeptide chain, which is released as the completed protein. The overall flow of genetic information is from DNA to mRNA to protein. For example, the polypeptide that is being formed in *Essential Biological Process 12A* began with the DNA nucleotide sequence TACGACTTA, which was transcribed into the mRNA sequence AUGCUGAAU. This sequence is then translated into a polypeptide composed of the amino acids methionine—leucine—asparagine.

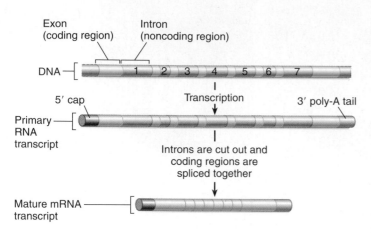

Figure 12.7 Processing eukaryotic RNA.

The gene shown here codes for a protein called ovalbumin. The ovalbumin gene and its primary transcript contain seven segments not present in the mRNA used by the ribosomes to direct the synthesis of the protein.

IMPLICATION Some mutations alter the sequence of DNA bases in a gene but do not produce a noticeable change in the gene's polypeptide product. Explain how this can happen.

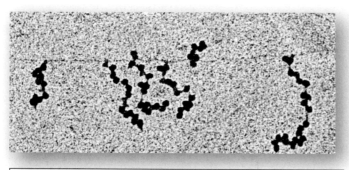

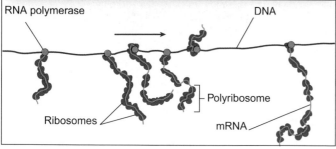

Figure 12.8 Transcription and translation in prokaryotes.

Ribosomes attach to an mRNA as it is formed, producing polyribosomes that translate the gene soon after it is transcribed.

12.3 Gene Expression

CONCEPT PREVIEW: The process of gene expression is similar in prokaryotes and eukaryotes, but differences exist in the architecture of the gene and the location in the cell where transcription and translation occur.

Architecture of the Gene

In prokaryotes, a gene is an uninterrupted stretch of DNA nucleotides. In eukaryotes, by contrast, genes are fragmented. In eukaryotic genes, the DNA nucleotide sequences encoding the amino acid sequence of a polypeptide are called **exons,** and the exons are interrupted frequently by extraneous nucleotides, "extra stuff" called **introns.** You can see them in the segment of DNA illustrated in figure 12.7; the exons are the blue areas and the introns are the orange areas. Imagine looking at an interstate highway from a satellite. Scattered randomly along the thread of concrete would be cars, some moving in clusters, others individually; most of the road would be bare. That is what a eukaryotic gene is like: scattered exons embedded within much longer sequences of introns. In humans, only 1.5% of the genome is devoted to exons, while 24% is devoted to the noncoding introns.

When a eukaryotic cell transcribes a gene, it first produces a **primary RNA transcript** of the entire gene, shown in figure 12.7 with the exons in green and the introns in orange. Enzymes add modifications called a *5´cap* and a *3´ poly-A* tail, which protect the RNA transcript from degradation. The primary transcript is then "processed," the introns removed and the exons joined together to form the shorter mRNA transcript that is actually translated into an amino acid chain. Notice that the mRNA transcript in figure 12.7 contains only exons (green segments).

Why have introns at all? It appears that many human genes exons are functional modules that can be spliced together in more than one way. One exon might encode a straight stretch of protein, another a curve, yet another a flat place. Like mixing Tinkertoy parts, you can construct quite different assemblies by employing the same exons in different combinations and orders. With this process called **alternative splicing,** the 20,000 to 25,000 genes of the human genome seem to encode as many as 120,000 different messenger RNAs.

Protein Synthesis

Prokaryotic cells lack a nucleus and so there is no barrier between where mRNA is synthesized during transcription and where proteins are formed during translation. Consequently, a gene can be translated as it is being transcribed (figure 12.8). In eukaryotic cells, a nuclear membrane separates the process of transcription from translation, making protein synthesis much more complicated. Figure 12.9 walks you through the entire process. Transcription (step ❶) and RNA processing (step ❷) occur within the nucleus. In step ❸, the mRNA travels to the cytoplasm where it binds to the ribosome. In step ❹, tRNAs bind to their appropriate amino acids, that correspond to their anticodons. In steps ❺ and ❻, the tRNAs bring the amino acids to the ribosome and the mRNA is translated into a polypeptide.

Concept Check

1. What process produces an mRNA molecule, transcription or translation?
2. Of the 64 codons, how many do not code for an amino acid?
3. How does an RNA transcript differ in prokaryotes and eukaryotes?

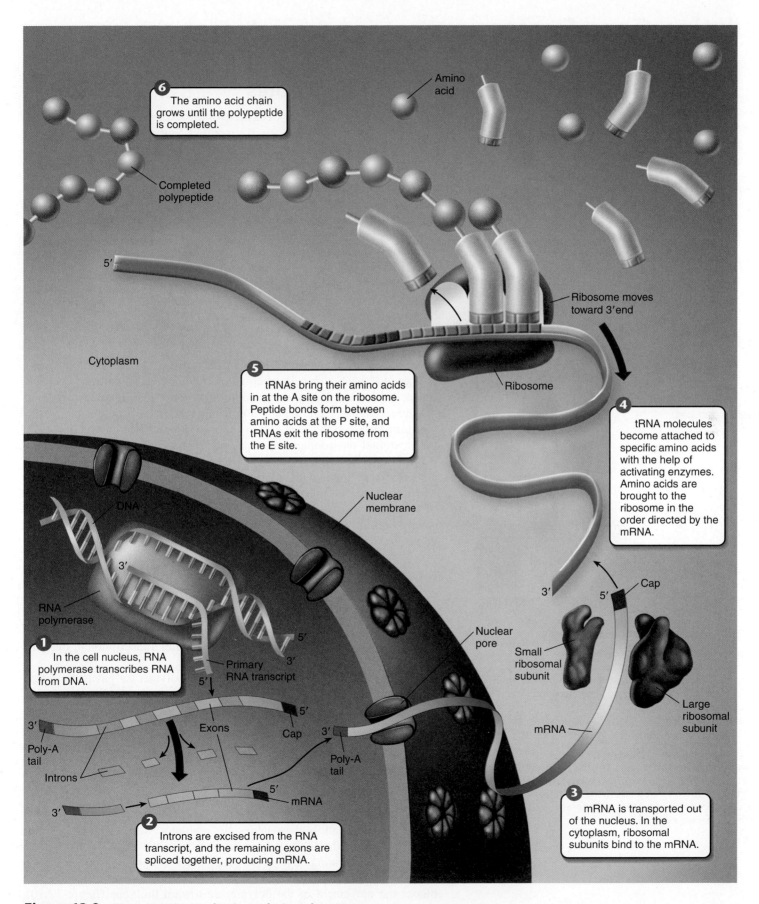

Figure 12.9 **How protein synthesis works in eukaryotes.**

Regulating Gene Expression

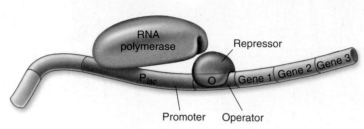

(a) *lac* operon is "repressed"

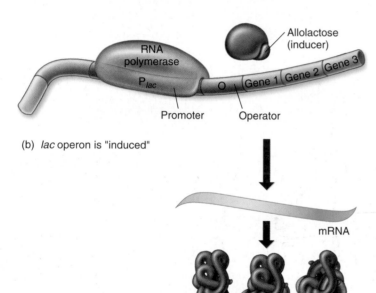

(b) *lac* operon is "induced"

Figure 12.10　**How the lac operon works.**

(a) The *lac* operon is shut down ("repressed") when the repressor protein is bound to the operator site. Because promoter and operator sites overlap, RNA polymerase and the repressor cannot bind at the same time. (b) The *lac* operon is transcribed ("induced") when allolactose binds to the repressor protein changing its shape so that it can no longer sit on the operator site and block polymerase binding.

IMPLICATION　Certain antibiotics act by interfering with transcription or translation in bacteria. The rifamycins are a group of drugs that are used in the treatment of tuberculosis. Unlike the repressor in the *lac* operon that binds to the operator site on the DNA, rifamycin binds to bacterial RNA polymerase. How can this effectively treat a bacterial infection?

12.4 Transcriptional Control in Prokaryotes

CONCEPT PREVIEW: Cells control the expression of genes by determining when they are transcribed. Some regulatory proteins block the binding of RNA polymerase, and others facilitate it.

Being able to translate a gene into a polypeptide is only part of gene expression. Every cell must also be able to regulate when particular genes are used. Imagine if every instrument in a symphony played at full volume all the time, all the horns blowing full blast and each drum beating as fast and loudly as it could! No symphony plays that way, because music is more than noise—it is the controlled expression of sound. In the same way, growth and development are due to the controlled expression of genes, each brought into play at the proper moment to achieve precise and delicate effects.

Scientist can determine which genes are expressed by measuring what proteins are present in the cell. Another way is to determine the identity of the mRNA molecules present in the cell. This can be accomplished with the use of gene microarrays, as discussed on page 224.

Control of gene expression is accomplished very differently in prokaryotes than in the cells of complex multicellular organisms. Prokaryotic cells have been shaped by evolution to grow and divide as rapidly as possible, enabling them to exploit transient resources. Proteins in prokaryotes turn over rapidly. This allows them to respond quickly to changes in their external environment by changing patterns of gene expression. In prokaryotes, the primary function of gene control is to adjust the cell's activities to its immediate environment. Changes in gene expression alter which enzymes are present in the cell in response to the quantity and type of available nutrients and the amount of oxygen present. Almost all of these changes are fully reversible, allowing the cell to adjust its enzyme levels up or down as the environment changes.

How Prokaryotes Turn Genes Off and On

Prokaryotes control the expression of their genes largely by saying *when* individual genes are to be transcribed. At the beginning of each gene are special regulatory sites that act as points of control. Specific regulatory proteins within the cell bind to these sites, turning transcription of the gene off or on.

For a gene to be transcribed, the RNA polymerase has to bind to a **promoter,** a specific sequence of nucleotides on the DNA that signals the beginning of a gene. In prokaryotes, gene expression is controlled by either blocking or allowing the RNA polymerase access to the promoter. Genes can be turned off by the binding of a **repressor,** a protein that binds to the DNA blocking the promoter. Genes can be turned on by the binding of an **activator,** a protein that makes the promoter more accessible to the RNA polymerase.

Repressors.　Many genes are "negatively" controlled: They are turned off except when needed. In these genes, the regulatory site is located between the place where the RNA polymerase binds to the DNA (the promoter site) and the beginning edge of the gene. When a regulatory protein called a repressor is bound to its regulatory site, called the *operator,* its presence blocks the movement of the polymerase toward the gene (figure 12.10).

Imagine if you went to sit down to eat dinner and someone was already sitting in your chair—you could not begin your meal until this person was removed from your chair. In the same way, the polymerase cannot begin transcribing the gene until the repressor protein is removed.

To turn on a gene whose transcription is blocked by a repressor, all that is required is to remove the repressor. Cells do this by binding special "signal" molecules to the repressor protein; the binding causes the repressor protein to contort into a shape that doesn't fit DNA, and it falls off, removing the barrier to transcription. A specific example demonstrating how repressor proteins work is the set of genes called the *lac* operon in the bacterium *Escherichia coli*. An **operon** is a segment of DNA containing a cluster of genes that are transcribed as a unit. The *lac* operon, shown in figure 12.10, consists of both polypeptide-encoding genes (labeled genes 1, 2, and 3, which code for enzymes involved in breaking down the sugar lactose) and associated regulatory elements—the operator (the purple segment) and promoter (the orange segment). Transcription is turned off when a repressor molecule binds to the operator such that RNA polymerase cannot bind to the promoter. When *E. coli* encounters the sugar lactose, a metabolite of lactose called allolactose binds to the repressor protein and induces a twist in its shape that causes it to fall from the DNA. As you can see in figure 12.10*b*, RNA polymerase is no longer blocked, so it starts to transcribe the genes needed to break down the lactose to get energy.

Activators.   Because RNA polymerase binds to a specific promoter site on one strand of the DNA double helix, it is necessary that the DNA double helix unzip in the vicinity of this site for the polymerase protein to be able to sit down properly. This unzipping requires the assistance of a regulatory protein called an activator that binds to the DNA nearby and helps it unwind. Cells can turn the genes on and off by binding "signal" molecules to this activator protein. In the *lac* operon, a protein called catabolite activator protein (CAP) acts as an activator. CAP has to bind a signal molecule, cAMP, before it can associate with the DNA. Once the CAP/cAMP complex forms, as seen in figure 12.11, it binds to the DNA and makes the promoter more accessible to RNA polymerase.

> Recall from the discussion on page 190 that the DNA is unwound by an enzyme called helicase. When the DNA is unwound, DNA polymerase has access to the DNA and can replicate it. Similarly, an activator allows RNA polymerase access to the gene.

Why bother with activators? Imagine if you had to eat every time you encountered food! Activator proteins enable a cell to cope with this sort of problem. Activators and repressors work together to control transcription. To understand how, let's consider the *lac* operon again, now shown in figure 12.12. When a bacterium encounters the sugar lactose, it may already have lots of energy in the form of glucose, as shown in panel ❶, and so does not need to break down more lactose. CAP can only bind and activate gene transcription when glucose levels are low. Because RNA polymerase requires the activator to function, the *lac* operon is not expressed. Also if glucose is present and lactose is absent, not only is the activator CAP unable to bind, but also a repressor blocks the promoter, as shown in panel ❷. In the absence of both glucose and lactose, cAMP, the "low glucose" signal molecule (the green pie-shaped piece in panels ❸ and ❹) binds to CAP, and CAP is able to bind to the DNA. However, the repressor is still blocking transcription, as shown in panel ❸. Only in the absence of glucose and in the presence of lactose, the repressor is removed, the activator (CAP) is bound, and transcription proceeds, as shown in panel ❹.

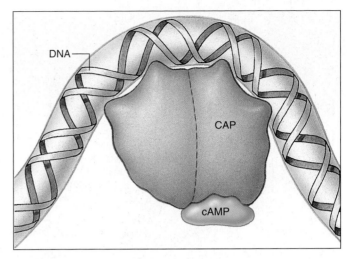

Figure 12.11   **How an activator works.**

Binding of the catabolite activator protein complex to DNA causes the DNA to bend around it. This increases the activity of RNA polymerase.

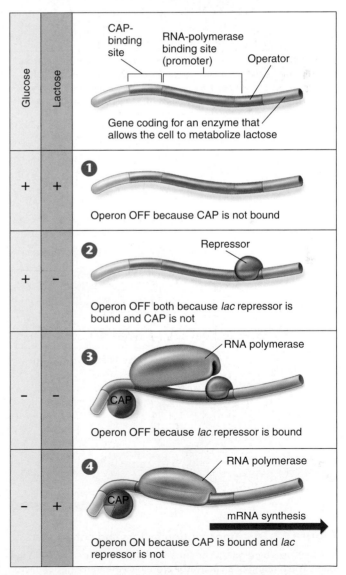

Figure 12.12   **Activators and repressors at the *lac* operon.**

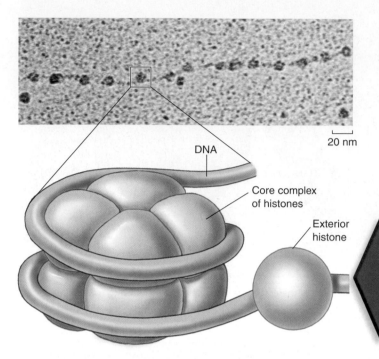

DNA

20 nm

Core complex
of histones

Exterior
histone

Figure 12.13 **DNA coils around histones.**

In chromatin, DNA is packaged into nucleosomes. In the electron micrograph (*top*), the DNA is partially unwound, and individual nucleosomes can be seen. In a nucleosome, the DNA double helix coils around a core complex of eight histones; one additional histone binds to the outside of the nucleosome.

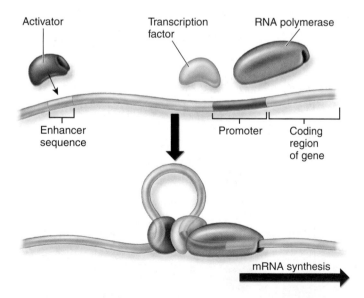

Activator

Transcription
factor

RNA polymerase

Enhancer
sequence

Promoter

Coding
region
of gene

mRNA synthesis

Figure 12.14 **Transcription factors and enhancers.**

The activator binding site, or enhancer, is often located far from the gene. The binding of an activator protein to the transcription factor is required for polymerase initiation.

12.5 Transcriptional Control in Eukaryotes

CONCEPT PREVIEW: Transcriptional control in eukaryotes can be effected by the tight packaging of DNA into nucleosomes. Transcription factors and enhancers also give these cells great flexibility in controlling gene expression.

Chromatin Structure

In multicellular organisms with relatively constant internal environments, the primary function of gene control in a cell is not to respond to that cell's immediate environment, like a prokaryote does, but rather to participate in regulating the body as a whole. Some of these changes in gene expression compensate for changes in the physiological condition of the body. Others mediate the decisions that ultimately produce the body, ensuring that the right genes are expressed in the right cells at the right time during development. In multicellular organisms, changes in gene expression serve the needs of the whole organism, rather than the survival of individual cells.

> An understanding of gene expression helps explain the gene theory, discussed on page 24. All cells in an organism have the same DNA, but not all of the cells express the same genes; different cells need different proteins to function. Many genes are never turned on in a given cell.

The first hurdle faced by RNA polymerase in transcribing a eukaryotic gene is gaining access to it. As described in chapter 8, eukaryotic chromosomes consist of chromatin, a complex of DNA and protein. In chromatin, the DNA is packaged with histone proteins into nucleosomes (figure 12.13). During mitosis, nucleosomes wind, twist, and coil into chromosomes, but even during interphase when transcription is occurring, some sections of DNA remain as part of nucleosomes. These nucleosomes seem to block the binding of RNA polymerase and other proteins called transcription factors to the promoter site. Histones can be chemically modified to result in a greater condensation of the chromatin, making it even less accessible. Another chemical modification of the DNA—adding a methyl group ($-CH_3$) to cytosine nucleotides—blocks accidental transcription of "turned-off" genes. DNA methylation ensures that once a gene is turned off, it stays off.

Transcription Factors and Enhancers

Eukaryotic transcription requires not only the RNA polymerase molecule, but also a variety of other proteins, called *transcription factors,* that interact with the polymerase to form an *initiation complex.* Transcription factors are necessary for the assembly of the initiation complex and recruitment of RNA polymerase to a promoter. While these factors, colored green in figure 12.14, are required for transcription to occur, they do not increase the rate of transcription.

While prokaryotic gene control regions, such as the operator, are positioned immediately upstream of the coding region, this is not true in eukaryotes, in which far away sites called **enhancers** can have a major impact on the rate of transcription. The ability of enhancers to act over large distances is accomplished by DNA bending to form a loop. In figure 12.14, the DNA loops, bringing the activator in contact with the RNA polymerase/initiation complex so that transcription can begin.

This complex transcriptional control is only one aspect of gene regulation in eukaryotic cells. Other processes associated with translation, such as gene silencing, protein synthesis, and post-translational modifications are described in *Essential Biological Process 12B* and provide the cell the ability to produce finely graded responses to environmental and developmental signals.

Essential Biological Process 12B

Control of Eukaryotic Gene Expression

1

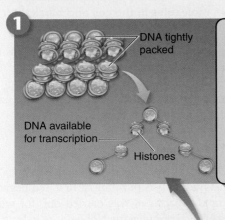

DNA tightly packed

DNA available for transcription

Histones

Chromatin structure
Access to many genes is affected by the packaging of DNA and by chemically altering the histone proteins.

2

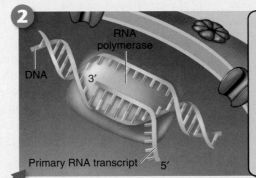

RNA polymerase

DNA

3'

Primary RNA transcript

5'

Initiation of transcription
Most control of gene expression is achieved by regulating the frequency of transcription initiation.

3

Intron   Exon

Cut intron

5' cap

3' poly-A tail   Mature RNA transcript

RNA splicing
Gene expression can be controlled by altering the rate of splicing in eukaryotes. Alternative splicing can produce multiple mRNAs from one gene.

4

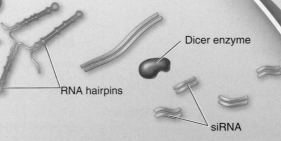

Dicer enzyme

RNA hairpins

siRNA

Gene silencing
Cells can silence genes with siRNAs, which are cut from inverted sequences that fold into double-stranded loops. siRNAs bind to mRNAs and block their translation.

6

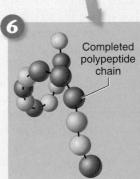

Completed polypeptide chain

Post-translational modification
Phosphorylation or other chemical modifications can alter the activity of a protein after it is produced.

5

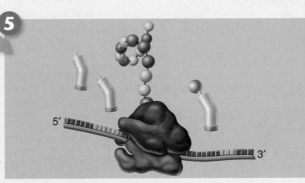

5'

3'

Protein synthesis
Many proteins take part in the translation process, and regulation of the availability of any of them alters the rate of gene expression by speeding or slowing protein synthesis.

IN THE NEWS

Silencing AIDS. After decades of searching unsuccessfully for an effective AIDS vaccine, researchers are turning to other approaches. One of the most promising centers on a 42-year-old American resident of Germany, who in 2008 was completely cured of AIDS. How? First, his immune system was wiped out with radiation and drugs, then he received blood stem cells (which make both circulating blood cells and immune system cells) from a person homozygous for a gene mutation known as *delta 32*. This mutation destroys the cell surface receptors that allow HIV to invade the immune system, so that people homozygous for it are immune to HIV infection and AIDS. Because the patient's own blood stem cells had been destroyed, these new ones reconstituted his blood supply and immune system with *delta 32* cells. He has been free of the HIV virus for 20 months and it appears he has been permanently cured of AIDS. Wiping out a person's immune system is far too dangerous as a general therapy (10%–30% die), but the success in this instance has encouraged researchers to try using RNA "scissors" as described on this page to cut out the *delta 32* gene from a patient's own blood stem cells and reinject them after wiping out only the patient's blood stem cells, not the entire immune system. Researchers report that the concept is working in monkeys.

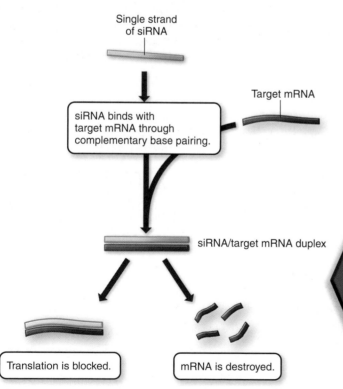

Figure 12.15   RNA interference.

RNA interference stops gene expression either by blocking the translation of the gene (on the left) or by targeting the mRNA for destruction before it can be translated.

12.6 RNA-Level Control

CONCEPT PREVIEW: Small interfering RNAs, called siRNAs, are formed from double-stranded sections of RNA molecules. These siRNAs bind to mRNA molecules in the cell and block their translation in a process called RNA interference.

Thus far we have discussed gene regulation entirely in terms of proteins that regulate the start of transcription by blocking or activating the "reading" of a particular gene by RNA polymerase. Within the last decade, however, it has become increasingly clear that RNA molecules also regulate gene expression, acting after transcription as a second equally important level of control.

Discovery of RNA Interference

As will be discussed in detail in chapter 13, the bulk of the eukaryotic genome is not translated into proteins. This finding was puzzling at first, but began to make sense in 1998, when a simple experiment was carried out, for which Americans Andrew Fire and Greg Mello later won the Nobel Prize in Physiology or Medicine in 2006. These investigators injected double-stranded RNA molecules into the nematode worm *Caenorhabditis elegans*. This resulted in the silencing of the gene whose sequence was complementary to the double-stranded RNA, and of no other gene. The investigators called this very specific effect **gene silencing,** or **RNA interference.** What is going on here? A group of viruses called RNA viruses, those that contain RNA as their hereditary storage molecule rather than DNA, replicate themselves through double-stranded intermediates. RNA interference may have evolved as a cellular defense mechanism against these viruses with double-stranded viral RNAs being targeted for destruction by RNA interference machinery. Without intending to do so, the nematode researchers had stumbled across this defense.

How RNA Interference Works

Investigating interference, researchers noted that in the process of silencing a gene, plants produced short RNA molecules (ranging in length from 21 to 28 nucleotides) that matched the gene being silenced. Earlier researchers focusing on far larger messenger RNA (mRNA), transfer RNA (tRNA), and ribosomal RNA (rRNA) had not noticed these far smaller bits, tossing them out during experiments. These bits of RNA are called "small interfering RNA" or siRNA. How do these small fragments of RNA silence the activity of specific genes? In a complex way we are just beginning to understand clearly, the small RNA fragments bind to any mRNA molecules in the cell that have a complementary sequence. Silencing of the gene that produced this mRNA is achieved in one of two ways (figure 12.15): either the expression of the mRNA is inhibited by blocking its translation into protein, or the mRNA is simply destroyed. In either case, the specific gene that produced that mRNA fails to be expressed—it is silenced.

> mRNA molecules are single-stranded, as described on page 52. During translation, the codon in the mRNA is exposed so it can bind to the complementary anticodon of the tRNA. If the mRNA codons are blocked by siRNA, the codons can't be read to make a polypeptide.

Concept Check

1. If lactose is added to their growth medium, bacteria begin to transcribe the *lac* operon. How is transcription possible now, when it wasn't before?
2. Enhancers are located far from the genes they regulate but in order to function they need to be in contact with the gene. How is that achieved?
3. How can RNA be used to regulate gene expression?

Silencing Genes to Treat Disease

The recent discovery that eukaryotes control their genes by selectively "silencing" particular gene transcripts has electrified biologists, as it opens exciting possibilities for treating disease and infection. Many diseases are caused by expression of one or more genes. AIDS for example requires the expression of several genes of the HIV virus. Many chronic human diseases result from excessively active genes. What if doctors could somehow shut these genes off? The expression of the diseases could be halted in their tracks.

The idea is simple. If you can isolate a gene involved in the disorder and determine its sequence, then in principle you could synthesize an RNA molecule with the sequence of the opposite or "antisense" strand. This RNA would thus have a sequence complementary to the messenger RNA produced by that gene. Introduced into cells, this synthesized RNA might be able to bind to the mRNA, creating a double-stranded RNA that could not be read by ribosomes. If an antisense therapy could be made to work and be delivered practically and inexpensively, the AIDS epidemic could be halted in its tracks. Indeed, any viral infection could be combatted in this way. Influenza is perhaps the greatest killer of all infectious diseases. A workable antisense therapy could provide a means of stamping out a bird flu epidemic before the virus spreads.

By far the most exciting promise of antisense gene silencing therapy is the possibility of a practical cancer therapy. Discussed in chapter 8, cancer kills more Americans than any other disease. We now know in considerable detail how cancer comes about. It results from damage to genes that regulate the cell cycle. The great promise of RNA gene silencing therapy comes from those cancer-causing gene mutations that increase the effectiveness of one or more "divide" signals. If these mutant genes could be silenced, the cancer could be shut down.

The prospect of using complementary RNA to silence troublesome genes has gotten a huge boost in the last few years from the discovery of a unique virus defense system in eukaryotes. In order to protect themselves from RNA virus infection, cells have a complex system for detecting, attacking, and destroying viral RNA. The system takes advantage of a subtle vulnerability of the infecting virus: at some point, in order to multiply within the infected cell, the virus must express its genes—it must make complementary copies of them that can serve as messenger RNAs to direct production of viral proteins. At that point, while the viral RNA molecule is double-stranded, the virus is vulnerable to attack: at no place in the cell is double-stranded RNA usually found, so by targeting double-stranded RNAs for immediate destruction, a cell can defeat viral infections.

Silencing genes with complementary RNA, dubbed "RNA interference," offers the exciting hope that successful treatment of many diseases may be literally at our doorstep. First, however, scientists must figure out how to make RNA interference therapies work. They are facing some formidable technical problems, not the least of which is to find a way to deliver the interfering RNA to, and into, the

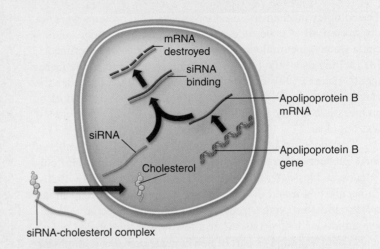

siRNA-cholesterol complex

target cells. The problem is that RNA is rapidly broken down in the bloodstream, and most of the body's cells don't readily absorb it, even if it does reach them. Some researchers are attempting to package the RNA into viruses, although gene therapies that have attempted this approach can trigger an immune response and could even cause cancer. Gene therapy researchers have been seeking safer virus gene-delivery vehicles; what they learn will surely be put to good effect.

One interesting alternative approach is to modify the RNA to protect it and make it more easily taken up by cells. This work focuses on the mRNA that encodes apolipoprotein B, a molecule involved in the metabolism of cholesterol. High levels of apolipoprotein are found in people with high levels of cholesterol, associated with increased risk of coronary heart disease. Interfering RNAs that target apolipoprotein B mRNA result in destruction of the mRNA, and lower levels of cholesterol. To effectively deliver it to the body's tissues, researchers simply attached a molecule of cholesterol to each interfering RNA molecule (see diagram above). Levels of apolipoprotein B were reduced 50–70%, and blood cholesterol levels plummeted downward, to the same levels seen in cells from which the apolipoprotein B gene had been deleted. It is not clear if this approach will work for many other RNAs, but it looks promising.

A second major problem confronting those seeking to develop successful therapies based on RNA silencing of troublesome genes is one of specificity. It is very important that only the target gene be silenced. Before carrying out clinical trials involving large numbers of people, it is imperative that we be sure the interfering RNA will not shut down vital human genes as well as the targeted virus or cancer genes. Some studies suggest this will not be a problem, while in others, a range of "off-target" genes seem to be affected. This possibility will have to be carefully evaluated for each new therapy being developed.

Building Proteins in a Test Tube

The complex mechanisms used by cells to build proteins were not discovered all at once, in a flash of insight. Our understanding came slowly, accumulating through a long series of experiments, each telling us a little bit more. To gain some sense of the incremental nature of this experimental journey, and to appreciate the excitement that each step gave, it is useful to step into the shoes of an investigator back when little was known and the way forward was not clear.

The shoes we will step into are those of Paul Zamecnik, an early pioneer in protein synthesis research. Working with colleagues at Massachusetts General Hospital in the early 1950s, Zamecnik first asked the most direct of questions: Where in the cell are proteins synthesized? To find out, they injected radioactive amino acids into rats. After a few hours, the labeled amino acids could be found as part of newly made proteins in the livers of the rats. And, if the livers were removed and checked only minutes after injection, radioactive-labeled proteins were found only associated with small particles in the cytoplasm. Composed of protein and RNA, these particles, later named ribosomes, had been discovered years earlier by electron microscope studies of cell components. This experiment identified them as the sites of protein synthesis in the cell.

After several years of trial-and-error tinkering, Zamecnik and his colleagues had worked out a "cell-free" protein-synthesis system that would lead to the synthesis of proteins in a test tube. It included ribosomes, mRNA, and ATP to provide energy. It also included a collection of required soluble "factors" isolated from homogenized rat cells that somehow worked with the ribosome to get the job done. When Zamecnik's team characterized these required factors, they found most of them to be proteins, as expected, but also present in the mix was a small RNA, very unexpected.

To see what this small RNA was doing, they performed the following experiment. In a test tube, they added various amounts of ^{14}C-leucine (that is, the radioactively labeled amino acid leucine) to the cell-free system containing the soluble factors, ribosomes, and ATP. After waiting a bit, they then isolated the small RNA from the mixture and checked it for radioactivity. You can see the results in graph (a) above.

In a follow-up experiment, they mixed the radioactive leucine-small RNA complex that this experiment had generated with cell extracts containing intact endoplasmic reticulum (that is, a cell system of ribosomes on membranes quite capable of making protein). Looking to see where the radioactive label now went, they then isolated the newly-made protein [red in graph (b)] as well as the small RNA [blue in graph (b)].

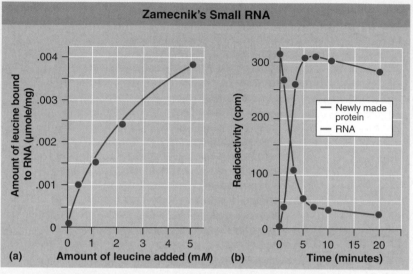

Analysis

EXPERIMENT A, shown in graph (a)

1. **Interpreting Data**   Does the amount of leucine added to the test tube have an effect on the amount of leucine found bound to the small RNA?

2. **Making Inferences**   Is the amount of leucine bound to small RNA proportional to the amount of leucine added to the mixture?

3. **Drawing Conclusions**   Can you reasonably conclude from this result that the amino acid leucine is binding to the small RNA?

EXPERIMENT B, shown in graph (b)

1. **Interpreting Data**
   a. Monitoring radioactivity for 20 minutes after the addition of the radioactive leucine-small RNA complex to the cell extract, what happens to the level of radioactivity in the small RNA (blue)?
   b. Over the same period, what happens to the level of radioactivity in the newly made protein (red)?

2. **Making Inferences**   Is the same amount of radioactivity being lost from the small RNA that is being gained by the newly made protein?

3. **Drawing Conclusions**
   a. Is it reasonable to conclude that the small RNA is donating its amino acid to the growing protein? [Hint: As a result of this experiment, the small RNA was called transfer RNA.]
   b. If you were to isolate the protein from this experiment made after 20 minutes, which amino acids would be radioactively labeled? Explain.

Concept Summary

From Gene to Protein

12.1 Transcription

- DNA is the storage site of genetic information in the cell. The process of gene expression, DNA to RNA to protein, is called the central dogma, shown here from **figure 12.1.**

- Gene expression occurs in two stages using three types of RNA: transcription, where an mRNA copy is made from the DNA, and translation, where the information on the mRNA is translated into a protein using rRNA and tRNA.

- In transcription, DNA serves as a template for mRNA synthesis by a protein called RNA polymerase. RNA polymerase binds to one strand of the DNA at a site called the promoter. RNA polymerase adds complementary nucleotides onto the growing mRNA in a way that is similar to the actions of DNA polymerase (**figure 12.2**).

12.2 Translation

- The message within the mRNA is coded in its sequence of nucleotides, which are read in three nucleotide units called codons. Codons correspond to particular amino acids. The rules that govern the translation of codons on the mRNA into amino acids is called the genetic code (**figure 12.3**).

- In translation, the mRNA carries the message to the cytoplasm. rRNA combines with proteins to form a structure called a ribosome (**figure 12.4**), the platform on which proteins are assembled.

- A ribosome contains a small and a large subunit. Translation begins when the mRNA binds to the small ribosomal subunit, which triggers the binding of the large subunit to the small subunit, forming a complete ribosome. The ribosome has three sites, called A, P, and E, to which a third type of RNA, called transfer RNA (tRNA) binds.

- tRNA molecules, like one shown here from **figure 12.5,** carry amino acids to the ribosome to build the polypeptide chain. The amino acid that attaches to the tRNA is determined by a three-nucleotide sequence on the tRNA called the anticodon.

- A ribosome moves along mRNA, while the tRNA molecules that contain anticodon sequences that are complementary to the codons bring the appropriate amino acids to the ribosome. An incoming tRNA first attaches to the A site, then moves to the P and E sites as the ribosome moves, leaving the A site open for the next tRNA. The amino acids add to the growing polypeptide chain (***Essential Biological Process 12A*** and **figure 12.6**).

- When a "stop" codon is reached, the ribosome dissociates and releases the polypeptide into the cell.

12.3 Gene Expression

- Prokaryotic genes are contained within a stretch of DNA that is transcribed and translated in its entirety.

- Eukaryotic genes are fragmented, containing coding regions called exons, and noncoding regions called introns. The entire eukaryotic gene is transcribed into RNA, but the introns are spliced out before translation (**figure 12.7**).

- Exons can be spliced together in different ways, a process called alternative splicing, producing different protein products from the same sections of DNA.

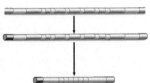

- The process of protein synthesis is different in eukaryotes compared to prokaryotes. In prokaryotes, transcription and translation can occur simultaneously. Ribosomes attach to the mRNA as it is transcribed, beginning translation before transcription is completed (**figure 12.8**). In eukaryotes, the RNA transcript is first produced and then processed (introns spliced out) in the nucleus. The mRNA then travels to the cytoplasm, where it is translated into a polypeptide (**figure 12.9**).

Regulating Gene Expression

12.4 Transcriptional Control in Prokaryotes

- In prokaryotic cells, genes are turned off when a repressor protein blocks the promoter, as shown here from **figure 12.10,** binding to a site called an operator.

- Some genes can be turned on only when a protein called an activator binds to the DNA and opens up the double helix so that RNA polymerase can bind to the promoter.

- The *lac* operon contains a cluster of genes that are involved in the breakdown of the sugar lactose. When the proteins produced by the *lac* operon genes are needed, an inducer molecule will bind to the repressor protein so it can't attach to the DNA, thereby freeing the promoter so that RNA polymerase can bind (**figure 12.10**).

- The *lac* operon is also controlled by an activator. The activator alters the shape of DNA, which allows the RNA polymerase to bind to the DNA (**figure 12.11**). It is only when the activator binds to the DNA and the repressor is removed that the RNA polymerase can bind to the promoter (**figure 12.12**).

12.5 Transcriptional Control in Eukaryotes

- The coiling of DNA around histones (**figure 12.13**) restricts RNA polymerase's access to the DNA. Controlling gene expression may involve chemical modification of histones, that make the DNA more accessible, or methylation of the DNA itself that keeps a gene turned off.

- In eukaryotic cells, transcription requires the binding of transcription factors before RNA polymerase can bind to the promoter. Eukaryotic genes are controlled from distant locations called enhancers. A regulatory protein binds to the enhancer region far from the gene and the DNA forms a loop, as shown here from **figure 12.14,** bringing the distant enhancer region into contact with the transcription factors. Eukaryotic cells have several levels of gene expression (***Essential Biological Process 12B***).

12.6 RNA-Level Control

- RNA interference blocks translation. Small sections of RNA, called siRNA, bind to mRNA in the cytoplasm, which blocks the expression of the gene (**figure 12.15**).

Self-Test

1. Which of the following is not a type of RNA?
   a. nRNA (nuclear RNA)   **c.** rRNA (ribosomal RNA)
   b. mRNA (messenger RNA) **d.** tRNA (transfer RNA)
2. Each amino acid in a polypeptide is specified by
   a. an enhancer.   **c.** an rRNA molecule.
   b. a promoter.   **d.** a codon.
3. Which of the following statements is correct about the genetic code?
   a. Every codon encodes an amino acid.
   b. Each amino acid is encoded by only one codon.
   c. A codon consists of three nuceotides.
   d. A codon and its complementary anticodon have the same sequences.
4. The process of obtaining a copy of the information in a gene as a strand of messenger RNA is called
   a. polymerase.   **c.** transcription.
   b. expression.   **d.** translation.
5. The site where RNA polymerase attaches to the DNA molecule to start the formation of an RNA molecule is called a(n)
   a. promoter.   **c.** intron.
   b. exon.   **d.** enhancer.
6. The process of taking the information on a strand of messenger RNA and building an amino acid chain, which will become all or part of a protein molecule, is called
   a. polymerase.   **c.** transcription.
   b. expression.   **d.** translation.

7. If an mRNA codon reads UAC, its complementary anticodon will be
   a. TUC.   **c.** AUG.
   b. ATG.   **d.** UAC.
8. Which of the following accurately describes gene expression in prokaryotic cells?
   a. All genes are on all the time in all cells, making the needed amino acid sequences.
   b. Some genes are always off unless a promoter turns them on.
   c. Some genes are always on unless a promoter turns them off.
   d. Some genes remain off as long as a repressor is bound.
9. Which of the following statements is correct about eukaryotic gene expression?
   a. mRNAs must have introns spliced out.
   b. mRNAs contain the transcript of only one gene.
   c. Enhancers act from a distance.
   d. All of these statements are correct.
10. Which of the following is *not* a mechanism of controlling gene expression in eukaryotic cells?
    a. Blocking translation with siRNA.
    b. Activating an enhancer.
    c. Translating a gene as it is being transcribed.
    d. Alternative splicing of the primary RNA transcript.

Visual Understanding

1. **Figure 12.1** This figure of the central dogma represents the general process of gene expression in prokaryotes and eukaryotes, however the process is slightly more complex in eukaryotic cells. What step is missing here that occurs in eukaryotic cells but not in prokaryotic cells?

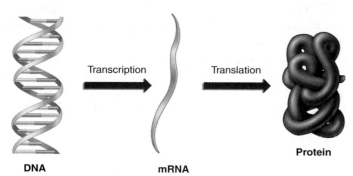

Transcription   Translation

Protein

DNA   **mRNA**

2. **Figure 12.10** Can genes 1, 2, and 3 be transcribed? What would happen if an inducer molecule was present in the cell?

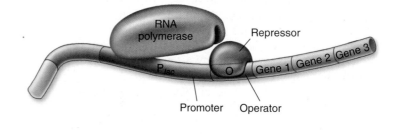

RNA polymerase   Repressor

P_lac   O   Gene 1 Gene 2 Gene 3

Promoter   Operator

Challenge Questions

1. Compare DNA to a cookbook. The book is kept in a library and cannot be checked out (removed). Start with the letters and words in the cookbook compared with the bases and codons in DNA; end with the amino acid chain being folded into a protein compared to a cake being baked.

2. What would happen if all the genes in a cell were always active?

3. If you travel to a foreign country where you don't speak the language, you might hire a translator to translate what you want to say from your language to the language used in the country you are visiting. The central dogma works in a similar way: The language of nucleic acids

(which uses nucleotides) is translated into the language of proteins (which uses amino acids). That being the case, which molecule serves as the translator?

4. The nucleotide sequence of a hypothetical gene is:

   TACATACTTAGTTACGTCGCCCGGAAATAT
   a. What will be the sequence on the mRNA when it is transcribed?
   b. What will be the amino acid sequence of the protein when it's translated?
   c. What would happen to the amino acid chain if the highlighted nucleotide underwent a mutation and was changed to an A nucleotide?

Chapter **13**

The New Biology

Sequencing Entire Genomes

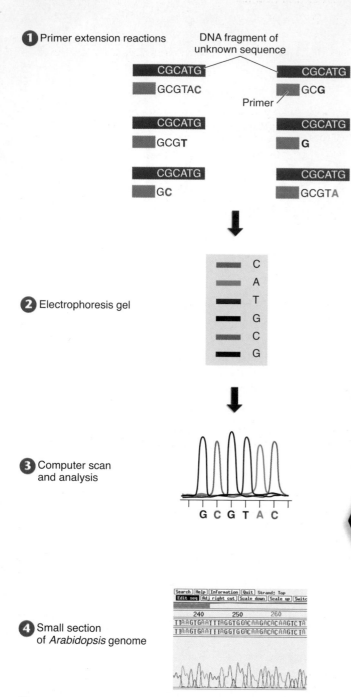

1 Primer extension reactions

DNA fragment of unknown sequence

CGCATG
GCGTAC
CGCATG
GCG

Primer

CGCATG
GCGT
CGCATG
G

CGCATG
GC
CGCATG
GCGTA

2 Electrophoresis gel

C
A
T
G
C
G

3 Computer scan and analysis

G C G T A C

4 Small section of *Arabidopsis* genome

Search Help Information Quit Strand: Top
Edit seq Adj right cut Scale down Scale up Switc

240   250   260
TTAAGTGAATTTAGGTGGACAAGACACAAGTCTA
TTAAGTGAATTTAGGTGGACAAGACACAAGTCTA

Figure 13.1   How to sequence DNA.

1 DNA is sequenced by adding complementary bases to a single-stranded fragment. DNA synthesis stops when a chemical tag is inserted instead of a nucleotide, resulting in different sizes of DNA fragments. **2** The DNA fragments of varying lengths are separated by gel electrophoresis, the smaller fragments migrating farther down the gel. (The letters indicate the chemical tags added in step 1 that stopped the replication process.) **3** Computers scan the gel, from smallest to largest fragments, and display the DNA sequence as a series of colored peaks. **4** Data from an automated DNA-sequencing run show the nucleotide sequence for a small section of the *Arabidopsis* (plant) genome.

13.1 Genomics

CONCEPT PREVIEW: Powerful automated DNA sequencing technology is now revealing the DNA sequences of entire genomes.

The full complement of genetic information of an organism—all of its genes and other DNA—is called its **genome.** To study a genome the DNA is first sequenced, a process that allows each nucleotide of a DNA strand to be read in order. The first genome to be sequenced was a very simple one: a small bacterial virus called φ-X174 (φ is the Greek letter phi). Frederick Sanger, inventor of the first practical way to sequence DNA, obtained the sequence of this 5,375-nucleotide genome in 1977. This was followed by the sequencing of dozens of prokaryotic genomes. The advent of automated DNA sequencing machines in recent years has made the DNA sequencing of much larger eukaryotic genomes practical, including our own (table 13.1).

Sequencing DNA

In sequencing DNA, the DNA of unknown sequence is first cut into fragments. Each DNA fragment is then copied (amplified), so there are thousands of copies of the fragment. The DNA fragments are then mixed with copies of DNA polymerase, copies of a primer (recall from chapter 11 that DNA polymerase can only add nucleotides onto an existing strand of nucleotides), a supply of the four nucleotide bases, and a supply of four different chain-terminating chemical tags. The chemical tags act as one of the four nucleotide bases in DNA synthesis, undergoing complementary base pairing. First, heat is applied to denature the double-stranded DNA fragments. The solution is then allowed to cool, allowing the primer (the lighter blue box in figure 13.1 **1**) to bind to a single strand of the DNA, and synthesis of the complementary strand proceeds. Whenever a chemical tag is added instead of a nucleotide base, the synthesis stops, as shown in the figure. For example, the terminating red "T" was added after three normal nucleotides and synthesis stopped. Because of the relatively low concentration of the chemical tags compared with the nucleotides, a tag that binds to G on the DNA fragment, for example, will not necessarily be added to the first G site. Thus, the mixture will contain a series of double-stranded DNA fragments of different lengths, corresponding to the different distances the polymerase traveled from the primer before a chain-terminating tag was incorporated (six are shown in **1**).

> To better understand DNA sequencing it would be helpful to review the process of DNA replication on page 190. For example, if the T nucleotide being added in figure 11.7 was a chain-terminating nucleotide, replication would stop before any more nucleotides were added.

The series of fragments are then separated according to size by gel electrophoresis. The fragments become arrayed like the rungs of a ladder, each rung being one base longer than the one below it. Compare the lengths of the fragments in **1** and their positions on the gel in **2**. The shortest fragment has only one nucleotide (G) added to the primer, so it is the lowest rung on the gel. In automated DNA sequencing, fluorescently colored chemical tags are used to label the fragments, one color for each type of nucleotide. Computers read the colors on the gel to determine the DNA sequence and display this sequence as a series of colored peaks (**3** and **4**). The development of automated sequencers in the mid-1990s made the sequencing of large eukaryotic genomes practical. A research institute with several hundred such instruments can sequence 100 million base pairs every day, with only 15 minutes of human attention!

TABLE 13.1	Some Eukaryotic Genomes			
Organism		**Estimated Genome Size (Mbp)**	**Number of Genes (×1,000)**	**Nature of Genome**
Vertebrates				
	Homo sapiens (human)	3,200	20–25	The first large genome to be sequenced; the number of transcribable genes is far less than expected; much of the genome is occupied by repeated DNA sequences.
	Pan troglodytes (chimpanzee)	2,800	20–25	There are few base substitutions between chimp and human genomes, less than 2%, but many small sequences of DNA have been lost as the two species diverged, often with significant effects.
	Mus musculus (mouse)	2,500	25	Roughly 80% of mouse genes have a functional equivalent in the human genome; importantly, large portions of the noncoding DNA of mouse and human have been conserved; overall, rodent genomes (mouse and rat) appear to be evolving more than twice as fast as primate genomes (humans and chimpanzees).
	Gallus gallus (chicken)	1,000	20–23	One-third the size of the human genome; genetic variation among domestic chickens seems much higher than the genetic variation seen in humans.
	Fugu rubripes (pufferfish)	365	35	The *Fugu* genome is only one-ninth the size of the human genome, yet it contains 10,000 more genes.
Invertebrates				
	Caenorhabditis elegans (nematode)	97	21	The fact that every cell of *C. elegans* has been identified makes its genome a particularly powerful tool in developmental biology.
	Drosophila melanogaster (fruit fly)	137	13	*Drosophila* telomere regions lack the simple repeated segments that are characteristic of most eukaryotic telomeres. About one-third of the genome consists of gene-poor centric heterochromatin.
	Anopheles gambiae (mosquito)	278	15	The extent of similarity between *Anopheles* and *Drosophila* is approximately equal to that between human and pufferfish.
	Nematostella vectensis (sea anemone)	450	18	The genome of this cnidarian is much more like vertebrate genomes than nematode or insect genomes that appear to have become streamlined by evolution.
Plants				
	Oryza sativa (rice)	430	33–50	The rice genome contains only 13% as much DNA as the human genome, but roughly twice as many genes; like the human genome, it is rich in repetitive DNA.
	Populus trichocarpa (cottonwood tree)	500	45	This fast-growing tree is widely used by the timber and paper industries. Its genome, fifty times smaller than the pine genome, is one-third heterochromatin.
Fungi				
	Saccharomyces cerevisiae (brewer's yeast)	13	6	*S. cerevisiae* was the first eukaryotic cell to have its genome fully sequenced.
Protists				
	Plasmodium falciparum (malaria parasite)	23	5	The *Plasmodium* genome has an unusually high proportion of adenine and thymine. Scarcely 5,000 genes contain the bare essentials of the eukaryotic cell.

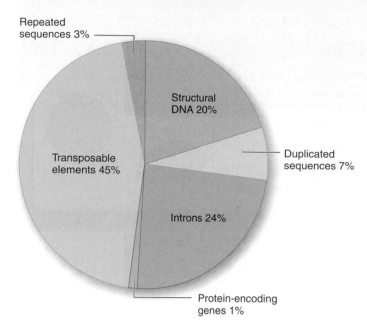

Repeated sequences 3%

Structural DNA 20%

Transposable elements 45%

Duplicated sequences 7%

Introns 24%

Protein-encoding genes 1%

Figure 13.2   The human genome.

Very little of the human genome is devoted to protein-encoding genes, indicated by the light blue section in this pie chart.

13.2 The Human Genome

CONCEPT PREVIEW: The entire 3.2-billion-base-pair human genome has been sequenced. Only about 1% to 1.5% of the human genome is devoted to protein-encoding genes. Much of the rest is composed of transposable elements.

On June 26, 2000, geneticists announced that the entire human genome had been sequenced. This effort presented no small challenge, as the human genome is huge—more than 3 billion base pairs, which is the largest genome sequenced to date. To get an idea of the magnitude of the task, consider that if all 3.2 billion base pairs were written down on the pages of this book, the book would be 500,000 pages long, and it would take you about 60 years, working eight hours a day, every day, at five bases a second, to read it all.

Reading the human genome for the first time, geneticists encountered three big surprises.

The Number of Genes Is Surprisingly Low

The human genome sequence contains only 20,000 to 25,000 protein-encoding genes, only 1% of the genome (figure 13.2). As you can see in table 13.1, this is scarcely more genes than in a nematode worm (21,000 genes), not quite double the number in a fruit fly (13,000 genes). Researchers had confidently anticipated at least four times as many genes, because over 100,000 unique messenger RNA (mRNA) molecules can be found in human cells—surely, they argued, it would take as many genes to make them.

How can human cells contain more mRNAs than genes? Recall from chapter 12 that in a typical human gene, the sequence of DNA nucleotides that specifies a protein is broken into many bits called exons, scattered among much longer segments of nontranslated DNA called introns. Imagine this paragraph was a human gene; all the occurrences of the letter "e" could be considered exons, while the rest would be noncoding introns, which make up 24% of the human genome.

When a cell uses a human gene to make a protein, it first manufactures mRNA copies of the gene, then splices the exons together, getting rid of the intron sequences in the process. Now here's the turn of events researchers had not anticipated: The transcripts of human genes are often spliced together in different ways, called *alternative splicing*. As we discussed in chapter 12, each exon is actually a module; one exon may code for one part of a protein, another for a different part of a protein. When the exon transcripts are mixed in different ways, very different protein shapes can be built.

> Alternative splicing, described on page 204 explains how fewer genes can encode many mRNA molecules. The mRNA shown in figure 12.7 consists of all the exons. You can imagine how a different mRNA would be produced if only half of the exons were used to make the mRNA.

With alternative splicing, it is easy to see how 25,000 genes can encode four times as many proteins. The added complexity of human proteins occurs because the gene parts are put together in new ways. Great music is made from simple tunes in much the same way.

Some Chromosomes Have Very Few Genes

In addition to the fragmenting of genes by the scattering of exons throughout the genome, there is another interesting "organizational" aspect of the genome. Genes are not distributed evenly over the genome. The small chromosome number 19 is packed densely with genes, transcription factors, and other functional elements.

> Transcription factors are proteins involved in eukaryotic gene expression as discussed on page 208. Transcription cannot occur without the involvement of transcription factors.

BIOLOGY & YOU

Sequencing the Genome of a Cancer Patient. In 2008 scientists at Washington University in Saint Louis sequenced the entire genome of a cancer patient, a woman who had acute myeloid leukemia (AML). This was the first time all the genes of a cancer patient had been examined—a very important milestone because for the first time medical researchers could see exactly what genes had been changed to cause the cancer and aid its progression. The researchers discovered genetic mutations in the cells of the patient's cancerous tumor; mutations present in every tumor cell but not present in any other body cells. Two of these mutations were expected, as they had been linked to AML in many other patients. Importantly, however, there were eight other mutations, rare among humans and never before linked to AML. Most work on cancer has focused on the few hundred genes known to be directly involved in the disease, not the 20,000 others that make up the full human genome. Using this new approach, scientists can identify which mutations are relevant to a particular patient's cancer and determine which therapy would be most suitable for that patient. Because one-third of the students reading these words will someday develop cancer, that's good news for all of us.

The much larger chromosome numbers 4 and 8, by contrast, have few genes, scattered like isolated hamlets in a desert. On most chromosomes, vast stretches of seemingly barren DNA fill the chromosomes between clusters rich in genes.

Most of Genome Is Noncoding DNA

The third notable characteristics of the human genome is the startling amount of noncoding DNA it possesses. Only 1% to 1.5% of the human genome is coding DNA, devoted to genes encoding proteins. Each of your cells has about six feet of DNA stuffed into it, but of that, less than one inch is devoted to genes! Nearly 99% of the DNA in your cells seems to have little or nothing to do with the instructions that make you who you are (table 13.2).

There are four major types of noncoding human DNA:

Noncoding DNA Within Genes.   As discussed earlier, a human gene is made up of numerous fragments of protein-encoding information (exons) embedded within a much larger matrix of noncoding DNA (introns). Introns make up 24% of the human genome—exons only 1%!

Structural DNA.   Some regions of the chromosomes remain highly condensed, tightly coiled, and untranscribed throughout the cell cycle. These portions—about 20% of the DNA—tend to be localized around the centromere, or located near the telomeres, or ends, of the chromosome.

Repeated Sequences.   Scattered about chromosomes are simple sequence repeats of two or three-nucleotides like CA or CGG, repeated like a broken record thousands and thousands of times. These make up about 3% of the human genome. An additional 7% is devoted to other sorts of duplicated sequences. Repetitive sequences with excess C and G tend to be found in the neighborhood of genes, while A- and T-rich repeats dominate the nongene deserts. The light bands on chromosome karyotypes now have an explanation—they are regions rich in GC and genes. Dark bands signal neighborhoods rich in A and T which are thin on genes. Chromosome 8 in figure 13.3 contains many nongene areas that are indicated by the dark bands, while chromosome 19 is dense with genes and so it has few dark bands.

Transposable Elements.   Fully 45% of the human genome consists of mobile parasitic bits of DNA called transposable elements. Discovered by Barbara McClintock in 1950 (she won the Nobel Prize in Physiology or Medicine in 1983 for her discovery), transposable elements are bits of DNA that are able to jump from one location on a chromosome to another—tiny molecular versions of Mexican jumping beans. Because they leave a copy of themselves behind when they jump, their numbers in the genome increase as generations pass. Nested within the human genome are over half a million copies of an ancient transposable element called *Alu*, composing fully 10% of the entire human genome. Often jumping right into genes, *Alu* transpositions cause many harmful mutations.

> Transposable elements (also called transposons) are most harmful when they insert right in the middle of a gene, as discussed on page 193 and shown in table 11.1.

Class	Frequency	Description
Protein-encoding genes	1–1.5%	Translated exons, within some 25,000 genes scattered about the chromosomes
Introns	24%	Noncoding DNA comprising the great majority of most genes
Structural DNA	20%	DNA localized near centromeres and telomeres
Repeated sequences	3%	Simple sequence repeats (SSRs) of a few nucleotides repeated thousands and thousands of times
Duplicated sequences	7%	Duplicated sequences, other than the SSRs
Transposable elements	45%	20% long interspersed nuclear elements (LINEs), active transposons; 15% other transposable elements, including long terminal repeats (LTRs);10% the parasite sequence *Alu*, present in half a million copies

TABLE 13.2 Types of DNA Sequences Found in the Human Genome

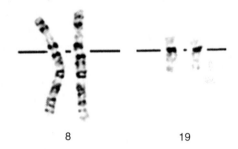

8   19

Figure 13.3   Banding on karyotype chromosomes.

The dark bands are areas rich in A and T nucleotides, which indicate nongene areas. Lighter gene-heavy areas of the chromosome indicate repetitive sequences containing C and G nucleotides.

Concept Check

1. How can DNA be sequenced by listing differences in fragment size?
2. How many genes do the chromosomes of one of your skin cells contain?
3. What fraction of your DNA is composed of protein-encoding genes?

Genetic Engineering

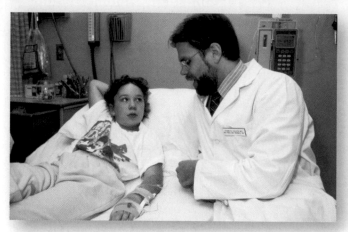

Curing disease. One of two young girls who were the first humans "cured" of a hereditary disorder by transferring into their bodies healthy versions of a defective gene. The transfer was successfully carried out in 1990, and the girls remain healthy.

Increasing yields. The genetically engineered salmon on the *right* have shortened production cycles and are heavier than the nontransgenic salmon of the same age on the *left*.

Pest-proofing plants. The genetically engineered cotton plants on the *right* have a gene that inhibits feeding by weevils; the cotton plants on the *left* lack this gene, and produce far fewer cotton bolls.

Figure 13.4   Examples of genetic engineering.

13.3 A Scientific Revolution

CONCEPT PREVIEW: Restriction enzymes, the key tools that make genetic engineering possible, bind to specific short sequences of DNA and cut the DNA. This produces fragments with "sticky ends," which can be rejoined in different combinations.

In recent years, **genetic engineering**—the ability to manipulate genes and move them from one organism to another—has led to great advances in medicine and agriculture (figure 13.4). Most of the insulin used to treat diabetes is now obtained from bacteria that contain a human insulin gene. In late 1990, the first transfers of genes from one human to another were carried out to correct the effects of defective genes in a rare genetic immune disorder called severe combined immunodeficiency (SCID)—often called the Bubble Boy disorder based on a young boy who lived out his life in an enclosed, germ-free environment. In addition, cultivated plants and animals can be genetically engineered to resist pests, grow bigger, or grow faster.

Restriction Enzymes

The first stage in any genetic engineering experiment is to chop up the "source" DNA to get a copy of the gene you wish to transfer. This first stage is the key to successful transfer of the gene, and learning how to do it is what has led to the genetic revolution. The trick is in how the DNA molecules are cut. The cutting must be done in such a way that the resulting DNA fragments have "sticky ends" that can later be joined with another molecule of DNA.

This special form of molecular surgery is carried out by **restriction enzymes,** also called *restriction endonucleases*, which are special enzymes that bind to specific short sequences (typically four to six nucleotides long) on the DNA. These sequences are very unusual in that they are symmetrical—the two strands of the DNA duplex have the same nucleotide sequence, running in opposite directions! The sequence in figure 13.5 for example, is GAATTC. Try writing down the sequence of the opposite strand: it is CTTAAG—the same sequence, written backward. This sequence is recognized by the restriction enzyme *Eco*RI. Other restriction enzyme recognize other sequences.

What makes the DNA fragments "sticky" is that most restriction enzymes do not make their incision in the center of the sequence; rather, the cut is made to one side. In the sequence in figure 13.5 **❶**, the cut is made on both strands between the G and A nucleotides, G/AATTC. This produces a break, with short, single strands of DNA dangling from each end. Because the two single-stranded ends are complementary in sequence, they could pair up and heal the break, with the aid of a sealing enzyme—*or they could pair with any other DNA fragment cut by the same enzyme,* because all would have the same single-stranded sticky ends. Figure 13.5 **❷** shows how DNA from another source (the orange DNA) also cut with *Eco*RI has the same sticky ends as the original source DNA. Any gene in any organism cut by the enzyme that attacks GAATTC sequences will have the same sticky ends, and can be joined to any other with the aid of a sealing enzyme called *DNA ligase* **❸**.

> As discussed on page 53, the two strands of DNA are held together through complementary base pairing of nucleotides. Base pairing is universal, whether the DNA resides in a bacterial cell or a human cell, and that is why restriction enzymes work on any DNA.

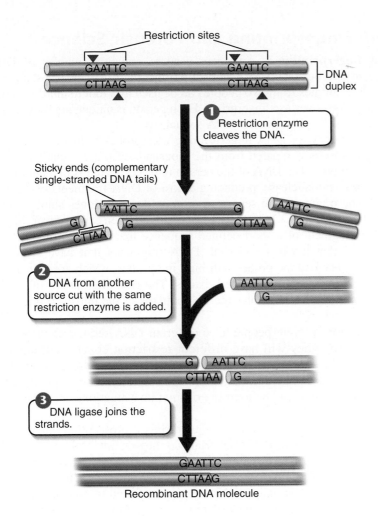

Restriction sites

GAATTC   GAATTC
CTTAAG   CTTAAG   } DNA duplex

1 Restriction enzyme cleaves the DNA.

Sticky ends (complementary single-stranded DNA tails)

AATTC   G   AATTC
G   CTTAA   G
CTTAA

2 DNA from another source cut with the same restriction enzyme is added.

AATTC
G

G   AATTC
CTTAA   G

3 DNA ligase joins the strands.

GAATTC
CTTAAG

Recombinant DNA molecule

Figure 13.5 How restriction enzymes produce DNA fragments with sticky ends.

The restriction enzyme *Eco*RI always cleaves the sequence GAATTC between G and A. Because the same sequence occurs on both strands, both are cut. However, the two sequences run in opposite directions on the two strands. As a result, single-stranded tails are produced that are complementary to each other, or "sticky."

IMPLICATION   Restriction enzymes evolved as a way for bacteria to destroy the DNA of invading viruses. Some 2,500 different restriction enzymes have been isolated from bacteria, using over 200 different 4–8 base recognition sequences. In each instance, the bacterium also contains an enzyme that adds a methyl group ($-CH_3$) to that same 4–8 base sequence. Why do you suppose bacteria always possess the pair of enzymes, and never just the restriction enzyme alone? *Hint:* There is a lot of viral DNA that the restriction enzyme has to "look through" to find the recognition sequence. Could something speed up this process?

Formation of cDNA

As you will recall from chapter 12, eukaryotic genes are encoded in segments called *exons* separated from one another by numerous non-translated sequences called *introns*. The entire gene is transcribed by RNA polymerase, producing what is called the *primary RNA transcript* (figure 13.6). Before a eukaryotic gene can be translated into a protein, the introns must be cut out of this primary transcript. The fragments that remain are then spliced together to form the *mRNA*, which is eventually translated in the cytoplasm. When transferring eukaryotic genes into bacteria (discussed in section 13.4), it is necessary to transfer DNA that has had the intron information removed because bacteria lack the enzymes to carry out this processing. Bacterial genes do not contain introns. To produce eukaryotic DNA without introns, genetic engineers first isolate from the cytoplasm the processed mRNA corresponding to a particular gene. The cytoplasmic mRNA has *only* exons, properly spliced together. An enzyme called *reverse transcriptase* is then used to make a complementary DNA strand of the mRNA. A double-stranded DNA molecule is then produced. Such a version of a gene is called complementary DNA, or **cDNA.**

cDNA technology is being used in other ways, such as determining the patterns of gene expression in different cells. Every cell in a organism contains the same DNA, but genes are selectively turned on and off in any given cell. Researchers can see what genes are actively expressed using cDNAs.

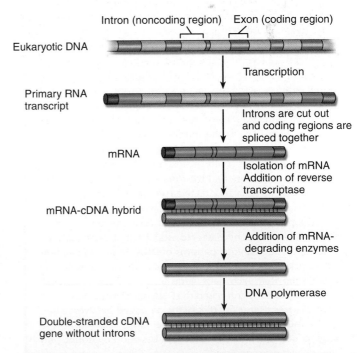

Intron (noncoding region)   Exon (coding region)

Eukaryotic DNA

Transcription

Primary RNA transcript

Introns are cut out and coding regions are spliced together

mRNA

Isolation of mRNA
Addition of reverse transcriptase

mRNA-cDNA hybrid

Addition of mRNA-degrading enzymes

DNA polymerase

Double-stranded cDNA gene without introns

Figure 13.6   cDNA: Producing an intron-free version of a eukaryotic gene for genetic engineering.

In eukaryotic cells, a primary RNA transcript is processed into the mRNA. The mRNA is isolated and converted into cDNA.

Probe 1 tags the same fragments of DNA in the rapist and the suspect but tags a different fragment of DNA in the victim. This probe indicates a match between the rapist and the suspect.

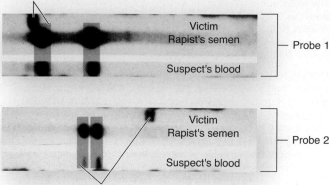

Like probe 1, probe 2 tags the same fragments of DNA in the rapist and the suspect but tags a different fragment of DNA in the victim. This probe indicates another match between the rapist and the suspect.

Figure 13.7 DNA fingerprints that led to conviction.

The two DNA probes seen here were used to characterize DNA isolated from the victim, the semen left by the rapist, and the suspect's blood. There is a clear match between the suspect's DNA and the DNA of the rapist's semen.

IN THE NEWS

Plant Witness for the Prosecution. The first case in which a murderer was convicted on plant DNA evidence was in Phoenix, Arizona. A pager left at the scene of the murder of a young woman led the police to a prime suspect. He admitted picking up the victim at a bar, but claimed she had robbed him of his wallet and pager. The crime scene squad examined the suspect's pickup truck and collected pods wedged by the front fender that were later identified as the fruits of the palo verde tree (*Cercidium spp.*). One squad member went back to the murder scene and found several palo verde trees, one recently damaged by a car. If the pods on the suspect's car could be linked to the damaged tree at the crime scene, they had their man. The detective heading the crime scene squad contacted a geneticist at the University of Arizona to establish evidence that would stand up in court that the pod indeed came from this very tree and no other. First, crucially, it was necessary to establish that individual palo verde trees have unique patterns of DNA. Sampling different palo verde trees at the murder scene and elsewhere quickly established that each palo verde tree is unique in its DNA pattern. Next, they compared the DNA pattern of a pod found on the suspect's truck to that of the damaged palo verde tree at the murder scene. They were the same. This placed the suspect at the scene of the murder beyond any doubt, and the suspect was convicted.

DNA Fingerprinting and Forensic Science

DNA fingerprinting is a process used to compare samples of DNA. Because each person differs genetically from others in many ways, comparing DNA samples from two people can be used as a tool to tell them apart, much as detectives use fingerprints. Each person is unique not only in their fingerprints, but also in their DNA sequences.

The process of DNA fingerprinting uses probes to fish out particular sequences to be compared from the thousands of other sequences in the human genome. The DNA of the person to be compared is first cut up by a restriction endonuclease, producing a host of DNA fragments of different sizes. The fragments are separated on a gel by techniques similar to that described in figure 13.1. The DNA fragments are then exposed to a "probe," a DNA fragment of a short defined sequence that has been radioactively tagged so that it can be viewed. DNA fragments that have sequences complementary to the probes will bind the probes, which are then visible on autoradiographic film as dark bands. The resulting pattern of parallel bars on X-ray film resembles the line patterns of the universal price code found on groceries.

Because different people have different DNA sequences throughout their genome, they will have different restriction enzyme cutting sites, and so produce different-sized fragments that migrate to different locations on the gel. The autoradiograph gel patterns are in essence DNA "fingerprints" that can be used in criminal investigations and other identification applications.

Figure 13.7 shows the DNA fingerprints a prosecuting attorney presented in a rape trial in 1987. This trial was the first time DNA evidence was used in a court of law. Usually six to eight probes are used to identify the source of a DNA sample, two such probes are shown as examples in figure 13.7. The probes are unique DNA sequences found in noncoding regions of human DNA that vary much more frequently from one individual to the next than do coding regions of the DNA. The chances of any two individuals, other than identical twins, having the same restriction pattern for these noncoding sequences varies from 1 in 800,000 to 1 in 1 billion, depending on the number of probes used.

The results of two probes are indicated in figure 13.7, probe 1 in the upper set of bars and probe 2 in the lower set of bars. A vaginal swab was taken from the victim within hours of her attack; from it semen was collected and the semen DNA analyzed for its restriction endonuclease patterns. The red bars compare the restriction endonuclease patterns of the semen with that of the suspect, Tommie Lee Andrews. You can see that the suspect's patterns match that of the rapist (and are not at all like those of the victim). Although a DNA fingerprint doesn't prove with 100% certainty that the semen came from the suspect, it is at least as reliable as traditional fingerprinting when several probes are used. The more probes used, the more reliable are the results. On November 6, 1987, the jury returned a verdict of guilty. Andrews became the first person in the United States to be convicted of a crime based on DNA evidence. Since the Andrews verdict, DNA fingerprinting has been admitted as evidence in thousands of court cases.

Just as fingerprinting revolutionized forensic evidence in the early 1900s, so DNA fingerprinting is revolutionizing it today. A hair, a minute speck of blood, a drop of semen, all can serve as sources of DNA to convict or clear a suspect. Of course, laboratory analyses of DNA samples must be carried out properly—sloppy procedures could lead to a wrongful conviction. After widely publicized instances of questionable lab procedures, national standards have been developed.

13.4 Genetic Engineering and Medicine

CONCEPT PREVIEW: Genetic engineering has facilitated the production of medically important proteins and led to novel vaccines.

Much of the excitement about genetic engineering has focused on its potential to improve medicine—to aid in curing and preventing illness. Major advances have been made in the production of proteins used to treat illness, and in the creation of new vaccines to combat infections.

Making "Magic Bullets"

Many genetic defects occur because our bodies fail to make critical proteins. Juvenile diabetes is such an illness. The body is unable to control levels of sugar in the blood because a critical protein, insulin, cannot be made. This failure can be overcome if the body can be supplied with the protein it lacks. The donated protein is in a very real sense a "magic bullet" to combat the body's inability to regulate itself.

> Insulin is a type of chemical called a hormone that is produced in very small quantities in one area of the body and is carried throughout the body in the bloodstream. Insulin, as discussed on pages 521 and 584, helps cells take up sugar from the blood.

Until recently, the principal problem with using regulatory proteins as drugs was in manufacturing the protein. Proteins that regulate the body's functions are typically present in the body in very low amounts, and this makes them difficult and expensive to obtain in quantity. With genetic engineering techniques, the problem of obtaining large amounts of rare proteins has been largely overcome. The cDNA of genes encoding medically important proteins are now introduced into bacteria. Because the host bacteria can be grown cheaply, large amounts of the desired protein can be easily isolated. In 1982, the U.S. Food and Drug Administration approved the use of human insulin produced from genetically engineered bacteria, the first commercial product of genetic engineering.

The use of genetic engineering techniques in bacteria has provided ample sources of therapeutic proteins, but the application extends beyond bacteria. Today hundreds of pharmaceutical companies around the world are busy producing other medically important proteins expanding the use of these genetic engineering techniques. A gene added to the DNA of the mouse on the right in figure 13.8 produces human growth hormone, allowing the mouse to grow larger than its twin.

The advantage of using genetic engineering is clearly seen with factor VIII, a protein that promotes blood clotting. A deficiency in factor VIII leads to hemophilia, an inherited disorder (discussed in chapter 10), which is characterized by prolonged bleeding. For a long time, hemophiliacs received blood factor VIII that had been isolated from donated blood. Unfortunately, some of the donated blood had been infected with viruses such as HIV and hepatitis B, which were then unknowingly transmitted to those people who received blood transfusions. Today the use of genetically engineered factor VIII produced in the laboratory eliminates the risks associated with blood products obtained from other individuals.

BIOLOGY & YOU

Eliminating Tooth Decay. The human mouth is home to billions of bacteria belonging to more than 300 species, but one species, *Streptococcus mutans*, is the major cause of tooth decay. It thrives on the organic film that coats tooth surfaces, and in the process of doing so produces lots of lactic acid, a corrosive chemical that gradually dissolves the protective enamel coating of the teeth. Genetic engineers have devised a way to block this path to the dentist's chair by replacing harmful *S. mutans* with a version of the species engineered to be more friendly to your mouth. To do this, they transferred into *S. mutans* a gene to increase production of an enzyme called urease. This enzyme converts urea, plentiful in the mouth, into ammonia, a base that neutralizes acid and creates conditions that enhance the buildup of tooth enamel. When the mouths of laboratory rats were colonized with this engineered form of *S. mutans*, they got far fewer cavities. Of course, this doesn't mean it will work with people. And even if it does, did anyone check the mouth odor of the rats?

Figure 13.8 Genetically engineered human growth hormone.

These two mice are genetically identical, but the large one has one extra gene: the gene encoding human growth hormone. The gene was added to the mouse's genome by genetic engineers and is now a stable part of the mouse's genetic makeup. In humans, growth hormone is used to treat various forms of dwarfism.

Putting Your Genome to Work

Microarrays

A **gene microarray** is a glass square smaller than a postage stamp, covered with hundreds of thousands of different single strands of DNA rising from the surface like blades of grass. At each position on the glass plate, a particular DNA sequence of a hundred or more nucleotides is assembled, forming the microarray. The gene microarray chip you see to the right, called a *GeneChip* by its manufacturer, contains all known human gene sequences and can be purchased for as little as $200.

How could you use such a microarray chip to delve into a person's genes? All you would have to do is to obtain a little of the person's DNA, say from a blood sample, and denature it to form single-stranded DNA. You would then flush fluid containing the person's denatured DNA over the chip surface with known DNA sequences. Wherever the DNA has a gene matching one of the microarray strands, it will stick to it in a way a computer can detect.

Gene microarrays can also be used to determine patterns of gene expression. To do this, mRNA isolated from the cells being studied is reverse transcribed, using fluorescently labeled nucleotides, to make complementary DNA (cDNA, see page 221). Because the cDNA contains fluorescently labeled nucleotides, it is easily recognized by a computer. When this labeled cDNA is mixed with a gene microarray representing many thousands of genes, spots light up on the computer screen corresponding to those genes being transcribed in the cells.

Similarly, two different sources of DNA can be compared, such as DNA from two different individuals, to determine their levels of genetic similarities. In this case, the DNA from the two sources are labeled with different-colored fluorescent labels, typically one labeled with a green fluorescent dye and the other with a red fluorescent dye. Spots that fluoresce are places where the samples of DNA bind to DNA on the microarray; the spots are reddish where one source binds and greenish where the other source binds. Where the two sources have similar DNA sequences, they bind to the same spot on the microarray, and it shows up as yellow spots. The more yellow spots there are, the more similar the source DNAs are.

Researchers are busily comparing the "reference sequence" of the human genome with the DNA of individual people, and noting any differences they detect. In this way, they are finding SNPs (single nucleotide polymorphisms), or spot differences in the identity of particular nucleotides, which record every way in which a particular individual differs from the reference sequence. Some SNPs are associated with disorders like cystic fibrosis or sickle-cell disease. Others may give you red hair or elevated cholesterol in your blood. The human genome tells us that SNPs can be expected to occur at a frequency of about 5 per 1,000 nucleotides, scattered about randomly over the chromosomes. Each of us can be expected to differ from the standard "type sequence" by thousands of nucleotide SNPs.

Microarrays Raise Critical Issues of Personal Privacy

Humans are thought to contain some 10 million different SNPs, all of which could reside on a small library of gene microarrays. When your

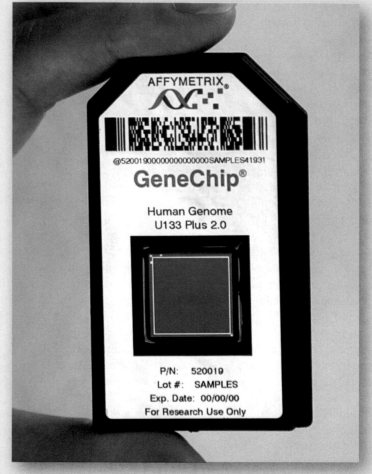

Humanity on a chip.

Microarrays such as this Affymetrix GeneChip now include all known human genes.

DNA is flushed over a SNP microarray, the sequences that light up will instantly reveal your SNP profile, the genetic characteristics that make up who you are. Genes that might affect your health, your behavior, your future potential—all are there to be read. Your SNP profile will reflect all of this variation: a table of contents of your chromosomes, a molecular window to who you are.

When millions of such SNP profiles have been gathered over the coming years, computers will be able to identify other individuals with profiles like yours, and, by examining health records, standard personality tests, and the like, correlate parts of your profile with particular traits. Even behavioral characteristics involving many genes, which until now have been thought too complex to ever analyze, cannot resist a determined assault by a computer comparing SNP profiles. This raises issues of privacy—who should have access to this information and how should it be used?

DNA and the Innocence Project

Every person's DNA is uniquely their own, a sequence of nucleotides found in no other person. Like a molecular social security number a billion digits long, the nucleotide sequence of an individual's genes can provide proof-positive identification of a rapist or murderer from DNA left at the scene of a crime—proof more reliable than fingerprints, more reliable than an eye witness, even more reliable than a confession by the suspect.

Never was this demonstrated more clearly than on May 16, 2006 in a Rochester, New York courtroom. There a judge freed convicted murderer Douglas Warney after ten years in prison.

The murder occurred on New Year's Day in 1996. The bloody body of one William Beason, a prominent community activist, was found in his bed. Police called in all the usual suspects, in this case everyone known to be an acquaintance of the victim. Warney, an unemployed 34-year-old who had dropped out of school in the eighth grade, committed robberies, and worked as a male hustler, learned that detectives wanted to speak to him about the killing, and went to the police station for questioning. Within hours he was charged with murder.

Warney's interrogation was not recorded, but it resulted in a signed confession based on the words that the detective sergeant said Warney uttered, a confession that contained accurate details about the murder scene that had not been made public. Warney said that the victim was wearing a nightgown and had been cooking chicken in a pot, and that the murderer had used a 12-inch serrated knife. The case against him in court rested almost entirely on these vivid details. Even though Warney recanted his confession at his trial, the accuracy of his confession was damning. The details that Warney provided to the sergeant could have come only from someone who was present at the crime scene.

There were problems with his confession, brought out at trial. Three elements of the signed statement's account of that night were clearly not true. It said Warney had driven to the victim's house in his brother's car, but his brother did not own a car; it said Warney disposed of his bloody clothes after the stabbing in the garbage can behind the house, but the can, buried in snow from the day of the crime, did not contain bloody clothes; it named a relative of Warney as an accomplice, but that relative was in a secure rehabilitation center on that day.

The most difficult bit of evidence to match to the written confession was blood found at the scene that was not that of the victim. Drops of a second person's blood were found on the floor and on a towel. The difficult bit was that the blood was a different blood type than Warney's.

At trial, prosecutors pointed out that the blood could have come from the accomplice mentioned in the confession, and pounded

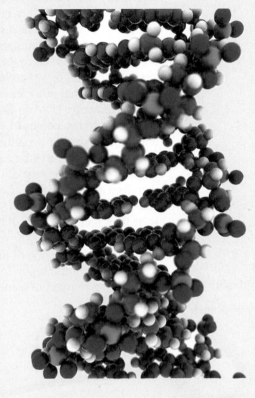

away on the point that the details in the confession could only have come from first-hand knowledge of the crime.

After a short trial, Douglas Warney was convicted of the murder of William Beason and sentenced to 25 years.

When appeals failed, Mr. Warney in 2004 sought help from the Innocence Project, a non-profit legal clinic that helps identify wrongly-convicted individuals and secure their freedom. Set up at the Benjamin N. Cardozo School of Law in New York by lawyers Barry Schech and Peter Neufeld, the Innocence Project specializes in using DNA technology to establish innocence.

The Innocence Project staff petitioned the court for additional DNA testing using new sensitive DNA probes, arguing that this might disclose evidence that would have resulted in a different verdict. The judge refused, ruling that the possibility that the blood found at the scene of the crime might match that of a criminal already in the state databank was "too speculative and improbable" to warrant the new tests.

Someone in the prosecutor's office must have been persuaded, however. Without notifying Warney's legal team, Project Innocence, or the court, this good Samaritan arranged for new DNA tests on the blood drops found at the crime scene. When compared to the New York State criminal DNA database, they hit a strong match. The blood was that of Eldred Johnson, in prison for slitting his landlady's throat in Utica two weeks before Beason was killed.

When confronted, the prisoner Johnson readily admitted to the stabbing of Beason. He said he was the sole killer, and had never met Douglas Warney.

This leaves the interesting question of how Warney's signed confession came to include such accurate information about the crime scene. It now appears he may have been fed critical details about the crime scene by the homicide detective leading the investigation, a Sergeant Gropp. It seems Gropp's partner had stepped out to get some papers when the confession was obtained. Sergeant Gropp died in March, 2006.

DNA testing has become a pillar of the American criminal justice system. It has provided key evidence that has established the guilt of thousands of suspects beyond any reasonable doubt—and, as you see here, also provided the evidence that our criminal justice system sometimes convicts and sentences innocent people. Over more than a decade, the Innocence Project and other similar efforts have cleared hundreds of convicted people, strong proof that wrongful convictions are not isolated or rare events. DNA testing opens a window of hope for the wrongly convicted.

Piggyback Vaccines

Another area of potential significance involves the use of genetic engineering to produce subunit vaccines against viruses such as those that cause herpes and hepatitis. Genes encoding part of the protein-polysaccharide coat of the herpes simplex virus or hepatitis B virus are spliced into a fragment of the vaccinia (cowpox) virus genome. The vaccinia virus, which is essentially harmless to humans and was used by British physician Edward Jenner more than 200 years ago in his pioneering vaccinations against smallpox, is now used as a vector to carry the herpes or hepatitis viral coat gene into cultured mammalian cells. The steps, highlighted in figure 13.9, begin with ❶ extracting the herpes simplex viral DNA and ❷ isolating a gene that codes for a protein on the surface of the virus. The cowpox viral DNA is extracted and cleaved in step ❸, and the herpes gene is combined with the cowpox DNA in ❹. The recombinant DNA is inserted into a cowpox virus. Many copies of the recombinant virus, which have the outside coat of a herpes (or hepatitis) virus, are produced. When this recombinant virus is injected into a human ❺, the immune system produces antibodies directed against the coat of the recombinant virus as in step ❻. The person therefore develops an immunity to the virus. Vaccines produced in this way, also known as **piggyback vaccines,** are harmless because the vaccinia virus is benign, and only a small fragment of the DNA from the disease-causing virus is introduced via the recombinant virus.

The great attraction of this approach is that it does not depend upon the nature of the viral disease. In the future, similar recombinant viruses may be injected into humans to confer resistance to a wide variety of viral diseases.

Figure 13.9 **Constructing a subunit, or piggyback, vaccine for the herpes simplex virus.**

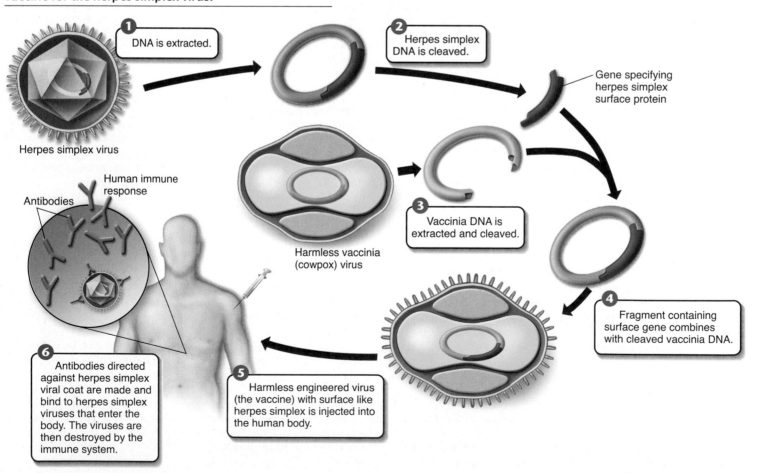

13.5 Genetic Engineering and Agriculture

CONCEPT PREVIEW: Genetic engineering affords great opportunities for progress in food production. On balance, there are risks, but the potential benefits are substantial.

Pest Resistance

An important effort of genetic engineers in agriculture has involved making crops resistant to insect pests without spraying with pesticides, a great saving to the environment. Consider cotton. Its fibers are a major source of raw material for clothing throughout the world, yet the plant itself can hardly survive in a field because many insects attack it. Over 40% of the chemical insecticides used today are employed to kill insects that eat cotton plants. The world's environment would greatly benefit if these thousands of tons of insecticide were not needed. Biologists are now in the process of producing cotton plants that are resistant to attack by insects.

One successful approach uses a kind of soil bacterium, *Bacillus thuringiensis* (Bt) that produces a protein that is toxic when eaten by crop pests, such as larvae (caterpillars) of butterflies. When the gene producing the Bt protein was inserted into the chromosomes of tomatoes, the plants began to manufacture Bt protein. While not harmful to humans, it makes the tomatoes highly toxic to hornworms (one of the most serious pests of commercial tomato crops).

Many important plant pests also attack roots. To combat these pests, genetic engineers are introducing the *Bt* gene into different kinds of bacteria, ones that colonize the roots of crop plants. Any insects eating such roots consume the bacteria and so are lethally attacked by the Bt protein.

Herbicide Resistance

A major advance has been the creation of crop plants that are resistant to the herbicide *glyphosate,* a powerful biodegradable herbicide that kills most actively growing plants. Glyphosate is used in orchards and agricultural fields to control weeds. Growing plants need to make a lot of protein, and glyphosate stops them from making protein by destroying an enzyme necessary for the manufacture of so-called aromatic amino acids (that is, amino acids that contain a ring structure). Humans are unaffected by glyphosate because we don't make aromatic amino acids—we obtain them from plants we eat! To make crop plants resistant to this powerful plant killer, genetic engineers screened thousands of organisms until they found a species of bacteria that could make aromatic amino acids in the presence of glyphosate. They then isolated the gene encoding the resistant enzyme and successfully introduced the gene into plants. They inserted the gene into the plants using DNA particle guns, also called gene guns. You can see in figure 13.10 how a DNA particle gun works. Small tungsten or gold pellets are coated with DNA (red in the figure) that contains the gene of interest and placed in the DNA particle gun. The DNA gun literally shoots the gene into plant cells in culture where the gene can be incorporated in the plant genome and then expressed. Plants that have been genetically engineered in this way are shown in figure 13.11. The two plants on top were genetically engineered to be resistant to glyphosate, the herbicide that killed the two plants at the bottom of the photo.

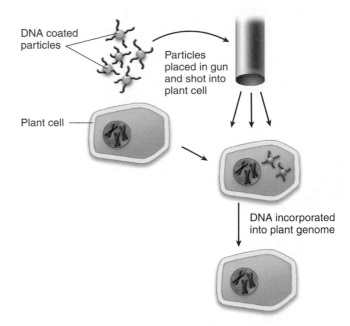

Figure 13.10   Shooting genes into cells.

A DNA particle gun, also called a gene gun, fires tungsten or gold particles coated with DNA into plant cells. The DNA coated particles pass through the cell wall and into the cell, where the DNA is incorporated into the plant cell's DNA. The gene encoded by the DNA is expressed.

DNA coated particles

Particles placed in gun and shot into plant cell

Plant cell

DNA incorporated into plant genome

Figure 13.11   Genetically engineered herbicide resistance.

All four of these petunia plants were exposed to equal doses of an herbicide. The two on *top* were genetically engineered to be resistant to glyphosate, the active ingredient in the herbicide, whereas the two dead ones on the *bottom* were not.

More Nutritious Crops

The cultivation of genetically modified (GM) crops of corn, cotton, soybeans, and other plants (table 13.3) has become commonplace in the United States. In 2004, 85% of soybeans in the United States were planted with seeds genetically modified to be herbicide resistant. The result has been that less tillage is needed and as a consequence soil erosion is greatly lessened. Pest-resistant GM corn in 2004 comprised over 50% of all corn planted in the United States, and pest-resistant GM cotton comprised 81% of all cotton. In both cases, the change greatly lessens the amount of chemical pesticide used on the crops.

The real promise of plant genetic engineering is to produce genetically modified plants with desirable traits that directly benefit the consumer. One recent advance, nutritionally improved "golden" rice, gives us a hint of what is to come. In developing countries large numbers of people live on simple diets that are poor sources of both iron, which affects 1.4 billion women, 24% of the world population, and vitamin A, affecting 40 million children, 7% of the world population. These two deficiencies are especially severe in developing countries where the major staple food is rice. To solve the problem of dietary iron deficiency among rice eaters, gene engineers first asked why rice is such a poor source of dietary iron. The problem, and the answer, proved to have three parts:

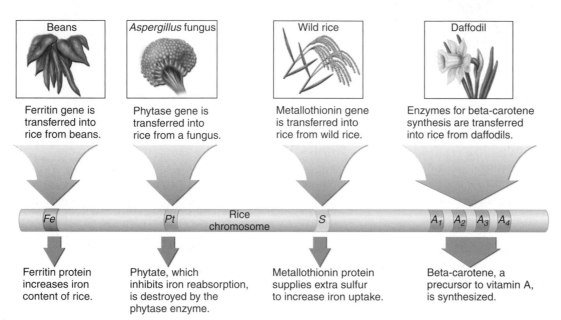

| Beans | *Aspergillus* fungus | Wild rice | Daffodil |

Ferritin gene is transferred into rice from beans.

Phytase gene is transferred into rice from a fungus.

Metallothionin gene is transferred into rice from wild rice.

Enzymes for beta-carotene synthesis are transferred into rice from daffodils.

Ferritin protein increases iron content of rice.

Phytate, which inhibits iron reabsorption, is destroyed by the phytase enzyme.

Metallothionin protein supplies extra sulfur to increase iron uptake.

Beta-carotene, a precursor to vitamin A, is synthesized.

1. *Too little iron.* The proteins of rice endosperm have unusually low amounts of iron. To solve this problem, a ferritin gene (abbreviated as Fe in figure 13.12) was transferred into rice from beans. Ferritin is a protein with an extraordinarily high iron content, and so greatly increased the iron content of the rice.

2. *Inhibition of iron absorption by the intestine.* Rice contains an unusually high concentration of a chemical called phytate, which inhibits iron absorption in the intestine—it stops your body from taking up the iron in the rice. To solve this problem, a gene encoding an enzyme called phytase (abbreviated as Pt) that destroys phytate was transferred into rice from a fungus.

3. *Too little sulfur for efficient iron absorption.* The human body requires sulfur for the uptake of iron, and rice has very little of it. To solve this problem, a gene encoding a sulfur-rich protein (abbreviated as S) was transferred into rice from wild rice.

To solve the problem of vitamin A deficiency, the same approach was taken. First, the problem was identified. It turns out rice only goes partway toward making vitamin A; there are no enzymes in rice to catalyze the last four steps. To solve the problem, genes encoding these four enzymes (abbreviated A_1 A_2 A_3 A_4) were added to rice from a flower, the daffodil.

The development of this transgenic rice is only the first step. Many years will be required to breed the genes into lines adapted to local conditions, but it is a promising start, hinting at the very real promise of genetic engineering.

Figure 13.12   Transgenic "golden" rice.

IMPLICATION   The "Free Rice Game" on the internet (www.freerice.com) asks you quiz questions. For each answer you get right, 20 grains of rice are donated to the UN World Food Program to help end hunger. Since its inception in October 2007, over 50 billion grains of rice have been donated, about 20,000,000 grains a day. For each answer you get right, you then get a harder question to attempt. How many questions in a row do you think you can get right (the average is four)?

How Do We Measure the Potential Risks of Genetically Modified Crops?

Is Eating Genetically Modified Food Dangerous? Many consumers worry that when genetic engineers introduce novel genes into genetically modified (GM) crops, there may be dangerous consequences for the food we eat. The introduction of glyphosate-resistance into soybeans is an example. Is the soybean that results nutritionally different? No. But could introduced proteins like the enzyme making the GM soybeans glyphosate tolerant cause a fatal immune reaction in some people? Because the potential danger of allergic reactions is quite real, every time a protein-encoding gene is introduced into a GM crop it is necessary to carry out extensive tests of the introduced protein's allergen potential. No GM crop currently being produced in the United States contains a protein that acts as an allergen to humans. On this score, then, the risk of genetic engineering to the food supply seems to be slight.

Are GM Crops Harmful to the Environment? Those concerned about the widespread use of GM crops raise three legitimate concerns:

1. *Harm to Other Organisms.* Results from a small laboratory experiment suggested that pollen from Bt corn could harm larvae from the Monarch butterfly. While this preliminary report received considerable publicity, subsequent studies suggest little possibility of harm. Monarch butterflies lay their eggs on milkweed, not corn, and there is little if any milkweed growing in or near cornfields.
2. *Resistance.* All insecticides used in agriculture share the problem that pests eventually evolve resistance to them, in much the same way that bacterial populations evolve resistance to antibiotics. However, despite the widespread use of Bt crops like corn, soybeans, and cotton since 1996, there are as of yet only a few cases of insects developing resistance to Bt plants in the field. Still, because of this possibility, farmers are required to plant at least 20% non-Bt crops alongside Bt crops to provide refuges where insect populations are not under selection pressure and in this way to slow the development of resistance.
3. *Gene Flow.* How about the possibility that introduced genes will pass from GM crops to their wild or weedy relatives? For the major GM crops, there is usually no potential relative around to receive the modified gene from the GM crop. There are no wild relatives of soybeans in Europe, for example. Thus there can be no gene escape from GM soybeans in Europe, any more than genes can flow from you to your pet dog or cat. However—and this is a big however—for secondary crops only now being genetically modified, studies suggest that it may be difficult to prevent GM crops from interbreeding with surrounding relatives to create new hybrids.

Concept Check

1. If a six-base restriction enzyme recognition site begins with T A G _ _ _, fill in the remaining bases.
2. Why doesn't the herbicide glyphosate hurt people?
3. The possibility of gene flow seems to be a potential problem with GM crops. With which crops? Why these?

TABLE 13.3	Genetically Modified Crops
Rice 	Genes have been added to commercial rice from daffodils for vitamin A, and from beans, fungi, and wild rice to supply dietary iron; transgenic strains that are cold-tolerant are under development.
Wheat 	New strains of wheat, resistant to the herbicide glyphosate, greatly reduce the need for tilling and so reduce loss of topsoil; these strains have not been made commercially available yet because of public resistance to GM crops.
Soybean 	A major animal feed crop, soybeans tolerant of the herbicide glyphosate were used in 85% of U.S. soybean acreage in 2004. Varieties are being developed that contain the *Bt* gene, to protect the crop from insect pests without chemical pesticides. The nutritional value of soybean crops is being improved by genetic engineers in several ways, including transgenic varieties with high tryptophan (soybeans are poor in this essential amino acid), reduced trans-fatty acids, and enhanced omega-3 (beneficial) fatty acids, common in fish oil but low in plants.
Corn 	Corn varieties resistant to insect pests (Bt corn) are widely planted (40% of U.S. acreage); varieties also tolerant of the herbicide glyphosate have been recently developed. Varieties that are drought-resistant are being developed, as well as nutritionally improved lines with high lysine, vitamin A, and high levels of the unsaturated fat oleic acid, which reduces harmful cholesterol and so prevents clogged arteries.
Cotton 	Cotton crops are attacked by cotton bollworm, budworm, and other lepidopteran insects; more than 40% of all chemical pesticide tonnage worldwide is applied to cotton. A form of the *Bt* gene toxic to all lepidopterans but harmless to other insects has transformed cotton to a crop that requires few chemical pesticides. 81% of U.S. acreage is Bt cotton.
Peanut	The lesser cornstalk borer causes serious damage to peanut crops. An insect-resistant variety is under development by gene engineers to control this pest.
Potato 	Verticillium wilt (a fungal disease) infects the water-conducting tissues of potatoes, reducing crop yields 40%. An antifungal gene from alfalfa reduces infections sixfold.
Canola 	Canola, a major vegetable oil and animal feed crop, is typically grown in narrow rows with little cultivation, requiring extensive application of chemical herbicides to keep down weeds. New glyphosate-tolerant varieties require far less chemical treatment. 80% of U.S. canola acreage planted is gene-modified canola.

The Revolution in Cell Technology

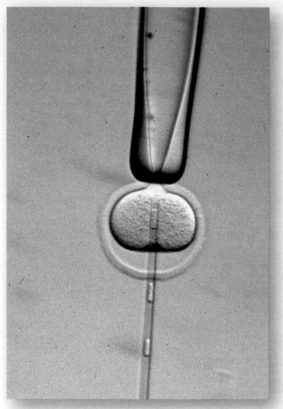

Figure 13.13   A cloning experiment.

In this photo, a nucleus is being injected from a micropipette *(bottom)* into an enucleated egg cell held in place by a pipette.

13.6 Reproductive Cloning

CONCEPT PREVIEW: Although recent experiments have demonstrated the possibility of cloning animals from adult tissue, the cloning of farm animals often fails for lack of proper genomic imprinting.

One of the most active and exciting areas of biology involves recently developed approaches to manipulating animal cells. In this section, you will encounter three areas where landmark progress is being made in cell technology: reproductive cloning of farm animals, stem cell research, and gene therapy. Advances in cell technology hold the promise of revolutionizing our lives.

The idea of cloning animals was first suggested in 1938 by German embryologist Hans Spemann (called the "father of modern embryology"), who proposed what he called a "fantastical experiment": remove the nucleus from an egg cell (an enucleated egg) and put in its place a nucleus from another cell. When attempted many years later (figure 13.13), this experiment actually succeeded in frogs, sheep, monkeys, and many other animals. However, only donor nuclei extracted from early embryos seemed to work. After repeated failures using nuclei from adult cells, many researchers became convinced that the nuclei of animal cells become irreversibly committed to a developmental pathway after the first few cell divisions of the developing embryo.

Wilmut's Lamb

Then, in the 1990s, a key insight was made in Scotland by geneticist Keith Campbell, a specialist in studying the cell cycle of agricultural animals. Recall from chapter 8 that the division cycle of eukaryotic cells progresses in several stages. Campbell reasoned, "Maybe the egg and the donated nucleus need to be at the same stage in the cell cycle." This proved to be a key insight. In 1994 researchers succeeded in cloning farm animals from advanced embryos by first starving the cells, so that they paused at the beginning of the cell cycle. Two starved cells are thus synchronized at the same point in the cell cycle.

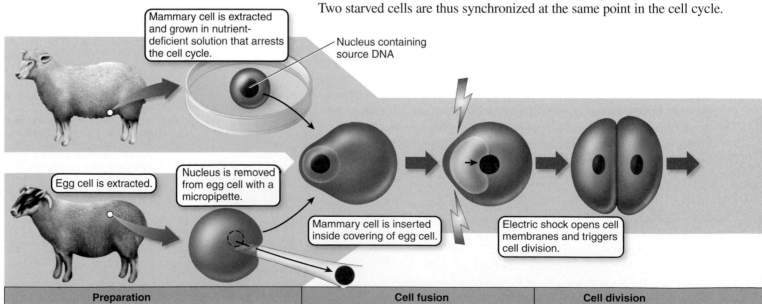

Mammary cell is extracted and grown in nutrient-deficient solution that arrests the cell cycle.

Nucleus containing source DNA

Egg cell is extracted.

Nucleus is removed from egg cell with a micropipette.

Mammary cell is inserted inside covering of egg cell.

Electric shock opens cell membranes and triggers cell division.

| Preparation | Cell fusion | Cell division |

Figure 13.14   Wilmut's animal cloning experiment.

Campbell's colleague Ian Wilmut then attempted the key breakthrough, the experiment that had been eluding researchers: He set out to transfer the nucleus from an adult differentiated cell into an enucleated egg, and to allow the resulting embryo to grow and develop in a surrogate mother, hopefully producing a healthy animal (figure 13.14). Approximately five months later, on July 5, 1996, the mother gave birth to a lamb. This lamb, "Dolly," was the first successful clone generated from a differentiated animal cell. Dolly grew into a healthy adult, and as you can see in the photo at the beginning of this chapter, she went on to have healthy offspring normal in every respect.

Progress with Reproductive Cloning

Since Dolly's birth in 1996, scientists have successfully cloned a wide variety of farm animals with desired characteristics, including cows, pigs, goats, horses, and donkeys, as well as pets like cats and dogs. Snuppy, the puppy in figure 13.15, was the first dog to be cloned. For most farm animals, cloning procedures have become increasingly efficient since Dolly was cloned. However, the development of clones into adults tends to go unexpectedly haywire. Almost none survive to live a normal life span. Even Dolly died prematurely in 2003, having lived only half a normal sheep life span.

The Importance of Genomic Imprinting

What is going wrong? It turns out that as mammalian eggs and sperm mature, their DNA is conditioned by the parent female or male, a process called reprogramming. Chemical changes are made to the DNA that alter when particular genes are expressed without changing the nucleotide sequences. In the years since Dolly, scientists have learned a lot about gene reprogramming, also called **genomic imprinting.** Genomic imprinting works by blocking the cell's ability to read certain genes. A gene is locked in the off position by adding a —CH$_3$ (methyl) group to some of its cytosine nucleotides. After a gene has been altered like this, the polymerase protein that is supposed to "read" the gene can no longer recognize it. The gene has been shut off.

We are only beginning to learn how to reprogram human DNA, so any attempt to clone a human is simply throwing stones in the dark, hoping to hit a target we cannot see. For this and many other reasons, any attempt at human reproductive cloning is regarded as highly unethical.

Figure 13.15   Cloning the family pet.

This puppy named "Snuppy" is the first dog cloned. Beside him to the left, is the adult male dog who provided the skin cell from which Snuppy was cloned. The dog in the photo on the right was Snuppy's surrogate mother.

IMPLICATION   Do you have a dog or cat? Do you think it would be ethical for you to have your dog or cat cloned when it grows old, so you can continue to enjoy its company for many more years? Then, in later years, do you clone the clone? What, if any, limits would you place on this process? Discuss.

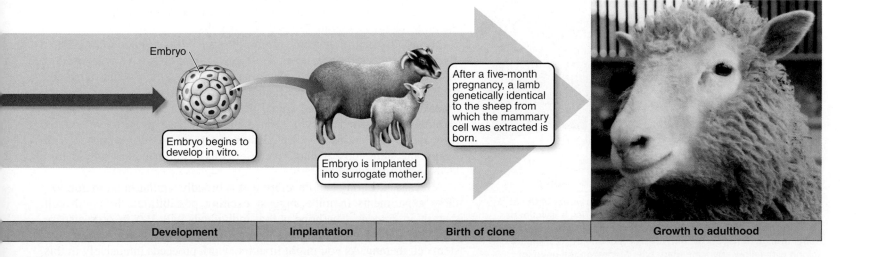

Embryo

Embryo begins to develop in vitro.

Embryo is implanted into surrogate mother.

After a five-month pregnancy, a lamb genetically identical to the sheep from which the mammary cell was extracted is born.

| Development | Implantation | Birth of clone | Growth to adulthood |

Figure 13.16   Human embryonic stem cells (×20).

This mass is a colony of undifferentiated human embryonic stem cells surrounded by fibroblasts (elongated cells) that serve as a "feeder layer." The image depicts stem cell research occurring at the University of Wisconsin–Madison.

Figure 13.17   Promoting a cure for Parkinson's.

Michael J. Fox, with whom you may be familiar as a star of the *Back to the Future* film series and the TV show *Family Ties*, is a victim of Parkinson's disease, and a prominent spokesman for those who suffer from it. Here you see him testifying before the U.S. Senate (along with fellow advocate Mary Tyler Moore) on the need for vigorous efforts to support research seeking a cure.

13.7   Stem Cell Therapy

CONCEPT PREVIEW: Human adult and embryonic stem cells offer the possibility of replacing damaged or lost human tissues.

You can see a colony of human embryonic stem cells in figure 13.16. Each is **totipotent**—able to form any body tissue, and even an entire adult animal. What is an embryonic stem cell, and why is it totipotent? To answer this question, we need to consider for a moment where an embryo comes from. At the dawn of a human life, a sperm fertilizes an egg to create a single cell destined to become a child. As development commences, that cell begins to divide, producing after five or six days a small ball of a few hundred cells called a blastocyst, enclosing an inner cell mass of up to 200 **embryonic stem cells.** Each embryonic stem cell has all of the genes needed to produce a normal individual. In cattle breeding, for example, these cells are frequently separated by the breeder and used to produce multiple clones of valuable offspring.

As development proceeds, some of these embryonic stem cells become committed to forming specific types of tissues, such as nerve tissues, and, after this step is taken, cannot ever produce any other kind of cell. The genes needed to produce those other types of cells are inactivated and cannot be expressed. For example, in nerve tissue they are then called *nerve stem cells.* Others become specialized to produce blood cells, others to produce muscle tissue, and still others to form the other tissues of the body. Each major tissue is formed from its own kind of tissue-specific **adult stem cell.** Because an adult stem cell forms only that one kind of tissue, it is not totipotent.

Using Stem Cells to Repair Damaged Tissues

Stem cells offer the exciting possibility of restoring damaged tissues. For some tissues like blood, it is possible to harvest adult stem cells for this use. For other tissues, like nerve cells, this is unfortunately not yet possible. However, it is in theory possible to use embryonic stem cells instead, because they can develop into any tissue. To understand how embryonic stem cells could be used to repair damaged tissue, follow along in figure 13.18. A few days after fertilization, the blastocyst forms ❶. Embryonic stem cells are harvested from its inner cell mass or from cells of the embryo at a later stage ❷. These embryonic stem cells can be grown in tissue culture (see figure 13.16) and in principle be induced to form any type of tissue in the body ❸. The resulting healthy tissue can then be injected into the patient where it will grow and replace damaged tissue ❹. Alternatively, where possible, adult stem cells can be isolated and when injected back into the body, can form certain types of tissue cells.

Both adult and embryonic stem cell transfer experiments have been carried out successfully in mice. Adult blood stem cells have been used to cure leukemia. Heart muscle cells grown from mouse embryonic stem cells have successfully replaced the damaged heart tissue of a living mouse. In other experiments, damaged spinal neurons have been partially repaired. DOPA-producing neurons of mouse brains, whose progressive loss is responsible for Parkinson's disease, have been successfully replaced with embryonic stem cells, as have the islet cells of the pancreas whose loss leads to juvenile diabetes.

Because the course of development is broadly similar in all mammals, these experiments in mice suggest exciting possibilities, for stem cell therapy in humans. The hope is that individuals with Parkinson's disease like Michael J. Fox (figure 13.17) might be partially or fully cured with stem cell therapy. As you might imagine, work proceeds intensively in this field of research.

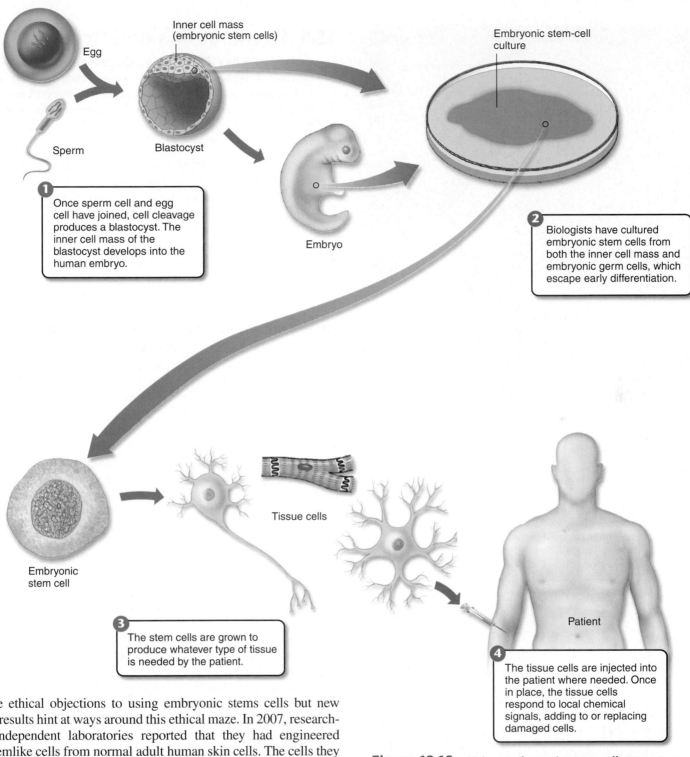

Egg

Sperm

Inner cell mass
(embryonic stem cells)

Blastocyst

Embryo

Embryonic stem-cell
culture

1 Once sperm cell and egg cell have joined, cell cleavage produces a blastocyst. The inner cell mass of the blastocyst develops into the human embryo.

2 Biologists have cultured embryonic stem cells from both the inner cell mass and embryonic germ cells, which escape early differentiation.

Embryonic
stem cell

Tissue cells

Patient

3 The stem cells are grown to produce whatever type of tissue is needed by the patient.

4 The tissue cells are injected into the patient where needed. Once in place, the tissue cells respond to local chemical signals, adding to or replacing damaged cells.

Figure 13.18   Using embryonic stem cells to restore damaged tissue.

Embryonic stem cells can develop into any body tissue. Methods for growing the tissue and using it to repair damaged tissue in adults, such as the brain cells of multiple sclerosis patients, heart muscle, and spinal nerves, are being developed.

There are ethical objections to using embryonic stems cells but new experimental results hint at ways around this ethical maze. In 2007, researchers in two independent laboratories reported that they had engineered embryonic stemlike cells from normal adult human skin cells. The cells they created were pluripotent—they could differentiate into many different cell types. Whether pluripotency extends to totipotency is still being investigated. How were these cells transformed? The essential clue came six years earlier, when fusing adult cells with embryonic stem cells transformed the adult cells into pluripotent cells, as if factors had been transmitted to the adult cells that conferred pluripotency. The 2007 researchers introduced genes for just four transcription factors into adult human skin cells piggyback on viruses, an approach discussed on page 226. Once inside, these four factors induced a series of events that led to pluripotency. From proof of principle in a laboratory culture dish to actual medical application is still a leap, but the possibility is exciting.

Figure 13.19 **Embryonic stem cells growing in cell culture.**

These embryonic stem cells, at the University of Wisconsin–Madison, are derived from early human embryos and will grow indefinitely in tissue culture. When transplanted, they can sometimes be induced to form new cells of the adult tissue into which they have been placed. This suggests exciting therapeutic uses.

IN THE NEWS

Turning One Cell Type Into Another. In 2008 doctors at Children's Hospital in Boston set out to cure type I diabetes, in which individuals lack the pancreatic beta cells needed to produce insulin. For two years they sifted through more than 1,000 transcription factors (proteins that tell cells which genes to turn on and off) to find ones that would turn the normal cells of the pancreas into beta cells. In the end they found that just three were needed to do the trick. When the three factors were injected into living mice whose islet cells had been destroyed, normal cells in the pancreas were turned into insulin-producing beta cells. The added insulin led to significant lowering of blood sugar levels, although not enough cells were transformed to cure diabetes. The new beta cells remained stable for many months. Although work needs to be done to improve efficiency, and a mouse is not a human, this result is proof in principle that adult cell transformation can work to cure tissue diseases.

13.8 Therapeutic Cloning

CONCEPT PREVIEW: Therapeutic cloning involves initiating blastocyst development from a patient's tissue using nuclear transplant procedures, then using the blastocyst's embryonic stem cells to replace the patient's damaged or lost tissue. Reprogramming adult tissue cells may allow a less controversial approach.

While exciting, the therapeutic uses of stem cells to cure leukemia, type I diabetes, Parkinson's disease, damaged heart muscle, and injured nerve tissue were all achieved in experiments carried out using strains of mice without functioning immune systems. Why is this important? Because had these mice possessed fully functional immune systems, they almost certainly would have rejected the implanted stem cells as foreign. Humans with normal immune systems might well refuse to accept transplanted stem cells simply because they are from another individual. For such stem cell therapy to work in humans, this problem needs to be addressed and solved.

Cloning to Achieve Immune Acceptance

Early in 2001, a research team at the Rockefeller University reported a way around this potentially serious problem. Their solution? They first isolated skin cells from a mouse, then using the same procedure that created Dolly, they created a 120-cell embryo from them. The embryo was then destroyed, its embryonic stem cells harvested and cultured (figure 13.19) for transfer to replace injured tissue. This procedure is called **therapeutic cloning.** Therapeutic cloning and the procedure that was used to create Dolly, called **reproductive cloning,** are contrasted in figure 13.20. You can see that steps ❶ through ❺ are essentially the same for both procedures, but the two methods proceed differently after that. In reproductive cloning, the blastocyst from step ❺ is implanted in a surrogate mother in step ❻ₐ, developing into a baby that is genetically identical to the nucleus donor, step ❼ₐ. In therapeutic cloning, by contrast, stem cells from the blastocyst of step ❺ are removed and grown in culture, step ❻. These stem cells develop into pancreatic islet cells in step ❼ and are injected or transplanted into the diabetic patient, where they begin producing insulin.

Therapeutic cloning, or, more technically, *somatic cell nuclear transfer,* successfully addresses the key problem that must be solved before embryonic stem cells can be used to repair damaged human tissues, which is immune acceptance. Because stem cells are cloned from the body's own tissues in therapeutic cloning, they pass the immune system's "self" identity check, and the body readily accepts them.

Reprogramming to Achieve Immune Acceptance

In therapeutic cloning, the cloned embryo is destroyed to obtain embryonic stem cells. What is the moral standing of a six-day human embryo? Considering it a living individual, many people regard therapeutic cloning to be ethically unacceptable. Recent research discussed on the previous page suggests an alternative approach that avoids this problem: reprogramming adult cells into embryonic stemlike cells by introducing just a few genes into the adult cells. The genes are so-called transcription factors, turning on key genes that act to reverse the "shut off" changes that have occurred during development of the adult cells. Human applications, if even possible, are probably far into the future, but the possibility of reprogramming adult cells is exciting.

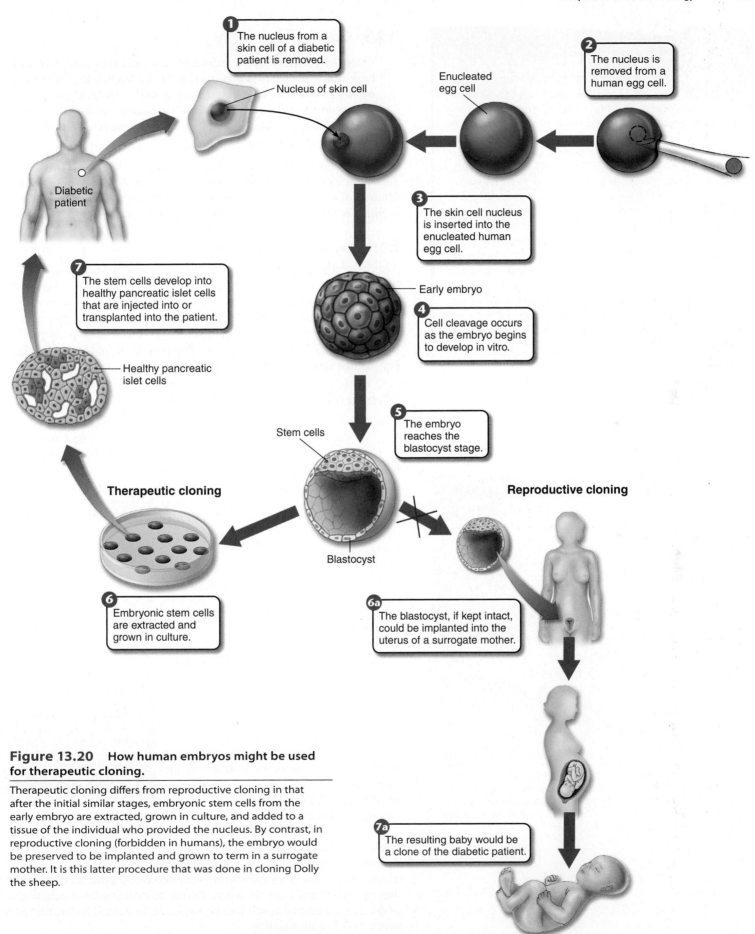

1 The nucleus from a skin cell of a diabetic patient is removed.

2 The nucleus is removed from a human egg cell.

Enucleated egg cell

Nucleus of skin cell

Diabetic patient

3 The skin cell nucleus is inserted into the enucleated human egg cell.

7 The stem cells develop into healthy pancreatic islet cells that are injected into or transplanted into the patient.

Early embryo

4 Cell cleavage occurs as the embryo begins to develop in vitro.

Healthy pancreatic islet cells

Stem cells

5 The embryo reaches the blastocyst stage.

Therapeutic cloning

Reproductive cloning

Blastocyst

6 Embryonic stem cells are extracted and grown in culture.

6a The blastocyst, if kept intact, could be implanted into the uterus of a surrogate mother.

7a The resulting baby would be a clone of the diabetic patient.

Figure 13.20 How human embryos might be used for therapeutic cloning.

Therapeutic cloning differs from reproductive cloning in that after the initial similar stages, embryonic stem cells from the early embryo are extracted, grown in culture, and added to a tissue of the individual who provided the nucleus. By contrast, in reproductive cloning (forbidden in humans), the embryo would be preserved to be implanted and grown to term in a surrogate mother. It is this latter procedure that was done in cloning Dolly the sheep.

IN THE NEWS

Shooting Genes into Cells. As you will learn on this page, one of the chief obstacles to successful gene therapy is getting the desired gene into a patient's cells without harming the patient. The usual virus vectors proved problematic, often leading to the development of cancer. A novel new approach has been taken in Germany that picks up where plant bioengineers left off (figure 13.10, page 227). These researchers literally shoot the genes into a patient's tissue. While plant bioengineers must use gold pellets to pierce the plant cell wall, the German researchers simply squirt the genes in across the plasma membranes of the patient's cells. Jet injection of "naked" genes (that is, without a protective virus coat) in a fine beam of water delivered at high pressure does not harm the genes and seems to be well tolerated by the tissue cells. Importantly, introducing genes into human tissues in this way is safe. Because the patient's DNA is not altered, there is no danger of cancer as there is with commonly used adenovirus vectors.

Figure 13.21 **Adenovirus and AAV vectors (×200,000).**
Adenovirus, the *red* virus particles above, has been used to carry healthy genes in clinical trials of gene therapy. Its use as a vector is problematic, however. AAV, the much smaller *bluish-green* virus particles seen in association with adenovirus here, lacks the many problems of adenovirus and is a much more promising gene transfer vector.

13.9 Gene Therapy

CONCEPT PREVIEW: In principle, it should be possible to cure hereditary disorders like cystic fibrosis by transferring a healthy gene into the cells of affected tissues. Early attempts using adenovirus vectors were not often successful. New virus vectors like AAV avoid the problems of earlier vectors and offer promise of gene transfer therapy cures.

The third major advance in cell technology involves introducing "healthy" genes into cells that lack them. For decades scientists have sought to cure often-fatal genetic disorders like cystic fibrosis, muscular dystrophy, and multiple sclerosis by replacing the defective gene with a functional one.

Early Success

A successful **gene transfer therapy** procedure was first demonstrated in 1990 (see section 13.3). Two girls were cured of a rare blood disorder due to a defective gene for the enzyme adenosine deaminase. Scientists isolated working copies of this gene and introduced them into bone marrow cells taken from the girls. The gene-modified bone marrow cells were allowed to proliferate, then were injected back into the girls. The girls recovered and stayed healthy. For the first time, a genetic disorder was cured by gene therapy.

The Rush to Cure Cystic Fibrosis

Researchers quickly set out to apply the new approach to one of the big killers, cystic fibrosis. The defective gene, labelled *cf*, had been isolated in 1989. Five years later, in 1994, researchers successfully transferred a healthy *cf* gene into a mouse that had a defective one—they in effect had cured cystic fibrosis in a mouse. They achieved this remarkable result by adding the *cf* gene to a virus that infected the lungs of the mouse, carrying the gene with it "piggyback" into the lung cells. The virus chosen as the "vector" was adenovirus (the red viruses in figure 13.21), a virus that causes colds and is very infective of lung cells. Very encouraged by these preliminary trials with mice, several labs set out to cure cystic fibrosis by transferring healthy copies of the *cf* gene into human patients the same way. But the gene-modified cells in the patients' lungs soon came under attack by the patients' own immune systems. The "healthy" *cf* genes were lost and with them any chance of a cure.

Problems with the Vector

In retrospect, although it was not obvious then, the problem with these early attempts seems predictable. Adenovirus causes colds. Do you know anyone who has never had a cold? When you get a cold, your body produces antibodies to fight off the infection, and so all of us have antibodies directed against adenovirus. We were introducing genes in a vector our bodies are primed to destroy.

> The human immune response, the subject of chapter 28, primes the body with a first exposure to a virus and so the immune response, called the secondary response, discussed on pages 536 to 541, can quickly fight subsequent infections by the same type of virus.

A second serious problem is that when the adenovirus infects a cell, it inserts its DNA into the human chromosome. Unfortunately, it does so at a random location. This means that the insertion events could cause mutations—if the viral DNA inserts into the middle of a gene, it could inactivate that gene. Because the spot where the adenovirus inserts is random, some of the mutations that result can be expected to cause cancer, certainly an unacceptable consequence.

A More Promising Vector

Researchers are now investigating a much more promising vector, a tiny virus called *adeno-associated virus* (AAV—the smaller bluish-green viruses in figure 13.21) that has only two genes. To create a vector for gene transfer, researchers remove both of the AAV genes. The shell that remains is still quite infective and can carry human genes into patients. AAV does not elicit a strong immune response—cells infected with AAV are not eliminated by a patient's immune system. Importantly, AAV enters human DNA far less frequently than adenovirus, and so is less likely to produce cancer-causing mutations.

A simple experiment using AAV cured dogs of a hereditary disorder leading to retinal degeneration and blindness. These dogs had a defective gene that produced a mutant form of a protein associated with the retina of the eye and were blind. Recombinant viral DNA was made using a healthy version of the gene, shown in steps ❶ and ❷ in figure 13.22. Injection of AAV bearing the needed gene into the fluid-filled compartment behind the retina in step ❸ restored their sight in step ❹. This procedure was recently tried on human patients with some success.

Human clinical trials are under way again. Scientists have performed the first gene therapy experiment for muscular dystrophy, injecting genes into a 35-year-old South Dakota man. He is an early traveler on what is likely to become a well-traveled therapeutic highway. Trials are also under way for cystic fibrosis, rheumatoid arthritis, hemophilia, and a wide variety of cancers. The way seems open, the possibility of progress tantalizingly close.

Concept Check

1. Explain why the development of cloned animals often goes wrong.
2. What is the difference between an embryonic and an adult stem cell?
3. What makes AAV safer than adenovirus as a gene therapy vector?

Figure 13.22   Using gene therapy to cure a retinal degenerative disease in dogs.

Researchers were able to use genes from healthy dogs to restore vision in dogs blinded by an inherited retinal degenerative disease. This disease also occurs in human infants and is caused by a defective gene that leads to early vision loss, degeneration of the retinas, and blindness. In the gene therapy experiments, genes from dogs without the disease were inserted into three-month-old dogs that were known to carry the defective gene and that had been blind since birth. Six weeks after the treatment, the dogs' eyes were producing the normal form of the gene's protein product, and by three months, tests showed that the dogs' vision was restored.

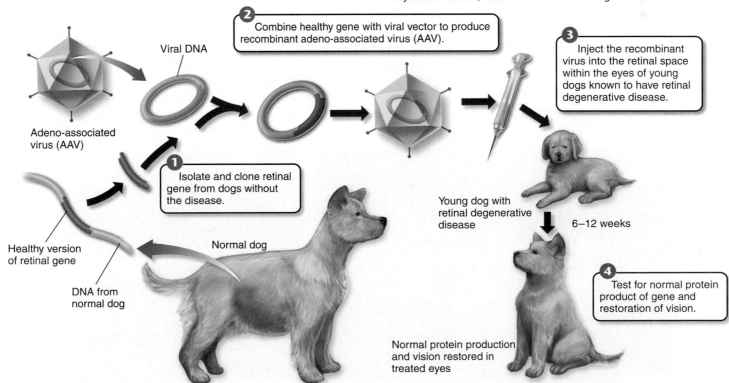

Inquiry & Analysis

Can Modified Genes Escape from GM Crops?

On page 229, the question of whether gene flow of GM crops posed a problem to the environment was discussed. A field experiment conducted in 2004 by the Environmental Protection Agency assessed the possibility that introduced genes could pass from genetically modified golf course grass to other plants. Investigators introduced a gene conferring herbicide resistance (the EPSP synthetase gene for resistance to glyphosate) into golf course bentgrass, *Agrostis stolonifera,* and then looked to see if the gene passed from the GM grass to other plants of the same species, and also if it passed to other related species.

The map at the bottom displays the setup of this elaborate field study. A total of 178 *A. stolonifera* plants were placed outside the golf course, many of them downwind. An additional 69 bentgrass plants were found to be already growing downwind, most of them the related species *A. gigantea.* Seeds were collected from each of these plants, and the DNA of resulting seedlings tested for the presence of the gene introduced into the GM golf course grass. In the graph, the upper red histogram (a **histogram** is a "bar graph" that sorts data into a series of discontinuous categories, the value of each bar representing the number of individuals in a category, or, as in this case, the average value of entries in that category) presents the relative frequency with which the gene was found in *A. stolonifera* plants located at various distances from the golf course. The lower blue histogram does the same for *A. gigantea* plants.

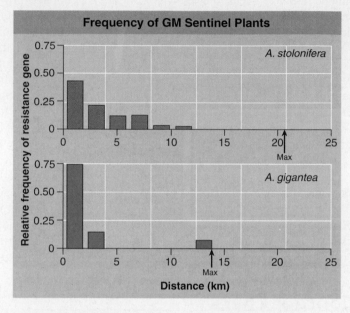

Analysis

1. Applying Concepts

a. **Reading a Histogram.** Does the gene conferring resistance to herbicide pass to other plants of this species, *A. stolonifera*? to individuals of the related species *A. gigantea*?

b. What is the maximal distance over which the herbicide resistance gene is transferred to other plants of this species? of the related species? what are these distances, expressed in miles (km × 0.62 = mile)?

2. Interpreting Data

a. What general statement can be made about the effect of distance on the likelihood that the herbicide resistance gene will pass to another plant?

b. Are there any significant differences in the gene flow to individuals of *A. stolonifera* and to individuals of the related species *A. gigantea*?

3. Making Inferences
What mechanism do you propose to account for this gene flow?

4. Drawing Conclusions
Is it fair to conclude that genetically modified traits can pass from crops to other plants? What qualifications would you place on your conclusion?

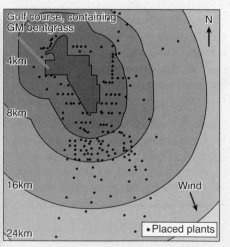

Concept Summary

Sequencing Entire Genomes

13.1 Genomics

- The genetic information of an organism, its genes and other DNA, is called its genome. The sequencing and study of genomes is an area of biology called genomics.

- The sequencing of entire genomes, a once long and tedious process, has been made faster and easier with automated systems (**figure 13.1**).

13.2 The Human Genome

- The human genome contains about 20,000 to 25,000 genes, not much more than other organisms (**table 13.1**) and far less than what was expected based on the number of unique mRNA molecules present in our cells.

- Genes are organized in different ways in the genome, with nearly 99% of the human genome containing noncoding segments of DNA (**figure 13.2** and **table 13.2**).

Genetic Engineering

13.3 A Scientific Revolution

- Genetic engineering is the process of moving genes from one organism to another. It is having a major impact on medicine and agriculture (**figure 13.4**).

- Restriction enzymes are a special kind of enzyme that binds to specific short sequences of DNA and cuts them. The cut made is often a staggered cut producing "sticky ends," which are sections of single-stranded DNA (**figure 13.5 ❷**). The significance of these sticky ends is that any two molecules of DNA cut with the same restriction enzyme will have the same sticky end sequences. This allows the two sources of DNA to combine through base pairing of their sticky ends (**figure 13.5 ❸**).

- Before transferring a eukaryotic gene into a bacterial cell, the intron regions must be removed. This is accomplished with the use of cDNA, a complementary copy of the gene using the processed mRNA to make double-stranded DNA that doesn't contain the introns (**figure 13.6**).

- DNA fingerprinting is a process using probes to compare two samples of DNA. The probes bind to the DNA samples, creating restriction patterns that can be compared. Two DNA samples with the same restriction patterns are likely from the same source (**figure 13.7**).

13.4 Genetic Engineering and Medicine

- Genetic engineering is used in the production of medically important proteins used to treat illnesses. Genes encoding the proteins are inserted into bacteria that produce large quantities of proteins that can then be administered to patients.

- Vaccines are developed using genetic engineering. A gene that encodes a viral protein of a pathogenic virus is inserted into the DNA of a harmless virus that serves as a vector, as shown here from **figure 13.9**. The vector carrying the recombinant DNA is injected into a human. The vector infects the body, replicates, and the recombinant DNA is translated producing the viral proteins. The body elicits an immune response against the proteins, which protects the person from an infection by the pathogenic virus in the future.

13.5 Genetic Engineering and Agriculture

- Genetic engineering has been used in crop plants to make them more cost effective to grow or more nutritious (**figures 13.10–13.12**).

- GM plants are a source of controversy because of potential dangers that may result from the genetic manipulation of crop plants.

The Revolution in Cell Technology

13.6 Reproductive Cloning

- The idea of cloning animals, placing the nucleus from an adult cell into an enucleated egg cell, is not new, but early cloning attempts had mixed results. In 1996, Ian Wilmut and colleagues succeeded in cloning a sheep by synchronizing the donated nucleus and the egg cell to the same stage of the cell cycle (**figure 13.14**).

- Other animals have been successfully cloned (**figure 13.15**), but problems and complications arise, often causing premature death. The problems with cloning appear to be caused by the lack of modifications that need to be made to the DNA, which turns certain genes on or off, a process called genomic imprinting.

13.7 Stem Cell Therapy

- Embryonic stem cells are totipotent cells, which are cells that are able to divide and develop into any type of cell in the body or develop into an entire individual. These cells are present in the early embryo. Because of the totipotent nature of embryonic stem cells, they could be used to replace tissues lost or damaged due to accident or disease (**figure 13.18**).

13.8 Therapeutic Cloning

- The use of embryonic stem cells to replace damaged tissue has one major drawback: tissue rejection. The embryonic stem cells are treated as foreign cells by the patient's body and are rejected. A process called therapeutic cloning could alleviate this problem.

- Therapeutic cloning is the process whereby a cell from an individual who has lost tissue function is cloned, producing an embryo that is genetically identical to the person. Embryonic stem cells are then harvested from the cloned embryo and injected into the same individual. The embryonic stem cells regrow the lost or damaged tissue without eliciting an immune response (**figure 13.20**). However, this procedure, like others using embryonic stem cells, is controversial. Adult cells reprogrammed to behave like embryonic stem cells might prove to be an acceptable treatment.

13.9 Gene Therapy

- Using gene therapy, a patient with a genetic disorder is cured by replacing a defective gene with a "healthy" gene. In theory, this should work—but early attempts to cure cystic fibrosis failed because of immunological reactions to the adenovirus vector used to carry the healthy genes into the patient.

- The recent focus of gene therapy has been to identify a vector that avoids the problems encountered with the adenovirus vector. Promising results in experiments using a virus called adeno-associated virus (AAV) has scientists hopeful that this vector will eliminate the problems seen with adenovirus (**figure 13.22**).

Self-Test

1. The total amount of DNA in an organism, including all of its genes and other DNA, is its
   a. heredity.   c. genome.
   b. genetics.   d. genomics.

2. A possible reason why humans have such a small number of genes as opposed to what was anticipated by scientists is that
   a. humans don't need more than 25,000 genes to function.
   b. the exons used to make a specific mRNA can be rearranged to form different proteins.
   c. the sample size used to sequence the human genome was not big enough, so the number of genes estimated could be low.
   d. the number of genes will increase as scientists find out what all of the noncoding DNA actually does.

3. A protein that can cut DNA at specific DNA base sequences is called a
   a. DNase.   c. restriction enzyme.
   b. DNA ligase.   d. DNA polymerase.

4. Complementary DNA or cDNA is produced by
   a. inserting a gene into a bacterial cell.
   b. exposing the mRNA of the desired eukaryotic gene to reverse transcriptase.
   c. exposing the source DNA to restriction enzymes.
   d. exposing the source DNA to a probe.

5. Which of the following statements is correct.
   a. DNA fingerprinting is not admissible in court.
   b. DNA fingerprinting can prove with 100% certainty that two samples of DNA are from the same person.
   c. DNA fingerprinting becomes more and more reliable as more probes are used.
   d. No two people will ever have the same restriction pattern.

6. Using drugs produced by genetically engineered bacteria allows
   a. the drug to be produced in far larger amounts than in the past.
   b. humans to permanently correct the effect of a missing gene from their own systems.
   c. humans to cure cystic fibrosis.
   d. All these answers are correct.

7. Some of the advantages to using genetically modified organisms in agriculture include
   a. increased yield.
   b. maintaining current nutritive value.
   c. mass producing proteins.
   d. curing genetic diseases.

8. Which of the following is *not* a concern about the use of genetically modified crops?
   a. possible danger to humans after consumption
   b. insecticide resistance developing in pest species
   c. gene flow into natural relatives of GM crops
   d. harm to the crop itself from mutations

9. One of the main biological problems with replacing damaged tissue through the use of embryonic stem cells is
   a. immunological rejection of the tissue by the patient.
   b. that stem cells may not target appropriate tissue.
   c. the time needed to grow sufficient amounts of tissue.
   d. that genetic mutation of chosen stem cells may cause future problems.

10. In gene therapy, healthy genes are placed into animal cells that have defective genes by using
    a. DNA particle gun.
    b. micropipettes (needles).
    c. viruses.
    d. Cells are not modified genetically. Instead, healthy tissue is grown and transplanted into the patient.

Visual Understanding

1. **Figure 13.1** Can you sequence the unknown section of DNA with the DNA fragments obtained from the following DNA analysis?

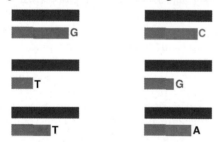

2. **Figure 13.14** Your friend Thomas wants to know why scientists can't just take the egg cell, with its own nucleus intact, and shock it to begin cell division. How do you answer him?

Challenge Questions

1. The goal behind therapeutic cloning is to replace tissue that is damaged due to an accident or nongenetic disease. Why wouldn't therapeutic cloning as shown in figure 13.20 work to replace tissue damage caused by genetic disorders?

2. If a person has a genetic disease such as cystic fibrosis, the hope is that we will be able to use gene therapy to cure him or her. When that happens, will the patient no longer be able to pass the *cf* gene on to his or her children?

3. Much of the technology for producing GM foods is owned by multinational corporations, which seek to maintain intellectual ownership of their creations. As one example, Monsanto Corporation requires farmers to sign contracts for glyphosate-tolerant soybeans that prevent the farmers from saving seed for replanting the next year. The company has aggressively brought suit against violators. On the one hand, companies need to be able to profit from their products, and the development costs of GM foods are enormous. Without potential profit, future GM crops will not be developed. On the other hand, in many highly populated regions of the world, people who face famine when their crops fail simply cannot afford to pay the price of seeds every year. How would you want to see this challenging issue handled?

Chapter 14

Evolution and Natural Selection

Evolution

Figure 14.1   **The theory of evolution by natural selection was proposed by Charles Darwin.**

This rediscovered photograph appears to be the last ever taken of the great biologist. It was taken in 1881, the year before Darwin died.

14.1 Darwin's Voyage on HMS *Beagle*

CONCEPT PREVIEW: Darwin was the first to propose natural selection as the mechanism of evolution that produced the diversity of life on earth.

The great diversity of life on earth—ranging from bacteria to elephants and roses—is the result of a long process of **evolution**, the change that occurs in organisms' characteristics through time. In 1859, the English naturalist Charles Darwin (1809–82; figure 14.1) first suggested an explanation for why evolution occurs, a process he called *natural selection*. Biologists soon became convinced Darwin was right and now consider evolution one of the central concepts of the science of biology. In this chapter we examine Darwin and evolution in detail, as the concepts we encounter will provide a solid foundation for your exploration of the living world.

The theory of evolution proposes that a population can change over time, sometimes forming a new species. A **species** is a population or group of populations that possess similar characteristics and can interbreed and produce fertile offspring. This famous theory provides a good example of how a scientist develops a hypothesis—in this case, a hypothesis of how evolution occurs—and how, after much testing, the hypothesis is eventually accepted as a theory.

Charles Robert Darwin was an English naturalist who, after 30 years of study and observation, wrote one of the most famous and influential books of all time. This book, *On the Origin of Species by Means of Natural Selection, or The Preservation of Favoured Races in the Struggle for Life,* created a sensation when it was published, and the ideas Darwin expressed in it have played a central role in the development of human thought ever since.

In Darwin's time, most people believed that the various kinds of organisms and their individual structures resulted from direct actions of the Creator. Species were thought to be specially created and unchangeable over the course of time. In contrast to these views, a number of earlier philosophers had presented the view that living things must have changed during the history of life on earth. Darwin proposed a concept he called natural selection as a coherent, logical explanation for this process. Darwin's book, as its title indicates, presented a conclusion that differed sharply from conventional wisdom. Although his theory did not challenge the existence of a Divine Creator, Darwin argued that this Creator did not simply create things and then leave them forever unchanged. Instead, Darwin's God expressed Himself through the operation of natural laws that produced change over time—evolution.

The story of Darwin and his theory begins in 1831, when he was 22 years old. The small British naval vessel HMS *Beagle* that you see in figure 14.2

Figure 14.2   **Cross section of HMS *Beagle*.**

HMS *Beagle*, a 10-gun brig of 242 tons, only 90 feet in length, had a crew of 74 people! After he first saw the ship, Darwin wrote to his college professor Henslow: "The absolute want of room is an evil that nothing can surmount."

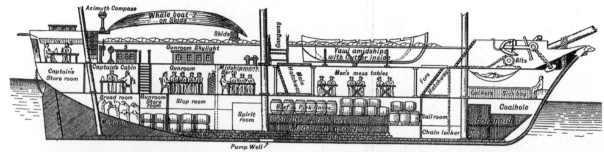

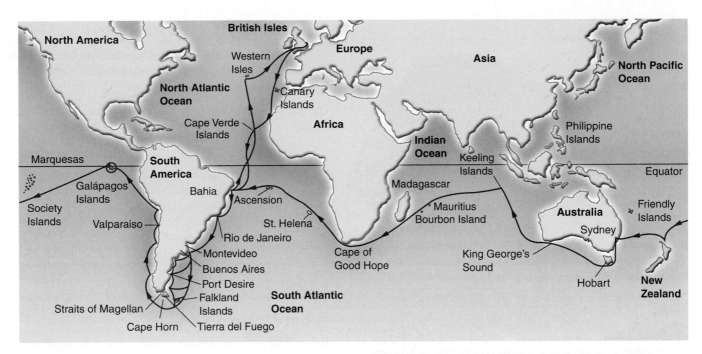

Figure 14.3 **The five-year voyage of HMS *Beagle*.**

Although the ship sailed around the world, most of the time was spent exploring the coasts and coastal islands of South America, such as the Galápagos Islands. Darwin's studies of the animals of these islands played a key role in the eventual development of his theory of evolution by means of natural selection.

IMPLICATION During Darwin's five years on HMS *Beagle*, he was confined to a ship that crammed 74 people into a very small space for months at a time. Have you ever lived in very tight quarters with others who were not members of your own family? Do you think the lack of privacy and forced interactions would have an effect on you? Do you think they influenced Darwin?

was about to set sail on a five-year navigational mapping expedition around the coasts of South America. The red arrows in figure 14.3 indicate the route taken by HMS *Beagle*. The young (26-year-old) captain of HMS *Beagle,* unable by British naval tradition to have social contact with his crew, and anticipating a voyage that would last many years, wanted a gentleman companion, someone to talk to. Indeed, the *Beagle*'s previous skipper had broken down and shot himself to death after three solitary years away from home.

On the recommendation of one of his professors at Cambridge University, Darwin, the son of a wealthy doctor and very much a gentleman, was selected to serve as the captain's companion, primarily to share his table at mealtime during every shipboard dinner of the long voyage. Darwin paid his own expenses, and even brought along a manservant.

Darwin took on the role of ship's naturalist (the official naturalist, a man named Robert McKormick, left the ship before the first year was out). During this long voyage, Darwin had the chance to study a wide variety of plants and animals on continents and islands and in distant seas. He was able to explore the biological richness of the tropical forests, examine the extraordinary fossils of huge extinct mammals in Patagonia at the southern tip of South America, and observe the remarkable series of related but distinct forms of life on the **Galápagos Islands.** Such an opportunity clearly played an important role in the development of his thoughts about the nature of life on earth.

When Darwin returned from the voyage at the age of 27, he began a long period of study and contemplation. During the next 10 years, he published important books on several different subjects, including the formation of oceanic islands from coral reefs and the geology of South America. He also devoted eight years of study to barnacles, a group of small shelled marine animals that inhabit rocks and pilings, eventually writing a four-volume work on their classification and natural history. In 1842, Darwin and his family moved out of London to a country home at Down, in the county of Kent. In these pleasant surroundings, Darwin lived, studied, and wrote for the next 40 years.

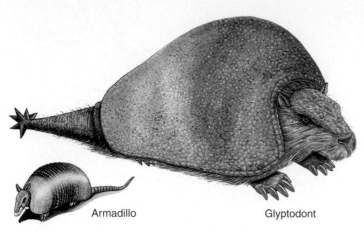

Figure 14.4   Fossil evidence of evolution.

The now-extinct glyptodont was a large 2,000-kilogram South American armadillo (about the size of a small car), much larger than the modern armadillo, which weighs an average of about 4.5 kilograms and is about the size of a house cat. The similarity of fossils such as the glyptodonts to living organisms found in the same regions suggested to Darwin that evolution had taken place.

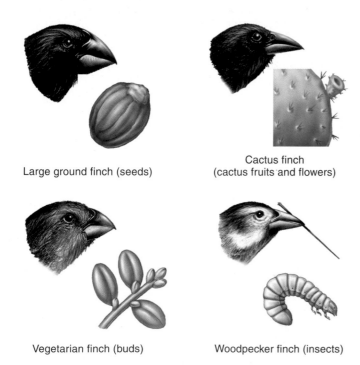

Figure 14.5   Four Galápagos finches and what they eat.

Darwin observed 14 different species of finches on the Galápagos Islands, differing mainly in their beaks and feeding habits. These four finches eat very different food items, and Darwin surmised that the very different shapes of their beaks represented evolutionary adaptations improving their ability to do so.

14.2 Darwin's Evidence

CONCEPT PREVIEW: The fossils and patterns of life that Darwin observed on the voyage of HMS *Beagle* eventually convinced him that evolution had taken place.

One of the obstacles that had blocked the acceptance of any theory of evolution in Darwin's day was the incorrect notion, widely believed at that time, that the earth was only a few thousand years old. The discovery of thick layers of rocks, evidences of extensive and prolonged erosion, and the increasing numbers of diverse and unfamiliar fossils discovered during Darwin's time made this assertion seem less and less likely. The great geologist Charles Lyell (1797–1875), whose *Principles of Geology* (1830) Darwin read eagerly as he sailed on HMS *Beagle*, outlined for the first time the story of an ancient world of plants and animals in flux. In this world, species were constantly becoming extinct while others were emerging. It was this world that Darwin sought to explain.

What Darwin Saw

When HMS *Beagle* set sail, Darwin was fully convinced that species were immutable, meaning that they were not subject to being changed. Indeed, it was not until two or three years after his return that he began to seriously consider the possibility that they could change. Nevertheless, during his five years on the ship, Darwin observed a number of phenomena that were of central importance to him in reaching his ultimate conclusion. For example, in the rich fossil beds of southern South America, he observed fossils of the extinct armadillo shown in figure 14.4 on the right. They were surprisingly similar in form to the armadillos that still lived in the same area, shown on the left. Why would similar living and fossil organisms be in the same area unless the earlier form had given rise to the other? Later, Darwin's observations would be strengthened by the discovery of other examples of fossils that show intermediate characteristics, pointing to successive change.

Repeatedly, Darwin saw that the characteristics of similar species varied somewhat from place to place. These geographical patterns suggested to him that organismal lineages change gradually as individuals move into new habitats. On the Galápagos Islands, 900 kilometers (540 miles) off the coast of Ecuador, Darwin encountered a variety of different finches on the islands. The 14 species, although related, differed slightly in appearance. Darwin felt it most reasonable to assume all these birds had descended from a common ancestor blown by winds from the South American mainland several million years ago. Eating different foods, on different islands, the species had changed in different ways, most notably in the size of their beaks. The larger beak of the ground finch in the upper left of figure 14.5 is better suited to crack open the large seeds it eats. As the generations descended from the common ancestor, these ground finches changed and adapted, what Darwin referred to as "descent with modification"—evolution.

In a more general sense, Darwin was struck by the fact that the plants and animals on these relatively young volcanic islands resembled those on the nearby coast of South America. If each one of these plants and animals had been created independently and simply placed on the Galápagos Islands, why didn't they resemble the plants and animals of islands with similar climates, such as those off the coast of Africa? Why did they resemble those of the adjacent South American coast instead?

14.3 The Theory of Natural Selection

CONCEPT PREVIEW: The fact that populations do not really expand geometrically implies that nature acts to limit population numbers. The traits of organisms that survive to produce more offspring will be more common in future generations—a process Darwin called natural selection.

It is one thing to observe the results of evolution but quite another to understand how it happens. Darwin's great achievement lies in his formulation of the hypothesis that evolution occurs because of natural selection.

Darwin and Malthus

Of key importance to the development of Darwin's insight was his study of Thomas Malthus's *Essay on the Principle of Population* (1798). In his book, Malthus pointed out that populations of plants and animals (including human beings) tend to increase geometrically, while their food supply increases only arithmetically. A geometric progression is one in which the elements increase by a constant factor; the blue line in figure 14.6 shows the progression 2, 6, 18, 54, . . . where each number is three times the preceding one. An arithmetic progression, in contrast, is one in which the elements increase by a constant difference; the red line shows the progression 2, 4, 6, 8, . . . where each number is two greater than the preceding one.

> Malthus's ideas about population growth will come into play again in chapter 19 on page 365 in relation to carrying capacity. Carrying capacity is the size at which a population stabilizes because of limited resources.

Because populations increase geometrically, virtually any kind of animal or plant would cover the entire surface of the world within a surprisingly short time, if it could reproduce unchecked. Instead, population sizes of species remain fairly constant year after year, because death limits population numbers. Malthus's conclusion provided the key ingredient that was necessary for Darwin to develop the hypothesis that evolution occurs by natural selection.

Natural Selection

Sparked by Malthus's ideas, Darwin saw that although every organism has the potential to produce more offspring than can survive, only a limited number actually do survive and produce further offspring. Many examples appear in nature. Sea turtles, for instance, will return to the beaches where they hatched to lay their eggs. Each female will lay about 100 eggs. The beach could be covered with thousands of hatchlings, like in figure 14.7, trying to make it to water's edge. Less than 10% will actually reach adulthood and return to this beach to reproduce. Darwin combined his observation with what he had seen on the voyage of HMS *Beagle,* as well as with his own experiences in breeding domestic animals, and made an important association:

> This variation in attributes enters the population through mutation, as discussed on pages 192 and 193. Without this variation, a population cannot evolve. Selection, artificial or natural, can only work on variation that exists in the population.

Those individuals that possess physical, behavioral, or other attributes that help them live in their environment are more likely to survive than those that do not have these characteristics. By surviving, they gain the opportunity to pass on their favorable characteristics to their offspring. As the frequency of these characteristics increases in the population, the nature of the population as a whole will gradually change. Darwin called this process **natural selection.** The driving force he identified has often

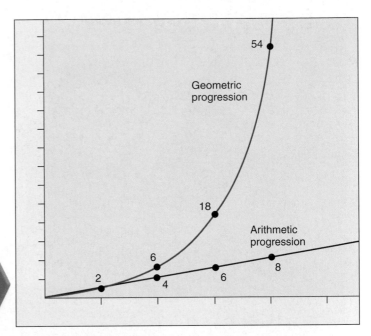

Figure 14.6 Geometric and arithmetic progressions.

An arithmetic progression increases by a constant difference (for example, units of 1 or 2 or 3), while a geometric progression increases by a constant factor (for example, by 2 or by 3 or by 4). Malthus contended that the human growth curve was geometric, but the human food production curve was only arithmetic. Can you see the problems this difference would cause?

Figure 14.7 Sea turtle hatchlings.

These newly hatched sea turtles make their way to the ocean from their nests on the beach. Thousands of eggs may be laid on a beach during a spawning, but less than 10% will survive to adulthood. Natural predators, human egg poachers, and environmental challenges prevent the majority of offspring from surviving. As Darwin observed, sea turtles produce more offspring than will actually survive to reproduce.

Figure 14.8   Darwin greets his monkey ancestor.

In his time, Darwin was often portrayed unsympathetically, as in this drawing from an 1874 publication.

IMPLICATION In 2008 the Spanish parliament approved resolutions granting to gorillas, chimpanzees, and orangutans statutory rights currently only applicable to humans. This was the first time a country has taken such action. The resolutions were based on the Great Ape Project, a framework designed by scientists and philosophers to provide apes' closest relatives with the right to life, liberty, and protection from torture. Zoos could still legally hold apes, but living conditions must be "optimal." Using apes in performances will be illegal. The law will also ban using apes in potentially useful research, if it might hurt the ape in any way. Do you think this last condition of the law is appropriate? Explain.

been referred to as survival of the fittest. However, this is not to say the biggest or the strongest always survive. These characteristics may be favorable in one environment but less favorable in another. The organisms that are "best suited" to their particular environment survive more often, and therefore produce more offspring than others in the population, and in this sense are the "fittest."

Darwin's theory provides a simple and direct explanation of biological diversity, or why animals are different in different places—because habitats differ in their requirements and opportunities, the organisms with characteristics favored locally by natural selection will tend to vary in different places. As we will discuss later in this chapter, section 14.9, there are five evolutionary forces that can affect biological diversity, although natural selection is the only evolutionary force that produces *adaptive* changes.

Darwin Drafts His Argument

Darwin drafted the overall argument for evolution by natural selection in a preliminary manuscript in 1842. After showing the manuscript to a few of his closest scientific friends, however, Darwin put it in a drawer and for 16 years turned to other research.

Wallace Has the Same Idea

The stimulus that finally brought Darwin's theory into print was an essay he received in 1858. A young English naturalist named Alfred Russel Wallace (1823–1913) sent the essay to Darwin from Malaysia; it concisely set forth the theory of evolution by means of natural selection, a theory Wallace had developed independently of Darwin. Like Darwin, Wallace had been greatly influenced by Malthus's 1798 book. After receiving Wallace's essay, Darwin arranged for a joint presentation of their ideas at a seminar in London. Darwin then completed his own book, expanding the 1842 manuscript that he had written so long ago, and submitted it for publication.

Publication of Darwin's Theory

Darwin's book appeared in November 1859 and caused an immediate sensation. Although people had long accepted that humans closely resembled apes in many characteristics, the possibility that there might be a direct evolutionary relationship was unacceptable to many. Darwin did not actually discuss this idea in his book, but it followed directly from the principles he outlined. In a subsequent book, *The Descent of Man*, Darwin presented the argument directly, building a powerful case that humans and living apes have common ancestors. Many people were deeply disturbed with the suggestion that human beings were descended from the same ancestor as apes, and Darwin's book on evolution caused him to become a victim of the satirists of his day—the cartoon in figure 14.8 is a vivid example. Darwin's arguments for the theory of evolution by natural selection were so compelling, however, that his views were almost completely accepted within the intellectual community of Great Britain after the 1860s.

Concept Check

1. How long after his original voyage on HMS *Beagle* did Darwin present his work and publish *On the Origin of Species*?
2. What did Darwin see on the Galápagos Islands that hinted at evolution?
3. How did Malthus influence both Darwin and Wallace?

Darwin's Finches: Evolution in Action

14.4 The Beaks of Darwin's Finches

CONCEPT PREVIEW: In Darwin's finches, natural selection adjusts the shape of the beak in response to the nature of the food supply, adjustments that are occurring even today.

Darwin's Galápagos finches played a key role in his argument for evolution by natural selection. He collected 31 specimens of finches from three islands when he visited the Galápagos Islands in 1835. Darwin, not an expert on birds, had trouble identifying the specimens. He believed by examining their beaks that his collection contained wrens, "gross-beaks," and blackbirds.

The Importance of the Beak

Upon Darwin's return to England, ornithologist John Gould examined the finches. Gould recognized that Darwin's collection was in fact a closely related group of distinct species, all similar to one another except for their beaks. In all, 14 species are now recognized, 13 from the Galápagos and one from far-distant Cocos Island. The ground finches with the larger beaks in figure 14.9 feed on seeds that they crush in their beaks, whereas those with narrower beaks eat insects, including the warbler finch (named for its resemblance to a mainland bird). Other species include fruit and bud eaters, and species that feed on cactus fruits and the insects they attract; some populations of the sharp-beaked ground finch even include "vampires" that creep up on seabirds and use their sharp beaks to drink their blood. Perhaps most remarkable are the tool users, like the woodpecker finch you see in the upper left of the figure, that picks up a twig, cactus spine, or leaf stalk, trims it into shape with its beak, and then pokes it into dead branches to pry out grubs.

The differences in the beaks of Darwin's finches are due to differences in the genes of the birds. When biologists compare the DNA of large ground finches (with stout beaks for cracking large seeds) to the

Figure 14.9   A diversity of finches on a single island.

Ten species of Darwin's finches from Isla Santa Cruz, one of the Galápagos Islands. The ten species show differences in beaks and feeding habits. These differences presumably arose when the finches arrived and encountered habitats lacking small birds. Scientists concluded that all of these birds derived from a single common ancestor.

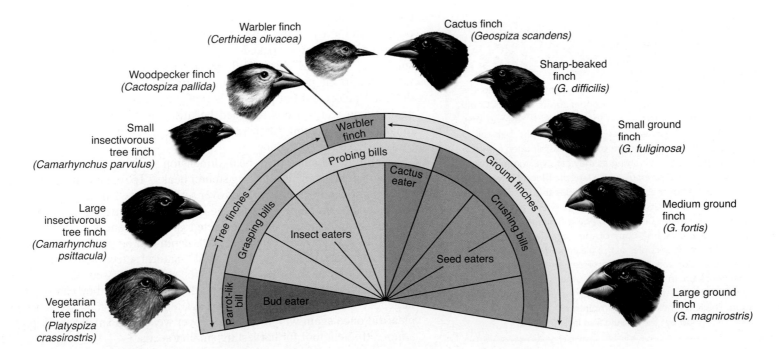

Figure 14.10 **A gene shapes the beaks of Darwin's finches.**

A cell signalling molecule called "bone morphogenic protein 4" (BMP4) has been shown by DNA researchers to tailor the shape of the beak in Darwin's finches.

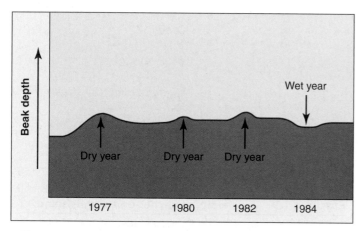

Figure 14.11 **Evidence that natural selection alters beak size in *Geospiza fortis*.**

In dry years, when only large, tough seeds were available, the mean beak size increased. In wet years, when many small seeds were available, smaller beaks became more common.

E V O L U T I O N

How Vampires Evolved. Anyone viewing the film *Twilight* could not help but wonder about vampires. Could vampires ever have evolved in the real world? Yes, and they have—in bats. There are three kinds of vampire bats, all of which employ a salivary enzyme called plasminogen activator to prevent blood from clotting as they eat. The common vampire bat, *Desmodus rotundus,* which laps the blood of mammals, has four copies of the gene that encodes this enzyme. There are other vampire bats: *Diaemus youngi,* which also feeds on mammals but prefers birds, and *Diphylla ecaudata,* which sticks to birds, have only one copy of the gene. DNA sequencing reveals that the four copies of the plasminogen activator gene that the common vampire bat possesses all lack a section called *Kringle 2* present in this gene in both other species. Its deletion in *Desmodus* may have aided a dietary switch to mammalian blood. There are no human vampires, of course, but if there were it is a good bet they would also lack *Kringle 2*.

DNA of small ground finches (with more slender beaks), the only growth factor gene that is different in the DNA of the two species is *BMP4* (figure 14.10 and figure 1.13). The difference is in how the gene is used. The large ground finches, with larger beaks, make more BMP4 protein than do the small ground finches.

The correspondence between the beaks of the 14 finch species and their food source immediately suggested to Darwin that evolution had shaped them:

"Seeing this gradation and diversity of structure in one small, intimately related group of birds, one might really fancy that from an original paucity of birds in this archipelago, one species has been taken and modified for different ends."

Checking to See if Darwin Was Right

If Darwin's suggestion that the beak of an ancestral finch had been "modified for different ends" is correct, then it ought to be possible to see the different species of finches acting out their evolutionary roles, each using its beak to acquire its particular food specialty. The four species that crush seeds within their beaks, for example, should feed on different seeds, with those with stouter beaks specializing on harder-to-crush seeds.

Starting in 1973, Peter and Rosemary Grant of Princeton University and generations of their students have studied the medium ground finch, *Geospiza fortis,* on a tiny island in the center of the Galápagos called Daphne Major. These finches (one of which appears in the photo at the beginning of this chapter) feed preferentially on small tender seeds, abundantly available in wet years. The birds resort to larger, drier seeds that are harder to crush when small seeds are hard to find. Such lean times come during periods of dry weather, when plants produce few seeds, large or small.

By carefully measuring the beak shape of many birds every year, the Grants were able to assemble for the first time a detailed portrait of evolution in action. The Grants found that beak depth changed from one year to the next in a predictable fashion. During droughts, plants produced few seeds, and all available small seeds quickly were eaten, leaving large seeds as the major remaining source of food. As a result, birds with large beaks survived better, because they were better able to break open these large seeds. Consequently, the average beak depth of birds in the population increased the next year because this next generation included offspring of the large-beaked birds that survived. The offspring of the surviving "dry year" birds had larger beaks, an evolutionary response which led to the peaks you see in the graph in figure 14.11. The reason there are peaks and not plateaus is that the average beak size decreased again when wet seasons returned because the larger beak size was no longer more favorable when seeds were plentiful and so smaller-beaked birds survived to reproduce.

Could these changes in beak dimension simply reflect a response to diet, with poorly fed birds having stouter beaks? To rule out this possibility, the Grants examined many broods over several years. The depth of the beak was passed down faithfully from one generation to the next, suggesting the differences in beak size indeed reflected gene differences.

If the year-to-year changes in beak depth can be predicted by the pattern of dry years, then Darwin was right—natural selection influences beak size based on available food supply. In the study discussed here, birds with stout beaks have an advantage during dry periods, for they can break the large, dry seeds that are the only food available. When small seeds become plentiful once again with the return of wet weather, a smaller beak proves a more efficient tool for harvesting smaller seeds.

14.5 How Natural Selection Produces Diversity

CONCEPT PREVIEW: Darwin's finches, all derived from one similar mainland species, have radiated widely on the Galápagos Islands, filling unoccupied niches in a variety of ways.

Darwin believed that each Galápagos finch species had adapted to the particular foods and other conditions on the island it inhabited. Because the islands presented different opportunities, a cluster of species resulted. Presumably, the ancestor of Darwin's finches reached these newly formed islands before other land birds, so that when it arrived, all of the niches where birds occur on the mainland were unoccupied. A *niche* is what a biologist calls the way a species makes a living—the biological (that is, other organisms) and physical (climate, food, shelter, etc.) conditions with which an organism interacts as it attempts to survive and reproduce. As the new arrivals to the Galápagos moved into vacant niches and adopted new lifestyles, they were subjected to diverse sets of selective pressures. Under these circumstances, the ancestral finches rapidly split into a series of populations, some of which evolved into separate species.

Competition between organisms that are utilizing the same niche cannot continue indefinitely. As discussed on pages 370-372, one of the participants dies out in competitive exclusion, or the niche is divided between them in resource partitioning.

The phenomenon by which a cluster of species change, as they occupy a series of different habitats within a region, is called *adaptive radiation*. Figure 14.12 shows how the 14 species of Darwin's finches on the Galápagos Islands and Cocos Island are thought to have evolved. The ancestral population, indicated by the base of the brackets, migrated to the islands about 2 million years ago and underwent adaptive radiation giving rise to the 14 different species. Such species clusters are often particularly impressive on island groups, in series of lakes, or in other sharply discontinuous habitats.

The descendants of the original finches that reached the Galápagos Islands now occupy many different kinds of habitats on the islands. The 14 species that inhabit the Galápagos Islands and Cocos Island occupy four types of niches:

1. **Ground finches.** There are six species of *Geospiza* ground finches. Most of the ground finches feed on seeds. The size of their beaks is related to the size of the seeds they eat. Some of the ground finches feed primarily on cactus flowers and fruits and have longer, larger, more pointed beaks.
2. **Tree finches.** There are five species of insect-eating tree finches. Four species have beaks that are suitable for feeding on insects. The woodpecker finch has a chisel-like beak. This unique bird carries around a twig or a cactus spine, which it uses to probe for insects in deep crevices.
3. **Vegetarian finch.** The very heavy beak of this bud-eating bird is used to wrench buds from branches.
4. **Warbler finches.** These birds play the same ecological role in Galápagos woods that warblers play on the mainland, searching continually over the leaves and branches for insects. They have a slender, warblerlike beak.

Concept Check

1. What evidence did Darwin use to formulate his hypothesis of evolution by natural selection?
2. Describe the different finches Darwin saw on the Galápagos Islands.
3. What four kinds of niches do the finches occupy?

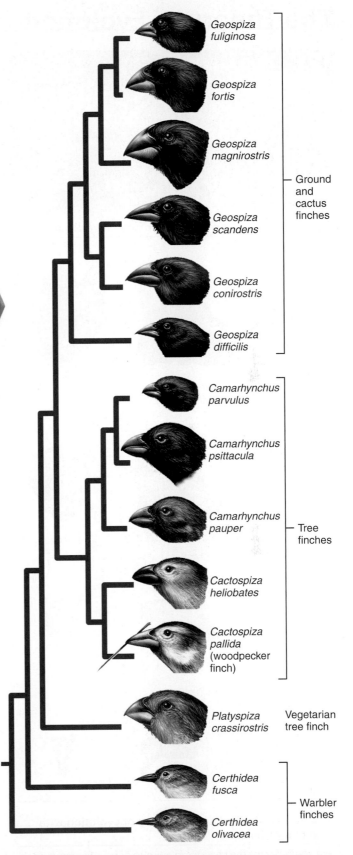

Figure 14.12   An evolutionary tree of Darwin's finches.

This family tree was constructed by comparing DNA of the 14 species. Their position at the base of the finch tree suggests that warbler finches were among the first adaptive types to evolve in the Galápagos.

Labels in figure:
Geospiza fuliginosa
Geospiza fortis
Geospiza magnirostris
Geospiza scandens
Geospiza conirostris
Geospiza difficilis
— Ground and cactus finches

Camarhynchus parvulus
Camarhynchus psittacula
Camarhynchus pauper
Cactospiza heliobates
Cactospiza pallida (woodpecker finch)
— Tree finches

Platyspiza crassirostris — Vegetarian tree finch

Certhidea fusca
Certhidea olivacea
— Warbler finches

The Theory of Evolution

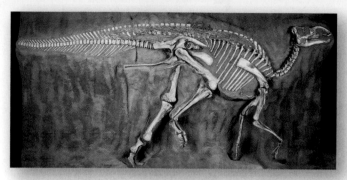

Figure 14.13   Dinosaur fossil of *Parasaurolophus*.

14.6 The Evidence for Evolution

CONCEPT PREVIEW: The fossil record provides a clear record of successive evolutionary change. Comparative anatomy also offers evidence that evolution has occurred. Finally, the genetic record exhibits successive evolution, the DNA of organisms accumulating increasing numbers of changes over time.

The evidence that Darwin presented in *The Origin of Species* to support his theory of evolution was strong. We will now examine other lines of evidence supporting Darwin's theory, including information revealed by examining fossils, anatomical features, and molecules such as DNA and proteins.

The Fossil Record

The most direct evidence of evolution is found in the fossil record. **Fossils** are the preserved remains, tracks, or traces of once-living organisms. Fossils are created when organisms become buried in sediment. The calcium in bone or other hard tissue mineralizes, and the surrounding sediment eventually hardens to form rock. Most fossils are, in effect, skeletons, as shown by the dinosaur fossil in figure 14.13. In the rare cases when fossils form in very fine sediment, feathers may also be preserved. When remains are frozen or become suspended in amber (fossilized plant resin), however, the entire body may be preserved. The fossils contained in layers of sedimentary rock reveal a history of life on earth.

> Isotopes are atoms that have the same number of protons but different numbers of neutrons. Some isotopes, called radioisotopes, are unstable and tend to break up into particles with lower atomic numbers. There are many uses for radioisotopes as discussed on page 34.

By dating the rock in which a fossil occurs, we can get an accurate idea of how old the fossil is. Rocks are dated by measuring the amount of certain radioisotopes in the rock. A radioisotope will break down, or decay, into other isotopes or elements. This occurs at a constant rate and so the amount of a radioisotope present in the rock is an indication of the rock's age.

Using Fossils to Test the Theory of Evolution

If the theory of evolution is correct, then the fossils we see preserved in rock should represent a history of evolutionary change. The theory makes the clear prediction that a parade of successive changes should be seen, as first one change occurs and then another. If the theory of evolution is not correct, on the other hand, then such orderly change is not expected.

To test this prediction, biologists follow a very simple procedure:

1. *Assemble a collection of fossils of a particular group of organisms.* You might for example gather together a collection of fossil titanotheres, a hoofed mammal that lived about 50–35 million years ago.
2. *Date each of the fossils.* In dating the fossils, it is important to make no reference to what the fossil looks like. Imagine it as being concealed in a black box of rock, with only the box being dated.
3. *Order the fossils by their age.* Without looking in the "black boxes," place them in a series, beginning with the oldest and proceeding to the youngest.
4. *Now examine the fossils.* Do the differences between the fossils appear jumbled, or is there evidence of successive change as evolution predicts? You can judge for yourself in figure 14.14, which traces the

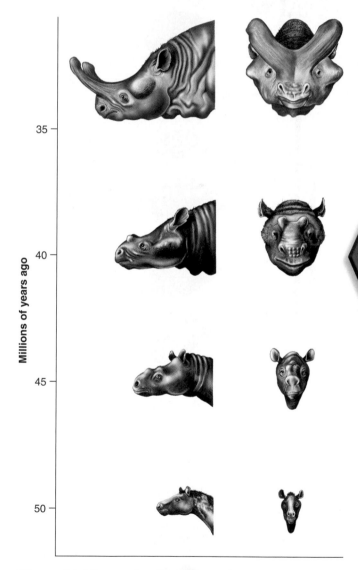

Figure 14.14   Testing the theory of evolution with fossil titanotheres.

Here you can trace the changes in a group of hoofed mammals known as titanotheres from about 50 million years ago (at the bottom) through 35 million years ago (at the top). During this time, the small, bony protuberance located above the nose 50 million years ago evolved into relatively large, blunt horns.

Darwin and Moby Dick

Moby Dick, the white whale hunted by Captain Ahab in Melville's novel, was a sperm whale. One of the ocean's great predators, a large sperm whale is a voracious meat-eater that may span over 60 feet and weigh 50 tons. A sperm whale is not a fish, though. Unlike the great white shark in *Jaws*, a whale has hairs (not many), and a female whale has milk-producing mammary glands with which it feeds its young. A sperm whale is a mammal, just as you are! This raises an interesting question. If Darwin is right about the fossil record reflecting life's evolutionary past, then fossils tell us mammals evolved from reptiles on land at about the time of the dinosaurs. How did they end up back in the water?

The evolutionary history of whales has long fascinated biologists, but only in recent years have fossils been discovered that reveal the answer to this intriguing question. A series of discoveries now allows biologists to trace the evolutionary history of the most colossal animals ever to live on earth back to their beginnings at the dawn of the Age of Mammals. Whales, it turns out, are the descendants of four-legged land mammals that reinvaded the sea some 50 million years ago, much as seals and walruses are doing today. It's pretty startling to realize that Moby Dick's evolutionary ancestor lived on the steppes of Asia and looked like a modest-sized pig a few feet long and weighing perhaps 50 pounds.

From what group of land mammals did whales arise? Researchers had long speculated that it might be a hoofed meat-eater with three toes known as a mesonychid, related to rhinoceroses. Subtle clues suggested this—the arrangement of ridges on the molar teeth, the positioning of the ear bones in the skull. But findings announced in 2001 reveal these subtle clues to have been misleading. Ankle bones from two newly described 50 million-year-old whale species discovered by Philip Gingerich of the University of Michigan are those of an artiodactyl, a four-toed mammal related to hippos, cattle, and pigs. Even more recently, Japanese researchers studying DNA have discovered unique genetic markers shared today only by whales and hippos.

Biologists now conclude that whales, like hippos, are descended from a group of early four-hoofed mammals called anthracotheres, modest-sized grazing animals with a piggish appearance abundant in Europe and Asia 50 million years ago.

In Pakistan in 1994, biologists discovered its descendant, the oldest known whale. The fossil was 49 million years old, had four legs, each with four-toed feet and a little hoof at the tip of each toe. Dubbed *Ambulocetus* (walking whale), it was sharp-toothed and about the size of a large sea lion. Analysis of the minerals in its teeth reveal it drank fresh water, so like a seal it was not yet completely a marine animal. Its nostrils were on the end of the snout, like a dog's.

Appearing in the fossil record a few million years later is *Rodhocetus,* also seal-like but with smaller hind limbs and the teeth of an ocean water drinker. Its nostrils are shifted higher on the skull, halfway towards the top of the head.

Almost 10 million years later, about 37 million years ago, we see the first representatives of *Basilosaurus*, a giant 60-foot-long serpent-like whale with shrunken hind legs still complete down to jointed knees and toes.

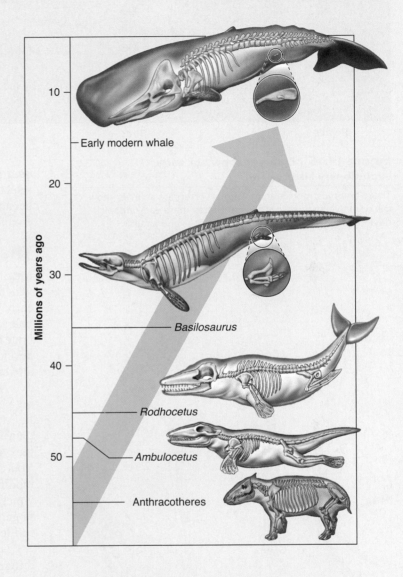

Early modern whale

Basilosaurus

Rodhocetus

Ambulocetus

Anthracotheres

Millions of years ago

The earliest modern whales appear in the fossil record 15 million years ago. The nostrils are now in the top of the head, a "blowhole" that allows it to break the surface, inhale, and resubmerge without having to stop or tilt the head up. The hind legs are gone, with vestigial tiny bones remaining that are unattached to the pelvis. Still, today's whales retain all the genes used to code for legs—occasionally a whale is born having sprouted a leg or two.

So it seems to have taken 35 million years to for a whale to evolve from the piglike ancestor of a hippopotamus—intermediate steps preserved in the fossil record for us to see. Darwin, who always believed that gaps in the vertebrate fossil record would eventually be filled in, would have been delighted.

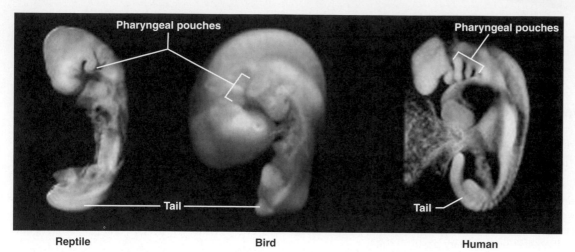

Figure 14.15 **Embryos show our early evolutionary history.**

These embryos, representing various vertebrate animals, show the primitive features that all vertebrates share early in their development, such as pharyngeal pouches and a tail.

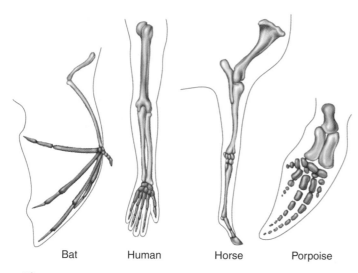

Bat   Human   Horse   Porpoise

Figure 14.16 **Homology among vertebrate limbs.**

Homologies among the forelimbs of four mammals show the ways in which the proportions of the bones have changed in relation to the particular way of life of each organism. Although considerable differences can be seen in form and function, the same basic bones are present in each forelimb.

fossils through time from the oldest at the bottom to the more recent at the top. During the 15 million years spanned by this collection of titanothere fossils, the small, bony protuberance located above the nose 50 million years ago evolved in a series of continuous changes into relatively large blunt horns.

It is important not to miss the key point of the result you see illustrated in figure 14.14: Evolution is an observation, not a conclusion. Because the dating of the samples is independent of what the samples are like, *successive change through time is a data statement.* While the statement that evolution is the result of natural selection is a theory advanced by Darwin, the statement that evolution has occurred is a factual observation. Many other examples illustrate this clear confirmation of the key prediction of Darwin's theory.

The Anatomical Record

Much of the evolutionary history of vertebrates can be seen in the way in which their embryos develop. Figure 14.15 shows three different embryos early in development, and as you can see, all vertebrate embryos have pharyngeal pouches (that develop into gill slits in fish); and every vertebrate embryo has a long bony tail, even if the tail is not present in the fully developed animal. These relict developmental forms strongly suggest that all vertebrates share a basic set of developmental instructions.

As vertebrates have evolved, the same bones are sometimes still there but put to different uses, their presence betraying their evolutionary past. For example, the forelimbs of vertebrates are all **homologous structures;** that is, although the structure and function of the bones have diverged, they are derived from the same body part present in a common ancestor. You can see in figure 14.16 how the bones of the forelimb have been modified for different functions. The yellow- and purple-colored bones, which correspond in humans to the bones of the forearm and wrist and fingers, respectively, are modified to make up the wings in the bat, the full leg of the horse, and the paddle in the fin of the porpoise.

Not all similar features are homologous. Sometimes features found in different lineages come to resemble each other as a result of parallel evolutionary adaptations to similar environments. This form of evolutionary change is referred to as *convergent evolution,* and these similar-looking features are called **analogous structures.** For example, the wings of birds, pterosaurs, and bats are analogous structures, with different bones modified through natural selection to serve the same function and therefore produce wings that look the same.

Sometimes structures are put to no use at all! In living whales, which evolved from hoofed mammals, the bones of the pelvis that formerly anchored the two hind limbs are all that remain of the rear legs, unattached to any other bones and serving no apparent purpose (the reduced pelvic bone can be seen in the figure in the "Today's Biology" reading on the previous page). Another example of what are called *vestigial organs* is the human appendix. In the great apes, our closest relatives, we find an appendix much larger than ours attached to the gut tube, which functions in digestion, holding bacteria used in digesting the cellulose cell walls of the plants eaten by these primates. The human appendix is a vestigial version of this structure that now serves no function in digestion (but it may have acquired an alternate function in the lymphatic system).

The Molecular Record

Traces of our evolutionary past are also evident at the molecular level. We possess the same set of color vision genes as our ancestors, only more complex, and we employ pattern formation genes during early development that all animals share. Indeed, if you think about it, the fact that organisms have evolved from a series of simpler ancestors implies that a record of evolutionary change is present in the cells of each of us, in our DNA. According to evolutionary theory, new alleles arise from older ones by mutation and come to predominance through favorable selection. A series of evolutionary changes thus implies a continual accumulation of genetic changes in the DNA. From this you can see that evolutionary theory makes a clear prediction: Organisms that are more distantly related should have accumulated a greater number of evolutionary differences than two species that are more closely related.

> Our knowledge of the molecular record is expanding with the advancements made in genomics. The sequencing of genomes, discussed on page 216, allows scientists to compare the DNA of closely and distantly related organisms.

This prediction is now subject to direct test. Recent DNA research, allows us to directly compare the genomes of different organisms. The result is clear: For a broad array of vertebrates, the more distantly related two organisms are, the greater their genomic difference. This research is described later in this chapter on pages 256 and 257.

This same pattern of divergence can be clearly seen at the protein level. Comparing the hemoglobin amino acid sequence of different species with the human sequence in figure 14.17, you can see that species more closely related to humans have fewer differences in the amino acid structure of their hemoglobin. Macaques, primates closely related to humans, have fewer differences from humans (only 8 different amino acids) than do more distantly related mammals like dogs (which have 32 different amino acids). Nonmammalian terrestrial vertebrates differ even more, and marine vertebrates are the most different of all. Again, the prediction of evolutionary theory is strongly confirmed.

Molecular Clocks. This same pattern is seen when the DNA sequence of an individual gene is compared over a much broader array of organisms. One well-studied case is the mammalian *cytochrome c* gene (cytochrome *c* is a protein that plays a key role in oxidative metabolism). Figure 14.18 compares the time when two species diverged on the *x* axis to the number of differences in their *cytochrome c* gene on the *y* axis. To practice using this data set, go back about 75 million years ago to find a common ancestor for humans and rodents—in that time there have been about 60 base substitutions in cytochrome *c*. This graph reveals a very important finding: Evolutionary changes appear to accumulate in cytochrome *c*

> The central dogma, DNA-RNA-protein, is discussed on page 200, and describes the information pathway used by cells. Mutations in DNA that occur over time can affect protein products and those changes can be traced in the DNA itself and in the proteins.

at a constant rate, as indicated by the straightness of the blue line connecting the points. This constancy is sometimes referred to as a molecular clock. Most proteins for which data are available appear to accumulate changes over time in this fashion, although different proteins can evolve at very different rates.

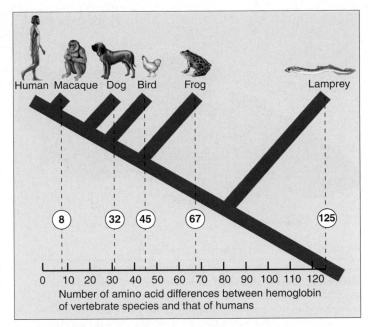

Figure 14.17   Molecules reflect evolutionary divergence.

The greater the evolutionary distance from humans (as revealed by the *blue* evolutionary tree based on the fossil record), the greater the number of amino acid differences in the vertebrate hemoglobin polypeptide.

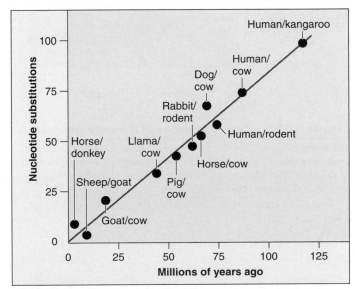

Figure 14.18   The molecular clock of cytochrome c.

When the time since each pair of organisms presumably diverged is plotted against the number of nucleotide differences in cytochrome *c*, the result is a straight line, suggesting that the *cytochrome c* gene is evolving at a constant rate.

Figure 14.19   **An intermediate fossil.**

The animal that produced this fossil is an extinct lobe-finned fish (genus *Tiktaalik*) that lived approximately 375 million years ago. Coined by its discoverer as a "fishopod," it clearly has some characteristics that are fishlike, similar to fish that lived about 380 million years ago, and others that are more like early tetrapods, which lived about 365 million years ago. *Tiktaalik* appears to be a transitional animal, between fish and amphibians.

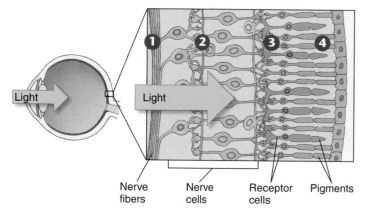

Figure 14.20   **The vertebrate eye is poorly designed.**

The visual pigments in a vertebrate eye that are stimulated by light are embedded in the retinal tissue, facing backward to the direction of the light. The light has to pass through nerve fibers ❶, nerve cells ❷, and receptor cells ❸, before reaching the pigments ❹.

14.7 Evolution's Critics

CONCEPT PREVIEW: Darwin's theory of evolution, while accepted overwhelmingly by scientists, has its objectors. Their criticisms are without scientific merit.

Critics of evolution have raised a variety of objections to Darwin's theory of evolution by natural selection:

1. **Evolution is not solidly demonstrated.** *"Evolution is just a theory,"* critics point out, as if theory meant lack of knowledge, some kind of guess. Scientists, however, use the word theory in a very different sense than the general public does (see section 1.6). Theories are the solid ground of science, supported with much experimental evidence and that of which we are most certain. Few of us doubt the theory of gravity because it is "just a theory."

2. **There are no fossil intermediates.** *"No one ever saw a fin on the way to becoming a leg,"* critics claim, pointing to the many gaps in the fossil record in Darwin's day. Since then, however, most fossil intermediates in vertebrate evolution have indeed been found. A clear line of fossils now traces the transition between whales and hoofed mammals, between reptiles and mammals, and between apes and humans. The fossil evidence of evolution between major forms is compelling (figure 14.19).

3. **The intelligent design argument.** *"The organs of living creatures are too complex for a random process to have produced."* This classic "argument from design" was first proposed nearly 200 years ago by William Paley in his book *Natural Theology*—the existence of a clock is evidence of the existence of a clockmaker, Paley argues. Similarly, Darwin's critics argue that organs like the mammalian ear are too complex to be due to blind evolution. There must have been a designer. Biologists do not agree. Complex structures like the mammalian ear evolved as a progression of slight improvements. The intermediates in the evolution of the mammalian ear are well documented in the fossil record, each favored by natural selection because they each had value—being able to amplify sound a little is better than not being able to amplify it at all. Nor is the solution always optimal, as your own eyes attest. As you can see in the blown-up image in figure 14.20, the receptor cells in the human eye are actually facing backward to the stimulus (light). No intelligent designer would design an eye backwards!

4. **Evolution violates the second law of thermodynamics.** *"A jumble of soda cans doesn't by itself jump neatly into a stack—things become more disorganized due to random events, not more organized."* Biologists point out that this argument ignores what the second law really says: Disorder increases in a closed system, which the earth most certainly is not. Energy enters the biosphere from the sun, fueling life and all the processes that organize it.

5. **Proteins are too improbable.** *"Hemoglobin has 141 amino acids. The probability that the first one would be leucine is 1/20, and that all 141 would be the ones they are by chance is $(1/20)^{141}$, an impossibly rare event."* You cannot use probability to argue backward. The probability that a student in a classroom has a particular birthday is 1/365; arguing this way, the probability that everyone in a class of 50 would have the birthdays they do is $(1/365)^{50}$, and yet there the class sits.

The Irreducible Complexity Fallacy

The century-and-a-half-old "intelligent design" argument of William Paley has been recently articulated in a new molecular guise. Today's proponents of intelligent design now argue that the intricate molecular machinery of our cells is so elaborate that it cannot be explained by evolution from simpler stages—it is "irreducibly complex." Each part of a molecular machine plays a vital role. Remove just one, they claim, and cell molecular machinery cannot function.

As an example of such an irreducibly complex system, intelligent design advocates point to the series of more than a dozen blood clotting proteins that act in our body to cause blood to clot around a wound. Take out any step in the complex cascade of reactions that leads to coagulation of blood, they say, and your body's blood would leak out from a cut like water from a ruptured pipe. If dozens of different proteins all must work correctly to clot blood, how could natural selection act to fashion any one of the individual proteins? No one protein does anything on its own, just as a portion of a watch doesn't tell time. Like Paley's watch, the blood clotting system must have been designed all at once, as a single functioning machine.

What's wrong with this argument, as evolutionary scientists have been quick to point out, is that evolution acts on the system, not its parts. Natural selection can evolve a complex system because at every stage of its evolution, the system functions. Parts that improve function are added, and, because of later changes, eventually become essential, in the same way that the second rung of a ladder becomes essential once you have added a third.

The mammalian blood clotting system, for example, has evolved in stages from much simpler systems (figure 14.21). The core of the vertebrate clotting system, called the "common pathway" (highlighted in blue), evolved at the dawn of the vertebrates approximately 600 million years ago, and is found today in lampreys, the most primitive fish. As vertebrates evolved, proteins were added to the clotting system, improving its efficiency. The so-called extrinsic pathway (highlighted in pink), triggered by substances released from damaged tissues, was added 500 million years ago. Each step in the pathway amplifies what goes before, so adding the extrinsic pathways greatly increases the amplification and thus the sensitivity of the system. Fifty million years later, a third component was added, the so-called intrinsic pathway (highlighted in tan). It is triggered by contact with the jagged surfaces produced by injury. Again, amplification and sensitivity were increased to ultimately end up with blood clots formed by the cross linking of fibrin (highlighted in green). At each stage as the clotting system evolved to become more complex, its overall performance came to depend on the added elements. Mammalian clotting, which utilizes all three pathways, no longer functions if any one of them is disabled. Blood clotting has become "irreducibly complex"—as the result of Darwinian evolution. Intelligent design proponents claim that complex cellular and molecular processes can't be explained by Darwinism. Indeed, examination of the human genome reveals that the cluster of blood clotting genes arose through duplication of genes, with increasing amounts of change. The evolution of the blood clotting system is an observation, not a surmise. Its irreducible complexity is a fallacy.

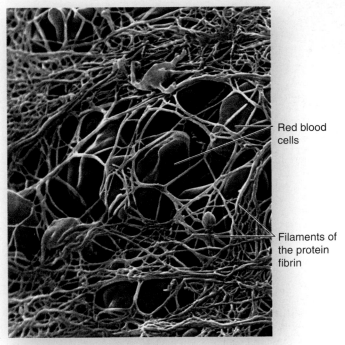

(a) A blood clot

Red blood cells

Filaments of the protein fibrin

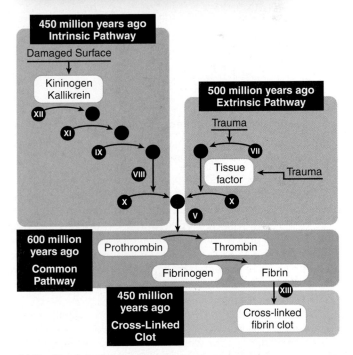

(b) The blood clotting system

Figure 14.21 **How blood clotting evolved.**

The blood clotting system evolved in steps, with new proteins adding on to the preceding step.

Concept Check

1. Describe how fossils are used to provide evidence of evolution.
2. Do genome comparisons establish evolution as a fact? Evaluate.
3. What is the theory of intelligent design? Is it a scientific theory?

Putting Intelligent Design to the Test

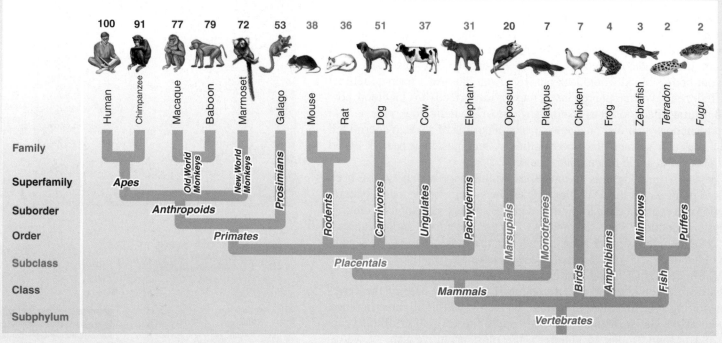

Genomic similarity reflects evolutionary relatedness.

The number above each organism is the percent of the nucleotides in selected regions of that organism's genome that match those of the same regions in the human genome.

In the spring of 2006 the South Carolina Board of Education rejected a state panel's proposal to change high school standards by calling on students to critically analyze evolution. The Board stated it felt the proposal was a ploy to promote the avoidance of teaching evolution. Similar proposals to add a requirement that students critically analyze evolution had been rejected earlier in the year by the Utah and Ohio Boards of Education, and are currently under consideration in several other states.

What are we to make of this? Surely no scientist can object to critically analyzing any theory. That is what science is all about, seeking explanations for what can be observed, tested, replicated, and possibly falsified. Indeed, biologists claim that Darwin's theory of evolution has been subjected to as much critical analysis as any theory in the history of science.

So why the objection to this change in high school standards? Because many scientists and teachers, apparently including the South Carolina Board of Education, feel the change is simply intended to promote the teaching of a non-scientific alternative to evolution in classrooms.

This distinction between an assertion that can be tested and one that cannot goes to the very nature of science. Actually, nothing makes this difference more clear cut than the critical analysis so sought after by South Carolina's critics of evolution. So let's do it. Let's put Darwin to the test.

As explained earlier in the chapter, if Darwin's assertion is correct, that organisms evolved from ancestral species, then we should be able to track evolutionary changes in our DNA. The variation that we see between species reflects adaptations to environmental challenges, adaptations that result from changes in DNA. Therefore, a series of evolutionary changes should be reflected in an accumulation of genetic changes in the DNA. This hypothesis, that evolutionary changes reflect accumulated changes in DNA, leads to the following prediction: Two species that are more distantly related (for example, humans and mice) should have accumulated a greater number of evolutionary differences than two species that are more closely related (say, humans and chimpanzees).

So have they? Let's compare vertebrate species to see. The "family tree" above shows how biologists believe 18 different vertebrate species are related. Apes and monkeys, because they are in the same order (primates), are considered more closely related to each other than either are to members of another order, such as mice and rats (rodents).

The wealth of genomes (a genome is all the DNA that an organism possesses) that have been sequenced since completion of the human genome project allows us to directly compare the DNA of these 18 vertebrates. To reduce the size of the task, investigators at the National Human Genome Research Institute working at the University of California, Santa Cruz, focused on 44 so-called ENCODE regions scattered around the vertebrate genomes. These regions, corresponding to 30 Mb (megabase, or million bases) or roughly 1% of the total human genome, were selected to be representative of the genome as a whole, containing protein-encoding genes as well as noncoding DNA.

For each vertebrate species, the investigators determined the similarity of its DNA to that of humans—that is, the percent of the nucleotides in that organism's 44 ENCODE regions which match those of the human genome.

You can see the result in each instance presented as a number above the picture of each organism on the vertebrate family tree. As Darwin's theory predicts, the closer the relatives, the less the genomic difference we see. The chimpanzee genome is more like the human genome (91% for these ENCODE regions) than the monkey genomes are (72 to 79%). Furthermore, these five genomes, all in the primate order, are more like each other than any are to those of another order, such as rodents (mouse and rat).

In general, as you proceed through the taxonomic categories of the vertebrate family tree from very distant relatives on the right (some in the same class as humans) to very close ones on the left (in the same family), you can see clearly that genomic similarity increases as taxonomic distance decreases —just as Darwin's theory predicts. The prediction of evolutionary theory is solidly confirmed.

The analysis does not have to stop here. The evolutionary history of the vertebrates is quite well known from fossils, and because many of these fossils have been independently dated using tools such as radioisotope dating, it is possible to recast the analysis in terms of concrete intervals of time, and assess directly whether or not vertebrate genomes accumulate more differences over longer periods of time as Darwin's theory predicts.

For each of the 18 vertebrates being analyzed, the graph shown here plots genomic similarity—how alike the DNA sequence of the vertebrate's ENCODE regions are to those of the human genome—against divergence time (that is, how many millions of years have elapsed since that vertebrate and humans shared a common ancestor in the fossil record). Thus the last common ancestor shared by chickens and humans was an early reptile called a dicynodont that lived some 250 million years ago; since then the genomes of the two species have changed so much that only 7% of their ENCODE sequences are still the same.

The result seen in the graph is striking and very clear: Over their more than 300 million year history, vertebrates have accumulated more and more genetic change in their DNA. "Descent with modification" was Darwin's definition of evolution, and that is exactly what we see in the graph. The evolution of the vertebrate genome is not a theory, but an observation.

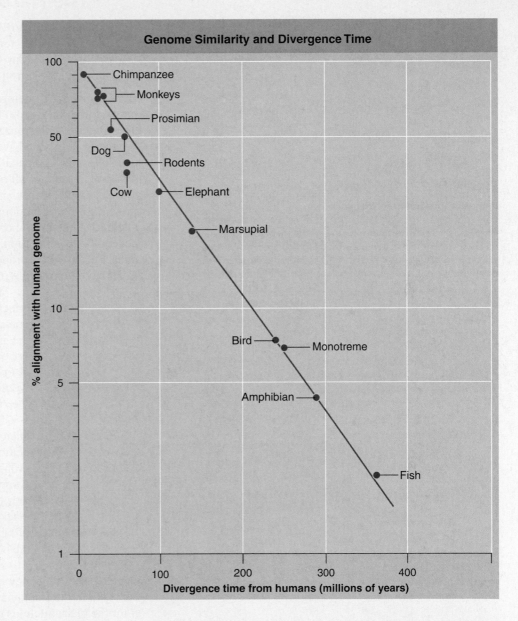

The wealth of data made available by the human genome project has allowed a powerful test of Darwin's prediction. The conclusion to which the test leads us is that evolution is an observed fact, clearly revealed in the DNA of vertebrates.

This is the sort of critical analysis that science requires, and that the theory of evolution has again passed. Anyone suggesting that a nonscientific alternative to evolution, such as Intelligent Design, offers an alternative scientific explanation to evolution is welcome to subject it to the same sort of critical analysis you have seen employed here. Can you think of a way to do so? It is precisely because the assertion of intelligent design cannot be critically analyzed—it does not make any testable prediction—that it is not science and has no place in science classrooms.

How Populations Evolve

14.8 Genetic Change in Populations: The Hardy-Weinberg Rule

CONCEPT PREVIEW: In a large, randomly mating population that fulfills the other Hardy-Weinberg assumptions, allele frequencies can be expected to be in Hardy-Weinberg equilibrium. If they are not, then the population is undergoing evolutionary change.

Darwin and his contemporaries were puzzled why dominant **alleles** (alternative forms of a gene) did not drive recessive alleles out of populations. The solution to this puzzle was explained in 1908 by G. H. Hardy and W. Weinberg. Hardy and Weinberg studied **allele frequencies** (the proportion of alleles of a particular type in a population) in a population's *gene pool*, which is the sum of all of the genes in a population, including all alleles in all individuals. Hardy and Weinberg pointed out that in a large population in which there is random mating, and in the absence of forces that change allele frequencies, the original genotype proportions remain constant from generation to generation. Dominant alleles do not, in fact, replace recessive ones. Because their proportions do not change, the genotypes are said to be in **Hardy-Weinberg equilibrium.**

The Hardy-Weinberg rule is viewed as a baseline to which the frequencies of alleles in a population can be compared. If the allele frequencies are not changing (they are in Hardy-Weinberg equilibrium), the population is not evolving. If, however, allele frequencies are sampled at one point in time and they differ greatly from what would be expected under Hardy-Weinberg equilibrium, then the population is undergoing evolutionary change.

Hardy and Weinberg came to their conclusion by analyzing the frequencies of alleles in successive generations. The **frequency** of something is defined as the proportion of individuals with a certain characteristic, compared to the entire population. Thus, in the population of 1,000 cats shown in figure 14.22, there are 840 black cats and 160 white cats. To determine the frequency of black cats, divide 840 by 1,000 (840/1,000), which is 0.84. The frequency of white cats is 160/1,000 = 0.16.

Knowing the frequency of the phenotype, one can calculate the frequency of the genotypes and alleles in the population. By convention, the frequency of the more common of two alleles (in this case *B* for the black allele) is designated by the letter *p* and that of the less common allele (*b* for the white allele) by the letter *q*. Because there are only two alleles, the sum of *p* and *q* must always equal 1 ($p + q = 1$).

In algebraic terms, the Hardy-Weinberg equilibrium is written as an equation. For a gene with two alternative alleles *B* (frequency *p*) and *b* (frequency *q*), the equation looks like this:

$$p^2 \quad + \quad 2pq \quad + \quad q^2 \quad = \quad 1$$

p^2	$2pq$	q^2
Individuals homozygous for allele **B**	Individuals heterozygous for alleles **B** and **b**	Individuals homozygous for allele **b**

Figure 14.22 Hardy-Weinberg equilibrium.

In the absence of factors that alter them, the frequencies of gametes, genotypes, and phenotypes remain constant generation after generation. The example shown here involves a population of 1,000 cats, in which 160 are white and 840 are black. White cats are *bb*, and black cats are *BB* or *Bb*. The potential crosses in this cat population can be determined using a Punnett square analysis.

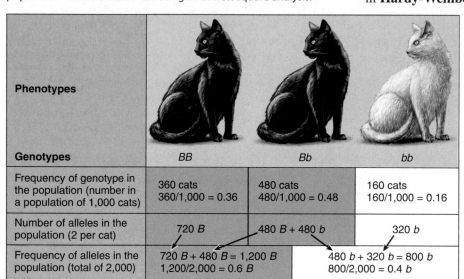

Phenotypes			
Genotypes	*BB*	*Bb*	*bb*
Frequency of genotype in the population (number in a population of 1,000 cats)	360 cats 360/1,000 = 0.36	480 cats 480/1,000 = 0.48	160 cats 160/1,000 = 0.16
Number of alleles in the population (2 per cat)	720 *B*	480 *B* + 480 *b*	320 *b*
Frequency of alleles in the population (total of 2,000)	720 *B* + 480 *B* = 1,200 *B* 1,200/2,000 = 0.6 *B*	480 *b* + 320 *b* = 800 *b* 800/2,000 = 0.4 *b*	

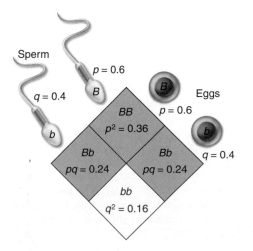

Sperm

$p = 0.6$

$q = 0.4$

Eggs

$p = 0.6$

$q = 0.4$

BB
$p^2 = 0.36$

Bb
$pq = 0.24$

Bb
$pq = 0.24$

bb
$q^2 = 0.16$

You will notice that not only does the sum of the allele frequencies p and q add up to 1, but so does the sum of the frequencies of genotypes.

Hardy-Weinberg Assumptions

The Hardy-Weinberg rule is based on certain assumptions. The equation on the facing page is true only if the following five assumptions are met:

1. The size of the population is very large or effectively infinite.
2. Individuals mate with one another at random.
3. There is no mutation.
4. There is no input of new copies of any allele from any extraneous source (such as migration from a nearby population) or losses of copies of alleles through emigration (individuals leaving the population).
5. All alleles are replaced equally from generation to generation (natural selection is not occurring).

Many populations, and most human populations, are large and randomly mating with respect to most traits (a few traits affecting appearance undergo strong sexual selection in humans). Thus, many populations are similar to the ideal population envisioned by Hardy and Weinberg. For some genes, however, the observed proportion of heterozygotes does not match the value calculated from the allele frequencies. When this occurs, it indicates that something is acting on the population to alter one or more of the genotypic frequencies, whether it is selection, nonrandom mating, migration, or some other factor. The factors that can affect the frequencies of alleles in a population are discussed in detail later in this chapter.

Case Study: Cystic Fibrosis in Humans

How valid are the predictions made by the Hardy-Weinberg equation? For many genes, they prove to be very accurate. As an example, consider the recessive allele responsible for the serious human disease cystic fibrosis. This allele (q) is present in Caucasians in North America at a frequency of 0.022. What proportion of Caucasian North Americans, therefore, is expected to express this trait? The frequency of double-recessive individuals (q^2) is expected to be:

$$q^2 = 0.022 \times 0.022 = 0.00048$$

which equals 0.48 in every 1,000 individuals or about 1 in every 2,000 individuals, very close to real estimates.

What proportion is expected to be heterozygous carriers? If the frequency of the recessive allele q is 0.022, then the frequency of the dominant allele p must be $p = 1 - q$ or:

$$p = 1 - 0.022 = 0.978$$

The frequency of heterozygous individuals ($2pq$) is thus expected to be:

$$2 \times 0.978 \times 0.022 = 0.043$$

It is estimated that 12 million individuals in the United States are carriers of the cystic fibrosis allele. In a population of 292 million people, that is a frequency of 0.041, very close to projections using the Hardy-Weinberg equation.

IN THE NEWS

Why Doesn't Natural Selection Eliminate Cystic Fibrosis? As you learned in this chapter, the frequency of the allele responsible for cystic fibrosis in the United States indicates that it is in Hardy Weinberg equilibrium, and thus not undergoing evolutionary change. Why hasn't natural selection acted to eliminate cystic fibrosis from the human population? Recent experiments on mice offer an answer. They suggest that the nearly 5% of all Caucasians who carry just one copy of the *cf* allele—and thus don't suffer from the disease—are protected against another deadly scourge: cholera. A geneticist in Barcelona, Spain, announced results in 2008 indicating that the *cf* mutation known as *delta-F508*—a deletion of three DNA base pairs from among the 250,000 that make up the gene of the allele form responsible for half of all cases of cystic fibrosis—arose at least 52,000 years ago. This was before the settlement of Europe by modern humans but after humans migrated to Asia (among modern-day East Asians, only 1 in 100,000 develop the disease). The survival of this harmful allele among human populations for so long suggests that there must be a selective advantage to the disorder. A few months later, a University of North Carolina researcher announced results indicating what that advantage might be: The *cf* mutation offers increased resistance to deadly epidemics of cholera, by lessening the diarrhea that kills people with this disease. The researcher found that laboratory mice with one copy of the *cf* allele had half as many functioning chloride channels as normal mice, and so lost only half as much body water to diarrhea when injected with cholera toxin. While not suffering cystic fibrosis, he suggests, early Europeans with one *cf* allele were better able to survive the killing dehydration of cholera. Why wasn't this an advantage in South East Asians? Because one *cf* allele also makes a person lose more salt in their sweat, and in hot climates excessive salt loss would result. It seems that chronic salt loss was a graver problem than occasional exposure to cholera, and because of this, natural selection in Asia favored the lesser of two evils.

TABLE 14.1	Agents of Evolution	
Factor		**Description**
Mutation	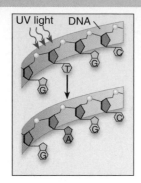	The ultimate source of genetic variation. Individual mutations occur so rarely that mutation alone does not change allele frequency much.
Migration		A very potent agent of change. Migration acts to promote evolutionary change by enabling populations that exchange members to converge toward one another. This bee carries pollen from one population of flowers to another.
Genetic drift		Chance events may result in the loss of individuals and therefore the loss of alleles in a population. Usually occurs only in very small populations. A small number of alleles can also impact a newly formed population, such as the founder effect on an island.
Nonrandom mating	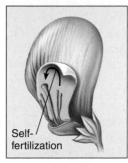	Inbreeding is the most common form of nonrandom mating. It does not alter allele frequency but decreases the proportion of heterozygotes ($2pq$). See text for explanation.
Selection		The only form that produces *adaptive* evolutionary changes. Only rapid for allele frequency greater than .01.

14.9 Agents of Evolution

CONCEPT PREVIEW: Five evolutionary forces have the potential to significantly alter allele and genotype frequencies in populations: mutation, migration, genetic drift, nonrandom mating, and selection. Selection on traits affected by many genes can favor intermediate values, or one or both extremes.

Many factors can alter allele frequencies. But only five alter the proportions of homozygotes and heterozygotes enough to produce significant deviations from the proportions predicted by the Hardy-Weinberg rule (table 14.1):

1. Mutation
2. Migration
3. Genetic drift
4. Nonrandom mating
5. Selection

Mutation

A **mutation** is a change in a nucleotide sequence in DNA. For example, a T nucleotide could undergo a mutation and be replaced with an A nucleotide. Mutation from one allele to another obviously can change the proportions of particular alleles in a population. But mutation rates are generally too low to significantly alter Hardy-Weinberg proportions of common alleles. Many genes mutate 1 to 10 times per 100,000 cell divisions. Some of these mutations are harmful, while others are neutral or, even rarer, beneficial. Also, the mutations must affect the DNA of the germ cells (egg and sperm), or the mutation will not be passed on to offspring. The mutation rate is so slow that few populations are around long enough to accumulate significant numbers of mutations. However, no matter how rare, mutation is the ultimate source of genetic variation in a population.

Migration

Migration is defined in genetic terms as the movement of individuals between populations. It can be a powerful force, upsetting the genetic stability of natural populations. Migration includes movement of individuals into a population, called *immigration,* or the movement of individuals out of a population, called *emigration.* If the characteristics of the newly arrived individuals differ from those already there, and if the newly arrived individuals adapt to survive in the new area and mate successfully, then the genetic composition of the receiving population may be altered.

Sometimes migration is not obvious. Subtle movements include the drifting of gametes of plants, or of the immature stages of marine organisms, from one place to another. For example, a bee can carry pollen from a flower in one population to a flower in another population. By doing this, the bee may be introducing new alleles into a population. However it occurs, migration can alter the genetic characteristics of populations and cause a population to be out of Hardy-Weinberg equilibrium. Thus, migration can cause evolutionary change. The magnitude of effects of migration is based on two factors: (1) the proportion of migrants in the population, and (2) the difference in allele frequencies between the migrants and the original population. The actual evolutionary impact of migration is difficult to assess, and depends heavily on the selective forces prevailing at the different places where the populations occur.

Genetic Drift

In small populations, the frequencies of particular alleles may be changed drastically by chance alone. In an extreme case, individual alleles of a given gene may all be represented in few individuals, and may be accidentally lost if those individuals fail to reproduce or die. This loss of individuals and their alleles is due to random events rather than the fitness of the individuals carrying those alleles. This is not to say that alleles are always lost with genetic drift, but allele frequencies appear to change randomly, as if the frequencies were drifting; thus, random changes in allele frequencies is known as **genetic drift.** A series of small populations that are isolated from one another may come to differ strongly as a result of genetic drift.

When one or a few individuals migrate and become the founders of a new, isolated population at some distance from their place of origin, the alleles that they carry are of special significance in the new population. Even if these alleles are rare in the source population, they will become a significant fraction of the new population's genetic endowment. This is called the **founder effect.** As a result of the founder effect, rare alleles and combinations often become more common in new, isolated populations. The founder effect is particularly important in the evolution of organisms that occur on oceanic islands, such as the Galápagos Islands, which Darwin visited. Most of the kinds of organisms that occur in such areas were probably derived from one or a few initial founders. In a similar way, isolated human populations are often dominated by the genetic features that were characteristic of their founders, particularly if only a few individuals were involved initially (figure 14.23).

Nonrandom Mating

Individuals with certain genotypes sometimes mate with one another either more or less frequently than would be expected on a random basis, a phenomenon known as **nonrandom mating.** One type of nonrandom mating is **sexual selection,** choosing a mate often based on certain physical characteristics. Another type of nonrandom mating is **inbreeding,** or mating with relatives, such as in the self-fertilization of a flower. Inbreeding increases the proportions of individuals that are homozygous because no individuals mate with any genotype but their own. As a result, inbred populations contain more homozygous individuals than predicted by the Hardy-Weinberg rule. For this reason, populations of self-fertilizing plants consist primarily of homozygous individuals, whereas outcrossing plants, which interbreed with individuals different from themselves, have a higher proportion of heterozygous individuals. Nonrandom mating alters genotype frequencies but not allele frequencies. The allele frequencies remain the same—the alleles are just distributed differently among the offspring.

Selection

As Darwin pointed out, some individuals leave behind more progeny than others, and the likelihood they will do so is affected by their inherited characteristics. The result of this process is called **selection** and was familiar even in Darwin's day to breeders of domestic animals. In so-called **artificial selection,** the breeder selects for the desired characteristics. For example, mating larger animals with each other produces offspring that are larger. In **natural selection,** Darwin suggested the environment plays this role, with conditions in nature determining which kinds of individuals in a population are the most fit (best suited to their environment) and so affecting the proportions of genes among individuals of future populations. The

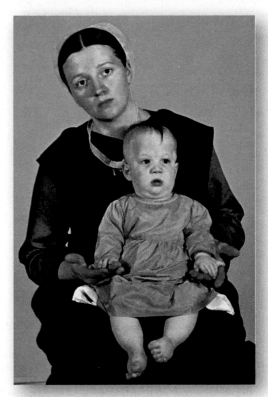

Figure 14.23   The founder effect.

This Amish woman is holding her child, who has Ellis-van Creveld syndrome. The characteristic symptoms are short limbs, dwarfed stature, and extra fingers. This disorder was introduced in the Amish community by one of its founders in the eighteenth century and persists to this day because of reproductive isolation.

EVOLUTION

Why Cheetahs Are All Alike. Even if organisms do not move from place to place, occasionally their populations may be drastically reduced in size from flooding, drought, or other natural catastrophes, a form of genetic drift. The surviving individuals constitute a random genetic sample of the original population. Such a restriction in genetic variability has been termed the bottleneck effect. The very low levels of genetic variability seen in African cheetahs (*Acinonyx jubatus*) today is thought to reflect a near-extinction event 10,000 years ago. Only a few individuals are thought to have survived, and among them they represented only a tiny portion of the species' genetic variation—less than that seen in deliberately inbred strains of laboratory mice. African cheetahs are so genetically similar that they don't reject each other's skin grafts! Unfortunately, this lack of genetic variability is considered to be a significant contributing factor to a lack of disease resistance in African cheetahs. In 1983 a productive and healthy Oregon breeding colony of cheetahs was nearly wiped out by an epidemic of corona virus, a virus which kills less than 1% of infected domestic cats. Cheetah populations have simply lost the alleles that make other cat groups resistant. This vulnerability of cheetahs to disease is leading to a modern decline that threatens both natural and captive populations with extinction.

environment imposes the conditions that determine the results of selection and, thus, the direction of evolution.

Selection operates in natural populations of a species as skill does in a football game. In any individual game, it can be difficult to predict the winner, because chance can play an important role in the outcome. But over a long season, the teams with the most skillful players usually win the most games. In nature, those individuals best suited to their environments tend to win the evolutionary game by leaving the most offspring, although chance can play a major role in the life of any one individual. While you cannot predict the fate of any one individual, just as you cannot predict "heads" or "tails" in any one coin toss, it is possible to predict which kind of individual will tend to become more common in populations of a species. For example, it is possible to predict the proportion of heads after many coin tosses.

In nature, many traits, perhaps most, are affected by more than one gene. The interactions between genes are typically complex, as you saw in chapter 10. For example, alleles of many different genes play a role in determining human height (see figure 10.12). In such cases, selection operates on all the genes, influencing most strongly those that make the greatest contribution to the phenotype. How selection changes the population depends on which genotypes are favored. Three types of natural selection have been identified: stabilizing selection, disruptive selection, and directional selection. Figure 14.24 shows the results of these three types of selection on body size.

Figure 14.24   Three kinds of natural selection.

In the top panels, the *blue* areas indicate the phenotypes that are being selected for and the *red* areas are the phenotypes that are not being preferentially selected for. The bottom panels show the phenotypic results of the selection. (a) In *stabilizing selection,* individuals with midrange phenotypes are favored, with selection acting against both ends of the range of phenotypes. (b) In *disruptive selection,* individuals in the middle of the range of phenotypes of a certain trait are selected against, and the extreme forms of the trait are favored. (c) In *directional selection,* individuals concentrated toward one extreme of the array of phenotypes are favored.

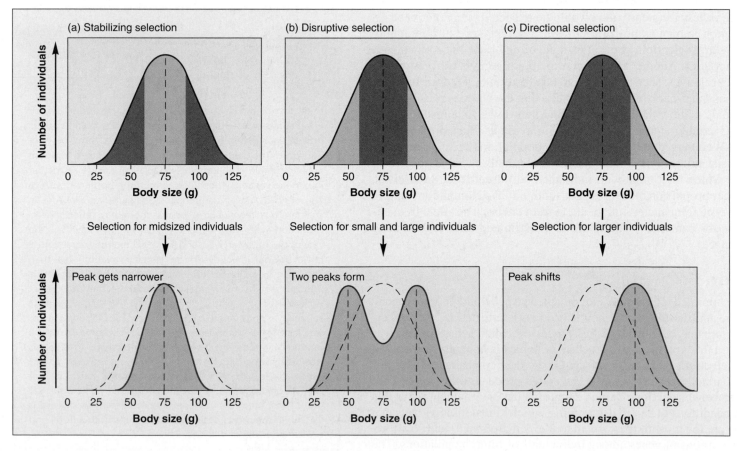

Stabilizing Selection

When selection acts to eliminate both extremes from an array of phenotypes—for example, eliminating the larger and smaller body sizes (figure 14.24a)—the result is an increase in the frequency of the already common intermediate phenotype (a midsized body). This is called **stabilizing selection.** In effect, selection is operating to prevent change away from the middle range of values. In a classic study carried out after an "uncommonly severe storm of snow, rain, and sleet" on February 1, 1898, 136 starving English sparrows were collected and brought to the laboratory of H. C. Bumpus at Brown University in Providence, Rhode Island. Of these, 64 died and 72 survived. Bumpus took standard measurements on all the birds. He found that among the female birds that perished were many more individuals that had extreme measurements, either very large or very small. Selection had acted most strongly against these "extreme-sized" birds. Stabilizing selection does not change which phenotype is the most common of the population— the average-sized birds were already the most common phenotype—but rather makes it even more common by eliminating extremes. Many examples similar to Bumpus's female sparrows are known. In humans, infants with intermediate weight at birth have the highest survival rate (figure 14.25a).

Disruptive Selection

In some situations, selection acts to eliminate the intermediate type (figure 14.24b), resulting in the two more extreme phenotypes becoming more common in the population. This type of selection is called **disruptive selection.** A clear example is the different beak sizes of the African black-bellied seedcracker finch *Pyrenestes ostrinus* (figure 14.25b). Populations of these birds contain individuals with large and small beaks, but very few individuals with intermediate-sized beaks. As their name implies, these birds feed on seeds, and the available seeds fall into two size categories: large and small. Only large-beaked birds, like the one on the left in the figure, can open the tough shells of large seeds, whereas birds with the smallest beaks, like the one on the right, are more adept at handling small seeds. Birds with intermediate-sized beaks are at a disadvantage with both seed types: unable to open large seeds and too clumsy to efficiently process small seeds. Consequently, selection acts to eliminate the intermediate phenotypes, partitioning the population into two distinct groups.

Directional Selection

When selection acts to eliminate one extreme from an array of phenotypes (figure 14.24c), the genes determining this extreme become less frequent in the population. This form of selection is called **directional selection.** An experiment set up to test directional selection is shown in figure 14.25c. In the *Drosophila* population, flies that flew toward light, a behavior called phototropism, were eliminated from the population. The remaining flies were mated and the experiment repeated. After 20 generations of selected mating, flies exhibiting phototropism were far less frequent in the population.

Concept Check

1. Which agent of evolution is the ultimate source of genetic variation?
2. If a population is in Hardy-Weinberg equilibrium, is it evolving? Explain.
3. Which type of selection eliminates intermediate phenotypes?

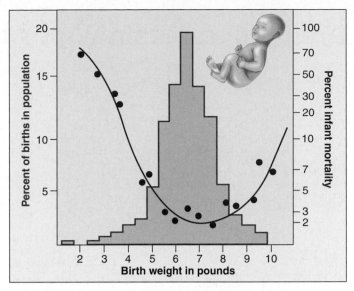

(a) Stabilizing selection for birth weight. The death rate among human babies is lowest at an intermediate birth weight between 7 and 8 pounds indicated by the *red* line. The intermediate weights are also the most common in the population, indicated by the *blue* area. Larger and smaller babies both occur less frequently and have a greater tendency to die at or near birth.

(b) Disruptive selection for large and small beaks. Differences in beak size in the black-bellied seedcracker finch of West Africa are the result of disruptive selection for two distinct food sources.

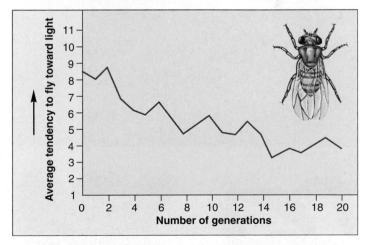

(c) Directional selection for negative phototropism in *Drosophila*. Individuals of the fly *Drosophila* were selectively bred. Flies that moved toward light were discarded; only flies that moved away from light were used as parents for the next generation. After 20 generations, the offspring had an ever greater tendency to avoid light.

Figure 14.25 Examples of selection.

Adaptation Within Populations

Figure 14.26 **The first known patient with sickle-cell disease.**

Dr. Ernest Irons's blood examination report on his patient Walter Clement Noel, December 31, 1904, described his oddly shaped red blood cells.

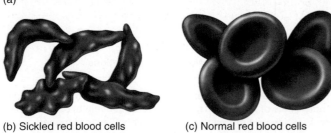

Val 6

(a)

(b) Sickled red blood cells   (c) Normal red blood cells

Figure 14.27 **Why the sickle-cell mutation causes hemoglobin to clump.**

14.10 Sickle-Cell Disease

CONCEPT PREVIEW: The prevalence of sickle-cell disease in African populations is thought to reflect the action of natural selection. Natural selection favors individuals carrying one copy of the sickle-cell allele, because they are resistant to malaria, which is common in Africa.

In the time since Darwin suggested the pivotal role of natural selection in evolution, many examples have been found in which natural selection is clearly acting to change the genetic makeup of species, just as Darwin predicted. Here we will examine three examples: sickle-cell disease (a defect in human hemoglobin proteins), industrial melanism in European moths, and color selection in South American guppy populations.

Sickle-cell disease (which used to be called sickle cell anemia) is a hereditary disease affecting hemoglobin molecules in the blood. It was first detected in 1904 in Chicago in a blood examination of an individual complaining of tiredness. You can see the original doctor's report in figure 14.26. The disorder arises as a result of a single nucleotide change in the gene encoding β-hemoglobin, one of the key proteins used by red blood cells to transport oxygen. The sickle-cell mutation changes the sixth amino acid in the β-hemoglobin chain (position B6) from glutamic acid (very polar) to valine (nonpolar). The unhappy result of this change is that the nonpolar *valine* at position B6, protruding from a corner of the hemoglobin molecule, fits nicely into a nonpolar pocket on the opposite side of another hemoglobin molecule; the nonpolar regions associate with each other. As the two-molecule unit that forms still has both a B6 valine and an opposite nonpolar pocket, other hemoglobins clump on, and long chains form as in figure 14.27a. The result is the deformed "sickle-shaped" red blood cell you see in figure 14.27b. In normal hemoglobin, by contrast, the polar amino acid *glutamic acid* occurs at position B6. This polar amino acid is not attracted to the nonpolar pocket, so no hemoglobin clumping occurs, and the cells are normal in shape, as in figure 14.27c.

Persons homozygous for the sickle-cell genetic mutation in the β-hemoglobin gene frequently have a reduced life span. This is because red blood cells that are sickled do not flow smoothly through the tiny capillaries but instead jam up and block blood flow, which interferes with oxygen transport. Heterozygous individuals, who have both a defective and a normal form of the gene, make enough functional hemoglobin to keep their red blood cells healthy.

The Puzzle: Why So Common?

The disorder is now known to have originated in Central Africa, where the frequency of the sickle-cell allele is about 0.12. One in 100 people is homozygous for the defective allele and develops the fatal disorder. Sickle-cell disease affects roughly two African Americans out of every thousand but is almost unknown among other racial groups.

If Darwin is right, and natural selection drives evolution, then why has natural selection not acted against the defective allele in Africa and eliminated it from the human population there? Why is this potentially fatal allele instead relatively common there?

The Answer: Stabilizing Selection

The defective allele has not been eliminated from Central Africa because people who are heterozygous for the sickle-cell allele are much less susceptible to malaria, one of the leading causes of death in Central Africa. Examine the maps in figure 14.28, and you will see the relationship between sickle-cell disease and malaria clearly. The upper map shows the frequency of the sickle-cell allele, the darker green areas indicating a 10% to 20% frequency of the allele. The map on the bottom indicates the distribution of malaria in dark orange. Clearly, the areas that are colored in darker green on the upper map overlap many of the dark orange areas in the map at the bottom. Although the population pays a high price—the many individuals in each generation who are homozygous for the sickle-cell allele die—the deaths are far fewer than would occur due to malaria if the heterozygous individuals were not malaria resistant. One in 5 individuals (20%) are heterozygous and survive malaria, while only 1 in 100 (1%) are homozygous and die of sickle-cell disease. Similar inheritance patterns of the sickle-cell allele are found in other countries frequently exposed to malaria, such as areas around the Mediterranean, India, and Indonesia. Natural selection has favored the sickle-cell allele in Central Africa and other areas hit by malaria because the payoff in survival of heterozygotes more than makes up for the price in death of homozygotes. This phenomenon is an example of **heterozygote advantage.**

Stabilizing selection (also called *balancing selection*) is thus acting on the sickle-cell allele: (1) selection tends to eliminate the sickle-cell allele because of its lethal effects on homozygous individuals, and (2) selection tends to favor the sickle-cell allele because it protects heterozygotes from malaria. Like a manager balancing a store's inventory, natural selection increases the frequency of an allele in a species as long as there is something to be gained by it, until the cost balances the benefit.

Stabilizing selection occurs because malarial resistance counterbalances lethal sickle-cell disease. Malaria is a tropical disease that has essentially been eradicated in the United States since the early 1950s, and stabilizing selection has not favored the sickle-cell allele here. Africans brought to America several centuries ago have not gained any evolutionary advantage in all that time from being heterozygous for the sickle-cell allele. There is no benefit to being resistant to malaria if there is no danger of getting malaria anyway. As a result, the selection against the sickle-cell allele in America is not counterbalanced by any advantage, and the allele has become less common among African Americans than among native Africans living in Central Africa.

Stabilizing selection is thought to have influenced many other human genes in a similar fashion. The recessive *cf* allele causing cystic fibrosis is unusually common in northwestern Europeans. As discussed in the "In the News" feature on page 259, heterozygotes for the *cf* allele are protected from the dehydration caused by cholera, and the *cf* allele provides other advantages as well. Apparently, the bacterium causing typhoid fever uses the healthy version of the CFTR protein (see page 82) to enter the cells it infects, but it cannot use the cystic fibrosis version of the protein. As with sickle-cell disease, heterozygotes are protected.

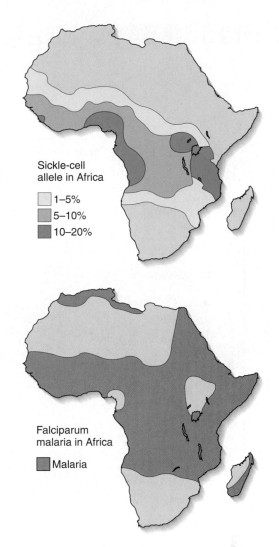

Sickle-cell allele in Africa
- 1–5%
- 5–10%
- 10–20%

Falciparum malaria in Africa
- Malaria

Figure 14.28   **How stabilizing selection maintains sickle-cell disease.**

The diagrams show the frequency of the sickle-cell allele (*top*) and the distribution of falciparum malaria (*bottom*). Falciparum malaria is one of the most devastating forms of the often fatal disease. As you can see, its distribution in Africa is closely correlated with that of the allele of the sickle-cell characteristic.

IMPLICATION   1 in 500 African Americans have sickle-cell disease (SCD). While there is no cure, the antitumor drug hydroxyurea can be used to block expression of the disorder. Administration of hydroxyurea reverses the course of hemoglobin development, inducing the reading of the fetal hemoglobin gene in place of the adult hemoglobin gene, and in so doing preventing sickling in SCD patients. Can you suggest a reason why the substitution of fetal for adult hemoglobin might prevent sickling and so block expression of the disorder?

Figure 14.29 Tutt's hypothesis explaining industrial melanism.

These photographs show color variants of the peppered moth (*Biston betularia*). Tutt proposed that the dark moth is more visible to predators on unpolluted trees (*top*), while the light moth is more visible to predators on bark blackened by industrial pollution (*bottom*).

IMPLICATION Peppered moths are creatures of the woodland. The moths you find in your house are much smaller, about the size of the end of your pinky. There are two very common types you may find in your home, Indian meal moths (called pantry moths) and clothes moths, whose larvae eat clothing. You can usually deal with clothes moths with ironing, vacuuming, moth balls, or by using chemical poisons called pyrethrins. Indian meal moths are a different story altogether, as you don't want the poison or smell of moth balls on your food. How can you rid your kitchen of a pantry moth infestation? After careful cleaning, you can use a commercial moth trap loaded with pheromones (powerful sexual attractants). It's a rare pantry moth that can resist their call. They fly in and can't fly back out and soon the trap will contain every moth in your kitchen. Why is this approach not effective in blocking the damage caused by an infestation of clothing moths in your closets?

14.11 Peppered Moths and Industrial Melanism

CONCEPT PREVIEW: Natural selection has favored the dark form of the peppered moth in areas subject to severe air pollution, perhaps because on darkened trees they are less easily seen by moth-eating birds. Selection has in turn favored the light form as pollution has abated.

The peppered moth, *Biston betularia,* is a European moth that rests on tree trunks during the day. Until the mid-nineteenth century, almost every captured individual of this species had light-colored wings. From that time on, individuals with dark-colored wings increased in frequency in the moth populations near industrialized centers, until they made up almost 100% of these populations. Dark individuals had a dominant allele that was present but very rare in populations before 1850. Biologists soon noticed that in industrialized regions where the dark moths were common, the tree trunks were darkened almost black by the soot of pollution. Dark moths were much less conspicuous resting on them than light moths were. In addition, air pollution that was spreading in the industrialized regions had killed many of the light-colored lichens on tree trunks, making the trunks darker.

Selection for Melanism

Can Darwin's theory explain the increase in the frequency of the dark allele? Why did dark moths gain a survival advantage around 1850? An amateur moth collector named J. W. Tutt proposed in 1896 what became the most commonly accepted hypothesis explaining the decline of the light-colored moths. He suggested that light forms were more visible to predators on sooty trees that have lost their lichens. Consequently, birds ate the peppered moths resting on the trunks of trees during the day. The dark forms, in contrast, were at an advantage because they were camouflaged (figure 14.29). Although Tutt initially had no evidence, British ecologist Bernard Kettlewell tested the hypothesis in the 1950s by rearing populations of peppered moths with equal numbers of dark and light individuals. Kettlewell then released these populations into two sets of woods: one, near heavily polluted Birmingham, the other, in unpolluted Dorset. Kettlewell set up traps in the woods to see how many of both kinds of moths survived. To evaluate his results, he had marked the released moths with a dot of paint on the underside of their wings, where birds could not see it.

In the polluted area near Birmingham, Kettlewell trapped 19% of the light moths, but 40% of the dark ones. This indicated that dark moths had a far better chance of surviving in these polluted woods, where the tree trunks were dark. In the relatively unpolluted Dorset woods, Kettlewell recovered 12.5% of the light moths but only 6% of the dark ones. This indicated that where the tree trunks were still light-colored, light moths had a much better chance of survival. Kettlewell later solidified his argument by placing dead moths on trees and filming birds looking for food. Sometimes the birds actually passed right over a moth that was the same color as its background.

Industrial Melanism

Industrial melanism is a term used to describe the evolutionary process in which darker individuals come to predominate over lighter individuals since the industrial revolution as a result of natural selection. The process is widely believed to have taken place because the dark organisms are better concealed from their predators in habitats that have been darkened by soot and other forms of industrial pollution, as suggested by Kettlewell.

Dozens of other species of moths have changed in the same way as the peppered moth in industrialized areas throughout Eurasia and North America, with dark forms becoming more common from the mid-nineteenth century onward as industrialization spread.

Selection against Melanism

As of the second half of the twentieth century, with the widespread implementation of pollution controls, these trends are reversing, not only for the peppered moth in many areas in England but also for many other species of moths throughout the northern continents. These examples provide some of the best-documented instances of changes in allelic frequencies of natural populations as a result of natural selection due to specific factors in the environment.

In England, the air pollution promoting industrial melanism began to reverse following enactment of Clean Air legislation in 1956. Beginning in 1959, the *Biston* population at Caldy Common outside Liverpool has been sampled each year. The frequency of the melanic (dark) form dropped from a high of 94% in 1960 to a low of 19% in 1995 (figure 14.30). Similar reversals have been documented at numerous other locations throughout England. The drop correlates well with a drop in air pollution, particularly with tree-darkening sulfur dioxide and suspended particulates.

Interestingly, the same reversal of industrial melanism appears to have occurred in America during the same time that it was happening in England. Industrial melanism in the American subspecies of the peppered moth was not as widespread as in England, but it has been well documented at a rural field station near Detroit. Of 576 peppered moths collected there from 1959 to 1961, 515 were melanic, a frequency of 89%. The American Clean Air Act, passed in 1963, led to significant reductions in air pollution. Resampled in 1994, the Detroit field station peppered moth population had only 15% melanic moths! The moths in Liverpool and Detroit, both part of the same natural experiment, exhibit strong evidence of natural selection.

Reconsidering the Target of Natural Selection

Tutt's hypothesis, widely accepted in the light of Kettlewell's studies, is currently being reevaluated. The problem is that the recent selection against melanism does not appear to correlate with changes in tree lichens. At Caldy Common, the light form of the peppered moth began its increase in frequency long before lichens began to reappear on the trees. At the Detroit field station, the lichens never changed significantly as the dark moths first became dominant and then declined over the last 30 years. In fact, investigators have not been able to find peppered moths on Detroit trees at all, whether covered with lichens or not. Wherever the moths rest during the day, it does not appear to be on tree bark. Some evidence suggests they rest on leaves on the treetops, but no one is sure.

The action of selection may depend on other differences between light and dark forms of the peppered moth as well as their wing coloration. Researchers report, for example, a clear difference in their ability to survive as caterpillars under a variety of conditions. Perhaps natural selection is also targeting the caterpillars rather than the adults. While we can't yet say exactly what the targets of selection are, researchers are actively investigating this, one of the best-documented instances of natural selection in action.

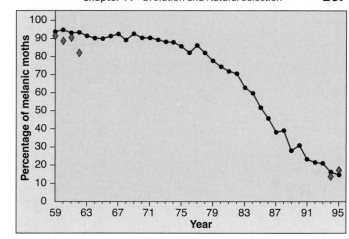

Figure 14.30 **Selection against melanism.**

The dots indicate the frequency of melanic *Biston betularia* moths at Caldy Common in England, sampled continuously from 1959 to 1995. Red diamonds indicate frequencies of melanic *B. betularia* in Michigan from 1959 to 1962 and from 1994 to 1995.

EVOLUTION

Natural Selection for Melanism in Mice. Melanism isn't restricted to just insects. Many mammals have melanic forms that are subject to natural selection in much the same way as moths. The coat color of desert pocket mice that live on differently colored rock habitats provides a clear-cut example of natural selection acting on melanism. In Arizona and New Mexico, these small wild pocket mice live in isolated black volcanic lava beds and the pale soils between them. Melanin synthesis during hair development of pocket mice is regulated by the receptor gene *MC1R*. Mutations that disable *MC1R* lead to melanism. Such mutations are dominant alleles, so whenever they are present in a population, dark pocket mice are seen. When wild populations of pocket mice were surveyed by biologists from the University of Arizona, there was a striking correlation between coat color and the color of the rock on which the population of pocket mice lived. As you can see in the upper photographs, the close match between coat color and background color gives the mice cryptic protection from avian predators, particularly owls. These mice are very visible when placed in the opposite habitats (lower photos).

14.12 Selection on Color in Guppies

CONCEPT PREVIEW: Experiments can be conducted in nature to test hypotheses about how evolution occurs. Such studies reveal that natural selection can lead to rapid evolutionary change.

To study evolution, biologists have traditionally investigated what has happened in the past, sometimes many millions of years ago, relying on observation rather than experiments to examine ideas about past events. Nonetheless, evolutionary biology is not entirely an observational science. In some circumstances evolutionary change can occur rapidly. Consequently, it is possible to establish experimental studies to test evolutionary hypotheses in real time. Although laboratory studies on fruit flies and other organisms have been common for more than 50 years, it has only been in recent years that scientists have started conducting experimental studies of evolution in nature. One excellent example of how observations of the natural world can be combined with rigorous experiments in the lab and in the field concerns research on the guppy, *Poecilia reticulata.*

Guppies Live in Different Environments

The guppy is a popular aquarium fish because of its bright coloration and prolific reproduction. In nature, guppies are found in small streams in northeastern South America and the nearby island of Trinidad. In Trinidad, guppies are found in many mountain streams. One interesting feature of several streams is that they have waterfalls. Amazingly, guppies and some other fish are capable of colonizing portions of the stream above the waterfall. The killifish, *Rivulus hartii,* is a particularly good colonizer; apparently on rainy nights, it will wriggle out of the stream and move through the damp leaf litter. Guppies are not so proficient, but they are good at swimming upstream. During flood seasons, rivers sometimes overflow their banks, creating secondary channels that move through the forest. During these occasions, guppies may be able to move upstream and invade pools above waterfalls. By contrast, not all species are capable of such dispersal and thus are only found in these streams below the first waterfall. One species whose distribution is restricted by waterfalls is the pike cichlid, *Crenicichla alta,* a voracious predator that feeds on other fish, including guppies.

Because of these barriers to dispersal, guppies can be found in two very different environments. The guppies you see living in pools just below the waterfalls in figure 14.31 are faced with predation by the pike cichlid. This substantial risk keeps rates of survival relatively low. By contrast, in similar pools just above the waterfall, the only predator present is the killifish, which only rarely preys on guppies. Guppy populations above and below waterfalls exhibit many differences. In the high-predation pools, male guppies exhibit the drab coloration you see in the guppies below the waterfall in figure 14.31. Moreover, they tend to reproduce at a younger age and attain relatively smaller adult sizes. By contrast, male fish above the waterfall in the figure display gaudy colors that they use to court females. Adults mature later and grow to larger sizes.

These differences suggest the function of natural selection. In the low-predation environment, males display gaudy colors and spots that help in mating. Moreover, larger males are most successful at holding territories and mating with females, and larger females lay more eggs. Thus, in the absence of predators, larger and more colorful fish may have produced more offspring, leading to the evolution of those traits. In pools below the waterfall, however, natural selection would favor different traits. Colorful males are likely to attract the attention of the pike cichlid, and high predation rates mean that most fish live short lives; thus, individuals that are more

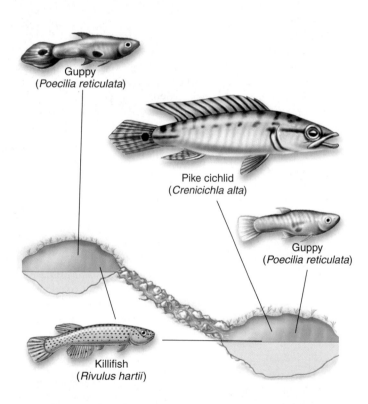

Figure 14.31 **The evolution of protective coloration in guppies.**

In pools below waterfalls where predation is high, male guppies (*Poecilia reticulata*) are drab colored. In the absence of the highly predatory pike cichlid (*Crenicichla alta*), male guppies in pools above waterfalls are much more colorful and attractive to females. The killifish (*Rivulus hartii*) is also a predator but only rarely eats guppies. The evolution of these differences in guppies can be experimentally tested.

drab and shunt energy into early reproduction, rather than into growth to a larger size, are likely to be favored by natural selection.

The Experiments

Although the differences between guppies living above and below the waterfalls suggest that they represent evolutionary responses to differences in the strength of predation, alternative explanations are possible. Perhaps, for example, only very large fish are capable of swimming upstream past the waterfall to colonize pools. If this were the case, then the new population would be established solely by individuals with genes for large size.

Laboratory Experiment. The only way to rule out such alternative possibilities is to conduct a controlled experiment. John Endler, now of the University of California, Santa Barbara, conducted the first experiments in large pools in laboratory greenhouses. At the start of the experiment, a group of 2,000 guppies was divided equally among 10 large pools. Six months later, pike cichlids were added to four of the pools and killifish to another four,

> A control experiment, as discussed on page 21 is an experiment in which no variable is altered. It is usually run in parallel with an experiment in which a variable has been changed in order to test a hypothesis.

with the remaining two pools left to serve as "no predator" controls. Fourteen months later (which corresponds to 10 guppy generations), the scientists compared the populations. You can see their results in figure 14.32a. The guppies in the killifish pool (the blue line) and control pools (the green line) were notably large, brightly colored fish with about 13 colorful spots per individual. In contrast, the guppies in the pike cichlid pools (the red line) were smaller and drab in coloration, with a reduced number of spots (about 9 per fish). These results clearly suggest that predation can lead to rapid evolutionary change, but do these laboratory experiments reflect what occurs in nature?

Field Experiment. To find out, Endler and colleagues—including David Reznick, now at the University of California, Riverside—located two streams that had guppies in pools below a waterfall, but not above it (see the photograph in figure 14.32b). As in other Trinidadian streams, the pike cichlid was present in the lower pools, but only the killifish was found above the waterfalls. The scientists then transplanted guppies to the upper pools and returned at several-year intervals to monitor the populations. Despite originating from populations in which predation levels were high, the transplanted populations rapidly evolved the traits characteristic of low-predation guppies: they matured late, attained greater size and had brighter colors. Control populations in the lower pools, by contrast, continued to be drab and matured early and at smaller sizes. Laboratory studies confirmed that the differences between the populations were the result of genetic differences. These results demonstrate that substantial evolutionary change can occur in less than 12 years. The results give strong support to the theory of evolution by natural selection.

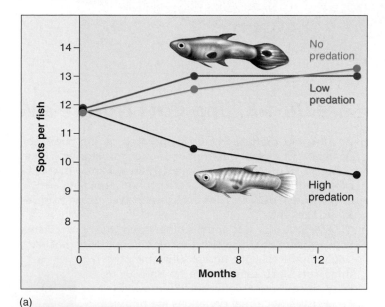

(a)

(b)

Figure 14.32 **Evolutionary change in spot number.**

(a) Guppies raised in low-predation or no predation environments in laboratory greenhouses had a greater number of spots, whereas selection in more dangerous environments, like the pools with the highly predatory pike cichlid, led to less conspicuous fish. (b) The same results are seen in field experiments conducted in pools above and below waterfalls.

IMPLICATION Many people raise guppies as a hobby. Purchasing a few fish at the local pet store can soon lead to a very large number of guppies in your fish tank, as they breed prodigiously. After several generations of such breedings, however, sickly and deformed individuals begin to be seen in increasing numbers. What might be happening?

Concept Check

1. What does malaria have to do with why the sickle-cell hemoglobin allele has not been eliminated from the human population by natural selection?
2. If birds aren't eating noncryptic moths, is there evidence of selection?
3. How do pike cichlids influence the evolution of color in guppies?

Are Bird-Killing Cats Nature's Way of Making Better Birds?

Death is not pretty, early in the morning on the doorstep. A small dead bird was left at our front door one morning, lying by the newspaper as if it might at any moment fly away. I knew it would not. Like other birds before it, it was a gift to our household by Feisty, a cat who lives with us. Feisty is a killer of birds, and every so often he leaves one for us, like rent.

We have four cats, and the other three, true housecats, would not know what to do with a bird. Feisty is different, a long-haired gray Persian with the soul of a hunter. While the other three cats sleep safely in the house with us, Feisty spends most nights outside, prowling.

Feisty's nocturnal donations are not well received by my family. More than once it has been suggested, as we donate the bird to the trash can, that perhaps Feisty would be happier living in the country.

As a biologist I try to take a more scientific view. I tell my girls that getting rid of Feisty is unwarranted, because hunting cats like Feisty actually help birds, in a Darwinian sort of way. Like an evolutionary quality control check, I explain, predators ensure that only those individuals of a population that are better suited to their environment contribute to the next generation, by the simple expedient of removing the lesser suited. By taking the birds who are least able to escape predation—the sick and the old—Feisty culls the local bird population, leaving it on average a little better off.

That's what I tell my girls. It all makes sense, from a biological point of view, and it is a story they have heard before, in movies like *Never Cry Wolf*, and *The Lion King*. So Feisty is given a reprieve, and survives to hunt another night.

What I haven't told my girls is how little evidence actually backs up this pretty defense of Feisty's behavior. My explanation may be couched in scientific language, but without proof this "predator-as-purifier" tale is no more than a hypothesis. It might be true, and then again it might not. By such thin string has Feisty's future with our family hung.

Recently the string became a strong cable. Two French biologists put the hypothesis I had been using to defend Feisty to the test. To my great relief, it was supported.

Drs. Anders Møller and Johannes Erritzoe of the Université Pierre et Marie Curie in Paris devised a simple way to test the hypothesis. They compared the health of birds killed by domestic cats like Feisty with that of birds killed in accidents such as flying into glass windows or moving cars. Glass windows do not select for the weak or infirm—a healthy bird flies into a glass window and breaks its neck just as easily as a sickly bird. If cats are actually selecting the less-healthy birds, then their prey should include a larger proportion of sickly individuals than those felled by flying into glass windows.

How can we know what birds are sickly? Drs. Møller and Erritzoe examined the size of the dead bird's spleens. The size of its spleen is a good indicator of how healthy a bird is. Birds experiencing a lot of infections, or harboring a lot of parasites, have smaller spleens than healthy birds.

They examined 18 species of birds, more than 500 individuals. In all but two species (robins and goldcrests) they found that the spleens of birds killed by cats were significantly smaller than those killed accidently. We're not splitting hairs here, talking about some minor statistical difference. Spleens were on average a third smaller in cat-killed birds. In five bird species (blackcaps, house sparrows, lesser whitethroats, skylarks, and spotted flycatchers), the spleens of birds pounced on by cats were less than half the size of those killed by flying at speed into glass windows or moving cars.

As a control to be sure that additional factors were not operating, the Paris biologists checked for other differences between birds killed by cats and birds killed accidentally. Weight, sex, and wing length, all of which you could imagine might be important, were not significant. Cat-killed birds had, on average, the same weight, proportion of females, and wing length as accident-killed birds.

One other factor did make a difference: age. About 50% of the birds killed accidently were young, while fully 70% of the birds killed by cats were. Apparently it's not quite so easy to catch an experienced old codger as it is a callow youth.

So Feisty was just doing Darwin's duty, I pleaded, informing my girls that the birds he catches would soon have died anyway. But a dead bird on a doorstep argues louder than any science, and they remained unconvinced.

They are my daughters, and thus not ones to give in without a fight. Scouring the Internet, they assembled this counter-argument: Predatory house cats not unlike Feisty, as well as feral cats (domesticated cats that have been abandoned to the wild), are causing major problems for native bird populations of England, New Zealand, and Australia, as well as here in the United States. Although house cats like Feisty have the predatory instincts of their ancestors, they seem to lack the restraint that their wild relatives have. Most wild cats hunt only when hungry, but pet and feral cats seem to "love the kill," not killing for food but for sport.

So Darwin and I lost this argument. It seems I must restrict Feisty's hunting expeditions after all. While a little pruning may benefit a bird population, wholesale slaughter only devastates it. I will always see a lion whenever I look at Feisty on the prowl, but it will be a lion restricted to indoor hunting.

How Species Form

14.13 The Biological Species Concept

CONCEPT PREVIEW: A species is generally defined as a group of similar organisms that does not exchange genes extensively with other groups in nature.

A key aspect of Darwin's theory of evolution is his proposal that small adaptations (what is called microevolution) lead ultimately to large-scale changes leading to species formation and higher taxonomic groups (macroevolution). The way natural selection leads to the formation of new species has been thoroughly documented by biologists, who have observed the stages of the species-forming process, or **speciation,** in many different plants, animals, and microorganisms. Speciation usually involves successive change: first, local populations become increasingly specialized; then, if they become different enough, natural selection may act to keep them that way.

Before we can discuss how one species gives rise to another, we need to understand exactly what a species is. The evolutionary biologist Ernst Mayr coined the **biological species concept,** which defines species as "groups of actually or potentially interbreeding natural populations which are reproductively isolated from other such groups."

In other words, the biological species concept says that a species is composed of populations whose members mate with each other and produce fertile offspring—or would do so if they came into contact. Conversely, populations whose members do not mate with each other or who cannot produce fertile offspring are said to be **reproductively isolated** and, thus, members of different species.

What causes reproductive isolation? If organisms cannot interbreed or cannot produce fertile offspring, they clearly belong to different species. However, some populations that are considered to be separate species can interbreed and produce fertile offspring, but they ordinarily do not do so under natural conditions. They are still considered to be reproductively isolated in that genes from one species generally will not be able to enter the gene pool of the other species. Table 14.2 summarizes the steps at which barriers to successful reproduction may occur. Examine this table carefully. We will return to it throughout our discussion of species formation. Such barriers are termed **reproductive isolating mechanisms** because they prevent genetic exchange between species. We will first discuss *prezygotic isolating mechanisms,* those that prevent the formation of zygotes. Then we will examine *postzygotic isolating mechanisms,* those that prevent the proper functioning of zygotes after they have formed.

Even though the definition of what constitutes a species is of fundamental importance to evolutionary biology, this issue has still not been completely settled and is currently the subject of considerable research and debate. For example, the biological species concept has had a number of problems. Plants of different species cross fertilize and produce fertile hybrids at much higher frequencies than first thought. Hybridization also occurs in animal populations. However, hybridization is not causing species to coalesce into one genetically variable population, but rather the species are maintaining their distinctiveness. This is not to say that hybridization is rampant, but it is common enough to cast doubt about whether reproductive isolation is the only force maintaining the integrity of species.

TABLE 14.2	Isolating Mechanisms	
Mechanism		**Description**
Prezygotic Isolating Mechanisms		
Geographic isolation		Species occur in different areas, which are often separated by a physical barrier such as a river or mountain range.
Ecological isolation		Species occur in the same area, but they occupy different habitats. Survival of hybrids is low because they are not adapted to either environment of their parents.
Temporal isolation		Species reproduce in different seasons or at different times of the day.
Behavioral isolation		Species differ in their mating rituals.
Mechanical isolation	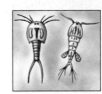	Structural differences between species prevent mating.
Prevention of gamete fusion		Gametes of one species function poorly with the gametes of another species or within the reproductive tract of another species.
Postzygotic Isolating Mechanisms		
Hybrid inviability or infertility		Hybrid embryos do not develop properly, hybrid adults do not survive in nature, or hybrid adults are sterile or have reduced fertility.

(a)

(b)

(c)

Figure 14.33   Lions and tigers are ecologically isolated.

The ranges of lions and tigers used to overlap in India. However, lions and tigers do not hybridize in the wild because they utilize different portions of the habitat. (a) Tigers are solitary animals that live in the forest, whereas (b) lions live in open grassland. (c) Hybrids, such as this liger, have been successfully produced in captivity, but hybridization does not occur in the wild.

14.14 Isolating Mechanisms

CONCEPT PREVIEW: Prezygotic isolating mechanisms lead to reproductive isolation by preventing the formation of hybrid zygotes. Postzygotic mechanisms lead to the failure of hybrid zygotes to develop normally, or they prevent hybrids from becoming established in nature.

Prezygotic Isolating Mechanisms

Geographical Isolation.   This mechanism is perhaps the easiest to understand; species that exist in different areas are not able to interbreed. The two populations of flowers in the first panel of table 14.2 are separated by a mountain range and so would not be capable of interbreeding.

Ecological Isolation.   Even if two species occur in the same area, they may utilize different portions of the environment and thus not hybridize because they do not encounter each other, like the lizards in the second panel of table 14.2. One lives on the ground and the other in the trees. Another example in nature is the ranges of lions and tigers in India. Their ranges overlapped until about 150 years ago. Even when they did overlap, however, there were no records of natural hybrids. Lions stayed mainly in the open grassland and hunted in groups called prides; tigers tended to be solitary creatures of the forest. Because of their ecological and behavioral differences, lions and tigers rarely came into direct contact with each other, even though their ranges overlapped thousands of square kilometers. Figure 14.33 shows that hybrids are possible; the liger shown in figure 14.33*c* is a hybrid of a male lion and a female tiger (a tigon is the hybrid of a male tiger and a female lion). These matings do not occur in the wild but can happen in artificial environments such as zoos.

Temporal Isolation.   *Lactuca graminifolia* and *L. canadensis,* two species of wild lettuce, grow together along roadsides throughout the southeastern United States. Hybrids between these two species are easily made experimentally and are completely fertile. But such hybrids are rare in nature because *L. graminifolia* flowers in early spring and *L. canadensis* flowers in summer. This is called temporal isolation and is shown in the third panel in table 14.2. When the blooming periods of these two species overlap, as they do occasionally, the two species do form hybrids, which may become locally abundant.

Behavioral Isolation.   In chapter 21, we will consider the often elaborate courtship and mating rituals of some groups of animals, which tend to keep these species distinct in nature even if they inhabit the same places. This behavioral isolation is discussed in the fourth panel of table 14.2. For example, mallard and pintail ducks are perhaps the two most common freshwater ducks in North America. In captivity, they produce completely fertile offspring, but in nature they nest side-by-side and rarely hybridize.

Mechanical Isolation. Structural differences that prevent mating between related species of animals and plants is called mechanical isolation and is shown in panel five of table 14.2. Flowers of related species of plants often differ significantly in their proportions and structures. Some of these differences limit the transfer of pollen from one plant species to another. For example, bees may pick up the pollen of one species on a certain place on their bodies; if this area does not come into contact with the receptive structures of the flowers of another plant species, the pollen is not transferred.

> As discussed on page 326, the flower contains the sexual structures in a group of plants called angiosperms. The male gametes, called pollen, are produced on one part of the flower and the female gametes, eggs, are produced on another part of the flower.

Prevention of Gamete Fusion. In animals that shed their gametes directly into water, eggs and sperm derived from different species may not attract one another. Many land animals may not hybridize successfully because the sperm of one species may function so poorly within the reproductive tract of another that fertilization never takes place. In plants, the growth of pollen tubes may be impeded in species hybrids. In both plants and animals, the operation of such isolating mechanisms prevents the union of gametes even after mating.

Postzygotic Isolating Mechanisms

If hybrid matings do occur, and zygotes are produced, many factors may still prevent those zygotes from developing into normally functioning, fertile individuals. In hybrids, the genetic complements of two species may be so different that they cannot function together normally in embryonic development. For example, hybridization between sheep and goats usually produces embryos that die in the earliest developmental stages.

Figure 14.34 shows four species of leopard frogs (genus *Rana*) and their ranges throughout North America. It was assumed for a long time that they constituted a single species. However, careful examination revealed that although the frogs appear similar, successful mating between them is rare because of problems that occur as the fertilized eggs develop. Many of the hybrid combinations cannot be produced even in the laboratory.

Even if hybrids survive the embryo stage, however, they may not develop normally. If the hybrids are weaker than their parents, they will almost certainly be eliminated in nature. Even if they are vigorous and strong, as in the case of the mule, a hybrid between a female horse and a male donkey, they may still be sterile and thus incapable of contributing to succeeding generations. Sterility may result in hybrids because the development of sex organs may be abnormal, because the chromosomes derived from the respective parents may not pair properly, or from a variety of other causes.

Concept Check

1. How is a biological species generally defined?
2. Why do ligers and tigons not occur in the wild?
3. What sort of isolating mechanism prevents the mule from becoming a separate true-breeding species?

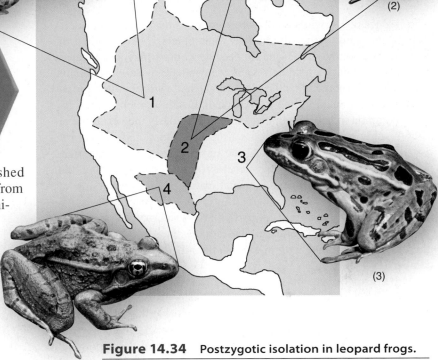

Figure 14.34   Postzygotic isolation in leopard frogs.

Numbers indicate the following species in the geographic ranges shown: (1) *Rana pipiens;* (2) *Rana blairi;* (3) *Rana sphenocephala;* (4) *Rana berlandieri*. These four species resemble one another closely in their external features. Their status as separate species was first suspected when hybrids between them were found to produce defective embryos in the laboratory. Subsequent research revealed that the mating calls of the four species differ substantially, indicating that the species have both pre- and postzygotic isolating mechanisms.

IN THE NEWS

Why Does a Wolf Not Cross the Road? Geographical isolation need not involve massive barriers like mountain ranges. Constructing a road across a forest creates a barrier that many animals cannot cross. Anyone traveling by car who counts the roadkill can see one reason why. A less obvious reason is that many animals simply don't like to be close to the humans that roads bring to the forest. A recent study of wolves in Wisconsin has revealed that road density has been a critical factor in the disappearance of breeding wolf populations in that state. The critical density, measured in several studies extending back a century, seems to be roughly 0.5 km of road per square km of forest. At higher road densities, radio-collared wolves simply move away to regions where they can live further from roads.

Does Natural Selection Act on Enzyme Polymorphism?

The essence of Darwin's theory of evolution is that, in nature, selection favors some gene alternatives over others. Many studies of natural selection have focused on genes encoding enzymes because populations in nature tend to possess many alternative alleles of their enzymes (a phenomenon called *enzyme polymorphism*). Often investigators have looked to see if weather influences which alleles are more common in natural populations. A particularly nice example of such a study was carried out on a fish, the mummichog (*Fundulus heteroclitus*), which ranges along the East Coast of North America. Researchers studied allele frequencies of the gene encoding the enzyme lactate dehydrogenase, which catalyzes the conversion of pyruvate to lactate. As you learned in chapter 7, this reaction is a key step in energy metabolism, particularly when oxygen is in short supply. There are two common alleles of lactate dehydrogenase in these fish populations, with allele *a* being a better catalyst at lower temperatures than allele *b*.

In an experiment, investigators sampled the frequency of allele *a* in 41 fish populations located over 14 degrees of latitude, from Jacksonville, Florida (31° North), to Bar Harbor, Maine (44° North). Annual mean water temperatures change 1° C per degree change in latitude. The survey is designed to test a prediction of the hypothesis that natural selection acts on this enzyme polymorphism. If it does, then you would expect that allele *a,* producing a better "low-temperature" enzyme, would be more common in the colder waters of the more northern latitudes. The graph on the right presents the results of this survey. The points on the graph are derived from pie chart data such as shown for 20 populations in the map (a **pie chart diagram** assigns a slice of the pie to each variable; the size of the slice is proportional to the contribution made by that variable to the total). The blue line on the graph is the line that best fits the data (a **"best-fit" line,** also called a **regression line,** is determined statistically by a process called *regression analysis*).

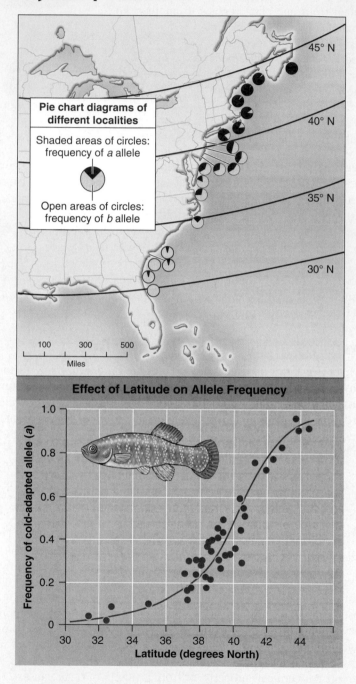

Pie chart diagrams of different localities

Shaded areas of circles: frequency of *a* allele

Open areas of circles: frequency of *b* allele

Effect of Latitude on Allele Frequency

Analysis

1. **Applying Concepts**
   a. In the fish population located at 35° N latitude, what is the frequency of the *a* allele? Locate this point on the graph.
   b. Compare the frequency of allele *a* among fish captured in waters at 44° N latitude with the frequency among fish captured at 31° N latitude. Is there a pattern? Describe it.
2. **Interpreting Data** At what latitude do fish populations exhibit the greatest variability in allele *a* frequency?
3. **Making Inferences**
   a. Are fish populations in cold waters at 44° N latitude more or less likely to contain heterozygous individuals than fish populations in warm waters at 31° N latitude?
   b. Where along this latitudinal gradient in the frequency of allele *a* would you expect to find the highest frequency of heterozygous individuals? Why?
4. **Drawing Conclusions** Are the differences in population frequencies of allele *a* consistent with the hypothesis that natural selection is acting on the alleles encoding this enzyme? Explain.

Concept Summary

Evolution

14.1 Darwin's Voyage on HMS *Beagle*

- The theory of evolution through natural selection, a theory proposed by Darwin, is overwhelmingly accepted by scientists and is considered to be the backbone of the science of biology.

14.2 Darwin's Evidence

- Darwin observed fossils in South America of extinct species that resembled living species (**figure 14.4**). On the Galápagos Islands, Darwin observed finches that differed slightly in appearance between islands but resembled finches found on the South American mainland (**figure 14.5**).

14.3 The Theory of Natural Selection

- Key to Darwin's hypothesis was the observation by Malthus that the food supply limits population growth. A population grows only as large as that which can be supported by available food which limits geometric growth of a population (**figure 14.6**).

- Using Malthus's observations and his own, Darwin proposed that individuals that are better suited to their environments survive to produce offspring, gaining the opportunity to pass their characteristics on to future generations, what Darwin called natural selection.

Darwin's Finches: Evolution in Action

14.4 The Beaks of Darwin's Finches

- By observing the different sizes and shapes of beaks in the closely related finches of the Galápagos Islands and correlating the beaks with the types of food consumed, Darwin concluded that the birds' beaks were modified from an ancestral species based on the food available, each suited to its food supply (**figure 14.9**). Scientists have identified a gene, *BMP4*, that is expressed differently in birds with different shaped beaks.

14.5 How Natural Selection Produces Diversity

- The 14 species of finches found on the islands off the coast of South America descended from a mainland species that adapted to different niches, a process called adaptive radiation (**figure 14.12**).

The Theory of Evolution

14.6 The Evidence for Evolution

- The evidence for evolution includes the fossil record. The titanothere, shown here from **figure 14.14,** and its ancestors are known only from the fossil record, which in many cases reveals organisms that are intermediate in form.

- The evidence for evolution includes the anatomical record, which reveals similarities in structures between species (**figure 14.15**). Homologous structures are similar in structure but differ in their functions (**figure 14.16**). Analogous structures are similar in function but differ in their underlying structure.

- The evidence for evolution includes the molecular record, which traces changes in the genomes of species (**figures 14.17** and **14.18**).

14.7 Evolution's Critics

- Darwin's theory of evolution through natural selection has always had its critics. Their criticisms of evolution however are without scientific merit (**figures 14.19** and **14.20**).

How Populations Evolve

14.8 Genetic Change in Populations: The Hardy-Weinberg Rule

- If a population follows the five assumptions of Hardy-Weinberg, the frequencies of alleles within the population, like the alleles that determine coat color in these cats from **figure 14.22,** will not change. However, if a population is small, has selective mating, experiences mutations or migrations, or is under the influence of natural selection, the allele frequencies will be different from those predicted by the Hardy-Weinberg rule.

14.9 Agents of Evolution

- Five factors act on populations to change their allele and genotype frequencies (**table 14.1**). Mutations are changes in DNA. Migrations are the movements of individuals or alleles into or out of a population. Genetic drift is the random loss of alleles in a population due to chance occurrences. In the founder effect, a small group of individuals establishes a new population such that their alleles become common in the newly established population. Nonrandom mating occurs when individuals seek out mates based on certain traits. Selection occurs when individuals with certain traits leave more offspring because their traits allow them to better respond to the challenges of their environment. Selection can act on the genes in that population in several different ways (**figures 14.24** and **14.25**). Stabilizing selection tends to reduce extreme phenotypes. Disruptive selection tends to reduce intermediate phenotypes. Directional selection tends to reduce one extreme phenotype from the population.

Adaptation Within Populations

14.10 Sickle-Cell Disease

- Sickle-cell disease is an example of heterozygote advantage, where individuals who are heterozygous for a trait tend to survive better. In areas with malaria, people who are heterozygous for the sickle-cell trait survive better than individuals with either of the two homozygous phenotypes (**figures 14.27** and **14.28**).

14.11 Peppered Moths and Industrial Melanism

- Natural selection favors dark-colored moths in areas of heavy pollution (**figures 14.29** and **14.30**).

14.12 Selection on Color in Guppies

- Experimentation, both in the laboratory and in the field, have shown selection pressure from predation resulted in evolutionary change in guppies—drab-colored fish where there is predation and brightly colored fish, like the fish here from **figure 14.31,** when there was no selection pressure (**figure 14.32**).

How Species Form

14.13 The Biological Species Concept

- The biological species concept states that a species is a group of organisms that are capable of mating with each other and produce fertile offspring. If they cannot mate, or mate but produce sterile offspring, they are said to be reproductively isolated (**table 14.2**).

14.14 Isolating Mechanisms

- There are two types of isolating mechanisms, prezygotic and postzygotic. Prezygotic mechanisms prevent the formation of a hybrid zygote. Postzygotic mechanisms prevent normal development of a hybrid zygote or result in sterile offspring.

Self-Test

1. Darwin was greatly influenced by Thomas Malthus who observed that
   a. food supplies increase geometrically.
   b. populations increase arithmetically.
   c. populations are capable of geometric increase, yet remain at constant levels.
   d. the food supply usually increases faster than the population that depends on it.

2. Darwin proposed that individuals with traits that help them live in their immediate environment are more likely to survive and reproduce than individuals without those traits. He called this
   a. natural selection.
   b. arithmetic progression.
   c. the theory of evolution.
   d. geometric progression.

3. A great deal of research has been done on Darwin's finches over the last 70 years. The research
   a. seems to contradict Darwin's original ideas.
   b. seems to agree with Darwin's original ideas.
   c. does not show any clear patterns that support or refute Darwin's original ideas.
   d. suggests a different explanation for evolution of finches.

4. One of the major sources of evidence for evolution is in the comparative anatomy of organisms. Features that look different but have similar structural origin are called
   a. homologous structures.
   b. analogous structures.
   c. vestigial structures.
   d. equivalent structures.

5. The Hardy-Weinberg equilibrium allows scientists to study
   a. natural selection.
   b. evolution.
   c. isolating mechanisms.
   d. punctuated equilibrium.

6. A large group of organisms lives in a large, stable ecosystem. There is no competition for resources. Individuals show no mate preferences. All organisms appear to be identical except for a few individuals in the most recent generation of offspring that exhibit a different fur coat color and pattern. The ecosystem and population are geographically isolated from other populations of the same organism. Which Hardy-Weinberg assumption has been violated?
   a. large population size
   b. random mating within the population
   c. no mutation within the population
   d. no input of new alleles from outside or loss of alleles

7. A population of 1,000 individuals has 200 individuals who show a homozygous recessive phenotype and 800 individuals who express the dominant phenotype. What is the frequency of homozygous recessive individuals in this population?
   a. 0.20
   b. 0.30
   c. 0.45
   d. 0.55

8. A chance event occurs that causes a population to lose some individuals (they died)—hence, a loss of alleles in the population results from
   a. mutation.
   b. migration.
   c. selection.
   d. genetic drift.

9. Selection that causes one extreme phenotype to be more frequent in a population is an example of
   a. disruptive selection.
   b. stabilizing selection.
   c. directional selection.
   d. equivalent selection.

10. A key element of Ernst Mayr's biological species concept is
   a. homologous isolation.
   b. divergent isolation.
   c. convergent isolation.
   d. reproductive isolation.

Visual Understanding

1. **Figure 14.24** Because of prolonged drought, the trees on an island are producing nuts that are much smaller with thicker and harder shells. What will happen to the birds that depend on the nuts for food? What type of selection will result?

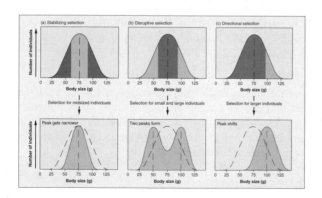

2. **Table 14.2** A very heavy rainstorm floods a mountain river, changing its course and digging a deep canyon through the soft soils of the meadow in the valley below. How could mice populations on either side of the valley be affected?

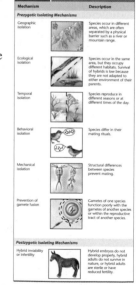

Challenge Questions

1. How likely is it that any naturally occurring population on a continent (not on an island) will meet all the assumptions of the Hardy-Weinberg rule? Explain your reasoning.

2. The evolutionary pathways of some groups of related organisms seem to have moved from larger to smaller organisms over time, such as the glyptodont to the armadillo, the mammoth to the elephant. Yet other groups of related organisms have exhibited a trend of increased sizes of species over time, such as the tiny eohippus to the horse. Explain how this can occur.

3. In a courtroom in 2005, biologist Ken Miller criticized the claims of intelligent design. After noting that 99.9% of the organisms that have ever lived on earth are now extinct, he said that "an intelligent designer who designed things, 99.9% of which didn't last, certainly wouldn't be very intelligent." Evaluate Miller's criticism.

Chapter **15**

Exploring Biological Diversity

The Classification of Organisms

Figure 15.1 Carolus Linnaeus (1707-1778).

This eighteenth-century Swedish professor, physician, and naturalist devised the binomial system for naming species of organisms and established the major categories that are used in the hierarchical system of biological classification. When he was 25, Linnaeus spent five months exploring Lapland for the Swedish Academy of Sciences; he is shown here wearing his Lapland collector's outfit.

15.1 The Invention of the Linnaean System

CONCEPT PREVIEW: Two-part (binomial) Latin names, first used by Linnaeus, are now universally employed by biologists to name organisms.

It is estimated that our world is populated by some 10 to 100 million different kinds of organisms. To talk about them and study them, it is necessary to give them names, just as it is necessary that people have names. Of course, no one can remember the name of every kind of organism, so biologists use a kind of multilevel grouping of individuals called **classification.**

Organisms were first classified more than 2,000 years ago by the Greek philosopher Aristotle, who categorized living things as either plants or animals. He classified animals as either land, water, or air dwellers, and he divided plants into three kinds based on stem differences. This simple classification system was expanded by the Greeks and Romans, who grouped animals and plants into basic units such as cats, horses, and oaks. Eventually, these units began to be called **genera** (singular, **genus**), the Latin word for "group." Starting in the Middle Ages, these names began to be systematically written down, using Latin, the language used by scholars at that time. Thus, cats were assigned to the genus *Felis,* horses to *Equus,* and oaks to *Quercus*—names that the Romans had applied to these groups. For genera that were not known to the Romans, new names were invented.

The classification system of the Middle Ages, called the *polynomial system,* was used virtually unchanged for hundreds of years, until it was replaced about 250 years ago by the **binomial system** introduced by Linnaeus.

The Polynomial System

Until the mid-1700s, biologists usually added a series of descriptive terms to the name of the genus when they wanted to refer to a particular kind of organism, which they called a **species.** These phrases, starting with the name of the genus, came to be known as **polynomials** (*poly,* many, and *nomial,* name), strings of Latin words and phrases consisting of up to 12 or more words. For example, the common wild briar rose was called *Rosa sylvestris inodora seu canina* by some and *Rosa sylvestris alba cum rubore, folio glabro* by others. This would be like the mayor of New York referring to a particular citizen as "Brooklyn resident: Democrat, male, Caucasian, middle income, Protestant, elderly, likely voter, short, bald, heavyset, wears glasses, works in the Bronx selling shoes." As you can imagine, these polynomial names were cumbersome. Even more worrisome, the names were altered at will by later authors, so that a given organism really did not have a single name that was its alone, as was the case with the briar rose.

The Binomial System

A much simpler system of naming animals, plants, and other organisms stems from the work of the Swedish biologist Carolus Linnaeus (1707–78). Linnaeus (figure 15.1) devoted his life to a challenge that had defeated many biologists before him—cataloging all the different kinds of organisms. Linnaeus, a botanist studying the plants of Sweden and from around the world, developed a plant classification system that grouped plants based on their reproductive structures. This system resulted in some seemingly unnatural groupings and therefore was never universally accepted. However, in the 1750s he produced several major works that, like his earlier books, employed the polynomial system. But as a kind of shorthand, Linnaeus also included in these books a two-part name for each species (others had also occasionally done this, but Linnaeus used these shorthand names consistently). These two-part names, or **binomials** (*bi* is the Latin prefix for "two"), have become our standard way of designating species. For example, he designated the willow oak (shown in figure 15.2*a* with its smaller, unlobed leaves) *Quercus phellos* and the red oak (with the larger deeply lobed leaves in figure 15.2*b*) *Quercus rubra,* even though he also included the polynomial name for these species. We also use binomial names for ourselves, our so-called given and family names. So, this naming system is like the mayor of New York calling the Brooklyn resident Sylvester Kingston.

Linnaeus took the naming of organisms a step further, grouping similar organisms into higher-level categories based on similar characteristics (discussed later). Although not intended to show evolutionary connections between different organisms, this hierarchical system acknowledged that there were broad similarities shared by groups of species that distinguished them from other groups.

15.2 Species Names

CONCEPT PREVIEW: By convention, the first part of a binomial species name identifies the genus to which it belongs, and the second part distinguishes that particular species from other species in the genus.

A group of organisms at a particular level in a classification system is called a **taxon** (plural, **taxa**), and the branch of biology that identifies and names such groups of organisms is called **taxonomy.** Taxonomists are in a real sense detectives, biologists who must use clues of appearance and behavior to identify and assign names to organisms.

By formal agreement among taxonomists throughout the world, no two organisms can have the same name. So that no one country is favored, a language spoken by no country—Latin—is used for the names. Because the scientific name of an organism is the same anywhere in the world, this system provides a standard and precise way of communicating, whether the language of a particular biologist is Chinese, Arabic, Spanish, or English. This is a great improvement over the use of common names, which often vary from one place to the next. As you can see in figure 15.3, in America corn refers to the plant in the upper left photo, but in Europe it refers to the plant Americans call wheat, the lower left photo. A bear is a large placental omnivore in the United States (the upper right photo), but in Australia it is a koala, a vegetarian marsupial (the lower right photo).

By convention, the first word of the binomial name is the genus to which the organism belongs. This word is always capitalized. The second word, called the *specific epithet,* refers to the particular species and is not capitalized. The two words together are called the **scientific name,** or species name, and are written in italics. The system of naming animals, plants, and other organisms established by Linnaeus has served the science of biology well for nearly 250 years.

> Throughout this text, you will often see scientific names provided for examples of organisms discussed or shown in photos. You will notice that the same system is used whether identifying a prokaryote, protist, fungus, plant, or animal.

(a) *Quercus phellos*
(Willow oak)

(b) *Quercus rubra*
(Red oak)

Figure 15.2 How Linnaeus named two species of oaks.

(a) Willow oak, *Quercus phellos.* (b) Red oak, *Quercus rubra.* Although they are clearly oaks (members of the genus *Quercus*), these two species differ sharply in the shapes and sizes of their leaves and in many other features, including their overall geographical distributions.

(a) (b)

Figure 15.3 Common names make poor labels.

The common names corn (a) and bear (b) bring clear images to our minds (photos on *top*), but the images would be very different to someone living in Europe or Australia (photos on *bottom*). There, the same common names are used to label very different species.

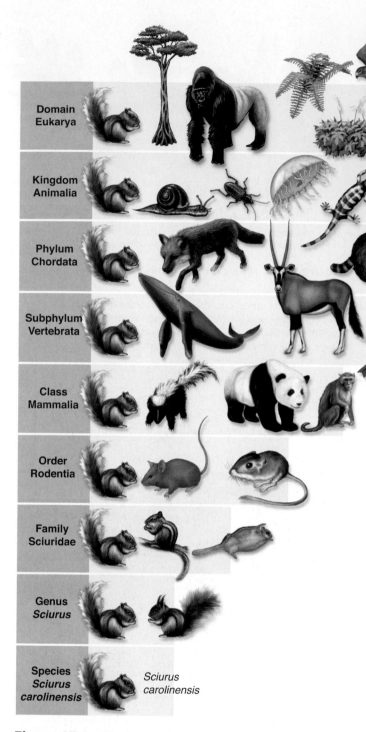

Domain
Eukarya

Kingdom
Animalia

Phylum
Chordata

Subphylum
Vertebrata

Class
Mammalia

Order
Rodentia

Family
Sciuridae

Genus
Sciurus

Species
*Sciurus
carolinensis*

*Sciurus
carolinensis*

Figure 15.4 **The hierarchical system used to classify a gray squirrel.**

In this example, the organism is first recognized as a eukaryote (domain: Eukarya). Second, within this domain, it is an animal (kingdom: Animalia). Among the different phyla of animals, it is a vertebrate (phylum: Chordata, subphylum: Vertebrata). The organism's fur characterizes it as a mammal (class: Mammalia). Within this class, it is distinguished by its gnawing teeth (order: Rodentia). Next, because it has four front toes and five back toes, it is a squirrel (family: Sciuridae). Within this family, it is a tree squirrel (genus: *Sciurus*), with gray fur and white-tipped hairs on the tail (species: *Sciurus carolinensis*, the eastern gray squirrel).

15.3 Higher Categories

CONCEPT PREVIEW: A hierarchical system is used to classify organisms, in which higher categories convey more general information about the group.

A biologist needs more than two categories to classify all the world's living things. Taxonomists group the genera with similar properties into a cluster called a **family.** For example, the Eastern gray squirrel at the bottom in figure 15.4 is placed in a family with other squirrel-like animals including prairie dogs, marmots, and chipmunks. Similarly, families that share major characteristics are placed into the same **order** (for example, squirrels placed in with other rodents). Orders with common properties are placed into the same **class** (squirrels in the class Mammalia), and classes with similar characteristics into the same **phylum** (plural, **phyla**) such as the Chordata. Botanists (that is, those who study plants) also call plant phyla "divisions." Finally, the phyla are assigned to one of several gigantic groups, the **kingdoms.** Biologists currently recognize six kingdoms: two kinds of prokaryotes (Archaea and Bacteria), a largely unicellular group of eukaryotes (Protista), and three multicellular groups (Fungi, Plantae, and Animalia). To remember the seven categories in their proper order, it may prove useful to memorize a phrase such as "**k**indly **p**ay **c**ash **o**r **f**urnish **g**ood **s**ecurity" or "**K**ing **P**hilip **c**ame **o**ver **f**or **g**reen **s**paghetti" (**k**ingdom–**p**hylum–**c**lass–**o**rder–**f**amily–**g**enus–**s**pecies).

In addition, an eighth level of classification, called *domains,* is sometimes used. Domains are the broadest and most inclusive taxa, and biologists recognize three of them, Bacteria, Archaea, and Eukarya—which are discussed later in this chapter.

Each of the categories in this Linnaean system of classification is loaded with information. For example, consider a honeybee:

Level 1: Its **species** name, *Apis mellifera,* identifies the particular species of honey bee.

Level 2: Its **genus** name, *Apis,* tells you it is a honey bee.

Level 3: Its **family,** Apidae, are all bees, some solitary, others living in hives as *A. mellifera* does.

Level 4: Its **order,** Hymenoptera, tells you that it is likely able to sting and may live in colonies.

Level 5: Its **class,** Insecta, says that *A. mellifera* has three major body segments, with wings and three pairs of legs attached to the middle segment.

Level 6: Its **phylum,** Arthropoda, tells us that it has a hard cuticle of chitin and jointed appendages.

Level 7: Its **kingdom,** Animalia, says that it is a multicellular heterotroph whose cells lack cell walls.

Level 8: An addition to the Linnaean system, its **domain,** Eukarya, says that its cells contain membrane-bounded organelles.

15.4 What Is a Species?

CONCEPT PREVIEW: Among animals, species are generally defined as reproductively isolated groups; among the other kingdoms such a definition is less useful, as their species typically have weaker barriers to hybridization.

The basic biological unit in the Linnaean system of classification is the species. John Ray (1627–1705), an English clergyman and scientist, was one of the first to propose a general definition of species. In about 1700, he suggested a simple way to recognize a species: all the individuals that belong to it can breed with one another and produce fertile offspring. By Ray's definition, the offspring of a single mating were all considered to belong to the same species, even if they contained different-looking individuals, as long as these individuals could interbreed. All domestic cats are one species (they can all interbreed), while carp are not the same species as goldfish (they cannot interbreed). The donkey you see in figure 15.5 is not the same species as the horse, because when they interbreed, the offspring—mules—are sterile.

The Biological Species Concept

Linnaeus used Ray's species concept, and it is still widely used today. The *biological species concept* defines species as groups that are reproductively isolated. Hybrids (offspring of different species that mate) are rare in nature, while individuals that belong to the same species are able to interbreed freely.

The biological species concept works fairly well for animals, where strong barriers to hybridization between species exist, but very poorly for members of the other kingdoms. The problem is that in prokaryotes and many protists, fungi, and plants, asexual reproduction (reproduction without sex) predominates. These species clearly cannot be characterized in the same way as animals—they do not breed with one another, much less with individuals of other species.

Complicating matters further, the reproductive barriers that are key to the biological species concept, although common among animal species, are not typical of other kinds of organisms. In fact, there are essentially no barriers to hybridization between the species in many groups of trees, such as oaks, and other plants, such as orchids. Even among animals, fish species are able to form fertile hybrids with one another, though they may not do so in nature.

In practice, biologists today recognize species in different groups in much the way they always did, as groups that differ from one another in their visible features. Molecular data are causing a reevaluation of traditional classification systems, and are changing the way scientists classify plants, protists, fungi, prokaryotes, and even animals.

How Many Kinds of Species Are There?

Since the time of Linnaeus, about 1.5 million species have been named. But the actual number of species in the world is undoubtedly much greater, judging from the very large numbers that are still being discovered. Some scientists estimate that at least 10 million species exist on earth, two-thirds in the tropics.

Concept Check

1. List the eight levels of classification in order of inclusiveness.
2. Explain why the biological species concept works so much better for animals than for other kinds of organisms.
3. Distinguish between what Linnaeus and Ray contributed to taxonomy.

Horse

Donkey

Mule

Figure 15.5   Ray's definition of a species.

According to Ray, donkeys and horses are not the same species. Even though they produce very hardy offspring (mules) when they mate, the mules are sterile, meaning that two mules that mate cannot produce offspring. The sterility arises because the two species have different numbers of chromosomes: donkeys have 62 chromosomes, whereas horses have 64.

IMPLICATION   Horses, donkeys, and zebras are all members of the same genus, *Equus,* and, when interbred, produce hardy but sterile offspring: a "zonky" is the offspring of a zebra and a donkey, while a "zorse" is the offspring of a zebra and a horse. In the film *Racing Stripes,* the son of Stripes (a zebra) and Sandy (a white filly horse) is a zorse. Do you think Stripes has the same number of chromosomes as Sandy? Explain your answer.

Inferring Phylogeny

15.5 How to Build a Family Tree

CONCEPT PREVIEW: A phylogeny may be represented as a cladogram based on the order in which groups evolved or as a traditional taxonomic tree that weights characters according to assumed importance.

By looking at the differences and similarities between organisms, biologists attempt to reconstruct the tree of life, inferring which organisms evolved from which other ones, in what order, and when. The evolutionary history of an organism and its relationship to other species is called **phylogeny.**

Cladistics

A simple and objective way to construct an evolutionary history, or **phylogenetic tree,** is to focus on key characters that some organisms share because they have inherited them from a common ancestor. A **clade** is a group of organisms related by descent, and this approach to constructing a phylogeny is called **cladistics.** Cladistics infers phylogeny (that is, builds family trees) according to similarities derived from a common ancestor, so-called **derived characters.** Derived characters are defined as characters that are present in a group of organisms that arose from a common ancestor that lacked the character. The key to the approach is being able to identify morphological, physiological, or behavioral traits that differ among the organisms being studied and can be attributed to a common ancestor. By examining the distribution of these traits among the organisms, it is possible to construct a **cladogram,** a branching diagram that represents the phylogeny. A cladogram of the vertebrates is shown in figure 15.6.

Cladograms are not true family trees, derived directly from data that document ancestors and descendants like the fossil record does. Instead, cladograms convey comparative information about *relative* relationships. Organisms that are closer together on a cladogram simply share a more recent common ancestor than those that are farther apart. Because the analysis is comparative, it is necessary to have something to anchor the comparison to, some solid ground against which the comparisons can be made. To achieve this, each cladogram must contain an **outgroup,** a rather different organism (but not *too* different) to serve as a baseline for comparisons among the other organisms being evaluated, called the **ingroup.** For example, in figure 15.6, the lamprey is the outgroup to the clade of animals that have jaws. Comparisons are then made up the cladogram, beginning with lampreys and sharks, based on the emergence of derived characters. For example, the shark differs from the lamprey in that it has jaws, the derived character missing in the lamprey. The derived characters are in the colored boxes along the main line of the cladogram. Salamanders differ from sharks in that they have lungs, and so on up the cladogram.

Sometimes cladograms are adjusted to "weight" characters, or take into account the variation in the "strength" (importance) of a character—the size or location of a fin, the effectiveness of a lung. For example, let's say that the following are five unique events that occurred on September 11, 2001: (1) my cat was declawed, (2) I had a wisdom tooth pulled, (3) I sold my first car, (4) terrorists attacked the United States using commercial airplanes, and (5) I passed physics. Without weighting the events, each one is assigned equal importance. In a nonweighted cladistic sense, they are equal (all happened only once, and on that day), but in a practical, real-world sense, they certainly are not. One event, the terrorist attack, had a far

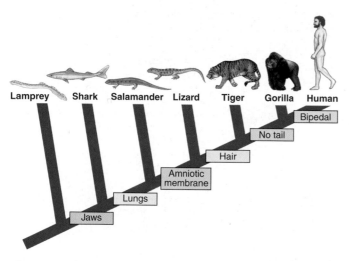

Figure 15.6   A cladogram of vertebrate animals.

The derived characters between the branch points are shared by all the animals to the right of each character and are not present in any organisms to the left of it.

IMPLICATION   Do you think you would obtain the same tree of relationships you see here if instead of looking at derived characters you instead simply counted the number of DNA bases different from humans in the genomes of each of the groups in the figure above? Explain your answer. [Hint: Look at the family tree on page 256].

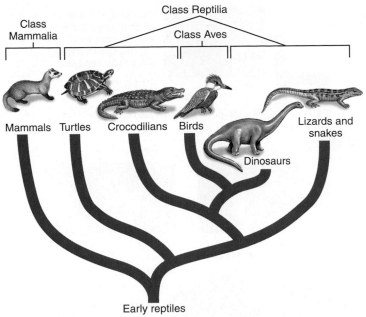

Traditional phylogeny and taxonomic classification

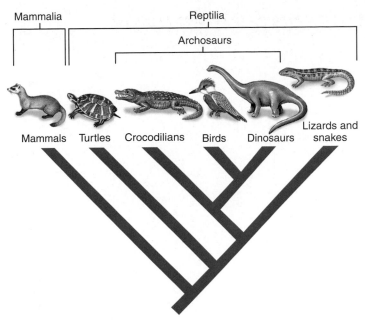

Cladogram and cladistic classification

Figure 15.7 Two ways to classify terrestrial vertebrates.

Traditional taxonomic analyses place birds in their own class (Aves) because birds have evolved several unique adaptations that separate them from the reptiles. Cladistic analyses, however, place crocodiles, dinosaurs, and birds together (as archosaurs) because they share many derived characters, indicating a recent shared ancestry. In practice, most biologists adopt the traditional approach and consider birds as members of the class Aves rather than Reptilia.

greater impact and importance than the others. Because evolutionary success depends so critically on just such high-impact events, these weighted cladograms attempt to assign extra weight to the evolutionary significance of key characters.

Traditional Taxonomy

Weighting characters lies at the core of **traditional taxonomy.** In this approach, phylogenies are constructed based on all available information about the morphology and biology of the organism. The large amount of information permits a knowledgeable weighting of characters according to their biological significance. For example, in classifying the terrestrial vertebrates, traditional taxonomists, shown by the phylogeny on the left in figure 15.7, place birds in their own class (Aves), giving great weight to the characters that made powered flight possible, such as feathers. However, a cladogram of vertebrate evolution, as shown on the right, lumps birds in among the reptiles with crocodiles and dinosaurs. This accurately reflects their ancestry but ignores the immense evolutionary impact of a derived character such as feathers.

Overall, phylogenetic trees based on traditional taxonomy are information-rich, while cladograms often do a better job of deciphering evolutionary histories. Traditional taxonomy is the better approach when a great deal of information is available to guide character weighting. However, cladistics is the preferred approach when less information is available about how the character affects the life of the organism. For example, the cat family tree in figure 15.8 on the next page is based on thousands of DNA differences among the different groups of felines, few of which affect the life of the cats in ways we know and understand.

Concept Check

1. Define cladistics.
2. Can a derived character be present in an outgroup?
3. Is a bird really a reptile? Justify your answer.

IN THE NEWS

The Biodiversity Crisis. Biologists, worried by the rapid rate of extinction of animals and plants, have recently announced completion of a comprehensive survey of threatened species. For plants, invertebrate animals, and the five major categories of vertebrates, they first determined the number of species that had been evaluated in the last decade, and then for each group counted how many of these had proven to be threatened. They concluded that our world now faces a biodiversity crisis, with over a third of all species living on it threatened today with extinction. Birds seemed to be doing better than the others, and plants worse. All mammalian species had been evaluated, and over 20% of them proved to be endangered, most by habitat loss or poaching.

Taxonomic Group	Threatened
Mammals	21%
Birds	12%
Reptiles	31%
Amphibians	30%
Fishes	37%
Invertebrates	41%
Plants	70%

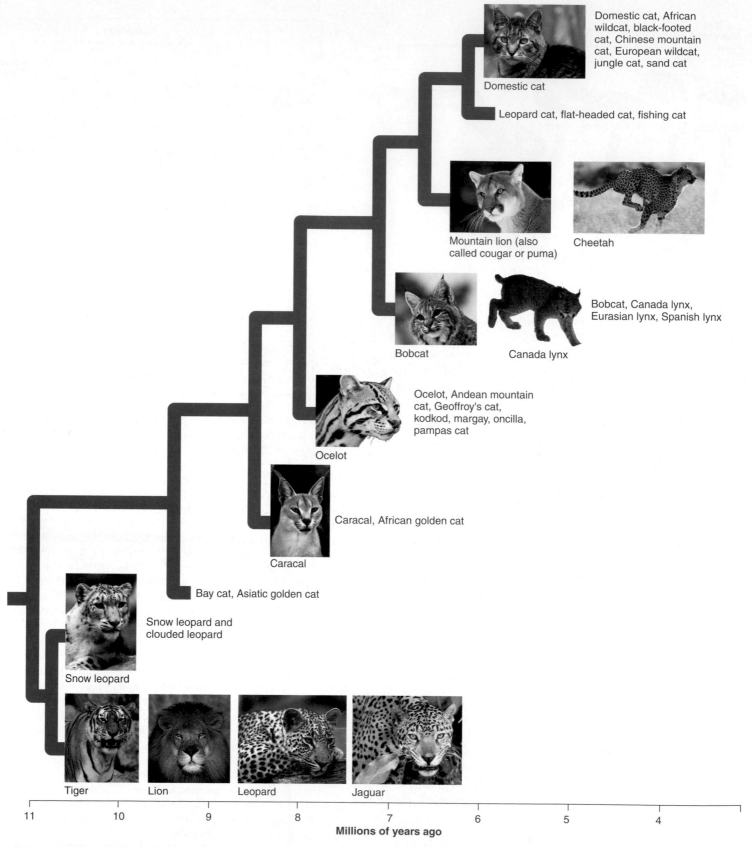

Domestic cat, African wildcat, black-footed cat, Chinese mountain cat, European wildcat, jungle cat, sand cat

Domestic cat

Leopard cat, flat-headed cat, fishing cat

Mountain lion (also called cougar or puma)

Cheetah

Bobcat, Canada lynx, Eurasian lynx, Spanish lynx

Bobcat

Canada lynx

Ocelot, Andean mountain cat, Geoffroy's cat, kodkod, margay, oncilla, pampas cat

Ocelot

Caracal, African golden cat

Caracal

Bay cat, Asiatic golden cat

Snow leopard and clouded leopard

Snow leopard

Tiger

Lion

Leopard

Jaguar

Millions of years ago

11 10 9 8 7 6 5 4

Figure 15.8 **The cat family tree.**

Recent studies of DNA similarities reported in 2006 have allowed biologists to construct this feline family tree of the eight major cat lineages and their individual species. Among the oldest of all cats are the four big panthers: tiger, lion, leopard, and jaguar. The other big cats, the cheetah and mountain lion, are members of a much younger lineage and are not close relatives of the big four. Domestic cats evolved most recently.

Race and Medicine

Few issues in biology have stirred more social controversy than race. Race has a deceptively simple definition, referring to groups of individuals related by ancestry that differ from other groups, but not enough to constitute separate species. The controversy arises because of the way people have used the concept of race to justify the abuse of humans. The African slave trade is but one obvious example. Perhaps in some measure responding to these sorts of injustice, scientists have largely abandoned the concept of race except as a social construct. However, scientists are again evaluating the biology of race. This reevaluation had its beginnings in 1972, when geneticists pointed out that if one looked at genes rather than faces, the differences between the genes of an African and a European would be hardly greater than the difference between those of any two Europeans. For thirty years, gene data has continually reinforced the validity of this observation, and the concept of genetic races has been largely abandoned.

Recent detailed comparisons of the genomes of people around the world reveal that particular alleles defining human features often occur in clusters on chromosomes. Because they are close together, the genes experience little recombination over the centuries. The descendants of a person who has a particular combination of alleles will almost always have that same combination. The set of alleles, technically called a "haplotype," reflects the common ancestry of these descendants from that ancestor.

Ignoring skin color and eye shape, and instead comparing the DNA sequences of hundreds of regions of the human genome, investigators at the University of Southern California showed in 2002 that when a large sample of people from around the world are compared, the computer sorts them into five large groups containing similar clusters of variation: Europe, East Asia, Africa, America, and Australasia. That these are more or less the major races of traditional anthropology is not the point. The point is that humanity evolved in these five regions in isolation from one another, that today each of these groups is composed of individuals with a shared ancestry, and that by analyzing the DNA of an individual we can deduce that ancestry.

Why bother making that rather arcane and socially controversial point? Because analysis of the human genome is revealing that many diseases are influenced by alleles that have arisen since the five major branches of the human family tree separated from one another, and are much more common in the human ancestral group within which the DNA mutation causing the disease first occurred. For example, a mutation causing hemochromatosis, a disorder of iron metabolism, is rare or absent among Indians or Chinese, but very common among northern Europeans (it occurs in 7.5% of Swedes), who also commonly possess an allele leading to adult lactose intolerance (inability to digest lactose) not common in many other groups. Similarly, the hemoglobin S mutation causing sickle cell disease is common among Africans of Bantu ancestry, but seems to have arisen only there.

This is a pattern we see again and again as we compare genomes of people living in different parts of the world—and it has a very important consequence. Because of common ancestry, genetic diseases (disorders that are inherited) have a lot to do with geography. The risk that an African American man will be afflicted with hypertensive heart disease or prostate cancer is nearly three times greater than that for a European American man, while the European American is far more likely to develop cystic fibrosis or multiple sclerosis.

These differences in medically important DNA variants carry over to genetic differences in how individuals respond to treatment. African Americans, for example, respond poorly to some of the main drugs used to treat heart conditions, such as beta-blockers and angiotensin enzyme inhibitors.

Scientists and doctors that recognize this unfortunate fact are not racists. They fully agree that while using the races of traditional anthropology to pin-point which therapeutic treatment to recommend is an improvement over treating all patients alike, it is still a very poor way of getting at these differences. It would be far better to simply ignore skin color and other single-gene differences and instead perform for each patient a broad "gene variation" analysis. Hopelessly difficult only a few years ago, this now seems an attractive avenue to improve medical treatment for all of us.

It is important to keep clearly in mind the goal of sorting out individual human ancestry, which is to identify common lines of descent that share common response to potential therapies. It is NOT to assign people to over-arching racial categories. This point is being made with great clarity in a course being taught at Pennsylvania State University, where about 90 students had their DNA sampled and compared with four of the five major human groups. Many of these students had thought of themselves as "100% white," but only a few were. One "white" student learned that 14% of his DNA came from Africa, and 6% from East Asia. Similarly, "black" students found that as much as half of their genetic material came from Europe, and a significant amount from Asia as well.

The point is, rigid ideas about the biological basis of identity are wrong. Humanity is much more complex than indicated by a few genes affecting skin color and eye shape. The more clearly we can understand that complexity, the better we can deal with the medical consequences, and the great potential that human diversity provides all of us.

Kingdoms and Domains

BIOLOGY & YOU

How Biodiversity Benefits You—Direct Economic Value. Many threatened species have direct value to you as sources of food, medicine, clothing, energy, and shelter:

1. **Gene variation.** Most of the world's food crops, such as corn, wheat, and rice, are derived from a small number of domesticated plants and so contain relatively little genetic variation; on the other hand, their wild relatives have great diversity. In the future, genetic variation from wild strains will be needed if we are to improve yields or find a way to breed resistance to new pests.

2. **Medicine.** About 70% of the world's population depends directly on wild plants as their source of medicine. In addition, some 40% of the prescription drugs used today have active ingredients extracted from animals or plants. Aspirin was first extracted from the leaves of a tropical willow tree. Vinblastine, the most effective drug for combatting childhood leukemia, is extracted from the rosy periwinkle vine. The cancer-fighting drug taxol is produced from the bark of the Pacific yew tree.

3. **Gene prospecting.** Genetic engineers are using the advances described in chapter 13 to transfer beneficial genes into crops. We have been able to examine only a minute proportion of the world's species to see if they contain genes that have useful properties for humans. When a species is extinct, its entire library of genes is lost to us forever.

15.6 The Kingdoms of Life

CONCEPT PREVIEW: Living organisms are grouped into three categories called domains. One of the domains, Eukarya, is divided into four kingdoms: Protista, Fungi, Plantae, and Animalia.

Classification systems have gone through their own evolution of sorts, as illustrated in figure 15.9. The earliest classification systems recognized only two kingdoms of living things: animals, shown in blue in (a), and plants, shown in green. But as biologists discovered microorganisms (the yellow-colored boxes in b) and learned more about other organisms like the protists (in teal) and the fungi (in light brown), they added kingdoms in recognition of fundamental differences. Most biologists now use a six-kingdom system (indicated by the six different-colored boxes in c) first proposed by Carl Woese of the University of Illinois.

In this system, four kingdoms consist of eukaryotic organisms. The two most familiar kingdoms, **Animalia** and **Plantae,** contain organisms that are multicellular during most of their life cycle. These groups of animals and plants are no doubt familiar to you. The kingdom **Fungi** contains multicellular forms, such as mushrooms and molds, and single-celled yeasts, which are thought to have multicellular ancestors. Fundamental differences divide these three kingdoms. Plants are mainly stationary, but some have motile sperm; fungi have no motile cells; animals are mainly motile. Animals ingest their food, plants manufacture it, and fungi digest it by means of secreted extracellular enzymes. Each of these kingdoms probably evolved from a different single-celled ancestor.

The large number of unicellular eukaryotes are arbitrarily grouped into a single kingdom called **Protista** (see chapter 16). They include the algae and many kinds of microscopic aquatic organisms. This kingdom is an artificial group in that many of these organisms are only distantly related, and the classification of the protists is in flux.

> Prokaryotic organisms, as described on page 65, are single-celled organisms that lack a nucleus and other internal membrane-bounded structures.

The remaining two kingdoms, **Archaea** and **Bacteria,** consist of prokaryotic organisms, which are vastly different from all other living things (also in chapter 16). The prokaryotes with which you are most familiar, those that cause disease or

Figure 15.9 Different approaches to classifying living organisms.

(a) Linnaeus popularized a two-kingdom approach, in which the fungi and the photosynthetic protists were classified as plants and the nonphotosynthetic protists as animals; when prokaryotes were described, they too were considered plants. (b) Whittaker in 1969 proposed a five-kingdom system that soon became widely accepted. (c) Woese has championed splitting the prokaryotes into two kingdoms for a total of six kingdoms or even assigning them separate domains, with a third domain containing the four eukaryotic kingdoms (d).

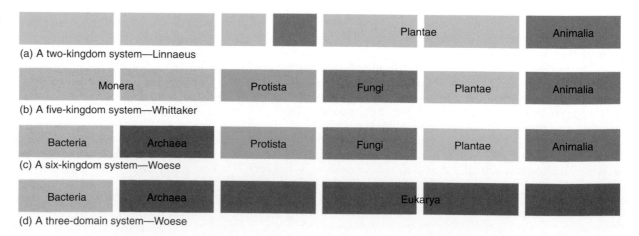

(a) A two-kingdom system—Linnaeus

(b) A five-kingdom system—Whittaker

(c) A six-kingdom system—Woese

(d) A three-domain system—Woese

are used in industry, are members of the kingdom Bacteria. Archaea are a diverse group including the methanogens and some that live in physically extreme conditions, such as the extreme thermophiles. They differ greatly from bacteria in many ways. The characteristics of these six kingdoms are presented in table 15.1.

As biologists have learned more about the archaea, it has become increasingly clear that this ancient group is very different from all other organisms. When the full genomic DNA sequences of an archaean and a bacterium were first compared in 1996, the differences proved striking. Based on comparing DNA and cell structures, archaea are as different from bacteria as bacteria are from eukaryotes. Recognizing this, biologists have in recent years adopted a taxonomic level higher than kingdom that recognizes three **domains** (figure 15.9*d*). Archaea (red-colored box) are in one domain, bacteria (yellow-colored box) in a second, and eukaryotes (the four purple boxes representing the four eukaryotic kingdoms) in the third. While the domain Eukarya contains four kingdoms of organisms, the domains Bacteria and Archaea contain only one kingdom in each. Because of this, the kingdom level of classification for Bacteria and Archaea is now often omitted, with biologists using just their domain and phyla names.

BIOLOGY & YOU

How Biodiversity Benefits You—Ecosystem Services. The most important value of biological diversity to you is indirect, as diversity is of vital importance to the health of ecosystems (communities of organisms and the places where they live). Diverse biological communities help maintain the chemical quality of natural water, buffer natural areas against storms and drought, preserve soils and prevent loss of minerals, moderate local and regional climate, absorb pollution, and promote the breakdown of organic wastes and the recycling of minerals. All of these benefits are grouped together by economists as so-called ecosystem services. The stability and productivity of the world's ecosystems depend critically on these services, and they in turn depend critically on the number of different species present in an area, called species richness. By destroying biodiversity we are creating conditions of instability and lessened productivity, and promoting desertification (the natural transformation of lands into deserts), tropical soil loss through mineralization, and many other undesirable outcomes throughout the world.

TABLE 15.1 Characteristics of the Six Kingdoms

Domain	Bacteria	Archaea	Eukarya			
Kingdom	Bacteria	Archaea	Protista	Plantae	Fungi	Animalia
Cell type	Prokaryotic	Prokaryotic	Eukaryotic	Eukaryotic	Eukaryotic	Eukaryotic
Nuclear envelope	Absent	Absent	Present	Present	Present	Present
Mitochondria	Absent	Absent	Present or absent	Present	Present or absent	Present
Chloroplasts	None (photosynthetic membranes in some types)	None (bacteriorhodopsin in one species)	Present in some forms	Present	Absent	Absent
Cell wall	Present in most; peptidoglycan	Present in most; polysaccharide, glycoprotein, or protein	Present in some forms; various types	Cellulose and other polysaccharides	Chitin and other noncellulose polysaccharides	Absent
Means of genetic recombination, if present	Conjugation, transduction, transformation	Conjugation, transduction, transformation	Fertilization and meiosis	Fertilization and meiosis	Fertilization and meiosis	Fertilization and meiosis
Mode of nutrition	Autotrophic (chemosynthetic, photosynthetic) or heterotrophic	Autotrophic (photosynthesis in one species) or heterotrophic	Photosynthetic or heterotrophic or combination of both	Photosynthetic, chlorophylls *a* and *b*	Absorption	Digestion
Motility	Bacterial flagella, gliding, or nonmotile	Unique flagella in some	9 + 2 cilia and flagella; amoeboid, contractile fibrils	None in most forms, 9 + 2 cilia and flagella in gametes of some forms	Nonmotile	9 + 2 cilia and flagella, contractile fibrils
Multicellularity	Absent	Absent	Absent in most forms	Present in all forms	Present in most forms	Present in all forms

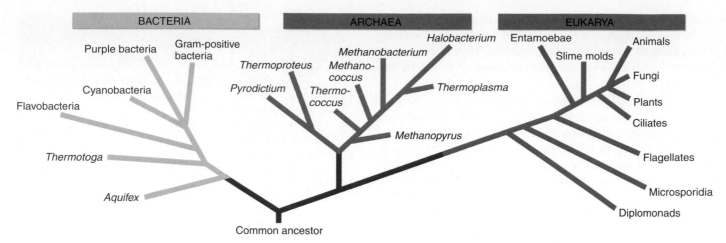

Figure 15.10   A tree of life.

This phylogeny, prepared from rRNA analyses, shows the evolutionary relationships among the three domains. The base of the tree was determined by examining genes that are duplicated in all three domains, the duplication presumably having occurred in the common ancestor. When one of the duplicates is used to construct the tree, the other can be used to root it. This approach clearly indicates that the root of the tree is within the bacterial domain. Archaea and eukaryotes diverged later and are more closely related to each other than either is to bacteria.

15.7 Domain: A Higher Level of Classification

CONCEPT PREVIEW: Organisms in the three domains differ in basic cellular structure. Bacteria and Archaea are prokaryotes but vary in several aspects. Eukarya contains single-celled and multicellular eukaryotic organisms.

Domain Bacteria

The domain Bacteria contains one kingdom of the same name, Bacteria. The bacteria are the most abundant organisms on earth. There are more living bacteria in your mouth than there are mammals living on earth. Although too tiny to see with the unaided eye, bacteria play critical roles throughout the biosphere. For example, they extract from the air all the nitrogen used by organisms, and they play key roles in cycling carbon and sulfur.

> All organisms need nitrogen, but only a few kinds of bacteria can transform the nitrogen in the air into a form that can be used by other organisms. This process, called nitrogen fixation, is discussed on page 394.

There are many different kinds of bacteria, and the evolutionary links between them are not well understood. The archaea and eukaryotes are more closely related to each other than to bacteria and are on a separate evolutionary branch of the phylogenetic tree in figure 15.10.

Domain Archaea

The domain Archaea contains one kingdom by the same name, the Archaea. The term *archaea* (Greek, *archaio,* ancient) refers to the ancient origin of this group of prokaryotes, which most likely diverged very early from the bacteria. Notice in figure 15.10 that the Archaea, in red, branched off from a line of prokaryotic ancestors that led to the evolution of eukaryotes. Though a diverse group, all archaea share certain key characteristics: They possess very unusual cell walls, lipids, and ribosomal RNA (rRNA) sequences. Some of their genes possess introns, unlike those of bacteria.

As the genomes of archaea have become better known, microbiologists have been able to identify signature sequences of DNA present in all archaea and in no other organisms. When samples from soil or seawater are tested for genes matching these signature sequences, many of the prokaryotes living there prove to be archaea. Clearly, archaea are not restricted to extreme habitats, as microbiologists used to think.

Domain Eukarya

For at least 1 billion years, prokaryotes ruled the earth. No other organisms existed to eat them or compete with them, and their tiny cells formed the world's oldest fossils. The third great domain of life, the eukaryotes, appear in the fossil record much later, only about 1.5 billion years ago.

Three Largely Multicellular Kingdoms.   Fungi, plants, and animals are well-defined evolutionary groups. They are largely multicellular, each group clearly stemming from a different single-celled eukaryotic ancestor. The ancestor of each distinct evolutionary line would be classified in the kingdom Protista.

The amount of diversity among the protists, however, is much greater than that within or between the three largely multicellular kingdoms derived from the protists. Because of the size and ecological dominance of plants, animals, and fungi, and because they are predominantly multicellular, we recognize them as kingdoms distinct from Protista.

A Fourth Very Diverse Kingdom.   When multicellularity evolved, the diverse kinds of single-celled organisms that existed at that time did not simply become extinct. A wide variety of unicellular eukaryotes and their relatives exist today. They are grouped together in the kingdom Protista solely because they are not fungi, plants, or animals. Protists are a fascinating group containing many organisms of intense interest and great biological significance.

Symbiosis and the Origin of Eukaryotes.   The hallmark of eukaryotes is complex cellular organization, highlighted by an extensive endomembrane system that subdivides the eukaryotic cell into functional compartments called organelles (see chapter 4). Not all of these organelles, however, are derived from the endomembrane system. Mitochondria and chloroplasts are both believed to have entered early eukaryotic cells by a process called endosymbiosis (*endo,* inside) where an organism such as a bacterium is taken into the cell and remains functional inside the cell.

With few exceptions, all modern eukaryotic cells possess energy-producing organelles, the mitochondria. Mitochondria are about the size of bacteria and contain DNA. Comparison of the nucleotide sequence of this mitochondrial DNA with that of a variety of organisms indicates clearly that mitochondria are the descendants of purple bacteria that were incorporated into eukaryotic cells early in the history of the group. Some protist phyla have also acquired chloroplasts during the course of their evolution and thus are photosynthetic. These chloroplasts are derived from cyanobacteria that became symbiotic in several groups of protists early in their history. Figure 15.11*a* shows how this could have happened, with the green cyanobacterium being engulfed by an early protist. Some of these photosynthetic protists gave rise to land plants. Endosymbiosis still happens today (the green structures inside the coral in figure 15.11*b* are endosymbiotic protists). We discussed the endosymbiotic origin of mitochondria and chloroplasts in chapter 4.

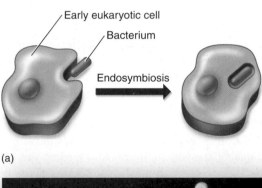

Early eukaryotic cell

Bacterium

Endosymbiosis

(a)

(b)

Figure 15.11   Endosymbiosis.

(a) This figure shows how an organelle could have arisen in early eukaryotic cells through a process called endosymbiosis. An organism, such as a bacterium, is taken into the cell through a process similar to endocytosis but remains functional inside the host cell. (b) Many corals contain endosymbionts, algae called zooxanthellae that carry out photosynthesis and provide the coral with nutrients. In this photograph, the zooxanthellae are the golden-brown spheres packed into the tentacles of a coral animal.

Concept Check

1. Which domain contains humans?
2. What is the difference between archaea and bacteria?
3. Which kingdom is the most diverse?

What Causes New Forms to Arise?

Biologists once presumed that new forms—genera, families, and orders—arose most often during times of massive geological disturbance, stimulated by the resulting environmental changes. But no such relationship exists. An alternative hypothesis was proposed by evolutionist George Simpson in 1953. He proposed that diversification followed new evolutionary innovations, "inventions" that permitted an organism to occupy a new "adaptive zone." After a burst of new orders that define the major groups, subsequent specialization would lead to new genera.

The early bony fishes, typified by the sturgeon below, had feeble jaws and long shark-like tails. They dominated the Devonian (the Age of Fishes), but were succeeded in the Triassic (the period when dinosaurs appeared) by fishes like the gar pike with a more powerful jaw that improved feeding and a shortened more maneuverable tail that improved locomotion. Gar pikes were in turn succeeded by teleost fishes like the perch, with an even better tail for fast, maneuverable swimming, and a complex mouth with a mobile upper jaw that slides forward as the mouth opens.

This history allows a clear test of Simpson's hypothesis. Was the appearance of these three orders followed by a burst of evolution as Simpson predicts, the new innovations in feeding and locomotion opening wide the door of opportunity? If so, many new genera should be seen in the fossil record soon after the appearance of each new order. If not, the pattern of when new genera appear should not track the appearance of new orders.

The graph shows the evolutionary history of the class Osteichthyes, the bony fishes, since they first appeared in the Silurian some 420 million years ago.

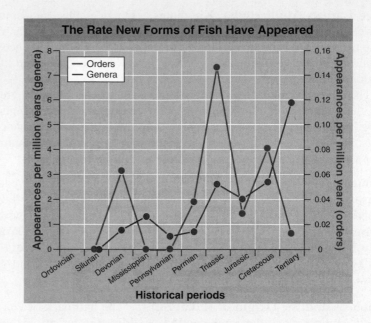

The Rate New Forms of Fish Have Appeared

Analysis

1. **Interpreting Data**   Three great innovations in jaw and tail occur during the history of the bony fishes, producing the superorders represented by sturgeons, then gars, and then teleost fishes. In what period did each innovation occur?
2. **Making Inferences**   Do bursts of new genera appear at these same three times, or later?
3. **Drawing Conclusions**   Does the data presented in the graph support Simpson's hypothesis? Explain.

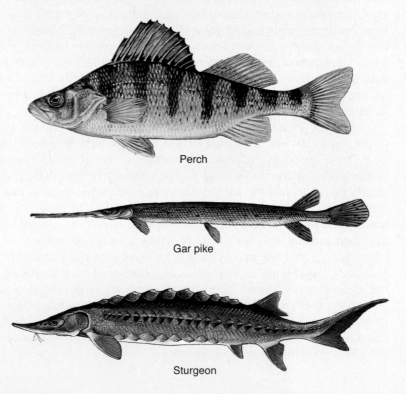

Perch

Gar pike

Sturgeon

Concept Summary

The Classification of Organisms

15.1 The Invention of the Linnaean System

- Scientists use a system of grouping similar organisms together, called classification. Latin is used because it was the language used by earlier philosophers and scientists.

- The polynomial system of classification named an organism by using a list of adjectives that described the organism. The binomial system, using a two-part name, was originally developed as a "shorthand" reference to the polynomial name. Linnaeus used this two-part naming system consistently and its use became widespread (**figure 15.2**).

15.2 Species Names

- Taxonomy is the area of biology involved in identifying, naming, and grouping organisms. Scientific names consist of two parts, the genus and species. The genus name is capitalized but the species name is not. The two parts of the name are italicized. Scientific names are standardized, universal names that are less confusing than common names (**figure 15.3**).

15.3 Higher Categories

- In addition to the genus and species names, an organism is also assigned to higher levels of classification. The higher categories convey more general information about the organisms in a particular group. The most general category, domain, is the largest grouping followed by ever-increasing specific information that is used to group organisms into a kingdom, phylum, class, order, family, genus, and species (**figure 15.4**).

15.4 What Is a Species?

- The biological species concept states that a species is a group of organisms that is reproductively isolated, meaning that the individuals mate and produce fertile offspring with each other but not with those of other species.

- This concept works well to define animal species because animals regularly outcross (mate with other individuals—**figure 15.5**) as do many kinds of plants. However, the concept does not apply to other organisms (fungi, protists, prokaryotes, and some plants) that regularly reproduce without mating through asexual reproduction. The classification of these organisms relies more on physical, behavioral, and molecular characteristics.

Inferring Phylogeny

15.5 How to Build a Family Tree

- In addition to organizing a great number of organisms, the study of taxonomy also gives us a glimpse of the evolutionary history of life on earth. Organisms with similar characteristics are more likely to be related to each other. The evolutionary history of an organism and its relationship to other species is called phylogeny, and relationships are often mapped out using phylogenetic trees.

- Phylogenetic trees can be created using key characteristics that are shared by some organisms, presumably having been inherited from a common ancestor. A group of organisms that are related by descent is called a clade, and a phylogenetic tree organized in this manner is called a cladogram (**figure 15.6**). A cladogram suggests the order in which evolutionary changes occurred.

- Cladograms can sometimes be misleading when the characteristics are weighted, placing more importance on a characteristic that seems to have a more significant impact on evolution. The problem with this system is that some characteristics may turn out to be less important than first thought. For this reason, cladograms work best when all characteristics are weighted equally.

- Traditional taxonomy focuses more on the significance or evolutionary impact of a characteristic and not just on the commonality of the characteristic. For example, in traditional taxonomy, birds are placed in their own class, as shown here from **figure 15.7,** even though, evolutionarily, birds fall within the reptile group. Traditional taxonomy is used when more information is available to weight characteristics that seem more significant (**figure 15.8**), while cladistics places more emphasis on the order or timing in which unique, or derived, characteristics appear.

Kingdoms and Domains

15.6 The Kingdoms of Life

- The designation of kingdoms, the second-highest category used in classification, has changed over the years as more and more information about organisms has been uncovered. Currently six kingdoms have been identified: Bacteria, Archaea, Protista, Fungi, Plantae, and Animalia, (**figure 15.9** and **table 15.1**).

- The domain level of classification was added in the mid-1990s, recognizing three fundamentally different types of cells: Eukarya (eukaryotic cells), Archaea (prokaryotic archaea), and Bacteria (prokaryotic bacteria) (**figure 15.10**).

15.7 Domain: A Higher Level of Classification

- The domain Bacteria contains prokaryotic organisms in the kingdom Bacteria. These single-celled organisms are the most abundant organisms on earth, and play key roles in ecology.

- The domain Archaea contains prokaryotic organisms in the kingdom Archaea. Although they are prokaryotes, archaea are as different from bacteria as they are from eukaryotes. These single-celled organisms are found in diverse environments including very extreme environments.

- The domain Eukarya contains very diverse organisms from four kingdoms but are similar in that they are all eukaryotes. Fungi, plants, and animals are multicellular organisms, and the protists are primarily single-celled but very diverse. Eukaryotes contain cellular organelles, some that were most likely acquired through endosymbiosis, the process shown here from **figure 15.11.**

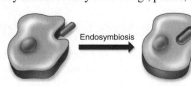

Self-Test

1. The wolf, domestic dog, and red fox are all in the same family, Canidae. The scientific name for the wolf is *Canis lupus,* the domestic dog is *Canis familiaris,* and the red fox is *Vulpes vulpes.* This means that
   a. the red fox is in the same family, but different genus than dogs and wolves.
   b. the dog is in the same family, but different genus than red fox and wolves.
   c. the wolf is in the same family, but different genus than dogs and red foxes.
   d. all three organisms are in different genera.

2. The evolutionary relationship of organisms, and their relationships to other species, is called
   a. taxonomy.   c. ontogeny.
   b. phylogeny.   d. systematics.

3. Organisms are classified based on
   a. physical, behavioral, and molecular characteristics.
   b where the organism lives.
   c. what the organism eats.
   d. the size of the organism.

4. Organisms that are closer together on a cladogram
   a. are in the same family.
   b. comprise an outgroup.
   c. share a more recent common ancestor than those organisms that are farther apart.
   d. share fewer derived characters than organisms that are farther apart.

5. The six kingdoms of organisms can be organized into three domains based on
   a. where the organism lives.
   b. what the organism eats.

   c. how the organism moves.
   d. cell structure and DNA sequence.

6. All of the extremophiles belong to the domain of
   a. Bacteria.   c. Prokarya.
   b. Archaea.   d. Eukarya.

7. Bacteria are similar to Archaea in that they
   a. arose through endosymbiosis.
   b. are multicellular.
   c. live in extreme environments.
   d. are prokaryotes.

8. It is theorized that the ancestral organisms that gave rise to the plants, animals, and fungi originated in the kingdom
   a. Bacteria.
   b. Archaea.
   c. Protista.
   d. All of the above, each giving rise to one of the three kingdoms listed.

9. One difference between the kingdom Protista and the other three kingdoms in the domain Eukarya is that the other kingdoms are mostly
   a. prokaryotic.   c. eukaryotic.
   b. multicellular.   d. unicellular.

10. It is thought that mitochondrion and chloroplast organelles found in eukaryotic cells came from
   a. development of the internal membrane system.
   b. protists.
   c. mutation.
   d. endosymbiosis of bacteria.

Visual Understanding

1. **Figure 15.6** Which of the organisms shown have amniotic membranes that surround the fetus? (If necessary, go back to the figure in the chapter to clearly see it.)

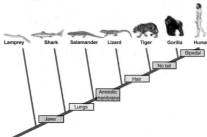

2. **Figure 15.10** Compared to the bacteria and archaea, how similar are animals and plants?

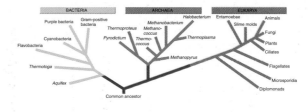

Challenge Questions

1. A friend wants to know what the big deal is—everyone knows that a rose is a rose, why bother with all the fancy Latin stuff, like *Rosa odorata?* What do you tell him?

2. Why are birds classified so differently in traditional phylogeny and in cladistics?

3. If we already have organisms divided into kingdoms, why do we also need domains?

Evolution of Microbial Life

Origin of Life

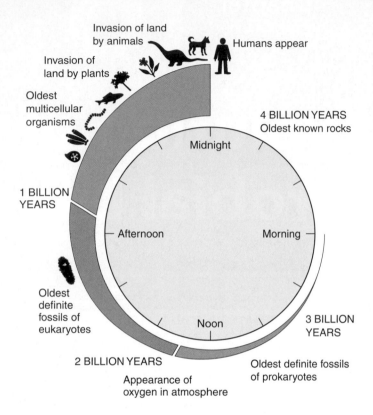

Figure 16.1 A clock of biological time.

A billion seconds ago, most students using this text had not yet been born. A billion minutes ago, Jesus was alive and walking in Galilee. A billion hours ago, the first modern humans were beginning to appear. A billion days ago, the ancestors of humans were beginning to use tools. A billion months ago, the last dinosaurs had not yet been hatched. A billion years ago, no creature had ever walked on the surface of the earth.

IMPLICATION On average, the heart of an American adult male beats 70 times a minute, that of a female slightly faster, at 75 beats a minute. How many times has your heart beat since you were born? How old will you be when it has beat 1 billion times?

16.1 How Cells Arose

CONCEPT PREVIEW: Life appeared on earth 2.5 billion years ago. It may have arisen spontaneously, although the nature of the process is not clearly understood. Little is known about how the first cells originated.

All living organisms are constructed of the same four kinds of macromolecules discussed in chapter 3. They are the building blocks of a cell, like the bricks and mortar of a building. Where the first macromolecules came from and how they came to be assembled together into cells are among the least understood questions in biology—questions that address the very origin of life itself.

No one knows for sure where the first organisms (thought to be like today's bacteria) came from. It is not possible to go back in time and watch how life originated, nor are there any witnesses.

In this chapter we will attempt to understand whether the forces of evolution could have led to the origin of life and, if so, how the process might have occurred. This is not to say that evolution is definitely the source of life on earth. Nothing rules out the possibility that life did not originate on earth at all but instead was carried to it, perhaps as an extraterrestrial infection of spores originating on a planet of a distant star. Or, life-forms may have been put on earth by supernatural or divine forces. In Biology we limit the scope of our inquiry to scientific matters. Of these possibilities, only evolution permits testable hypotheses to be constructed and so provides the only scientific explanation—that is, one that could potentially be disproved by experiment.

Forming Life's Building Blocks

If we look at the development of living organisms as a 24-hour clock of biological time shown in figure 16.1, with the formation of the earth 4.5 billion years ago being midnight, the first cells appear just before noon, but humans do not appear until the day is almost all over, only minutes before its end. How then can we learn about the origin of the first cells? One way is to try to reconstruct what the earth was like when life originated 2.5 billion years ago. We know from rocks that were forming at that time that there was little or no oxygen in the earth's atmosphere then and more of the hydrogen-rich gases hydrogen sulfide (H_2S), ammonia (NH_3), and methane (CH_4). Electrons in these gases would have been frequently pushed to higher energy levels by photons crashing into them from the sun or by electrical energy in lightning. Today, high-energy electrons are quickly soaked up by the oxygen in earth's atmosphere (the atmosphere or air is 21% oxygen, all of it contributed by photosynthesis) because oxygen atoms have a great "thirst" for such electrons. But in the absence of oxygen, high-energy electrons would have been free to help form biological molecules.

Electrons circulate around the nucleus of an atom in energy levels, as discussed on page 33. When an electron moves to a higher energy level its potential energy increases. This potential energy can be used to build molecules through the formation of covalent bonds.

When the scientists Stanley Miller and Harold Urey reconstructed the oxygen-free atmosphere of the early earth in their laboratory and subjected it to the lightning and UV radiation it would have experienced then, they found that many of the building blocks of organisms, such as amino acids and nucleotides, formed spontaneously. They concluded that life may have evolved in a "primordial soup" of biological molecules formed in the ancient earth's oceans.

Recently, concerns have been raised regarding the "primordial soup" hypothesis as the origin of life on earth. If the earth's atmosphere had no oxygen soon after it was formed, as Miller and Urey assumed (and most evidence supports this assumption), then there would have been no protective layer of ozone to shield the earth's surface from the sun's damaging UV radiation. Without an ozone layer, scientists think UV light would have destroyed any ammonia and methane in the atmosphere. When these gases are missing, the Miller-Urey experiment does not produce key biological molecules such as amino acids. If the necessary ammonia and methane were not in the atmosphere, where were they?

The bubble model, shown in figure 16.2, proposes that the key chemical processes generating the building blocks of life took place not in a primordial soup but rather within bubbles on the ocean's surface. The bubble model solves a key problem with the primordial soup hypothesis. Inside the bubbles, methane and ammonia would have been protected from destruction by UV radiation.

The First Cells

We don't know how the first cells formed, but most scientists suspect they aggregated spontaneously. When complex carbon-containing macromolecules are present in water, they tend to gather together, sometimes forming aggregations big enough to see without a microscope. Try vigorously shaking a bottle of oil-and-vinegar salad dressing—tiny bubbles called **microspheres** form spontaneously, suspended in the vinegar. Similar microspheres might have represented the first step in the evolution of cellular organization. Such microspheres have many cell-like properties—their outer boundary resembles the membranes of a cell in that it has two layers (see figure 4.9), and the microspheres can increase in size and divide. Over millions of years, those microspheres better able to incorporate molecules and energy would have tended to persist longer than others.

Scientists suspect that the first macromolecules to form were RNA molecules, which can behave as enzymes and catalyze their own assembly. Eventually DNA may have taken the place of RNA as the storage molecule for genetic information because the double-stranded DNA would have been more stable than single-stranded RNA.

As you can see, the scientific vision of life's origin is at best a hazy outline. Many different scenarios seem possible, and some have solid support from experiments. How life might have originated naturally and spontaneously remains a subject of intense interest, research, and discussion among scientists.

Concept Check

1. Name three possible sources for the origin of life.
2. What objection has been raised to the "primordial soup" hypothesis?
3. Which is thought to have come first, DNA or RNA?

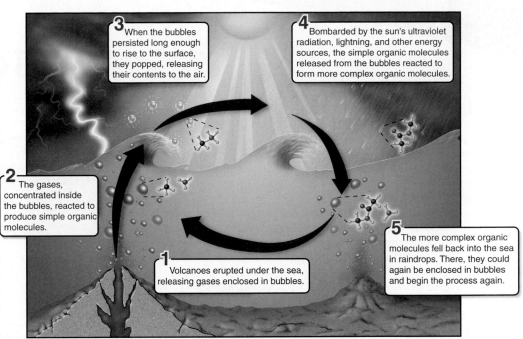

Figure 16.2   A chemical process involving bubbles may have preceded the origin of life.

In 1986, geophysicist Louis Lerman proposed that the chemical processes leading to the evolution of life took place within bubbles on the ocean's surface.

EVOLUTION

Where do new kinds of genes come from?

However life's first cell originated, it probably happened only once—the first living cell was the ancestor to all of life on earth. It seems a fair question, then, to ask where genes came from that appear only later in evolutionary history. What is the origin of novel genes? A very clear example of this issue is seen with nylon-eating bacteria. Nylon is a tough, flexible, and durable chemical invented by a DuPont chemist named Wallace Carothers in 1935. This chemical had never been present on earth before, and yet the factory ponds that held leftover nylon fragments soon formed scummy mats of bacteria. Containing only water and nylon, nothing should have been able to live and grow in these ponds—but somehow these bacteria were able to "eat" nylon! Researchers discovered the bacteria possessed an enzyme capable of breaking down nylon. They had evolved "nylonase." Where did the enzyme come from? It turns out the nylonase enzyme was created in two steps: first, a gene was duplicated that had another, quite different function; then second, a single extra base was inserted into one copy. This one-base insertion shifted the reading frame of the gene, creating an entirely new protein able to bind and break the chemical bonds of nylon. Bacteria with the new gene could thrive on the wastewater ponds, living on this novel source of food.

Has Life Evolved Elsewhere?

We should not overlook the possibility that life processes might have evolved in different ways on other planets. A functional genetic system, capable of accumulating and replicating changes and thus of adaptation and evolution, could theoretically evolve from molecules other than carbon, hydrogen, nitrogen, and oxygen in a different environment. Silicon, like carbon, needs four electrons to fill its outer energy level, and ammonia is even more polar than water. Perhaps under radically different temperatures and pressures, these elements might form molecules as diverse and flexible as those carbon has formed on earth.

The universe has 10^{20} (100,000,000,000,000, 000,000) stars similar to our sun. We don't know how many of these stars have planets, but it seems increasingly likely that many do. Since 1996, astronomers have been detecting planets orbiting distant stars, more than 300 at last count. At least 10% of stars are thought to have planetary systems. If only 1 in 10,000 of these planets is the right size and at the right distance from its star to duplicate the conditions in which life originated on earth, the "life experiment" will have been repeated 10^{15} times (that is, a million billion times). It does not seem likely that we are alone.

A dull gray chunk of rock collected in 1984 in Antarctica ignited an uproar about ancient life on Mars with the report that the rock contains evidence of possible life. Analysis of gases trapped within small pockets of the rock indicate it is a meteorite from Mars. It is, in fact, the oldest rock known to science—fully 4.5 billion years old. Evidence collected by the 2004 NASA Mars mission (the photo below of the Martian surface was taken by the rover Spirit) suggests that the surface, now cold and arid, was much warmer when the Antarctic meteorite formed 4.5 billion years ago, that water flowed over its surface, and that it had a carbon dioxide atmosphere—conditions not too different from those that spawned life on earth.

When examined with powerful electron microscopes, carbonate patches within the meteorite exhibit what look like microfossils, some 20 to 100 nanometers in length. One hundred times smaller than any known bacteria, it is not clear they actually are fossils, but the resemblance to bacteria is striking.

Viewed as a whole, the evidence of bacterial life associated with the Mars meteorite is not compelling. Clearly, more painstaking research remains to be done before the discovery can claim a scientific consensus. However, while there is no conclusive evidence of bacterial life associated with this meteorite, it seems very possible that life has evolved on other worlds in addition to our own.

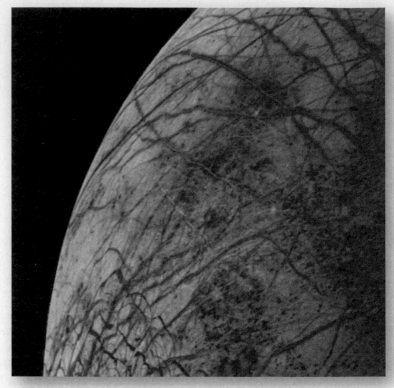

Europa—a large moon of Jupiter.

There are planets other than ancient Mars with conditions not unlike those on earth. Europa, a large moon of Jupiter, is a promising candidate (photo above). Europa is covered with ice, and photos taken in close orbit in the winter of 1998 reveal seas of liquid water beneath a thin skin of ice. Additional satellite photos taken in 1999 suggest that a few miles under the ice lies a liquid ocean of water larger than earth's, warmed by the push and pull of the gravitational attraction of Jupiter's many large satellite moons. The conditions on Europa now are far less hostile to life than the conditions that existed in the oceans of the primitive earth. In coming decades, satellite missions are scheduled to explore this ocean for life.

Mars landscape.

Prokaryotes

16.2 The Simplest Organisms

CONCEPT PREVIEW: Prokaryotes are the smallest and simplest organisms, single cells with no internal compartments or organelles. They divide by binary fission.

Judging from fossils in ancient rocks, prokaryotes have been plentiful on earth for over 2.5 billion years. From the diverse array of early living forms, a few became the ancestors of the great majority of organisms alive today. Several ancient forms including cyanobacteria have survived; others gave rise to different members of the diverse prokaryotic group Bacteria, and still others to the second group of prokaryotes, the Archaea. The fossil record indicates that eukaryotic cells, being much larger than prokaryotes and exhibiting elaborate shapes in some cases, did not appear until about 1.5 billion years ago. For at least 1 billion years, prokaryotes were the only organisms that existed.

Today, prokaryotes are the simplest and most abundant form of life on earth. In a spoonful of farmland soil, 2.5 billion bacteria may be present. In one hectare (about 2.5 acres) of wheat land in England, the weight of bacteria in the soil is approximately equal to that of 100 sheep!

It is not surprising, then, that prokaryotes occupy a very important place in the web of life on earth. They play a key role in cycling minerals within the earth's ecosystems. In fact, photosynthetic bacteria were in large measure responsible for the introduction of oxygen into the earth's atmosphere. Pathogenic bacteria are responsible for some of the most deadly animal and plant diseases, including many human diseases. Bacteria and archaea are our constant companions, present in everything we eat and on everything we touch.

The Structure of a Prokaryote

The essential character of prokaryotes can be conveyed in a simple sentence: **Prokaryotes** are small, simply organized, single cells that lack an organized nucleus. Therefore, bacteria and archaea are prokaryotes; their single circle of DNA is not confined by a nuclear membrane in a nucleus, as in the cells of eukaryotes. Table 16.1 presents other ways in which prokaryotes differ from eukaryotes. Too tiny to see with the naked eye, bacterial cells are simple in form, either rod-shaped (bacilli), spherical (cocci), or spirally coiled (spirilla). A few kinds of bacteria aggregate into stalked structures or filaments.

In most cases, the prokaryotic cell's plasma membrane is surrounded by a cell wall. The cell wall of bacteria are different than the cell walls of archaea, and those found in eukaryotic cells (protists, fungi, and plants). Bacterial cell walls are all composed of the same material, chemicals called peptidoglycan, but some bacteria cell walls are thicker, having more layers of peptidoglycan, than others. Those with thin cell walls have an outer membrane surrounding the cell that appears to protect them from attack by chemicals, such as the antibiotic penicillin. Those with thick cell walls lack an outer membrane. Many bacteria also have a gelatinous layer outside the cell wall, called a capsule.

Bacterial cell wall thickness and complexity was first discovered by a Danish researcher, Hans Gram, who developed a staining process that stains the cell types differently. Bacteria that possess a thinner cell wall stain red and are called Gram-negative bacteria. Those with thicker cell walls stain purple and are called Gram-positive bacteria. Today, scientists and medical doctors alike still refer to bacteria as either Gram-positive or

TABLE 16.1	Prokaryotes Compared to Eukaryotes

Internal compartmentalization. Unlike eukaryotic cells, prokaryotic cells contain no internal compartments, no internal membrane system, and no cell nucleus.

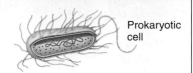
Prokaryotic cell

Cell size. Most prokaryotic cells are only about 1 micrometer in diameter, whereas most eukaryotic cells are well over 10 times that size.

 Prokaryotic cell   Eukaryotic cell

Unicellularity. All prokaryotes are fundamentally single-celled. Even though some may adhere together in a matrix or form filaments, their cytoplasms are not directly interconnected, and their activities are not integrated and coordinated, as is the case in multicellular eukaryotes.

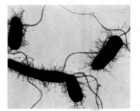
Unicellular bacteria

Chromosomes. Prokaryotes do not possess chromosomes in which proteins are complexed with the DNA, as eukaryotes do. Instead, their DNA exists as a single circle in the cytoplasm.

Prokaryotic chromosome   Eukaryotic chromosomes

Cell division. Cell division in prokaryotes takes place by binary fission (see chapter 8). The cells simply pinch in two. In eukaryotes, microtubules pull chromosomes to opposite poles during the cell division process, called mitosis.

Binary fission in prokaryotes   Mitosis in eukaryotes

Flagella. Prokaryotic flagella are simple, composed of a single fiber of protein that spins like a propeller. Eukaryotic flagella are more complex structures, with a 9 + 2 arrangement of microtubules, that whip back and forth rather than rotate.

Simple bacterial flagellum

Metabolic diversity. Prokaryotes possess many metabolic abilities that eukaryotes do not: prokaryotes perform several different kinds of anaerobic and aerobic photosynthesis; prokaryotes can obtain their energy from oxidizing inorganic compounds (so-called chemoautotrophs); and prokaryotes can fix atmospheric nitrogen.

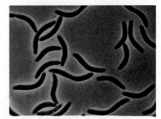
Chemoautotrophs

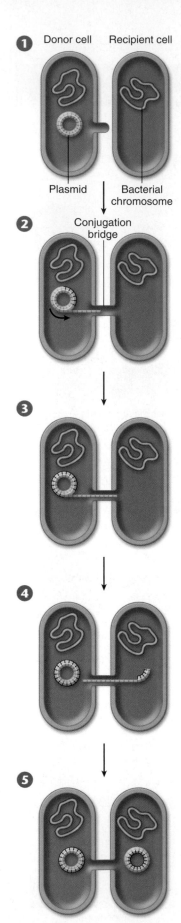

Figure 16.3 **Bacterial conjugation.**

Donor cells contain a plasmid that recipient cells lack. The pilus of a donor cell contacts a recipient cell and draws the cells closer. The plasmid replicates itself and transfers the copy across a conjugation bridge. The remaining strand of the plasmid serves as a template to build a replacement. When the single strand enters the recipient cell, it serves as a template to assemble a double-stranded plasmid. When the process is complete, both cells contain a complete copy of the plasmid.

Gram-negative which implies certain relationships among the bacterial species and indicates the type of antibiotic that might be an effective treatment and the environmental conditions in which they may be found.

Many kinds of bacteria possess threadlike **flagella,** long strands of protein that may extend out several times the length of the cell body. Bacteria swim by twisting these flagella in a corkscrew motion as shown below. Some bacteria also possess shorter outgrowths called **pili** (singular, **pilus**), which act as docking cables, helping the cell to attach to surfaces or to other cells.

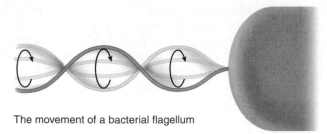

The movement of a bacterial flagellum

When exposed to harsh conditions (dryness or high temperature), some bacteria form thick-walled **endospores** around their DNA and a small bit of cytoplasm. These endospores are highly resistant to environmental stress and may germinate to form new active bacteria even after centuries. The formation of endospores is the reason that the bacterium responsible for botulism, *Clostridium botulinum,* sometimes persists in cans and bottles if the containers have not been heated at a high enough temperature to kill the spores.

How Prokaryotes Reproduce

Prokaryotes, like all other living cells, grow and divide. Prokaryotes reproduce using a process called **binary fission,** in which an individual cell simply increases in size and divides in two. Following replication of the cell's DNA, the plasma membrane and cell wall grow inward and eventually divide the cell by forming a new wall from the outside toward the center of the old cell.

Some bacteria can exchange genetic information by passing plasmids from one cell to another in a process called **conjugation.** A plasmid is a small, circular fragment of DNA that replicates outside the main bacterial chromosome. In bacterial conjugation, seen in figure 16.3, the pilus of a donor cell contacts a recipient cell ❶; the pilus draws the two cells close together. A passageway called a *conjugation bridge* forms between the two cells. The plasmid in the donor cell begins to replicate its DNA ❷, passing the replicated copy out across the bridge into the recipient cell ❸, where a complementary strand is synthesized ❹. The remaining strand in the donor cell serves as a template for the synthesis of its complement. Both the recipient cell and the donor cell now contain a complete copy of the genetic material found in the plasmid of the donor cell ❺. Genes that produce antibiotic resistance in bacteria are often transferred from one bacterial cell to another through conjugation.

Concept Check

1. How old are the earliest fossils in the fossil record? Might life be older?
2. Name four of the nine ways prokaryotes differ from eukaryotes.
3. Are recipient and donor copies of the plasmid identical after conjugation?

Viruses

16.3 Viruses Infect Organisms

CONCEPT PREVIEW: Viruses are genomes of DNA or RNA, encased in a protein shell, that can infect cells and replicate within them. They are chemical assemblies, not cells, and are not alive. Viruses are responsible for some of the most lethal diseases of humans.

The border between the living and the nonliving is very clear to a biologist. Living organisms are cellular and able to grow and reproduce independently, guided by information encoded within DNA. As discussed earlier in this chapter, the simplest creatures living on earth today that satisfy these criteria are prokaryotes. Viruses, on the other hand, do not satisfy the criteria for "living" because they possess only a portion of the properties of organisms. **Viruses** are literally "parasitic" chemicals, segments of DNA (or sometimes RNA) wrapped in a protein coat. They cannot reproduce on their own, and for this reason they are not considered alive by biologists. They can, however, replicate within cells, using the host cell's DNA replication and gene expression enzymes and ribosomes.

> Viruses don't contain cellular structures needed for proteins synthesis (such as certain enzymes, ribosomes, nucleotides, amino acids). As discussed in chapter 12, all of these are necessary for cell function. Viruses reproduce by using the host cell's structures and processes.

Viruses are very small. The smallest viruses are only about 17 nanometers in diameter; the largest ones are long filamentous viruses that measure up to 1,000 nanometers in length. Viruses are so small that they are smaller than many of the molecules in a cell. Most viruses can be detected only by using the higher resolution of an electron microscope.

Each virus is in fact a mixture of two chemicals: nucleic acid and protein. The tobacco mosaic virus (TMV) pictured in figure 16.4*a* has the structure of a Twinkie, a tube made of an RNA core (the green springlike structure) surrounded by a coat of protein (the purple structures that encircle the RNA). Researchers were able to separate the RNA from the protein and purify and store each chemical. Later, when they mixed the protein and RNA, the two components reassembled into an infectious unit fully able to infect healthy tobacco plants. Because it cannot reproduce on its own, a virus is best regarded as a chemical assembly rather than a living organism.

Viruses infect all organisms, from bacteria to humans, and in every case the basic structure of the virus is the same, a core of nucleic acid surrounded by protein. All viruses reproduce by first disintegrating into their component nucleic acid and protein molecules within a cell. Viruses then use the cell's machinery to manufacture new components and reform them into a new generation of virus particles.

There is considerable difference, however, in the details of virus structure. In figure 16.4 you can compare the structure of plant, bacterial, and animal viruses—they are clearly quite different from one another and there is even a wide variety of shapes and structures within each group of viruses. Bacterial viruses, called *bacteriophages,* can have elaborate structures like the bacteriophage in figure 16.4*b*, that looks more like a lunar module than a virus. Many plant viruses like TMV have a core of RNA, and some animal viruses like HIV (figure 16.4*c*) do too. One or more different segments of DNA may be present in other animal virus particles, along with many different kinds of protein. Like TMV, most viruses form a protein sheath, or *capsid,* around their nucleic acid core. In addition, many viruses form a membranelike *envelope,* rich in proteins and lipids, around the capsid.

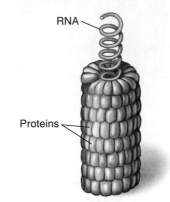

RNA

Proteins

(a) Tobacco mosaic virus

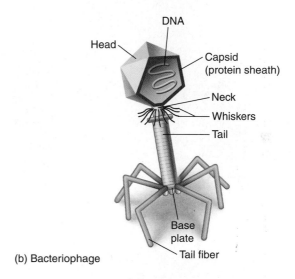

DNA

Head

Capsid (protein sheath)

Neck

Whiskers

Tail

Base plate

(b) Bacteriophage

Tail fiber

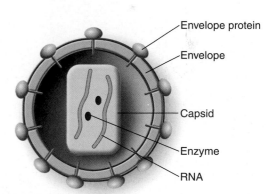

Envelope protein

Envelope

Capsid

Enzyme

RNA

(c) Human immunodeficiency virus

Figure 16.4 **The structure of plant, bacterial, and animal viruses.**

(a) TMV infects plants and consists of 2,130 identical protein molecules (*purple*) that form a cylindrical coat around the single strand of RNA (*green*). The RNA backbone determines the shape of the virus and is protected by the identical protein molecules packed tightly around it. (b) Bacterial viruses, called bacteriophages, often have a complex structure. (c) In the human immunodeficiency virus (HIV), the RNA core is held within a capsid that is encased by a protein envelope.

Birds are a natural reservoir for the influenza virus, which can then spread to humans and possibly between humans.

Chimpanzees passed a virus like HIV to humans, but the virus most likely originated in monkeys.

The Chinese horseshoe bat spreads the SARS virus without being affected itself.

Birds are the natural reservoir for the West Nile virus, but it spreads to humans through mosquitoes.

The Origin of Viral Diseases

Sometimes viruses that originate in one organism pass to another, causing a disease in the new host. For example, influenza is fundamentally a bird virus, and smallpox is thought to have passed from cattle to humans when cows were first domesticated.

Influenza.   Perhaps the most lethal virus in human history has been the influenza virus. Between 40 and 100 million worldwide died of flu within 19 months in 1918 and 1919—an astonishing number. The natural reservoir of influenza virus is in ducks, chickens, and pigs in Central Asia.

Influenza virus

AIDS (HIV, human immunodeficieny virus).   The virus that causes AIDS first entered humans from chimpanzees somewhere in Central Africa, probably between 1910 and 1950, and spread rapidly, mostly through sexual contact with an infected person. Where did chimpanzees acquire a virus like HIV? These types of viruses are rampant in African monkeys, and from studies of the nucleotide sequences of monkey immunodeficieny viruses, it seems certain that chimpanzees acquired the viruses from monkeys they ate.

Ebola virus.   Among the most lethal of emerging viruses are a collection of filamentous Ebola viruses arising in Central Africa that attack human connective tissue. With lethality rates in excess of 50%, these so-called filoviruses cause some of the most lethal infectious diseases known. One strain of Ebola virus has exhibited lethality rates in excess of 90% in isolated outbreaks in Central Africa. Luckily, victims die too fast to spread the disease very far. Extensive searches among wild and domestic animals have failed to reveal for certain the identity of the natural host of Ebola virus, although fruit bats are likely suspects.

Ebola virus

SARS.   A recently emerged strain of coronavirus was responsible for a worldwide outbreak in 2003 of *severe acute respiratory syndrome* (SARS), a respiratory infection with pneumonia-like symptoms that in over 8% of cases is fatal. Virologists in 2005 identified the Chinese horseshoe bat as the natural host of the SARS virus. Because these bats are healthy carriers not sickened by the virus, and occur commonly throughout Asia, it will be difficult to prevent future outbreaks.

SARS virus

West Nile Virus.   The mosquito-borne West Nile virus, carried by infected crows and other birds, spread across the country a few years ago, with 4,156 cases at its peak in 2002, 284 of whom died. By 2005 the wave of infection had greatly lessened. First detected in humans in Uganda, Africa, in 1937, the virus is common among birds, and is thought to have been transmitted to humans by mosquitoes that had previously bitten infected birds, much as it is being transmitted now. Earlier spread of the virus through Europe also abated after several years.

Concept Check

1. Name the two types of macromolecules found in TMV virus?
2. Is a virus alive? Can it evolve? Defend your answer.

Biology and Staying Healthy

Bird and Swine Flu

The influenza virus has been one of the most lethal viruses in human history. Flu viruses are animal RNA viruses containing 11 genes. An individual flu virus resembles a sphere studded with spikes composed of two kinds of protein. Different strains of flu virus, called subtypes, differ in their protein spikes. One of these proteins, hemagglutinin (H), aids the virus in gaining access to the cell interior. The other, neuraminidase (N), helps the daughter virus break free of the host cell once virus replication has been completed. Flu viruses are currently classified into 13 distinct H subtypes and 9 distinct N subtypes, each of which requires a different vaccine to protect against infection. The influenza virus subtype that caused the Hong Kong flu epidemic of 1968 has type 3 H molecules and type 2 N molecules, and is called H3N2.

How New Flu Strains Arise

Worldwide epidemics of the flu in the last century have been caused by shifts in flu virus H-N combinations as mutation creates new versions of H and N. The "killer flu" of 1918, H1N1, thought to have passed directly from birds to humans, killed between 40 and 100 million people worldwide. The Asian flu of 1957, H2N2, killed over 100,000 Americans, and the Hong Kong flu of 1968, H3N2, killed 70,000 Americans.

It is no accident that new strains of flu usually originate in the Far East. The most common hosts of influenza virus are ducks, chickens, and pigs, which in Asia often live in close proximity to each other and to humans. Pigs are subject to infection by both bird and human strains of the virus, and individual animals are often simultaneously infected with multiple strains. This creates conditions favoring genetic recombination between strains, as illustrated above, sometimes putting together novel combinations of H and N spikes unrecognizable by human immune defenses specific for the old configuration. The Hong Kong flu, for example, arose from recombination between H3N8 from ducks and H2N2 from humans. The new strain of influenza, in this case H3N2, then passed back to humans, creating an epidemic because the human population had never experienced that H-N combination before.

Conditions for a Pandemic

Not every new strain of influenza creates a worldwide flu epidemic. Three conditions are necessary: (1) The new strain must contain a novel combination of H and N spikes, so that the human population has no significant immunity from infection; (2) The new strain must be able to replicate in humans and cause death—many bird influenza viruses are harmless to people because they cannot multiply in human cells; (3) The new strain must be efficiently transmitted between humans. The H1N1 killer flu of 1918 spread in water droplets exhaled from infected individuals and subsequently inhaled into the lower respiratory tract of nearby people.

The new strain need not be deadly to every infected person in order to produce a pandemic—the H1N1 flu of 1918 had an overall mortality rate of only 2%, and yet killed 40–100 million people.

Recombination within humans A person infected with a flu virus can become infected with another type of flu virus by direct contact with birds. The two viruses can undergo genetic recombination to produce a third type of virus, which can spread from human to human.

Recombination within pigs Pigs can contract flu viruses from both birds and humans. The flu viruses can undergo genetic recombination in the pig, to produce a new kind of flu virus, which can spread from pigs to humans.

Why did so many die? Because so much of the world's population was infected.

Bird Flu

A potentially deadly new strain of flu virus emerged in Hong Kong in 1997, H5N1. Like the 1918 pandemic strain, H5N1 passes to humans directly from infected birds, usually chickens or ducks, and for this reason has been dubbed "bird flu." Bird flu has gotten an unusual amount of public attention, because it satisfies the first two conditions of a pandemic. Bird flu is a novel combination of H and N spikes for which humans have little immunity, and the resulting strain is particularly deadly. Of 256 individuals infected by handling birds by the end of 2006, 151 died of the infection, a mortality of 59% (recall, the mortality of the 1918 strain was only 2%). Fortunately, the third condition for a pandemic is not yet met: The H5N1 strain of flu virus does not spread easily from person to person, and the number of human infections remains small.

Swine Flu

A second potentially pandemic form of flu virus emerged in Mexico in 2009, H1N1, first passed to humans from infected pigs. A descendant of the 1918 H1N1 virus, this flu (dubbed "swine flu") seems to pass just as easily from person to person and has quickly spread around the world. Also like the 1918 virus, the initial wave of 2009 H1N1 swine flu infection triggers only mild symptoms in most people. The 1918 H1N1 virus only became deadly in a second wave of infection, after the virus had better adapted to living in the human body. Public health officials are watching the 2009 H1N1 flu carefully, for fear that a subsequent wave of infection may too become more lethal.

The Protists

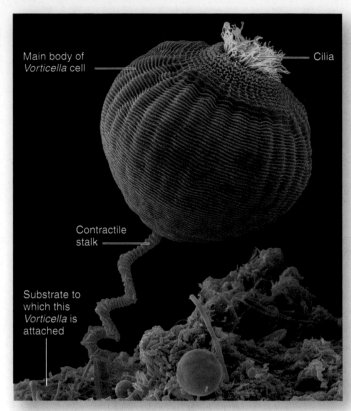

Main body of *Vorticella* cell

Cilia

Contractile stalk

Substrate to which this *Vorticella* is attached

Figure 16.5   A unicellular protist.

The protist kingdom is a catch-all kingdom for many different groups of unicellular organisms, such as this *Vorticella* (phylum Ciliophora), which is heterotrophic, feeding on bacteria, and has a contractible stalk.

BIOLOGY & YOU

Traveler Beware. If you travel to an under-developed country, particularly one where sanitation is poor, you would be well-advised to drink only bottled water. Over 10 million people each year get "traveler's diarrhea" from drinking fecally contaminated water. Usually this means a few days of cramps and frequent trips to the bathroom, but 80% of the time antibiotics quickly eliminate the bacterial infection. In some 20% of the cases, however, the problem is not so simply solved, because the infecting agent is a protist, not a bacterium, and a lot harder to kill. Giardia attaches to the inside of the small intestine, while cryptosporidium (a relative of malaria) and amoebic parasites each form dormant cysts. All three protists leave the body in feces which, when contaminating drinking water, spread the infection to others. All three protists cause serious diseases that are more difficult to treat than bacterial infections. Cysts make treatment particularly difficult, because dormant cysts are resistant to antibiotics (most of which work by interfering with metabolism or genome replication). In a sign of the medical importance of amoebic dysentery, which infects 50 million people worldwide, the amoebic parasite (*Entamoeba histolytica*) was one of the first protists to have its genome completely sequenced.

16.4 General Biology of Protists

CONCEPT PREVIEW: Protists exhibit a wide range of forms, locomotion, nutrition, and reproduction. Their cells form clusters with varying degrees of specialization.

As discussed in chapter 15, animals, plants, fungi, and protists are all eukaryotes. **Protists** are the most ancient eukaryotes and are united on the basis of a single negative characteristic: They are not fungi, plants, or animals. In all other respects, they are highly variable with no uniting features. Many are unicellular, like the *Vorticella* you see in figure 16.5 with its contractible stalk, but there are numerous colonial and multicellular groups. Most are microscopic, but some are as large as trees.

The Cell Surface

Protists possess varied types of cell surfaces. All protists have plasma membranes, but some protists, like algae and molds, are additionally encased within strong cell walls. Still others, like diatoms and radiolarians, secrete glassy shells of silica.

Locomotor Organelles

Movement in protists is also accomplished by diverse mechanisms. Protists move by flagella, cilia, pseudopods, or gliding mechanisms. Many protists wave one or more flagella to propel themselves through the water, whereas others use banks of short, flagella-like structures called cilia to create water currents for their feeding or propulsion. Among amoeba large, blunt extensions of the cell body called pseudopodia are the chief means of locomotion (see figure 16.8). Other related protists extend thin, branching protrusions, sometimes supported by axial rods of microtubules. These so-called axopodia can be extended or retracted. Because the tips can adhere to adjacent surfaces, the cell can move by a rolling motion, shortening the axopodia in front and extending those in the rear.

Cyst Formation

Many protists with delicate surfaces are successful in quite harsh habitats. How do they manage to survive so well? They survive inhospitable conditions by forming **cysts.** A cyst is a dormant form of a cell with a resistant outer covering in which cell metabolism is more or less completely shut down. Amoebic parasites in humans form cysts that are quite resistant to gastric acidity.

Nutrition

Protists employ every form of nutritional acquisition except chemoautotrophy (the ability of an organism to use energy from chemical reactions to make its own food), which has so far been observed only in prokaryotes. Some protists are photosynthetic autotrophs (using energy from the sun to make food as plants do) and are called **phototrophs.** Others are heterotrophs that obtain energy from organic molecules synthesized by other organisms. Among heterotrophic protists, those that ingest visible particles of food are called phagotrophs, or holozoic feeders. Those ingesting food in soluble form are called osmotrophs, or saprozoic feeders.

Phagotrophs ingest food particles into intracellular vesicles called *food vacuoles,* or phagosomes. Lysosomes fuse with the food vacuoles, introducing enzymes that digest the food particles within. The food vacuoles shrink as the digested molecules are absorbed across their membranes.

Reproduction

Protists typically reproduce asexually, reproducing sexually only in times of stress. Asexual reproduction involves mitosis, but the process is often somewhat different from the mitosis that occurs in multicellular animals, discussed in chapter 8. The nuclear membrane, for example, often persists throughout mitosis, with the microtubular spindle forming within it. In some groups, asexual reproduction involves spore formation, and fission in others (figure 16.6*a*). Sexual reproduction occurs only rarely by exchanging nuclei (figure 16.6*b*)

Multicellularity

A single cell has limits. It can only be so big without encountering serious surface-to-volume problems. Said simply, as a cell becomes larger, there is too little surface area for so much volume. The evolution of multicellular individuals composed of many cells solved this problem. **Multicellularity** is a condition in which an organism is composed of many cells, permanently associated with one another, that integrate their activities. This allows the organism to function on a scale and with a complexity that would have been impossible for its unicellular ancestors. The key advantage of multicellularity is that it allows specialization—distinct types of cells, tissues, and organs can be differentiated within an individual's body, each with a different function. With such functional "division of labor" within its body, a multicellular organism can possess cells devoted specifically to protecting the body, others to moving it about, still others to seeking mates and prey, and yet others to carry on a host of other activities. In just this way, a small city of 50,000 inhabitants is vastly more complex and capable than a crowd of 50,000 people in a football stadium—each city dweller is specialized in a particular activity that is interrelated to everyone else's, rather than just being another body in a crowd.

Colonies. A colonial organism is a collection of cells that are permanently associated but in which little or no integration of cell activities occurs. Many protists form colonial assemblies, consisting of many cells with little differentiation or integration. In some protists, the distinction between colonial and multicellular is blurred. For example, in the green algae *Volvox* shown in figure 16.7, individual motile cells aggregate into a hollow ball of cells that moves by a coordinated beating of the flagella of the individual cells—like scores of rowers all pulling their oars in concert. A few cells near the rear of the moving colony are reproductive cells, but most are relatively undifferentiated.

Multicellular Individuals. True multicellularity, in which the activities of the individual cells are coordinated and the cells themselves are in contact, occurs only in eukaryotes and is one of their major characteristics. Three groups of protists have independently attained true but simple multicellularity—the brown algae (phylum Phaeophyta), green algae (phylum Chlorophyta), and red algae (phylum Rhodophyta). In multicellular organisms, individuals are composed of many cells that interact with one another and coordinate their activities.

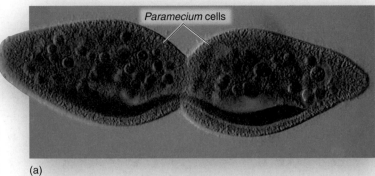

(a)

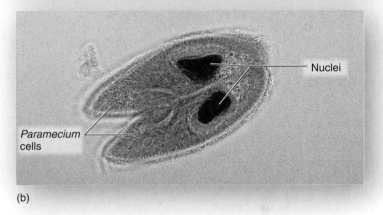

(b)

Figure 16.6 **Reproduction among paramecia.**

(a) When *Paramecium* reproduces asexually, a mature individual divides, and two genetically identical individuals result. (b) In sexual reproduction, two mature cells fuse in a process called conjugation (×100) and exchange haploid nuclei.

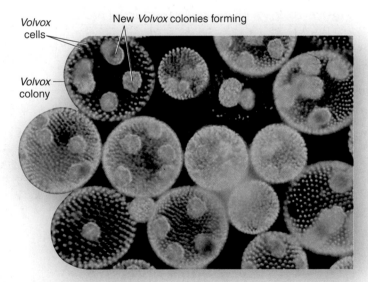

Figure 16.7 **A colonial protist.**

Individual, motile, unicellular green algae are united in the protist *Volvox* as a hollow colony of cells that moves by the beating of the flagella of its individual cells. Some species of *Volvox* have cytoplasmic connections between the cells that help coordinate colony activities. The *Volvox* colony is a highly complex form that has many of the properties of multicellular life.

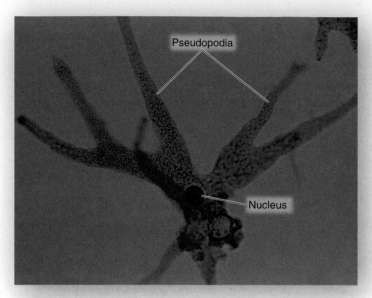

Figure 16.8 An amoeba.

Amoeba proteus is a relatively large amoeba (×45). The projections are pseudopodia; an amoeba moves simply by flowing cytoplasm into them. The nucleus is plainly visible.

16.5 Kinds of Protists

CONCEPT PREVIEW: Protista is an artificially created kingdom containing groups as different from each other as they are from plants or animals.

Protists are the most diverse of the four kingdoms in the domain Eukarya. The 200,000 different forms in the kingdom Protista include many unicellular (figure 16.8), colonial, and multicellular groups (figure 16.9 and table 16.2).

Probably the most important statement we can make about classifying the kingdom Protista is that it is an artificial group; as a matter of convenience, single-celled eukaryotic organisms have typically been grouped together into this kingdom. This lumps many very different and only distantly related forms together, as different from one another as any are to plants or animals.

How Protists Impact Your Life

Protists have major impacts on humanity. Some of humanity's most important diseases are caused by protists. Sleeping sickness, which causes extreme fatigue, is caused by trypanosomes and affects millions of people each year in tropical areas, and makes it impossible to raise domestic cattle in a large portion of Africa. Malaria, caused by the *Plasmodium* parasite, is one of humanity's greatest scourges, killing over a million people each year.

Protists have great economic impact, both negative and positive. The poisonous and destructive "red tides" in coastal areas are caused by population explosions, or "blooms," of red-pigmented dinoflagellates. Diatoms, which are photosynthetic protists with shells of silica, have many industrial applications as grinding materials. Algae are particularly useful. Massive marine algae called kelp that occur in shallow ocean waters throughout the world provide commercial fiber. Sushi rolls are wrapped in nori, a red algae. Polysaccharides from red algae are used to thicken ice cream and cosmetics.

Concept Check

1. Name four different modes of protist locomotion. Are any unique?
2. Explain how some protists survive harsh conditions.
3. Which if any protists are multicellular?

Figure 16.9 A phylogenetic tree for the protists.

This tree shows seven major groups of protists and a tentative array of their relationship to each other. As more information becomes available, these relationships, and the groups, may change.

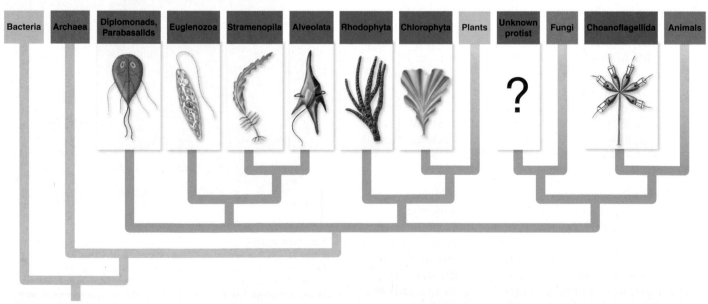

TABLE 16.2 | Kinds of Protists

Group	Phylum	Typical Examples	Key Characteristics
DIPLOMONADS			
Diplomonads	Diplomonadida	*Giardia intestinalis*	Move by flagella; have two nuclei
PARABASALIDS			
Parabasalids	Parabasalida	*Trichomonas vaginalis*	Undulating membrane; some are pathogens, others digest cellulose in gut of termites
EUGLENOZOA			
Euglenoids	Euglenozoa	*Euglena*	Unicellular; some photosynthetic; others heterotrophic; contain chlorophylls *a* and *b* or none
Trypanosomes	Euglenozoa	Trypanosomes	Heterotrophic; unicellular
STRAMENOPILA			
Brown algae	Phaeophyta	Kelp	Multicellular; contain chlorophylls *a* and *c*
Diatoms	Chrysophyta	*Diatoma*	Unicellular; manufacture the carbohydrate chrysolaminarin; unique double shells of silica; contain chlorophylls *a* and *c*
Water molds	Oomycota	*Phytophthora infestans*	Terrestrial and freshwater
ALVEOLATA			
Ciliates	Ciliophora	*Paramecium*	Heterotrophic unicellular protists with cells of fixed shape possessing two nuclei and many cilia; many contain highly complex and specialized organelles
Dinoflagellates	Pyrrhophyta	Red tides	Unicellular; two flagella; contain chlorophylls *a* and *b*
Sporozoans	Apicomplexa	*Plasmodium*	Nonmotile; unicellular; the apical end of the spores contains a complex mass of organelles
RHODOPHYTA			
Red algae	Rhodophyta	Coralline algae	Most multicellular; contain chlorophylls *a* and a red pigment
CHLOROPHYTA			
Green algae	Chlorophyta	*Chlamydomonas, Ulva*	Unicellular or multicellular; contain chlorophylls *a* and *b*; ancestor of plants
CHOANOFLAGELLIDA			
Choanoflagellates	Choanozoa	Choanoflagellates	Flagellated feeding funnel; ancestor of animals
PHYLOGENY NOT YET DETERMINED			
Amoebas	Rhizopoda	*Amoeba*	Move by pseudopodia
Forams	Foraminifera	Forams	Rigid shells; move by protoplasmic streaming
Radiolarians	Actinopoda	Radiolarians	Glassy skeletons; needlelike pseudopods
Cellular slime molds	Acrasiomycota	*Dictyostelium*	Colonial aggregations of individual cells; most closely related to amoebas
Plasmodial slime molds	Myxomycota	*Fuligo*	Stream along as a multinucleate mass of cytoplasm

Fungi

(a)   **(b)**

Figure 16.10   **Mushrooms.**

Most of the body of a fungus is belowground, a network of fine threads that penetrate the ground, often over great distances. Mushrooms are aboveground reproductive structures that release spores into the air, allowing the fungus to invade new habitats. (a) Yellow morel. (b) *Amanita* mushroom.

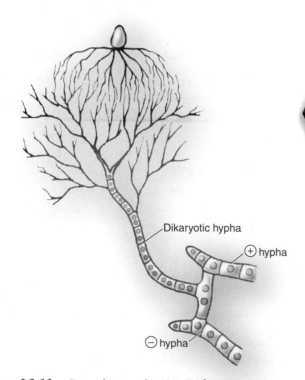

Dikaryotic hypha

⊕ hypha

⊖ hypha

Figure 16.11   **Sexual reproduction in fungi.**

In some fungi, different mating strains ("+" and "–") fuse and form a dikaryotic hypha that grows into a mycelium. The nuclei in certain cells will eventually fuse, forming a diploid zygote.

16.6 A Fungus Is Not a Plant

CONCEPT PREVIEW: Although they are anchored to their substrates, fungi are not at all like plants. The fungal body is basically long strings of cells, often interconnected. Fungi reproduce both asexually and sexually. They obtain nutrients by secreting digestive enzymes into their surroundings and then absorbing the digested molecules back into the fungal body.

The fungi are a distinct kingdom of organisms, comprising about 74,000 named species, some microbial, most multicellular. Mycologists, scientists who study fungi, believe there may be many more species in existence. Although fungi were at one time included in the plant kingdom, they lack chlorophyll and resemble plants only in their general appearance and lack of mobility. Significant differences between fungi and plants include the following:

Fungi are heterotrophs.   Perhaps most obviously, a mushroom is not green because it does not contain chlorophyll. Virtually all plants are photosynthesizers, whereas no fungi carry out photosynthesis. Instead, fungi obtain their food by secreting digestive enzymes onto whatever they are attached to and then absorbing into their bodies the organic molecules that are released by the enzymes.

Fungi have filamentous bodies.   A plant is built of groups of functionally different cells called tissues, with different parts typically made of several different tissues. Fungi by contrast are basically filamentous in their growth form (that is, their bodies consist entirely of cells organized into long, slender filaments called *hyphae*), even though these filaments may be packed together to form a mass, called a *mycelium*.

Fungi have nonmotile sperm.   Some plants have motile sperm with flagella. The majority of fungi do not.

Fungi have cell walls made of chitin.   The cell walls of fungi contain chitin, the same tough material that a crab shell is made of. The cell walls of plants are made of cellulose, also a strong building material. Chitin, however, is far more resistant to microbial degradation than is cellulose.

> The cellulose that makes up plant cell walls and the chitin that makes up fungal cell walls are both polysaccharides, made of glucose subunits. Refer back to table 3.1 on page 55 to see the structural difference between cellulose and chitin.

Fungi have nuclear mitosis.   Mitosis in fungi is different from mitosis in plants, animals, and most protists in one key aspect: The nuclear envelope does not break down and re-form; instead, all of mitosis takes place *within* the nucleus. A spindle apparatus forms there, dragging chromosomes to opposite poles of the *nucleus* (not the cell).

You could build a much longer list, but already the take-home lesson is clear: Fungi are not like plants at all! Their many unique features are strong evidence that fungi are not closely related to any other group of organisms.

The Body of a Fungus

Fungi exist mainly in the form of slender filaments, barely visible with the naked eye, called **hyphae** (singular, **hypha**). A hypha is basically a long string of cells. Many hyphae then associate with each other to form a much larger structure, called a **mycelium** (plural, **mycelia**), like the mushrooms in figure 16.10.

The main body of a fungus is not the mushroom, which is a temporary reproductive structure, but rather the extensive network of fine hyphae that penetrate the soil, wood, or flesh in which the fungus is growing. A mycelium may contain many meters of individual hyphae.

Fungal cells are able to exhibit a high degree of communication within such structures, because although most cells of fungal hyphae are separated by cross-walls called *septa* (singular, *septum*), these septa rarely form a complete barrier. Openings in the septa allow cytoplasm to flow throughout the hyphae from one cell to another.

Because of such cytoplasmic streaming, proteins synthesized throughout the hyphae can be carried to the hyphal tips. This novel body plan is perhaps the most important innovation of the fungal kingdom. As a result of it, fungi can respond quickly to environmental changes, growing very rapidly when food and water are plentiful and the temperature is optimal. All parts of the fungal body work together, secreting digestive enzymes and actively attempting to digest and absorb any organic material with which the fungus comes in contact.

Also due to cytoplasmic streaming, many nuclei may be connected by the shared cytoplasm of a fungal mycelium. None of them (except for reproductive cells) are isolated in any one cell; all of them are linked cytoplasmically with every cell of the mycelium. Indeed, the entire concept of multicellularity takes on a new meaning among the fungi, the ultimate communal sharers among the multicellular organisms.

Figure 16.12   Many fungi produce spores.

Spores explode from the surface of a puffball fungus.

How Fungi Reproduce

Fungi reproduce both asexually and sexually. All fungal nuclei except for the zygote are haploid. Often in the sexual reproduction of fungi, individuals of genetically different "mating types" must participate, much as two sexes are required for human reproduction. Sexual reproduction is initiated when two hyphae of different mating types (indicated by "+" and "–") come in contact, and the hyphae fuse (figure 16.11). What happens next? In animals and plants, when the two haploid gametes fuse, the two haploid nuclei immediately fuse to form the diploid nucleus of the zygote. As you might by now expect, fungi handle things differently. In most fungi, the two nuclei do not fuse immediately. Instead, they remain unmarried inhabitants of the same house, coexisting in a common cytoplasm for most of the life of the fungus! A fungal hypha that has two nuclei is called **dikaryotic.** The nuclei in certain cells will fuse, forming a zygote.

Spores are a common means of reproduction among the fungi. The puffball fungus in figure 16.12 is releasing spores in a somewhat explosive manner. Spores are well suited to the needs of an organism anchored to one place. They are so small and light that they may remain suspended in the air for long periods of time and may be carried great distances. When a spore lands in a suitable place, it germinates and begins to divide, giving rise to a new fungal hypha.

How Fungi Obtain Nutrients

All fungi obtain food by secreting digestive enzymes into their surroundings and then absorbing back into the fungus the organic molecules produced by this *external digestion*. Many fungi are able to break down the cellulose in wood, cleaving the linkages between glucose subunits and then absorbing the glucose molecules as food. That is why fungi are so often seen growing on trees (figure 16.13).

Figure 16.13   The oyster mushroom.

This species, *Pleurotus ostreatus*, not only uses cellulose from the tree as a food source, but it is also a predator and attracts tiny roundworms called nematodes. It secretes a substance that immobilizes nematodes, which the fungus then uses as a source of food.

(a)   (b)

Figure 16.14   **Edible and poisonous mushrooms.**

Edible mushrooms include (a) button mushrooms (*Agaricus bisporus*). Poisonous mushrooms, including (b) *Amanita muscaria*, can cause a range of symptoms, from slight allergic and digestive reactions, to organ failure and death.

16.7 Kinds of Fungi

CONCEPT PREVIEW: The fungal phyla are distinguished primarily by their modes of sexual reproduction. Fungi are key decomposers within almost all terrestrial ecosystems. In mycorrhizae and lichens, fungi form ecologically important associations with autotrophic organisms.

Fungi are an ancient group of organisms at least 400 million years old. There are nearly 74,000 described species, in five groups (described in table 16.3), and many more awaiting discovery. Many fungi are harmful because they decay, rot, and spoil many different materials as they obtain food and because they cause serious diseases in animals and particularly in plants. Some mushrooms are edible, but others are poisonous (figure 16.14). Other fungi are used for commercial purposes. The manufacture of both bread and beer depends on the biochemical activities of yeasts, single-celled fungi that produce abundant quantities of carbon dioxide and ethanol. Fungi are used in industry to convert one complex organic molecule into another; many commercially important steroids are synthesized in this way.

TABLE 16.3	**Fungi**			
Phylum	**Typical Examples**		**Key Characteristics**	**Approximate Number of Living Species**
Zygomycota	*Rhizopus* (black bread mold)		Reproduce sexually and asexually; multinucleate hyphae lack septa except for reproductive structures; fusion of hyphae leads directly to formation of a zygote, in which meiosis occurs just before it germinates	1,050
Ascomycota	Yeasts, truffles, morels		Reproduce by sexual means; ascospores are formed inside a sac called an ascus; asexual reproduction is also common	32,000
Basidiomycota	Mushrooms, toadstools, rusts		Reproduce by sexual means; basidiospores are borne on club-shaped structures called basidia; the terminal hyphal cell that produces spores is called a basidium; asexual reproduction occurs occasionally	22,000
Chytridiomycota	*Allomyces*		Produce flagellated gametes (zoospores); predominately aquatic, some freshwater and some marine; oldest group of fungi	1,500
Imperfect fungi (not a phylum)	*Aspergillus, Penicillium*		Sexual reproduction has not been observed; most are thought to be ascomycetes that have lost the ability to reproduce sexually	17,000

The four major fungal phyla, distinguished from one another primarily by their mode of sexual reproduction, are the zygomycetes, the ascomycetes, the basidiomycetes, and the chytridiomycetes. A fifth group, the imperfect fungi, is an artificial grouping of fungi in which sexual reproduction has not been observed. Molecular data is contributing to our understanding of the fungal phylogeny and as additional molecular evidence is acquired, a new fungal phylogeny is likely to appear. However, it already seems clear that fungi are more closely related to animals than to plants.

Ecological Roles as Decomposers

Fungi, together with bacteria, are the principal decomposers in the biosphere. They break down organic materials and return the substances that had been locked in those molecules to circulation in the ecosystem. Fungi are virtually the only organisms capable of breaking down lignin, one of the major constituents of wood. By breaking down such substances, fungi release critical building blocks, such as carbon, nitrogen, and phosphorus, from the bodies of dead organisms and make them available to other organisms.

In breaking down organic matter, some fungi attack living plants and animals, acting as disease-causing organisms. Pathogenic fungi are responsible for billions of dollars in agricultural losses every year. Not only are fungi the most harmful pests of living plants, but they also attack food products once they have been harvested and stored. In addition, fungi often secrete substances into the foods they are attacking that make these foods unpalatable or poisonous.

Commercial Uses

The same aggressive metabolism that makes fungi ecologically important has been put to commercial use in many ways. The manufacture of both bread and beer depends on the biochemical activities of yeasts, single-celled fungi mentioned on the previous page. Cheese and wine achieve their delicate flavors because of the metabolic processes of certain fungi, and others make possible the manufacture of soy sauce. Vast industries depend on the biochemical manufacture of organic substances such as citric acid by fungi in culture, and yeasts are now used on a large scale to produce protein for the enrichment of animal food. Many antibiotics, including the first one that was used on a wide scale, penicillin, are derived from fungi.

Fungal Associations

Two kinds of mutualistic associations between fungi and autotrophic organisms are ecologically important: mycorrhizae and lichens. In each case, a photosynthetic organism fixes atmospheric carbon dioxide and thus makes organic material available to the fungi. The metabolic activities of the fungi, in turn, enhance the overall ability of the symbiotic association to exist in a particular habitat. Mycorrhizae are symbiotic associations between fungi and the roots of plants where the fungal partner expedites the plant's absorption of essential nutrients such as phosphorus (figure 16.15). Lichens are symbiotic associations between fungi and either green algae or cyanobacteria. They are prominent nearly everywhere in the world, especially in unusually harsh habitats such as bare rock (figure 16.16).

Concept Check

1. How do materials made in the hyphae reach the growing tip?
2. When two haploid hyphae of different mating type fuse, is the resulting cell diploid? Explain.
3. All fungi are heterotrophs. Where do fungi get their food?

BIOLOGY & YOU

Making Your Own Beer. Single-celled fungi called yeasts are used to make beer and wine. Beer is made from four basic ingredients: water, malted barley, hops, and yeast. To start, you soak barley grain in water until it sprouts, then dry it, a process called malting which develops the sugars and starches needed for fermentation. You must then "mash" the malt to extract the sugars and starches. You can also bypass these initial steps by buying malt extract. To the extract you now add hops, which are green flowers that grow on a vine and look like little pine cones. Since not all the sugars of the malt will ferment, the hops balances out the sweetness with a degree of bitterness, and adds distinctive taste and aroma to the brew. The malt, hops, and water are now boiled, producing a mixture called wort (pronounced *wert*) which is poured into a fermentor and allowed to cool. The yeast is then added and the container is sealed to keep air out. After about 7 to 10 days, the yeast will have fermented all it can, and falls to the bottom of the container. Time to bottle the beer and wait a couple of weeks for carbonation to complete. Then your "home brew" is ready to drink!

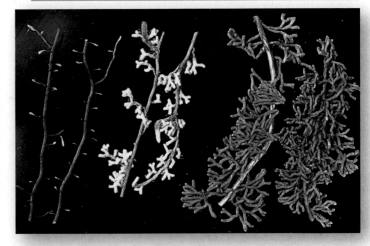

Figure 16.15 **Mycorrhizae on the roots of pines.**

From left to right are pine roots not associated with a fungus, white mycorrhizae formed by *Rhizopagon*, and yellow-brown mycorrhizae formed by *Pisolithus*. The mycorrhizae act to expand the surface area of the root, increasing water and nutrient absorption in the root.

Figure 16.16 **Lichens growing on a rock.**

Defining a Treatment Window for Malaria

While malaria kills more people each year than any other infectious disease, the combination of mosquito control and effective treatment has virtually eliminated this disease from the United States. In 1941 more than 4,000 Americans died of malaria; in the year 2006, by contrast, fewer than five people died of malaria.

The key to controlling malaria has come from understanding its life cycle. The first critical advance came in 1897 in a remote field hospital in Secunderabad, India, when English physician Ronald Ross observed that hospital patients who did not have malaria were more likely to develop the disease in the open wards (those without screens or netting) than in wards with closed windows or screens. Observing closely, he saw that patients in the open wards were being bitten by mosquitoes of the genus *Anopheles*. Dissecting mosquitoes who had bitten malaria patients, he found the plasmodium parasite. Newly-hatched mosquitoes who had not yet fed, when allowed to feed on malaria-free blood, did not acquire the parasite. Ross reached the conclusion that mosquitoes were spreading the disease from one person to another, passing along the parasite while feeding. In every country where it has been possible to eliminate the *Anopheles* mosquitoes, the incidence of the disease malaria has plummeted.

The second critical advance came with the development of drugs to treat malaria victims. The British had discovered in India in the mid-1800s that a bitter substance called quinine taken from the bark of cinchona trees was useful in suppressing attacks of malaria. The boys in the photograph are being treated with an intravenous solution of quinine. Quinine also reduces the fever during attacks, but does not cure the disease. Today physicians use instead the synthetic drugs chloroquine and primaquine, which are much more effective than quinine, with fewer side effects.

Unlike quinine these two drugs can cure patients completely, because they attack and destroy one of the phases of the plasmodium life cycle, the merozoites released into the bloodstream several days after infection—but only if the drugs are administered soon enough after the bite that starts the infection.

In order to determine the time frame for successful treatment, doctors have carefully studied the time course of a malarial infection. The graph above presents what they have found. Numbers of merozoites are presented on the y axis on a log scale—each step reflects a 10-fold increase in numbers. The infection becomes life-threatening if 1% of red blood cells become infected, and death is almost inevitable if 20% of red blood cells are infected.

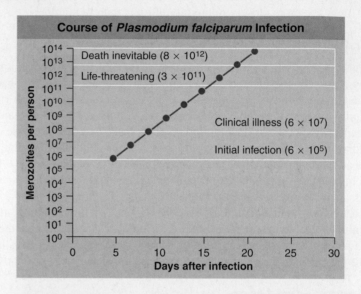

Course of *Plasmodium falciparum* Infection

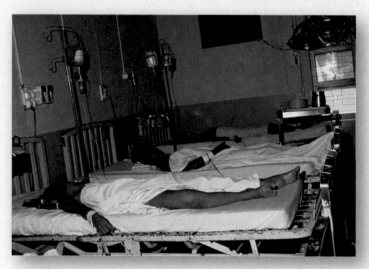

Analysis

1. **Making Inferences**
   a. How long after infection is it before the liver releases merozoites into the blood stream (that is, initial infection by merozoites)? before the disease becomes life-threatening? before death is inevitable?
   b. How long does it take merozoites to multiply 10-fold?

2. **Drawing Conclusions** After the first appearance of clinical illness symptoms, how many days do doctors have to treat the disease before it becomes life-threatening? before treatment has little or no chance of saving the patient's life?

Concept Summary

Origin of Life

16.1 How Cells Arose

- Life on earth may have arisen in bubbles in the ocean. The "bubble model" suggests that biological molecules were captured in bubbles where they underwent chemical reactions, leading to the origin of life (**figure 16.2**). The first cells may have formed from bubble-enclosed molecules, such as RNA.

Prokaryotes

16.2 The Simplest Organisms

- Prokaryotes have very simple internal structures, lacking nuclei and other membrane-bound compartments. The plasma membrane of prokaryotes is encased in a cell wall. The cell wall of bacteria is made of peptidoglycan, which is different than cell walls found in archaea, or in certain eukaryotic cells. Bacteria are divided into two groups, gram-positive and gram-negative, based on the construction of their cell walls.

- Many bacteria have flagella or pili and may form endospores.

- Bacteria reproduce by splitting in two, called binary fission, and may exchange genetic information through conjugation. In conjugation, bacterial cells exchange genetic information by transferring copies of plasmids to another bacterium (**figure 16.3**).

Viruses

16.3 Viruses Infect Organisms

- Viruses are not living organisms, but are parasitic chemicals that enter and replicate inside cells. They contain a nucleic acid core surrounded by a protein coat. Some viruses have an outer protein capsid or membranelike envelope. Viruses infect bacteria, plants, and animals, and while all viruses have the same general structure, they vary greatly in shape and size (**figure 16.4**). Viruses cause many diseases, often spreading from animals to humans.

The Protists

16.4 General Biology of Protists

- Protists are a very diverse group of eukaryotes, having diverse cell surfaces, modes of locomotion, and modes of acquiring nutrients. Some protists form cysts to protect themselves in harsh environmental conditions. Most protists reproduce asexually except during times of stress, when they engage in sexual reproduction. Some exist as single cells (**figure 16.5**), whereas others form colonies (**figure 16.7**) or aggregates, but only a few exhibit true multicellularity.

16.5 Kinds of Protists

- The classification of organisms into the kingdom Protista is mainly a matter of convenience; many different forms have been lumped together. Some forms of protists are as different from one another as they are from plants or animals (**figure 16.9** and **table 16.2**).

- There are many different types of protists, including some that cause diseases. Sleeping sickness and malaria are caused by protists.

- Population explosions of protists called dinoflagellates produce "red tides," and protists called diatoms have many industrial uses. Another type of protist called algae has many uses and are key components of marine ecosystems.

Fungi

16.6 A Fungus Is Not a Plant

- Fungi are nonmobile heterotrophs. The body of a fungus is composed of long slender filaments called hyphae that pack together to form a mycelium (**figure 16.10**). Most fungi have nonmotile sperm, unlike some plants. Fungi have cell walls made of chitin, which is different from the cellulose found in plant cell walls. Fungi also undergo nuclear mitosis, where the nuclei divide but the cell does not. Most fungal cells are separated by an incomplete wall called a septum, which allows cytoplasm to pass between cells for intercellular communication.

- Fungi reproduce both sexually and asexually. Sexual reproduction takes place between two genetically different "mating types." In fungi, two hyphae of different mating types fuse. Their haploid nuclei may stay separate in the cell, this is called a dikaryotic cell (**figure 16.11**).

- Many fungi form spores. Spores are released and may be carried by wind to considerable distances away where they will land and germinate new environments.

- Fungi obtain nutrients through external digestion. They grow on their food, releasing enzymes that digest the food. The products of digestion are then absorbed by the fungus.

16.7 Kinds of Fungi

- There are four major phyla of fungi classified according to their mode of sexual reproduction. A fifth group, the imperfect fungi, is a "catch-all" category of fungi in which sexual reproduction has not been observed (**table 16.3**).

- Fungi play key roles in the environment as decomposers. They breakdown the bodies of dead organisms, returning substances back to the ecosystem. However, fungi can also cause diseases and agricultural damage.

- Certain fungi have commercial uses. Some are edible, while others are poisonous. The manufacturing of bread, beer, wine, and cheese use single-celled yeast. Other types of fungi are used to make antibiotics, and still others are used to breakdown contaminants in the environment.

- They are involved in symbiotic associations with the roots of some plants, called mycorrhizae (**figure 16.15**), and with cyanobacteria or algae, called lichens (**figure 16.16**).

Self-Test

1. While it is still unknown how the first cells formed, scientists suspect that the first active biological macromolecule was
   a. protein.
   c. RNA.
   b. DNA.
   d. carbohydrates.

2. Bacteria
   a. are prokaryotic.
   b. have been on the earth for at least 2.5 billion years.
   c. are the most abundant life-form on earth.
   d. All answers are correct.

3. Bacterial cell walls are composed of
   a. proteins.
   c. chitin.
   b. peptidoglycan.
   d. cellulose.

4. Viruses are
   a. protein coats that contain DNA or RNA.
   b. simple eukaryotic cells.
   c. simple prokaryotic cells.
   d. alive.

5. Which is a natural host for the influenza virus?
   a. mosquitoes
   c. birds
   b. monkeys
   d. bats

6. Many protists survive unfavorable environmental conditions by forming
   a. gametes.
   c. cysts.
   b. zygotes.
   d. aggregations.

7. Protists do *not* include
   a. algae.
   c. multicellular organisms.
   b. amoebas.
   d. mushrooms.

8. A fungus is composed of filaments called
   a. hyphae.
   c. flagella.
   b. septa.
   d. mycelia.

9. Fungi reproduce
   a. both sexually and asexually.
   b. sexually only.
   c. asexually only.
   d. by fragmentation.

10. Lichens are mutualistic associations between
    a. plants and fungi.
    c. animals and fungi.
    b. algae and fungi.
    d. all of the above.

Visual Understanding

1. **Figure 16.4** Viruses have dramatically different forms and shapes. What are the consistent features of all the viruses shown?

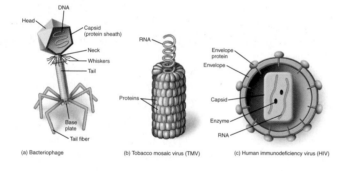

(a) Bacteriophage   (b) Tobacco mosaic virus (TMV)   (c) Human immunodeficiency virus (HIV)

2. **Figure 16.7** Are the *Volvox* shown here multicellular organisms, or numerous single-celled organisms working together?

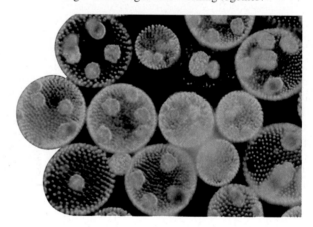

Challenge Questions

1. Why can't we prove how life began on earth? (Hint: What would be necessary to prove how life began—what types of evidence would have to be gathered?)

2. Why do we say that bacteria are alive but viruses are not?

3. Compare the methods of movement found in some of the groups of protists.

4. Describe some of the ways that fungi cooperate with other organisms that protists and animals cannot manage to do. Why can only fungi manage these forms of cooperation?

Evolution of Plants

Plants

Figure 17.1 The importance of plants.

Plants are astonishingly diverse and are key components in the biosphere. Photosynthesis conducted by plants and algae creates the majority of oxygen in the atmosphere. Also, through photosynthesis, plants convert energy from the sun into organic material that can be used by the many members of an ecosystem. Plants play key roles in the cycling of water and many nutrients. They are the main source of food for humans and many other organisms (directly and indirectly), and plants provide us with wood, cloth, paper, and many other nonfood products.

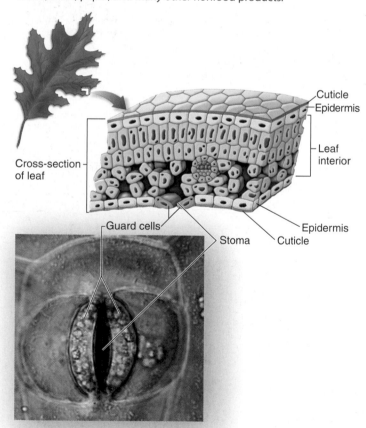

Figure 17.2 A stoma.

A stoma is a passage through the cuticle covering the epidermis of a leaf (×400). Water and oxygen pass out through the stoma, and carbon dioxide enters by the same portal. The cells flanking the stoma are called guard cells and control the opening and closing of the stoma.

17.1 Adapting to Terrestrial Living

CONCEPT PREVIEW: Plants are multicellular terrestrial photosynthesizers that evolved from green algae. Plants adapted to life on land by developing ways to absorb minerals in partnership with fungi, to conserve water with watertight coverings, and to reproduce on land.

Plants are complex multicellular organisms that are terrestrial **autotrophs**—that is, they occur almost exclusively on land and feed themselves by photosynthesis. The name *autotroph* comes from the Greek, *autos*, self, and *trophos*, feeder. Today, plants are the dominant organisms on the surface of the earth and are key components of ecosystems (figure 17.1). Nearly 300,000 species are now in existence, covering almost every part of the terrestrial landscape. In this chapter, we see how plants adapted to life on land.

The green algae that were probably the ancestors of today's plants are aquatic organisms that are not well adapted to living on land. Before their descendants could live on land, they had to overcome many environmental challenges. For example, they had to absorb minerals from the rocky surface. They had to find a means of conserving water. They had to develop a way to reproduce on land.

Absorbing Minerals

Plants require relatively large amounts of six inorganic minerals: nitrogen, potassium, calcium, phosphorus, magnesium, and sulfur. Each of these minerals constitutes 1% or more of a plant's dry weight. Algae absorb these minerals from water, but where is a plant on land to get them? From the soil. Soil is the weathered outer layer of the earth's crust. It is composed of a mixture of ingredients, which may include sand, rocks, clay, silt, humus (partly decayed organic material), and various other forms of mineral and organic matter. The soil is also rich in microorganisms that break down and recycle organic debris. Plants absorb these materials, along with water, through their *roots* (described in chapter 32). Most roots are found in topsoil, which is a mixture of mineral particles, living organisms, and humus. When topsoil is lost due to erosion, the soil loses its ability to hold water and nutrients.

Top soil is a nonreplaceable resource, as discussed on page 441. The success of the U.S. agricultural industry depends on the quality of the top soil, and so preserving the top soil is of major concern for the U.S. and world economies.

The first plants seem to have developed a special relationship with fungi, which was a key factor in their ability to absorb minerals in terrestrial habitats. Within the roots or underground stems of many early fossil plants like *Cooksonia* (see figure 17.7) and *Rhynia*, fungi can be seen living intimately within and among the plant's cells. As you may recall from chapter 16, these kinds of symbiotic associations are called **mycorrhizae.** In plants with mycorrhizae, the fungi enable the plant to take up phosphorus and other nutrients from rocky soil, while the plant supplies organic molecules to the fungus.

Conserving Water

One of the key challenges to living on land is to avoid drying out. To solve this problem, plants have a watertight outer covering called a **cuticle.** The covering is formed from a waxy substance that is impermeable to water. Like the wax on a shiny car, the cuticle prevents water from entering or leaving the stem or leaves. Water enters the plant only through the roots, while the cuticle prevents water loss to the air. Passages do exist through the cuticle, in the form of specialized pores called **stomata** (singular, **stoma**) in the leaves and sometimes the green portions of the stems. Figure 17.2 shows a stoma on the underside of a leaf. The cutaway view allows you to see the placement of the stoma in relation to other cells in the leaf. Stomata, which occur on at least some portions of all plants except liverworts, allow carbon dioxide to pass into the plant bodies (by diffusion) for photosynthesis and allow water and oxygen gas to pass out of them. The two *guard cells* that border the stoma swell (opening the stoma) and shrink (closing the stoma) when water moves into and out of them by osmosis. This controls the loss of water from the leaf while allowing the entrance of carbon dioxide (see also figure 32.21). In most plants, water enters through the roots as a liquid and exits through the stomata as water vapor. Water movement within a plant is discussed in more detail in chapter 32, section 32.6.

Reproducing on Land

To reproduce sexually on land, it is necessary to pass gametes from one individual to another, which is a challenge because plants cannot move about. In the first plants, the eggs were surrounded by a jacket of cells, and a film of water was required for the sperm to swim to the egg and fertilize it. In later plants, pollen evolved, providing a means of transferring gametes without drying out. The pollen grain is protected from drying and allows plants to transfer gametes by wind or insects.

> Recall that the process of meiosis produces haploid cells. Haploid cells, as discussed on page 144, have only half the number of chromosomes as a diploid cell from the same organism. Each haploid cell receives only one copy of each chromosome.

Changing the Life Cycle. Among many algae, haploid cells occupy the major portion of the life cycle. The zygote formed by the fusion of gametes is the only diploid cell, and it immediately undergoes meiosis to form haploid cells again. In early plants, by contrast, meiosis was delayed and the cells of the zygote divided to produce a multicellular diploid structure. Thus for a significant portion of the life cycle, the cells were diploid. This change resulted in an **alternation of generations,** in which a diploid generation alternates with a haploid one. Figure 17.3 shows a life cycle that exhibits an alternation of generation. Botanists call the haploid generation (indicated by the upper yellow area) the **gametophyte** because it forms haploid *gametes* by mitosis ❶. The diploid generation (indicated by the lower blue area) is called the **sporophyte** because it forms haploid *spores* by meiosis ❹.

When you look at primitive plants such as liverworts and mosses, you see mostly gametophyte tissue, the green leafy structures in figure 17.4a—the sporophytes are smaller brown structures (not visible in the photo) attached to or enclosed within the tissues of the larger gametophyte. When you look at plants that evolved later, such as a pine tree, you see mostly sporophyte tissue. The gametophytes of these plants are always much smaller than the sporophytes and are often enclosed within sporophyte tissues. Figure 17.4b shows the male gametophyte (pollen grains) of a pine tree. They are not photosynthetic cells and are dependent upon the sporophyte tissue in which it is usually enclosed.

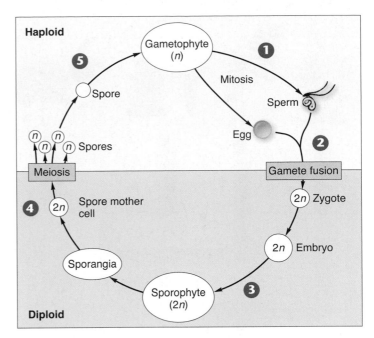

Figure 17.3 Generalized plant life cycle.

In a plant life cycle, a diploid generation alternates with a haploid one. Gametophytes, which are haploid (n), alternate with sporophytes, which are diploid (2n). ❶ Gametophytes give rise by mitosis to sperm and eggs. ❷ The sperm and egg ultimately come together to produce the first diploid cell of the sporophyte generation, the zygote. ❸ The zygote undergoes cell division, ultimately forming the sporophyte. ❹ Meiosis takes place within the sporangia, the spore-producing organs of the sporophyte, resulting in the production of the spores, which are haploid and are the first cells ❺ of gametophyte generations.

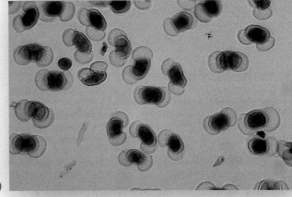

Figure 17.4 Two types of gametophytes.

(a) This gametophyte of a moss (a more primitive plant) is green and free-living. (b) Male gametophytes (pollen) of a pine tree (a more advanced plant) are barely large enough to see with the naked eye (×200).

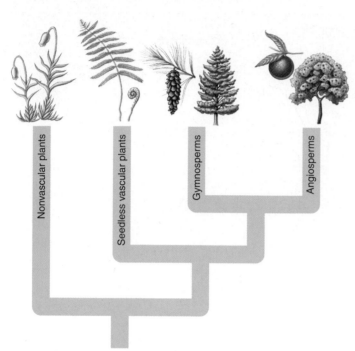

Figure 17.5 The evolution of plants.

17.2 Plant Evolution

CONCEPT PREVIEW: Plants evolved from green algae and eventually developed more dominant diploid phases of the life cycle, conducting systems of vascular tissue, seeds that protected the embryo, and flowers and fruits that aided in fertilization and distribution of seeds.

Once plants became established on land, they gradually developed many other features that aided their evolutionary success in this new, demanding habitat. For example, among the first plants there was no fundamental difference between the aboveground and the belowground parts. Later, roots and shoots with specialized structures evolved, each suited to its particular below- or aboveground environment.

As we explore plant diversity, we will examine several key evolutionary innovations that have given rise to the wide variety of plants we see today. Table 17.1 provides you with an overview of the plant phyla and their characteristics.

Four key evolutionary innovations serve to trace the evolution of plants:

1. **Alternation of generations.** Although algae exhibit a haploid and diploid phase, the diploid phase is not a significant portion of their life cycle. By contrast, even in early plants (the nonvascular plants indicated by the first vertical bar in figure 17.5) the diploid sporophyte is a larger structure and offers protection for the egg and developing embryo. The dominance of the sporophyte, both in size and the proportion of time devoted to it in the life cycle, becomes greater throughout the evolutionary history of plants.

2. **Vascular tissue.** A second key innovation was the emergence of vascular tissue. Vascular tissue transports water and nutrients throughout the plant body and provides structural support. With the evolution of vascular tissue, plants were able to more efficiently supply the upper portions of their bodies with water absorbed from the soil and had some rigidity, allowing the plants to grow larger and in drier conditions. The first vascular plants were the seedless vascular plants, the second vertical bar in figure 17.5.

3. **Seeds.** The evolution of seeds was a key innovation that allowed plants to dominate their terrestrial environments. Seeds provide nutrients and a tough, durable cover that protects the embryo until it encounters favorable growing conditions. The first plants with seeds were the gymnosperms, the third vertical bar in figure 17.5.

4. **Flowers and fruits.** The evolution of flowers and fruits were key innovations that improved the chances of successful mating in sedentary organisms and facilitated the dispersal of their seeds. Flowers protected the egg and improved the odds of its fertilization, allowing plants that were located at considerable distances to mate successfully. Fruit, which surrounds the seed and aids in its dispersal, allowed plant species better opportunities to invade new and possibly more favorable environments. The angiosperms, the fourth vertical bar in figure 17.5, are the only plants to produce flowers and fruits.

Concept Check

1. What advantage did the first plants have in forming mycorrhizae?
2. Is a sporophyte haploid or diploid?
3. Do all plants with fruits contain seeds? Explain.
4. A pine tree is a gymnosperm. Do pine trees have flowers at any point in their life cycle? If so, when?

TABLE 17.1	PLANT PHYLA			
Phylum	**Typical Examples**		**Key Characteristics**	**Approximate Number of Living Species**
Nonvascular Plants				
Hepaticophyta (liverworts)	*Marchantia*		Without true vascular tissues; lack true roots and leaves; live in moist habitats and obtain water and nutrients by osmosis and diffusion; require water for fertilization; gametophyte is dominant structure in the life cycle; the three phyla were once grouped together	15,600
Anthocerophyta (hornworts)	*Anthoceros*			
Bryophyta (mosses)	*Polytrichum, Sphagnum* (hairy cap and peat moss)			
Seedless Vascular Plants				
Lycophyta (lycopods)	*Lycopodium* (club mosses)		Seedless vascular plants, some are similar in appearance to mosses but diploid; require water for fertilization; sporophyte is dominant structure in life cycle; found in moist woodland habitats	1,150
Pterophyta (ferns)	*Azolla, Sphaeropteris* (water and tree ferns)		Seedless vascular plants; require water for fertilization; sporophytes diverse in form and dominate the life cycle	11,000
	Equisetum (horsetails)			
	Psilotum (whisk ferns)			
Seed Plants				
Coniferophyta (conifers)	Pines, spruce, fir, redwood, cedar		Gymnosperms; wind pollinated; ovules partially exposed at time of pollination; flowerless; seeds are dispersed by the wind; sperm lack flagella; sporophyte is dominant structure in life cycle; leaves are needlelike or scalelike; most species are evergreens and live in dense stands; among the most common trees on earth	601
Cycadophyta (cycads)	Cycads, sago palms		Gymnosperms; wind pollination or possibly insect-pollination; very slow growing, palmlike trees; sperm have flagella; trees are either male or female; sporophyte dominant in the life cycle	206
Gnetophyta (shrub teas)	Mormon tea, *Welwitschia*		Gymnosperms; nonmotile sperm; shrubs and vines; wind pollination and possibly insect pollination; plants are either male or female; sporophyte is dominant in the life cycle	65
Ginkgophyta (ginkgo)	Ginkgo trees		Gymnosperms; fanlike leaves that are dropped in winter (deciduous); seeds fleshy and ill-scented; motile sperm; trees are either male or female; sporophyte is dominant in the life cycle	1
Anthophyta (flowering plants, also called angiosperms)	Oak trees, corn, wheat, roses		Flowering; pollination by wind, animal, and water; characterized by ovules that are fully enclosed by the carpel; fertilization involves two sperm nuclei: one forms the embryo, the other fuses with the polar body to form endosperm for the seed; after fertilization, carpels and the fertilized ovules (now seeds) mature to become fruit; sporophyte is dominant in life cycle	250,000

Seedless Plants

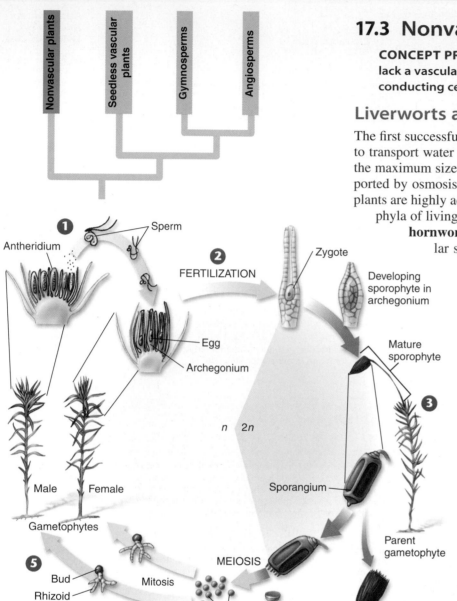

17.3 Nonvascular Plants

CONCEPT PREVIEW: While liverworts and hornworts totally lack a vascular system, mosses have simple soft strands of conducting cells.

Liverworts and Hornworts

The first successful land plants had no vascular system—no tubes or pipes to transport water and nutrients throughout the plant. This greatly limited the maximum size of the plant body because all materials had to be transported by osmosis and diffusion (see section 4.9). However, these simple plants are highly adapted to a variety of terrestrial environments. Only two phyla of living plants, the **liverworts** (phylum Hepaticophyta) and the **hornworts** (phylum Anthocerophyta), completely lack a vascular system. The word *wort* meant *herb* in medieval Anglo-Saxon when these plants were named. Liverworts are the simplest of all living plants. About 6,000 species of liverworts and 100 species of hornworts survive today, usually growing in moist and shady places.

Mosses Use Primitive Conducting Systems

Another phylum of plants, the **mosses** (phylum Bryophyta), were the first plants to evolve strands of specialized cells that conduct water and carbohydrates up the stem of the gametophyte. The conducting cells do not have specialized wall thickenings; instead they are like nonrigid pipes and cannot carry water very high. Because these conducting cells could at the most be considered a primitive vascular system, mosses are usually grouped by botanists with the liverworts and hornworts as "nonvascular" plants.

Today about 9,500 species of mosses grow in moist places all over the world. In the Arctic and Antarctic, mosses are the most abundant plants. "Peat moss" (genus *Sphagnum*) can be used as a fuel or a soil conditioner. The moss *Physcomitrella patens,* whose genome was sequenced in 2006, has been the subject of many genetic studies. The life cycle of a moss is illustrated in figure 17.6. You can see that the majority of the life cycle consists of the haploid gametophyte generation (the green part of the plant), which exists as male or female plants. The diploid sporophyte (the brown stalk with the swollen head) grows out of the gametophyte of the female plant ❸, from the zygote, which is the fertilized egg. Cells within the sporophyte undergo meiosis to produce haploid spores ❹ that grow into gametophytes ❺.

Figure 17.6 **The life cycle of a moss.**

On the haploid gametophytes, sperm are released from each antheridium (sperm-producing structure) ❶. They then swim through free water to an archegonium (egg-producing structure) and down to the egg. Fertilization takes place there ❷; the resulting zygote develops into a diploid sporophyte. The sporophyte grows out of the archegonium, forming a structure called a sporangium at its apex ❸. The sporophyte grows on the gametophyte, as shown in the photo, and eventually produces spores as a result of meiosis. The spores are shed from the sporangium ❹. The spores germinate, giving rise to gametophytes ❺.

A hair-cup moss, *Polytrichum*

17.4 The Evolution of Vascular Tissue

CONCEPT PREVIEW: Vascular plants have specialized vascular tissue that conducts water and food throughout the plant. Plants grow tall through primary growth and expand in diameter through secondary growth.

The remaining seven phyla of plants, which have efficient vascular systems made of highly specialized cells, are called **vascular plants.** The first vascular plants appeared approximately 430 million years ago, but only incomplete fossils have been found. The earliest vascular plants for which we have relatively complete fossils, the extinct phylum Rhyniophyta, lived 410 million years ago. Among them is the oldest known vascular plant, *Cooksonia*. The fossil of *Cooksonia* in figure 17.7 clearly shows that the plant had branched, leafless shoots that formed spores at the tips in structures called *sporangia.*

Cooksonia and the other early plants that followed became successful colonizers of the land through the development of efficient water- and food-conducting systems known as **vascular tissues** (Latin, *vasculum,* vessel or duct). These tissues consist of strands of specialized cylindrical or elongated cells that form a network of tubelike structures throughout a plant. This network extends from near the tips of the roots (when present), through the stems, and into the leaves (when present; figure 17.8). One type of vascular tissue, *xylem,* conducts water and dissolved minerals upward from the roots; another type, *phloem,* conducts carbohydrates throughout the plant. The presence of a cuticle and stomata are also characteristic of vascular plants.

> The vascular system in plants consists of xylem and phloem. As described on page 618, xylem is made of dead cells that form continuous tubelike structures that carry water and minerals, while phloem, the food-conducting tissue, is made of living cells that also form tubelike structures.

Most early vascular plants seem to have grown by cell division at the tips of the stem and roots. Imagine stacking dishes—the stack can get taller but not wider! This sort of growth is called **primary growth** and was quite successful. During the so-called Coal Age (between 350 and 290 million years ago, when much of the world's fossil fuel was formed) the lowland swamps that covered Europe and North America were dominated by an early form of seedless tree called a lycophyte. Lycophyte trees grew to heights of 10 to 35 meters (33 to 115 ft), and their trunks did not branch until they attained most of their total height. The pace of evolution was rapid during this period, for the world's climate was changing, growing dryer and colder. As the world's swamplands began to dry up, the lycophyte trees vanished, disappearing abruptly from the fossil record. They were replaced by tree-sized ferns, a form of vascular plant that will be described in detail in section 17.5. Tree ferns grew to heights of more than 20 meters (66 ft) with trunks 30 centimeters (12 in) thick. Like the lycophytes, the trunks of tree ferns were formed entirely by primary growth.

About 380 million years ago, vascular plants developed a new pattern of growth, in which a cylinder of cells beneath the bark divides, producing new cells in regions around the plant's periphery. This growth is called **secondary growth.** Secondary growth makes it possible for a plant stem to increase in diameter. Only after the evolution of secondary growth could vascular plants become thick-trunked and therefore considerably taller. Redwood trees today reach heights of up to 117 meters (384 ft) and trunk diameters in excess of 11 meters (36 ft). This evolutionary advance made possible the dominance of the tall forests that today cover northern North America. You are familiar with the product of plant secondary growth as **wood.** The growth rings that are so visible in cross sections of trees are zones of secondary growth (spring–summer) spaced by zones of little growth (fall–winter).

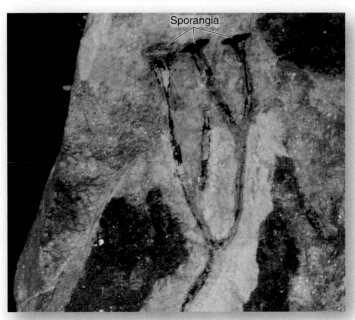

Figure 17.7   The earliest vascular plant.

The earliest vascular plant of which we have complete fossils is *Cooksonia.* This fossil shows a plant that lived some 410 million years ago; its upright branched stems terminated in spore-producing sporangia at the tips.

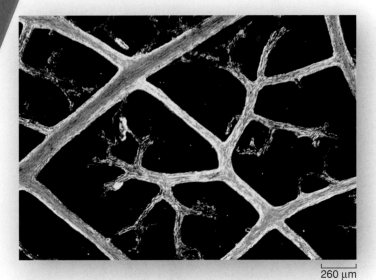

260 µm

Figure 17.8   The vascular system of a leaf.

The veins of a vascular plant contain strands of specialized cells for conducting food and water. The veins from a leaf are shown in this magnified view of a leaf with the rest of the tissue removed.

17.5 Seedless Vascular Plants

CONCEPT PREVIEW: Ferns are among the vascular plants that lack seeds, reproducing with spores as nonvascular plants do.

The earliest vascular plants lacked seeds, and two of the seven phyla of modern-day vascular plants do not have them. The two phyla of living seedless vascular plants include the ferns, phylum Pterophyta and the club mosses, phylum Lycophyta. In addition to the typical ferns seen growing in forests, like those shown in figure 17.9*a*, *b*, the phylum Pterophyta also contains the whisk ferns (figure 17.9*c*) and the horsetails (figure 17.9*d*). The other phylum, Lycophyta, contains the club mosses (figure 17.9*e*). Both phyla have free-swimming sperm that require the presence of free water for fertilization.

(a)

(b)

(c)

(d)

(e)

Figure 17.9   Seedless vascular plants.

(a) A tree fern in the forests of Malaysia (phylum Pterophyta). The ferns are by far the largest group of spore-producing vascular plants. (b) Ferns on the floor of a redwood forest. (c) A whisk fern. Whisk ferns have no roots or leaves. (d) A horsetail, *Equisetum telmateia*. This species forms two kinds of erect stems; one is green and photosynthetic, and the other, which terminates in a spore-producing "cone," is mostly light brown. (e) A club moss (phylum Lycophyta). Although superficially similar to the gametophytes of mosses, the conspicuous club moss plants shown here are sporophytes.

By far the most abundant of seedless vascular plants are the **ferns,** with about 11,000 living species. Ferns are found throughout the world, although they are much more abundant in the tropics than elsewhere. Many are small, with stems only a few centimeters in diameter, but some of the largest plants that live today are also ferns. Descendants of ancient tree ferns, they can have trunks more than 20 meters (66 ft) tall and leaves up to 5 meters (16 ft) long!

The Life of a Fern

In ferns, the life cycle of plants begins a revolutionary change that culminates later with seed plants. Nonvascular plants like mosses are made largely of gametophyte (haploid) tissue. Vascular seedless plants like ferns have both gametophyte and sporophyte individuals, each independent and self-sufficient. The haploid gametophyte is a small structure, 1 to 2 cm in diameter, (the heart-shaped plant in figure 17.10, not drawn to scale) which produces eggs and sperm. Rhizoids (anchoring structures) project from their lower surface. After sperm swim through water and fertilize the egg, the zygote grows into a sporophyte. Eventually, the sporophyte, becomes much larger than the gametophyte. Most ferns have horizontal stems, called rhizomes, that creep along below ground. The sporophyte bears haploid spores on the underside of their leaves, in the brown clusters called sori (singular, sorus). The spores are released from the sorus and float to the ground where they germinate, growing into haploid gametophytes. The fern gametophytes are small, thin photosynthetic plants that live in moist places.

The fern sporophytes are much larger and more complex, with long vertical leaves called **fronds.** When you see a fern, you are almost always looking at the sporophyte.

Figure 17.10 **The life cycle of a fern.**

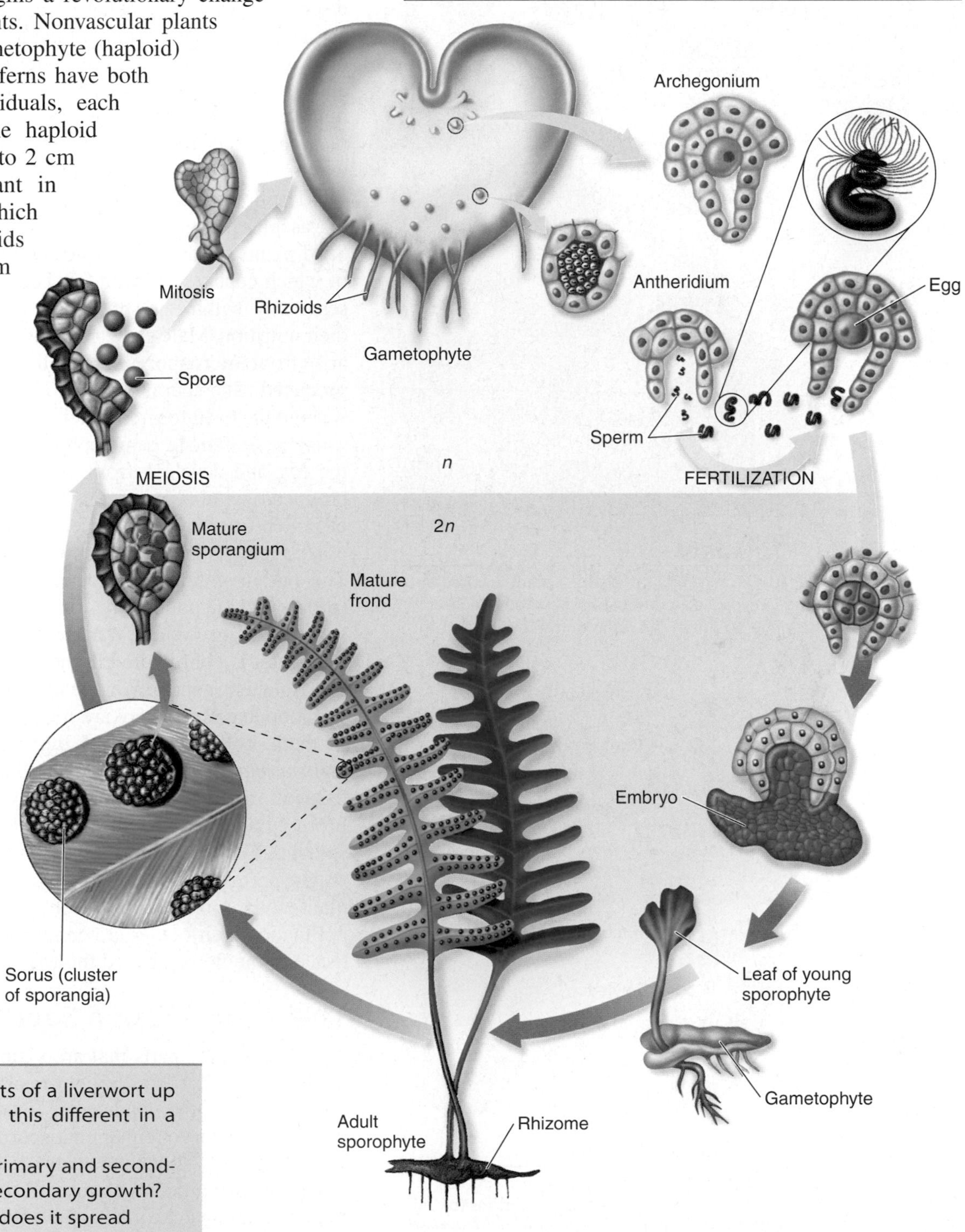

Concept Check

1. How does water get from the roots of a liverwort up through the plant body? How is this different in a fern plant?
2. What is the difference between primary and secondary growth? Does a moss show secondary growth?
3. If a fern doesn't have seeds, how does it spread offspring to distant locations?

The Advent of Seeds

Figure 17.11   **A seed plant.**

The seeds of this cycad, like all seeds, consist of a plant embryo and a protective covering. A cycad is a gymnosperm (naked-seeded plant), and its seeds develop out in the open on the edges of the cone scales.

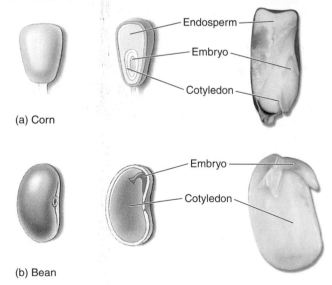

(a) Corn

(b) Bean

Figure 17.12   **Basic structure of seeds.**

A seed contains a sporophyte (diploid) embryo and a source of food, either endosperm (a) or food stored in the cotyledons (b). A seed coat surrounds the seed and protects the embryo. The seed structures are visible in the cutaway photos of seeds on the *right*.

17.6 Evolution of Seed Plants

CONCEPT PREVIEW: A seed is a dormant diploid embryo encased with food reserves in a hard protective coat. Seeds play a critical role improving a plant's chances of successfully reproducing in a varied environment.

When we think of plants, we often think of seed plants, but none of the plants discussed up to this point have seeds. Seeds were a very important evolutionary adaptation in the history of plants. The **seed** is a crucial adaptation to life on land because it protects the embryonic plant when it is at its most vulnerable stage. The plant in figure 17.11 is a cycad, and its seeds (the green balls in the photo) develop on the edges of the scales of the cone. The embryonic plants are inside the seeds where they are protected. The evolution of the seed was a critical step in allowing plants to dominate life on land.

The dominance of the sporophyte (diploid) generation in the life cycle of vascular plants reaches its full force with the advent of the seed plants. Seed plants produce two kinds of gametophytes—male and female, each of which consists of just a few cells. Both kinds of gametophytes develop separately within the sporophyte and are completely dependent on it for their nutrition. Male gametophytes, commonly referred to as **pollen grains,** arise from **microspores.** The pollen grains become mature when sperm are produced. The sperm are carried to the egg in the female gametophyte without the need for free water in the environment. A female gametophyte contains the egg and develops from a **megaspore** produced within an **ovule.** The transfer of pollen to an ovule by insects, wind, or other agents is referred to as **pollination.** The pollen grain then cracks open and sprouts and the pollen tube, containing the sperm cells, grows out, transporting the sperm directly to the egg. Thus free water is not required in the pollination and fertilization process.

Pollination, fertilization, and seed formation are discussed in detail on pages 634–637, but it is important to note here that unlike haploid spores in the seedless vascular plants that develop into gametophytes, seeds contain diploid cells that develop into sporophytes.

Botanists generally agree that all seed plants are derived from a single common ancestor. There are five phyla with living representatives. In four of them, collectively called the **gymnosperms** (Greek, *gymnos,* naked, and *sperma,* seed), the ovules are not completely enclosed by sporophyte tissue at the time of pollination. Gymnosperms were the first seed plants. From gymnosperms evolved the fifth group of seed plants, called **angiosperms** (Greek, *angion,* vessel, and *sperma,* seed), phylum Anthophyta. Angiosperms, or flowering plants, are the most recently evolved of all the plant phyla. Angiosperms differ from all gymnosperms in that at the time of pollination their ovules are completely enclosed by a vessel of sporophyte tissue in the flower, called the **carpel.**

The Structure of a Seed

A seed has three parts that are visible in the corn and bean seeds shown in figure 17.12: (1) a drought-resistant protective cover called the seed coat (visible in the whole seeds on the left), (2) a source of food for the developing embryo (either endosperm or cotyledons), and (3) a sporophyte plant embryo (visible in the cutaway views in the middle and on the right). The **endosperm** makes up most of the seed in corn and is the white part of popcorn, but in some plants, the endosperm is used up during the development of the embryo and is stored as food by the embryo in thick "leaflike"

Figure 17.13   Seeds allow plants to bypass the dry season.

Seeds can remain dormant until conditions are favorable for growth. For example, when it rains, seeds can germinate, and plants can grow rapidly to take advantage of the relatively short periods when water is available. This palo verde desert tree (*Cercidium floridum*) has tough seeds (*inset*) that germinate only after they are cracked. Rains leach out the chemicals in the seed coats that inhibit germination, and the hard coats of the seeds may be cracked when they are washed down gullies in temporary floods.

structures called **cotyledons.** For example, the endosperm is replaced by the cotyledon in the bean seed.

Seeds are one way in which plants, anchored by their roots to one place in the ground, are able to disperse their progeny to new locations. The hard cover of the seed protects the seed while it travels to a new location. The seed travels by many means such as by air, water, and animals. Many airborne seeds have devices to aid in carrying them farther. Most species of pine, for example, have seeds with thin flat wings attached. These wings help catch air currents, which carry the seeds to new areas. Seed dispersal will be discussed in chapter 33.

Once a seed has fallen to the ground, it may lie there, dormant, for many years. When conditions are favorable, however, and particularly when moisture is present, the seed germinates and begins to grow into a young plant (figure 17.13). Most seeds have abundant food stored in them to provide a ready source of energy for the new plant as it starts its growth.

The advent of seeds had an enormous influence on the evolution of plants. Seeds are particularly adapted to life on land in at least four respects:

1. **Dispersal.** Most important, seeds facilitate the migration and dispersal of plant offspring into new habitats.
2. **Dormancy.** Seeds permit plants to postpone development when conditions are unfavorable, as during a drought, keeping the embryonic plant protected until conditions improve.
3. **Germination.** By making the reinitiation of development dependent upon environmental factors such as temperature, seeds permit the course of embryonic development to be synchronized with critical aspects of the plant's habitat, such as the season of the year.
4. **Nourishment.** Seeds offer a young plant nourishment during the critical period just after germination, when the seedling must establish itself.

> Seed germination, as discussed on page 639, occurs before the plant is able to carry out photosynthesis. However, the embryo must have energy to start growing. The food stored in the seed provides a source of energy for this early growth.

IN THE NEWS

The Oldest Living Thing Is a Seed. Seeds permit a plant to take a time-out when the going gets rough, but rarely is the wait for improved conditions as long as that of an Israeli date palm. Popular as food in Palestine in the time of Jesus, dates were grown extensively several thousand years ago. Several ancient date seeds were recovered in 1973 from an excavation at Masada, a historic mountain fortress. Carbon dating indicated that the seeds were about 2,000 years old. The seeds sat untouched in a drawer in Jerusalem for many years, until recently a researcher asked permission to try and germinate one of them. A little water, and presto! A three-foot sapling with nine leaves soon was growing vigorously, none the worse for wear for its 2,000 year nap.

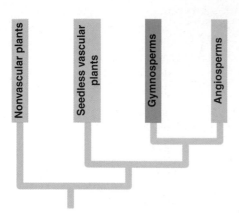

17.7 Gymnosperms

CONCEPT PREVIEW: Gymnosperms are seed plants in which pollen grains, the male gametophytes, are produced on pollen cones and fertilization occurs in the seed cones that contain the female gametophytes. Gymnosperms do not have flowers.

Four phyla constitute the gymnosperms (figure 17.14): the conifers (Coniferophyta), the cycads (Cycadophyta), the gnetophytes (Gnetophyta), and the ginkgo (Ginkgophyta). The conifers are the most familiar of the four phyla of gymnosperms and include pine, spruce, hemlock, cedar, redwood, yew, cypress, and fir trees (figure 17.14*a*). Conifers are trees that produce their seeds in cones. The seeds (ovules) of conifers develop on scales within the cones and are exposed at the time of pollination. Most of the conifers have needlelike leaves, an evolutionary adaptation for retarding water loss. Conifers are often found growing in moderately dry regions of the world, including the vast taiga forests of the northern latitudes. Many are very important as sources of timber and pulp.

There are about 600 living species of conifers. The tallest living vascular plant, the coastal sequoia (*Sequoia sempervirens*), found in coastal California and Oregon, is a conifer and reaches over 100 meters (328 ft). The biggest redwood, however, is the mountain sequoia redwood species (*Sequoiadendron gigantea*) of the Sierra Nevadas. The largest individual tree is nicknamed after General Sherman of the Civil War, and it stands more than 83 meters (274 ft) tall while measuring 31 meters (102 ft) around its base. Another much smaller type of conifer, the bristlecone pines in Nevada, may be the oldest trees in the world—about 5,000 years old.

The other three gymnosperm phyla are much less widespread. Cycads (figure 17.14*b*), the predominant land plant in the golden age of dinosaurs, the Jurassic period (213–144 million years ago), have short stems and palmlike leaves. They are still widespread throughout the tropics. The

(a)

(b)

(c)

(d)

Figure 17.14 Gymnosperms.

(a) These Douglas fir trees, a type of conifer, often occur in vast forests. (b) An African cycad, *Encephalartos transvenosus,* phylum Cycadophyta. The cycads have fernlike leaves and seed-forming cones, like the ones shown here. (c) *Welwitschia mirabilis,* phylum Gnetophyta, is found in the extremely dry deserts of southwestern Africa. In *Welwitschia,* two enormous, beltlike leaves grow from a circular zone of cell division that surrounds the apex of the carrot-shaped root. (d) Maidenhair tree, *Ginkgo biloba,* the only living representative of the phylum Ginkgophyta, a group of plants that was abundant 200 million years ago. Among living seed plants, only the cycads and ginkgo have swimming sperm.

gnetophytes, phylum Gnetophyta, contains only three kinds of plants, all unusual. One of them is perhaps the most bizarre of all plants, *Welwitschia,* shown in figure 17.14*c,* which grows on the exposed sands of the harsh Namibian Desert of southwestern Africa. *Welwitschia* acts like a plant standing on its head! Its two beltlike, leathery leaves are generated continuously from their base, splitting as they grow out over the desert sands. There is only one living species of ginkgo, which has fan-shaped leaves (shown in figure 17.14*d*) that are shed in the autumn. Because ginkgos are resistant to air pollution, they are commonly planted along city streets.

The fossil record indicates that members of the ginkgo phylum were once widely distributed, particularly in the Northern Hemisphere; today, only one living species, the maidenhair tree (*Ginkgo biloba*), remains. The reproductive structures of ginkgos are produced on separate trees. The fleshy outer coverings of the seeds of female ginkgo plants exude the foul smell of rancid butter caused by butyric and isobutyric acids.

The Life of a Gymnosperm

We will examine conifers as a typical gymnosperm. The conifer life cycle is illustrated in figure 17.15. Conifer trees form two kinds of cones. Pollen cones ❶ contain the male gametophytes, pollen grains; seed cones ❸ contain the female gametophytes, with their egg cells. Conifer pollen grains ❷ are small and light and are carried by the wind to seed cones. Because it is very unlikely that any particular pollen grain will succeed in being carried to a seed cone (the wind can take it anywhere), a great many pollen grains are produced to be sure that at least a few succeed in pollinating seed cones. For this reason, pollen grains are shed from their cones in huge quantities, often appearing as a sticky yellow layer on the surfaces of ponds and lakes—and even on windshields.

When a grain of pollen settles down on a scale of a female cone, a slender tube grows out of the pollen cell up into the scale, delivering the male gamete to the female gametophyte containing the egg, or ovum. Fertilization occurs when the sperm cell fuses with the egg ❺, forming a zygote that develops into an embryo. This zygote is the beginning of the sporophyte generation. What happens next is the essential improvement in reproduction achieved by seed plants. Instead of the zygote growing immediately into an adult sporophyte—just as you grew directly into an adult from a fertilized zygote—the fertilized ovule forms a seed ❻. The pine seed contains a sail-like structure that helps carry the seed by the wind. The seed can then be dispersed into new habitats. If conditions are favorable where the seed lands, it will germinate and begin to grow, forming a new sporophyte plant ❼.

Concept Check

1. Does a megaspore give rise to male or female gametes, or to both?
2. Name the three components of a seed and their functions.
3. Is the plant embryo within a seed a gametophyte or a sporophyte?

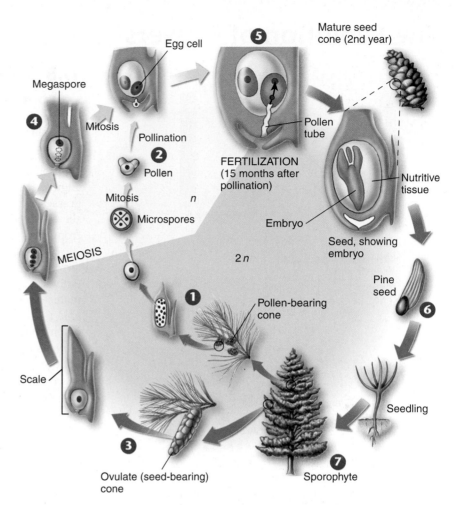

Figure 17.15   The life cycle of a conifer.

In all seed plants, the gametophyte generation is greatly reduced. In conifers, the relatively delicate pollen-bearing cones contain microspores ❶, which give rise to pollen grains ❷, which are the male gametophytes. The familiar seed-bearing cones of pines ❸ are much heavier and more substantial structures than the pollen-bearing cones. Two ovules, and ultimately two seeds, are borne on the upper surface of each scale, which contains the megaspores that give rise to the female gametophytes ❹. After a pollen grain reaches a scale, it germinates, and a slender pollen tube grows toward the egg. When the pollen tube grows to the vicinity of the female gametophyte, sperm are released ❺, fertilizing the egg and producing a zygote. The development of the zygote into an embryo takes place within the ovule, which matures into a seed ❻. Eventually, the seed falls from the cone and germinates, the embryo resuming growth, becoming a new pine tree ❼.

IMPLICATION   Each year Americans spend some $1.5 billion on 36 million Christmas trees grown on 22,000 tree farms occupying half a million acres. Almost all American Christmas trees are Douglas or Balsam firs, as fir trees don't shed their leaves when they dry out as other evergreens do like spruce, hemlock, and pine. American trees are sheared during growth to achieve a more conical shape, while Europeans favor the open aspect of unsheared trees. Which do you prefer?

The Evolution of Flowers

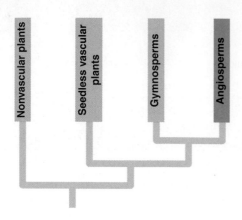

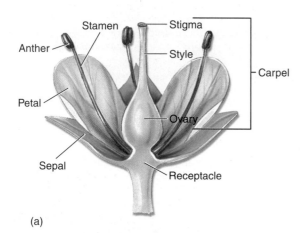

(a)

(b)

Figure 17.16   An angiosperm flower.

(a) The basic structure of a flower is a series of four concentric circles, or whorls: the sepals, petals, stamens, and the carpel (ovary, style, and stigma), also seen in (b) the wild woodland plant *Geranium*.

17.8 Rise of the Angiosperms

CONCEPT PREVIEW: Angiosperms are flowering seed plants in which the ovule is completely enclosed by diploid tissue at pollination and which undergo double fertilization.

Angiosperms are plants in which the ovule is completely enclosed by sporophyte tissue when it is fertilized. Ninety percent of all living plants are angiosperms, about 250,000 species, including many trees, shrubs, herbs, grasses, vegetables, and grains—in short, nearly all of the plants that we see every day. Virtually all of our food is derived, directly or indirectly, from angiosperms.

Angiosperms successfully meet the last difficult challenge posed by terrestrial living: the inherent conflict between the need to obtain nutrients (solved by roots, which anchor the plant to one place) and the need to find mates (solved by accessing plants of the same species). Angiosperms are able to deliver their pollen directly, as if in an addressed envelope, from one individual of a species to another. How? *By inducing insects and other animals to carry it for them!* The tool that makes this animal-dictated pollination possible, the great advance of the angiosperms, is the flower. While some very successful later-evolving angiosperms like grasses have reverted to wind pollination of their flowers, the directed pollination of flowering plants has led to phenomenal evolutionary success.

The Flower

Flowers are the reproductive organs of angiosperm plants. A flower is a sophisticated pollination machine. It may employ bright colors to attract the attention of insects (or birds or small mammals), nectar to induce the insect to enter the flower, and structures that coat the insect with pollen grains while it is visiting. Then, when the insect visits another flower, it carries the pollen with it.

The basic structure of a flower consists of four concentric circles, or **whorls,** connected to a base called the **receptacle:**

1. The outermost whorl, called the **sepals** of the flower, typically serves to protect the flower from physical damage. These are the green leaf-like structures in figure 17.16*a* and are in effect modified leaves that protect the flower while it is a bud. The sepals can be seen surrounding the flower bud to the lower left of the flower in figure 17.16*b*.
2. The second whorl, called the **petals** of the flower, serves to attract particular pollinators. Petals have particular pigments, often vividly colored like the light purple color in figure 17.16.
3. The third whorl, called the **stamens** of the flower, contains the "male" parts that produce the pollen grains. Stamens are the slender, thread-like filaments in figure 17.16 with a swollen **anther** at the tip containing pollen.
4. The fourth and innermost whorl, called the **carpel** of the flower, contains the "female" parts that produce eggs. The carpel is the vase-shaped structure in figure 17.16. The carpel is sporophyte tissue that completely encases the ovules within which the egg cell develops. The ovules occur in the bulging lower portion of the carpel, called the **ovary;** usually there is a slender stalk rising from the ovary called the **style,** with a sticky tip called a **stigma,** which receives pollen. When the flower is pollinated, a pollen tube grows down from the pollen grain on the stigma through the style to the ovary to fertilize the egg.

Improving Seeds: Double Fertilization

The seeds of gymnosperms often contain food to nourish the developing plant in the critical time immediately after germination, but the seeds of angiosperms have greatly improved on this aspect of seed function. Angiosperms produce a special, highly nutritious tissue called **endosperm** within their seeds. Here is how it happens. The angiosperm life cycle is presented in figure 17.17, but there are actually two parts to the cycle, a male and a female part, indicated by the two sets of arrows in the top half of the cycle. We'll begin with the flower on the left side of the cycle ❶. The development of the male gametophyte (the pollen grain) occurs in the anthers and is indicated in the upper set of arrows. The anthers are shown in cross section in ❷ so you can see the microspore mother cells that develop into pollen grains. The pollen grain contains two haploid sperm. Upon adhering to the stigma at the top of the carpel (the female organ where the egg cell is produced in the ovary ❸), the pollen begins to form a pollen tube ❹. The pollen tube grows down into the carpel until it reaches the ovule in the ovary ❺. The two sperm (the small purplish cells) travel down the pollen tube and into the ovary. The first sperm fuses with the egg (the green cell at the base of the ovary) forming the zygote that develops into the embryo. The other sperm cell fuses with two other products of meiosis, called polar nuclei, to form a triploid (three copies of the chromosomes, 3*n*) endosperm cell. This cell divides much more rapidly than the zygote, giving rise to the nutritive endosperm tissue (the tan material surrounding the embryo ❻). This process of fertilization with two sperm to produce both a zygote and endosperm is called **double fertilization.** Double fertilization is exclusive to angiosperms. As described earlier, in some angiosperms, the endosperm is used up during development, producing leaflike structures called cotyledons that store the food.

Some angiosperm embryos have two cotyledons, and are called *dicotyledons,* or **dicots.** Dicot plants typically have leaves with netlike branching of veins and flowers with four or five parts per whorl. Oak and maple trees are dicots. The embryos of other angiosperms, which evolved somewhat later, have a single cotyledon and are called *monocotyledons,* or **monocots.** Monocots typically have leaves with parallel veins and flowers with three parts per whorl. Grasses are wind-pollinated monocots. Monocots and dicots are compared in more detail in chapter 32.

Concept Check

1. Name three things unique to angiosperms.
2. What is the role of nectar in a flower?
3. Describe the four whorls of a flower. Which of them tend to be brightly colored? Why?

Figure 17.17 Life cycle of an angiosperm.

In angiosperms, as in gymnosperms, the sporophyte is the dominant generation. However, in angiosperms the reproductive structures are the flowers ❶. The male gametophytes, the pollen grains, are formed within the anthers ❷ and complete their differentiation to the mature, three-celled stage either before or after grains are shed. Meanwhile, eggs form within the embryo sac, inside the ovules ❸, which, in turn, are enclosed in the carpels. The carpel is differentiated in most angiosperms into a slender portion, or style, ending in a stigma, the surface on which the pollen grains germinate ❹. Fertilization is distinctive in angiosperms, being a double process ❺. A sperm and an egg come together, producing a zygote; at the same time, another sperm fuses with the two polar nuclei, producing the primary endosperm nucleus, which is triploid. The zygote and the primary endosperm nucleus divide mitotically, giving rise, respectively, to the embryo and the endosperm ❻. The endosperm is the tissue, almost exclusive to angiosperms, that nourishes the embryo and young plant.

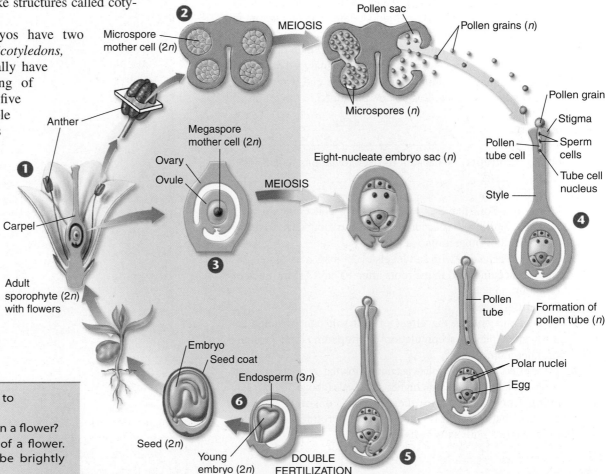

How Does Arrowgrass Tolerate Salt?

Plants grow almost everywhere on earth, thriving in many places where exposure, drought, and other severe conditions challenge their survival. In deserts a common stress is the presence of high levels of salt in the soils. Soil salinity is also a problem for millions of acres of abandoned farmland because the accumulation of salt from irrigation water restricts growth. Why does excess salt in the soil present a problem for a plant? For one thing, high levels of sodium ions that are taken up by the roots are toxic. For another, a plant's roots cannot obtain water when growing in salty soil. Osmosis (the movement of water molecules to areas of higher solute concentrations, see page 78) causes water to move in the opposite direction, drawn out of the roots by the soil's high levels of salt. And yet plants do grow in these soils. How do they manage?

To investigate this, researchers have studied seaside arrowgrass (*Triglochin maritima*), the plant you see to the right. Arrowgrass plants are able to grow in very salty seashore soils, where few other plants survive. How are they able to survive? Researchers found that their roots do not take up salt, and so do not accumulate toxic levels of salt.

However, this still leaves the arrowgrass plant the challenge of preventing its root cells from losing water to the surrounding salty soil. How then do the roots achieve osmotic balance? In an attempt to find out, researchers grew arrowgrass plants in nonsalty soil for two weeks, then transferred them to one of several soils which differed in salt level. After 10 days, shoots were harvested and analyzed for amino acids, because accumulating amino acids could be one way that the cells maintain osmotic balance. Results are presented in the graph.

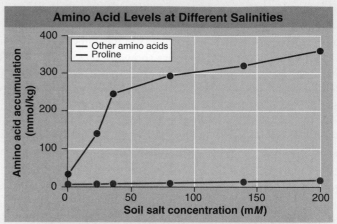

Analysis

1. **Applying Concepts** What do the abbreviations "mM" and "mmol/kg" mean?
2. **Interpreting Data**
   a. In salt-free soil (that is, the mM soil salt concentration = 0), how much proline has accumulated in the roots after 10 days? How much of other amino acids?
   b. In salty beach soils with salt levels of 35 mM, how much proline has accumulated in the roots after 10 days? How much of other amino acids?
3. **Making Inferences**
   a. In general, what is the effect of soil salt concentration on arrowgrass plant's accumulation of the amino acid proline? of other amino acids?
   b. Is the effect of salt on proline accumulation the same at lower salt levels (below 50 mM) as at higher salt levels (above 50 mM)?
4. **Drawing Conclusions** Are these results consistent with the hypothesis that arrowgrass accumulates proline to achieve osmotic balance with salty soils?

Concept Summary

Plants

17.1 Adapting to Terrestrial Living

- Plants are complex multicellular autotrophs, producing their own food through photosynthesis.

- Plants acquire water and minerals from the soil. They control water loss through a watertight layer called the cuticle. Openings, called stomata, allow for gas exchange with the air (**figure 17.2**).

- The adaptation of pollen allowed land plants to transfer gametes on land under dry conditions. The pollen grain protects the gamete from drying out.

- Plant life cycles involve an alternation of generations where a haploid gametophyte alternates with a diploid sporophyte, with the sporophyte dominating the life cycle more and more as plants evolved (**figure 17.3**).

17.2 Plant Evolution

- Plants evolved from green algae, helped by four innovations. These innovations include an alternation of generations life cycle, vascular tissue, seeds, and flower and fruits (**figure 17.5** and **table 17.1**).

- In an alternation of generations, the plant spends a portion of its life in a multicellular haploid phase that produces gametes and another portion in a multicellular diploid phase. As plants evolved, the diploid sporophyte became a more dominant structure, both in size and in its duration of time in the life cycle.

- The evolution of vascular tissue allowed plants to transport water and minerals from the soil and up through their bodies, distributing food throughout the plant and allowing them to grow taller.

- Seeds provide protection and nutrients for the developing embryo. Flowers and fruits improve the chances of successful mating and the distribution of seeds.

Seedless Plants

17.3 Nonvascular Plants

- Liverworts and hornworts don't have vascular tissue. Mosses have specialized cells for conducting water and carbohydrates in the plant, but not rigid vascular tissue. These three types of plants are grouped together as nonvascular plants. Their life cycles are dominated by the haploid gametophyte (**figure 17.6**).

17.4 The Evolution of Vascular Tissue

- Vascular tissue consists of specialized cylindrical or elongated cells that form a network throughout the plant. Xylem conducts water from the roots up the stem to the leaves. Phloem conducts carbohydrates manufactured in the leaves throughout the plant (**figure 17.8**).

- The evolution of vascular tissue gave plants a rigidity that allowed them to grow taller.

- Plants grow in two directions. All plants extend roots and stems through primary growth by adding cells to the growing tips. Plants that exhibit secondary growth become thicker as new cells are added around the periphery of the plant.

17.5 Seedless Vascular Plants

- Among the most primitive vascular plants are the Pterophyta (ferns) and Lycophyta (**figure 17.9**). These plants can grow very tall due to vascularization, but they lack seeds and require water for fertilization. The sporophyte, shown here from **figure 17.10,** is larger than its gametophyte.

The Advent of Seeds

17.6 Evolution of Seed Plants

- In seed plants, the sporophyte becomes the dominant structure in the life cycle. The plants contain separate male and female gametophytes. The male gametophyte, called a pollen grain, produces sperm which are carried to the egg, which is contained in the small female gametophyte. Pollination in seed plants doesn't require water.

- The seed was an evolutionary innovation that provided protection for the plant embryo. The seed contains the embryo and food inside a drought-resistant cover (**figure 17.12**). Seeds are a key adaptation to land-dwelling plants because they improve the dispersal of embryos, allow dormancy when needed, germinate when conditions are favorable, and provide nourishment during germination.

17.7 Gymnosperms

- Gymnosperms are nonflowering seed plants and include conifers, cycads, gnetophytes, and the ginkgo. The seeds are produced in cones where pollination occurs. Pollen, produced in smaller cones, travels by wind to the seed-bearing cones, where fertilization of the egg takes place. The seed, like the one shown here from **figure 17.15,** disperses the embryo.

The Evolution of Flowers

17.8 Rise of the Angiosperms

- Angiosperms, flowering plants, are plants in which the ovule is completely enclosed by sporophyte tissue.

- The flowers are the reproductive structures. The basic structure of a flower includes four concentric whorls, shown here from **figure 17.16.** The outermost green sepals protect the flower. Next, the colorful petals attract pollinators. The third whorl contains the stalklike stamen with the pollen-bearing anthers. The innermost whorl is a vaselike structure called the carpel that contains the egg within the ovary. Flowers improved the efficiency of mating.

- Double fertilization provides food for the germinating plant. The pollen grain produces two sperm cells that travel through the pollen tube, as shown here from **figure 17.17.** One sperm cell fuses with the egg forming the plant embryo. The other sperm cell fuses with two polar nuclei, producing triploid endosperm. Both structures become enclosed in the seed. The angiosperms are divided into two groups: dicots and monocots. Dicots have two cotyledons, netlike veins in the leaves, and flower parts in numbers of four or five. Monocots have only one cotyledon, parallel veins in the leaves, and flower parts in threes or multiples of three.

Self-Test

1. A major consideration for the evolution of terrestrial plants is the problem of
   a. too much sunlight.
   c. dehydration.
   b. predators.
   d. not enough carbon.

2. Which of the following structures or systems is not unique to plant evolution?
   a. chloroplasts
   c. seeds
   b. vascular tissue
   d. flowers and fruits

3. Mosses, liverworts, and hornworts do not reach a large size because
   a. they lack chlorophyll.
   b. they do not have specialized vascular tissue to transport water very high.
   c. photosynthesis does not take place at a very fast rate.
   d. alternation of generations does not allow the plants to grow very tall before reproduction.

4. One characteristic that separates ferns from complex vascular plants is that ferns do not have
   a. a vascular system.
   b. chloroplasts.
   c. alternation of generations in their life cycle.
   d. seeds.

5. As seed plants evolved, the _____ form became more dominant in the life cycle.
   a. gametophyte
   c. sporophyte
   b. gymnosperm
   d. angiosperm

6. In seeds, the endosperm helps with
   a. fertilization.
   c. nourishment.
   b. photosynthesis.
   d. dispersal.

7. What separates the gymnosperms from other seed plants?
   a. a vascular system
   b. wind dispersion of pollen
   c. fertilization occurs in seed cones
   d. fruits and flowers

8. What separates the angiosperms from other seed plants?
   a. a vascular system
   b. wind dispersion of pollen
   c. fruits and flowers
   d. all of the above

9. Flower shape and color can be linked to the process of
   a. pollination.
   c. germination.
   b. photosynthesis.
   d. secondary growth.

10. Monocots and dicots differ in all of the following ways except
    a. vein patterns in the leaves.
    b. the process of double fertilization.
    c. the number of flower parts.
    d. the number of cotyledons.

Visual Understanding

1. **Figure 17.5** Explain which key innovation led to the evolution of each group and why the innovation was significant.

2. **Figure 17.12** When farmers harvest corn, soybeans, wheat, and rice, they collect the seeds and get rid of the rest of the plant. Why do we eat the seeds rather than the stems or roots of these plants? *Hint:* Think about the purpose of a seed in the life cycle of the plant.

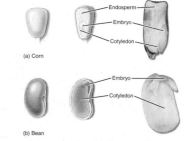

3. **Figure 17.13** The seeds that fall from this tree are scattered about on the ground nearby. What are the benefits or drawbacks to having: (a) chemicals in the seed to inhibit sprouting, and (b) hard seed coats?

Challenge Questions

1. Explain why mycorrhizae are advantageous to both the plant and fungal partners.

2. Why are mosses limited to moist places?

3. Why does a pine tree, which might live more than 100 years, put so much energy and resources into producing hundreds of pine cones each year, each with 30 to 50 seeds?

4. Grass species are angiosperms and so they have flowers, but their flowers are usually small and lack the bright colors and fragrances. Explain why grasses don't have more conspicuous flowers.

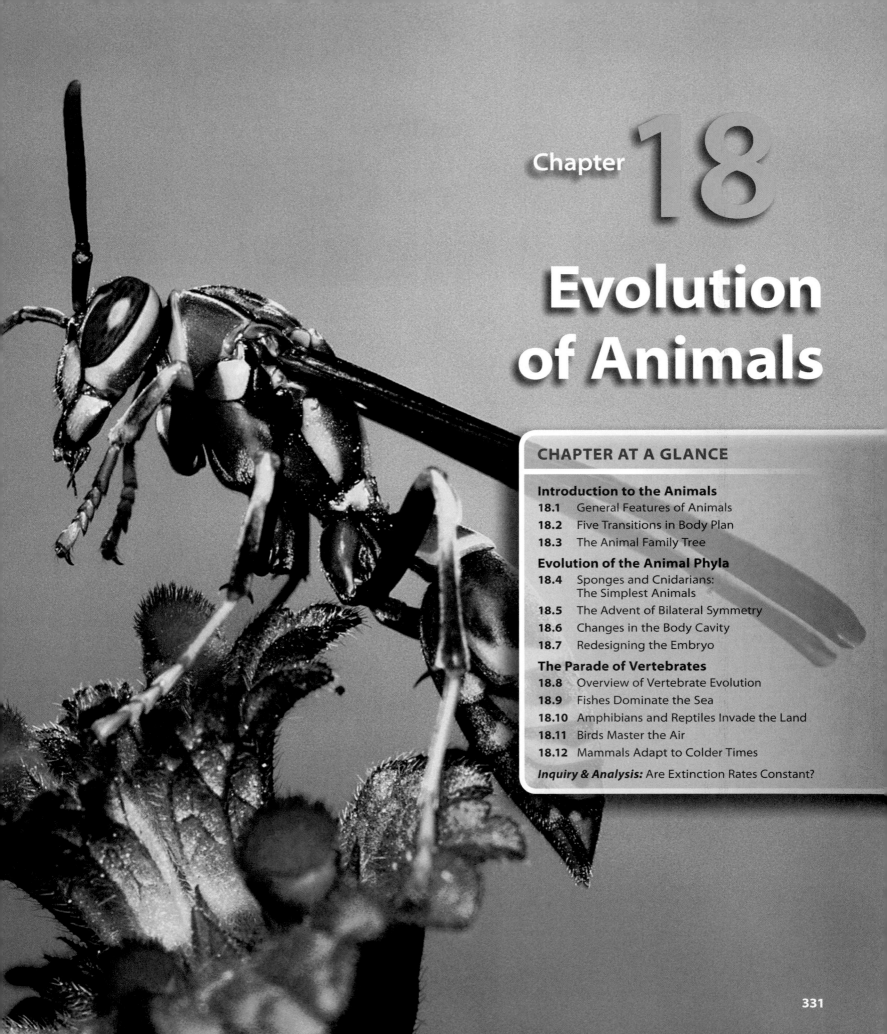

CHAPTER AT A GLANCE

Introduction to the Animals

18.1 General Features of Animals

CONCEPT PREVIEW: Animals are complex, multicellular, heterotrophic organisms.

From protist ancestors, an astonishing diversity of animals has evolved. All animals, however, have certain key features in common (table 18.1): (1) Animals are heterotrophs and must ingest plants, algae, or other animals for nourishment. (2) All animals are multicellular, but unlike protists, fungi, and plants, animal cells lack cell walls. (3) Animals are able to move from place to place. (4) Animals are very diverse in form and habitat. (5) Most animals reproduce sexually. (6) Animals have characteristic tissues and patterns of development.

TABLE 18.1	General Features of Animals

Heterotrophs. Unlike autotrophic plants and algae, animals cannot construct organic molecules from inorganic chemicals. All animals are heterotrophs—that is, they obtain energy and organic molecules by ingesting other organisms. Some animals (herbivores) consume autotrophs, other animals (carnivores) consume heterotrophs; others, like the bear to the right, are omnivores that eat both autotrophs and heterotrophs, and still others (detritivores) consume decomposing organisms.

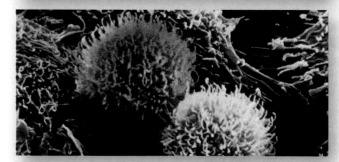

Multicellular. All animals are multicellular, often with complex bodies like that of this brittlestar (*right*). The unicellular heterotrophic organisms called protozoa, which were at one time regarded as simple animals, are now considered members of the large and diverse kingdom Protista, discussed in chapter 16.

No Cell Walls. Animal cells are distinct among those of multicellular organisms because they lack rigid cell walls and are usually quite flexible, like these cancer cells. The many cells of animal bodies are held together by extracellular lattices of structural proteins such as collagen. Other proteins form a collection of unique intercellular junctions between animal cells.

Active Movement. The ability of animals to move more rapidly and in more complex ways than members of other kingdoms is perhaps their most striking characteristic, one that is directly related to the flexibility of their cells and the evolution of nerve and muscle tissues. A remarkable form of movement unique to animals is flying, an ability that is well developed among vertebrates and insects like this butterfly. The only terrestrial vertebrate group never to have evolved flight is amphibians.

TABLE 18.1 | General Features of Animals *(continued)*

Diverse in Form.   Almost all animals (99%) are invertebrates, which, like this millipede, lack a backbone. Of the estimated 10 million living animal species, only 42,500 have a backbone and are referred to as vertebrates. Animals are very diverse in form, ranging in size from organisms too small to see with the unaided eye to enormous whales and giant squids.

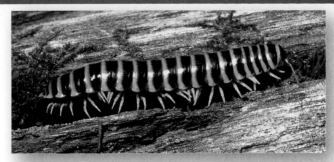

Diverse in Habitat.   The animal kingdom includes about 35 phyla, most of which, like these jellyfish (phylum Cnidaria), occur in the sea. Far fewer phyla occur in freshwater and fewer still occur on land. Members of three successful marine phyla, Arthropoda (insects), Mollusca (snails), and Chordata (vertebrates), dominate animal life on land.

Sexual Reproduction.   Most animals reproduce sexually, as these tortoises are doing. Animal eggs, which are nonmotile, are much larger than the small, usually flagellated sperm. In animals, cells formed in meiosis function directly as gametes. The haploid cells do not divide by mitosis first, as they do in plants and fungi, but rather fuse directly with each other to form the zygote. Consequently, with a few exceptions, there is no counterpart among animals to the alternation of haploid (gametophyte) and diploid (sporophyte) generations characteristic of plants.

Embryonic Development.   Most animals have a similar pattern of embryonic development. The zygote first undergoes a series of mitotic divisions, called *cleavage*, and, like this dividing frog egg, becomes a solid ball of cells, the morula, then a hollow ball of cells, the blastula. In most animals, the blastula folds inward at one point to form a hollow sac with an opening at one end called the blastopore. An embryo at this stage is called a gastrula. The subsequent growth and movement of the cells of the gastrula differ widely from one phylum of animals to another.

Unique Tissues.   The cells of all animals except sponges are organized into structural and functional units called tissues, collections of cells that have joined together and are specialized to perform a specific function. Animals are unique in having two tissues associated with movement: (1) muscle tissue, which powers animal movement, and (2) nervous tissue, which conducts signals among cells. Neuromuscular junctions, where nerves connect with muscle tissue, are shown here.

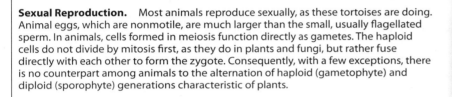

Radial symmetry

Bilateral symmetry

Segmentation

18.2 Five Transitions in Body Plan

CONCEPT PREVIEW: Five key transitions in body design are responsible for most of the differences we see among the major animal phyla.

The features described in the previous section evolved over millions of years. And indeed the great diversity of animals today is the result of this long evolutionary journey. The multicellular animals, or metazoans, are traditionally divided into about 35 distinct and very different phyla. These phyla are characterized by the evolution of five key innovations: tissues, bilateral symmetry, a body cavity, deuterostome development, and segmentation. These five body transitions have traditionally been used by biologists as benchmarks in classifying animals, as indicated at the branch-points of the animal evolutionary tree in figure 18.1. New molecular data suggests a somewhat different tree that takes into account another novel innovation, molting. In this text we will keep with the traditional classification, although this evolutionary tree will most likely change in the future, as discussed in section 18.3.

1. Evolution of Tissues

The simplest animals, the Parazoa, lack both defined tissues and organs. Characterized by the sponges, these animals exist as aggregates of cells with minimal intercellular coordination.

2. Evolution of Bilateral Symmetry

Sponges also lack any definite symmetry, growing as irregular masses. Virtually all other animals have a definite shape and symmetry that can be defined along an imaginary axis through the animal's body. **Radial symmetry** is where body parts are arranged around a central axis, while **bilateral symmetry** is where the body has a right and a left half that are mirror images of each other.

3. Evolution of a Body Cavity

A third key transition in the evolution of the animal body plan was the evolution of the body cavity. The evolution of efficient organ systems within the animal body was not possible until a body cavity evolved for supporting organs, distributing materials, and fostering complex developmental interactions.

4. The Evolution of Deuterostome Development

All but two of the major phyla of bilateral animals are members of a group called the **protostomes** (from the Greek words *protos,* first, and *stoma,* mouth) which includes the flatworms, nematodes, mollusks, annelids, and arthropods. They all undergo similar embryonic development. Two outwardly dissimilar groups, the echinoderms and the chordates, comprise a second group, the **deuterostomes** (Greek, *deuteros,* second, and *stoma,* mouth), with key differences in their pattern of embryonic development compared to that of the protostomes.

5. The Evolution of Segmentation

The fifth key transition in the animal body plan involved the subdivision of the body into **segments,** clearly seen in the earthworm to the left. Just as it is efficient for workers to construct a tunnel from a series of identical prefabricated parts, so segmented animals are assembled from a succession of identical segments.

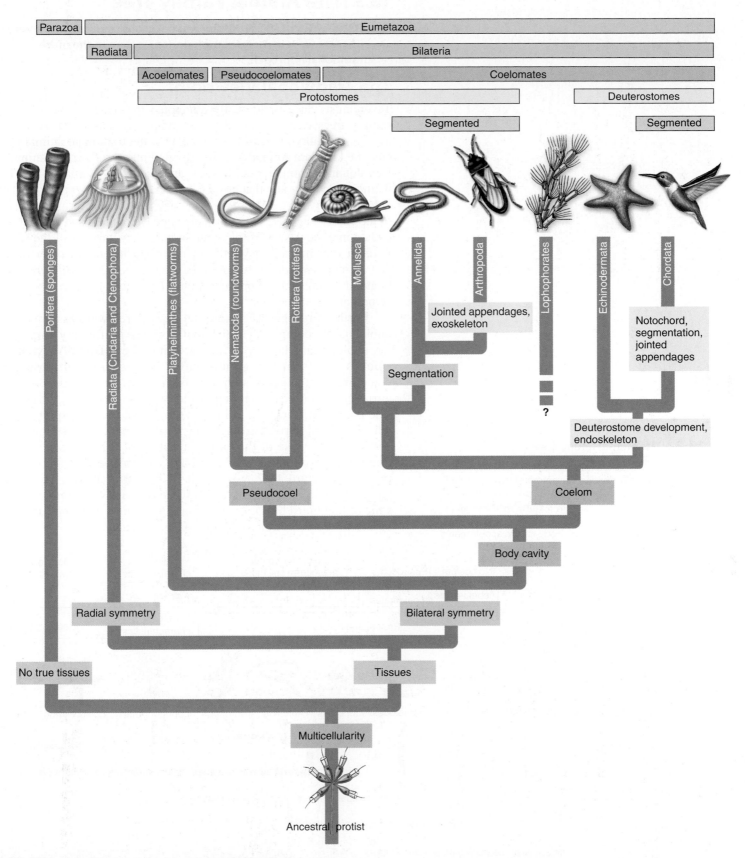

Figure 18.1   Evolutionary trends among the animals.

In this chapter, we examine a series of key evolutionary innovations in the animal body plan, shown here along the branches. Some of the major animal phyla are shown on this tree. Lophophorates is a group that contains two phyla that exhibit a mix of protostome and deuterostome characteristics.

18.3 The Animal Family Tree

CONCEPT PREVIEW: Major animal groups are related in very different ways in molecular phylogenies than in the more traditional approach.

The Traditional Viewpoint

Taxonomists have traditionally attempted to create animal *phylogenies* (family trees; see section 15.5) by comparing anatomical features and aspects of embryological development. In the traditional animal phylogeny, the kingdom Animalia is first divided into the *Parazoa*, animals that lack definite symmetry and tissues, and *Eumetazoa*, animals that have a definite symmetry and that possess tissues. The eumetazoan branch of the animal family tree then has two principal branches, differing in the nature of the embryonic layers that form during development and go on to differentiate into the tissues of the adult animal. Eumetazoans of the subgroup *Radiata* (having radial symmetry) have two layers, an outer *ectoderm* and an inner *endoderm*. All other eumetazoans are the *Bilateria* (having bilateral symmetry) and produce a third layer, the *mesoderm,* between the ectoderm and endoderm.

Further branches of the animal family tree were assigned by taxonomists by comparing traits that seemed profoundly important to the evolutionary history of phyla. Thus, the bilaterally symmetrical animals were split into groups with a body cavity and those without (acoelomates); animals with a body cavity were split into those with a true coelom (body cavity enclosed by mesoderm) and those without (the pseudocoelomates); animals with a coelom were split into those whose coelom derived from the digestive tube and those that did not, and so on. Because of the either-or nature of the categories set up by traditional taxonomists, this approach has produced a family tree like the one in figure 18.2, with a lot of paired branches. The arbitrary nature of the divisions has always been obvious to biologists, but in spite of that, most biologists feel that it faithfully represents the general nature of the evolutionary history of metazoans.

Figure 18.2 The animal family tree: the traditional viewpoint.

Biologists have traditionally divided the animals into 35 distinct phyla. This diagram illustrates the relationships among some of the major animal phyla. The bilaterally symmetrical animals (those to the right of Radiata in the figure) are sorted into three groups that differ with respect to their body cavity: acoelomates, pseudocoelomates, and coelomates.

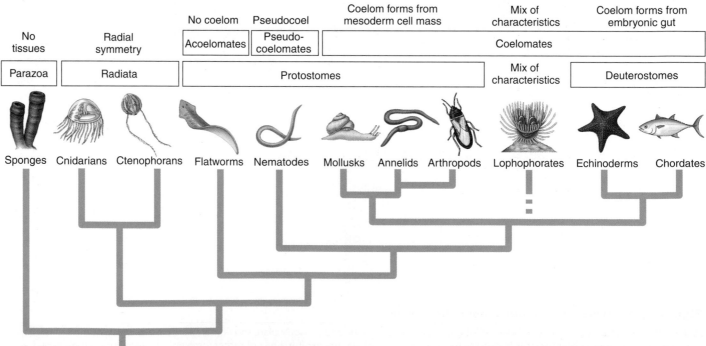

A New Look at the Animal Family Tree

The last decade has seen a wealth of new molecular RNA and DNA sequence data on the various animal groups. Using these sorts of molecular data, a variety of molecular phylogenies have been produced. While differing from one another in many important respects, the new molecular phylogenies have the same deep branch structure as the traditional animal family tree. However, most agree on one revolutionary difference from the traditional phylogeny: The protostomes (which have a different pattern of development than the deuterostomes—a topic that will be discussed later in this chapter) are broken into two distinct clades. Figure 18.3 is a consensus molecular phylogeny developed from DNA, ribosomal RNA, and protein studies. In it, the traditional protostome group is broken up into Lophotrochozoa and Ecdysozoa.

Lophotrochozoans are animals that grow by adding mass to an existing body. They are named for a distinctive feeding apparatus called a lophophore found in some phyla of the molecularly-defined group. These animals—which usually live in water, have ciliary locomotion, and trochophore larvae—include flatworms, mollusks, and annelids. **Ecdysozoans** have exoskeletons that must be shed for the animal to grow. This sort of molting process is called ecdysis, which is why these animals are called ecdysozoans. They include the roundworms (nematodes) and arthropods.

Molecular phylogenetic analysis of the animal kingdom is in its infancy, but the childhood of the new molecular approach is likely to be short. Over the next few years, a mountain of additional molecular data can be anticipated. As more data are brought to bear, the confusion can be expected to lessen and the relationships within groupings more confidently resolved.

Concept Check

1. Name at least six key features that all animals share.
2. Are you bilaterally symmetrical? Explain.
3. What is different between protostomes and deuterostomes?

Figure 18.3   The animal family tree: a new look.

New phylogenies suggest that the protostomes might be better grouped according to whether they grow by adding mass to an existing body (Lophotrochozoa) or by molting (Ecdysozoa).

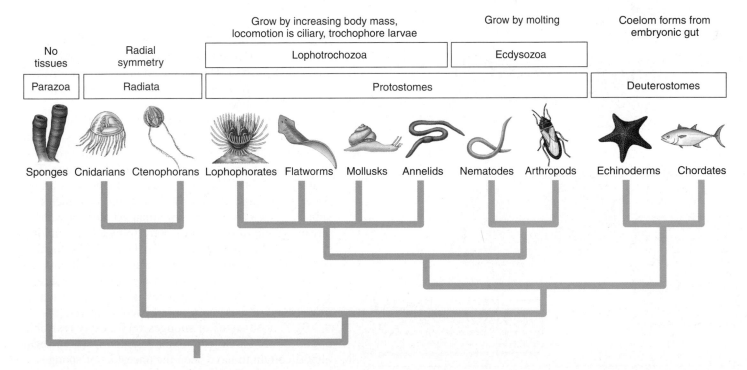

Evolution of the Animal Phyla

(a)   (b)

Figure 18.4 Diversity in sponges.

These two marine sponges are barrel sponges. They are among the largest of sponges, with well-organized forms. Many are more than 2 meters in diameter (a), while others are smaller (b).

18.4 Sponges and Cnidarians: The Simplest Animals

CONCEPT PREVIEW: Sponges have a multicellular body with specialized cells but lack definite symmetry and organized tissues. Cnidarians possess radial symmetry and specialized tissues and carry out extracellular digestion.

The kingdom Animalia consists of two subkingdoms: (1) *Parazoa,* animals of phylum Porifera that lack a definite symmetry, and (2) *Eumetazoa,* 35 phyla of animals that have a definite shape and symmetry

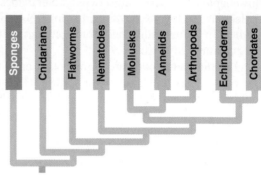

Sponges: Multicellularity

Sponges, members of the phylum Porifera, are the simplest animals. Most sponges completely lack symmetry, and although some of their cells are highly specialized, they are not organized into tissues. The bodies of sponges consist of little more than masses of specialized cells embedded in a gel-like matrix, like chopped fruit in Jell-O. However, sponge cells do possess a key property of animal cells: cell recognition. For example, when a sponge is passed through a fine silk mesh, individual cells separate and then reaggregate on the other side to re-form the sponge. Clumps of cells disassociated from a sponge can give rise to entirely new sponges.

About 5,000 species exist, almost all in the sea. Some are tiny, and others are more than 2 meters in diameter (figure 18.4*a*). The body of an adult sponge is anchored in place on the seafloor and is shaped like a vase (figure 18.4*b*). The outside of the sponge is covered with a skin of flattened cells called epithelial cells that protect the sponge.

The body of the sponge is perforated by tiny holes. The name of the phylum, Porifera, refers to this system of pores. Unique flagellated cells called **choanocytes,** or collar cells, line the body cavity of the sponge. The beating of the flagella of the many choanocytes draws water in through the pores. One cubic centimeter of sponge tissue can propel more than 20 liters of water a day in and out of the sponge body! Why all this moving of water? The sponge is a "filter-feeder." The beating of each choanocyte's flagellum draws water through its collar, made of small hairlike projections resembling a picket fence. Any food particles in the water, such as protists and tiny animals, are trapped in the fence.

The choanocytes of sponges very closely resemble a kind of protist called choanoflagellates, which seem almost certain to have been the ancestors of sponges.

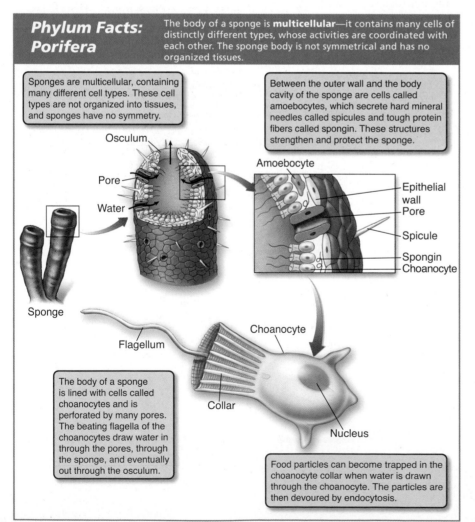

Phylum Facts: Porifera

The body of a sponge is **multicellular**—it contains many cells of distinctly different types, whose activities are coordinated with each other. The sponge body is not symmetrical and has no organized tissues.

Sponges are multicellular, containing many different cell types. These cell types are not organized into tissues, and sponges have no symmetry.

Between the outer wall and the body cavity of the sponge are cells called amoebocytes, which secrete hard mineral needles called spicules and tough protein fibers called spongin. These structures strengthen and protect the sponge.

Osculum
Pore
Water
Sponge
Flagellum

Amoebocyte
Epithelial wall
Pore
Spicule
Spongin
Choanocyte

Choanocyte
Collar
Nucleus

The body of a sponge is lined with cells called choanocytes and is perforated by many pores. The beating flagella of the choanocytes draw water in through the pores, through the sponge, and eventually out through the osculum.

Food particles can become trapped in the choanocyte collar when water is drawn through the choanocyte. The particles are then devoured by endocytosis.

Cnidarians: Symmetry and Tissues

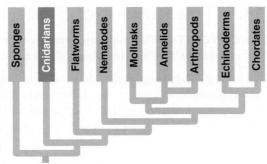

All animals other than sponges have both symmetry and tissues. The most primitive are two radially symmetrical phyla whose bodies are organized around a central axis like the petals of a daisy. Radial symmetry offers advantages to these animals, which remain attached to the surface or are free-floating, as their bodies don't pass through the environment, but rather interact with it on all sides. These two phyla are Cnidaria (pronounced ni-DAH-ree-ah) and Ctenophora (pronounced ten-NO-fo-rah). Cnidaria includes hydra (figure 18.5*a*), jellyfish (figure 18.5*b*), corals, and sea anemones. Ctenophora is a minor phylum of comb jellies. These two phyla together are called the Radiata. The bodies of all other animals with tissues, the Bilateria, are marked by a fundamental bilateral symmetry (discussed in next section).

Cnidarians (phylum Cnidaria) are carnivores that capture prey such as fishes and shellfish with tentacles that ring their mouths. These tentacles, and sometimes the body surface, bear unique stinging cells called **cnidocytes** that give the phylum its name. Within each cnidocyte is a small but powerful harpoon called a **nematocyst** that cnidarians use to spear their prey and then draw the harpooned prey back. The cnidocyte builds up a very high internal osmotic pressure and uses it to push the nematocyst outward so explosively that the barb can penetrate the hard shell of a crab.

A major evolutionary innovation that arose in the radiates is *extracellular digestion* of food. In sponges, food is taken directly into cells by endocytosis and digested. In radiates, food is digested in a cavity, the *gastrovascular cavity,* where digestive enzymes break down the food, and the products of digestion are absorbed by cells that line the cavity. This evolutionary advance has been retained by all of the more advanced animal groups. For the first time, it became possible to digest an animal larger than oneself.

Cnidarians have two basic body forms. **Medusae** are free-floating, gelatinous, umbrella-shaped forms (see figure 18.5*b*). Their mouths point downward, with a ring of tentacles hanging down around the edges. Medusae are commonly called "jellyfish" because of their gelatinous interior or "stinging nettles" because of their nematocysts. **Polyps** are cylindrical, pipe-shaped forms that usually attach to a rock. *Hydra* as seen in figure 18.5*a*, sea anemones, and corals are examples of polyps. In polyps, the mouth faces away from the rock, often directed upward. For shelter and protection, corals deposit an external "skeleton" of calcium carbonate within which they live. This is the structure usually identified as coral. Many cnidarians exist only as medusae, others only as polyps, and still others alternate between these two phases during the course of their life cycles.

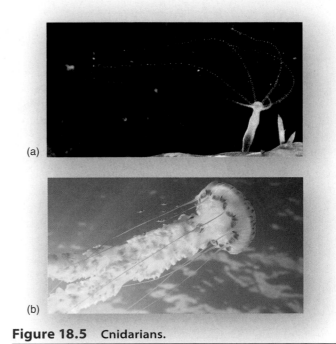

Figure 18.5   Cnidarians.

(a) Hydroids are a group of cnidarians that are mostly marine and colonial. However, *Hydra,* shown above, is a freshwater genus whose members exist as solitary polyps. (b) Jellyfish are translucent, marine cnidarians.

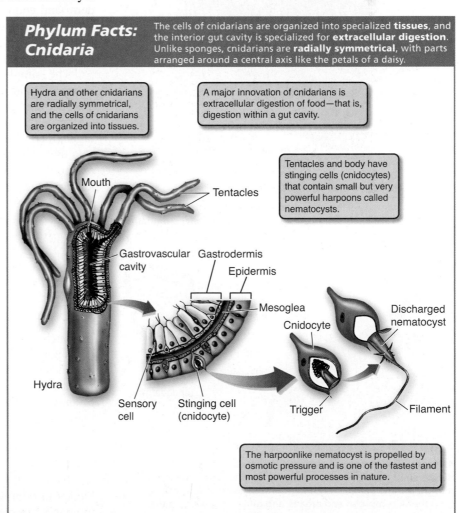

Phylum Facts: Cnidaria

The cells of cnidarians are organized into specialized **tissues**, and the interior gut cavity is specialized for **extracellular digestion**. Unlike sponges, cnidarians are **radially symmetrical**, with parts arranged around a central axis like the petals of a daisy.

Hydra and other cnidarians are radially symmetrical, and the cells of cnidarians are organized into tissues.

A major innovation of cnidarians is extracellular digestion of food—that is, digestion within a gut cavity.

Tentacles and body have stinging cells (cnidocytes) that contain small but very powerful harpoons called nematocysts.

Mouth

Tentacles

Gastrovascular cavity

Gastrodermis

Epidermis

Mesoglea

Cnidocyte

Discharged nematocyst

Hydra

Sensory cell

Stinging cell (cnidocyte)

Trigger

Filament

The harpoonlike nematocyst is propelled by osmotic pressure and is one of the fastest and most powerful processes in nature.

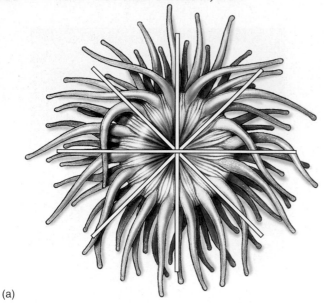

(a)

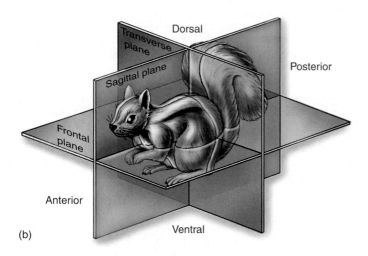

(b)

Figure 18.6   **Radial and bilateral symmetry.**

(a) Radial symmetry is the regular arrangement of parts around a central axis. (b) Bilateral symmetry is reflected in a body form that has a left and right half.

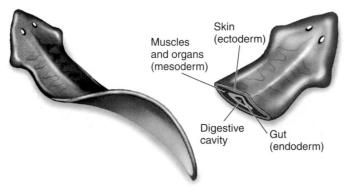

Figure 18.7   **Body plan of a solid worm.**

All bilaterally symmetrical eumetazoans produce three layers during embryonic development: an outer ectoderm, a middle mesoderm, and an inner endoderm. These layers differentiate to form the skin, muscles and organs, and gut, respectively, in the adult animal.

18.5 The Advent of Bilateral Symmetry

CONCEPT PREVIEW: Flatworms have internal organs, bilateral symmetry, and a distinct head. They do not have a body cavity.

All animals with tissues other than cnidarians and ctenophores are **bilaterally symmetrical**—that is, they have a right half and a left half that are mirror images of each other. This is apparent when you compare the radially symmetrical sea anemone in figure 18.6 with the bilaterally symmetrical squirrel. Any of the three planes that cut the sea anemone in half produce mirror images, but only one plane, the green sagittal plane, produces mirror images of the squirrel. In looking at a bilaterally symmetrical animal, you refer to the top half of the animal as **dorsal** and the bottom half as **ventral.** The front is called **anterior** and the back **posterior.** Bilateral symmetry was a major evolutionary advance among the animals because it allows different parts of the body to become specialized in different ways. For example, most bilaterally symmetrical animals have evolved a definite head end, a process called **cephalization.** Animals that have heads are often active and mobile, moving through their environment headfirst, with sensory organs concentrated in front so the animal can test for food, danger, and mates, as it enters new surroundings.

The bilaterally symmetrical animals produce three embryonic layers that develop into the tissues of the body: an outer **ectoderm** (colored blue in the drawing of a flatworm in figure 18.7), an inner **endoderm** (colored yellow), and a third layer, the **mesoderm** (colored red), between the ectoderm and endoderm. In general, the outer coverings of the body and the nervous system develop from the ectoderm, the digestive organs and intestines develop from the endoderm, and the skeleton and muscles develop from the mesoderm.

The simplest of all bilaterally symmetrical animals are the solid worms. By far the largest phylum of these, with about 20,000 species, is Platyhelminthes (pronounced plat-ee-hel-MIN-theeze), which includes the flatworms (figure 18.7). Flatworms are the simplest animals in which organs occur. An organ is a collection of different tissues that function as a unit. The testes and uterus of flatworms are reproductive organs, for example. The dark spots on the head are eyespots that can detect light, although they cannot focus an image like your eyes can.

Solid worms lack any internal cavity other than the digestive tract. Flatworms are soft-bodied animals flattened from top to bottom, like a piece of tape or ribbon. If you were to cut a flatworm in half across its body, as in figure 18.7, you would see that the gut is completely surrounded by tissues and organs. This solid body construction is called **acoelomate,** meaning without a body cavity.

Flatworms

Although flatworms have a simple body design, they do have a definite head at the anterior end and they possess organs. Flatworms range in size from a millimeter or less to many meters long, as in some tapeworms. Most species of flatworms are parasitic, occurring within the bodies of many other kinds of animals. Other flatworms are free-living, occurring in a wide variety of marine and freshwater habitats (figure 18.8), as well as moist places on land. Free-living flatworms are carnivores and scavengers; they eat various small animals and bits of organic debris. They move from place to place by means of ciliated epithelial cells concentrated on their ventral surfaces.

Parasitic flatworms such as the human liver fluke *Clonorchis sinensis* and blood flukes of the genus *Schistosoma* afflict more than 200 million people throughout tropical Asia, Africa, Latin America, and the Middle East.

Characteristics of Flatworms

Flatworms that have a digestive cavity have a gut with only one opening, a muscular tube called the pharynx. Through this opening, food is taken in as well as waste material expelled. Thus, flatworms cannot feed continuously, as more advanced animals can. The gut is branched and extends throughout the body, functioning in both digestion and the transport of food. Cells that line the gut engulf most of the food particles by phagocytosis and digest them. Parasitic flatworms lack digestive systems and absorb their food directly through their body walls.

Unlike cnidarians, flatworms have an excretory system, which consists of a network of fine tubules (little tubes) that runs throughout the body. Cilia line the hollow centers of bulblike **flame cells,** which are located on the side branches of the tubules. Cilia in the flame cells move water and excretory substances into the tubules and then out through exit pores located between the epidermal cells. More metabolic wastes diffuse directly into the gut. Flame cells were named because of the flickering movements of the tuft of cilia within them. They also play a major role in regulating the body's water balance. Like sponges, cnidarians, and ctenophorans, flatworms lack a circulatory system. Consequently, all flatworm cells must be within diffusion distance of oxygen and food. Flatworms have thin bodies and highly branched digestive cavities, making such a relationship possible.

The nervous system of flatworms is very simple, a longitudinal nerve cord that acts as a simple central nervous system. Between two longitudinal cords are cross connections, so the flatworm nervous system resembles a ladder.

Free-living flatworms use sensory pits or tentacles along the sides of their heads to detect food, chemicals, or movements of the fluid in which they are living. Free-living members of this phylum also have eyespots on their heads. These are inverted, pigmented cups containing light-sensitive cells connected to the nervous system (see figure 29.25). These eyespots enable the worms to distinguish light from dark. Flatworms are far more active than cnidarians or ctenophores. Such activity is characteristic of bilaterally symmetrical animals.

The reproductive systems of flatworms are complex. Most flatworms are **hermaphroditic,** with each individual containing both male and female sexual structures. In some parasitic flatworms, there is a complex succession of distinct larval forms. Some genera of flatworms are also capable of asexual regeneration; when a single individual is divided into two or more parts, each part can regenerate an entirely new flatworm.

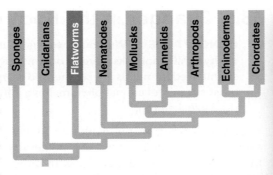

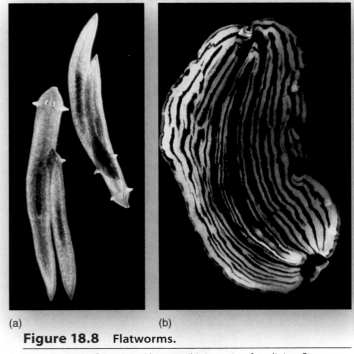

(a) (b)

Figure 18.8 Flatworms.

(a) A common flatworm, *Planaria*. (b) A marine free-living flatworm.

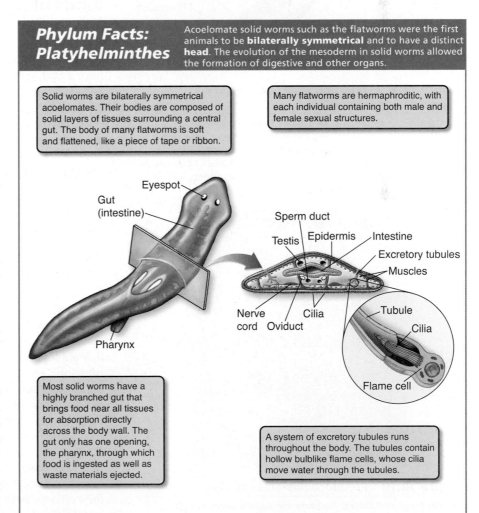

Phylum Facts: Platyhelminthes

Acoelomate solid worms such as the flatworms were the first animals to be **bilaterally symmetrical** and to have a distinct **head.** The evolution of the mesoderm in solid worms allowed the formation of digestive and other organs.

Solid worms are bilaterally symmetrical acoelomates. Their bodies are composed of solid layers of tissues surrounding a central gut. The body of many flatworms is soft and flattened, like a piece of tape or ribbon.

Many flatworms are hermaphroditic, with each individual containing both male and female sexual structures.

Eyespot

Gut (intestine)

Sperm duct

Testis Epidermis Intestine

Excretory tubules

Muscles

Nerve cord Cilia Oviduct

Tubule

Cilia

Pharynx

Flame cell

Most solid worms have a highly branched gut that brings food near all tissues for absorption directly across the body wall. The gut only has one opening, the pharynx, through which food is ingested as well as waste materials ejected.

A system of excretory tubules runs throughout the body. The tubules contain hollow bulblike flame cells, whose cilia move water through the tubules.

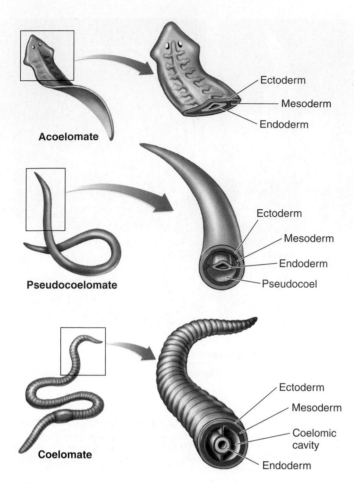

Figure 18.9 Three body plans for bilaterally symmetrical animals.

Acoelomates, such as flatworms, have no body cavity between the digestive tract (endoderm) and the outer body layer (ectoderm). Pseudocoelomates have a body cavity, the pseudocoel, between the endoderm and the mesoderm. Coelomates have a body cavity, the coelom, that develops entirely within the mesoderm, and so is lined on both sides by mesoderm tissue.

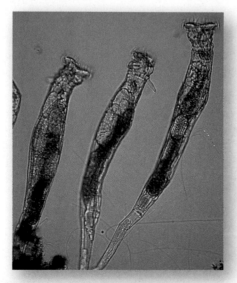

Figure 18.10 Pseudocoelomates: rotifers.

Rotifers (phylum Rotifera) are common aquatic animals that depend on their crown of cilia for feeding and locomotion.

18.6 Changes in the Body Cavity

CONCEPT PREVIEW: Roundworms have a pseudocoel body cavity, while mollusks have a coelom body cavity; neither is segmented. Annelids are segmented worms. Arthropods, the most successful animal phylum, have jointed appendages and a rigid exoskeleton.

A key innovation in the evolution of the animal body plan was the evolution of the body cavity. All bilaterally symmetrical animals other than solid worms have a cavity within their body. The evolution of an internal body cavity was an important improvement in body design for three reasons:

1. **Circulation.** Fluids that move within the body cavity can serve as a circulatory system, permitting the rapid passage of materials from one part of the body to another and opening the way to larger bodies.
2. **Movement.** Fluid in the cavity makes the animal's body rigid, permitting resistance to muscle contraction and thus opening the way to muscle-driven body movement.
3. **Organ function.** In a fluid-filled enclosure, body organs can function without being deformed by surrounding muscles. For example, food can pass freely through a gut suspended within a cavity, at a rate not controlled by when the animal moves.

Kinds of Body Cavities

There are three basic kinds of body plans found in bilaterally symmetrical animals (figure 18.9). Acoelomates, such as solid worms that we discussed in the previous section, have no body cavity. **Pseudocoelomates** have a body cavity called the **pseudocoel** located between the mesoderm and endoderm. A third way of organizing the body is one in which the fluid-filled body cavity develops entirely within the mesoderm. Such a body cavity is called a **coelom,** and animals that possess such a cavity are called **coelomates.**

Pseudocoelomates

Seven phyla are characterized by their possession of a pseudocoel. In all pseudocoelomates, the pseudocoel serves as a hydrostatic skeleton—one that gains its rigidity from being filled with fluid under pressure. The animal's muscles can work against this "skeleton," thus making the movements of pseudocoelomates far more efficient than those of the acoelomates.

Only one of the seven pseudocoelomate phyla includes a large number of species. This phylum, Nematoda, includes some 20,000 recognized species of nematodes, eelworms, and other roundworms. Scientists estimate that the actual number might approach 100 times that many. Members of this phylum are found everywhere. **Nematodes** are abundant and diverse in marine and freshwater habitats, and many members of this phylum are parasites of animals and plants. Many nematodes are microscopic and live in soil. It has been estimated that a spadeful of fertile soil may contain, on the average, a million nematodes.

Another phylum, Rotifera, contains about 2,000 species of rotifers. **Rotifers** are common, small, basically aquatic animals that have a crown of cilia at their heads (figure 18.10); they range from 0.04 to 2 millimeters long. Bilaterally symmetrical and covered with chitin, rotifers depend on their cilia for both locomotion and feeding, ingesting bacteria, protists, and small animals.

Phylum Nematoda: The Roundworms

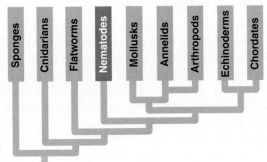

Nematodes are bilaterally symmetrical, cylindrical, unsegmented worms (figure 18.11). They are covered by a flexible, thick cuticle, which they shed as they grow, a process call **molting.** Their muscles constitute a layer beneath the epidermis and extend along the length of the worm, rather than encircling its body. These longitudinal muscles attach to the outer layer of the body (see the *Phylum Facts* below) and pull against the cuticle and the pseudocoel, which forms a type of fluid skeleton. When nematodes move, their bodies whip about from side to side.

All pseudocoelomates lack a defined circulatory system; this role is performed by the fluids that move within the pseudocoel. Most pseudocoelomates have a complete, one-way digestive tract that acts like an assembly line. Food is broken down, absorbed, and then treated and stored.

Near the mouth of a nematode, at its anterior end, are usually 16 raised, hairlike sensory organs. The mouth is often equipped with piercing organs called *stylets.* Food passes through the mouth as a result of the sucking action of the pharynx. After passing through a short corridor into the pharynx, food continues through the other portions of the digestive tract, where it is broken down and then digested. Some of the water with which the food has been mixed is reabsorbed near the end of the digestive tract, and material that has not been digested is eliminated through the anus.

Nematodes completely lack flagella or cilia, even on sperm cells. Reproduction in nematodes is sexual, with sexes usually separate. Their development is simple, and the adults consist of very few cells. For this reason, nematodes have become extremely important subjects for genetic and developmental studies. The 1-millimeter-long *Caenorhabditis elegans* matures in only three days, its body is transparent, and it has only 959 cells. It is the only animal whose complete developmental cellular anatomy is known, and the first animal whose genome (97 million DNA bases encoding over 21,000 different genes) was fully sequenced.

Some nematodes are parasitic in humans, cats, dogs, and animals of economic importance, such as cows and sheep. Heartworm infections in dogs and cats are caused by a parasitic nematode that infest the heart of the animal. About 50 species of nematodes, including several that are rather common in the United States, regularly parasitize human beings. Trichinosis, a nematode-caused disease in temperate regions, is caused by worms of the genus *Trichinella.* A more prevalent human parasitic nematode is *Ascaris lumbricoides.* This intestinal worm infects approximately one in six people worldwide but is rare in areas with modern plumbing.

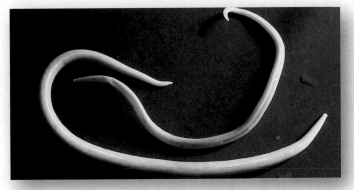

Figure 18.11   Pseudocoelomates: nematodes.

These nematodes (phylum Nematoda) are intestinal roundworms that infect humans and some other animals. Their fertilized eggs pass out with feces and can remain viable in soil for years.

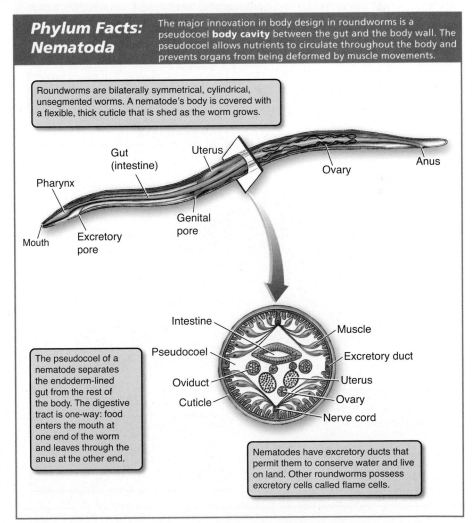

Phylum Facts: Nematoda

The major innovation in body design in roundworms is a pseudocoel **body cavity** between the gut and the body wall. The pseudocoel allows nutrients to circulate throughout the body and prevents organs from being deformed by muscle movements.

Roundworms are bilaterally symmetrical, cylindrical, unsegmented worms. A nematode's body is covered with a flexible, thick cuticle that is shed as the worm grows.

Gut (intestine)
Uterus
Ovary
Anus
Pharynx
Mouth
Excretory pore
Genital pore

The pseudocoel of a nematode separates the endoderm-lined gut from the rest of the body. The digestive tract is one-way: food enters the mouth at one end of the worm and leaves through the anus at the other end.

Intestine
Muscle
Pseudocoel
Excretory duct
Oviduct
Uterus
Cuticle
Ovary
Nerve cord

Nematodes have excretory ducts that permit them to conserve water and live on land. Other roundworms possess excretory cells called flame cells.

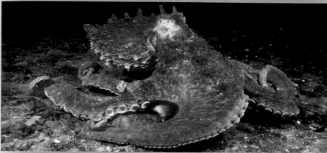

Figure 18.12 Mollusks.

Mollusks include the bivalves (*top*), cephalopods (*bottom*), and the gastropod snails (*illustrated below*).

Coelomates

Even though acoelomates and pseudocoelomates have proven very successful, the bulk of the animal kingdom consists of coelomates. A major advantage of the coelomate body plan is that it allows contact between mesoderm and endoderm during development, permitting localized portions of the digestive tract to develop into complex, highly specialized regions like the stomach. Coelomates also have a *circulatory system,* a network of vessels that carries fluids, oxygen, and food molecules to all parts of the body. The circulating fluid is usually pushed through the circulatory system by contraction of one or more muscular hearts.

Mollusks

The **mollusks** (phylum Mollusca) are the second largest animal phylum, with over 110,000 species. Mollusks, while mostly marine, are found nearly everywhere.

There are three kinds of mollusks with very

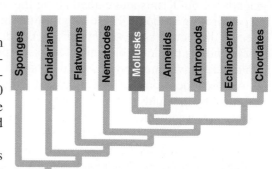

different body plans: *gastropods* (snails and slugs), *bivalves* (clams, oysters, scallops, and mussels), and *cephalopods* (octopuses, squids, and nautiluses) (figure 18.12). The differences hide a basically similar body design. The body of all mollusks is composed of three distinct parts: a head-foot, a central section called the visceral mass that contains the body's organs, and a mantle. The foot of a mollusk is muscular and may be adapted for locomotion, attachment, food capture, or various combinations of these functions. The **mantle** is a heavy fold of tissue wrapped around the visceral mass like a cape, with the respiratory organs (gills or lungs) positioned on its inner surface like the lining of a coat. Mollusks were among the first animals to develop an efficient excretory system. Tubular structures called *nephridia* (a type of kidney) gather wastes from the coelom and discharge them into the mantle cavity.

One of the most characteristic features of gastropods and cephalopods is the **radula,** a rasping, tongue-like organ. With rows of pointed, backward-curving teeth, the radula is used by some snails to scrape algae off rocks. The small holes often seen in oyster shells are produced by gastropods that have bored holes to kill the oyster and extract its body.

Pearls are formed when a foreign object, such as a grain of sand, becomes lodged between the mantle and the inner shell layer of a bivalve, including clams and oysters. The mantle coats the foreign object with layer upon layer of shell material to reduce irritation. The shell serves primarily for protection, with some mollusks withdrawing into their shells when threatened.

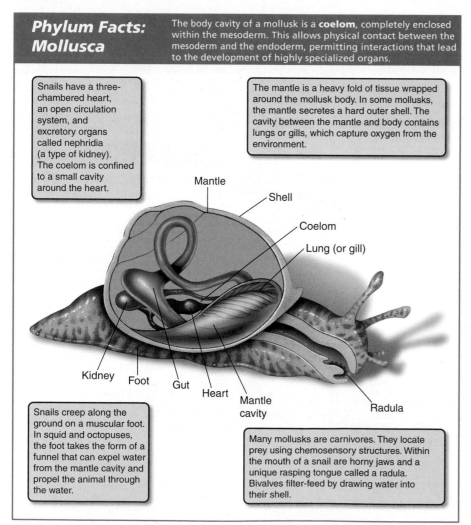

Phylum Facts: Mollusca

The body cavity of a mollusk is a **coelom**, completely enclosed within the mesoderm. This allows physical contact between the mesoderm and the endoderm, permitting interactions that lead to the development of highly specialized organs.

Snails have a three-chambered heart, an open circulation system, and excretory organs called nephridia (a type of kidney). The coelom is confined to a small cavity around the heart.

The mantle is a heavy fold of tissue wrapped around the mollusk body. In some mollusks, the mantle secretes a hard outer shell. The cavity between the mantle and body contains lungs or gills, which capture oxygen from the environment.

Mantle
Shell
Coelom
Lung (or gill)
Kidney
Foot
Gut
Heart
Mantle cavity
Radula

Snails creep along the ground on a muscular foot. In squid and octopuses, the foot takes the form of a funnel that can expel water from the mantle cavity and propel the animal through the water.

Many mollusks are carnivores. They locate prey using chemosensory structures. Within the mouth of a snail are horny jaws and a unique rasping tongue called a radula. Bivalves filter-feed by drawing water into their shell.

Annelids

A key innovation in body plan to arise among the coelomates was **segmentation,** the building of a body from a series of similar segments. The first segmented animals to evolve were the annelid worms, phylum Annelida (figure

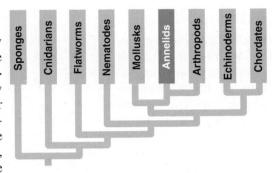

18.13). These advanced coelomates are assembled as a chain of nearly identical segments, like the boxcars of a train. The great advantage of such segmentation is the evolutionary flexibility it offers—a small change in an existing segment can produce a new kind of segment with a different function. Thus, segments are modified for reproduction, feeding, and eliminating wastes.

Two-thirds of all annelids live in the sea (about 8,000 species), but some live in freshwater, and most of the rest—some 3,100 species—are earthworms. The basic body plan of an annelid is a tube within a tube: The digestive tract is suspended within the coelom, which is itself a tube running from mouth to anus. The body segments of an annelid are visible as a series of ringlike structures running the length of the body, looking like a stack of doughnuts. The segments are divided internally from one another by partitions. In each of the cylindrical segments, the excretory and locomotor organs are repeated. The body fluid within the coelom of each segment creates a hydrostatic (liquid-supported) skeleton that gives the segment rigidity, like an inflated balloon. Because each segment is separate, each is able to expand or contract independently. This lets the worm's body move in ways that are quite complex. When an earthworm crawls on a flat surface, for example, it lengthens the front half of its body while shortening the back half, pulling itself along.

> A hydrostatic skeleton provides an effective means of locomotion in soft-bodied animals, as discussed on page 467. The fluid-filled cavity is surrounded by muscles, and when the muscles contract, the fluid is displaced, which moves the body.

The anterior (front) segments of annelids contain the sensory organs. Elaborate eyes with lenses and retinas have evolved in some annelids. One anterior segment contains a well-developed cerebral ganglion, or brain. The digestive tract, circulatory system, and nervous system are connected between segments. In annelids, a closed circulatory system carries blood from one segment to another through a network of blood vessels. Nerve cords connect the nerve centers located in each segment with each other and the brain. The brain can then coordinate the worm's activities.

Segmentation underlies the body organization of all complex coelomate animals, not only annelids but also arthropods (crustaceans, spiders, and insects) and chordates (vertebrates). For example, vertebrate muscles develop from repeated blocks of tissue called somites that occur in the embryo. Vertebrate segmentation is also seen in the vertebral column, which is a stack of very similar vertebrae.

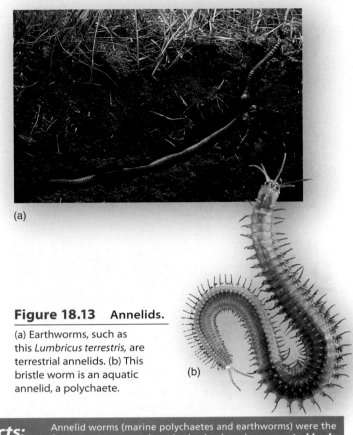

(a)

Figure 18.13 Annelids.

(a) Earthworms, such as this *Lumbricus terrestris*, are terrestrial annelids. (b) This bristle worm is an aquatic annelid, a polychaete.

(b)

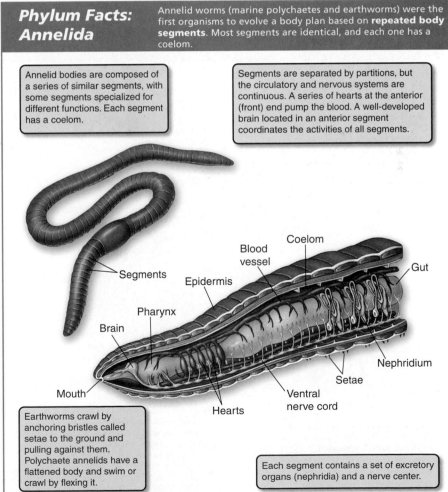

Phylum Facts: Annelida

Annelid worms (marine polychaetes and earthworms) were the first organisms to evolve a body plan based on **repeated body segments**. Most segments are identical, and each one has a coelom.

Annelid bodies are composed of a series of similar segments, with some segments specialized for different functions. Each segment has a coelom.

Segments are separated by partitions, but the circulatory and nervous systems are continuous. A series of hearts at the anterior (front) end pump the blood. A well-developed brain located in an anterior segment coordinates the activities of all segments.

Segments

Brain

Mouth

Pharynx

Epidermis

Blood vessel

Coelom

Gut

Nephridium

Setae

Hearts

Ventral nerve cord

Earthworms crawl by anchoring bristles called setae to the ground and pulling against them. Polychaete annelids have a flattened body and swim or crawl by flexing it.

Each segment contains a set of excretory organs (nephridia) and a nerve center.

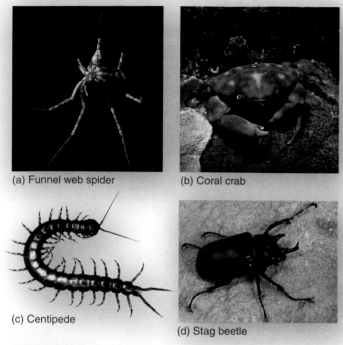

(a) Funnel web spider

(b) Coral crab

(c) Centipede

(d) Stag beetle

Figure 18.14 Arthropods.

Arthropods are a large group which includes (a) arachnids, (b) crustaceans, (c) centipedes, and (d) insects.

Arthropods

The evolution of segmentation among the annelids marked a major innovation in body structure among the coelomates. An even more profound innovation is found in the body plan of the most successful of all animal

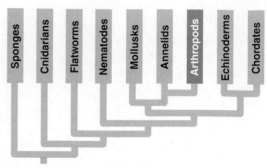

groups, the **arthropods,** phylum Arthropoda (figure 18.14). This innovation was the development of jointed appendages. Two-thirds of all named species on earth are arthropods, which include the arachnids (spiders, ticks, mites, scorpions, and daddy longlegs), crustaceans (lobsters, crabs, shrimp, and barnacles), centipedes and millipedes, and insects. About 80% of all arthropod species are insects, and about half of these are beetles!

Jointed Appendages. The name *arthropod* comes from two Greek words, *arthros,* jointed, and *podes,* feet. All arthropods have jointed appendages. Some are legs, and others may be modified for other uses. To gain some idea of the importance of jointed appendages, imagine yourself without them—no hips, knees, ankles, shoulders, elbows, wrists, or knuckles. Without jointed appendages, you could not walk or grasp an object. Arthropods use jointed appendages as legs and wings for moving, as antennae to sense their environment, and as mouthparts for sucking, ripping, and chewing prey.

Rigid Exoskeleton. The arthropod body plan has a second great innovation: a rigid external skeleton, or **exoskeleton,** made of chitin. As the body grows, its hard exoskeleton is discarded in the process of molting, to be replaced by a new larger one. In arthropods the muscles attach to the interior surface of this hard chitin shell, which also protects the animal from predators and impedes water loss.

However, while chitin is hard and tough, it is also brittle and cannot support great weight. As a result, the exoskeleton must be much thicker to bear the pull of the muscles in large insects than in small ones, so there is a limit to how big an arthropod body can be. The great majority of arthropod species consist of small animals—mostly about a millimeter in length—but members of the phylum range in adult size from about 80 micrometers long (some parasitic mites) to 3.6 meters across (a gigantic crab found in the sea off Japan). Some lobsters are nearly a meter in length. The largest living insects are about 33 centimeters long, but the giant dragonflies that lived 300 million years ago had wingspans of as much as 60 centimeters (2 feet)!

Arthropod bodies are segmented like those of annelids, from which they almost certainly evolved.

Phylum Facts: Arthropoda

Insects and other arthropods have a coelom, segmented bodies, and **jointed appendages**. During development, segments often fuse to form different body regions. Arthropods also have a strong **exoskeleton** made of chitin; most insects also have wings.

Arthropods have jointed appendages, a chitinous exoskeleton, and a segmented body. Segments can fuse to form different body regions; in insects, there are three body regions: head, thorax, and abdomen.

The jointed appendages of insects may be modified into antennae, mouthparts, legs, or wings.

Thorax

Head

Eye

Antenna

Mouthparts

Air sac

Malpighian tubules

Abdomen

Rectum

Sting

Spiracles

Midgut

Poison sac

Insects have complex sensory organs located on the head, including a single pair of antennae and compound eyes composed of many independent visual units.

Insects eliminate wastes by collecting circulatory fluid osmotically in Malpighian tubules. Insects breathe through small tubes called tracheae that pass throughout the body and open externally through spiracles.

18.7 Redesigning the Embryo

CONCEPT PREVIEW: Echinoderms and chordates are deuterostomes. Deuterostome eggs cleave radially, and the blastopore of their embryo becomes the animal's anus. Echinoderms have an endoskeleton of hard plates, often fused together. Adults are radially symmetrical. Chordates have a notochord at some stage of their development. In adult vertebrates, the notochord is replaced by a backbone.

There are two major kinds of coelomate animals representing two distinct evolutionary lines. All the coelomates we have met so far have essentially the same kind of embryonic development. Cell divisions of the fertilized egg produce a hollow ball of cells, a blastula, which indents to form a two-layer-thick ball with a blastopore opening to the outside. In mollusks, annelids, and arthropods, the mouth (stoma) develops from or near the blastopore. This same pattern of development, in a general sense, is seen in all noncoelomate animals. An animal whose mouth develops in this way is called a **protostome** (figure 18.15, *column on left*). If such an animal has a distinct anus or anal pore, it develops later in another region of the embryo.

A second distinct pattern of embryological development occurs in the echinoderms and the chordates. In these animals, the anus forms from or near the blastopore, and the mouth forms subsequently on another part of the blastula. This group of phyla consists of animals that are called the **deuterostomes** (figure 18.15, *column on right*).

Deuterostomes represent a revolution in embryonic development. In addition to the fate of the blastopore, deuterostomes differ from protostomes in two other features:

1. The progressive division of cells during embryonic growth is called *cleavage.* The cleavage pattern relative to the embryo's polar axis determines how the cells array. In nearly all protostomes, each new cell buds off at an angle oblique to the polar axis. As a result, a new cell nestles into the space between the older ones in a closely packed array. This pattern is called **spiral cleavage** because a line drawn through a sequence of dividing cells spirals outward from the polar axis, (indicated by the curving blue arrow at the 32-cell stage in figure 18.15).

   In deuterostomes, the cells divide parallel to and at right angles to the polar axis. As a result, the pairs of cells from each division are positioned directly above and below one another; this process gives rise to a loosely packed array of cells. This pattern is called **radial cleavage** because a line drawn through a sequence of dividing cells describes a radius outward from the polar axis (indicated by the straight blue arrow at the 32-cell stage).

2. In protostomes, the developmental fate of each cell in the embryo is fixed when that cell first appears. Even at the four-celled stage, each cell is different, containing different chemical developmental signals and no one cell, if separated from the others, can develop into a complete animal. In deuterostomes, on the other hand, the first cleavage divisions of the fertilized egg produce identical daughter cells, and any single cell, if separated, can develop into a complete organism.

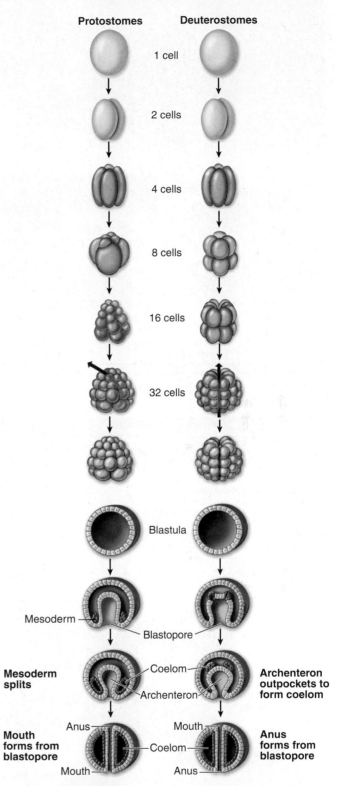

Figure 18.15 Embryonic development in protostomes and deuterostomes.

Cleavage of the egg produces a hollow ball of cells called the blastula. Invagination, or infolding, of the blastula produces the blastopore. In protostomes, embryonic cells cleave in a spiral pattern and become tightly packed. The blastopore becomes the animal's mouth. In deuterostomes, embryonic cells cleave radially and form a loosely packed array. The blastopore becomes the animal's anus.

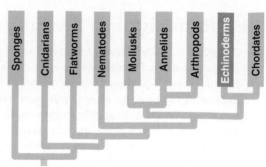

Echinoderms

The first deuterostomes, marine animals called **echinoderms** in the phylum Echinodermata, appeared more than 650 million years ago. The term *echinoderm* means "spiny skin" and refers to an **endoskeleton** composed of hard, calcium-rich plates called ossicles that lie just beneath the delicate skin. When they are first formed, the plates are enclosed in living tissue, and so are truly an endoskeleton, although in adults they fuse, forming a hard shell. About 6,000 species of echinoderms are living today, almost all of them on the ocean bottom (figure 18.16). Many of the most familiar animals seen along the seashore are echinoderms, including sea stars (starfish), sea urchins, sand dollars, and sea cucumbers.

The body plan of echinoderms undergoes a fundamental shift during development: All echinoderms are bilaterally symmetrical as larvae but become radially symmetrical as adults. Many biologists believe that early echinoderms were sessile and evolved adult radial symmetry as an adaptation to the sessile existence. Bilateral symmetry is of adaptive value to an animal that travels through its environment, whereas radial symmetry is of value to an animal whose environment meets it on all sides. Adult echinoderms have a five-part body plan, easily seen in the five arms of a sea star. Some echinoderms like feather stars have 10 or 15 arms, but always multiples of five. The nervous system consists of a central ring of nerves from which radial branches arise—while the animal is capable of complex response patterns, there is no centralization of function, no "brain."

A key evolutionary innovation of echinoderms is the development of a hydraulic system to aid movement. Called a **water vascular system,** this fluid-filled system is composed of a central ring canal from which radial canals extend out. From each radial canal, tiny vessels extend through short side branches into thousands of tiny, hollow tube feet. At the base of each tube foot is a fluid-filled muscular sac that acts as a valve. When a sac contracts, its fluid is prevented from reentering the radial canal and instead is forced into the tube foot, thus extending it. When extended, the tube foot attaches itself to the ocean bottom, often aided by suckers. An echinoderm can then pull against these tube feet and so haul itself over the seafloor.

Most echinoderms reproduce sexually, but they have the ability to regenerate lost parts, which can lead to asexual reproduction. In a few sea stars, asexual reproduction takes place by splitting, and the broken parts of the sea star can sometimes regenerate whole animals.

Figure 18.16 Echinoderms.

(a) Sea star, *Oreaster occidentalis.* (b) Sand dollar, *Echinarachnius parma.* (c) Giant red sea urchin, *Strongylocentrotus franciscanus.*

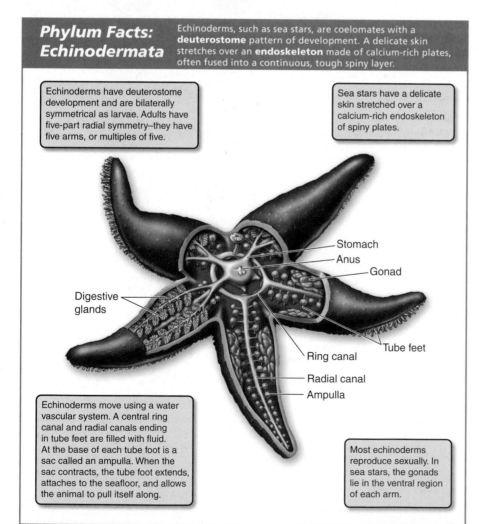

Phylum Facts: Echinodermata

Echinoderms, such as sea stars, are coelomates with a **deuterostome** pattern of development. A delicate skin stretches over an **endoskeleton** made of calcium-rich plates, often fused into a continuous, tough spiny layer.

Echinoderms have deuterostome development and are bilaterally symmetrical as larvae. Adults have five-part radial symmetry—they have five arms, or multiples of five.

Sea stars have a delicate skin stretched over a calcium-rich endoskeleton of spiny plates.

Echinoderms move using a water vascular system. A central ring canal and radial canals ending in tube feet are filled with fluid. At the base of each tube foot is a sac called an ampulla. When the sac contracts, the tube foot extends, attaches to the seafloor, and allows the animal to pull itself along.

Most echinoderms reproduce sexually. In sea stars, the gonads lie in the ventral region of each arm.

- Stomach
- Anus
- Gonad
- Digestive glands
- Tube feet
- Ring canal
- Radial canal
- Ampulla

Chordates

Chordates (phylum Chordata) evolved an endoskeleton that is truly internal, based initially on a flexible rod called a **notochord** that develops along the back of the embryo. Muscles attached to this rod allowed early chordates to

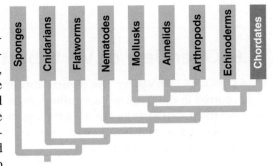

swing their bodies back and forth, swimming through the water. This key evolutionary innovation, attaching muscles to an internal element, started chordates along an evolutionary path that leads to the vertebrates and, for the first time, to truly large animals.

The approximately 56,000 species of chordates are distinguished by four principal features, which all chordates have at some time in their lives (figure 18.17):

1. **Notochord.** A stiff, but flexible, rod that forms beneath the nerve cord in the early embryo.
2. **Nerve cord.** A single hollow dorsal nerve cord, to which the nerves that reach the different parts of the body are attached.
3. **Pharyngeal pouches.** A series of pouches behind the mouth that develop into slits in some animals.
4. **Postanal tail.** A tail that extends beyond the anus. The pharyngeal pouches and postanal tail disappear during human development, and the notochord is replaced with the vertebral column.

Vertebrates

With the exception of tunicates and lancelets, all chordates are **vertebrates.** Vertebrates differ from tunicates and lancelets in two important respects:

1. **Backbone.** The notochord becomes surrounded and then replaced during the course of development by a bony vertebral column. The vertebral column is a stack of hollow bones called *vertebrae* that encloses the dorsal nerve cord like a sleeve and protects it.
2. **Head.** All vertebrates except the earliest fishes have a distinct and well-differentiated head, with a skull and brain.

All vertebrates have an internal skeleton made of bone or cartilage against which the muscles work. This endoskeleton makes possible the great size and extraordinary powers of movement that characterize the vertebrates.

Concept Check

1. What three embryonic layers do all bilaterally symmetrical animals produce?
2. Compare a coelom to a pseudocoel.
3. What phylum of coelomates is not segmented?

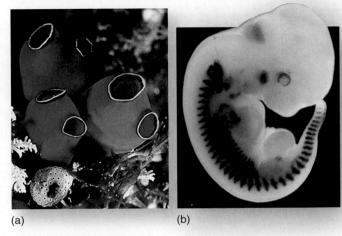

Figure 18.17 Chordates.

(a) Beautiful blue and gold tunicates. (b) In this mouse embryo, the muscle segments (stained *dark* in this photo) can be seen, reflecting the fundamentally segmented nature of all chordates.

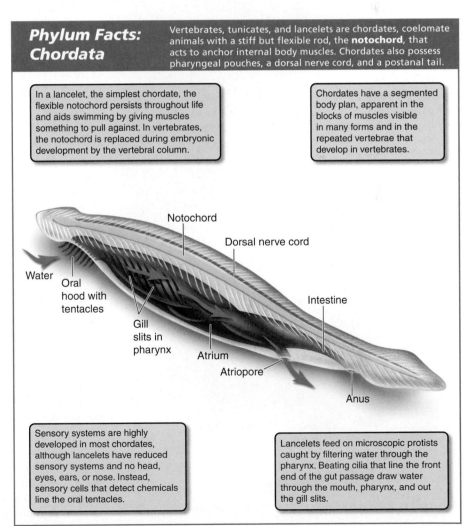

Phylum Facts: Chordata

Vertebrates, tunicates, and lancelets are chordates, coelomate animals with a stiff but flexible rod, the **notochord**, that acts to anchor internal body muscles. Chordates also possess pharyngeal pouches, a dorsal nerve cord, and a postanal tail.

In a lancelet, the simplest chordate, the flexible notochord persists throughout life and aids swimming by giving muscles something to pull against. In vertebrates, the notochord is replaced during embryonic development by the vertebral column.

Chordates have a segmented body plan, apparent in the blocks of muscles visible in many forms and in the repeated vertebrae that develop in vertebrates.

Notochord

Dorsal nerve cord

Water

Oral hood with tentacles

Gill slits in pharynx

Atrium

Atriopore

Intestine

Anus

Sensory systems are highly developed in most chordates, although lancelets have reduced sensory systems and no head, eyes, ears, or nose. Instead, sensory cells that detect chemicals line the oral tentacles.

Lancelets feed on microscopic protists caught by filtering water through the pharynx. Beating cilia that line the front end of the gut passage draw water through the mouth, pharynx, and out the gill slits.

The Parade of Vertebrates

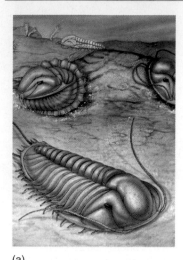

(a)   (b)

Figure 18.18   Life in the Cambrian.

(a) Trilobites are shown in this reconstruction of a community of marine organisms in the Cambrian period, 545 to 490 million years ago. (b) A fossil trilobite.

Figure 18.19
Ammonite fossil.

This shelled mollusk evolved during the Paleozoic era and became the most abundant creature on earth, until the end of the Mesozoic era (65 M.Y.A.).

Figure 18.20   An early reptile: the pelycosaur.

18.8 Overview of Vertebrate Evolution

CONCEPT PREVIEW: The diversification of the animal phyla occurred in the sea. The two animal phyla that have invaded the land most successfully are the arthropods and the chordates. Mass extinctions, where the loss of species outpaces species formation, have occurred five times so far.

When scientists first began to study and date fossils, they had to find some way to organize the different time periods from which the fossils came. They divided the earth's past into large blocks of time called **eras.** Eras are further subdivided into smaller blocks of time called **periods,** and some periods, in turn, are subdivided into **epochs,** which can be divided into **ages.**

Virtually all of the major groups of animals that survive at the present time originated in the sea at the beginning of the **Paleozoic era,** during or soon after the Cambrian period (545–490 million years ago—M.Y.A.). Thus, the major diversification of animal life on earth occurred largely in the sea, and fossils from the early Paleozoic era are found in the marine fossil record. Many of the animal phyla that appeared in the Cambrian period, like the bizarre trilobites you see in figure 18.18, have no surviving close relatives. Ammonites, shelled, cephalopod mollusks that originated in the Paleozoic era, were among the most abundant creatures on earth 100 million years ago (figure 18.19).

The first vertebrates evolved about 500 million years ago in the oceans—fishes without jaws. They didn't have paired fins either—many of them looked something like a flattened hotdog with a hole at one end and a fin at the other. For over 100 million years, a parade of different kinds of fishes were the only vertebrates on earth. They became the dominant creatures in the sea, some as large as 10 meters, larger than most cars.

Invasion of the Land

The first organisms to colonize the land were fungi and plants, over 500 million years ago. The ancestors of plants were specialized members of a group of photosynthetic protists known as the green algae. It seems probable that plants first occupied the land in symbiotic association with fungi, as discussed in chapter 17.

Only a few of the animal phyla that evolved in the Cambrian seas have invaded the land successfully. The first invasion of the land by animals, and perhaps the most successful invasion of the land, was accomplished by the arthropods, a phylum of hard-shelled animals with jointed legs and a segmented body. This invasion occurred about 410 million years ago.

Vertebrates invaded the land during the Carboniferous period (360–280 M.Y.A.). The first vertebrates to live on land were the amphibians, represented today by frogs, toads, salamanders, and caecilians (legless amphibians). The earliest amphibians known are from the Devonian period. Then about 300 million years ago, the first reptiles appeared. Within 50 million years, the reptiles, better suited than amphibians to living out of water, replaced them as the dominant land animal on earth. The pelycosaur, the fan-backed animal pictured in figure 18.20, was an early reptile. By the Permian period (280 M.Y.A.), the major evolutionary lines on land had been established and were expanding.

The **Mesozoic era** (248–65 M.Y.A.) followed the Paleozoic era and has traditionally been divided into three periods: the Triassic, the Jurassic, and the Cretaceous. The Mesozoic era was a time of intensive evolution of terrestrial plants and animals. With the success of the reptiles, vertebrates truly came to dominate the surface of the earth. Many kinds of reptiles evolved, from those smaller than a chicken to others even larger than a semitrailer truck (figure 18.21). Some flew, and others swam. From among reptile ancestors, dinosaurs, birds, and mammals evolved. Although dinosaurs and mammals appear at about the same time in the fossil record, 200 to 220 million years ago, the dinosaurs, especially large dinosaurs, dominated the face of the earth for over 150 million years. And then, about 65 million years ago at the end of the Cretaceous period, dinosaurs disappeared, along with the flying reptiles called pterosaurs (figure 18.22), the great marine reptiles, and other animals, such as ammonites. This extinction marks the end of the Mesozoic era. During the **Cenozoic era** (65 M.Y.A. to present), new forms of life were able to invade new habitats. Mammals diversified from earlier, small nocturnal forms to many new forms. Most present-day orders of mammals appeared at this time, a period of great diversity.

Mass Extinctions

The history of life on earth has been marked by periodic episodes of extinction, where the loss of species outpaces the formation of new species. Particularly sharp declines in species diversity are called **mass extinctions.** Five mass extinctions have occurred, the first of them near the end of the Ordovician period about 438 million years ago. At that time, most of the existing families of trilobites (see figure 18.18) became extinct. Another mass extinction occurred about 360 million years ago at the end of the Devonian period.

The third and most drastic mass extinction in the history of life on earth happened during the last 10 million years of the Permian period (250 M.Y.A.), marking the end of the Paleozoic era. It is estimated that 96% of all species of marine animals that were living at that time became extinct! All of the trilobites disappeared forever. Brachiopods, marine animals resembling mollusks but with a different filter-feeding system, were extremely diverse and widespread during the Permian; only a few species survived. Bryozoans, marine filter-feeders that formed coral-like colonies throughout the world, became rare afterward.

Mass extinctions left vacant many ecological opportunities, and for this reason they were followed by rapid evolution among the relatively few plants, animals, and other organisms that survived the extinction. Little is known about the causes of major extinctions. In the case of the Permian mass extinction, some scientists argue that the extinction was brought on by a gradual accumulation of carbon dioxide in ocean waters, the result of large-scale volcanism due to the collision of the earth's landmasses during formation of the single large "supercontinent" of Pangaea. Such an increase in carbon dioxide would have severely disrupted the ability of animals to carry out metabolism and form their shells.

The most famous and well-studied extinction, though not as drastic, occurred at the end of the Cretaceous period (65 M.Y.A), at which time the dinosaurs and a variety of other organisms went extinct. Recent findings have supported the hypothesis that this fifth mass extinction event was triggered when a large asteroid slammed into the earth, perhaps causing global forest fires and obscuring the sun for months by throwing particles into the air.

We are living during a new sixth mass extinction event. The number of species in the world is greater today than it has ever been. Unfortunately, that number is decreasing at an alarming rate due to human activity. Some estimate that as many as one-fourth of all species will become extinct in the near future, a rate of extinction not seen on earth since the Cretaceous mass extinction.

Figure 18.21   **Some dinosaurs were truly enormous.**

Here the paleontologist Jim Jensen is standing by a reconstruction of the leg of a sauropod fossil he found. Sauropods were herbivores that had enormous barrel-shaped bodies with heavy columnlike legs and very long necks and tails. Some weighed 55 tons, stood 10 meters tall, and were over 30 meters long.

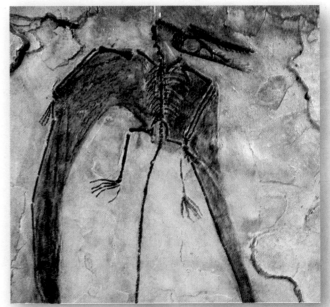

Figure 18.22   **An extinct flying reptile.**

Pterosaurs, such as the fossil pictured here, became extinct with the dinosaurs about 65 million years ago. Flight has evolved three separate times among vertebrates; however, birds and bats are the only representatives still in existence.

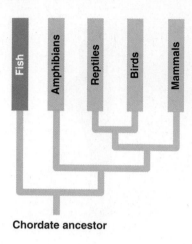

Chordate ancestor

18.9 Fishes Dominate the Sea

CONCEPT PREVIEW: Fishes are characterized by gills, a vertebral column, and a single-loop circulatory system. Sharks are fast swimmers, whereas the very successful bony fishes have unique characteristics such as swim bladders and lateral line systems.

A series of key evolutionary advances allowed vertebrates to first conquer the sea, and then the land. Figure 18.23 shows a phylogeny of the vertebrates. Branch points in the family tree indicate key adaptations. About half of all vertebrates are **fishes.** The most diverse and successful vertebrate group, they provided the evolutionary base for the invasion of land by amphibians.

Characteristics of Fishes

From whale sharks that are 12 meters long to tiny cichlids no larger than your fingernail, fishes vary considerably in size, shape, color, and appearance. However varied, all fishes have four important characteristics in common:

1. **Gills.** Fish are water-dwelling creatures, and they use gills to extract dissolved oxygen gas from the water around them. Gills are fine filaments of tissue rich in blood vessels. When water passes over the gills at the back of the mouth, oxygen gas diffuses from the water into the fish's blood.

The gills of bony fish are the most efficient respiratory machines that have ever evolved in animals. Fishes' gills use a countercurrent flow system, described on page 493, that allows for constant diffusion of oxygen into the blood as it passes through the gills.

Figure 18.23   **Vertebrate family tree.**

Primitive amphibians arose from lobe-finned fishes. Primitive reptiles arose from amphibians and gave rise in turn to mammals and to dinosaurs, which are the ancestors of today's birds.

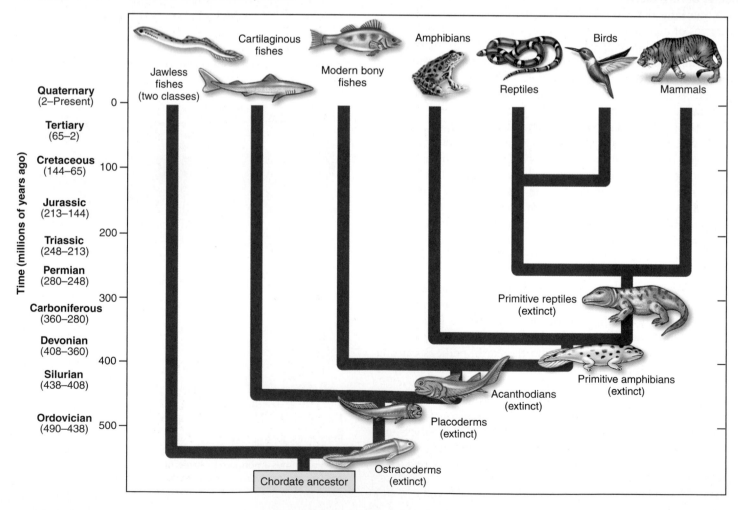

2. **Vertebral column.** All fishes have an internal skeleton made of bone or cartilage with a spine surrounding the dorsal nerve cord. The brain is fully encased within a protective case, called the skull or cranium.
3. **Single-loop blood circulation.** Blood is pumped from the heart to the gills. From the gills, the oxygenated blood passes to the rest of the body and then returns to the heart.
4. **Nutritional deficiencies.** Fishes are unable to synthesize the aromatic amino acids and must consume them in their diet. This inability has been inherited by all their vertebrate descendants.

The first backboned animals were jawless fishes that appeared in the sea about 500 million years ago. Only their head-shields were made of bone; their elaborate internal skeletons were constructed of cartilage. Jawless and toothless, these fishes sucked up small food particles from the ocean floor. One group of jawless fishes, the agnathans, survive today as hagfish and parasitic lampreys. These first fishes were replaced by larger, heavier jawed fishes that were predators. Early forms of jawed fishes were eventually replaced by fishes that moved through the water faster: the sharks and the bony fishes.

Sharks

Sharks are fast and maneuverable swimmers due to their light and flexible cartilaginous skeleton. Members of this group, the class Chondrichthyes, consist of sharks, skates, and rays. Sharks are very powerful swimmers, with a back fin, a tail fin, and two sets of paired side fins for controlled thrusting through the water (figure 18.24). Skates and rays are flattened sharks that are bottom-dwellers. Today there are about 750 species of sharks, skates, and rays.

Some of the largest sharks filter their food from the water like jawless fishes, but most are predators, their mouths armed with rows of hard, sharp teeth. Sharks are well-adapted to their predatory life due to their sophisticated sensory systems. Reproduction among the Chondrichthyes is the most advanced of any fish. About 40% of sharks, skates, and rays lay fertilized eggs. The eggs of other species develop within the female's body, and the pups are born alive.

Bony Fishes

Bony fish are also fast and maneuverable swimmers, but instead of gaining speed through lightness, as sharks did, bony fishes adopted a heavy internal skeleton made completely of bone. This internal skeleton is very strong, providing a base against which very strong muscles can pull, but it is also heavy. Bony fishes are buoyant because they possess a **swim bladder.** The swim bladder (figure 18.25) is a gas-filled sac that allows fish to regulate their buoyant density and so remain effortlessly suspended at any depth in the water.

Bony fish are the most successful of all fishes, indeed of all vertebrates. Bony fishes (class Osteichthyes) consist of about 30,000 species. The remarkable success of the bony fishes has resulted from a series of significant adaptations. In addition to the swim bladder, they have a highly developed **lateral line system,** a system of sensory cells that run the length of the body on each side that enables them to detect changes in water pressure and thus the movement of predators and prey in the water. Also, most bony fishes have a hard plate called the **operculum** that covers the gills on each side of the head. Flexing the operculum permits bony fishes to pump water over their gills. Using the operculum as very efficient bellows, bony fishes can pass water over the gills while stationary in the water. That is what a goldfish in a fish tank is doing when it seems to be gulping.

Figure 18.24 Chondrichthyes.

The Galápagos shark is a member of the class Chondrichthyes, which are mainly predators or scavengers and spend most of their time in graceful motion. As they move, they create a flow of water past their gills, from which they extract oxygen.

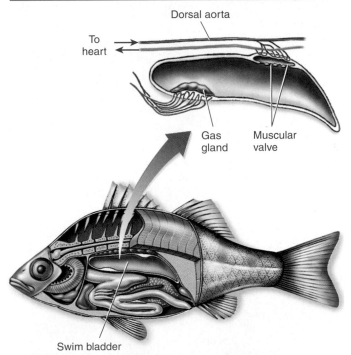

Figure 18.25 Diagram of a swim bladder.

The bony fishes use a swim bladder to control their buoyancy in water. The swim bladder can be filled or drained of gas taken from the blood. The gas gland secretes the gases into the swim bladder; gas is released from the bladder by a muscular valve.

Figure 18.26   **A key adaptation of amphibians: the evolution of legs.**

In ray-finned fish, the fins contain only bony rays. (a) In lobe-finned fish, the fins have a central core of bones (inside a fleshy lobe) in addition to rays. Some lobe-finned fishes could move out onto land. (b) In the incomplete fossil of *Tiktaalik* (which did not contain the hindlimbs), the shoulder, forearm, and wrist bones were like those of amphibians, but the end of the limb was like that of lobe-finned fishes. (c) In primitive amphibians, the positions of the limb bones are shifted, and bony "toes" are present.

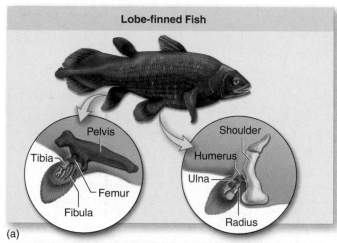
(a)

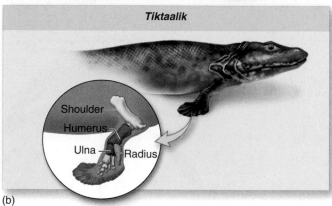
(b)

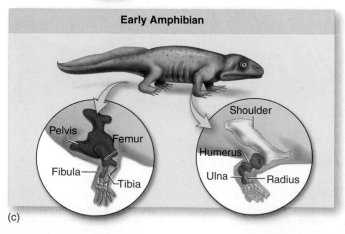
(c)

18.10 Amphibians and Reptiles Invade the Land

CONCEPT PREVIEW: Amphibians were the first vertebrates to successfully invade land, helped by legs, lungs, and two pathways of blood circulation. Reptiles have three characteristics that make them well-suited for life on land: a watertight (amniotic) egg, dry skin, and thoracic breathing.

Frogs, salamanders, and caecilians (damp-skinned and legless) are direct descendants of fishes. They are the sole survivors of a very successful group, the **amphibians,** the first vertebrates to walk on land. Amphibians likely evolved from a type of bony fish called the lobe-finned fishes, fish with paired fins that consist of a long fleshy muscular lobe supported by a central core of bones that form fully articulated joints with one another. In 2006, the discovery of a new fossil fish (genus *Tiktaalik*) exhibited the transition between fish and amphibians (figure 18.26*c*).

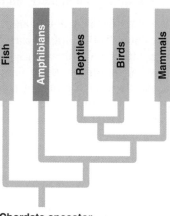

Chordate ancestor

Amphibians

Amphibians have five key characteristics that allowed them to successfully invade the land:

1. **Legs.** Frogs and salamanders have four legs and can move about on land quite well. The way in which legs are thought to have evolved from fins is illustrated in figure 18.26.
2. **Lungs.** Most amphibians possess a pair of lungs, necessary because the delicate structure of fish gills requires the buoyancy of water for support.
3. **Cutaneous respiration.** Frogs, salamanders, and caecilians all supplement the use of lungs by respiring directly across their skin, which is kept moist and provides an extensive surface area (figure 18.27).

Figure 18.27
Cutaneous respiration.

This red-eyed tree frog, *Agalychnis callidryas,* is a member of the group of amphibians that includes frogs and toads (order Anura). Frogs live in moist environments allowing for cutaneous respiration.

4. **Pulmonary veins.** After blood is pumped through the lungs, two large veins called pulmonary veins return the aerated blood to the heart for repumping. This allows aerated blood to be pumped to tissues at a much higher pressure than when it leaves the lungs.

5. **Partially divided heart.** Greater amounts of oxygen are required by muscles for movement and support on land. The chambers of the amphibian heart are separated by a dividing wall that helps prevent aerated blood from the lungs from mixing with nonaerated blood being returned to the heart from the rest of the body. The separation is incomplete, however, and some mixing does occur.

Amphibians were the dominant land vertebrates for 100 million years, when 40 families existed. Sixty percent of them were fully terrestrial, with bony plates and armor covering their bodies. Many of these terrestrial amphibians grew to be very large—some as big as a pony! Only a tiny fraction of these have living descendants, all descended from three aquatic families of amphibians that survived competition with reptiles by reinvading the water. Today's 4,850 species of amphibians are classified in three orders: Anura (frogs and toads), Urodela (salamanders and newts), and Apoda (caecilians). Most of today's amphibians are tied to moist if not aquatic environments and must reproduce in water. Many have a tadpole or intermediate larval phase in water when development pauses for growth before adult limbs develop. In moist habitats, particularly in the tropics, amphibians are often the most abundant vertebrates.

Reptiles

If we think of amphibians as the "first draft" of a manuscript about survival on land, then **reptiles** were the published book. All living reptiles share certain fundamental characteristics, features they retained from the time when they replaced amphibians as the dominant terrestrial vertebrates. Among the most important are:

1. **Amniotic egg.** Amphibians never succeeded in becoming fully terrestrial because amphibian eggs must be laid in water to avoid drying out. Most reptiles lay watertight, leathery **amniotic eggs** that offer various layers of protection from drying out (figure 18.28).

2. **Dry skin.** Today's amphibians have a moist skin and must remain in moist places to avoid drying out. In reptiles, a layer of scales or armor covers their bodies, preventing water loss.

3. **Thoracic breathing.** Amphibians breathe by squeezing their throat to pump air into their lungs; this limits their breathing capacity to the volume of their mouth. Reptiles developed thoracic breathing, expanding and contracting the rib cage to suck air into the lungs and then force it out.

Reptiles first evolved about 300 million years ago when the world was entering a long, dry period. Early reptiles rose to prominence and also gave rise to the ancestor of mammals. Reptiles continued their diversification over millions of years, giving rise to many different forms, including marine reptiles, flying reptiles (pterosaurs), dinosaurs, and their descendants the birds (figure 18.29). Dinosaurs became the most successful of all land vertebrates, until their abrupt extinction around 65 million years ago.

Today some 7,000 species in the class are found in practically every wet and dry habitat on earth. Modern reptiles include four groups: turtles and tortoises, crocodiles and alligators, snakes and lizards, and tuataras. Extinct reptiles include marine forms, pterosaurs, dinosaurs, and the ancestors of mammals and birds.

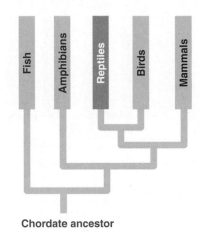

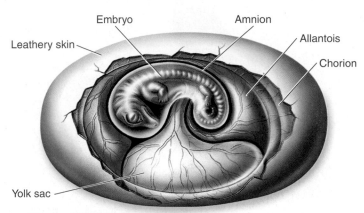

Figure 18.28 A key adaptation of reptiles: watertight eggs.

The watertight amniotic egg allows reptiles to live in a wide variety of terrestrial habitats.

Figure 18.29 An extinct reptile.

Euparkeria, a thecodont reptile, had rows of bony plates along the sides of the backbone, as seen in modern crocodiles and alligators. Thecodonts likely gave rise to crocodiles, dinosaurs, pterosaurs (flying reptiles), and birds.

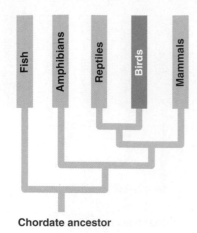

Chordate ancestor

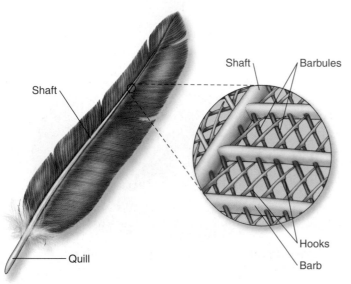

Figure 18.30   A key adaptation of birds: feathers.

The barbs off the main shaft of a feather have secondary branches called barbules that attach to one another by microscopic hooks.

Figure 18.31   *Archaeopteryx.*

Archaeopteryx lived about 150 million years ago and is the oldest fossil bird.

18.11 Birds Master the Air

CONCEPT PREVIEW: Birds are descendents of dinosaurs. Feathers and a strong, light skeleton make flight possible.

Birds evolved from small bipedal dinosaurs about 150 million years ago. Birds are so structurally similar to dinosaurs that some biologists consider them "feathered dinosaurs." Most biologists, however, continue to place birds in their own class, Aves, rather than lumping them in with the reptiles.

Characteristics of Birds

Modern birds lack teeth and have only vestigial tails, but they still retain many reptilian characteristics. For instance, birds lay amniotic eggs, although the shells of bird eggs are hard rather than leathery. Also, reptilian scales are present on the feet and lower legs of birds. What makes birds unique? What distinguishes them from living reptiles?

1. **Feathers.**   Derived from reptilian scales, feathers serve two functions: providing lift for flight and conserving heat. Feathers consist of a center shaft with barbs extending out (figure 18.30). The barbs are held together with secondary branches called barbules that hook over each other. This reinforces the structure of the feather without adding much weight to it. Like scales, feathers can be replaced. Among living animals, feathers are unique to birds. Several types of dinosaurs also possessed feathers, however these feathers were most likely used for insulation or display.
2. **Flight skeleton.**   The bones of birds are thin and hollow. Many of the bones are fused, making the bird skeleton more rigid than a reptilian skeleton and forming a sturdy frame that anchors muscles during flight. The power for active flight comes from large breast muscles that can make up 30% of a bird's total body weight. They stretch down from the wing and attach to the breastbone, which is greatly enlarged and bears a prominent keel for muscle attachment. They also attach to the fused collarbones that form the so-called wishbone. No other living vertebrates have a fused collarbone or a keeled breastbone.

Birds are endothermic. Like mammals, they generate enough heat through metabolism to maintain a high body temperature. Birds maintain body temperatures significantly higher than most mammals. This permits a faster metabolism, necessary to satisfy the large energy requirements of flight.

History of Birds

The oldest bird of which there is a clear fossil is ***Archaeopteryx*** (meaning "ancient wing"; figure 18.31), which was about the size of a crow and shared many features with small, bipedal, carnivorous dinosaurs. For example, it had teeth and a long reptilian tail. Its bones were solid, unlike the hollow bones of today's birds, and it had feathers (recently discovered fossils in China show that several species of dinosaurs possessed feathers). By the early Cretaceous, only a few million years after *Archaeopteryx,* a diverse array of birds had evolved, with many of the features of modern birds. The diverse birds of the Cretaceous shared the skies with the flying reptiles called pterosaurs for 70 million years. Today about 8,600 species of birds (class Aves) occupy a variety of habitats all over the world.

18.12 Mammals Adapt to Colder Times

CONCEPT PREVIEW: Mammals are endotherms that nurse their young with milk and have three bones in their middle ear, helping to amplify sound waves. All mammals have at least some hair.

Characteristics of Mammals

The **mammals** (class Mammalia) that evolved about 220 million years ago side by side with the dinosaurs would look strange to you, not at all like modern-day lions and tigers and bears. They shared three key characteristics with mammals today:

1. **Mammary glands.**   Female mammals have mammary glands, which produce milk to nurse the newborns. Even baby whales are nursed by their mother's milk. Milk is a very-high-calorie food (human milk has 750 kcal per liter), important because of the high energy needs of a rapidly growing newborn mammal.

2. **Hair.**   Among living vertebrates, only mammals have hair (even whales and dolphins have a few sensitive bristles on their snout). A hair is a filament composed of dead cells filled with the protein keratin. The primary function of hair is insulation. The insulation provided by fur may have ensured the survival of mammals when the dinosaurs perished.

3. **Middle ear.**   All mammals have three middle-ear bones, which evolved from bones in the reptile jaw. These bones play a key role in hearing by amplifying vibrations created by sound waves that beat upon the eardrum.

History of Mammals

The first mammals arose from therapsids, in the late Triassic about 220 million years ago, just as the first dinosaurs evolved from thecodont archosaurs. Tiny, shrewlike creatures that ate insects, most mammals were only a minor element in a land that quickly came to be dominated by dinosaurs. Fossils reveal that these early mammals had large eye sockets, evidence that they may have been active at night. At the end of the Cretaceous period, 65 million years ago, when dinosaurs and many other land and marine animals became extinct, mammals rapidly diversified, taking over areas vacated by the extinction of dinosaurs. Mammals reached their maximum diversity about 15 million years ago.

Rapid diversification of animals to fill empty niches was discussed on page 249 regarding the diversification of finches on the Galápagos Islands.

Today, over 4,500 species of mammals occupy all the areas that dinosaurs once claimed, among many others (figure 18.32). Present day mammals range in size from 1.5-gram shrews to 100-ton whales. Almost half of all mammals are rodents—mice and their relatives. Almost one-quarter of all mammals are bats! Mammals have even invaded the seas, as plesiosaur and ichthyosaur reptiles did so successfully millions of years earlier—79 species of whales and dolphins live in today's oceans.

Concept Check

1. What three phyla of animals successfully invaded the land?
2. How do bony fishes stay afloat? Compare this to sharks.
3. Which are more like reptiles, mammals or birds?

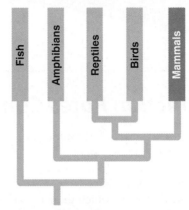

Fish Amphibians Reptiles Birds Mammals

Chordate ancestor

(a)   (b)

(c)

Figure 18.32   Today's mammals.

Today mammals include the **monotremes** (egg-laying mammals), the **marsupials** (pouched mammals), and the **placental mammals** (mammals that produce a true placenta). (a) Monotremes include the duck-billed platypus and the echidna, or spiny anteater, like this *Tachyglossus aculeatus*. (b) Marsupials include kangaroos, like this adult with young in its pouch. In marsupials, the tiny embryo is born and crawls into the marsupial pouch, where it completes its development. (c) This female African lion, *Panthera leo* (order Carnivora), is a placental mammal. In placental mammals, a specialized organ called the placenta nourishes the embryo throughout its development. Most species of mammals living today, including humans, are placental mammals.

The placenta contains tissue that originates from the fetus as well as from the mother, as discussed on page 606. Materials such as nutrients, gases, and waste products pass between the baby and the mother, but blood in the two circulatory systems never come in contact.

Inquiry & Analysis

Are Extinction Rates Constant?

Since the time of the dinosaurs, the number of living species has risen steadily. Today, for example, there are over 700 families of marine animals, containing thousands of described species.

Interspersed, however, have been a number of major setbacks, termed mass extinctions, in which the number of species has greatly decreased. Five major mass extinctions have been identified, the most severe of which occurred at the end of the Permian Period, approximately 225 million years ago, at which time more than half of all families and as many as 96% of all species may have perished.

The most famous and well-studied extinction, though not as drastic, occurred at the end of the Cretaceous Period (65 million years ago), at which time the dinosaurs and a variety of other organisms went extinct. Recent studies have provided support for the hypothesis that this extinction event was triggered by a large asteroid that slammed into the earth. This mass extinction did have one positive effect, though: With the disappearance of dinosaurs, mammals, which previously had been relatively small and inconspicuous, quickly experienced a vast evolutionary radiation, which ultimately produced a wide variety of organisms, including elephants, tigers, whales, and humans. Indeed, a general observation is that biological diversity tends to rebound quickly after mass extinctions, reaching comparable levels of species richness, even if the organisms making up that diversity are not the same.

The number of species in the world in recent times is greater than it has ever been. Unfortunately, that number is decreasing at an alarming rate, a rate of extinction not seen on earth since the Cretaceous mass extinction.

One thing that the Cretaceous mass extinction and the present-day mass extinction share is that species are becoming extinct for reasons that have nothing to do with what they themselves are like. Is this generally true of all extinctions, or are mass extinctions a special case?

Evolutionist Lee Van Valen put forth the hypothesis in 1973 that extinction is indeed usually due to random events unrelated to a species's particular adaptations. If this is in fact the case, then the likelihood that a species will go extinct would be expected to be virtually constant, when viewed over long periods of time.

Van Valen's hypothesis has been tested by evolutionary biologists for a variety of animal groups. One of the most complete fossil records available for such a test is that of marine echinoids (sea urchins and sand dollars). The fossil echinoid you see in the photo, genus *Cidaris*, is from the Cretaceous some 75 million years ago. In the graph above it, you see an examination of the 200 million year fossil record of echinoids. Data are presented as the number of echinoid families that have survived for over a period of 200 million years (the blue dots on the graph), estimated from the fossil record. The red dashed line shows a theoretical constant extinction rate, as postulated by Van Valen. The blue line is a "best-fit" curve determined by statistical regression analysis of the family number estimates.

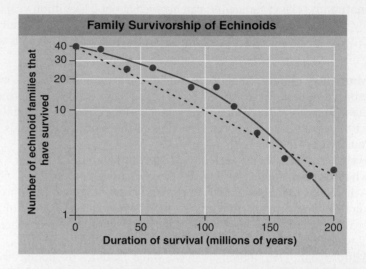

Family Survivorship of Echinoids

Analysis

1. **Interpreting Data** Estimate how many echinoid families were surviving at the time of the fossil in the photo?
2. **Making Inferences** Over the 200 million year fossil record of echinoids, which of the two lines best represents the data?
3. **Drawing Conclusions** Is Van Valen's hypothesis supported by this analysis?

Concept Summary

Introduction to the Animals

18.1 General Features of Animals

- Animals are complex multicellular heterotrophs. They are mobile and reproduce sexually. Animals cells do not have cell walls, and animal embryos have a similar pattern of development (**table 18.1**).

18.2 Five Transitions in Body Plan

- The large diversity of animals can be traced to five key evolutionary changes in body plan—the evolution of tissues, bilateral symmetry, a body cavity, deuterostome pattern of development, and body segmentation. The beetle shown here from **figure 18.1** exhibits four of these five transitions, lacking only deuterostome development.

18.3 The Animal Family Tree

- Traditionally, animals have been classified based on anatomical and embryological features (**figure 18.2**). But, molecular methods that compare RNA and DNA of animals are producing new phylogenies (**figure 18.3**).

Evolution of the Animal Phyla

18.4 Sponges and Cnidarians: The Simplest Animals

- Sponges, in the subkingdom Parazoa, are aquatic, have specialized cells, but lack tissues. The vase-shaped adult is anchored to the substrate (**figure 18.4**). Sponges are filter-feeders. Specialized cells called choanocytes trap food particles that are filtered from the water (*Phylum Facts,* **page 338**).

- Cnidarians (**figure 18.5**) have radially symmetrical bodies and tissues. They are carnivores, capturing their prey using harpoonlike nematocysts that are launched from cells called cnidocytes (*Phylum Facts,* **page 339**). They digest their food extracellularly in the gastrovascular cavity.

18.5 The Advent of Bilateral Symmetry

- The simplest bilaterally symmetrical animals are the solid worms, including flatworms (**figure 18.7**). Solid worms have three embryonic tissue layers and a digestive cavity but lack a body cavity, so they are called acoelomates (*Phylum Facts,* **page 341**).

18.6 Changes in the Body Cavity

- The evolution of a body cavity improved circulation, movement, and organ function. The roundworms have a body cavity but it is not a true body cavity, so they are called pseudocoelomates (*Phylum Facts,* **page 343**).

- Mollusks have a true body cavity and so they are called coelomates. Although a diverse group, they all contain a head-foot, a visceral mass, and a mantle (*Phylum Facts,* **page 344**).

- Segmentation first evolved in annelid worms, which allowed for specialization of body parts (*Phylum Facts,* **page 345**).

- Arthropods are the most successful animal phylum. Jointed appendages first evolved in this group, as well as the evolution of a rigid exoskeleton (*Phylum Facts,* **page 346**).

18.7 Redesigning the Embryo

- In the coelomates, there are two different developmental patterns that are reflected in the organization of cells during cleavage, the future of the blastopore, and the formation of mesoderm. A deuterostome developmental pattern evolved in echinoderms and chordates, suggesting that they share a common ancestor. All other coelomates are protostomes (**figure 18.15**).

- Echinoderms have an endoskeleton made up of bony plates that lie under the skin. The adults are radially symmetrical, but that appears to be an adaptation to their environment (*Phylum Facts,* **page 348**).

- The chordates have a truly internal endoskeleton and are distinguished by the presence of a notochord, a dorsal nerve cord, pharyngeal pouches, and a postanal tail (*Phylum Facts,* **page 349**). The notochord is replaced with the backbone in vertebrates.

The Parade of Vertebrates

18.8 Overview of Vertebrate Evolution

- Animal life began in the sea primarily during the Paleozoic era with only two animal groups having successfully invaded the land, arthropods and vertebrates.

- Mass extinctions, where the loss of species outpaces new species formation, have occurred five times throughout earth's history.

18.9 Fishes Dominate the Sea

- Fishes are the ancestors of all vertebrates (**figure 18.23**). Although they are a very diverse group, all fishes have certain characteristics in common such as gills, a vertebral column, a single-loop circulatory system, and nutritional deficiencies.

- Sharks have a flexible cartilaginous skeleton and are very fast swimmers (**figure 18.24**). Bony fish have swim bladders to control buoyancy in the water (**figure 18.25**), and they are the most successful group of vertebrates.

18.10 Amphibians and Reptiles Invade the Land

- Amphibians were the first vertebrates to invade the land. They most certainly evolved from lobe-finned fishes (**figure 18.26**). Adaptations to a terrestrial environment included the development of legs, lungs, cutaneous respiration, pulmonary circulatory system, and a partially divided heart.

- The key adaptations found in reptiles that made them better suited to a terrestrial environment were the evolution of a watertight amniotic egg (**figure 18.28**), watertight skin, and thoracic breathing.

18.11 Birds Master the Air

- Birds evolved from dinosaurs but were a minor group until the mass extinction of dinosaurs. Two key adaptations allowed birds to dominate the skies: feathers (**figure 18.30**) and a skeleton modified for flight with its sturdy but lightweight, hollow bones.

18.12 Mammals Adapt to Colder Times

- Mammals (**figure 18.32**), like birds, expanded and diversified after the mass extinction of the dinosaurs. Mammals are distinguished by the presence of mammary glands for feeding their young, hair for insulation, and bones in the middle ear that amplify sound.

Self-Test

1. Sponges possess unique, collared flagellated cells called
   a. cnidocytes.
   b. choanocytes.
   c. choanoflagellates.
   d. epithelial cells.

2. Which of the following characteristics is *not* seen in the phylum Platyhelminthes (flatworms)?
   a. cephalization
   b. the presence of a mesoderm
   c. specialization of digestive tract
   d. bilateral symmetry

3. One difference between the pseudocoel found in the phylum Nematoda (roundworms) and the coelom found in the phylum Annelida (segmented worms) is that the pseudocoel develops between the mesoderm and the _____ in roundworms, and the coelom develops in the _____ in segmented worms.
   a. ectoderm; mesoderm
   b. endoderm; mesoderm
   c. ectoderm; endoderm
   d. endoderm; ectoderm

4. Which of the following is *not* found in arthropods?
   a. jointed appendages
   b. a coelom
   c. deuterostome development
   d. an exoskeleton

5. Echinoderms take on radial symmetry as adults. Some biologists explain this change in symmetry as an adaptation for
   a. an animal traveling through the environment, rather than a sessile lifestyle.
   b. an animal with a sessile lifestyle, rather than one that moves through the environment.
   c. a predator, rather than a filter-feeder.
   d. an animal living in a marine environment, rather than a freshwater environment.

6. All fish species share all of the following characteristics *except*
   a. gills.
   b. jaws.
   c. internal skeleton with dorsal nerve cord.
   d. single loop circulatory system.

7. Chondrichthyes (sharks) and Osteichthyes (bony fish) have evolved anatomical solutions to increase swimming speed and maneuverability. Which of the following modifications is *not* found in Osteichthyes?
   a. a lateral line system
   b. buoyancy control through swim bladders
   c. an internal skeleton made of cartilage
   d. an operculum

8. Adaptations in reptiles do *not* include
   a. an amniotic egg.
   b. a layer of scales on the skin.
   c. middle ear bones.
   d. modifications to the respiratory system.

9. Characteristics that evolved in birds to allow for flight include
   a. reptilian-like scales on the legs.
   b. a hard-shelled amniotic egg.
   c. internal fertilization.
   d. thin, hollow bones in the skeleton.

10. A characteristic unique to most species of mammals and no other vertebrates is
    a. a notochord.
    b. skin that offers insulation and protection from dehydration.
    c. hair.
    d. flight.

Visual Understanding

1. About two-thirds of all the named species on our planet are arthropods. From the arthropod pie chart shown here, estimate about what percentage of the named species on our planet are beetles.

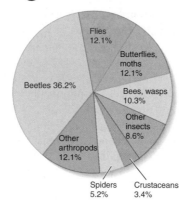

Flies 12.1%
Butterflies, moths 12.1%
Beetles 36.2%
Bees, wasps 10.3%
Other insects 8.6%
Other arthropods 12.1%
Spiders 5.2%
Crustaceans 3.4%

2. **Figure 18.28** What are the advantages for reptiles by having their eggs covered with a leathery outer shell?

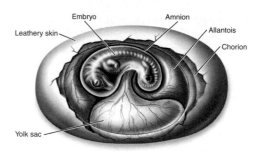

Leathery skin
Embryo
Amnion
Allantois
Chorion
Yolk sac

Challenge Questions

1. You have discovered a new organism deep in the jungle. You are trying to decide if it is a very slow-moving animal or a plant that responds to light and touch. What characteristics can you investigate to help you decide?

2. Why is the body cavity such a useful innovation in the animal kingdom?

3. Why is the internal skeleton of vertebrates so useful, compared to the external skeleton of many other animals?

4. A friend tells you that his father can't understand why people get so upset about the possibility of a few species dying out because of global warming or rain forest destruction when you consider the fact that there have already been five mass extinctions on our planet. What do you tell him?

Chapter **19**

Populations
and Communities

Ecology

19.1 What Is Ecology?

CONCEPT PREVIEW: Ecology is the study of how the organisms that live in a place interact with each other and with their physical environment. An ecosystem is a dynamic ecological system that challenges organisms to adjust to its changing conditions.

Ecology is the study of how organisms interact with each other and with their environment. Ecology also encompasses the study of the distribution and abundance of organisms, which includes population growth and the limits and influences on population growth. The word ecology was coined in 1866 by the great German biologist Ernst Haeckel and comes from the Greek words *oikos* (house, place where one lives) and *logos* (study of). Our study of ecology, then, is a study of the house in which we live. Do not forget this simple analogy built into the word ecology—most of our environmental problems could be avoided if we treated the world in which we live the same way we treat our own homes. Would you pollute your own house?

Levels of Ecological Organization

> The discussion of the hierarchical organization of life discussed on pages 16 and 17 began at the cellular level and traced the increasing complexity of life up through the levels to ecosystems. In this chapter, we pick up the hierarchy at the population level.

Ecologists consider groups of organisms at six progressively more encompassing levels of organization. As mentioned in chapter 1, new characteristics called *emergent properties* arise at each higher level, resulting from the way components of each level interact.

1. **Populations.** Individuals of the same species that live together are members of a population. They potentially interbreed with one another, share the same habitat, and use the same pool of resources the habitat provides.
2. **Species.** All populations of a particular kind of organism form a species. Populations of the species can interact and affect the ecological characteristics of the species as a whole.
3. **Communities.** Populations of different species that live together in the same place are called communities. Different species typically use different resources within the habitat they share (figure 19.1).
4. **Ecosystems.** A community and the nonliving factors with which it interacts is called an **ecosystem.** An ecosystem is affected by the flow of energy, ultimately derived from the sun, and the cycling of the essential elements on which the lives of its constituent organisms depend. The redwood forest community pictured in figure 19.1 is part of an ecosystem, where the giant trees and other organisms interact with each other and with their physical surroundings.
5. **Biomes.** Biomes are major terrestrial assemblages of plants, animals, and microorganisms that occur over wide geographical areas that have distinct physical characteristics. Examples include deserts, tropical rain forests, and grasslands. Similar types of groupings occur in marine and freshwater habitats.
6. **The biosphere.** All the world's biomes, along with its marine and freshwater assemblages, together constitute an interactive system we call the biosphere. Changes in one biome can have profound consequences for others.

Figure 19.1   A redwood community.

(a) The redwood forest of coastal California and southwestern Oregon is dominated by the population of redwoods *(Sequoia sempervirens)*. Other species in the redwood community include (b) redwood sorrel *(Oxalis oregana)*, (c) sword ferns *(Polystichum munitum),* and (d) ground beetles *(Scaphinotus velutinus),* this one feeding on a slug on a sword fern leaf.

Although we include biomes and the biosphere as higher levels of ecological organization in this list of organizational levels, the *ecosystem* is viewed as the "basic functional unit," in much the same way the cell rather than tissues or organs is considered the basic unit of living organisms.

Some ecologists, called *population ecologists,* focus on a particular species and how its populations grow. Other ecologists, called *community ecologists,* study how the different species living in a place interact with one another. Still other ecologists, called *systems ecologists,* are interested in how biological communities interact with their physical environment.

We will begin our study of ecology at the basic levels, by examining populations and communities. We will then work our way up the hierarchy by examining ecosystems, biomes, and ending with a critical look at the conditions of the biosphere. Although we break these topics into separate chapters, we should not overlook the fact that an organism does not live in a vacuum. Individuals interact with each other and with their physical environment and these interactions introduce challenges and obstacles to survival.

The Environmental Challenge

The nature of the physical environment determines to a great extent which organisms live in a particular climate or region. Key elements of the environment include:

Temperature.   Most organisms are adapted to live within a relatively narrow range of temperatures and will not thrive if temperatures are colder or warmer. The growing season of plants, for example, is strongly influenced by temperature.

Water.   All organisms require water. On land, water is often scarce, so patterns of rainfall have a major influence on life.

Sunlight.   Almost all ecosystems rely on energy captured by photosynthesis and so the availability of sunlight influences the amount of life an ecosystem can support, particularly below the surface in marine environments.

Soil.   The physical consistency, pH, and the availability of minerals in the soil often severely limit plant growth, particularly the amount of nitrogen and phosphorus present in the soil.

During the course of a day, a season, or a lifetime, an individual organism must cope with a range of living conditions. Many organisms are able to adapt to environmental change by making physiological, morphological, or behavioral adjustments. For example, you sweat when it is hot, increasing heat loss through evaporation and thus preventing overheating. Morphological adaptations in some mammals may include growing a thicker coat of fur in winter (figure 19.2). And, many animals deal with variations in the environment through behavior, such as moving from one place to another, thereby avoiding areas that are unsuitable. For example, a tropical lizard manages to maintain a fairly uniform body temperature by basking in the sunlight but then retreating to the shade when it becomes too hot (figure 19.3). These physiological, morphological, and behavioral abilities are a product of natural selection acting in a particular environmental setting over time, which explains why an individual organism that is moved to a different environment may not survive.

Concept Check

1. What is the difference between a community and an ecosystem?
2. Are deep-sea ecosystems affected by the availability of sunlight?

Figure 19.2   Wolf in winter.

This gray wolf grows a thicker coat of fur in the winter to insulate its body. Escaping body heat is trapped in the air surrounding the hairs of its coat, holding in heat and thus helping to maintain the wolf's body temperature in the cold winter.

Figure 19.3   Costa Rican lizard.

This green iguana escapes to the shade in the heat of the day, helping to keep its body cooler as the temperature outside rises.

Population Dynamics

(a)

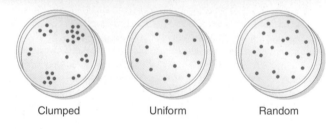

Clumped   Uniform   Random

(b)

Figure 19.4   **Population dispersion.**

(a) Starlings dispersed uniformly along telephone wires, and
(b) different arrangements of bacterial colonies.

BIOLOGY & YOU

Living Exponentially. Throughout most of human history, the size of the world's human population has been limited by food availability. Two thousand years ago, some 130 million people populated the earth. It took a thousand years for that number to double, and it was 1650 before it doubled again, to about 500 million. In all these years, the human population grew only slowly, limited by the earth's ability to support it. That all changed in the early 1700s with the advent of the industrial revolution. For the first time, technology gave humans control over the food supply, removing the limit to earth's carrying capacity and unleashing exponential growth. Although the human birthrate has remained unchanged for the last 300 years at about 30 per 1,000 per year, the death rate has fallen from 20 per thousand per year to its present level of 13. The difference between birth and death rates means the human population grows unchecked at 1% to 2% a year, leading to today's population of 6.7 billion. 80 million more will be added this year, and the population will double in the next 58 years. How long do you think this exponential growth can continue? What do you think will happen then?

19.2 Population Growth

CONCEPT PREVIEW: The size at which a population stabilizes in a particular place is defined as its carrying capacity. Populations increase in size to the carrying capacity of their environment.

Populations are groups of individuals of a species that live together and influence each other's survival. In this section, we will explore the properties of populations, focusing on the factors that influence whether a population will grow or shrink, and at what rate. Although we humans picture ourselves as different from populations of animals living in the wild, factors that affect wild populations, such as population densities, dispersion, growth, competition, and sharing resources, affect human populations in similar ways.

One of the critical properties of any population is its **population size**—the number of individuals in the population. For example, if an entire species consists of only one or a few small populations, that species is likely to become extinct, especially if it occurs in areas that have been or are being radically changed. In addition to population size, **population density**—the number of individuals that occur in a unit area, such as per square kilometer—is often an important characteristic. The density of a population, how closely individuals associate with each other, is an indication of how they live. Animals that live in large groups, such as herds of wildebeests, may find safety in numbers.

A third significant property is **population dispersion,** the scatter of individual organisms within the population's range. As you can see in figure 19.4, individuals may be spaced in clumps (due to uneven distribution of resources or in response to social interaction), uniformly (as a result of competition for resources), or randomly, which is less common in nature. In addition to size, density, and dispersion, another key characteristic of any population is its capacity to grow. To understand populations we must consider what factors in nature promote or limit **population growth.**

The Exponential Growth Model

The simplest model of population growth defines a population's growth rate (the change in its numbers over time) as the difference between the birthrate and the death rate, corrected for any movement of individuals into (immigration) or out of (emigration) the population. Movements of individuals can have a major impact on population growth rates. For example, the increase in human population in the United States during the closing decades of the twentieth century was mostly due to immigrants. Less than half of the increase came from the reproduction of the people already living there.

Even when a population's growth rate remains constant, the actual increase in the *number* of individuals in the population accelerates rapidly; the larger a population is, the faster it grows. This sort of rapid growth is called *exponential growth.* Rapid exponential growth is indicated by the red line in figure 19.5. This sort of growth pattern is similar to that obtained by compounding interest on an investment. In practice, such patterns prevail only for short periods, usually when an organism reaches a new habitat with abundant resources. Natural examples include dandelions reaching the lawns of North America from Europe for the first time; algae colonizing a newly-formed pond; or the first plants arriving on an island recently thrust up from the sea. Figure 19.6a shows the establishment of tree populations in the British Isles after the glaciers receded in the Northern Hemisphere 20,000 years ago. The tree population grew at an exponential rate.

Carrying Capacity

No matter how rapidly populations grow, they eventually reach a limit imposed by shortages of important environmental factors such as space, light, water, or nutrients and will experience *logistic growth*. A population usually ultimately stabilizes at a certain size, called the **carrying capacity** of the particular place where it lives, and the size of the population levels off, like the blue line in figure 19.5. The carrying capacity is the maximum number of individuals that an area can support.

The human population, as discussed on page 443, is experiencing exponential growth but like other populations of organisms, humans are affected by environmental conditions which will ultimately act to stabilize population size.

The Logistic Growth Model

As a population approaches its carrying capacity, its rate of growth slows greatly as fewer and fewer resources remain for each new individual to use, until the population size levels off at or near the carrying capacity. The growth curve of such a population, which is always limited by one or more factors in the environment, is approximated by an S-shaped sigmoid growth curve, characteristic of most biological populations. The curve is called "sigmoid" because its shape has a double curve like the letter *S*. As the size of a population stabilizes at the carrying capacity, its rate of growth slows down, eventually coming to a halt. The fur seal population in figure 19.6*b* has a carrying capacity of about 10,000 breeding male seals.

Processes such as competition for resources, emigration, and the accumulation of toxic wastes all tend to increase as a population approaches its carrying capacity for a particular habitat. The resources for which the members of the population are competing may be food, shelter, light, mating sites, mates, or any other factor needed to survive and reproduce.

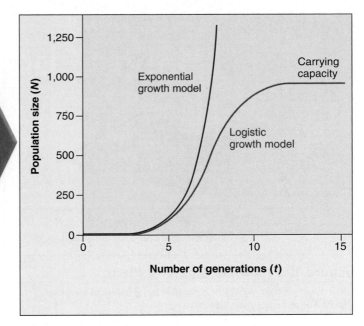

Figure 19.5 **Two models of population growth.**

The *red line* illustrates the exponential growth model for a population. The *blue line* illustrates the logistic growth model. At first, logistic growth accelerates exponentially, and then, as resources become limiting, the birth rate decreases or the death rate increases, and growth slows. Growth ceases when the death rate equals the birthrate. The carrying capacity ultimately depends on the resources available in the environment.

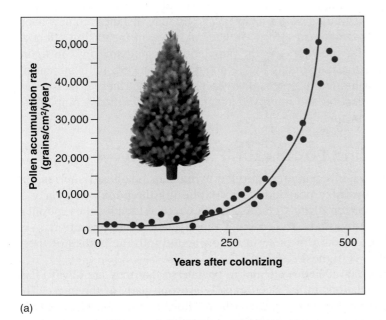

(a)

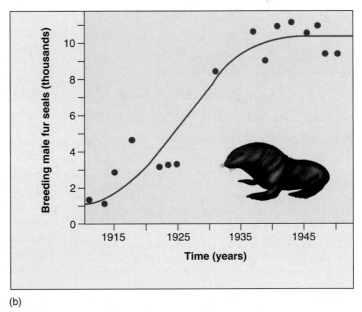

(b)

Figure 19.6 **Examples of population growth in two natural populations.**

Most natural populations exhibit logistic growth except when a species is inhabiting a new environment. (a) These data present the exponential growth of Scotch pine trees after glaciers receded from the Norfolk region of Great Britain. The size of the population was estimated based on the rate of pollen accumulation in lake sediments. (b) These data present the history of a fur seal (*Callorhinus ursinus*) population on St. Paul Island, Alaska. Driven almost to extinction by hunting in the late 1800s, the fur seal made a comeback after hunting was banned in 1911. Today the number of breeding males with "harems" oscillates around 10,000 individuals, presumably the carrying capacity of the island.

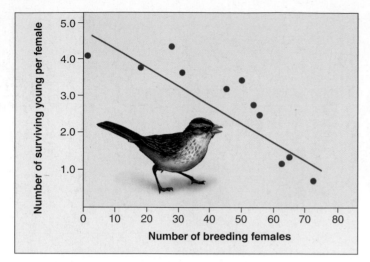

Figure 19.7   Density-dependent effects.

Reproductive success of the song sparrow (*Melospiza melodia*) decreases as population size increases.

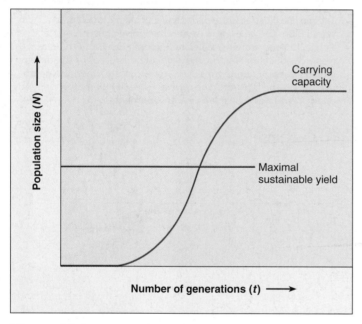

Figure 19.8   Maximal sustainable yield.

The goal of harvesting organisms for commercial purposes is to harvest just enough organisms to maximize current yields but also to sustain the population for future yields. Harvesting the organisms when the population is in the rapid growth phase of the sigmoidal curve, but not overharvesting, will result in sustained yields.

IMPLICATION   Do you like tuna fish? How about tuna steaks? Atlantic bluefin tuna (*Thunnus thynnus*) are among the largest bony fishes in the ocean (reaching over 10 feet long and weights of 1,200 pounds) and are prized catches of commercial fishers. Not surprisingly, their numbers have fallen dramatically due to the calamitous overexploitation of Atlantic populations of this species. The bluefins are now protected in the western Atlantic, but not in the eastern Atlantic, even though tagging of fish confirms that they go back and forth. Do you think the Atlantic bluefin population will be sustainable under this policy?

19.3 The Influence of Population Density

CONCEPT PREVIEW: Density-independent effects are controlled by factors that operate regardless of population size. Density-dependent effects are caused by factors that come into play particularly when the population size is larger.

Many factors act to regulate the growth of populations in nature. Some of these factors act independently of the size of the population; others do not.

Density-Independent Effects

Effects that are independent of the size of a population and act to regulate its growth are called **density-independent effects.** A variety of factors may affect populations in a density-independent manner. Most of these are aspects of the external environment, such as weather (extremely cold winters, droughts, storms, floods) and physical disruptions (volcanic eruptions and fire). Individuals often will be affected by these activities regardless of the size of the population. Populations that occur in areas in which such events occur relatively frequently will display erratic population growth patterns, increasing rapidly when conditions are relatively good, but suffering extreme reductions whenever the environment turns hostile.

Density-Dependent Effects

Effects that are dependent on the size of the population and act to regulate its growth are called **density-dependent effects.** Among animals, these effects may be accompanied by hormonal changes that can alter behavior that will directly affect the ultimate size of the population. One striking example occurs in migratory locusts ("short-horned" grasshoppers). When they become crowded, the locusts produce hormones that cause them to enter a migratory phase; the locusts take off as a swarm (as pictured at the beginning of this chapter) and fly long distances to new habitats. Density-dependent effects, in general, have an increasing effect as population size increases. As the population of song sparrows in figure 19.7 grows, the individuals in the population compete with increasing intensity for limited resources. Darwin proposed that these effects result in natural selection and improved adaptation as individuals compete for the limiting factors.

Maximizing Population Productivity

In natural systems that are exploited by humans, such as fisheries, the aim is to maximize productivity by exploiting the population early in the rising portion of its sigmoid growth curve. At such times, populations and individuals are growing rapidly, and net productivity—in terms of the amount of material incorporated into the bodies of these organisms—is highest.

Commercial fisheries attempt to operate so that they are always harvesting populations in the steep, rapidly growing parts of the curve. The point of *maximal sustainable yield* (the red line in figure 19.8) lies partway up the sigmoid curve. Harvesting the population of an economically desirable species near this point will result in the best sustained yields. Overharvesting a population that is smaller than this critical size can destroy its productivity for many years or even drive it to extinction.

19.4 Life History Adaptations

CONCEPT PREVIEW: Some life history adaptations favor near exponential growth, while others favor the more competitive logistic growth.

Populations of many species, including annual plants, some insects, and most bacteria, can have very fast rates of growth when not limited by dwindling environmental resources. Habitats with more available resources than the population requires favor very rapid reproduction rates, which often approximate the exponential growth model discussed earlier. In mathematical formulas used to calculate population growth rates, the maximum growth rate is indicated by *r*.

Populations of most animals have much slower rates of growth with numbers limited by available resources. Growth slows as available resources become limiting, producing a sigmoid growth curve approximating the logistic growth model discussed earlier. Habitats with limited resources lead to more intense competition for resources, and favor individuals that can survive and successfully reproduce more efficiently. The number of individuals that can survive at this limit is the carrying capacity of the environment. In mathematical formulas used to calculate population growth rates, the carrying capacity is indicated by **K.**

The complete life cycle of an organism constitutes its life history. Life histories are very diverse, with different organisms having different adaptations in response to their environments. Some life history adaptations of a population favor very rapid growth in a habitat with abundant resources, or in unpredictable or volatile environments in which organisms have to take advantage of the resources when they are available. In these situations reproducing early, producing many small offspring that mature quickly, and engaging in other aspects of "big bang" reproduction are favored. Using the terms of the exponential model, these adaptations that favor a high rate of increase, approaching the maximum growth rate or *r,* are called ***r*-selected adaptations.** Examples of organisms displaying *r*-selected life history adaptations include dandelions, aphids, mice, and cockroaches (figure 19.9).

Other life history adaptations favor survival in an environment in which individuals are competing for limited resources. These features include reproducing late, having small numbers of larger-sized offspring that mature slowly and receive intensive parental care, and other aspects of "carrying capacity" reproduction. In terms of the logistic model, these adaptations favoring reproduction near the carrying capacity of the environment or *K,* are called **K-selected adaptations.** Examples of organisms displaying *K*-selected life history adaptations include coconut palms, whooping cranes, and whales.

Although the *r/K* concept of life histories is often used to compare different taxa, it also provides a powerful way to examine more closely related organisms living in different types of habitats. In general, populations living in rapidly changing habitats tend to exhibit *r*-selected adaptations, whereas populations of closely related organisms living in more stable and competitive habitats exhibit more *K*-selected adaptations.

Selection in guppies, discussed on pages 268 and 269, illustrates how two populations of the same species may evolve different life histories. High-predation populations reproduced at a younger age (*r*-selected), while low-predation populations reproduced later (*K*-selected).

Most natural populations show life history adaptations that exist along a continuum, ranging from completely *r*-selected traits to completely *K*-selected traits. Table 19.1 outlines the adaptations at the extreme ends of the continuum.

Figure 19.9   The consequences of exponential growth.

All organisms have the potential to produce populations larger than those that actually occur in nature. The German cockroach *(Blatella germanica),* a major household pest, produces 80 young every six months. If every cockroach that hatched survived for three generations, kitchens might look like this theoretical culinary nightmare concocted by the Smithsonian Museum of Natural History.

TABLE 19.1	*r*-Selected and *K*-Selected Life History Adaptations	
Adaptation	***r*-Selected Populations**	***K*-Selected Populations**
Age at first reproduction	Early	Late
Homeostatic capability	Limited	Often extensive
Life span	Short	Long
Maturation time	Short	Long
Mortality rate	Often high	Usually low
Number of offspring produced per reproductive episode	Many	Few
Number of reproductions per lifetime	Usually one	Often several
Parental care	None	Often extensive
Size of offspring or eggs	Small	Large

Figure 19.10   **Age structure in different types of populations.**

The grass you see in this photo is an annual plant. The individual plants are all the same age because they die every year. By contrast, the herd of bison contains individuals of varying ages because they can survive from year to year.

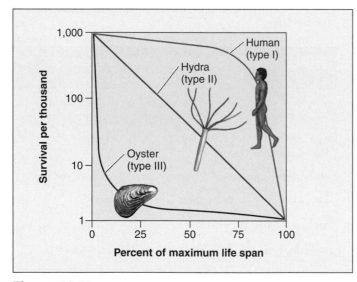

Figure 19.11   **Survivorship curves.**

By convention, survival (the *vertical axis*) is plotted on a log scale. In hydra, animals related to jellyfish, individuals are equally likely to die at any age, as indicated by the straight survivorship curve (the blue line, type II). Oysters, like plants, produce vast numbers of offspring, only a few of which live to reproduce. However, once they become established as reproductive individuals, their mortality is extremely low (red line, type III survivorship curve). Even though human babies are susceptible to death at relatively high rates, mortality in humans, as in many animals and protists, rises in the postreproductive years (green line, type I survivorship curve).

19.5 Population Demography

CONCEPT PREVIEW: The growth rate of a population is a sensitive function of its age structure. In some species, mortality is higher among the young, and in others, among the old; in only a few is mortality independent of age.

Demography is the statistical study of populations. Demography is the science that helps predict how population sizes will change in the future. Populations grow if births outnumber deaths and shrink if deaths outnumber births. Because birth and death rates also depend on age and sex, the future size of a population depends on its present age structure and sex ratio.

Age Structure

Many annual plants and insects time their reproduction to particular seasons of the year and then die. All members of these populations are the same age. Perennial plants and longer-lived animals contain individuals of more than one generation, so that in any given year individuals of different ages are reproducing within the population (figure 19.10). A group of individuals of the same age is referred to as a **cohort.**

Within a population, every cohort has a characteristic birthrate, or **fecundity,** defined as the number of offspring produced in a standard time (for example, per year), and a characteristic death rate, or **mortality,** the number of individuals that die in that period. The rate of a population's growth depends on the difference between these two rates.

The relative number of individuals in each cohort defines a population's age structure. A population with a large proportion of young individuals tends to grow rapidly because an increasing proportion of its individuals are reproductive. A population with more females usually also tend to grow more rapidly. However, among monogamous species like many birds, pairs often form long-lasting reproductive relationships and a reduction in the number of males can directly reduce the number of births. The proportion of males to females in a population is its **sex ratio.**

Mortality and Survivorship Curves

A population's intrinsic rate of increase depends on the ages of the organisms in it and the reproductive performance of the individuals in the various age groups. When a population lives in a constant environment for a few generations, its **age distribution**—the proportion of individuals in different age categories—tends to stabilize. This distribution differs greatly from species to species and even, to some extent, from population to population within a given species. A population whose size remains fairly constant through time is called a stable population. In such a population, births plus immigration must balance deaths plus emigration.

One way to express the age distribution characteristics of populations is through a **survivorship curve.** Survivorship is defined as the percentage of an original population that survives to a given age. Examples of different kinds of survivorship curves are shown in figure 19.11.

Concept Check

1. What has to happen to convert logistic growth into exponential growth?
2. Do density-independent effects result in natural selection? Explain.
3. What is fecundity? If it increases, what happens to the population growth rate?

How Competition Shapes Communities

19.6 Communities

CONCEPT PREVIEW: A community consists of all species that occur at a site. Their interactions shape ecological and evolutionary patterns.

The term **community** refers to the mix of species that occur at any particular locality. Some communities contain many species, like the array of plants and animals that you see in the savanna photo in figure 19.12*a* as well as the ones you can't see (like fungi, protists, and microbes). Some communities contain only a few species, such as in the near-boiling waters of Yellowtone's geysers (where a limited number of microbial species live). Communities can be characterized either by their constituent species, a list of all species present in the community, or by their properties, such as species richness (the number of different species present) or primary productivity (the amount of solar energy captured through photosynthesis and stored as organic compounds).

Interactions among community members govern many ecological and evolutionary processes. These interactions, such as predation (figure 19.12*b*) and competition (figure 19.12*c*), affect the population biology of particular species—whether a population increases or decreases in abundance for example—as well as the ways in which energy and nutrients are used in the ecosystem. As discussed further in chapter 20, an *ecosystem* includes a community of living organisms and the nonliving components that surround them.

Scientists study biological communities in many ways, ranging from detailed observations to elaborate, large-scale experiments. In some cases, such studies focus on the entire community, whereas in other cases only a subset of species that are likely to interact with each other are studied. Regardless of how they are studied, two views exist on the makeup and functioning of communities.

The *individualistic concept* of communities, first championed by H. A. Gleason of the University of Chicago early in the twentieth century, holds that a community is nothing more than an aggregation of species that happen to coexist in one place. By contrast, the *holistic concept* of communities, which can be traced to the work of F. E. Clements, also about a century ago, views communities as an integrated unit. In this sense, the community could be viewed as a superorganism whose constituent species have coevolved to the extent that they function as a part of a greater whole, just as the kidneys, heart, and lungs all function together within an animal's body. In this view, then, a community would amount to more than the sum of its parts.

Most ecologists today favor the individualistic concept. For the most part, species seem to respond independently to changing environmental conditions. As a result, community composition changes gradually across landscapes as some species appear and become more abundant, while others decrease in abundance and eventually disappear. Competition is an important factor that affects individuals and in so doing affects the community.

(a)

(b)

(c)

Figure 19.12   A Tanzanian savanna community.

A community consists of all the species—plants, animals, fungi, protists, and prokaryotes—that occur at a locality. (a) A savanna community in Lake Manyara National Park in Tanzania. Species within a community interact with each other, such as through (b) predation or (c) competition for a resource.

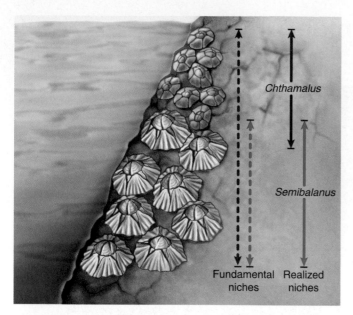

Figure 19.13   Competition among two species of barnacles limits niche use.

Chthamalus can live in both deep and shallow zones (its fundamental niche), but *Semibalanus* forces *Chthamalus* out of the part of its fundamental niche that overlaps the realized niche of *Semibalanus*.

IN THE NEWS

Beavers Back in Britain. The highlands of Scotland, once heavily timbered, were lumbered so intensively centuries ago that the mountains were denuded of trees. Starting a few decades ago, the highlands began the long road back. A major program of tree replanting was started, and today woodlands have begun to again pepper the highlands. Many of the deer and other forest inhabitants that used to frequent Scottish highland forests are now seen there again—but not beavers. For 400 years there have been no beavers in Scotland, or indeed anywhere in Great Britain—they disappeared in the seventeenth century, at the dawn of the Industrial Revolution. This raises a very interesting question: Does the absence of beavers imply that beavers cannot compete successfully for habitat in Great Britain—that the new replanted Scottish highlands are outside a beaver's fundamental niche? Or might the reborn Scottish highlands simply be outside today's realized niche for beavers, unoccupied only because there are no beavers around in the highlands these days to do the job? To answer this question, British government conservationists in 2009 released four wild beaver families into the highlands of western Scotland. Trapped in Norway and quarantined for six months, these intrepid colonizers will soon settle the issue, creating new wetland habitats or disappearing again into the highland mists.

19.7 The Niche and Competition

CONCEPT PREVIEW: A niche may be defined as the way in which an organism uses its environment. No two species can occupy the same niche indefinitely without competition driving one to extinction if resources are limiting. Sympatric species partition available resources, reducing competition between them.

Within a community, each organism occupies a particular biological role, or **niche.** The niche an organism occupies is the sum total of all the ways it uses the resources of its environment, including space, food, and many other factors of the environment. A niche may be described in terms of space utilization, food consumption, temperature range, appropriate conditions for mating, requirements for moisture, and other factors. *Niche* is not synonymous with **habitat,** the place where an organism lives. *Habitat* is a place, and *niche* is a pattern of living. Many species can share a habitat, but as we shall see, no two species can long occupy exactly the same niche.

Sometimes species are not able to occupy their entire niche because of the presence or absence of other species. Species can interact with each other in a number of ways, and these interactions can either have positive or negative effects. **Competition** describes the interaction when two organisms attempt to use the same resource when there is not enough of the resource to satisfy both.

Competition between individuals of different species is called **interspecific competition.** Interspecific competition is often greatest between organisms that obtain their food in similar ways and between organisms that are more similar. Another type of competition, called **intraspecific competition,** is competition between individuals of the same species.

The Realized Niche

Because of competition, organisms may not be able to occupy the entire niche they are theoretically capable of using, called the **fundamental niche** (or theoretical niche). The actual niche the organism is able to occupy in the presence of competitors is called its **realized niche.**

In a classic study, J. H. Connell of the University of California, Santa Barbara, investigated competitive interactions between two species of barnacles that grow together on rocks along the coast of Scotland. Barnacles are marine animals (crustaceans) that have free-swimming larvae. The larvae eventually settle down, cementing themselves to rocks and remaining attached for the rest of their lives. Of the two species Connell studied, *Chthamalus stellatus* (the smaller barnacle in figure 19.13) lives in shallower water, where tidal action often exposes it to air, and *Semibalanus balanoides* (the larger barnacle) lives at lower depths, where it is rarely exposed to the atmosphere. In the deeper zone, *Semibalanus* could always outcompete *Chthamalus* by crowding it off the rocks, undercutting it, and replacing it even where it had begun to grow. When Connell removed *Semibalanus* from the area, however, *Chthamalus* was easily able to occupy the deeper zone, indicating that no physiological or other general obstacles prevented it from becoming established there. In contrast, *Semibalanus* could not survive in the shallow-water habitats where *Chthamalus* normally occurs; it evidently does not have the special physiological and morphological adaptations that allow *Chthamalus* to occupy this zone. Thus, the fundamental niche of the barnacle *Chthamalus* in Connell's experiments in Scotland included that of *Semibalanus* (the red dashed arrow), but its realized niche was much narrower (the red solid arrow) because *Chthamalus* was outcompeted by *Semibalanus* in its fundamental niche.

Predators, as well as competitors, can limit the realized niche of a species. For example, a plant called the St. John's-wort was introduced and became widespread in open rangeland habitats in California. It occupied all of its fundamental niche until a species of beetle that feeds on the plant was introduced into the habitat. Populations of the plant then quickly decreased, and it is now only found in shady sites where the beetle cannot thrive.

Competitive Exclusion

In classic experiments carried out between 1934 and 1935, Russian ecologist G. F. Gause studied competition among three species of *Paramecium*, a tiny protist. All three species grew well alone in culture tubes (figure 19.14*a*), preying on bacteria and yeasts that fed on oatmeal suspended in the culture fluid. However, when Gause grew *P. aurelia* together with *P. caudatum* in the same culture tube (figure 19.14*b*), the numbers of *P. caudatum* (the green line) always declined to extinction, leaving *P. aurelia* the only survivor. Why? Gause found *P. aurelia* was able to grow six times faster than its competitor, *P. caudatum*, because it was able to better use the limited available resources.

From experiments such as this, Gause formulated what is now called the *principle of competitive exclusion*. This principle states that if two species are competing for a resource, the species that uses the resource more efficiently will eventually eliminate the other locally—no two species with the same niche can coexist.

Niche Overlap

In a revealing experiment, Gause challenged *P. caudatum*—the defeated species in his earlier experiments—with a third species, *P. bursaria*. Because he expected these two species to also compete for the limited bacterial food supply, Gause thought one would win out, as had happened in his previous experiments. But that's not what happened. Instead, both species survived in the culture tubes (figure 19.14*c*); the paramecia found a way to divide the food resources. How did they do it? In the upper part of the culture tubes, where the oxygen concentration and bacterial density were high, *P. caudatum* dominated because it was better able to feed on bacteria. However, in the lower part of the tubes, the lower oxygen concentration favored the growth of a different potential food, yeast, and *P. bursaria* was better able to eat this food. The fundamental niche of each species was the whole culture tube, but the realized niche of each species was only a portion of the tube. This graph also demonstrates the negative effect competition had on the participants: Competition was always detrimental to both species involved. Both species more than doubled their densities when grown without a competitor as when grown together.

Gause's principle of competitive exclusion can be restated to say that no two species can occupy the same niche indefinitely when resources are limiting. When two species are able to coexist on a long-term basis, either resources are not limited or their niches differ in one or more features; otherwise, one species outcompetes the other and the extinction of the second species inevitably results through competitive exclusion.

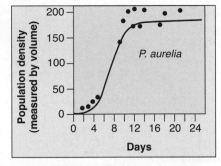
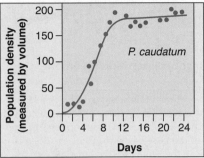
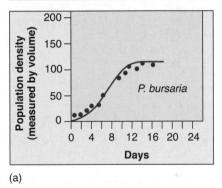

(a)

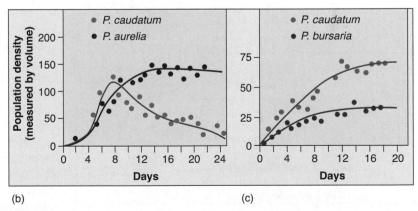

(b)   (c)

Figure 19.14 **Competitive exclusion among three species of *Paramecium*.**

In the microscopic world, *Paramecium* is a ferocious predator. Paramecia eat by ingesting their prey; their plasma membranes surround bacterial or yeast cells, forming a food vacuole containing the prey cell. In his experiments, (a) Gause found that three species of *Paramecium* grew well alone in culture tubes. (b) However, *P. caudatum* declined to extinction when grown with *P. aurelia* because they shared the same realized niche, and *P. aurelia* outcompeted *P. caudatum* for food resources. (c) *P. caudatum* and *P. bursaria* were able to coexist, although in smaller populations, because the two have different realized niches and thus avoid competition.

Figure 19.15 Resource partitioning among lizard species.

Species of *Anolis* lizards in the Caribbean partition their tree habitats in a variety of ways. Some species of anoles occupy the canopy of trees (a), others use twigs on the periphery (b), and still others are found at the base of the trunk (c). In addition, some use grassy areas in the open (d). This same pattern of resource partitioning has evolved independently on different Caribbean islands.

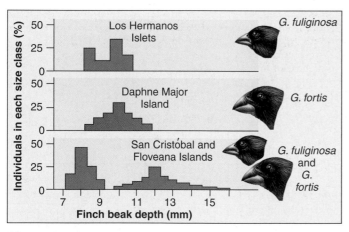

Figure 19.16 Character displacement.

These two species of Galápagos finches (genus *Geospiza*) have beaks of similar sizes when living apart, but different sizes when living together.

Resource Partitioning

Gause's exclusion principle has a very important consequence: Persistent and intense competition between two species is rare in natural communities. Either one species drives the other to extinction, or natural selection reduces the competition between them, such as through **resource partitioning** (dividing up resources to create two realized niches). In resource partitioning, species that live in the same geographical area avoid competition by living in different portions of the habitat or by using different food or other resources. A clear example of this is seen in *Anolis* lizards (figure 19.15), where species may live in different parts of a tree habitat to avoid competition for food and space with other species that may live on the twigs, trunks, or grass.

Resource partitioning can often be seen in closely related species that occupy the same geographical area. Called **sympatric species** (Greek, *syn,* same, and *patria,* country), these species avoid competition by evolving different adaptations to use different portions of the habitat, food or other resources. Closely related species that do not live in the same geographical area, called **allopatric species** (Greek, *allos,* other, and *patria,* country), often use the same habitat locations and food resources—because they are not in competition, natural selection does not favor evolutionary changes that subdivide their niche.

When a pair of closely related species occur in the same place, they tend to exhibit greater differences in morphology and behavior than the same two species do when living in different areas. Called **character displacement,** the differences between sympatric species are thought to have been favored by natural selection as a mechanism to facilitate resource partitioning and thus reduce competition. Character displacement can be seen clearly among Darwin's finches. The two Galápagos finches in figure 19.16 have beaks of similar size when each is living on an island where the other does not occur. On islands where they are found living together, the two species have evolved beaks of different sizes, one adapted to larger seeds, the other to smaller ones. In essence, the two finches have subdivided the food niche, creating two new smaller niches. By partitioning the available food resources, the two species have avoided direct competition with each other, and so are able to live together in the same habitat.

Concept Check

1. How does a holistic community differ from an individualistic one?
2. What is the difference between fundamental and realized niches?
3. Does character displacement occur between allopatric species?

Species Interact in Many Ways

19.8 Coevolution and Symbiosis

CONCEPT PREVIEW: Coevolution is a term that describes the long-term evolutionary adjustments of species to one another. In symbiosis, two or more species live together. Mutualism involves cooperation between species, to the mutual benefit of both. Commensalism is the benign use of one organism by another. In parasitism, one organism serves as a host to another organism, usually to the host's disadvantage.

The previous section described the "winner take all" results of competition between two species whose niches overlap. Other relationships in nature are less competitive and more cooperative.

Coevolution

The plants, animals, protists, fungi, and prokaryotes that live together in communities have changed and adjusted to one another continually over millions of years. For example, many features of flowering plants have evolved in relation to the dispersal of the plant's gametes by animals (figure 19.17). These animals, in turn, have evolved a number of special traits that enable them to obtain food or other resources efficiently from the plants they visit, often from their flowers. In addition, the seeds of many flowering plants have features that make them more likely to be dispersed to new areas of favorable habitat.

Such interactions, which involve the long-term, mutual evolutionary adjustment of the characteristics of the members of biological communities, are examples of **coevolution.** Coevolution is the adaptation of two or more species to each other. In this section, we consider the many ways species interact, some of which involve coevolution.

Symbiosis Is Widespread

In symbiotic relationships, two or more kinds of organisms live together in often elaborate and more or less permanent relationships. All symbiotic relationships carry the potential for coevolution between the organisms involved, and in

> As you will recall from the discussion on page 309, mycorrhizae are symbiotic relationships between fungi and the roots of plants; lichens are symbiotic relationships between fungi and a photosynthetic partner such as cyanobacteria or green algae.

many instances the results of this coevolution are fascinating. Examples of symbiosis include lichens, which are associations of certain fungi with green algae or cyanobacteria. Another important example are mycorrhizae, the association between fungi and the roots of most kinds of plants. The fungi expedite the plant's absorption of certain nutrients, and the plants in turn provide the fungi with carbohydrates. Similarly, root nodules that occur in legumes and certain other kinds of plants contain bacteria that fix atmospheric nitrogen and make it available to their host plants.

In the tropics, leaf-cutter ants are often so abundant that they can remove a quarter or more of the total leaf surface of the plants in a given area. They do not eat these leaves directly; rather, they take them to underground nests, where they chew them up and inoculate them with the spores of particular fungi. These fungi are cultivated by the ants and brought from one specially prepared bed to another, where they grow and reproduce. In turn, the fungi constitute the primary food of the ants and their larvae. The relationship between leaf-cutter ants and these fungi is an excellent example of symbiosis.

Figure 19.17   Pollination by bat.

Many flowers have coevolved with other species to facilitate pollen transfer. Insects are widely known as pollinators, but they're not the only ones. Notice the cargo of pollen on the bat's snout.

BIOLOGY & YOU

Your Little Friends. The 10 trillion bacteria that live in the human gut have evolved over millions of years to live in symbiosis with the human digestive system. One such bacterium, *Bacteroides thetaiotaomicron*, salvages energy from complex polysaccharide sugars that are otherwise nondigestible by humans, benefiting both itself and its host. Another, *Bacteroides fragilis*, offers protection from inflammatory bowel diseases, estimated to affect one million Americans. When *B. fragilis* detects the presence of the pathogenic bacterium *Helicobacter* that causes the disease, it releases a polysaccharide "symbiosis factor" which signals the human immune system to begin producing a chemical defense in the gut against *Helicobacter*. So while it is useful to eliminate disease-causing bacteria from our bodies with antibiotics, in some instances disease may result from the absence of beneficial bacteria and their symbiotic good effects. The frequency of inflammatory bowl diseases has skyrocketed in the last 50 years, as antibiotic use became prevalent. Perhaps there is a downside to living without our little friends.

Figure 19.18   The pistol shrimp defends the coral, which he calls home—an example of mutualism.

Figure 19.19   Oxpeckers eat insects off an impala—an example of commensalism.

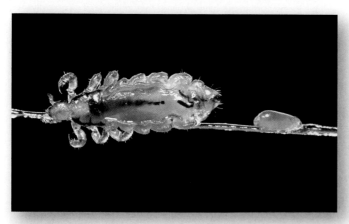

Figure 19.20   The head louse, shown here with its egg, feeds on its host and is an example of parasitism.

The major kinds of symbiotic relationships include (1) **mutualism,** in which both participating species benefit; (2) **commensalism,** in which one species benefits while the other neither benefits nor is harmed; and (3) **parasitism,** in which one species benefits but the other is harmed. Parasitism can also be viewed as a form of predation, although the organism that is preyed upon does not necessarily die.

Mutualism

Mutualism is a symbiotic relationship in which both species benefit. The pistol shrimp patrolling the surface of the coral in figure 19.18 is defending its homestead from sea stars, which prey on coral. When it encounters a sea star on the coral, the shrimp attacks, pinching the sea star's spines and tube feet and making loud snapping sounds with its enlarged pincers. The loud popping sounds, which have given the shrimp its name, are so intense they stun small fish. Protection from being eaten by sea stars certainly provides a benefit to the coral. The shrimp also clearly benefit, obtaining food and shelter from the coral. Because both parties benefit, their relationship is an example of mutualism.

Another example of mutualism involves ants and aphids. Aphids, also called greenflies, are small insects that suck fluids with their piercing mouthparts from the phloem of living plants. They extract a certain amount of the sucrose and other nutrients from this fluid, but they excrete much of it in an altered form through their anus. Certain ants have taken advantage of this—in effect, domesticating the aphids. The ants carry the aphids to new plants, where they come into contact with new sources of food, and then consume as food the "honeydew" that the aphids excrete.

Commensalism

Commensalism is a symbiotic relationship that benefits one species and neither hurts nor helps the other. The best-known examples of commensalism involve the relationships between certain small tropical fishes and sea anemones, marine animals that have stinging tentacles (see chapter 18). These fish have evolved the ability to live among the tentacles of sea anemones, even though these tentacles would quickly paralyze other fishes that touched them. The fishes feed on the detritus left from the meals of the host anemone, remaining uninjured under remarkable circumstances. In another example, birds called oxpeckers eat ticks and other insects off of grazing animals (figure 19.19). In this symbiotic relationship, the oxpeckers receive a clear benefit in the form of nutrition. If the removal of the ticks benefits the impala, then the relationship is mutually beneficial and the relationship would be considered a form of mutualism. However, there is no evidence that this is so. In this instance, as in most examples of commensalism, it is difficult to be certain whether the partner receives a benefit or not.

Parasitism

Parasitism is a symbiotic relationship which benefits one species at the expense of the other. Typically the parasite is much smaller than its host, and remains closely associated with it. Parasitism is sometimes considered a special kind of predator-prey relationship in which the predator is much smaller than the prey, but unlike a predator, the parasite typically does not kill its host. Parasites are very common among animals. The head louse you see in figure 19.20, for example, is one of two types of sucking lice that parasitize humans.

19.9 Predation

CONCEPT PREVIEW: Predators and their prey often show similar cyclic oscillations. This pattern of oscillating population sizes is promoted by refuges that prevent prey populations from being driven to extinction.

Species that live together in a community interact in many ways, one of which is to eat one another. **Predation** is the consuming of one organism by another. The organism doing the eating is called the *predator*, while the organism being eaten is the *prey*. Examples of predation include a leopard capturing and eating an antelope, a whale grazing on millions of microscopic ocean plankton, and locusts eating the leaves of plants.

In nature, predators often have large effects on prey populations. Some of the most dramatic examples involve situations in which humans have either added or eliminated predators from an area. For example, the elimination of large carnivores from much of the eastern United States has led to population explosions of white-tailed deer, which strip the habitat of all edible plant life within their reach. Similarly, when sea otters were hunted to near extinction on the western coast of the United States, populations of sea urchins, a principal prey item of the otters, exploded. Appearances, however, sometimes can be deceiving. On Isle Royale in Lake Superior, moose reached the island by crossing over ice in an unusually cold winter and multiplied freely there in isolation. When wolves later reached the island by crossing over the ice, naturalists widely assumed that the wolves were playing a key role in controlling the moose population. More careful studies have demonstrated that this is not in fact the case. The moose that the wolves eat are, for the most part, old or diseased animals that would not survive long anyway. In general, the moose are controlled by food availability, disease, and other factors rather than by the wolves (figure 19.21).

Figure 19.21   Wolves chasing a moose—what will the outcome be?

On Isle Royale, Michigan, a large pack of wolves pursue a moose. They chased this moose for almost 2 kilometers; it then turned and faced the wolves, who by that time were exhausted from running through chest-deep snow. The wolves lay down, and the moose walked away.

Predator/Prey Cycles

Why doesn't a predator exterminate its prey, and then become extinct itself, having nothing left to eat. This is just what happens in laboratory experiments, such as the experiment shown in figure 19.22, where the predator, *Didinium* (red line) often exterminates its prey, *Paramecium* (blue line). However, if refuges are provided for the prey, its population drops to low levels but not to extinction. Low prey population levels then provide inadequate food for the predators, causing the predator population to decrease. When this occurs, the prey population can recover. In this way, predator and prey populations cycle in their abundance.

Cycles in Hare Populations: A Case Study

Population cycles are characteristic of some species of small mammals, such as lemmings, and they appear to be stimulated, at least in some situations, by their predators. Ecologists have studied cycles in hare populations since the 1920s. They have found that the North American snowshoe hare, *Lepus americanus,* follows a "10-year cycle" (in reality, it varies from 8 to 11 years). Its numbers fall 10-fold to 30-fold in a typical cycle, and 100-fold changes can occur. Two factors appear to be generating the cycle: food plants and predators.

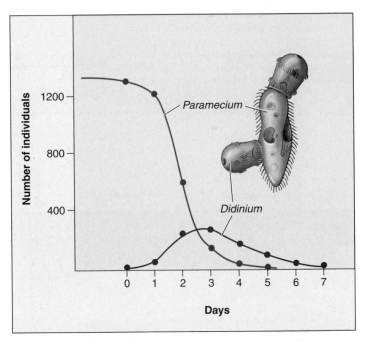

Figure 19.22   Predator-prey in the microscopic world.

When the predatory *Didinium* is added to a *Paramecium* population, the numbers of *Didinium* initially rise, while the numbers of *Paramecium* steadily fall. As the *Paramecium* population is depleted, however, the *Didinium* individuals also die.

(a)

Figure 19.23 A predator-prey cycle.

(a) A snowshoe hare being chased by a lynx. (b) The numbers of lynxes and snowshoe hares oscillate in tune with each other in northern Canada. The data are based on numbers of animal pelts from 1845 to 1935. As the number of hares grows, so does the number of lynxes, with the cycle repeating about every 10 years. Both predators (lynxes) and available food resources control the number of hares. The number of lynxes is controlled by the availability of prey (snowshoe hares).

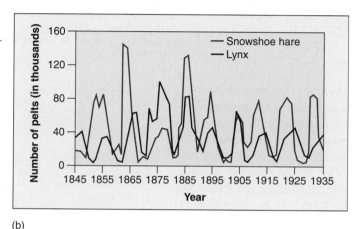

(b)

1. **Food plants.** The preferred foods of snowshoe hares are willow and birch twigs. As hare density increases, the quantity of these twigs decreases, leading to a precipitous decline in willow and birch twig abundance. The result is that hares are forced to feed on high-fiber (low-quality) food, causing lower birthrates, low juvenile survivorship, low growth rates, and a corresponding fall in hare abundance. It takes two to three years for the quantity of mature twigs to recover.

2. **Predators.** A key predator of the snowshoe hare is the Canada lynx, *Lynx canadensis*. The Canada lynx shows a "10-year cycle" of abundance that seems remarkably entrained to the hare abundance cycle (compare the hare population cycle, the blue line in figure 19.23 with the lynx population, the red line). As hare numbers increase, lynx numbers do, too, rising in response to the increased availability of lynx food. When hare numbers fall, so do lynx numbers, their food supply depleted.

Which factor is responsible for the predator-prey oscillations? Do increasing numbers of hares lead to overharvesting of plants (a hare-plant cycle), or do increasing numbers of lynx lead to overharvesting of hares (a hare-lynx cycle)? Field experiments carried out by C. Krebs and coworkers in 1992 provide an answer. Krebs set up experimental plots containing hare populations in Canada's Yukon. If food is added and predators excluded from an experimental area, hare numbers increase 10-fold and stay there—the cycle is lost. However, the cycle is retained if either of the factors is allowed to operate alone: exclude predators but don't add food, or add food in presence of predators. Thus, both factors can affect the cycle, which, in practice, seems to be generated by the interaction between the two factors.

Predation Reduces Competition

Predator-prey interactions are an essential factor in the maintenance of communities that are rich and diverse in species. The predators prevent or greatly reduce competitive exclusion by reducing the numbers of individuals of competing species. For example, in preying selectively on bivalves in marine intertidal habitats, sea stars prevent bivalves from monopolizing such habitats, opening up space for many other organisms. When sea stars are removed, species diversity falls precipitously, the sea floor community coming to be dominated by a few species of bivalves (figure 19.24). Because predation tends to reduce competition in natural communities, it is usually a mistake to attempt to eliminate a major predator such as sea stars, wolves, or mountain lions from a community. The result is to decrease rather than increase the biological diversity of the community, the opposite of what is intended.

> The predatory-prey relationship is not the only factor that decreases biodiversity. As you will see in chapter 22, page 437, several factors are responsible for the loss of biodiversity, including habitat loss, overexploitation, and introduced species.

Figure 19.24 Predation reduces competition.

When a key predator, starfish (*Pisaster*), is removed from a coastal ecosystem, fiercely competitive mussels explode in growth, effectively crowding out seven other indigenous species.

19.10 Plant and Animal Defenses

CONCEPT PREVIEW: The members of many groups of plants are protected from most herbivores by synthesizing protective chemicals. Animals defend themselves against predators with chemical defenses such as poisons, warning coloration, and camouflage.

Plant Defenses

Plants have evolved many mechanisms to defend themselves from their predators, called herbivores. The most obvious are thorns, spines, and prickles. Chemical defenses are even more crucial, and are widespread in plants. Mustard oils, which give the sharp pungent taste to mustard, capers, cabbage, and horseradish, are toxic to many groups of insects. Interestingly, certain groups of herbivores have developed the ability to feed on these plants without harm, thereby avoiding competition with other herbivores. For example, cabbage butterfly caterpillars (subfamily Pierinae) feed almost exclusively on plants of the mustard and caper families (figure 19.25). Similarly, caterpillars of monarch butterflies and their relatives (subfamily Danainae) feed on plants of the milkweed and dogbane families that produce a milky sap that deters predators.

Animal Defenses

Some animals that feed on chemically protected plants can concentrate and store the protective chemicals. Monarch butterflies contain protective chemicals because the caterpillars feed on plants of the milkweed family. In addition, animals also manufacture and use a startling array of defensive substances. Bees, wasps, predatory bugs, scorpions, spiders, and many other arthropods use chemicals to defend themselves and to kill their prey. Various chemical defenses have also evolved among both marine animals, such as jellyfish, and vertebrates, including venomous snakes, lizards, fishes, and some birds. The poison-dart frogs of the family Dendrobatidae produce toxic alkaloids in the mucus that covers their brightly colored skin (figure 19.26). Some of these toxins are so powerful that a few micrograms will kill a person if injected into the bloodstream.

Defensive Coloration. Many animals that use chemical defenses are brightly colored (see figure 19.26), advertising their poisonous nature with an ecological strategy known as **warning coloration** or **aposematic coloration.** How does a predator know not to eat brightly colored organisms? By learning. One attempt is rarely repeated. Organisms that lack specific chemical defenses are seldom brightly colored. In fact, many have **cryptic coloration**—color that blends with the surroundings and thus hides the individual from predators (figure 19.27). Camouflaged animals usually do not live together in groups because a predator that discovers one individual gains a valuable clue to the presence of others.

Coevolution of Predator and Prey

Predation can exert strong selective pressures on prey populations. Any feature that acts to decrease the probability of capture should thus be strongly favored by natural selection. In turn, the evolution of such features by prey would be expected to encourage natural selection to then favor counteradaptations in their predator populations. In this way, a coevolutionary arms race may ensue, in which predators and prey are continually evolving better defenses and better means of circumventing these defenses. Herbivores that are able to feed on mustard, milkweed, and dogbane are predators that have acquired counteradaptations to the chemicals produced by the plants.

Figure 19.25 **Insect herbivores are well suited to their hosts.**

Although mustard oils protect plants in the mustard family against most herbivores, the caterpillars of the cabbage butterfly (*Pieris rapae*) are able to break down the mustard oil compounds.

Figure 19.26 **Warning coloration serves as a defense mechanism in Dendrobatidae.**

The skin coloration of this poison dart frog warns potential predators of the toxic mucus covering the skin, and predators keep their distance.

Figure 19.27 **Cryptic coloration.**

An inchworm caterpillar (family Geometridae) closely resembles a twig.

(a) Model

(b) Batesian mimic

Figure 19.28   A Batesian mimic.

(a) The model. Monarch butterflies (*Danaus plexippus*) are protected from birds and other predators by the cardiac glycosides they incorporate from the milkweeds and dogbanes they feed on as larvae. Adult monarch butterflies advertise their poisonous nature with warning coloration. (b) The mimic. Viceroy butterflies, *Limenitis archippus*, are Batesian mimics of the poisonous monarch. Although the viceroy is not related to the monarch, it looks a lot like it, so predators that have learned not to eat distasteful monarchs avoid viceroys, too.

EVOLUTION

Wandering Mimics. Batesian mimicry also occurs in vertebrates. Probably the most famous case is the nonvenomous scarlet kingsnake, whose red, black, and yellow bands mimic those of the venomous eastern coral snake. Paradoxically, the scarlet kingsnake is found hundreds of miles away from the range of its coral snake model. Why is this a paradox? Theory tells us that mimics should not occur in areas where their model is absent, because predators there would not be under selection to avoid the dangerous banding pattern. So how can scarlet kingsnakes get away with disobeying the rules? It turns out that they don't—male kingsnakes have simply dispersed far from regions where both snakes are present. Researchers have found that, once separated, natural selection promotes the evolution of scarlet kingsnakes that look less like their model in the model-free area.

19.11 Mimicry

CONCEPT PREVIEW: In Batesian mimicry, unprotected species resemble others that are distasteful. Both species exhibit aposematic coloration. In Müllerian mimicry, two or more unrelated but protected species resemble one another, thus achieving a kind of group defense. In self-mimicry, one body part resembles another, in some cases helping the prey and in other cases helping the predators.

During the course of their evolution, many unprotected (nonpoisonous) species have come to resemble distasteful ones that exhibit aposematic coloration. Also, protected species can mimic each other, or organisms can have adaptations that mimic body parts. These types of mimicry are discussed below.

Batesian Mimicry

Batesian mimicry is named for Henry Bates, the nineteenth-century British naturalist who first brought this type of mimicry to general attention in 1857. In his journeys to the Amazon region of South America, Bates discovered many palatable insects that resembled brightly colored, distasteful species. He reasoned that the mimics are avoided by predators, who are fooled by the disguise into thinking the mimic actually is the distasteful model.

Many of the best-known examples of Batesian mimicry occur among butterflies and moths. Obviously, predators in systems of this kind must use visual cues to hunt for their prey; otherwise, similar color patterns would not matter to potential predators. There is also increasing evidence indicating that Batesian mimicry can also involve nonvisual cues, such as olfaction, although such examples are less obvious to humans.

The kinds of butterflies that provide the models in Batesian mimicry are, not surprisingly, members of groups whose caterpillars feed only on one or a few closely related plant families. The plant families on which they feed are strongly protected by toxic chemicals. The model butterflies incorporate the poisonous molecules from these plants into their bodies. The mimic butterflies, in contrast, belong to groups in which the feeding habits of the caterpillars are not so restricted. As caterpillars, these butterflies feed on a number of different plant families unprotected by toxic chemicals.

One often-studied mimic among North American butterflies is the viceroy, *Limenitis archippus* (figure 19.28*b*). This butterfly, which resembles the poisonous monarch (in figure 19.28*a*), ranges from central Canada through much of the United States and into Mexico. The caterpillars feed on willows and cottonwoods, and neither caterpillars nor adults were thought to be distasteful to birds, although recent findings may dispute this. Interestingly, the Batesian mimicry seen in the adult viceroy butterfly does not extend to the caterpillars. Viceroy caterpillars are camouflaged on leaves, resembling bird droppings, whereas the monarch's distasteful caterpillars are very conspicuous.

Müllerian Mimicry

Another kind of mimicry, **Müllerian mimicry,** was named for German biologist Fritz Müller, who first described it in 1878. In Müllerian mimicry, several unrelated but protected animal species come to resemble one another (figure 19.29). Thus, different kinds of stinging wasps have yellow-and-black-striped abdomens, but they may not all be descended from a common yellow-and-black-striped ancestor. In general, yellow and black and bright red tend to be

(a)

(b)

(c)

(d)

Figure 19.29 Müllerian mimics.

Because the color patterns of these insects are very similar, and because they all sting, they are Müllerian mimics. The yellow jacket (a), the masarid wasp (b), the sand wasp (c), and the anthidiine bee (d) all act as models for each other, strongly reinforcing the color pattern that they all share.

IMPLICATION A close relative of the yellow jacket is the European hornet, *Vespa crabro*, introduced to the eastern United States in 1840. It is the only true hornet in North America. It also has a yellow and black banded body. Do you think it is a Müllerian mimic? How would you decide?

common color patterns that warn predators relying on vision. If animals that resemble one another are all poisonous or dangerous, they gain an advantage because a predator learns more quickly to avoid them. In both Batesian and Müllerian mimicry, mimic and model must not only look alike but also act alike if predators are to be deceived. For example, the members of several families of insects that resemble wasps behave surprisingly like the wasps they mimic, flying often and actively from place to place.

Self-Mimicry

Another type of mimicry, called **self-mimicry,** involves adaptations in which one animal body part comes to resemble another body part. This type of mimicry is used by both prey and predators. In prey, it is used to increase survival during an attack. For example, many moths, butterflies, and fish have coloration patterns with "eyespots" (figure 19.30). In some cases, these eyespots startle a predator, allowing the prey time to escape, or they present a false target for the predator's attack, for example attacking the tail where the eyespots appear instead of the head. Predators may also use mimicry to simulate bait to lure prey in. For example, some predators have accessory body parts that look like food, such as the tongue of the alligator snapping turtle that looks like a wriggly worm. The wormlike tongue attracts prey and brings them in close enough for a successful attack.

Figure 19.30 Self-mimicry.

The eyespots on the wings of this moth may startle predators or may be a pattern that predators are instinctively afraid of, because the eyespots resemble the eyes of predators' predators.

Concept Check

1. Compare coevolution to evolution.
2. Name and describe three kinds of symbiotic relationships.
3. Removing a predator often causes loss of biodiversity. Why?

Invasion of the Killer Bees

One of the harshest lessons of environmental biology is that the unexpected does happen. Precisely because science operates at the edge of what we know, scientists sometimes stumble over the unexpected. This lesson has been brought clearly to mind in recent years, with reports of killer bees being discovered east of the Mississippi River.

In the continental United States, we are used to the mild-mannered European honeybee, a subspecies called *Apis mellifera mellifera*. The African variety, subspecies *A. m. scutelar*, looks very much like them, but they are hardly mild mannered. They are in fact very aggressive critters with a chip on their shoulder, and when swarming do not have to be provoked to start trouble.

A few individuals may see you at a distance, "lose it," and lead thousands of bees in a concerted attempt to do you in. The only thing you can do is run—fast. They will keep after you for up to a mile. It doesn't do any good to duck under water, as they just wait for you to surface.

They are nicknamed "killer" bees not because any one sting is worse than the European kind, but rather because so many of the bees try to sting you. An average human can survive no more than 300 bee stings. A horrible total of more than 8,000 bee stings is not unusual for a killer bee attack. An American graduate student attacked and killed in a Costa Rican jungle had 10,000 stings.

Killer bees remind us of the "law of unintended consequences" because their invasion of this continent is the direct result of scientists stumbling over the unexpected.

Killer bees were brought from Africa to Brazil 47 years ago by a prominent Brazilian scientist, Warwick Estevan Kerr. A famous geneticist, Kerr is the only Brazilian to be a member of the U.S. National Academy of Sciences. What he was doing in 1956, at the request of the Brazilian government, was attempting to establish tropical bees in Brazil to expand the commercial bee industry (bees pollinate crops and make commercial honey in the process). African bees seemed ideal candidates, better adapted to the tropics than European bees and more prolific honey producers.

Kerr established a quarantine colony of African bees at a remote field station outside the city of Rio Claro, several hundred miles from his university at São Paulo. Although he had brought back many

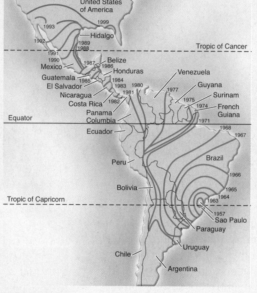

queens from South Africa, the colony came to be dominated by the offspring of a single very productive queen from Tanzania. Kerr noted at the time that she appeared unusually aggressive.

In the fall of 1957, the field station was visited by a beekeeper. As no one else was around that day, the visitor performed the routine courtesy of tending the hives. A hive with a queen in it has a set of bars across the door so the queen can't get out. Called a "queen excluder," the bars are far enough apart that the smaller worker bees can squeeze through. Once a queen starts to lay eggs, she never leaves the hive, so there is little point in slowing down entry of workers to the hive, and the queen excluder is routinely removed. On that day, the visitor saw the Tanzanian queen laying eggs in the African colony hive, and so removed the queen excluder.

One and a half days later, when a staff member inspected the African colony, the Tanzanian queen and 26 of her daughter queens had decamped. Out into the neighboring forest they went. And that's what was unexpected. European queen bees never leave the hive after they have started to lay eggs. No one could have guessed that the Tanzanian queen would behave differently. But she did.

By 1970 the superaggressive African bees had blanketed Brazil, totally replacing local colonies. They reached Central America by 1980, Mexico by 1986, and Texas by 1990, having conquered 5 million square miles in 33 years. In the process they killed an estimated 1,000 people and over 100,000 cows. The first American to be killed, a rancher named Lino Lopez, died of multiple stings in Texas in 1993.

All during the 1990s, the bees continued their invasion of the United States. All of Arizona and much of Texas has been occupied. A few years ago, Los Angeles county was officially declared colonized, and it looks like African bees will eventually move at least halfway up the state of California.

Soon, however, the invasion is predicted to cease, on a line roughly from San Francisco, California, to Richmond, Virginia. Winter cold is expected to limit any further northward advance of the invading hoards.

For the states below this line, sure as the sun rises, the bees are coming, not caring one bit that they were unanticipated. The deep lesson—that unexpected things do happen—is being driven home by millions of tiny aggressive teachers.

Community Stability

19.12 Ecological Succession

CONCEPT PREVIEW: In succession, communities change through time, often in a predictable sequence.

Marked changes in habitat can result in the orderly replacement of one community with another, from simple to complex, in a process known as **succession.** This process is familiar to anyone who has seen a vacant lot or cleared woods slowly become occupied by an increasing number of plants, or a pond become dry land as it is filled with vegetation encroaching from the sides.

Secondary Succession

If a wooded area is cleared and left alone, plants slowly reclaim the area. Eventually, traces of the clearing disappear and the area is again woods. Similarly, intense flooding may clear a stream bed of many organisms, leaving mostly sand and rock; afterward, the bed is progressively reinhabited by protists, invertebrates, and other aquatic organisms. This kind of succession, which occurs in areas where an existing community has been disturbed, is called **secondary succession.** Humans are often responsible for initiating secondary succession, such as when a fire has burned off an area or in abandoned agricultural fields.

Primary Succession

In contrast, **primary succession** occurs on bare, lifeless substrate, such as rocks. Primary succession occurs in lakes left behind after the retreat of glaciers, on volcanic islands that rise above the sea, and on land exposed by retreating glaciers. Primary succession on glacial moraines provides an example. The graph in figure 19.31 shows how nitrogen concentrations change in the soil as primary succession occurs. On bare, mineral-poor soil, lichens grow first, forming small pockets of soil. Acidic secretions from the lichens help to break down the substrate and add to the accumulation of soil. Mosses then colonize these pockets of soil (indicated by *a* in the graph), eventually building up enough nutrients in the soil for alder shrubs to take hold (*b*). These first plants to appear form a **pioneering community.** Over 100 years, the alders (*c* and in the *photo*) build up the soil nitrogen levels until spruce are able to thrive, eventually crowding out the alder and forming a dense spruce forest (*d*).

Primary successions end with a community called a **climax community,** whose populations remain relatively stable and are characteristic of the region as a whole.

Why Succession Happens

Succession happens because species alter the habitat and the resources available in it, often in ways that favor other species. As ecosystems mature, and more *K*-selected species replace *r*-selected ones, species richness and total biomass increase but net productivity decreases.

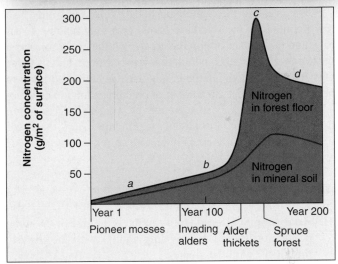

Figure 19.31 Plant succession produces progressive changes in the soil.

Initially the glacial moraine at Glacier Bay, Alaska, had little soil nitrogen, but nitrogen-fixing alders (shown in the photo) led to a buildup of nitrogen in the soil, encouraging the subsequent growth of the conifer forest. The graph breaks down the steps of primary succession. (a) The first invaders after the glaciers retreat are pioneering mosses with nitrogen-fixing mutualistic microbes. (b) Within 20 years, young alder shrubs take hold. (c) Rapidly fixing nitrogen, they soon form dense thickets (in photo). (d) As soil nitrogen levels rise, spruce crowd out the mature alders, forming a forest.

IMPLICATION Plant succession occurs wherever habitats are disturbed, even in your own backyard. If you are wildly interested in this, you can demonstrate it for yourself. In the fall, dig up an area of your yard about the size of a dining room table, turning all the soil over and removing as many plants as possible. The following summer, lay out a 1-meter wide path across the disturbed area and onward an equal distance into the untouched yard; this is called a transect line. Inventory the plants you find within the transect, classifying them as either abundant, occasional, or absent. How do the disturbed and untouched areas compare?

Concept Check

1. Is reforestation of an abandoned field primary succession? Explain.
2. Does (*d*) in the graph indicate a pioneering or climax community?
3. What is the effect of succession on net productivity in the area?

Are Island Populations of Song Sparrows Density Dependent?

When island populations are isolated, receiving no visitors from other populations, they provide an attractive opportunity to test the degree to which a population's growth rate is affected by its size. A population's size can influence the rate at which it grows because increased numbers of individuals within a population tend to deplete available resources, leading to an increased risk of death by deprivation. Also, predators tend to focus their attention on common prey, resulting in increasing rates of mortality as populations grow. However, simply knowing that a population is decreasing in numbers does not tell you that the decrease has been caused by the size of the population. Many factors such as severe weather, volcanic eruption, and human disturbance can influence island population sizes too.

The graph to the right displays data collected from 13 song sparrow populations on Mandarte Island (see map below). In an attempt to gauge the impact of population size on the evolutionary success of these populations, each population was censured, and its juvenile mortality rate estimated. On the graph, these juvenile mortality rates have been plotted against the number of breeding adults in each population. Although the data appear scattered, the "best-fit" regression line is statistically significant (**statistically significant** means that there is a less than 5% chance that there is in fact no correlation between dependent and independent variables).

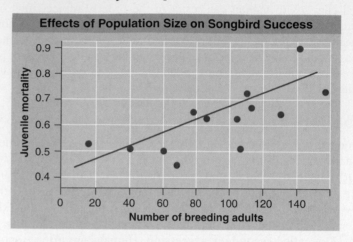

Effects of Population Size on Songbird Success

(y-axis: Juvenile mortality; x-axis: Number of breeding adults)

Analysis

1. **Analyzing Scattered Data**   What is the size of the song sparrow population (based on breeding adults) with the lowest juvenile mortality? with the greatest?

2. **Interpreting Data**
   a. What is the average juvenile mortality of all 13 populations, estimated from the 13 points on the graph?
   b. How many populations were observed to have juvenile mortality rates *below* this average value? What is the average size of these populations?
   c. How many populations were observed to have juvenile mortality rates *above* this average value? What is the average size of these populations?

3. **Making Inferences**   Are the populations with lower juvenile mortality bigger or smaller than the populations with higher juvenile mortality?

4. **Drawing Conclusions**   Do the population sizes of these song sparrows appear to exhibit density dependence?

Concept Summary

Ecology

19.1 What Is Ecology?

- Ecology is the study of how organisms interact with each other and with their physical environment. There are six levels of ecological organization: populations, species, communities, ecosystems, biomes, and the biosphere.

Population Dynamics

19.2 Population Growth

- A population is a group of individuals of the same species that live together and influence each other's survival. Several factors affect population growth, including population size, density, and dispersion, as shown here in **figure 19.4.**

- Exponential growth occurs in a population when no factors are limiting its growth. As resources are used up, a population's growth slows and stabilizes at a size called the carrying capacity. At this point, the population experiences logistic growth and exhibits a sigmoid growth curve (**figures 19.5** and **19.6**).

19.3 The Influence of Population Density

- Factors such as weather and physical disruptions are density-independent effects and act on population growth, regardless of population size. Density-dependent effects are factors, such as resources, that are affected by increases in population size. As resources are used up, individuals die off and reduce the size of the population (**figure 19.7**). A population is less affected by losses during the rising portion of the sigmoid growth curve, a point called the maximal sustainable yield (**figure 19.8**).

19.4 Life History Adaptations

- Populations with abundant resources and little competition reproduce rapidly and exhibit *r*-selected adaptations. Populations that experience competition over limited resources tend to be more reproductively efficient and exhibit *K*-selected adaptations (**table 19.1**).

19.5 Population Demography

- Population demography is the statistical study of populations. Using demography, predictions can be made about population sizes in the future. Survivorship curves illustrate the impact of mortality rates among different age groups in a population (**figure 19.11**).

How Competition Shapes Communities

19.6 Communities

- The array of organisms that live together in an area is called a community. These individuals compete and cooperate with each other to make the community stable (**figure 19.12**).

19.7 The Niche and Competition

- A niche is the way an organism uses all available resources in its environment. Competition limits an organism from using its entire niche, as shown here from **figure 19.13.** Two species cannot use the same niche; one will either outcompete the other, driving it to extinction, called competitive exclu-

sion (**figure 19.14***a,b*), or they will divide the niche into two smaller niches, called resource partitioning (**figures 19.14***c* and **19.15**).

- As each species adapts to its portion of the niche, resource partitioning can affect morphological characteristics, called character displacement (**figure 19.16**).

Species Interact in Many Ways

19.8 Coevolution and Symbiosis

- Coevolution is the adaptation of two or more species to each other. Symbiotic relationships involve two or more organisms of different species that live together and form a somewhat permanent relationship. Symbiotic relationships, such as in lichen and mycorrhizae, can lead to coevolution. The major kinds of symbioses include mutualism (**figure 19.18**), commensalism (**figure 19.19**), and parasitism (**figure 19.20**).

19.9 Predation

- In predatory-prey relationships, the predator kills and consumes the prey. Predator and prey populations often exhibit cyclic oscillations, the prey population being hunted to a low number, which begins to negatively affect predator population size. When the predator population decreases in numbers, the prey population rebounds (**figures 19.22** and **19.23**). Predators also act to maintain diversity in a community. In the absence of a predator, a prey population can grow rapidly, taking over the habitat and reducing diversity.

19.10 Plant and Animal Defenses

- Plants have evolved defense mechanisms to protect themselves from predation by herbivores. Thorns, prickles, and distasteful or toxic chemical discourage animals from eating the plants.

- The evolution of chemical defenses in plants has led to counter-adaptations in some herbivores that allow them to consume the plants (**figures 19.25**).

- Animals have also evolved chemical defense mechanisms and advertise this with warning coloration to let potential predators know that they should "stay away." The bright coloring of this frog from **figure 19.26** lets predators know that a toxic chemical covers its skin. Cryptic coloration is a defense mechanism in which animals use camouflage to avoid predation (**figure 19.27**).

19.11 Mimicry

- Mimicry is where one organism takes advantage of the warning coloration of another organism. Batesian mimicry is where a harmless species has come to resemble a harmful species and so is avoided (**figure 19.28**). Müllerian mimicry is where a group of harmful species has a similar warning coloration pattern (**figure 19.29**). Self mimicry is where one animal's body part resembles another, confusing the predator (**figure 19.30**).

Community Stability

19.12 Ecological Succession

- Succession is the replacement of one community with another. Secondary succession occurs following the disturbance of an existing community, and primary succession is the emergence of a pioneering community where no life existed before (**figure 19.31**).

Self-Test

1. In the levels of ecological organization, the lowest level, composed of individuals of a single species who live near each other, share the same resources, and can potentially mate is called a
   a. population.
   b. community
   c. ecosystem.
   d. biome.

2. When the number of organisms in a population remains more or less the same over time in the specific place where these organisms live, it is said that this population of organisms has reached its _____.
   a. dispersion
   b. biotic potential
   c. carrying capacity
   d. population density

3. Which of the following is a density-dependent effect on a population?
   a. earthquake
   b. increased competition for food
   c. habitat destruction by humans
   d. seasonal flooding

4. Which of the following traits is *not* a characteristic of an organism that has *K*-selected adaptations?
   a. short life span
   b. few offspring per breeding season
   c. extensive parental care of offspring
   d. low mortality rate

5. All the organisms that live in the same location (fungi, protists, bacteria, archaea, animals, and plants) make up a(n)
   a. biome.
   b. population.
   c. ecosystem.
   d. community.

6. For similar species to occupy the same space, their niches must be different in some way. One way for these species to both survive is
   a. competitive exclusion.
   b. interspecific competition.
   c. resource partitioning.
   d. intraspecific competition.

7. A relationship between two species where one species benefits and the other is neither hurt nor helped is known as
   a. parasitism.
   b. commensalism.
   c. mutualism.
   d. competition.

8. Predators can assist in maintaining the species diversity of an area by
   a. increasing competitive exclusion between prey species.
   b. decreasing competitive exclusion between prey species.
   c. not affecting competitive exclusion between prey species.
   d. decreasing resource partitioning between prey species.

9. The bright colors of poison-dart frogs is example of
   a. aposematic coloration.
   b. Müllerian mimicry.
   c. cryptic coloration.
   d. Batesian mimicry.

10. Succession that occurs on abandoned agricultural fields is best described as
    a. coevolution.
    b. primary succession.
    c. secondary succession.
    d. prairie succession.

Visual Understanding

1. **Figure 19.5** What factors in an environment could cause the carrying capacity of a population to decrease? Explain.

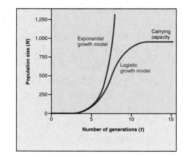

2. **Figure 19.13** How would the niche change for *Chthamalus* if *Semibalanus* was removed from the community? How would the niche change for *Semibalanus* if *Chthamalus* was removed from the community?

Challenge Questions

1. Give at least two examples, from your own area, of limiting factors on a population that are density-dependent factors, and two that are density-independent factors on the same population.

2. Briefly discuss and give examples of the three types of symbiosis.

3. There are many examples of coevolution between flowers and their pollinators. A group of flowers, often referred to as carrion flowers, smell like a dead, rotting animal. What insects do you think pollinate these flowers and why?

4. Many U.S. communities struggle with issues of deer overpopulation. Explain how human activities have created this situation.

Chapter

20

Ecosystems

The Energy in Ecosystems

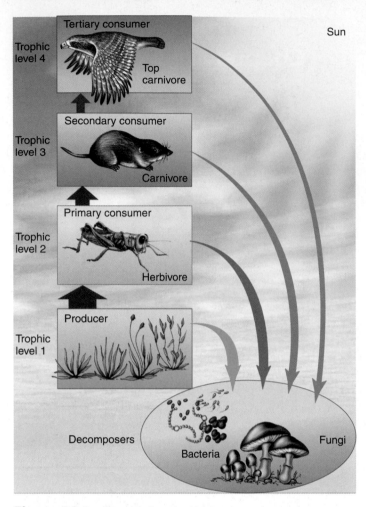

Figure 20.1   **Trophic levels within an ecosystem.**

Ecologists assign all the members of a community to various trophic levels based on feeding relationships.

BIOLOGY & YOU

Your Own Personal Ecosystem. You may not realize it, but you are a walking ecosystem. Tiny, eight-legged mites nestle head down inside the follicles of your eye lashes, feasting unnoticed on skin cells. Microscopic bacteria live on your tongue, teeth, and skin. There are about 3.3 pounds of bacteria living in your gut. Most of the time we share our bodies harmoniously with these other inhabitants, a balanced ecological community. A few examples of the neighbors with whom you share your body: **1.** Over 500 species of bacteria live inside your gut, breaking down carbohydrates for you, making essential vitamins like K and B$_{12}$, and crowding out harmful bacteria. **2.** Bacteria that inhabit the vagina secrete lactic acid, which fends off hostile invaders like pathogenic *Candida* yeast. **3.** A flat, wingless insect called a head louse has been found attached to a strand of human hair 10,000 years old. Less than a tenth of an inch long, head lice suck on your blood, cementing their eggs, or nits, to your hair.

20.1 Energy Flows Through Ecosystems

CONCEPT PREVIEW: Energy moves through ecosystems from producers, to herbivores, to carnivores, and finally to detritivores and decomposers, which consume the dead bodies of all the others. Much energy is lost at each stage of a food chain.

What Is an Ecosystem?

The ecosystem is the most complex level of biological organization. The biosphere includes all the ecosystems on earth. The earth is a closed system with respect to chemicals but an open system in terms of energy. The organisms in ecosystems regulate the capture and expenditure of energy and the cycling of chemicals. All organisms depend on the ability of photosynthetic organisms to recycle the basic components of life.

Ecologists, the scientists who study ecology, view the world as a patchwork quilt of different environments, all bordering on and interacting with one another. Consider for a moment a patch of forest, the sort of place a deer might live. Ecologists call the collection of creatures that live in a particular place a **community**—all the animals, plants, fungi, and microorganisms that live together in a forest, for example, are the forest community. Ecologists call the place where a community lives its **habitat**—the soil, and the water flowing through it, are key components of the forest habitat. The sum of these two, community and habitat, is an ecological system, or **ecosystem.** An ecosystem is a largely self-sustaining collection of organisms and their physical environment. An ecosystem can be as large as a forest or as small as a tidepool.

The Path of Energy: Who Eats Whom in Ecosystems

Energy flows into the biological world from the sun, which shines a constant beam of light on our earth. Life exists on earth because some of that light energy can be captured and transformed into chemical energy through the process of photosynthesis and used to make organic molecules such as carbohydrates, nucleic acids, proteins, and fats. These organic molecules are what we call food. Living organisms use the energy in food to make new materials for growth, to repair damaged tissues, to reproduce, and to do a myriad of other things requiring energy.

You can think of all the organisms in an ecosystem as chemical machines fueled by energy captured in photosynthesis. The organisms that first capture the energy, the **producers,** are plants, algae and some bacteria, which produce their own energy-storing molecules by carrying out photosynthesis. They are also referred to as *autotrophs.* All other organisms in an ecosystem are **consumers,** obtaining energy-storing molecules by consuming plants or other animals, and are referred to as *heterotrophs.* Ecologists assign every organism in an ecosystem to a trophic (or feeding) level, depending on the source of its energy. A **trophic level** is composed of those organisms within an ecosystem whose source of energy is the same number of consumption "steps" away from the sun. Thus, as shown in figure 20.1, a plant's trophic level is 1, while *herbivores* (animals that graze on plants) are in trophic level 2, and *carnivores* (animals that eat these grazers) are in trophic level 3.

> Heterotrophs consume food to acquire energy stored in the chemical bonds of the food's molecules. As described on page 121, this energy is converted to ATP, which the cell uses to fuel metabolism. If ATP isn't needed, the energy is converted into fat for long term storage.

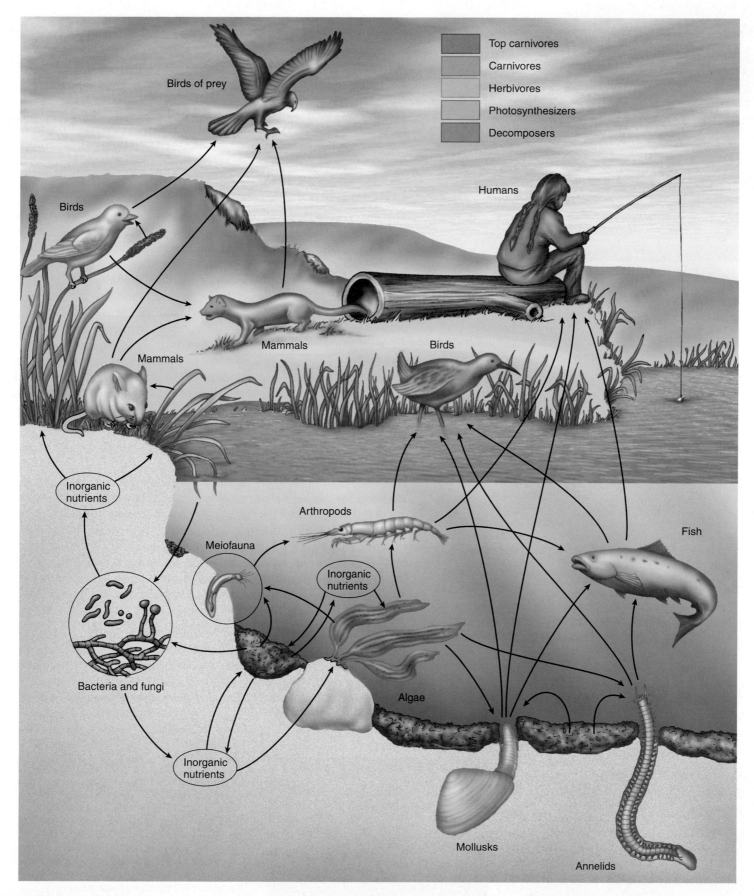

Figure 20.2   A food web.

A food web is much more complicated than a linear food chain. The path of energy passes from one trophic level to another and back again in complex ways.

(a) Producers and herbivores

Higher trophic levels exist for animals that eat other carnivores (like the top carnivore in figure 20.1). Food energy passes through an ecosystem from one trophic level to another. When the path is a simple linear progression, like the links of a chain, it is called a **food chain.** The chain ends with *decomposers,* who break down dead organisms, or their excretions, and return the organic matter to the soil.

In most ecosystems, however, the path of energy is not a simple linear one, because individual animals often feed at several trophic levels. This creates a more complicated path of energy flow called a **food web** (figure 20.2).

Producers

The lowest trophic level of any ecosystem is occupied by the producers (figure 20.3*a*)—green plants in most land ecosystems (and, usually, algae in aquatic ecosystems). Plants use the energy of the sun to build energy-rich sugar molecules. They also absorb carbon dioxide from the air, and nitrogen and other key substances from the soil, and use them to build biological molecules. It is important to realize that plants consume as well as produce. The roots of a plant, for example, do not carry out photosynthesis—there is no sunlight underground. Roots obtain their energy the same way you do, by using energy-storing molecules produced elsewhere (in this case, in the leaves of the plant).

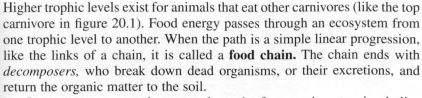

The Calvin cycle of photosynthesis uses the energy from the sun to build carbohydrate molecules, as discussed on pages 100 to 103. Plants use energy from the sun, carbon dioxide from the atmosphere, and water and nutrients from the soil to make organic molecules.

(b) Carnivores

Herbivores

At the second trophic level are **herbivores,** animals that eat plants (figure 20.3*a*). They are the *primary consumers* of ecosystems. Deer and horses are herbivores, and so are rhinoceroses, chickens (primarily herbivores), and caterpillars. Most herbivores rely on "helpers" to aid in the digestion of cellulose, a structural material found in plants. A cow, for instance, has a thriving colony of bacteria in its gut that digests cellulose. So does a termite. Humans cannot digest cellulose because we lack these bacteria— that is why a cow can live on a diet of grass and you cannot.

Carnivores

At the third trophic level are animals that eat herbivores, called **carnivores** (meat-eaters). They are the *secondary consumers* of ecosystems. Tigers and wolves are carnivores (figure 20.3*b*), and so are mosquitoes and blue jays. Some animals, like bears and humans, eat both plants and animals and are

(c) Omnivore

(d) Detritivore

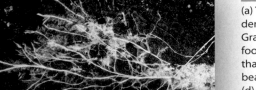

(e) Decomposer

Figure 20.3 **Members of the food chain.**

(a) The East African grasslands are covered by a dense growth of grasses, the primary producers. Grazing herbivores like these zebras obtain their food from plants. (b) These wolves are carnivores that live in North American forests. (c) This grizzly bear is an omnivore, this one fishing for salmon. (d) This crab is a detritivore. (e) Fungi, such as this basidiomycete growing through the soil, and bacteria are the primary decomposers of terrestrial ecosystems.

called **omnivores** (figure 20.3*c*). They use the simple sugars and starches stored in plants as food but not the cellulose. Many complex ecosystems contain a fourth trophic level, composed of animals that consume other carnivores. They are called *tertiary consumers,* or *top carnivores.* A weasel that eats a blue jay is a tertiary consumer. Only rarely do ecosystems contain more than four trophic levels, for reasons we will discuss later.

Detritivores and Decomposers

In every ecosystem there is a special class of consumers that include **detritivores** (figure 20.3*d*), organisms that eat dead organisms (also referred to as scavengers) and **decomposers,** organisms that break down organic substances making the nutrients available to other organisms (figure 20.3*e*). Worms, arthropods, and vultures are examples of detritivores. Bacteria and fungi are the principal decomposers in land ecosystems.

Energy Flows Through Trophic Levels

How much energy passes through an ecosystem? **Primary productivity** is the total amount of light energy converted by photosynthetic organisms into organic compounds in a given area per unit of time. An ecosystem's **net primary productivity** is the total amount of energy fixed by photosynthesis per unit of time, minus that which is expended by photosynthetic organisms to fuel metabolic activities. In short, it is the energy stored in organic compounds that is available to heterotrophs. The total weight of all an ecosystem's organisms, called its **biomass,** increases as a result of the ecosystem's net productivity.

When a plant uses the energy from sunlight to make structural molecules such as cellulose, it loses a lot of the energy as heat. In fact, only about half of the energy captured by the plant ends up stored in its molecules. The other half of the energy is lost. This is the first of many such losses as the energy passes through the ecosystem. When the energy flow through an ecosystem is measured at each trophic level, we find that 80% to 95% of the energy available at one trophic level is not transferred to the next. In other words, only 5% to 20% of the available energy passes from one trophic level to the next. For example, the amount of energy that ends up in the beetle's body in figure 20.4 is approximately only 17% of the energy present in the plant molecules it eats. Similarly, when a carnivore eats the herbivore, a comparable amount of energy is lost from the amount of energy present in the herbivore's molecules. This is why food chains generally consist of only three or four steps. So much energy is lost at each step that little usable energy remains after it has been incorporated into the bodies of organisms at four successive trophic levels.

Lamont Cole of Cornell University studied the flow of energy in a freshwater ecosystem in Cayuga Lake in upstate New York. Figure 20.5 shows how much energy flowed through the ecosystem, indicated by the red arrows. Each block represents the energy obtained by a different trophic level, with the producers, the algae and cyanobacteria, being the largest block. He calculated that about 150 of each 1,000 calories of potential energy fixed by algae and cyanobacteria are transferred into the bodies of animal plankton (small heterotrophs). Of these, about 30 calories are incorporated into the bodies of a type of small fish called a smelt, the principal secondary consumers of the system. If humans eat the smelt, they gain about 6 of the 1,000 calories that originally entered the system. If trout eat the smelt and humans eat the trout, humans gain only about 1.2 of the 1,000 calories.

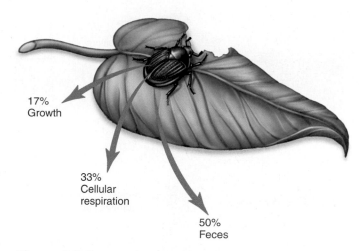

Figure 20.4 **How heterotrophs use food energy.**

A heterotroph assimilates only a fraction of the energy it consumes. For example, if a "bite" is composed of 500 Joules of energy (1 Joule = 0.239 calories), about 50%, 250 J, is lost in feces, about 33%, 165 J, is used to fuel cellular respiration, and about 17%, 85 J, is converted into consumer biomass. Only this 85 J (or roughly 20 calories) is available to the next trophic level.

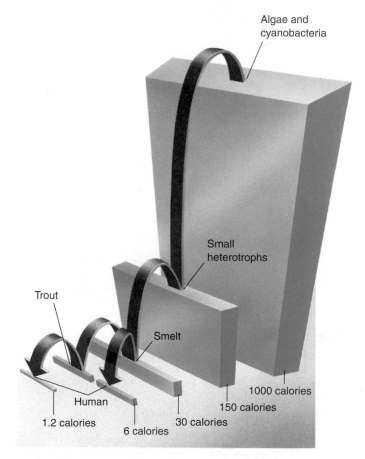

Figure 20.5 **Energy loss in an ecosystem.**

In a classic study of Cayuga Lake in New York, the path of energy was measured precisely at all points in the food web.

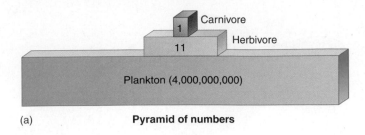

(a)   **Pyramid of numbers**

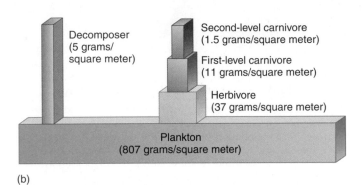

(b)

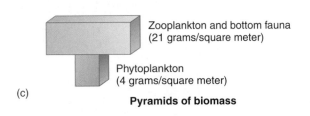

(c)   **Pyramids of biomass**

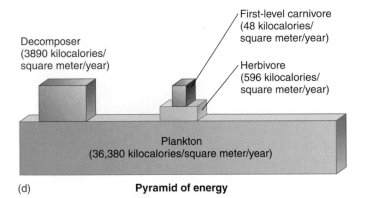

(d)   **Pyramid of energy**

Figure 20.6   **Ecological pyramids.**

Ecological pyramids measure different characteristics of each trophic level. In these aquatic ecosystems, plankton are the primary producers. (a) Pyramid of numbers. Pyramids of biomass, both normal (b) and inverted (c). (d) Pyramid of energy.

20.2 Ecological Pyramids

CONCEPT PREVIEW: Because energy is lost at every step of a food chain, the biomass of primary producers (photosynthesizers) tends to be greater than that of the herbivores that consume them, and herbivore biomass tends to be greater than that of the carnivores that consume them.

A plant fixes about 1% of the sun's energy that falls on its green parts. The successive members of a food chain, in turn, process into their own bodies on average about 10% of the energy available in the organisms on which they feed. For this reason, there are generally far more individuals at the lower trophic levels of any ecosystem than at the higher levels. Similarly, the biomass of the primary producers present in a given ecosystem is greater than the biomass of the primary consumers, with successive trophic levels having a lower and lower biomass and so less and less potential energy.

These key ecological relationships appear as pyramids when expressed as diagrams. Ecologists speak of "pyramids of numbers," where the sizes of the blocks reflect the number of individuals at each trophic level. The producers in the green box of figure 20.6*a* represent the largest number of individuals. Similarly, the producers (plankton) represent the largest group in "pyramids of biomass," shown in figure 20.6*b*. The inverted pyramid in figure 20.6*c* is an exception and is discussed below. The "pyramid of energy" in figure 20.6*d* shows the producers as the largest block.

Inverted Pyramids

Some aquatic ecosystems have inverted biomass pyramids, like in figure 20.6*c*. In a planktonic ecosystem—dominated by small organisms floating in water—the turnover of photosynthetic phytoplankton at the lowest level is very rapid, with zooplankton consuming phytoplankton so quickly that the phytoplankton (the producers at the base of the food chain) can never develop a large population size. Because the phytoplankton reproduce very rapidly, the community can support a population of heterotrophs that is larger in biomass and more numerous than the phytoplankton. However, don't confuse the sizes of these bars with the energy present at each level. The zooplankton that eat the phytoplankton are present in greater numbers but contain about only 10% of the energy.

Top Carnivores

The loss of energy that occurs at each trophic level places a limit on how many top-level carnivores a community can support. As we have seen, only about one-thousandth of the energy captured by photosynthesis passes all the way through a three-stage food chain to a tertiary consumer such as a snake or hawk. This explains why there are no predators that subsist on lions—the biomass of these animals is simply insufficient to support another trophic level.

In the pyramid of numbers, top-level predators tend to be fairly large animals. Thus, the small residual biomass available at the top of the pyramid is concentrated in a relatively small number of individuals.

Concept Check

1. What does an ecosystem include that a habitat does not?
2. What trophic level do you occupy in a food chain? Explain.
3. Explain how the pyramid of biomass can be inverted in oceans when the pyramid of energy is not.

Materials Cycle Within Ecosystems

20.3 The Water Cycle

CONCEPT PREVIEW: Water cycles through ecosystems in the atmosphere via precipitation and evaporation, some of it passing through plants on the way.

Unlike energy, which flows through the earth's ecosystems in one direction (from the sun to producers to consumers), the physical components of ecosystems are passed around and reused within ecosystems. Ecologists speak of such constant reuse as recycling or, more commonly, cycling. Materials that are constantly recycled include all the chemicals that make up the soil, water, and air. While many are important, the proper cycling of four materials is particularly critical to the health of any ecosystem: water, carbon, and the soil nutrients nitrogen and phosphorus.

The paths of water, carbon, and soil nutrients as they pass from the environment to living organisms and back form closed circles, or cycles. In each cycle, the chemical resides for a time in an organism and then returns to the nonliving environment, often referred to as a *biogeochemical cycle*.

Of all the nonliving components of an ecosystem, water has the greatest influence on the living portion. The availability of water and the way in which it cycles in an ecosystem in large measure determines the biological richness of that ecosystem—the kinds of creatures that live there and how many of each.

Water cycles within an ecosystem in two ways: the environmental water cycle and the organismic water cycle. Both cycles are shown in figure 20.7.

BIOLOGY & YOU

You and the Water Cycle. How many gallons of water do you personally consume in a day? Take a guess. Here are some numbers to help you make an estimate: *bath*: 40 gallons; *shower*: 2 gallons per minute; *teeth brushing*: 1 gallon; *hands/face washing*: 1 gallon; *dishwasher*: 15 gallons/load; *clothes machine load*: 25 gallons/load; *toilet flush*: 3 gallons; *glasses of water drunk*: 1/16 gallon/glass. Your per-capita water use (*per* is Latin for "by" and *capita* is Latin for "head") can rapidly become quite alarming—and the number you get from a calculation like this is quite likely an underestimation. It does not take into account everything you use water for, such as cooking, washing your dog, or cleaning muddy shoes. You might water your lawn, wash your car, or leave the water running while you brush your teeth. It all adds up. Even using water conservation measures such as low-flow shower heads and toilets, few of us have a per-capita water use less than 100 gallons a day.

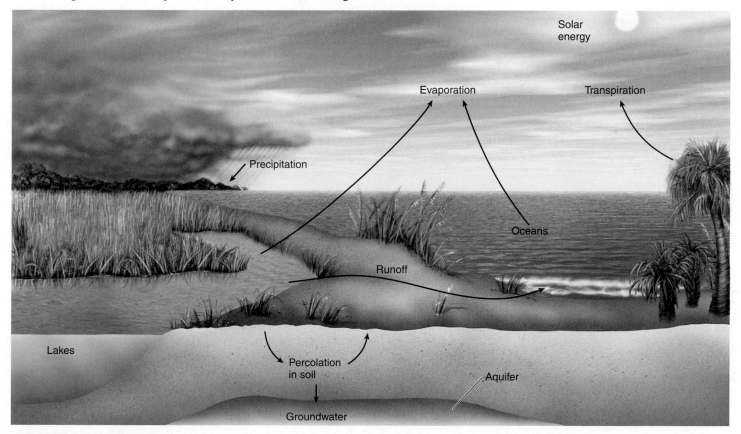

Figure 20.7   The water cycle.

Precipitation on land eventually makes its way to the ocean via groundwater, lakes, and rivers. Solar energy causes evaporation, adding water to the atmosphere. Plants give off excess water through transpiration, also adding water to the atmosphere. Atmospheric water falls as rain or snow over land and oceans, completing the water cycle.

The Environmental Water Cycle

In the environmental water cycle, water vapor in the atmosphere condenses and falls to the earth's surface as rain or snow (called precipitation in figure 20.7). Heated there by the sun, it reenters the atmosphere by **evaporation** from lakes, rivers, and oceans, where it condenses and falls to the earth again.

The Organismic Water Cycle

In the organismic water cycle, surface water does not return directly to the atmosphere. Instead, it is taken up by the roots of plants. After passing through the plant, the water reenters the atmosphere through tiny openings (stomata) in the leaves, evaporating from their surface. This evaporation from leaf surfaces is called **transpiration.** Transpiration is also driven by the sun: The sun's heat creates wind currents that draw moisture from the plant by passing air over the leaves.

> The mechanism of transpiration is described in more detail on pages 627 and 628. Air moving across the stomata in the leaves causes water to evaporate from the leaf tissue. The water in the leaves is replaced by water passing into the roots.

Breaking the Cycle

In very dense forest ecosystems, such as tropical rain forests, more than 90% of the moisture in the ecosystem is taken up by plants and then transpired back into the air. Because so many plants in a rain forest are doing this, the vegetation is the primary source of local rainfall. In a very real sense, these plants create their own rain: The moisture that travels up from the plants into the atmosphere falls back to earth as rain.

Where forests are cut down, the organismic water cycle is broken, and moisture is not returned to the atmosphere. Water drains off to the sea instead of rising to the clouds and falling again on the forest. During his expeditions from 1799–1805, the great German explorer Alexander von Humboldt reported that stripping the trees from a tropical rain forest in Colombia prevented water from returning to the atmosphere and created a semiarid desert. It is a tragedy of our time that just such a transformation is occurring in many tropical areas, as rain forests are being clear-cut or burned in the name of "development" (figure 20.8).

Figure 20.8   Burning or clear-cutting forests breaks the water cycle.

The high density and large size of plants in a forest translate into great quantities of water being transpired to the atmosphere, creating rain over the forests. In this way rain forests perpetuate the wet climate that supports them. Tropical deforestation permanently alters the climate in these areas, creating arid zones.

IMPLICATION   Proponents of clear-cutting suggest that this practice, if followed by planting neat rows of lumber tree seedlings, actually improves the forest, as it allows "plantation forestry," making future lumbering much more efficient. What do you think of this argument?

Groundwater

Much less obvious than the surface waters seen in streams, lakes, and ponds is the groundwater, which occurs in permeable, saturated, underground layers of rock, sand, and gravel called *aquifers*. In many areas, groundwater is the most important water reservoir; for example, in the United States, more than 96% of all freshwater is groundwater. In the United States, groundwater provides about 25% of the water used for all purposes and provides about 50% of the population with drinking water.

Because of the greater rate at which groundwater is being used, the increasing chemical pollution of groundwater is a very serious problem. Pesticides, herbicides, and fertilizers are key sources of groundwater pollution. Because of the large volume of water, its slow rate of turnover, and its inaccessibility, removing pollutants from aquifers is virtually impossible.

20.4 The Carbon Cycle

CONCEPT PREVIEW: Carbon captured from the atmosphere by photosynthesis is returned to it through respiration, combustion, and erosion.

The earth's atmosphere contains plentiful carbon, present as carbon dioxide (CO_2) gas. This carbon cycles between the atmosphere and living organisms, often being locked up for long periods of time in organisms or deep underground. The cycle is begun by plants that use CO_2 in photosynthesis to build organic molecules—in effect, they trap the carbon atoms of CO_2 within the living world. The carbon atoms are returned to the atmosphere's pool of CO_2 through respiration, combustion, and erosion, as shown in figure 20.9.

Carbon dioxide is generated at two points during cellular respiration. First, two molecules of CO_2 are produced in the oxidation of pyruvate to acetyl-CoA, discussed on page 117. Another four molecules of CO_2 are produced in the Krebs cycle, discussed on page 118.

Respiration

All organisms in ecosystems respire—that is, they extract energy from organic food molecules, which involves stripping away the carbon atoms and combining them with oxygen to form CO_2. This end product of respiration is released into the atmosphere.

Combustion

Plants that become buried in sediment may be transformed by pressure into coal or oil. The carbon originally trapped by these plants is only released back into the atmosphere when the coal or oil (called **fossil fuels**) is burned.

Human activity is causing the carbon cycle to become unbalanced. As described on page 436, the influx of large amounts of carbon dioxide into the atmosphere by the burning of fossil fuels is causing atmospheric temperatures to rise, a process called global warming.

Erosion

Very large amounts of carbon are present in limestone, formed from calcium carbonate shells of marine organisms. When the limestone becomes exposed to weather and erodes, the carbon washes back and is dissolved again in oceans.

BIOLOGY & YOU

You and the Carbon Cycle. You and every American make a contribution to global warming, because the way we live each day adds a lot of carbon dioxide to the earth's atmosphere. Your "carbon footprint" is a measure of the amount of carbon dioxide produced to fuel your daily life. In the diagram below, your carbon footprint is the sum of two parts: the *primary footprint* over which you have direct control, and the *secondary footprint* which measures the CO_2 emissions associated with the manufacture and breakdown of products we use. The average worldwide carbon footprint is about 4 tons per year. The average footprint of people in the United States is five times that, over 20 tons per year! The worldwide target to combat global warming is 2 tons per year. Imagine what your life would be like with your carbon footprint reduced 90%. The world's future depends upon it.

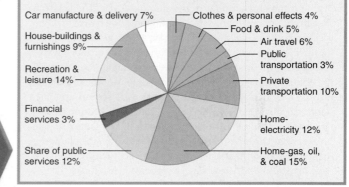

Car manufacture & delivery 7%
Clothes & personal effects 4%
Food & drink 5%
Air travel 6%
Public transportation 3%
House-buildings & furnishings 9%
Private transportation 10%
Recreation & leisure 14%
Home-electricity 12%
Financial services 3%
Share of public services 12%
Home-gas, oil, & coal 15%

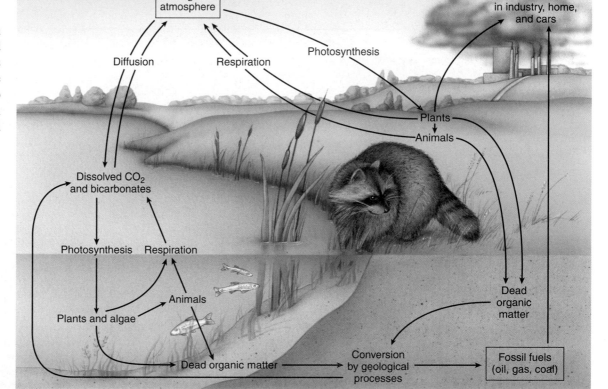

Figure 20.9
The carbon cycle.

Carbon from the atmosphere and from water is fixed by photosynthetic organisms and returned through respiration, combustion, and erosion.

You and the Nitrogen Cycle. As a way to step back and see the larger picture of how the nitrogen cycle affects your life, imagine setting up an aquarium tank with many fish. To care for the fish, plants, and other aquatic life you add to the tank, you have to provide light (for the plants) and fish food (for the critters). If that is all you do, however, the fish will soon die. Why? While fish get along fine in rivers and lakes, a fish tank is a confined body of water, and that turns out to make a critical difference. Fish food contains a lot of protein. A combination of uneaten fish food, decaying plant life, and urine and feces excreted by the fish will soon cause a toxic buildup of ammonia in the tank. *Nitrosomonas* bacteria in the water convert the ammonia into the chemical compound nitrite, and other *Nitrobacter* bacteria oxidize the nitrite, adding an oxygen to produce nitrate, much less harmful to the fish and the end product of the nitrogen cycle. The key to a healthy aquarium, then, is to start with only one or two fish. The desired bacteria are present in the air, and so will start to grow naturally in the tank water as the fish begin to produce waste products. Over time the bacteria will grow in numbers and will be able to handle more waste products. In 6–8 weeks the tank can be fully populated with fish. What does setting up a fish tank tell you about your nitrogen cycle? Our nitrogen wastes are "cleaned" by bacteria in sewage treatment plants, and like in a fish tank the ammonia concentrations within the plants are kept low enough that bacteria can convert toxic ammonia to safe nitrate. In a very real way, we too live in fish tanks.

20.5 The Nitrogen and Phosphorus Cycles

CONCEPT PREVIEW: Certain bacteria are able to convert atmospheric nitrogen gas into usable ammonia. Phosphorus, also critical to organisms, is available in soil and dissolved in water. These two chemicals are often the limiting factors determining what lives in an ecosystem.

The Nitrogen Cycle

Earth's atmosphere is 78.08% nitrogen gas (N_2), but most living organisms are unable to use the N_2 so plentifully available in the air surrounding them. The two nitrogen atoms of N_2 are bound together by a particularly strong "triple" covalent bond that is very difficult to break. Luckily, a few kinds of bacteria can break this triple bond and bind nitrogen atoms to hydrogen, forming "fixed" nitrogen, ammonia (NH_3), in a process called **nitrogen fixation.**

> Recall from the discussion of covalent bonds on page 36, that a triple covalent bond involves the sharing of three pairs of electrons. This is a lot of energy holding together two atoms, which is why the bond holding the nitrogen atoms together in nitrogen gas is so difficult to break.

Bacteria evolved the ability to fix nitrogen early in the history of life, before photosynthesis had introduced oxygen gas into the earth's atmosphere, and that is still the only way the bacteria are able to do it—even a trace of oxygen poisons the process. In today's world, awash with oxygen, these bacteria live encased within bubbles called cysts that admit no oxygen or within special airtight cells in nodules of tissue on the roots of beans, aspen trees, and a few other plants. Figure 20.10 shows the workings of the nitrogen cycle.

The growth of plants in ecosystems is often severely limited by the availability of "fixed" nitrogen in the soil, which is why farmers fertilize fields. Today most fixed nitrogen added to soils by farmers is produced in factories by industrial rather than bacterial nitrogen fixation—this industrial process today accounts for a prodigious 30% of the entire nitrogen cycle.

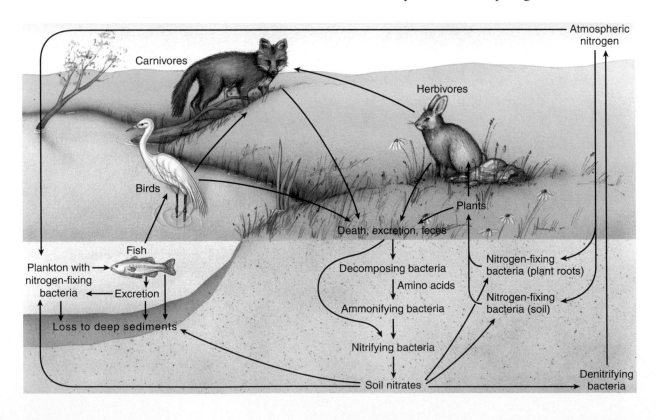

Figure 20.10
The nitrogen cycle.

Relatively few kinds of organisms—all of them bacteria—can convert atmospheric nitrogen into forms that can be used for biological processes.

Figure 20.11
The phosphorus cycle.

Phosphorus plays a critical role in plant nutrition; next to nitrogen, phosphorus is the element most likely to be so scarce that it limits plant growth.

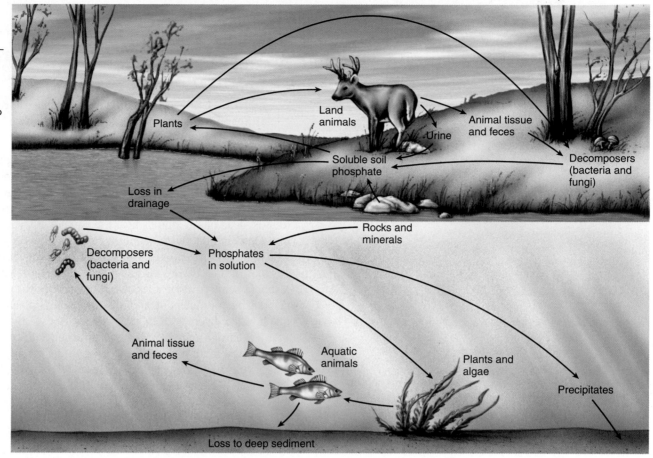

The Phosphorus Cycle

Phosphorus is an essential element in all living organisms, a key part of both ATP and DNA. Phosphorus is often in very limited supply in the soil of particular ecosystems, and because phosphorus does not form a gas, none is available in the atmosphere. Most phosphorus exists in soil and rock as the mineral calcium phosphate, which dissolves in water to form phosphate ions, as shown in figure 20.11. These phosphate ions are absorbed by the roots of plants and used by them to build organic molecules like ATP and DNA. When the plants and animals die and decay, bacteria in the soil convert the organic phosphorus back into phosphorus ions, completing the cycle.

The phosphorus level in freshwater lake ecosystems is often quite low, preventing much growth of photosynthetic algae in these systems. Pollution of a lake by the inadvertent addition of phosphorus to its waters (agricultural fertilizers and many commercial detergents are rich in phosphorus) first produces a green scum of algal growth on the surface of the lake, and then proceeds to "kill" the lake: As aging algae die, bacteria feeding on the dead algae cells use up so much of the lake's dissolved oxygen that fish and invertebrate animals suffocate. Such rapid, uncontrolled growth caused by excessive nutrients in an aquatic ecosystem is called **eutrophication.**

Concept Check

1. Explain how clear-cutting breaks the environmental water cycle.
2. How is carbon released into the atmosphere from plants?
3. If air is 78% nitrogen gas, why is plant growth often severely limited by available nitrogen?

BIOLOGY & YOU

You and the Phosphorus Cycle. Most of us have little direct impact on the phosphorus cycle, with one glaring exception: fertilizing lawns and gardens. Soils in most states already have an adequate amount of phosphorus to grow healthy lawns and gardens. Adding more phosphorus in fertilizers does not benefit the plants. Instead, the extra phosphorus runs off in rainwater and is washed down the nearest street or storm drain, ending up in your local lake or river. Soon the phosphorus produces algal blooms, a green scum over what was once clear water. States like Minnesota have passed laws that meet this problem head on. Starting in 2005, it became illegal to use fertilizers containing phosphorus on lawns anywhere in Minnesota. If you want to help keep phosphorus levels down in your community, look at the bag or box of fertilizer before you buy it. On its label there is a string of three numbers that list its percent nitrogen, phosphorus, and potassium content, in that order. A "0" in the middle means a phosphorus-free fertilizer.

How Weather Shapes Ecosystems

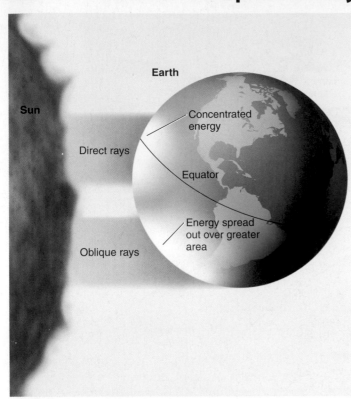

Figure 20.12   **Latitude affects climate.**

The relationship between the earth and sun is critical in determining the nature and distribution of life on earth. The tropics are warmer than the temperate regions because the sun's rays strike at a direct angle, providing more energy per unit of area.

20.6 The Sun and Atmospheric Circulation

CONCEPT PREVIEW: The sun drives circulation of the atmosphere, causing rain in the tropics and a band of deserts at 30 degrees latitude.

The world contains a wide variety of ecosystems because its climate varies a great deal from place to place. On a given day, Miami and Boston often have very different weather. There is no mystery about this. The tropics are warmer than the temperate regions because the sun's rays arrive almost perpendicular (that is, dead on) at regions near the equator. As you move from the equator into temperate latitudes, sunlight strikes the earth at more oblique angles, which spreads it out over a much greater area, thus providing less energy per unit of area (figure 20.12). This simple fact—that because the earth is a sphere some parts of it receive more energy from the sun than others—is responsible for much of the earth's different climates and thus, indirectly, for much of the diversity of its ecosystems.

The earth's annual orbit around the sun and its daily rotation on its own axis are also both important in determining world climate. Because of the daily cycle, the climate at a given latitude is relatively constant. Because of the annual cycle and the inclination of the earth's axis, all parts away from the equator experience a progression of seasons. In summer in the Southern Hemisphere, the earth is tilted toward the sun as shown in figure 20.12, and rays hit more directly leading to higher temperatures; in the winter, the tilt of the earth is opposite, with the Northern Hemisphere nearer to the sun, experiencing summer.

The major atmospheric circulation patterns result from the interactions between six large air masses. These great air masses (shown as circulating arrows in figure 20.13) occur in pairs, with one air mass of the pair occurring in the northern latitudes and the other occurring in the southern latitudes. These air masses affect climate because the rising and falling of an air mass influence its temperature, which, in turn, influences its moisture-holding capacity.

Near the equator, warm air rises and flows toward the poles (indicated by arrows at the equator that rise and circle toward the poles). As it rises and cools, this air loses most of its moisture because cool air holds less water vapor than warm air. (This explains why it rains so much in the tropics where the air is warm.) When this air has traveled to about 30 degrees north and south latitudes, the cool, dry air sinks and becomes reheated, soaking up water like a sponge as it warms, producing a broad zone of low rainfall. It is no accident that all of the great deserts of the world lie near 30 degrees north or 30 degrees south latitude. Air at these latitudes is still warmer than it is in the polar regions, and thus it continues to flow toward the poles. At about 60 degrees north and south latitudes, air rises and cools and sheds its moisture, and such are the locations of the great temperate forests of the world. Finally, this rising air descends near the poles, producing zones of very low precipitation.

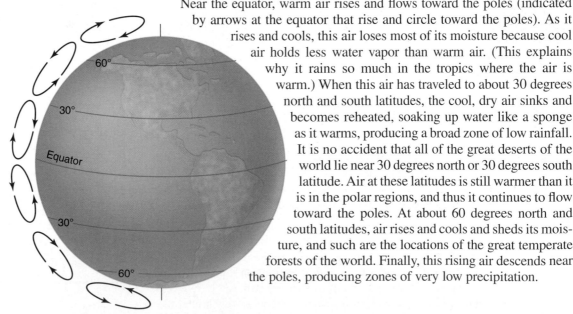

Figure 20.13   **Air rises at the equator and then falls.**

The pattern of air movement out from and back to the earth's surface forms three pairs of great cycles.

20.7 Latitude and Elevation

CONCEPT PREVIEW: Temperatures fall with increasing latitude and also with increasing elevation. Rainfall is higher on the windward side of mountains, with air losing its moisture as it rises up the mountain; descending on the far side, the dry air warms and sucks up moisture, creating deserts.

Temperatures are higher in tropical ecosystems for a simple reason: more sunlight per unit area falls on tropical latitudes (see figure 20.12). Solar radiation is most intense when the sun is directly overhead, and this occurs only in the tropics, where sunlight strikes the equator perpendicularly. Temperature also varies with elevation, with higher altitudes becoming progressively colder. At any given latitude, air temperature falls about 6°C for every 1,000-meter increase in elevation. The ecological consequences of temperature varying with elevation are the same as temperature varying with latitude. Figure 20.14 illustrates this principle comparing changes in ecosystems that occur with increasing latitudes in North America with the ecosystem changes that occur with increasing elevation at the tropics. A 1,000-meter increase in elevation on a mountain in southern Mexico (figure 20.14*b*) results in a temperature drop equal to that of an 880-kilometer increase in latitude on the North American continent (figure 20.14*a*). This is why "timberline" (the elevation above which trees do not grow) occurs at progressively lower elevations as one moves farther from the equator.

Rain Shadows

When a moving body of air encounters a mountain (figure 20.15), it is forced upward, and as it is cooled at higher elevations the air's moisture-holding capacity decreases, producing the rain you see on the windward side of the mountains—the side from which the wind is blowing. The effect on the other side of the mountain—the leeward side—is quite different. As the air passes the peak and descends on the far side of the mountains, it is warmed, so its moisture-holding capacity increases. Sucking up all available moisture, the air dries the surrounding landscape, often producing a desert. This effect, called a **rain shadow,** is responsible for deserts such as Death Valley, which is in the rain shadow of Mount Whitney, the tallest mountain in the Sierra Nevada.

Similar effects can occur on a larger scale. Regional climates are areas that are located on different parts of the globe but share similar climates because of similar geography. A so-called Mediterranean climate results when winds blow from a cool ocean onto warm land during the summer. As a result, the air's moisture-holding capacity is increased and precipitation is blocked, similar to what occurs on the leeward side of mountains. This effect accounts for dry, hot summers and cool, moist winters in areas with a Mediterranean climate such as portions of southern California or Oregon, central Chile, southwestern Australia, and the Cape region of South Africa. Such a climate is unusual on a world scale. In the regions where it occurs, many unusual kinds of endemic (local in distribution) plants and animals have evolved.

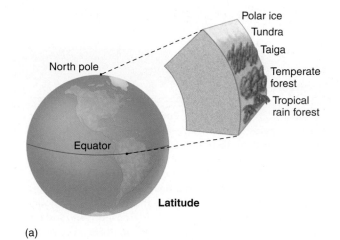
(a)

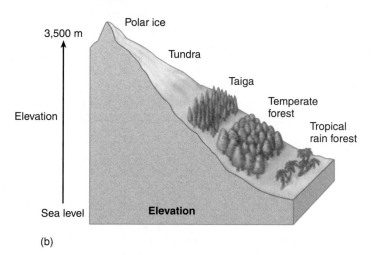
(b)

Figure 20.14 **How elevation affects ecosystems.**

The same land ecosystems that normally occur as latitude increases north and south of the equator at sea level (a) can occur in the tropics as elevation increases (b).

Figure 20.15 **The rain shadow effect.**

Moisture-laden winds from the Pacific Ocean rise and are cooled when they encounter the Sierra Nevada. As they cool, their moisture-holding capacity decreases and precipitation occurs. As the air descends on the east side of the range, it warms, its moisture-holding capacity increases, and the air picks up moisture from its surroundings. As a result, arid conditions prevail on the east side of these mountains.

20.8 Patterns of Circulation in the Ocean

CONCEPT PREVIEW: The world's oceans circulate in huge gyres deflected by continental landmasses. Disturbances in ocean currents like an El Niño can have profound influences on world climate.

Patterns of ocean circulation are determined by the patterns of atmospheric circulation, which means that indirectly, the currents are driven by solar energy. The radiant input of heat from the sun sets the atmosphere in motion as already described, and then the winds set the ocean in motion. Oceanic circulation is dominated by the movement of surface waters in huge spiral patterns called gyres, which move around the subtropical zones of high pressure between approximately 30 degrees north and south latitudes. These gyres, indicated by the red and blue arrows in figure 20.16, move clockwise in the Northern Hemisphere and counterclockwise in the Southern Hemisphere. The ways they redistribute heat profoundly affects life not only in the oceans but also on coastal lands. For example, the Gulf Stream, in the North Atlantic (the red-colored arrow, meaning it carries warm waters), swings away from North America near Cape Hatteras, North Carolina, and reaches Europe near the southern British Isles. Because of the Gulf Stream, western Europe is much warmer and more temperate than eastern North America at similar latitudes.

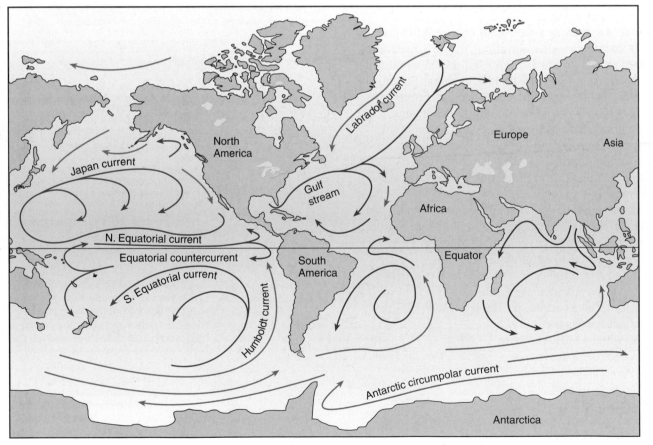

Figure 20.16
Oceanic circulation.

The circulation in the oceans moves in great surface spiral patterns called gyres; oceanic circulation affects the climate on adjacent lands.

→ Cold water current

→ Warm water current

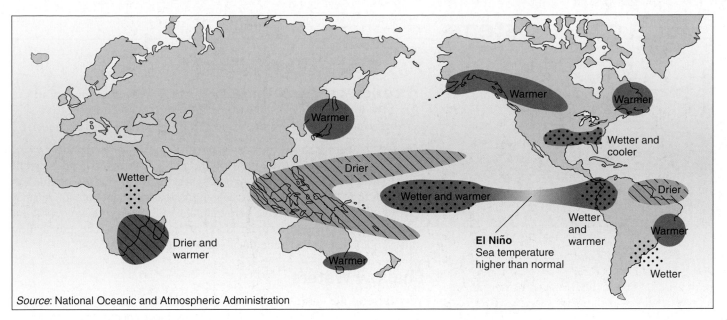

Figure 20.17   An El Niño winter.

El Niño currents produce unusual weather patterns all over the world as warm waters from the western Pacific move eastward.

El Niño Southern Oscillations and Ocean Ecology

Every Christmas a warm current sweeps down the coast of Peru and Ecuador from the tropics, reducing the fish population slightly and giving local fishers some time off. The local fishers named this Christmas current *El Niño* (literally, "the child," after the Christ Child). A dramatic version of the same phenomenon occurs every two to seven years, felt on a global scale.

Scientists now have a pretty good idea of what goes on in an El Niño. Normally the Pacific Ocean is fanned by constantly blowing east-to-west trade winds. These winds push warm surface water away from the ocean's eastern side (Peru, Ecuador, and Chile) and allow cold water to well up from the depths in its place, carrying nutrients that feed plankton and hence fish. This warm surface water piles up in the west, around Australia and the Philippines, making it several degrees warmer and a meter or so higher than the eastern side of the ocean. But if the winds slacken briefly, warm water begins to slosh back across the ocean (indicated by the darker red band stretching between Australia and the northern coast of South America in figure 20.17), causing an El Niño.

The end result is to shift the weather systems of the western Pacific Ocean 6,000 kilometers eastward. The tropical rainstorms that usually drench Indonesia and the Philippines soak the western edge of South America, leaving the previously rainy areas in drought (indicated by the light pink, hatched areas in figure 20.17).

Concept Check

1. Explain why the earth's great deserts all lie near 30° latitude.
2. Death Valley, within 160 miles of the Pacific Ocean, is one of the driest places on earth. How could it be so dry, if it is so close to so much water?
3. What sets the water in the earth's oceans into motion?

IN THE NEWS

El Niño and Global Warming. El Niño, the unusual rise in sea-surface temperatures in the equatorial Pacific Ocean, visited three times from 1990 to 1995, an unwanted guest that wouldn't go away. In between, temperatures never went quite back to normal, as if it were all one long El Niño. This has led some scientists like Kevin Trenberth at the National Center for Atmospheric Research in Boulder, Colorado, to speculate that the apparent increase in El Niño events may be linked to warmer ocean temperatures resulting from global warming. He believes that as global warming drives temperatures higher, ocean currents and weather systems are not able to release all the extra heat getting pumped into tropical seas. El Niño, in his view, acts as a kind of pressure release valve, expelling the excess heat. As global warming becomes more pronounced, El Niño events become more intense and frequent. Other scientists suspect that the recent upturn in El Niño events are a statistical blip—we would have expected to see one to two events between 1990 and 1995 anyway. Without better computer models or decades of data, we are not going to know with any certainty who is right.

Major Kinds of Ecosystems

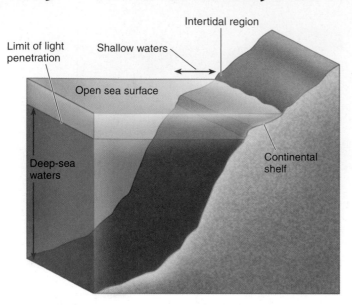

Figure 20.18 Ocean ecosystems.

There are three primary ecosystems found in the earth's oceans. Shallow water ecosystems occur along the shoreline and at areas of coral reefs. Open sea surface ecosystems occur in the upper 100 meters where light can penetrate. Finally, deep-sea water ecosystems are bottom areas below 300 meters.

20.9 Ocean Ecosystems

CONCEPT PREVIEW: Three principal ocean ecosystems occur: shallow waters, open-sea surface, and deep-sea waters. Both intertidal shallows and deep-sea communities are very diverse.

Most of the earth's surface—nearly three-quarters—is covered by water. The seas have an average depth of more than 3 kilometers, and they are, for the most part, cold and dark. Photosynthetic organisms are confined to the upper few hundred meters because light does not penetrate any deeper. Almost all organisms that live below this level feed on organic debris that rains downward. The three main kinds of marine ecosystems are shallow waters, open-sea surface, and deep-sea waters (figure 20.18).

Shallow Waters

Very little of the earth's ocean surface is shallow—mostly that along the shoreline—but this small area contains many more species than other parts of the ocean (figure 20.19*a*). The world's great commercial fisheries occur on banks in the coastal zones, where nutrients derived from the land are more abundant than in the open ocean. Part of this zone consists of the **intertidal region,** which is exposed to the air whenever the tides recede. Partly enclosed bodies of water, such as those that often form at river mouths and in coastal bays, where the salinity is intermediate between that of seawater and freshwater, are called **estuaries.** Estuaries are among the most naturally fertile areas in the world, often containing rich stands of submerged and emergent plants, algae, and microscopic organisms. They provide the breeding grounds for most of the coastal fish and shellfish that are harvested both in the estuaries and in open water.

Open-Sea Surface

Drifting freely in the upper, better-illuminated waters of the ocean is a diverse biological community of microscopic organisms. Most of the plankton occurs in the top 100 meters of the sea. Many fish swim in these waters as well, feeding on the plankton and one another (figure 20.19*b*). Some members of the plankton, including algae and some bacteria, are

(a)   (b)

Figure 20.19 Shallow waters and open sea surface.

(a) Fish and many other kinds of animals find food and shelter among the coral in the coastal waters of some regions. (b) The upper layers of the open ocean contain plankton and large schools of fish, like these bigeye snappers.

photosynthetic and are called phytoplankton. Collectively, these organisms are responsible for about 40% of all photosynthesis that takes place on earth. Over half of this is carried out by organisms less than 10 micrometers in diameter—at the lower limits of size for organisms—and almost all of it near the surface of the sea, in the zone into which light from the surface penetrates freely.

Deep-Sea Waters

In the deep waters of the sea, below 300 meters, little light penetrates. Very few organisms live there, compared to the rest of the ocean, but those that do include some of the most bizarre organisms found anywhere on earth. Many deep-sea inhabitants have bioluminescent (light-producing) body parts that they use to communicate or to attract prey (figure 20.20a).

The supply of oxygen can often be critical in the deep ocean, and as water temperatures become warmer, the water holds less oxygen. For this reason, the amount of available oxygen becomes an important limiting factor for deep-sea organisms in warmer marine regions of the globe. Carbon dioxide, in contrast, is almost never limited in the deep ocean. The distribution of minerals is much more uniform in the ocean than it is on land, where individual soils reflect the composition of the parent rocks from which they have weathered.

Frigid and bare, the floors of the deep sea have long been considered a biological desert. Recent close-up looks taken by marine biologists, however, paint a different picture (figure 20.20b). The ocean floor is teeming with life. Often kilometers deep, thriving in pitch darkness under enormous pressure, crowds of marine invertebrates have been found in hundreds of deep samples from the Atlantic and Pacific. Rough estimates of deep-sea diversity have soared to hundreds of thousands of species. Many appear endemic (local). The diversity of species is so high it may rival that of tropical rain forests! This profusion is unexpected. The formation of new species usually require some kind of barrier to diverge (see chapter 14), and the ocean floor seems boringly uniform. However, little migration occurs among deep populations, and this lack of movement may encourage local specialization and species formation. A patchy environment may also contribute to species formation there; deep-sea ecologists find evidence that fine but nonetheless formidable resource barriers arise in the deep sea.

No light falls in the deep ocean. From where do deep-sea organisms obtain their energy? While some utilize energy falling to the ocean floor as debris from above, other deep-sea organisms are autotrophic, gaining their energy from **hydrothermal vent systems,** areas in which seawater circulates through porous rock surrounding fissures where molten material from beneath the earth's crust comes close to the surface. Hydrothermal vent systems, also called deep-sea vents, support a broad array of heterotrophic life (figure 20.20c). Water in the area of these hydrothermal vents is heated to temperatures in excess of 350°C, and contains high concentrations of hydrogen sulfide. Prokaryotes that live by these deep-sea vents obtain energy and produce carbohydrates through chemosynthesis instead of photosynthesis. Like plants, they are autotrophs; they extract energy from hydrogen sulfide to manufacture food, much as a plant extracts energy from the sun to manufacture its food. These prokaryotes live symbiotically within the tissues of heterotrophs that live around the deep-sea vents. The animals provide a place for the prokaryotes to live and obtain nutrients, and in turn the prokaryotes supply the animal with organic compounds to use as food.

(a)

(c)

(b)

Figure 20.20   Deep-sea waters.

(a) The luminous spot below the eye of this deep-sea fish results from the presence of a symbiotic colony of luminous bacteria. (b) Looking for all the world like some undersea sunflower, these two sea anemones (actually animals) use a glass-sponge stalk to catch "marine snow," food particles raining down on the ocean floor from the ocean surface several kilometers above. (c) These giant beardworms live along vents where water jets from fissures at 350°C and then cools to the 2°C of the surrounding water.

IMPLICATION   No sunlight penetrates to the bottom of the deep sea. Why do you think that fish in this subphotic region have eyes, often extremely large, rather than evolution having selected for no eyes, as it has done for freshwater fish that inhabit deep dark caves?

Figure 20.21 Freshwater organism.

Some organisms, such as this giant waterbug with eggs on its back, can only live in freshwater habitats.

20.10 Freshwater Ecosystems

CONCEPT PREVIEW: Freshwater ecosystems cover only about 2% of the earth's surface, and all are strongly tied to adjacent terrestrial ecosystems. In some, organic materials are common, and in others, scarce. The temperature zones in lakes overturn twice a year, in spring and fall.

Freshwater ecosystems (lakes, ponds, rivers, and wetlands) are distinct from both ocean and land ecosystems, and they are very limited in area. Inland lakes cover about 1.8% of the earth's surface and rivers, streams, and wetlands about 0.4%. All freshwater habitats are strongly connected to land habitats, with marshes and swamps (wetlands) constituting intermediate habitats. In addition, a large amount of organic and inorganic material continually enters bodies of freshwater from communities growing on the land nearby. Many kinds of organisms are restricted to freshwater habitats (figure 20.21). When they occur in rivers and streams, they must be able to attach themselves in such a way as to resist or avoid the effects of current or risk being swept away.

Like the ocean, ponds and lakes have three zones in which organisms live (figure 20.22*a*): a shallow "edge" zone (the littoral zone), an open-water surface zone (the limnetic zone), and a deep-water zone where light does not penetrate (the profundal zone). Also, lakes can be divided into two categories, based on their production of organic material. In **oligotrophic lakes** (figure 20.22*b*), organic matter and nutrients are relatively scarce. Such lakes are often deep, and their deep waters are always rich in

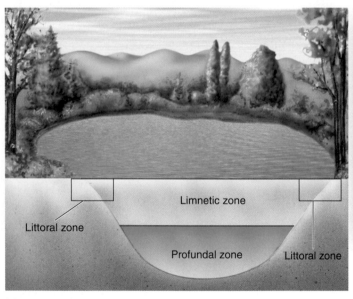

(a)

(b) Oligotrophic lake

(c) Eutrophic lake

Figure 20.22 Characteristics of ponds and lakes.

(a) Ponds and lakes can be divided into three zones based on the types of organisms that live in each. A shallow "edge" (littoral) zone lines the periphery of the lake where attached algae and their insect herbivores live. An open-water surface (limnetic) zone lies across the entire lake and is inhabited by floating algae, zooplankton, and fish. A dark, deep-water (profundal) zone overlies the sediments at the bottom of the lake. The profundal zone contains numerous bacteria and wormlike organisms that consume dead debris settling at the bottom of the lake. Lakes can be oligotrophic (b), containing scarce amounts of organic material, or eutrophic (c), containing abundant amounts of organic material.

oxygen. Oligotrophic lakes are highly susceptible to pollution from excess phosphorus from such sources as fertilizer runoff, sewage, and detergents. **Eutrophic lakes,** on the other hand, have an abundant supply of minerals and organic matter (figure 20.22c). Oxygen is depleted at the lower depths in the summer because of the abundant organic material and high rate at which aerobic decomposers in the lower layer use oxygen. These stagnant waters circulate to the surface in the fall (during the fall overturn, as discussed below) and are then infused with more oxygen.

> Aerobic decomposers consume dead organisms as a means of acquiring energy. Because they carry out aerobic respiration, they need oxygen as the final electron acceptor in the electron transport chain, as discussed on page 121. Their active metabolism uses up a lot of oxygen.

Thermal stratification, characteristic of the larger lakes in temperate regions, is the process whereby water at a temperature of 4°C (which is when water is most dense) sinks beneath water that is either warmer or cooler. Follow through the changes in a large lake in figure 20.23 beginning in winter ❶, where water at 4°C sinks beneath cooler water that freezes at the surface at 0°C. Below the ice, the water remains between 0° and 4°C, and plants and animals survive there. In spring ❷, as the ice melts, the surface water is warmed to 4°C and sinks below the cooler water, bringing the cooler water to the top with nutrients from the lake's lower regions. This process is known as the *spring overturn.*

In summer ❸, warmer water forms a layer over the cooler water that lies below. In the area between these two layers, called the thermocline, temperature changes abruptly. You may have experienced the existence of these layers if you have dived into a pond in temperate regions in the summer. Depending on the climate of the particular area, the warm upper layer may become as much as 20 meters thick during the summer. In autumn ❹, its surface temperature drops until it reaches that of the cooler layer underneath—4°C. When this occurs, the upper and lower layers mix—a process called the fall overturn. Therefore, colder waters reach the surfaces of lakes in the spring and fall, bringing up fresh supplies of dissolved nutrients.

BIOLOGY & YOU

Putting Thermal Stratification to Work: Ice Fishing. When a large lake freezes in winter, it doesn't freeze solid like a huge ice cube. Only the surface freezes. Beneath its hard skin lies cold, dense water where game fish like pickerel, walleye, and pike thrive all winter long. Ice fishing is the activity of catching these fish with lines and fish hooks through a small opening augered through the ice. Some hardy ice anglers sit on a stool in the open on a frozen lake; other less adventurous souls await a catch in a heated cabin on the ice, some with bunks, kitchens, and portable bathrooms! Ice fishing has long been a popular solitary pastime in Finland, Norway, and Sweden, with fish houses a rare occurrence. In the United States, by contrast, ice fishing is often a social activity, with fish houses rented out by the day in states like Minnesota or Wisconsin—indeed, any state with lakes and long cold winters. Winter fishing makes up nearly 1/4 the annual catch in Wisconsin. Some confident but reckless ice fishers will walk out on 2.5 inches of ice, but because ice thickness can vary across an area, a full 4 inches of good ice is recommended—"thick and blue, tried and true." If you are on a snowmobile (it can be a long way from a lake's edge out to its center) you need 6 inches of ice; in a pickup or car, you must have 12–14 inches if you don't want your vehicle to join the fishes you seek to catch.

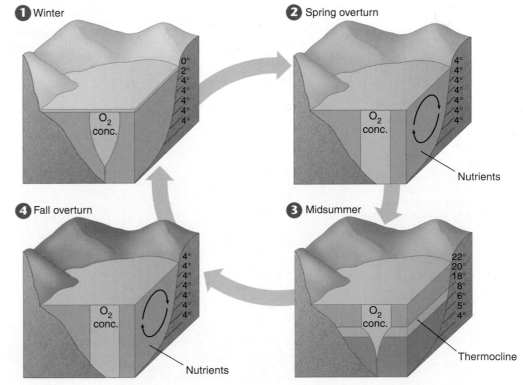

Figure 20.23 Spring and fall overturns in freshwater ponds or lakes.

The pattern of stratification in a large pond or lake in temperate regions is upset in the spring and fall overturns. Of the three layers of water shown in midsummer (*lower right*), the densest water occurs at 4°C. The warmer water at the surface is less dense. The thermocline is the zone of abrupt change in temperature that lies between them. In summer and winter, oxygen concentrations are lower at greater depths, whereas in the spring and fall, they are more similar at all depths.

IN THE NEWS

California Wildfires. Each autumn the Los Angeles region is plagued by wildfires. During a two-week period in October 2003, 721,791 acres burned and in 2007, some 426,000 acres. In 2008 a single fire in the Los Padres National Forest burned 134,000 acres, nearly 210 square miles. Every year there are fires—fall is the "fire season" in Southern California. Ever wonder why? Turns out it is no mystery. Every fall strong, extremely dry winds called Santa Ana winds sweep down across Southern California from the Mohave Desert, bringing with them hot dry air, often very hot and very dry—the perfect incubator for a forest fire. What sort of biome would you expect to find in the path of annual Santa Ana winds? One that has evolved to tolerate fire—chaparral. Indeed, many chaparral plants can germinate only when exposed to the extreme heat of a forest fire. Fall forest fires are a normal and necessary part of the chaparral biome. It is only because we humans insist on building so many homes within this chaparral biome that damage is done and life lost. Unlike chaparral plants, human homes are not a natural part of this biome.

Figure 20.24   **Distribution of the earth's biomes.**

The seven primary types of biomes are tropical rain forest, savanna, desert, temperate grassland, temperate deciduous forest, taiga, and tundra. In addition, seven less widespread biomes are shown.

20.11 Land Ecosystems

CONCEPT PREVIEW: Biomes are major terrestrial communities defined largely by temperature and rainfall patterns.

Biomes are major types of ecosystems that occur on land. Each biome occurs over a broad area and is characterized by a particular climate and a defined group of organisms. While biomes can be classified in a number of ways, the seven most widely occurring biomes are (1) tropical rain forest, (2) savanna, (3) desert, (4) temperate grassland, (5) temperate deciduous forest, (6) taiga, and (7) tundra (figure 20.24). The reason that there are seven primary biomes, and not one or 80, is that they have evolved to suit the climate of the region, and the earth has seven principal climates. The seven biomes differ remarkably from one another but show many consistencies within; a particular biome often looks similar, with many of the same types of creatures living there, wherever it occurs on earth.

There are seven other less widespread biomes also shown in figure 20.24: chaparral; polar ice; mountain zone; temperate evergreen forest; warm, moist evergreen forest; tropical monsoon forest; and semidesert.

If there were no mountains and no climatic effects caused by the irregular outlines of the continents and by different sea temperatures, each biome would form an even belt around the globe. In fact, their distribution is greatly affected by these factors, especially by elevation. Thus, the summits of the Rocky Mountains are covered with a vegetation type that resembles tundra, whereas other forest types that resemble taiga occur farther down. It is for reasons such as these that the distributions of the biomes are so irregular. One trend that is apparent is that those

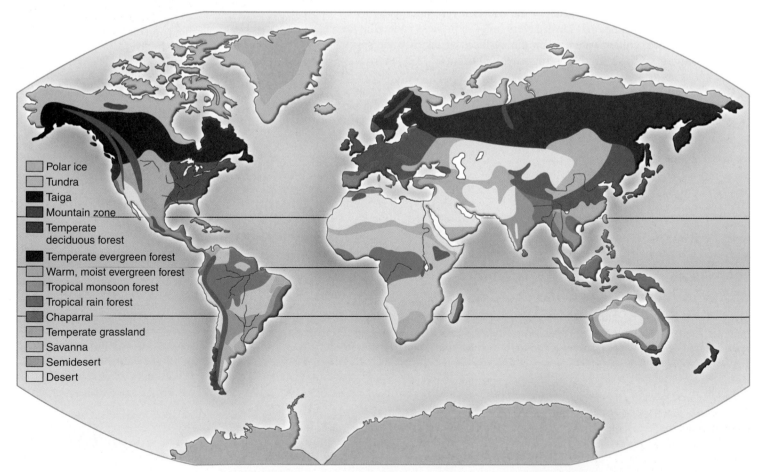

Polar ice
Tundra
Taiga
Mountain zone
Temperate deciduous forest
Temperate evergreen forest
Warm, moist evergreen forest
Tropical monsoon forest
Tropical rain forest
Chaparral
Temperate grassland
Savanna
Semidesert
Desert

biomes that normally occur at high latitudes also follow an altitudinal gradient along mountains. That is, biomes found far north and far south of the equator at sea level also occur in the tropics but at high mountain elevations (see figure 20.14).

Lush Tropical Rain Forests

Rain forests, which experience over 250 centimeters of rain a year, are the richest ecosystems on earth (figure 20.25). They contain at least half of the earth's species of terrestrial plants and animals—more than 2 million species! In a single square mile of tropical forest in Rondonia, Brazil, there are 1,200 species of butterflies—twice the total number found in the United States and Canada combined. The communities that make up tropical rain forests are diverse in that each kind of animal, plant, or microorganism is often represented in a given area by very few individuals. There are extensive tropical rain forests in South America, Africa, and Southeast Asia. But the world's tropical rain forests are being destroyed, and with them, countless species, many of them never seen by humans. Perhaps a quarter of the world's species will disappear with the rain forests during the lifetime of many of us.

Figure 20.25 Tropical rain forest.

Savannas: Dry Tropical Grasslands

In the dry climates that border the tropics are found the world's great grasslands, called **savannas.** Landscapes are open, often with widely spaced trees, and rainfall (75 to 125 cm annually) is seasonal. Many of the animals and plants are active only during the rainy season. The huge herds of grazing animals that inhabit the African savanna are familiar to all of us (figure 20.26). Such animal communities occurred in the temperate grasslands of North America during the Pleistocene epoch but have persisted mainly in Africa. On a global scale, the savanna biome is transitional between tropical rain forest and desert. As these savannas are increasingly converted to agricultural use to feed rapidly expanding human populations in subtropical areas, their inhabitants are finding it difficult to survive. The elephant and rhino are now endangered species; the lion, giraffe, and cheetah will soon follow them.

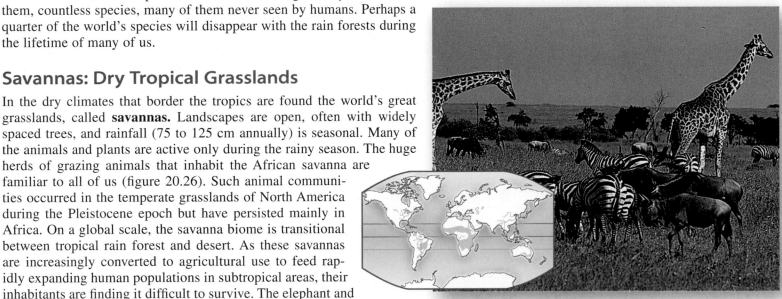

Figure 20.26 Savanna.

Deserts: Burning Hot Sands

In the interior of continents are found the world's great deserts, especially in Africa (the Sahara), Asia (the Gobi), and Australia (the Great Sandy Desert). **Deserts** are dry places where less than 25 centimeters of rain falls in a year—an amount so low that vegetation is sparse and survival depends on water conservation (figure 20.27). One quarter of the world's land surface is desert. The plants and animals that live in deserts may restrict their activity to favorable times of the year, when water is present. To avoid high temperatures, most desert vertebrates live in deep, cool, and sometimes even somewhat moist burrows. Those that are active over a greater portion of the year emerge only at night, when temperatures are relatively cool. Some, such as camels, can drink large quantities of water when it is available and then survive long, dry periods. Many animals simply migrate to or through the desert, where they exploit food that may be abundant seasonally.

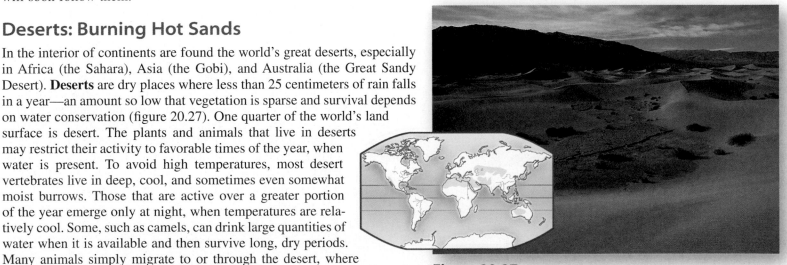

Figure 20.27 Desert.

Figure 20.28   Temperate grassland.

Figure 20.29   Temperate deciduous forest.

Figure 20.30   Taiga.

Grasslands: Seas of Grass

Halfway between the equator and the poles are temperate regions where rich **grasslands** grow. These grasslands once covered much of the interior of North America, and they were widespread in Eurasia and South America as well. Such grasslands are often highly productive when converted to agriculture. Many of the rich agricultural lands in the United States and southern Canada were originally occupied by **prairies,** another name for temperate grasslands. These natural temperate grasslands are one of the biomes adapted to periodic fire.

The roots of perennial grasses characteristically penetrate far into the soil, and grassland soils tend to be deep and fertile. Temperate grasslands are often populated by herds of grazing mammals. In North America, the prairies were once inhabited by huge herds of bison and pronghorns (figure 20.28). The herds are almost all gone now, with most of the prairies having been converted to the richest agricultural region on earth.

Deciduous Forests: Rich Hardwood Forests

Mild climates (warm summers and cool winters) and plentiful rains promote the growth of **deciduous** ("hardwood") **forests** in Eurasia, the northeastern United States, and eastern Canada (figure 20.29). A deciduous tree is one that drops its leaves in the winter. Deer, bears, beavers, and raccoons are the familiar animals of the temperate regions. Because the temperate deciduous forests represent the remnants of more extensive forests that stretched across North America and Eurasia several million years ago, these remaining areas—especially those in eastern Asia and eastern North America—share animals and plants that were once more widespread. Alligators, for example, are found only in China and in the southeastern United States. The deciduous forest in eastern Asia is rich in species because climatic conditions have remained constant.

Taiga: Trackless Conifer Forests

A great ring of northern forests of coniferous trees (spruce, hemlock, larch, and fir) extends across vast areas of Asia and North America. Coniferous trees are ones with leaves like needles that are kept all year long. This ecosystem, called **taiga,** is one of the largest on earth (figure 20.30). Here, the winters are long and cold. Rain, often as little as in hot deserts, falls in the summer. Because it has too short a growing season for farming, few people live there. Many large mammals, including elk, moose, deer, and such carnivores as wolves, bears, lynx, and wolverines, live in the taiga.

Traditionally, fur trapping has been extensive in this region. Lumber production is also important. Marshes, lakes, and ponds are common and are often fringed by willows or birches. Most of the trees occur in dense stands of one or a few species.

Tundra: Cold Boggy Plains

In the far north, above the great coniferous forests and below the polar ice, there are few trees. There the grassland, called **tundra,** is open, windswept, and often boggy (figure 20.31). Enormous in extent, this ecosystem covers one-fifth of the earth's land surface. Very little rain or snow falls. When rain does fall during the brief arctic summer, it sits on frozen ground, creating a sea of boggy ground. **Permafrost,** or permanent ice, usually exists within a meter of the surface. Trees are small and are mostly confined to the margins of streams and lakes. Large grazing mammals, including musk-oxen, caribou, reindeer, and carnivores such as wolves, foxes, and lynx, live in the tundra. Lemming populations rise and fall on a long-term cycle, with important effects on the animals that prey on them.

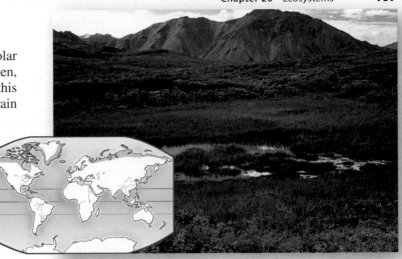

Figure 20.31   Tundra.

Chaparral

Chaparral consists of evergreen, often spiny shrubs and low trees that form communities in regions with what is called a "Mediterranean," dry summer climate. These regions include California, central Chile, the Cape region of South Africa, southwestern Australia, and the Mediterranean area itself (figure 20.32). Many plant species found in chaparral can germinate only when they have been exposed to the hot temperatures generated during a fire. The chaparral of California and adjacent regions is historically derived from deciduous forests.

Figure 20.32   Chaparral.

Polar Ice Caps

Polar ice caps lie over the Arctic Ocean in the north and Antarctica in the south (figure 20.33). The poles receive almost no precipitation, so although ice is abundant, freshwater is scarce. The sun barely rises in the winter months. Life in Antarctica is largely limited to the coasts. Because the Antarctic ice cap lies over a landmass, it is not warmed by the latent heat of circulating ocean water and becomes very cold, so only prokaryotes, algae, and some small insects inhabit the vast Antarctic interior.

Figure 20.33   Polar ice.

Tropical Monsoon Forest

Tropical upland forests (figure 20.34) occur in the tropics and semitropics at slightly higher latitudes than rain forests or where local climates are drier. Most trees in these forests are deciduous, losing many of their leaves during the dry season. This loss of leaves allows sunlight to penetrate to the understory and ground levels of the forest, where a dense layer of shrubs and small trees grow rapidly. Rainfall is typically very seasonal, measuring several inches daily in the monsoon season and approaching drought conditions in the dry season, particularly in locations far from oceans, such as in central India.

Concept Check

1. How can bacteria living near deep sea vents be auto-trophs, when no light penetrates so deep?
2. Why are there 7 primary biomes, and not 12?
3. In what biome do you live?

Figure 20.34   Tropical monsoon forest.

Does Clear-Cutting Forests Cause Permanent Damage?

The lumber industry practice called "clear-cutting" has been common in many states. Loggers find it more efficient to simply remove all trees from a watershed, and sort the logs out later, than to selectively cut only the most desirable mature trees. While the open cuts seem a desolation to the casual observer, the loggers claim that new forests can become established more readily in the open cut as sunlight now more easily reaches seedlings at ground level. Ecologists counter that clear-cutting fundamentally changes the forest in ways which cannot be easily reversed.

Who is right? The most direct way to find out is to try it, clear-cut an area and watch it very carefully. Just this sort of massive field test was carried out in a now-classic experiment at the Hubbard Brook Experimental Forest in New Hampshire. Hubbard Brook is the central stream of a large watershed that drains a region of temperate deciduous forest in northern New Hampshire. The research team, led by then-Dartmouth College professors Herbert Bormann and Gene Likens, first gathered a great deal of information about the forest watershed. Starting in 1963, they censused the trees, measured the flow of water through the watershed, and carefully documented the levels of minerals and other nutrients in the water leaving the eco-system via Hubbard Brook. To keep track, they constructed concrete dams across each of the six streams that drain the forest and moni-tored the runoff, chemically analyzing samples. The undisturbed forest proved very efficient at retaining nitrogen and other nutrients. The small amounts of nutrients that entered the ecosystem in rain and snow were approximately equal to the amounts of nutrients that ran out of the valleys into Hubbard Brook.

Now came the test. In the winter of 1965 the investigators felled all the trees and shrubs in 48 acres drained by one stream (as shown in the photo), and examined the water running off. The immediate effect was dramatic: the amount of water running out of the valley increased

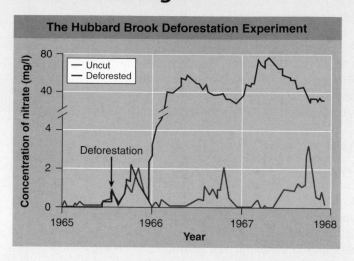

The Hubbard Brook Deforestation Experiment

Concentration of nitrate (mg/l) — *Year*

Legend: Uncut / Deforested

Deforestation

by 40%. Water that otherwise would have been taken up by vegetation and released into the atmosphere through evaporation was now simply running off. It was clear that the forest was not retaining water as well, but what about the soil nutrients, the key to future forest fertility?

The red line in the graph above shows nitrogen minerals leaving the ecosystem in the runoff water of the stream draining the clear-cut area; the blue line shows the nitrogen runoff in a neighboring stream draining an adjacent uncut portion of the forest.

Analysis

1. **Applying Concepts**
   Scale. What is the significance of the break in the vertical axis between 4 and 40?
2. **Interpreting Data**
   a. What is the approximate concentration of nitrogen in the runoff of the uncut valley before cutting? of the cut valley before cutting?
   b. What is the approximate concentration of nitrogen in the run-off of the uncut valley one year after cutting? of the clear-cut valley one year after cutting?
3. **Making Inferences**
   a. Is there any yearly pattern to the nitrogen runoff in the uncut forest? Can you explain it?
   b. How does the loss of nitrogen from the ecosystem in the clear-cut forest compare with nitrogen loss from the uncut forest?
4. **Drawing Conclusions**
   a. What is the impact of this forest's trees upon its ability to retain nitrogen?
   b. Has clear-cutting harmed this ecosystem? Explain.

Concept Summary

The Energy in Ecosystems

20.1 Energy Flows Through Ecosystems

- An ecosystem includes the community and the habitat present in a particular area. Energy constantly flows into an ecosystem from the sun, and is passed among organisms in a food chain. Energy from the sun is captured by photosynthetic producers, which are eaten by herbivores, which are in turn eaten by carnivores, as shown here from **figure 20.1.** Organisms at all trophic levels die and are consumed by detritivores and decomposers. A food chain is organized linearly, but in nature the flow of energy is often more complex, and is called a food web (**figure 20.2**).

- An ecosystem's net primary productivity is the total amount of energy that is captured by producers. Energy is lost at every level in a food chain, such that only about 5% to 20% of available energy is passed on to the next trophic level (**figures 20.4** and **20.5**).

20.2 Ecological Pyramids

- Because energy is lost as it passes up through the trophic levels of the food chain, there tends to be more individuals at the lower trophic levels. Similarly, the amount of biomass is also less at the higher trophic levels, as is the amount of energy. Ecological pyramids illustrate this distribution of number of individuals, biomass, and energy (**figure 20.6**).

Materials Cycle Within Ecosystems

20.3 The Water Cycle

- Physical components of the ecosystem cycle through the ecosystem, being used then recycled and reused. This cycling of materials often involves living organisms and is referred to as a biogeochemical cycle.

- Water availability can limit the number of organisms in an ecosystem. There are two water cycles: the environmental cycle and the organismic cycle. In the environmental cycle, shown here from **figure 20.7,** water cycles from the atmosphere as precipitation, where it falls to the earth and reenters the atmosphere through evaporation. In the organismic cycle, water cycles through plants, entering through the roots and leaving as water vapor by transpiration. Groundwater held in underground aquifers cycles more slowly through the water cycle.

20.4 The Carbon Cycle

- Carbon cycles from the atmosphere through plants, via carbon fixation of CO_2 in photosynthesis. Carbon then returns to the atmosphere as CO_2 via cellular respiration, but some carbon is also stored in the tissues of the organisms. Eventually, this carbon reenters the atmosphere by the burning of fossil fuels and by diffusion after erosion (**figure 20.9**).

20.5 The Nitrogen and Phosphorus Cycles

- Nitrogen gas in the atmosphere cannot be readily used by organisms and needs to be fixed by certain types of bacteria into ammonia and nitrate that is then used by plants. Animals eat plants that have taken up the fixed nitrogen. Nitrogen reenters the ecosystem through decomposition and animal excretions (**figure 20.10**).

- Phosphorus also cycles through the ecosystem and may limit growth when not available (**figure 20.11**) or it can cause problems in aquatic ecosystems when found in excess amounts. Phosphorus is not a gas and so does not cycle through the atmosphere. Instead it cycles from soil and rock through plants and back to the soil through decomposition.

How Weather Shapes Ecosystems

20.6 The Sun and Atmospheric Circulation

- The heating power of the sun and air currents affect evaporation, causing certain parts of the globe, such as the tropics, to have larger amounts of precipitation (**figures 20.12** and **20.13**).

20.7 Latitude and Elevation

- Temperature and precipitation are similarly affected by elevation and latitude. Changes in ecosystems from the equator to the poles are similarly reflected in the changes in ecosystems from sea level to mountaintops, as shown here from **figure 20.14.** Changes in temperature cause the rain shadow effect, where precipitation is deposited on the windward side of mountains, and deserts form on the leeward side (**figure 20.15**).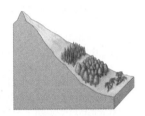

20.8 Patterns of Circulation in the Ocean

- The earth's oceans circulate in patterns that distribute warmer and cooler waters to different areas of the world. These ocean patterns affect climates across the globe (**figures 20.16** and **20.17**).

Major Kinds of Ecosystems

20.9 Ocean Ecosystems

- There are three primary ocean ecosystems: shallow waters, open-sea surfaces, and deep-sea bottoms (**figure 20.18**). Each is affected by light and temperature.

20.10 Freshwater Ecosystems

- Freshwater ecosystems are closely tied to the terrestrial environments that surround them. Freshwater ecosystems are affected by light, temperature, and nutrients. The penetration of light divides a lake into three zones with varying amounts of light (**figure 20.22**). Temperature variations in a lake, called thermal stratification, also bring about an overturning of the lake that distributes nutrients (**figure 20.23**).

20.11 Land Ecosystems

- Biomes are terrestrial communities found throughout the world. Each biome contains its own characteristic group of organisms based on temperature and rainfall patterns (**figure 20.24**).

Self-Test

1. Energy from the sun is captured and converted into chemical energy by
   a. herbivores.
   b. carnivores.
   c. producers.
   d. detritivores.

2. As energy is transferred from one trophic level to the next, substantial amounts of energy are lost to/as
   a. undigestible biomass.
   b. heat.
   c. metabolism.
   d. All answers are correct.

3. The number of carnivores found at the top of an ecological pyramid is limited by the
   a. number of organisms below the top carnivores.
   b. number of trophic levels below the top carnivores.
   c. amount of biomass below the top carnivores.
   d. amount of energy transferred to the top carnivores.

4. Hydrologists, scientists who study the movements and cycles of water, refer to the return of water from the ground to the air as evapotranspiration. The first part of the word refers to evaporation. The second part of the word refers to transpiration, which is evaporation of water
   a. from plants.
   b. through animal perspiration.
   c. off the ground shaded by plants.
   d. from the surface of rivers.

5. The carbon cycle includes a store of carbon as fossil fuels that is released through
   a. respiration.
   b. combustion.
   c. erosion.
   d. all of the above.

6. The element phosphorus is needed in organisms to build
   a. proteins.
   b. carbohydrates.
   c. ATP.
   d. steroids.

7. A rain shadow results in
   a. extremely wet conditions due to the lack of wind over a mountain range.
   b. dry air moving toward the poles that cools and sinks in regions 15 to 30 degrees north/south latitude.
   c. global polar regions that rarely receive moisture from the warmer, tropical regions, and are therefore dryer.
   d. desert conditions on the downwind side of a mountain due to increased moisture-holding capacity of the winds as the air heats up.

8. As one travels from northern Canada south to the United States, the timberline increases in elevation. This is because as latitude
   a. increases, temperature increases.
   b. decreases, temperature increases.
   c. increases, humidity decreases.
   d. decreases, humidity increases.

9. In freshwater lakes during the summer, layers of sudden temperature change called _____ form.
   a. eutrophy
   b. thermal restratification
   c. oligotrophy
   d. thermocline

10. Which of the following biomes is *not* found south of the equator?
    a. polar ice cap
    b. savanna
    c. tundra
    d. tropical monsoon forest

Visual Understanding

1. **Figure 20.5** A. One thousand calories of energy is harvested by a certain amount of algae. Explain the probable efficiency of having humans eat, respectively, (1) the algae itself, (2) the small heterotrophs, (3) the smelt, or (4) the trout. B. How many more people could be fed if we all ate algae? C. Why can't we survive just by eating algae?

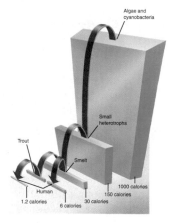

2. **Figure 20.7** Imagine that you are a molecule of water. Create a journey for yourself, starting with falling as part of a drop of rain onto the earth, and ending in a cloud ready to fall once again. Take a trip through a plant along the way.

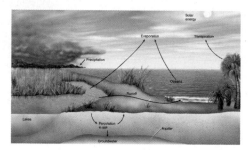

Challenge Questions

1. Given the amount of sunlight that hits the plants on our planet, and the ability of plants for rapid growth and reproduction, how come we aren't all hip deep in dead plants?

2. Plants need both carbon (C) and nitrogen (N) in large amounts. Compare and contrast how plants obtain these two important elements.

3. Why do increasing latitude and increasing elevation affect, in the same way, which plant species grow?

4. Pick two very different biomes and research them. Compare and contrast the sunshine, rainfall, major temperature features, and something about the plants and animals.

Chapter 21

Behavior and the Environment

Some Behavior Is Genetically Determined

Figure 21.1 Two ways to look at a behavior.

This male songbird can be said to be singing because his levels of testosterone are elevated, triggering innate "song" programs in his brain. Viewed in a different light, he is singing to defend his territory and attract a mate, behaviors that have evolved to increase his reproductive fitness.

EVOLUTION

Do Humans Have Instincts? While researchers agree that human reflexes like the knee jerk are instinctive because they are universal and free of environmental influences, what about complex behaviors like sex, laughing, and sleeping? Do humans have instinctive behaviors? How this question is answered depends on who you ask. Biologists tend to say "Yes," arguing that there are strong evolutionary advantages for our species to have adopted these instinctive reflexive behaviors. Sociologists reach the opposite conclusion, arguing that humans have no behavioral instincts. They define instincts as "complex behaviors present in all humans that are innate and cannot be overridden." Sex, laughing, sleeping, and other human behaviors can all be overridden, they say, and so cannot be considered instincts. This semantic argument is present in many introductory sociology textbooks but in few biology texts, and is still hotly debated.

21.1 Approaches to the Study of Behavior

CONCEPT PREVIEW: Animal behavior is the way an animal responds to stimuli in its environment. Some biologists study the physiological mechanisms producing the behavior, others the evolutionary forces responsible for its development.

Biologists define behavior as the way an animal responds to stimuli in its environment. A bacterial cell "behaves" by moving toward higher concentrations of sugar. In animals, behaviors are far more complex than this simple bacterial response. Using eyes, ears, and a variety of other sense organs, animals are able to perceive environmental stimuli, process the information, and dictate appropriate body responses, which can be both complex and subtle.

Explaining Behavior

When we observe animal behavior, we can examine it in two different ways. First, we might ask *how* it all works. How do the animal's senses, nerve networks, and internal state act together physiologically to produce the behavior? Like a mechanic studying the behavior of a car, we are asking how the machine works. A psychologist would say we would be asking a question of proximate causation. To analyze the proximate cause of behavior, we might measure hormone levels, or record the impulse activity of particular neurons in the brain. The field of psychology often focuses on proximate causes.

We might also ask *why* it all works the way it does. Why did the behavior evolve in this way? What is the adaptive value of this particular response to the environment? This is a question of ultimate causation. To study the ultimate cause of a behavior, we would attempt to find how it influenced the animal's survival or reproductive success. The field of animal behavior typically focuses on ultimate causes.

Any behavior can be looked at either way. For example, a male songbird may sing during the breeding season (figure 21.1). Why? One explanation is that longer spring days have induced in his body elevated levels of the steroid sex hormone testosterone, which at these high levels bind to hormone receptors in the songbird's brain and trigger the production of song. Elevated testosterone would be the proximate cause of the male songbird's song.

Another explanation is that the male is exhibiting a pattern of behavior produced by natural selection to better adapt it to its environment. Seen in this light, the male songbird sings to defend a territory from other males, and to attract a female to reproduce. These reproductive motives are the ultimate, or evolutionary, explanation of the male songbird's behavior.

A Controversial Field of Biology

The study of behavior has had a long history of controversy. One source of controversy has been the question of whether an animal's behavior is determined more by an individual's genes, or by its learning and experience. In other words, is behavior the result of nature (instinct) or nurture (learning)? In the past, this question has been considered an "either-or" proposition, but we now know that instinct and learning both play significant roles, often interacting in complex ways to produce the final behavior.

21.2 Instinctive Behavioral Patterns

CONCEPT PREVIEW: The ethological approach to studying animal behavior has emphasized innate, instinctive behaviors that are the result of preset pathways in the nervous system.

Early research in the field of animal behavior focused on behavioral patterns that appeared to be instinctive (or "innate"). Because behavior in animals is often stereotyped (that is, appearing in the same way in different individuals of a species), behavioral scientists argued that the behaviors must be based on preset paths in the nervous system. In their view, these neural paths are structured from genetic blueprints and cause animals to show essentially the same behavior from the first time it is produced throughout their lives.

This study of the instinctive nature of animal behavior was typically carried out in the field rather than on laboratory animals. The study of animal behavior in natural conditions is called **ethology.** The three scientists most responsible for founding the field of ethology were Karl von Frisch, Konrad Lorenz, and Niko Tinbergen. These scientists were awarded the Nobel Prize in Physiology or Medicine in 1973 for their path-making contributions.

An Example of Innate Behavior

The study by Konrad Lorenz of the egg retrieval behavior of geese provides a clear example of what ethologists mean by innate behavior. The drawings in figure 21.2*a* illustrate how when a goose is incubating its eggs in a nest and it notices that an egg has been knocked out of the nest, it will extend its neck toward the egg, get up, and roll the egg back into the nest with a side-to-side motion of its neck while the egg is tucked beneath its bill. Even if the egg is removed during retrieval, the goose completes the behavior, as if driven by a program released by the initial sight of the egg outside the nest.

According to ethologists like Lorenz, egg retrieval behavior is triggered by a *sign stimulus* (also called a key stimulus), which in this case is the appearance of an egg outside the nest. The way in which nerves are connected in the goose's brain, the *innate releasing mechanism,* responds to the sign stimulus by providing the neural instructions for the motor program, or *fixed action pattern,* which causes the goose to carry out the intricate egg retrieval behavior.

More generally, the sign stimulus is a "signal" in the environment that triggers a behavior. The innate releasing mechanism is the hard-wired element of the brain, and the fixed action pattern is the stereotyped act.

Studying fixed action patterns in birds and other animals, ethologists have discovered that in some situations, a wide variety of objects will trigger a fixed action pattern. For example, geese will attempt to roll baseballs, and even beer cans, back into their nests!

A clear example of the general nature of some sign stimuli is seen in the mating behavior of male stickleback fish studied by Niko Tinbergen. During the breeding season, males develop bright red coloration on their undersides. The males are very territorial, reacting aggressively to the approach of other males. They first perform an aggressive display (shown on the right in figure 21.2*b*), and if the invading male is not deterred, they attack it. However, when Niko Tinbergen observed a male stickleback in a laboratory aquarium display aggressively when a red firetruck passed by the window, he realized that the red color was the sign stimulus. In experiments using models shown on the left in figure 21.2*b,* he was able to produce the aggressive display in males by challenging them with the many unfishlike models, as long as the models had a red strip.

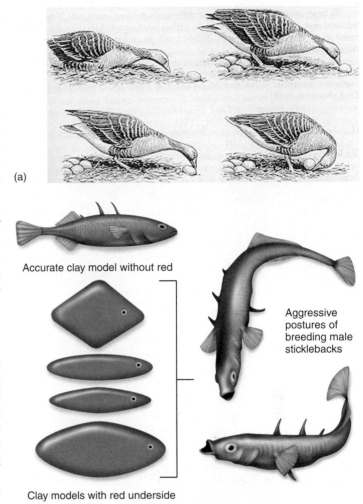

(a)

Accurate clay model without red

Aggressive postures of breeding male sticklebacks

Clay models with red underside

(b)

Figure 21.2 Sign stimulus and fixed action pattern.

(a) The series of movements used by a goose to retrieve an egg is a fixed action pattern. Once it detects the sign stimulus (in this case, an egg outside the nest), the goose goes through the entire set of movements: it will extend its neck toward the egg, get up, and roll the egg back into the nest with a side-to-side motion of its neck while the egg is tucked underneath its bill. (b) In stickleback fish, a red color acts as a sign stimulus to trigger a fixed action pattern in males: aggressive threat displays or postures. When the models above are presented to a male stickleback, he will display more often to the last four models than to the first model, which looks like a male stickleback but lacks the red belly characteristic of males.

IMPLICATION Like the goose egg retrieval described above, human laughter is also considered to be a reflex action, the same fixed action pattern in every human. But is it really the same? Congenitally deaf people laugh out loud, but do they produce the same sounds as those who hear normally? How would you go about determining if they do?

Figure 21.3   Genetics of lovebird behavior.

Some species of lovebirds carry nest material, such as paper strips, under their flank feathers. When mated with species that carry material in their beaks, the hybrids show intermediate behavior, attempting to carry the material in both ways.

(a)

(b)

Figure 21.4   A gene alters maternal care.

In mice, normal mothers (a) take very good care of their offspring, retrieving them if they move away and crouching over them. Mothers with the mutant *fosB* allele (b) perform neither of these behaviors, leaving their pups exposed.

21.3 Genetic Effects on Behavior

CONCEPT PREVIEW: The conclusion that genes play a key role in many behaviors is supported by a broad range of studies in many animals, including humans.

Although most animal behaviors are not "hard-wired" instincts, animal behaviorists have clearly demonstrated that many of them are strongly influenced by genes passed from parent to offspring. It appears "nature" plays a key role.

If genes determine behavior, then it should be possible to study their inheritance, much as Mendel studied the inheritance of flower color in garden peas. This sort of investigation is called behavioral genetics.

Studies of Genetic Hybrids

Behavioral genetic studies have revealed many examples of behaviors that seem to be inherited in a Mendelian manner. William Dilger of Cornell University examined two species of lovebirds that differ in the way they carry twigs, paper, and other materials used to build a nest. One species, Fisher's lovebird, holds nest materials in its beak, while another, the peachfaced lovebird, carries material tucked beneath its flank (tail) feathers (figure 21.3). When Dilger crossed the two species to produce hybrids, he found that the hybrids carry nest material in a way that is intermediate between that of the parents: They repeatedly shift material between the bill and the flank feathers.

Studies of Twins

The influence of genes on behavior can be clearly seen in humans. Identical twins are genetically identical. In some instances identical twins have been separated at birth, and raised apart. A recent study of 50 such sets of twins revealed many similarities in personality, temperament, and even leisure-time activities, even though the twins were raised in different circumstances. These results show that genes can play a key role in determining human behavior.

A Detailed Look at How One Gene Affects a Behavior

One well-studied gene mutation in mice provides a clear look at how a particular gene influences a behavior. In 1996 behavioral geneticists discovered a gene, *fosB*, that determines whether or not female mice will nurture their young. Females with both *fosB* alleles knocked out (experimentally removed) will initially investigate their newborn babies, but then ignore them, in stark contrast to the maternal care provided by normal females (figure 21.4).

How does *fosB* work? When mothers of new babies initially inspect them, *fosB* alleles are activated in an area of the brain called the hypothalamus, producing a protein signal which activates neural circuitry within the hypothalamus directing the female to react maternally toward her offspring. In a general way, the information gained from inspecting the newborn babies can be viewed as acting like a sign stimulus, the *fosB* gene as an innate releasing mechanism, and the maternal behavior as the resulting action pattern.

Concept Check

1. Suggest proximate and ultimate causes for a songbird singing.
2. List the three components of an instinctive behavior.
3. How does the *fosB* gene influence the proximate causes of maternal behavior in mice?

Behavior Can Also Be Influenced by Learning

21.4 How Animals Learn

CONCEPT PREVIEW: Nonassociative learning is simple learning in which there is no association between stimulus and response. In contrast, associative learning involves the formation of an association between two stimuli or between a stimulus and a response.

Many of the behavioral patterns displayed by animals are not the result solely of instinct. In many cases, animals alter their behavior as a result of previous experiences, a process termed **learning.** The simplest type of learning, **nonassociative learning,** does not require an animal to form an association between two stimuli, or between a stimulus and a response. One form of nonassociative learning is *habituation,* learning not to respond to a repeated stimulus. As an everyday example, are you still conscious of the chair you are sitting on? Being able to ignore unimportant stimuli is critical when facing a barrage of stimuli in a complex environment.

Classical Conditioning

A change in behavior that involves an association between two stimuli, or between a stimulus and a response, is called **associative learning.** The behavior is modified, or *conditioned,* through the association. In **classical conditioning,** the paired presentation of two kinds of stimuli causes the animal to form an association between the stimuli. When the Russian psychologist Ivan Pavlov presented meat powder, an *unconditioned stimulus,* to a dog, the dog responded by salivating. If an unrelated stimulus, such as the ringing of a bell, was present at the same time as the meat powder, then over repeated trials the dog would come to salivate in response to the sound of the bell alone. The dog had learned to associate the unrelated sound stimulus with the meat powder stimulus. Its response to the sound stimulus had become conditioned, and the bell was now a *conditioned stimulus.*

Operant Conditioning

In **operant conditioning,** an animal learns to associate its behavioral response with a reward or punishment. Psychologist B. F. Skinner studied operant conditioning in rats by placing them in an experimental cage nicknamed a "Skinner box." As the rat explored the box, it would occasionally press a lever by accident, causing a pellet of food to appear. At first, the rat would ignore the lever, eat the food pellet, and continue to move about. Soon, however, it would learn to associate pressing the lever (the behavioral response) with food (the reward). When a conditioned rat was hungry, it would spend all its time pressing the lever. This sort of trial-and-error learning is common among vertebrates.

Imprinting

As an animal matures, it may form preferences or social attachments to other individuals that will profoundly influence behavior later in life, a process called **imprinting.** In *filial imprinting,* social attachments form between parents and offspring. For example, young birds of some species begin to follow their mother within a few hours after hatching, forming a strong bond between mother and young. Ethologist Konrad Lorenz raised geese from eggs, and when he offered himself as a model for imprinting, the goslings treated him as if he were their parent, following him dutifully (figure 21.5).

Figure 21.5 An unlikely parent.

The eager goslings follow ethologist Konrad Lorenz as if he were their mother. He is the first object they saw when they hatched, and they used him as a model for imprinting.

IMPLICATION It appears that humans also exhibit imprinting. One of the most widely studied examples of human imprinting is reverse sexual imprinting, also known as the Westermarck effect. When people live in close proximity during the first few years of their lives, they are desensitized to later sexual attraction. In Israeli kibbutz collective farms, children of both sexes were reared together. Of 3,000 marriages of couples both raised during their first six years in kibbutz, none of the couples had been reared in the same kibbutz. Why do you think evolution might have favored the development of this form of imprinting in humans?

IN THE NEWS

Counting Bee. When people say that animals can think, they are usually referring to chimps, the family dog, or perhaps birds like the raven on the facing page—all animals with backbones. But what about all the other kinds of animals? Do spiders think? What about bees? In a remarkable study reported in 2008, Australian researchers discovered that honeybees (*Apis mellifera*) can count up to four! The bees were trained to fly down a tunnel in search of food placed beside one of five identical landmarks positioned at intervals. When trained bees flew into a tunnel that had no food, they spent most of the time searching previously rewarding landmarks—unless it was landmark number five. Moving the landmarks closer to or further away from each other did not fool the bees, showing they were not relying on distance, but were counting the number of landmarks before the food. Changing landmarks from stripes to spots had no effect either, suggesting that bees can use numbers in an abstract way.

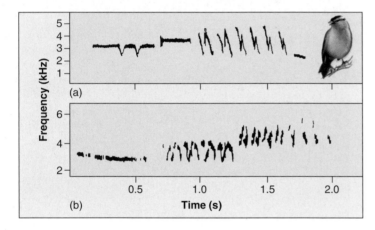

Figure 21.6 **Song development in birds involves both instinct and learning.**

The sonograms of songs produced by male white-crowned sparrows that had been exposed to their own species' song during development (a) are different from those of male sparrows that heard no song during rearing (b). This difference indicates that the genetic program itself is insufficient to produce a normal song.

21.5 Instinct and Learning Interact to Determine Behavior

CONCEPT PREVIEW: Behavior is both instinct (influenced by genes) and learned through experience. Genes are thought to limit the extent to which behavior can be modified and the types of associations that can be made.

Some animals have innate predispositions toward forming certain associations. Certain pairs of stimuli can be linked by operant conditioning, others not. For example, pigeons can learn to associate food with colors but not with sounds; on the other hand, they can associate danger with sounds, but not with colors. This sort of learning preparedness demonstrates that what an animal can learn is biologically influenced—that is, learning is possible only within the boundaries set by instinct.

The innate programs that make up an animal's instincts have evolved because each of them reinforces an adaptive response. The seed a pigeon eats may have a distinctive color that the pigeon can see, but it makes no sound the pigeon can hear. The approach of a predator a pigeon fears may generate noise, but involves no distinctive color.

The way in which white-crowned sparrows first acquire their courtship songs provides an excellent example of the interaction between instinct and learning in the development of behavior. Courtship songs are sung by mature birds and are species specific. By rearing male birds in soundproof incubators provided with speakers and microphones, animal behaviorist Peter Marler could control what a bird heard as it matured, and then record the song it produced as an adult, a recording called a sonogram. When compared to a normal sonogram, in figure 21.6*a*, he found that white-crowned sparrows that heard no song at all during development had a poorly developed song as adults, shown in the sonogram in figure 21.6*b*. The same thing happened if they heard only the song of a different species, the song sparrow. But birds that heard the song of their own species sang a fully developed white-crowned sparrow song as adults. This was true even if the young birds also heard the song sparrow song as well.

Marler's results suggest that these birds have a genetic template, or instinctive program, that guides them to learn the appropriate song. During a critical period in development, the template will accept the correct song as a model. Thus, song acquisition depends on learning, but only the song of the correct species can be learned.

Although the song template is genetically determined, Marler found that learning also plays a prominent role in song development. If a young white-crowned sparrow becomes deaf *after* it hears its species' song during the critical period, it will sing a poorly developed song as an adult. The bird must "practice" listening to himself, matching what he hears to the template he has accepted.

The males of some bird species have no opportunity to hear the song of their own species. In such cases, it appears that the males instinctively "know" their own species song. For example, cuckoos are brood parasites; females lay their eggs in the nest of another species of bird. The young are hatched and reared by the foster parents. When the cuckoos become adults, they sing the song of their own species rather than that of the foster parents. Natural selection has provided the birds with a genetically programmed song guided by instinct.

Parasitism, as described on page 374, is a symbiotic relationship where one species benefits at the expense of the other. Given this definition, the cuckoo is a parasite. Its offspring are raised by the nest "host," taking food and parenting time from the legitimate offspring.

21.6 Animal Cognition

CONCEPT PREVIEW: Research on the cognitive abilities of animals is in its infancy, but evidence suggests that some animals can reason.

In recent years, serious attention has been given by researchers to the topic of animal awareness. The central question is whether animals other than humans show cognitive behavior—that is, do they process information and respond in a manner that suggests thinking?

Evidence of Conscious Planning

What kinds of behavior would demonstrate cognition? Some birds remove the foil caps from nonhomogenized milk bottles to get at the cream beneath. Japanese macaques (a kind of monkey) learn to float grain on water to separate it from sand, and teach other macaques to do it. A chimpanzee pulls the leaves off of a tree branch and uses the stick to probe the entrance to a termite nest to gather termites, suggesting that the ape is consciously planning ahead, with full knowledge of what it intends to do. A sea otter will use a rock as an "anvil," against which it bashes a clam to break it open, often keeping a favorite rock for a long time, as though it had a clear idea of its future use (figure 21.7).

Problem Solving

Some instances of problem solving by animals are hard to explain in any other way than as a result of some sort of cognitive process—what, if we were doing it, we would call reasoning. For example, in a series of classic experiments conducted in the 1920s, a chimpanzee was left in a room with a banana hanging from the ceiling out of reach. Also in the room were several boxes, each lying on the floor. After unsuccessful attempts to jump up and grab the banana, the chimpanzee suddenly looked at the boxes and immediately proceeded to move them underneath the banana, placing one on top of another, and climbed up the boxes to claim its prize. Would you have solved the problem as well?

It is not surprising to find obvious intelligence in animals so closely related to us as chimpanzees. Perhaps more surprising, however, are recent studies finding that other animals also show evidence of cognition. Ravens have always been considered among the most intelligent of birds. Researchers placed a piece of meat on the end of a string and hung it from a branch in an outdoor aviary containing six ravens. The birds like to eat meat, but had never seen string before and were unable to get at the meat. After several hours, during which time the birds periodically looked at the meat but did nothing else, one bird flew to the branch, reached down, grabbed the string with its beak, pulled it up, and placed it under his foot. He then reached down and pulled up another length of the string, repeating this action over and over, each time bringing the meat closer (figure 21.8). Eventually the meat was within reach, and was grasped and eaten by the bird. The raven, presented with a completely novel problem, had devised a solution. Eventually, three of the other five ravens also figured out how to get the meat. This result can leave little doubt that ravens have advanced cognitive abilities.

Concept Check

1. Distinguish between classical and operant conditioning.
2. Compare the singing of an adult white-crowned sparrow that heard only the song of its own species when young with the singing of an adult that also heard the song of a different sparrow species when it was young.

Figure 21.7
Conscious planning.

This sea otter is using a rock as an "anvil" to break open a clam.

Figure 21.8   Problem solving by a raven.

Confronted with a problem it had never previously encountered, the raven figures out how to get the meat at the end of the string by repeatedly pulling up a bit of string and stepping on it.

Evolutionary Forces Shape Behavior

Figure 21.9 **The adaptive value of eggshell removal.**

Niko Tinbergen painted chicken eggs to resemble the mottled brown camouflage of gull eggs. Mottled eggs look like the rocky ground around a gull nest. The eggs were used to test the hypothesis that camouflaged eggs are more difficult for a predator to find and thus increase the young's chance of survival. He then placed broken eggshells near the camouflaged ones to test the hypothesis that the white interior of broken eggshells attracts predators.

IMPLICATION When you go to the grocery store to buy eggs, do you look at them before selecting a box to purchase? If you did, you would discover that not all commercial chicken eggs look alike. While many are pure white, others are pigmented tan (some even seem blue-grey, although these are usually culled by the producer and not sold in markets). How might you test whether or not egg color influences which eggs a customer selects to purchase?

21.7 Behavioral Ecology

CONCEPT PREVIEW: Behavioral ecology is the study of how natural selection shapes behavior.

The investigation of animal behavior can be conveniently divided into three sorts of questions: (1) *The study of its development.* Lorenz's study of imprinting in geese was a study of this sort. (2) *The study of its physiological basis.* Analysis of the impact of the *fosB* gene on maternal behavior in mice was a study of this sort. (3) *The study of its function* (that is, its evolutionary significance). This third sort of question is addressed by biologists working in the field of **behavioral ecology.** Behavioral ecology is the study of how natural selection shapes behavior.

Behavioral ecology examines the survival value of behavior. How does an animal's behavior allow it to stay alive and reproduce, or keep its offspring alive to reproduce? Research in behavioral ecology thus focuses on a behavior's adaptive significance—that is, on the contribution a behavior makes to an animal's reproductive success, or fitness.

> As discussed on page 245, natural selection acts on variation in a population such that certain characteristics become more prevalent in future populations. While it is easier to follow the evolution of physical traits, the process is the same with behavioral traits.

It is important to remember that all genetic differences in behavior need not have survival value. Many genetic differences in natural populations are the result of random mutations that become common by accident, a process called genetic drift. It is only by experiment that we can learn if a particular behavior has been favored by natural selection.

> Genetic drift was discussed as an agent of evolution, on page 261, where random events can alter allele frequencies in a population. An allele may be lost from a population, or a rarer allele may become more common, such as in the founder effect in an isolated population.

Nobel laureate Niko Tinbergen's pioneering study of seagull nesting provides an excellent example of how a behavioral ecologist investigates the potential evolutionary significance of a behavior. Tinbergen observed that after gull nestlings hatched from their eggs, the parent birds quickly remove the eggshells from the nest. Why? What possible evolutionary advantage would this behavior confer on the birds?

To investigate this, Tinbergen camouflaged chicken eggs by painting them to resemble gulls' eggs that blend in with the natural background where the gull nests were located (figure 21.9), and distributed them on the ground throughout the nesting area. He placed broken eggshells next to some of the eggs, and, as a control, he left other camouflaged eggs alone without eggshells. He then watched to see which eggs were found more easily by crows. Because the crows could use the white interior of a broken eggshell as a cue, they repeatedly ate the camouflaged eggs that were near broken eggshells, and tended to ignore solitary camouflaged eggs that sat on the ground in plain sight. Tinbergen concluded that eggshell removal behavior is adaptive, that it does confer an evolutionary advantage on birds. Removing broken eggshells from the nest reduces predation of unhatched eggs (and probably of newborn chicks) and thus increases the chance that offspring will survive.

It is not always so easy to learn how an adaptive trait confers its evolutionary advantage. Some behaviors, like eggshell removal, reduce predation. Other behaviors enhance energy intake, allowing an increased number of offspring to be supported. Still others increase resistance to disease, or in some other way increase an individual's ability to contribute offspring to the next generation.

21.8 A Cost-Benefit Analysis of Behavior

CONCEPT PREVIEW: Natural selection tends to favor the evolution of foraging and territorial behaviors that maximize energy gain, although other factors such as avoiding predators are also important.

One important way to examine the evolutionary advantage of a behavior is to ask if it provides an evolutionary benefit greater than its cost.

Foraging Behavior

For many animals, the food that they eat can be found in many sizes, and in many places. An animal must choose what food to select, and how far to go in search of it. These choices are called the animal's **foraging behavior.** Each choice involves benefits and associated costs. Thus, although a larger food may contain more energy, it may be harder to capture and less abundant. In addition, more desirable foods may be farther away than other types. Hence, an animal's foraging behavior involves a trade-off between a food's energy content and the cost of obtaining it.

The net energy (in calories) gained by feeding on each kind of food available to a foraging animal is simply the energy content of the food minus the energy costs of pursuing and handling it. At first glance, one might expect that evolution would favor foraging behaviors that are as energetically efficient as possible. This sort of reasoning has led to what is known as the **optimal foraging theory,** which predicts that animals will select food items that maximize their net energy intake per unit of foraging time (figure 21.10).

Is the optimal foraging theory correct? Many foragers do preferentially use food items that maximize the energy return per unit time. Shore crabs, for example, tend to feed primarily on intermediate-sized mussels, which provide the greatest energy return (see page 430). Larger mussels provide more energy, but also take considerably more energy to crack open. In other cases, however, the costs of foraging seem to outweigh the benefits. An animal in danger of being eaten itself is often better off to minimize the amount of time it spends foraging. Many animals alter their foraging behavior when predators are present, reflecting this trade-off between food and risk.

Territorial Behavior

Animals often move over a large area, or *home range*. In many species, the home range of several individuals overlap, but each individual uses a portion of its home range exclusively. This behavior is called **territoriality.**

Territories are defended by displays that advertise that the territories are occupied, and by overt aggression. A bird sings from its perch within its territory to prevent invasion of its territory by a neighboring bird. If the intruding bird is not deterred by the song, the territory owner may attack and attempt to drive the invader away.

Why aren't all animals territorial? The answer involves a cost-benefit analysis. Territoriality offers clear benefits, including increased food intake from nearby resources (figure 21.11), access to refuges from predators, and exclusive access to mates. The costs of territorial behavior, however, may also be significant. The singing of a bird, for example, is energetically expensive, and advertisement through song or visual display can reveal one's location to a predator. In many instances, particularly when food sources are abundant, defending easily obtained resources is simply not worth the cost.

Figure 21.10   Optimal foraging.

Optimal foraging in this golden mantled ground squirrel pays off with increased reproductive success.

Figure 21.11   The benefit of territoriality.

Sunbirds, found in Africa and ecologically similar to hummingbirds, increase nectar availability by defending flowers. A sunbird will expend 3,000 calories per hour chasing intruders away.

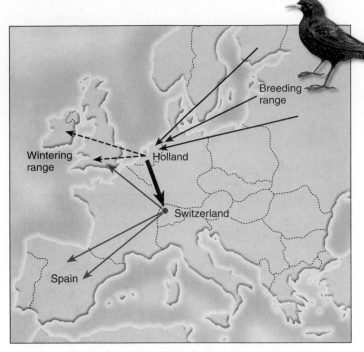

Figure 21.12 **Starlings learn how to navigate.**

The navigational abilities of inexperienced birds differ from those of adults that have made the migratory journey before. Starlings were captured in Holland, halfway along their full migratory route from Baltic breeding grounds to wintering grounds in the British Isles. These captured birds were transported to Switzerland and released. Experienced older birds compensated for the displacement and flew toward the normal wintering grounds (*blue* arrow). Inexperienced young birds kept flying in the same direction as before, on a course that took them toward Spain (*red* arrows). These observations indicate that inexperienced birds migrate with an innate compass sense, while experienced birds utilize a learned map sense to aid their navigation.

BIOLOGY & YOU

Fly Away Home. Birds that are hatched in captivity have no mentor birds to teach them their traditional migratory routes. Italian hang glider pilot Angelo d'Arrigo noted that migratory birds fly in much the same way as hang gliders: both use updrafts of hot air to gain altitude which then permits soaring flight over long distances. With this in mind, he set out to reintroduce threatened species of eagles into the wild. The chicks hatched under the wings of his glider and imprinted on him. The young birds followed him not only on the ground (like Lorenz's geese in figure 21.5), but also in the air. Taking the path of their migratory routes, he flew his eagles across the Sahara and over the Mediterranean to Sicily. Now the educated birds were able to reenter the wild. In subsequent flights, d'Arrigo flew from Siberia to Iran (5,500 km) leading a flock of Siberian cranes, and over Mount Everest with Nepalese eagles. Learning from d'Arrigo, Canadian ultralight enthusiast Bill Lishman trained orphaned Canada Geese to fly their normal migration route. You may have seen his story shown in the fact-based movie *Fly Away Home*.

21.9 Migratory Behavior

CONCEPT PREVIEW: Many animals migrate in predictable ways, navigating by looking at the sun and stars, and in some cases by detecting magnetic fields. In many instances, young individuals learn the route by following experienced ones.

Many animals breed in one part of the world, and spend the rest of the year in another. Long-range two-way annual movements like this are called **migrations.** Migratory behavior is particularly common in birds. Ducks and geese migrate southward along flyways from northern Canada across the United States each fall, overwinter, and then return northward each spring to nest. Warblers and many other songbirds winter in the tropics and breed in the United States and Canada in spring and summer, when insects are plentiful.

Biologists have studied migration with great interest. In attempting to understand how animals are able to navigate accurately over such long distances, it is important to understand the difference between *compass sense* (an innate ability to move in a particular direction, called following a bearing) and *map sense* (a learned ability to adjust a bearing depending on the animal's location). Experiments on starlings shown in figure 21.12 indicate that inexperienced birds migrate using a compass sense, and older birds that have migrated previously also employ a map sense to help them navigate—in essence, they learn the route. Migrating birds were captured in Holland, the halfway point of their migration, and were taken to Switzerland where they were released. Inexperienced birds (the red arrows) kept flying in their original direction, while experienced birds (the blue arrow) were able to adjust course and reached their normal wintering grounds.

The Compass Sense

We now have a good understanding of how the compass sense is achieved in birds. Many migrating birds have the ability to detect the earth's magnetic field and to orient themselves with respect to it. In a closed indoor cage, they will attempt to move in the correct geographical direction, even though there are no visible external clues. However, the placement of a powerful magnet near the cage can alter the direction in which the birds attempt to move.

The first migration of young birds appears to be innately guided by the earth's magnetic field. Inexperienced birds also use the sun and particularly the stars to orient themselves (migrating birds fly mainly at night).

The Map Sense

Much less is known about how migrating birds and other animals acquire their map sense. During their first migration, young birds move with a flock of experienced older birds that know the route, and during the course of the journey they appear to learn to recognize certain cues, such as the position of mountains and coastline.

Animals that migrate through featureless terrain present more of a puzzle. Consider the green sea turtle. Every year great numbers of these large 400-pound turtles migrate with incredible precision from Brazil halfway across the Atlantic to Ascension Island, over 1,400 miles of open ocean, where females lay their eggs. Plowing head down through the waves, how do they find this tiny rocky island, over the horizon, more than a thousand miles away? No one knows for sure, although recent studies suggest that the direction of wave action provides an important navigational clue.

The Great Pigeon Race Disaster of 1997 Suggests an Answer to an Enduring Mystery

Homing pigeons (*Columba livia*) are small gray birds with a remarkable ability to find their way home from distant, unfamiliar places. A homing pigeon released hundreds of miles from where it lives, at a location it has never seen, will unerringly set out on a beeline homeward, flying through darkness and storms at speeds that often average 50 miles per hour, without stopping, until it arrives at its home loft hours or even days later. How can an unassuming little bird carry out a feat a normal person could only match with a satellite GPS navigation system?

Navigating accurately over unfamiliar terrain as homing pigeons do demands of the birds both a sense of direction (compass sense) and a sense of location (map sense).

The compass sense of migrating birds is pretty well understood by biologists. Their usual daytime compass is the sun, but they can also navigate on cloudy days because they possess as well a magnetic sense that they can use to locate "north."

However, how does a homing pigeon determine its location, so it can select the correct homeward bearing? This has until now remained a mystery. They don't do it by memorizing landmarks—birds fitted with frosted contact lenses find their way home just fine. Nor do they sense local variations in the earth's magnetic field, or smell particular odors.

How then do homing pigeons figure out where they are? A researcher at the U.S. Geological Survey, Jonathan Hagstrum, has come up with a novel suggestion. It involves, of all things, pigeon races.

In Europe, and to a lesser extent in the United States, pigeon racing has become an international sport for which birds are carefully bred and trained. Birds from many lofts are taken to a common distant location, released together, and their return speeds timed. The most rapid return wins. Some birds are a little tardy, but typically over 90% of the birds return within a few days, and eventually almost all do.

On Sunday, June 29, 1997, a great race was held to celebrate the centenary of the Royal Pigeon Racing Association. More than 60,000 homing pigeons were released at 6:30 A.M. from a field in Nantes (southern France), flying to lofts all over southern England 400 to 500 miles away. By 11:00 A.M., the majority of the racing birds had made it out of France and were over the English Channel. They should have arrived at their lofts by early afternoon. They didn't.

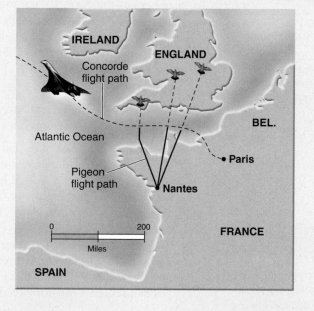

A few thousand of the birds straggled in over the next few days. Most were never seen again. In pigeon racing terms, the loss of so many birds was practically unheard of, a disaster. Any one bird could get lost, but tens of thousands?

Hagstrum, in studying this event, noticed an odd fact. At the very same time the racing pigeons were crossing the Channel, 11:00 A.M., the Concorde supersonic transport (SST) airliner was flying along the Channel on its morning flight from Paris to New York (a flight discontinued in 2004). In flight, the SST generates a shock wave that pounds down toward the earth, a carpet of sound almost a hundred miles wide. The racing pigeons flying below the Concorde could not have escaped the intense wave of sound. The birds that did eventually arrive at their lofts were lucky enough to be very slow racers—they were still south of the Channel when the SST passed over, ahead of them.

Perhaps, Hagstrum suggests, racing pigeons locate where they are using atmospheric infrasounds that the SST obliterated. These very-low-frequency sounds travel thousands of miles from their sources. That's why you can hear distant thunder. Pigeons can hear infrasounds very well, because a pigeon's ear is particularly good at detecting very-low-frequency sounds.

What sort of infrasounds are available to guide pigeons? All over the world, there is one infrasound pigeons should all be able to hear—the very-low-frequency acoustic shock waves generated by ocean waves banging against one another! Like an acoustic beacon, a constant stream of these tiny seismic waves would always say where the ocean is. Even more valuable to a racing pigeon looking for home, infrasounds reflect from cliffs, mountains, and other steep-sided features of the earth's surface. Ocean wave infrasounds reflecting off of local terrain could provide a pigeon with a detailed sound picture of its surroundings, near and far.

Hagstrum's suggestion explains neatly what happened to the lost pigeons in the great race of 1997. The enormous wave of infrasound generated by the SST's sonic boom would have blotted out all of the normal oceanic infrasound information. Any bird flying in its path would lose all orientation.

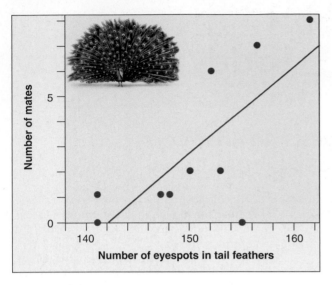

Figure 21.13 **The male peacock's tail feathers are a product of sexual selection.**

While female tail feathers are drab, the male tail feathers are decorated with colorful "eyespots." Experiments show that female peahens prefer to mate with males with greater numbers of eyespots in their tail feathers.

BIOLOGY & YOU

The Dormitory Effect. In 1968, Martha McClintock, an undergraduate student at Wellesley College, noticed that women in her dorm were menstruating at the same time. She wrote a paper about her finding that was proved to be very controversial. Subsequent studies have tended to confirm this "McClintock effect." In a representative study, groups of women were brought in to live together in close quarters, with researchers monitoring their menstrual cycles. Within three months, 30% of the women had started to cycle with each other. In another study, by the end of the summer, groups of female lifeguards were menstruating around the same time. What causes menstrual synchrony? Most current studies implicate pheromones, odorless chemicals picked up by the human sense of smell. Not all researchers are convinced, however. Some argue that synchrony often does not occur, and that when it does, other factors, such as birth control pills, may be responsible.

21.10 Reproductive Behaviors

CONCEPT PREVIEW: Natural selection has favored the evolution of mate choice, mating systems, and parenting behaviors that maximize reproductive success.

Among the many different types of animal behaviors (table 21.1), reproductive behaviors are among the most complex. The reproductive success of an animal is directly affected by its reproductive behaviors, because these behaviors influences how long the individual lives, how frequently it mates, and how many offspring it produces per mating. The second of these factors, competition for mating opportunities, has been termed **sexual selection.**

Sexual selection among males leads to the evolution of structures used in combat with other males (such as a deer's antlers or a ram's horns). Sexual selection by females, also called **mate choice,** leads to the evolution of complex courtship behaviors, and of ornaments used to "persuade" members of the opposite sex to mate, such as long tail feathers or bright plumage. The male peacock will parade in front of females with tail feathers displayed. Figure 21.13 shows that the more eyespots on a male's tail feathers, the more mates he will attract.

The Benefits of Mate Choice

Why did mating preferences evolve? What is their adaptive value? Biologists have proposed several reasons:

1. In many species of birds and mammals, males help raise the offspring. In these cases, females would benefit by choosing the male that can provide the best care—the better the male parent, the more offspring she is likely to rear successfully.

2. In other species, males provide no care, but maintain territories that provide food, nesting sites, and predator refuges. In such species, females that choose males with the best territories will maximize reproductive success.

3. Even if males provide no direct benefits of any kind to the female, a more vigorous male is more likely to have a good genetic makeup, ensuring that a female's offspring receive good genes from their father.

Mating Systems

Some animals mate with many partners during the breeding season, others with only one. The typical number of mates an animal has during its breeding season is called the **mating system.** Among animals, there are three principal mating systems: *monogamy* (one male mates with one female), *polygyny* (one male mates with more than one female), and *polyandry* (one female mates with more than one male).

Natural selection favors the evolution of polygyny in territorial animals. A male holding a high-quality territory may already have a mate, but it is still more advantageous for a female to breed with that male than with an unmated male that defends a low-quality territory.

Polyandrous systems—in which one female mates with several males—also occur. In the spotted sandpiper, a shore bird, males take care of all incubation and parenting, and females mate and leave eggs with two or more males.

Concept Check

1. Why do gulls remove egg shells from the nest after the eggs hatch?
2. What does optimal foraging theory predict?
3. What's the difference between a compass sense and a map sense?

TABLE 21.1 | Animal Behaviors

Foraging Behavior

Select, obtain, and consume food

Oystercatchers forage for food by stabbing the ground or rocks in search of arthropods or mollusks.

Territorial Behavior

Defend portion of home range and use it exclusively

Male elephant seals fight with each other for possession of territories. Only the largest males can hold territories, which contain many females.

Migratory Behavior

Move to a new location for part of the year

Wildebeests undergo an annual migration in search of new pastures and sources of water. Migratory herds can contain up to a million individuals and can stretch thousands of miles.

Courtship

Attract and communicate with potential mates

This male frog is vocalizing, producing a call that attracts females.

Parental Care

Produce and rear offspring

Female lions share the responsibility of raising the pride's young, increasing the probability that the young will survive into adulthood.

Social Behavior

Communicate information and interact with members of a social group

These leaf-cutter ants are members of different castes (or worker classes) in an insect society. The large ant is a worker, carrying leaves to the nest, and the smaller ants are protecting the worker from attack.

Social Behavior

Figure 21.14   Ants following a pheromone trail.

Trail pheromones organize cooperative foraging. The trails taken by the first ants to travel to a food source are soon followed by most of the other ants due to the release of pheromones by the first ants.

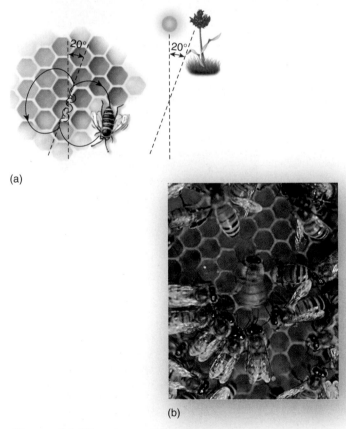

(a)

(b)

Figure 21.15   The waggle dance of honeybees.

(a) The angle between the food source, the nest, and the sun is represented by a dancing bee as the angle between the straight part of the dance and vertical. Here the food is 20 degrees to the right of the sun, and the straight part of the bee's dance on the hive is 20 degrees to the right of vertical. (b) A scout bee dancing in the hive.

21.11   Communication Within Social Groups

CONCEPT PREVIEW: The study of animal communication involves analysis of the specificity of signals, their information content, and the methods used to produce and receive them.

Many insects, fish, birds, and mammals live in social groups in which information is communicated between group members. For example, some individuals in mammalian societies serve as "guards." When a predator appears, the guards give an alarm call, and group members respond by seeking shelter. Social insects, such as ants and honeybees, secrete chemicals called pheromones that can trigger defensive or foraging behaviors (figure 21.14). Honeybees have an extremely complex dance language that directs nestmates to rich nectar sources.

The Dance Language of the Honeybee

The European honeybee, *Apis mellifera,* lives in hives consisting of 30,000 to 40,000 individuals whose behaviors are integrated into a complex colony. Worker bees may forage miles from the hive, collecting nectar and pollen from a variety of plants on the basis of how energetically rewarding their food is. The food sources used by bees tend to occur in patches, and each patch offers much more food than a single bee can transport to the hive. A colony is able to exploit the resources of a patch because of the behavior of scout bees, which locate patches and communicate their location to hivemates through a *dance language.* Over many years, Nobel laureate Karl von Frisch was able to unravel the details of this communication system.

After a successful scout bee returns to the hive, she performs a remarkable behavior pattern called a *waggle dance* on a vertical comb (figure 21.15). The path of the bee during the dance resembles a figure-eight. On the straight part of the path, the bee vibrates or waggles her abdomen while producing bursts of sound. She may stop periodically to give her hivemates a sample of the nectar she has carried back to the hive in her crop. As she dances, she is followed closely by other bees, which soon appear as foragers at the new food source.

Von Frisch and his colleagues claimed that the other bees use information in the waggle dance to locate the food source. According to their explanation, the scout bee indicates the *direction* of the food source by representing the angle between the food source, the hive, and the sun as the deviation from vertical of the straight part of the dance performed on the hive wall (that is, if the bee moved straight up, then the food source would be in the direction of the sun, but if the food were at a 30 degree angle relative to the sun's position, then the bee would move upward at a 30 degree angle from vertical). The *distance* to the food source is indicated by the tempo, or degree of vigor, of the dance.

Adrian Wenner, a scientist at the University of California, did not believe that the dance language communicated anything about the location of food, and he challenged von Frisch's explanation. Wenner maintained that flower odor was the most important cue allowing recruited bees to arrive at a new food source. A heated controversy ensued as the two groups of researchers published articles supporting their positions.

Such controversies can be very beneficial, because they often generate innovative experiments. In this case, the "dance language controversy" was resolved (in the minds of most scientists) in the mid-1970s by the creative research of James L. Gould. Gould devised an experiment in which hive members were tricked into misinterpreting the directions given by the scout bee's dance. As a result, Gould was able to manipulate where the hive members would go if they were using visual signals. If odor were the cue they were using, hive members would have appeared at the food source, but instead they appeared exactly where Gould predicted. This confirmed von Frisch's ideas.

Recently, researchers have extended the study of the honeybee dance language by building robot bees whose dances can be completely controlled. Their dances are programmed by a computer and perfectly reproduce the natural honeybee dance; the robots even stop to give food samples! The use of robot bees has allowed scientists to determine precisely which cues direct hivemates to food sources.

Primate Language

Some primates have a "vocabulary" that allows individuals to communicate the identity of specific predators. The vocalizations of African vervet monkeys, for example, distinguish eagles, leopards, and snakes. The two sonograms in figure 21.16b show the alarm calls for an eagle and a leopard, each eliciting different responses in other members of the troop. Chimpanzees and gorillas can learn to recognize a large number of symbols and use them to communicate abstract concepts.

The complexity of human language would at first appear to defy biological explanation, but closer examination suggests that the differences are in fact superficial—all languages share many basic structural similarities. All of the roughly 3,000 human languages draw from the same set of 40 consonant sounds (English uses two dozen of them), and any human can learn them. Researchers believe these similarities reflect the way our brains handle abstract information, a genetically determined characteristic of all humans.

Language develops at an early age in humans. Human infants are capable of recognizing the 40 consonant sounds characteristic of speech, including those not present in the particular language they will learn, while they ignore other sounds. In contrast, individuals who have not heard certain consonant sounds as infants can only rarely distinguish or produce them as adults. That is why English speakers have difficulty mastering the throaty French "r," French speakers typically replace the English "th" with "z," and native Japanese often substitute "r" for the unfamiliar English "l." Children go through a "babbling" phase, in which they learn by trial and error how to make the sounds of language. Even deaf children go through a babbling phase using sign language. Next, children quickly and easily learn a vocabulary of thousands of words. Like babbling, this phase of rapid learning seems to be genetically programmed. It is followed by a stage in which children form simple sentences that, though they may be grammatically incorrect, can convey information. Learning the rules of grammar constitutes the final step in language acquisition.

While language is the primary channel of human communication, odor and other nonverbal signals (such as "body language") may also convey information. However, it is difficult to determine the relative importance of these other communication channels in humans.

(a)

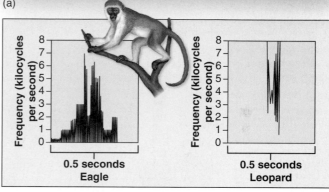
(b)

Figure 21.16 **Primate semantics.**

(a) This leopard, an efficient predator of primates, has attacked and will eat a vervet monkey. (b) Escaping a leopard attack presents a very different challenge to a vervet monkey than avoiding an eagle, another key predator of monkeys. Vervet monkeys give different alarm calls when leopards are sighted by troop members than they do when a member sees an eagle. Each distinctive call elicits a different and adaptive escape behavior.

EVOLUTION

Can Apes Use Language? Nonhuman primates can learn sign language. Kanzi, a bonobo (pygmy chimpanzee), is believed to understand more human language than any nonhuman animal in the world, apparently by eavesdropping on keyboard lessons a researcher was giving his mother. One day, using the computer, the researcher asked Kanzi, "Can you make the dog bite the snake?" Kanzi searched among the objects present until he found a dog toy and a snake toy, put the snake in the dog's mouth, and used his thumb and finger to close the dog's mouth over the snake! But at best chimps like Kanzi learn 125 signs; you have a vocabulary of greater than 100,000 words. Further, linguistic analysis has demonstrated that the "language" of Kanzi and other apes is symbolic, and lacks grammar of the conceptual kind that humans use. So the answer is "No." Although some apes can communicate quite well, apes cannot use language.

Figure 21.17 An altruistic act—or is it?

A meerkat sentinel on duty. Meerkats, *Suricata suricata*, are a species of highly social mongoose living in the semiarid sands of the Kalahari Desert in southern Africa. This meerkat is taking its turn to act as a lookout for predators. Under the security of its vigilance, the other members of the group can focus their attention on foraging. The sentinel puts his own life at risk when he gives an alarm, an apparent example of altruistic behavior.

IMPLICATION   This meerkat looks as if it is listening for trouble. Just what sounds do you suppose it might be listening for? A miniature antelope called a dik-dik that also lives in Africa has been shown to listen to nearby birds, reacting to alarm calls while ignoring nonalarmist calls of the same volume. How would you test if meerkats are doing what dik-diks do?

21.12 Altruism and Group Living

CONCEPT PREVIEW: Many factors could be responsible for the evolution of altruistic behaviors. Individuals may benefit directly if altruistic acts are reciprocated; kin selection explains how alleles for altruism can increase in frequency if altruistic acts are directed toward relatives.

Altruism—the performance of an action that benefits another individual at a cost to the actor—occurs in many guises in the animal world. In many bird species, for example, parents are assisted in raising their young by other birds, which are called *helpers at the nest*. In species of both mammals and birds, individuals that spy a predator will give an alarm call, alerting other members of their group, even though such an act would seem to call the predator's attention to the caller. Finally, lionesses with cubs will allow all cubs in the pride to nurse, including cubs of other females.

The existence of altruism has long perplexed evolutionary biologists. If altruism imposes a cost to an individual, how could an allele for altruism be favored by natural selection? One would expect such alleles to be at a disadvantage, and thus their frequency in the gene pool to decrease through time.

A number of explanations have been put forward to explain the evolution of altruism. One suggestion often heard on television documentaries is that such traits evolve for the good of the species. The problem with such explanations is that natural selection operates on individuals within species, not on species themselves. Thus, natural selection will not favor alleles that lead an individual to benefit others at a cost to itself; it is even possible for traits to evolve that are detrimental to the species as a whole, as long as they benefit the individual. Consequently, the "good of the species" cannot explain the evolution of altruistic traits.

Another possibility is that seemingly altruistic acts aren't altruistic after all. For example, helpers at the nest are often young and gain valuable parenting experience by assisting established breeders. Moreover, by hanging around an area, such individuals may inherit the territory when the established breeders die. Similarly, alarm callers (figure 21.17) may actually benefit by causing other animals to panic. In the ensuing confusion, the caller may be able to slip off undetected. Detailed field studies in recent years have demonstrated that some acts truly are altruistic, but others are not as they seemed.

Reciprocity

Robert Trivers, now of Rutgers University, proposed that individuals may form "partnerships" in which mutual exchanges of altruistic acts occur because it benefits both participants to do so. In the evolution of such reciprocal altruism, "cheaters" (nonreciprocators) are discriminated against and are cut off from receiving future aid. According to Trivers, if the altruistic act is relatively inexpensive, the small benefit a cheater receives by not reciprocating is far outweighed by the potential cost of not receiving future aid. Under these conditions, cheating should not occur.

For example, vampire bats roost in hollow trees in groups of 8 to 12 individuals. Because these bats have a high metabolic rate, individuals that have not fed recently may die. Bats that have found a host imbibe a great deal of blood; giving up a small amount presents no great energy cost to the donor, and it can keep a roostmate from starvation. Vampire bats tend to share blood with past reciprocators. If an individual fails to give blood to a bat from which it had received blood in the past, it will be excluded from future bloodsharing.

Kin Selection

The most influential explanation for the origin of altruism was presented by William D. Hamilton in 1964. It is perhaps best introduced by quoting a passing remark made in a pub in 1932 by the great population geneticist J. B. S. Haldane. Haldane said that he would willingly lay down his life for two brothers or eight first cousins. Evolutionarily speaking, Haldane's statement makes sense, because for each allele Haldane received from his parents, his brothers each had a 50% chance of receiving the same allele. Consequently, it is statistically expected that two of his brothers would pass on as many of Haldane's particular combination of alleles to the next generation as Haldane himself would. Similarly, Haldane and a first cousin would share an eighth of their alleles. Their parents, which are siblings, would each share half their alleles, and each of their children would receive half of these, of which half on the average would be in common: $1/2 \times 1/2 \times 1/2 = 1/8$. Eight first cousins would therefore pass on as many of those alleles to the next generation as Haldane himself would. Hamilton saw Haldane's point clearly: Natural selection will favor any strategy that increases the net flow of an individual's alleles to the next generation.

Hamilton showed that by directing aid toward kin, or close genetic relatives, an altruist may increase the reproductive success of its relatives enough to compensate for the reduction in its own fitness. Because the altruist's behavior increases the propagation of its own alleles in relatives, it will be favored by natural selection. Selection that favors altruism directed toward relatives is called **kin selection.** Although the behaviors being favored are cooperative, the genes are actually "behaving selfishly," because they encourage the organisms to support copies of themselves in other individuals. In other words, if an individual has a dominant allele that causes altruism, any action that increases the fitness of relatives and thus increases the frequency of this allele in future generations will be favored, even if that action is detrimental to the particular individual taking the action.

Examples of Kin Selection

Hamilton's kin selection model predicts that altruism is likely to be directed toward close relatives. The more closely related two individuals are, the greater the potential genetic payoff.

Many examples of kin selection are known from the animal world. Belding's ground squirrels give alarm calls when they spot a predator such as a coyote or a badger. Such predators may attack a calling squirrel, so giving a signal places the caller at risk. The social unit of a ground squirrel colony consists of a female and her daughters, sisters, aunts, and nieces. When they mature, males disperse long distances from where they are born; thus, adult males in the colony are not genetically related to the females. By marking all squirrels in a colony with an individual dye pattern on their fur and by recording which individuals gave calls and the social circumstances of their calling, researchers found that females who have relatives living nearby are more likely to give alarm calls than females with no kin nearby. Males tend to call much less frequently, as would be expected because they are not related to most colony members.

Another example of kin selection is seen in a bird called the white-fronted bee-eater, which lives along rivers in Africa in colonies of 100 to 200 birds. Many male bee-eaters do not raise their own offspring, but rather help others. The presence of a single helper, on average, doubles the number of offspring that survive. Helper males almost invariably choose to help the parents to which they are most closely related (figure 21.18).

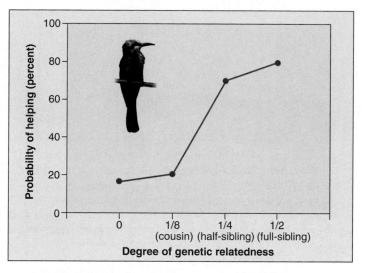

Figure 21.18   Kin selection is common among vertebrates.

A clear example of kin selection comes from a bird called the white-fronted bee-eater (*Merops bullockoides*). Male birds often help other individuals raise offspring at the nest. This graph compares the probability of a bird helping in a nest (on the *y* axis) and the relationship of the helper (on the *x* axis). The more closely related the helper (toward the right side of the graph), the higher the probability that it will be a helper in the nest.

BIOLOGY & YOU

Is The Golden Rule Scientific?   The golden rule *"Do unto others as you would have them do unto you"* is found in most, if not all, of the world's major religions. From a biological point of view, the evolution of altruism described in this chapter raises an interesting issue: Is there an evolutionary justification for morality? Is the golden rule altruism or enlightened self-interest? What do you think? It is important as you study biology to appreciate that science need not be divorced from other lines of inquiry, that biologists, philosophers, and theologians can have much to say to each other.

Figure 21.19 Savanna-dwelling African weaver birds form colonial nests.

21.13 Animal Societies

CONCEPT PREVIEW: Animal societies, mainly insects, exhibit a division of labor and altruism. The extent of vertebrate sociality is affected by environmental conditions.

Organisms as diverse as bacteria, cnidarians, insects, fish, birds, prairie dogs, whales, and chimpanzees exist in social groups. To encompass the wide variety of social phenomena, we can broadly define a **society** as a group of organisms of the same species that are organized in a cooperative manner.

Insect Societies

In insects, sociality has chiefly evolved in two orders, the Hymenoptera (ants, bees, and wasps) and the Isoptera (termites), although a few other insect groups include social species. These social insect colonies are composed of different **castes,** groups of individuals that differ in size and morphology and that perform different tasks, such as workers and soldiers.

Honeybees. In honeybees, the queen maintains her dominance in the hive by secreting a pheromone, called "queen substance," that suppresses development of the ovaries in other females, turning them into sterile workers. Drones (male bees) are produced only for purposes of mating. When the colony grows larger in the spring, some members do not receive a sufficient quantity of queen substance, and the colony begins preparations for swarming. Workers make several new queen chambers, in which new queens begin to develop. Scout workers look for a new nest site and communicate its location to the colony. The old queen and a swarm of female workers then move to the new site. Left behind, a new queen emerges, kills the other potential queens, flies out to mate, and returns to assume "rule" of the hive.

Leaf-Cutter Ants. The leafcutter ants provide another fascinating example of the remarkable lifestyles of social insects. Leafcutters live in colonies of up to several million individuals, their mound-like nests covering more than 100 square meters, with hundreds of entrances and chambers as deep as 5 meters beneath the ground. The division of labor among the worker ants is related to their size. Every day, workers travel along trails from the nest to a tree or a bush, cut its leaves into small pieces, and carry the pieces back to the nest. Smaller workers chew the leaf fragments into a mulch, which they spread like a carpet in the underground fungus chambers. Even smaller workers implant fungal hyphae in the mulch (recent molecular studies suggest that ants have been cultivating these fungi for more than 50 million years!). Soon a luxuriant garden of fungi is growing. While other workers weed out undesirable kinds of fungi, nurse ants carry the larvae of the nest to choice spots in the garden, where the larvae graze. Some of these larvae grow into reproductive queens that will disperse from the parent nest and start new colonies, repeating the cycle.

Vertebrate Societies

In contrast to the highly structured and integrated insect societies and their remarkable forms of altruism, vertebrate social groups are usually less rigidly organized and cohesive. Each social group of vertebrates has a certain size, stability of members, number of breeding males and females, and type of mating system. Behavioral ecologists have learned that the way a vertebrate group like the African weaver birds seen in figure 21.19 is organized is influenced most often by ecological factors such as food type and predation.

21.14 Human Social Behavior

CONCEPT PREVIEW: In humans, just as in other animals, behavior is molded by experience within limits set by our genes. Both heredity and learning play key roles in determining how we behave.

One of the most profound lessons of biology is that we human beings are animals, quite close relatives of the chimpanzee, and not some special form of life set apart from the earth's other creatures. This biological view of human life raises an important issue: To what degree are the animal behaviors we have described in this chapter characteristic of the social behavior of humans?

Genes and Human Behavior

As we have seen repeatedly, genes play a key role in determining many aspects of animal behavior. From maternal behavior in mice to migratory behavior in songbirds, changes in genes have been shown to have a profound impact. This leads to an important prediction. If behaviors have a genetic basis, and if behaviors affect an animal's ability to survive and raise young, then surely behaviors must be subject to natural selection, and the complex social behaviors of animals, including humans, should evolve. The study of this facet of animal behavior, sometimes called **sociobiology,** has proven highly controversial.

Genes certainly affect human behavior in important ways. Human facial expressions are similar the world over, no matter the culture or language, arguing that they have a deep genetic basis. Infants born blind still smile and frown, even though they have never seen these expressions on another face.

It is also true, however, that learning has a huge impact on how humans behave. Humans babies in the United States quickly lose the ability to make the consonant sounds they do not hear. Just as birds learn songs from experienced adults, so human babies learn to speak from listening to adults around them.

Diversity Is the Hallmark of Human Culture

When a group of biologists that studied social behavior in chimpanzees among seven widely separated areas compared notes, they identified 39 behaviors involving such things as social behavior, courtship, and tool use that were common in some groups but absent in others. Each population seemed to have a distinct repertoire of behaviors. It seems that each population has its own collection of customary behaviors, which each generation teaches its children. In a word, chimpanzees have culture.

No animal, not even our close relative the chimpanzee, exhibits cultural differences to the degree seen in human populations (figure 21.20). To what degree human culture is the product of evolution acting on behavior-determining genes is a question that is hotly debated by behavioral biologists. However one chooses to answer this question, it is perfectly clear that much of our behavior is molded by experience. Human cultures, and the behaviors that produce them, change very rapidly, far too rapidly to reflect genetic evolution. Human cultural diversity, a hallmark of our species, is certainly strongly influenced, if not largely determined, by learning and experience.

Concept Check

1. Regarding honeybees, did Gould prove Wenner right or wrong?
2. The English language uses 2 dozen consonant sounds. How many others can a child learn?
3. Discuss how natural selection might favor altruism.

Figure 21.20 A New York City street scene.

EVOLUTION

Evolutionary Psychology. The research into the evolutionary aspects of behavior, once dubbed sociobiology, has been largely replaced by narrower, less controversial fields of investigation. Perhaps the most important of these, evolutionary psychology, studies if human cognitive behaviors have been shaped by natural selection to serve survival and reproduction, and if so how. As you will learn in chapter 29, the human brain is functionally organized, each part specialized to perform a specific function. It is the central assumption of evolutionary psychology that this organization has evolved, like the other parts of the body, to improve the odds of survival and reproduction—stated simply, that the human brain is shaped by evolution. This seemingly simple assumption can take you to surprising and sometimes unexpected places. Evolutionary psychologists argue that much of human behavior, shaped by the structure of our brains, can be considered to be psychological adaptations that evolved to solve recurrent problems in ancestral human environments. The universal fear of spiders and snakes is seen as an evolved adaptation, and so is the ability to learn a language. These conclusions remain controversial among most biologists, who worry that human behaviors are shaped by many factors other than brain structure, and in particular that learning plays a huge part in shaping each person's behavior. The controversy has many aspects of the classic "nature versus nurture" argument, written in modern terms.

Do Crabs Eat Sensibly?

Many behavioral ecologists claim that animals exhibit so-called optimal foraging behavior. The idea is that because an animal's choice in seeking food involves a trade-off between the food's energy content and the cost of obtaining it, evolution should favor foraging behaviors that optimize the trade-off.

While this all makes sense, it is not at all clear that this is what animals would actually do. This optimal foraging approach makes a key assumption, that maximizing the amount of energy acquired will lead to increased reproductive success. In some cases this is clearly true. In ground squirrels, zebra finches, and orb-weaving spiders, researchers have found a direct relationship between net energy intake and the number of offspring raised successfully.

However, animals have other needs besides energy, and sometimes these needs conflict. One obvious "other need," important to many animals, is to avoid predators. It makes little sense for you to eat a little more food if doing so greatly increases the probability that you will yourself be eaten. Often the behavior that maximizes energy intake increases predation risk. A shore crab foraging for mussels on a beach exposes itself to predatory gulls and other shore birds with each foray. Thus the behavior that maximizes fitness may reflect a trade-off, obtaining the most energy with the least risk of being eaten. Not surprisingly, a wide variety of animals use a more cautious foraging behavior when predators are present—becoming less active and staying nearer to cover.

So what does a shore crab do? To find out, an investigator looked to see if shore crabs in fact feed on those mussels which provide the most energy, as the theory predicts. He found that the mussels on the beach he studied come in a range of sizes, from small ones less than 10 mm in length that are easy for a crab to open but yield the least amount of energy, to large mussels over 30 mm in length that yield the most energy but also take considerably more energy to pry open. To obtain the most net energy, the optimal approach, described by the blue curve in the graph above, would be for shore crabs to feed primarily on intermediate-sized mussels about 22 mm in length. Is this in fact what shore crabs do? To find out, the researcher carefully monitored the size of the mussels eaten each day by the beach's population of shore crabs. The results he obtained—the numbers of mussels of each size actually eaten—are presented in the red histogram.

Analysis

1. **Applying Concepts**   What variable do the curve and the histogram have in common?

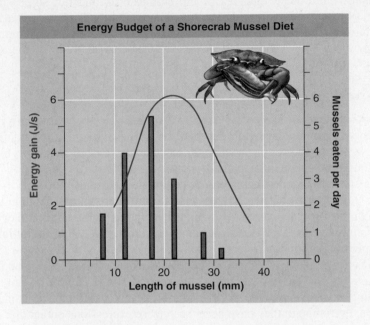

Energy Budget of a Shorecrab Mussel Diet

2. **Making Inferences**
   a. What is the most energetically optimal mussel size for the crabs to eat, in mm?
   b. What size mussel is most frequently eaten by crabs, in mm?
3. **Drawing Conclusions**   Do shore crabs tend to feed on those mussels that provide the most energy?
4. **Further Analysis**   What factors might be responsible for the slight difference in peak prey length relative to the length optimal for maximal energy gain?

Concept Summary

Some Behavior Is Genetically Determined

21.1   Approaches to the Study of Behavior

- The study of animal behavior, how animals respond to stimuli in their environment, includes examining how and why the behavior occurs. Both instinct (nature) and learning (nurture) play significant roles in behavior.

21.2   Instinctive Behavioral Patterns

- Instinctive, or innate, behaviors are those that are the same in all individuals of a species and appear to be controlled by preset pathways in the nervous system. A sign stimulus triggers the behavior, called a fixed action pattern, such as egg retrieval in geese shown here from **figure 21.2.**

21.3   Genetic Effects on Behavior

- Most behaviors are not "hard-wired" instincts, but they are strongly influenced by genes and so they can be studied as inherited traits. Hybrids, twins, and genetically altered animals (**figure 21.4**) have been used to study genetically influenced behaviors.

Behavior Can Also Be Influenced by Learning

21.4   How Animals Learn

- Many behaviors are learned, having been formed or altered based on previous experiences. Classical conditioning results when two stimuli are paired such that the animal learns to associate the two stimuli. Operant conditioning results when an animal associates a behavior with a reward or punishment. Imprinting is when an animal forms social attachments, usually during a critical window of time (**figure 21.5**).

21.5   Instinct and Learning Interact to Determine Behavior

- Behavior is often both genetically determined (instinctive) and modified by learning. Genes can limit the extent to which a behavior can be modified through learning. Ecology has a lot to do with behavior, and knowing an animal's ecological niche can reveal much about its behavior (**figure 21.6**).

21.6   Animal Cognition

- While humans have evolved a great capacity for cognitive thought, studies show that other animals possess varying degrees of cognitive abilities. Some behaviors in animals show conscious planning ahead. Still other animals show problem solving abilities. When presented with a novel situation, such as a piece of meat dangling out of reach of a raven, shown here from **figure 21.8,** these animals respond to the situation in a problem-solving capacity.

Evolutionary Forces Shape Behavior

21.7   Behavioral Ecology

- Behavioral ecology is the study of how natural selection shapes behavior. Only behaviors that have a genetic basis and offer some advantage for survival or reproduction can be acted upon through natural selection (**figure 21.9**).

21.8   A Cost-Benefit Analysis of Behavior

- For every behavior that offers an individual an advantage for survival, there is usually an associated cost. For example, foraging and territorial behaviors, exhibited by this sunbird from **figure 21.11,** offer benefits by providing food and shelter for the individual and their offspring but may endanger the parents through predation or expenditure of energy. The benefits have to outweigh the costs in order for the behaviors to be favored by natural selection.

21.9   Migratory Behavior

- Migration is a behavior that changes throughout the life of an animal. Inexperienced animals seem to rely on compass sense (following a direction), while experienced animals may rely more on map sense (learned ability to alter the path based on location) (**figure 21.12**).

21.10   Reproductive Behaviors

- Behaviors that maximize reproduction are favored by natural selection. Often these behaviors involve mate choice, mating systems, and parenting behaviors. Mate choice has lead to the evolution of complex courtship behaviors and ornate physical characteristics (**figure 21.13**).

Social Behavior

21.11   Communication Within Social Groups

- Communication is a behavior found in animals that live in groups or societies. Some animals secrete chemical pheromones to communicate information to others (**figure 21.14**). Others use movements, like the waggle dance of the honeybee shown here from **figure 21.15.** Although not as complex as human language, other animals are able to communicate a great deal of information through auditory signals (**figure 21.16**).

21.12   Altruism and Group Living

- Altruistic behaviors evolved in animals that live in groups. The reason why may involve the reciprocation of altruistic acts or benefits for relatives, called kin selection (**figure 21.18**).

21.13   Animal Societies

- Many types of animals live in social groups, or societies. Some insect societies are highly structured.

21.14   Human Social Behavior

- Both genetics and learning play key roles in human behaviors and establishing cultures, but the extent of each is hotly debated.

Self-Test

1. Innate behavior patterns
   a. can be modified if the stimulus changes.
   b. cannot be modified, as these behaviors seem built into the brain and nervous system.
   c. can be modified if environmental conditions begin to vary over a long period, a year or more.
   d. cannot be modified, as these behaviors are learned while very young.
2. The study of mouse maternal behavior shows a clearer genetic effect on behavior than lovebird or human twin studies because
   a. there is a clear link between presence or absence of a specific gene, a specific metabolic pathway, and a specific behavior.
   b. the behavior of mice was less complex and easier to study than lovebirds or human twins.
   c. parental care has a larger influence on mice than on lovebirds and humans.
   d. No answer is correct.
3. Training a dog to perform tricks using verbal commands and treats is an example of
   a. nonassociative learning.   c. classical conditioning.
   b. operant conditioning.   d. imprinting.
4. Behavior can be tied to ecology and evolution by considering what the behavior does to increase
   a. body size.   c. reproductive fitness.
   b. number of breeding sites. d. territory size.

5. The selection of foods and the journey to seek those foods is called
   a. territoriality.   c. migratory behavior.
   b. imprinting.   d. foraging behavior.
6. Courtship rituals are thought to have come about through
   a. intrasexual selection.   c. intersexual selection.
   b. agonistic behavior.   d. kin selection.
7. A mating system where the female mates with more than one male is
   a. protandry.   c. polygyny.
   b. polyandry.   d. monogamy.
8. Sharing of blood meals between fed and hungry vampire bats and the shunning of a bat who takes but doesn't share are examples of
   a. helpers.   c. reciprocity.
   b. kin selection.   d. group selection.
9. Bird offspring who help their parents care for younger offspring are showing
   a. brood parasitism.   c. reciprocity.
   b. kin selection.   d. group selection.
10. Which of the following is *not* a caste in an insect society?
    a. worker bees.   c. drones
    b. queens   d. helpers at the nest

Visual Understanding

1. **Figure 21.2b** What does the experiment illustrated here imply about certain types of fixed or innate behaviors?

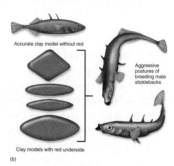

Accurate clay model without red

Aggressive postures of breeding male sticklebacks

Clay models with red underside

(b)

2. **Figure 21.16b** What does this graph imply about communication in nonhuman primates?

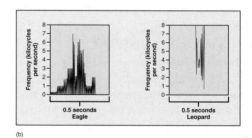

(b)

Challenge Questions

1. "Nature versus nurture" is an old argument regarding which factor predominates in behavior. How have studies of hybrids and of human twins helped us learn more?

2. Some behavior requires both genetic (nature) and environmental (nurture) input to be carried out. Explain how these two forces interact to affect the song of the white-crowned sparrow.

3. Optimal foraging theory balances costs and benefits of the various methods of finding and hoarding food items, and the danger of being exposed to predators. Discuss the probable differences in foraging behavior of a skunk, a porcupine, and a field mouse.

4. What are some of the reasons advanced for altruism among groups of animals? Give some examples.

How Humans Influence the Living World

Global Change

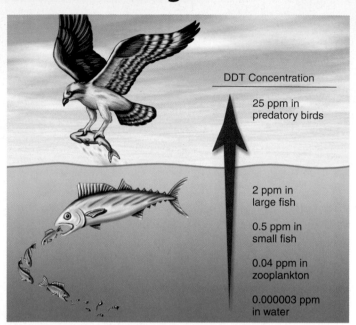

DDT Concentration

25 ppm in predatory birds

2 ppm in large fish

0.5 ppm in small fish

0.04 ppm in zooplankton

0.000003 ppm in water

Figure 22.1 **Biological magnification of DDT.**

Because DDT accumulates in animal fat, the compound becomes increasingly concentrated in higher levels of the food chain.

BIOLOGY & YOU

Do You Use Bottled Water? One sort of chemical pollution that many of us contribute to without thinking about it is drinking bottled water packaged in plastic. Bottled water is purified water or spring water, popular in the United States both because water is a healthy choice instead of soft drinks, and because bottles can be purchased conveniently in stores or vending machines. Americans buy some 28 billion water bottles a year, consuming more bottled water than beer, even though in the United States tap water quality is as high as bottled water, and an awful lot cheaper. Bottled water typically costs $1 to $2 per bottle (90% of the cost is making the bottle, label, and cap), while tap water is essentially free. Why do the Sierra Club, the World Wildlife Fund, and other conservation groups urge you to consume less bottled water—because the bottles, made of plastic, do not degrade and 80% are not recycled. They simply accumulate by the billions in refuse dumps, landfills and the world's oceans. Adding to the environmental damage, producing the bottles creates more than 2.5 million tons of CO_2 each year, an estimated 250 grams for each bottle. That's like filling a quarter of every bottle with oil and burning it. If you want to do your bit to curb global warming and pollution, drink more tap water.

22.1 Pollution

CONCEPT PREVIEW: All over the globe, increasing industrialization is leading to higher levels of pollution.

Our world is one ecological continent, one highly interactive biosphere, and damage done to any one ecosystem can have ill effects on many others. Burning high-sulfur coal in Illinois kills trees in Vermont, while dumping refrigerator coolants in New York destroys atmospheric ozone over Antarctica and leads to increased skin cancer in Madrid. Biologists call such widespread effects on the worldwide ecosystem **global change.** The pattern of global change that has become evident within recent years, including chemical pollution, acid precipitation, the ozone hole, the greenhouse effect, and the loss of biodiversity, is one of the most serious problems facing humanity's future.

Chemical Pollution

The problem posed by chemical pollution has grown very serious in recent years, both because of the growth of heavy industry and because of an overly casual attitude in industrialized countries. Chemicals are released into both the air and into water; therefore, their effects are far reaching.

Air Pollution. Air pollution is a major problem in the world's cities. In Mexico City, oxygen is sold routinely on corners for patrons to inhale. Cities such as New York, Boston, and Philadelphia are known as gray-air cities because the pollutants in the air are usually sulfur oxides emitted by industry. Cities such as Los Angeles, however, are called brown-air cities because the pollutants in the air undergo chemical reactions in the sunlight to form smog.

Water Pollution. Disposing of chemicals into the water, a "flushing it down the sink" approach, doesn't work in today's crowded world. There is simply not enough water available to dilute the many substances that the enormous human population produces continuously. Lakes and rivers throughout the world are becoming increasingly polluted with sewage, and fertilizers and insecticides that wash from the land to the water in great quantities.

Agricultural Chemicals

The spread of "modern" agriculture has caused very large amounts of many kinds of new chemicals to be introduced into the global ecosystem, particularly pesticides, herbicides, and fertilizers.

The chlorinated hydrocarbons, compounds that include DDT, chlordane, lindane, and dieldrin, have been banned in the United States but are still widely used in many tropical countries. Chlorinated hydrocarbon molecules accumulate in animal fat tissue, so as they pass through a food chain they become increasingly concentrated, a process called **biological magnification.** Figure 22.1 shows how a minute concentration of DDT in plankton increases to significant levels as it is passed up through this aquatic food chain. DDT caused serious ecological problems, leading to the production of thin, fragile eggshells in many predatory bird species such as peregrine falcons, bald eagles, osprey, and brown pelicans until the late 1960s, when it was banned in time to save the birds from extinction.

Food chains, as discussed on page 388, track the path of energy through an ecosystem from the sun through the trophic levels, based on who eats whom. Available energy decreases as it passes up through the food chain, but chlorinated hydrocarbons increase in concentration as they travel through the food chain.

22.2 Acid Precipitation

CONCEPT PREVIEW: Pollution-acidified precipitation—loosely called acid rain—is destroying forest and lake ecosystems in Europe and North America. The solution is to clean up the emissions.

The smokestacks you see in figure 22.2 are those of the Four Corners power plant in New Mexico. This facility burns coal, sending the smoke high into the atmosphere through these tall stacks. The smoke contains high concentrations of sulfur dioxide and other sulfates, which produce acid when they combine with water vapor in the air. The first tall stacks were introduced in Britain in the mid-1950s, and the design rapidly spread through Europe and the United States. The intent of having tall smokestacks was to release the sulfur-rich smoke high in the atmosphere, where winds would disperse and dilute it.

However, in the 1970s, scientists began noticing that the acids from the sulfur-rich smoke were having devastating effects. Throughout northern Europe, lakes were reported to have suffered drastic drops in biodiversity, some even becoming devoid of life. The trees of the great Black Forest of Germany were dying—and the damage was not limited to Europe. In the eastern United States and Canada, many of the forests and lakes have been seriously damaged.

It turns out that when the sulfur introduced into the upper atmosphere combined with water vapor to produce sulfuric acid, the acid was taken far from its source, but it later fell along with water as acidic rain and snow. This pollution-acidified precipitation is called **acid rain** (but the term acid precipitation is actually more correct). Natural rainwater rarely has a pH lower than 5.6; however, rain and snow in many areas of the United States have pH values less than 5.3, and in the northeastern United States, pHs of 4.2 or below have been recorded, with occasional storms as low as 3.0.

> The pH scale, discussed on page 40, is a measurement of the amount of hydrogen ions (H^+) in a solution on a scale from 0 to 14. The lower the value, the more acidic the solution. A pH of 7 is a neutral solution. A pH of 3.0, as recorded in some storms in the United States, is equal to the pH of vinegar or carbonated drinks, such as cola.

Acid precipitation destroys life. Many of the forests of the northeastern United States and Canada have been seriously damaged. In fact, it is now estimated that at least 1.4 million acres of forests in the Northern Hemisphere have been adversely affected by acid precipitation (figure 22.3). In addition, thousands of lakes in Sweden and Norway no longer support fish—these lakes are now eerily clear. In the northeastern United States and Canada, tens of thousands of lakes are dying biologically as their pH levels fall to below 5.0, so acidic that many fish species and other aquatic animals die, unable to reproduce.

Acidification of the world's oceans driven by rising levels of carbon dioxide in the atmosphere is creating an even more serious problem. CO_2 dissolves in water to form carbonic acid, lowering the pH. Ocean acidification will lower ocean water pH by 0.5 pH units by 2100 at current rates. Marine organisms like shellfish and coral that use calcium carbonate to form structures will be disastrously affected, their shells dissolving at the lower pH.

The solution seems like it would be easy—clean up the sulfur and CO_2 emissions. But there have been serious problems with implementing this solution. First, it is expensive. Estimates of the cost of installing and maintaining the necessary "scrubbers" in the United States are on the order of $5 billion a year. An additional difficulty is that the polluter and the recipient of the pollution are far from one another, and neither wants to pay so much for what they view as someone else's problem. The Clean Air Act revisions of 1990 have begun to address this problem by mandating some cleaning of emissions in the United States, although much still remains to be done worldwide.

Figure 22.2   Tall stacks export pollution.

Tall stacks like those of the Four Corners coal-burning power plant in New Mexico send pollution far up into the atmosphere.

Figure 22.3   Acid precipitation.

Acid precipitation is killing many of the trees in North American and European forests. Much of the damage is done to the mycorrhizae, fungi growing within the cells of the tree roots. Trees need mycorrhizae in order to extract nutrients from the soil.

IMPLICATION   Humans when they breathe are exposed to the same acidified moisture that is killing the trees in North American forests. Why do you suppose it doesn't kill us?

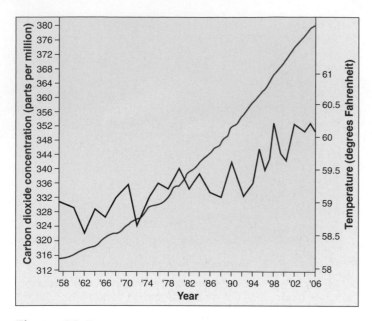

Figure 22.4 **The greenhouse effect.**

The concentration of carbon dioxide in the atmosphere has shown a steady increase for many years (*blue line*). The *red line* shows the average global temperature for the same period of time. Note the general increase in temperature since the 1950s and, specifically, the sharp rise beginning in the 1980s. Data from the National Center for Atmospheric Research and other sources.

IN THE NEWS

Clean Coal. The key problem we must face in combating global warming is the world's growing thirst for cheap energy. While we in the United States begin to search for effective ways to obtain large amounts of energy without burning fossil fuels, the problem is spiraling out of control. India and China are initiating large-scale commercial projects that would tap enormous but inaccessible coal reserves by burning the coal where it lies, deep below the earth's surface. The amount of carbon dioxide released by such burning would be staggering. Such underground coal gasification was pioneered by the Russians in the 1930s. Because underground gasification produces no sulfur oxide or nitrogen oxide, and the ash stays underground, the U.S. Coal Industry has taken to calling the process "clean coal." They suggest that the carbon dioxide released by the burning of underground coal can be captured and then sequestered (that is, locked away from the atmosphere) by pumping it back into the underground void left by burning the coal. The problem, ignored in television ads promoting governmental incentives for clean coal development in the United States, is that carbon sequestration costs a lot of money. This makes the energy obtained from underground coal gasification much more expensive, a strong incentive to skip this step. A clear-sighted government environmental policy would tie any financial incentive for coal gasification to caps on carbon emissions. Otherwise, governmental support of coal gasification may accelerate rather than slow the speed of climate change.

22.3 Global Warming

CONCEPT PREVIEW: Humanity's burning of fossil fuels has greatly increased atmospheric levels of CO_2 leading to global warming.

For over 150 years, the growth of our industrial society has been fueled by cheap energy, much of it obtained by burning fossil fuels—coal, oil, and gas. Coal, oil, and gas are the remains of ancient plants, transformed by pressure and time into carbon-rich "fossil fuels." When such fossil fuels are burned, this carbon is combined with oxygen atoms, producing carbon dioxide (CO_2). Industrial society's burning of fossil fuels has released huge amounts of carbon dioxide into the atmosphere. No one paid any attention to this because the carbon dioxide was thought to be harmless and because the atmosphere was thought to be a limitless reservoir, able to absorb and disperse any amount. It turns out neither assumption is true, and in recent decades, the levels of carbon dioxide in the atmosphere have risen sharply and are continuing to rise.

What is alarming is that the carbon dioxide doesn't just sit in the air doing nothing. The chemical bonds in carbon dioxide molecules transmit radiant energy from the sun but trap the longer wavelengths of infrared light, or heat, that are reflected off the earth's surface and prevent them from radiating back into space. This creates what is known as the **greenhouse effect.** Planets that lack this type of "trapping" atmosphere are much colder than those that possess one. If the earth did not have a "trapping" atmosphere, the average earth temperature would be about –20°C, instead of the actual +15°C.

Global Warming Due to Greenhouse Gases

The rise in average global temperatures during recent decades, a profound change in the earth's atmosphere referred to as **global warming** (shown as the red line in figure 22.4), is correlated with increased carbon dioxide concentrations in the atmosphere (the blue line). The suggestion that global warming might in fact be caused by the accumulation of greenhouse gases (carbon dioxide, CFCs, nitrogen oxides, and methane) in the atmosphere has been controversial, and is examined in detail in the Inquiry & Analysis feature at the end of this chapter. After serious examination of the evidence, the overwhelming consensus among scientists is that indeed greenhouse gases are causing global warming.

Increases in the amounts of greenhouse gases increase average global temperatures from 1° to 4°C, which could have serious impact on rain patterns in prime agricultural lands, and in changes in sea levels.

Effects on Rain Patterns. Global warming is predicted to have a major effect on rainfall patterns. Areas that have already been experiencing droughts may see even less rain, contributing to even greater water shortages.

Effects on Agriculture. Both positive and negative effects of global warming on agriculture are predicted. Warmer temperatures and increased levels of carbon dioxide in the atmosphere would be expected to increase the yields of some crops, while having a negative impact on others. Droughts that may result from global warming will also negatively affect crops.

Rising Sea Levels. Much of the water on earth is locked into ice in glaciers and polar ice caps. As global temperatures increase, these large stores of ice have begun to melt. Most of the water from the melted glaciers ends up in the oceans, causing water levels to rise. Higher water levels can be expected to cause increased flooding of low-lying lands.

22.4 Loss of Biodiversity

CONCEPT PREVIEW: The loss of biodiversity can usually be attributed to one of a few main causes, including habitat loss, overexploitation, and introduced species.

Just as death is as necessary to a normal life cycle as reproduction, so extinction is as normal and necessary to a stable world ecosystem as species formation. Most species, probably all, go extinct eventually. More than 99% of species known to science (most from the fossil record) are now extinct. However, current rates of extinctions are alarmingly high. The extinction rate for birds and mammals was about one species every decade from 1600 to 1700, but it rose to one species every year during the period from 1850 to 1950, and four species per year between 1986 and 1990. It is this increase in the rate of extinction that is the heart of the **biodiversity** crisis.

Factors Responsible for Extinction

What factors are responsible for extinction? Studying a wide array of recorded extinctions, and many species currently threatened with extinction, biologists have identified three factors that seem to play a key role in many extinctions: habitat loss, species overexploitation, and introduced species (figure 22.5).

Habitat Loss. Habitat loss is the single most important cause of extinction. Given the tremendous amounts of ongoing destruction of all types of habitat, from rain forest to ocean floor, this should come as no surprise. Natural habitats may be adversely affected by human influences in four ways: (1) destruction, (2) pollution, (3) human disruption, and (4) habitat fragmentation (dividing up the habitat into small isolated areas). Habitat destruction is rapidly endangering species on Madagascar. You can see the loss of rain forest habitat, colored in red on the overlain maps in figure 22.6.

Species Overexploitation. Species that are hunted or harvested by humans have historically been at grave risk of extinction, even when the species populations are initially very abundant. There are many examples in our recent history of overexploitation: passenger pigeons, bison, many species of whales, commercial fish such as Atlantic bluefin tuna, and mahogany trees in the West Indies are but a few.

Introduced Species. Occasionally, a new species will enter a habitat and colonize it, usually at the expense of native species. Colonization occurs in nature, but it is rare; however, humans have made this process more common with devastating ecological consequence. The introduction of exotic species has wiped out or threatened many native populations. Species introductions occur in many ways, usually unintentionally. Plants and animals can be transported in nursery plants, in the ballast of large ocean vessels, as stowaways in boats, cars, and planes, and as beetle larvae within wood products. These species enter new environments where they have no native predators to keep their population sizes in check. Free to populate the habitat, they crowd out native species.

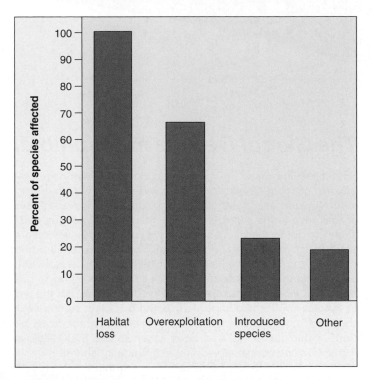

Figure 22.5   **Factors responsible for animal extinction.**

These data represent known extinctions of mammals in Australia, Asia, and the Americas. Some extinctions have more than one cause.

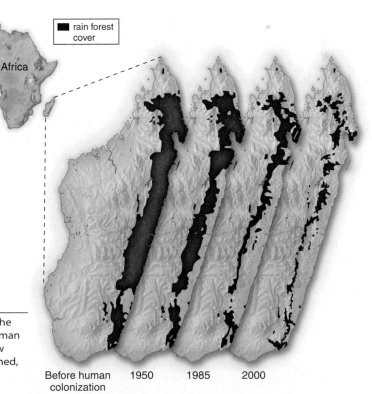

Figure 22.6   **Extinction and habitat destruction.**

The rain forest covering the eastern coast of Madagascar, an island off the coast of East Africa, has been progressively destroyed as the island's human population has grown. Ninety percent of the original forest cover is now gone. Many species have become extinct, and many others are threatened, including 16 of Madagascar's 31 primate species.

The Global Decline in Amphibians

Sometimes important things happen, right before our eyes, without anyone noticing. That thought occurred to David Bradford as he stood looking at a quiet lake high in the Sierra Nevada Mountains of California in the summer of 1988. Bradford, a biologist, had hiked all day to get to the lake, and when he got there his worst fears were confirmed. The lake was on a list of mountain lakes that Bradford had been visiting that summer in Sequoia-Kings Canyon National Parks while looking for a little frog with yellow legs. The frog's scientific name was *Rana muscosa,* and it had lived in the lakes of the parks for as long as anyone had kept records. But this silent summer evening, the little frog was gone. The last major census of the frog's populations within the parks had been taken in the mid-1970s, and *R. muscosa* had been everywhere, a common inhabitant of the many freshwater ponds and lakes within the parks. Now, for some reason Bradford did not understand, the frogs had disappeared from 98% of the ponds that had been their homes.

After Bradford reported this puzzling disappearance to other biologists, an alarming pattern soon became evident. Throughout the world, local populations of amphibians (frogs, toads, and salamanders) were becoming extinct. Waves of extinction have swept through high-elevation amphibian populations in the western United States, and have also cut through the frog populations of Central America and coastal Australia.

Amphibians have been around for 350 million years, since long before the dinosaurs. Their sudden disappearance from so many of their natural homes sounded an alarm among biologists. What are we doing to our world? If amphibians cannot survive the world we are making, can we?

In 1998 the U.S. National Research Council brought scientists together from many disciplines in a serious attempt to address the problem. After years of intensive investigation, they have begun to sort out the reasons for the global decline in amphibians. Like many important questions in science, this one does not have a simple answer.

Five factors seem to be contributing in a major way to the worldwide amphibian decline: (1) habitat deterioration and destruction, particularly clear-cutting of forests, which drastically lowers the humidity (water in the air) that amphibians require; (2) the introduction of exotic species that outcompete local amphibian populations; (3) chemical pollutants that are toxic to amphibians; (4) fatal infections by pathogens; and (5) global warming, which is making some habitats unsuitable.

Infection by parasites appears to have played a particularly important role in the western United States and coastal Australia. Amphibian ecology expert James Collins of Arizona State University has reported one clear instance of infection leading to amphibian decline. When Collins examined populations of salamanders living on the Kaibab Plateau along the Grand Canyon rim, he found many sick salamanders. Their skin was covered with white pustules, and most infected ones died, their hearts and spleens collapsed. The infectious agent proved to be a virus common in fish called a ranavirus. Ranavirus isolated by Collins from one sick salamander would cause the disease in a healthy salamander, so there was no doubt that ranavirus was the culprit responsible for the salamander decline on the Kaibab Plateau.

Ranavirus outbreaks eliminate small populations, but in larger ones a few individuals survive infection, sloughing off their pustule-laden skin. These populations slowly recover.

A second kind of infection, very common in Australia but also seen in the United States, is having more widespread effects. Populations infected with this microbe, a kind of fungus called a chytrid (pronounced "kit-rid," see chapter 16), do not recover. Usually a harmless soil fungus that decomposes plant material, this particular chytrid (with the Latin name of *Batrachochytrium dendrobatidis*) is far from harmless to amphibians. It dissolves and absorbs the chitinous mouthparts of amphibian larvae, killing them.

This killer chytrid appeared in Australia near Melbourne in the early 1980s. Now almost all Australia is affected. How did the disease spread so rapidly? Apparently it traveled by truck. Infected frogs moved all across Australia in wooden boxes with bunches of bananas. In one year, 5,000 frogs were collected from banana crates in one Melbourne market alone.

In other parts of the world, infection does not seem to play as important a role as acid precipitation, habitat loss, and introduction of exotic species. This complex pattern of cause and effect only serves to emphasize the take-home lesson: Worldwide amphibian decline has no one culprit. Instead, all five factors play important roles. It is their total impact that has shifted the worldwide balance toward extinction.

To reverse the trend toward extinction, we must work to lessen the impact of all of these factors. It is important that we not get discouraged at the size of the job, however. Any progress we make on any one factor will help shift the balance back toward survival. Extinction is only inevitable if we let it be.

22.5 The Ozone Hole

CONCEPT PREVIEW: CFCs are catalytically destroying the ozone in the upper atmosphere, exposing the earth's surface to dangerous radiation. International efforts to solve the problem are succeeding.

For 2 billion years, life was trapped in the oceans because radiation from the sun seared the earth's surface unchecked. Nothing could survive that bath of destructive energy. Living things were able to leave the oceans and colonize the surface of the earth only after a protective shield of ozone (O_3) had been added to the atmosphere by photosynthesis. Imagine if that shield were taken away. Alarmingly, it appears that we are destroying it ourselves. Starting in 1975, the earth's ozone shield began to disintegrate. Over the South Pole in September of that year, satellite photos revealed that the ozone concentration was unexpectedly low. It was as if some "ozone eater" were chewing it up in the Antarctic sky, leaving a mysterious zone of lower-than-normal ozone concentration, an **ozone hole.** Every year after that, more of the ozone has been depleted, the hole growing bigger and deeper. The satellite image in figure 22.7 shows lower levels of ozone as purple—the ozone hole completely covers Antarctica.

What is eating the ozone? Scientists soon discovered that the culprit was a class of chemicals that everyone had thought to be harmless: **chlorofluorocarbons (CFCs).** Throughout the world, CFCs are used in large amounts as coolants in refrigerators and air conditioners, as the gas in aerosol dispensers, and as the foaming agent in Styrofoam containers. All of these CFCs eventually escape into the atmosphere, but no one worried about this until recently, because CFCs were thought to be chemically inert.

It turned out that the CFCs were causing mischief the chemists had not imagined. High over the South and North Poles, nearly 50 kilometers up, where it is very, very cold, the CFCs stick to frozen water vapor and act as catalysts of a chemical reaction. Just as an enzyme carries out a reaction in your cells without being changed itself, so the CFCs catalyze the conversion of ozone (O_3) into oxygen (O_2) without being used up themselves. Oxygen gas, unlike ozone, does not provide a protective shield. The drop in ozone worldwide is now over 3%.

Ultraviolet radiation is a serious human health concern. Every 1% drop in the atmospheric ozone content is estimated to lead to a 6% increase in the incidence of skin cancers. At middle latitudes, the drop of approximately 3% that has occurred worldwide is estimated to have led to an increase of perhaps as much as 20% in lethal melanoma skin cancers.

Experts generally agree that levels of ozone-killing chemicals in the upper atmosphere are leveling off since more than 180 countries in the 1980s signed an international agreement phasing out the manufacture of CFCs. Current computer models suggest the Antarctic ozone hole should disappear by 2065, and the lesser-damaged ozone layer over the Arctic should recover by about 2023.

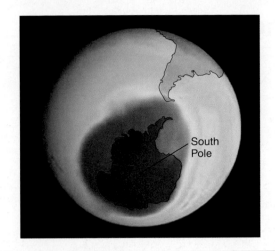

South Pole

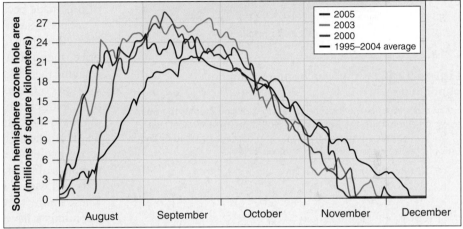

Figure 22.7   The ozone hole over Antarctica.

For decades NASA satellites have tracked the extent of ozone depletion over Antarctica. Every year since 1975 an ozone "hole" has appeared in August when sunlight triggers chemical reactions in cold air trapped over the South Pole during Antarctic winter. The hole intensifies during September before tailing off as temperatures rise in November–December. In 2000, the 28.4-million-square-kilometer hole (*purple* in the satellite image) covered an area larger than the United States, Canada, and Mexico combined. In September 2000, the hole extended over Punta Arenas, a city of about 120,000 people in southern Chile, exposing residents to very high levels of UV radiation.

IMPLICATION   The ozone hole reached its peak size in 2006, at 11.4 million square miles (largest single-day recording) and the largest average recording at 10.6 million square miles. The ozone hole was still nearly as large in 2008, at 10.5 million square miles average (27 million square kilometers). However, levels of CFCs over Antarctica in 2008 had decreased 3.8% from their peak in 2000. Based on these CFC levels, would you expect the size of the ozone hole to continue to decrease over the next few years? Explain your reasoning.

Concept Check

1. Why is DDT, widely used in the tropics, banned in the USA?
2. Name the three factors most responsible for loss of biodiversity.
3. How can burning coal be responsible for acid rain? Coal has been burned for centuries, while acid rain was unknown before 1950. Explain.

Saving Our Environment

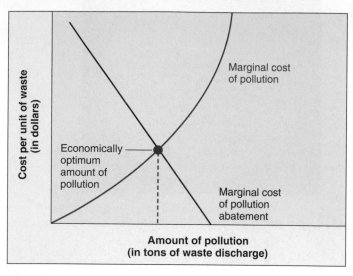

Figure 22.8   **Is there an optimum amount of pollution?**

Economists identify the "optimum" amount of pollution as the point at which eliminating the next unit of pollution (the marginal cost of pollution abatement) equals the cost in damages caused by that unit of pollution (the marginal cost of pollution).

22.6 Reducing Pollution

CONCEPT PREVIEW: Free-market economies often foster pollution when prices do not include environmental costs. Laws and taxes are being designed in an attempt to compensate.

To solve the problem of industrial and atmospheric pollution, it is first necessary to understand the cause of the problem. In essence, it is a failure of our economy to set a proper price on environmental health. The economy of the United States (and much of the rest of the industrial world) is based on a simple feedback system of supply and demand. As a commodity gets scarce, its price goes up, and this added profit acts as an incentive for more of the item to be produced; if too much is produced, the price falls and less of it is made because it is no longer as profitable to produce it.

This system works very well and is responsible for the economic strength of our nation, but it has one great weakness. If demand is set by price, then it is very important that all the costs be included in the price. Imagine that the person selling the item were able to pass off part of the production cost to a third person. The seller would then be able to set a lower price and sell more of the item! Driven by the lower price, the buyer would purchase more than if all the costs had been factored into the price.

Unfortunately, that sort of pricing error is what has driven the pollution of the environment by industry. The true costs of energy and of the many products of industry are composed of direct production costs, such as materials and wages, and of indirect costs, such as pollution of the ecosystem. These indirect costs are not always taken into account when setting prices. Economists have identified an "optimum" amount of pollution based on how much it costs to reduce pollution versus the social and environmental cost of allowing pollution. The economically optimum amount of pollution is indicated by the blue dot in figure 22.8. If more pollution than the optimum is allowed, the social cost is too high, but if less than the optimum is allowed, the economic cost is too high.

Antipollution Laws

Two effective approaches have been devised to curb pollution in this country. The first is to pass laws forbidding it. In the last 20 years, laws have begun to significantly curb pollution by setting stiff standards for what can be released into the environment. For example, all cars are required to have effective catalytic converters to eliminate automobile smog. Similarly, the Clean Air Act of 1990 requires that power plants eliminate sulfur emissions, typically by installing scrubbers on their smokestacks. The cost of the scrubbers increases the price of the energy. The new, higher costs are closer to the true costs, lowering consumption to more appropriate levels.

Pollution Taxes

A second approach to curbing pollution has been to increase the consumer costs directly by placing a tax on the pollution, in effect an artificial price hike imposed by the government as a tax added to the price of production. This added cost lowers consumption too, but by adjusting the tax, the government can attempt to balance the conflicting demands of environmental safety and economic growth. Such taxes, often imposed as "cap-and-trade pollution permits," are becoming an increasingly important part of antipollution laws.

BIOLOGY & YOU

Green Living. Few of us think of ourselves as sources of pollution, although all of us are. The production of the food we eat often involves the release into the environment of pesticides, herbicides, and other agricultural chemicals. The plastic products we use expose us to bisphenol A and other chemicals. As important as these sources of pollution are, however, our most serious individual impact comes from our use of fossil fuel energy. Burning fossil fuels pollutes the atmosphere with carbon dioxide and other greenhouse gases that are the root cause of global climate change. You can lessen your impact on environmental pollution and global warming in a variety of ways that are collectively coming to be called "green living." One simple way is to favor organic foods, those produced without extensive use of agricultural chemicals. Also, don't drive when you can walk or share a ride. Another very important contribution you can make is to go on a home energy diet. Replace incandescent light bulbs with compact fluorescent bulbs. Turn up the thermostat on your air conditioner (air conditioners account for 16% of residential electricity consumption), and turn heating down at night in the winter. Don't take overlong showers or run the dishwasher half-full. Each aspect of green living seems small, but if all of us were to live this way, it could have a major impact. We will not succeed unless every one of us gets involved.

22.7 Preserving Nonreplaceable Resources

CONCEPT PREVIEW: Nonreplaceable resources are being consumed at an alarming rate all over the world; key among them are topsoil, groundwater, and biodiversity.

Among the many ways ecosystems are being damaged, one problem stands out as more serious than all the rest: consuming or destroying resources that we all share in common but cannot replace (figure 22.9). Although a polluted stream can be cleaned, no one can restore an extinct species.

Topsoil

Soil is composed of a mixture of rocks and minerals with partially decayed organic matter called humus. Plant growth is strongly affected by soil composition. Minerals like nitrogen and phosphorus are critical to plant growth, and are abundant in humus-rich soils. The United States is one of the most productive agricultural countries on earth, largely because much of it is covered with particularly fertile soils. Our Midwestern farm belt sits astride what was once a great prairie. The soil of that ecosystem accumulated bit by bit from countless generations of animals and plants until, by the time humans came to plow, the humus-rich soil extended down several feet.

We cannot replace this rich **topsoil,** the capital upon which our country's greatness is built, yet we are allowing it to be lost at a rate of centimeters every decade. Our country has lost one-quarter of its topsoil since 1950! By repeatedly tilling (turning the soil over) to eliminate weeds, we permit rain to wash more and more of the topsoil away, into rivers and eventually out to sea. New approaches are desperately needed to lessen the reliance on intensive cultivation. Some possible solutions include using genetic engineering to make crops resistant to weed-killing herbicides, and terracing to recapture lost topsoil.

BIOLOGY & YOU

Organic Farming. Organic farming is a form of agriculture that avoids the use of synthetic chemicals. The use of fertilizers and pesticides is strictly limited, as are plant growth regulators and livestock feed additives. Organic farming relies upon crop rotation and green manure to provide nitrogen and nutrients to crops, instead of applying synthetic fertilizers. Insects are fought using biological pest control rather than synthetic pesticides. Weeds are eliminated from fields by mechanical cultivation rather than herbicides. Genetically modified (GM) crops are avoided, primarily because most GM crops like Roundup-ready soy beans have been created so that synthetic herbicides can be used more extensively (there is nothing inherently nonorganic about a DNA modification as such). Approximately 2% of total world farmland is now farmed organically. However, while ecologically very desirable, organic farming is controversial. It produces as little as half the output of high-yield conventional farming, suggesting that organic farming may be incapable of feeding a rapidly growing world population.

Figure 22.9 The tragedy of the commons.

In a now-famous essay, ecologist Garrett Hardin argues that destruction of the environment is driven by freedom without responsibility.

Reprinted with permission from "The Tragedy of the Commons," by G. Hardin, Science, 162, p. 1244. Copyright 1968 AAAS.

"Freedom in a Commons Brings Ruin to All"

The essence of Hardin's original essay:

Picture a pasture open to all. It is expected that each herdsman will try to keep as many cattle as possible on [this] commons....What is the utility...of adding one more animal?...Since the herdsman receives all the proceeds from the sale of the additional animal, the positive utility [to the herdsman] is nearly +1.... Since, however, the effects of overgrazing are shared by all the herdsmen, the negative utility for any particular decision-making herdsman is only a fraction of -1. Adding together the...partial utilities, the rational herdsman concludes that the only sensible course for him to pursue is to add another animal to [the] herd. And another; and another.... Therein is the tragedy. Each man is locked into a system that [causes] him to increase his herd without limit—in a world that is limited....Freedom in a commons brings ruin to all.

—G. Hardin, "The Tragedy of the Commons,"
Science **162,** 1243 (1968), p. 1244

(a)

(b)

(c)

Figure 22.10 **Tropical rain forest destruction.**

(a) These fires are destroying the rain forest in Brazil, which is being cleared for cattle pasture. (b) The flames are so widespread and so high that their smoke can be viewed from space. (c) The consequences of deforestation can be seen on these middle-elevation slopes in Ecuador. The slopes now support only low-grade pastures where they used to support highly productive forest.

Groundwater

A second resource that we cannot replace is **groundwater,** water trapped beneath the soil within porous rock reservoirs called aquifers. This water seeped into its underground reservoir very slowly during the last ice age over 12,000 years ago. We should not waste this treasure, for we cannot replace it.

In most areas of the United States, local governments exert relatively little control over the use of groundwater. As a result, a very large portion is wasted watering lawns, washing cars, and running fountains. A great deal more is inadvertently being polluted by poor disposal of chemical wastes—and once pollution enters the groundwater, there is no effective means of removing it. Some cities, like Phoenix and Las Vegas, may completely deplete their groundwater within several decades.

Biodiversity

The number of species in danger of extinction during your lifetime is far greater than the number that became extinct with the dinosaurs. This disastrous loss of biodiversity is important to every one of us, because as these species disappear, so does our chance to learn about them and their possible benefits for ourselves. The fact that our entire supply of food is based on 20 kinds of plants, out of the 250,000 available, should give us pause. Like burning a library without reading the books, we don't know what it is we waste. All we can be sure of is that we cannot retrieve it. Extinction is forever.

Over the last 20 years, about half of the world's tropical rain forests have been either burned to make pasture land or cut for timber (figure 22.10). Over 6 million square kilometers have been destroyed. Every year the rate of loss increases as the human population in the tropics grows. About 160,000 square kilometers were cut each year in the 1990s, a rate greater than 0.6 hectares (1.5 acres) per second! At this rate, all the rain forests of the world will be gone in your lifetime. In the process, it is estimated that one-fifth or more of the world's species of animals and plants will become extinct—more than a million species. This would be an extinction event unparalleled for at least 65 million years, since the Age of Dinosaurs.

You should not be lulled into thinking that loss of biodiversity is a problem limited to the tropics. The ancient forests of the Pacific Northwest are being cut at a ferocious rate today. At the current rate, very little will remain in a decade. Nor is the problem restricted to one area. Throughout our country, natural forests are being "clear-cut," replaced by pure stands of lumber trees planted in rows like so many lines of corn. It is difficult to scold those living in the tropics when we do such a poor job of preserving our own country's biodiversity.

But what is so bad about losing species? What is the value of biodiversity? Loss of a species entails three costs: (1) the direct economic value of the products we might have obtained from species; (2) the indirect economic value of benefits produced by species without our consuming them, such as nutrient recycling in ecosystems; and (3) their ethical and aesthetic value. It is not difficult to see the value in protecting species that we use to obtain food, medicine, clothing, energy, and shelter, but other species are vitally important to maintaining healthy ecosystems; by destroying biodiversity, we are creating conditions of instability and lessened productivity. Other species add beauty to the living world, no less crucial because it is hard to set a price upon.

22.8 Curbing Population Growth

CONCEPT PREVIEW: The problem at the core of all other environmental concerns is the rapid growth of the world's human population. Serious efforts are being made to slow its growth.

If we were to solve all the problems mentioned in this chapter, we would merely buy time to address the fundamental problem: there are getting to be too many of us.

Humans first reached North America at least 12,000 to 13,000 years ago, crossing the narrow straits between Siberia and Alaska and moving swiftly to the southern tip of South America. By 10,000 years ago, when the continental ice sheets withdrew and agriculture first developed, about 5 million people lived on earth, distributed over all the continents except Antarctica. With the new and much more dependable sources of food that became available through agriculture, the human population began to grow more rapidly. By the time of Christ, 2,000 years ago, an estimated 130 million people lived on earth. By the year 1650, the world's population had doubled, and doubled again, reaching 500 million. Starting in the early 1700s, changes in technology have given humans more control over their food supply, led to the development of cures for many diseases, and have produced improvements in shelter and storage capabilities that make humans less vulnerable to climatic uncertainties. Recall from chapter 19 that populations grow exponentially until they reach the limits of their environment, called the carrying capacity. These changes that occurred since the 1700s allowed humans to expand the carrying capacity of the habitats in which they lived and thus to escape the confines of logistic growth and reenter the exponential phase of the sigmoidal growth curve, shown by the explosive growth in figure 22.11.

Although the human population has grown explosively for the last 300 years, the average human birthrate has stabilized at about 21 births per year per 1,000 people worldwide. However, with the spread of better sanitation and improved medical techniques, the death rate has fallen steadily, to its present level of 9 per 1,000 per year. The difference between birth and death rates amounts to a population growth rate of 1.2% per year, which seems like a small number, but it is not, given the large population size (figure 22.12).

The world population reached 6.7 billion people in 2008, and the annual increase now amounts to about 80 million people, which leads to a doubling of the world population in about 58 years. Put another way, more than 219,000 people are added to the world population each day, or almost 152 every minute. At this rate, the world's population will continue to grow and perhaps stabilize at a figure around 10 billion. Such growth cannot continue, because our world cannot support it. Just as a cancer cannot grow unabated in your body without eventually killing you, so humanity cannot continue to grow unchecked in the biosphere without killing it.

Most countries are devoting considerable attention to slowing the growth rate of their populations and there are genuine signs of progress, but the world population may still gain another 1 to 4 billion people before it stabilizes. No one knows whether the world can support so many people indefinitely. Finding a way to do so is the greatest task facing humanity. The quality of life that will be available for your children in this new century will depend to a large extent on our success.

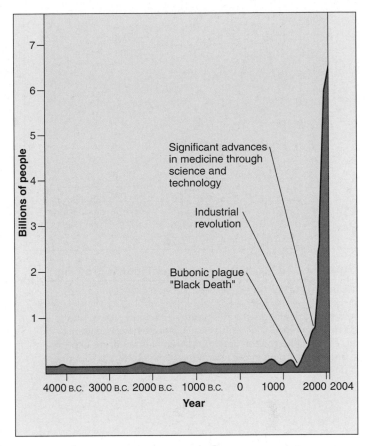

Figure 22.11   Growth curve of the human population.

Over the past 300 years, the world population has been growing steadily. Currently, there are over 6 billion people on the earth.

Figure 22.12   Mexico City has a population of over 19 million people.

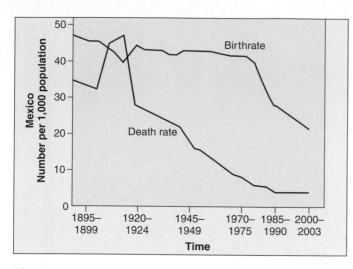

Figure 22.13   **Why Mexico's population is growing.**

The death rate (*red line*) in Mexico has been falling, while the birthrate (*blue line*) remained fairly steady until 1970. The difference between birth and death rates has fueled a high growth rate. Efforts begun in 1970 to reduce the birthrate have been quite successful. Although the growth rate remains rapid, it is expected to begin leveling off in the near future as the birthrate continues to drop.

Figure 22.14   **Population pyramids.**

Population pyramids are graphed according to a population's age distribution. Kenya's pyramid has a broad base because of the great number of individuals below child-bearing age. When all of the young people begin to bear children, the population will experience rapid growth. The 2005 U.S. pyramid demonstrates a larger number of individuals in the "baby boom" cohort—the pyramid bulges because of an increase in births between 1945 and 1964, as shown at the base of the 1964 pyramid. The 25 to 34 cohort in the 1964 pyramid represents people born during the Depression and is smaller in size than the cohorts in the preceding and following years.

Population Growth Rate Has Been Declining

The world population growth rate has been declining, from a high of 2.0% in the period 1965–70 to 1.2% in 2007. Nonetheless, because of the larger population, this amounts to an increase of 80 million people per year to the world population, compared to 53 million per year in the 1960s.

The United Nations attributes the decline to increased family planning efforts and the increased economic power and social status of women. As family size decreases in developing countries, education programs improve, leading to increased education levels for women, which in turn tends to result in further decreases in family size.

No one knows whether the world can sustain today's population of over 6.7 billion people, much less the far greater numbers expected in the future. If we are to avoid catastrophic increases in death rates, such as the tragedy we are seeing in sub-Saharan Africa, the birthrates must continue to fall dramatically.

Population Pyramids

While the human population as a whole continues to grow rapidly, this growth is not occurring uniformly over the planet. Some countries, like Mexico, are currently growing rapidly. Figure 22.13 shows how Mexico's birthrate, while declining (the blue line), still greatly exceeds its death rate (the red line). There is often a correlation in how developed a country is and how rapidly its population grows. Table 22.1, on the next page, compares three countries that differ in their levels of development. Ethiopia, a developing country, has a higher fertility rate, which results in a higher birthrate than either Brazil or the United States. But Ethiopia also has a much higher infant mortality rate and a lower life expectancy. Overall, the population in Ethiopia will double much more quickly than the population of Brazil or the United States. The rate at which a population can be expected to grow in the future can be assessed graphically by means of a population pyramid—a bar graph displaying the numbers of people in each age category (some examples are shown in figure 22.14). Males are conventionally shown to the left of the vertical age axis (colored blue here) and females to the right (colored red). In most human population pyramids, the number of older females is disproportionately large compared with the number of older males, because females in most regions have a longer life expectancy than males. This is apparent in the upper portion of the 2005 U.S. pyramid.

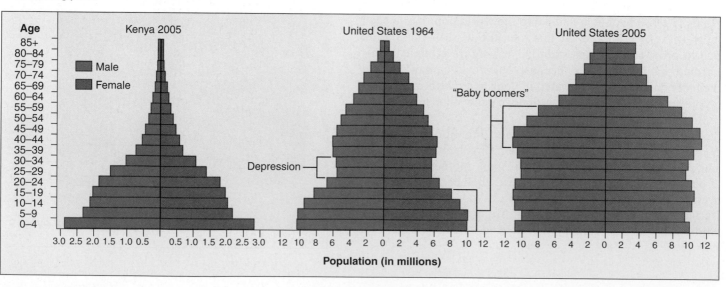

TABLE 22.1	A Comparison of 2006 Population Data in Developed and Developing Countries		
	United States (highly developed)	Brazil (moderately developed)	Ethiopia (developing)
Fertility rate	2.1	2.3	5.4
Doubling time at current rate (yr)	72.2	55.5	27.9
Infant mortality rate (infant deaths/1,000 births)	6.5	27	77
Life expectancy (yr)	78	72	49
Per capita income (U.S. dollar equivalent)	$44,260	$8,800	$1,190

Viewing such a pyramid, one can predict demographic trends in births and deaths. In general, rectangular pyramids are characteristic of countries whose populations are stable; their numbers are neither growing nor shrinking. A triangular pyramid, like the 2005 Kenya pyramid, is characteristic of a country that will exhibit rapid future growth, as most of its population has not yet entered the child-bearing years. Inverted triangles indicate shrinking populations.

Compare the differences in the population pyramids for the United States and Kenya in figure 22.14. In the somewhat more rectangular population pyramid for the United States in 2005, the cohort (group of individuals) 40 to 59 years old represents the "baby boom," the large number of babies born following World War II. The very triangular pyramid of Kenya, by contrast, predicts explosive future growth. The population of Kenya is predicted to double in less than 20 years.

It is important to note that these estimates do not take into account the huge impact that natural disasters and epidemics such as AIDS will have on population sizes. In sub-Saharan Africa, the AIDS epidemic has reduced the life expectancy at birth by 20 years. Figure 22.15 shows two population pyramid projections for Botswana, Africa, where over 36% of the population is living with HIV or AIDS. The uncolored portions of the bars indicate the population projections in 2025 without the effect of the AIDS epidemic, and the colored bars are projections with AIDS.

The Level of Consumption in the Developed World Is Also a Problem

The wealthiest 20% of the world's population accounts for 86% of the world's consumption of resources and produces 53% of the world's carbon dioxide emissions, whereas the poorest 20% of the world is responsible for only 1.3% of consumption and 3% of CO_2 emissions.

One way of quantifying this disparity is by calculating what has been termed the **ecological footprint,** which is the amount of productive land required to support an individual at the standard of living of a particular population through the course of his or her life. As figure 22.16 illustrates, the ecological footprint of an individual in the United States is more than 10 times greater than that of someone in India. Based on these measurements, researchers have calculated that resource use by humans is now one-third greater than the amount that nature can sustainably replace; if all humans lived at the standard of living in the developed world, two additional planet earths would be needed.

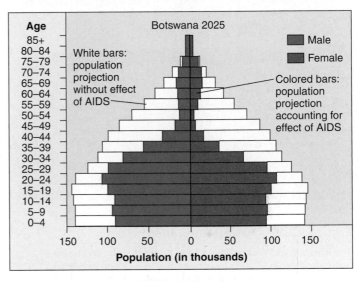

Figure 22.15 Projected AIDS effect on Botswana population (year 2025).

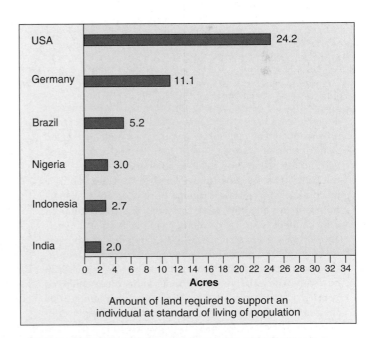

Figure 22.16 Ecological footprint of individuals in different countries in 2003.

Concept Check

1. Explain why the "optimum amount of pollution" is not zero.
2. What is the human population growth rate? Is this fast? Explain.
3. What is the downside of a triangular population pyramid?

Solving Environmental Problems

(a)

(b)

Figure 22.17   Habitat restoration.

The University of Wisconsin-Madison Arboretum has pioneered restoration ecology. (a) The restoration of the prairie was at an early stage in November, 1935. (b) The prairie as it looks today. This picture was taken at approximately the same location as the 1935 photograph.

BIOLOGY & YOU

Endangered Species Fur. Got $50,000 to spare? That's a typical penalty for breaking the Endangered Species Act of 1973, which outlaws trade of any product from an endangered animal. Suppose you have a genuine leopard skin fur coat and want to sell it. You list the coat on eBay with no restrictions and sell it to someone in another state or country. Not only does this violate eBay policies, you have just broken a federal law. Although once widely available, fur from leopards and other large cats like tigers and cheetahs is now banned—even fur garments made as long as 30 years ago. Virtually every bear species is endangered, as are many types of seal and otter. Endangered plants are protected too. Digging up a cactus plant you encounter on a desert hike can be a crime! The list of protected species under the Endangered Species Act is lengthy and complex, containing about 2,000 species of mammals, birds, fish, insects and other invertebrates, and plants. So unless you are positive the fur garment, purse, or throw rug you want to buy or sell is not from an animal listed as endangered or threatened, you should first check to be certain.

22.9 Preserving Endangered Species

CONCEPT PREVIEW: Recovery programs at the species level must deal with habitat loss and fragmentation, and often with a marked reduction in genetic diversity. Captive breeding programs that stabilize genetic diversity and pay attention to habitat preservation and restoration are typically involved in successful recoveries.

Once you understand the reasons why a particular species is endangered, it becomes possible to think of designing a recovery plan. If the cause is commercial overharvesting, regulations can be designed to lessen the impact and protect the threatened species. If the cause is habitat loss, plans can be instituted to restore lost habitat. Loss of genetic variability in isolated subpopulations can be countered by transplanting individuals from genetically different populations. Populations in immediate danger of extinction can be captured, introduced into a captive breeding program, and later reintroduced to other suitable habitats.

Habitat Restoration

Conservation biology typically concerns itself with preserving populations and species in danger of decline or extinction. However, in many situations conservation is no longer an option. The clear-cutting of the temperate forests of Washington State leaves little behind to conserve, nor does converting a piece of land into a wheat field or an asphalt parking lot. Redeeming these situations requires restoration rather than conservation. Three quite different sorts of habitat restoration programs might be undertaken, depending very much on the cause of the habitat loss.

Pristine Restoration.   In situations where all species have been effectively removed, one might attempt to restore the plants and animals that are believed to be the natural inhabitants of the area, when such information is available. Restoring abandoned farmland to prairie (figure 22.17) requires that you know the identity of all of the original inhabitants, and the ecologies of each of the species. We rarely have this much information, so no restoration is truly pristine.

Removing Introduced Species.   Sometimes the habitat of a species has been destroyed by a single introduced species. In such a case, habitat restoration involves removal of the introduced species. For example, Lake Victoria, Africa, was home to over 300 species of cichlid fishes, small perchlike fishes that display incredible diversity. However, in 1954, the Nile perch, a commercial fish with a voracious appetite, was introduced into Lake Victoria. This resulted in the loss of over 70% of cichlid species, including all open-water species.

Restoration of the once-diverse cichlid fishes to Lake Victoria will require that the Nile perch population (as well as other introduced species such as water hyacinth plants), be brought under control or removed, as well as breeding and restocking the endangered species.

Cleanup and Rehabilitation.   Habitats seriously degraded by chemical pollution cannot be restored until the pollution is cleaned up. The successful restoration of the Nashua River in New England, discussed later in this chapter, is one example of how a concerted effort can succeed in restoring a heavily polluted habitat to a relatively pristine condition.

Captive Propagation

Recovery programs, particularly those focused on one or a few species, often must involve direct intervention in natural populations to avoid an immediate threat of extinction. Introducing wild-caught individuals into captive breeding programs is being used in an attempt to save the black-footed ferret and California condor populations in immediate danger of disappearing. Several other such captive propagation programs have had success.

Case History: The Peregrine Falcon.   U.S. populations of birds of prey such as the peregrine falcon (*Falco peregrinus*) began an abrupt decline shortly after World War II. Of the approximately 350 breeding pairs east of the Mississippi River in 1942, all had disappeared by 1960. The culprit proved to be the chemical pesticide DDT and related organochlorine pesticides. Birds of prey are particularly vulnerable to DDT because they feed at the top of the food chain, where DDT becomes concentrated (see the discussion of biological magnification in section 22.1). DDT interferes with the deposition of calcium in the bird's eggshells, causing most eggs to break before they hatch.

The use of DDT was banned by federal law in 1972, causing levels in the eastern United States to fall quickly. Because there were no peregrine falcons left in the eastern United States to reestablish a natural population, falcons from other parts of the country were used to establish a captive breeding program at Cornell University in 1970. The intent of this program was to reestablishing the peregrine falcon in the eastern United States by releasing offspring of these birds. By the end of 1986, over 850 birds had been released in 13 eastern states, producing an astonishingly strong recovery (figure 22.18).

Figure 22.18   Captive propagation.

The reestablishment of peregrine falcon populations in the eastern U.S. is a success story for captive propagation programs.

Sustaining Genetic Diversity

One of the chief obstacles to a successful species recovery program is that a species is generally in serious trouble by the time a recovery program is instituted. When populations become very small, much of their genetic diversity is lost. If a program is to have any chance of success, every effort must be made to sustain as much genetic diversity as possible.

Case History: The Black Rhino.   All five species of rhinoceros are critically endangered. The three Asian species live in a forest habitat that is rapidly being destroyed, while the two African species are illegally killed for their horns. Fewer than 11,000 individuals of all five species survive today. The problem is intensified by the fact that many of the remaining animals live in very small, isolated populations. The 2,400 wild-living individuals of the black rhino, *Diceros bicornis* (figure 22.19), live in approximately 75 small, widely separated groups

> Small populations are at greater risk of extinction. Genetic drift, as discussed on page 261, is caused by random events that eliminate individuals, and therefore alleles, from a population and can be devastating if that population is small to begin with.

that are adapted to local conditions throughout the species' range. All six subspecies appear to have low genetic variability; in three of the subspecies, only a few dozen animals remain. Analysis of mitochondrial DNA suggests that in these populations most individuals are genetically very similar.

This lack of genetic variability represents one of the greatest challenges to the future of the species. Much of the range of the black rhino is still open and not yet subject to human encroachment. To have any

Figure 22.19   Sustaining genetic diversity.

The black rhino is highly endangered, living in 75 small, widely separated populations. Only about 2,400 individuals survive in the wild. Conservation biologists have the difficult job of finding ways to preserve genetic diversity in small, isolated populations.

IMPLICATION   A basic problem faced by black rhinos is that the species has lost much of its genetic diversity as its numbers have decreased. Do you think that expanding the size of today's small rhino populations would fix this problem?

Figure 22.20 Preserving keystone species.

The flying fox is a keystone species in many Old World tropical islands. It pollinates many of the plants, and is a key disperser of seeds. Its elimination by hunting and habitat loss is having a devastating effect on the ecosystems of many South Pacific islands.

IMPLICATION Can you think of a keystone species that plays a key role in natural ecosystems in your area? How would you go about confirming that it is indeed a keystone species?

Figure 22.21 A mega-reserve.

The Baviaanskloof mega-reserve in South Africa's Eastern Cape Province includes 270,000 hectares of unspoiled terrain, habitat for many species of plants and animals.

significant chance of success, a species recovery program will have to find a way to sustain the genetic diversity that remains in this species. Heterozygosity could be best maintained by bringing all black rhinos together in a single breeding population, but this is not a practical possibility. A more feasible solution would be to move individuals between populations. Managing the black rhino populations for genetic diversity could prevent the loss of genetic variation, which might prove fatal to this species.

Preserving Keystone Species

Keystone species are species that exert a particularly strong influence on the structure and functioning of their ecosystem. Their removal can have disastrous consequences.

Case History: Flying Foxes. The severe decline of many species of pteropodid bats, or "flying foxes," in the Old World tropics is an example of how the loss of a keystone species can have dramatic effects on the other species living within an ecosystem, sometimes even leading to a cascade of further extinctions (figure 22.20). These bats have very close relationships with important plant species on the islands of the Pacific and Indian Oceans. Widespread on the islands of the South Pacific, flying foxes are the most important—and often the only—pollinators and seed dispersers. A study in Samoa found that 80% to 100% of the seeds landing on the ground during the dry season were deposited by flying foxes. Many species are entirely dependent on these bats for pollination.

Flying foxes are being driven to extinction by human hunting. They are hunted for food and by orchard farmers, who consider them pests. Flying foxes are particularly vulnerable because they live in large, easily seen groups of up to a million individuals. Because they move in predictable patterns and can be tracked to their home roost, hunters can easily bag thousands at a time.

In Guam, where the two local species of flying fox have recently been driven extinct or nearly so, the impact on the ecosystem appears to be substantial. Many plant species are not fruiting, or are doing so only marginally, with fewer fruits than normal. Attempts to preserve flying fox populations are underway. In each instance, legal protection, captive breeding programs, and habitat preservation are key ingredients to successful programs.

Conservation of Ecosystems

Habitat fragmentation is one of the most pervasive enemies of biodiversity conservation efforts. Some species simply require large patches of habitat to thrive, and conservation efforts that cannot provide suitable habitat of such a size are doomed to failure. It has become clear that species placed in isolated patches of habitat are lost far more rapidly than species placed in large preserves. As a result, conservation biologists have promoted the creation, particularly in the tropics, of so-called megareserves, large areas of land containing a core of one or more undisturbed habitats (figure 22.21).

However, ample land to form mega-reserves may not be available. In these cases, the linking together of habitat islands through the use of corridors is proving an option. Corridors are areas with suitable vegetation and topography between habitats that allow animals to wander outside of their home ranges. Corridors allow for seasonal migrations, the dispersal of juveniles into new territories, and the expansion of growing populations.

22.10 Finding Cleaner Sources of Energy

CONCEPT PREVIEW: Renewable energy, particularly auto fuels from biomass, are becoming increasingly important as cleaner energy sources.

Modern society's propensity to burn fossil fuels has recycled an enormous amount of CO_2 back into the earth's atmosphere. To gain some idea, focus for a moment on your own personal contribution to the carbon cycle. Every mile you drive your car releases about a pound of CO_2 into the air. How many miles do you drive in a year? You see the point? Think about the natural gas that heats your home, the electricity that lights it (mostly generated by the burning of fossil fuels). Your life is having a significant impact on the earth's carbon balance. And you are not alone. Three hundred million other Americans are having a similar impact. In 2006 we saw reductions in carbon dioxide emission in the United States, but Americans still released nearly 6 billion metric tons of carbon dioxide into earth's atmosphere from the burning of fossil fuels (5,973,000,000,000,000 pounds!).

This massive flow of carbon dioxide into the atmosphere is having an unintended and very grave consequence—the earth is getting warmer. What can we do about it? For now the best way to avoid contributing to global warming will be to switch to *alternative energy sources*, and in particular to a different way of powering automobiles that doesn't add to the atmosphere's load of CO_2.

Many countries are turning to nuclear power for their energy needs. Over 70% of France's electricity is now produced by nuclear power plants. In theory nuclear power can provide plentiful, cheap energy, but nuclear power presents several problems—safety, waste disposal, security—that must be overcome if it is to provide a significant portion of the world's energy (figure 22.22).

Alternative Energy Sources

A variety of other sources of cleaner energy can help reduce our use of fossil fuels. Many of these are *renewable energy*—sources of energy such as solar power that are naturally replenished. The solar panels in figure 22.23*a* capture the energy of sunlight to heat water or other fluids to make steam that turns a turbine, generating electricity. Smaller applications use solar panels connected to photovoltaic cells, which convert solar energy directly into electricity. Other sources of renewable energy include wind (figure 22.23*b)*, and biomass, plants like corn and sugar cane that can be used to produce ethanol, replacing gasoline in cars.

Looking Closer at Ethanol

Ethanol is a simple two-carbon alcohol, CH_3CH_2OH—the same alcohol found in beer and wine. Rich in energy-storing C—H bonds, ethanol makes a good fuel. Burning a gallon of ethanol in your car releases about 80% as much car-powering energy as burning a gallon of gasoline.

How can we burn ethanol and not add more CO_2 to the atmosphere? Focus on the word "more." Ethanol is produced through yeast fermentation of sugars found in plants, the same fermentation process that is used to make beer and wine. If our automobiles burn carbon molecules recently produced via photosynthesis by living plants, then they are simply returning to the atmosphere the CO_2 recently extracted from it! No net increase in atmospheric CO_2 occurs.

Figure 22.22   Three Mile Island nuclear power plant.

Since a nuclear accident here in 1979, the building of nuclear power stations in the United States has slowed dramatically.

(a)

(b)

Figure 22.23   Alternate energy sources.

(a) Solar energy uses large mirrors to collect energy from the sun. These solar panels absorb heat from the sun that is used to boil water (or other fluids), which creates steam. The steam turns large turbines (not pictured), generating electricity. (b) Wind-powered energy is an old technology modernized for large-scale use. Large wind fields harness the kinetic energy in wind, converting it into electricity.

Figure 22.24 **The starch in corn kernels is used in ethanol production.**

To understand this, focus on where the carbon atoms come from. Burning a fossil fuel like gasoline releases into the atmosphere stores of carbon that had been trapped for thousands of years as oil buried deep in the earth. Burning ethanol also releases carbon, but in this case the carbon dioxide released into the atmosphere has just been taken from it. Think of the atmosphere as a fountain. A fountain recycles the water it shoots into the air. The level of water in the pool stays the same because the water added to the pool by the falling spray is recycled back to the pump to shoot up again. That is how ethanol works with regards to carbon dioxide emissions. Carbon dioxide is taken from the atmosphere and used by the plant to build plant tissue; that tissue is then used to make ethanol. Now imagine if there is a nearby tank of water (representing fossil fuels), and water from the tank was pumped into the fountain's pool. Not only would the tank get depleted, but the pool would overflow (too much carbon dioxide). That is what happens when fossil fuels are burned.

When used as fuel, ethanol is typically added to gasoline rather than burned by itself. New cars called Flexible Fuel Vehicles (FFV) have a redesigned engine that can burn 100% gasoline, but can also burn a gasoline blend called E85 that is 85% ethanol and 15% gasoline.

In the United States, commercial ethanol fuel is traditionally produced by fermenting sugars obtained from starch stored in corn kernels (figure 22.24). How might we get ethanol from the rest of the corn plant—the stalk, leaves, and cob? They are made largely of three kinds of organic molecules, other than starch: 40% cellulose, 40% hemicellulose, and 10% lignins.

Focus first on the cellulose. Fully one-half of all the organic carbon in the living world is cellulose. Like starch, cellulose consists of chains of glucose sugars linked together. Why not use the sugars in cellulose to make ethanol? Because there is a subtle but very important chemical difference between starch and cellulose. Each of the glucose sugars in a starch molecule has six carbon atoms arranged in a ring like a group of children holding hands in a circle. In a cellulose molecule, the rings of the glucose molecules are inside-out, as if the children are facing with their backs to the center of the circle. Yeast enzymes do not attack links between these kinds of sugars. To use cellulose to produce ethanol, bioengineers must find a way to teach yeasts to break these links.

There are microbes with enzymes that can do this, otherwise cows could not survive by eating grass, nor termites wood. Using the sort of genetic engineering technology discussed in chapter 13, it is possible to "bioengineer" a yeast to be able to ferment cellulose. Researchers in Spain have succeeded in commercially producing ethanol from cellulose biomass. They added DNA taken from plant-digesting bacteria in the gut of termites to yeast—DNA containing the genes these bacteria use to break down cellulose and free sugars.

Nor should we forget hemicellulose, which contains one-fifth of the corn plant's carbon. Hemicellulose is like cellulose, but with five-carbon sugars. Genetically-engineered bacteria containing the enzymes necessary to break down hemicellulose into free five-carbon sugars have already been constructed, so this approach seems rich with promise.

In addition to corn, there are other fast-growing plants that could be dedicated to fuel production, including switchgrass and trees such as hybrid poplar and willows. These plants can be grown on lands that are not suitable for row crops like corn, especially erosion-prone soils, leaving fertile soil for consumable crops such as corn, soybeans, and wheat. Other sources of cellulose from industrial and commercial waste, such as sawdust and paper pulp, could also be used to produce ethanol, and so could leaves and yard waste and the paper and cardboard that make up the bulk of municipal dumps.

IN THE NEWS

Plug-in Hybrids. Increased use of biofuels is not the only way to reduce American automobile carbon emissions. A very promising alternative is to power our cars with electricity. Although today's electricity comes largely from the burning of coal, this will be less true in the future. Wind, solar, and nuclear power are all becoming increasingly important components of the nation's power grid. The first electrically powered cars, due to arrive on American highways in about 2010, are so-called "plug-in hybrids." Like traditional hybrid vehicles, plug-in hybrid cars have both electric and internal combustion motors. In traditional hybrids, the internal combustion motor is the primary source of power, with the electrical motor providing supplementary power. The battery, used during idling, is charged by the car's internal combustion engine, not by plug-in. The plug-in hybrids have high-capacity lithium ion batteries that are capable of powering the car for about 40 miles of continuous driving using only the electric motor (the length of the daily commute of 70% of Americans is 35 miles). If you drive further, the internal combustion engine kicks in to spin the shaft of a generator that recharges the car's batteries as you continue to drive. The car uses no fossil fuel at all within its all-electric range, and its batteries can be conveniently recharged at home simply by plugging it in like a lamp or kitchen appliance. As future batteries improve, it seems likely that plug-in hybrids will evolve into all-electric vehicles. If such cars become widely used, we can expect that the need for gasoline or biofuels to power American cars will greatly diminish.

22.11 Individuals Can Make the Difference

CONCEPT PREVIEW: In solving environmental problems, the commitment of one person can make a critical difference. Biological literacy is no longer a luxury for scientists—it has become a necessity for all of us.

It is important not to lose sight of the key role often played by informed individuals in solving environmental problems. Often one person has made the difference; two examples serve to illustrate the point.

The Nashua River

Running through the heart of New England, the Nashua River was severely polluted by mills established in Massachusetts in the early 1900s. By the 1960s, the river was clogged with pollution and declared ecologically dead. When Marion Stoddart moved to a town along the river in 1962, she was appalled. She approached the state about setting aside a "greenway" (trees running the length of the river on both sides), but the state wasn't interested in buying land along a filthy river. So Stoddart organized the Nashua River Cleanup Committee and began a campaign to ban the dumping of chemicals and wastes into the river. The committee presented bottles of dirty river water to politicians, spoke at town meetings, recruited businesspeople to help finance a waste treatment plant, and began to clean garbage from the Nashua's banks. This citizen's campaign, coordinated by Stoddart, greatly aided passage of the Massachusetts Clean Water Act of 1966. Industrial dumping into the river is now banned, and the river has largely recovered (figure 22.25).

Lake Washington

A large, 86-square-kilometer freshwater lake east of Seattle, Lake Washington became surrounded by Seattle suburbs in the building boom following the Second World War. Between 1940 and 1953, a ring of 10 municipal sewage plants discharged their treated effluent into the lake. Safe enough to drink, the effluent was believed "harmless." By the mid-1950s a great deal of effluent had been dumped into the lake (try multiplying 80 million liters/day × 365 days/year × 10 years). In 1954, an ecology professor at the University of Washington in Seattle, W. T. Edmondson, noted that his research students were reporting filamentous blue-green algae growing in the lake. Such algae require plentiful nutrients, which deep freshwater lakes usually lack—the sewage had been fertilizing the lake! Edmondson, alarmed, began a campaign in 1956 to educate public officials to the danger: Bacteria decomposing dead algae would soon so deplete the lake's oxygen that the lake would die. After five years, joint municipal taxes financed the building of a sewer to carry the effluent out to sea. The lake is now clean (figure 22.26).

Concept Check

1. In what way does each of the three sorts of habitat restoration depend on the cause of the habitat loss?
2. Name three sources of renewable energy.
3. Burning a gallon of ethanol releases CO_2 into the air just as burning a gallon of gasoline does, so why does ethanol contribute less to global warming?

Figure 22.25   Cleaning up the Nashua River.

The Nashua River, seen on the left in the 1960s, was severely polluted because factories set up along its banks dumped their wastes directly into the river. Seen on the right today, the river is mostly clean.

Figure 22.26   Lake Washington, Seattle.

Lake Washington in Seattle is surrounded by residences, businesses, and industries. By the 1950s, the dumping of sewage and the runoff of fertilizers had caused an algal bloom in the lake, which would eventually deplete the lake's oxygen. Efforts to reverse this effect and clean up the lake were started by W. T. Edmondson of the University of Washington in 1956. The lake is now clean.

IMPLICATION   It is easy to become discouraged when considering the world's many environmental problems, but do not lose track of the single most important conclusion that emerges from our examination of these problems—the fact that each is solvable. A polluted lake can be cleaned; a dirty smokestack can be altered to remove noxious gas; waste of key resources can be stopped. What is required is a clear understanding of the problem and a commitment to doing something about it. The extent to which U.S. families recycle aluminum cans and newspapers is evidence of the degree to which people want to become part of the solution, rather than part of the problem. What do you recycle? What other items could you be recycling?

How Real Is Global Warming?

The controversy over global warming has two aspects. The first contentious issue is the claim that global temperatures are rising significantly, a profound change in the earth's atmosphere and oceans referred to as "global warming." The second contentious issue is the assertion that global warming is the consequence of elevated concentrations of carbon dioxide in the atmosphere as a consequence of the widespread burning of fossil fuels.

Resolution of the second issue requires detailed science and is only now reaching consensus acceptance. Resolution of the first issue is a simpler proposition, because it is, in essence, a data statement. The graph to the right displays the data in question, global air temperatures for the last century and a half. Temperature data is collected from measuring stations across the globe, as shown in the image below, and averaged. The bars of the histogram represent mean yearly global air temperatures for each year since 1850. In order to dampen the effects of random year-to-year variations and so better reveal accumulating influences, the data are presented as an anomaly histogram (in an **anomaly histogram**, each bar presents the deviation of the value during that period from the average value determined for some standard period). In this instance, the anomaly histogram shows the deviation of each year's global mean air temperature from the mean of these values observed over a standard 30-year period between 1961 and 1990.

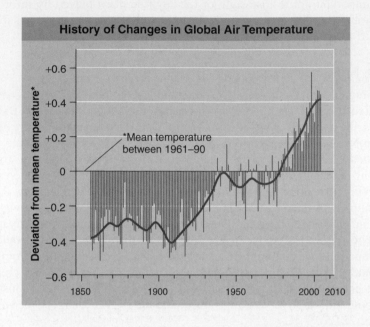

History of Changes in Global Air Temperature

*Mean temperature between 1961–90

Deviation from mean temperature*

Analysis

1. **Applying Concepts** What fraction of the 155 years do not deviate from the 1961–1990 mean value? What fraction deviates more than +0.2°C? more than −0.2°C? more than +0.4°C? more than −0.4°C?

2. **Interpreting Data**
   a. Of the years that deviate more than +0.2°C, how many are before 1940? between 1940 and 1980? after 1980? What fraction occur after 1980?
   b. Of the years that deviate more than +0.4°C, how many are before 1980? after 2000? What fraction occur after 2000?
   c. Of the years that deviate more than −0.2°C, how many are before 1940? between 1940 and 1980? after 1980? What fraction occur before 1940?
   d. Of the years that deviate more than −0.4°C, how many are before 1940? 1900? What fraction occur before 1900?

3. **Making Inferences** If you were to pick a year at random between 1850 and 1900, would it be most likely to deviate +0.2, +0.4, 0, −0.2, or −0.4? a year between 1900 and 1940? a year after 1980?

4. **Drawing Conclusions** Has the global air temperature been warming progressively over the last century and a half?

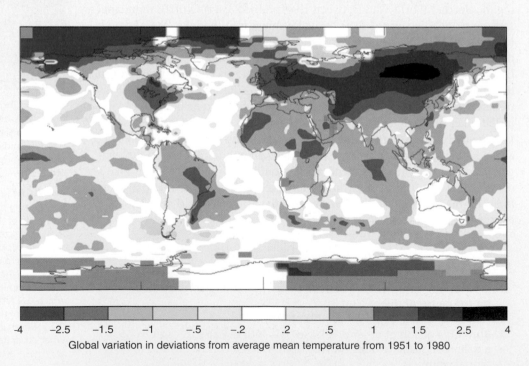

Global variation in deviations from average mean temperature from 1951 to 1980

Concept Summary

Global Change

22.1 Pollution

- Pollution leads to global change because its effects can spread far from the source. Air and water become polluted when chemicals that are harmful to organisms are released into the ecosystem. The use of agricultural chemicals, such as pesticides, herbicides, and fertilizers, has been widespread with devastating effects on animals. Biological magnification occurs when harmful chemicals become more concentrated as they pass up through the food chain, as shown here from **figure 22.1.**

22.2 Acid Precipitation

- The burning of coal releases sulfur into the atmosphere where it combines with water vapor to form sulfuric acid. This acid falls back to earth in rain and snow, commonly called acid rain, far from the source of the pollution, killing animals and vegetation (**figures 22.2** and **22.3**).

22.3 Global Warming

- The burning of fossil fuels releases carbon dioxide into the atmosphere. It remains in the atmosphere, where it traps infrared light (heat) from the sun, a phenomenon known as the greenhouse effect. As a result, the average global temperatures have been steadily increasing as CO_2 levels increase in the atmosphere (**figure 22.4**), a process known as global warming. Global warming is predicted to have major impacts on global rain patterns, agriculture, and rising sea levels.

22.4 Loss of Biodiversity

- Extinction is a fact of life, but the current rate of species loss is alarmingly high. Three factors mostly responsible for present-day extinctions include loss of habitat, overexploitation, and the introduction of new species (**figure 22.5**). Loss of habitat is the most devastating.

22.5 The Ozone Hole

- Ozone (O_3) forms a protective shield in the earth's upper atmosphere that blocks out harmful UV rays from the sun. In the mid-1970s, scientists determined that ozone was being depleted. The culprit, chlorofluorocarbons (CFCs), used in refrigeration systems, reacts with ozone, converting it to oxygen gas (O_2), which doesn't block UV rays. Termed the ozone hole, this reduction in ozone over and extending from the South Pole, as shown here from **figure 22.7,** is resulting in dangerously high levels of radiation reaching the earth. This drop in O_3 levels is resulting in an increase in the incidence of skin cancer.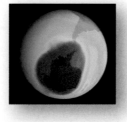

Saving Our Environment

22.6 Reducing Pollution

- Human activities are placing severe stress on the biosphere. Reducing pollution requires examining the costs associated with pollution (**figure 22.8**). Antipollution laws and pollution taxes are ways to begin factoring in the costs of pollution.

22.7 Preserving Nonreplaceable Resources

- The consumption or destruction of non-replaceable resources is perhaps the most serious problem humans face. Topsoil, necessary for agriculture, is being depleted rapidly. Groundwater, which percolates through the soil to underground reservoirs,  is our primary source of drinking water, but it is being wasted and polluted. Biodiversity is being reduced through extinctions, due primarily to loss of habitat, such as rain forests (**figure 22.10**).

22.8 Curbing Population Growth

- Technology has allowed the human population to grow exponentially for the last 300 years to a current population of over 6.7 billion (**figure 22.11**).

- Human populations grow at different rates, with developing countries' populations growing more rapidly than developed countries' populations as tracked using population pyramids (**figures 22.13** and **22.14** and **table 22.1**). Disease and natural disasters can affect regional population growth projections (**figure 22.15**).

- Although population growth rates are lower for developed countries, it takes more resources to support populations in developed countries (**figure 22.16**).

Solving Environmental Problems

22.9 Preserving Endangered Species

- In an attempt to slow the loss of biodiversity, recovery programs are under way, designed to save endangered species. These programs include habitat restoration,  breeding in captivity, sustaining genetic diversity, preserving keystone species, and conservation of ecosystems (**figures 22.17–22.21**).

22.10 Finding Cleaner Sources of Energy

- The burning of fossil fuels leads to pollution, depletion of valuable resources, and global warming. Alternative sources of energy are needed, and some countries have looked to nuclear power, but nuclear power has its own drawbacks. Renewable energy sources such as solar and wind power and ethanol are promising alternatives (**figure 22.23**).

22.11 Individuals Can Make the Difference

- There are environmental success stories, where one or a few individuals made a difference and reversed an ecological disaster (**figures 22.25** and **22.26**).

Self-Test

1. "Gray-air cities" are the result of
   a. biological magnification of air pollutants.
   b. chlorinated hydrocarbons as a major air pollutant.
   c. pesticides as a major air pollutant.
   d. sulfur oxides as a major air pollutant.

2. The main cause of acid rain is
   a. car and truck exhaust.   c. chlorofluorocarbons.
   b. coal-powered industry.   d. chlorinated hydrocarbons.

3. Global warming effects all of the following except
   a. rain patterns.   c. ozone levels.
   b. rising sea levels.   d. agriculture.

4. The factor most responsible for present day extinctions is
   a. habitat loss.
   b. overexploitation of species.
   c. introduction of new species.
   d. All are equally responsible.

5. Destruction of the ozone layer is due to
   a. car and truck exhaust.
   b. coal-powered industry.
   c. chlorofluorocarbons.
   d. chlorinated hydrocarbons.

6. Free market economies often promote pollution. This is because
   a. environmental costs are hardly ever recognized as part of the economy.
   b. supply never keeps up with demand, so industry must increase output to address the demand.
   c. the costs of energy and raw materials are so variable.
   d. laws controlling pollution are unenforceable.

7. Preserving biodiversity is
   a. needed to preserve possible direct value from species, such as new medicines.
   b. not needed as extinction is a "natural" cycle and should not be disturbed.
   c. needed to make sure all niches are filled.
   d. not needed as it interferes with industrial development.

8. Which factor is *not* responsible for the large increase in the human population over the last 300 or so years?
   a. larger and more reliable food reserves from the modernization of farming techniques
   b. decreasing mortality rate due to improvements in medicine
   c. increasing amounts of open space as countries develop
   d. increased sanitation practices

9. Removal of the endangered black-footed ferret and California condor populations from the wild for breeding programs in zoos and field laboratories are examples of preservation through
   a. pristine restoration.   c. habitat rehabilitation.
   b. habitat restoration.   d. captive propagation.

10. If the removal of a species causes an ecosystem to collapse, that species is known as a(n)
    a. keystone species.   c. threatened species.
    b. endangered species.   d. No answer is correct.

Visual Understanding

1. **Figure 22.5** Your friends tell you that their father complains that environmentalists are trying to save "every confounded bug and weed on the planet." He says that things have always gone extinct, why should now be any different. Use the graph to help you respond to your friends.

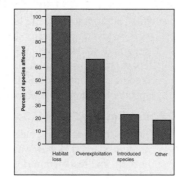

2. **Figures 19.5** and **22.11** Discuss the human growth curve from this chapter in terms of the graph from chapter 19.

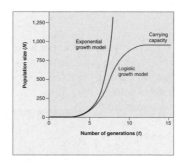
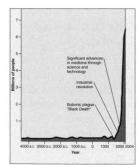

Challenge Questions

1. Explain why being exposed to even very tiny amounts of some chemical pollutants can, over time, be hazardous to your health.

2. Economists tend to look at the world in terms of costs. They say, "If you increase pollution, then you increase the social costs to people's health, but if you try to decrease pollution, then you increase the economic costs of cleaning it up." Discuss whether you think these two costs are paid equally by the same set of people.

3. Consider the stories of Marion Stoddart, W. T. Edmondson, and the following statement: "Never doubt that a small group of thoughtful, committed citizens can change the world, indeed it's the only thing that ever has."—Margaret Mead, Anthropologist. What *could* you do to make your neighborhood/area/community a better, healthier place? What *will* you do?

Chapter **23**

The Animal Body and How It Moves

The Animal Body Plan

23.1 Organization of the Vertebrate Body

CONCEPT PREVIEW: Groups of cells of the same type are organized in the vertebrate body into tissues. Organs are body structures composed of several different tissues. An organ system is a group of organs that work together to carry out an important function.

All vertebrates and other coelomates have the same general architecture: a long internal tube (the gut or digestive system) that extends from mouth to anus, which is suspended within an internal body cavity called the *coelom*. The coelom of many terrestrial vertebrates is divided into two parts: the *thoracic cavity*, which contains the heart and lungs, and the *abdominal cavity*, which contains the stomach, intestines, and liver. The vertebrate body is supported by an internal scaffold, or skeleton, made up of jointed bones or cartilage.

Like all animals, the vertebrate body is composed of cells—over 10 to 100 trillion of them in your body. It's difficult to picture how large this number actually is. A line of 10 trillion cars would stretch from the earth to the sun and back 50 million times! Not all of these cells in your body are the same, of course. If they were, we would not be bodies but amorphous blobs. Vertebrate bodies contain over 100 different kinds of cells.

Tissues

Groups of cells of the same type are organized within the body into **tissues,** which are the structural and functional units of the vertebrate body. A tissue is a group of cells of the same type that performs a particular function. It is possible to assemble many different kinds of tissue from 100 cell types, but biologists have traditionally grouped adult tissues into four general classes: *epithelial, connective, muscle,* and *nerve tissue.* The bird pictured in figure 23.1 contains all four classes of tissues, and you can see by the circled enlargements, each class of tissue contains different types of cells.

Organs

Organs are body structures composed of several different tissues grouped together into a larger structural and functional unit, just as a factory is a group of people with different jobs who work together to make something. The heart is an organ. It contains cardiac muscle tissue wrapped in connective tissue and joined to many nerves. All of these tissues work together to pump blood through the body: The cardiac muscles contract, which squeezes the heart to push the blood; the connective tissues act as a bag to hold the heart in the proper shape and ensure that the different chambers of the heart squeeze in the proper order; and the nerves control the rate at which the heart beats. No single tissue can do the job of the heart, any more than one piston can do the job of an automobile engine. Organs are the machines of the vertebrate body, each built from several different tissues and each doing a particular job.

Figure 23.1 **Vertebrate tissue types.**

The four basic classes of tissue are epithelial, nerve, connective, and muscle.

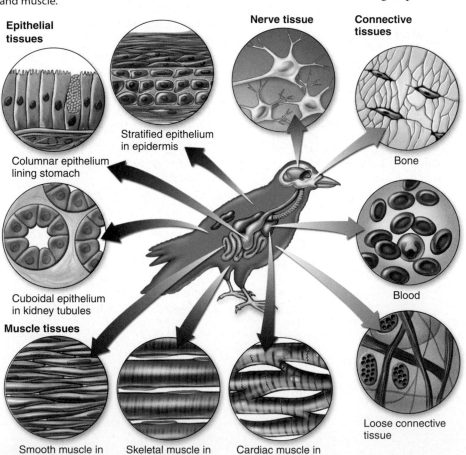

Epithelial tissues

Columnar epithelium lining stomach

Stratified epithelium in epidermis

Cuboidal epithelium in kidney tubules

Muscle tissues

Smooth muscle in intestinal wall

Skeletal muscle in voluntary muscles

Cardiac muscle in heart

Nerve tissue

Connective tissues

Bone

Blood

Loose connective tissue

Organ Systems

An **organ system** is a group of organs that work together to carry out an important function (figure 23.2). For example, the vertebrate digestive system is an organ system composed of individual organs that break up the food (beaks or teeth), pass the food to the stomach (esophagus), break down the food (stomach and intestine), absorb the food (intestine), and expel the solid residue (rectum). If all of these organs do their job right, the body obtains energy and necessary building materials from food. The digestive system is a particularly complex organ system with many different organs consisting of many different types of cells, all working together to carry out a complex function.

The vertebrate body contains 11 principal organ systems (Figure 23.3):

1. **Skeletal.** The most distinguishing feature of the vertebrate body is its internal skeleton made of cartilage or bone. The skeletal system protects the body and provides support for loco-motion and movement. Its principal components are bones, cartilage, and ligaments. Like arthropods, vertebrates have jointed appendages—arms, hands, legs, and feet.

> Jointed appendages, as discussed on page 346, was a key evolutionary innovation in arthropods that is maintained in the vertebrates. The term arthropods comes from two Greek words, *arthros*, meaning jointed, and *podes*, meaning feet.

2. **Circulatory.** The circulatory system transports oxygen, nutrients, and chemical signals to the cells of the body and removes carbon dioxide, chemical wastes, and water. Its principal components are the heart, blood vessels, and blood.

3. **Endocrine.** The endocrine system coordinates and integrates the activities of the body through the release of hormones. Its principal components are the pituitary, adrenal, thyroid, and other ductless glands.

4. **Nervous.** The activities of the body are coordinated by the nervous system. Its principal components are the nerves, sense organs, brain, and spinal cord.

5. **Respiratory.** The respiratory system captures oxygen and exchanges gases and is composed of the lungs, trachea, and other air passageways.

6. **Immune.** The immune system removes foreign bodies from the blood-stream using special cells, such as lymphocytes and macrophages.

7. **Digestive.** The digestive system captures soluble nutrients from ingested food. Its principal components are the mouth, esophagus, stomach, intestines, liver, and pancreas.

8. **Urinary.** The urinary system removes metabolic wastes from the bloodstream. Its principal components are the kidneys, bladder, and associated ducts.

9. **Muscular.** The muscular system produces movement, both within the body and of its limbs. Its principal components are skeletal muscle, cardiac muscle, and smooth muscle.

10. **Reproductive.** The reproductive system carries out reproduction. Its principal components are the testes in males, ovaries in females, and associated reproductive structures.

11. **Integumentary.** The integumentary system covers and protects the body. Its principal components are the skin, hair, nails, and sweat glands.

Concept Check

1. Name a human organ other than the heart. What tissues is it made of?
2. Which organ systems extend throughout all of the major parts of the human body?
3. How many human cells are in your body? How many different tissues?

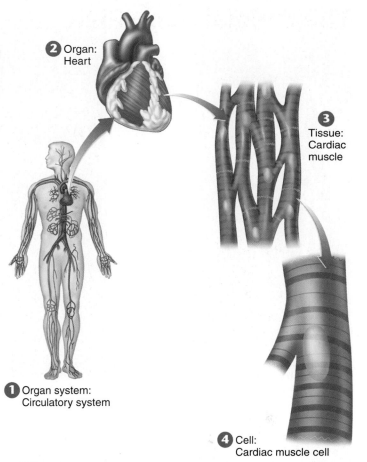

2 Organ: Heart

3 Tissue: Cardiac muscle

1 Organ system: Circulatory system

4 Cell: Cardiac muscle cell

Figure 23.2   Levels of organization within the vertebrate body.

1 The circulatory system you see illustrated here is an example of an organ system. **2** The heart is one of several organs that work together to carry out the function of circulation for the body. **3** Within an organ like the heart, individual tissues like cardiac muscle function together with other tissues. **4** Particular cell types like cardiac muscle cells operate together with other similar types of cells to form each kind of tissue.

EVOLUTION

Regrowing Complex Organs. Many verte-brates such as newts can renew damaged parts of their bodies, but not mammals like us. The only example of a mammal being able to regrow a large complex organ is antler growth. Deer antlers are large structures made of living bone, cartilage, blood vessels, and fibrous tissue, covered in skin. Each year antlers grow in the summer (they are said to be "in velvet"), die in the fall (the hard bone is used for fighting), are shed in the winter, and then regenerate in the spring. Unlike the regenerative processes of newts, antler growth does not involve reversal of the differentiated state, but rather local acti-vation of stem cells. Researchers are seeking to understand this process, hoping it may someday be possible to activate organ renewal of damaged human tissues.

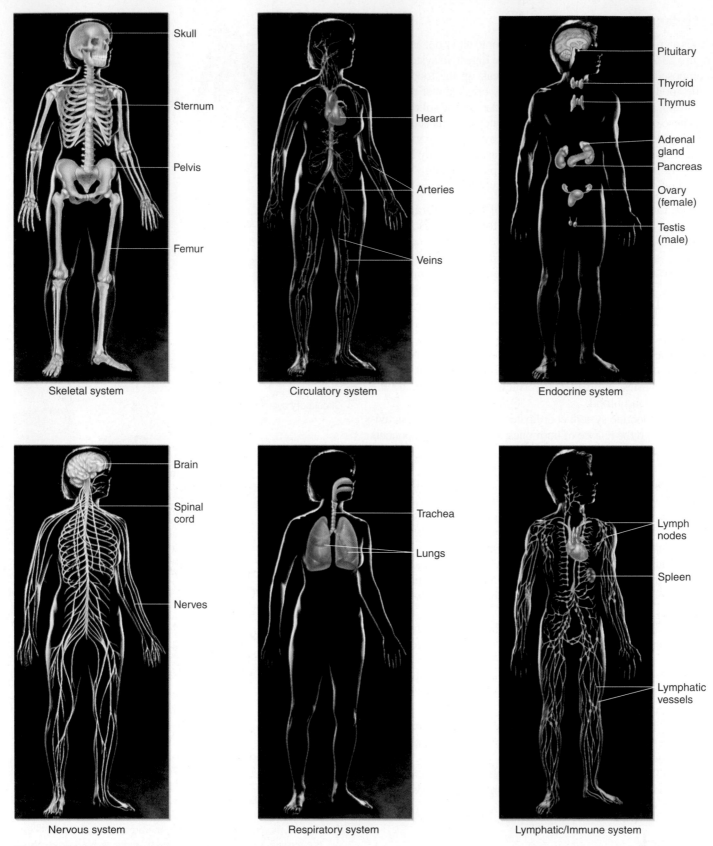

Figure 23.3 **Vertebrate body organ systems.**

The 11 principal organ systems of the human body are shown, including both male and female reproductive systems.

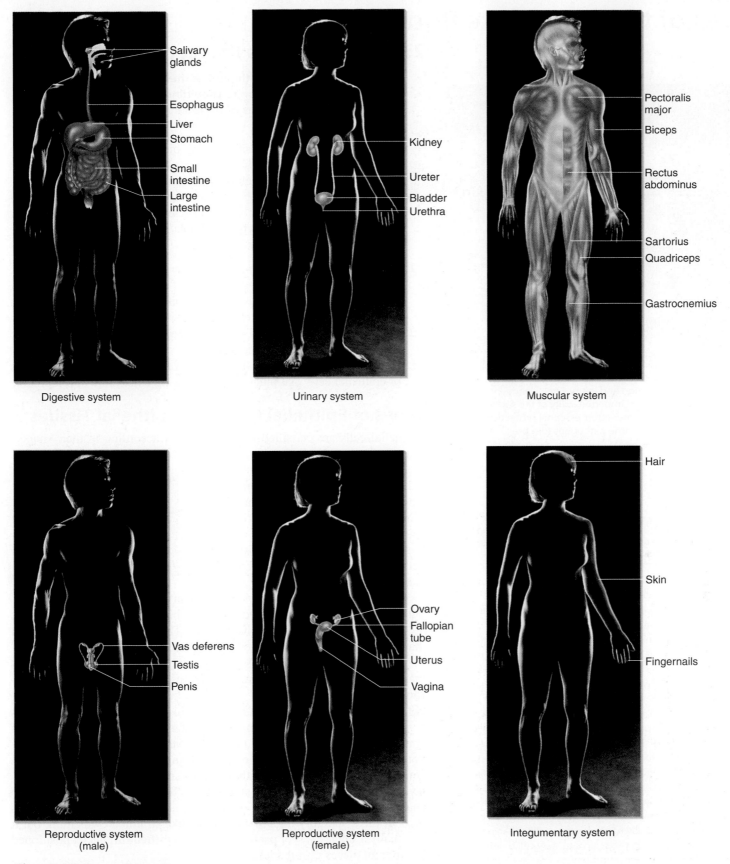

Digestive system

Urinary system

Muscular system

Reproductive system
(male)

Reproductive system
(female)

Integumentary system

Figure 23.3   (continued)

Tissues of the Vertebrate Body

Figure 23.4   The epithelium prevents dehydration.

The tough, scaly skin of this gila monster provides a layer of protection against dehydration and injury. For all land-dwelling vertebrates, the relative impermeability of the surface epithelium (the epidermis) to water offers essential protection from dehydration and from airborne pathogens (disease-causing organisms).

BIOLOGY & YOU

Dandruff. Have you ever had dandruff? You know, the white flaky stuff on your collar and shoulders. If so, you are not alone. Many, if not most, Americans experience dandruff at one time or another. Dandruff is the shedding of excessive dead skin cells from the scalp. It is normal for skin cells to die and flake off, producing small flakes too tiny to see—the epidermal layer continually replaces itself from its inside, pushing cells outward where they eventually die and flake off. However, in people with dandruff this process is greatly accelerated, with skin cells maturing and shedding in two to seven days as opposed to around a month in people without dandruff. Dead skin cells build up so fast that they are shed in large, oily clumps which appear as white flakes. What causes dandruff? A scalp fungus called *Malassezia globosa*. It metabolizes skin oils, producing a by-product that triggers an inflammatory response in susceptible persons, leading to accelerated division of skin cells. Shampoos control dandruff in different ways: Sebulex uses salicylic acid to remove dead skin cells from the scalp and slow cell division; Head and Shoulders uses zinc pyrithione to kill the *Malassezia* fungus; Selsun Blue uses selenium sulfide to achieve the results of both salicylic acid and zinc pyrithione. All reduce flaking dramatically.

23.2 Epithelium Is Protective Tissue

CONCEPT PREVIEW: Epithelium is the protective tissue of the vertebrate body. In addition to providing protection and support, vertebrate epithelial tissues provide sensory surfaces and secrete key materials.

Tissues are the basic building blocks of the animal body. We begin our discussion of tissues with one located on the body's surfaces, epithelium. Epithelial cells are the guards and protectors of the body. They cover both internal and external surfaces of the body, and determine which substances enter the body and which do not. The body's epithelial layers function in three ways:

1. They *protect underlying tissues* from water loss and mechanical damage. Because epithelium encases all the body's surfaces, every substance that enters or leaves the body must cross an epithelial layer, even one as thick as the gila monster's in figure 23.4.
2. They *provide sensory surfaces.* Many of a vertebrate's sense organs are in fact modified epithelial cells.
3. They *secrete materials.* Most secretory glands are derived from pockets of epithelial cells that pinch together during embryonic development.

Types of Epithelial Cells and Epithelial Tissues

Epithelial cells are classified into three types according to their shapes: squamous, cuboidal, and columnar. Layers of epithelial tissue are usually only one or a few cells thick. Individual epithelial cells possess only a small amount of cytoplasm and have a relatively low metabolic rate. A characteristic of all epithelia is that sheets of cells are tightly bound together, with very little space between them. This forms the barrier that is key to their functioning.

> Like animals, plants also have a surface layer of cells that provides a physical barrier that protects underlying tissues, as well as reducing water loss. As described on pages 617 and 618, the plant epidermis contains numerous specialized cells and structures.

The cells of epithelial layers are constantly being replaced throughout the life of the organism. The cells lining the digestive tract, for example, are continuously replaced every few days. The epidermis, the epithelium that forms the skin, is renewed every two weeks.

There are two general kinds of epithelial tissue. First, the membranes that line the lungs and the major cavities of the body are a **simple epithelium** only a single cell layer thick. The first three entries in table 23.1 (❶, ❷, ❸) are simple epithelium. These single-celled layers are surfaces across which many materials must pass, entering and leaving the body's compartments—it is important that the "road" into and out of the body not be too long. Second, the skin, or epidermis, is a **stratified epithelium** composed of more complex epithelial cells several layers thick ❹. Multilayers are necessary to provide adequate cushioning and protection and to enable the skin to continuously replace its cells. The epithelium that lines parts of the respiratory tract ❺ also looks like stratified epithelium, but it is actually a single layer of **pseudostratified epithelium;** nuclei positioned in different places give it a layered appearance.

A type of simple epithelial tissue that has a secretory function is cuboidal epithelium, which is found in the **glands** of the body. Endocrine glands secrete hormones into the blood. Exocrine glands (those with ducts that open to the body's outside) secrete sweat, milk, saliva, and digestive enzymes.

TABLE 23.1	Epithelial Tissue		
Tissue		**Typical Location**	**Tissue Function**
Simple Epithelium			
❶ Squamous	Simple squamous epithelial cell — Nucleus —	Lining of lungs, capillary walls, and blood vessels	Flat and thin cells; provides a thin layer across which diffusion can readily occur; the cells when viewed from the surface look like tiles on a floor
❷ Cuboidal	Cuboidal epithelial cells — Nucleus — Cytoplasm —	Lining of some glands and kidney tubules; covering of ovaries	Cells rich in specific transport channels; functions in secretion and specific absorption
❸ Columnar	Columnar epithelial cells — Nucleus — Goblet cell —	Surface lining of stomach, intestines, and parts of respiratory tract	Thicker cell layer; provides protection and functions in secretion and absorption
Stratified Epithelium			
❹ Squamous	Stratified squamous cells — Nuclei —	Outer layer of skin; lining of mouth	Layers of cells; provide protection
Pseudostratified Epithelium			
❺ Columnar	Cilia — Pseudo-stratified columnar cell — Goblet cell —	Lining of parts of respiratory tract	Functions in secretion of mucus; dense with cilia (small, hairlike projections) that aid in movement of mucus; provides protection

BIOLOGY & YOU

Getting Fat. It will come as no surprise to you that overeating can make you fat. Is this because overeating triggers the production of more fat cells, or because the fat cells get larger? It turns out, both situations occur. You are born with a predetermined number of fat cells, with women generally inheriting more than men. The number then grows through childhood until late puberty, after which it is pretty much set. It was generally thought that the only difference between obese people and nonobese people was that obese people had their fat cells filled up to capacity. When overweight people lose weight, they empty their fat cells, but the fat cells don't go away. They are still there, like little sponges waiting to refill. We now know that we can—and do—"grow" more fat cells in adulthood, and that overweight and obese people do in fact have more fat cells. It turns out that when fat cells expand to their maximum size, this triggers cell division, producing an increase in the actual number of fat cells. It is one of the curses of dieters, that these new cells do not go away when you lose weight. While dieting may shrink them, they tend to regain lost weight rapidly. If you wind up with a lot of extra fat cells, you have to work harder than lean people to keep excess weight off. It does not seem fair, and it is not fun, but it is important to good health. You CAN lose weight and keep it off with commitment and determination.

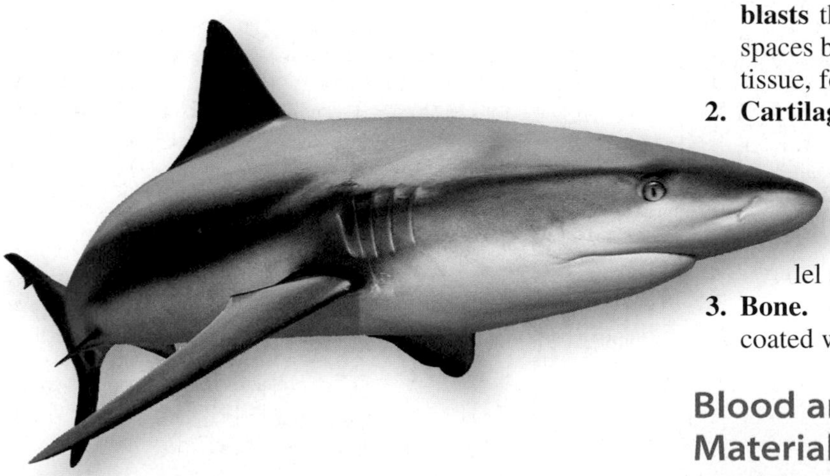

Figure 23.5 Cartilage: skeletal connective tissue.

The entire skeleton of this Caribbean Reef shark is made of cartilage, a strong but flexible type of connective tissue. In other vertebrates, cartilage is found in joints, noses, and external ears.

23.3 Connective Tissue Carries Out Various Functions

CONCEPT PREVIEW: Connective tissues support the vertebrate body and consist of cells embedded in an extracellular matrix. They include cells of the immune system, bone and other cells of the skeletal system, and cells found throughout the body like blood and fat cells.

The cells of **connective tissue** provide the vertebrate body with its structural building blocks and also with its most potent defenses. These cells are sometimes densely packed together, and sometimes widely dispersed. Although very diverse, all connective tissues share a key common structural feature: They all have abundant extracellular **matrix** material between widely spaced cells.

Immune Connective Tissue Defends the Body

The cells of the immune system, the many kinds of so-called "white blood cells" (❶ in table 23.2) roam the body within the bloodstream. They are mobile hunters of invading microorganisms and cancer cells. The two principal kinds of immune system cells are **macrophages,** which engulf and digest invading microorganisms (as shown in table 23.2), and **lymphocytes,** which attack virus-infected cells or make antibodies that tag cells for destruction. Immune cells are carried through the body in a fluid matrix, called *plasma.*

Skeletal Connective Tissue Supports the Body

Three kinds of connective tissue are the principal components of the skeletal system: fibrous connective tissue, cartilage, and bone. Although composed of similar cells, they differ in the nature of the matrix that is laid down between individual cells.

1. **Fibrous connective tissue.** The most common kind of connective tissue in the vertebrate body is fibrous connective tissue, ❷ in table 23.2. It is composed of flat, irregularly branching cells called **fibroblasts** that secrete structurally strong proteins like collagen into the spaces between the cells. Fibroblasts are active in wound healing; scar tissue, for example, possesses a collagen matrix.
2. **Cartilage.** In cartilage ❸, a collagen matrix between cartilage cells (technically called *chondrocytes*) forms in long parallel arrays along the lines of mechanical stress. What results is a firm and flexible tissue of great strength (figure 23.5), just as strands of nylon molecules laid down in long, parallel arrays produce strong, flexible ropes.
3. **Bone.** Bone ❹ is similar to cartilage, except the collagen fibers are coated with a calcium phosphate salt, making the tissue rigid.

Blood and Fat Cells Store and Transport Materials

The third general class of connective tissue is made up of cells that are specialized to accumulate and transport particular molecules. They include the fat-accumulating cells of **adipose tissue** ❺. The large seemingly empty areas in the adipose tissue in table 23.2 are actually fat-containing vacuoles within the adipose cells. Red blood cells ❻, called **erythrocytes,** also function in transport and storage. About 5 billion erythrocytes are present in every milliliter of your blood. Erythrocytes transport oxygen and carbon dioxide in blood.

TABLE 23.2 | Connective Tissue

Tissue		Typical Location	Tissue Function	Characteristic Cell Types
Immune				
❶ White blood cells	White blood cell Invading micro-organism .15 μm	In plasma of the circulatory system	Attack invading microorganisms and virus-infected cells	Macrophages; lymphocytes
Skeletal				
❷ Fibrous connective tissue				
Loose	Loose connective tissue (fibroblasts)	Beneath skin and other epithelial tissues	Support; provide a fluid reservoir for epithelium	Fibroblasts
Dense		Tendons; sheath around muscles; kidney; liver; dermis of skin	Provide flexible, strong connections	Fibroblasts
Elastic		Ligaments; large arteries; lung tissue; skin	Enable tissues to expand and then return to original size	Fibroblasts
❸ Cartilage	Chondrocytes (cartilage cells)	Spinal disks; knees and other joints; ear; nose; tracheal rings, and the skeleton in some vertebrates	Provides flexible support; functions in shock absorption and reduction of friction on load-bearing surfaces	Chondrocytes (specialized fibroblast-like cells)
❹ Bone (see figure 23.6)		Most vertebrate skeletons	Protects internal organs; provides rigid support for muscle attachment	Osteocytes (specialized fibroblast-like cells)
Storage and Transport				
❺ Adipose tissue	Adipose tissue	Beneath skin	Stores fat	Specialized fibroblasts (adipocytes)
❻ Red blood cells	Red blood cells	In plasma of the circulatory system	Transport oxygen	Red blood cells (erythrocytes)

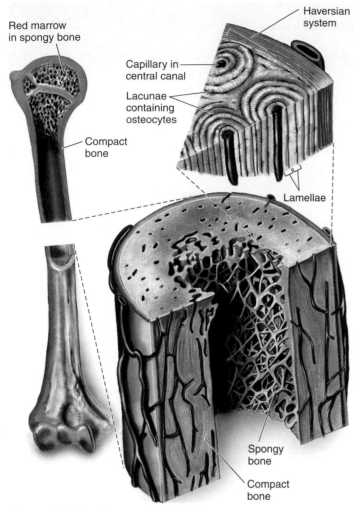

Red marrow
in spongy bone

Haversian
system

Capillary in
central canal

Lacunae
containing
osteocytes

Compact
bone

Lamellae

Spongy
bone

Compact
bone

Figure 23.6   The structure of bone.

Some parts of bones are dense and compact, giving the bone strength. Other parts are spongy, with a more open lattice; red blood cells form in the bone marrow. Osteocytes, which are mature osteoblasts, lie in tight spaces called lacunae.

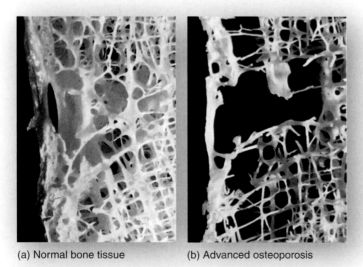

(a) Normal bone tissue

(b) Advanced osteoporosis

Figure 23.7   Osteoporosis.

Common in older women, osteoporosis is a bone disorder in which bones progressively lose minerals.

A Closer Look at Bone

Bone consists of living bone cells embedded within an inert matrix composed of the structural protein collagen coated with a calcium phosphate salt called *hydroxyapatite*. Why coat the collagen fibers with calcium salts? The construction of bone is similar to that of fiberglass, a composite composed of glass fibers embedded in epoxy glue: Small, needle-shaped crystals of hydroxyapatite surround and impregnate collagen fibers within bone. No crack can penetrate far into bone because any stress that breaks a hard hydroxyapatite crystal passes into the collagenous matrix, which dissipates the stress before it encounters another crystal. Hydroxyapatite provides rigidity (like the glass in fiberglass), the collagen "glue" provide flexibility.

Most of us think of bones as solid and rocklike. But actually, bone is a dynamic tissue that is constantly being reconstructed. The cross section through a bone in figure 23.6 shows the outer layer of bone is very dense and compact and so is called **compact bone.** The interior is less compact, with a more open lattice structure, and is called **spongy bone.** Red blood cells form in the red marrow of spongy bone. New bone is formed in two stages: First, collagen is secreted by cells called **osteoblasts,** which lay down a matrix of fibers along lines of stress. Then calcium minerals impregnate the fibers. Bone is laid down in thin, concentric layers, like layers of paint on an old pipe. The layers form as a series of tubes around a narrow channel called a **central canal,** also called a *Haversian canal,* which runs parallel to the length of the bone. You can see the central canals and the concentric rings surrounding the canals in the enlargement in figure 23.6. The many central canals within a bone are all interconnected and contain blood vessels and nerves that provide a lifeline to its living, bone-forming cells.

When bone is first formed in the embryo, osteoblasts use a cartilage skeleton as a template for bone formation. During childhood, bones grow actively. The total bone mass in a healthy young adult, by contrast, does not increase much from one year to the next. This does not mean change is not occurring. Large amounts of calcium and thousands of **osteocytes** (mature osteoblasts) are constantly being removed and replaced, but total bone mass does not change because deposit and removal take place at about the same rate.

Two cell types are responsible for this dynamic bone "remodeling": *Osteoblasts* deposit bone, and **osteoclasts** secrete enzymes that digest the organic matrix of bone, liberating calcium for reabsorption by the bloodstream. The dynamic remodeling of bone adjusts bone strength to workload, new bone being formed along lines of stress. When a bone is subjected to compression, mineral deposition by osteoblasts exceeds withdrawals by osteoclasts. That is why long-distance runners must slowly increase the distances they attempt, to allow their bones to strengthen along lines of stress; otherwise, stress fractures can cripple them.

> Bone contains large stores of calcium ions, and as you will see on pages 586 and 587, the body draws on these calcium stores when needed for body functions. When calcium levels in the blood drop, the body will actually break down bone tissue, releasing calcium.

As a person ages, the backbone and other bones tend to decline in mass. Excessive bone loss is a condition called **osteoporosis.** After the onset of osteoporosis, the replacement of calcium and other minerals lags behind withdrawal, causing the bone tissue to gradually erode. Compare the normal bone in figure 23.7*a* with bone from a person with osteoporosis in figure 23.7*b*. Eventually the bones become brittle and easily broken. Women are four times more likely to develop osteoporosis compared to men.

23.4 Muscle Tissue Lets the Body Move

CONCEPT PREVIEW: Muscle tissue is the tool the vertebrate body uses to move its limbs, contract its organs, and pump the blood through its circulatory system.

Muscle cells are the motors of the vertebrate body. The distinguishing characteristic of muscle cells, the thing that makes them unique, is the abundance of contractible protein fibers within them. These fibers, called **myofilaments,** are made of the proteins actin and myosin. Crammed in like the fibers of a rope, they take up practically the entire volume of the muscle cell. When actin and myosin slide past each other, the muscle contracts. Like slamming a spring-loaded door, the shortening of all of these fibers together within a muscle cell can produce considerable force. The process of muscle contraction will be discussed later in this chapter.

Smooth Muscle

Smooth muscle cells are long and spindle-shaped, each containing a single nucleus. Smooth muscle tissue is organized into sheets of cells. In some tissues, smooth muscle cells contract only when they are stimulated by a nerve or hormone. Examples are the muscles that line the walls of many blood vessels and those that make up the iris of the vertebrate eye. In other smooth muscle tissue, such as that found in the wall of the gut, the individual cells contract spontaneously, leading to a slow, steady contraction of the tissue.

Skeletal Muscle

Skeletal muscles are attached to and move the bones of the skeleton when they contract. Skeletal muscle cells are produced during development by the fusion of several cells at their ends to form a very long fiber still containing all the original nuclei. Looking at figure 23.8, you can see how the nuclei are positioned to the outside of the muscle fiber. Each **muscle fiber** consists of many elongated **myofibrils** composed of myofilaments of the proteins actin and myosin. The arrangement of actin and myosin give the muscle fibers a banded appearance, called *striations*.

Cardiac Muscle

The vertebrate heart is composed of striated **cardiac muscle** in which the fibers are arranged very differently from the fibers of skeletal muscle. Instead of very long, multinucleate cells running the length of the muscle, heart muscle is composed of chains of single cells, organized into fibers that branch and interconnect, forming a latticework. This lattice structure is critical to the way heart muscle functions, as electrical impulses pass from cell to cell through small openings called gap junctions, causing the heart to contract in an orderly pulsation.

Figure 23.8   A skeletal muscle fiber, or muscle cell.

Each muscle is composed of bundles of muscle cells, or fibers. Each fiber is composed of many myofibrils, which are each, in turn, composed of myofilaments. Muscle cells have a modified endoplasmic reticulum called the sarcoplasmic reticulum that is involved in the regulation of calcium ions in muscles.

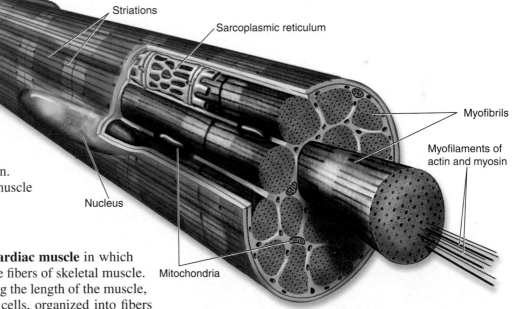

Striations

Sarcoplasmic reticulum

Myofibrils

Myofilaments of actin and myosin

Nucleus

Mitochondria

TABLE 23.3	Types of Neurons	
Neuron	**Typical Location**	**Function**
❶ Sensory neurons 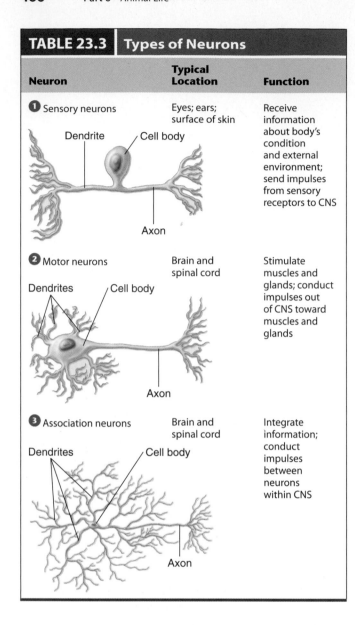	Eyes; ears; surface of skin	Receive information about body's condition and external environment; send impulses from sensory receptors to CNS
❷ Motor neurons	Brain and spinal cord	Stimulate muscles and glands; conduct impulses out of CNS toward muscles and glands
❸ Association neurons	Brain and spinal cord	Integrate information; conduct impulses between neurons within CNS

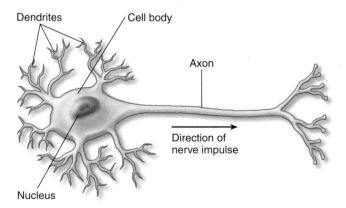

Figure 23.9 Neurons carry nerve impulses.

Neurons carry nerve impulses, which are electrical signals, from their initiation in dendrites, to the cell body, and down the length of the axon, where they may pass the signal to a neighboring cell.

23.5 Nerve Tissue Conducts Signals Rapidly

CONCEPT PREVIEW: Nerve tissue provides the vertebrate body with a means of communication and coordination. Three different types of neurons carry out different functions in the nervous system.

Nerve tissue, the fourth major class of vertebrate tissue, is composed of two kinds of cells: (1) **neurons,** which are specialized for the rapid transmission of nerve impulses from one organ to another, and (2) supporting **glial cells,** which supply the neurons with nutrients, support, and insulation.

Neurons have a highly specialized cell architecture that enables them to conduct signals rapidly throughout the body. Their plasma membranes are rich in ion-selective channels that maintain a voltage difference between the interior and the exterior of the cell, the equivalent of a battery. When ion channels in a local area of the membrane open, ions flood in from the exterior, temporarily wiping out the charge difference. This process, called depolarization, tends to open nearby voltage-sensitive channels in the neuron membrane, resulting in a wave of electrical activity that travels down the entire length of the neuron as a nerve impulse.

> Ions pass in and out of cells. As discussed on page 554, Na^+ and K^+ ion gradients are established by the actions of the Na^+/K^+ pump. Na^+ is pumped out of the cell and K^+ is pumped into the cell. When specific ion channels open, the ions flow down their concentration gradients.

Each neuron is composed of three parts, as illustrated in figure 23.9: (1) a **cell body,** which contains the nucleus; (2) threadlike extensions called **dendrites** extending from the cell body, which act as antennae, bringing nerve impulses to the cell body from other cells or sensory systems; and (3) a single, long extension called an **axon,** which carries nerve impulses away from the cell body. Axons often carry nerve impulses for considerable distances: The axons that extend from the skull to the pelvis in a giraffe are about 3 meters long!

The body contains neurons of various sizes and shapes. Some are tiny and have only a few projections, others are bushy and have more projections, and still others have extensions that are meters long. However, all fit into one of three general categories of neurons as shown in table 23.3. *Sensory neurons* ❶ generally carry electrical impulses from the body to the central nervous system (CNS), the brain and spinal cord. *Motor neurons* ❷ generally carry electrical impulses from the central nervous system to the muscles. *Association neurons* ❸ occur within the central nervous system and act as a "connector" between sensory and motor neurons. These will be discussed in more detail in chapter 29. Neurons are not normally in direct contact with one another. Instead, a tiny gap called a **synapse** separates them. Neurons communicate with other neurons by passing chemical signals called **neurotransmitters** across the gap.

Vertebrate nerves appear as fine white threads when viewed with the naked eye, but they are actually composed of bundles of axons. Like a telephone trunk cable, nerves include large numbers of independent communication channels—bundles composed of hundreds of axons, each connecting a nerve cell to a muscle fiber or other type of cell. It is important not to confuse a nerve with a neuron. A nerve is made up of the axons of many neurons, just as a cable is made of many wires.

Concept Check

1. What are the three principal functions of the body's epithelium?
2. What is the common structural feature of all connective tissues?
3. Describe how bones adjust strength to workload.

The Skeletal and Muscular Systems

23.6 Types of Skeletons

CONCEPT PREVIEW: The animal skeletal system provides a framework against which the body's muscles can pull. Many soft-bodied invertebrates employ a hydraulic skeleton, whereas arthropods have a rigid, hard exoskeleton surrounding their body. Echinoderms and vertebrates have an internal endoskeleton to which muscles attach.

With muscles alone, the animal body could not move—it would simply pulsate as its muscles contracted and relaxed in futile cycles. For a muscle to produce movement, it must direct its force against another object. Animals are able to move because the opposite ends of their muscles are attached to a rigid scaffold, or **skeleton,** so that the muscles have something to pull against. There are three types of skeletal systems in the animal kingdom: hydraulic skeletons, exoskeletons, and endoskeletons.

Hydraulic skeletons are found in soft-bodied invertebrates such as earthworms and jellyfish. In this case, a fluid-filled cavity is encircled by muscle fibers that raise the pressure of the fluid when they contract. The earthworm in figure 23.10 moves forward by a wave of contractions of circular muscles that begins anteriorly and compresses the body, so that the fluid pressure pushes it forward. Contractions of longitudinal muscles then pull the rest of the body along.

Exoskeletons surround the body as a rigid hard case to which muscles attach internally. When a muscle contracts, it moves the section of exoskeleton to which it is attached. Arthropods, such as crustaceans (figure 23.11) and insects, have exoskeletons made of the polysaccharide *chitin.* An animal with an exoskeleton cannot grow too large because its exoskeleton would have to become thicker and heavier to prevent collapse. If an insect were the size of an elephant, its exoskeleton would have to be so thick and heavy it would hardly be able to move.

Endoskeletons, found in vertebrates and echinoderms, are rigid internal skeletons to which muscles are attached. Vertebrates have a soft, flexible exterior that stretches to accommodate the movements of their skeleton. The endoskeleton of some vertebrates is made of cartilage but in many others it is composed of bone (figure 23.12). Unlike chitin, bone is a cellular, living tissue capable of growth, self-repair, and remodeling in response to physical stresses.

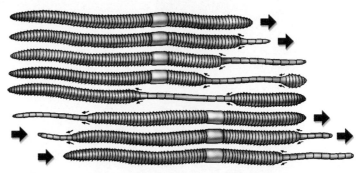

Figure 23.10   **Earthworms have a hydraulic skeleton.**

When an earthworm's circular muscles contract, the internal fluid presses on the longitudinal muscles, which then stretch to elongate segments of the earthworm. A wave of contractions down the body of the earthworm produces forward movement.

Figure 23.11   **Crustaceans have an exoskeleton.**

The exoskeleton of this rock crab is bright orange.

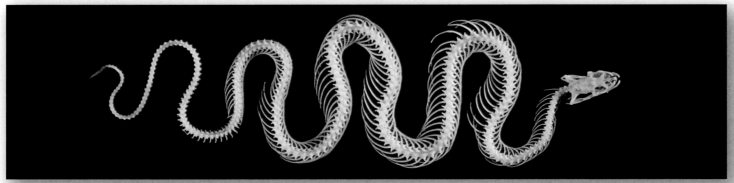

Figure 23.12   **Snakes have an endoskeleton.**

The endoskeleton of most vertebrates is made of bone. A snake's skeleton is specialized for quick lateral movement.

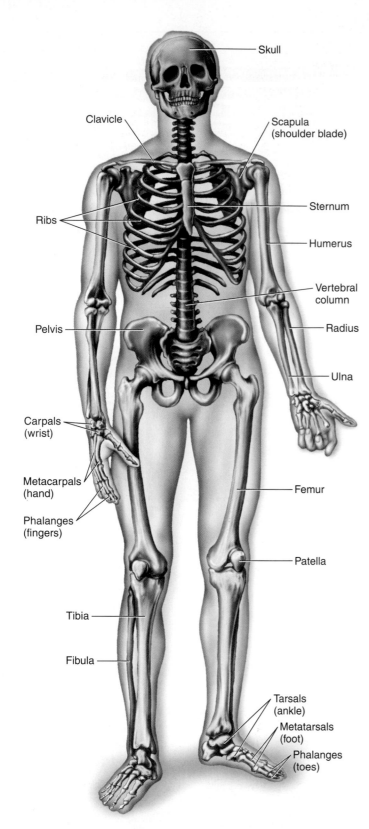

Figure 23.13   Axial and appendicular skeletons.

The axial skeleton is shown in *purple,* and the appendicular skeleton is shown in *tan.* Some of the joints are shown in *green.*

A Vertebrate Endoskeleton: The Human Skeleton

The human skeleton is made up of 206 individual bones. If you saw them as a pile of bones jumbled together, it would be hard to make any sense of them. To understand the skeleton, it is necessary to group the 206 bones according to their function and position in the body. The 80 bones of the **axial skeleton,** the purple-colored bones in figure 23.13, support the main body axis, while the remaining 126 bones of the **appendicular skeleton,** the tan-colored bones, support the arms and legs. These two skeletons function more or less independently—that is, the muscles controlling the axial skeleton (postural muscles) are managed by the brain separately from those controlling the appendages (manipulatory muscles).

The Axial Skeleton

The axial skeleton is made up of the skull, vertebral column (backbone), and rib cage that includes the sternum. Of the skull's 28 bones, only 8 form the cranium, which encases the brain; the rest are facial bones and middle ear bones.

The skull is attached to the upper end of the vertebral column, which is also called the spine. The spine is made up of 26 individual bones called vertebrae, stacked one on top of the other to provide a flexible column surrounding and protecting the spinal cord. Curving forward from the vertebrae are 12 pairs of ribs, attached at the front to the sternum, or breastbone, and forming a protective cage around the heart and lungs.

The Appendicular Skeleton

The 126 bones of the appendicular skeleton are attached to the axial skeleton at the shoulders and hips. The shoulder, or **pectoral girdle,** is composed of two large, flat shoulder blades, each connected to the top of the sternum by a slender, curved collarbone (clavicle). The arms are attached to the pectoral girdle; each arm and hand contains 30 bones. The clavicle is the most frequently broken bone of the body. Can you guess why? Because if you fall on an outstretched arm, a large component of the force is transmitted to the clavicle.

The **pelvic girdle** forms a bowl that provides strong connections for the legs, which must bear the weight of the body. Each leg and foot contains a total of 30 bones.

Joints are points where two bones come together. They confer flexibility to the rigid endoskeleton, allowing a range of motion determined by the type of joint. There are three main classes of joints based on mobility. *Immovable joints,* such as the sutures of the skull, are capable of little to no movement. *Slightly movable joints,* such as the joints between vertebrae in the spine, allow the bones some movement. And *freely movable joints* allow a range of motion and are seen in the joints of the limbs (for example, the shoulder, elbow, hip, knee), the jaw, and fingers and toes. Depending on the type of joint, bones are held together at the joint by cartilage, fibrous connective tissue, or a fibrous capsule filled with a lubricating fluid.

23.7 Muscles and How They Work

CONCEPT PREVIEW: Muscles are made of many tiny threadlike filaments of actin and myosin called myofilaments. Muscles work by the sliding of myosin along actin, causing the myofibrils to contract. Muscle contraction is powered by ATP, which is needed to reposition the myosin heads for reattachment to the actin filaments.

Three kinds of muscle together form the vertebrate muscular system. As we have discussed, the vertebrate body is able to move because *skeletal muscles* pull the bones with considerable force. The heart pumps because of the contraction of *cardiac muscle*. Food moves through the intestines because of the rhythmic contractions of *smooth muscle*.

Actions of Skeletal Muscle

Skeletal muscles move the bones of the skeleton. Some of the major human muscles are labeled on the right in figure 23.14. Muscles are attached to bones by straps of dense connective tissue called **tendons.** Bones pivot about flexible connections called joints, pulled back and forth by the muscles attached to them. Each muscle pulls on a specific bone. One end of the muscle, the *origin,* is attached by a tendon to a bone that remains stationary during a contraction. This provides an object against which the muscle can pull. The other end of the muscle, the *insertion,* is attached to a bone that moves if the muscle contracts. For example, the origin and insertion for the sartorius muscle are labeled on the left in figure 23.14. This muscle helps bend the leg at the hip, bringing the knee to the chest. The origin of the muscle is at the hip and stays stationary. The insertion is just below the knee, such that when the muscle contracts (gets shorter) the knee is pulled up toward the chest.

Muscles can only pull, not push, because myofibrils contract rather than expand. For this reason, the muscles that extend across movable joints of vertebrate are attached in opposing pairs, called flexors and extensors, which when contracted, move the bones in different directions. As you can see in figure 23.15, when the **flexor** muscle at the back of your upper leg contracts, the lower leg is moved closer to the thigh. When the **extensor** muscle at the front of your upper leg contracts, the lower leg is moved in the opposite direction, away from the thigh.

All muscles contract, but there are two types of muscle contractions, isotonic and isometric contractions. In **isotonic contractions,** the muscle shortens, moving the bones as just described. In **isometric contractions,** a force is exerted by the muscle, but the muscle does not shorten. This occurs when you try to lift something very heavy. Eventually, if your muscles generate enough force and you are able to lift the object, the isometric contraction becomes isotonic.

Muscle Contraction

Recall from figure 23.8 that myofibrils are composed of bundles of myofilaments. Far too fine to see with the naked eye, the individual myofilaments of vertebrate muscles are only 8 to 12 nanometers thick. Each is a long, threadlike filament of the proteins actin or myosin. An **actin filament** consists of two strings of actin molecules wrapped around one another, like two strands of pearls loosely wound together. A **myosin filament** is about twice as long as an actin filament and is composed of bundles of myosin molecules. A myosin molecule consists of two polypeptide chains

> Actin isn't found only in muscle cells. All animal cells contain some actin filaments as components of the cytoskeleton, discussed on page 76. The cytoskeleton helps maintain the shape of animal cells.

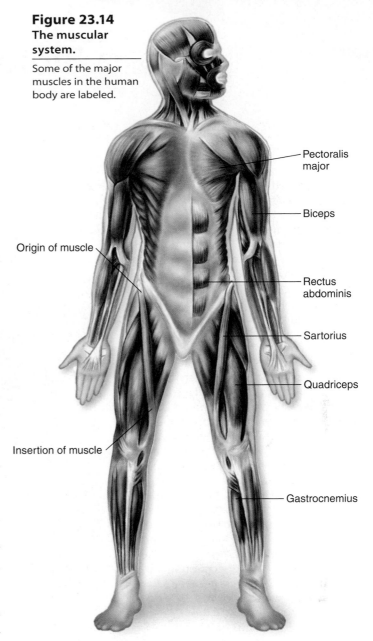

**Figure 23.14
The muscular system.**

Some of the major muscles in the human body are labeled.

- Pectoralis major
- Biceps
- Origin of muscle
- Rectus abdominis
- Sartorius
- Quadriceps
- Insertion of muscle
- Gastrocnemius

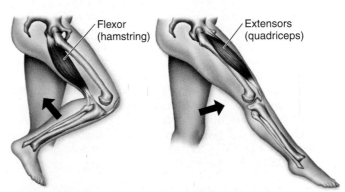

- Flexor (hamstring)
- Extensors (quadriceps)

Figure 23.15   Flexor and extensor muscles.

Limb movement is always the result of muscle contraction, never muscle extension. Muscles that retract limbs are called flexors; those that extend limbs are called extensors.

The Author Works Out

No one seeing the ring of fat decorating my middle would take me for a runner. Only in my memory do I get up with the robins, lace up my running shoes, bounce out the front door, and run the streets around Washington University before going to work. Now my 5-K runs are 30-year-old memories. Any mention I make of my running in a race only evokes screams of laughter from my daughters, and an arch look from my wife. Memory is cruelest when it is accurate.

I remember clearly the day I stopped running. It was a cool fall morning in 1978, and I was part of a mob running a 5-K (that's 5 kilometers for the uninitiated) race, winding around the hills near the university. I started to get flashes of pain in my legs below the knees—like shin splints, but much worse. Imagine fire pouring on your bones. Did I stop running? No. Like a bonehead I kept going, "working through the pain," and finished the race. I have never run a race since.

I had pulled a muscle in my thigh, which caused part of the pain. But that wasn't all. The pain in my lower legs wasn't shin splints, and didn't go away. A trip to the doctor revealed multiple stress fractures in both legs. The X rays of my legs looked like tiny threads had been wrapped around the shaft of each bone, like the red stripe on a barber's pole. It was summer before I could walk without pain.

What went wrong? Isn't running supposed to be GOOD for you? Not if you run improperly. In my enthusiasm to be healthy, I ignored some simple rules and paid the price. The biology lesson I ignored had to do with how bones grow. The long bones of your legs are not made of stone, solid and permanent. They are dynamic structures, constantly being re-formed and strengthened in response to the stresses to which you subject them.

To understand how bone grows, we first need to recall a bit about what bone is like. Bone, as you have learned in this chapter, is made of fibers of a flexible protein called collagen stuck together to form cartilage. While an embryo, all your bones are made of cartilage. As your body develops, the collagen fibers become impregnated with tiny needle-shaped crystals of calcium phosphate, turning the cartilage into bone. The crystals are brittle but rigid, giving bone great strength. Collagen is flexible but weak, but like the epoxy of fiberglass, it acts to spread any stress over many crystals, making bone resistant to fracture. As a result, bone is both strong and flexible.

When you subject a bone in your body to stress—say, by running—the bone grows so as to withstand the greater workload. How does the bone "know" just where to add more material? When stress deforms the collagen fibers of a leg bone, the interior of the collagen fibers becomes exposed, like opening your jacket and exposing your shirt. The fiber interior produces a minute electrical charge. Cells called fibroblasts are attracted to the electricity like bugs to night lights, and secrete more collagen there. As a result, new collagen fibers are laid down on a bone along the lines of stress. Slowly, over months, calcium phosphate crystals convert the new collagen to new bone. In your legs, the new bone forms along the long stress lines that curve down along the shank of the bone.

Now go back 30 years, and visualize me pounding happily down the concrete pavement each morning. I had only recently begun to run on the sidewalk, and for an hour or more at a stretch. Every stride I took those mornings was a blow to my shinbones, a stress to which my bones no doubt began to respond by forming collagen along the spiral lines of stress. Had I run on a softer surface, the daily stress would have been far less severe. Had I gradually increased my running, new bone would have had time to form properly in response to the added stress. I gave my leg bones a lot of stress, and no time to respond to it. I pushed them too hard, too fast, and they gave way.

Nor was my improper running limited to overstressed leg bones. Remember that pulled thigh muscle? In my excessive enthusiasm, I never warmed up before I ran. I was having too much fun to worry about such details. Wiser now, I am sure the pulled thigh muscle was a direct result of failing to properly stretch and warm up before running.

I was reminded of that pulled muscle recently, listening to a good friend of my wife's describe how she sets out early each morning for a long run without stretching or warming up. I can see her in my mind's eye, bundled up warmly on the cooler mornings, an enthusiastic gazelle pounding down the pavement in search of good health. Unless she uses more sense than I did, she may fail to find it.

wound about each other. Each polypeptide chain has a very unusual shape, like that of a golf club. One end consists of a very long rod, while the other end consists of a globular region, or "head." The two polypeptide chains wound around each other looks a bit like a two-headed snake. This odd structure is the key to how muscles work.

How Myofilaments Contract

The **sliding filament model** of muscle contraction, illustrated in figure 23.16, describes how actin and myosin cause muscles to contract. Focus on the knob-shaped myosin head in panel 1. When a muscle contraction begins the heads of the myosin filaments, which are attached to actin, move first. Like flexing your hand downward at the wrist, the heads bend backward and inward as in panel 2. This moves the myosin heads closer to their rod-like backbones in the direction of the flex. In itself, this myosin head-flex accomplishes nothing—but because the myosin head is attached to the actin filament, the actin filament is pulled along with the myosin head as it flexes. This causes the actin filament to slide by the myosin filament in the direction of the flex (indicated by the arrows in panel 2). As one myosin head after another flexes, the myosin in effect "walks" step by step along the actin. Each step uses a molecule of ATP to recock the myosin head (in panel 3) before it attaches to the actin again, ready for the next flex.

> ATP is the energy currency of the cell, as discussed on page 94. An ATP molecule contains high-energy bonds that hold the phosphate groups together. The breaking of the outermost phosphate bond releases a burst of energy that resets the myosin heads.

How does this sliding of actin past myosin lead to myofibril contraction and muscle cell movement? The actin filament is anchored at one end, at a position in striated muscle called the Z line, indicated by the lavender-colored bars toward the edges in figure 23.17. Two Z lines with the actin and myosin filaments in between make up a contractile unit called a **sarcomere.** Because it is tethered like this, the actin cannot simply move off. Instead, the actin pulls the anchor with it! As actin moves past myosin, it drags the Z line toward the myosin. The secret of muscle contraction is that each myosin is interposed between two pairs of actin filaments, which are anchored at both ends to Z lines, as shown in panel 1 of figure 23.17. One moving to the left and the other to the right, the two pairs of actin molecules drag the Z lines toward each other as they slide past the myosin core, shown in panel 2. As the Z lines are pulled closer together, the plasma membranes to which they are attached move toward one another, and the cell contracts.

Concept Check

1. Describe three kinds of animal skeletons.
2. What is the difference between your axial and your appendicular skeleton?
3. What precisely is the difference between isotonic and isometric muscle contraction?

Figure 23.17   The sliding filament model of muscle contraction.

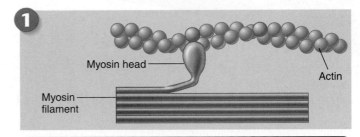

1

Myosin head ——   —— Actin

Myosin filament

The myosin head is attached to actin.

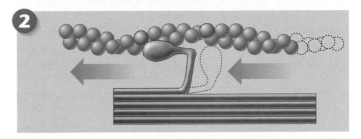

2

The myosin head flexes, advancing the actin filament.

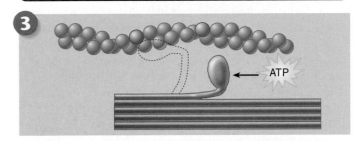

3

← ATP

The myosin head releases and unflexes, powered by ATP. The myosin head is then able to reattach to actin, farther along the fiber.

Figure 23.16   How myofilament contraction works.

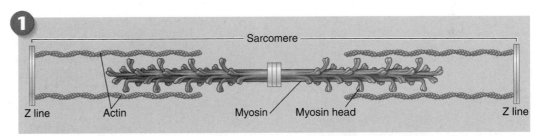

1

Sarcomere

Z line   Actin   Myosin   Myosin head   Z line

The heads on the two ends of the myosin filament are oriented in opposite directions.

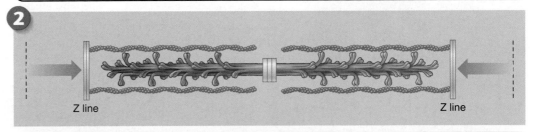

2

Z line   Z line

Thus, as the right-hand end of the myosin filament "walks" along the actin filaments, pulling them and their attached Z line leftward toward the center, the left-hand end of the same myosin filament "walks" along the actin filaments, pulling them and their attached Z line rightward toward the center. The result is that both Z lines move toward the center—and contraction occurs.

Which Mode of Locomotion Is the Most Efficient?

Running, flying, and swimming require more energy than sitting still, but how do they compare? The greatest differences between moving on land, in the air, and in water result from the differences in support and resistance to movement provided by water and air. The weight of a swimming animal is fully supported by the surrounding water, and no effort goes into supporting the body, while running and flying animals must support the full weight of their bodies. On the other hand, water presents considerable resistance to movement, air much less, so that flying and running require less energy to push the medium out of the way.

A simple way to compare the costs of moving for different animals is to determine how much energy it takes to move. The energy cost to run, fly, or swim is in each case the energy required to move one unit of body mass over one unit of distance with that mode of locomotion. (Energy is measured in the **metric system** as a **kilocalorie** (**kcal**) or technically 4.184 kilojoules [note that the Calorie measured in food diets and written with a capital C is equivalent to 1 kcal]; body mass is measured in kilograms, where one kilogram [**kg**] is 2.2 pounds; distance is measured in kilometers, where one kilometer [**km**] is 0.62 miles). The graph to the right displays three such "cost-of-motion" studies. The blue squares are running, the red circles are flying, and the green triangles are swimming. In each study, the line is drawn as the statistical "best-fit" for the points. Some animals like humans have data in two lines, as they both run (well) and swim (poorly). Ducks have data in all three lines, as they not only fly (very well), but also run and swim (poorly).

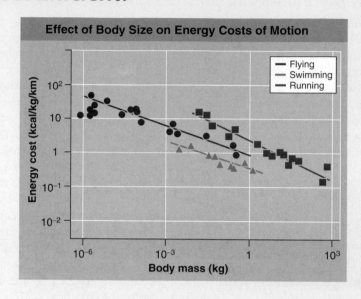

Effect of Body Size on Energy Costs of Motion

Energy cost (kcal/kg/km) vs. Body mass (kg)

Legend: Flying, Swimming, Running

Analysis

1. **Applying Concepts**   Do the three modes of locomotion have the same or different costs?
2. **Interpreting Data**
   a. For any given mode of locomotion, what is the impact of body size on cost of moving?
   b. Is the impact of body mass the same for all three modes of locomotion? If not, which mode's cost is least affected by body mass? Why do you think this is so?
3. **Making Inferences**
   a. Comparing the energy costs of running versus flying for animals of the same body size, which mode of locomotion is the most expensive? Why would you expect this to be so?
   b. Comparing the energy costs of swimming to flying, which uses the least energy? Why would you expect this to be so?
4. **Drawing Conclusions**   In general, which mode of locomotion is the most efficient? the least efficient? Why do you think this is so?

Concept Summary

The Animal Body Plan

23.1 Organization of the Vertebrate Body

- All vertebrates and some invertebrates have the same general architecture: a tube (the gut or digestive system) suspended in a cavity (the coelom) that in many terrestrial vertebrates is divided into a thoracic and an abdominal cavity.
- The vertebrate body is composed of cells that exhibit increasing levels of structural and functional complexity.
- Cells that group together into tissues act as functional units (**figure 23.1**). Organs of the body are composed of several different tissues that act together to perform a higher level of function. Organs work together in an organ system to perform larger-scale body functions (**figure 23.2**) The skeletal system, shown here from **figure 23.3**, is an example of an organ system in the body.

Tissues of the Vertebrate Body

23.2 Epithelium Is Protective Tissue

- Epithelial tissue is composed of different types of epithelial cells. It covers internal and external surfaces of the body providing protection.
- The structure of the epithelium determines its function (**table 23.1**). Some types of epithelium are a single layer of cells through which substances can pass. Some epithelium is stratified, providing protection. Pseudostratified epithelium appears to have several layers, but is actually just a single layer of cells with nonuniform placement of nuclei. Cuboidal epithelium lines glands in the body and have a secretory function. They produce and release hormones into the blood.

23.3 Connective Tissue Carries Out Various Functions

- The connective tissues of the body are very diverse in structure and function, but all are composed of cells embedded in an extracellular matrix. The matrix may be hard as in bone, flexible as in fibrous connective tissue, adipose tissue, and cartilage, or fluid as in blood, like the red blood cells shown here from **table 23.2** floating in the fluid matrix called plasma.
- Immune connective tissue contains white blood cells that float in the blood plasma and protect the body from infection.
- Skeletal connective tissue, such as fibrous connective tissue, cartilage, and bone, provide structural support of the body.
- Adipose tissue stores fat deposits and red blood cells, erythrocytes, transport oxygen and carbon dioxide throughout the body.
- Bone is a living tissue. Bone cells, called osteoblasts, lay down a fibrous matrix. Calcium minerals then impregnate the fibers causing the matrix to harden into compact bone (**figure 23.6**).

23.4 Muscle Tissue Lets the Body Move

- Muscle tissue is dynamic tissue; it contracts, causing the body to move. There are three types of muscle tissue: smooth, skeletal, and cardiac muscle. All three types of muscle contain actin and myosin myofilaments but differ in the organization of the myofilaments.
- Smooth muscle cells are long, spindle-shaped cells organized into sheets. Smooth muscle is found in the walls of blood vessels, in the iris of the vertebrate eye, and in the walls of the digestive system.
- Skeletal muscle cells are fused into long fibers (**figure 23.8**). Skeletal muscle is attached to the skeleton, so when the muscle contracts, the skeleton moves.
- Cells of cardiac muscle found in the heart are interconnected so that they contract together in an orderly pulsation.

23.5 Nerve Tissue Conducts Signals Rapidly

- Nerve tissue is composed of neurons and supporting glial cells. Neurons have three parts: branching dendrites, the cell body that contains the nucleus, and a long axon (**figure 23.9**). Neurons carry electrical impulses from one area of the body to another. The electrical impulse travels down the length of the neuron's axon by the movement of ions across the plasma membrane. There are three categories of neurons: sensory neurons that carry impulses to the central nervous system, motor neurons that carry impulses from the central nervous system to muscles, and association neurons that act as connectors within the central nervous system (**table 23.3**).

The Skeletal and Muscular Systems

23.6 Types of Skeletons

- The skeletal system provides a framework on which muscles act to move the body. Soft-bodied invertebrates have hydraulic skeletons, where muscles act on a fluid-filled cavity (**figure 23.10**).
- Arthropods have exoskeletons, where muscles attach from within to the hard outer covering of the body (**figure 23.11**).
- Vertebrates and echinoderms have endoskeletons, where muscles attach to bones or cartilage inside the body (**figure 23.12**). The human skeleton has 206 individual bones that make up the axial and appendicular skeletons (**figure 23.13**).

23.7 Muscles and How They Work

- Skeletal muscles attach to the skeleton at two points. The end that attaches to the stationary bone is called the origin. The muscle passes over a joint and attaches to another bone at a point called the insertion (**figure 23.14**). As the muscle contracts, the insertion is brought closer to the origin and the joint flexes. Muscles act in opposing pairs to flex or extend a joint (**figure 23.15**).
- Myofibrils are composed of bundles of actin and myosin. During muscle contraction, myosin attaches to actin (as shown here from **figure 23.16**) and pulls the actin along its length, causing the myofilaments to slide past  each other. Actin myofilaments are anchored at each end to a structure called the Z line. As the actin slides along the myosin the anchor points are brought closer together, resulting in a shortening of the muscle (**figure 23.17**). Energy from ATP causes the myofilaments to dissociate and reset, triggering the sliding motion again.

Self-Test

1. Which of the following contains the heart and lungs in a vertebrate body?
   a. the coelom
   b. the gut
   c. the thoracic cavity
   d. the abdominal cavity

2. Which of the following is the correct organization sequence from smallest to largest in animals?
   a. cells, tissues, organs, organ systems, organism
   b. organism, organ systems, organs, tissues, cells
   c. tissues, organs, cells, organ systems, organism
   d. organs, tissues, cells, organism, organ systems

3. Which of the following is *not* a function of the epithelial tissue?
   a. secrete materials
   b. provide sensory surfaces
   c. move the body
   d. protect underlying tissue from damage and dehydration

4. An example of connective tissue is
   a. nerve cells in your fingers.
   b. skin cells.
   c. brain cells.
   d. red blood cells.

5. When a person has osteoporosis, the work of _____ falls behind the work of _____.
   a. osteoclasts; osteoblasts
   b. osteoclasts; collagen
   c. osteoblasts; osteoclasts
   d. osteoblasts; collagen

6. Nerve impulses pass from one nerve cell to another through the use of
   a. hormones.
   b. neurotransmitters.
   c. pheromones.
   d. calcium ions.

7. The type of muscle used to move the leg when walking is
   a. skeletal.
   b. cardiac.
   c. smooth.
   d. all of the above.

8. The vertebral column is part of the
   a. appendicular skeleton.
   b. axial skeleton.
   c. hydraulic skeleton.
   d. exoskeleton.

9. Movement of a limb in two directions requires a pair of muscles because
   a. a single muscle can only pull and not push.
   b. a single muscle can only push and not pull.
   c. moving a limb requires more force than one muscle can generate.
   d. No answer is correct.

10. ATP is required for muscle contraction to
    a. pull the actin along the myosin fiber.
    b. pull the myosin along the actin fiber.
    c. to reset the myosin head so it can reattach to the actin.
    d. All answers are correct; ATP is required at several different stages in the contraction of the muscle.

Visual Understanding

1. **Figure 23.7a** Areas of normal bone are composed of an open lattice framework of minerals, including calcium. Why wouldn't it be more sensible for bones to be almost solid mineral?

2. **Figure 23.13** The bones of your head and trunk form three cagelike structures; these are the skull, the rib cage, and the pelvic girdle. What is the importance of these structures?

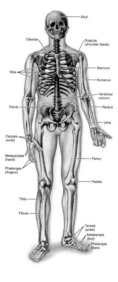

Challenge Questions

1. Your friends are having a friendly debate about which organ systems are the most important to them. Lucas says it's his muscles and bones so that he can play soccer. Brigit insists that her nervous system and respiratory system are the most important so that she can think, feel, and talk. Joseph is sure that it's his digestive system; otherwise he couldn't enjoy foods like pizza, tacos, and hamburgers. Pick one or two systems and explain why you think they are the most important to your life.

2. Imagine that you are designing a living organism, some type of vertebrate. Explain briefly how the four types of tissue are all necessary to your design.

3. When you are exercising rapidly, such as playing tennis, dancing to fast music, or doing aerobics, you begin to breathe rapidly and your heart rate increases. If you continue, you might even say that you are "out of breath." Why is rapid breathing and heart rate important when you are giving your muscles a lot of work to do? (Hint: rapid breathing and heart rate speeds oxygen to your muscles, why is this important to muscle function?)

Chapter **24**

Circulation

Circulation

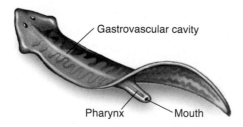

(a) *Planaria:* gastrovascular cavity

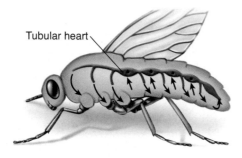

(b) Insect: open circulation

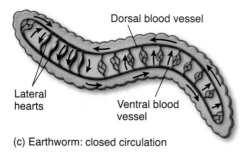

(c) Earthworm: closed circulation

Figure 24.1 **Three types of circulatory systems found in the animal kingdom.**

(a) The gastrovascular cavity of *Planaria* serves as both a digestive and circulatory system, delivering nutrients and oxygen directly to the tissue cells by diffusion from the digestive cavity. (b) In the open circulation of an insect, hemolymph is pumped from a tubular heart into cavities in the insect's body; the hemolymph then returns to the blood vessels so that it can be recirculated. (c) In the closed circulation of the earthworm, blood pumped from the hearts remains within a system of vessels that returns it to the hearts. All vertebrates also have closed circulatory systems.

24.1 Open and Closed Circulatory Systems

CONCEPT PREVIEW: Circulatory systems may be open or closed. All vertebrates have a closed circulatory system in which blood circulates away from the heart in arteries and back to the heart in veins. The circulatory system serves a variety of functions, including transportation, regulation, and protection.

Among the unicellular protists, oxygen and nutrients are obtained directly by simple diffusion from the aqueous external environment. Cnidarians, such as *Hydra,* and flatworms, such as *Planaria,* have cells that are directly exposed to either the external environment or to a body cavity called the **gastrovascular cavity** that functions in both digestion and circulation. The gastrovascular cavity of *Hydra* extends even into the tentacles, and that of *Planaria* (colored green in figure 24.1*a*) branches extensively to supply every cell with oxygen and the nourishment obtained by digestion. Larger animals, however, have tissues that are several cell layers thick, so many cells are too far away from the body surface or digestive cavity to exchange materials directly with the environment. Instead, oxygen and nutrients are transported from the environment and digestive cavity to the body cells by an internal fluid within a **circulatory system.**

> The gastrovascular cavity in cnidarians was described on page 339. Water brings oxygen into the cavity, which passes into the cells that line the cavity by diffusion. Food enters the cavity and is broken down with enzymes. The products of digestion are also absorbed by cells.

There are two main types of circulatory systems: *open* and *closed.* In an **open circulatory system,** such as that found in arthropods and many mollusks, there is no distinction between the circulating fluid (blood) and the extracellular fluid of the body tissues. This fluid is thus called **hemolymph.** Insects, like the fly in figure 24.1*b,* have a muscular tube that serves as a heart to pump the hemolymph through a network of open-ended channels that empty into the cavities in the body (downward-pointing arrows). There, the hemolymph delivers nutrients to the cells of the body. It then reenters the circulatory system through pores in the heart (upward-pointing arrows). The pores close when the heart pumps to keep the hemolymph from flowing back out into the body cavity.

In a **closed circulatory system,** the circulating fluid, or *blood,* is always enclosed within blood vessels that transport blood away from and back to the heart. Annelids and all vertebrates have a closed circulatory system. In annelids such as the earthworm in figure 24.1*c,* a dorsal blood vessel contracts rhythmically to function as a pump. Blood is pumped through five small connecting vessels called lateral hearts, which also function as pumps, to a ventral blood vessel, which transports the blood posteriorly (lower arrows) until it eventually reenters the dorsal blood vessel (upper arrows). Smaller vessels branch between the ventral and dorsal blood vessels to supply the tissues of the earthworm.

In vertebrates, blood vessels form a tubular network that permits blood to flow from the heart to all the cells of the body and then back to the heart. *Arteries* carry blood away from the heart, whereas *veins* return blood to the heart. Blood passes from the arterial to the venous system in *capillaries,* which are the thinnest and most numerous of the blood vessels.

As blood passes through capillaries, the pressure of the blood forces some fluid out of the capillary walls. Fluid derived this way is called

interstitial fluid. Some of this fluid returns directly to capillaries, and some enters into lymph vessels, located in the connective tissues around the blood vessels. This fluid, now called *lymph,* is returned to the venous blood at specific sites. The lymphatic system is considered a part of the circulatory system and is discussed later in this chapter.

The Functions of Vertebrate Circulatory Systems

The functions of the circulatory system can be divided into three areas: transportation, regulation, and protection.

1. **Transportation.** Substances essential for cellular functions are transported by the circulatory system, and can be categorized as follows:
   Respiratory. Red blood cells, or erythrocytes, transport oxygen to the tissue cells. In the capillaries of the lungs or gills, oxygen attaches to hemoglobin molecules within the erythrocytes and is transported to the cells for aerobic respiration. Carbon dioxide produced by cell respiration is carried by the blood to lungs or gills for elimination.

   > Cells need oxygen to make ATP. Oxygen, as discussed on page 120 is the final electron acceptor in the electron transport chain. Carbon dioxide is a waste product of cellular respiration, being produced during the oxidation of pyruvate and during the Krebs cycle.

   Nutritive. The digestive system is responsible for the breakdown of food so that nutrients can be absorbed through the intestinal wall and into the blood vessels of the circulatory system. The blood then carries these absorbed products of digestion to the cells of the body.
   Excretory. Metabolic wastes, excessive water and ions, and other molecules in plasma (the fluid portion of blood) are filtered through the capillaries of the kidneys and excreted in urine.
   Endocrine. The blood carries hormones from the endocrine glands, where they are secreted, to the distant target organs they regulate.

2. **Regulation.** The cardiovascular system regulates body temperature.
   Temperature regulation. In warm-blooded vertebrates, or homeotherms, a constant body temperature is maintained, regardless of the surrounding temperature. This is accomplished in part by blood vessels located just under the epidermis. When the ambient temperature is cold, the superficial vessels constrict to divert the warm blood to deeper vessels, reducing heat loss. When the ambient temperature is warm, the superficial vessels dilate so that the warmth of the blood can be lost by radiation.

   Some vertebrates also retain heat in a cold environment by using a **countercurrent heat exchange.** Figure 24.2 shows how a countercurrent heat exchange system works in the flipper of a killer whale. In this process, the warm blood going out heats the cold blood returning from the body surface, helping to maintain a stable core body temperature.

3. **Protection.** The circulatory system protects against injury and foreign microbes or toxins introduced into the body.
   Blood clotting. The clotting mechanism protects against blood loss when vessels are damaged. This clotting mechanism involves both proteins from the blood plasma and blood cell structures called platelets.
   Immune defense. The blood contains proteins and white blood cells, or leukocytes, that provide immunity against many disease-causing agents. Some white blood cells are phagocytic, some produce antibodies, and some act by other mechanisms to protect the body.

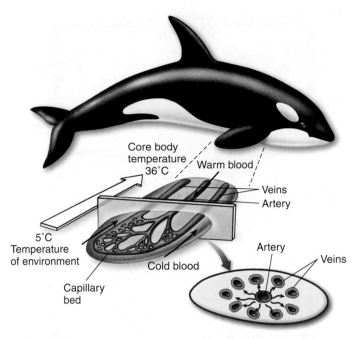

Figure 24.2 Countercurrent heat exchange.

Many marine mammals, such as this killer whale, limit heat loss in cold water by countercurrent flow that allows heat exchange between arteries and veins. The warm blood pumped from within the body in arteries warms the cold blood returning from the skin in veins, so that the core body temperature can remain constant in cold water. The cut-away portion in the figure shows how the veins surround the artery, maximizing the heat exchange between the artery and the veins.

E V O L U T I O N

Why Bother with Blood? Despite the many advantages listed on this page, many animals with simple bodies seem to get along just fine without circulatory systems. The body cavity of a flatworm (phylum *Platyhelminthes*) has no lining or internal fluid to circulate. Think of its body as having the structure of a drinking straw. The tissue wall between the inside cavity and the body's exterior is only a few cells thick, so that oxygen can diffuse from the interior cavity or the surrounding water directly into each of the body's cells, and carbon dioxide can diffuse out just as easily. Nutrients are forced into the gastrovascular cavity by a muscular pharynx, passing directly to all the body's cells by simple diffusion. Because every cell can obtain food, water, and oxygen directly, the animal has no need of a transport system. It gives up a few niceties, of course: Without circulating fluid, there is no way to regulate body temperature, so the animal must always adopt the temperature of its immediate surroundings. Nor can it protect itself as we do with a circulating army of defensive cells, or control its development with circulating hormones as we do. Circulatory systems like ours have evolved to overcome the limits imposed by diffusion.

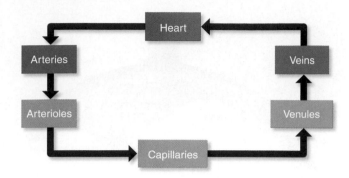

Figure 24.3 **The flow of blood through the circulatory system.**

24.2 Architecture of the Vertebrate Circulatory System

CONCEPT PREVIEW: The vertebrate circulatory system is composed of arteries and arterioles, which are elastic and carry blood away from the heart; a fine network of capillaries across whose thin walls the exchange of gases and nutrients takes place; and venules and veins, which return blood from the capillaries to the heart.

The vertebrate circulatory system, also called the **cardiovascular system,** is made up of three elements: (1) the *heart,* a muscular pump that pushes blood through the body; (2) the *blood vessels,* a network of tubes through which the blood moves; and (3) the *blood,* which circulates within these vessels.

Blood moves through the body in a cycle, from the heart, through a system of vessels: from the arteries and arterioles, into the capillaries, and then back to the heart through the venules and veins, as shown in figure 24.3. The designation of a blood vessel as a vein or artery is determined, in part, by the direction of blood flow in relation to the heart (veins carry blood toward the heart, arteries carry blood away from the heart), not whether the vessel is carrying oxygenated or deoxygenated blood.

Blood leaves the heart through vessels known as **arteries.** From the arteries, the blood passes into a network of smaller arteries called **arterioles,** shown in figure 24.4*a* ❶. From these, it is eventually forced through a capillary bed ❷, a fine latticework of very narrow tubes called **capillaries** (from the Latin, *capillus,* "a hair"). While passing through the capillaries, the blood exchanges gases and metabolites (glucose, vitamins, hormones) with the cells of the body. After traversing the capillaries, the blood passes into a fourth kind of vessel, the **venules,** or small veins, ❸. A network of venules empties into larger **veins** that collect the circulating blood and carry it back to the heart.

Capillary beds can be opened or closed, based on the physiological needs of the tissues. Smooth muscles can control blood flow to the arterioles (described later), but in addition, flow through the capillaries can be controlled by the relaxation or contraction of small circular muscles called *precapillary sphincters.* The closing of the precapillary sphincters by contraction is shown in figure 24.4*b* with the blood being diverted from the capillary bed.

The capillaries have a much smaller diameter than the other blood vessels of the body. Blood leaves the mammalian heart through a large artery, the aorta, a tube that in your body that has a diameter of about 2 centimeters (about the width of your thumb), but when it reaches the capillaries it passes through vessels with an average diameter of only 8 micrometers, a reduction in radius of some 1,250 times!

This decrease in size of blood vessels has a very important consequence. Although each capillary is very narrow, there are so many of them that the capillaries have the greatest *total* cross-sectional area of any other type of vessel. Consequently, this allows more time for blood to exchange materials with the surrounding extracellular fluid. By the time the blood reaches the end of a capillary, it has released some of its oxygen and nutrients and picked up carbon dioxide and other waste products. Blood loses most of its pressure and velocity in passing through the vast capillary networks, and so is under very low pressure when it enters the veins.

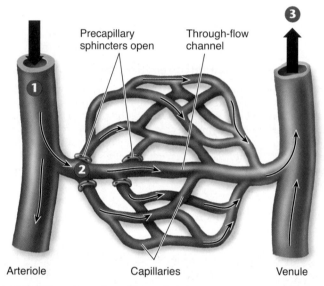

(a) Blood flows through capillary network

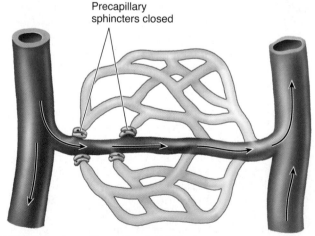

(b) Blood flow in capillary network is limited

Figure 24.4 **The capillary network connects arteries with veins.**

Through-flow channels connect arterioles directly to venules. Branching from these through-flow channels are the capillaries. Most of the exchange between the body tissues and red blood cells occurs in this capillary network. The flow of blood into the capillaries is controlled by bands of muscle called precapillary sphincters. (a) When a sphincter is open, blood flows through that capillary. (b) When a sphincter contracts, it closes off the capillaries.

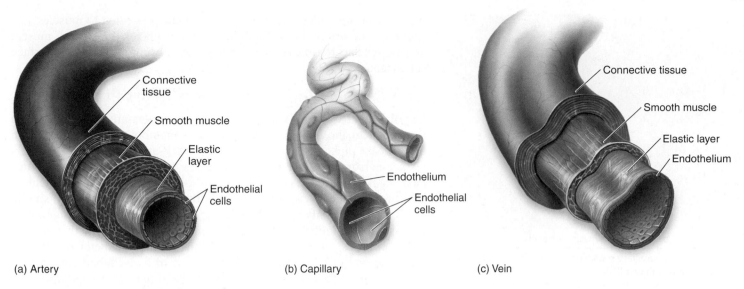

Figure 24.5   The structure of blood vessels.

(a) Arteries, which carry blood away from the heart, are expandable and are composed of layers of tissue. (b) Capillaries are simple tubes whose thin walls facilitate the exchange of materials between the blood and the cells of the body. (c) Veins, which transport blood back to the heart, do not need to be as sturdy as arteries. The walls of veins have thinner muscle layers than arteries, and they collapse when empty. Note, this drawing is not to scale; as stated in the text, arteries can be up to 2 centimeters in diameter, the capillaries are only about 8 micrometers in diameter, and the largest veins can be up to 3 centimeters in diameter.

The blood flow through the capillaries is like water flowing out of the sprinkler head of a watering can, the stream of blood spreads out to many small streams. These smaller streams don't flow with as much force, nor as quickly, as the larger stream that entered the capillary bed.

Arteries: Highways from the Heart

The arterial system, composed of arteries and arterioles, carries blood away from the heart. An artery is more than simply a pipe. Blood comes from the heart in pulses rather than in a smooth flow, forcing blood into the artery in big slugs as the heart ejects its contents with each contraction. The artery has to be able to *expand* to withstand the pressure caused by each contraction of the heart. An artery, then, is designed as an expandable tube, with its walls made up of four layers of tissue. Figure 24.5*a* shows some of these layers pulled out from the artery like a telescope so they are more easily seen. The innermost thin layer is composed of endothelial cells. Surrounding them is a layer of elastic fibers and then a thick layer of smooth muscle, which in turn is encased within an envelope of protective connective tissue. Because this sheath and envelope are elastic, the artery is able to expand its volume considerably when the heart contracts, pumping a new volume of blood into the artery—just as a tubular balloon expands when you blow more air into it. The steady contraction of the smooth muscle layer strengthens the wall of the vessel against overexpansion.

Arterioles differ from arteries in two ways. They are smaller in diameter, and the muscle layer that surrounds an arteriole can be relaxed under the influence of hormones to enlarge the diameter. When the diameter increases, the blood flow also increases, an advantage during times of high body activity. Most arterioles are also in contact with nerve fibers. When stimulated by these nerves, the muscle lining of the arteriole contracts, constricting the diameter of the vessel. Such a contraction limits the flow of blood to the extremities during periods of low temperature or stress. You turn pale when you are scared or cold because the arterioles in your skin are constricting. You blush for just the opposite reason. When you

BIOLOGY & YOU

Figuring Out Arteries. Among the ancient Greeks, the arteries of the body were considered to be "air holders" responsible for transporting air and connected to the windpipe. The Greeks came to this conclusion because they found the arteries of the dead to be empty. In medieval times, it was recognized that arteries carried a fluid. Galen taught that the bright arterial fluid, what he called "vital spirits," was different from the darker blood of the veins, and that it was continually produced rather than recycled. It wasn't until the 17th century that the British physician William Harvey first understood that the body's blood recycles between arteries and veins, driven by the beating of the heart. The modern concept of the circulatory system as two separate closed loops, one for the lungs and another for the body's organs and tissues, was written down in Latin in 1628.

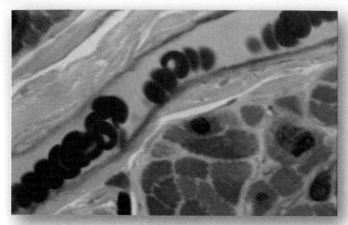

Figure 24.6   Red blood cells within a capillary.

The red blood cells in this capillary pass along in single file. Red blood cells can pass through capillaries even narrower than their own diameter, pushed along by the pressure of the pumping heart.

IMPLICATION   The volume of blood in your body is about 5 liters, if you are a typical adult. When you donate a pint of blood, what percentage of your total blood are you giving away?

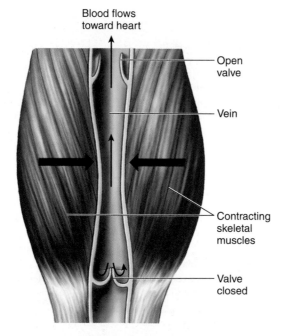

Blood flows
toward heart

Open
valve

Vein

Contracting
skeletal
muscles

Valve
closed

Figure 24.7   Flow of blood through veins.

Venous valves ensure that blood moves through the veins in only one direction back to the heart. This movement of blood is aided by the contraction of skeletal muscles surrounding the veins.

overheat or are embarrassed, the nerve fibers connected to muscles surrounding the arterioles are inhibited, which relaxes the smooth muscle and causes the arterioles in the skin to expand, bringing heat to the surface for escape.

Capillaries: Where Exchange Takes Place

Capillaries are where oxygen and food molecules are transferred from the blood to the body's cells and where metabolic waste and carbon dioxide are picked up. To facilitate this back-and-forth traffic, capillaries are narrow and have thin walls across which gases and metabolites pass easily. Capillaries have the simplest structure of any element in the cardiovascular system. They are built like a soft-drink straw, simple tubes with walls only one cell thick (see figure 24.5b). The average capillary is about 1 millimeter long and connects an arteriole with a venule. All capillaries are very narrow, with an internal diameter of about 8 micrometers, just bigger than the diameter of a red blood cell (5 to 7 micrometers). This design is critical to the function of capillaries. By bumping against the sides of the vessel as they pass through (like the cells in figure 24.6), the red blood cells are forced into close contact with the capillary walls, making exchange easier.

Almost all cells of the vertebrate body are no more than 100 micrometers from a capillary. At any one moment, about 5% of the circulating blood is in capillaries, a network that amounts to several thousand miles in overall length. If all the capillaries in your body were laid end to end, they would extend across the United States! Individual capillaries have high resistance to flow because of their small diameters. However, the total cross-sectional area of the extensive capillary network (that is, the sum of all the diameters of all the capillaries, expressed as area) is greater than that of the arteries leading to it. As a result, the blood pressure is actually far lower in the capillaries than in the arteries. This is important, because the walls of capillaries are not strong, and they would burst if exposed to the pressures that arteries routinely withstand.

Veins: Returning Blood to the Heart

Veins are vessels that return blood to the heart. Veins do not have to accommodate the pulsing pressures that arteries do, because much of the force of the heartbeat is weakened by the high resistance and great cross-sectional area of the capillary network. For this reason, vein walls have much thinner layers of muscle and elastic fiber (figure 24.5c). An empty artery will stay open, like a pipe, but when a vein is empty, its walls collapse like an empty balloon.

Because the pressure of the blood flowing within veins is low, it becomes important to avoid any further resistance to flow, lest there not be enough pressure to get the blood back to the heart. Because a wide tube presents much less resistance to flow than a narrow one, the internal passageway of veins is often quite large, requiring only a small pressure difference to return blood to the heart. The diameters of the largest veins in the human body, the venae cavae, which lead into the heart, are fully 3 centimeters; this is wider than your thumb! Pressure alone cannot force the blood in the veins back to the heart but several features provide help. Most significantly, when skeletal muscles surrounding the veins contract, they move blood by squeezing the veins. Veins also have unidirectional valves (the small flaps within the vein in figure 24.7) that ensure the return of this blood by preventing it from flowing backward. These structural features keep the blood flowing in a cycle through the circulatory system.

24.3 The Lymphatic System: Recovering Lost Fluid

CONCEPT PREVIEW: The lymphatic system returns fluids and proteins to the circulatory system that leak through the capillaries.

The cardiovascular system is very leaky. Fluids are forced out across the thin walls of the capillaries by the pumping pressure of the heart. Although this loss is unavoidable—the circulatory system could not do its job of gas and metabolite exchange without tiny vessels with thin walls—it is important that the loss be made up. In your body, about 4 liters of fluid leave your cardiovascular system in this way each day, more than half the body's total supply of about 5.6 liters of blood! To collect and recycle this fluid, the body uses a second circulatory system called the **lymphatic system.** Like the blood circulatory system, the lymphatic system is composed of a network of vessels; their distribution throughout the body is shown in figure 24.8.

Fluid is filtered from the capillaries near their arteriole ends where the blood pressure is higher, as shown by the red arrows in figure 24.9. Most of the fluid returns to the blood capillaries by osmosis (the light purple arrows); however, excess fluid enters the lymphatic capillaries (indicated by the green arrow). These capillaries gather up fluid from the spaces surrounding cells and carry it through a series of progressively larger vessels to two large lymphatic vessels, which drain into veins in the lower part of the neck through one-way valves. Once within the lymphatic system, this fluid is called **lymph.**

Fluid is driven through the lymphatic system when its vessels are squeezed by the movements of the body's muscles. The lymphatic vessels contain a series of one-way valves that permit movement only in the direction of the neck. In some cases, the lymphatic vessels also contract rhythmically. In many fishes, all amphibians and reptiles, bird embryos, and some adult birds, movement of lymph is propelled by *lymph hearts.*

The lymphatic system has three other important functions:

1. **It returns proteins to the circulation.** So much blood flows through the capillaries that there is significant leakage of blood proteins, even though the capillary walls are not very permeable to proteins (proteins are very large molecules). By recapturing the lost fluid, the lymphatic system also returns this lost protein. Whenever this protein is not returned to the blood and instead remains in the tissues, fluid also remains in the tissues and the body becomes swollen, a condition known as *edema.*

   > Without enough protein, water will leave the blood through osmosis. As discussed on page 78, water passes from areas of low solute concentrations to areas of higher concentrations. Proteins increase the solute concentration of the blood, and so keep the water from leaving.

2. **It transports fats absorbed from the intestine.** Lymph capillaries called *lacteals* permeate the lining of the small intestine. These lacteals absorb fats from the digestive tract and eventually transport them to the circulatory system by way of the lymphatic system.

3. **It aids in the body's defense.** Along a lymph vessel are swellings called *lymph nodes,* seen in figure 24.8. Inside each node is a honeycomb of spaces filled with white blood cells specialized for defense. The lymphatic system carries bacteria and dead blood cells to the lymph nodes and spleen for destruction. The thymus gland, also shown in the figure, produces white blood cells.

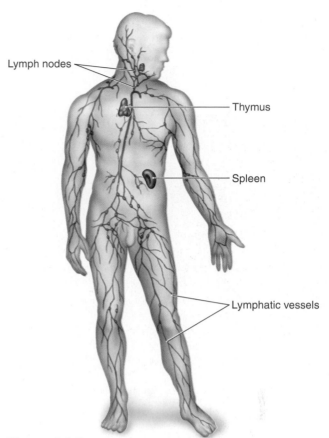

Figure 24.8   The human lymphatic system.

The lymphatic system consists of lymphatic vessels and capillaries, lymph nodes, and lymphatic organs, including the spleen and thymus gland.

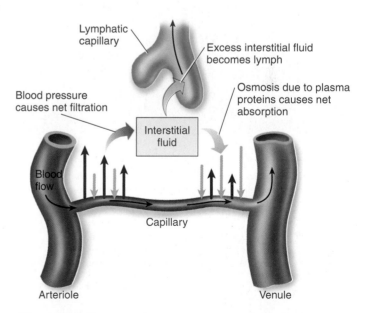

Figure 24.9   Lymphatic capillaries reclaim fluid from interstitial fluid.

Blood pressure forces fluid from capillaries into the interstitial fluid surrounding cells, where it reenters the blood or drains into lymphatic capillaries.

24.4 Blood

CONCEPT PREVIEW: Blood is a collection of cells that circulate within a protein-rich salty fluid called plasma. Some of the cells circulating in the blood carry out gas transport; others are engaged in defending the body from infection.

About 5% of your body mass is composed of the blood circulating through the arteries, veins, and capillaries of your body. This blood is composed of a fluid called **plasma,** together with several different kinds of cells that circulate within that fluid.

> Blood is a connective tissue in the body, and as explained on page 462, connective tissue is composed of cells within a matrix. The cells in blood tissue are red and white blood cells and the matrix is the fluid plasma.

Blood Plasma: The Blood's Fluid

Blood plasma is a complex solution of water with three very different sorts of substances dissolved within it:

1. **Metabolites and wastes.**   If the circulatory system is viewed as a highway of the vertebrate body, the blood contains the traffic traveling on that highway. Dissolved within its plasma are glucose, vitamins, hormones, and wastes that circulate between the cells of the body.
2. **Salts and ions.**   Plasma is a dilute salt solution. The chief plasma ions are sodium, chloride, and bicarbonate. In addition, trace amounts of other salts, such as calcium and magnesium, as well as metallic ions, including copper, potassium, and zinc, are present in plasma. The composition of the plasma is not unlike that of seawater.
3. **Proteins.**   Blood plasma is 90% water. Passing by all the cells of the body, blood would soon lose most of its water to them by osmosis if it did not contain as high a concentration of proteins as the cells it passes. Some of the proteins in blood plasma are antibodies that are active in the immune system. More than half the amount of protein that is necessary to balance the protein content of the cells of the body consists of a single protein, serum albumin, which circulates in the blood as an osmotic counterforce. Human blood contains 46 grams of serum albumin per liter—that's over half a pound of it in your body. Starvation and protein deficiency result in reduced levels of protein in the blood. This lack of plasma proteins produces swelling of the body because the body's cells, which now have a higher level of solutes than the blood, take up water from the albumin-deficient blood. A symptom of protein deficiency diseases such as kwashiorkor is edema, a swelling of tissues, although other factors can also result in edema.

The liver produces most of the plasma proteins, including albumin; the alpha and beta globulins, which serve as carriers of lipids and steroid hormones; and *fibrinogen,* which is required for blood clotting. When blood in a test tube clots, the fibrinogen is converted into insoluble threads of *fibrin* that become part of the clot. You can see a blood clot forming in figure 24.10. Red blood cells, the disk-shaped cells, are becoming trapped in and among the threads of fibrin. The fluid that's left, which lacks fibrinogen and so cannot clot, is called serum.

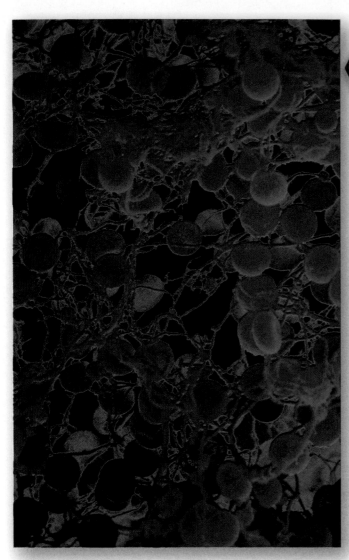

Figure 24.10   Threads of fibrin.

This scanning electron micrograph (×1,430) shows fibrin threads among red blood cells. Fibrin is formed from a soluble protein, fibrinogen, in the plasma to produce a blood clot when a blood vessel is damaged.

IMPLICATION   In 2007, Rockefeller University researchers reported that levels of fibrin are elevated in the brains of Alzheimer's patients. What effect do you think this could have on these patients?

Blood Cells: Cells That Circulate Through the Body

Although blood is liquid, nearly half of its volume is actually occupied by cells. The three principal cellular components of blood are erythrocytes (red blood cells), leukocytes (white blood cells), and cell fragments called platelets. The fraction of the total volume of the blood that is occupied by red blood cells is referred to as the blood's hematocrit. In humans, the hematocrit is usually about 45%.

Erythrocytes Carry Hemoglobin.   Each microliter (1 μl) of blood contains about 5 million **erythrocytes.** Each human erythrocyte (pictured at the top of figure 24.11) is a flat disk with a central depression on both sides, something like a doughnut with a hole that doesn't go all the way through. Erythrocytes carry oxygen to the cells of the body. Almost the entire interior of an erythrocyte is packed with hemoglobin, a protein that binds oxygen in the lungs and delivers it to the cells of the body.

> Hemoglobin is a large molecule made up of four subunits. Each subunit, as described on page 496, contains a central heme group that binds oxygen; therefore each hemoglobin molecule can transport four molecules of oxygen.

Mature mammalian erythrocytes function like boxcars rather than trucks. Like a vehicle without an engine, erythrocytes contain neither a nucleus nor the machinery to make proteins. Because they lack a nucleus, these cells are unable to repair themselves and therefore have a rather short life; any one human erythrocyte lives only about four months. New erythrocytes are constantly being synthesized and released into the blood by cells within the soft interior marrow of bones.

Leukocytes Defend the Body.   Less than 1% of the cells in mammalian blood are **leukocytes,** also called **white blood cells.** There are many different kinds of leukocytes, which are shown in figure 24.11. They are the somewhat transparent cells with large or odd-shaped nuclei. Most are larger than erythrocytes. They contain no hemoglobin and are essentially colorless. Each type of leukocyte has a different function. Neutrophils are the most numerous of the leukocytes, followed in order by lymphocytes, monocytes, eosinophils, and basophils. *Neutrophils* attack like kamikazes, responding to foreign cells by releasing chemicals that kill all the cells in the neighborhood—including themselves. Monocytes give rise to *macrophages,* which attack and kill foreign cells by ingesting them (the name means "large eater"). Lymphocytes include *B cells* which make antibodies, and *T cells* which kill infected cells.

All of these white blood cell types, and others, help defend the body against invading microorganisms and other foreign substances, as you will see in chapter 28. Unlike other blood cells, leukocytes are not confined to the bloodstream; they are mobile soldiers that migrate out into the surrounding fluid.

Platelets Help Blood to Clot.   Certain large cells within the bone marrow, called megakaryocytes, regularly pinch off bits of their cytoplasm. These cell fragments, called **platelets** (shown at the bottom of figure 24.11), contain no nuclei. Entering the bloodstream, they play a key role in blood clotting. In a clot, a gluey mesh of fibrin protein fibers (shown in figure 24.10) sticks platelets together to form a mass that plugs the rupture in the blood vessel. The clot provides a tight, strong seal, much as the inner lining of a tubeless tire seals punctures. The fibrin that forms the clot is made in a series of reactions that start when circulating platelets first encounter the site of an injury. Responding to chemicals released by the damaged blood vessel, platelets release a protein factor into the blood that starts the clotting process.

Blood cell	Life span in blood	Function
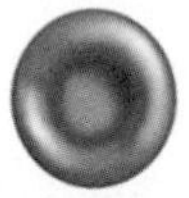 Erythrocyte	120 days	O_2 and CO_2 transport
 Neutrophil	7 hours	Immune defenses
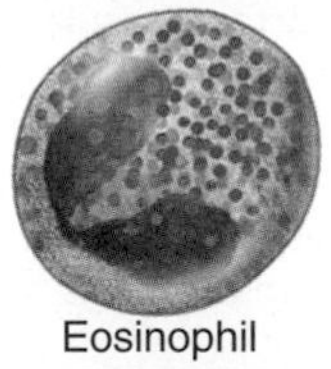 Eosinophil	Unknown	Defense against parasites
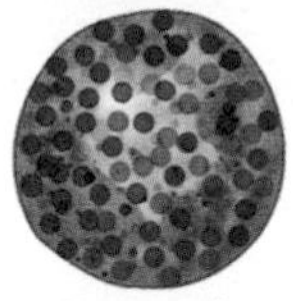 Basophil	Unknown	Inflammatory response
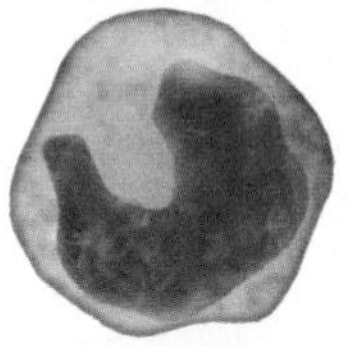 Monocyte	3 days	Immune surveillance (precursor of tissue macrophage)
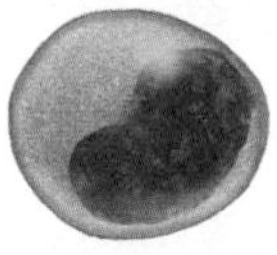 B lymphocyte	Unknown	Antibody production (precursor of plasma cells)
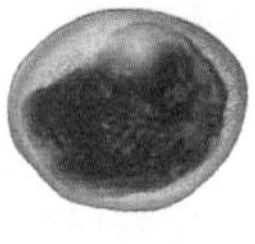 T lymphocyte	Unknown	Cellular immune response
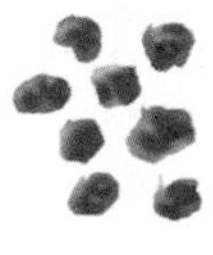 Platelets	7–8 days	Blood clotting

Figure 24.11   Types of blood cells.

Erythrocytes, leukocytes (neutrophils, eosinophils, basophils, monocytes, and lymphocytes), and platelets are the three principal cellular components of blood in vertebrates.

Figure 24.12   The heart and circulation in humans.

(a) Some of the major arteries and veins in the human circulatory system are shown. (b) In mammals (and birds), the heart has a septum dividing the ventricle into left and right ventricles. Oxygenated blood from the lungs enters the left atrium of the heart by way of the pulmonary veins. This blood then enters the left ventricle, from which it passes into the aorta to circulate throughout the body and deliver oxygen to the tissues. When gas exchange has taken place at the tissues, veins return blood to the heart. After entering the right atrium by way of the superior and inferior venae cavae, deoxygenated blood passes into the right ventricle and then through the pulmonary valve to the lungs by way of the pulmonary artery.

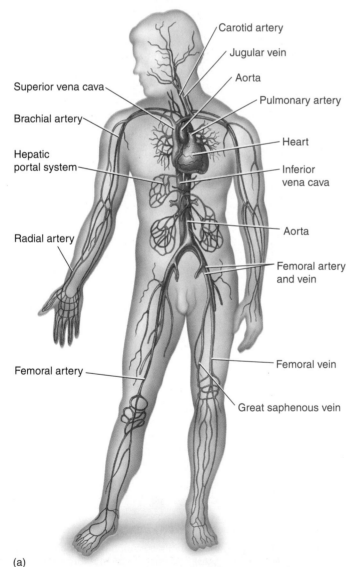

Carotid artery

Jugular vein

Aorta

Superior vena cava

Pulmonary artery

Brachial artery

Heart

Hepatic portal system

Inferior vena cava

Aorta

Radial artery

Femoral artery and vein

Femoral artery

Femoral vein

Great saphenous vein

(a)

24.5 Human Circulatory System

CONCEPT PREVIEW: The mammalian heart is a two-cycle pump. The left side pumps oxygenated blood to the body's tissues, while the right side pumps O_2-depleted blood to the lungs.

Humans and other mammals have a four-chambered heart that is really two separate pumping systems operating together within a single unit. One system pumps blood to the lungs, while the other pumps blood to the rest of the body. The left side has two connected chambers, and so does the right, but the two sides are not connected with one another.

Circulation Through the Heart

Let's follow the journey of blood through the human heart in figure 24.12*b*, starting with the entry of oxygen-rich blood into the heart from the lungs. Oxygenated blood (the pink arrows) from the lungs enters the left side of the heart (which is on the right as you look at the figure), emptying directly into the **left atrium** through large vessels called the **pulmonary veins.** From the atrium, blood flows down through an opening into the adjoining chamber, the **left ventricle.** Most of this flow, roughly 70%, occurs while the heart is relaxed. When the heart starts to contract, the atrium contracts first, pushing the remaining 30% of its blood into the ventricle.

After a slight delay, the ventricle contracts. The walls of the ventricle are far more muscular than those of the atrium (as seen in this cross section), and thus this contraction is much stronger. It forces most of the blood out of the ventricle in a single strong pulse. The blood is prevented from going back into the atrium by a large, one-way valve, the **bicuspid (mitral)**

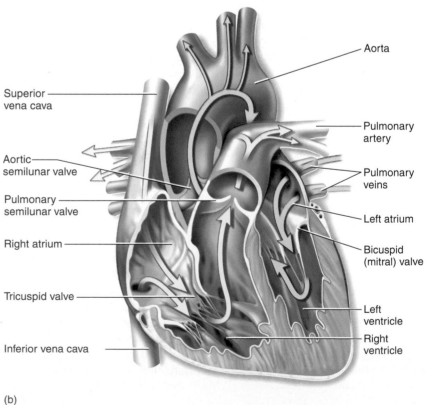

Aorta

Superior vena cava

Pulmonary artery

Aortic semilunar valve

Pulmonary veins

Pulmonary semilunar valve

Left atrium

Right atrium

Bicuspid (mitral) valve

Tricuspid valve

Left ventricle

Inferior vena cava

Right ventricle

(b)

valve, or left atrioventricular valve, whose two flaps are pushed shut as the ventricle contracts.

Prevented from reentering the atrium, the blood within the contracting left ventricle takes the only other passage out (partially covered by the large blue vessel). It moves through a second opening that leads into a large blood vessel called the **aorta.** The aorta is separated from the left ventricle by a one-way valve, the **aortic semilunar valve.** The aortic valve is oriented to permit the flow of the blood out of the ventricle. Once this outward flow has occurred, the aortic valve closes, preventing the reentry of blood into the heart. The aorta and its many branches carry oxygen-rich blood to all parts of the body.

Eventually, this blood returns to the heart after delivering its cargo of oxygen to the cells of the body. In returning, it passes through a series of progressively larger veins, ending in two large veins that empty into the right atrium.

The right side of the heart is similar in organization to the left side. Blood (the blue-colored arrows) passes from the **right atrium** into the **right ventricle** through a one-way valve, the **tricuspid valve** or right atrioventricular valve. It passes out of the contracting right ventricle through a second valve, the **pulmonary semilunar valve,** into the **pulmonary arteries,** which carry the deoxygenated blood to the lungs. The blood then returns from the lungs to the left side of the heart with a new cargo of oxygen, which is pumped to the rest of the body. Figure 24.12*a* shows the major veins and arteries in the human body. Veins carry blood to the heart and arteries carry blood away from the heart.

Monitoring the Heart's Performance

The simplest way to monitor heartbeat is to listen to the heart at work, using a stethoscope. The first sound you hear, a low-pitched *lub,* is the closing of the bicuspid and tricuspid valves at the start of ventricular contraction. A little later, you hear a higher-pitched *dub,* the closing of the pulmonary and aortic valves at the end of ventricular contraction. If the valves are not closing fully, or if they open incompletely, a turbulence is created within the heart. This turbulence can be heard as a liquid sloshing sound, called a *heart murmur.*

A second way to examine the events of the heartbeat is to monitor blood pressure. This is done using a device called a sphygmomanometer (figure 24.13). A cuff wrapped around the upper arm is tightened enough to stop the flow of blood to the lower part of the arm ❶. As the cuff is loosened, blood begins pulsating through the arm's arteries and can be detected using a stethoscope. Two measurements are recorded: The systolic pressure ❷ is recorded when a pulse is heard, and the diastolic pressure ❸ is recorded when the pressure in the cuff is so low that the sound stops.

During the first part of the heartbeat, the atria are filling. At this time the pressure in the arteries leading from the left side of the heart out to the tissues of the body decreases slightly. This low pressure is referred to as the **diastolic** pressure. During the contraction of the left ventricle, a pulse of blood is forced into the systemic arterial system, immediately raising the blood pressure within these vessels. The high blood pressure produced in this pushing period, which ends with the closing of the aortic valve, is referred to as the **systolic** pressure. Normal blood pressure values are 70 to 90 diastolic and 110 to 130 systolic. When the inner walls of the arteries accumulate fats, as they do in the condition known as *atherosclerosis,* the diameters of the passageways are narrowed. If this occurs, the systolic blood pressure is elevated.

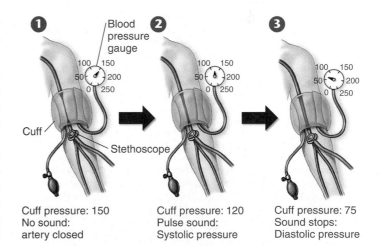

❶ Blood pressure gauge ❷ ❸

Cuff

Stethoscope

Cuff pressure: 150
No sound:
artery closed

Cuff pressure: 120
Pulse sound:
Systolic pressure

Cuff pressure: 75
Sound stops:
Diastolic pressure

Figure 24.13 **Measuring blood pressure.**

The blood pressure cuff is tightened to stop the blood flow through the brachial artery ❶. As the cuff is loosened, the systolic pressure is recorded as the pressure at which a pulse is heard through a stethoscope ❷. The diastolic pressure is recorded as the pressure at which a sound is no longer heard ❸.

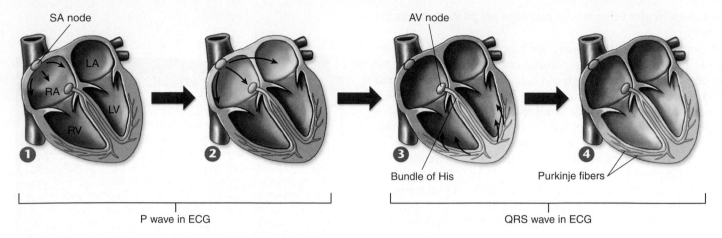

P wave in ECG

QRS wave in ECG

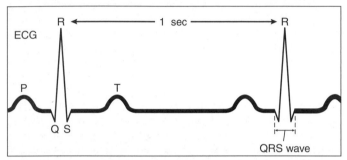

Figure 24.14 How the mammalian heart contracts.

A cardiac cycle consists of one sequence of contraction and relaxation of the heart. ❶ Contraction of the mammalian heart is initiated by an electrical signal (yellow highlighting) that begins at the SA node. ❷After passing over the right and left atria and causing their contraction, ❸the signal reaches the AV node, from which it passes to the ventricles by the bundle of His. ❹The signal is then conducted rapidly over the surface of the ventricles by a set of finer fibers called Purkinje fibers, causing the ventricles to contract. The electrical signal passing over the surface of the heart generates an electrical current that passes in a wave throughout the body. The magnitude of this electrical pulse is tiny, but it can be detected by sensors placed on the skin. The recording shown above, called an electrocardiogram (ECG or EKG), shows how the cells of the heart respond electrically during the cardiac cycle. The letters on the ECG correspond to the electrical stimulation of different parts of the heart. The P wave is the stimulation of the atria and the QRS wave is the stimulation of the ventricles. The T wave on the ECG corresponds to the relaxation of the ventricles. The ECG is showing a heart rate of 60 beats per minute. The time between peaks varies with changes in heart rates.

IMPLICATION To improve your cardiovascular fitness, you need to raise your heart rate from a resting rate of some 60 to 70 beats per minute to an exercise target heart rate of 100–150 beats per minute, and keep it there for 20 minutes. How would you expect the ECG shown above to change if measured during this period of exercise?

How the Heart Contracts

Contraction of the heart consists of a carefully orchestrated series of muscle contractions. First, both atria contract simultaneously. Then, after a brief time interval, both ventricles contract simultaneously. Contraction is initiated by the **sinoatrial (SA) node** (indicated in panel ❶ of figure 24.14). Its membranes spontaneously depolarize (that is, admit ions that cause it to become more positively charged), creating electrical signals with a regular rhythm that determines the rhythm of the heart's beating. In this way, the SA node acts as a *pacemaker* for the heart. Each electrical signal initiated within this pacemaker passes quickly from one heart muscle cell to another in a wave that envelops the left and the right atria at almost the same time (indicated by the yellow coloring in figure 24.14, panels ❶ and ❷).

But the electrical signals do not immediately spread to the ventricles. There is a pause before the lower half of the heart starts to contract. The reason for the delay is that the atria of the heart are separated from the ventricles by connective tissue that does not propagate the electrical signal. The signal would not pass to the ventricles at all except for a slender connection of cardiac muscle cells known as the **atrioventricular (AV) node** (labeled in panel ❸), which connects across the gap to a strand of specialized muscle known as the atrioventricular bundle, or **bundle of His.** Bundle branches divide into fast-conducting **Purkinje fibers,** which initiate the almost simultaneous contraction of all the cells of the right and left ventricles about 0.1 seconds after the atria contract. This delay permits the atria to finish emptying their contents into the corresponding ventricles before those ventricles start to contract. The contraction of the ventricles begins at the apex (the bottom of the heart) where depolarization of Purkinje fibers begins the electrical signal. The contraction then spreads up toward the atria (the yellow coloring in panels ❸ and ❹). This results in a "wringing out" of the ventricles, forcing blood up and out of the heart.

Because the human body contains so much water, it conducts electrical currents rather well. These depolarizations generate an electrical current that can be detected using an **electrocardiogram** (**ECG** or also known as **EKG**). The electrical impulses are recorded as deflections on the ECG, such as the one shown in figure 24.14).

Concept Check

1. Name and describe the types of blood vessels in humans.
2. Name and describe four functions of the lymphatic system.
3. What is the difference between erythrocytes and leukocytes?

Heart Disease

Heart disease and other diseases of the circulatory system are the leading cause of death in the United States, accounting for over 32% of all deaths in 2005. More than 80 million Americans have some form of cardiovascular disease. Heart attacks, the main killer, result from an insufficient supply of blood reaching one or more parts of the heart muscle, which causes myocardial cells in those parts to die. Heart attacks, also called myocardial infarctions, are often caused by a blood clot forming somewhere in the coronary arteries (the arteries that supply the heart muscle with blood); like a rock stuck in a pipe, the clot blocks the passage of blood through the artery to the muscles of the heart. Without blood-born oxygen, the affected cardiac muscles are damaged and the heart ceases to function properly. If enough cardiac muscle is affected, the heart stops beating altogether and the individual dies.

A heart attack may also result if a coronary artery is blocked by atherosclerosis (discussed later). Recovery from a heart attack is possible if the portion of the heart that was damaged is small. **Angina pectoris,** which literally means "chest pain," can occur when the supply of blood to the heart is reduced. The pain may occur in the heart and often also in the left arm and shoulder. Angina pectoris is a warning sign that the blood supply to the heart is inadequate but still sufficient to avoid myocardial cell death.

Strokes are caused by an interference with the blood supply to the brain. They may occur when a blood vessel bursts in the brain or when blood flow in a cerebral artery is blocked by a thrombus (blood clot) or by atherosclerosis. The effects of a stroke depend on how severe the damage is and where in the brain the stroke occurs.

Atherosclerosis is an accumulation within the arteries of fatty materials, abnormal amounts of smooth muscle, deposits of cholesterol or fibrin, or various kinds of cellular debris. These accumulations cause blood flow to be reduced. The lumen (interior) of the artery may be further reduced in size by a clot that forms as a result of the atherosclerosis. Three different coronary arteries are shown below with different degrees of blockage. The coronary artery on the left shows only minor blockage, while the artery you see in the middle exhibits severe atherosclerosis—much of the passage is blocked by buildup on the interior walls of the artery. In the severest cases (as seen on the right),

the artery may be blocked completely. Atherosclerosis is promoted by genetic factors, smoking, hypertension (high blood pressure), and high blood cholesterol levels. Diets low in cholesterol and saturated fats (from which cholesterol can be made) can help lower the level of blood cholesterol, and therapy for hypertension can reduce that risk factor. To stop smoking, however, is the single most effective action a smoker can take to reduce the risk of atherosclerosis.

Atherosclerosis is treated both with medications and with invasive procedures. Medications include enzymes, which help dissolve clots; anticoagulants, which prevent clots from forming (aspirin works as a weak anticoagulant); and nitroglycerin, which dilates blood vessels.

Invasive treatments include reducing the blockage with *angioplasty*. Angioplasty is a procedure where a small balloon is threaded into a partially blocked coronary artery. Once in the blocked artery, the balloon is inflated, flattening the atherosclerosis deposit against the side of the artery. In some cases, a small metal mesh sleeve called a *stent* may also be inserted to prop the artery open. More aggressive treatments include *coronary bypass surgery*, where healthy segments of blood vessels are patched into a coronary artery, which diverts the flow of blood around a blocked section of artery; and *heart transplants*, where the damaged heart is replaced by a donor heart.

Arteriosclerosis, or hardening of the arteries, occurs when calcium is deposited in arterial walls. It tends to occur when atherosclerosis is severe. Not only do such arteries have restricted blood flow, but they also lack the ability to expand as normal arteries do to accommodate the volume of blood pumped out by the heart. This inflexibility forces the heart to work harder. Treatments include those used in alleviating the symptoms of atherosclerosis and high blood pressure.

The great tragedy of heart disease is that it is largely a life style disease. People who do not smoke are at much lower risk of developing heart disease. People who exercise regularly and eat a "heart-healthy" low-fat diet tend to have lower blood cholesterol levels and so are far less likely to develop atherosclerosis. Many people with a genetic tendency to over-produce cholesterol can maintain acceptable levels with statin drugs like Lipitor that inhibit the initial enzyme carrying out the body's cholesterol biosynthesis.

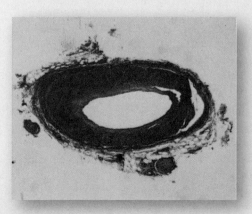  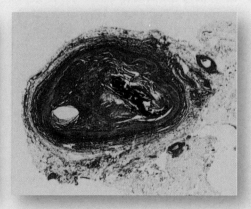

Inquiry & Analysis

Do Big Hearts Beat Faster?

Small animals live at a much faster pace than large animals. They reproduce more quickly, and live shorter lives. As a rule, they tend to move about more quickly, and so consume more oxygen per unit body weight. Interestingly, small and large mammals have about the same size heart, relative to body size (about 0.6% of body mass). It is interesting to ask whether all mammalian hearts beat at the same rate. The heart of a 7,000-kilogram (a kilogram is 1,000 grams) African bull elephant must push a far greater volume of blood through its body than the heart of a 3-gram mouse, but the elephant is able to do it through much-larger-diameter arteries, which impose far less resistance to the blood's flow. Does the elephant's heart beat faster? Or does the mouse's, in order to deliver more oxygen to its muscles? Or perhaps the mouse's heart beats more slowly, because of increased resistance to flow through narrower blood vessels?

The graph to the right displays the pulse rate of a number of mammals of different body sizes (the **pulse rate** is the number of heartbeats counted per minute, a measure of how rapidly the heart is beating). Note that both the *x* and *y* axes use log scales (see page 10). For comparison, the pulse rate of an adult human at rest is about 70 beats per minute. The largest mammal is the blue whale, as big as a supersized moving van with a body mass as great as 136,000 kilograms; the smallest is the pygmy shrew, smaller than a cockroach with a body mass of a few hundredths of a gram.

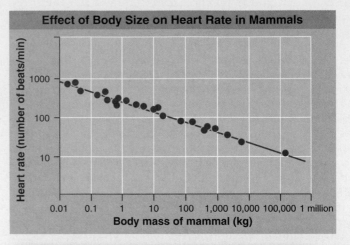

Analysis

1. **Applying Concepts** All mammals have the same size hearts relative to their body size. Do their hearts beat at the same rate?
2. **Interpreting Data**
   a. What is the resting pulse rate of a 7,000-kilogram African bull elephant?
   b. What is the resting pulse rate of a 3-gram shrew? How many complete heartbeats is that per second?
   c. What general statement can be made regarding the effect of body size on heart rate in mammals?
3. **Making Inferences**
   a. The data in the graph, plotted on logarithmic coordinates (that is, the scale rises in powers of 10), fall nicely upon a straight line. How would you expect them to look, plotted on linear coordinates?
   b. As you walk through the graph from left to right, the line slopes down; this is called a negative slope. What does the negative slope of the line signify?
4. **Drawing Conclusions** If you plotted data for an experiment measuring body mass versus resting oxygen consumption, you would get exactly the same slope of the line as shown in this graph. What does this tell us about why body size affects heart rate in mammals as it does?

488

Concept Summary

Circulation

24.1 Open and Closed Circulatory Systems

- Animal cells acquire oxygen from the environment and nutrients from the food they eat, but not all animals do this in the same way. Cnidarians and flatworms have a gastrovascular cavity that circulates the products of gas exchange and digestion (**figure 24.1a**).

- Mollusks and arthropods have open circulatory systems. Hemolymph is pumped by tubular hearts through blood vessels that open into a body cavity, as shown here from **figure 24.1b**. The hemolymph delivers oxygen and nutrients to the cells and then reenters the circulatory system through pores in the tubular heart.

- Annelids and all vertebrates have closed circulatory systems. The blood stays within closed vessels that extend throughout the body. The blood is propelled through the body by the pumping of a heart (**figure 24.1c**).

- In vertebrates, arteries carry blood away from the heart and veins bring blood back to the heart. Capillaries are small vessels that connect the arteries to the veins. As blood is forced through the capillaries the blood exchanges gases and delivers nutrients to the cells. Fluid is also forced out across the walls of the capillaries where it is deposited in the tissues as interstitial fluid. The fluid returns back to the blood directly or by means of the lymphatic system.

- The vertebrate circulatory system has several functions. It functions in the transportation of substances throughout the body, this includes gases in respiration, nutrients in digestion, metabolic wastes in excretion, and hormones in endocrine functions. It also functions in the regulation of body temperature, bringing warm blood to the surface where heat can be dissipated or using warm blood to heat colder blood (**figure 24.2**). The circulatory system also functions in the protection of the body through blood clotting and immunity. Platelets circulate in the blood so they are available to form blood clots to seal injured vessels and white blood cells circulate in the blood to protect against infection.

24.2 Architecture of the Vertebrate Circulatory System

- In vertebrates, blood circulates from the heart through arteries and arterioles to capillaries. Blood flows from the capillaries back to the heart through venules and larger veins (**figure 24.3**).

- Capillaries are the sites of gas, nutrient, and waste exchange. A capillary bed can be closed by the contraction of precapillary sphincter muscles that restricts the flow of blood to that area (**figure 24.4**).

- Blood vessels vary in size and structure from thick-walled arteries to capillaries that are a single cell layer (**figure 24.5**). Arteries have a thicker layer of smooth muscle that absorbs the force generated by the strong contraction of the heart. A layer of elastic tissue allows the artery to expand with the force of the blood flow. The walls of capillaries are only a single cell layer thick and small in diameter, which aids in the exchange of gases and nutrients with surrounding tissue (**figure 24.6**).

- Veins bring blood back to the heart. The blood moves through the veins with the help of contractions of skeletal muscles and one-way valves inside the veins that keeps the blood from backing up in the vessel (**figure 24.7**).

24.3 The Lymphatic System: Recovering Lost Fluid

- Needed fluid is lost to the body tissues through leaky capillary walls. This interstitial fluid is recovered by the lymphatic system (**figure 24.8**). The lymphatic system consists of a network of vessels called lymphatic capillaries (**figure 24.9**). The fluid, now called lymph, drains into lymphatic capillaries, which drain into larger lymphatic vessels. The lymphatic vessels return the fluid and lost proteins back to the circulatory system. The lymph travels through the lymphatic system by the contraction of skeletal muscles and one-way valves that keep the lymph from backing up in the system. The lymphatic system has other functions related to digestion and immunity.

24.4 Blood

- Blood is a fluid that circulates through the body in the circulatory system. Blood consists of a fluid called plasma that contains metabolites, waste products, salts and ions, and proteins. These substances help maintain the osmotic balance in the blood. Various cells also circulate in the plasma, including red blood cells (like the one pictured here from **figure 24.11**) that are involved in gas exchange, and various types of white blood cells, called leukocytes, that defend the body against infection. There are many different types of leukocytes in the blood including neutrophils, as shown here, macrophages, B lymphocytes, T lymphocytes, and others. Platelets are fragments of cells that circulate in the blood and are involved in blood clotting.

24.5 Human Circulatory System

- The human heart has four chambers that functions as a two-cycle pump (**figure 24.12**), pumping blood from the heart to the lungs, and pumping oxygenated blood from the heart to the rest of the body.

- Oxygenated blood from the lungs enters the left side of the heart through the pulmonary vein, emptying into the left atrium. From there the blood passes into the left ventricle where it is pumped out to the body. After delivering oxygen to the tissues of the body, the deoxygenated blood flows back to the heart and enters the right atrium. The blood then passes into the right ventricle, and when the ventricle contracts, the blood is pumped to the lungs.

- A relatively easy way to monitor the activity of the heart is with a stethoscope. The lub-dub sounds heard are the closing of heart valves. The activity of the heart can also be monitored by measuring blood pressure. Measurements of the systolic and diastolic pressures indicate how hard the heart is having to work to pump blood through the body (**figure 24.13**).

- A pacemaker, called the sinoatrial (SA) node, controls heartbeat rate. An electrical impulse that begins in the SA node triggers muscles in the atrium to contract. The electrical impulse passes through the atria, as shown in this panel from **figure 24.14,** and stimulates the AV node. An electrical impulse then travels from the AV node down to the apex of the heart, where it initiates another wave of muscle contraction of the ventricles. The electrical impulses of the heart can be detected and recorded as an electrocardiogram (ECG).

Self-Test

1. Which of the following describes the circulatory system found in insects?
   a. Fluids containing oxygen and nutrients are brought into a gastrovascular cavity where all body cells are exposed to the fluid.
   b. The blood is pumped throughout the body in a system of vessels, where the blood never leaves the blood vessels.
   c. A series of tubular hearts pump hemolymph into cavities of the body.
   d. The blood is pumped into the tissues of the body and is returned to the circulatory system through lymph capillaries.

2. Which of the following is *not* a function of the circulatory system?
   a. regulation of body temperature
   b. protection against injury and disease-causing microbes
   c. transportation of materials in the body
   d. All of the above are functions of the circulatory system.

3. How do some vertebrates maintain body temperature in cold environments?
   a. by mixing cold blood with warm blood in the heart
   b. by pumping more warm blood to the extremities
   c. by passing warm blood near cold blood in the extremities to warm the blood
   d. All of the above are used by vertebrates in cold environments.

4. Exchange of waste material, oxygen, carbon dioxide, and metabolites such as salts and food molecules, takes place in the
   a. capillaries.   c. arterioles.
   b. venules.   d. arteries.

5. The lymphatic system is like the circulatory system in that they both
   a. have nodes that filter out pathogens.
   b. are made up of arteries.
   c. deliver blood to the heart.
   d. carry fluids.

6. The most numerous type of blood cell is the
   a. macrophage.   c. platelet.
   b. leukocyte.   d. erythrocyte.

7. What would happen if the pulmonary semilunar valve in the heart wasn't closing completely?
   a. The person would suffer a heart attack.
   b. There would be a reduction in the amount of blood flowing to the body.
   c. Too much blood would be pumped out of the left ventricle.
   d. Blood would flow back into the right ventricle.

8. Which of the following correctly describes a cycle of blood through the body?
   a. heart->veins->venules->capillaries->arterioles->arteries->heart
   b. lungs->veins->heart->arteries->arteriole->capillaries->venules ->veins->heart->arteries->lungs
   c. lungs->capillaries->arterioles->arteries->veins->venules ->capillaries->heart
   d. heart->veins->lungs->arteries->heart->arteries->arterioles ->capillaries->venules->veins->heart

9. Which of the following statements is *false?*
   a. Only arteries carry oxygenated blood.
   b. Both arteries and veins have a layer of smooth muscle.
   c. Capillary beds lie between arteries and veins.
   d. Sphincters regulate the flow of blood through capillaries.

10. Which piece of equipment is the simplest way to monitor the beating of the heart?
    a. an electrocardiogram (ECG)
    b. a stethoscope
    c. a sphygmomanometer
    d. a blood pressure cuff

Visual Understanding

1. **Figure 24.7** Can being a dedicated "couch potato" or "video game addict" ever cause circulatory problems? Explain.

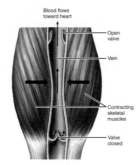

2. **Figure 24.12***b* A mnemonic is a learning tool to help a person remember or recall information. The mnemonic "VAVA lung, VAVA body" is sometimes used to recall the pathway of the blood in humans. Now that you know the anatomy of the heart, can you explain this mnemonic?

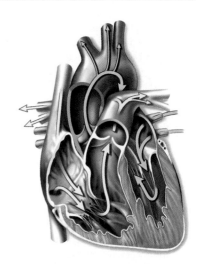

Challenge Questions

1. Your friend, James, has a painful paper cut on his finger and it's bleeding. Disgusted, because it already stained his shirt, he asks you why the blood is so important, anyway. What do you tell him?

2. Explain why the muscular wall of the left ventricle is thicker, resulting in stronger, more forceful contractions compared to the right ventricle.

Respiration

Respiration

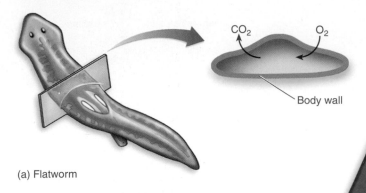

(a) Flatworm

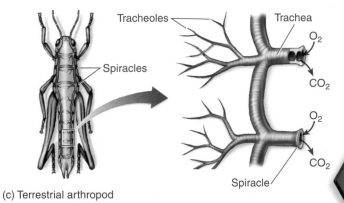

(b) Fish

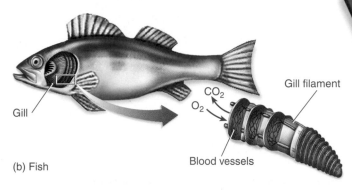

(c) Terrestrial arthropod

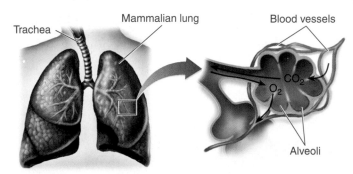

(d) Mammal

Figure 25.1   Gas exchange in animals.

(a) Gases diffuse directly across the body wall in many invertebrates and in some species of amphibians. (b) Fish gills provide a very large respiratory surface. (c) Terrestrial arthropods respire through tracheae, which open to the outside through spiracles. (d) The alveoli of mammalian lungs provide a large respiratory surface area.

25.1 Types of Respiratory Systems

CONCEPT PREVIEW: Aquatic animals extract oxygen dissolved in water, some directly across the body surfaces, others with gills. Terrestrial animals use tracheae or lungs.

All animals obtain the energy that fuels their lives by consuming other organisms, harvesting energy-rich electrons from the organic molecules of these creatures, and then using these electrons to drive the synthesis of ATP and other molecules. Afterward, the spent electrons are donated to oxygen gas (O_2) to form water (H_2O). The carbon atoms left over after the electrons were stripped away combine with oxygen to form carbon dioxide (CO_2), which is released from the cell as a "waste" by-product. The process that animals use to capture energy utilizes oxygen and produces carbon dioxide. The uptake of oxygen and the release of carbon dioxide together are called **respiration,** neatly defining one of the principle evolutionary challenges facing all animals—how to obtain oxygen and dispose of carbon dioxide.

> The process of breaking down organic molecules to transfer their energy into ATP molecules is called oxidative respiration (the subject of chapter 7). Oxygen is needed for this process as it is the final electron acceptor in the electron transport chain, discussed on page 120.

Most of the primitive phyla of organisms obtain oxygen by direct diffusion from their aquatic environments, which contains about 10 milliliters of dissolved oxygen per liter. Sponges, cnidarians, many flatworms and roundworms, and some annelid worms obtain their oxygen by diffusion from surrounding water. Oxygen and carbon dioxide diffuse across the surface of the body as shown in the flatworm in figure 25.1a. Similarly, some members of the vertebrate class Amphibia conduct gas exchange by direct diffusion through their moist skin.

The more advanced marine invertebrates (mollusks, arthropods, and echinoderms) and fishes possess special respiratory organs called gills. Gills increase the surface area available for diffusion of oxygen. A **gill** is basically a thin sheet of tissue that waves through the water. Gills can be simple or complex, as in the highly convoluted gills of fish (figure 25.1b). In fish, the gills are protected by a covering called an operculum (not shown in figure 25.1b so that the underlying gills are visible). Because of this, their gills do not wave in the water. Instead, water is pumped over

> The pumping action of the operculum, a key adaptation in bony fish as discussed on page 353, allows a fish to remain stationary in the water and still have the water pass over the gills.

the gills and gas exchange occurs across the walls of capillaries contained in the gills.

Terrestrial arthropods do not have a single major respiratory organ like a gill. Instead, a network of air ducts, beginning with the **trachea** (the purple tubes in figure 25.1c) that branches into smaller and smaller tubes, carry air to every part of the body. The tracheae open to the outside of the body through structures called **spiracles,** which can be closed and opened.

Terrestrial vertebrates, except for some amphibians, have a pair of respiratory organs called the **lungs.** Gas exchange in the lungs occurs across the walls of capillaries and air sacs, which in mammals are called *alveoli* (shown in the enlargement in figure 25.1d).

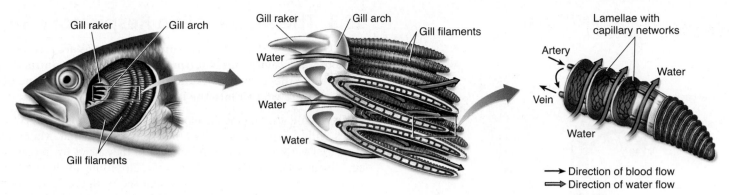

→ Direction of blood flow
⟹ Direction of water flow

Figure 25.2 Structure of a fish gill.

Water passes from the gill arch over the filaments (from *left* to *right* in the center diagram). Water always passes the lamellae in the same direction, which is opposite to the direction the blood circulates across the lamellae (the diagram on the *far right*).

25.2 Respiration in Aquatic Vertebrates

CONCEPT PREVIEW: Fish gills achieve countercurrent flow, making them very efficient at extracting oxygen from water.

Have you ever seen the face of a swimming fish up close? A swimming fish continuously opens and closes its mouth, pushing water through the mouth cavity and out a slit at the rear of the mouth—and (this is the whole point) past the gills on its one-way journey.

This swallowing process, which seems so awkward, is at the heart of a great advance in gill design in fishes. What is important about the swallowing is that it causes the water to move past the fish's gills *always in the same direction.* Moving the water past the gills in the same direction permits **countercurrent flow,** which is a very efficient way of extracting oxygen.

Here is how it works: Each gill is composed of two rows of gill filaments (two sections of gills are shown in the middle panel in figure 25.2, each with two rows of gill filaments). The gill filaments are made of thin membranous plates stacked one on top of the other and projecting out into the flow of water. As water flows past the filaments (indicated by the blue arrows in the enlargements), oxygen diffuses from the water into blood circulating within the gill filament. Within each filament the blood flows in the direction opposite the movement of the water (through the capillaries from artery to vein). The advantage of the countercurrent flow system is that blood in the capillaries of the gill filaments always encounters water that has a higher oxygen concentration, resulting in the diffusion of oxygen into the blood. To understand this, compare the countercurrent exchange system in figure 25.3a to a concur-

> A similar countercurrent system is found in the extremities of some mammals to control heat loss. A countercurrent heat exchange system is discussed on page 477 in relation to maintaining body temperature in a killer whale where warm blood is used to warm cold blood.

rent exchange system in figure 25.3b. In the countercurrent system, when blood and water flow in opposite directions, the initial oxygen concentration difference at the bottom is not large (10% in the blood and 15% in water), but it is sufficient for oxygen diffusion. As the blood flows through the vessel (upward in the figure), its oxygen concentration increases, but it continually encounters water with a higher oxygen concentration. Even at 85% oxygen concentration in the blood, it is still encountering oxygen concentrations in water of 100%. In the concurrent exchange system, oxygen diffusion is rapid at first, because of the large difference in oxygen concentrations between the blood and water (0% versus 100%), but quickly slows as the difference approaches equilibrium at 50% saturation.

Because of the countercurrent flow, the blood in the fish's gills can build up oxygen concentrations as high as those of the water entering the gills. The gills of bony fishes are the most efficient respiratory machines that have ever evolved among organisms.

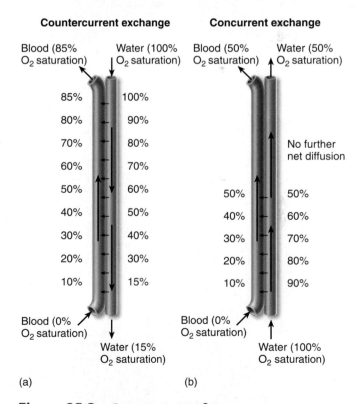

Figure 25.3 Countercurrent flow.

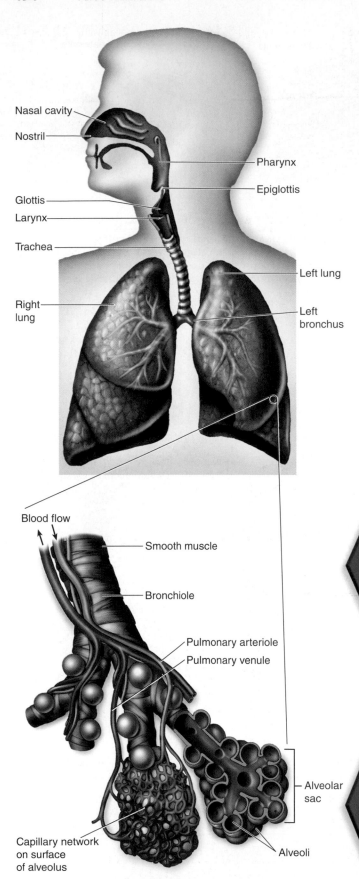

Figure 25.4 **The human respiratory system.**

The respiratory system consists of the lungs and the passages that lead to them.

Labels (top figure): Nasal cavity, Nostril, Glottis, Larynx, Trachea, Right lung, Pharynx, Epiglottis, Left lung, Left bronchus

Labels (bottom figure): Blood flow, Smooth muscle, Bronchiole, Pulmonary arteriole, Pulmonary venule, Alveolar sac, Capillary network on surface of alveolus, Alveoli

25.3 The Mammalian Respiratory System

CONCEPT PREVIEW: In mammals, the lungs are located within a thoracic cavity that is surrounded by muscles. Contracting and relaxing these muscles expands and reduces the volume of the cavity, drawing air into the lungs or forcing it out.

The oxygen-gathering mechanism of mammals, although less efficient than fishes' gills, adapts them well to their terrestrial habitat. Mammals, like all terrestrial vertebrates, obtain the oxygen they need for metabolism from air, which is about 21% oxygen gas. A pair of lungs is located in the chest within the **thoracic** cavity. As you can see in figure 25.4, the two lungs hang free within the cavity, connected to the rest of the body only at one position, where the lung's blood vessels and air tube enter. This air tube is called a **bronchus** (plural, bronchi). It connects each lung to a long tube called the **trachea,** which passes upward and opens into the rear of the mouth. The trachea and both the right and left bronchi are supported by C-shaped rings of cartilage.

Air normally enters through the nostrils into the nasal cavity where it is moistened and warmed. In addition, the nostrils are lined with hairs that filter out dust and other particles. As the air passes through the nasal cavity, an extensive array of cilia further filters it. The air then passes to the back of the mouth, through the pharynx (the common passage of food and air), and then through the larynx (voice box) and the trachea. Because the air crosses the path of food at the back of the throat, a special flap called the epiglottis covers the trachea whenever food is swallowed, to keep it from "going down the wrong pipe." From the trachea, air passes down through several branchings of bronchi in the lungs and

The epiglottis is a very important structure. The mechanism which closes the epiglottis over the trachea when swallowing is described on page 509.

eventually to bronchioles that lead to alveoli. Mucous secretory ciliated cells in the trachea and bronchi also trap foreign particles and carry them upward to the pharynx, where they can be swallowed. The lungs contain millions of alveoli, tiny sacs clustered like grapes. The alveoli are surrounded by an extremely extensive capillary network. All gas exchange between the air and blood takes place across the walls of the capillaries and the alveoli via diffusion.

The walls of capillaries consist of a single cell layer, as discussed on page 480 and shown in figure 24.5. They are also very narrow, just wider than the diameter of a red blood cell. Because of their structure, capillaries are ideal for the process of gas exchange.

The mammalian respiratory apparatus is simple in structure and functions as a one-cycle pump. The thoracic cavity is bounded on its sides by the ribs and on the bottom by a thick layer of muscle, the **diaphragm,** which separates the thoracic cavity from the abdominal cavity. Each lung is covered by a very thin, smooth membrane called the pleural membrane.

The adhesion of the lungs to the inner wall of the thoracic cavity is due to hydrogen bonding in the water that bathes the pleural membranes. This type of hydrogen bonding leads to cohesion and adhesion, discussed on page 39, which keeps the lungs attached to the thoracic cavity.

This membrane also folds back on itself to line the interior wall of the thoracic cavity, into which the lungs hang. The space between these two layers of membrane is very small and filled with fluid. This fluid causes the two membranes to adhere to each other in the same way a thin film of water can hold two plates of glass together, effectively coupling the lungs to the walls of the thoracic cavity. Similar to the way a layer of fluid causes the pleural membrane of the lungs to adhere to the inside of the thoracic cavity, fluid inside the alveoli can cause these air sacs to collapse. This doesn't occur because the epithelial cells that line the alveoli secrete onto their surface

Essential Biological Process 25A

Breathing

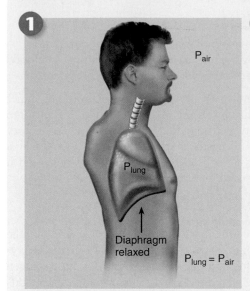

1 P_{air}

P_{lung}

Diaphragm relaxed

$P_{lung} = P_{air}$

Before inhalation, the air pressure in the lungs (P_{lung}) is equal to the atmospheric pressure (P_{air}).

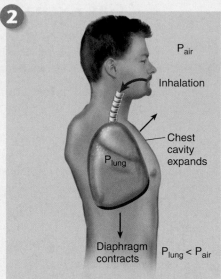

2 P_{air}

Inhalation

Chest cavity expands

P_{lung}

Diaphragm contracts

$P_{lung} < P_{air}$

During inhalation, the diaphragm contracts, and the chest cavity expands downward and outward. This increases the volume of the chest cavity and lungs, which reduces the air pressure inside the lungs, and the air from outside the body flows into the lungs.

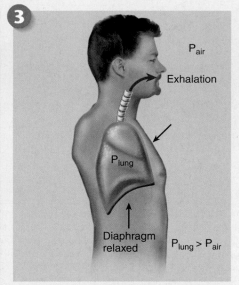

3 P_{air}

Exhalation

P_{lung}

Diaphragm relaxed

$P_{lung} > P_{air}$

During exhalation, the diaphragm relaxes, decreasing the volume of the chest cavity. The pressure increases in the lungs, forcing air out of the lungs.

---

a mixture of lipoprotein molecules called *surfactant*, reducing the surface tension that would otherwise cause the alveoli to collapse.

Air is drawn into the lungs by the creation of negative pressure—that is, pressure in the lungs is less than atmospheric pressure. The pressure in the lungs is reduced when the volume of the lungs is increased. This is similar to how a bellow pump or accordion works. In both cases, when the bellow is extended the volume inside increases, causing air to rush in. How does this occur in the lungs? The volume in the lungs increases when muscles that surround the thoracic cavity contract, causing the thoracic cavity to increase in size. Because the lungs adhere to the thoracic cavity, they also expand in size. This creates negative pressure in the lungs, and air rushes in.

The Mechanics of Breathing

The active pumping of air in and out of the lungs is called breathing. During *inhalation,* muscular contractions cause the rib cage to move outward and upward. The diaphragm, the red-colored lower border of the lung in *Essential Biological Process 25A*, is dome-shaped when relaxed (panel 1) but moves downward and flattens during contraction (panel 2). These rib cage movements expand the chest cavity, which decreases the air pressure in the lungs compared to the atmosphere ($P_{lung} < P_{air}$). When this happens, air flows into the lungs. In effect, inhalation sucks air into the lungs.

During *exhalation* (panel 3) the ribs and diaphragm return to their original resting position. In doing so, they exert pressure on the lungs. This pressure is transmitted uniformly over the entire surface of the lung, forcing air from the inner cavity back out to the atmosphere. Because the diffusion surfaces of the lungs are not exposed to fully oxygenated air, but rather to a mixture of fresh and partly oxygenated air, the respiratory efficiency of mammalian lungs is far from maximal.

BIOLOGY & YOU

Asthma. Over 9% of American children are affected by asthma, a chronic disease of the respiratory system in which the airways occasionally constrict, causing shortness of breath, coughing, and in some instances a life-threatening difficulty in breathing. Asthma attacks can be triggered by a variety of events, including viral illnesses like colds, stress, airborne allergens, smoke, and exercise. Researchers do not yet fully understand the cause of asthma, or why it seems to be more common. Many genetic and environmental factors appear to interact in complex ways to produce this common disorder. At least 25 different genes have been implicated, some of which may only cause asthma when combined with specific environmental triggers. The symptoms can be treated by inhaling chemicals called bronchodilaters that expand the constricted passageways. Longer-term preventive medications like inhaled corticosteroids reduce inflammation of the airway lining and so suppress attacks. The most effective treatment for asthma sufferers is to identify their triggers, such as pets or aspirin, and limit exposure to them. As children mature into adults, half of asthma sufferers (54%) lose their sensitivity to asthma triggers; the less fortunate suffer from asthma all their lives.

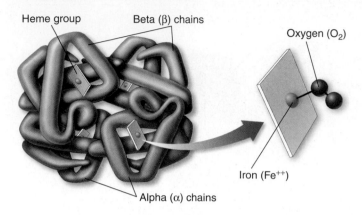

Figure 25.5 **The hemoglobin molecule.**

The hemoglobin molecule is actually composed of four protein chain subunits: two copies of the "alpha chain" and two copies of the "beta chain." Each chain is associated with a heme group, and each heme group has a central iron atom, which can bind to a molecule of oxygen.

BIOLOGY & YOU

Hiccups. A hiccup (pronounced "HICK-up") is a spasmodic contraction of the diaphragm that repeats several times a minute. The abrupt rush of air into your lungs causes your epiglottis to close, making the "hic" noise. Hiccups are caused by irritation of the phrenic and vagus nerves, which activates reflexive motor pathways to the diaphragm muscles. Hiccups often occur after drinking carbonated soda or alcohol, but can be initiated by any of a host of other reasons, like eating too fast or taking a cold drink while eating a hot meal. What do you do when you get the hiccups? In most cases they can be stopped simply by forgetting about them—the basis of the common home remedy of "scaring them away" with a surprise or fright. Increasing respired CO_2 by breathing into a paper bag also often works, for the interesting reason that hiccups may be an evolutionary remnant of earlier amphibian respiration. Air gulping in frogs is inhibited by CO_2, just as your hiccuping is. Frogs and other amphibians don't have a diaphragm, and instead gulp air via a simple motor reflex much like your hiccuping reflex. In humans the hiccuping motor pathway forms early in fetal development, well before the motor pathways that drive normal breathing. Premature infants born before their lung motor pathways are fully functional spend 2.5% of their time hiccuping, gulping air just like amphibians.

25.4 How Respiration Works: Gas Exchange

CONCEPT PREVIEW: Oxygen molecules move through the circulatory system carried by the protein hemoglobin within red blood cells. Most CO_2 is transported in the blood plasma as bicarbonate.

When oxygen has diffused from the air into the moist cells lining the inner surface of the lung, its journey has just begun. Passing from these cells into the bloodstream, the oxygen travels throughout the body in the circulatory system, described in chapter 24. It has been estimated that it would take a molecule of oxygen three years to diffuse from your lung to your toe if it moved only by diffusion, unassisted by a circulatory system.

O_2 Transport

Oxygen (O_2) moves within the circulatory system carried piggyback on the protein **hemoglobin.** Hemoglobin molecules contain iron, which binds oxygen, as shown in figure 25.5. Hemoglobin molecules act like little oxygen sponges, soaking up oxygen within red blood cells and causing more to diffuse in from the blood plasma. The oxygen binds in a reversible way, which is necessary so that the oxygen can unload when it reaches the tissues of the body. Hemoglobin is manufactured within red blood cells and never leaves these cells, which circulate in the bloodstream like ships bearing cargo.

At the high O_2 levels that occur in the lung (panel 1 in *Essential Biological Process 25B*), most hemoglobin molecules carry a full load of oxygen atoms. In tissue, the presence of carbon dioxide (CO_2) causes the hemoglobin molecule to assume a different shape, one that gives up its oxygen more easily (as in panel 3). The effect of CO_2 on oxygen unloading is important, because CO_2 is produced by the tissues through cell metabolism. For this reason, the blood unloads oxygen more readily within tissues undergoing metabolism.

CO_2 Transport

At the same time the red blood cells are unloading oxygen in panel 3, they are also absorbing CO_2 from the tissue. About 8% of the CO_2 in blood is simply dissolved in plasma. Another 20% is bound to hemoglobin; however, this CO_2 does not bind to the heme group but rather to another site on the hemoglobin molecule, and so it does not compete with oxygen binding. The remaining 72% of the CO_2 diffuses into the red blood cells. To keep CO_2 from diffusing out of the red blood cells back to the plasma where CO_2 levels are low, the enzyme *carbonic anhydrase* combines CO_2 molecules with water molecules (panel 4) to form **carbonic acid** (H_2CO_3) in the cell. This acid dissociates into **bicarbonate** (HCO_3^-) and hydrogen (H^+) ions. The H^+ binds to hemoglobin. A transporter protein in the membrane of the red blood cell moves the bicarbonate out of the red blood cell into the plasma. This transporter protein exchanges one chloride ion (Cl^-) for a bicarbonate, a process called the *chloride shift*. This reaction keeps the levels of CO_2 in the blood plasma low, facilitating the diffusion of more CO_2 into it from the surrounding tissue. The facilitation is critical to CO_2 removal, because the difference in CO_2 concentration between blood and tissue is not large (only 5%). The formation of carbonic acid and bicarbonate is also important in maintaining the acid-base balance of the blood because the molecules act as a buffering system.

As discussed on page 40 a buffer is an acid-base pair of molecules that adjusts the levels of H^+ in a solution, thereby controlling pH levels. The reversible reaction between carbonic acid and bicarbonate regulates the amount of H^+ in the blood, thereby adjusting pH levels in the body.

The blood plasma carries bicarbonate ions back to the lungs. The lower CO_2 concentration in the air inside the lungs causes the carbonic anhydrase reaction to proceed in

Essential Biological Process 25B

Gas Exchange During Respiration

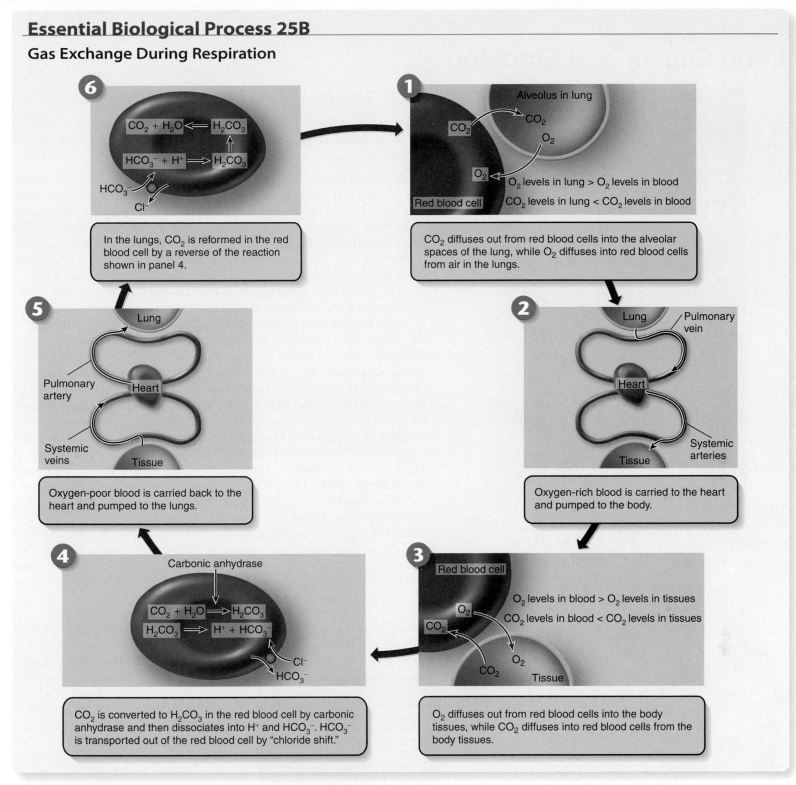

6 $CO_2 + H_2O \Longleftarrow H_2CO_3$

$HCO_3^- + H^+ \Longrightarrow H_2CO_3$

HCO_3^-

Cl^-

In the lungs, CO_2 is reformed in the red blood cell by a reverse of the reaction shown in panel 4.

1 Alveolus in lung

CO_2

CO_2

O_2

O_2

Red blood cell

O_2 levels in lung > O_2 levels in blood

CO_2 levels in lung < CO_2 levels in blood

CO_2 diffuses out from red blood cells into the alveolar spaces of the lung, while O_2 diffuses into red blood cells from air in the lungs.

5 Lung

Pulmonary artery

Heart

Systemic veins

Tissue

Oxygen-poor blood is carried back to the heart and pumped to the lungs.

2 Lung

Pulmonary vein

Heart

Systemic arteries

Tissue

Oxygen-rich blood is carried to the heart and pumped to the body.

4 Carbonic anhydrase

$CO_2 + H_2O \longrightarrow H_2CO_3$

$H_2CO_3 \Longrightarrow H^+ + HCO_3^-$

Cl^-

HCO_3^-

CO_2 is converted to H_2CO_3 in the red blood cell by carbonic anhydrase and then dissociates into H^+ and HCO_3^-. HCO_3^- is transported out of the red blood cell by "chloride shift."

3 Red blood cell

O_2

CO_2

O_2

CO_2

Tissue

O_2 levels in blood > O_2 levels in tissues

CO_2 levels in blood < CO_2 levels in tissues

O_2 diffuses out from red blood cells into the body tissues, while CO_2 diffuses into red blood cells from the body tissues.

the reverse direction (panel 6), releasing gaseous CO_2, which diffuses outward from the blood into the alveoli. With the next exhalation, this CO_2 leaves the body. The diffusion of CO_2 out from the red blood cells causes the hemoglobin within these cells to release its bound CO_2 and take up O_2 instead. The cells with a new load of O_2 then start their next respiratory journey.

Concept Check

1. Why is it key that water flow across fish gills in one direction only?
2. Why is there water in the space between pleural membrane layers?
3. What stops CO_2 from diffusing out of red blood cells into plasma?

Lung Cancer and Smoking

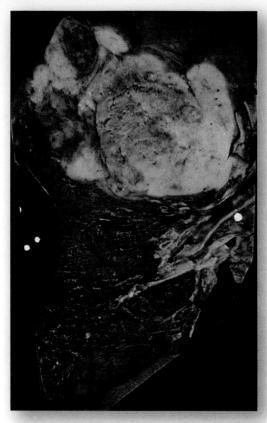

Figure 25.6 Lung cancer.

The bottom half of this lung contains normal tissue, but a cancerous tumor has completely taken over the top half.

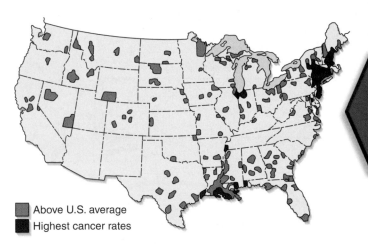

■ Above U.S. average
■ Highest cancer rates

Figure 25.7 Cancer in the United States.

The incidence of cancer per 1,000 people is not uniform throughout the United States. It is centered in cities where chemical manufacturing is common and in the Mississippi Delta.

25.5 The Nature of Lung Cancer

CONCEPT PREVIEW: Cancer results from the destruction of genes by mutations that, when healthy, enable the cell to regulate cell division. To avoid lung cancer, don't smoke.

Of all human diseases, none is more feared than cancer (see chapter 8). Nearly one in every four deaths in the United States was caused by cancer in 2005. The American Cancer Society estimates that 559,650 people died of cancer in the United States in 2006. About 29% of these—160,390 people—died of lung cancer (figure 25.6). About 140,000 cases of lung cancer were diagnosed each year in the 1980s, and 90% of these persons died within three years. What has caused lung cancer to become a major killer of Americans?

The search for a cause of cancers such as lung cancer has uncovered a host of environmental factors that appear to be associated with cancer. For example, the incidence of cancer per 1,000 people is not uniform throughout the United States. Rather, it is centered in cities, like the heavily populated Northeast indicated by the red areas in figure 25.7, and in the Mississippi Delta, indicated by the red and brown areas, suggesting that environmental factors such as pollution and pesticide runoff may contribute to cancer. When the many environmental factors associated with cancer are analyzed, a clear pattern emerges: Most cancer-causing agents, or carcinogens, share the property of being potent mutagens. Recall from chapter 11 that a mutagen is a chemical that damages DNA, destroying or changing genes (a change in DNA is called a mutation). That cancer is caused by mutation is now supported by an overwhelming body of evidence.

What sort of genes are being mutated? In the last several years, researchers have found that mutation of only a few genes is all that is needed to transform normally dividing cells into cancerous ones. Identifying and isolating these cancer-causing genes, investigators have learned that all are involved with regulating cell proliferation (how fast cells grow and divide). A key element in this regulation are so-called tumor suppressors, genes that actively prevent tumors from forming.

How do tumor suppressor genes work? The **p53 protein,** a tumor suppressor sometimes called the "guardian angel" of the cell, inspects the DNA before the cell divides. When p53 detects damaged or foreign DNA, it stops cell division and activates the cell's DNA repair systems. If the damage isn't repaired in a certain amount of time, p53 pulls the plug, triggering events that kill the cell. In this way, mutations such as those that cause cancer are either repaired or the cells containing them eliminated. If the gene that produces p53 is itself destroyed by mutation, future damage accumulates unrepaired. Among this damage are mutations that lead to cancer, mutations that would have been repaired by healthy p53. Fully 50% of all cancers have a disabled *p53* gene.

> The protein product of the *p53* gene functions as a tumor suppressor. It keeps cells that have damaged DNA from dividing and destroys them if their DNA can't be corrected. The actions of *p53* are discussed in the featured reading "Curing Cancer" on pages 138 and 139.

Smoking Causes Lung Cancer

If cancer is caused by damage to growth-regulating genes, what then has led to the rapid increase in lung cancer in the United States? Two lines of evidence are particularly telling. The first consists of detailed information about cancer rates among smokers. The annual incidence of lung cancer among nonsmokers is only a few per 100,000 but increases with the number of cigarettes smoked per day to a staggering 300 per 100,000 for 30 cigarettes a day smokers.

A second line of evidence consists of changes in the incidence of lung cancer that mirror changes in smoking habits. Look carefully at the data presented in figure 25.8. The upper graph is compiled from data on men and shows the incidence of smoking (blue line) and of lung cancer (red line) in the United States since 1900. As late as 1920, lung cancer was a rare disease. About 30 years after the incidence of smoking began to increase among men, lung cancer also started to become more common. Now look at the lower graph, which presents data on women. Because of social mores, a significant number of women in the United States did not smoke until after World War II (see blue line), when many social conventions changed. As late as 1963, only 6,588 women had died of lung cancer. But as women's frequency of smoking has increased, so has their incidence of lung cancer (red line), again with a lag of about 30 years. Women today have achieved equality with men in the number of cigarettes they smoke, and their lung cancer death rates are now rapidly approaching those for men. In 2008, an estimated 71,000 American women died of lung cancer.

How does smoking cause cancer? Cigarette smoke contains many powerful mutagens, among them benzo[*a*]pyrene, and smoking introduces these mutagens to the lung tissues. Benzo[*a*]pyrene, for example, binds to three sites on the *p53* gene and causes mutations at these sites that inactivate the gene. In 1997, scientists studying this tumor-suppressor gene demonstrated a direct link between cigarettes and lung cancer. They found that the *p53* gene is inactivated in 70% of all lung cancers. When these inactivated *p53* genes are examined, they prove to have mutations at just the three sites where benzo[*a*]pyrene binds! Clearly, the chemical in cigarette smoke is responsible for the lung cancer.

> The link between certain chemicals and cancer was first made over 200 years ago, as discussed in the feature "Protecting Your Genes" on page 194. The specific connection between lung cancer and smoking was proposed in 1949 and has since been supported by much evidence.

In the face of these facts, why do so many people continue to smoke? Because the nicotine in cigarette smoke is an addictive drug. Researchers have identified the receptor on brain neurons that it binds to, and they have demonstrated that smoking leads to a decrease in the brain's population of these receptors, leading to a craving for more cigarettes. Once a person becomes addicted, it is difficult to quit. Only half of those who try eventually succeed.

Clearly, an effective way to avoid lung cancer is not to smoke. Life insurance companies have computed that, on a statistical basis, smoking a single cigarette lowers your life expectancy 10.7 minutes (more than the time it takes to smoke the cigarette!). Every pack of 20 cigarettes bears an unwritten label: *The price of smoking this pack of cigarettes is 3 1/2 hours of your life.* Smoking a cigarette is very much like going into a totally dark room with a person who has a gun and standing still. The person with the gun cannot see you, does not know where you are, and shoots once in a random direction. A hit is unlikely, and most shots miss. As the person keeps shooting, however, the chance of eventually scoring a hit becomes more likely. Similarly, every time an individual smokes a cigarette, mutagens are being shot at his or her genes. The more cigarettes smoked, the more likely that genes will be damaged.

Concept Check

1. What is the function of *p53* in a normal cell? What happens if it is mutated?
2. Explain how chemicals in cigarette smoke can cause lung cancer.
3. How does nicotine cause cigarette addiction?

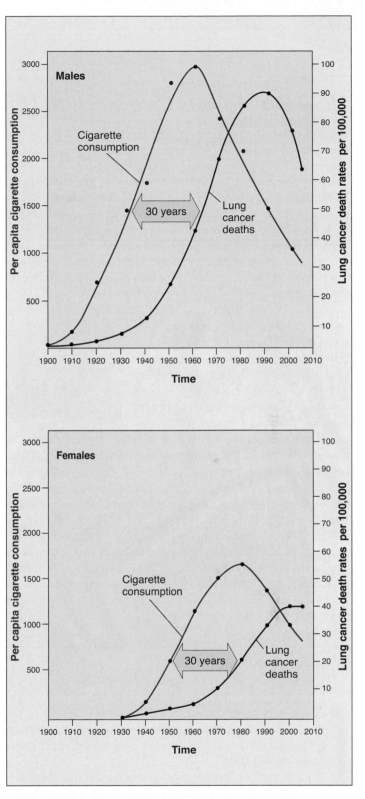

Figure 25.8   Incidence of lung cancer in men and women.

Lung cancer was a rare disease a century ago. As men in the United States increased smoking in the early 1900s, the incidence of lung cancer also increased. Women followed suit years later, and in 2008, more than 71,000 women died of lung cancer, a rate of 40 per 100,000. In that year, about 20% of women smoked and about 25% of men.

How Do Llamas Live so High Up?

Because of mixing, the air animals breathe is 21% oxygen everywhere, even way up into the sky 100 km above earth's surface. However, the amount of air (the number of molecules in a unit volume) decreases sharply with altitude, as shown in the upper graph. Air pressure at 5,000 meters is half of that at sea level. This lack of air presents a serious problem to humans, as mountain climbers know. The amount of oxygen in the air (measured as oxygen partial pressure) is lower, so there is simply too little oxygen to fuel a climber's muscles. To combat this problem, high-altitude climbers typically spend months acclimating to high altitudes, a period in which their bodies greatly increase the amount of hemoglobin in their red blood cells and so increase the amount of oxygen the red blood cells can capture. Many mammals live their entire lives at high altitudes. The llama and the vicuna (pictured here) both live in the high Andes of South America, often above 5,000 meters. Do they deal with the problem of low oxygen in the same way, with elevated hemoglobin levels, or is there another answer?

The graph on the lower right displays three "oxygen loading curves" that reveal the effectiveness with which hemoglobin binds oxygen. The more effective the binding, the less oxygen required before hemoglobin becomes fully loaded. In the graph, the percent hemoglobin saturation (that is, how much of the hemoglobin is bound to oxygen) is presented on the *y* axis, and the oxygen partial pressure (a measure of the amount of oxygen available to the hemoglobin molecules) is presented on the *x* axis. Oxygen-loading curves are presented for three mammalian species: humans living at sea level, and llamas and vicunas, each living in the Andes above 5,000 meters.

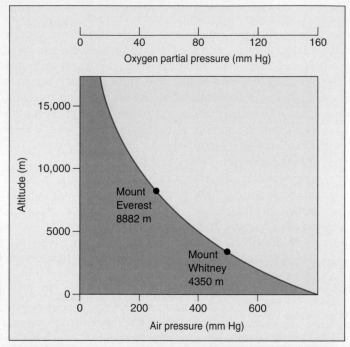

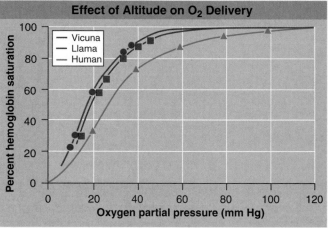

Analysis

1. **Applying Concepts**
   Comparing Curves. Extrapolating on the lower graph, which of the three species possesses hemoglobin able to load oxygen well at sea-level partial pressures (160 mm Hg)? Which of the three possesses hemoglobin better able to load oxygen on Mount Everest (get oxygen partial pressure value from the upper graph)?

2. **Interpreting Data**
   a. The partial pressure of oxygen in human muscle tissue at sea level is about 40 mm Hg. What is the percent hemoglobin bound to O_2 for each of the three species at this partial pressure? What percent of the human hemoglobin has released its oxygen? of the llama? of the vicuna?
   b. Are there any significant differences in the hemoglobin saturation values for the two high-altitude species?

3. **Making Inferences** At an elevation of 5,000 meters, the partial pressure of oxygen is 80 mm Hg (half of what it is at sea level). At this elevation, how much of human hemoglobin has succeeded in binding oxygen? How much of llama hemoglobin? of vicuna?

4. **Drawing Conclusions** What is the effect of shifting the oxygen loading curve to the left? What general statement can be made regarding the affinity of hemoglobin for oxygen in the three species? What saturation values would you expect in llamas raised from birth in the National Zoo at Washington, D.C.? Why would you expect this?

Concept Summary

Respiration

25.1 Types of Respiratory Systems

- To acquire energy, all animals need oxygen to fuel cellular respiration and need to expel carbon dioxide, a waste product produced during cellular respiration. However, animals carry out gas exchange in different ways (**figure 25.1**).

- Some aquatic animals extract oxygen and release carbon dioxide across the skin.

- The evolution of gills increased the surface area for gas exchange in higher aquatic animals. The complexity of gills varies greatly in these animal groups from simple structures to the highly convoluted gills of fish.

- Terrestrial animals extract oxygen from the air with tracheae or lungs. The tracheae system in arthropods is a network of air ducts with openings to the outside, called spiracles. Other terrestrial vertebrates use lungs where gas exchange occurs across the walls of capillaries and air sacs called alveoli.

25.2 Respiration in Aquatic Vertebrates

- Fish gills, shown here from **figure 25.2,** are very efficient at extracting oxygen from water because of a countercurrent flow system. Water is pumped passed the gill filaments in a one-way direction that is opposite the flow of blood in the capillaries, called countercurrent flow. Because of this flow pattern, the water passing over the gills always has a higher concentration of oxygen compared to that in the underlying blood. This concentration gradient drives the diffusion of oxygen into the blood (**figure 25.3**).

25.3 The Mammalian Respiratory System

- Mammalian lungs are positioned within an internal cavity, called the thoracic cavity, as shown here from **figure 25.4.** An airway called a bronchus connects each lung to the trachea. The trachea is a tube that extends up to the rear of the mouth.

- Air is warmed and filtered as it enters through the nostrils and passes through the nasal cavity and down the trachea. At the back of the throat the air crosses the path of food. A flap called the epiglottis will close over the trachea when food is swallowed so that the food doesn't go down the trachea.

- The airways of the mammalian lung end in alveoli, which are surrounded by capillaries. Gas exchange occurs across the single cell layers of the alveoli and capillaries walls.

- Pleural membranes line the outside of the lungs and the inner layer of the thoracic cavity. Fluid between these two membranes causes the lungs to stick to the walls of the thoracic cavity. Contraction of the muscles that line the thoracic cavity expands the space in the lungs, causing air to rush into the lungs. Relaxation of the muscles causes air to be exhaled (***Essential Biological Process 25A***).

25.4 How Respiration Works: Gas Exchange

- Hemoglobin, found in red blood cells and shown here from **figure 25.5,** carries oxygen from the lungs to the cells of the body. In the lungs where oxygen concentrations are high, oxygen diffuses into the red blood cells in the blood. Oxygen binds to iron atoms contained within the heme groups of hemoglobin. The oxygen is then carried to distant areas of the body that are low in oxygen. The oxygen diffuses down its concentration gradient entering cells in the surrounding tissues (***Essential Biological Process 25B***).

- As red blood cells unload oxygen, they take up carbon dioxide. CO_2 is primarily carried as bicarbonate and hydrogen ions. CO_2 enters red blood cells and is converted into carbonic acid by carbonic anhydrase. Carbonic acid dissociates into bicarbonate and hydrogen ions. The bicarbonate is transported out of the cell into the plasma with the counter transport of a chloride ion. This is called the chloride shift (***Essential Biological Process 25B***).

- The conversion of CO_2 to bicarbonate increases the rate of diffusion from the cells into the blood. In the lungs, the carbonic anhydrase reaction is reversed, converting bicarbonate and hydrogen ions back to carbon dioxide where it passes out of the blood and into the alveoli traveling down its concentration gradient. Once in the lungs, the CO_2 is exhaled.

Lung Cancer and Smoking

25.5 The Nature of Lung Cancer

- Cancer results from mutations of the DNA, often caused by chemicals in the environment. Areas of the country that have a higher amount of environmental contaminants also have a higher incidence of cancer as shown in this map from **figure 25.7**.

- Many cancers have been linked to mutations of a key tumor suppressor gene *p53*. The protein products of this gene controls cell division, keeping the cell from dividing when it is not suppose to. If mutations damage this gene, the cells are able to divide uncontrollably.

- Lung cancer results from mutations caused by chemicals in cigarette smoke. The incidence of cigarette smoking and lung cancer mirror each other with a separation of about 30 years, time for the accumulation of mutations that lead to cancer (**figure 25.8**).

Self-Test

1. Which of the following respiratory structures would you expect to find in aquatic animals?
   - **a.** gills
   - **b.** tracheae
   - **c.** alveoli
   - **d.** spiracles
2. Which respiratory structure is the most efficient?
   - **a.** amphibian skin
   - **b.** fish gills
   - **c.** insect tracheae
   - **d.** mammalian lungs
3. In the gills of fish, countercurrent flow allows
   - **a.** the animal's blood to be continually exposed to water of lower oxygen concentration.
   - **b.** the animal's blood to be continually exposed to water of equal oxygen concentration.
   - **c.** the animal's blood to be continually exposed to water of higher oxygen concentration.
   - **d.** for a larger surface area for gas exchange to occur.
4. Which of the following paths reflects the correct flow of air into the lungs?
   - **a.** nasal cavity->bronchus->trachea->alveoli->capillary
   - **b.** trachea->bronchus->capillaries->alveoli
   - **c.** nostrils->nasal cavity->trachea->bronchus->alveoli
   - **d.** notrils->bronchus->trachea->alveoli
5. Which of the following is *not* involved in mammalian respiration?
   - **a.** diffusion of oxygen
   - **b.** active transport of carbon dioxide
   - **c.** the creation of negative pressure
   - **d.** the chloride shift
6. When you take a deep breath, your chest moves out because
   - **a.** the incoming air expands the thoracic cavity.
   - **b.** contracting your abdominal muscles forces the chest cavity outward.
   - **c.** contracting the muscles around the thoracic cavity pulls your chest out.
   - **d.** positive pressure builds up in the lungs, expanding the chest cavity and forcing the chest out.
7. Oxygen is transported by
   - **a.** hemoglobin in red blood cells.
   - **b.** dissolving it in the blood plasma.
   - **c.** binding to proteins in the blood plasma.
   - **d.** platelets in the blood plasma.
8. Most of the carbon dioxide is transported
   - **a.** by hemoglobin in red blood cells.
   - **b.** as bicarbonate in the blood plasma.
   - **c.** by proteins in the blood plasma.
   - **d.** as carbonic acid in red blood cells.
9. Which of the following is *not* carried by hemoglobin?
   - **a.** bicarbonate
   - **b.** oxygen
   - **c.** carbon dioxide
   - **d.** All are carried by hemoglobin.
10. Which of the following can lead to cancer?
    - **a.** smoking
    - **b.** pollution
    - **c.** mutations of *p53*
    - **d.** all of the above

Visual Understanding

1. *Essential Biological Process 25B* Explain what would happen to carbon dioxide transport if a person was poisoned with a chemical that blocked the actions of carbonic anhydrase.

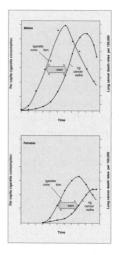

2. **Figure 25.8** What conclusions can you draw about cigarette smoking and women's health? Explain.

Challenge Questions

1. Sometimes when people are eating, they take a bite that is too big, or is not completely chewed, and when they swallow it becomes stuck partway down the esophagus near the epiglottis (a flap that covers the trachea when swallowing). When food is stuck in this location, a person usually can't breathe. In this case, people have been instructed to do the Heimlich maneuver, which is a method of pushing up rapidly on the diaphragm, compressing the lungs. How might this help?

2. How is it that cigarette smoking can be linked to an increased incidence of many kinds of cancer, not just lung cancer?

Chapter 26

The Path of Food Through the Animal Body

CHAPTER AT A GLANCE

Food Energy and Essential Nutrients

Figure 26.1   **The pyramid of nutrition.**

The width of each section indicates how much you should consume of that food group. The gold section is grains, green is vegetables, red is fruits, yellow is fats, blue is dairy, and purple is meats and beans. However, one size does not fit all; go to www.mypyramid.gov to customize the food pyramid that is right for you.

IN THE NEWS

The Peanut Butter Debate. Severe malnutrition occurs in some 20 million children in Africa and South Asia every year. A new type of ready-to-use food is changing the way severe malnutrition is treated. At therapeutic feeding centers, children are given a quick health checkup and, if malnourished, a silvery packet. Open and squeeze the packet and out pours 92 grams of a brown paste that looks like dark peanut butter. Called Plumpy'nut, it's made of roasted ground peanuts combined with vegetable oil, milk powder, sugar, and a mix of minerals and vitamins. One serving has 550 calories and plenty of proteins, vitamins, and minerals. Plumpy'nut has a long shelf life, does not need to be mixed with water as powdered milk treatments do (local water may be unavailable or unhealthy) and is simple for mothers to give children at home—and children love the sweet, sticky stuff. Eighty percent of severely malnourished children recover when fed Plumpy'nut. But should peanut butter-like pastes, like Plumpy'nut, be distributed far more widely, given to children at risk of malnutrition? The cost of attempting to *prevent* malnutrition in this way would be high ($55 per child in 2007). Many argue that the approach is not cost-effective and widespread use would make poor countries totally dependent on foreign aid. One solution, discussed by the World Health Organization in 2008, is to produce Plumpy'nut locally, and more cheaply—for instance, by replacing the powdered milk (the most expensive ingredient) with soy.

26.1 Food for Energy and Growth

CONCEPT PREVIEW: Food is an essential source of calories. It is important to maintain a proper balance of carbohydrate, protein, and fat. Individuals with a body mass index of 25 or more are considered overweight. Food also provides key components that the body cannot manufacture for itself.

The food animals eat provides both a source of energy and essential molecules such as certain amino acids and fats that the animal body is not able to manufacture for itself. An optimal diet contains more carbohydrates than fats and also a significant amount of protein, as recommended by the federal government's "pyramid of nutrition" in figure 26.1. The pyramid is intended as a general guideline of what a person should eat. A healthy diet should include more of the foods indicated by the larger sections—for example, the gold section indicates grains and cereals with an emphasis on whole grains. Fats (the yellow section) are recommended in much smaller amounts because they have a far greater number of energy-rich carbon–hydrogen bonds and thus a much higher energy content per gram than carbohydrates or proteins.

Carbohydrates are obtained primarily from cereals, grains, breads, fruits (the red section), and vegetables (the green section). On average, carbohydrates contain 4.1 calories per gram; fats, by comparison, contain 9.3 calories per gram, over twice as much. Dietary fats are obtained from oils, margarine, and butter and are abundant in fried foods, meats, and processed snack foods, such as potato chips and crackers. Like carbohydrates, proteins have 4.1 calories per gram and can be obtained from many foods, including dairy products, poultry, meat (the blue and purple sections), and grains.

In wealthy countries, such as those of North America and Europe, being significantly overweight is common, the result of habitual overeating and high-fat diets, in which fats constitute over 35% of the total caloric intake. The international standard measure of appropriate body weight is the body mass index (BMI), estimated as your body weight in kilograms, divided by your height in meters squared. A BMI chart is presented in figure 26.2. To determine your BMI, find your height in the left hand column (in feet and inches) and trace it across to the column with your weight (in pounds). A BMI value of 25 (dark blue boxes) and above is considered overweight and 30 or over is considered obese. In the United States, the National Institutes of Health estimated in 2004 that 66% of adults, 133.6 million Americans, were overweight, with a body mass index of 25 or more. Of those individuals, 63.6 million were considered obese with a body mass index of 30 or greater. Being overweight is highly correlated with coronary heart disease, diabetes, and many other disorders. However, starving yourself is also not the answer. A BMI of less than 18.5 is also unhealthy, often resulting from eating disorders including anorexia nervosa.

All animals must eat. Even an animal that is completely at rest requires energy to support its metabolism. This minimum rate of energy consumption, called the *basal metabolic rate* (BMR), is relatively constant for a given individual. Exercise raises the metabolic rate above the basal levels, so the amount of energy the body requires per day is determined not only by the BMR but also by the level of physical activity. Energy that is not used for metabolism or exercise is stored as fat. Therefore, energy needs can be altered by the choice of diet (caloric intake) and the amount of energy expended in exercise.

	25 OVERWEIGHT LIMIT										OVERWEIGHT											
WEIGHT	**100**	**105**	**110**	**115**	**120**	**125**	**130**	**135**	**140**	**145**	**150**	**155**	**160**	**165**	**170**	**175**	**180**	**185**	**190**	**195**	**200**	**205**
HEIGHT																						
5' 0"	20	21	21	22	23	24	25	26	27	28	29	30	31	32	33	34	35	36	37	38	39	40
5' 1"	19	20	21	22	23	24	25	26	26	27	28	29	30	31	32	33	34	35	36	37	38	39
5' 2"	18	19	20	21	22	23	24	25	26	27	27	28	29	30	31	32	33	34	35	36	37	37
5' 3"	18	19	19	20	21	22	23	24	25	26	27	27	28	29	30	31	32	33	34	35	35	36
5' 4"	17	18	19	20	21	21	22	23	24	25	26	27	27	28	29	30	31	32	33	33	34	35
5' 5"	17	17	18	19	20	21	22	22	23	24	25	26	27	27	28	29	30	31	32	32	33	34
5' 6"	16	17	18	19	19	20	21	22	23	23	24	25	26	27	27	28	29	30	31	31	32	33
5' 7"	16	16	17	18	19	20	20	21	22	23	23	24	25	26	27	27	28	29	30	31	31	32
5' 8"	15	16	17	17	18	19	20	21	21	22	23	24	24	25	26	27	27	28	29	30	30	31
5' 9"	15	16	16	17	18	18	19	20	21	21	22	23	24	24	25	26	27	27	28	29	30	30
5' 10"	14	15	16	17	17	18	19	19	20	21	22	22	23	24	24	25	26	27	27	28	29	29
5' 11"	14	15	15	16	17	17	18	19	20	20	21	22	22	23	24	24	25	26	26	27	28	29
6' 0"	14	14	15	16	16	17	18	18	19	20	20	21	22	22	23	24	24	25	26	26	27	28
6' 1"	13	14	15	15	16	16	17	18	18	19	20	20	21	22	22	23	24	24	25	26	26	27
6' 2"	13	13	14	15	15	16	17	17	18	19	19	20	21	21	22	22	23	24	24	25	26	26
6' 3"	12	13	14	14	15	16	16	17	17	18	19	19	20	21	21	22	22	23	24	24	25	26
6' 4"	12	13	13	14	15	16	16	16	17	18	18	19	19	20	21	21	22	23	23	24	24	25

Figure 26.2   Are you overweight?

This chart presents the body mass index (BMI) values used by federal health authorities to determine who is overweight. Your body mass index is at the intersection of your height and weight.

Essential Substances for Growth

Over the course of their evolution, many animals have lost the ability to manufacture certain substances they need. Mosquitoes, for example, cannot manufacture cholesterol, and must obtain it in their diet—human blood is rich in cholesterol. Humans are unable to manufacture 8 of the 20 amino acids used to make proteins: lysine, tryptophan, threonine, methionine, phenylalanine, leucine, isoleucine, and valine. These amino acids, called **essential amino acids,** must therefore be obtained from proteins in the food we eat.

Trace Elements. In addition to supplying energy, food that is consumed must also supply the body with a wide variety of **trace elements,** which are minerals required in very small amounts. Among the trace elements are iodine (a component of thyroid hormone), cobalt (a component of vitamin B_{12}), zinc and molybdenum (components of enzymes), manganese, and selenium.

Vitamins. Essential organic substances that are used in trace amounts are called **vitamins.** Humans require at least 13 different vitamins. Humans, monkeys, and guinea pigs, for example, lack the ability to synthesize ascorbic acid (vitamin C) and will develop the potentially fatal disease called scurvy—characterized by weakness, spongy gums, and bleeding of the skin and mucous membranes—if vitamin C is not supplied in their diets.

Concept Check

1. About what portion of the recommended food pyramid is fat?
2. What is your body mass index?
3. Why are eight of the 20 amino acids called essential amino acids?

BIOLOGY & YOU

Vitamin Supplements: Nutrition in a Pill? It seems like wherever we go these days, we are bombarded with advertisements touting the health benefits of food supplements. Megadoses of vitamin C are said to aid in avoiding colds, and supplements of antioxidant vitamins (A, C, and E) are said to help prevent heart attacks and cancer. Is any of this true? Do dietary supplements make us healthier? According to most health professionals, the answer is "No." The American Heart Association states that healthy people get adequate nutrients by eating a healthy diet, and recommend no supplements. Clinical trials are underway to see if increased vitamin antioxidant intake may have an overall benefit, but in early results, a large placebo-controlled, randomized study failed to show any benefit from vitamin E on heart disease. While dietary supplements may be necessary if you are a vegetarian or consume less than 1,600 calories a day, it appears they are wasted on most of us. The lone exception: omega-3 fatty acid supplements, which have been associated with decreased risk of heart disease. If you don't eat fish (salmon, herring, and trout are rich in omega-3), the American Heart Association suggests this supplement may be of value.

Digestion

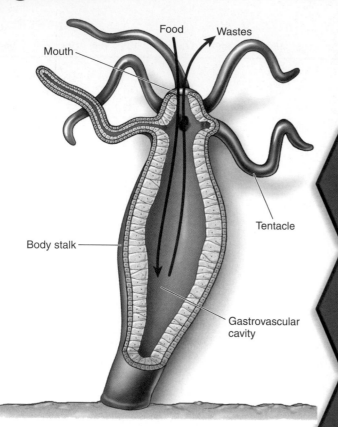

Figure 26.3 The gastrovascular cavity of *Hydra*.

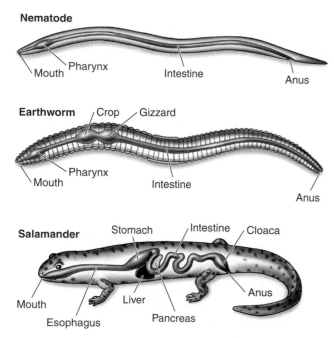

Figure 26.4 One-way digestive tracts.

One-way movement through the digestive tract allows different regions of the digestive system to become specialized for different functions.

26.2 Types of Digestive Systems

CONCEPT PREVIEW: Some animals digest their food intracellularly, but most digest their food extracellularly. A one-way digestive tract allows specialization of regions for different functions.

Heterotrophs are divided into three groups on the basis of their food sources. Animals that eat plants exclusively are classified as **herbivores;** common examples include cows, horses, rabbits, and sparrows. Animals that are meat eaters, such as cats, eagles, trout, and frogs, are **carnivores.** **Omnivores** are animals that eat both plants and other animals. We humans are omnivores, as are pigs, bears, and crows.

Sponges digest their food intracellularly, breaking down food particles with digestive enzymes inside their cells. Other animals digest their food extracellularly, within a digestive cavity. In this case, the digestive enzymes are released into a cavity that is continuous with the animal's external environment. In flatworms (such as *Planaria*) and cnidarians, like the hydra in figure 26.3, the digestive cavity has only one opening at the top that serves as both mouth (the red arrow) and anus (the blue arrow). There can be no specialization within this type of digestive system, called a *gastrovascular cavity,* because every cell is exposed to all stages of food digestion.

Specialization occurs when the digestive tract, or alimentary canal, has a separate mouth and anus, so that transport of food is one way. Three examples are shown in figure 26.4. The most primitive digestive tract is seen in nematodes (phylum Nematoda), where it is simply a tubular *gut* lined by an epithelial membrane. Earthworms (phylum Annelida) have a digestive tract specialized in different regions for the ingestion, storage (crop), fragmentation (gizzard), digestion, and absorption of food (intestine). All higher animals, like the salamander, show similar specializations.

The ingested food may be stored in a specialized region of the digestive tract or may first be subjected to physical fragmentation through the chewing action of teeth (in the mouth of many vertebrates) or the grinding action of pebbles (in the gizzard of earthworms and birds). The process of chemical digestion occurs primarily in the intestine, breaking down the larger food molecules of polysaccharides, fats, and proteins into smaller subunits. Chemical digestion involves hydrolysis reactions that liberate the subunits—primarily monosaccharides, amino acids, and fatty acids—from the food. These products of chemical digestion pass through the epithelial lining of the gut and ultimately into the blood, in a process known as absorption. Any molecules in the food that are not absorbed cannot be used by the animal. These waste products are excreted from the body through the anus.

Energy flows from the sun through the trophic levels of a food chain, discussed on page 386. Solar energy is absorbed through photosynthesis by producers and passes up through the food chain, from herbivores to carnivores. Omnivores are able to fill different trophic levels.

The type of cell in sponges that absorbs food and breaks it down is a choanocyte, described in the Phylum Facts on page 338. The beating of its flagellum brings water and food toward the cell where food particles are trapped, brought into the cell, and digested.

The gastrovascular cavity also functions in circulation, as described on page 476. The water that is brought into the gastrovascular cavity contains oxygen that diffuses into the cells that line the cavity. CO_2 diffuses into the cavity from cells and is carried away.

The process of digestion involves breaking the chemical bonds that hold the atoms of a food molecule together. Whether carbohydrates, proteins, or fats, the process of breaking chemical bonds is through hydrolysis, as discussed on page 47.

26.3 Vertebrate Digestive Systems

CONCEPT PREVIEW: The vertebrate digestive system consists of a tubular gastrointestinal tract composed of a series of tissue layers. The digestive system is modified in animals based on what they eat.

In humans and other vertebrates, the digestive system consists of a tubular gastrointestinal tract and accessory digestive organs (figure 26.5). Working through the figure from the top down, the initial components of the gastrointestinal tract are the mouth and the pharynx, which is the common passage of the oral and nasal cavities. The pharynx leads to the esophagus, a muscular tube that delivers food to the stomach. From the stomach, where some preliminary digestion occurs, food passes to the first part of the small intestine, where a battery of digestive enzymes continues the digestive process. Accessory digestive organs, such as the liver, gallbladder, and pancreas aid in digestion. The products of digestion pass across the wall of the small intestine into the bloodstream. The small intestine empties what remains into the large intestine, also called the colon, where water and minerals continue to be absorbed. In most vertebrates other than mammals, the waste products emerge from the large intestine into a cavity called the cloaca (see the salamander in figure 26.4), which also receives the products of the urinary and reproductive systems. In mammals, the urogenital products are separated from the fecal material in the large intestine; the fecal material enters the rectum and is expelled through the anus.

The tubular gastrointestinal tract of a vertebrate has a characteristic layered structure (figure 26.6). Working from the inside (the lumen) outward, the innermost layer is the mucosa, an epithelium that lines the lumen. The next major tissue layer, composed of connective tissue, is called the submucosa. Just outside the submucosa is the muscularis, which consists of a double layer of smooth muscles. The muscles in the inner layer have a circular orientation, and those in the outer layer are arranged longitudinally. An outer connective tissue layer, the serosa, covers the external surface of the tract. Nerves, intertwined in regions called plexuses, are located in the submucosa and help regulate the gastrointestinal activities.

In general, carnivores have shorter intestines for their size than do herbivores. A short intestine is adequate for a carnivore, but herbivores ingest a large amount of plant cellulose, which resists digestion. These animals have a long, convoluted small intestine. In addition, mammals called *ruminants* (such as cows) that consume grass and other vegetation have stomachs with multiple chambers, where microorganisms aid in the digestion of cellulose. The first chamber is the *rumen*, which serves as a fermentation vat. Prokaryotes and protists convert cellulose into simpler compounds. These animals will then regurgitate and rechew the contents of the rumen, breaking down more and more of the cellulose. This activity is called rumination, or "chewing the cud." The cud is then swallowed into another chamber of the stomach where digestion continues. Other herbivores, including rabbits and horses, also digest cellulose with the aid of bacteria, in a blind pouch called the *cecum* located at the beginning of the large intestine. Because the cecum is located beyond the stomach, regurgitation of its contents is impossible. However, rodents and rabbits have evolved another way to use the products of cellulose digestion. They eat their feces, thus passing the food through their digestive tract a second time where the nutrients can be absorbed. Eating their feces is required to maintain their health.

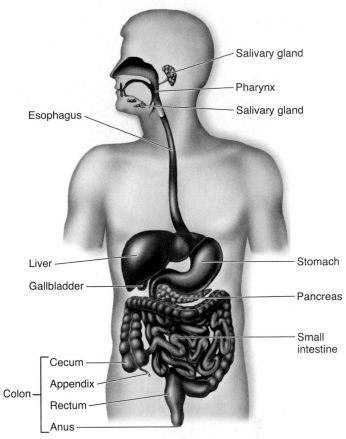

Figure 26.5   The human digestive system.

The tubular gastrointestinal tract and accessory digestive organs are shown. The colon extends from the cecum to the anus.

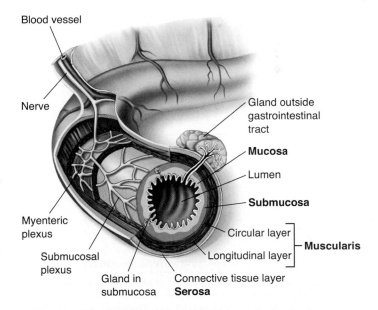

Figure 26.6   The layers of the gastrointestinal tract.

The mucosa contains a lining epithelium, the submucosa is composed of connective tissue (as is the outer serosa layer), and the muscularis consists of smooth muscles.

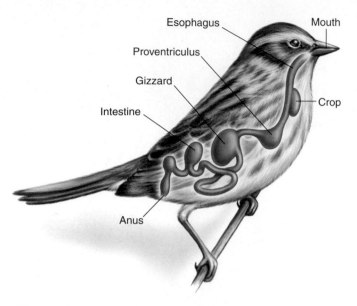

Figure 26.7 The digestive tract of birds.

In birds, food enters the mouth and is stored in the crop. Because birds lack teeth, they swallow gritty objects or pebbles, which lodge in the gizzard, to help pulverize food. Digestive enzymes produced in the proventriculus are churned up with the food and gritty objects in the gizzard before passing into the intestine.

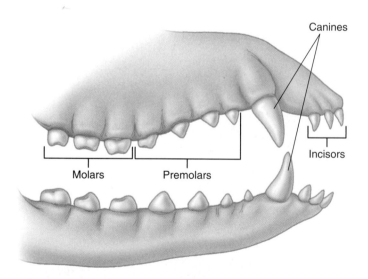

Figure 26.8 Diagram of heterodont dentition.

Different mammals have specific variations of heterodont dentition, depending on whether the mammal is an herbivore, carnivore, or omnivore. In this carnivore, the canines are prominent, and the premolars and molars are pointed—adaptations for tearing and ripping food. In herbivores, some of the incisors are large, the canines are reduced or absent, and the premolars and molars are flattened—adaptations for nipping and grinding vegetation.

26.4 The Mouth and Teeth

CONCEPT PREVIEW: Many vertebrates fragment ingested food through the tearing or grinding action of specialized teeth; birds accomplish this through the grinding action of pebbles in the gizzard. Food mixed with saliva is swallowed and enters the esophagus.

Specializations of the digestive systems in different kinds of vertebrates reflect differences in the way these animals live. Fishes have a large pharynx with gill slits, whereas air-breathing vertebrates have a greatly reduced pharynx. Many vertebrates have teeth (discussed below), and chewing (mastication) breaks up food into small particles and mixes it with fluid secretions. Birds, which lack teeth, break up food in their two-chambered stomachs. The first chamber, the proventriculus (figure 26.7), produces digestive enzymes, which are passed along with the food into the second chamber, the gizzard. The gizzard contains small pebbles ingested by the bird, which are churned together with the food by muscular action. This churning grinds up the seeds and other hard plant material into smaller chunks that can be digested more easily in the intestine.

Vertebrate Teeth

Reptiles and fish have homodont dentition (teeth that are all the same), whereas most mammals have heterodont dentition, teeth of different specialized types. Up to four different types of teeth are seen: Incisors are chisel-shaped teeth used for nipping and biting; "canines" are sharp, pointed teeth used for tearing food; and premolars (bicuspids) and molars usually have flattened, ridged surfaces used for grinding and crushing food. The front teeth in the upper and lower jaws of mammals are incisors. On each side of the incisors are the canines. Behind the canines are premolars and then molars.

This general pattern of heterodont dentition is modified in different mammals depending on their diet (figure 26.8). For example, in carnivorous mammals the canines are prominent, and the premolars and molars are more blade-like, with sharp edges adapted for cutting and shearing. Carnivores often tear off pieces of their prey but have little need to chew them, because digestive enzymes can act directly on animal cells. (Have you ever noticed how a cat or dog gulps down its food?) By contrast, grass-eating herbivores, such as cows and horses, must pulverize the cellulose cell walls of plant tissue before digesting it. In these mammals, the incisors can be well-developed and are used to cut grass and other plants. The canines are reduced or absent, and the premolars and molars are large, flat teeth with complex ridges well suited for grinding.

Humans are omnivores, and human teeth are adapted for eating both plant and animal food. Viewed simply, humans are carnivores in the front of the mouth and herbivores in the back. Children have only 20 teeth, but these deciduous teeth are lost during childhood and are replaced by 32 adult teeth. The third molars are the wisdom teeth, which usually grow in during the late teens or early twenties, when a person is assumed to have gained some "wisdom."

As you can see in figure 26.9, the tooth is a living organ, composed of connective tissue, nerves, and blood vessels, held in place by cementum, a bonelike substance that anchors the tooth in the jaw. The interior of the tooth contains connective tissue called pulp that extends into the root canals and contains nerves and blood vessels. A layer of calcified tissue called dentin surrounds the pulp cavity. The portion of the tooth that projects above the gums is called the crown and is covered with an extremely hard, nonliving substance called enamel. Enamel protects the tooth against abrasion and acids that are produced by bacteria living in the mouth. Cavities form when bacterial acids break down the enamel, allowing bacteria to infect the inner tissues of the tooth.

Processing Food in the Mouth

Inside the mouth, the tongue mixes food with a mucous solution, called **saliva.** In humans, three pairs of salivary glands secrete saliva into the mouth through ducts in the mouth's mucosal lining. Saliva moistens and lubricates the food so that it is easier to swallow and does not abrade the tissue it passes on its way through the esophagus. Saliva also contains the hydrolytic enzyme *salivary amylase*, which initiates the breakdown of the polysaccharide starch into the disaccharide maltose. This digestion is usually minimal in humans, however, because most people don't chew their food very long.

The secretions of the salivary glands are controlled by the nervous system, which in humans maintains a constant flow of about half a milliliter per minute when the mouth is empty of food. This continuous secretion keeps the mouth moist. The presence of food in the mouth triggers an increased rate of secretion, as taste-sensitive neurons in the mouth send impulses to the brain, which responds by stimulating the salivary glands. The most potent stimuli are acidic solutions; lemon juice, for example, can increase the rate of salivation eightfold. The sight, sound, or smell of food can stimulate salivation markedly in dogs, but in humans, these stimuli are much less effective than thinking or talking about food.

Swallowing

When food is ready to be swallowed, the tongue moves it to the back of the mouth. In mammals, the process of swallowing begins when the soft palate elevates, pushing against the back wall of the pharynx (figure 26.10). Elevation of the soft palate seals off the nasal cavity and prevents food from entering it ❶. Pressure against the pharynx stimulates neurons within its walls, which send impulses to the swallowing center in the brain. In response, muscles are stimulated to contract and raise the *larynx* (voice box). This pushes the *glottis,* the opening from the larynx into the trachea (windpipe), against a flap of tissue called the *epiglottis* ❷. These actions keep food out of the respiratory tract, directing it instead into the esophagus ❸.

> Recall from the discussion on page 494 that the respiratory system also uses the mouth and pharynx. The elevation of the soft palate and the movement of the epiglottis over the trachea helps to keep the processes of breathing and swallowing separate.

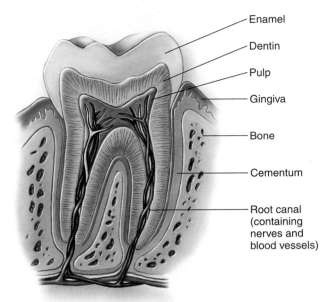

Figure 26.9 Human teeth.

Each vertebrate tooth is alive, with a central pulp containing nerves and blood vessels. The actual chewing surface is a hard enamel layered over the softer dentin, which forms the body of the tooth.

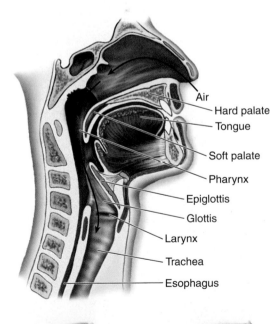

Figure 26.10 The human pharynx, palate, and larynx.

Swallowing triggers the closing of the epiglottis over the trachea, which keeps food and liquids from going down the windpipe.

 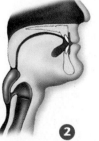 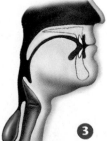

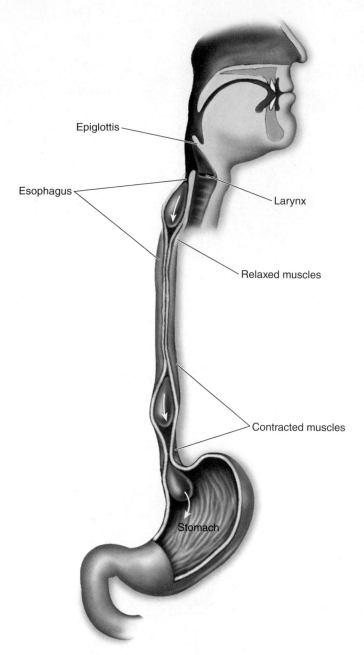

Figure 26.11 **The esophagus and peristalsis.**

Labels on figure: Epiglottis, Esophagus, Larynx, Relaxed muscles, Contracted muscles, Stomach

26.5 The Esophagus and Stomach

CONCEPT PREVIEW: Peristaltic waves of contraction propel food along the esophagus to the stomach. Gastric juice contains HCl and the protein-digesting enzyme pepsin, which begins the digestion of proteins into shorter polypeptides. The chyme then passes to the small intestine.

Structure and Function of the Esophagus

Swallowed food enters a muscular tube called the **esophagus,** which connects the pharynx to the stomach. In adult humans, the esophagus is about 25 centimeters long; the upper third is enveloped in skeletal muscle, for voluntary control of swallowing, while the lower two-thirds is surrounded by involuntary smooth muscle. The swallowing center stimulates successive waves of contraction in these muscles that move food along the esophagus to the stomach. The muscles relax ahead of the food, allowing it to pass freely, and contract behind the food to push it along, as shown in figure 26.11. These rhythmic waves of muscular contraction are called **peristalsis;** they enable humans and other vertebrates to swallow even if they are upside down.

Stomach contents can be brought back up during vomiting, when the **sphincter** (a ring of circular smooth muscle between the stomach and esophagus) is relaxed and the contents of the stomach are forcefully expelled through the mouth. The relaxing of this sphincter can also result in the movement of stomach acid into the esophagus, causing an irritation called *heartburn.* Chronic and severe heartburn is a condition known as *acid reflux.*

Structure and Function of the Stomach

The **stomach** is a saclike portion of the digestive tract (figure 26.12). Its inner surface is highly convoluted, enabling it to fold up when empty and open out like an expanding balloon as it fills with food. Thus, while the human stomach has a volume of only about 50 milliliters when empty, it may expand to contain 2 to 4 liters of food when full.

The stomach contains an extra layer of smooth muscle for churning food and mixing it with *gastric juice,* an acidic secretion of the tubular gastric glands of the mucosa. The gastric glands lie at the bottom of deep depressions, the gastric pits shown in the enlargement in figure 26.12. These exocrine glands contain two kinds of secretory cells: *parietal cells,* which secrete hydrochloric acid (HCl); and *chief cells,* which secrete pepsinogen, a weak protease (protein-digesting enzyme) that requires a very low pH to be active. This low pH is provided by the HCl. Activated pepsinogen molecules then cleave each other at specific sites, producing a much more active protease, pepsin. This process of secreting a relatively inactive enzyme that is then converted into a more active enzyme outside the cell prevents the chief cells from digesting themselves. It should be noted that only proteins are partially digested in the stomach—there is no significant digestion of carbohydrates or fats there.

Action of Acid

A solution with a low pH contains a higher concentration of hydrogen ions (H^+), as discussed on page 40. The HCl in the stomach dissociates creating a high concentration of H^+, which disrupts hydrogen bonding and causes proteins to denature, as described on page 51.

The human stomach produces about 2 liters of HCl and other gastric secretions every day, creating a very acidic solution inside the stomach. The concentration of HCl in this solution is about 10 millimolar, corresponding to a pH of 2. Thus, gastric

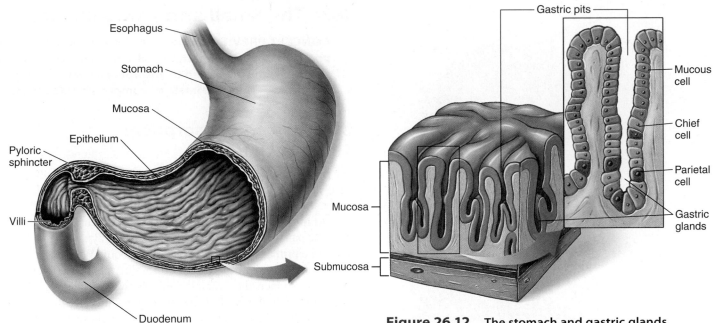

Figure 26.12   The stomach and gastric glands.

Food enters the stomach from the esophagus. The epithelial walls of the stomach are dotted with gastric pits, which contain glands that secrete hydrochloric acid (HCl) and the enzyme pepsinogen. The gastric glands consist of mucous cells, chief cells that secrete pepsinogen, and parietal cells that secrete HCl. Gastric pits are the openings of the gastric glands.

juice is about 250,000 times more acidic than blood, whose normal pH is 7.4. The low pH in the stomach helps denature food proteins, making them easier to digest, and keeps pepsin maximally active. Active pepsin hydrolyzes food proteins into shorter chains of polypeptides that are not fully digested until the mixture enters the small intestine. The mixture of partially digested food and gastric juice is called **chyme.**

Ulcers

It is important that the stomach not produce too much acid. If it did, the body could not neutralize the acid later in the small intestine, a step essential for the final stage of digestion. Production of acid is controlled by hormones. These hormones are produced by endocrine cells scattered within the walls of the stomach. The hormone gastrin regulates the synthesis of HCl by the parietal cells of the gastric pits, permitting HCl to be made only when the pH of the stomach is higher than about 1.5.

Overproduction of gastric acid can occasionally eat a hole through the wall of the stomach. Such **gastric ulcers** are rare, however, because epithelial cells in the mucosa of the stomach are protected by a layer of alkaline mucus, and because those cells are rapidly replaced by cell division if they become damaged (gastric epithelial cells are replaced every two to three days). Over 90% of gastrointestinal ulcers are **duodenal ulcers,** which are ulcers of the small intestine. These may be produced when the mucosal barriers to self-digestion are weakened by an infection of the bacterium *Helicobacter pylori.* Modern antibiotic treatments can reduce symptoms and often cure the ulcer.

> The mucus that protects the lining of the stomach is an alkaline substance. As discussed on page 40, an alkaline substance is a base and has a higher pH. The alkaline mucus combines with H⁺, which reduces the acidity of the stomach and protects the lining of the stomach.

Leaving the Stomach

Chyme leaves the stomach at its base through the *pyloric sphincter*, as shown in figure 26.12, and enters the small intestine. This is where all terminal digestion of carbohydrates, fats, and proteins occurs, and where the products of digestion—amino acids, glucose, and fatty acids—are absorbed into the blood.

IN THE NEWS

Are Microbes Making Americans Fatter?   The number of overweight Americans is skyrocketing. What is fueling this trend? Surely some of it is lifestyle. Today's "super-size-it" fast foods and "big gulp" drinks, coupled with lack of physical exercise, leave us consuming more calories without burning them off. It now appears, however, that this isn't the whole story. Researchers have recently stumbled onto a second potential cause of our collective weight problem. Your digestive system is teeming with bacteria, some 100 trillion of them, and it now appears that overweight people have a different assortment than lean people. There are two basic phyla of intestinal bacteria, *Firmicutes* and *Bacteriodetes*, and it turns out that obese people have a much higher proportion of firmicutes than lean people. When obese people are put on a diet and lose weight, the balance of bacteria shifts to that seen in lean people. The researchers discovered that overweight people gained weight not because they always ate more, but because their firmicutes are more efficient at breaking down polysaccharides into simple sugars, which the body stores as fat. In addition, firmicutes suppress your ability to produce a substance called "fasting-induced adipose factor" (Fiaf) that keeps fat cells from storing fat. By suppressing Fiaf, firmicutes cause the gates to your fat cells to stay open, so the excess sugars produced can be channeled directly into fat. Fast-food diets and lack of exercise will make anyone overweight, but firmicute intestinal microbes appear to accelerate the process. It would seem we are not only what we eat, but also who helps us eat it.

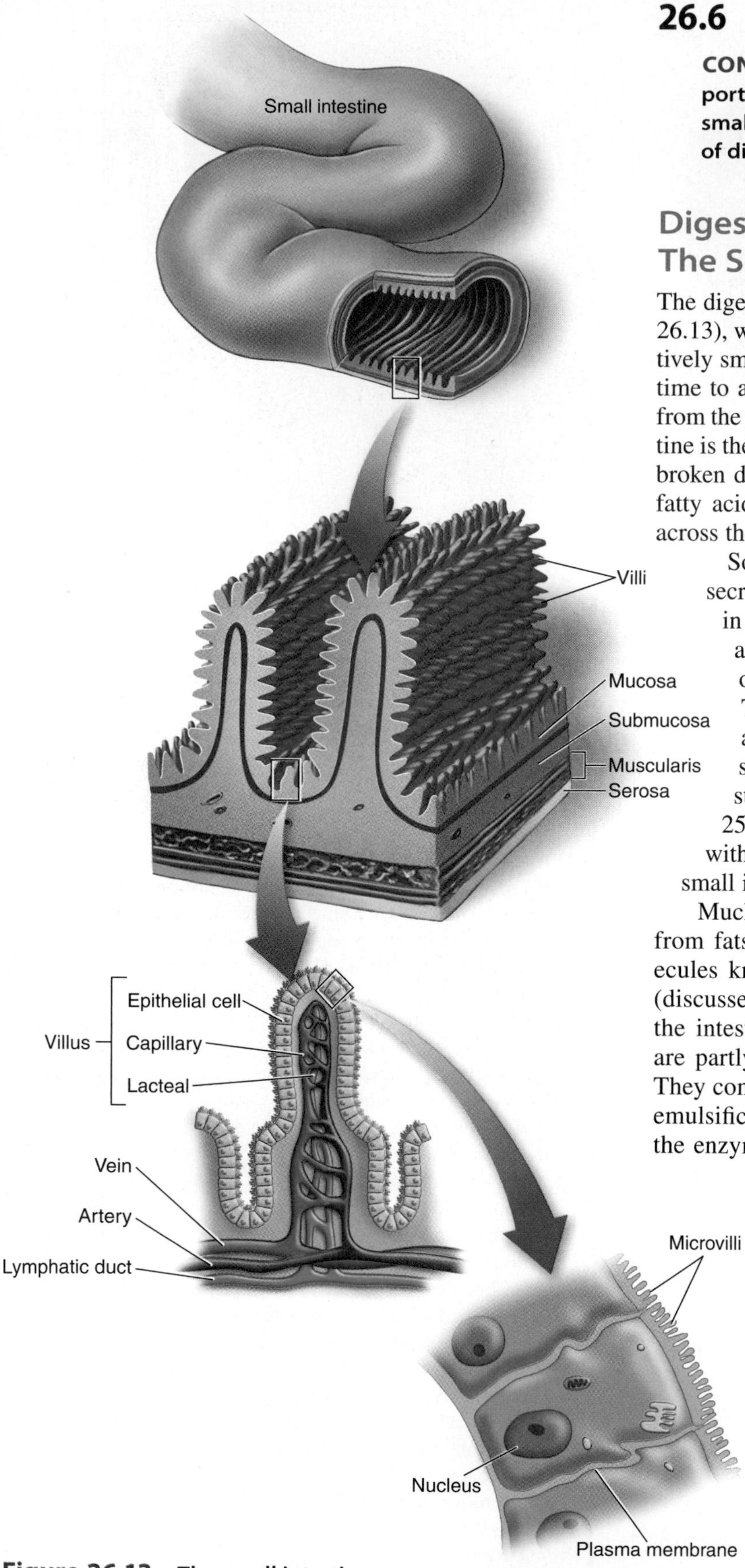

26.6 The Small and Large Intestines

CONCEPT PREVIEW: Most digestion occurs in the initial upper portion of the small intestine, called the duodenum. The rest of the small intestine is devoted to absorption of water and the products of digestion. The large intestine compacts residual solid wastes.

Digestion and Absorption: The Small Intestine

The digestive tract exits from the stomach into the **small intestine** (figure 26.13), where large molecules are broken down into small ones. Only relatively small portions of food are introduced into the small intestine at one time to allow time for acid to be neutralized (using bicarbonate released from the pancreas and discussed later) and enzymes to act. The small intestine is the primary digestive organ of the body. Within it, carbohydrates are broken down into simple sugars, proteins into amino acids, and fats into fatty acids. Once these small molecules have been produced, they pass across the epithelial wall of the small intestine into the bloodstream.

Some of the enzymes necessary for these digestive processes are secreted by the cells of the intestinal wall. Most, however, are made in a large gland called the *pancreas* (discussed in section 26.7), situated near the junction of the stomach and the small intestine. It is one of the body's major exocrine glands (secreting through ducts). The pancreas sends its secretions into the small intestine through a duct that empties into its initial segment, the **duodenum.** Your small intestine is approximately 6 meters long—unwound and stood on its end, it would be far taller than you are! Only the first 25 centimeters, about 4% of the total length, is the duodenum. It is within this initial segment, where the pancreatic enzymes enter the small intestine, that the majority of digestion occurs.

Much of the food energy the vertebrate body harvests is obtained from fats. The digestion of fats is carried out by a collection of molecules known as *bile salts* secreted into the duodenum from the *liver* (discussed in section 26.7). Because fats are insoluble in water, they enter the intestine as drops within the watery chyme. The bile salts, which are partly lipid-soluble and partly water-soluble, work like detergents. They combine with fats to form microscopic droplets in a process called emulsification. These tiny droplets have greater surface areas upon which the enzyme that breaks down fats, called *lipase*, can work. This allows the digestion of fats to proceed more rapidly. The digested fats are first absorbed into lymphatic vessels called lacteals before they later enter the bloodstream.

Two areas make up the rest of the small intestine (96% of its length), the **jejunum** and the **ileum.** Digestion continues into the jejunum, but the ileum is devoted to absorbing water and the products of digestion into the bloodstream. The lining of the small intestine is folded into ridges, as shown in figure 26.13. The ridges are covered with fine fingerlike projections called **villi** (singular, **villus**), each too small to see with the naked eye. In turn, each of the cells covering a villus is covered on its outer surface by a field of cytoplasmic projections called **microvilli.** The enlargement of the villus shows epithelial cells lining the villus, and the further enlargement of these cells shows the microvilli on the surface side of the cells. Scanning and transmission electron micro-

Figure 26.13   The small intestine.

A cross section of the small intestine shows the structure of the villi and microvilli.

graphs in figure 26.14 give you different perspectives of the microvilli. Both villi and microvilli greatly increase the absorptive surface of the lining of the small intestine. The average surface area of the small intestine of an adult human is about 300 square meters, more than the surface of many swimming pools!

The amount of material passing through the small intestine is startlingly large. An average human consumes about 800 grams of solid food, and 1,200 milliliters of water per day, for a total volume of about 2 liters. To this amount is added about 1.5 liters of fluid from the salivary glands, 2 liters from the gastric secretions of the stomach, 1.5 liters from the pancreas, 0.5 liters from the liver, and 1.5 liters of intestinal secretions. The total adds up to a remarkable 9 liters—more than 10% of the total volume of your body! However, although the flux is great, the *net* passage is small. Almost all these fluids and solids are reabsorbed during their passage through the small intestine—about 8.5 liters across the walls of the small intestine and 0.35 liters across the wall of the large intestine. Of the 800 grams of solids and 9 liters of liquids that enter the digestive tract each day, only about 50 grams of solids and 100 milliliters of liquids leave the body as feces. The fluid absorption efficiency of the digestive tract thus approaches 99%, very high indeed.

Concentration of Solids: The Large Intestine

The **large intestine,** or **colon,** is much shorter than the small intestine, approximately 1 meter long, but it is called the large intestine because of its larger diameter. The small intestine empties directly into the large intestine at a junction where the cecum and the appendix are located, which are two structures no longer actively used in humans (see figure 26.5). No digestion takes place within the large intestine, and only about 6% to 7% of fluid absorption occurs there. The large intestine is not convoluted, lying instead in three relatively straight segments, and its inner surface does not possess villi. As a consequence, the large intestine has only one-thirtieth the absorptive surface area of the small intestine. Although some water, sodium, and vitamin K are absorbed across its walls, the primary function of the large intestine is to act as a refuse dump. Within it, undigested material, including large amounts of plant fiber and cellulose, is compacted and stored. Many bacteria live and actively divide within the large intestine, where they play a role in the processing of undigested material into the final excretory product, **feces.** Bacterial fermentation produces gas within the human colon at a rate of about 500 milliliters per day. This rate increases greatly after the consumption of beans or other vegetable matter because the passage of undigested plant material (fiber) into the large intestine provides substrates for fermentation.

The final segment of the digestive tract is a short extension of the large intestine called the **rectum.** Compact solids within the colon pass through the rectum as a result of the peristaltic contractions of the muscles encasing the large intestine, and then out of the body through the **anus.**

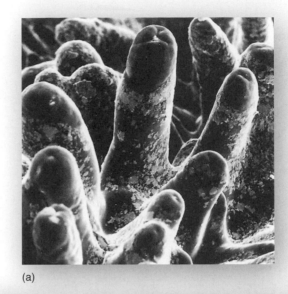
(a)

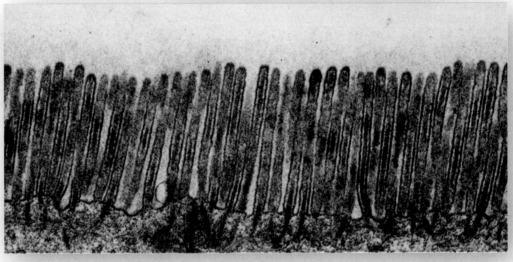
(b)

Figure 26.14 **Microvilli in the small intestine.**

(a) Microvilli, shown in a scanning electron micrograph, are very densely clustered, giving the small intestine an enormous surface area, which is very important for efficient absorption. (b) Intestinal microvilli as shown in a transmission electron micrograph.

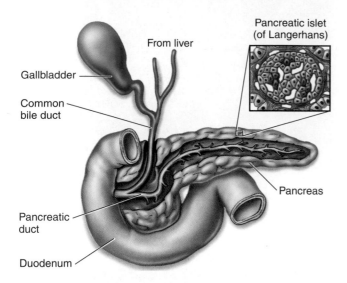

Pancreatic islet (of Langerhans)

From liver

Gallbladder

Common bile duct

Pancreatic duct

Duodenum

Pancreas

Figure 26.15 **The pancreatic and bile ducts empty into the duodenum.**

26.7 Accessory Digestive Organs

CONCEPT PREVIEW: The pancreas secretes digestive enzymes and bicarbonate into the pancreatic duct. The liver produces bile that is stored in the gallbladder. The liver and the pancreatic hormones regulate blood glucose concentration.

The Pancreas

The **pancreas,** a large gland situated near the junction of the stomach and the small intestine (see figure 26.5), is one of the accessory organs that contribute secretions to the digestive tract. Fluid from the pancreas is secreted into the duodenum through the *pancreatic duct* shown in figure 26.15. This fluid contains a host of enzymes which digest proteins, starch, and fats. Pancreatic enzymes digest proteins into smaller polypeptides, polysaccharides into shorter chains of sugars, and fat into free fatty acids and other products.

Pancreatic fluid also contains bicarbonate, which neutralizes the HCl from the stomach and gives the chyme in the duodenum a slightly alkaline pH. In addition to its exocrine role in digestion, the pancreas also functions as an endocrine gland, secreting several hormones into the blood that control the blood levels of glucose and other nutrients. These hormones are produced in the **islets of Langerhans,** clusters of endocrine cells scattered throughout the pancreas and shown in the enlarged view in figure 26.15. Two pancreatic hormones, insulin and glucagon, are discussed in chapters 27 and 30.

Endocrine glands release their products, hormones, directly into the bloodstream as described on page 578, while exocrine glands release their products into ducts that connect to nearby structures. The pancreatic hormones are discussed on page 584.

The Liver and Gallbladder

The **liver** is the largest internal organ of the body. In an adult human, the liver weighs about 1.5 kilograms and is the size of a football. The main exocrine secretion of the liver is **bile,** a fluid mixture consisting of *bile pigments* and *bile salts* that is delivered into the duodenum during the digestion of a meal. The bile salts work like detergents, dispersing large drops of fat into a fine suspension of smaller droplets. This breaking up, or emulsification, of the fat into droplets produces a greater surface area of fat upon which the lipase enzymes can act, and thus allows the digestion of fat to proceed more rapidly.

After it is produced in the liver, bile is stored and concentrated in the gallbladder (the green organ in figure 26.15). The arrival of fatty food in the duodenum stimulates the gallbladder to contract, causing bile to be injected into the duodenum through the common bile duct.

Regulatory Functions of the Liver

A large vein carries blood from the stomach and intestine directly to the liver, which is the first stop for substances absorbed from the gastrointestinal tract. The liver acts as a kind of purification plant. Ingested alcohol and other drugs are taken into liver cells and metabolized; this is why the liver is often damaged as a result of alcohol and drug abuse. The liver also removes toxins, pesticides, carcinogens, and other poisons, converting them into less toxic forms. The liver and all of the digestive organs work together (figure 26.16).

Concept Check

1. What is the function of a rumen? a cecum?
2. Distinguish between a parietal cell and a chief cell?
3. In what part(s) of the gastrointestinal tract does digestion occur for proteins, carbohydrates, and fats?

BIOLOGY & YOU

Vegans. We humans are omnivores, meaning we can eat a broad range of plant and animal tissues—but not all of us choose to do so. Some people don't like spinach and love steak, while others called vegetarians choose not to eat meat. Some become vegetarians because they judge it a more healthy diet—plants are low in saturated fats linked to heart disease. Others make the choice for ethical reasons, sensitive to the animal rights issues associated with livestock agriculture. Still others simply don't like meat. The most extreme form of vegetarian diet is the "vegan" diet. Vegans avoid all animal proteins. They don't eat red meat, poultry, fish, eggs, or milk. Instead, they obtain all protein and nutrients from grains, vegetables, fruits, legumes, nuts, and seeds. The vegan diet, mirroring that of our early human ancestors, is very challenging, because no single fruit, vegetable, or grain contains all the essential amino acids which humans require in their diet. Vegetal foods must be eaten in particular combinations to provide this necessary balance. Beans and rice together provide a balanced diet, but neither food does so when eaten alone. For calcium, which is usually obtained from milk, vegans must eat green leafy vegetables like broccoli or spinach. In practice it is not difficult to achieve this balance if a vegan eats a variety of plants.

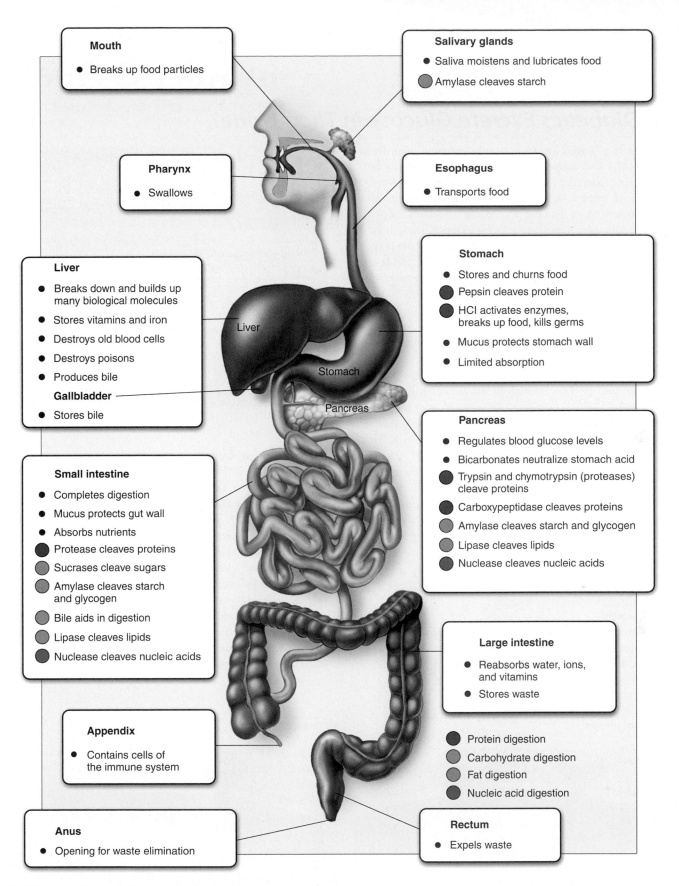

Mouth
- Breaks up food particles

Salivary glands
- Saliva moistens and lubricates food
- Amylase cleaves starch

Pharynx
- Swallows

Esophagus
- Transports food

Liver
- Breaks down and builds up many biological molecules
- Stores vitamins and iron
- Destroys old blood cells
- Destroys poisons
- Produces bile

Gallbladder
- Stores bile

Stomach
- Stores and churns food
- Pepsin cleaves protein
- HCl activates enzymes, breaks up food, kills germs
- Mucus protects stomach wall
- Limited absorption

Pancreas
- Regulates blood glucose levels
- Bicarbonates neutralize stomach acid
- Trypsin and chymotrypsin (proteases) cleave proteins
- Carboxypeptidase cleaves proteins
- Amylase cleaves starch and glycogen
- Lipase cleaves lipids
- Nuclease cleaves nucleic acids

Small intestine
- Completes digestion
- Mucus protects gut wall
- Absorbs nutrients
- Protease cleaves proteins
- Sucrases cleave sugars
- Amylase cleaves starch and glycogen
- Bile aids in digestion
- Lipase cleaves lipids
- Nuclease cleaves nucleic acids

Large intestine
- Reabsorbs water, ions, and vitamins
- Stores waste

- Protein digestion
- Carbohydrate digestion
- Fat digestion
- Nucleic acid digestion

Appendix
- Contains cells of the immune system

Anus
- Opening for waste elimination

Rectum
- Expels waste

Liver
Stomach
Pancreas

Figure 26.16 **The organs of the digestive system and their functions.**

The digestive system contains some dozen different organs that act on the food that is consumed, starting with the mouth and ending with the anus. All of these organs must work properly for the body to effectively obtain nutrients.

Why Do Diabetics Excrete Glucose in Their Urine?

Late-onset diabetes is a serious and increasingly common disorder in which the body's cells lose their ability to respond to insulin, a hormone which is needed to trigger their uptake of glucose. As illustrated below, the binding of insulin to a receptor in the plasma membrane causes the rapid insertion of glucose transporter channels into the plasma membrane, allowing the cell to take up glucose. In diabetics, however, glucose molecules accumulate in the blood while the body's cells starve for the lack of them. In mild cases, blood glucose levels rise to several times the normal value of 4 mM; in severe untreated cases, blood glucose levels may become enormously elevated, up to 25 times the normal value. A characteristic symptom of even mild diabetes is the excretion of large amounts of glucose in the urine. The name of the disorder, *diabetes mellitus*, means "excessive secretion of sweet urine." In normal individuals, by contrast, only trace amounts of glucose are excreted. The kidney very efficiently reabsorbs glucose molecules from the fluid passing through it. Why doesn't it do so in diabetic individuals?

The graph on the upper right displays so-called glucose tolerance curves for a normal person (*blue line*) and a diabetic (*red line*). After a night without food, each individual drank a test dose of 100 grams of glucose dissolved in water. Blood glucose levels were then monitored at 30-minute and one-hour intervals. The dotted line indicates the kidney threshold, the maximum concentration of blood glucose molecules (about 10 mM) that the kidney is able to retrieve from the fluid passing through it when all of its glucose-transporting channels are being utilized full-bore.

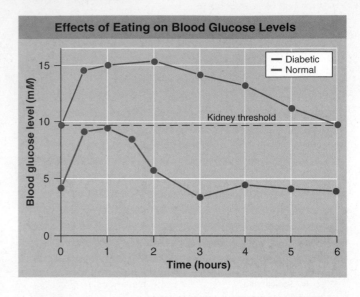

Effects of Eating on Blood Glucose Levels

Analysis

1. **Applying Concepts**
   Reading a Curve. What is the immediate impact on the normal individual's blood glucose levels of consuming the test dose of glucose? How long does it take for the normal person's blood glucose level to return to the level before the test dose?

Comparing Curves. Is the impact any different for the diabetic person? How long does it take for the diabetic person's blood glucose levels to return to the level before the test dose?

2. **Interpreting Data**
   a. Is there any point at which the normal individual's blood glucose levels exceed the kidney threshold?
   b. Is there any point at which the diabetic individual's blood glucose levels do *not* exceed the kidney threshold?

3. **Making Inferences**
   a. Why do you suppose the diabetic individual took so much longer to recover from the test dose?
   b. Would you expect the normal individual to excrete glucose? Explain. The diabetic individual? Explain.

4. **Drawing Conclusions** Why do diabetic individuals secrete sweet urine?

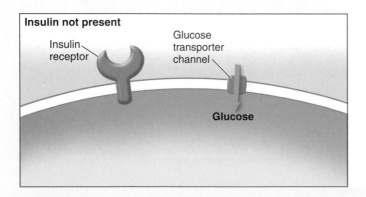

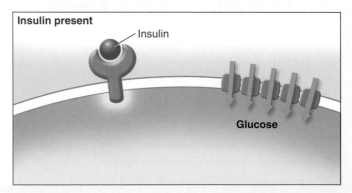

Concept Summary

Food Energy and Essential Nutrients

26.1   Food for Energy and Growth

- Animals consume food as a source of energy and as a source of essential molecules and minerals. A balanced diet that is higher in complex carbohydrates, fruits, and vegetables and low in fats is recommended, as indicated by the pyramid of nutrition shown here from **figure 26.1.**

- The energy from food is either used up through metabolic activity (the BMR and exercise) or is stored as fat in fat cells. Fats have a higher energy content per gram, about 9.3 calories per gram, compared to 4.1 calories for carbohydrates and proteins. If a person consumes a lot of calories, especially in the form of fats, but doesn't burn off those calories through exercise, the energy is stored as fat. A measurement called a body mass index (BMI) is an easy guide to determining whether a person is overweight or obese (**figure 26.2**).

- Animals must consume proteins, fruits, and vegetables to obtain essential amino acids, minerals, and vitamins that the body needs but cannot produce itself.

Digestion

26.2   Types of Digestive Systems

- All animals are heterotrophs. Heterotrophs are divided into three groups based on what they eat. Herbivores eat exclusively plants, carnivores are meat eaters, and omnivores have diets that contains both plants and meat.

- Single-celled organisms and lower animal phyla, such as sponges, digest food intracellularly. Food is taken up by individual cells and is broken down inside the cells.

- All other animals digest food extracellularly. Digestive enzymes are released into a cavity or tract where they break down food. The products of digestion are then absorbed by cells in the body. The digestive cavity in flatworms and cnidarians, like the *Hydra* shown here from **figure 26.3** is called the gastrovascular cavity.

- Specialization of the digestive tract, where different regions of the tract are involved in different digestive functions, was possible with the evolution of a one-way digestive tract (**figure 26.4**).

26.3   Vertebrate Digestive Systems

- Vertebrate digestion occurs in a gastrointestinal tract with specialized areas for different digestive functions (**figures 26.5** and **26.6**). The size or structure of specialized digestive areas vary in different animal groups depending on the animal's diet.

26.4   The Mouth and Teeth

- In vertebrates, food is first brought into the mouth. Birds break up food in the gizzard, a compartment of

their digestive system (**figure 26.7**). In the gizzard, food is churned and ground up with pebbles that the bird has swallowed. In other vertebrates, teeth are used to chew up food, breaking it into smaller pieces (**figures 26.8** and **26.9**).

- The chewed food mixes with saliva in the mouth. Saliva moistens and lubricates the food and it contains the enzyme salivary amylase that begins the digestion of starches. The moistened food is then swallowed passing from the mouth into the esophagus (**figure 26.10**). A flap called the epiglottis closes over the trachea so that the food passes down the esophagus and not into the windpipe.

26.5   The Esophagus and Stomach

- Food is moved along the esophagus to the stomach by peristaltic waves of muscle contractions (**figure 26.11**). A ring of smooth muscle, called a sphincter, closes off the esophagus from the stomach, keeping food from coming back up.

- In the stomach, muscle contractions churn up the food with gastric juice, which contains hydrochloric acid and pepsin, a protein-digesting enzyme activated by HCl (**figure 26.12**). Gastric juice is produced by two gastric glands. These exocrine glands in the stomach lining contain two types of secretory cells: parietal cells that secrete HCl and chief cells that secrete pepsinogen. The low pH in the stomach due to the HCl activates pepsinogen to pepsin.

- Proteins are partially digested in the stomach by pepsin. The acidic conditions in the stomach help denature proteins so they are easier to digest in the small intestine. The partially digested food and gastric juice that leaves the stomach is called chyme.

26.6   The Small and Large Intestines

- The acidic chyme from the stomach passes into the upper portion of the small intestine where it is neutralized and mixed with other digestive enzymes. Some enzymes are secreted by the cells that line the walls of the intestine, but most enzymes and other digestive substances are produced in the pancreas or other accessory organs. The rest of the small intestine is involved in absorption of food molecules and water. The lining of the small intestine is folded into ridges which are covered with fingerlike projections called villi, shown here from **figure 26.13**. The surface of the cells that line the villi are themselves covered with cytoplasmic projections called microvilli. Villi and microvilli increase the surface area of the absorptive surface.

- The large intestine collects and compacts solid waste, releasing it through the rectum and anus.

26.7   Accessory Digestive Organs

- The pancreas produces protein-digesting enzymes, a starch-digesting enzyme, and a fat-digesting enzyme. The pancreatic fluid also contains bicarbonate, which neutralizes the acidic chyme.

- The liver produces bile (a mixture of bile pigments and bile salts), which breaks down fats. Bile is stored in the gallbladder and released into the small intestine (**figure 26.15**). All of the organs of digestion work together (**figure 26.16**).

Self-Test

1. One-way passage of food through the digestive system of many animal groups allows
   a. intracellular digestion.
   b. specialization of different regions of the digestive system.
   c. release of digestive enzymes into the gut.
   d. extracellular digestion.

2. Organisms with longer digestive systems, which help break down difficult to digest food, are usually
   a. herbivores.   c. omnivores.
   b. carnivores.   d. detritivores.

3. The purpose of a gizzard, like teeth, is to
   a. hold on to prey.
   b. begin the chemical digestion of food.
   c. release enzymes.
   d. begin the physical digestion of food.

4. When a mammal swallows food, the food is prevented from going up into the nasal cavity by the
   a. esophagus.   c. soft palate.
   b. tongue.   d. epiglottis.

5. The first site of protein digestion in the digestive system occurs in the
   a. mouth.   c. stomach.
   b. esophagus.   d. small intestine.

6. Which of the following statements is *false*?
   a. Villi and microvilli improve the efficiency of absorption in the small intestine.
   b. The surface area of the large intestine is greater than that in the small intestine.

   c. Digestion occurs primarily in the duodenum.
   d. Ruminants are able to regurgitate food.

7. Most of the absorption of food molecules takes place in the
   a. stomach.   c. small intestine.
   b. liver.   d. large intestine.

8. The purpose of the villi and microvilli in the small intestine is to
   a. neutralize stomach acid.
   b. produce bile.
   c. produce digestive enzymes.
   d. increase the surface area of the small intestine for absorption of nutrients.

9. The primary function of the large intestine is
   a. the breakdown and absorption of fats.
   b. the absorption of water.
   c. the concentration of solid wastes.
   d. the absorption of vitamin C.

10. The _____ secretes digestive enzymes and bicarbonate solution into the small intestine to aid digestion.
    a. pancreas   c. gallbladder
    b. liver   d. All of these are correct.

Visual Understanding

1. **Figure 26.10** "Don't talk with your mouth full" is parental advice that is important for not only social reasons (manners) but for medical reasons. Talking while eating can result in choking. Explain how this could occur.

2. **Figure 26.15** What are the functions of the tan-colored organ and the green-colored organ shown here?

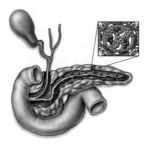

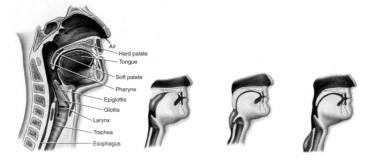

Challenge Questions

1. Why is it so important for Popeye, and you, to eat your green leafy vegetables (spinach, chard, turnip and mustard greens, broccoli, cauliflower, cabbage)?

2. You're going on a backpacking trip with three friends. Lisa wants to pack trail snacks of chocolate cupcakes; Chris wants to take beef jerky.

Andre insists that you should all pack GORP (*good old r*aisins and *p*eanuts—sometimes known as trail mix). They turn to you to decide. Explain which one is best for a quick snack to keep up your energy on a long hike, and why.

Maintaining the Internal Environment

Homeostasis

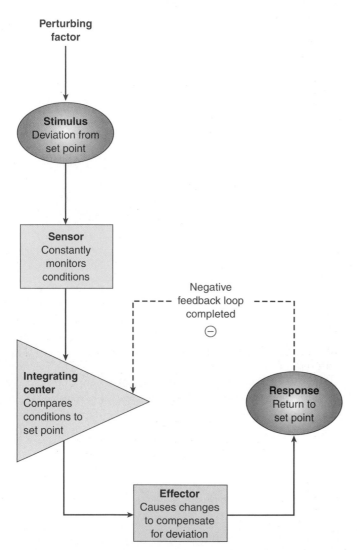

Figure 27.1 **A generalized diagram of a negative feedback loop.**

Negative feedback loops maintain a state of homeostasis, or dynamic constancy of the internal environment, by correcting deviations from a set point.

27.1 How the Animal Body Maintains Homeostasis

CONCEPT PREVIEW: Negative feedback mechanisms correct deviations from a set point for different internal variables. In this way, body conditions, such as temperature and blood glucose, are kept within a normal range.

Each animal cell is a sophisticated machine, finely tuned to carry out a precise role within the body. Such specialization of cell function is possible only when extracellular conditions are kept within narrow limits. Temperature, pH, the concentration of glucose and oxygen, and many other factors must be held constant for cells to function efficiently and interact properly with one another.

Homeostasis may be defined as the dynamic constancy of the internal environment. The term *dynamic* is used because conditions are never absolutely constant, but fluctuate continuously within narrow limits. Homeostasis is essential for life, and most of the regulatory mechanisms of the vertebrate body are concerned with maintaining homeostasis.

Negative Feedback Loops

To maintain internal constancy, the vertebrate body must have **sensors** that are able to measure each condition of the internal environment (indicated by the light green box in figure 27.1). Sensors constantly monitor the extracellular conditions and relay this information (usually via nerve signals) to an **integrating center** (the yellow triangle), which contains the *set point,* which is the proper value for that condition. This set point is analogous to the temperature setting on a house thermostat. In a similar manner, there are set points for body temperature, blood glucose concentration, the tension on a tendon, and so on. The integrating center is often a particular region of the brain or spinal cord, but in some cases it can also be cells of endocrine glands. It receives messages from several sensors, weighs the relative strengths of each sensor input, and then determines whether the value of the condition is deviating from the set point. When a deviation in a condition occurs, sensors detect it, and the integrating center sends a message to increase or decrease the activity of particular effectors. **Effectors** (the blue box in the figure) are generally muscle or glands, and can change the value of the condition in question back toward the set point value, which is "the response" (the purple oval).

In humans, if the body temperature exceeds the set point of 37°C, sensors in the brain detect this deviation. Acting via an integrating center (also in the brain), these sensors stimulate effectors, such as sweat glands, that act to lower the temperature. One can think of the effectors as "defending" the set points of the body against deviations. Because the activity of the effectors is influenced by the effects they produce, and because this regulation is in a negative, or reverse, direction, this type of control system is known as a **negative feedback loop** (figure 27.1).

The nature of the negative feedback loop becomes clear when we again refer to the analogy of the thermostat in a house. After the air conditioner has been on for some time, the room temperature may fall significantly below the set point of the thermostat. When this occurs, the air conditioner will be turned off. The effector (air conditioner) is turned on by high temperature; and when activated, it produces a negative change (lowering of the temperature) that ultimately causes the effector to be turned off, maintaining constancy.

Regulating Body Temperature

Humans, together with other mammals and with birds, are endothermic; they can maintain relatively constant body temperatures independent of the environmental temperature. When the temperature of your blood exceeds 37°C (98.6°F), neurons in a part of the brain called the hypothalamus (discussed in chapter 29) detect the temperature change. Acting through the control of neurons, the hypothalamus responds by promoting the dissipation of heat through sweating, dilation of blood vessels in the skin, and other mechanisms. These responses tend to counteract the rise in body temperature. When body temperature falls, the hypothalamus coordinates a different set of responses, such as shivering and the constriction of blood vessels in the skin, which help to raise body temperature and correct the initial challenge to homeostasis.

> The hypothalamus is an area of the brain that functions as an integration center. As discussed on page 559, the hypothalamus integrates internal activities of the body involved in regulating body temperature, blood pressure, respiration, and heartbeat.

Vertebrates other than mammals and birds are ectothermic; their body temperatures are more or less dependent on the environmental temperature. However, many ectothermic vertebrates attempt to maintain constant body temperature through behavioral means—by placing themselves in varying locations of sun and shade. That's why often you see lizards basking in the sun.

Most invertebrates modify behaviors to adjust their body temperature. Many butterflies, for example, must reach a certain body temperature before they can fly. In the cool of the morning, they orient their bodies to maximize their absorption of sunlight. Moths and other insects use a shivering reflex to warm their flight muscles.

Regulating Blood Glucose

When you digest a meal containing carbohydrates, you absorb glucose into your blood. This causes a temporary rise in the blood glucose concentration, which is brought back down in a few hours. What counteracts the rise in blood glucose following a meal?

Glucose levels within the blood are constantly monitored by a sensor, cells called the islets of Langerhans in the pancreas. When glucose levels increase (figure 27.2a), the islets secrete the hormone *insulin*, which stimulates the uptake of blood glucose into muscles, liver, and adipose tissue. The muscles and liver convert the glucose into the polysaccharide glycogen. Adipose cells convert glucose into fat. These actions lower the blood glucose. When enough blood glucose is absorbed, reaching the set point, the release of insulin stops. When blood glucose levels decrease below the set point, as they do between meals, during periods of fasting, and during exercise, the liver secretes glucose into the blood (the center arrow in figure 27.2b). This glucose is obtained in part from the breakdown of liver glycogen. The breakdown of liver glycogen is stimulated in two ways: by the hormone *glucagon*, which is also secreted by the islets of Langerhans, and the hormone adrenaline, which is secreted from the adrenal gland, as will be discussed in chapter 30.

> The pancreas also functions as an exocrine gland in the digestive system, as discussed on page 514, secreting digestive enzymes into the small intestine through the pancreatic duct. These enzymes are not produced in the islets of Langerhans, but in other cells of the pancreas.

Concept Check

1. In what sense is homeostasis dynamic?
2. How does negative feedback act to maintain body temperature?
3. Contrast the action of insulin with that of glucagon.

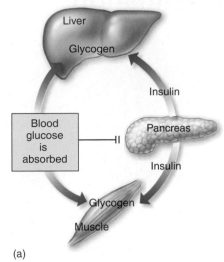

HIGH BLOOD SUGAR

(a)

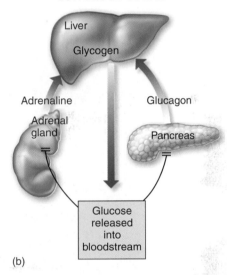

LOW BLOOD SUGAR

(b)

Figure 27.2   Control of blood glucose levels.

(a) When blood glucose levels are high, cells within the pancreas produce the hormone insulin, which stimulates the liver and muscles to convert blood glucose into glycogen. (b) When blood glucose levels are low, other cells within the pancreas release the hormone glucagon into the bloodstream; in addition, cells within the adrenal gland release the hormone adrenaline into the bloodstream. When they reach the liver, these two hormones act to increase the liver's breakdown of glycogen to glucose.

IMPLICATION   In some people, the pancreas produces very low levels of insulin or the body's cells do not respond to insulin. These people suffer from a condition called diabetes mellitus. A test that is used to detect this condition is the "oral glucose tolerance test." The test is administered to a person who has fasted for 12 to 16 hours. At the beginning of the test, the blood is drawn and tested for glucose. The person is given a high-glucose drink and the blood is retested for glucose five times over a three-hour period. What do you think happens to the blood glucose readings of a normal person over the three-hour period? What about those of a person suffering from diabetes?

Osmoregulation

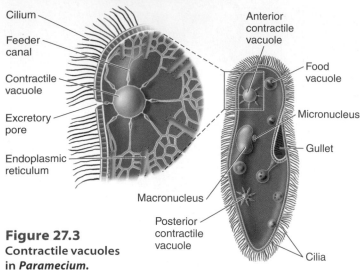

Figure 27.3
Contractile vacuoles in *Paramecium*.

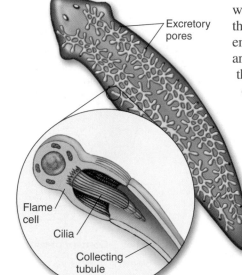

Figure 27.4
The protonephridia of flatworms.

A branching system of tubules, bulblike flame cells, and excretory pores make up the protonephridia of flatworms.

Figure 27.5
The nephridia of annelids.

Most invertebrates, such as the annelid shown here, have nephridia. These consist of tubules that receive a filtrate of coelomic fluid, which enters the funnel-like nephrostomes. Salt can be reabsorbed from these tubules, and the fluid that remains, urine, is released from pores into the external environment.

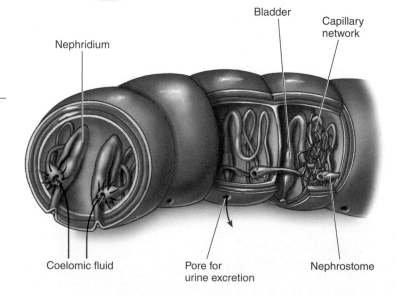

27.2 Regulating the Body's Water Content

CONCEPT PREVIEW: Many invertebrates filter fluid into a system of tubules and then reabsorb ions and water, leaving waste products for excretion. Insects create an excretory fluid by secreting K^+ into tubules, which draws in water osmotically. The vertebrate kidney produces a filtrate from which water is reabsorbed.

Animals must carefully monitor the water content of their bodies. The first animals evolved in seawater, and the physiology of all animals reflects this origin. Approximately two-thirds of every vertebrate's body is water. If the amount of water in the body of a vertebrate falls much lower than this, the animal dies. Animals use various mechanisms for **osmoregulation,** the regulation of the body's osmotic composition, or how much water and salt it contains. The proper operation of many vertebrate organ systems requires that the osmotic concentration of the blood—the concentration of solutes dissolved within it—be kept within a narrow range.

Animals have evolved a variety of mechanisms to cope with problems of water balance. In many animals and single-celled organisms, the removal of water or salts from the body is coupled with the removal of metabolic wastes through the excretory system. Protists, like the *Paramecium* in figure 27.3, employ contractile vacuoles for this purpose, as do the cells of sponges. Water and metabolic wastes are collected by the endoplasmic reticulum and pass through feeder canals to the contractile vacuole. The water and wastes are expelled when the vacuole contracts and releases its contents out a pore.

Multicellular animals have a system of excretory tubules (little tubes) that expel fluid and wastes from the body. In flatworms, these tubules are called *protonephridia* (the green structures you see in figure 27.4). They branch throughout the body into bulblike **flame cells,** shown in the enlargement. While these simple excretory structures open to the outside of the body, they do not open to the inside of the body. Rather, the beating action of cilia within the flame cells draw in fluid from the body, which is passed into a collecting tube. Water and metabolites are then reabsorbed, and the substances to be excreted are expelled through excretory pores.

Other invertebrates have a system of tubules that open both to the inside and to the outside of the body. In the earthworm, these tubules are known as *nephridia,* the blue structures you see in figure 27.5. The nephridia obtain fluid from the body cavity through a process of filtration into funnel-shaped structures called *nephrostomes.* The term *filtration* is used because the fluid is formed under pressure and passes through small openings, so that molecules larger than a certain size are excluded. This filtered fluid is isotonic (having the same osmotic concentration) to the fluid in the coelom, but as it passes through the tubules of the nephridia, NaCl is removed by active transport processes. Because salt is reabsorbed from the filtrate, the urine excreted is more dilute than the body fluids (meaning it is hypotonic).

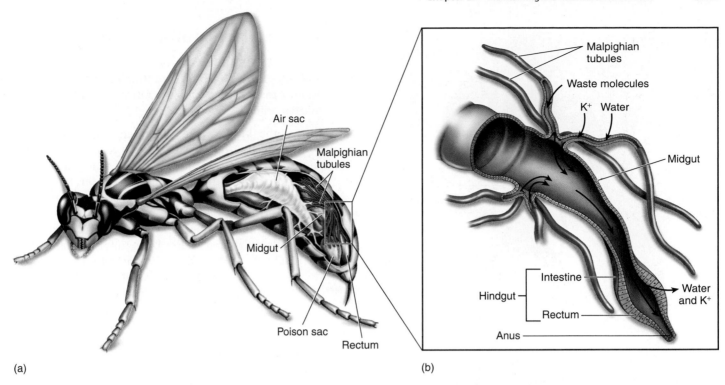

(a)

(b)

Figure 27.6 **The Malpighian tubules of insects.**

(a) The Malpighian tubules of insects are extensions of the digestive tract that collect water and wastes from the body's circulatory system. (b) K⁺ is secreted into these tubules, drawing water with it osmotically. Much of this water and K⁺ is reabsorbed across the wall of the hindgut.

The excretory organs in insects are called **Malpighian tubules,** the green structures you see in figure 27.6. Malpighian tubules are extensions of the digestive tract that branch off anterior to the hindgut. Urine is not formed by filtration in these tubules, because there is no pressure difference between the blood in the body cavity and the tubule. Instead, waste molecules and potassium ions (K⁺) are secreted into the tubules by active transport. In *secretion,* ions or molecules are transported from the body fluid into the tubule. The secretion of K⁺ creates an osmotic gradient that causes water to enter the tubules by osmosis from the body's open circulatory system. Most of the water and K⁺ is then reabsorbed into the circulatory system through the epithelium of the hindgut, leaving only small molecules and waste products to be excreted from the rectum along with feces. Malpighian tubules thus provide a very efficient means of water conservation.

Kidneys are the excretory organs in vertebrates and is the topic for the rest of the chapter. Unlike the Malpighian tubules of insects, kidneys create a tubular fluid by filtration of the blood under pressure. In addition to containing waste products and water, the filtrate contains many small molecules that are of value to the animal, including glucose, amino acids, and vitamins. These molecules and most of the water are reabsorbed from the tubules into the blood, while wastes remain in the filtrate. Additional wastes may be secreted by the tubules and added to the filtrate, and the final waste product, urine, is eliminated from the body.

It may seem odd that the vertebrate kidney should filter out almost everything from blood plasma (except proteins, which are too large to be filtered) and then spend energy to take back or reabsorb what the body needs. But selective reabsorption provides great flexibility; various vertebrate groups have evolved the ability to reabsorb different molecules that are especially valuable in particular habitats. This flexibility is a key factor underlying the successful colonization of many diverse environments by the vertebrates.

Putting Malpighian Tubules to Other Uses.
Although primarily involved in excretion and osmoregulation, Malpighian tubules have evolved in some insects to serve quite different functions. Larvae of the New Zealand glowworm *Arachnocampa luminosa* (a kind of fly) have Malpighian tubules in which the swollen ends have been modified to produce a blue-green light. The larvae produce mucus-coated strings that dangle from trees or ceilings of caves; they use the light, produced with the expenditure of ATP, to attract prey toward mucus-coated trap lines. The larvae then attack and consume any inquisitive prey insects unfortunate enough to become stuck onto the lines.

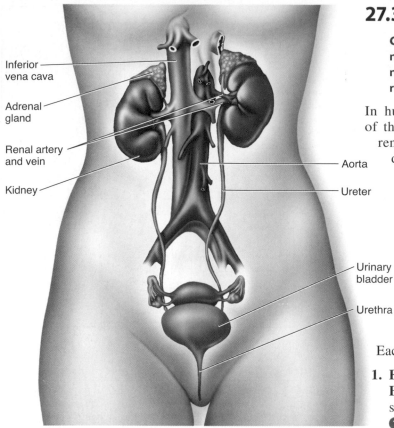

Inferior vena cava

Adrenal gland

Renal artery and vein

Kidney

Aorta

Ureter

Urinary bladder

Urethra

Figure 27.7 **The mammalian urinary system contains two kidneys.**

The urinary system consists of the kidneys, the ureters, which transport urine from the kidneys to the urinary bladder, and the urethra, which carries urine out of the body.

BIOLOGY & YOU

Kidney Stones. Your kidneys filter metabolic wastes and minerals from your circulatory system. Unfortunately, some of the minerals in urine are able to crystallize, forming solid deposits that attach to the kidney walls. Called kidney stones, these solids typically leave the body in the urine stream without causing symptoms. However, if the stone grows large (over 3 millimeters, about the size of a grain of sand) it can obstruct the ureter, and the resulting backup behind the blockage causes quite severe pain. Stones less than 6 mm usually pass in a few days; larger stones must be broken up with sound waves or surgically removed. In the United States, about 15% of adults will be diagnosed with a kidney stone, three-quarters of them men. What causes these kidney stones? Eighty percent of kidney stones are composed of calcium oxalate crystals. It seems that excessive oxalate in your diet promotes calcium oxalate precipitation in your kidneys. Thus to prevent the reoccurrence of kidney stones, affected individuals are urged to restrict their intake of foods that are high in oxalate, such as tea, spinach, and beets.

27.3 The Mammalian Kidney

CONCEPT PREVIEW: The mammalian kidney forces small molecules through a filter and then reclaims water and useful metabolites and ions from the filtrate before eliminating the residual urine.

In humans, the kidneys are fist-sized organs located in the region of the lower back (figure 27.7). Each kidney receives blood from a renal artery, and it is from this blood that urine is produced. Urine drains from each kidney through a **ureter,** which carries the urine to a **urinary bladder.** Urine passes out of the body through the **urethra.**

Within the kidney, shown in cross section in figure 27.8*a*, the mouth of the ureter flares open to form a funnel-like structure, the renal pelvis. The renal pelvis, in turn, has cup-shaped extensions that receive urine from the renal tissue. This tissue is divided into an outer **renal cortex** and an inner **renal medulla.** The structure in the kidney that performs the functions of filtration, reabsorption, and secretion is called the **nephron.**

The mammalian kidney is composed of roughly 1 million nephrons (an enlarged view of a nephron is shown in figure 27.8*b*). Each nephron is composed of four regions:

1. **Filter.** The filtration device at the top of each nephron is called a **Bowman's capsule.** Within each capsule an arteriole enters and splits into a fine network of capillaries called a **glomerulus** (labeled ❶ in figure 27.8*b*). The walls of these capillaries act as a filtration device. Blood pressure forces fluid through the capillary walls. These walls withhold proteins and other large molecules in the blood, while passing water, small molecules, ions, and urea, which is the primary waste product of metabolism. This filtrate enters the Bowman's capsule.

2. **Proximal tubule.** The Bowman's capsule empties into the proximal tubule, which reclaims most of the water (75%), as well as molecules useful to the body, such as glucose and a variety of ions.

3. **Renal tube.** The proximal tubule is connected to a long narrow tube called a renal tubule (labeled ❷ through ❹), which is bent back on itself in its center. This long hairpin loop, called the **loop of Henle,** is a reabsorption device. As the filtrate passes, the renal tubule extracts another 10% of water in the descending loop.

4. **Collecting duct.** The renal tubule empties into a large collection tube called a **collecting duct** (❺). The collecting duct operates as a water conservation device, reclaiming another 14% of water from the filtrate so that it is not lost from the body. Humans produce a urine that is four times as concentrated as blood plasma—that is, the collecting ducts remove much of the water from the filtrate passing through the kidney. Your kidneys achieve this remarkable degree of water conservation by a simple but superbly designed mechanism: The duct bends back alongside the renal tubule and is permeable to urea. Urea passes out of the collecting duct by diffusion. This greatly increases the local salt (urea) concentration in the tissue surrounding the tube, causing water in the filtrate to pass out of the tube by osmosis. The salty tissue sucks up water from the filtrate like blotting paper, passing it on to blood vessels that carry it out of the kidneys and back to the bloodstream. The filtrate that is left in the collecting duct is called urine.

The Kidney at Work

The formation of urine within the mammalian kidney involves the movement of several kinds of molecules between nephrons and the capillaries that surround them. Five steps are involved and indicated in figure 27.8b: pressure filtration ❶, reabsorption of water ❷, selective reabsorption of ions ❸, tubular secretion ❹, and further reabsorption of water ❺.

Pressure Filtration. Driven by the blood pressure, small molecules are pushed across the thin walls of the glomerulus to the inside of the Bowman's capsule ❶. Blood cells and large molecules like proteins cannot pass through, and as a result the blood that enters the glomerulus is divided into two paths: nonfilterable blood components that are retained and leave the glomerulus in the bloodstream, and filterable components that pass across and leave the glomerulus. This filterable stream is called the **glomerular filtrate.** It contains water, nitrogenous wastes (principally urea), nutrients (principally glucose and amino acids), and a variety of ions.

> The cardiovascular system is leaky, as you learned on page 481. Capillaries are only one cell layer thick and so fluid is easily forced out by the pumping pressure of the heart. Fluid forced out of the capillaries in the glomerulus is collected in the Bowman's capsule.

Reabsorption of Water. Filtrate from the glomerulus passes down the proximal tubule into the descending arm of the loop of Henle. The walls of this descending arm are impermeable to either salts or urea but are freely permeable to water. Because (for reasons we discuss later) the surrounding tissue has a high concentration of urea, water passes out of the descending arm by osmosis ❷, leaving behind a more concentrated filtrate.

> The body retains water through osmosis. Recall on page 78 that water will pass from an area of low solute concentration to an area of high solute concentration by the process of osmosis. The high solute concentration in the tissue surrounding the nephron drives osmosis.

Selective Reabsorption. At the turn in the loop, the walls of the tubule become permeable to salts but much less permeable to water. As the concentrated filtrate passes up this ascending arm, these nutrients pass out into the surrounding tissue ❸, where they are carried away by blood vessels. In the upper region of the ascending arm are active transport channels that pump out salt (NaCl). Left behind in the filtrate is the urea that initially passed through the glomerulus as nitrogenous waste. At this point in the tubule, the urea concentration is becoming very high.

Tubular Secretion. In the distal tubule, substances are also added to the urine by a process called tubular secretion ❹. This active transport process secretes into the urine other nitrogenous wastes such as uric acid and ammonia.

Further Reabsorption of Water. The tubule then empties into a collecting duct that passes back through the tissue of the kidney. Unlike the tubule, the lower portions of the collecting duct are permeable to urea, some of which diffuses out into the surrounding tissue (that is why the tissue surrounding the descending arm of the loop of Henle has a high urea concentration indicated by the darker pink color in the figure). The high urea concentration in the tissue causes even more water to pass outward from the filtrate by osmosis ❺. The filtrate that is left in the collecting duct after salts, nutrients, and water have been removed is urine. This urine can have a higher osmotic concentration than the fluids in the body. This ability to produce a hypertonic urine allows mammals, as well as birds, to excrete their waste products in a small volume of water, so that more water can be retained in the body.

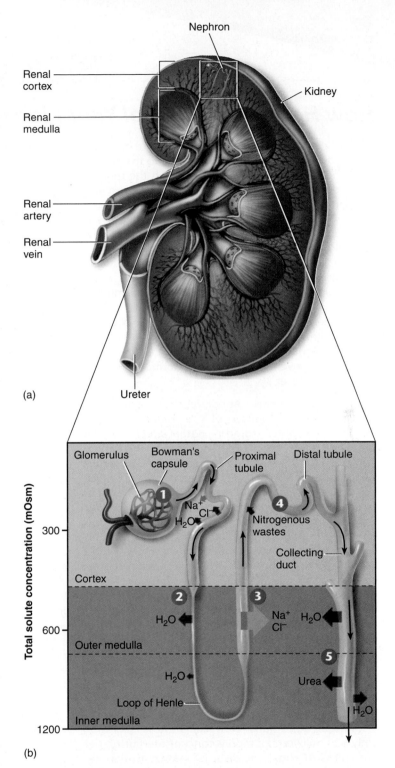

(a)

(b)

Figure 27.8 Structure of the kidney.

(a) About 1 million nephrons lie in the renal cortex. (b) The glomerulus is enclosed within a collection device called a Bowman's capsule. Blood pressure forces liquid from blood through the glomerulus and into the proximal tubule of the nephron, where glucose and small proteins are reabsorbed from the filtrate. The filtrate then passes through a loop arrangement consisting of the proximal tubule, the loop of Henle, and the collecting duct, all of which act to remove water from the filtrate. The water is then collected by blood vessels and transported out of the kidney to the systemic (body) circulation.

Biology and Staying Healthy

How Hormones Control Your Kidney's Functions

Like all mammals and birds, the amount of water excreted in your urine varies according to the changing needs of your body. Acting through the mechanisms described in this chapter, your kidneys excrete a hypertonic urine when your body needs to conserve water. If you drink a lot of water, your kidneys will excrete a hypotonic urine. As a result, the volume of your blood, your blood pressure, and the salt levels of your blood plasma are all maintained relatively constant by the kidneys, no matter how much water you drink. The kidneys also regulate the plasma Na^+ and K^+ concentrations and blood pH within narrow limits. These homeostatic functions of the kidneys are coordinated primarily by hormones, which are chemical signals produced in one part of the body but influence events in other parts. Hormones are the subject of chapter 30.

Antidiuretic Hormone. Antidiuretic hormone (ADH) is produced by a portion of your brain called the hypothalamus and is stored in another structure in the brain called the posterior pituitary gland. The primary stimulus for ADH secretion into your blood stream is an increase in the osmolality (concentration of salt) of the blood plasma. The salt in each milliliter of plasma increases when you are dehydrated or when you eat salty food. Consider the stimulus of dehydration shown in the figure here. Dehydration ❶ causes an increase in the solute concentrations in the blood (called the *osmolality*) ❷. Osmoreceptors in the hypothalamus ❸ respond to the elevated blood osmolality by triggering a sensation of thirst ❹ and by an increase in the secretion of ADH ❺.

ADH causes the walls of the distal tubules and collecting ducts in your kidneys (see figure 27.8b) to become more permeable to water, thus increasing the amount of water they absorb from the urine as it passes through your kidneys ❻. The increased absorption of water from the kidneys feeds back to the osmoreceptors (the dashed line), causing a reduction in the secretion of ADH. When the secretion of ADH is reduced, the walls become less permeable, and you excrete more water in your urine.

Under conditions of maximal ADH secretion, you excrete only 600 milliliters of highly concentrated urine per day. A person who lacks ADH has the disorder known as diabetes insipidus and constantly excretes a large volume of dilute urine. Such a person may become severely dehydrated and succumb to dangerously low blood pressure.

Aldosterone. Sodium ion is the major solute in your blood plasma. When the blood concentration of Na^+ falls, the blood

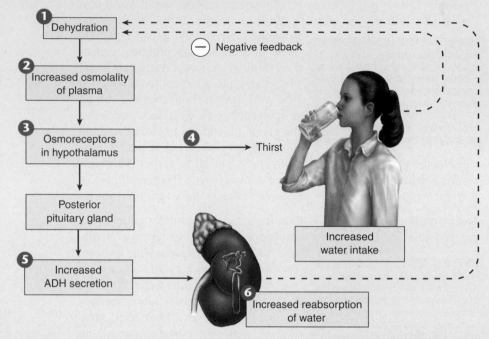

Antidiuretic hormone stimulates the reabsorption of water by the kidneys.

This action completes a negative feedback loop and helps to maintain homeostasis of blood volume and osmolality.

osmolality also falls. This drop in osmolality inhibits ADH secretion, causing more water to remain in the collecting duct for excretion in your urine. As a result, your blood volume and blood pressure decrease. A decrease in extracellular Na^+ also causes more water to be drawn into your cells by osmosis, partially offsetting the drop in plasma osmolality but further decreasing your blood volume and blood pressure. If Na^+ deprivation is severe, the blood volume may fall so low that there is insufficient blood pressure to sustain your life. For this reason, salt is necessary for life. Many animals have a "salt hunger" and actively seek salt. This is why a "salt lick" will attract deer.

A drop in blood Na^+ concentration is normally compensated by the kidneys under the influence of another hormone, aldosterone, which is also secreted by your brain. Indeed, under conditions of maximal aldosterone secretion, Na^+ may be completely absent from the urine. The reabsorption of Na^+ is followed by Cl^- and by water, so aldosterone has the net effect of promoting the retention of both salt and water. It thereby helps to maintain your blood volume, osmolality, and pressure.

27.4 Eliminating Nitrogenous Wastes

CONCEPT PREVIEW: The metabolic breakdown of amino acids and nucleic acids produces ammonia as a by-product. Ammonia is excreted by bony fish, but other vertebrates convert nitrogenous wastes into urea and uric acid.

Amino acids and nucleic acids are nitrogen containing molecules. When animals catabolize these molecules for energy or convert them into carbohydrates or lipids, they produce nitrogen containing by-products called *nitrogenous wastes* that must be eliminated from the body.

The first step in the metabolism of amino acids and nucleic acids is the removal of the amino (—NH_2) group and its combination with H^+ to form **ammonia** (NH_3) in the liver, ❶ in figure 27.9. Ammonia is quite toxic to cells and therefore is safe only in very dilute concentrations. The excretion of ammonia is not a problem for the bony fish and tadpoles, which eliminate most of it by diffusion through the gills and the rest by excretion in very dilute urine ❷. In sharks, adult amphibians, and mammals, the nitrogenous wastes are eliminated in the far less toxic form of **urea** ❸. Urea is water-soluble and so can be excreted in large amounts in the urine. It is carried in the bloodstream from its place of synthesis in the liver to the kidneys, where it is excreted in the urine.

Reptiles, birds, and insects excrete nitrogenous wastes in the form of **uric acid** ❹, which is only slightly soluble in water. As a result of its low solubility, uric acid precipitates and thus can be excreted using very little water. Uric acid forms the pasty white material in bird droppings.

EVOLUTION

The Dry Life of the Kangaroo Rat. The kangaroo rat (genus *Dipodomys*) is native to the deserts of North America, where it has evolved to live with little water. Indeed, a kangaroo rat obtains and conserves water so efficiently that it doesn't have to drink at all. An adult kangaroo rat may never have a single sip of water its whole life, but its body has about the same water content as other desert mammals. What is its secret? Water management. It gains water where there is no water, from the food it eats. Recall from chapter 7 that water is a chemical by-product of cellular respiration. This "metabolic water," which we expel when we breathe, the kangaroo rat collects on cool nasal passages and returns to the body by swallowing. Another crucial adaptation to its dry lifestyle is the kangaroo rat kidney, which is engineered to recapture much of the water passing through the kidneys before it is expelled in urine. In kangaroo rats, like in humans, water is selectively reabsorbed by the body in the loop of Henle and the collecting duct of the nephron. But the loop of Henle in the kangaroo rat is about five times longer than that found in humans and other mammals! As urine travels deeper into the kidney, more and more water passes out of the nephron into the surrounding tissues and back into the bloodstream, and the urine becomes highly concentrated, 4 to 5 times more concentrated than a human's.

Figure 27.9 Nitrogenous wastes.

When amino acids and nucleic acids are metabolized, the immediate nitrogen by-product is ammonia ❶, which is quite toxic but can be eliminated through the gills of bony fish ❷. Mammals convert ammonia into urea ❸, which is less toxic. Birds and terrestrial reptiles convert it instead into uric acid ❹, which is insoluble in water.

Concept Check

1. Contrast flame cells, nephridia, and Malpighian tubules.
2. What does a Bowman's capsule do?
3. How does the loop of Henle act to concentrate urine?

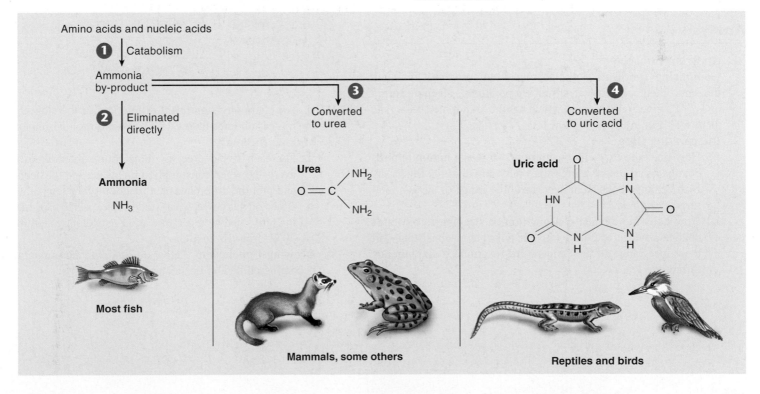

How Do Sleeping Birds Stay Warm?

Mammals and birds are endothermic—they maintain relatively constant body temperatures regardless of the temperature of their surroundings. This lets them reliably run their metabolism even when external temperatures fall—the rates of most enzyme-catalyzed reactions slow two- to threefold for every 10°C temperature drop. Your body keeps its temperature within narrow bounds at 37°C (98.6°F), and birds maintain even higher temperatures. To stay warm like this, mammals and birds continuously carry out oxidative metabolism, which generates heat. This requires a several-fold increase in metabolic rate, which is expensive, particularly when the animal is not active. The logical solution is to give up the struggle to keep warm and let the body temperature drop during sleep, a condition known as *torpor*. While humans don't adopt this approach, many other mammals and birds do. This raises an interesting question: What prevents a sleeping bird in torpor from freezing? Does its body simply adopt the temperature of its surroundings, or is there a body temperature below which metabolic heating kicks in to avoid freezing?

The graph to the right displays an experiment examining this issue in the tropical hummingbird *Eulampis*. The study examines the effect on metabolic rate (measured as oxygen consumption) of decreasing air temperature. Oxygen consumption was assessed over a range of air temperatures from 3° to 37°C, for two contrasting physiological states: The blue data were collected from birds that were awake, the red data from sleeping torpid birds. The blue and red lines, called regression lines, were plotted using curve-fitting statistics that provide the best fit to the data.

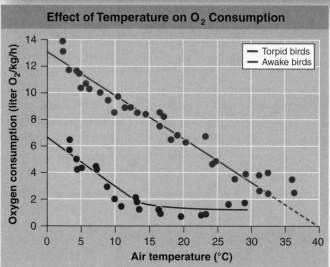

Effect of Temperature on O_2 Consumption

Analysis

1. **Applying Concepts**
   Comparing Two Data Sets. Do awake hummingbirds maintain the same metabolic rate at all air temperatures? sleeping torpid ones? At a given temperature, which has the higher metabolic rate, an awake bird or a sleeping one?

2. **Interpreting Data**
   a. How does the oxygen consumption of awake hummingbirds change as air temperature falls? Why do you think this is so? Is the change consistent over the entire range of air temperatures examined?
   b. How does the oxygen consumption of sleeping torpid birds change as air temperature falls? Is this change consistent over the entire range of air temperatures examined? Explain any difference you detect.

   c. Are there any significant differences in the slope of the two regression lines below 15°C? What does this suggest to you?

3. **Making Inferences**
   a. For each five-degree air temperature interval, estimate the average oxygen consumption for awake and for sleeping birds, and plot the difference as a function of air temperature.
   b. Based on this curve, what would you expect to happen to a sleeping bird's body temperature as air temperatures fall from 30° to 20°C? from 15° to 5°C?

4. **Drawing Conclusions** How do *Eulampis* hummingbirds avoid becoming chilled while sleeping on cold nights?

Concept Summary

Homeostasis

27.1 How the Animal Body Maintains Homeostasis

- Animals maintain relatively constant internal conditions, a process called homeostasis. Homeostasis refers to a dynamic constancy of the internal environment within narrow ranges.

- The body uses sensors to measure internal conditions. The measurements are sent to an integrating center. The integrating center compares the measurement to a set point for that particular condition. If the measurement is deviating from the set point the body will make adjustments through effectors, muscles or glands, to bring the body back to the set point (**figure 27.1**).

- This type of control system is called a negative feedback loop because the response to correct the deviation is in a reverse or negative direction, bringing the condition back to the set point.

- Examples of homeostasis include body temperature and blood glucose levels. Mammals and birds are endothermic, meaning that they maintain a relatively constant body temperature. When temperature sensors detect a change in body temperature, the body responds through sweating and the dilation of surface blood vessels in the skin to decrease body temperature or shivering and a constriction of surface blood vessels to raise body temperature. Ectothermic animals can't control body temperatures in this way, but can make some adjustments through behavioral means, such as sitting in the sun to warm muscles.

- The human body maintains a relatively constant blood glucose level by the actions of hormones. If the blood glucose levels rise, such as after eating, the pancreas releases the hormone insulin, which stimulates the uptake of glucose by the cells in muscles, the liver, and adipose tissue. When blood glucose levels drop, such as in between meals or during exercise, the hormone glucagon is released from the pancreas or adrenaline from the adrenal glands. These hormones trigger the release of glucose stores from the liver (**figure 27.2**).

Osmoregulation

27.2 Regulating the Body's Water Content

- Organisms have evolved various mechanisms to control water balance. Some single-celled organisms, like *Paramecium*, use contractile vacuoles that collect fluid and expels it through pores (**figure 27.3**). Many invertebrates use systems of tubules that collect fluid and reabsorb ions and water. Flatworms use a network of tubules called protonephridia (**figure 27.4**). Nephridia, such as those found in earthworms and shown here from **figure 27.5**, filter fluids and allow the body to reabsorb NaCl. Waste and fluids are excreted as urine through pores. Insects create an osmotic gradient by secreting K^+ into Malpighian tubules, causing water to enter the tubules by osmosis. Water and K^+ are reabsorbed through the epithelium of the hindgut (**figure 27.6**).

- Kidneys are the excretory organs in vertebrates. They filter fluid under pressure and then reabsorb molecules that are important to the body.

27.3 The Mammalian Kidney

- Each kidney receives blood from a renal artery and produces urine, which then drains from each kidney through a ureter and is carried to the urinary bladder (**figure 27.7**).

- Each kidney (**figure 27.8a**) is composed of roughly 1 million nephrons. Each nephron contains a Bowman's capsule, a loop of Henle, and a collecting duct. The function of the nephron is to collect the filtrate from the blood, selectively reabsorb ions and water, and expel the waste from the body.

- The blood is first filtered from the glomerulus, a network of blood capillaries encapsulated by a structure called Bowman's capsule. Water, small molecules, ions, and urea pass into the capsule. The Bowman's capsule is connected to a long tubule where selected molecules and ions are reabsorbed by the body. The tubule forms a hairpin loop called the loop of Henle. The tubule empties into a collecting duct, which recovers additional water from the filtrate (**figure 27.8b**).

- Different areas of the loop of Henle are permeable to different substances. In the proximal tubule, much water and important molecules are reclaimed. In the descending arm of the loop of Henle, water passes back into the surrounding tissue by osmosis, but the loop at this point is impermeable to either salts or urea. At the turn of the loop, the walls become permeable to salts and other molecules, which pass out of the tubules and into the surrounding tissue. This results in a higher concentration of solutes in the tissue surrounding the nephron tubule.

- As the filtrate passes into the collecting duct, whose walls are permeable to water, water is absorbed back into the tissues through osmosis. Osmosis occurs because the surrounding tissue is highly concentrated with solutes such as ions and other molecules that were reabsorbed from the ascending loop.

- The fluid that remains in the collecting duct is a highly hypertonic urine that passes to the ureter and is excreted from the body.

27.4 Eliminating Nitrogenous Wastes

- The metabolic breakdown of amino acids and nucleic acids produces ammonia in the liver. Ammonia, a nitrogenous waste product, is toxic and must be eliminated from the body. Different animals use different means of excretion (**figure 27.9**).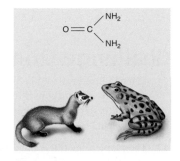

- Ammonia is excreted directly in bony fish and tadpoles through gills and in dilute urine.

- In sharks, adult amphibians, and mammals, ammonia is converted to the less toxic urea. Urea is water-soluble and is carried in the bloodstream to the kidneys where it is excreted in the urine.

- Reptiles, birds, and insects excrete nitrogenous wastes as uric acid, which has low toxicity and is only slightly soluble in water and so can be excreted using only small amounts of water. Uric acid is the pasty white substance in bird droppings.

Self-Test

1. The monitoring and adjusting of the body's internal conditions, such as temperature and pH, is known as
   a. exothermy.   c. osmoregulation.
   b. homeostasis.   d. ectothermy.

2. Which of the following describes the method of regulation whereby the response of an effector returns conditions to a set point?
   a. inhibitory regulation   c. positive feedback loop
   b. homeostasis   d. negative feedback loop

3. If your blood sugar is too low, the hormone glucagon is released by the pancreas. This hormone will cause
   a. the release of insulin.   c. glycogen to be formed.
   b. glycogen to break down.   d. fat to be formed.

4. Which of the following is *not* involved in osmoregulation?
   a. pancreas   c. Malpighian tubules
   b. nephridia   d. flame cells

5. Which of the following animals use Malpighian tubules for excretion?
   a. birds   c. kangaroo rats
   b. ants   d. earthworms

6. The structure in the kidneys that carries out osmoregulation is called the
   a. ureter.   c. urethra.
   b. nephron.   d. filtrate.

7. Where does filtration occur in the nephron?
   a. glomerulus   c. loop of Henle
   b. renal tube   d. collecting duct

8. Selective reabsorption of components of the filtrate occurs where?
   a. Bowman's capsule   c. loop of Henle
   b. glomerulus   d. ureter

9. Water is removed from kidney filtrate by the process of
   a. diffusion.   c. facilitated diffusion.
   b. active transport.   d. osmosis.

10. Humans excrete their excess nitrogenous wastes as
    a. uric acid crystals.
    b. compounds containing protein.
    c. ammonia.
    d. urea.

Visual Understanding

1. **Figure 27.1** Explain the regulation of blood glucose levels using figure 27.1. First indicate what the various components are, assuming the stimulus is high blood glucose after eating a meal. Then describe what happens to each of the components after the stimulus is detected.

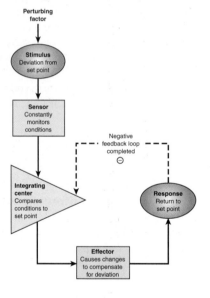

2. **Figure 27.8***b* When the body becomes dehydrated it produces a hormone called ADH (see "Biology and Staying Healthy" on page 526). ADH affects the body by making the collecting duct of the kidney more permeable to water. How would this help a dehydrated person? How would the person's urine be affected?

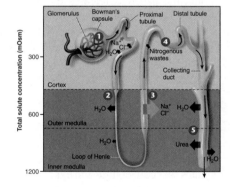

Challenge Questions

1. In a condition called type I diabetes, or juvenile diabetes, the pancreas doesn't release insulin. If left untreated, the person may begin losing weight, practically starving, even though they eat. Explain why this may occur.

2. The glomerulus filters out a tremendous amount of water and molecules needed by the body, which must be reabsorbed by the body in the kidney. The Malpighian tubules of insects might seem to function more logically, secreting mainly waste products that need to be excreted. What advantages might a filtration-reabsorption process found in humans and other mammals provide over a strictly secretion process of elimination found in insects?

Chapter 28

How the Animal Body Defends Itself

CHAPTER AT A GLANCE

Three Lines of Defense

Figure 28.1   **Skin is the vertebrate body's first line of defense.**

This young elephant has tough, leathery skin thicker than a belt, allowing it to follow the herd through dense thickets without injury.

BIOLOGY & YOU

Pimples.   Many of us as teenagers develop acne, our face pocked with blemishes called pimples. A pimple is the result of a blockage of one of the skin's pores. What causes them? Not stress, despite what you might read in magazines. Inside the pores of your face's skin are sebaceous glands which become active around puberty, stimulated by male hormones from the adrenal glands of both boys and girls. Activated, these glands begin to produce an oil called sebum which protects your aging skin. Unfortunately, when the outer layers of your skin shed (as they do continuously), sometimes the dead skin cells left behind become "glued" together by the sebum, causing a blockage in the pore. When this happens, the sebum continually being produced by the sebaceous gland builds up behind the blockage, causing a pimple to form. Bacteria grow in the oil within the pimple, causing the surrounding tissues to become inflamed. If the oil breaks through to the surface, the result is a "whitehead." If the oil accumulates melanin pigment or becomes oxidized, the oil changes from white to black, and the result is a "blackhead." Blackheads are not dirt, and do not reflect poor hygiene. How do you treat acne? Lots of scrubbing to keep the affected area clean, plus over-the-counter medications like benzoyl peroxide and salicylic acid that help the skin slough off more easily, plus patience. In time, acne tends to fade, although severe cases may require consulting a physician.

Multicellular bodies offer a feast of nutrients for tiny, single-celled creatures, as well as a warm, sheltered environment in which they can grow and reproduce. We live in a world awash with bacteria and viruses (microbes). Some are **pathogens,** causing disease and wreaking havoc in the body; no animal can long withstand their onslaught unprotected. Animals survive because they have a variety of very effective defenses against this constant attack.

Overview of the Three Lines of Defense

The vertebrate body is defended from infection the same way knights defended medieval cities. "Walls and moats" make entry difficult; "roaming patrols" attack strangers; and "guards" challenge anyone wandering about and signal for an attack if a proper "ID" is not presented.

1. **Walls and moats.** The outermost layer of the vertebrate body, the skin, is the first barrier to penetration by microbes. Mucous membranes in the respiratory and digestive tracts are also important barriers that protect the body from invasion.
2. **Roaming patrols.** If the first line of defense is penetrated, the response of the body is to mount a *cellular counterattack,* using a battery of cells and chemicals that kill microbes. These defenses act very rapidly after the onset of infection.
3. **Guards.** Lastly, the body is also policed by cells that circulate in the bloodstream and scan the surfaces of every cell they encounter. They are part of the *specific immune response.* One kind of immune cell aggressively attacks and kills any cell identified as foreign or infected with viruses, whereas the other type marks the foreign cell or virus for elimination by the roaming patrols.

28.1   Skin: The First Line of Defense

CONCEPT PREVIEW: Skin is the body's first line of defense against invading microbes. Multiple cell layers provide a barrier to entry.

The Skin

Skin, like the thick, tough skin of the elephants in figure 28.1, is the outermost layer of the vertebrate body and provides the first defense against invasion by microbes. Skin is our largest organ, comprising some 15% of our total weight. One square centimeter of skin from your forearm (about the size of a dime) contains 200 nerve endings, 10 hairs and muscles, 100 sweat glands, 15 oil glands, 3 blood vessels, 12 heat-sensing organs, 2 cold-sensing organs, and 25 pressure-sensing organs. The section of skin you see in figure 28.2 has two distinct layers: an outer **epidermis** and a lower **dermis.** A subcutaneous layer lies underneath the dermis. Cells of the outer epidermis are continually being worn away and replaced by cells moving up from below—in one hour your body loses and replaces approximately 1.5 million skin cells!

The epidermis of skin is from 10 to 30 cells thick, about as thick as this page. The outer layer, called the **stratum corneum,** is the one you see when you look at your arm or face. Cells from this layer are continuously subjected to damage. They are

The outer epidermis of the skin is composed of stratified epithelial cells as described on page 460. The layering of cells protects the skin and also provides cushioning.

abraded, injured, and worn by friction and stress during the body's many activities. They also lose moisture and dry out. The body deals with this damage not by repairing cells but by replacing them. Cells from the stratum corneum are shed continuously, replaced by new cells produced deep within the epidermis (the dark layer of cells at the border of the epidermis and dermis). The cells of this inner *basal layer* are among the most actively dividing cells of the vertebrate body. New cells formed there migrate upward, and as they move they manufacture keratin protein, which makes them tough. Each cell resides at the surface for about a month before it is shed and replaced by a newer cell.

The skin not only defends the body by providing a nearly impermeable barrier, but it also reinforces this defense with chemical weapons housed within structures in the dermis layer of the skin. For example, the oil glands that occur along the shaft of the hair and sweat glands secrete chemicals that make the skin's surface very acidic (pH of between 4 and 5.5), which inhibits the growth of many microbes. Sweat also contains the enzyme lysozyme, which attacks and digests the cell walls of many bacteria.

In addition to the skin, other surfaces, such as the eyes are exposed to the outside. Like sweat, the tears that bathe the eyes contain lysozyme to fight bacterial infections. The digestive tract and the respiratory tract are two other potential routes of entry by bacteria and viruses. Microbes present in food are killed by saliva (that contains lysozyme), the acid environment in the stomach (pH of 2), and digestive enzymes in the intestines. Microbes present in inhaled air are trapped by a layer of sticky mucus secreted by cells that line the bronchi and bronchioles before they can reach the lungs. Other cells that line these passages have cilia that continually sweeps the mucus up to the throat, like an escalator, where they can be swallowed and killed.

The surface defense provided by skin is very effective, but it is occasionally breached. Through breathing, eating, or cuts and nicks, bacteria and viruses now and then enter our bodies. When these invaders reach deeper tissue, a second line of defense comes into play, a cellular counterattack.

Figure 28.2 A section of human skin.

The skin defends the body by providing a barrier with sweat and oil glands, whose secretions make the skin's surface acidic enough to inhibit the growth of microorganisms.

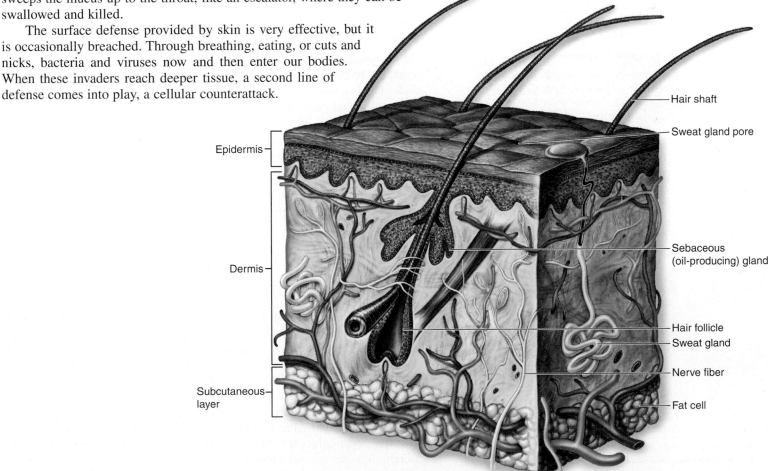

Hair shaft

Sweat gland pore

Epidermis

Sebaceous (oil-producing) gland

Dermis

Hair follicle

Sweat gland

Nerve fiber

Subcutaneous layer

Fat cell

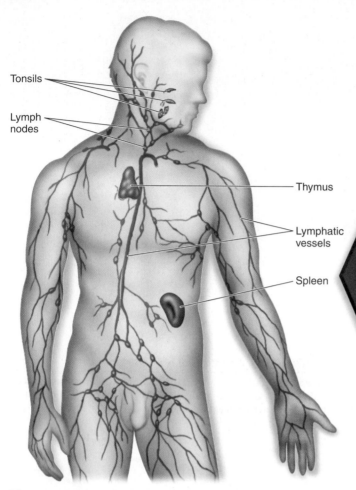

Tonsils

Lymph
nodes

Thymus

Lymphatic
vessels

Spleen

> The vessels of the lymphatic system also serve a circulatory function. As described on page 481, fluid that leaks out of blood capillaries accumulates in tissues as interstitial fluid. It is collected by lymphatic vessels and is returned to the circulatory system.

Figure 28.3   **The lymphatic system.**

Immune system cells and proteins roam through the body and are stored and distributed by the lymphatic system. The lymphatic system consists of lymph nodes, lymphatic organs, and a network of lymphatic capillaries that drain into lymphatic vessels emptying back into the circulatory system.

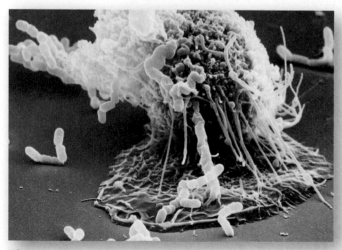

Figure 28.4   **A macrophage in action.**

In this scanning electron micrograph, a macrophage is "fishing" for rod-shaped bacteria.

28.2 Cellular Counterattack: The Second Line of Defense

CONCEPT PREVIEW: Vertebrates respond to infection with a battery of cellular and chemical weapons, including cells and proteins that kill invading microbes, and inflammatory and temperature responses.

When the body's interior is invaded, a host of cellular and chemical defenses swing into action. Four are of particular importance: (1) cells that kill invading microbes; (2) proteins that kill invading microbes; (3) the temperature response, which elevates body temperature to slow the growth of invading bacteria; and (4) the inflammatory response, which speeds defending cells to the point of infection.

Although these cells and proteins roam throughout the body, they are stored in and distributed from the lymphatic system. The lymphatic system consists of the structures shown in figure 28.3.

Cells That Kill Invading Microbes

The most important counterattack to infection is mounted by white blood cells, which attack invading microbes. These cells patrol the bloodstream and await invaders within the tissues. The three basic kinds of killing cells are macrophages and neutrophils, which are phagocytes, and natural killer cells. Natural killer cells can distinguish the body's cells (self) from foreign cells (nonself) because the body's cells contain self-identifying MHC proteins (discussed in more detail later in this chapter).

Macrophages.   White blood cells called **macrophages** (Greek, "big eaters") kill bacteria by ingesting them, much as an amoeba ingests a food particle. The macrophage in figure 28.4 is sending out long, sticky cytoplasmic extensions that catch the sausage-shaped bacteria and draw them back to the cell where they are engulfed. Although some macrophages are anchored within particular organs most patrol the byways of the body, circulating as precursor cells called **monocytes** in the blood, lymph, and fluid between cells. Macrophages are among the most actively mobile cells of the body.

Neutrophils.   Other white blood cells called **neutrophils** act like kamikazes. In addition to ingesting microbes, they release chemicals (identical to household bleach) to "neutralize" the entire area, killing any bacteria in the neighborhood—and themselves in the process. A neutrophil is like a grenade tossed into an infection. It kills everything in the vicinity. Macrophages, by contrast, kill only one or a few invading cells at a time but live to keep on doing it.

Natural Killer Cells.   A third kind of white blood cell, called a **natural killer cell,** does not attack invading microbes but rather the body cells that are infected by them. Natural killer cells puncture the membrane of the infected target cell with a molecule called *perforin.* The natural killer cell in figure 28.5 is releasing several perforin molecules that insert in the membrane of the infected cell, like boards on a picket fence, forming a pore that allows water to rush in, causing the cell to swell and burst. Natural killer cells are particularly effective at detecting and attacking body cells that have been infected with viruses. They are also one of the body's most potent defenses against cancer.

Proteins That Kill Invading Microbes

The cellular defenses of vertebrates are complemented by a very effective chemical defense called the **complement system.** This system consists of approximately 20 different proteins that circulate freely in the blood plasma in an inactive state. Their defensive activity is triggered when they encounter the cell walls of bacteria or fungi. The complement proteins then aggregate to form a *membrane attack complex* that inserts itself into the foreign cell's plasma membrane, forming a pore like that produced by natural killer cells. Like the perforin pore, the membrane attack complex in figure 28.6 allows water to enter the foreign cell, causing the cell to swell and burst. The difference between the two is that perforin released from natural killer cells attacks host cells that have become infected, while free-floating complement attacks the foreign cell directly. Aggregation of the complement proteins is also triggered by the binding of antibodies to invading microbes, as we'll see in a later section.

The Temperature Response

Human pathogenic bacteria do not grow well at high temperatures. Thus, when macrophages initiate their counterattack, they increase the odds in their favor by sending a message to the brain to raise the body's temperature. The cluster of brain cells that serves as the body's thermostat responds to the chemical signal by boosting the body temperature several degrees above the normal 37°C (98.6°F). The higher-than-normal temperature that results is called a *fever*. Although a fever is quite effective at inhibiting microbial growth, very high fevers are dangerous because excessive heat can inactivate critical cellular enzymes. In general, temperatures greater than 39.4°C (103°F) are dangerous, those greater than 40.6°C (105°F) are often fatal.

The Inflammatory Response

The aggressive cellular and chemical counterattacks to infection are made more effective by the **inflammatory response.** The inflammatory response can be broken down into three stages:

1. First, infected or injured cells release chemical alarm signals, most notably histamine and prostaglandins.
2. These chemical alarm signals cause blood vessels to expand, both increasing the flow of blood to the site of infection or injury and, by stretching their thin walls, making the capillaries more permeable. This produces the redness and swelling so often associated with infection.
3. Now the increased blood flow through larger, leakier capillaries promotes the migration of phagocytes from the blood to the site of infection, squeezing between cells in the capillary walls. Neutrophils arrive first, spilling out chemicals that kill the microbes (as well as tissue cells in the vicinity and themselves). Monocytes follow and become macrophages, which engulf the pathogens as well as the remains of all the dead cells. This counterattack takes a considerable toll; the pus associated with infections is a mixture of dead neutrophils, cells, and pathogens.

The second line of defense, with both chemical and cellular weapons, provides a sophisticated defense against microbial infection. Only occasionally do bacteria or viruses overwhelm this defense. When this happens, they face yet a third line of defense, more difficult to evade than any they have encountered. It is the specific immune response, the most elaborate of the body's defenses.

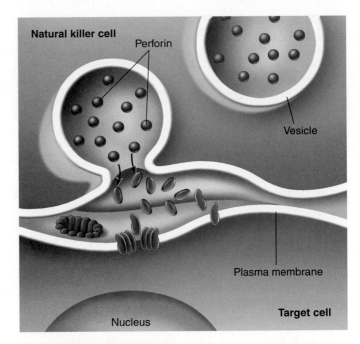

Figure 28.5   Natural killer cells attack target cells.

Tight binding of the natural killer cell to the target cell initiates a chain of events within the natural killer cell in which vesicles loaded with perforin molecules move to the outer plasma membrane and expel their contents into the intercellular space over the target. The perforin molecules insert into the membrane, forming a pore that admits water and ruptures the cell.

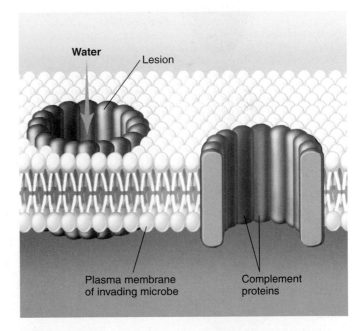

Figure 28.6   Complement proteins attack invaders.

Complement proteins form a transmembrane channel resembling the perforin-lined lesion made by natural killer cells, but complement proteins are free-floating, and they attach to the invading microbe directly, while perforin molecules insert into infected body cells.

TABLE 28.1	Cells of the Immune System
Cell Type	**Function**
Helper T cell	Commander of the immune response; detects infection and sounds the alarm, initiating both T cell and B cell responses
Memory T cell	Provides a quick and effective response to an antigen previously encountered by the body
Cytotoxic T cell	Detects and kills infected body cells; recruited by helper T cells
Suppressor T cell	Dampens the activity of T and B cells, scaling back the defense after the infection has been checked
B cell	Precursor of plasma and memory cells; specialized to recognize specific foreign antigens
Memory B cell	Provides a quick and effective response to an antigen previously encountered by the body
Neutrophil	Engulfs invading bacteria and releases chemicals that kill neighboring bacteria
Plasma cell	Biochemical factory devoted to the production of antibodies directed against specific foreign antigens
Mast cell	Initiator of the inflammatory response, which aids the arrival of leukocytes at a site of infection; secretes histamine and is important in allergic responses
Monocyte	Precursor of macrophage
Macrophage	The body's first cellular line of defense; also serves as antigen-presenting cell to B and T cells and engulfs antibody-covered cells
Natural killer cell	Recognizes and kills infected body cells; natural killer cell detects and kills cells infected by a broad range of invaders

28.3 Specific Immunity: The Third Line of Defense

CONCEPT PREVIEW: T cells attack cells that carry molecules called antigens. When a B cell encounters a specific antigen, it gives rise to plasma cells that produce antibodies. Antibodies tag the antigen for destruction by other white blood cells.

Specific immune defense mechanisms of the body involve the actions of white blood cells, or leukocytes. They are very numerous—of the 10 to 100 trillion cells of your body, two in every 100 are white blood cells! Macrophages are white blood cells, as are neutrophils and natural killer cells. In addition, there are T cells, B cells, plasma cells, mast cells, and monocytes (table 28.1). T cells and B cells are called **lymphocytes** and are critical to the specific immune response.

After their origin in the bone marrow, **T cells** migrate to the thymus (hence the designation "T"), a gland just above the heart (see figure 28.3). There they develop the ability to identify cells that have been infected by microorganisms or viruses. These infected cells display antigens on their surfaces that are produced by the infecting microorganism or virus. An **antigen** is a molecule that provokes a specific immune response. Anti-

> Antigens are proteins that are inserted in the plasma membrane, as described on page 69. Antigen proteins project out from the surface of the cell and contain carbohydrates or lipids that are different from the body's cell surface proteins, marking the antigen as foreign.

gens are large, complex molecules, such as proteins, and they are generally foreign to the body, usually belonging to bacteria and viruses. Tens of millions of different T cells are made, each specializing in the recognition of one particular antigen. No invader can escape being recognized by at least a few T cells. There are four principal kinds of T cells: Helper T cells (symbolized T_H) initiate the immune response; memory T cells (T_M) provide a quick response to a previously encountered antigen; cytotoxic T cells (T_C) lyse cells that have been infected by viruses; and suppressor T cells (T_S) terminate the immune response.

Unlike T cells, **B cells** do not travel to the thymus; they complete their maturation in the bone marrow. (B cells are so named because they were originally characterized in a region of chickens called the bursa.) From the bone marrow, B cells are released to circulate in the blood and lymph. Individual B cells, like T cells, are specialized to recognize particular foreign antigens. When a B cell encounters the antigen to which it is targeted, it begins to divide rapidly, and its progeny differentiate into *plasma cells* and memory cells. Each plasma cell is a miniature factory producing markers called *antibodies*. These antibodies are large proteins that stick like flags to that antigen wherever it occurs in the body, marking any cell bearing the antigen for destruction. So, B cells don't kill foreign invaders directly, but rather they mark these cells so that they are more easily recognized by the other white blood cells that do the dirty work.

B cells and T cells also produce memory cells that provide the body with the ability to recall a previous exposure to an antigen and mount an attack against that antigen very quickly. As described later in this chapter, the initial specific immune response to an antigen encountered for the first time is delayed, which allows the pathogen time to infect the body. A second infection is halted much earlier due to the presence of memory cells, which respond to the pathogen more quickly.

Concept Check

1. Explain the difference between chemical and cellular responses.
2. Compare and contrast perforin and complement.
3. Which cells provide specific immunity in the body?

The Immune Response

28.4 Initiating the Immune Response

CONCEPT PREVIEW: When macrophages encounter cells without the proper MHC proteins, they secrete a chemical alarm that initiates the immune defense.

> Influenza is caused by a virus that invades the body. As described in the "Biology and Staying Healthy" reading on page 301, the flu virus can have a unique combination of antigens on its surface that is not immediately detected by the body and so the immune response is delayed.

To understand how this third line of defense works, imagine you have just come down with the flu. Influenza viruses enter your body in small water droplets inhaled into your respiratory system. If they are able to pass across the epithelium lining the respiratory system (first line of defense), and avoid consumption by macrophages (second line of defense), the viruses infect and kill mucous membrane cells.

At this point, T cells and macrophages initiate the immune response. When a foreign particle infects the body, it is taken into the cells by endocytosis and partially digested by lysosomes. Within the cells, the viral antigens are processed and moved to the surface of the plasma membrane, as shown in figure 28.7c. The cells that perform this function are called **antigen-presenting cells** and are usually macrophages.

> The foreign microbe is taken into the cell by endocytosis, described on page 80. Once inside, the microbe is destroyed by the actions of lysosome enzymes. A lysosome fuses with the vesicle containing the microbe, and enzymes attack the microbe, as discussed on page 73.

T cells inspect the surfaces of all cells they encounter. Every cell in the body carries special marker proteins on its surface, called major histocompatibility proteins, or **MHC proteins.** The MHC proteins are different for each individual, much as fingerprints are. The MHC protein on the cell in figure 28.7a is exactly the same on all cells in that person's body. As a result, the MHC proteins on the tissue cells serve as "self" markers that enable the individual's immune system to distinguish its cells from foreign cells. For example, the foreign microbe in figure 28.7b has different surface proteins that are recognized as antigens.

At the membrane, the processed antigens of an infecting microbe are complexed with the MHC proteins. This process is critical for the function of T cells because T cell receptors can be called into action only when antigens are presented in this way (figure 28.7d). B cells can interact with free antigens directly.

Macrophages that encounter pathogens—either a foreign cell such as a bacterial cell, which lacks proper MHC proteins, or a virus-infected body cell with telltale viral proteins stuck to its surface—respond by secreting a chemical alarm signal. The alarm signal is a protein called **interleukin-1** (Latin for "between white blood cells"). This protein stimulates **helper T cells.** The helper T cells respond

> Recall from the discussions of bacteria and viruses on pages 297 to 300, that some are pathogens, agents that cause disease. Not all bacteria are pathogens, but viruses, by their very nature, are pathogens because they must infect cells in order to replicate themselves.

to the interleukin-1 alarm by simultaneously initiating two different parallel lines of immune system defense: the cellular immune response carried out by T cells and the antibody or humoral response carried out by B cells. The immune response carried out by T cells is called the *cellular response* because the T lymphocytes attack the cells that carry antigens. The B cell response is called the *humoral response* because antibodies are secreted into the blood and body fluids (*humor* refers to a body fluid).

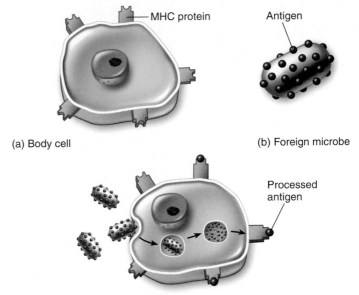

(a) Body cell

(b) Foreign microbe

(c) Antigen-presenting cell

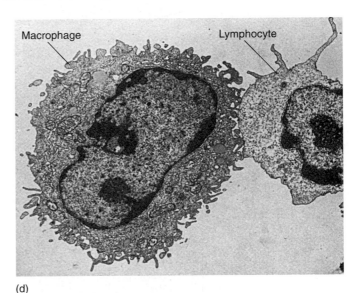

(d)

Figure 28.7   How antigens are presented.

(a) Cells of the body have MHC proteins on their surfaces that identify them as "self" cells. Immune system cells do not attack these cells. (b) Foreign cells or microbes have antigens on their surfaces. (c) T cells can bind to the antigens to initiate an attack only after the antigens are processed and complexed with MHC proteins on the surface of an antigen-presenting cell. B cells recognize the antigens directly, not requiring an antigen-presenting cell. (d) In this electron micrograph, a lymphocyte (*right*) contacts a macrophage (*left*), an antigen-presenting cell.

28.5 T Cells: The Cellular Response

CONCEPT PREVIEW: The cellular immune response is carried out by T cells, which mount an immediate attack on infected cells, killing any that present foreign surface antigens.

When macrophages process foreign antigens, they trigger the **cellular immune response,** illustrated in figure 28.8. Macrophages secrete interleukin-1 ❶, which stimulates cell division and proliferation of T cells. Helper T cells become activated when they bind to the complex of MHC proteins and antigens presented to them by the macrophages. The helper T cells then secrete interleukin-2 ❷, which stimulates the proliferation of cytotoxic T cells ❸, which recognize and destroy infected body cells displaying the foreign antigen together with their MHC proteins ❹.

The body makes millions of different versions of T cells. Each version bears a single, unique kind of receptor protein on its membrane, a receptor able to bind to a particular antigen-MHC protein complex on the surface of an antigen-presenting cell. Any cytotoxic T cell whose receptor fits the particular antigen-MHC protein complex present in the body begins to multiply rapidly. Large numbers of infected cells can be quickly eliminated, because the single T cell able to recognize the invading virus is amplified in number to form a large clone of identical T cells, all able to carry out the attack. Any of the body's cells that bear traces of viral infection are destroyed. The method used by cytotoxic T cells to kill infected body cells is similar to that used by natural killer cells and complement—they puncture the plasma membrane of the infected cell. Following an infection, some activated T cells give rise to memory T cells ❺ that remain in the body, ready to mount an attack quickly if the antigen is encountered again.

> The binding of T cells with the MHC-antigen complex on macrophages or antigen-presenting cells is very specific. You encountered this same type of specific binding between enzymes and their substrates on page 91.

Figure 28.8   The T cell immune defense.

After a macrophage has processed an antigen, it releases interleukin-1, signaling helper T cells to bind to the antigen-MHC protein complex. This triggers the helper T cell to release interleukin-2, which stimulates the multiplication of cytotoxic T cells. In addition, proliferation of cytotoxic T cells is stimulated when a T cell with a receptor that fits the antigen displayed by an antigen-presenting cell binds to the antigen-MHC protein complex. Body cells that have been infected by the antigen are destroyed by the cytotoxic T cells.

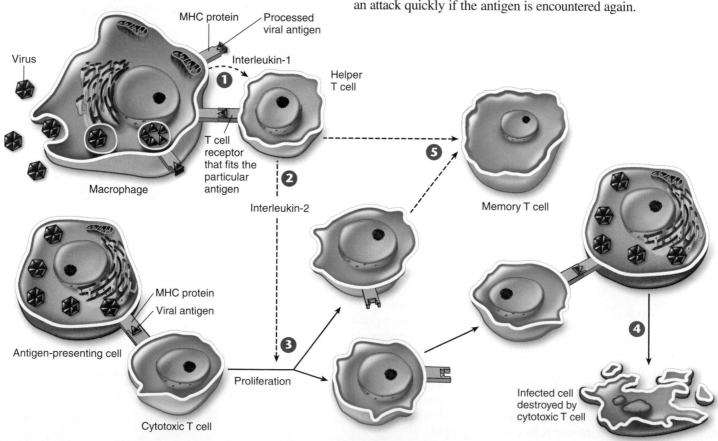

28.6 B Cells: The Humoral Response

CONCEPT PREVIEW: In the humoral immune response, B cells label antigens on pathogens or infected cells with antibodies to tag them for destruction by complement proteins, natural killer cells, and macrophages.

B cells also respond to helper T cells activated by interleukin-1. Like cytotoxic T cells, B cells have receptor proteins on their surfaces, a different receptor for each version of B cell. B cells recognize invading microbes much as cytotoxic T cells recognize infected cells, but unlike cytotoxic T cells, they do not go on the attack themselves. Rather, they mark the pathogen for destruction by mechanisms that have no "ID check" system of their own. Early in the immune response known as the **humoral immune response,** the markers placed by B cells alert complement proteins to attack the cells carrying them. Later in the response, the markers placed by B cells activate macrophages and natural killer cells.

The way B cells do their marking is simple and foolproof. Unlike the receptors on T cells, which bind only to antigen-MHC protein complexes on antigen-presenting cells, B cell receptors can bind to free, unprocessed antigens, shown in step ❶ of figure 28.9. When a B cell encounters an antigen, antigen particles enter the B cell by endocytosis and get processed and placed on the surface complexed with MHC proteins. Helper T cells recognize the specific antigen, bind to the antigen-MHC protein complex on the B cell ❷, and release interleukin-2, which stimulates the B cell to divide. In addition, free, unprocessed antigens stick to antibodies (the green Y-shaped structures) on the B cell surface. This antigen exposure triggers even more B cell proliferation. B cells divide to produce plasma

BIOLOGY & YOU

Passive Immunity. Your humoral immune response protects you all your life—except at its beginning. As a newborn infant, you had no prior exposure to microbes and so were particularly vulnerable to infection. Lucky for you, several layers of passive protection were provided by your mother. During pregnancy, a particular type of antibody, called IgG, was transported from your mother to you across the placenta, so that you had high levels of protective antibodies even at birth, offering you the same protection that your mom's humoral immune response offered her. After birth, this protection is continued with breast milk, which also contains antibodies that protected you as a newborn until you could synthesize your own antibodies. During the first several months of your dangerous entrance to this world, you did not actually make any protective antibodies—you only borrowed them.

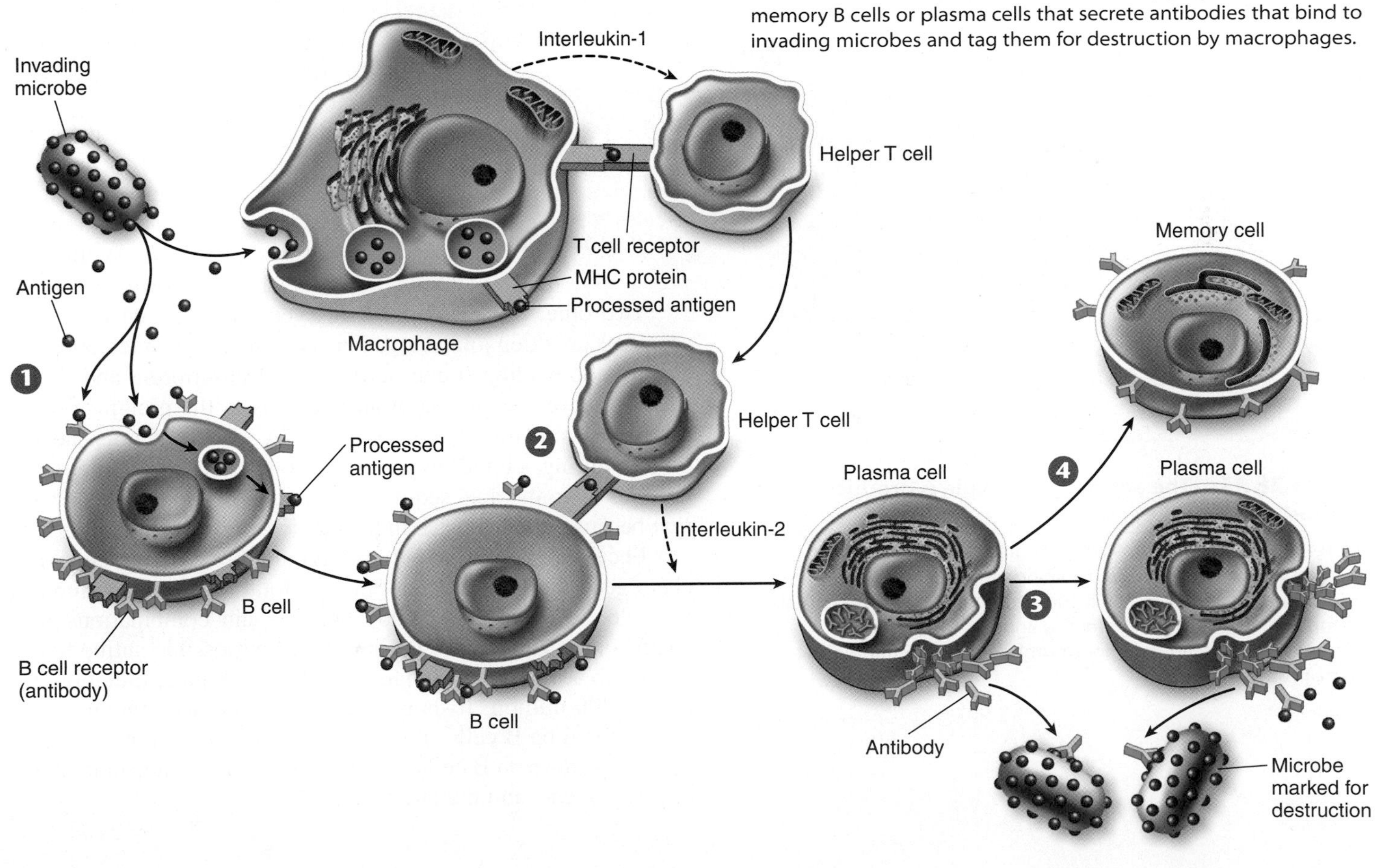

Figure 28.9 The B cell immune defense.

Invading particles are bound by B cells, which interact with helper T cells and are activated to divide. The multiplying B cells produce memory B cells or plasma cells that secrete antibodies that bind to invading microbes and tag them for destruction by macrophages.

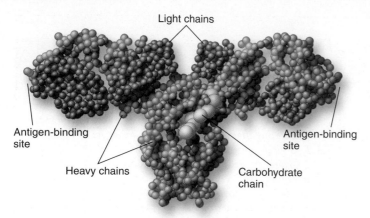

Figure 28.10   An antibody molecule.

In this molecular model of an antibody molecule, each amino acid is represented by a small sphere. Each molecule consists of four protein chains, two "light" (*red*) and two "heavy" (*blue*). The four protein chains wind around one another to form a Y shape. Foreign molecules, called antigens, bind to the arms of the Y.

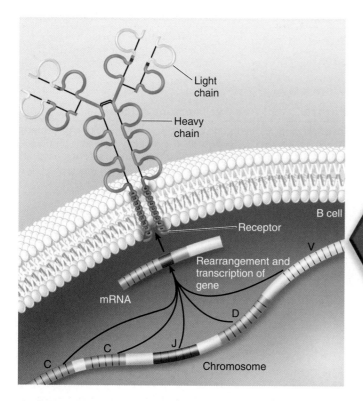

Figure 28.11   The antibody molecule is produced by a composite gene.

Different regions of the DNA code for different regions of the antibody (C, constant regions; J, joining regions; D, diversity regions; and V, variable regions) and are brought together to make a composite gene that codes for the antibody. By combining different segments, an enormous number of different antibodies can be produced.

cells that serve as short-lived antibody factories ❸ and long-lived memory B cells ❹ that remain in the body after the initial infection and are available to mount a quick attack if the antigen enters the body again.

The **plasma cells** that are derived from B cells produce lots of the same specific antibody that was able to bind the antigen in the initial immune response. Flooding through the bloodstream, these antibody proteins (figure 28.10) are able to stick to antigens on any cells or microbes that present them, flagging those cells and microbes for destruction. Complement proteins, macrophages, or natural killer cells then destroy the antibody-tagged cells and microbes.

The B cell defense is very powerful because it amplifies the reaction to an initial pathogen encounter a millionfold. It is also a very long-lived defense in that a few of the multiplying B cells do not become antibody producers. Instead they become a line of **memory B cells** that continue to patrol your body's tissues, circulating through your blood and lymph for a long time—sometimes for the rest of your life.

Antibody Diversity

The vertebrate immune system is capable of recognizing as foreign practically any nonself molecule presented to it—literally millions of different antigens. Although vertebrate chromosomes contain only a few hundred receptor-encoding genes, it is estimated that human B cells can make between 10^6 and 10^9 different antibody molecules. How do vertebrates generate millions of different antigen receptors when their chromosomes contain only a few hundred versions of the genes encoding those receptors?

The answer to this question is that the millions of immune receptor genes do not exist as single sequences of nucleotides. Rather, they are assembled by stitching together three or four DNA segments that code for different parts of the receptor molecule. When an antibody is assembled, different sequences of DNA are transcribed and brought together to form a composite gene. The antibody molecule in figure 28.11 was produced by a composite gene, different sections of DNA were used to produce the constant regions (green), the joining regions (red), the diversity regions (blue), and the variable regions (yellow). This process is called **somatic rearrangement.**

> Somatic rearrangement is similar to the process of alternative splicing discussed on page 204 where exons are spliced together in different combinations resulting in different protein products.

Two other processes generate even more sequences. First, the DNA segments are often joined together with one or two nucleotides off-register, shifting the reading frame during gene transcription and so generating a totally different sequence of amino acids in the protein. Second, random mistakes occur during successive DNA replications as the lymphocytes divide during clonal expansion. Both mutational processes produce changes in amino acid sequences, a phenomenon known as *somatic mutation* because it takes place in a somatic cell rather than in a gamete.

Because a cell may end up with any heavy chain gene and any light chain gene during its maturation, the total number of different antibodies possible is staggering: 16,000 heavy chain combinations × 1,200 light chain combinations = 19 million different possible antibodies. If one also takes into account the changes induced by somatic mutation, the total can exceed 200 million! It should be understood that although this discussion has centered on B cells and their receptors, the receptors on T cells are as diverse as those on B cells because they also are subject to similar somatic rearrangements and mutations.

28.7 Active Immunity Through Clonal Selection

CONCEPT PREVIEW: The primary immune response prepares the body for a strong secondary immune response because infecting microbes stimulate B cells and T cells to divide repeatedly, forming clones of responding cells. Memory cells are maintained in the body and are available to elicit a faster, stronger response with a second and subsequent infections.

As we discussed earlier, B cells and T cells have receptors on their cell surfaces that recognize and bind to specific antigens. When a particular antigen enters the body, it must, by chance, encounter the specific lymphocyte with the appropriate receptor to provoke an immune response. The first time a pathogen invades the body, there are only a few B cells or T cells that may have the receptors that can recognize the invader's antigens. Binding of the antigen to its receptor on the lymphocyte surface, however, stimulates cell division and produces a *clone* (a population of genetically identical cells). This process is known as **clonal selection.** For example, in the first encounter with a chicken pox virus in figure 28.12 there are only a few cells that can mount an immune response, and the response is relatively weak. This is called a **primary immune response** and is indicated by the first curve, which shows the initial amount of antibody produced upon exposure to the virus.

> A clone is produced through mitotic cell division. As described on pages 134 to 135, daughter cells that are produced through mitosis are genetically identical to the parent cell. In this way, clonal selection results in a large group of genetically identical cells.

The primary immune response involves B cells, and so some of the cells become plasma cells that secrete antibodies (taking 10 to 14 days to clear the chicken pox virus from the system), and some become memory cells. Some of the T cells involved in the primary response also become memory cells. Because a clone of memory cells specific for that antigen develops after the primary response, the immune response to a second infection by the same pathogen is swifter and stronger, as shown by the second curve. Many memory cells can be produced following the primary response, providing a jump start for the production of antibodies should a second exposure occur. The next time the body is invaded by the same pathogen, the immune system is ready. As a result of the first infection, there is now a large clone of lymphocytes that can recognize that pathogen. This more effective response, elicited by subsequent exposures to an antigen, is called a **secondary immune response.** The "Inquiry & Analysis" feature at the end of this chapter further explores the nature of the secondary immune response.

Memory cells can survive for several decades, which is why people rarely contract chicken pox a second time after they have had it once. Memory cells are also the reason that vaccinations are effective. The viruses causing childhood diseases have surface antigens that change little from year to year, so the same antibody is effective for decades. Other diseases, such as influenza, are caused by viruses whose genes that encode surface proteins mutate rapidly. This rapid genetic change causes new strains to appear every year or so that are not recognized by memory cells from previous infections.

Although the cellular and humoral immune responses were discussed separately, they occur simultaneously in the body. Figure 28.13 on the next page follows the steps of a viral infection and shows how the cellular and humoral lines of defense work together to produce the body's specific immune response.

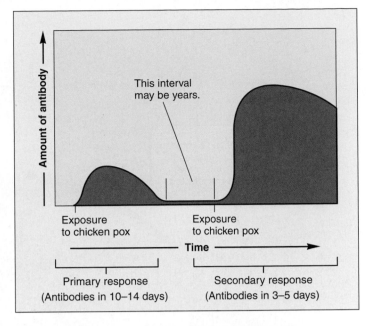

Figure 28.12 **The development of active immunity.**

Immunity to chicken pox occurs because the first exposure stimulated the development of lymphocyte clones with receptors for the chicken pox virus. As a result of clonal selection, a second exposure stimulates the immune system to produce large amounts of the antibody more rapidly than before, keeping the person from getting sick again.

IN THE NEWS

But Is It Safe? New Jersey became the first state in the nation to require flu shots for young schoolchildren. As of January 2009, all children between six months and five years who are attending licensed day care and preschool programs must be vaccinated for influenza. This protects the entire population, health officials say, because children can inadvertently infect each other at school, then carry the infection home and elsewhere. Mandatory vaccinations are not new. New Jersey preschoolers are already required to get vaccinated against pneumonia. Later, as sixth graders, they get another battery of vaccinations, protecting them against meningitis, tetanus, diphtheria, and pertussis. A lot of New Jersey parents are not happy about the new law, fearing that the vaccination shots are not safe. Many of these parents say they believe vaccines cause autism, even though multiple studies have found no such link. Should children of parents who feel this way be given an exemption from the vaccination requirement? New Jersey health officials say "No," pointing out that if you don't vaccinate your child, you put other children at risk.

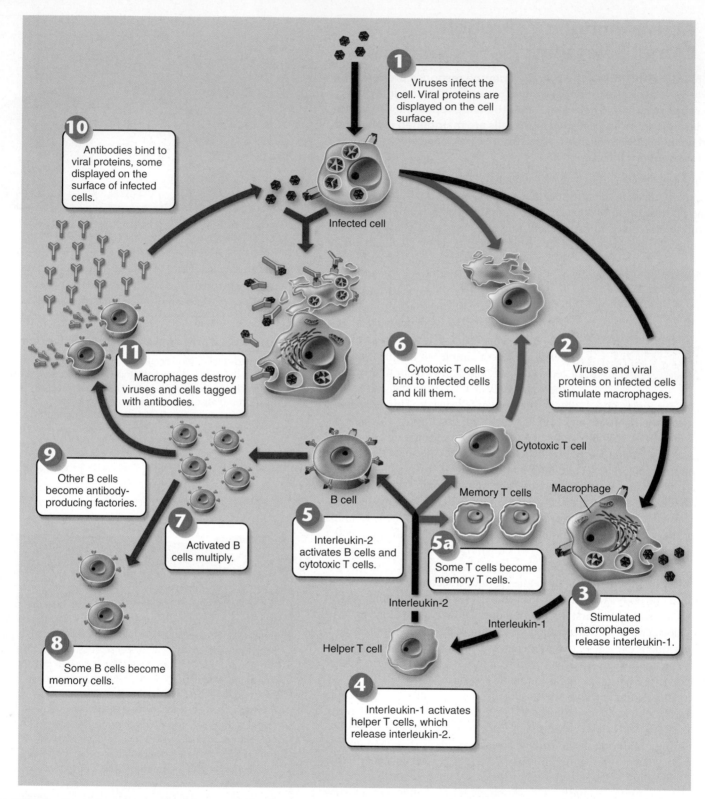

Figure 28.13 How the immune response works.

28.8 Vaccination

CONCEPT PREVIEW: Vaccines introduce antigens into the body that are similar or identical to those of pathogens. This elicits an immune response that defends against the pathogen should it enter the body later.

In 1796, an English country doctor named Edward Jenner carried out an experiment that marks the beginning of the study of immunology. Smallpox was a common and deadly disease in those days. Jenner observed, however, that milkmaids who had caught a much milder form of "the pox" called cowpox (presumably from cows) rarely caught smallpox. Jenner set out to test the idea that cowpox conferred protection against smallpox. He infected people with mild cowpox (figure 28.14), and as he predicted, many of them became immune to smallpox.

We now know that smallpox and cowpox are caused by two different but similar viruses. Jenner's patients who were injected with the cowpox virus mounted a defense that was also effective against a later infection of the smallpox virus. Jenner's procedure of injecting a harmless microbe to confer resistance to a dangerous one is called vaccination. **Vaccination** is the introduction into your body of a dead or disabled pathogen or, more commonly these days, of a harmless microbe with pathogen proteins displayed on its surface. The vaccination triggers an immune response against the pathogen, without an infection ever occurring. Afterward the vaccinated person has memory cells (B and T cells) directed against that specific pathogen. The vaccinated person is said to have been "immunized" against the disease.

Through genetic engineering, scientists are now routinely able to produce "piggyback," or subunit, vaccines. These vaccines are made of harmless viruses that contain in their DNA a single gene cut out of a pathogen, a gene encoding a protein normally exposed on the pathogen's surface. By splicing the pathogen gene into the DNA of the harmless host, that host is induced to display the protein on its surface. The harmless virus displaying the pathogen protein is like a sheep in wolf's clothing, unable to hurt you but raising alarm as if it could. Your body responds to its presence by making an antibody directed against the pathogen protein that acts like an alarm to the immune system, should that pathogen ever enter your body.

If the activities of memory cells provide such an effective defense against future infection, why can you catch some diseases like flu more than once? The reason you don't stay immune to flu is that the flu virus has evolved a way to evade the immune system—it changes. The genes encoding the surface proteins of the flu virus mutate very rapidly. Thus, the shapes of these surface proteins alter swiftly. Your memory cells do not recognize viruses with altered surface proteins as being the same viruses they have already successfully defeated or been vaccinated against, because the memory cells' receptors no longer "fit" the new shape of the flu surface proteins. When the new version of flu virus invades your body, you need to mount an entirely new immune defense.

Some new versions of the flu can be quite deadly. When mutations arose in a bird flu in 1918 that allowed this flu virus to pass easily from one infected human to another, over 20 million Americans and Europeans died in 18 months (figure 28.15). Less profound changes in flu virus surface proteins occur periodically, resulting in new strains of flu for which we are not immune. The annual flu shots are vaccines against these new strains. Researchers are able to predict the current year's strain of the flu by examining pre-season flu reports from across the globe and determining which strain seems to be dominant. Researchers then prepare a vaccine against this year's strain.

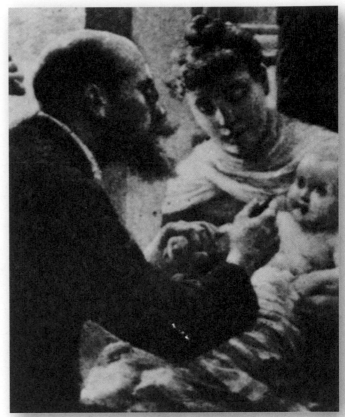

Figure 28.14 The birth of immunology.

This famous painting shows Edward Jenner inoculating patients with cowpox in the 1790s and thus protecting them from smallpox.

Figure 28.15 The flu epidemic of 1918 killed over 20 million in 18 months.

With 25 million Americans alone infected during the influenza epidemic, it was hard to provide care for everyone. The Red Cross often worked around the clock.

Recipient's blood

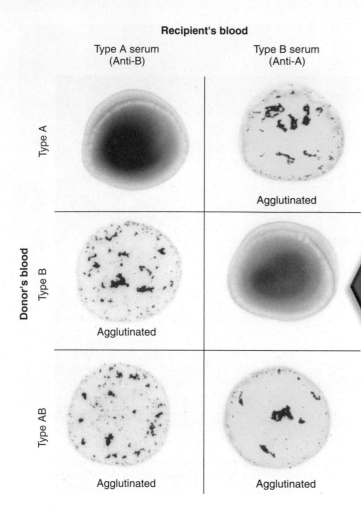

Figure 28.16   Blood typing.

Agglutination of the red blood cells is seen when blood types are mixed with sera containing antibodies against the A and B antigens. Note that no agglutination would be seen if type O blood (not shown) was used.

BIOLOGY & YOU

Erythroblastosis. Because fetal and maternal blood are normally kept separate across the placenta (see chapter 31), the Rh-negative mother is not usually exposed to the Rh antigen of an Rh-positive fetus during pregnancy. At the time of birth, however, a varying degree of exposure may occur, and the Rh-negative mother's immune system may become sensitized and produce antibodies against the Rh antigen. If so, these antibodies can cross the placenta in subsequent pregnancies and cause hemolysis of the Rh-positive red blood cells of the fetus, a condition called erythroblastosis fetalis, or hemolytic disease of the newborn. Erythroblastosis fetalis can be prevented by injecting the Rh-negative mother with an antibody preparation against the Rh factor soon after the birth of each Rh-positive baby. The injected antibodies inactivate the Rh antigens and thus prevent the mother from becoming actively immunized to them.

28.9 Antibodies in Medical Diagnosis

CONCEPT PREVIEW: Blood cells carry different surface proteins. Agglutination occurs because different antibodies exist for the ABO and Rh factor antigens on the surface of red blood cells. For this reason, donors and recipients must be of the same blood type. Antibodies are also used in other lab tests.

Blood Typing

A person's blood type indicates the class of antigens found on the red blood cell surface. Red blood cell antigens are clinically important because their types must be matched between donors and recipients for blood transfusions. There are several groups of red blood cell antigens, but the major group is known as the **ABO system.** In terms of the antigens present on the red blood cell surface, a person may be *type A* (with only A antigens), *type B* (with only B antigens), *type AB* (with both A and B antigens), or *type O* (with neither A nor B antigens).

> Human ABO blood types are genetically determined, as discussed on page 169. The inheritance of the ABO blood type is an example of codominance. A person who inherits both the A and B genes, expresses both antigens on their red blood cells.

The immune system is tolerant to its own red blood cell antigens but produces antibodies against nonself blood antigens. For example, a person who is type A does not produce anti-A antibodies, but does make antibodies against the B antigen. People who are type AB develop tolerance to both antigens and thus do not produce either anti-A or anti-B antibodies. Those who are type O make both anti-A and anti-B antibodies.

If type A blood is mixed on a glass slide with serum from a person with type B blood, the anti-A antibodies in the serum cause the type A blood cells to clump together, or agglutinate (this is shown in the upper right panel of figure 28.16). These tests allow the blood types to be matched prior to transfusions, so that agglutination will not occur in the blood vessels, where it could lead to inflammation and organ damage.

Rh Factor.   Another group of antigens found on most red blood cells is the *Rh factor*. People who have these antigens are said to be Rh-positive, whereas those who do not are Rh-negative. The Rh-negative allele is less common in the human population. The Rh factor is of particular significance when Rh-negative mothers give birth to Rh-positive babies (see the "Biology & You" App to the left).

Monoclonal Antibodies

Monoclonal antibodies are commercially prepared antibodies for use in clinical laboratory tests. Modern pregnancy tests, for example, use monoclonal antibodies produced against a pregnancy hormone (hCG—see chapter 31). hCG in a pregnant woman's urine binds to the monoclonal antibodies within the testing strip and indicates a positive result.

Concept Check

1. What is the role of interleukins in initiating the immune response?
2. Contrast T cells with B cells.
3. How is an invading cell killed after being tagged with an antibody?
4. Describe the process of somatic rearrangement.
5. Distinguish between primary and secondary immune responses.
6. Why do people need flu shots each year—doesn't the vaccine make them immune to flu?

Defeat of the Immune System

28.10 Overactive Immune System

CONCEPT PREVIEW: Autoimmune diseases are inappropriate responses to "self" cells. Allergies are inappropriate responses to harmless antigens.

Although the immune system is one of the most sophisticated systems of the vertebrate body, it is still not perfect. Several major diseases we face, and some minor irritations as well, reflect an overactive immune system.

Autoimmune Diseases

The ability of T cells and B cells to distinguish cells of your own body—"self" cells—from nonself cells is the key ability of the immune system that makes your body's third line of defense so effective. In certain diseases, this ability breaks down, and the body attacks its own tissues. Such diseases are called **autoimmune diseases.** For reasons we don't understand, autoimmune diseases are far more common among women than men.

Multiple sclerosis is an autoimmune disease that usually strikes people between the ages of 20 and 40. In multiple sclerosis, the immune system attacks and destroys a sheath of fatty material called myelin that insulates motor nerves (like the rubber covering electrical wires). Degeneration of the myelin sheath interferes with transmission of nerve impulses until eventually they cannot travel at all. Voluntary functions, such as movement of limbs, and involuntary functions, such as bladder control, are lost, leading finally to paralysis and death. Scientists do not know what stimulates the immune system to attack myelin.

Another autoimmune disease is type I diabetes, thought to result from an immune attack on the insulin-manufacturing cells of the pancreas. No one knows why the attack occurs. Other autoimmune diseases are rheumatoid arthritis (an immune system attack on the tissues of the joints), lupus (in which the connective tissue and kidneys are attacked), and Graves' disease (in which the thyroid is attacked).

Allergies

Although your immune system provides very effective protection against pathogens, sometimes it does its job too well, mounting an immune response that is greater than necessary to eliminate an antigen. The antigen in this case is called an **allergen,** and such an immune response is called an **allergy.** Hay fever, sensitivity to even tiny amounts of plant pollen, is a familiar example of an allergy. Many people are allergic to proteins released from the feces of a minute mite that lives on grains of house dust (figure 28.17).

What makes an allergic reaction uncomfortable, and sometimes dangerous, is the involvement of antibodies attached to a kind of white blood cell called a *mast cell.* It is the job of the mast cells in an immune response to initiate an inflammatory response. Figure 28.18 shows what happens when a mast cell encounters something that matches its antibody. Mast cells release histamines and other chemicals that cause capillaries to swell. Histamines also increase mucus production by cells of the mucous membranes, resulting in runny noses and nasal congestion. Most allergy medicines relieve these symptoms with antihistamines, chemicals that block the action of histamines.

Asthma is a form of allergic response in which histamines cause the narrowing of air passages in the lungs. People who have asthma have trouble breathing when exposed to substances to which they are allergic.

Figure 28.17   The house dust mite.

This tiny animal, *Dermatophagoides*, causes an allergic reaction in many people. The dust mite lives in mattresses and pillows and feeds on the large quantities of dead skin cells that we all shed.

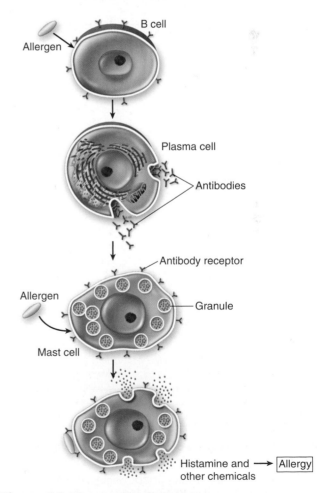

Figure 28.18   An allergic reaction.

In an allergic response, B cells secrete antibodies that attach to the plasma membranes of mast cells, which secrete histamine in response to antigen-antibody binding.

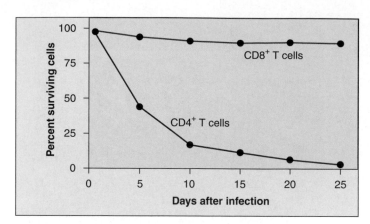

Figure 28.19   Survival of T cells in culture after exposure to HIV.

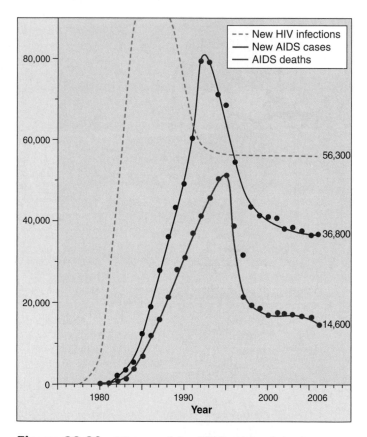

Figure 28.20   History of the AIDS epidemic in the United States.

The U.S. Centers for Disease Control (CDC) reports that 36,552 new AIDS cases were reported in 2005 and 36,800 new cases in 2006 in the United States, with a total of over one million cases and over 600,000 deaths since the beginning of the epidemic. Over one million other individuals are thought to be infected with the HIV virus in the United States and over 33 million worldwide.

28.11 AIDS: Immune System Collapse

CONCEPT PREVIEW: HIV cripples the immune defense by infecting and killing key lymphocytes, namely helper T cells.

AIDS (acquired immunodeficiency syndrome) was first recognized as a disease in 1981. By the end of 2006 in the United States, more than 545,000 individuals had died of AIDS, and more than one million others were thought to be infected with **HIV** (human immunodeficiency virus), the virus that causes the disease. Worldwide, 33 million are infected, and 25 million have died. HIV apparently evolved from a very similar virus that infects chimpanzees in Africa when a mutation arose that allowed the virus to recognize a human cell surface receptor called CD4. This receptor is present in the human body on certain immune system cells, notably macrophages and helper T cells. It is the identity of these immune system cells that leads to the devastating nature of the disease.

How HIV Attacks the Immune System

HIV attacks and cripples the immune system by targeting cells that have CD4 receptors (CD4+ cells), including helper T cells. The significance of killing helper T cells is that it leaves the immune system unable to mount a response to *any* foreign antigen. AIDS is a deadly disease for just this reason.

HIV's attack on CD4+ T cells progressively cripples the immune system, because HIV-infected cells die after releasing replicated viruses that proceed to infect other CD4+ T cells. Over time, the body's entire population of CD4+ T cells is destroyed (demonstrated by the experiment in figure 28.19). In a normal individual, CD4+ T cells make up 60% to 80% of circulating T cells; in AIDS patients, CD4+ T cells often become too rare to detect, wiping out the human immune defense. With no defense against infection, any of a variety of commonplace infections proves fatal. Also, with no ability to recognize and destroy cancer cells when they arise, death due to cancer becomes far more likely. More AIDS victims die of cancer than from any other cause.

The fatality rate of AIDS is 100%; no patient exhibiting the symptoms of AIDS has recovered. The disease is *not* highly contagious, because it is only transmitted from one individual to another through the transfer of internal body fluids, typically in semen or vaginal fluid during sexual intercourse and in blood transmitted by needles during drug use. However, symptoms of AIDS do not usually show up for several years after infection with HIV, apparent in the delay of the onset of AIDS in the United States (the red line in figure 28.20) after initial infection with HIV (the green line). Because of this symptomless delay, infected individuals often unknowingly spread the virus to others. Awareness campaigns have helped reduce the numbers of new cases.

A variety of drugs inhibit HIV in the test tube. These include AZT and its analogs (which inhibit virus nucleic acid replication) and protease inhibitors (which inhibit the production of functional viral proteins). A combination of a protease inhibitor and two AZT analog drugs entirely eliminates the HIV virus from many patients' bloodstreams. Widespread use of this *combination therapy* has cut the U.S. AIDS death rate by almost two-thirds since its introduction in the mid-1990s, from 51,414 AIDS deaths in 1995 to 38,074 in 1996, and 10 years later in 2006, deaths remained low, at approximately 14,600.

Concept Check

1. Distinguish between allergies and autoimmune diseases.
2. How does HIV target T cells?
3. How is HIV spread from one person to another?

Biology and Staying Healthy

AIDS Drugs Target Different Phases of the HIV Infection Cycle

AIDS is so devastating because it dismantles the very process that the body uses to fight infections. While the search for an AIDS vaccine continues, scientists are also trying to find ways to slow down or halt the process whereby the virus spreads throughout the body, infecting other cells. This doesn't cure the person, they still have an HIV infection that can be spread to others, but by slowing down HIV's ability to replicate inside a person, these drugs might help curb the effects of HIV that lead to the development of AIDS.

Many new drugs are being developed that target different phases of the viral infection cycle. To date there are six classes of drugs, described below and keyed to the diagram, that disrupt the ability of HIV to enter the cell, to replicate its DNA, and to form new viral particles. As of yet, there has been little drug development targeting the final phase of the infection cycle, viral exit (**7** where new viruses leave the cell).

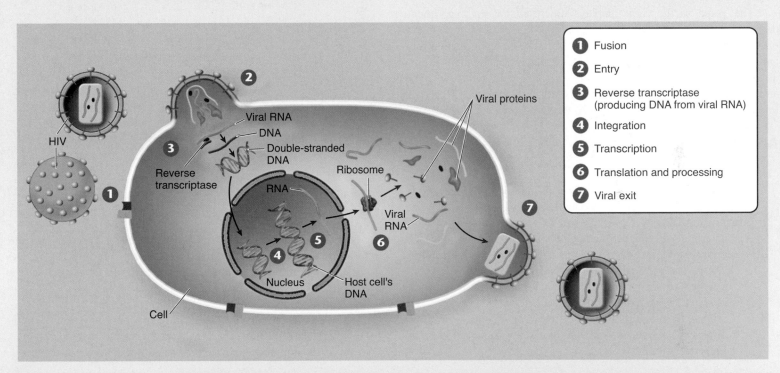

Class	Example	Number of Approved Drugs	Mechanisms of Action	Coming Soon...
❶ Fusion inhibitors	Fuzeon	1	Peptides that bind to a certain receptor protein on the outside of the cell and so prevent fusion of HIV particles to the cell surface	Other experimental drugs that target other receptor proteins
❷ Uptake inhibitors	Maraviroc	1	Bind to a necessary coreceptor protein on the cell surface and so prevent cell uptake of HIV particles	Other experimental drugs that target another coreceptor or the CD4 receptor
❸ Reverse transcriptase inhibitors	Atripla	4	Bind to HIV's "reverse transcriptase" enzyme's active site, gumming it up so that the enzyme cannot function, and so blocking HIV replication.	—
❹ Integrase inhibitors	Raltegravir	2	Bind to HIV "integrase" enzyme that inserts HIV genes into the cell's DNA, and so block HIV replication	A second drug is in the testing phase
❺ Nucleoside analogs	ACT	10	Faulty nucleotides cannot be assembled into functional DNA, and so block HIV replication	Nucleotide analogs that do not require phosphorylation within the cell as nucleosides do
❻ Protease inhibitors	Prezista	10	Bind to HIV "protease" enzyme that cuts primary transcript into functional segments, and so block HIV replication	—

Is Immunity Antigen-Specific?

The immune response provides a valuable protection against infection because it can remember prior experiences. We develop lifelong immunity to many infectious diseases after one childhood exposure. This long-term immunity is why vaccines work. A key question about immune protection is whether or not it is specific. Does exposure to one pathogen confer immunity to only that one, or is the immunity you acquire a more general response, protecting you from a range of infections?

The graph to the right displays the results of an experiment designed to answer this question. A colony of rabbits is immunized once with antigen A, and the level of antibody directed against this antigen monitored in each individual. After 40 days, each rabbit is reinjected, some with antigen A and others with antigen B, and the level of antibody directed against the reinjected antigens is monitored. The red line is typical of results for antigen A, the blue line for antigen B.

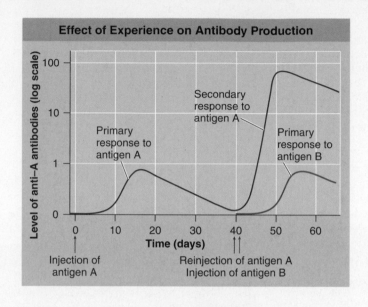

Analysis

1. **Applying Concepts**
   Reading a Continuous Curve. Does each injection of antigen A result in detectable antibody production? antigen B?
2. **Interpreting Data**
   a. The initial response to antigen A is called the "primary" response, and the second response to antigen A administered 40 days later is called the "secondary" response. Compare the speed of the primary and secondary responses—which reaches maximal antibody response quicker?
   b. Compare the magnitude of the primary and secondary immune responses to antigen A—are they similar, or is one response of greater magnitude?

3. **Making Inferences**
   a. Why is the secondary response induced by a second exposure to antigen A different from the primary response?
   b. Is the response to antigen B more similar to the primary or secondary response of antigen A?
4. **Drawing Conclusions**
   a. Does the prior exposure to antigen A have any impact on the speed or magnitude of the primary response to antigen B?
   b. Is the immune response to these antigens antigen-specific?

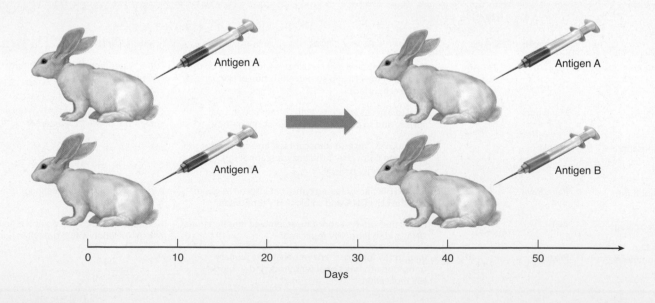

Concept Summary

Three Lines of Defense

28.1 Skin: The First Line of Defense

- The body has three lines of defense against infection, the first being skin (**figure 28.2**). Skin forms a barrier to pathogens.

28.2 Cellular Counterattack: The Second Line of Defense

- The second line of defense is a nonspecific cellular attack. The cells and chemicals used in this line of defense attack all foreign agents they encounter.

- Macrophages and neutrophils attack the invading pathogen by engulfing them before they can infect cells of the body (**figure 28.4**). Natural killer cells attack infected cells, killing them before the pathogen can spread to other cells. They kill the cells by inserting perforin proteins into the plasma membrane, causing the cell to burst (**figure 28.5**).

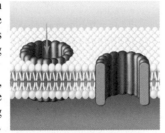

- Free-floating proteins in the blood, called complement, insert into the membranes of foreign cells, killing them, as shown here from **figure 28.6.**

- The temperature and inflammatory responses are also part of the second line of defense. Increasing body temperature slows bacterial growth and inflammation brings immune cells to the site of infection.

28.3 Specific Immunity: The Third Line of Defense

- The third line of defense is a specific immune response and involves cells other than macrophages, neutrophils, and natural killer cells (**table 28.1**). Lymphocytes called T cells and B cells are "programmed" by exposure to specific antigens and once programmed, will seek out the antigens or cells carrying the antigens and destroy them.

The Immune Response

28.4 Initiating the Immune Response

- Invading microorganisms display proteins on their surfaces that are different from the person's MHC proteins. Macrophages survey cells for "self" MHC proteins. A cell or virus that exhibits "nonself" proteins is engulfed by a macrophage and is partially digested. Antigens from the microbe are inserted into the surface of the macrophage and it is then called an antigen-presenting cell (**figure 28.7**). This antigen-presenting cell presents these antigens to T cells, activating the T cell response. This cell also secretes interleukin-1. Interleukin-1 stimulates helper T cells that trigger immune responses.

28.5 T Cells: The Cellular Response

- The cellular response involves T cells. Antigen-presenting cells activate helper T cells that release interleukin-2. Interleukin-2 stimulates the cloning of cytotoxic T cells that recognize and kill infected cells that display the specific antigen. Following the infection, memory T cells form and remain in the body to fight subsequent infections (**figure 28.8**).

28.6 B Cells: The Humoral Response

- The humoral response involves B cells. Helper T cells release interleukin-2 which activates the B cells. Activated B cells multiply and "mark" the foreign invaders with specific antibody proteins. The "marked" cells are then attacked by the nonspecific immune response. Activated B cells divide to form plasma cells and memory cells. Plasma cells produce and release antibodies and memory cells circulate in the blood. Memory cells become plasma cells in the event of reinfection by the same antigen (**figure 28.9**).

- The vertebrate immune system can produce some 10^9 different antibodies from only a few hundred versions of the genes encoding the antibody receptor proteins. This is accomplished by somatic rearrangement of the genes encoding different parts of the antibody molecule. The various antibodies are produced from composite genes (**figure 28.11**).

28.7 Active Immunity Through Clonal Selection

- The initial immune response triggered by infection is called the primary response. B and T cells are stimulated to begin dividing, producing a clone of cells. This is called clonal selection. The primary immune response is a delayed and somewhat weak response. But through clonal selection, a large population of memory cells is present in the body, such that a second infection by the same antigen will trigger a quicker and larger response, called the secondary response (**figure 28.12**).

- Memory cells can survive for several decades, conferring disease immunity to a person carrying the memory cells for that particular pathogen.

- Cellular and humoral immune responses occur simultaneously in the body. The activation of helper T cells triggers both immune responses, which also include nonspecific cellular attacks (**figure 28.13**).

28.8 Vaccination

- Vaccination takes advantage of the mechanism of the primary and secondary immune responses. Vaccines introduce pathogenic antigens into the body in a way that doesn't cause disease. This triggers the primary immune response. Later with an actual infection, the body elicits a swift and large secondary immune response.

28.9 Antibodies in Medical Diagnosis

- Antibodies are keen detectors of antigens, and so are used in various medical diagnostic applications such as blood typing (**figure 28.16**).

Defeat of the Immune System

28.10 Overactive Immune System

- Sometimes the immune system attacks antigens that are not foreign or pathogenic. In autoimmune responses, the body attacks its own cells. In allergic reactions, the body mounts an attack against a harmless substance, called an allergen. The allergic response includes activating mast cells that release histamines (**figure 28.18**). Histamines trigger the inflammatory response that results in allergy symptoms.

28.11 AIDS: Immune System Collapse

- AIDS is a fatal disease caused by infection with the HIV virus. HIV attacks macrophages and helper T cells, eventually destroying T cells that protect the body from other infections (**figures 28.19** and **28.20**).

Self-Test

1. Skin is a physical barrier but also provides chemical protection generated from
   a. sweat and oil glands.
   b. mucus.
   c. the skin's basal layer.
   d. the stratum corneum of the skin.

2. The immune system can identify foreign cells in the bloodstream because these foreign cells
   a. are observed destroying other cells by the immune system.
   b. have cell surface proteins that are different from the body's own cell surface receptors.
   c. have CD4 receptors similar to T cells.
   d. All of the above.

3. The purpose of the inflammatory response is to
   a. increase the temperature of an infected area.
   b. reduce pain of an infected area.
   c. increase the number of immune system cells in an infected area.
   d. form a barrier around the infected area.

4. Increasing human body temperature—that is, causing a fever— assists the immune system because
   a. increased temperature speeds up the chemical reactions used by the immune system.
   b. pathogenic bacteria do not grow well at high temperatures.
   c. increased temperature causes body cell proteins to denature.
   d. All of these answers are correct.

5. Immune responses tailored to specific antigens involve
   a. T cells.
   b. macrophages.
   c. complement.
   d. neutrophils.

6. Antibody production takes place in
   a. T cells.
   b. natural killer cells.
   c. B cells.
   d. mast cells.

7. Cytotoxic T cells
   a. produce antibodies.
   b. destroy pathogens directly.
   c. destroy foreign antigens floating freely in the bloodstream.
   d. destroy cells infected by pathogens.

8. Immunity to future invasion of a specific pathogen is accomplished by production of
   a. plasma cells.
   b. memory T and B cells.
   c. helper T cells.
   d. monocytes.

9. When a body's immune system attacks the body's own cells, this is called
   a. an inflammatory response.
   b. a temperature response.
   c. an autoimmune response.
   d. an allergic response.

10. HIV-infected people with advanced AIDS usually die of an infectious disease or cancer. This is because HIV attacks
    a. helper T cells.
    b. neutrophils.
    c. memory T and B cells.
    d. mast cells.

Visual Understanding

1. **Figure 28.12** Certain diseases are considered to be primarily "childhood" diseases—measles, mumps, chicken pox, for instance—and once someone has had these illnesses as a child, they don't catch them again when, as parents, they take care of their own children who are sick. Use the graph to help explain the reason.

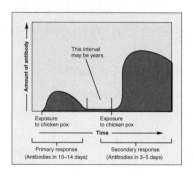

2. **Figure 28.20** What might account for the large increase in the number of AIDS cases, from the early1980s to 1992, then the decline, and now the essentially steady state, neither growing nor diminishing by very much, in the past decade?

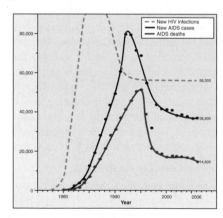

Challenge Questions

1. Your friend fell yesterday while she was skateboarding and has a deep cut on her forearm from a piece of wire. She shows it to you and complains about how sore it is. You can see the skin is red and swollen around the small puncture wound. Explain to her what is happening.

2. A doctor will give you an antibiotic to treat a bacterial infection, but you are given a vaccine to keep a virus (and some bacteria) from making you ill. What is the difference between these approaches?

3. There are two ways to avoid getting chicken pox when you are older, by catching it when you're young or by receiving a vaccination. How do each of these methods work and are there any similarities?

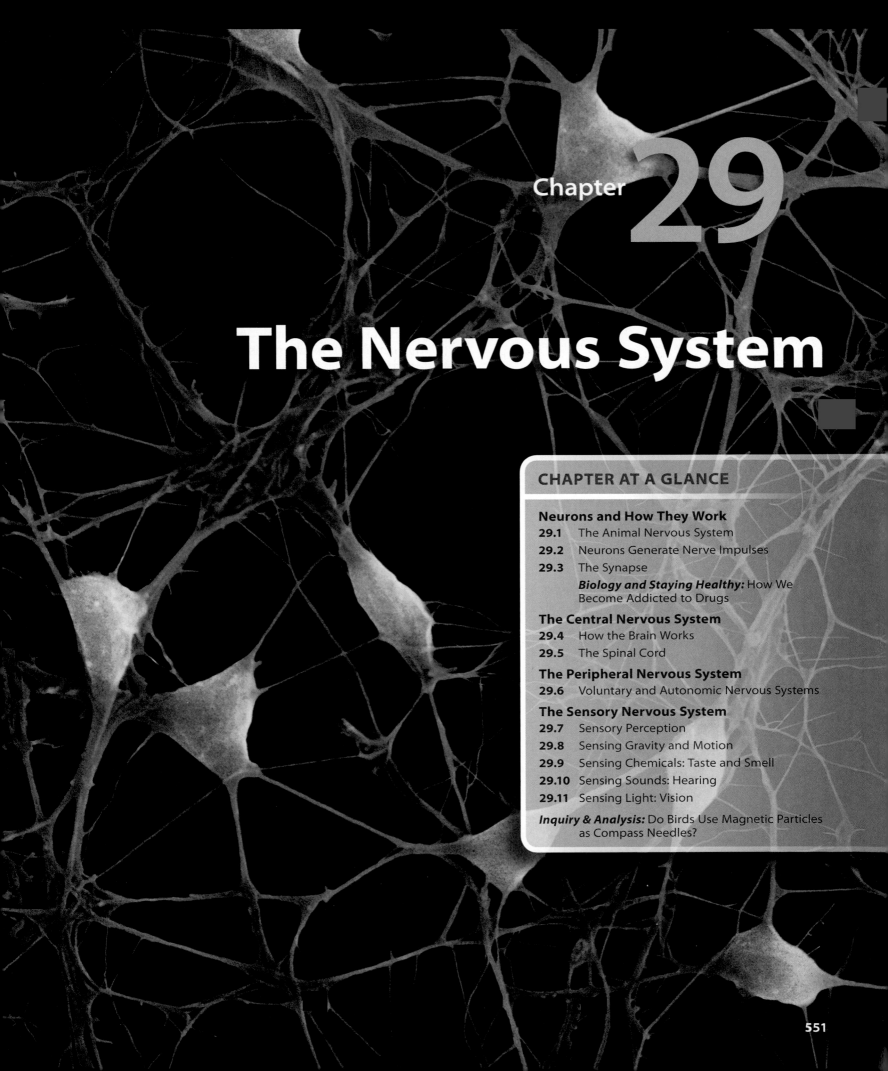

Chapter **29**

The Nervous System

Neurons and How They Work

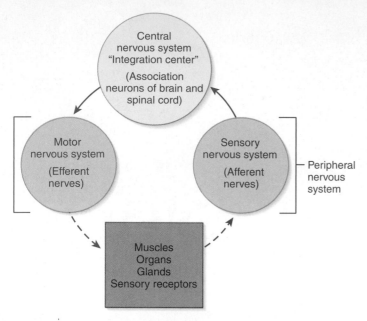

Figure 29.1 Organization of the vertebrate nervous system.

The central nervous system, consisting of the brain and spinal cord, issues commands via the motor nervous system and receives information from the sensory nervous system. The motor and sensory nervous systems together make up the peripheral nervous system.

29.1 The Animal Nervous System

CONCEPT PREVIEW: The nervous system of animals is organized into a central nervous system that includes a brain and spinal cord and contains association neurons, and the peripheral nervous system that contains sensory and motor neurons.

An animal must be able to respond to environmental stimuli. To do this, it must have sensory receptors that can detect the stimulus and motor effectors that can respond to it. In most invertebrate phyla and in all vertebrate classes, sensory receptors and motor effectors are linked by way of the **nervous system** (figure 29.1). As described in chapter 23, the nervous system consists of neurons and supporting cells. One type of neuron, called **association neurons** (or **interneurons**), is present in the nervous systems of most invertebrates and all vertebrates. These neurons are located in the brain and spinal cord of vertebrates, together called the **central nervous system (CNS),** represented by the yellow circle in figure 29.1. They help provide more complex reflexes and in the case of the brain, higher associative functions, including learning and memory, which require integration of many sensory inputs.

There are two other types of neurons. **Sensory** (or **afferent**) **neurons** (❶ in figure 29.2) carry impulses from sensory receptors to the CNS. **Motor** (or **efferent**) **neurons** ❸ carry impulses away from the CNS to effectors—muscles and glands. The association neurons ❷ link these two types of neurons together in the CNS. Together, motor and sensory neurons constitute the **peripheral nervous system (PNS)** of vertebrates (the bracketed tan circles in figure 29.1).

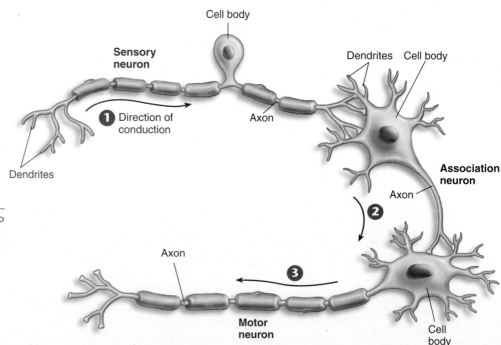

Figure 29.2 Three types of neurons.

Sensory neurons carry information about the environment to the brain and spinal cord. *Association neurons* are found in the brain and spinal cord and often provide links between sensory and motor neurons. *Motor neurons* carry impulses to muscles and glands (effectors).

29.2 Neurons Generate Nerve Impulses

CONCEPT PREVIEW: Neurons are cells specialized to conduct impulses. Nerve impulses result from ion movements across the neuron plasma membrane through special protein channels that open and close in response to chemical or electrical stimulation.

Neurons

The basic structural unit of the nervous system, whether central, motor, or sensory, is the nerve cell, or **neuron.** All neurons have the same basic structure as you can see by comparing the three cell types in figure 29.2 and the generalized cell in figure 29.3*a*. The *cell body* in figure 29.3*a* is the flat region of the neuron containing the nucleus. Short, slender branches called *dendrites* extend from one end of a neuron's cell body. Dendrites are input channels. Nerve impulses travel inward along them, toward the cell body. Motor and association neurons possess a profusion of highly branched dendrites, enabling those cells to receive information from many different sources simultaneously. Projecting out from the other end of the cell body is a single, long, tubelike extension called an *axon*. Axons are output channels. Nerve impulses travel outward along them, away from the cell body, toward muscles or glands, or to other neurons.

Most neurons are unable to survive alone for long; they require the nutritional support provided by supporting cells. More than half the volume of the human nervous system is composed of supporting cells and two of the most important kinds are the **Schwann cells** and **oligodendrocytes.** These cells envelop the axons of neurons with a sheath of fatty material called myelin, which acts as an electrical insulator. Schwann cells produce myelin in the PNS, while oligodendrocytes produce myelin in the CNS. During development, these cells associate with the axon, as shown at the top in figure 29.3*b*, and begin to wrap themselves around the axon several times to form a **myelin sheath,** an insulating covering consisting of multiple layers of membrane. Axons that have myelin sheaths are said to be myelinated, and those that don't are unmyelinated. The myelin sheath is interrupted at intervals, leaving uninsulated gaps called **nodes of Ranvier** (the nodes are where the yellow underlying axon can be seen). At the node regions, the axon is in direct contact with the surrounding fluid. The nerve impulse jumps from node to node, speeding its travel down the axon. Multiple sclerosis, discussed in chapter 28, is a debilitating clinical disorder that results from the degeneration of the myelin sheath.

The Nerve Impulse

The Na⁺/K⁺ pump pumps Na⁺ out of the cell against its concentration gradient and K⁺ into the cell, also against its concentration gradient. The Na⁺/K⁺ pump is a form of active transport and requires the expenditure of energy to function, as discussed on page 83.

When a neuron is "at rest," not carrying an impulse, active transport channels in the neuron's plasma membrane transport sodium ions (Na⁺) out of the cell and potassium ions (K⁺) in. Sodium ions cannot easily move back into the cell once they are pumped out, so the concentration of sodium ions builds up outside the

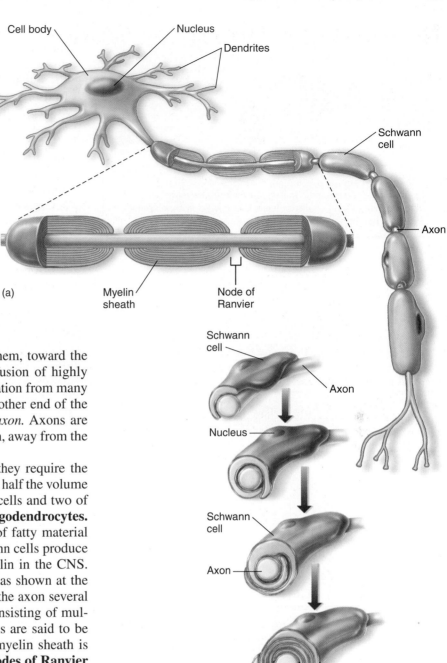

(a)

Figure 29.3 **Structure of a typical neuron and formation of the myelin sheath.**

(a) Extending from the cell body are many dendrites, which receive information and carry it to the cell body. A single axon transmits impulses away from the cell body. Many axons are encased by a myelin sheath, whose multiple membrane layers facilitate a more rapid conduction of impulses. The sheath is interrupted at regular intervals by small gaps called nodes of Ranvier. In the peripheral nervous system, myelin sheaths are formed by supporting Schwann cells. (b) The myelin sheath is formed by successive wrappings of Schwann cell membranes around the axon.

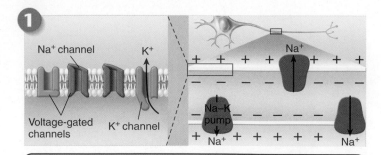

At the resting membrane potential, the inside of the axon is negatively charged because the sodium-potassium pump keeps a higher concentration of Na⁺ outside. Voltage-gated ion channels are closed, but there is some leakage of K⁺.

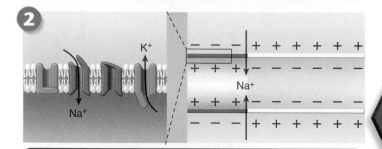

In response to a stimulus, the membrane depolarizes: voltage-gated Na⁺ channels open, Na⁺ flows into the cell, and the inside becomes more positive.

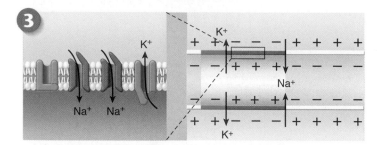

The local change in voltage opens adjacent voltage-gated Na⁺ channels, and an action potential is produced.

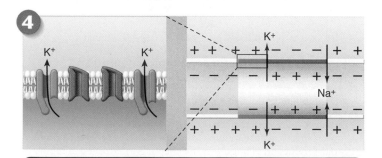

As the action potential travels farther down the axon, voltage-gated Na⁺ channels close and K⁺ channels open, allowing K⁺ to flow out of the cell and restoring the negative charge inside the cell. Ultimately, the sodium-potassium pump restores the resting membrane potential.

Figure 29.4 How an action potential works.

cell. Similarly, potassium ions accumulate inside the cell, although they are not as highly concentrated because many potassium ions are able to diffuse out through open channels. The result is to make the outside of the neuron more positive than the inside, a condition called the *resting membrane potential*, indicated by the yellow coloring in panel 1 of figure 29.4 The resting plasma membrane is said to be "polarized."

Neurons are constantly expending energy to pump sodium ions out of the cell, in order to maintain the resting membrane potential. The net negative charge of most proteins within the cell also adds to this charge difference. Using sophisticated instruments, scientists have been able to measure the voltage difference between the neuron interior and exterior as −70 millivolts (thousandth of a volt). The resting membrane potential is the starting point for a nerve impulse.

A nerve impulse travels along the axon and dendrites as an electrical current caused by ions moving in and out of the neuron through **voltage-gated channels** (that is, protein channels in the neuron membrane that open and close in response to an electrical voltage). The impulse starts when pressure or other sensory inputs disturb a neuron's plasma membrane, causing sodium channels on a dendrite to open (the purple channels in panel 2). As a result, sodium ions flood into the neuron from outside, down its concentration gradient, and for a brief moment a localized area inside of the membrane is "depolarized," becoming more positively charged in that immediate area of the axon (indicated by the pink coloring in panel 2).

> Although it takes a change in electrical voltage across the membrane to open the sodium and potassium channels, once they are open, Na⁺ travels into the cell and K⁺ travels out of the cell due to selective diffusion, discussed on page 81.

The sodium channels in the small patch of depolarized membrane remain open for only about a half a millisecond. However, if the change in voltage is large enough, called the **threshold potential,** it causes nearby voltage-gated sodium and potassium channels to open (panel 3). The sodium channels open first, which starts a wave of depolarization moving down the neuron. The opening of the gated channels causes nearby voltage-gated channels to open, like a chain of falling dominoes. This local reversal of voltage moving along the axon is called an **action potential.** An action potential follows an all-or-none law: A large enough depolarization produces either a full action potential or none at all, because the voltage-gated Na⁺ channels open completely or not at all. Once they open, an action potential occurs. After a slight delay, potassium voltage-gated channels open and K⁺ flows out of the cell down its concentration gradient, making the inside of the cell more negative. The increasingly negative membrane potential (colored green in panel 4) causes the voltage-gated sodium channels to snap closed again. This period of time after the action potential has passed and before the resting membrane potential is restored, is called the *refractory period*. A second action potential cannot fire during the refractory period, not until the resting potential is restored by the actions of the sodium-potassium pump.

The depolarization and restoration of the resting membrane potential takes only about 5 milliseconds. Fully 100 such cycles could occur, one after another, in the time it takes to say the word *nerve.*

29.3 The Synapse

CONCEPT PREVIEW: A synapse is the junction of an axon with another cell. The cells are separated by a gap across which neurotransmitters carry a signal that has either an excitatory or inhibitory effect, depending on which ion channels are opened.

A nerve impulse travels along a neuron until it reaches the end of the axon, usually positioned very close to another neuron, a muscle cell, or gland. Axons, however, do not actually make direct contact with other cells. Instead, a narrow gap, 10 to 20 nanometers across, called the *synaptic cleft,* separates the axon tip and the target cell. This junction of an axon with another cell is called a **synapse.** A synapse is shown in figure 29.5. The cell on the axon side of the synapse (on the left in the photo) is called the **presynaptic neuron;** the cell on the receiving side of the synapse (on the right) is called the **postsynaptic cell.**

Neurotransmitters

When a nerve impulse reaches the end of an axon, its message must cross the synapse if it is to continue. Messages do not "jump" across synapses. Instead, they are carried across by chemical messengers called **neurotransmitters.** These chemicals are packaged in tiny sacs, or vesicles, at the tip of the axon. When a nerve impulse arrives at the tip, it causes the sacs to release their contents into the synapse, as shown in figure 29.6a. The neurotransmitters diffuse across the synaptic cleft and bind to receptors (the purple structures) in the postsynaptic membrane. The signal passes to the postsynaptic cell when the binding of the neurotransmitter opens special ion channels, allowing ions to enter the postsynaptic cell and cause a change in electrical charge across its membrane. The enlarged view of figure 29.6a shows how the channel opens and the ion (the yellow ball) enters the cell. Because these channels open when stimulated by a chemical, they are said to be *chemically gated.*

Why go to all this trouble? Why not just wire the neurons directly together? For the same reason that the wires of your house are not all connected but instead are separated by a host of switches. When you turn on one light switch, you don't want every light in the house to go on, the toaster to start heating, and the television to come on! If every neuron in your body were connected to every other neuron, it would be impossible to move your hand without moving every other part of your body at the same time. Synapses are the control switches of the nervous system. However, the control switch must be turned off at some point by getting rid of the neurotransmitter, or the postsynaptic cell would stay stimulated and keep firing action potentials. In some cases, the neurotransmitter molecules diffuse away from the synapse. In other cases, the neurotransmitter molecules are either reabsorbed by the presynaptic cell, or are degraded in the synaptic cleft.

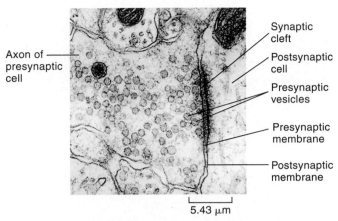

Figure 29.5   A synapse between two neurons.

This micrograph clearly shows the space between the presynaptic and postsynaptic membranes, which is called the synaptic cleft.

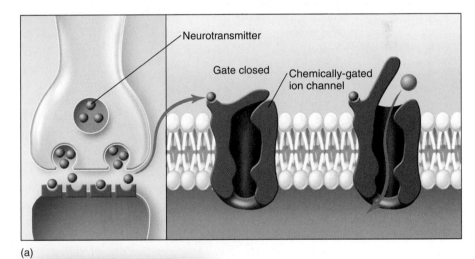

(a)

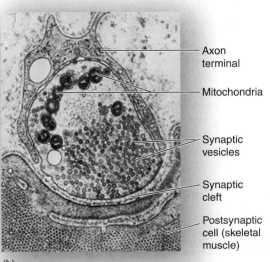

(b)

Figure 29.6   Events at the synapse.

(a) When a nerve impulse reaches the end of an axon, it releases a neurotransmitter into the synaptic cleft. The neurotransmitter molecules diffuse across the synapse and bind to receptors on the postsynaptic cell, opening ion channels. (b) A transmission electron micrograph of the tip of an axon filled with synaptic vesicles.

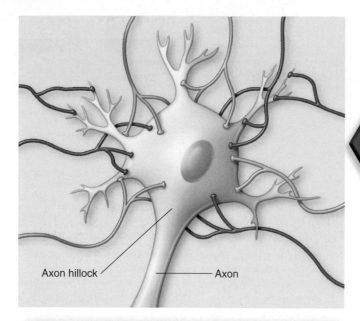

Axon hillock   Axon

> In this case, the sodium channel opens due to the binding of a neurotransmitter, but once open, Na⁺ travels into the cell by diffusion, down its concentration gradient as discussed on page 81.

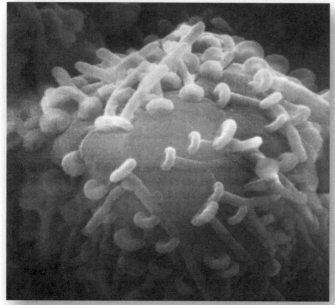

Figure 29.7   Integration.

Many different axons synapse with the cell body and dendrites of the postsynaptic neuron illustrated here. Excitatory synapses are shown in *red* and inhibitory synapses are shown in *blue*. The summed influence of their input at the base of the axon, called the axon hillock, determines whether or not a nerve impulse will be sent down the axon extending below. The scanning electron micrograph shows a neuronal cell body with numerous synapses.

Kinds of Synapses

The vertebrate nervous system uses dozens of different kinds of neurotransmitters, each recognized by specific receptors on receiving cells. They fall into two classes depending on whether they excite or inhibit the postsynaptic cell.

In an *excitatory synapse,* the receptor protein is usually a chemically gated sodium channel. On binding with a neurotransmitter whose shape fits it, the sodium channel opens, allowing sodium ions to flood inward. If enough sodium ion channels are opened by neurotransmitters, an action potential begins.

In an *inhibitory synapse,* the receptor protein is a chemically gated potassium or chloride channel. Binding with its neurotransmitter opens these channels, leading to the exit of positively charged potassium ions or the influx of negatively charged chloride ions, resulting in a more negative interior in the receiving cell. This inhibits the start of an action potential, because the negative voltage change inside means that even more sodium ion channels must be opened to get a domino effect started among voltage-gated sodium channels to start an action potential.

An individual nerve cell, like the neuron in figure 29.7, can possess both kinds of synaptic connections to other nerve cells. In the drawing, the excitatory synapses are colored red and the inhibitory synapses are colored blue. When signals from both excitatory and inhibitory synapses reach the cell body of a neuron, the excitatory effects (which cause less internal negative charge) and the inhibitory effects (which cause more internal negative charge) interact with one another. The result is a process of **integration** in which the various excitatory and inhibitory electrical effects tend to cancel or reinforce one another. If the result of the integration is a large enough depolarization (where the inside of the cell becomes more positive), an action potential will fire. Neurons often receive many inputs. A single motor neuron in the spinal cord may have as many as 50,000 synapses on it!

Neurotransmitters and Their Functions

Acetylcholine (ACh) is the neurotransmitter released at the neuromuscular junction, the synapse that forms between a neuron and a muscle fiber. ACh forms an excitatory synapse with skeletal muscle but has the opposite effect on cardiac muscle, causing an inhibitory synapse.

> The reason why ACh can have opposite effects depending on the type of muscle is because the protein channels found in the different muscles have different structures, allowing different ions through. The structure of a protein affects its function, as discussed on page 51.

Glycine and *GABA* are inhibitory neurotransmitters. This inhibitory effect is very important for neural control of body movements and other brain functions. Interestingly, the drug diazepam (Valium) causes its sedative and other effects by enhancing the binding of GABA to its receptors.

Biogenic amines are a group of neurotransmitters that include *dopamine* important in controlling body movements, *norepinephrine* and the hormone *epinephrine* both involved in the autonomic nervous system, and *serotonin* involved in sleep and emotional states.

Concept Check

1. Do sensory nerves carry impulses toward or away from the CNS?
2. What is a threshold potential?
3. Which type of synapse, excitatory or inhibitory, uses sodium channels?

Biology and Staying Healthy

How We Become Addicted to Drugs

The body sometimes deliberately prolongs the transmission of a signal across a synapse by slowing the destruction of neurotransmitters. It does this by releasing into the synapse special long-lasting chemicals called **neuromodulators** that inhibit the reabsorption of neurotransmitters or delay their breakdown after reabsorption, leaving them to be released back into the synapse when the next signal arrives.

Mood, pleasure, pain, and other mental states are determined by particular groups of neurons in the brain that use special sets of neurotransmitters and neuromodulators. Mood, for example, is strongly influenced by the neurotransmitter serotonin. Many researchers think that depression results from a shortage of serotonin. Prozac, the world's best-selling antidepressant, inhibits the reabsorption of serotonin, thus increasing the amount in the synapse.

When a nerve cell is exposed to a chemical signal for a prolonged period, it tends to lose its ability to respond to the stimulus with its original intensity. You are familiar with this loss of sensitivity—when you sit in a chair, how long are you aware of the chair? If receptor proteins within a synapse are exposed to high levels of neurotransmitter molecules for prolonged periods, the nerve cell often responds by inserting fewer receptor proteins into the membrane. This feedback is a normal part of the functioning of all neurons, a simple mechanism that has evolved to make cells more efficient by adjusting the number of "tools" (receptor proteins) in the membrane "workshop" to suit the workload. How does a drug like cocaine affect nerve cells in the brain's pleasure pathways to produce addiction? Cells of the limbic system (an area of the brain controlling strong emotions and drives) transmit pleasure messages using the neurotransmitter dopamine. Cocaine acts as a neuromodulator, blocking transporter proteins in limbic system synapses and, as a consequence, leaving the neurotransmitter active in these synapses for a longer time. Eventually the receptor number is reduced because of this extra activity.

The illustration below walks you through the addiction process:

1 In a normal synapse, dopamine molecules (the red balls), are reabsorbed by transporters in the presynaptic cell.

2 Cocaine binds tightly to the transporter proteins on presynaptic membranes. These proteins normally remove the neurotransmitter dopamine after it has acted. Eventually there are no unoccupied transporter proteins available to the dopamine molecules, so the dopamine stays in the synapse, firing the receptors again and again. As new signals arrive, more and more dopamine is added, firing the pleasure pathway more and more often.

3 When receptor proteins on limbic system nerve cells are exposed to high levels of dopamine neurotransmitter molecules for prolonged periods of time, the nerve cells "turn down the volume" of the signal by decreasing the number of receptor proteins on their surfaces, reducing the number of targets available for dopamine molecules to hit.

4 The cocaine user is now addicted. **Addiction** occurs when chronic exposure to a drug induces the nervous system to adapt physiologically. With so few receptors, normal levels of dopamine are not able to trigger an action potential in the postsynaptic cell and so the user needs the drug to maintain even normal levels of limbic activity.

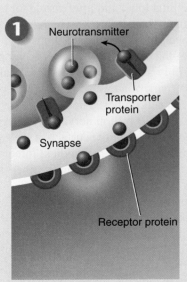

Neurotransmitter

Transporter protein

Synapse

Receptor protein

At a normal synapse, neurotransmitters are quickly recycled by transporter proteins, so the firing rate of receptor proteins stays low.

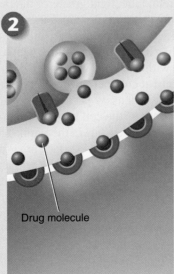

Drug molecule

Drug molecules like cocaine bind to the transporters and block recycling, so the level of neurotransmitters rises, and the firing rate increases.

The receiving neuron "turns down the volume" by lowering the number of receptors, so the firing rate returns to normal.

If the cocaine is removed, the level of neurotransmitters falls to normal, too low to fire the reduced number of receptors.

The Central Nervous System

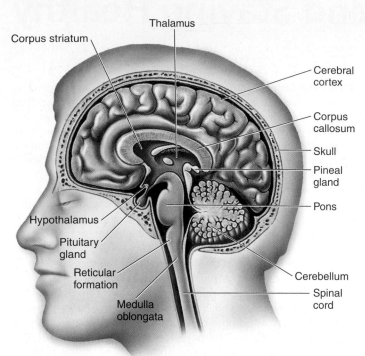

Corpus striatum

Thalamus

Cerebral cortex

Corpus callosum

Skull

Pineal gland

Pons

Hypothalamus

Pituitary gland

Reticular formation

Cerebellum

Spinal cord

Medulla oblongata

Figure 29.8   A section through the human brain.

The cerebrum occupies most of the brain. Only its outer layer, the cerebral cortex, is visible on the surface.

IN THE NEWS

A Taste Test for Depression? Depression is a mental disorder characterized by a pervasive low mood, low self-esteem, anxiety, and loss of pleasure in daily life. Our understanding of the nature and causes of depression remains incomplete, and recommended treatments vary. A body of biological evidence points to low levels of the neurotransmitter serotonin in certain regions of the brain: the hippocampus (which regulates mood) and the anterior cingulate (which governs emotions). The most commonly prescribed antidepressants (Zoloft, Prozac, and Paxil) all work by inhibiting serotonin uptake from neural synapses in the brain—serotonin concentrations within brain synapses rise when this neurotransmitter is not removed, compensating for its low levels and so relieving depression. Researchers at University of Bristol reported in 2008 that when patients taking Prozac got a serotonin jump, they were suddenly able to taste a tiny dab of faint sweet flavor placed on their tongue—depressed patients had blunted taste sensitivity that the antidepressant relieved. This simple taste test may provide a very useful tool for treating depression. It is very difficult to predict beforehand which antidepressant drug will work for a particular individual, and it takes weeks before these drugs have an effect on behavior. The taste test may provide a useful way to determine which drug to use much more quickly. Clinical trials are underway.

29.4 How the Brain Works

CONCEPT PREVIEW: The associative activity of the brain is centered in the wrinkled cerebral cortex, which lies over the cerebrum. Beneath, the thalamus and hypothalamus process information and integrate body activities. At the base of the brain, the cerebellum coordinates muscle movements. The brain stem controls unconscious activities such as breathing, swallowing, and other life-sustaining processes.

The structure and function of the vertebrate brain have long been the subject of scientific inquiry. Despite ongoing research, scientists are still not sure how the brain performs many of its functions. For instance, scientists continue to look for the mechanism the brain employs to store memories, and they do not understand how some memories can be "locked away," only to surface in times of stress. The brain is the most complex vertebrate organ ever to evolve, and it can perform a bewildering variety of complex functions.

The Cerebrum Is the Control Center of the Brain

Although vertebrate brains differ in the relative importance of different components, the human brain is a good model of how vertebrate brains function. About 85% of the weight of the human brain is made up of the cerebrum, the tan convoluted area in figure 29.8. The cerebrum is a large rounded area of the brain divided by a groove into right and left halves called cerebral hemispheres. The sectioned brain in figure 29.8 is cut along the center groove, with the left hemisphere removed, showing the right hemisphere. The cerebrum functions in language, conscious thought, memory, personality development, vision, and a host of other activities we call "thinking and feeling." Figure 29.9 shows general areas of the brain color-coded for easy identification (yellow for the frontal lobe, orange for the parietal lobe, light green for the occipital lobe, and light purple for temporal lobe) and the functions they control. The cerebrum, which looks like a wrinkled mushroom, is positioned over and surrounding the rest of the brain, like a hand holding a fist. Much of the neural activity of the cerebrum occurs within a thin, gray outer layer only a few millimeters thick called the **cerebral cortex** (*cortex* is Latin for "bark of a tree"). This layer is gray because it is densely packed with neuron cell bodies. The human cerebral cortex contains the cell bodies of more than 10 billion nerve cells, roughly 10% of all the neurons in the brain. The wrinkles in the surface of the cerebral cortex increase its surface area (and number of cell bodies) threefold. Underneath the cortex is a solid white region of myelinated nerve fibers that shuttle information between the cortex and the rest of the brain.

The right and left cerebral hemispheres are linked by bundles of neurons called tracts. These tracts serve as information highways, telling each half of the brain what the other half is doing. Because these tracts cross over, in the area of the brain called the *corpus callosum* (the blue-colored band in figure 29.8), each half of the brain controls muscles and glands on the opposite side of the body. In general, the left brain is associated with language, speech, and mathematical abilities, whereas the right brain is associated with intuitive, musical, and artistic abilities.

Researchers have found that the two sides of the cerebrum can operate as two different brains. For instance, in some people the tract between the two hemispheres has been cut by accident or surgery. In laboratory experi-

Figure 29.9 **The major functional regions of the human brain.**

Specific areas of the cerebral cortex are associated with different regions and functions of the body.

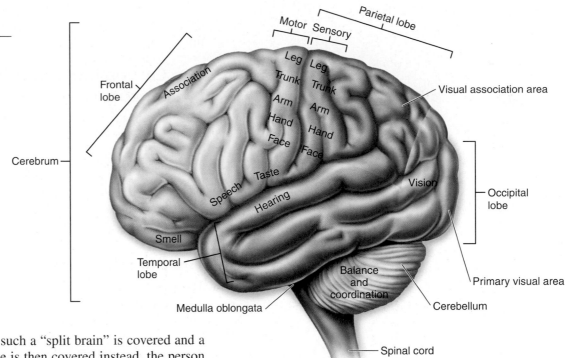

ments, one eye of an individual with such a "split brain" is covered and a stranger is introduced. If the other eye is then covered instead, the person does not recognize the stranger who was just introduced!

Sometimes blood vessels in the brain are blocked by blood clots, causing a disorder called a *stroke*. During a stroke, circulation to an area in the brain is blocked and the brain tissue dies. A severe stroke in one side of the cerebrum often causes paralysis of the other side of the body.

The Thalamus and Hypothalamus Process Information

Beneath the cerebrum are the thalamus and hypothalamus, important centers for information processing. The thalamus is the major site of sensory processing in the brain. Auditory (sound), visual, and other information from sensory receptors enter the thalamus and then are passed to the sensory areas of the cerebral cortex (indicated in figure 29.9). The thalamus also controls balance. Information about posture, derived from the muscles, and information about orientation, derived from sensors within the ear, combine with information from the cerebellum and pass to the thalamus. The thalamus processes the information and channels it to the appropriate motor center on the cerebral cortex.

The hypothalamus integrates all the internal activities of the body. It controls centers in the brain stem that in turn regulate body temperature, blood pressure, respiration, and heartbeat. It also directs the secretions of the brain's major hormone-producing gland, the pituitary gland. The hypothalamus is linked by an extensive network of neurons to other areas of the cerebral cortex. This network, along with parts of the hypothalamus and areas of the brain called the *hippocampus* and *amygdala*, make up the **limbic system.** The areas

> The role of the hypothalamus in the endocrine system is quite significant, as discussed on page 583. The hypothalamus directs the release of hormones from the pituitary gland, some of which are also produced in the hypothalamus.

highlighted in green in figure 29.10 indicate the components of the limbic system. The operations of the limbic system are responsible for many of the most deep-seated drives and emotions of vertebrates, including pain, anger, sex, hunger, thirst, and pleasure, centered in the amygdala.

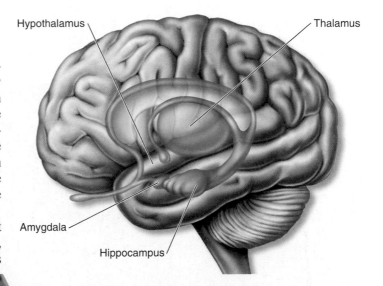

Figure 29.10 **The limbic system.**

The hippocampus and the amygdala are the major components of the limbic system, which controls our most deep-seated drives and emotions.

You'll recall on page 557 that the limbic system is the area of the brain affected by cocaine. It is also involved in memory, which is centered in the hippocampus.

The Cerebellum Coordinates Muscle Movements

Extending back from the base of the brain is a structure known as the **cerebellum.** The cerebellum controls balance, posture, and muscular coordination. This small, cauliflower-shaped structure, while well developed in humans and other mammals, is even more developed in birds. Birds perform more complicated feats of balance than we do, because they move through the air in three dimensions. Imagine the kind of balance and coordination needed for a bird to land on a branch, stopping at precisely the right moment without crashing into it.

The Brain Stem Controls Vital Body Processes

The **brain stem,** a term used to collectively refer to the midbrain, pons, and medulla oblongata, connects the rest of the brain to the spinal cord. This stalklike structure contains nerves that control your breathing, swallowing, and digestive processes, as well as the beating of your heart and the diameter of your blood vessels. A network of nerves called the *reticular formation* runs through the brain stem and connects to other parts of the brain. Their widespread connections make these nerves essential to consciousness, awareness, and sleep. One part of the reticular formation filters sensory input, enabling you to sleep through repetitive noises such as traffic yet awaken instantly when a telephone rings.

Language and Other Higher Functions

Although the two cerebral hemispheres seem structurally similar, they are responsible for different activities. The most thoroughly investigated example of this lateralization of function is language. The left hemisphere is the "dominant" hemisphere for language—the hemisphere in which most neural processing related to language is performed—in 90% of right-handed people and nearly two-thirds of left-handed people. There are two language areas in the dominant hemisphere: One is important for language comprehension and the formulation of thoughts into speech, and the other is responsible for the generation of motor output needed for language communication. Different language activities shown in figure 29.11 confirm that different parts of the brain are involved.

While the dominant hemisphere for language is adept at sequential reasoning, like that needed to formulate a sentence, the nondominant hemisphere (the right hemisphere in most people) is adept at spatial reasoning, the type of reasoning needed to assemble a puzzle or draw a picture. It is also the hemisphere primarily involved in musical ability—a person with damage to the speech area in the left hemisphere may not be able to speak but may retain the ability to sing! Damage to the nondominant hemisphere may lead to an inability to appreciate spatial relationships and may impair musical activities such as singing. Reading, writing, and oral comprehension remain normal. The

Figure 29.11   Different brain regions control various activities.

The colors indicate how the brain reacts in human subjects asked to listen to a spoken word, to read that same word silently, to repeat the word out loud, and then to speak a word related to the first. Regions of white, red, and yellow show the greatest activity.

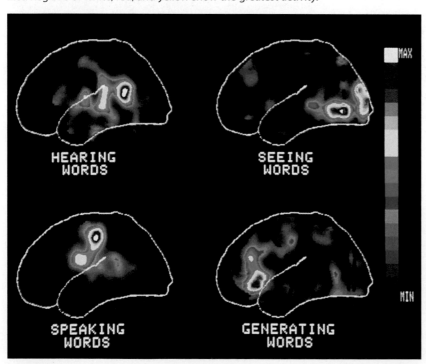

nondominant hemisphere is also important for the consolidation of memories of nonverbal experiences.

One of the great mysteries of the brain is the basis of memory and learning. There is no one part of the brain in which all aspects of a memory appear to reside. Although memory is impaired if portions of the brain, particularly the temporal lobes, are removed, it is not lost entirely. Many memories persist in spite of the damage, and the ability to access them gradually recovers with time. Therefore, investigators who have tried to probe the physical mechanisms underlying memory often have felt that they were grasping at a shadow. Although we still do not have complete understanding, we have learned a good deal about the basic processes in which memories are formed.

There appear to be fundamental differences between short-term and long-term memory. Short-term memory is transient, lasting only a few moments. Such memories can readily be erased by the application of an electrical shock, leaving previously stored long-term memories intact. This result suggests that short-term memories are stored electrically in the form of a transient neural excitation. Long-term memory, in contrast, appears to involve structural changes in certain neural connections within the brain. Two parts of the temporal lobes, the hippocampus and the amygdala, are involved in both short-term memory and its consolidation into long-term memory. Damage to these structures impairs the ability to process recent events into long-term memories.

The Mechanism of Alzheimer's Disease Is Still a Mystery

In the past, little was known about Alzheimer's disease, a condition in which the memory and thought processes of the brain become dysfunctional. As many as 5 million Americans are living with Alzheimer's disease. It is most often seen in adults over the age of 65 and increases in severity with age. It is fatal and is now the 6th leading cause of death in the United States (figure 29.12). Drug companies are eager to develop new products for the treatment of Alzheimer's disease, but they have little concrete evidence to go on. Scientists disagree about the biological nature of the disease and its cause. Two hypotheses have been proposed: One suggests that nerve cells in the brain are killed from the outside in, the other from the inside out.

In the first hypothesis, external proteins called β-amyloid peptides kill nerve cells. A mistake in protein processing produces an abnormal form of the peptide, which then forms aggregates, or plaques. The plaques begin to fill-in the brain, which then damages and kills nerve cells. However, these amyloid plaques have been found in autopsies of people who did not have Alzheimer's disease.

The second hypothesis maintains that the nerve cells are killed by an abnormal form of an internal protein. This protein, called tau (τ), normally functions to maintain protein transport microtubules. Abnormal forms of τ assemble into helical segments that form tangles, which interfere with the normal functioning of the nerve cells. Researchers continue to study whether tangles and plaques are causes or effects of Alzheimer's disease.

Progress has been made in identifying genes that increase the likelihood of developing Alzheimer's disease and genes that, when mutated, can cause the disorder. Most Alzheimer's patients do not have these mutated genes, but for those that do, the symptoms of Alzheimer's are expressed much earlier in life.

Figure 29.12 Alzheimer's disease.

In 1994, former president Ronald Reagan announced he had been diagnosed with Alzheimer's disease. He lived for another 10 years dying at the age of 93 (shown here in 1996).

IN THE NEWS

Testing for Alzheimer's. Medical scientists expect that the number of elderly people with Alzheimer's disease will climb to more than 80 million by 2040, when students like you will be old enough to be at risk. Neuroscientists are pretty sure they know what causes Alzheimer's disease—plaques and tangles of amyloid protein that eat away at the brain. However, clinical trials of therapies to remove amyloid plaques have not been successful. By the time people are diagnosed with dementia, their brains have been irreversibly damaged. The challenge is to recognize the disease earlier, before dementia sets in. Some researchers are using brain imaging to detect tell-tale changes in brain volume; others are measuring concentrations of amyloid proteins in the brain using special radioactive peptides that bind to the amyloid and can be seen in a PET scan. Many Alzheimer's patients carry a high-risk variant of a gene called *APOE* that makes them 15 times more likely to develop Alzheimer's. A private company sold a simple genetic test for the Alzheimer's-linked *APOE* variant for a few months in 2008, but sales of the test have been stopped, for fear that widespread commercial availability of the test may simply cause needless worry, as there is no prevention or treatment.

Figure 29.13 **A view down the human spinal cord.**

Pairs of spinal nerves can be seen extending out from the spinal cord. Along these nerves, the brain and spinal cord communicate with the body.

29.5 The Spinal Cord

CONCEPT PREVIEW: The spinal cord, protected in vertebrates by a back bone, contains nerves that extend from the brain out to the body.

The **spinal cord** is a cable of neurons extending from the brain down through the backbone, which is the view in figure 29.13. The cross section through the spinal cord in figure 29.14 shows a darker gray area in the center that consists of neuron cell bodies, which form a column down the length of the cord. This column is surrounded by axons and dendrites, which make the outer edges of the cord white because they are coated with myelin. The spinal cord is surrounded and protected by a series of bones called the vertebrae. Spinal nerves pass out to the body from between the vertebrae. Messages between the body and the brain run up and down the spinal cord, like an information highway.

> The vertebrae, individual bones stacked one on top of the other, make up the vertebral column as discussed on page 468. Because the vertebral column is made up of a series of bones, it is flexible, allowing the animal to flex its back, but it also protects the spinal cord.

In each segment of the spine, motor nerves extend out of the spinal cord to the muscles. Motor nerves from the spine control most of the muscles below the head. This is why injuries to the spinal cord often paralyze the lower part of the body. A muscle is paralyzed and cannot move if its motor neurons are damaged.

Spinal Cord Regeneration

In the past, scientists have tried to repair severed spinal cords by installing nerves from another part of the body to bridge the gap and act as guides for the spinal cord to regenerate. But most of these experiments have failed because the nerve bridges did not go from white matter to gray matter. Also, there is a factor that inhibits nerve growth in the spinal cord. After discovering that fibroblast growth factor stimulates nerve growth, neurobiologists tried gluing on the nerves, from white to gray matter, with fibrin that had been mixed with the fibroblast growth factor. Three months later, rats with the nerve bridges began to show movement in their lower bodies. In further analyses of the experimental animals, dye tests indicated that the spinal cord nerves had regrown from both sides of the gap. Many scientists are encouraged by the potential to use a similar treatment in human medicine. However, most spinal cord injuries in humans do not involve a completely severed spinal cord; often, nerves are crushed. Also, although the rats with nerve bridges did regain some locomotory ability, tests indicated that they were barely able to walk or stand.

Figure 29.14 **The vertebrate nervous system.**

The spinal cord (colored *yellow*) connects to the base of the brain (colored *tan*) and nerves extend out from the spinal cord to all parts of the body.

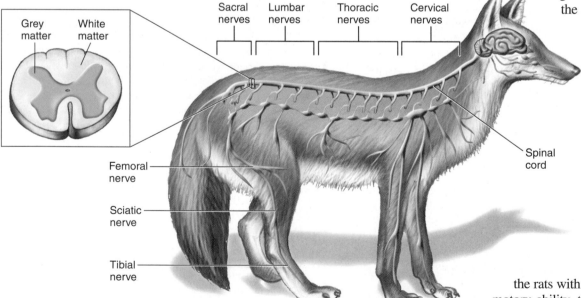

Grey matter   White matter

Sacral nerves   Lumbar nerves   Thoracic nerves   Cervical nerves

Spinal cord

Femoral nerve

Sciatic nerve

Tibial nerve

Concept Check

1. Where in the cerebrum is the cerebral cortex located?
2. What does the limbic system do?
3. What area of the brain is involved with balance and coordination?

The Peripheral Nervous System

29.6 Voluntary and Autonomic Nervous Systems

CONCEPT PREVIEW: The voluntary nervous system relays commands to skeletal muscles and can be controlled by conscious thought. The autonomic nervous system relays commands to muscles and glands that cannot be controlled by conscious thought. The autonomic nervous system consists of two opposing sets of neurons making up the sympathetic and parasympathetic systems.

As you learned in the opening discussion in section 29.1, the nervous system is divided into two main parts: the central nervous system (the pink boxes in figure 29.15) and the peripheral nervous system (the blue boxes). The motor pathways of the peripheral nervous system of a vertebrate can be further subdivided into the **somatic (voluntary) nervous system,** which relays commands to skeletal muscles, and the **autonomic (involuntary) nervous system,** which stimulates glands and relays commands to the smooth muscles of the body and to cardiac muscle. The voluntary nervous system can be controlled by conscious thought. You can, for example, command your hand to move. The autonomic nervous system, by contrast, cannot be controlled by conscious thought. You cannot, for example, tell the smooth muscles in your digestive tract to speed up their action. The central nervous system issues commands over both voluntary and autonomic systems, but you are conscious of only the voluntary commands.

Voluntary Nervous System

Motor neurons of the voluntary nervous system stimulate skeletal muscles to contract in two ways. First, motor neurons may stimulate the skeletal muscles of the body to contract in response to conscious commands. For example, if you want to bounce a basketball, your CNS sends messages through motor neurons to the muscles in your arms and hands. However, skeletal muscle can also be stimulated as part of reflexes that do not require conscious control.

Reflexes Enable Quick Action. The motor neurons of the body have been wired to enable the body to act particularly quickly in time of danger—even before the animal is consciously aware of the threat. These sudden, involuntary movements are called reflexes. A **reflex** produces a rapid motor response to a stimulus because the sensory neuron bringing information about the threat passes the information directly to a motor neuron. One of the most frequently used reflexes in your body is blinking, a reflex that protects your eyes. If anything, such as an insect or a cloud of dust, approaches your eye, the eyelid blinks closed even before you realize what has happened. The reflex occurs before the cerebrum is aware the eye is in danger.

Because they involve passing information between few neurons, reflexes are very fast. Many reflexes never reach the brain. The "danger" nerve impulse travels only as far as the spinal cord and then comes right back as a motor response. Most reflexes involve a single connecting interneuron between the sensory neuron and the motor neuron. A few, like the knee-jerk reflex in figure 29.16, are

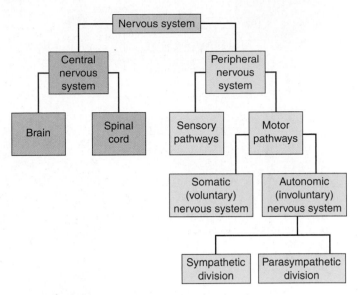

Figure 29.15 **The divisions of the vertebrate nervous system.**

The motor pathways of the peripheral nervous system are the somatic (voluntary) and autonomic nervous systems.

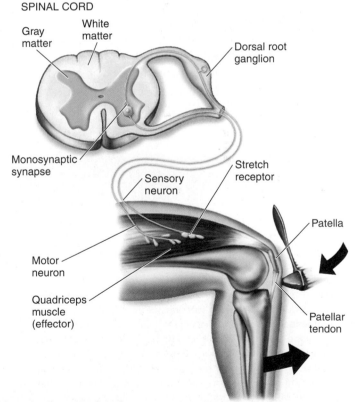

Figure 29.16 **The knee-jerk reflex.**

The most famous involuntary response, the knee jerk, is produced by activating stretch receptors in the quadriceps muscle. When a rubber mallet taps the patellar tendon, the muscle and stretch receptors in the muscle are stretched. A signal travels up a sensory neuron to the spinal cord, where the sensory neuron stimulates a motor neuron, which sends a signal to the quadriceps muscle to contract.

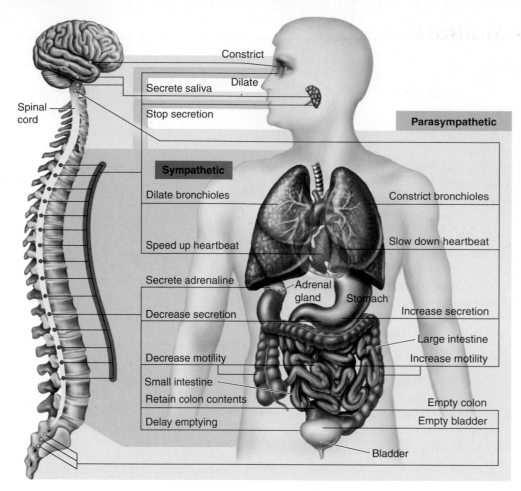

Figure 29.17 How the sympathetic and parasympathetic nervous systems interact.

A nerve path runs from both systems to every organ indicated except the adrenal gland, which is only innervated by the sympathetic nervous system.

![Earth icon] **BIOLOGY & YOU**

Biofeedback. Biofeedback is a technique in which you are trained to influence certain involuntary body processes under autonomic nervous system control, such as heart rate, blood pressure, muscle tension, and skin temperature. These processes are measured with electrodes and displayed on a monitor, providing you with "real-time" feedback on the internal workings of your body. How does biofeedback work? Scientists are not sure. It doesn't work for everyone, and most people who do benefit from biofeedback treatments have conditions that are brought on or made worse by stress. For this reason, many scientists believe that relaxation is key to successful biofeedback therapy. What is biofeedback good for? It shows considerable promise in treating urinary incontinence, which affects over 15 million Americans. Biofeedback is also becoming a popular treatment for attention deficit/hyperactivity disorder (ADHD), although its efficacy in moderating ADHD is controversial.

monosynaptic reflex arcs. You see in the figure that the sensory neuron, a stretch receptor embedded in a muscle, "senses" the stretching of the muscle when the tendon is tapped. This stretching could harm the muscle and so a nerve impulse is sent to the spinal cord where it synapses directly with a motor neuron—there is no interneuron between them. Similarly, if you step on something sharp, your leg jerks away from the danger; the prick causes nerve impulses in sensory neurons, which pass to the spinal cord and then to motor neurons, which cause your leg muscles to contract, jerking your leg up.

Autonomic Nervous System

Some motor neurons are active all the time. These neurons carry messages from the CNS that keep the body going even when it is not active. These neurons make up the autonomic nervous system. The autonomic nervous system carries messages to muscles and glands that work without the animal noticing.

The autonomic nervous system is the command network used by the CNS to maintain the body's homeostasis. Using it, the CNS regulates heartbeat and controls muscle contractions in the walls of the blood vessels. It directs the muscles that control blood pressure, breathing, and the movement of food through the digestive system. It also carries messages that help stimulate glands to secrete tears, mucus, and digestive enzymes.

The autonomic nervous system is composed of two elements that act in opposition to one another. One division, the **sympathetic nervous system,** dominates in times of stress. It controls the "fight-or-flight" reaction, increasing blood pressure, heart rate, breathing rate, and blood flow to the muscles. The sympathetic nervous system is colored pink in figure 29.17, with neurons extending from the middle section of the spinal cord. Long motor neurons extend from the ganglia directly to each target organ. Another division, the **parasympathetic nervous system,** has the opposite effect. It conserves energy by slowing the heartbeat and breathing rate and by promoting digestion and elimination. The parasympathetic nervous system is colored in blue in the figure, with neurons extending from the upper and lower sections of the spinal cord.

Most glands, smooth muscles, and cardiac muscles get constant input from *both* systems. The CNS controls activity by varying the ratio of the two signals to either stimulate or inhibit the organ.

Concept Check

1. How many neurons are involved in the knee-jerk reflex?
2. What is the functional role of the autonomic nervous system?
3. Which element of the autonomic nervous system acts to slow the heartbeat, the sympathetic or parasympathetic?

The Sensory Nervous System

29.7 Sensory Perception

CONCEPT PREVIEW: Sensory receptors initiate nerve impulses in response to stimulation. All sensory nerve impulses are the same, differing only in the stimulus that fires them and their destination in the brain. A variety of different sensory receptors inform the hypothalamus about different aspects of the body's internal environment, enabling it to maintain the body's homeostasis.

Did you ever wonder what it would be like not to know anything about what is going on around you? Imagine if you couldn't hear, or see, or feel, or smell. After a while, a human goes mad if completely deprived of sensory input. The senses are the bridge to experience and perceive the way the body relates to everything around it.

Sensory Receptors

The **sensory nervous system** tells the central nervous system what is happening. Sensory neurons carry impulses to the CNS from more than a dozen different types of sensory cells that detect changes outside and inside the body. Called **sensory receptors,** these specialized sensory cells detect many different things, including changes in blood pressure, strain on ligaments, and smells in the air. Particularly complex sensory receptors, made up of many cell and tissue types, are called sensory organs. The eyes and ears (figure 29.18) are sensory organs, and so are the taste buds in your mouth.

How does the brain know whether an incoming nerve impulse is light, sound, or pain? This information is built into the "wiring"—into which neurons interact while passing the information to the CNS and into the location in the brain where the information is sent. The brain "knows" it is responding to light because the message from a sensory neuron is wired to light receptor cells.

The Path of Sensory Information

Sensory receptors are able to initiate nerve impulses by opening or closing *stimulus-gated channels* within sensory neuron membranes. If the stimulus is large enough, the depolarization will trigger an action potential. The size of the stimulus determines the frequency of action potentials, not the size of the action potentials—all action potentials are the same size. The greater the sensory stimulus, the greater the depolarization of the sensory receptor and the higher the frequency of action potentials. Stimulus-gated channels are opened by chemical or mechanical stimulation, often a disturbance such as touch, heat, or cold. The receptors differ from one another in the nature of the environmental input that triggers the opening of the channel. The body contains many sorts of receptors, each sensitive to a different aspect of the body's condition or to a different quality of the external environment.

Exteroceptors are receptors that sense stimuli that arise in the external environment. Almost all of a vertebrate's exterior senses evolved in water before vertebrates invaded the land. Consequently, many senses of terrestrial vertebrates emphasize stimuli that travel well in water, using receptors that have been retained in the transition from the sea to the land. Hearing, as you will discover in section 29.10, converts an airborne stimulus into a waterborne one, using receptors similar to those that originally evolved in aquatic animals.

Figure 29.18   Kangaroo rats have specialized ears.

The ears of kangaroo rats (*Dipodomys*) are adapted to nocturnal life and allow them to hear the low-frequency sounds of their predators, such as an owl's wingbeats or a sidewinder rattlesnake's scales rubbing against the ground. Also, the ears seem to be adapted to the poor sound-carrying quality of dry, desert air.

© Wendy Shattil/Bob Rozinski

E V O L U T I O N

"Seeing" Heat. Visible light is by no means the only part of the electromagnetic spectrum that vertebrates use to sense their environment. Those parts of the spectrum with wavelengths longer than visible light ("below red," or *infrared*)—what we normally think of as heat—are too low in energy to be detected by photoreceptors. The only vertebrates known to have the ability to sense infrared radiation are rattlesnakes and other pit vipers. These snakes possess a pair of heat-detecting pit organs located beneath their eyes. Each pit organ is composed of two chambers separated by a membrane. Temperature-sensitive neurons compare the temperatures of the inner and outer chambers, providing stereoscopic information about the environment to the brain in much the same way as two eyes do. Pit organs allow a blindfolded rattlesnake to accurately strike at a warm, dead rat.

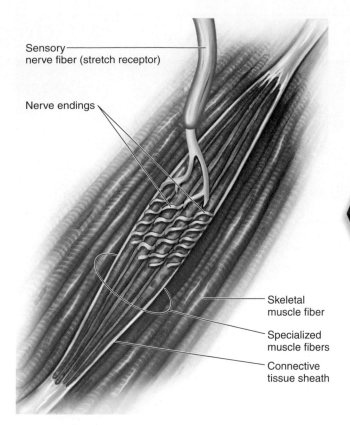

Sensory nerve fiber (stretch receptor)

Nerve endings

Skeletal muscle fiber

Specialized muscle fibers

Connective tissue sheath

Figure 29.19 **A stretch receptor embedded within skeletal muscle.**

Stretching the muscle elongates the specialized muscle fibers, which deforms the nerve endings, causing them to send a nerve impulse out along the nerve fiber.

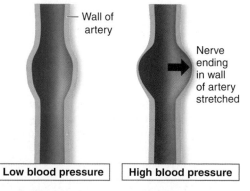

Wall of artery

Nerve ending in wall of artery stretched

Low blood pressure

High blood pressure

Figure 29.20 **How a baroreceptor works.**

A network of nerve endings covers a region where the wall of the artery is thin. High blood pressure causes the wall to balloon out there, stretching the nerve endings and causing them to fire impulses.

Interoceptors sense stimuli that arise from within the body. These internal receptors detect stimuli related to muscle length and tension, limb position, pain, blood chemistry, blood volume and pressure, and body temperature. Many of these receptors are simpler than those that monitor the external environment and are believed to bear a closer resemblance to primitive sensory receptors.

Sensing the Internal Environment

Sensory receptors inside the body inform the CNS about the condition of the body. Much of this information passes to a coordinating center in the brain, the hypothalamus, the part of the brain responsible for maintaining the body's homeostasis—that is, keeping the body's internal environment constant. The vertebrate body uses a variety of different sensory receptors to respond to different aspects of its internal environment.

> Homeostasis is the state of dynamic constancy of internal conditions, as discussed on page 520. Negative feedback loops maintain a state of homeostasis by correcting deviations from a set point. The deviation is the stimulus that is detected by the sensory receptors and acted on.

Temperature change. Two kinds of nerve endings in the skin are sensitive to changes in temperature, one stimulated by cold, the other by warmth. By comparing information from the two, the CNS can learn what the temperature is and if it is changing.

Blood chemistry. Receptors in the walls of arteries sense CO_2 levels in the blood. The brain uses this information to regulate the body's respiration rate, increasing it when CO_2 levels rise above normal.

Pain. Damage to tissue is detected by special nerve endings within tissues, usually near the surface, where damage is most likely to occur. When these nerve endings are physically damaged or deformed, the CNS responds by reflexively withdrawing the body segment and often by changing heartbeat and blood pressure as well.

Muscle contraction. Buried deep within muscles are sensory receptors called stretch receptors. You heard about these receptors in the discussion on reflexes. In each, the end of a sensory neuron is wrapped around a muscle fiber, like the receptor shown in figure 29.19. When the muscle is stretched, the fiber elongates, stretching the spiral nerve ending (like stretching a spring) and causing repeated nerve impulses to be sent to the brain. From these signals the brain can determine the rate of change of muscle length at any given moment. The CNS uses this information to control movements that require the combined action of several muscles, such as those that carry out breathing or locomotion.

Blood pressure. Blood pressure is sensed by neurons called baroreceptors with highly branched nerve endings within the walls of major arteries. When blood pressure increases, the stretching of the arterial wall, like the expansion of the artery in figure 29.20, causes the sensory neuron to increase the rate at which it sends nerve impulses to the CNS. When the wall of the artery is not stretched, the rate of firing of the sensory neuron goes down. Thus, the frequency of impulses provides the CNS with a continuous measure of blood pressure.

Touch. Touch is sensed by pressure receptors buried below the surface of the skin. There are a variety of different types, some specialized to detect rapid changes in pressure, others to measure the duration and extent to which pressure is applied, and still others sensitive to vibrations.

29.8 Sensing Gravity and Motion

CONCEPT PREVIEW: The body senses gravity and acceleration by the deflection of cilia caused by shifting otoliths and shifting fluid in a cupula, respectively. The body cannot sense motion at a constant velocity and direction.

Two types of receptors in the ear inform the brain where the body is in three dimensions. This knowledge enables an animal to move freely and maintain its balance. Figure 29.21 shows the anatomy of the inner ear and the locations of these receptors.

Balance. To keep the body's balance the brain needs a frame of reference, and the reference point it uses is gravity. The sensory receptors that detect gravity are hair cells that extend from the floor of the utricle and the saccule (see the enlarged view of the inner ear for these chambers). The tips of the hair cells project into a gelatinous matrix with embedded particles called **otoliths** (figure 29.21*b*). To illustrate how these receptors work, imagine a pencil standing in a glass. No matter which way you tip the glass, the pencil rolls along the rim due to the pull of gravity, applying pressure to the lip of the glass. If you want to know the direction the glass is tipped, you need only ask where on the rim pressure is being applied. Similarly, the otoliths will shift in the matrix in response to the pull of gravity and stimulate hair cells. The brain uses the information from the hair cells to determine vertical positioning.

Motion. The brain senses motion by employing a receptor in which fluid deflects cilia of hair cells in a direction opposite that of the motion. Within the inner ear are three fluid-filled **semicircular canals,** each oriented in a different plane at right angles to the other two (see the enlarged view of the inner ear) so that motion in any direction can be detected. Protruding into the canal are groups of cilia from sensory hair cells. The cilia from each cell are arranged in a tentlike assembly called a *cupula*, shown in figure 29.21*c*, which is pushed when fluid in the semicircular canals moves in a direction opposite that of the head's movement. Because the three canals are oriented in all three planes, movement in any plane is sensed by at least one of them, and the brain is able to analyze complex movements by comparing the sensory inputs from each canal.

The semicircular canals do not react if the body moves in a straight line because the fluid in the canals does not move. That is why traveling in a car or airplane at a constant speed in one direction gives no sense of motion.

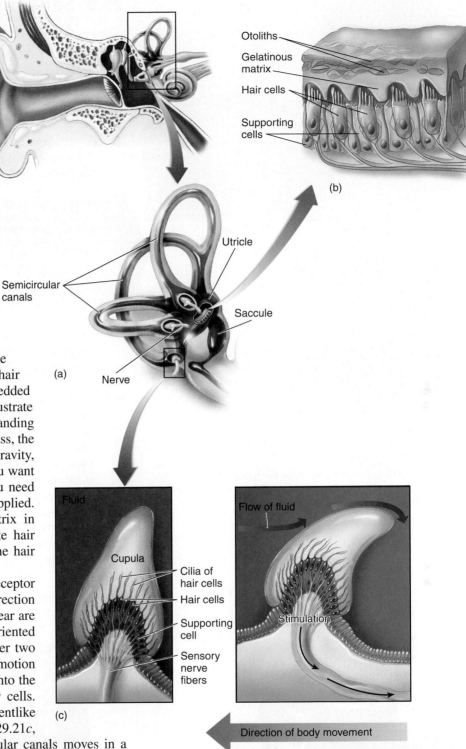

Figure 29.21 How the inner ear senses gravity and motion.

(a) The semicircular canals are part of the inner ear. (b) Enlargement of a section of the utricle or saccule. Otoliths embedded in the gelatinous matrix move in response to the pull of gravity. (c) The cupula within the semicircular canals are surrounded by fluid and contain hair cells. Movement in a particular direction causes fluid in the semicircular canal of that plane to move; the cupula is displaced, thereby stimulating the hair cells.

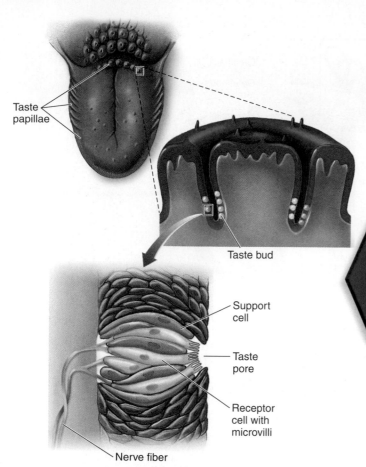

Figure 29.22 Taste.

Taste buds on the human tongue are typically grouped into projections called papillae. Individual taste buds are bulb-shaped collections of taste receptor cells that open out into the mouth through a taste pore.

29.9 Sensing Chemicals: Taste and Smell

CONCEPT PREVIEW: Taste and smell are chemical senses. The receptors are stimulated by chemicals in the environment. In many vertebrates, the sense of smell is very well developed.

Vertebrates are able to detect many of the chemicals in air and in food.

Taste. Embedded within the surface of the tongue are *taste buds* located within *papillae,* which are the raised areas on the tongue in figure 29.22. Taste buds are onion-shaped structures in the enlarged view of the papillae that contain many taste receptor cells, each of which has fingerlike microvilli that project into an opening called the taste pore. Chemicals from food dissolve in saliva and contact the taste cells through the taste pore. Salty, sour, sweet, bitter, and umami (a "meaty" taste) are perceived because chemicals in food are detected in different ways by taste buds. When the tongue encounters a chemical, information from the taste cells passes to sensory neurons, which transmit the signals to the brain.

> Microvilli are hairlike extensions of the plasma membrane that increase the surface area of the cell. Epithelial cells that line the small intestine, as discussed on page 512, contain microvilli on the lumen surface of the cells that increase the surface area for absorption.

Smell. The nose contains chemically sensitive neurons whose cell bodies are embedded within the epithelium of the nasal passage, shown in cross section in figure 29.23. When they detect chemicals, these sensory neurons (the red cells in the enlarged view) transmit information to a location in the brain where smell information is processed and analyzed. Although humans can only sense five different tastes, they can detect thousands of different smells. It appears that as many as a thousand different genes may code for different receptor proteins for smell. The particular set of neurons that respond to a given odor might serve as a "fingerprint" that the brain can use to identify the odor. For discovering how this works, American scientists Richard Axel and Linda Buck were awarded the Nobel Prize in Physiology or Medicine in 2004. In many vertebrates (dogs are a familiar example), these neurons are far more sensitive than in humans.

Smell and taste are very important senses in telling an animal about its food. That is why when you have a bad cold and your nose is stuffed up, your food has little taste. Other receptors also play a role. For example, the "hot" sensation of foods such as chili peppers is detected by pain receptors, not chemical receptors.

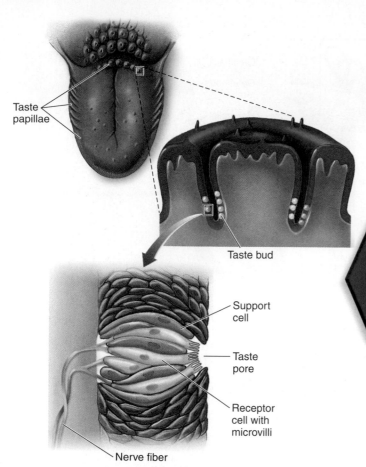

Figure 29.23 Smell.

Humans smell by using receptor cells located in the lining of the nasal passage. The receptor cells are neurons. Axons from these sensory neurons project back through the olfactory nerve directly to the brain.

29.10 Sensing Sounds: Hearing

CONCEPT PREVIEW: Sound receptors detect the vibrations of air as waves of pressure pushing against the membrane covering the ear. Inside, these waves are amplified and press down hair cells that send signals to the brain.

When you hear a sound, you are detecting the air vibrating—waves of pressure in the air beating against your ear, pushing a membrane called the eardrum in and out. As you can see in figure 29.24, on the inner side of the eardrum are three small bones, called ossicles, that act as a lever system to increase the force of the vibration. They transfer the amplified vibration across a second membrane to fluid within the inner ear. The fluid-filled chamber of the inner ear is shaped like a tightly coiled snail shell and is called the cochlea, from the Latin name for "snail." The middle ear, where the ossicles are located, is connected to the throat by the eustachian tube in such a way that there is no difference in air pressure between the middle ear and the outside. That is why your ears sometimes "pop" when landing in an airplane—the pressure is equalizing between the two sides of the eardrum. This equalized pressure is necessary for the eardrum to work.

The sound receptors within the cochlea are hair cells that rest on a membrane that runs up and down the middle of the spiraling chamber, separating it into two halves like a wall, into the upper and lower fluid-filled canals in the enlarged view. The hair cells do not project into the fluid-filled canals of the cochlea; instead, they are covered by a second membrane (the darker blue membrane in the figure). When a sound wave enters the cochlea, it causes the fluid in the chambers to move. The moving fluid causes this membrane "sandwich" to vibrate, bending the hairs pressed against the upper membrane and causing them to send nerve impulses to sensory neurons that travel to the brain.

Sounds of different frequencies travel different distances down the length of the cochlea and cause different parts of the membrane to vibrate. Each area of the membrane fires a different set of sensory neurons—the identity of the sensory neuron being fired tells the CNS the frequency of the sound. Sound waves of higher frequencies, about 20,000 vibrations (or cycles) per second, also called hertz (Hz), don't travel very far into the cochlea and move the membrane in the area closest to the middle ear. Medium-length frequencies, about 2,000 Hz, travel farther and move the membrane in the area about midway down the length of the cochlea. The lowest-frequency sound waves, about 500 Hz, move the membrane near the tip of the cochlea.

The intensity of the sound is determined by how *often* the neurons fire. Our ability to hear depends upon the flexibility of the membranes within the cochlea. Humans cannot hear low-pitched sounds, below 20 Hz, although some vertebrates can. As children, we can hear high-pitched sounds, up to 20,000 cycles per second, but this ability decreases as we get older. Other vertebrates can hear sounds at far higher frequencies. Dogs readily hear sounds of 40,000 cycles per second and so respond to a high-pitched dog whistle that seems silent to a human.

Frequent or prolonged exposure to loud noises can result in damage of the hair cells and membrane, especially in the high-frequency area of the cochlea. The loss of the ability to detect high-frequency sounds affects a person's ability to hear certain sounds, especially in a noisy setting.

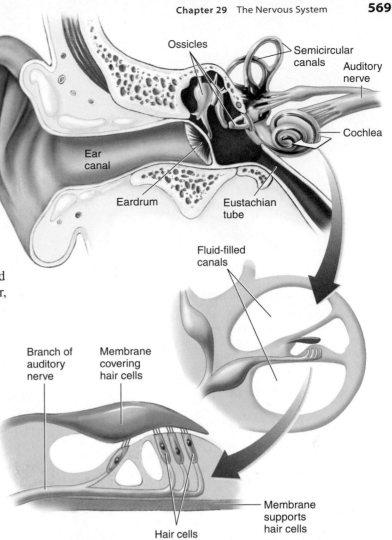

Figure 29.24   Structure and function of the human ear.

Sound waves passing through the ear canal beat on the eardrum, pushing a set of three small bones, or ossicles, against an inner membrane. This sets up a wave motion in the fluid filling the canals within the cochlea. The wave causes the membrane covering the hair cells to move back and forth against the hair cells, which causes associated neurons to fire impulses.

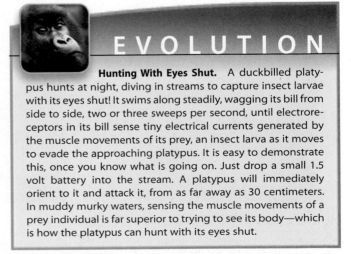

EVOLUTION

Hunting With Eyes Shut. A duckbilled platypus hunts at night, diving in streams to capture insect larvae with its eyes shut! It swims along steadily, wagging its bill from side to side, two or three sweeps per second, until electroreceptors in its bill sense tiny electrical currents generated by the muscle movements of its prey, an insect larva as it moves to evade the approaching platypus. It is easy to demonstrate this, once you know what is going on. Just drop a small 1.5 volt battery into the stream. A platypus will immediately orient to it and attack it, from as far away as 30 centimeters. In muddy murky waters, sensing the muscle movements of a prey individual is far superior to trying to see its body—which is how the platypus can hunt with its eyes shut.

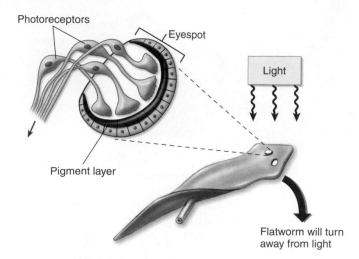

Figure 29.25 Simple eyespots in the flatworm.

Eyespots will detect the direction of light because a pigmented layer on one side of the eyespot screens out light coming from the back of the animal. Light is thus detected more readily coming from the front of the animal; flatworms will respond by turning away from the light.

29.11 Sensing Light: Vision

CONCEPT PREVIEW: Vision receptors called rods and cones contain pigment molecules that detect reflected light. Rods detect various shades of gray and are effective in dim light. Cones detect different colors of light. Binocular vision allows the brain to form three-dimensional images of objects.

No other stimulus provides as much detailed information about the environment as light. Vision, the perception of light, is carried out by a special sensory apparatus called an eye. All the sensory receptors described to this point have been chemical or mechanical ones. Eyes contain sensory receptors called rods and cones that respond to photons of light. The light energy is absorbed by pigments in the rods and cones, which respond by triggering nerve impulses in sensory neurons.

Evolution of the Eye

Vision begins with the capture of light energy by photoreceptors. Because light travels in a straight line and arrives virtually instantaneously, visual information can be used to determine both the direction and the distance of an object. No other stimulus provides as much detailed information.

Many invertebrates have simple visual systems with photoreceptors clustered in an eyespot. The flatworm in figure 29.25 has an eyespot, consisting of pigment molecules that are stimulated by light, triggering a nerve impulse in photoreceptor cells. Although an eyespot can perceive the direction of light,

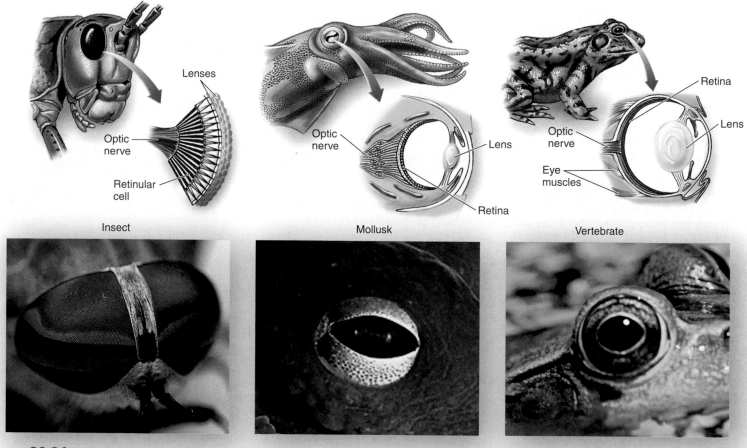

Figure 29.26 Eyes in three phyla of animals.

Although they are superficially similar, these eyes differ greatly in structure and are not homologous. Each has evolved separately and, despite the apparent structural complexity, has done so from simpler structures.

it cannot be used to construct a visual image. Well-developed, image-forming eyes have evolved in the members of four phyla—annelids, mollusks, arthropods, and vertebrates. True image-forming eyes in these phyla, though they at first seem similar (compare the eyes of an arthropod, mollusk, and vertebrate in figure 29.26), are believed to have evolved independently. Interestingly, the photoreceptors in all of them use the same light-capturing molecule, suggesting that not many alternative molecules are able to play this role.

Structure of the Vertebrate Eye

The vertebrate eye works like a lens-focused camera. Light first passes through a transparent protective covering, the **cornea** (the light blue layer in figure 29.27), which begins to focus the light onto the rear of the eye. The beam of light then passes through the **lens,** which completes the focusing. The lens is attached by stringlike *suspensory ligaments* to **ciliary muscles.** When these muscles contract and relax, they change the shape of the lens and thus allow the eye to view objects that are far and near. The amount of light entering the eye is controlled by a shutter, called the **iris** (the colored part of your eye), between the cornea and the lens. The transparent zone in the middle of the iris, the **pupil,** gets larger in dim light and smaller in bright light. The pupil also gets smaller when the eye is viewing close objects.

The light that passes through the pupil is focused by the lens onto the back of the eye. An array of light-sensitive receptor cells lines the back surface of the eye, called the **retina.** The retina is the light-sensing portion of the eye. The vertebrate retina contains two kinds of photoreceptors, called **rods** and **cones,** which, when stimulated by light, generate nerve impulses that travel to the brain along a short, thick nerve pathway called the optic nerve. Rods, the taller, flat-topped cell in figure 29.28, are receptor cells that are extremely sensitive to light, and they can detect various shades of gray even in dim light. However, they cannot distinguish colors, and because they do not detect edges well, they produce poorly defined images. Cones, the pointed-topped cells, are receptor cells that detect color and are sensitive to edges so that they produce sharp images. The center of the vertebrate retina contains a tiny pit, called the **fovea,** densely packed with some 3 million cones. This area produces the sharpest image, which is why we tend to move our eyes so that the image of an object we want to see clearly falls on this area.

The lens of the vertebrate eye is constructed to filter out short-wavelength light. This solves a difficult optical problem: Any uniform lens bends short wavelengths more than it does longer ones, a phenomenon known as chromatic aberration. Consequently, these short wavelengths cannot be brought into focus simultaneously with longer wavelengths. Unable to focus the short wavelengths, the vertebrate eye eliminates them. Insects, whose eyes do not focus light, are able to see these lower, ultraviolet wavelengths quite well and often use them to locate food or mates.

How Rods and Cones Work

A rod or cone cell in the eye is able to detect a single photon of light. How can it be so sensitive? The primary sensing event of vision is the absorption of a photon of light by a pigment. The pigments in rods and cones are made from plant pigments called carotenoids. That is why eating carrots is said to be good for night vision—the orange color of carrots is due to the presence of carotenoids called carotenes. The visual pigment in the human eye is a fragment of carotene called

> Pigments absorb certain wavelengths of light, as discussed on page 104. The primary photosynthetic pigment in plants is chlorophyll, which absorbs red and blue wavelengths, so that plants appear green.

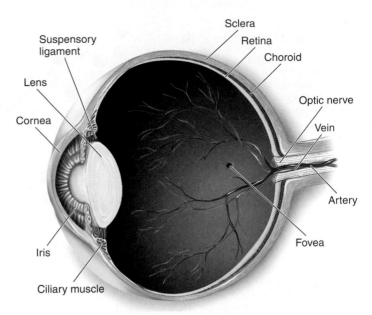

Figure 29.27 **The structure of the human eye.**

Light passes through the transparent cornea and is focused by the lens on the rear surface of the eye, the retina. The retina is rich in photoreceptors, with a high concentration in an area called the fovea.

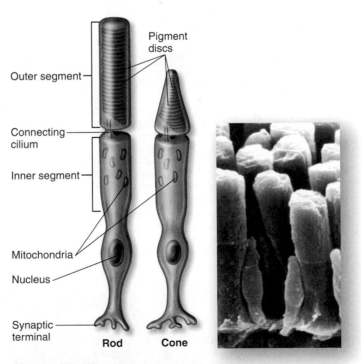

Figure 29.28 **Rods and cones.**

The broad tubular cell on the *left* is a rod. The shorter, tapered cell next to it is a cone. An electron micrograph of rods and cones is also shown.

Figure 29.29
Absorption of light.

When light is absorbed by *cis*-retinal, the pigment undergoes a change in shape and becomes *trans*-retinal.

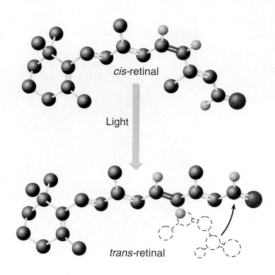

cis-retinal

Light

trans-retinal

cis-retinal. The pigment is attached to a protein called opsin to form a light-detecting complex called **rhodopsin.**

When it receives a photon of light, the pigment undergoes a change in shape. This change in shape must be large enough to alter the shape of the opsin protein attached to it. When light is absorbed by the *cis*-retinal pigment (the upper molecule in figure 29.29), the linear end of the molecule rotates sharply upward, straightening out that end of the molecule. The new form of the pigment is referred to as *trans*-retinal and the dashed outline in the figure shows the shape before it was stimulated by light. This radical change in the pigment's shape induces a change in the shape of the protein opsin to which the pigment is bound, initiating a chain of events that leads to the generation of a nerve impulse.

Each rhodopsin activates several hundred molecules of a protein called transducin. Each of these activates several hundred molecules of an enzyme whose product stimulates sodium channels in the photoreceptor membrane at a rate of about 1,000 per second. This cascade of events allows a single photon to have a large effect on the receptor.

Color Vision

Three kinds of cone cells provide us with color vision. Each possesses a different version of the opsin protein (that is, one with a distinctive amino acid sequence and thus a different shape). These differences in shape affect the flexibility of the attached retinal pigment, shifting the wavelength at which it absorbs light. The absorption spectrum in figure 29.30 shows the wavelength of light that is absorbed by each cone and rod cell. In rods, light is absorbed at 500 nanometers. In cones, the three versions of opsin absorb light at 420 nanometers (blue-absorbing), 530 nanometers (green-absorbing), or 560 nanometers (red-absorbing). By comparing the relative intensities of the signals from the three types of cones, the brain can calculate the intensity of other colors.

Some people are not able to see all three colors, a condition referred to as *color blindness*. Color blindness is typically due to an inherited lack of one or more types of cones. People with normal vision have all three types of cones. People with only two types of cones lack the ability to detect the third color. For example, people with red-green color blindness lack red cones and have difficulty distinguishing red from green (figure 29.31). Color blindness is a sex-linked trait, and so men are far more likely to be color blind than women.

> A sex-linked trait is controlled by a gene that is located on the X chromosome, as discussed on page 171. Because males carry only one X chromosome (the other sex chromosome is the Y chromosome), any gene that is carried on the X chromosome is expressed.

Most vertebrates, particularly those that are diurnal (active during the day), have color vision, as do many insects. Indeed, honeybees can see light in the near-ultraviolet range, which is invisible to the human eye. Color vision requires the presence of more than one photopigment in different receptor cells, but not all animals with color vision have the three-cone system characteristic of humans and other primates. Fish, turtles, and birds, for example, have four or five kinds of cones; the "extra" cones enable these

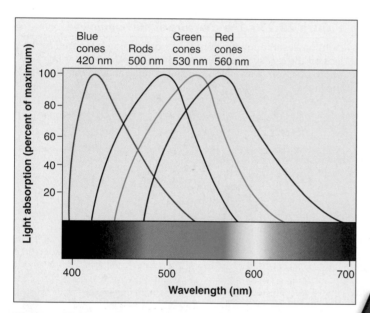

Figure 29.30   Color vision.

The absorption spectrum of *cis*-retinal is shifted in cone cells from the 500 nanometers characteristic of rod cells. The amount of the shift determines what color the cone absorbs: 420 nanometers yields blue absorption; 530 nanometers yields green absorption; and 560 nanometers yields red absorption. Red cones do not peak in the red part of the spectrum, but they are the only cones that absorb light in the far end of the spectrum.

Figure 29.31   Test for color blindness.

People with normal color vision see the number 16, but people that are red-green color blind see just spots and no discernible number.

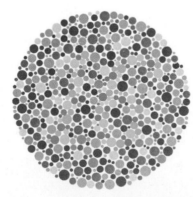

Source: This image has been reproduced from Ishihara's Tests for Color Deficiency published by KANEHARA TRADING INC., located in Tokyo, Japan. But tests for color deficiency cannot be conducted with this material. For accurate testing, the original plates should be used.

Conveying the Light Information to the Brain

The path of light through each eye is the reverse of what you might expect. The rods and cones are at the rear of the retina, not the front. If you track the path that light would take in figure 29.32, you will see that light passes through several layers of ganglion and bipolar cells before it reaches the rods and cones. Once the photoreceptors are activated, they stimulate bipolar cells, which in turn stimulate ganglion cells. The direction of nerve impulses in the retina is thus opposite to the direction of light.

Action potentials propagated along the axons of ganglion cells are relayed through structures called the *lateral geniculate nuclei* of the thalamus and projected to the occipital lobe of the cerebral cortex. There the brain interprets this information as light in a specific region of the eye's receptive field. The pattern of activity among the ganglion cells across the retina encodes a point-to-point map of the receptive field, allowing the retina and brain to image objects in visual space. In addition, the frequency of impulses in each ganglion cell provides information about the light intensity at each point, while the relative activity of ganglion cells connected (through bipolar cells) with the three types of cones provides color information.

Binocular Vision

Primates (including humans) and most predators have two eyes, one located on each side of the face. When both eyes are trained on the same object, the image that each sees is slightly different because each eye views the object from a different angle. This slight displacement of the images permits **binocular vision,** the ability to perceive three-dimensional images and to sense depth or the distance to an object. Having their eyes facing forward maximizes the field of overlap in which this stereoscopic vision occurs, as seen by the overlapping blue triangles in the human in figure 29.33. The triangles are the field of view for each eye.

In contrast, prey animals generally have eyes located to the sides of the head, preventing binocular vision but enlarging the overall receptive field. Depth perception is less important to prey than detection of potential enemies from any angle. The eyes of the American woodcock, for example, are located at exactly opposite sides of its skull so that it has a 360-degree field of view without turning its head! Most birds have laterally placed eyes and, as an adaptation, have two foveas in each retina. One fovea provides sharp frontal vision, like the single fovea in the retina of mammals, and the other fovea provides sharper lateral vision.

Concept Check

1. How is blood pressure sensed? How is pain sensed?
2. What is an otolith? What does it do?
3. How does the ear distinguish a sound's frequency?
4. Why can't rods produce color vision?

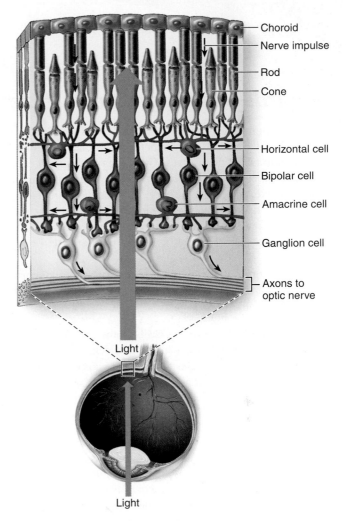

Figure 29.32 Structure of the retina.

The rods and cones are at the rear of the retina. Light passes over four other types of cells in the retina before it reaches the rods and cones. Nerve impulses then travel through the bipolar cells to the ganglion cells and on to the optic nerve (as indicated by the black arrows).

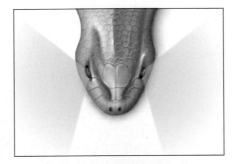

Figure 29.33 Binocular vision.

When the eyes are located on the sides of the head (as on the *left*), the two vision fields do not overlap and binocular vision does not occur. When both eyes are located toward the front of the head (as on the *right*) so that the two fields of vision overlap, depth can be perceived.

Do Birds Use Magnetic Particles as Compass Needles?

Although vision is the primary sense used by all vertebrates, birds also sense their environment using other cues. Some migrating birds use infrasound to orient themselves. Others may use visual cues, like the angle of polarizing light or the direction of sunset. Many birds that migrate long distances use the earth's geomagnetic field as a source of compass information. If the magnetic field of a blind "orientation cage" (see photo below) is deflected by 120 degrees clockwise by an artificial magnet, a bird that normally orients to the north will orient toward the southeast.

The sensory system underlying the magnetic compass of these birds is one of the great mysteries of sensory biology. There are two competing hypotheses.

The magnetite hypothesis. One hypothesis is that crystals of the magnetic mineral magnetite within brain cells of migrating birds act as miniature compass needles. While trace amounts of magnetite are indeed present in some brain cells, intensive research has failed to confirm that information about the orientation of magnetite particles within these cells is transmitted to any other cells of the brain.

The photoreceptor hypothesis. An alternative hypothesis is that the primary process underlying the compass is instead a magnetically-sensitive chemical reaction within the photoreceptors of the bird's eyes. The alignment of photopigment molecules with the earth's magnetic field might alter the visual pattern in a way that could be used to obtain directional information.

Which hypothesis is correct? Experiments have shown that the magnetic detector used by birds in blind cages is light sensitive, as a photoreceptor compass should be—but this would also be true of a light-activated magnetite compass.

In 2004, University of California Irvine researchers devised a clever experiment to distinguish between the two hypotheses. They studied the way in which migrating European robins held in orientation cages use the magnetic field as a source of compass information to hop in the appropriate migratory direction. They found that the robins oriented 16 degrees north during the spring migration, the appropriate direction. To distinguish between the magnetite and photoreceptor hypotheses, the robins in the cages were exposed to oscillating low-level radio frequencies (7 MHz) that would disrupt the energy state of any light-absorbing photoreceptor molecules involved in sensing the magnetic field, but would not affect the alignment of magnetite particles.

The chart above presents the results of this study. Each data entry is the mean of three recordings. For each recording, the bird was placed in a 35-inch conical orientation cage

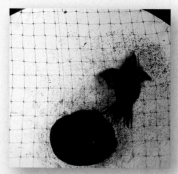

Effect of Radio-Disruption on Orientation

| | Mean Heading (degrees) | |
Bird	Geomagnetic field only	Radio-disrupted
1	26	110
2	20	126
3	4	86
4	350	17
5	15	162
6	1	330
7	18	297
8	20	220
9	354	58
10	24	261
11	358	278
12	37	3

lined with coated paper, and the vector (the directional position of first contact with the paper) recorded relative to magnetic North (North = 360 degrees; East = 90 degrees; South = 180 degrees; West = 270 degrees).

Analysis

1. **Interpreting Data** Plot each column on a circle. For birds orienting to the geomagnetic field without radio interference, what is the greatest difference (expressed in degrees) between recorded vectors and the mean vector of 16 degrees? for birds orienting with 7 MHz radio interference?

2. **Making Inferences**
   a. For birds orienting to the geomagnetic field without radio interference, how many of the 12 birds oriented with an accuracy of +/– 30 degrees relative to the mean vector of 16 degrees? for birds orienting with 7 MHz radio interference, how many were +/– 30 degrees?
   b. If you were to select a bird at random, what is the probability that it would orient within +/– 30 degrees of the appropriate migration direction (16 degrees North) without radio interference? with radio interference?

3. **Drawing Conclusions** Is the ability of European robins to orient correctly with respect to geomagnetic fields disrupted by 7 MHz radio frequencies? Is it fair to conclude that the birds' compass sense involves a molecule sensitive to radio disruption, such as a photoreceptor? that it does not involve particles not sensitive to radio disruption?

Concept Summary

Neurons and How They Work

29.1 The Animal Nervous System

- The nervous system is the communication network in the body (**figure 29.1**). Three types of neurons are found in the nervous system: sensory neurons, motor neurons, and association neurons (**figure 29.2**).

29.2 Neurons Generate Nerve Impulses

- Neurons are cells that conduct electrical impulses but are supported by neuroglial cells, such as Schwann cells and oligodendrocytes. These cells closely associate with the axons, wrapping them in a fatty material called myelin (**figure 29.3**).

- Electrical signals begin in dendrites and travel down an axon. Nerve impulses result from the movement of Na^+ and K^+ ions across the plasma membranes through voltage-gated channels. The movement of Na^+ in one area of the membrane causes a change in electrical properties, called depolarization, which causes the opening of adjacent ion channels. If the depolarization is large enough it will trigger an action potential that will spread down the axon (**figure 29.4**). The action of the Na^+/K^+ pumps restores the resting membrane potential.

29.3 The Synapse

- When a nerve impulse reaches the end of the axon, it triggers the release of neurotransmitters that pass across a small gap, called a synaptic cleft, between the presynaptic neuron and the postsynaptic cell. Neurotransmitter molecules bind to receptors on the postsynaptic cell causing chemically-gated ion channels to open. Ions flow across the plasma membrane creating electrical impulses in the postsynaptic cell (**figure 29.6**). Depending on the type of ion that flows into the cell, the synapse is excitatory or inhibitory. All neural inputs are integrated in the postsynaptic cell, shown here from **figure 29.7**, producing an overall positive or negative change in membrane potential.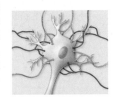

The Central Nervous System

29.4 How the Brain Works

- The cerebral cortex lies over the cerebrum and is the site of neural activities such as language, conscious thought, memory, personality development, vision, and many other higher-level activities (**figures 29.8** and **29.9**).

- The thalamus and hypothalamus, which lie underneath the cerebrum, process information and integrate bodily functions. Areas of the hypothalamus are also part of the limbic system, which is involved in deep-seated drives and emotions, such as pain, anger, sex, hunger, thirst, and pleasure (**figure 29.10**).

- The cerebellum controls balance, posture, and muscular coordination. The brain stem controls vital functions, such as breathing, swallowing, heart beat, and digestion. Language, memory, and learning are localized in the cerebrum. The left hemisphere is dominant for language in the majority of people (**figure 29.11**). The processes of memory and learning are not as well understood.

29.5 The Spinal Cord

- The spinal cord is a cable of neurons that extends from the brain down the back and is encased in the bony vertebrae of the backbone (**figure 29.14**). Motor nerves carry impulses from the brain and spinal cord out to the body, and sensory nerves carry impulses from the body to the brain and spinal cord.

The Peripheral Nervous System

29.6 Voluntary and Autonomic Nervous Systems

- The voluntary nervous system relays commands between the CNS and skeletal muscles and can be consciously controlled; however, reflexes, such as the knee-jerk reflex shown here from **figure 29.16,** work without conscious control. The autonomic nervous system consists of opposing sympathetic and parasympathetic divisions that unconsciously relay commands between the CNS and muscles and glands (**figure 29.17**).

The Sensory Nervous System

29.7 Sensory Perception

- Neurons called sensory receptors initiate and carry nerve impulses from throughout the body to the CNS. Different sensory cells are stimulated by different stimuli. Exteroceptors sense stimuli from the external environment, and interoceptors, such as stretch receptors and baroreceptors, sense stimuli within the body (**figures 29.19** and **29.20**).

29.8 Sensing Gravity and Motion

- Sensory receptors in the ear sense gravity and acceleration. The otolith sensory receptors detect gravity by the deflection of hair cells caused by the movement of otoliths in a gelatin-like matrix (**figure 29.21b**). Motion is detected by the deflection of hair cells in the cupula of the semicircular canals (**figure 29.21c**).

29.9 Sensing Chemicals: Taste and Smell

- Chemicals are detected through the senses of taste, using taste buds on the tongue and smell, using olfactory receptors that line the nasal passages (**figures 29.22** and **29.23**). Different tastes and smells are detected because different chemicals stimulate different receptors on the tongue and the nasal passages.

29.10 Sensing Sounds: Hearing

- Sound receptors detect vibrations of air through the deflection of hair cells in the inner ear. Sound waves cause the eardrum to vibrate. That vibration is amplified by bones in the middle ear that displaces fluid in the inner ear. This movement of fluid causes the deflection of hair cells; different frequencies of sound stimulate different cells (**figure 29.24**).

29.11 Sensing Light: Vision

- Sensory receptors in the eye use a pigment to detect light. Light receptors evolved in several different animal phyla (**figures 29.25–29.27**). Rod cells detect the intensity of light, while cone cells detect different colors of light (**figures 29.28–29.31**). Binocular vision allows for depth perception, but has a more limited field of view (**figure 29.33**).

Self-Test

1. Which of the following is *not* found in the peripheral nervous system?
   - **a.** Schwann cells
   - **c.** association neurons
   - **b.** sensory neurons
   - **d.** motor neurons
2. An action potential is caused by a quick depolarization of the membrane in a nerve cell resulting from the
   - **a.** influx of sodium ions.
   - **b.** actions of the Na$^+$/K$^+$ pump.
   - **c.** influx of potassium ions.
   - **d.** all of the above.
3. Excitatory neurotransmitters initiate an action potential in a postsynaptic neuron by opening
   - **a.** sodium ion gates in the postsynaptic cell.
   - **b.** potassium ion gates in the postsynaptic cell.
   - **c.** chloride ion gates in the postsynaptic cell.
   - **d.** calcium ion gates in the postsynaptic cell.
4. Neurotransmitters are released
   - **a.** from the postsynaptic cell.
   - **b.** into the synaptic cleft.
   - **c.** and bind to receptors on the presynaptic cell.
   - **d.** all of the above.
5. The integration of internal activities of the body are controlled by the
   - **a.** cerebrum.
   - **c.** hypothalamus.
   - **b.** cerebellum.
   - **d.** brain stem.

6. The purpose of the autonomic nervous system is to do all of the following *except*
   - **a.** stimulate glands.
   - **b.** relay messages to skeletal muscles.
   - **c.** relay messages to cardiac and smooth muscles.
   - **d.** regulate the body's homeostasis.
7. When arm muscles hurt after heavy exercise, the pain is detected by
   - **a.** neurotransmitters.
   - **c.** associative neurons.
   - **b.** interoceptors.
   - **d.** exteroceptors.
8. The ear senses different stimuli. Which of the following structures of the ear is associated with sensing motion and gravity?
   - **a.** cochlea
   - **c.** semicircular canals
   - **b.** ear bones (the ossicles)
   - **d.** eardrum
9. Which of the following is a chemical sensory system?
   - **a.** motion
   - **c.** taste
   - **b.** stretch receptors
   - **d.** vision
10. Which of the following statements is incorrect?
    - **a.** Vertebrates focus the eye by changing the shape of the lens.
    - **b.** The eyes of arthropods and vertebrates use the same light-capturing molecule.
    - **c.** Rod cells detect different colors and cones cells detect different shades of gray, working to see in dim light.
    - **d.** Light changes *cis*-retinal into *trans*-retinal.

Visual Understanding

1. **Page 557** Describe the feelings that might be experienced by someone who goes through the four stages related to drug addiction that are shown in the drawings.

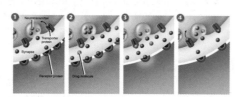

2. **Figure 29.26** Although they evolved separately, what do the visual sensory systems of annelids, mollusks, arthropods, and vertebrates have in common?

Challenge Questions

1. There is a lot of concern today about high sodium (Na) levels in our diet, and the resulting high blood pressure for many people. A friend suggests that if a low-sodium diet is healthy, then a no-sodium diet should be even better. What do you tell him?

2. Why is it important that the neurons in the autonomic nervous system are always active, not shutting down at night while you're asleep?

3. If you only taste a few things (salt, bitter, sweet, sour, umami), why do foods all seem so distinct?

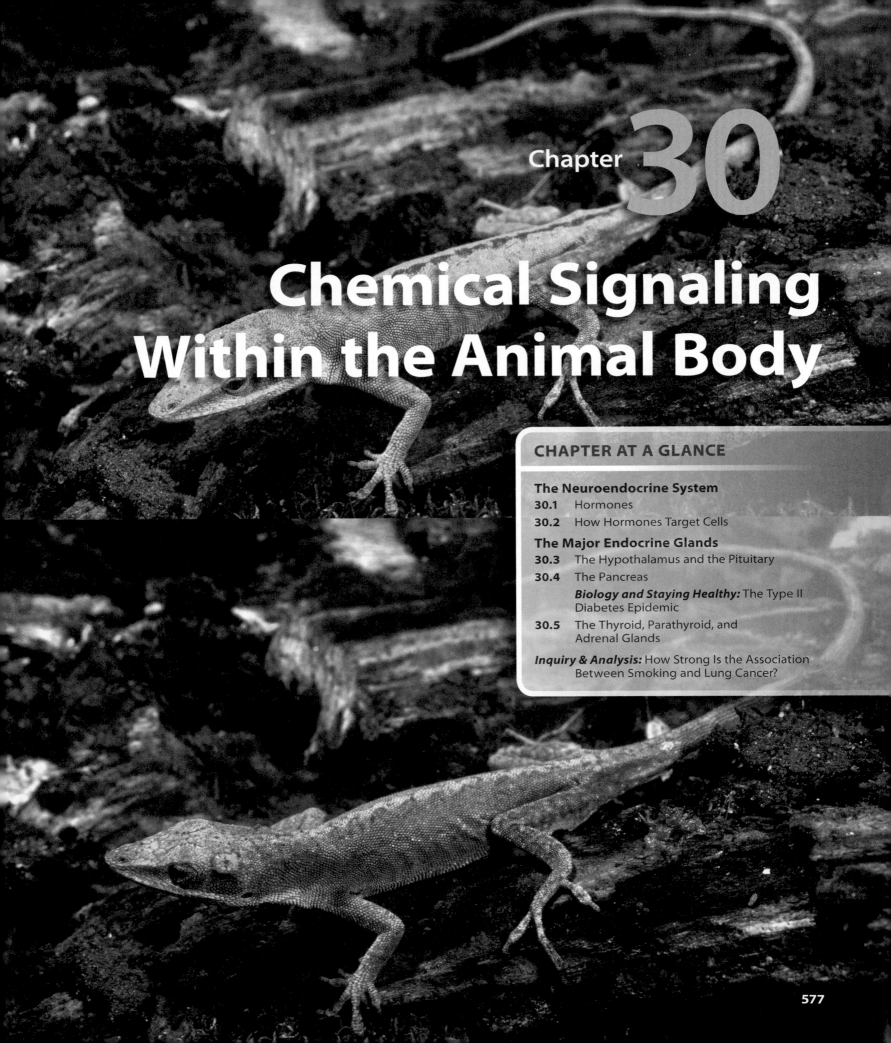

Chapter 30

Chemical Signaling Within the Animal Body

CHAPTER AT A GLANCE

The Neuroendocrine System

30.1 Hormones

30.2 How Hormones Target Cells

The Major Endocrine Glands

30.3 The Hypothalamus and the Pituitary

30.4 The Pancreas

Biology and Staying Healthy: The Type II Diabetes Epidemic

30.5 The Thyroid, Parathyroid, and Adrenal Glands

Inquiry & Analysis: How Strong Is the Association Between Smoking and Lung Cancer?

The Neuroendocrine System

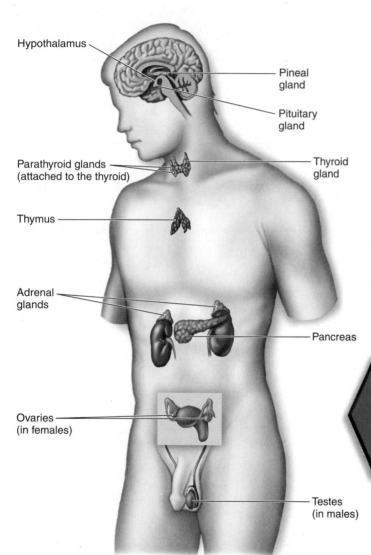

Hypothalamus

Pineal gland

Pituitary gland

Parathyroid glands (attached to the thyroid)

Thyroid gland

Thymus

Adrenal glands

Pancreas

Ovaries (in females)

Testes (in males)

Figure 30.1 **Major glands of the human endocrine system.**

Hormone-secreting cells are clustered in endocrine glands. The pituitary and adrenal glands are each composed of two glands.

30.1 Hormones

CONCEPT PREVIEW: Hormones are effective because they are recognized by specific receptors. Thus, only cells possessing the appropriate receptor will respond to a particular hormone.

A **hormone** is a chemical signal produced in one part of the body that is stable enough to be transported in active form far from where it is produced and that typically acts at a distant site. There are three big advantages to using chemical hormones as messengers rather than speedy electrical signals (like those used in nerves) to control body organs. First, chemical molecules can spread to all tissues via the blood (imagine trying to wire every cell with its own nerve!) and are usually required in only small amounts. Second, chemical signals can persist much longer than electrical ones, a great advantage for hormones controlling slow processes like growth and development. Third, many different kinds of chemicals can act as hormones, so different hormone molecules can be targeted at different tissues. For all these reasons, hormones are excellent messengers for signaling widespread, slow-onset, long duration responses.

Hormones, in general, are produced by glands, most of which are controlled by the central nervous system. Because these glands are completely enclosed in tissue rather than having ducts that empty to the outside, they are called **endocrine glands** (from the Greek, *endon,* within). Hormones are secreted from them directly into the bloodstream (this is in contrast to **exocrine glands,** like sweat glands, that have ducts). Your body has a dozen principal endocrine glands (figure 30.1) that together make up the endocrine system.

The *endocrine system* and the *motor nervous system* are the two main routes the central nervous system (CNS) uses to issue commands to the organs of the body. The two are so closely linked that they are often considered a single system—the **neuroendocrine system.** The **hypothalamus** can be considered the main switchboard of the neuroendocrine system. The hypothalamus is continually checking conditions inside the body to maintain a constant internal environment, a condition known as homeostasis. Is the body too hot or too cold? Is it running out of fuel? Is the blood pressure too high? If homeostasis is no longer maintained, the hypothalamus has several ways to set things right again. For example, if the hypothalamus needs to speed up the heart rate, it can send a nerve signal to the medulla oblongata, or it can use a chemical command, causing the adrenal gland to produce the hormone epinephrine, which also speeds up the heart rate. Which command the hypothalamus uses depends on the desired duration of the effect. A chemical message is typically far longer lasting than a nerve signal.

The hypothalamus was discussed on page 559 as part of the CNS, where it integrates all internal activities of the body through nerve impulses. As you will discover in this chapter it also has an endocrine function to maintain homeostasis, introduced on page 520.

The Chain of Command

The hypothalamus issues commands to a nearby gland, the pituitary, which in turn sends chemical signals to the various hormone-producing glands of the body. The CNS regulates the body's hormones through a chain of command. The "releasing" hormones made by the hypothalamus cause the pituitary to synthesize a corresponding pituitary hormone, which travels to a distant endocrine gland and causes that gland to begin

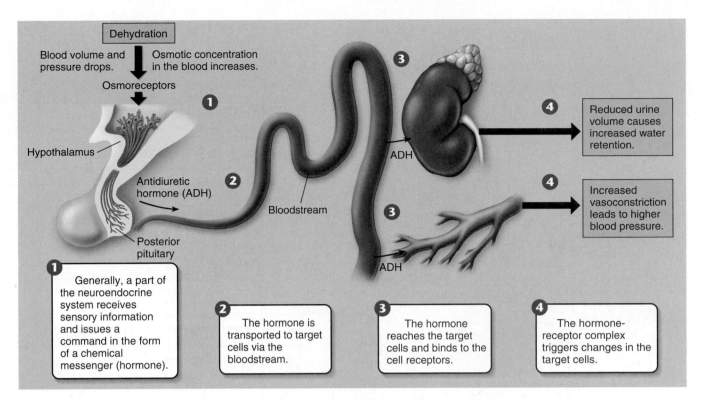

Figure 30.2   How hormonal communication works.

producing its particular endocrine hormone. The hypothalamus also secretes inhibiting hormones that keep the pituitary from secreting specific pituitary hormones.

How Hormones Work

The key reason why hormones are effective messengers within the body is because a particular hormone can influence a specific target cell. How does the target cell recognize that hormone, ignoring all others? Embedded in the plasma membrane or within the target cell are receptor proteins that match the shape of the potential signal hormone like a hand fits a glove. As you recall from chapter 29, nerve cells have highly specific receptors within their synapses, each receptor shaped to "respond" to a different neurotransmitter molecule. Cells that the body has targeted to respond to a particular hormone have receptor proteins shaped to fit that hormone and no other. Thus, chemical communication within the body involves *two* elements: a molecular signal (the hormone) and a protein receptor on or in target cells. The system is highly specific because each protein receptor has a shape that only a particular hormone fits.

The path of communication taken by a hormonal signal can be visualized as the series of simple steps shown in the example in figure 30.2:

❶ **Issuing the command.** Hormones produced in cells in the hypothalamus are stored in the posterior pituitary and are released into the bloodstream in response to a signal from the brain.

❷ **Transporting the signal.** The hormones travel through the bloodstream to the target cells.

❸ **Hitting the target.** When a hormone encounters a cell with a matching receptor, called a target cell, the hormone binds to that receptor.

❹ **Having an effect.** When the hormone binds to the receptor protein, the protein responds by changing shape, triggering a change in cell activity.

IN THE NEWS

Love-Enhancing Pharmaceuticals? Neuroscientists have developed many drugs that influence emotional states such as depression and anxiety, but how about love? They now report that dopamine-related reward regions of the brain are activated in people looking at photos of their lovers, the same regions stimulated by drugs such as nicotine and cocaine to produce euphoria, and by the hormone oxytocin to stimulate pair bonding. Other female mammals become attached to the nearest male if their brain is infused with oxytocin. Internet entrepreneurs have begun marketing oxytocin-laced sprays that claim to enhance dating success. While unlikely to do so, they point to a future when advances in neuroscience may transform love to an adjustable condition. What do you think of this?

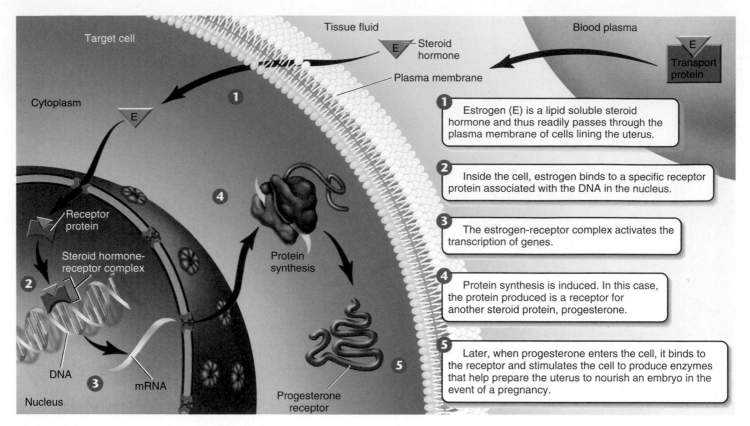

Figure 30.3 How steroid hormones work.

30.2 How Hormones Target Cells

CONCEPT PREVIEW: Steroid hormones pass through the cell's plasma membrane and bind to receptors inside the cell that alter the transcription of specific genes. Peptide hormones bind to cell surface receptors, triggering an enzyme cascade within the cell.

Steroid Hormones Enter Cells

Some protein receptors designed to recognize hormones are located in the cytoplasm or nucleus of the target cell. The hormones in these cases are typically lipid-soluble **steroid hormones.** The chemical shapes of these molecules are multi-ring-structures resembling chicken wire. All steroid hormones are manufactured from cholesterol, a complex molecule composed of four rings. They include the hormones that promote the development of secondary sexual characteristics such as testosterone, estrogen, and progesterone, discussed in detail in chapter 31.

Steroid hormones like estrogen, "E" in figure 30.3, can pass across the lipid bilayer of the cell plasma membrane ❶, and bind to receptors within the cell and often, as in the case with estrogen, within the nucleus. The hormone-receptor complex then binds to the DNA in the nucleus ❷, and activates the gene for a progesterone receptor protein, which is transcribed ❸. The protein is synthesized ❹, and the receptor is available to bind progesterone when it enters the cell ❺, which itself activates another set of genes.

Anabolic steroids are synthetic compounds that resemble the male sex hormone testosterone. The injection of anabolic steroids into muscles activates growth genes and causes the muscle cells to produce more pro-

> Steroids, discussed on page 56, are a type of lipid. Because lipids are hydrophobic, steroids are able to cross the hydrophobic inner layer of the lipid bilayer, passing directly into the cell.

BIOLOGY & YOU

Performance-Enhancing Sports Supplements. Many athletes, coaches, and fans feel that the use of anabolic steroids and other substances to improve performance is unethical in sports. Steroids are actually illegal. On the other hand, no one would argue that taking vitamin supplements was unethical. There is a large array of products between these two extremes. Many substances widely marketed as "performance-enhancing dietary supplements" are discouraged by most coaches and have limited research on their safety or effectiveness. For example, creatin is claimed to improve sprinting and weight lifting, ephedrine to improve strength and endurance, and glucosamine to stimulate cartilage formation, but there is little evidence that they do what they claim. On balance, when in doubt, don't take it.

tein, resulting in bigger muscles and increased strength. However, anabolic steroids have many dangerous side effects, including liver damage, heart disease, and high blood pressure. Anabolic steroids are illegal, and athletes in many sports are tested for their use.

Peptide Hormones Act at the Cell Surface

Other hormone receptors are embedded within the plasma membrane, with their recognition regions directed outward from the cell surface. **Peptide hormones** like the one binding to the receptor in figure 30.4 ❶, are typically short peptide chains (although some are full-sized proteins). The binding of the peptide hormone to the receptor triggers a change in the cytoplasmic end of the receptor protein. This change then triggers events within the cell cytoplasm, usually through intermediate within-cell signals called **second messengers** ❷, which greatly amplify the original signal and results in changes in the cell ❸.

How does a second messenger amplify a hormone's signal? Second messengers activate enzymes. One of the most common second messengers is cyclic AMP (cAMP), and its actions are shown in figure 30.5. A single hormone molecule binding to a receptor in the plasma membrane can result in the formation of many second messengers in the cytoplasm. Each second messenger can activate many molecules of a certain enzyme, and sometimes each of these enzymes can in turn activate many other enzymes. Thus, second messengers enable each hormone molecule to have a tremendous effect inside the cell, far greater than if the hormone had simply entered the cell and sought out a single target.

Concept Check

1. What does a steroid hormone do after it enters a cell?
2. Do anabolic steroids occur naturally in your body?
3. How do second messengers amplify hormone signals?

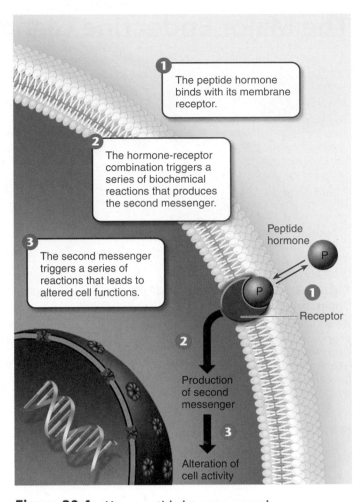

1. The peptide hormone binds with its membrane receptor.

2. The hormone-receptor combination triggers a series of biochemical reactions that produces the second messenger.

3. The second messenger triggers a series of reactions that leads to altered cell functions.

Peptide hormone

Receptor

2. Production of second messenger

3. Alteration of cell activity

Figure 30.4 How peptide hormones work.

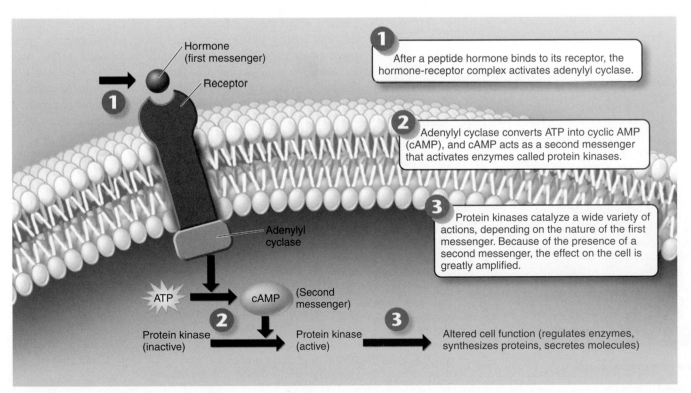

Hormone (first messenger)

Receptor

1. After a peptide hormone binds to its receptor, the hormone-receptor complex activates adenylyl cyclase.

2. Adenylyl cyclase converts ATP into cyclic AMP (cAMP), and cAMP acts as a second messenger that activates enzymes called protein kinases.

3. Protein kinases catalyze a wide variety of actions, depending on the nature of the first messenger. Because of the presence of a second messenger, the effect on the cell is greatly amplified.

Adenylyl cyclase

ATP

cAMP (Second messenger)

Protein kinase (inactive)

Protein kinase (active)

Altered cell function (regulates enzymes, synthesizes proteins, secretes molecules)

Figure 30.5 How second messengers work.

The Major Endocrine Glands

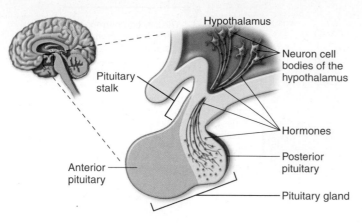

Figure 30.6   **The posterior pituitary contains cells that originate in the hypothalamus.**

A tract of nerve cells originates in the hypothalamus and extends down along the pituitary stalk and ends in the posterior pituitary. The cell bodies of the neurons produce hormones, which travel down the axons and are stored in the posterior pituitary. Thus, the hormones released from the posterior pituitary are actually synthesized in neurons in the hypothalamus.

30.3   The Hypothalamus and the Pituitary

CONCEPT PREVIEW: The posterior pituitary gland contains axons originating from neurons in the hypothalamus that produce hormones. The anterior pituitary responds to hormonal signals from the hypothalamus and produces a family of pituitary hormones that are carried to distant glands that in turn produce specific hormones.

The hypothalamus, the "control center" of the neuroendocrine system, exerts its control by releasing hormones that influence the nearby **pituitary gland,** located in a bony recess in the brain just below the hypothalamus. The pituitary in turn produces hormones that influence the body's other endocrine glands. Hormones produced by the back portion of the pituitary, or *posterior lobe,* regulate water conservation, as well as milk letdown and uterine contraction in women; hormones produced by the front portion, or *anterior lobe,* regulate the other endocrine glands.

The Posterior Pituitary

The posterior pituitary contains axons that originate in cell bodies within the hypothalamus. The hormones released from the posterior pituitary are actually produced by neuron cell bodies located in the hypothalamus. The hormones are transported to the posterior pituitary through axon tracts and are stored and released from the posterior pituitary (figure 30.6).

The role of the posterior pituitary first became evident in 1912, when a remarkable medical case was reported: A man who had been shot in the head developed a surprising disorder—he began to urinate every 30 minutes, unceasingly. The bullet had lodged in his pituitary gland, and subsequent research demonstrated that surgical removal of the pituitary also produces these unusual symptoms. Pituitary extracts were shown to contain a substance that makes the kidneys conserve water, and in the early 1950s the peptide hormone vasopressin (now more commonly called **antidiuretic hormone, ADH**) was isolated. As you learned in chapter 27, ADH regulates the kidney's retention of water. When ADH is missing, the kidneys cannot retain water, which is why the bullet caused excessive urination. Excessive alcohol and caffeine consumption, which inhibit ADH secretion, have a similar effect.

The posterior pituitary also releases a second hormone, oxytocin, of very similar structure—both are short peptides composed of nine amino acids—but very different function. Oxytocin initiates uterine contraction during childbirth and milk release in mothers. Here is how milk release works: Sensory receptors in the mother's nipples, when stimulated by sucking, send messages to the hypothalamus, causing the hypothalamus to stimulate the release of oxytocin from the posterior pituitary. The oxytocin travels in the bloodstream to the breasts, where it stimulates contraction of the muscles around the ducts into which the mammary glands secrete milk.

Both oxytocin and ADH are produced in the cell bodies of the hypothalamus but stored and released from the posterior pituitary.

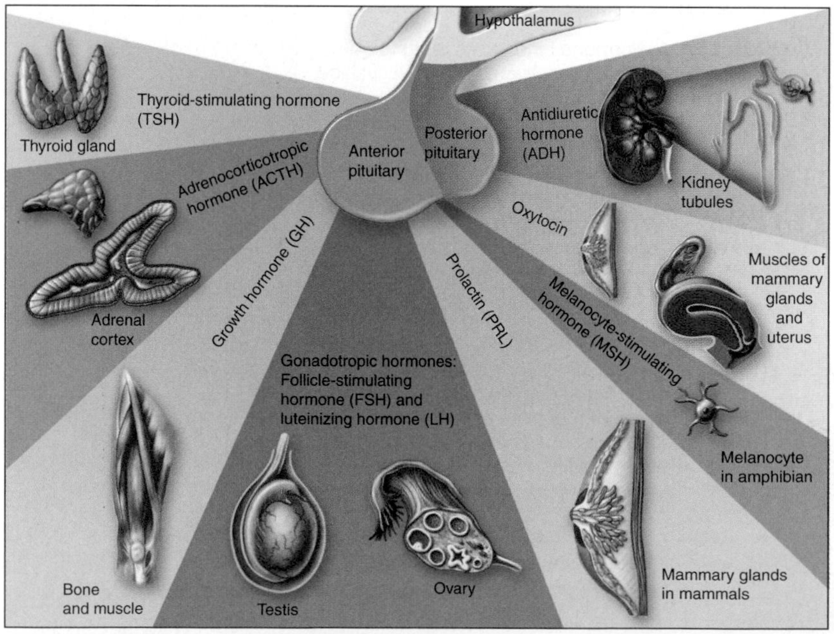

Figure 30.7   **The role of the pituitary.**

The Anterior Pituitary

The anterior pituitary gland produces seven major peptide hormones (highlighted in blue in figure 30.7), each controlled by a particular releasing signal secreted from the hypothalamus. All seven of them have major endocrine roles, as well as poorly understood roles in the central nervous system:

1. **Thyroid-stimulating hormone (TSH).** TSH stimulates the thyroid gland to produce the thyroid hormone thyroxine, which in turn stimulates oxidative respiration.

2. **Adrenocorticotropic hormone (ACTH).** ACTH stimulates the adrenal gland to produce a variety of steroid hormones. Some regulate the production of glucose; others regulate the balance of sodium and potassium ions in the blood.

3. **Growth hormone (GH).** GH stimulates the growth of muscle and bone throughout the body.

4. **Follicle-stimulating hormone (FSH).** FSH is significant in the female menstrual cycle by triggering the maturation of egg cells and stimulating the release of estrogen. In males, it stimulates cells in the testes, regulating development of the sperm.

5. **Luteinizing hormone (LH).** LH plays an important role in the female menstrual cycle by triggering ovulation, which is the release of a mature egg. It also stimulates the male gonads to produce testosterone.

> Follicle-stimulating hormone and luteinizing hormone (along with estrogen and progesterone that are produced elsewhere in the body) coordinate the female reproductive cycle, discussed in more detail on page 600.

6. **Prolactin (PRL).** Prolactin stimulates the breasts to produce milk, which is released in response to oxytocin.

7. **Melanocyte-stimulating hormone (MSH).** In reptiles and amphibians, MSH stimulates color changes in the epidermis, which caused the lizard on the opening page of this chapter to turn brown. The function of this hormone in humans is still poorly understood.

How the Hypothalamus Controls the Anterior Pituitary

As noted earlier, the hypothalamus controls production and secretion of the anterior pituitary hormones by means of a family of special hormones. Neurons in the hypothalamus secrete these releasing and inhibiting hormones into blood capillaries at the base of the hypothalamus. Figure 30.8 shows the relationship of two groups of neurons in the hypothalamus. As discussed earlier, neurons (colored blue in the figure) extend from the hypothalamus into the posterior pituitary, where axons deliver the hormones for storage and release. Other neurons of the hypothalamus (colored yellow in the figure) produce releasing and inhibiting hormones and release them into capillaries. These capillaries drain into small veins that run within the stalk of the pituitary to a second bed of capillaries in the anterior pituitary. This unusual system of vessels is known as the *hypothalamohypophyseal portal system*. It is called a portal system because it has a second capillary bed downstream from the first; the only other body location with a similar system is the liver.

Each releasing hormone delivered to the anterior pituitary by this portal system regulates the secretion of a specific anterior pituitary hormone. For example, thyrotropin-releasing hormone (TRH) stimulates the release of TSH, corticotropin-releasing hormone (CRH) stimulates the release of ACTH, gonadotropin-releasing hormone (GnRH) stimulates the release of FSH and LH, and growth-hormone-releasing hormone (GHRH) stimulates the release of GH.

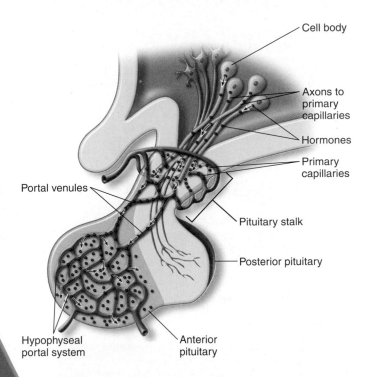

Figure 30.8 **Hormonal control of the anterior pituitary gland by the hypothalamus.**

Neurons in the hypothalamus secrete hormones that are carried by short blood vessels to the anterior pituitary gland, where they either stimulate or inhibit the secretion of anterior pituitary hormones.

BIOLOGY & YOU

Polycystic Ovary Syndrome. Sometimes disorders with odd-sounding names can play a very important role in the lives of many of us. A case in point is polycystic ovary syndrome (PCOS), an endocrine disorder that affects approximately 5% of all women. It is the most common hormonal disorder among women of reproductive age, and is a leading cause of infertility. The principal symptoms are weight problems, lack of regular ovulation, and excessive amounts of androgenic (masculinizing) hormones. What causes PCOS? Polycystic ovaries develop when the ovaries are stimulated to produce excessive amounts of male hormones, typically by high levels of insulin in the blood. A majority of patients with PCOS have insulin resistance, leading to the elevated insulin levels that trigger the condition. This explains why type II diabetes and obesity are both strongly correlated with PCOS. While there is considerable debate as to optimal treatment for PCOS (largely because of a lack of large-scale clinical trials comparing different treatments), general interventions to reduce weight and insulin resistance are recommended, because they address the underlying cause of the syndrome.

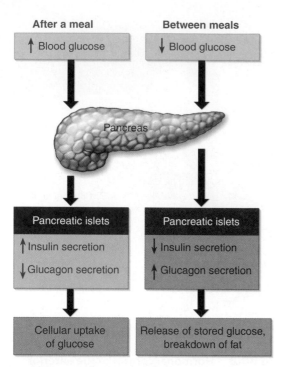

After a meal
↑ Blood glucose

Between meals
↓ Blood glucose

Pancreas

Pancreatic islets
↑ Insulin secretion
↓ Glucagon secretion

Pancreatic islets
↓ Insulin secretion
↑ Glucagon secretion

Cellular uptake of glucose

Release of stored glucose, breakdown of fat

Figure 30.9   Insulin and glucagon secreted by the pancreas regulate blood glucose levels.

After a meal, an increased secretion of insulin by the beta cells of the pancreatic islets of Langerhans promotes the movement of glucose from blood into tissue cells. Between meals, an increased secretion of glucagon by the alpha cells of the pancreatic islets and decreased secretion of insulin cause the release of stored glucose and the breakdown of fat.

IN THE NEWS

Pancreatic Cancer. Among the most deadly of all cancers are malignant tumors of the pancreas. Each year in the United States, about 37,700 individuals are diagnosed with this condition, and 34,300 die from the disease. Less than 5% of those diagnosed with pancreatic cancer are still alive five years after diagnosis; 50% are dead within six months. Pancreatic cancer is sometimes called a "silent killer" because early pancreatic cancer often does not cause symptoms, so a tumor is often not diagnosed until it is far advanced. Unfortunately, we don't have much of an idea of what causes pancreatic cancer. It is more common in men over 60, and slightly more common in cigarette smokers, diabetics, and obese people, but there are no established guidelines for preventing pancreatic cancer.

30.4 The Pancreas

CONCEPT PREVIEW: Clusters of cells within the pancreas secrete the hormones insulin and glucagon. Insulin stimulates the storage of glucose as glycogen, while glucagon stimulates glycogen breakdown to glucose. Working together, these hormones keep blood glucose levels within a narrow range.

The **pancreas** gland is located behind the stomach and is connected to the front end of the small intestine by a narrow tube. It secretes a variety of digestive enzymes into the digestive tract through this tube, and for a long time it was thought to be solely an exocrine gland. In 1869, however, a German medical student named Paul Langerhans described some unusual clusters of cells scattered throughout the pancreas. In 1893, doctors concluded that these clusters of cells, which came to be called islets of Langerhans, produced a substance that prevented diabetes mellitus. **Diabetes mellitus** is a serious disorder in which affected individuals' cells are unable to take up glucose from the blood, even though their levels of blood glucose become very high. Some individuals lose weight and literally starve; others develop poor circulation, sometime resulting in amputation of limbs with restricted circulation. Diabetes is the leading cause of blindness among adults, and it accounts for one-third of all kidney failures. It is the seventh leading cause of death in the United States.

> The pancreas is a major exocrine gland, as discussed on page 514. Pancreatic enzymes released into the small intestine catalyze the breakdown of proteins, carbohydrates, and fats. In addition, the pancreas also secretes bicarbonate, which neutralizes stomach acids.

The islets of Langerhans in the pancreas produce two hormones that interact to govern the levels of glucose in the blood. These hormones are *insulin* and *glucagon.* Insulin is a storage hormone, designed to put away nutrients for leaner times. It promotes the accumulation of glycogen in the liver and triglycerides in fat cells. When food is consumed (left side of figure 30.9), beta cells in the islets of Langerhans secrete insulin, causing the cells of the body to take up and store glucose as glycogen and triglycerides to be used later. When body activity causes the level of glucose in the blood to fall as it is used as fuel (right side of figure 30.9), other cells in the islets of Langerhans, called alpha cells, secrete glucagon, which causes liver cells to release stored glucose and fat cells to break down triglycerides for energy use. The two hormones work together to keep glucose levels in the blood within a narrow range.

Over 23 million people in the United States, and over 246 million people worldwide, have **diabetes.** There are *two* kinds of diabetes mellitus. About 5% to 10% of affected individuals suffer from type I diabetes, an autoimmune disease in which the immune system attacks the islets of Langerhans, resulting in abnormally low insulin secretion. Called juvenile-onset diabetes, this type usually develops before age 20. Affected individuals can be treated by daily injections of insulin. Active research on the possibility of transplanting islets of Langerhans holds promise as a lasting treatment for type I diabetes.

In type II diabetes, the level of insulin in the blood is often higher than normal, but cells don't respond to insulin. This form of diabetes usually develops in people over 40 years of age. It is almost always a consequence of excessive weight; in the United States, 80% of those who develop type II diabetes are obese. The cells of some type II diabetics, overwhelmed with food, adjust their appetite for glucose downward, reducing their sensitivity to insulin by reducing their number of insulin receptors. To compensate, the pancreas pumps out ever more insulin. Type II diabetes is usually treatable with diet and exercise, and most affected individuals do not need daily injections of insulin.

The Type II Diabetes Epidemic

We Americans love to eat, but in 2007 the Centers for Disease Control and Prevention released a report warning we are eating ourselves into a diabetes epidemic. Diabetes affected 7 million Americans in 1991. At the end of 2007, the number was over 23 million, more than 7% of all Americans, which is an alarming increase in just 16 years!

The same explosion of diabetes is being seen worldwide. Diabetes now effects 246 million people, and kills 3.8 million each year. Every 10 seconds one person dies of diabetes. In the same 10 seconds two more people develop the disease.

Diabetes is a disorder in which the body's cells fail to take up glucose from the blood. Tissues waste away as glucose-starved cells are forced to consume their own proteins. Diabetes is the leading cause of kidney failure, blindness, and amputation in adults. Almost all the increase in diabetes in the last decade is in the 90% of diabetics who suffer from type II, or "adult-onset," diabetes. These individuals lack the ability to use the hormone insulin.

Your body manufactures insulin after a meal as a way to alert cells that higher levels of glucose are coming soon. The insulin signal attaches to special receptors on the cell surfaces, which respond by causing the cell to turn on its glucose-transporting machinery. Some individuals who suffer from type II diabetes have normal or even elevated levels of insulin in their blood, and normal insulin receptors, but for some reason the binding of insulin to their cell receptors does not turn on the glucose-transporting machinery like it is supposed to do. For 30 years researchers have been trying to figure out why not.

How does insulin act to turn on a normal cell's glucose transporting machinery? Proteins called IRS proteins (the names refer not to tax collectors, but to *insulin receptor substrate*) snuggle up against the insulin receptor inside the cell. When insulin attaches to the receptor protein, the receptor responds by adding a phosphate group onto the IRS molecules. Like being touched by a red-hot poker, this galvanizes the IRS molecules into action. Dashing about, they activate a variety of processes, including an enzyme that turns on the glucose-transporting machinery.

When the IRS genes are deliberately taken out of action in so-called knockout mice, type II diabetes results. Are defects in the genes for IRS proteins responsible for type II diabetes? Probably not. When researchers look for IRS gene mutations in inherited type II diabetes, they don't find them. The IRS genes are normal.

This suggests that in type II diabetes something is interfering with the action of the IRS proteins. What might it be? An estimated 80% of those who develop type II diabetes are obese, a tantalizing clue. Look at the graph. Over the same 16 years that diabetes has undergone its explosive increase, the obesity rate increased from 12% of the U.S. population to over 34%.

What is the link between diabetes and obesity? Recent research suggests an answer to this key question. A team of scientists at the University of Pennsylvania School of Medicine had been investigating why a class of drugs called thiazolidinediones (TZDs) helped combat diabetes. They found that TZDs cause the body's cells to use insulin

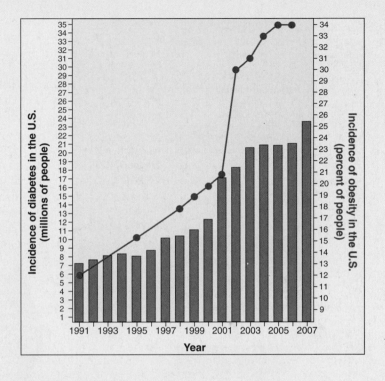

more effectively, and this suggested to them that the TZD drug might be targeting a hormone.

The researchers then set out to see if they could find such a hormone in mice. In search of a clue, they started by looking to see which mouse genes were activated or deactivated by TZD. Several were. Examining them, they were able to zero in on the hormone they sought. Dubbed *resistin,* the hormone is produced by fat cells and prompts tissues to resist insulin. The same *resistin* gene is present in humans too. The researchers speculate that resistin may have evolved to help the body deal with periods of famine.

Mice given resistin by the researchers lost much of their ability to take up blood sugar. When given a drug that lowers resistin levels, these mice recovered the lost glucose-transporting ability.

Researchers don't yet know how resistin acts to lower insulin sensitivity, although blocking the action of IRS proteins seems a likely possibility.

Importantly, dramatically high levels of the hormone were found in mice obese from overeating. Finding this sort of result is like ringing a dinner bell to diabetes researchers. If obesity is causing high resistin levels in humans, leading to type II diabetes, then resistin-lowering drugs might offer a diabetes cure!

On the scent of something important, resistin researchers are now shifting their efforts from mice to humans. Much needs to be checked, as there are no guarantees that what works in a mouse will do so in the same way in a human. Still, the excitement is tangible.

30.5 The Thyroid, Parathyroid, and Adrenal Glands

CONCEPT PREVIEW: The thyroid acts as a metabolic thermostat, secreting hormones that adjust metabolic rate. Parathyroid hormone regulates calcium levels in the blood. The adrenal hormones epinephrine, norepinephrine, and cortisol are released in response to stress and aldosterone promotes the uptake of sodium.

The Thyroid: A Metabolic Thermostat

The **thyroid gland** is shaped like a shield (its name comes from *thyros,* the Greek word for "shield") and lies just below the Adam's apple in the front of the neck. The thyroid makes several hormones, the two most important of which are **thyroxine,** which increases metabolic rate and promotes growth, and **calcitonin,** which inhibits the release of calcium from bones.

Thyroxine regulates the level of metabolism in the body in several important ways. Without adequate thyroxine, growth is retarded. For example, children with underactive thyroid glands are not able to carry out carbohydrate breakdown and protein synthesis at normal rates, a condition called cretinism, which results in stunted growth. Mental retardation can also result, because thyroxine is needed for normal development of the central nervous system.

The thyroid is stimulated to produce thyroxine by the hypothalamus, which is inhibited by thyroxine via negative feedback. The dashed lines in figure 30.10*a* illustrate how thyroxine inhibits the release of TRH and TSH from the hypothalamus and anterior pituitary, respectively.

Thyroxine contains iodine, and if the amount of iodine in the diet is too low, the thyroid cannot make adequate amounts of thyroxine to keep the hypothalamus inhibited. The hypothalamus will then continue to stimulate the thyroid, which will grow larger in a futile attempt to manufacture more thyroxine. The greatly enlarged thyroid gland that results is called a goiter (figure 30.10*b*). This need for iodine in the diet is why iodine is added to table salt.

Calcitonin, which is also produced by the thyroid and will be discussed later, plays a key role in maintaining proper calcium levels in the body.

> Negative feedback loops control many body functions, as discussed on page 520, maintaining them around a set point. In the case of hormone regulation, the hormone itself can be the response that feeds back to the effector (gland), increasing or decreasing its own production.

The Parathyroids: Regulating Calcium

The **parathyroid glands** are four small glands attached to the thyroid. Small and unobtrusive, they were ignored by researchers until well into the last century. The first suggestion that the parathyroids produce a hormone came from experiments in which they were removed from dogs: The concentration of calcium in the dogs' blood plummeted to less than half the normal level. However, if an extract of the parathyroid gland was administered, calcium levels returned to normal. If an excess was administered, calcium levels in the blood became *too* high, and the bones of the dogs were literally dismantled by the extract. It was clear that the parathyroid glands were producing a hormone that acted on calcium, both its uptake into bones and release from bones.

The hormone produced by the parathyroids is **parathyroid hormone (PTH).** It is one of only two hormones in the body that is absolutely essential for survival (the other is aldosterone, a hormone produced by the adrenal glands, discussed on the facing page). PTH regulates the level of calcium

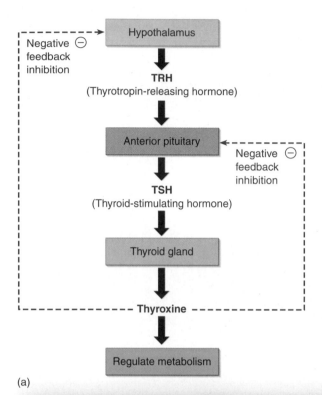

(b)

Figure 30.10 **The thyroid gland secretes thyroxine.**

(a) Thyroxine exerts negative feedback control of the hypothalamus and anterior pituitary. (b) A goiter is caused by a lack of iodine in the diet, which causes thyroxine secretion to decrease. As a result, there is less negative feedback, TSH is not inhibited from stimulating the thyroid gland, and the thyroid gland becomes enlarged.

IMPLICATION Do you think you would be able to treat this woman's goiter by administering iodine? Explain.

in blood. Calcium ions are needed in muscle contraction—by initiating calcium release, nerve impulses cause muscles to contract. A vertebrate cannot live without the muscles that pump the heart and drive the body, and these muscles cannot function if calcium levels are not kept within narrow limits.

PTH acts as a fail-safe to make sure calcium levels never fall too low. If they do, as in figure 30.11*a*, PTH is released into the bloodstream, travels to the bones, and acts on the osteoclast cells (the blue cells) within bones, stimulating them to dismantle bone tissue and release calcium into the bloodstream. PTH also acts on the kidneys to reabsorb calcium ions from the filtrate and leads to activation of vitamin D, necessary for calcium absorption

> Osteoblasts are bone cells that build new bone tissue and osteoclasts are bone cells that secrete enzymes that break down bone tissue. Both cell types are involved in bone remodeling as discussed on page 464.

by the intestine. A diet deficient in vitamin D leads to poor bone formation, a condition called rickets. The hormone calcitonin is released from the thyroid gland and acts in reverse of PTH. When calcium levels in the blood rise (figure 30.11*b*), calcitonin activates osteoblast cells (the orange cells) to take up calcium, and rebuild bone.

The Adrenals: Two Glands in One

Mammals have two **adrenal glands,** one located just above each kidney (see table 30.1 on the next page). Each adrenal gland is composed of two parts: (1) an inner core, the **medulla,** which produces the hormones epinephrine (also called adrenaline) and norepinephrine; and (2) an outer shell, the **cortex,** which produces the steroid hormones cortisol and aldosterone.

The Adrenal Medulla: Emergency Warning Siren. The adrenal medulla releases **epinephrine** and **norepinephrine** in times of stress. These hormones act as emergency signals that stimulate rapid deployment of body fuel. The "alarm" response these hormones produce throughout the body is identical to the individual effects achieved by the sympathetic nervous system, but it is much longer lasting. Among their effects are accelerated heartbeat, increased blood pressure, higher levels of blood sugar, and increased blood flow to the heart.

The Adrenal Cortex: Inflammation, Stress, and Maintaining the Proper Amount of Salt. The adrenal cortex produces the steroid hormone cortisol. **Cortisol** (also called hydrocortisone) acts on many different cells in the body to maintain nutritional well-being. It stimulates carbohydrate metabolism and reduces inflammation. Synthetic derivatives of this hormone, such as prednisone, have widespread medical use as anti-inflammatory agents. Cortisol is also called the *stress hormone,* released in times of stress to help the body deal with acute stress. Problems arise when the body experiences chronic stress and cortisol levels remain high in the body. This can lead to problems with maintaining blood sugar, high blood pressure, reduced immune function, fat accumulation, among others. These chronic effects of cortisol are unhealthy.

The adrenal cortex also produces **aldosterone.** Aldosterone acts primarily in the kidney to promote the uptake of sodium and other salts from the urine, which also increases the reabsorption of water. Aldosterone is, with PTH, one of the two endocrine hormones essential for survival. Removal of the adrenal glands is invariably fatal.

Concept Check

1. Why does pituitary injury lead to excessive urination?
2. What is the relationship between the hypothalamus and the pituitary gland?
3. Distinguish between type I and type II diabetes.

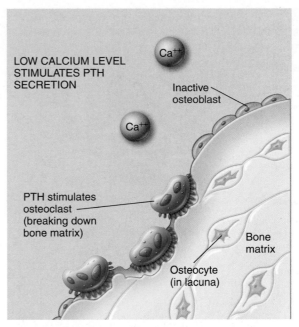

(a)

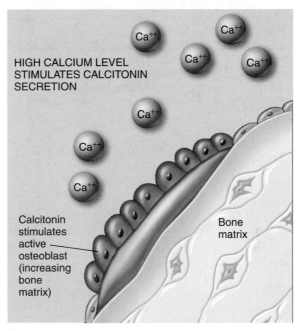

(b)

Figure 30.11 Maintenance of proper calcium levels in the blood.

(a) When calcium levels in the blood become too low, the parathyroid gland produces additional amounts of PTH, which stimulates the breakdown of bone, releasing calcium. (b) Conversely, abnormally high levels of calcium in the blood trigger the thyroid gland to secrete calcitonin, which inhibits the release of calcium from bone, and promotes the activity of osteoblasts to remove calcium from the blood and deposit it in bone.

TABLE 30.1 | The Principal Endocrine Glands

Endocrine Gland and Hormone	Target	Principal Actions
Adrenal Cortex		
Aldosterone	Kidney tubules	Maintains proper balance of sodium and potassium ions
Cortisol	General	Adaptation to long-term stress; raises blood glucose level; mobilizes fat
Adrenal Medulla		
Epinephrine (adrenaline) and norepinephrine (noradrenaline)	Smooth muscle, cardiac muscle, blood vessels, skeletal muscle	Initiate stress responses; increase heart rate, blood pressure, metabolic rate; dilate blood vessels; mobilize fat; raise blood glucose level
Hypothalamus		
Thyrotropin-releasing hormone (TRH)	Anterior pituitary	Stimulates TSH release from anterior pituitary
Corticotropin-releasing hormone (CRH)	Anterior pituitary	Stimulates ACTH release from anterior pituitary
Gonadotropin-releasing hormone (GnRH)	Anterior pituitary	Stimulates FSH and LH release from anterior pituitary
Prolactin-releasing factor (PRF)	Anterior pituitary	Stimulates PRL release from anterior pituitary
Growth-hormone-releasing hormone (GHRH)	Anterior pituitary	Stimulates GH release from anterior pituitary
Prolactin-inhibiting hormone (PIH)	Anterior pituitary	Inhibits PRL release from anterior pituitary
Growth-hormone-inhibiting hormone (somatostatin)	Anterior pituitary	Inhibits GH release from anterior pituitary
Melanotropin-inhibiting hormone (MIH)	Anterior pituitary	Inhibits MSH release from anterior pituitary
Ovary		
Estrogen	General; female reproductive structures	Stimulates development of secondary sex characteristics in females and growth of sex organs at puberty; prompts monthly preparation of uterus for pregnancy
Progesterone	Uterus, breasts	Completes preparation of uterus for pregnancy; stimulates development of breasts
Pancreas		
Insulin	General	Lowers blood glucose level; increases storage of glycogen in liver
Glucagon	Liver, adipose tissue	Raises blood glucose level; stimulates breakdown of glycogen in liver
Parathyroid Glands		
Parathyroid hormone (PTH)	Bone, kidneys, digestive tract	Increases blood calcium level by stimulating bone breakdown; stimulates calcium reabsorption in kidneys; activates vitamin D

TABLE 30.1	(continued)		
Endocrine Gland and Hormone		**Target**	**Principal Actions**
Pineal Gland			
Melatonin		Hypothalamus	Function not well understood; may help control onset of puberty in humans and help regulate sleep cycle
Posterior Lobe of Pituitary			
Oxytocin (OT)		Uterus	Stimulates contraction of uterus
		Mammary glands	Stimulates ejection of milk
Antidiuretic hormone (ADH) (vasopressin)		Kidneys	Conserves water; increases blood pressure
Anterior Lobe of Pituitary			
Growth hormone (GH)		General	Stimulates growth by promoting protein synthesis and breakdown of fatty acids
Prolactin (PRL)		Mammary glands	Sustains milk production after birth
Thyroid-stimulating hormone (TSH)		Thyroid gland	Stimulates secretion of thyroid hormones
Adrenocorticotropic hormone (ACTH)		Adrenal cortex	Stimulates secretion of adrenal cortical hormones
Follicle-stimulating hormone (FSH)		Gonads	Stimulates ovarian follicle growth and secretion of estrogen in females; stimulates production of sperm cells in males
Luteinizing hormone (LH)		Ovaries and testes	Stimulates ovulation and corpus luteum formation in females; stimulates secretion of testosterone in males
Melanocyte-stimulating hormone (MSH)		Skin	Stimulates color change in reptiles and amphibians; unknown function in mammals
Testes			
Testosterone		General; male reproductive structures	Stimulates development of secondary sex characteristics in males and growth spurt at puberty; stimulates development of sex organs; stimulates sperm production
Thyroid Gland			
Thyroid hormone (thyroxine, T4, and others)		General	Stimulates metabolic rate; essential to normal growth and development
Calcitonin		Bone	Lowers blood calcium level by inhibiting release of calcium from bone
Thymus			
Thymosin		White blood cells	Promotes production and maturation of white blood cells

How Strong Is the Association Between Smoking and Lung Cancer?

About a third of all cases of cancer in the United States are directly attributable to cigarette smoking. The association between smoking and cancer is particularly striking for lung cancer. The lung you see in the photograph below, riddled with cancer, is that of a smoker. The bottom half of the lung is normal, while a cancerous tumor has completely taken over the top half. The cancer cells eventually broke through into the lymph and blood vessels and spread through the body. This smoker died of secondary tumors that formed in his brain, not an uncommon result. Over half a million people died of cancer in the United States in 2006; about 29% of them died of lung cancer.

All Americans die. The tragedy of this statistic is that so many die unnecessarily soon—fully 87% of the lung cancer deaths were cigarette smokers. Smoking is a popular pastime among Americans. In the United States, 21% of adults and 23% of teens smoke, and U.S. smokers consumed 389 billion cigarettes in 2005. The smoke emitted from these cigarettes contains some 3,000 chemical components, including vinyl chloride, benzo[a]pyrenes, and N-nitrosonornicotine, all potent mutagens. Smoking places these mutagens into direct contact with the tissues of the lungs, with cancer the result.

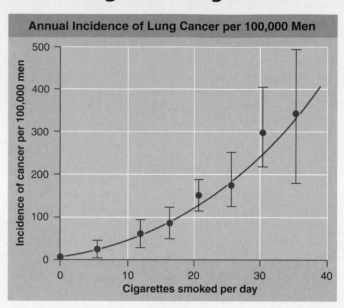

Annual Incidence of Lung Cancer per 100,000 Men

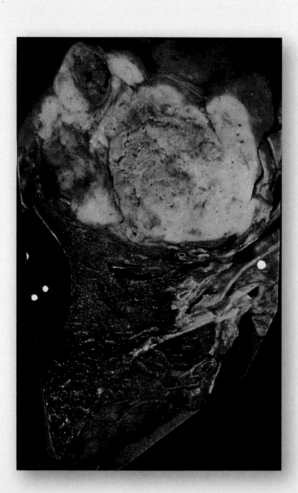

How strong is the correlation between the number of cigarettes smoked per day and the incidence of lung cancer? To find out, a detailed study was made of the incidence of lung cancer among American men, and of the cigarettes smoked per day. The results are presented in the graph above.

Analysis

1. **Applying Concepts** The vertical lines drawn through the points on the graph are "error bars." How much estimation error is associated with the estimate of cancer incidence among men smoking 20 cigarettes a day? 30 cigarettes?
2. **Interpreting Data**
   a. If the incidence of cancer per 100,000 men is 100, what is the percent of men with cancer?
   b. Do you see a trend in the magnitude of the error bars in the graph? What might account for this?
3. **Making Inferences**
   a. There are 20 cigarettes in a pack. What is the incidence of cancer among "a-pack-a-day" smokers?
   b. What is the incidence of lung cancer among nonsmokers?
   c. Compare the risk of contracting lung cancer among individuals who smoke one pack a day to the risk among nonsmokers.
   d. Is the relationship between cigarettes smoked per day and incidence of lung cancer linear? Why do you think the relationship is this way? What do you think this says about the risks of heavier smoking?
4. **Drawing Conclusions** Do these results support the hypothesis that cigarette smoking causes cancer? Do they prove it? Explain.

Concept Summary

The Neuroendocrine System

30.1 Hormones

- Hormones are chemical signals produced in glands or other endocrine tissues (**figure 30.1**) and transported to distant sites in the body. Endocrine glands produce hormones and release them into the bloodstream.

- Many endocrine glands and tissues are under the control of the central nervous system. The hypothalamus causes the release of hormones from the pituitary gland. Only cells that have receptors for the hormone respond and are called "target cells." The hormone binds to the receptor and elicits a response in the cell, often a change in cellular activity or genetic expression (**figure 30.2**).

30.2 How Hormones Target Cells

- Steroid hormones are lipid-soluble molecules. They pass through the plasma membrane of the target cell and bind to receptors in the cytoplasm or nucleus. The hormone-receptor complex binds to DNA, causing a change in gene expression that alters cell function (**figure 30.3**).

- Peptide hormones are unable to pass through the plasma membrane. Instead, they bind to membrane protein receptors (**figure 30.4**). The binding of the hormone, shown here from **figure 30.5,** causes a change in the internal side of the receptor which activates a second messenger. Second messengers, such as cyclic AMP, activate enzymes in the cell. The enzymes then trigger a change in cellular activity. The second messenger system is a cascade of reactions that amplifies the signal and facilitates the change in cellular activity.

The Major Endocrine Glands

30.3 The Hypothalamus and the Pituitary

- The pituitary gland is actually two glands: the posterior and anterior pituitary glands. The posterior pituitary develops as an extension of the hypothalamus and contains axons that extend from cell bodies in the hypothalamus (**figure 30.6**).

- The hormones released from the posterior pituitary are actually produced in the hypothalamus and transported by the axons to the posterior pituitary for storage and release. The hormones of the posterior pituitary include antidiuretic hormone (ADH), which regulates water retention in the kidneys, and oxytocin, which initiates uterine contractions during childbirth and milk release in the mother (**figure 30.7**).

- The anterior pituitary originates from epithelial tissue and produces the hormones it releases. Seven hormones are produced in the anterior pituitary (**figure 30.7**). They are thyroid-stimulating hormone (TSH), adrenocorticotropic hormone (ACTH), growth hormone (GH), follicle-stimulating hormone (FSH), luteinizing hormone (LH), prolactin (PRL), and melanocyte-stimulating hormone (MSH). The hypothalamus controls the anterior pituitary. The hypo-

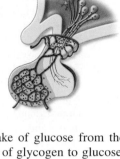

thalamus produces hormones that are released into blood capillaries that surround the pituitary stalk, called the hypothalamo-hypophyseal portal system. They travel a short distance to the anterior pituitary, as shown here from **figure 30.8.**

30.4 The Pancreas

- The pancreas secretes two hormones—insulin and glucagon into the blood. These hormones interact to maintain stable blood glucose levels. Insulin stimulates cell uptake of glucose from the blood. Glucagon stimulates the breakdown of glycogen to glucose. Two different types of cells in the islets of Langerhans produce insulin and glucagon.

- These hormones work opposite to each other (**figure 30.9**). An increase in blood glucose levels triggers the release of insulin and a decrease in blood glucose levels triggers the release of glucagon. When insulin is not available or cells fail to respond to insulin, diabetes mellitus can result.

30.5 The Thyroid, Parathyroid, and Adrenal Glands

- The thyroid gland is a shield-shaped organ that lies just beneath the Adam's apple in the front of the neck. The thyroid makes several hormones, but the two most important hormones produced by the thyroid are thyroxine, which increases metabolism and growth, and calcitonin, which stimulates calcium uptake by bones. Thyroxine is controlled by negative feedback. When enough of the hormone has been released, it feeds back to inhibit the hormone-production process (**figure 30.10***a*).

- Underproduction of thyroxine can lead to serious health problems. The underproduction of thyroxine in children can stunt growth and lead to mental retardation. A lack of iodine inhibits the production of thyroxine. In the absence of adequate levels of thyroxine, the hypothalamus will continue to stimulate the thyroid gland, leading to the formation of a goiter (**figure 30.10***b*).

- The parathyroid glands are four small glands that are attached to the thyroid. The parathyroids produce parathyroid hormone. PTH regulates the levels of calcium in the blood. Low calcium ion concentrations stimulate the release of PTH from the parathyroid glands. PTH acts on the bones to dismantle bone tissue, releasing Ca^{++} into the blood, shown here from **figure 30.11.** When Ca^{++} levels again increase, calcitonin is released from the thyroid and stimulates the uptake of Ca^{++} by bone cells and the synthesis of new bone tissue.

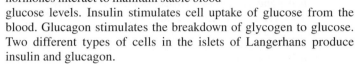

- The adrenal gland is actually two glands: The adrenal medulla is the inner core, and the adrenal cortex is the outer shell. The adrenal medulla secretes epinephrine and norepinephrine that stimulate an emergency response in the body. The adrenal cortex secretes cortisol, which is involved in inflammation and the regulation of glucose and stress responses. It also secretes aldosterone that promotes the uptake of sodium and water from urine.

Self-Test

1. One advantage chemical signaling has over electrical signaling is that
   a. reaction to stimuli can happen very quickly.
   b. although it takes large amounts of chemicals, the chemical signals are efficient.
   c. chemical signals stick around longer than electrical signals and can be used for slow processes.
   d. chemical signals are used in response to external and internal stimuli.

2. A coordination center for some of the endocrine system is the
   a. hypothalamus.   c. thyroid gland.
   b. adrenal gland.   d. pancreas.

3. Hormones and neurotransmitters are similar because they
   a. fit into receptors specifically shaped for them.
   b. are proteins.
   c. are released into the bloodstream.
   d. all of the above.

4. The action of steroid hormones is different from peptide hormones because
   a. peptide hormones must enter the cell to begin action, whereas steroid hormones must begin action on the external surface of the cell membrane.
   b. steroid hormones must enter the cell to begin action, whereas peptide hormones must begin action on the external surface of the cell membrane.
   c. peptide hormones produce a hormone-receptor complex that works directly on the DNA, whereas steroid hormones cause the release of a secondary messenger that triggers enzymes.
   d. No answer is correct.

5. Regulation of water concentration in the urine is done by a hormone released from the
   a. thyroid gland.   c. anterior pituitary gland.
   b. thymus.   d. posterior pituitary gland.

6. _____ is the hormone that stimulates the adrenal gland to produce a number of steroid hormones.
   a. ACTH   c. TSH
   b. LH   d. MSH

7. Type I diabetes is caused by an abnormality in endocrine cells of the
   a. pancreas.   c. adrenal glands.
   b. thymus.   d. hypothalamus.

8. The release of the thyroid hormone calcitonin is triggered by
   a. too much glucose in the blood.
   b. too much sodium in the blood.
   c. too much calcium in the blood.
   d. too much iodine in the blood.

9. Epinephrine mimics the effects of the
   a. somatic nervous system.
   b. central nervous system.
   c. parasympathetic nervous system.
   d. sympathetic nervous system.

10. Which of the following hormones is released from the adrenal cortex during periods of stress?
    a. cortisol   c. epinephrine
    b. Aldosterone   d. norepinephrine

Visual Understanding

1. **Figure 30.1** Some of the body systems are located primarily in one area of the body, or are obviously connected. The respiratory system, for instance, is located in the head and upper portion of the body. The skeletal system is articulated, almost every bone is connected to others. The endocrine system, however, is spread out, a batch of glands that do not appear connected with one another. Speculate on why this is so.

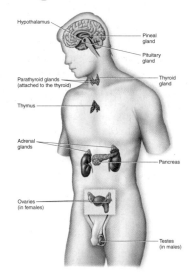

2. **Figure 30.7** A hypothetical patient has a disorder of the hypothalamus in which it can no longer secrete its "inhibiting" hormones. Which of the anterior pituitary hormones will be affected?

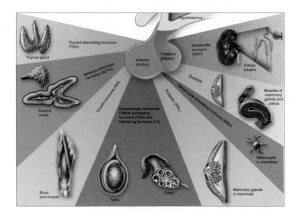

Challenge Questions

1. The younger brother of a friend wants to be a sports star in high school. Only in the eighth grade, he brags that he is taking steroids he gets from a friend in order to "bulk up" for next year. He asks if you have ever heard of any problems for kids his age—he only wants to take them "a couple of years" to get a football scholarship so his family can afford for him to go to college. How would you advise him?

2. Since your bones are so very important, why would your body ever need a system that includes osteoclasts, which literally break down your bone tissue?

Chapter 31

Reproduction and Development

Modes of Reproduction

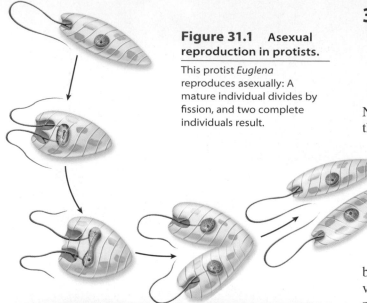

Figure 31.1   Asexual reproduction in protists.

This protist *Euglena* reproduces asexually: A mature individual divides by fission, and two complete individuals result.

(a)

(b)

31.1 Asexual and Sexual Reproduction

CONCEPT PREVIEW: While sexual reproduction is common among animals, many reproduce asexually by fission, budding, or parthenogenesis. Some animal species are hermaphroditic, while others change sex.

Not all reproduction involves two parents. Asexual reproduction, in which the offspring are genetically identical to the parent, is the primary means of reproduction among protists, cnidarians, and tunicates, and also occurs in some more complex animals.

Through mitosis, genetically identical cells are produced from a single parent cell. This permits asexual reproduction to occur in the *Euglena* in figure 31.1 by division of the organism, or **fission.** The DNA replicates and organs, such as the flagellum, duplicate. The nucleus divides with identical nuclei going to each daughter cell. Cnidaria commonly reproduce by **budding,** where a part of the parent's body becomes separated from the rest and differentiates into a new individual. The new individual may become an independent animal or may remain attached to the parent, forming a colony.

Unlike asexual reproduction, sexual reproduction occurs when a new individual is formed by the union of *two* cells. These cells are called **gametes,** and the two kinds that combine are generally called *sperm* and *eggs* (or ova). The union of a sperm and an egg produces a fertilized egg,

> Most animals undergo sexual reproduction. In the animals, diploid cells make up the majority of the life cycle; the only haploid cells in the life cycle are the gametes, as discussed on page 145.

or **zygote,** that develops by mitotic division into a new multicellular organism. The zygote and the cells it forms by mitosis are diploid; they contain both members of each homologous pair of chromosomes. The gametes, formed by meiosis in the sex organs, or **gonads**—the *testes* and *ovaries*—are haploid (see chapter 9). The processes of spermatogenesis (sperm formation) and oogenesis (egg formation) are described in later sections.

Different Approaches to Sex

Parthenogenesis, a type of reproduction in which offspring are produced from unfertilized eggs, is common in many species of arthropods. Some species are exclusively parthenogenic, whereas others switch between sexual reproduction and parthenogenesis in different generations. In honeybees, for example, a queen bee mates only once and stores the sperm. She then can control the release of sperm. If no sperm are released, the eggs develop parthenogenetically into drones, which are males; if sperm are allowed to fertilize the eggs, the fertilized eggs develop into other queens or worker bees, which are female. Parthenogenesis also occurs among populations of some lizard genera.

Figure 31.2   Hermaphroditism and protogyny.

(a) The hamlet bass (genus *Hypoplectrus*) is a deep-sea fish that is a hermaphrodite. In the course of a single pair-mating, one fish may switch sexual roles as many as four times. Here, the fish acting as a male curves around its motionless partner, fertilizing the upward-floating eggs. (b) The bluehead wrasse *Thalassoma bifasciatium* is protogynous. Here a large male, or sex-changed female, is seen among females, which are typically much smaller.

Another variation in reproductive strategies is **hermaphroditism,** when one individual has both testes and ovaries and so can produce both sperm and eggs. The hamlet bass in figure 31.2a are hermaphroditic, producing both eggs and sperm. During mating each fish switches from producing eggs that are fertilized by its partner, to producing sperm that fertilizes its partner's eggs. A tapeworm is hermaphroditic and can fertilize itself as well as cross fertilize, a useful strategy because it is unlikely to encounter another tapeworm living inside its host. Most hermaphroditic animals, however, require another individual to reproduce. Two earthworms, for example, are required for reproduction—like the hamlet bass, each functions as both male and female. Each leaves the encounter with fertilized eggs.

Numerous fish genera include species in which individuals can change their sex in response to social or environmental conditions, a process called *sequential hermaphroditism.* Among coral reef fish, for example, both **protogyny** ("first female," a change from female to male) and **protandry** ("first male," a change from male to female) occur. In the protogynous bluehead wrasse in figure 31.2b, the sex change appears to be under social control. These fish commonly live in large groups, or schools, where successful reproduction is typically limited to one or a few large, dominant males. If those males are removed, the largest female rapidly changes sex and becomes a dominant male (the blue-headed fish in the photo).

Sex Determination

In mammals, the sex is determined early in embryonic development. The reproductive systems of human males and females appear similar for the first 40 days after conception. During this time, the cells that will give rise to ova or sperm migrate to the embryonic gonads, which have the potential to become either ovaries in females or testes in males. If the embryo is XY, it is a male and will carry a gene on the Y chromosome whose protein product converts the gonads into testes (as on the left in figure 31.3). In females, who are XX, this Y chromosome gene and the protein it encodes are absent, and the gonads become ovaries (as on the right). Recent evidence suggests that the sex-determining gene may be one known as **SRY** (for "sex-determining region of the Y chromosome"). The *SRY* gene appears to have been highly conserved during the evolution of different vertebrate groups.

Once testes form in the embryo, they secrete testosterone and other hormones that promote the development of the male external genitalia and accessory reproductive organs (indicated in the blue box in the figure). If testes do not form, the embryo develops female external genitalia and accessory reproductive organs. The ovaries do not promote this development of female organs because the ovaries are nonfunctional at this stage. In other words, all mammalian embryos will develop female sex accessory organs and external genitalia by default unless they are masculinized by the secretions of the testes.

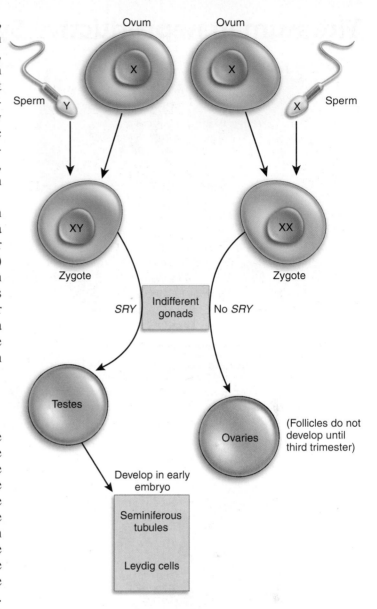

Figure 31.3 Sex determination.

Sex determination in mammals is made by a gene of the Y chromosome designated *SRY.* Testes are formed when the Y chromosome and *SRY* are present; ovaries are formed when they are absent.

IMPLICATION   Imagine if genetic engineering could be used to transfer the *SRY* gene into an X chromosome. Would a zygote carrying a normal X and this *SRY*-X be a female or a male? Explain why you think so. Do you think carrying out this sort of experiment in humans should be legal?

Concept Check

1. Explain the key difference between asexual and sexual reproduction.
2. What is parthenogenesis? Give an example of a parthenogenic animal.
3. How is sex determined among mammals?

The Human Reproductive System

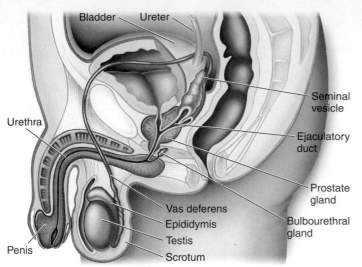

Figure 31.4 **The male reproductive organs.**

The testis is where sperm are formed. Cupped above the testis is the epididymis, a highly coiled passageway within which sperm complete their maturation. Extending away from the epididymis is a long tube, the vas deferens.

31.2 Males

CONCEPT PREVIEW: Male testes continuously produce large numbers of male gametes, sperm, which mature in the epididymis, are stored in the vas deferens, and are delivered through the penis into the female.

The human male gamete, or **sperm,** is highly specialized for its role as a carrier of genetic information. Produced by meiosis, sperm cells have 23 chromosomes instead of the 46 found in other cells of the male body. Sperm do not successfully complete their development at 37°C (98.6°F), the normal human body temperature. The sperm-producing organs, the **testes** (singular, **testis**), move during the course of fetal development into a sac called the scrotum (figure 31.4), which hangs between the legs of the male, maintaining the two testes at a temperature about 3°C cooler than the rest of the body. The testes contain cells that secrete the male sex hormone **testosterone.**

Male Gametes Are Formed in the Testes

An internal view of a testis in figure 31.5 ❶ shows that it is composed of several hundred compartments, each packed with large numbers of tightly coiled tubes called **seminiferous tubules** (seen in cross section in ❷). Sperm production, *spermatogenesis,* takes place inside the tubules. The process of spermatogenesis begins in germi-

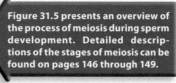

Figure 31.5 presents an overview of the process of meiosis during sperm development. Detailed descriptions of the stages of meiosis can be found on pages 146 through 149.

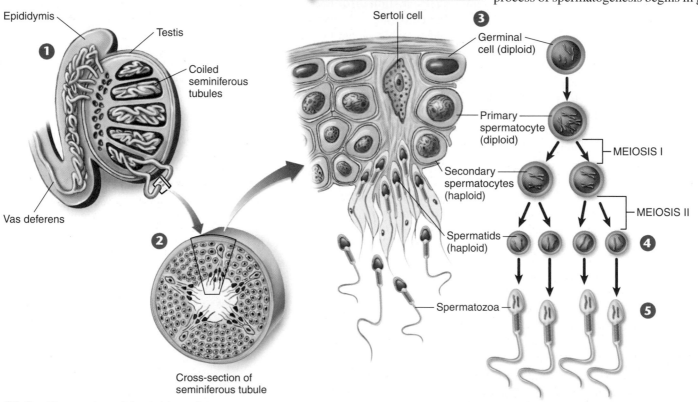

Figure 31.5 **The testis and formation of sperm.**

Inside the testis ❶, the seminiferous tubules ❷ are the sites of sperm formation. Germinal cells in the seminiferous tubules ❸ give rise to primary spermatocytes (diploid), which undergo meiosis to form haploid spermatids ❹. Spermatids develop into mobile spermatozoa, or sperm ❺. Sertoli cells are nongerminal cells within the walls of the seminiferous tubules. They assist spermatogenesis in several ways, such as helping to convert spermatids into spermatozoa.

nal cells toward the outside of the tubule (shown in the enlarged view in ❸). As the cells undergo meiosis, they move toward the lumen of the tubule, with spermatozoa being released into the tubule. The number of sperm produced is truly incredible. A typical adult male produces several hundred million sperm each day of his life! Those that are not ejaculated from the body are broken down, and their materials are reabsorbed and recycled.

After a sperm cell is manufactured within the testes through intermediate stages of meiosis, it is delivered to a long, coiled tube called the **epididymis** (see figure 31.5), where it matures. A sperm cell is not motile when it arrives in the epididymis, and it must remain there for at least 18 hours before its motility develops. Mature sperm are relatively simple cells, consisting of a head, body, and tail (figure 31.6). The head encloses a compact nucleus and is capped by a vesicle called an *acrosome*. The acrosome contains enzymes that aid in the penetration of the protective layers surrounding the egg. The body and tail provide a propulsive mechanism: Within the tail is a flagellum, and inside the body are centrioles, which act as a basal body for the flagellum, and mitochondria, which generate the energy needed for flagellar movement.

From the epididymis, the sperm is delivered to another long tube, the **vas deferens.** When sperm are released during intercourse, they travel through a tube from the vas deferens to the urethra, where the reproductive and urinary tracts join, emptying through the penis. Sperm is released in a fluid called **semen,** which also contains secretions mostly from the seminal vesicles and the prostate gland that provide metabolic energy sources for the sperm. Benign enlargement of the prostate occurs in 90% of men by age 70, but it can be cancerous. Prostate cancer is the second most common cancer in men and can be treated effectively if detected early during physical examinations, before it spreads.

Male Gametes Are Delivered by the Penis

In the case of humans and some other mammals, the **penis** is an external tube containing two long cylinders of spongy tissue side by side (seen in longitudinal and cross sections in figure 31.7). A third cylinder of spongy tissue contains in its center a tube called the urethra, through which both semen (during ejaculation) and urine (during urination) pass. Why this unusual design? The penis is designed to inflate. The spongy tissues that make up the three cylinders are riddled with small spaces between the cells, and when nerve impulses from the CNS cause the arterioles leading into this tissue to expand, blood collects within these spaces. Like blowing up a balloon, this causes the penis to become erect and rigid. Continued stimulation by the CNS is required for erection to continue.

Erection can be achieved without any physical stimulation of the penis, but physical stimulation is required for semen to be delivered. Stimulation of the penis, as by repeated thrusts into the vagina of a female, leads first to the mobilization of the sperm. In this process, muscles encircling the vas deferens contract, moving the sperm along the vas deferens into the urethra. The bulbourethral glands also secrete a clear, slippery fluid that neutralizes the acidity of any residual urine and lubricates the head of the penis. Further penis stimulation then leads to the strong contraction of the muscles at the base of the penis. The result is **ejaculation,** the forceful ejection of 2 to 5 milliliters of semen. Within this small 5-milliliter volume are several hundred million sperm. Because the odds against any one individual sperm cell successfully completing the long journey to the egg and fertilizing it are extraordinarily high, successful fertilization requires a high sperm count. Males with fewer than 20 million sperm per milliliter are generally considered sterile.

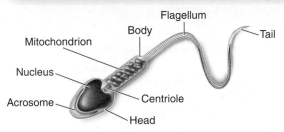

Figure 31.6 **Human sperm cells.**

Each sperm possesses a long tail that propels the sperm and a head that contains the nucleus. The tip, or acrosome, contains enzymes to help the sperm cell digest a passageway into the egg for fertilization.

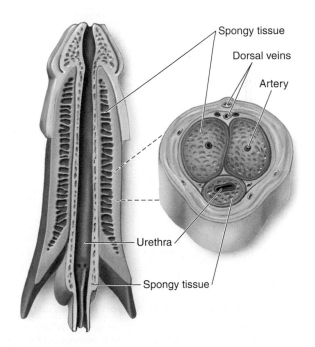

Figure 31.7 **Structure of the penis.**

(*Left*) longitudinal section; (*right*) cross section.

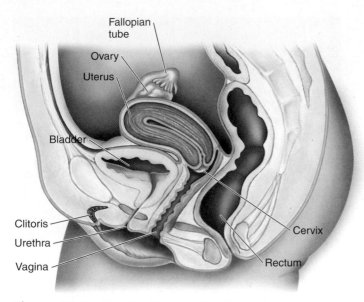

Figure 31.8   **The female reproductive system.**

The organs of the female reproductive system are specialized to produce gametes and to provide a site for embryonic development if the gamete is fertilized.

31.3 Females

CONCEPT PREVIEW: In human females, hormones trigger the development of one or a few oocytes once a month (about every 28 days). After ovulation, an egg cell travels down the fallopian tube and, if fertilized during its journey, implants in the wall of the uterus.

In females, eggs develop from cells called **oocytes,** located in the outer layer of compact masses of cells called **ovaries** within the abdominal cavity (figure 31.8). Recall that in males the gamete-producing cells are constantly dividing. In females all of the oocytes needed for a lifetime are already present at birth. During each reproductive cycle, one or a few of these oocytes are initiated to continue their development in a process called **ovulation;** the others remain in developmental holding patterns.

Only One Female Gamete Matures Each Month

At birth, a female's ovaries contain some 2 million oocytes, all of which have begun the first meiotic division. At this stage they are called *primary oocytes* (❶ in figure 31.9). Each primary oocyte waits to receive the proper developmental "go" signal before continuing on with meiosis. Until then, its meiosis remains arrested in prophase of the first meiotic division. Very few primary

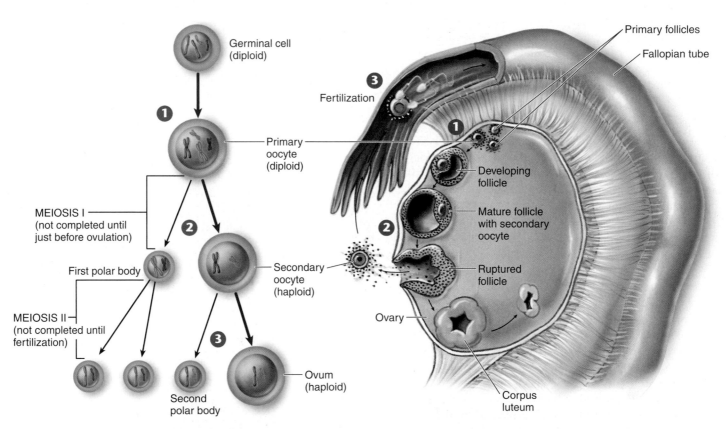

Figure 31.9   **The ovary and formation of an ovum.**

In this figure, the maturation of the ovum through meiosis is shown on the left, and the developmental journey of the ovum is on the right, with corresponding stages numbered on each. At birth, a human female's ovaries contain about 2 million egg-forming cells called oocytes, which have begun the first meiotic division and stopped. At this stage, they are called primary oocytes ❶, and their further development is halted until they receive the proper developmental signals, which are the hormones FSH and LH. Beginning at puberty, a monthly cycle of hormone secretion is established. When the hormones FSH and LH are released, meiosis resumes in a few oocytes, but only one oocyte usually continues to mature while the others regress. The primary oocyte (diploid) completes the first meiotic division, and one division product becomes a nonfunctional polar body. The other product, the secondary oocyte, is released during ovulation ❷, along with the polar body. The secondary oocyte does not complete the second meiotic division until fertilization ❸; that division yields two more nonfunctional polar bodies and a single haploid egg, or ovum. Fusion of the haploid egg with a haploid sperm during fertilization produces a diploid zygote.

oocytes ever receive the awaited signal, which turns out to be the pituitary hormones FSH and LH, which were discussed in chapter 30.

> FSH and LH are hormones that are released from the anterior pituitary gland in response to the release of the gonadotropin-releasing hormone (GnRH) from the hypothalamus, as discussed on page 583.

With the onset of puberty, females mature sexually. At this time, the release of FSH and LH initiates the resumption of the first meiotic division in a few oocytes. The first meiotic division produces the *secondary oocyte* and a nonfunctional *polar body* ❷. In humans, usually only a single oocyte is ovulated and the others regress. In some instances more than one oocyte develops; if both are fertilized, they become fraternal twins. Approximately every 28 days after that, another oocyte matures and is ovulated, although the exact timing may vary from month to month. Only about 400 of the approximately 2 million oocytes a woman is born with mature and are ovulated during her lifetime.

Fertilization Occurs in the Oviducts

The **oviducts** (also called **fallopian tubes** or uterine tubes) transport eggs from the ovaries to the **uterus.** In humans, the uterus is a muscular, pear-shaped organ about the size of a fist that narrows to a muscular ring called the **cervix,** which leads to the vagina (figure 31.10*a*). Mammals other than primates have more complex female reproductive tracts, where part of the uterus divides to form uterine "horns." The uterus is lined with a stratified epithelial membrane, called the **endometrium.** The surface of the endometrium is shed approximately once a month during menstruation, while the underlying portion remains to generate a new surface during the next cycle.

After ovulation, smooth muscles lining the fallopian tubes contract

> The movement of food through the digestive system is by peristaltic muscle contractions, discussed on page 510 and illustrated in figure 26.11.

rhythmically, moving the egg down the tube to the uterus in much the same way that food is moved down through your digestive system, pushing it along by squeezing the tube behind it. The journey of the egg through the fallopian tube is a slow one, taking from five to seven days to complete. However, if the egg is not fertilized within 24 hours of ovulation, it loses its capacity to develop.

During sexual intercourse, sperm are deposited within the vagina, a thin-walled muscular tube about 7 centimeters long that leads to the mouth of the uterus. Using their flagella, sperm entering the uterus swim up to and enter the fallopian tubes. Sperm can remain viable within the female reproductive tract for up to six days. If sexual intercourse takes place five days before ovulation or one day after, a viable egg will be present high up in the fallopian tubes. Of the several hundred million sperm that are ejaculated, only a few dozen make it to the egg. Once they reach the egg, the sperm must penetrate through two protective layers that surround the secondary oocyte (figure 31.10*b* ❶): a layer of granulosa cells and a protein layer called the zona pellucida. Enzymes within the acrosome cap of the sperm help digest the second of these layers ❷. The first sperm to make it through the second layer stimulates the oocyte to block the entry of other sperm ❸ and to complete meiosis II. Meiosis II produces the **ovum** (plural, **ova**) and two more nonfunctional polar bodies (see figure 31.9). When the female haploid nucleus within the ovum combines with the male haploid nucleus, the egg is fertilized and becomes a zygote. The zygote then begins a series of cell divisions while traveling down the fallopian tube. After about six days, it reaches the uterus, attaches itself to the endometrial lining, and continues the long developmental journey that eventually leads to the birth of a child.

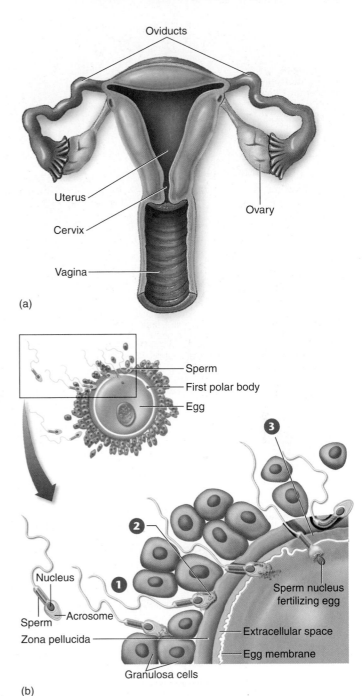

Figure 31.10   Fertilization occurs in the oviducts.

(a) The oviducts extend out from the uterus. Sperm are deposited in the vagina and travel to the oviducts. (b) Fertilization occurs in the oviduct when a sperm cell penetrates the outer layers of the egg cell.

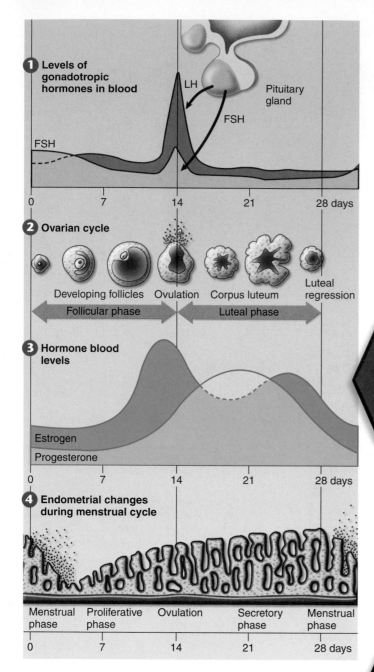

1 Levels of gonadotropic hormones in blood

LH

Pituitary gland

FSH

FSH

0   7   14   21   28 days

2 Ovarian cycle

Developing follicles   Ovulation   Corpus luteum   Luteal regression

Follicular phase   Luteal phase

3 Hormone blood levels

Estrogen

Progesterone

0   7   14   21   28 days

4 Endometrial changes during menstrual cycle

Menstrual phase | Proliferative phase | Ovulation | Secretory phase | Menstrual phase

0   7   14   21   28 days

Figure 31.11 The human menstrual cycle.

Ovulation and the preparation of the uterine lining for implantation is controlled by a group of four hormones during the menstrual cycle.

IMPLICATION Men do not exhibit a reproductive cycle, and are able to donate sperm at any time of the month. Why do you think evolution has favored development of a menstrual cycle in women? Do you think it might have been possible to have evolved human females without a menstrual cycle? What is the basis of your conclusion?

31.4 Hormones Coordinate the Reproductive Cycle

CONCEPT PREVIEW: Humans and apes have menstrual cycles driven by cyclic patterns of hormone secretion and ovulation. The cycle is composed of two distinct phases, follicular and luteal, coordinated by a family of hormones.

The female reproductive cycle, called a **menstrual cycle,** is composed of two distinct phases, the *follicular phase,* in which an egg reaches maturation and is ovulated, and the *luteal phase,* where the body continues to prepare for pregnancy. These phases are coordinated by a family of hormones. Hormones play many roles in human reproduction. Sexual development is initiated by hormones, released from the anterior pituitary and ovary, that coordinate simultaneous sexual development in many kinds of tissues. The production of gametes is another closely orchestrated process, involving a series of carefully timed developmental events. Successful fertilization initiates yet another developmental "program," in which the female body continues its preparation for the many changes of pregnancy.

> The hypothalamus is the control center of the neuroendocrine system. As discussed on page 559, the hypothalamus integrates all internal activities in the body through neural connections. It also exerts control over the pituitary, as discussed on page 583.

Production of the sex hormones that coordinate all these processes is coordinated by the hypothalamus, which sends releasing hormones to the pituitary, directing it to produce particular sex hormones. Negative feedback, discussed in chapters 27 and 30, plays a key role in regulating these activities of the hypothalamus. When target organs receive a pituitary hormone, they begin to produce a hormone of their own, which circulates back to the hypothalamus, shutting down production of the pituitary hormone. In addition, *positive feedback* mechanisms play a role too. In these cases, a hormone circulates back to the hypothalamus and increases the production of a pituitary hormone.

Triggering the Maturation of an Egg

The first phase of the menstrual cycle, the follicular phase, corresponds to days 0 through 14 in figure 31.11. During this time, several follicles (an oocyte and its surrounding tissue is called a *follicle*) are stimulated to develop. This development is carefully regulated by hormones. The anterior pituitary, after receiving a chemical signal (GnRH) from the hypothalamus, starts the cycle by secreting small amounts of **follicle-stimulating hormone (FSH)** and **luteinizing hormone (LH)** **1**. These hormones stimulate follicular growth **2** and the secretion of the female sex hormone **estrogen** **3**, more technically known as *estradiol,* from the developing follicles. Several follicles are stimulated to grow under FSH stimulation.

> Estrogen is a steroid hormone, and as shown in figure 30.3 on page 580, steroid hormones are able to pass through the plasma membrane and bind to receptors inside the cell. The binding of estrogen to its receptor activates the transcription of a receptor protein for progesterone.

Initially, the relatively low levels of estrogen have a negative-feedback effect on FSH and LH secretion. The low but rising levels of estrogen in the bloodstream feed back to the hypothalamus, which responds to the rising estrogen by commanding the anterior pituitary to decrease production of FSH and LH. As FSH levels fall, usually only one follicle achieves maturity. Late in the follicular phase, estrogen levels in the blood have increased drastically, and these higher levels of estrogen begin to have a positive-feedback effect on FSH and LH secretion. The rise in estrogen levels signals the completion of the follicular phase of the menstrual cycle.

Preparing the Body for Fertilization

The second phase of the cycle, the luteal phase (days 14 through 28), follows smoothly from the first. In a positive-feedback response to high levels of estrogen, the hypothalamus causes the anterior pituitary to rapidly secrete large amounts of LH and FSH (see figure 31.11 ❶). The surge of LH is larger than the surge of FSH and can last up to 24 hours. The peak in LH secretion triggers ovulation: LH causes the wall of the follicle to burst, and the egg within the follicle is released into one of the fallopian tubes extending from the ovary to the uterus (see ❶ in figure 31.12).

After the egg's release and departure, estrogen levels decrease, and LH directs the repair of the ruptured follicle, which fills in and becomes yellowish. In this condition, it is called the *corpus luteum,* which is simply the Latin phrase for "yellow body." The corpus luteum soon begins to secrete the hormone **progesterone** (the light green curve in figure 31.11 ❸), in addition to small levels of estrogen. Increased levels of progesterone and estrogen have a negative-feedback effect on the secretion of FSH and LH, preventing further ovulations. Progesterone completes the body's preparation of the uterus for fertilization including the thickening of the endometrium (figure 31.11 ❹). If fertilization does *not* occur soon after ovulation, however, production of progesterone slows and eventually ceases, marking the end of the luteal phase. The decreasing levels of progesterone cause the thickened layer of blood-rich tissue to be sloughed off, a process that results in the bleeding associated with menstruation. **Menstruation,** or "having a period," usually occurs about midway between successive ovulations (shown in figure 31.11 at 28 days).

At the end of the luteal phase, neither estrogen nor progesterone is being produced. In their absence, the anterior pituitary can again initiate production of FSH and LH, thus starting another reproductive cycle. Each cycle begins immediately after the preceding one ends. A cycle usually occurs every 28 days, or a little more frequently than once a month, although this varies in individual cases. The Latin word for "month" is *mens,* which is why the reproductive cycle is called the menstrual cycle, or monthly cycle.

If fertilization does occur high in the fallopian tube (❷ in figure 31.12a), the zygote undergoes a series of cell divisions called cleavage ❸, while traveling toward the uterus. At the blastocyst stage, it implants in the lining of the uterus ❹. The tiny embryo secretes human chorionic gonadotropin (hCG), an LH-like hormone, which maintains the corpus luteum. By maintaining the corpus luteum, hCG keeps the levels of estrogen and progesterone high, thereby preventing menstruation, which would terminate the pregnancy. Because hCG comes from the embryo and not from the mother, it is hCG that is tested in all pregnancy tests.

> As discussed on page 544, pregnancy tests use monoclonal antibodies which are produced using hCG as the antigen. The test strip is coated with the hCG monoclonal antibodies. If hCG is present in the urine, it will react with the antibodies and produce a positive result.

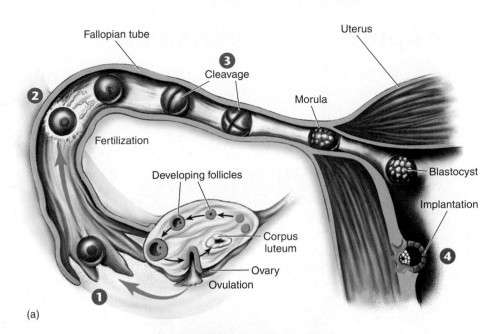

(a)

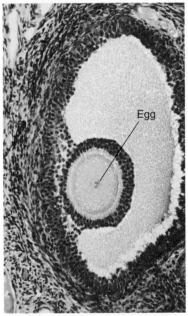

(b)

Figure 31.12   The journey of an ovum.

(a) Produced within a follicle and released at ovulation, an ovum is swept up into a fallopian tube ❶ and carried down by waves of contraction of the tube walls. Fertilization occurs within the tube ❷ by sperm journeying upward. Several mitotic divisions occur while the fertilized ovum undergoes cleavage and continues its journey down the fallopian tube ❸, becoming first a morula then a blastocyst. The blastocyst implants itself within the wall of the uterus ❹, where it continues its development. (b) A mature egg within an ovarian follicle. In each menstrual cycle, a few follicles are stimulated to grow under the influence of FSH and LH, but usually only one achieves full maturity and ovulation.

Concept Check

1. What is the function of the prostate gland?
2. Where is the endometrium found?
3. When in the menstrual cycle does luteinizing hormone peak?

The Course of Development

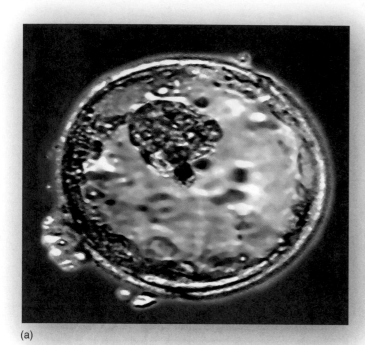

(a)

Ectoderm	Epidermis, central nervous system, sense organs, neural crest
Mesoderm	Skeleton, muscles, blood vessels, heart, gonads
Endoderm	Lining of digestive and respiratory tracts; liver, pancreas

(b)

Figure 31.13   The beginnings of human development.

(a) A human blastocyst. The formation of the blastocyst occurs when the zygote undergoes cleavage producing a hollow ball of cells. An inner cell mass will later differentiate into the different tissues of the embryo. (b) The fate of the three primary germ layers.

31.5 Embryonic Development

CONCEPT PREVIEW: The vertebrate embryo develops in three stages: cleavage, a hollow ball of cells forms; gastrulation, cells move into the interior, forming the primary tissues; and neurulation, organs form.

Cleavage: Setting the Stage for Development

Fertilization begins a carefully orchestrated series of developmental events. Table 31.1 traces the major stages of mammalian development, beginning with fertilization. Follow down the table as the stages of development are discussed here.

The first major event in human embryonic development is the rapid division of the zygote into a larger and larger number of smaller and smaller cells, becoming first 2 cells, then 4, then 8, and so on. The first of these divisions occurs about 30 hours after union of the egg and the sperm, and the second, 30 hours later. During this period of division, called **cleavage,** the overall size does not increase from that of the zygote. The resulting tightly packed mass of about 32 cells is called a **morula,** and each individual cell in the morula is referred to as a **blastomere.** The cells of the morula continue to divide, each cell secreting a fluid into the center of the cell mass. Eventually, a hollow ball of 500 to 2,000 cells is formed. This is the **blastocyst,** which contains a fluid-filled cavity called the **blastocoel** (figure 31.13a) Within the ball is an *inner cell mass* concentrated at one pole that goes on to form the developing embryo. The outer sphere of cells, called the *trophoblast,* releases the hCG hormone, discussed earlier.

During cleavage, the morula journeys down the mother's fallopian tube. On about the sixth day, the blastocyst has formed and reaches the uterus; it attaches to the uterine lining, and penetrates into the tissue of the lining. The blastocyst now begins to grow rapidly, initiating the formation of the membranes that will later surround, protect, and nourish it. One of these membranes, the **amnion,** will enclose the developing embryo, whereas another, the **chorion,** which forms from the trophoblast, will interact with uterine tissue to form the **placenta,** which will nourish the growing embryo (see figure 31.15). The placenta connects the developing embryo to the blood supply of the mother. Fully 61 of the cells at the 64-celled stage develop into the trophoblast and only 3 into the embryo proper.

Gastrulation: The Onset of Developmental Change

Ten to 11 days after fertilization, certain groups of cells move inward from the surface of the cell mass in a carefully orchestrated migration called **gastrulation.** First, the lower cell layer of the blastocyst cell mass differentiates into **endoderm,** one of the three primary embryonic tissues, and the upper layer into **ectoderm.** Just after this differentiation, much of the **mesoderm** arises by the invagination of cells that move from the upper layer of the cell mass *inward,* along the edges of a furrow that appears at the embryo midline, the primitive streak.

> These three germ layers appear early in the evolution of the animal kingdom. As discussed on page 340, the development of the endoderm, ectoderm, and mesoderm layers first appears in the lower invertebrate group of solid worms, called flatworms.

During gastrulation, about half of the cells of the blastocyst cell mass move into the interior of the human embryo. This movement largely determines the future development of the embryo. By the end of gastrulation, distribution of cells into the three primary germ layers has been completed. The ectoderm is destined to form the epidermis and neural tissue. The mesoderm is destined to form the connective tissue, muscle, and vascular elements. The endoderm forms the lining of the gut and its derivative organs (figure 31.13*b*).

Neurulation: Determination of Body Architecture

In the third week of embryonic development, the three primary cell types begin their development into the tissues and organs of the body. This stage in development is called **neurulation.**

The first characteristic vertebrate feature to form is the **notochord,** a flexible rod. Soon after gastrulation is complete, it forms from mesoderm tissue along the midline of the embryo, below its dorsal surface. After the notochord has been formed, the second characteristic vertebrate feature, the **neural tube,** forms from the region of the ectoderm that is located above the notochord and later differentiates into the spinal cord and brain. Just before the neural tube closes, two strips of cells break away and form the **neural crest.** These neural crest cells give rise to neural structures found in the vertebrate body.

While the neural tube is forming from ectoderm, the rest of the basic architecture of the human body is being rapidly determined by changes in the mesoderm. On either side of the developing notochord, segmented blocks of tissue form. Ultimately, these blocks, or **somites,** give rise to the muscles, vertebrae, and connective tissues. As development continues, more and more somites are formed. Within another strip of mesoderm that runs alongside the somites, many of the significant glands of the body, including the kidneys, adrenal glands, and gonads, develop. The remainder of the mesoderm layer moves out and around the inner endoderm layer of cells and eventually surrounds it entirely. As a result, the mesoderm forms two layers. The outer layer is associated with the body wall and the inner layer is associated with the gut. Between these two layers of mesoderm is the **coelom,** which becomes the body cavity of the adult.

By the end of the third week, over a dozen somites are evident, and the blood vessels and gut have begun to develop. At this point the embryo is about 2 millimeters (less than a tenth of an inch) long.

TABLE 31.1	Stages of Mammalian Development		
		Stage (age)	**Description**
	Sperm / Ovum	Fertilization (day 1)	The haploid male and female gametes fuse to form a diploid zygote.
	Blastocyst / Inner cell mass / Trophoblast / Blastocoel	Cleavage (days 2–10)	The zygote rapidly divides into many cells, with no overall increase in size. These divisions affect future development, because different cells receive different portions of the egg cytoplasm and, hence, different regulatory signals.
	Amniotic cavity / Ectoderm / Endoderm	Gastrulation (days 11–15)	The cells of the embryo move, forming three primary germ layers: ectoderm and endoderm form first, followed by the formation of mesoderm.
	Primitive streak / Ectoderm / Mesoderm / Endoderm / Formation of extraembryonic membranes		
	Neural groove / Notochord	Neurulation (days 16–25)	In all chordates, the first organ to form is the notochord; the second is the neural tube.
	Neural crest / Neural tube / Notochord		During neurulation, the neural crest is produced as the neural tube is formed. The neural crest gives rise to several uniquely vertebrate structures such as sensory neurons, sympathetic neurons, Schwann cells, and other cell types.
		Organogenesis (days 26+)	Cells from the three primary cell layers combine in various ways to produce the organs of the body.

31.6 Fetal Development

CONCEPT PREVIEW: Most of the key events in fetal development occur early. Organs begin to form in the fourth week, and by the end of the second month the development of the embryo is essentially complete. What remains in the second and third trimesters is primarily growth.

The Fourth Week: Organ Formation

In the fourth week of pregnancy, the body organs begin to form, a process called **organogenesis** (figure 31.14*a*, the drawing helps identify the structures in the photo). The eyes form, and the heart begins a rhythmic beating and develops four chambers. At 70 beats per minute, the little heart is destined to beat more than 2.5 billion times during a lifetime of about 70

Figure 31.14 **The developing human.**

(a) Four weeks; (b) seven weeks; (c) three months; and (d) four months.

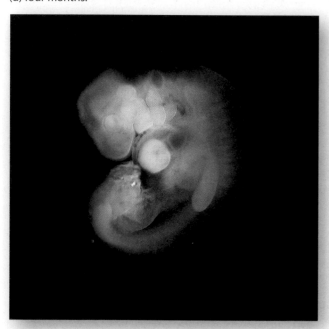

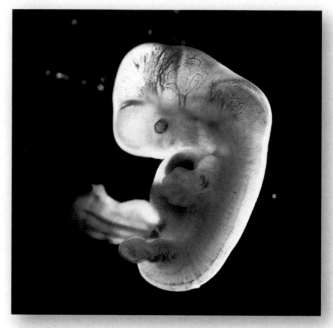

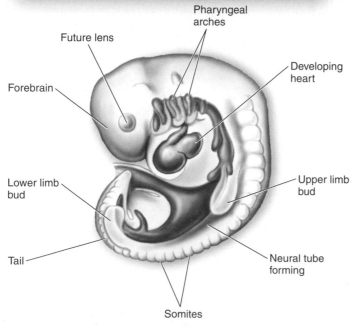

Pharyngeal arches

Future lens

Forebrain

Developing heart

Lower limb bud

Upper limb bud

Tail

Neural tube forming

Somites

(a)

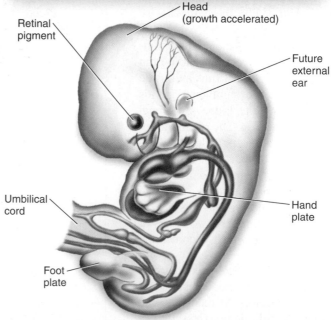

Head (growth accelerated)

Retinal pigment

Future external ear

Umbilical cord

Hand plate

Foot plate

(b)

years. More than 30 pairs of somites are visible by the end of the fourth week, and the arm and leg buds have begun to form. The embryo more than doubles in length during this week, reaching about 5 millimeters.

By the end of the fourth week, the developmental scenario is far advanced, although most women are not yet aware that they are pregnant. This is a crucial time in development because the proper course of events can be interrupted easily. For example, alcohol use by pregnant women during the first months of pregnancy is one of the leading causes of birth defects, producing *fetal alcohol syndrome,* in which the baby is born with a deformed face and often severe mental retardation. One in 250 newborns in the United States is affected with fetal alcohol syndrome.

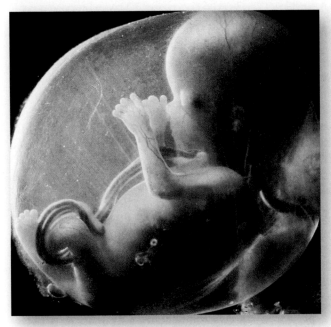

Development is essentially complete.

The developing human is now referred to as a fetus.

Facial expressions and primitive reflexes are carried out.

All of the major body organs have been established.

Arms and legs begin to move.

Bones actively enlarge.

Mother can feel baby kicking.

Following a period of rapid growth, the fetus is born.

Neurological growth continues after birth.

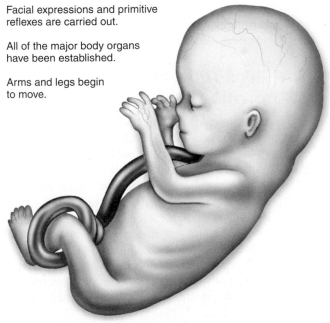

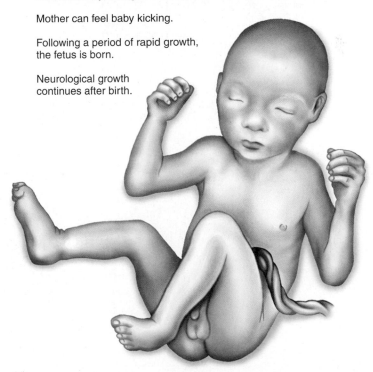

(c)

(d)

The Second Month: The Embryo Takes Shape

During the second month of pregnancy, great changes in morphology occur as the embryo takes shape (figure 31.14b). The miniature limbs of the embryo assume their adult shapes. The arms, legs, knees, elbows, fingers, and toes can all be seen as well as a short, bony tail. The bones of the embryonic tail, an evolutionary reminder of our past, later fuse to form the coccyx, or tailbone. Within the body cavity, the major internal organs are evident, including the liver and pancreas. By the end of the second month, the embryo has grown to about 25 millimeters in length—it is 1 inch long. It weighs perhaps a gram and is beginning to look distinctly human.

The Third Month: Completion of Development

Development of the embryo is essentially complete except for the lungs and brain. The lungs don't complete development until the third trimester, and the brain continues to develop even after birth. From this point on, the developing human is referred to as a **fetus** rather than an embryo. What remains is essentially growth. The nervous system and sense organs develop during the third month. The fetus begins to show facial expressions and carries out primitive reflexes such as the startle reflex and sucking. By the end of the third month, all of the major organs of the body have been established and the arms and legs begin to move (figure 31.14c).

The Second Trimester: The Fetus Grows in Earnest

The second trimester is a time of growth. In the fourth (figure 31.14d) and fifth months of pregnancy, the fetus grows to about 175 millimeters in length (almost 7 in long), with a body weight of about 225 grams. Bone formation occurs actively during the fourth month. During the fifth month, the head and body become covered with fine hair. This downy body hair, called *lanugo,* is another evolutionary relic and is lost later in development. By the end of the fourth month, the mother can feel the baby kicking; by the end of the fifth month, she can hear its rapid heartbeat with a stethoscope. In the sixth month, growth accelerates. By the end of the sixth month, the baby is over 0.3 meters (1 ft) long and weighs 0.6 kilograms (about 1.5 lb)—and most of its prebirth growth is still to come. At this stage, the fetus cannot yet survive outside the uterus without special medical intervention.

The Third Trimester: The Pace of Growth Accelerates

The third trimester is a period of rapid growth. In the seventh, eighth, and ninth months of pregnancy, the weight of the fetus more than doubles. This increase in bulk is not the only kind of growth that occurs. Most of the major nerve tracts are formed within the brain during this period, as are new brain cells.

All of this growth is fueled by nutrients provided by the mother's bloodstream, passing into the fetal blood supply within the placenta. The placenta in figure 31.15 contains blood vessels that extend from the umbilical cord into tissues that line the uterus (the *decidua basalis* in the figure). The mother's blood

Figure 31.15   Structure of the placenta.

The placenta contains a fetal component, the chorionic frondosum, and a maternal component, the decidua basalis. Oxygen and nutrients enter the fetal blood from the maternal blood by diffusion. Waste substances enter the maternal blood from the fetal blood, also by diffusion.

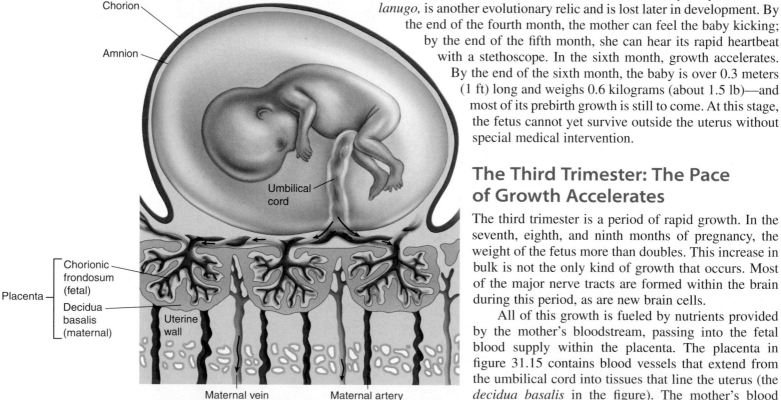

bathes this tissue so that nutrients can pass from the mother's blood into the blood vessels that carry it back to the fetus without the two blood systems ever mixing blood.

By the end of the third trimester, the neurological growth of the fetus is far from complete, but by this time the fetus is about as large as it can get and still be delivered through the pelvis without damage to mother or child. As any woman who has had a baby can testify, it is a tight fit. Birth takes place as soon as the probability of survival is high.

Birth

At approximately 40 weeks from the last menstrual cycle, the process of birth begins as hormonal changes in the mother initiate the onset of **labor.** During labor and delivery the cervix gradually dilates (the opening becomes larger), the amnion ruptures causing amniotic fluid to flow out through the vagina (sometimes referred to as the "water breaking"), and uterine contractions become strong and regular, usually resulting in the expulsion of the fetus from the uterus. Hormones called **oxytocin** and **prostaglandins** work in a positive-feedback mechanism to stimulate and increase uterine contractions. The fetus is usually in a head-down position near the end of pregnancy. In a vaginal birth, the fetus is pushed down through the cervix and out through the vagina (figure 31.16). After the birth of the fetus, continuing uterine contractions expel the placenta and associated membranes, collectively called the "afterbirth." In some cases, such as when a vaginal delivery would cause harm to the fetus or the mother, the fetus and placenta are surgically removed from the uterus in a procedure called a *caesarian section (C-section).* The umbilical cord is still attached to the baby, and a doctor, nurse, or parent clamps and cuts the cord. The baby transitions from living in a fluid environment to a gaseous one, and many of its organ systems undergo major changes.

In the mother, hormones during late pregnancy prepare the *mammary glands* for nourishing the baby after birth. For the first couple of days after childbirth, the mammary glands produce a fluid called *colostrum,* which contains protein (including antibodies) and lactose but little fat. Then milk production is stimulated by the anterior pituitary hormone **prolactin,** usually by the third day after delivery. When the infant suckles at the breast, the posterior pituitary hormone oxytocin is released, initiating milk release, or milk "letdown."

Postnatal Development

Growth continues rapidly after birth. Babies typically double their birth weight within a few months. Different organs grow at different rates, however, and the body proportions of infants are different than that of adults. The head, for example, is disproportionately large in newborns, but after birth it grows more slowly than the rest of the body. Such a pattern of growth, in which different components grow at different rates, is referred to as *allometric growth.*

The fact that the human brain continues to grow significantly for the first few years of postnatal life means that adequate nutrition and a safe environment are particularly crucial during this period.

Concept Check

1. Describe the events of gastrulation.
2. What tissues do mesoderm cells give rise to in the adult body?
3. What is the role of the hormone oxytocin? Prolactin?

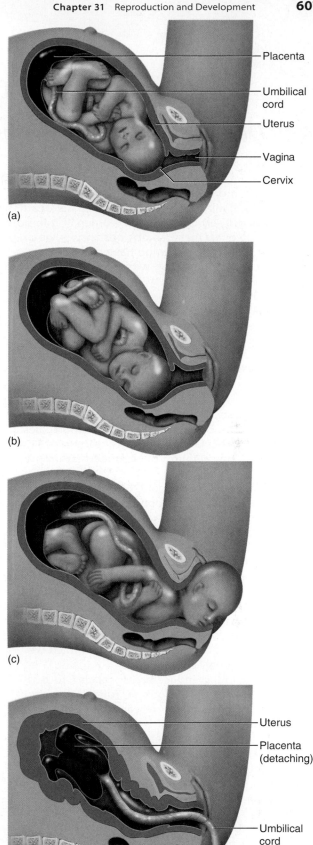

(a)

(b)

(c)

(d)

Figure 31.16   Stages of childbirth.

Why Don't Men Get Breast Cancer?

In 2007, an estimated 240,510 new cases of breast cancer were expected among women in the United States and 2,030 new cases among men—over 99% of the new breast cancer victims were women. Similarly, of 40,460 breast cancer deaths anticipated that year, all but 450 of them were women. It is impossible not to wonder, why so few men? An obvious answer would be that men don't have breasts, but men do have breast tissue, it just isn't as developed as a woman's. So we are back to the same question: why so few men?

While 5% of breast cancers are due to inherited genetic mutations (*BRCA1* and *BRCA2*), the cause of 95%—the overwhelming majority—remains a mystery. Hormone differences seem the most promising place to start, as the female sex hormone estrogen controls breast development in women; men, by contrast, lack physiologically significant amounts of estrogen. A logical suggestion is that in breast cancer patients, the effect of estrogen on breast cells is being altered by exposure to a so-called endocrine disrupter. Endocrine disrupters are man-made chemicals that mimic hormones. By sheer chance, their molecules are perfectly shaped to fit particular hormone receptors. In this case, the culprit would be a chemical mimic of estrogen that promotes cancerous growth in breast cells.

One candidate is bisphenol A (BPA), a molecule that is structurally similar to estrogen, with carbon rings at each end tipped with OH groups. Used to form the plastic packaging of many foods and drinks, as well as the clear plastic liners of metal food and beverage cans, BPA is a chemical to which all of us are exposed daily. Six billion pounds are produced worldwide each year.

It has been known since 1938 that BPA promotes excess estrogen production in rats. Alarmingly, in 1993 BPA was shown to have the same effects on human breast cancer cells growing in culture. What was alarming was that the effect could be measured at concentrations as low as two parts per billion, not much above the levels to which we humans are routinely exposed.

Does BPA in fact induce breast cancer? To test this possibility, Dr. Ana Soto of Tufts University School of Medicine and her research team in 2006 tested its effects on rats, which get breast cancer in much the same way as humans do. They exposed pregnant female rats to a range of BPA concentrations, and after 50 days sacrificed them for examination of their breast tissue. The researchers looked in particular for aberrant cell growth patterns in breast tissue called ductal hyperplasias, which in both rats and people are considered to be the precursors of breast cancer. BPA was administered to four groups of rats. Some received low doses not unlike what humans are exposed to, while others received much higher doses. In a fifth group, which served as a control, no BPA was administered.

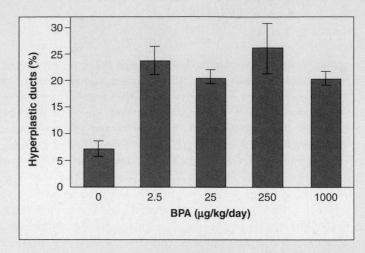

Effect of bisphenol A on breast cancer.

The histogram above shows what the researchers found. Amounts of BPA administered to the rats are reported in micrograms of BPA per kilogram of body weight per day. The incidence of breast cancer detected is presented as the percent of examined breast tissue ducts which were hyperplastic (that is, precancerous). Vertical black lines (called error bars) represent 5% confidence intervals (a measure of scatter among the data—95% of replicate experiments would be expected to fall within the ranges of the error bars).

What are we to make of the data the researchers obtained? Did any of the four doses of BPA they administered result in a percent of hyperplastic ducts significantly higher than that seen in the BPA control group? (Hint: If their error bars overlap with the control's, the difference is not significant.) The answer—yes, all four doses are significant. Do higher doses of BPA increase the incidence of hyperplastic ducts? Do the error bars of the four doses fail to overlap? No, all doses yielded similar results. It is difficult to avoid the conclusion that exposure to even low levels of bisphenol A induces ductal hyperplasias in laboratory rats.

These results suggest the rather alarming conclusion that BPA, a chemical to which we are all exposed every day, may cause breast cancer. The study is small, and a rat is not a human, but the possibility that BPA is a carcinogen certainly merits further investigation. Most government-funded breast cancer research has focused on the search for more effective breast cancer treatments; far less money is spent on searching for the causes of breast cancer. The most encouraging aspect of this study is its suggestion that such a search, if it became a priority, might prove fruitful.

Bisphenol A

Estrogen

Birth Control and Sexually Transmitted Diseases

31.7 Contraception and Sexually Transmitted Diseases

CONCEPT PREVIEW: A variety of birth control methods are available, many quite effective. STDs are diseases spread through sexual contact.

Contraception

Not all couples want to initiate a pregnancy every time they have sex, yet sexual intercourse may be a necessary and important part of their emotional lives together. The solution to this dilemma is to find a way to avoid reproduction without avoiding sexual intercourse, an approach that is commonly called **birth control,** or **contraception.** Table 31.2 on the next page provides a summary of birth control methods commonly available.

Abstinence. The simplest and most reliable way to avoid pregnancy is not to have sex at all. Of all methods of birth control, this is the most certain—and the most limiting, because it denies a couple the emotional support of a sexual relationship. A variant of this approach is to avoid sex only on the days which successful fertilization is likely to occur. The rest of the sexual cycle is considered relatively "safe" for intercourse. This approach, called the rhythm method, or natural family planning when other indicators are also monitored, is satisfactory in principle but difficult in application because ovulation is not easy to predict and may occur unexpectedly. Failure rate are as high as 20% to 30%.

Prevention of Egg Maturation. A widespread form of birth control in the United States has been the daily ingestion of hormones, or **birth control pills** (figure 31.17*a*). These pills contain estrogen and progesterone which shut down production of the pituitary hormones FSH and LH. The ovarian follicles do not ripen in the absence of FSH, and ovulation does not occur in the absence of LH. Other methods of hormone delivery include medroxy progesterone (Depo-Provera), which is injected every one to three months, the weekly birth control patch, which releases the hormones through the skin, and surgically implanted capsules that release hormones. Failure rates are less than 2%.

Emergency contraception, called **Plan B** or the "morning after pill," is a hormone manipulation therapy that blocks ovulation. It can be used as backup contraception after unprotected sex (either birth control wasn't used or is suspected of failure). Plan B is a high-dose progesterone pill taken in two doses 12 hours apart. It is most effective when taken within the first 24 hours, but can still offer some protection up to five days. The high levels of progesterone will block ovulation or may also inhibit fertilization or implantation. It will have no effect if pregnancy has occurred. Failure rates vary depending how quickly it is used following intercourse, from 5% to 40%.

Prevention of Embryo Implantation. The insertion of a coil or other irregularly shaped object into the uterus is an effective means of birth control. The irritation in the uterus prevents the implantation of the descending embryo within the uterine wall. Such **intrauterine devices (IUDs)** are very effective because once inserted, they can be forgotten. They have a failure rate of less than 2%.

A chemical means of preventing embryo implantation or ending an early pregnancy is **RU486.** This pill blocks the action of progesterone, causing the endometrium to slough off. RU486 must be administered under a doctor's care because of potentially serious side effects.

BIOLOGY & YOU

RU486. Emergency contraception like the "morning after" pill Plan B should not be confused with the highly publicized RU486 (Mifepristone). RU486 is not an emergency contraceptive—it is not a contraceptive at all. It ends an unwanted pregnancy after conception. To be effective, RU486 has to be administered within 49 days (or seven weeks) following the first day of the last menstrual period. An RU486-terminated pregnancy involves two drugs given two days apart. The first drug is RU486 (Mifepristone), a synthetic steroid compound that works by blocking progesterone receptors in the uterus. Without progesterone stimulation, the endometrium degenerates and is sloughed off, ejecting the embryo with the shed lining of the uterus. Because RU486 by itself is only effective 60% of the time in inducing removal of the embryo, a second medication, a prostaglandin called misoprostol, is given two days later to induce strong contractions of the uterus, forcing the embryo's expulsion. Because RU486 has numerous side effects, some potentially severe (including sepsis, a severe infection of the bloodstream, and excessive bleeding), it must be administered by a doctor.

Sperm Blockage.   If sperm cannot reach the uterus, fertilization cannot occur. One way to prevent the delivery of sperm is to encase the penis within a thin sheath, or condom (figure 31.17b). In principle, this method is easy to apply and foolproof, but in practice it has a failure rate of 3 to 15% because of incorrect use or inconsistent use. Nevertheless, it is the most commonly employed form of birth control in the United States. Condoms are also widely used to prevent the transmission of AIDS and other sexually transmitted diseases (STDs).

A second way to prevent the entry of sperm into the uterus is to place a cover over the cervix. The cover may be a relatively tight-fitting cervical cap, which is worn for days at a time, or a rubber dome called a diaphragm (figure 31.17c), which is inserted immediately before intercourse. Because the dimensions of individual cervices vary, a cervical cap or diaphragm must be fitted by a physician. Failure rates average 4 to 25% for dia-

TABLE 31.2	Methods of Birth Control				
Device	**Action**	**Failure Rate (percent)**	**Advantages**	**Disadvantages**
Oral contraceptive	Hormones (progesterone analogue alone or in combination with other hormones) primarily prevent ovulation	1–5, depending on type	Convenient; highly effective; provides significant noncontraceptive health benefits, such as protection against ovarian and endometrial cancers	Must be taken regularly; possible minor side effects which new formulations have reduced; not for women with cardiovascular risks (mostly smokers over age 35)
Condom	Thin sheath for penis that collects semen; "female condoms" sheath vaginal walls	3–15	Easy to use; effective; inexpensive; protects against some sexually transmitted diseases	Requires male cooperation; may diminish spontaneity; may deteriorate on the shelf
Diaphragm	Soft rubber cup covers entrance to uterus, prevents sperm from reaching egg, holds spermicide	4–25	No dangerous side effects; reliable if used properly; provides some protection against sexually transmitted diseases and cervical cancer	Requires careful fitting; some inconvenience associated with insertion and removal; may be dislodged during intercourse
Intrauterine device (IUD)	Small plastic or metal device placed in the uterus; prevents implantation; some contain copper, others release hormones	1–5	Convenient; highly effective; infrequent replacement	Can cause excess menstrual bleeding and pain; risk of perforation, infection, expulsion, pelvic inflammatory disease, and infertility; not recommended for those who eventually intend to conceive or are not monogamous; dangerous in pregnancy
Cervical cap	Miniature diaphragm covers cervix closely, prevents sperm from reaching egg, holds spermicide	Probably similar to that of diaphragm	No dangerous side effects; fairly effective; can remain in place longer than diaphragm	Problems with fitting and insertion; comes in limited number of sizes
Foams, creams, jellies, vaginal suppositories	Chemical spermicides inserted in vagina before intercourse that prevent sperm from entering uterus	10–25	Can be used by anyone who is not allergic; protects against some sexually transmitted diseases; no known side effects	Relatively unreliable; sometimes messy; must be used 5–10 minutes before each act of intercourse
Implant (levonorgestrel; Norplant)	Capsules surgically implanted under skin slowly release hormone that blocks ovulation	.03	Very safe, convenient, and effective; very long lasting (five years); may have nonreproductive health benefits like those of oral contraceptives	Irregular or absent periods; minor surgical procedure needed for insertion and removal; some scarring may occur
Injectable contraceptive (medroxy-progesterone; Depo-Provera)	Injection every 3 months of a hormone that is slowly released and prevents ovulation	1	Convenient and highly effective; no serious side effects other than occasional heavy menstrual bleeding	Animal studies suggest it may cause cancer, though new studies in humans are mostly encouraging; occasional heavy menstrual bleeding
Plan B ("morning after pill")	Emergency contraception taken within one to five days following intercourse. High levels of progesterone taken in two doses blocks ovulation and may prevent fertilization or implantation	5–40	Can prevent ovulation or possibly pregnancy following unprotected sex (if another method of birth control wasn't used or is suspected of failing, such as condom breaking or irregular use of oral contraception)	Available over the counter for women 18 and over but under 18 requires a prescription; effectiveness decreases after the first 24 hours; is emergency contraception, shouldn't be used in place of other methods of birth control

phragms, perhaps because of the propensity to insert them carelessly when in a hurry. Failure rates for cervical caps are lower.

Sperm Destruction. A final general approach to birth control is to eliminate the sperm after ejaculation. This can be achieved in principle by washing out the vagina immediately after intercourse, before the sperm have a chance to enter the uterus. Such a procedure is called a douche (French, "wash"). It involves a rapid dash to the bathroom immediately after ejaculation and a very thorough washing. Its failure rate is as high as 40%. Alternatively, sperm delivered to the vagina can be destroyed there with spermicidal jellies or foams. These treatments generally require application immediately before intercourse. Their failure rates vary from 10 to 25%. The use of a spermicide with a condom or diaphragm increases the effectiveness over each method used independently.

Sexually Transmitted Diseases

Sexually transmitted diseases (STDs) are diseases that spread from one person to another through sexual contact. AIDS, discussed in chapter 28, is a deadly viral STD. Other significant sexually transmitted diseases include:

Gonorrhea. Caused by the bacterium *Neisseria gonorrhoeae*. The primary symptom of this disease is discharge from the penis or vagina. It can be treated with antibiotics. If left untreated in women, gonorrhea can cause pelvic inflammatory disease (PID), a condition in which the fallopian tubes become scarred and blocked. PID can eventually lead to sterility.

Chlamydia. Caused by the bacterium *Chlamydia trachomatis,* this disease is sometimes called the "silent STD" because women usually experience no symptoms until after the infection has become established. Like gonorrhea, chlamydia can cause PID in women if left untreated.

Syphilis. Caused by the bacterium *Treponema pallidum,* this disease is one of the most potentially devastating STDs. Left untreated, the disease progresses to heart disease, mental deficiency, and nerve damage that may include loss of motor function or blindness.

Genital herpes. Caused by the herpes simplex virus type 2 (HSV-2), this disease is the most common STD in the United States. The virus causes red blisters on the penis or on the labia, vagina, or cervix that scab over.

Cervical Cancer. About 70% of cervical cancer is caused by HPV (human papillomavirus), a sexually transmitted virus. Gardasil, a newly developed vaccine which blocks HPV in women not yet exposed to the virus, could cut worldwide deaths by about 290,000 women each year.

This summary of STDs may give the impression that sexual activity is fraught with danger, and when practiced with unknown partners it is. In light of this, it is folly not to take precautions to avoid STDs. Anyone responsible enough to have sex should be responsible enough to protect themselves.

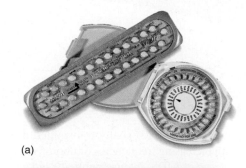
(a)

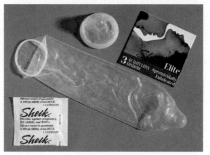
(b)

(c)

Figure 31.17   Three common birth control methods.

(a) Oral contraceptives; (b) condom; and (c) diaphragm and spermicidal jelly.

Concept Check

1. What is the failure rate of condoms? of birth control pills?
2. What are the active agents in birth control pills?
3. Which STD is the most common in the United States?

Why Do STDs Vary in Frequency?

As a general rule, the incidence of a sexually transmitted disease is expected to increase with increasing frequencies of unprotected sexual contact. With the emergence of AIDS, intense publicity and education has lessened such dangerous behavior. Both the number of sexual partners and the frequency of unprotected sex have fallen significantly in the United States in the last decade. It would follow, then, that the frequencies of sexually transmitted diseases (STDs) like syphilis, gonorrhea, and chlamydia should also be falling.

However, the level of one STD sometimes rises while another falls. What are we to make of this? The simplest explanation of such a difference is that the two STDs are occurring in different populations, and one population has rising levels of sexual activity, while the other has falling levels. However, nationwide statistics encompass all population subgroups, and there is no reason to expect subgroups to contain different STDs. Certainly each major subgroup contains all three major STDs mentioned above. So this would seem an unlikely explanation for the frequency of one STD to be rising while another falls.

A second possible explanation would be a change in the infectivity of one of the STDs. A less infective STD would tend to fall in frequency in the population, for the simple reason that fewer sexual contacts result in infection. To assess this possibility, we must examine the individual STDs more closely.

Syphilis is most infective in its initial stage, but this stage lasts only about a month. Most transmissions occur during the much longer second stage, marked by a pink rash and sores in the mouth. The bacteria can be transmitted at this stage by kissing or shared liquids. Any drop in infectivity of this STD would be expected to shorten this stage—but no such shortening has been observed.

Gonorrhea can be transmitted by various forms of sexual contact with an infected individual at any time during the infection. There has been no drop in infectivity per sexual contact reported.

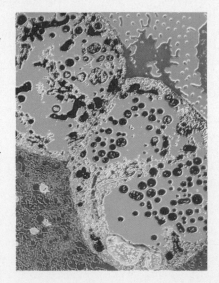

Chlamydia offers the most interesting possibility of changes in infectivity, because of its unusual nature. *Chlamydia trachomatis* is genetically a bacterium but is an obligate intracellular parasite, much like a virus in this respect—it can reproduce only

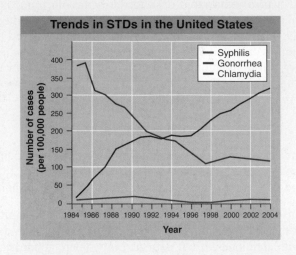

Trends in STDs in the United States

inside human cells. The red structures in the photo are chlamydia bacteria inside human cells. Like gonorrhea, chlamydia is transmitted through vaginal, anal, or oral intercourse with an infected person. With chlamydia, the person may show no symptoms. Because the disease agent lives inside cells, its infectivity would not be expected to change unless the number of cells of an infected individual to which his or her sex partner would be exposed during intercourse were to change, a very unlikely possibility.

So a drop in infectivity doesn't seem very likely. There is, however, a third possible explanation for why the frequency of one STD in a population might rise while the frequency of another STD in that same population falls. To grasp this third possible explanation, we will need to examine carefully the trends in the incidence in the United States of the gonorrhea, chlamydia, and syphilis. Detailed yearly statistics are reported in the graph above.

Analysis

1. **Making Inferences**
   a. Gonorrhea: What is the incidence in 1985? in 1995? Has the frequency declined or increased? In general, are individuals aware they are infected when they transmit the STD?
   b. Chlamydia: What is the incidence in 1985? in 1995? Has the frequency declined or increased? In general, are individuals aware they are infected when they transmit the STD?
2. **Drawing Conclusions** How might heightened public awareness explain why the trend in levels of gonorrhea differs from that of chlamydia?

Concept Summary

Modes of Reproduction

31.1 Asexual and Sexual Reproduction

- Asexual reproduction through fission (**figure 31.1**) or budding is the primary means of reproduction among protists and some animals, but most animals reproduce sexually.

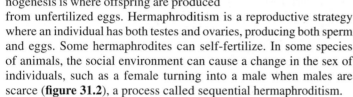

- Sexual reproduction is most common in animals, but parthenogenesis and hermaphroditism are two variations. Parthenogenesis is where offspring are produced from unfertilized eggs. Hermaphroditism is a reproductive strategy where an individual has both testes and ovaries, producing both sperm and eggs. Some hermaphrodites can self-fertilize. In some species of animals, the social environment can cause a change in the sex of individuals, such as a female turning into a male when males are scarce (**figure 31.2**), a process called sequential hermaphroditism.

- In mammals, sex is genetically determined and appears during embryonic development. An embryo that is XY develops into a male and an XX embryo develops into a female. The Y chromosome carriers a gene, possibly the *SRY* gene that converts the gonads into testes. The lack of this gene in females results in the formation of ovaries (**figure 31.3**).

The Human Reproductive System

31.2 Males

- Sperm is produced in the testes (**figure 31.4**). The testes contain a large number of tightly coiled tubes called seminiferous tubules, where sperm develop in a process called spermatogenesis (**figure 31.5**). As sperm develop and undergo meiosis, they move toward the lumen of the tubules, and from there they pass into the epididymis. Once matured, they are stored in the vas deferens. During sexual intercourse, sperm are delivered through the penis (**figure 31.7**) into the female.

31.3 Females

- Female gametes, eggs or ova, develop in the ovary from oocytes. The ovaries are located in the lower abdomen (**figure 31.8**). A female is born with some 2 million oocytes, all arrested during the first meiotic division. The hormones FSH and LH initiate the resumption of meiosis I in a few oocytes, but usually only one oocyte completes development.

- The egg ruptures from the ovary, called ovulation, and enters the fallopian tube (**figure 31.9**). Sperm deposited in the vagina travel up through the cervix and uterus, shown here from **figure 31.10a,** and into the fallopian tube, or oviduct, to reach the egg. Usually only one sperm cell penetrates the egg's protective layers (**figure 31.10b**). At this point, the oocyte completes meiosis II, and fertilization occurs. The fertilized egg, called a zygote, is transported to the uterus through the oviduct. The zygote attaches to the endometrial lining of the uterus where it completes development.

31.4 Hormones Coordinate the Reproductive Cycle

- The human reproductive cycle, called a menstrual cycle, is divided into two phases, the follicular and luteal phases (**figure 31.11**). The follicular phase begins with the secretion of FSH and LH, which stimulates the resumption of oocyte development and the secretion of estrogen. The estrogen acts as a negative-feedback signal to stop the secretion of FSH from the anterior pituitary.

- As the level of estrogen increases it begins to have a positive-feedback effect on FSH and LH. The luteal phase begins with a surge of LH, which causes ovulation and the formation of the corpus luteum (**figure 31.11**). The corpus luteum begins secreting progesterone, which acts to prepare the uterus for implantation of the zygote.

- If fertilization does not occur, estrogen and progesterone levels drop and the endometrial lining of the uterus sloughs off, a process called menstruation, and a new cycle begins.

- If a zygote implants in the lining of the uterus (**figure 31.12**), estrogen and progesterone levels remain high due to the release of human chorionic gonadotropin (hCG) from the embryo. The uterus is maintained and no further egg maturation occurs.

The Course of Development

31.5 Embryonic Development

- The vertebrate embryo develops in three stages (**table 31.1**). The first stage, called cleavage, involves hundreds of cell divisions that eventually produce a hollow ball of cells called a blastocyst. The second stage, called gastrulation, involves the orchestrated movement of cells, forming the three germ layers: endoderm, ectoderm, and mesoderm (**figure 31.13**). The third stage is neurulation, where the notochord and neural tube form.

31.6 Fetal Development

- Organs begin forming by the fourth week, and by the end of the second month the embryo looks distinctly human. By the end of the third month, all major organs except the brain and lungs are developed (**figure 31.14**).

- The second and third trimesters are periods of considerable growth, with the fetus receiving nourishment from the placenta (**figure 31.15**).

- During labor and delivery, the fetus and placenta are expelled from the uterus (**figure 31.16**). Hormones coordinate the production of milk in the mother for nourishing the newborn. Brain development continues in the baby after birth.

Birth Control and Sexually Transmitted Diseases

31.7 Contraception and Sexually Transmitted Diseases

- Various birth control methods are available and work by preventing egg maturation, preventing embryo implantation, and blocking or killing sperm (**table 31.2**).

- Sexually transmitted diseases are spread through sexual contact. AIDS is a deadly STD. Other STDs may not be as fatal as AIDS but are quite destructive, especially if left untreated.

Self-Test

1. If offspring are not genetically identical to each other or to the parent, then the organism reproduces through
   a. fission.
   c. budding.
   b. sexual reproduction.
   d. all of the above.

2. An animal that contains both ovaries and testes and produces both eggs and sperm reproduce by
   a. parthenogenesis.
   b. asexual reproduction.
   c. hermaphroditism.
   d. They cannot reproduce; they are sterile.

3. In mammals, the embryonic gonads will develop into ovaries
   a. if the *SRY* gene is expressed.
   b. if both sex chromosomes are X.
   c. if the sex chromosomes are X and Y.
   d. within the first 40 days.

4. Temperature regulation of spermatogenesis in human males is controlled by the location of the
   a. seminiferous tubules.
   c. vas deferens.
   b. epididymis.
   d. scrotum.

5. Oocyte development in human females requires the hormones
   a. estrogen and testosterone.
   b. FSH and LH.
   c. progesterone and testosterone.
   d. oxytocin and prolactin.

6. When pregnancy occurs, the endometrium is maintained by the
   a. embryo releasing hCG.
   b. decrease in levels of progesterone.
   c. hypothalamus releasing GnRH.
   d. increasing levels of FSH.

7. A human embryo has formed the three germ layers from which all tissues form by the time
   a. the blastula forms.
   c. the blastocyst forms.
   b. neurulation is complete.
   d. gastrulation is complete.

8. In a developing human, the first tissues to begin forming are the
   a. skeletal.
   c. neural.
   b. muscular.
   d. digestive.

9. Contractions of the uterus during labor are stimulated by the hormone
   a. estrogen.
   c. oxytocin.
   b. prolactin.
   d. progesterone.

10. Which of the following is *not* a method of contraception?
   a. destruction of the egg
   b. prevention of egg maturation
   c. sperm blockage
   d. prevention of embryo implantation

Visual Understanding

1. **Figure 31.3** Up until 40 days after conception, an embryo is neither male nor female. Explain what happens after that point in an embryo that carries the *SRY* gene.

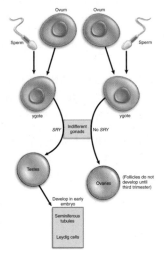

2. **Figure 31.10** Sometimes a zygote implants in the fallopian tube rather than in the uterus. This is called an ectopic pregnancy (ectopic meaning "out of place"). It is necessary to terminate an ectopic pregnancy because it endangers the mother and the fetus cannot survive. Explain why the mother is in danger and why the fetus can't survive.

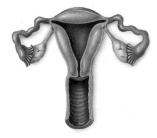

Challenge Questions

1. Why are all the parents of parthenogenic offspring female?

2. Speculate on why males produce so many sperm and females produce, comparatively, so few viable eggs.

3. Explain some of the ways that a pregnant woman's use of drugs—alcohol, cocaine, prescription drugs, over-the-counter drugs—can affect the embryo or fetus that she is carrying.

4. A friend tells you that the most reliable method of contraception and prevention of sexually transmitted disease is communication. Explain what she means.

Chapter **32**

Plant Form and Function

Structure and Function of Plant Tissues

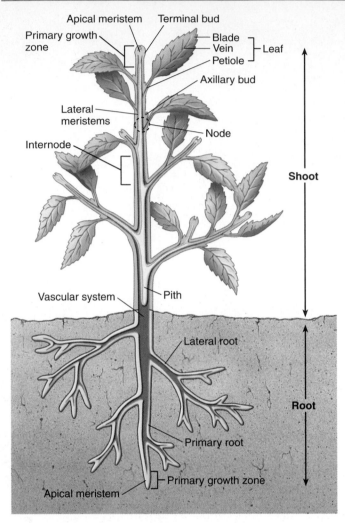

Figure 32.1   **The body of a plant.**

The body of this dicot plant consists of an aboveground portion called the shoot (stems and leaves) and a belowground portion called the root. Elongation of the plant, so-called primary growth, takes place when clusters of cells called the apical meristems (*lime green areas*) divide at the ends of the roots and the stems. Thickening of the plant, so-called secondary growth, takes place in the lateral meristems (*yellow areas*) of the stem, allowing the plant to increase in girth like letting out a belt.

IMPLICATION   Imagine that your body grew like a plant does. Where on your body do you think the apical meristems would be located? Why there?

32.1 Organization of a Vascular Plant

CONCEPT PREVIEW: The body of a vascular plant is a continuous structure, a grouping of tubes connecting roots to leaves, with growth zones called meristems.

While the plants you encountered in chapter 17 are remarkably diverse, most possess the same fundamental architecture and the same three major groups of organs—roots, stems, and leaves. The organs of vascular plants have an outer covering of protective tissue and an inner matrix of tissue. Within the inner matrix is embedded vascular tissue, which conducts water, minerals, and food throughout the plant. The cells and tissues of vascular plants, and how the plant body carries out the functions of living, are the focus of this chapter.

A vascular plant is organized along a vertical axis (figure 32.1). The part belowground is called the **root,** and the part aboveground is called the **shoot,** made up of the stem and leaves (however, in some instances roots may extend above the ground and stems can extend below it). Although roots and shoots differ in their basic structure, growth at the tips throughout the life of the individual is characteristic of both. The root penetrates the soil and absorbs water and various minerals, which are crucial for plant nutrition. It also anchors the plant. The **stem** serves as a framework for the positioning of the **leaves,** where most photosynthesis takes place. The arrangement, size, and other characteristics of the leaves are critically important in the plant's production of food.

Meristems

When an animal grows taller, all parts of its body lengthen—when you grew taller as a child, your arms and legs lengthened and so did your torso. A plant doesn't grow this way. Instead, it adds tissues to the tips of its roots and shoots. If you grew like this, your legs would get longer, and your head taller, while the central portion of your body would not change.

Why do plants grow in this way? The plant body contains growth zones of unspecialized cells called **meristems.** Meristems are areas with actively dividing cells that result in plant growth but also continually replenish themselves. That is, one cell divides to give rise to two cells. One remains meristematic, while the other is free to differentiate and contribute to the plant body, resulting in plant growth. In this way, meristem cells function much like "stem cells" in animals (see chapter 13). Molecular genetic evidence supports the hypothesis that animal stem cells and plant meristem cells may also share some common pathways of gene expression.

In plants, **primary growth** is initiated at the tips by the **apical meristems,** regions of active cell division that occur at the tips of roots and stems, colored lime green in figure 32.1. The growth of these meristems produces extension of the plant body. As the body tip elongates, it forms what is known as the primary plant body, which is made up of the primary tissues.

Growth in thickness, **secondary growth,** involves the activity of the **lateral meristems,** which are cylinders of meristematic tissue, colored yellow in figure 32.1. The continued division of their cells results primarily in the thickening of the plant body. There are two kinds of lateral meristems: the *vascular cambium,* which ultimately gives rise to thick accumulations of secondary xylem and phloem, and the *cork cambium,* from which arise the outer layers of bark on both roots and stems.

32.2 Plant Tissue Types

CONCEPT PREVIEW: Plants contain a variety of ground, dermal, and vascular tissues.

The organs of a plant—the roots, stem, leaves, and in some cases, flowers and fruits—are composed of different combinations of tissues. A tissue is a group of similar cells—cells that are specialized in the same way and are organized into a structural and functional unit. Each major tissue type is composed of distinctive kinds of cells, whose structures are related to the functions of the tissues in which they occur. Most plants have three major tissue types: (1) *ground tissue,* in which the vascular tissue is embedded; (2) *dermal tissue,* the outer protective covering of the plant; and (3) *vascular tissue,* which conducts water and dissolved minerals up the plant and conducts the products of photosynthesis throughout.

> Animal bodies are also made up of different types of cells organized into functional groups called tissues. Animal tissues are introduced on page 456, and include epithelial, connective, muscle, and nerve tissues, described on pages 460 through 466.

Ground Tissue

Parenchyma cells are the least specialized and the most common of all plant cell types; they form masses in leaves, stems, and roots. Parenchyma cells, unlike some other cell types, are characteristically alive at maturity, with fully functional cytoplasm and a nucleus. They are the cells that carry out the basic functions of living including photosynthesis, cellular respiration, and food and water storage. The edible parts of most fruits and vegetables are composed of parenchyma cells. They are capable of cell division and are important in cell regeneration and wound healing. Most parenchyma cells have only thin cell walls, as seen in figure 32.2, called **primary cell walls,** which are mostly cellulose that is laid down while the cells are still growing.

Collenchyma cells, which are also living at maturity, form strands or continuous cylinders beneath the epidermis of stems or leaf stalks and along veins in leaves. They are usually elongated, with unevenly thickened primary walls, which are clearly visible in figure 32.3, and are their distinguishing feature. Strands of collenchyma provide much of the support for plant organs in which secondary growth has not occurred.

In contrast to parenchyma and collenchyma cells, **sclerenchyma cells** have tough, thick cell walls called **secondary cell walls;** they usually do not contain living cytoplasm when mature. The secondary cell wall is laid down inside of the primary cell wall after the cell has stopped growing and expanding in size. The secondary cell wall provides cells with strength and rigidity. There are two types of sclerenchyma: *fibers,* which are long, slender cells that usually form strands, and *sclereids,* which are variable in shape but often branched. Sclereids, the reddish-colored cells in figure 32.4, are sometimes called "stone cells." Clusters of sclereids form the gritty texture you feel in the flesh of pears. Both fibers and sclereids are thick-walled and strengthen the tissues in which they occur. Compare the thickness of the cell walls in the sclereid cells in figure 32.4 with the green-stained parenchyma cells that surround them.

Dermal Tissue

All parts of the outer layer of a primary plant body are covered by flattened epidermal cells. These are the most abundant cells in the plant *epidermis,* or outer covering. The epidermis is one cell layer thick and is a protective

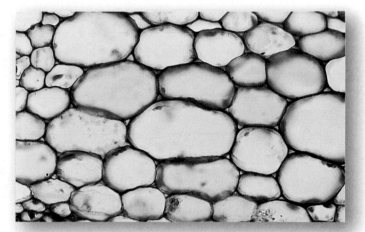

Figure 32.2 Parenchyma cells.

Cross section of parenchyma cells from grass. Only thin primary cell walls are seen in this living tissue.

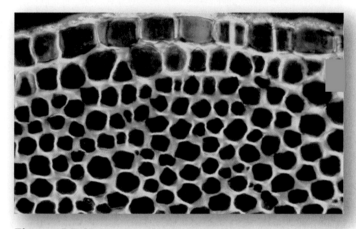

Figure 32.3 Collenchyma cells.

Cross section of collenchyma cells, with thickened side walls, from a young branch of elderberry *(Sambucus).* In some collenchyma cells, the thickened areas may occur at the corners of the cells or in others, as strands.

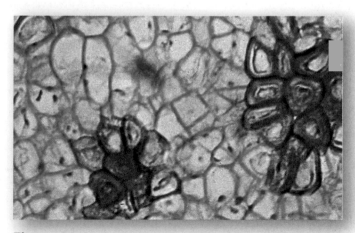

Figure 32.4 Sclerenchyma cells in sclereids.

Clusters of sclereids ("stone cells"), stained *red* in this preparation, in the pulp of a pear. These sclereid clusters give pears their gritty texture. The surrounding thin-walled cells, stained *green,* are parenchyma cells.

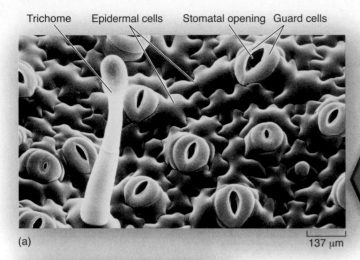

Trichome Epidermal cells Stomatal opening Guard cells

When the temperature is high or the humidity is low, plants will close their stomata to reduce water loss. As explained on page 109 and in figure 6.6, this can result in photorespiration, where oxygen levels increase in the cell and inhibit photosynthesis.

(a)   137 μm

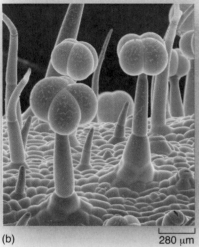

(b)   280 μm

Figure 32.5   Guard cells and trichomes.

(a) Numerous stomata occur among the leaf epidermal cells of this tobacco (*Nicotiana tabacum*) leaf. A trichome is also visible.
(b) Trichomes on a tomato plant (*Lycopersicon lycopersicum*).

layer that provides an effective barrier against water loss. The epidermis is often covered with a thick, waxy layer called the **cuticle,** which protects against ultraviolet light damage and water loss. In some cases, the dermal tissue is more extensive and forms the bark of trees.

Specialized cells and outgrowths can occur in the epidermis. **Guard cells** (figure 32.5*a*) are paired cells with an opening that lies between them called a **stoma** (plural, **stomata**). Guard cells and stomata occur frequently in the epidermis of leaves and occasionally on other parts of the shoot, such as on stems or fruits. Oxygen, carbon dioxide, and water vapor pass across the epidermis almost exclusively through the stomata, which open and shut in response to external factors such as moisture, temperature, and light.

Trichomes are outgrowths of the epidermis that occur on the surfaces of stems and leaves. Trichomes vary greatly in form in different kinds of plants, from the rounded tip trichome in figure 32.5*a* to the pointed and globular-tipped ones in figure 32.5*b*. Trichomes play an important role in regulating the heat and water balance of the leaf, much as the hairs of an animal's coat provide insulation.

Other outgrowths of the epidermis occur belowground, on the surface of roots near their tips. Called **root hairs,** these tubular extensions of single epidermal cells keep the root in intimate contact with the particles of soil (see figure 32.22). Root hairs play an important role in the absorption of water and minerals from the soil by increasing the surface area of the root.

Vascular Tissue

Vascular plants contain two kinds of conducting, or vascular, tissue: xylem and phloem. **Xylem** is the plant's principal water-conducting tissue, forming a continuous system that runs throughout the plant body. Within this system, water (and minerals dissolved in it) passes from the roots up through the shoot in an unbroken stream. When water reaches the leaves, much of it passes into the air as water vapor, through the stomata.

The two principal types of conducting cells in the xylem are **tracheids** and **vessel elements,** both of which have thick secondary walls that are laid down inside the primary cell wall. These cells are elongated, and they have no living cytoplasm (they are dead) at maturity. Tracheids are elongated cells that overlap at the ends as shown in figure 32.6*a*. In conducting elements composed of tracheids, water flows from tracheid to tracheid through openings called *pits* in the secondary walls. In contrast, vessel elements are elongated cells that line up end-on-end as shown in figure 32.6*b*, *c*. The end walls of vessel elements may be almost completely open or may have bars or strips of wall material perforated by pores through which water flows. A linked row of vessel elements forms a vessel. Primitive angiosperms and other vascular plants have only tracheids, but the majority of angiosperms have vessels. Vessels conduct water much more efficiently than do strands of tracheids.

Phloem is the principal food-conducting tissue in vascular plants. Food conduction in phloem is carried out through two kinds of elongated cells: **sieve cells** and **sieve-tube members.** The cells differ in the extent of the perforations between the cells, with the sieve cells having smaller perforations between cells. Seedless vascular plants and gymnosperms have only sieve cells; most angiosperms have sieve-tube members. Clusters of pores known as *sieve areas* occur on both kinds of cells and connect the cytoplasms of adjoining cells. Both cell types are living, but their nuclei are lost during maturation.

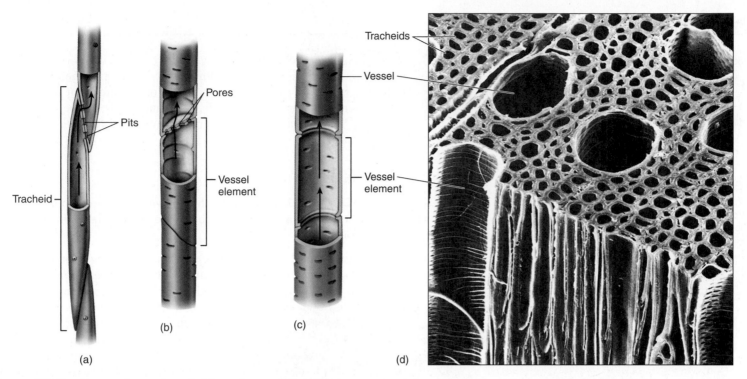

Figure 32.6 Comparison of tracheids and vessel elements.

(a) In tracheids the water passes from cell to cell by means of pits. (b) In vessel elements, pores in the end walls allow water to move from cell to cell, or (c) the end walls of vessel elements may be almost completely open. (d) A scanning electron micrograph of the red maple (*Acer rubrum*) wood, showing the xylem.

In sieve-tube members, some sieve areas have larger pores and are called *sieve plates*. Sieve-tube members occur end to end, as shown in the drawing and photo of figure 32.7, forming longitudinal series called **sieve tubes.** Specialized parenchyma cells known as **companion cells** occur regularly in association with sieve-tube members. The companion cells can be seen in figure 32.7 associated with the left side of the sieve-tube members. Companion cells apparently carry out some of the metabolic functions that are needed to maintain the associated sieve-tube member; their cytoplasms are connected to the sieve-tube members through openings called *plasmodesmata*.

In addition to conducting cells, xylem and phloem also contain fibers (sclerenchyma cells) and parenchyma cells. The parenchyma cells function in storage, and the fibers provide support and some storage. Xylem fibers are used in the production of paper.

Concept Check

1. Distinguish between primary and secondary growth.
2. Contrast the cell walls of parenchyma, collenchyma, and sclerenchyma.
3. Compare tracheids to vessel elements.

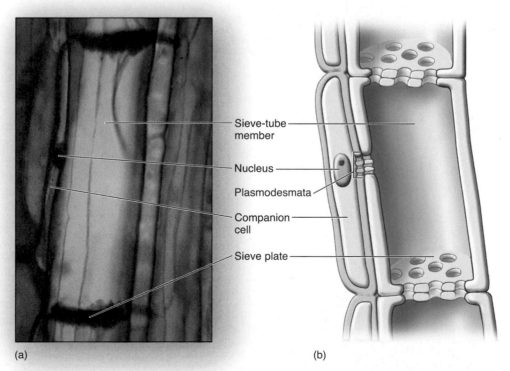

Figure 32.7 Sieve tubes.

(a) Sieve-tube member from the phloem of squash (*Cucurbita*), connected with the cells above and below to form a sieve tube. (b) In this diagram, note the thickened end walls, which are at right angles to the sieve tube. The narrow cell with the nucleus at the left of the sieve-tube member is a companion cell.

The Plant Body

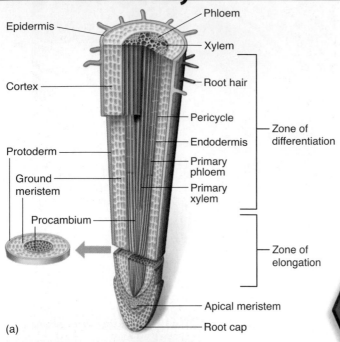

Epidermis

Phloem

Xylem

Root hair

Cortex

Pericycle

Zone of differentiation

Endodermis

Protoderm

Primary phloem

Ground meristem

Primary xylem

Procambium

Zone of elongation

Apical meristem

(a)

Root cap

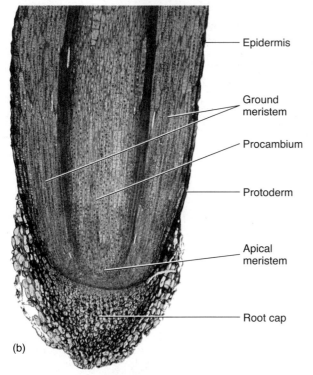

Epidermis

Ground meristem

Procambium

Protoderm

Apical meristem

Root cap

(b)

Figure 32.8 Root structure.

(a) Diagram of primary meristems in a dicot root, showing their relation to the apical meristem. The three primary meristems are the protoderm, which differentiates further into epidermis; the procambium, which differentiates further into primary vascular strands; and the ground meristem, which differentiates further into ground tissue. (b) Median longitudinal section of a monocot root tip in corn, *Zea mays*, showing the differentiation of protoderm, procambium, and ground meristem.

We now consider the three kinds of vegetative organs that form the body of a plant: roots, stems, and leaves. While we will examine their basic structures, it is important to understand that these organs are often also modified for different functions. For example, roots and stems can be modified for water and food storage, and leaves can be modified for defenses such as cactus spines.

32.3 Roots

CONCEPT PREVIEW: Roots, the belowground portion of the plant body, are adapted to absorb water and minerals from the soil.

Roots have a simpler pattern of organization and development than do stems, and we will examine them first. Although different patterns exist, the kind of root described here and shown in figure 32.8*a* is found in many dicots. Roots contain xylem (the pink areas) and phloem (the light green areas). As you can see, roots have a central column of xylem with radiating arms. Alternating with the radiating arms of xylem are strands of primary phloem. Surrounding the column of vascular tissue, and forming its outer boundary, is a cylinder of cells one or more cell layers thick called the **pericycle.** Branch, or lateral, roots are formed from cells of the pericycle. The outer layer of the root is the epidermis (the outer tan area). The mass of parenchyma in which the root's vascular tissue is located is the cortex (the cream-colored area). Its innermost layer—the endodermis—consists of specialized cells that regulate the flow of water between the vascular tissues and the root's outer portion. The **endodermis** is a single layer of cells (figure 32.9*b*) that lies just outside of the pericycle. Endodermis cells are encircled by a thickened, waxy band called the **Casparian strip.** Figure 32.9*c* is a drawing of endodermal cells showing how the wax substance that makes up the Casparian strip surrounds each cell. As the black arrows indicate, the Casparian strip blocks the movement of water *between* cells and instead directs the movement of water *through* the plasma membrane of the endodermis cells. In this way, the Casparian strip controls the passage of minerals into the xylem because transport through the endodermis cells is regulated by special channels embedded in the plasma membrane.

The apical meristem of the root (shown as a group of cells at the base of the root in figure 32.8*a, b*) divides and produces cells both inwardly, back toward the body of the plant, and outwardly. The three primary meristems, shown in the cross section of the root in figure 32.8*a* and in the photo of a monocot root in figure 32.8*b*, are the **protoderm,** which becomes the epidermis; the **procambium,** which produces primary vascular tissues (primary xylem and primary phloem); and the **ground meristem,** which differentiates further into ground tissue, which is composed of parenchyma cells. Outward cell division results in the formation of a thimblelike mass of relatively unorganized cells, called the **root cap,** which you can clearly see in figure 32.8*b*. The root cap covers and protects the root's apical meristem as it grows through the soil.

Dicots were the first type of angiosperms to evolve and, as described on page 327, differ from monocots in the number of cotyledons, in the structure of their flowers, and as discussed later in this chapter, in the distribution of vascular tissue in their roots, stems, and leaves.

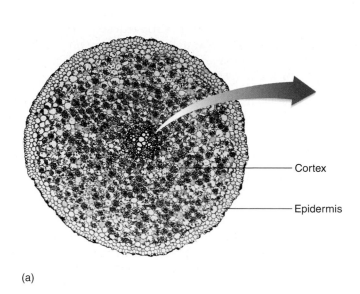

(a)

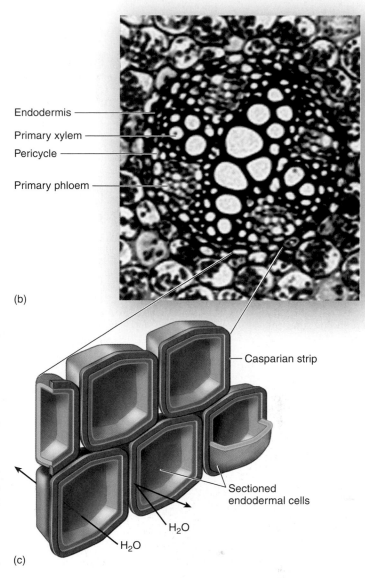

Endodermis

Primary xylem

Pericycle

Primary phloem

(b)

Casparian strip

Sectioned
endodermal cells

H_2O

H_2O

(c)

Figure 32.9 A root cross section.

(a) Cross section through a root of a buttercup (*Ranunculus*), a dicot (×10). (b) The enlargement shows the various tissues present. (c) Endodermal cells are surrounded by a water-proofing, waxy band, called a Casparian strip, that forces water and minerals to pass through the cell rather than between two cells.

The root elongates relatively rapidly just behind its tip in the area called the *zone of elongation.* Abundant root hairs (see figure 32.22), extensions of single epidermal cells, form above that zone, in the area called the *zone of differentiation.* Virtually all water and minerals are absorbed from the soil through the root hairs, which greatly increase the root's surface area and absorptive powers. In plants with symbiotic mycorrhizae, the root hairs are often greatly reduced in number, and the fungal filaments of the mycorrhizae play a role similar to that of the root hairs, increasing the surface area for absorption. Another symbiotic relationship involving plant roots is often key to the health of ecosystems. The roots of some plants, specifically plants of the pea family, called legumes, form symbiotic relationships with bacteria that are able to break down atmospheric nitrogen into a source that can be taken up and used by plants. These plants are a key component of the nitrogen cycle, cycling nitrogen back into the ecosystem in a form that can be used by other organisms.

The role of mycorrhizae was important in the evolution of plants and is described in more detail on page 314. The relationship of bacteria with the roots of legumes is important to the recycling of nitrogen in ecosystems and is described in more detail on page 394.

One of the fundamental differences between roots and stems has to do with the nature of their branching. In stems, branching occurs from buds on the stem surface; in roots, branching is initiated well back of the root tip as a result of cell divisions in the pericycle. The developing lateral roots (the red-stained mass of cells in figure 32.10) grow out through the cortex toward the surface of the root, eventually breaking through and becoming established as lateral roots. In some plants, roots may arise along a stem, or in some place other than the root of the plants. These roots are called *adventitious roots.* Adventitious roots occur in ivy, bulb plants such as onions, perennial grasses, and other plants that produce rhizomes, which are horizontal stems that grow underground.

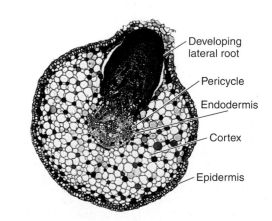

Developing
lateral root

Pericycle

Endodermis

Cortex

Epidermis

Figure 32.10 Lateral roots.

A lateral root growing out through the cortex of the black willow, *Salix nigra.* Lateral roots originate beneath the surface of the main root, whereas lateral stems originate at the surface.

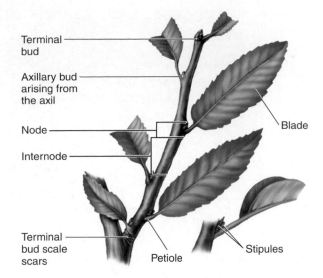

Figure 32.11 A woody twig.

This twig shows key stem structures, including the node and internode areas, the axillary bud in the axil, and leaves.

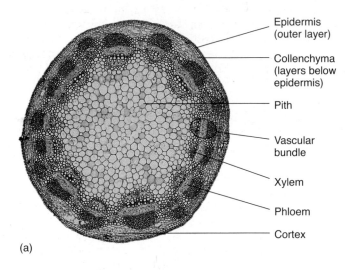

(a)

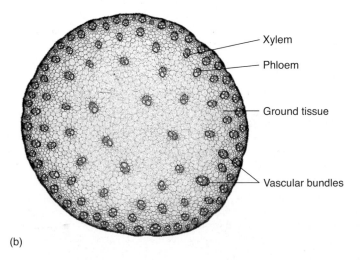

(b)

32.4 Stems

CONCEPT PREVIEW: Stems, the aboveground framework of the plant body, grow both at their tips, primary growth, and in circumference, secondary growth. Wood, the product of secondary growth, is the accumulation of secondary xylem.

Stems serve as the main structural support of the plant and the framework for the positioning of the leaves. Often experiencing both primary and secondary growth, stems are the source of an economically important product—wood.

Primary Growth

In the primary growth of a stem, leaves first appear as leaf primordia (singular, primordium), rudimentary young leaves that cluster around the apical meristem, unfolding and growing as the stem itself elongates. The places on the stem at which leaves form are called nodes (indicated by the small bracket in figure 32.11). The portions of the stem between these attachment points are called the internodes (the larger bracket). As the leaves expand to maturity, a bud, a tiny undeveloped side shoot, develops in the **axil** of each leaf, the angle between a leaf and the stem from which it arises. These buds, which have their own immature leaves called stipules (shown in figure 32.11), may elongate and form lateral branches, or they may remain small and dormant. A hormone moving downward from the terminal bud of the stem continually suppresses the expansion of the lateral buds in the upper portions of the stem. Lateral buds begin forming lower down when the stem lengthens to the point where the amount of hormone reaching the lower portion of the stem is reduced, or if the terminal bud is removed, such as when you prune a plant.

Lateral bud growth in plants is influenced by two different hormones, as discussed on page 644. The hormone auxin from the apical meristem inhibits lateral bud growth; when the apical meristem is removed by pruning, the hormone cytokinin promotes lateral bud growth.

Within the soft, young stems, the strands of vascular tissue, xylem and phloem, are arranged differently in dicots versus monocots. In dicots, the vascular bundles contain xylem and phloem and are arranged around the outside of the stem as a cylinder (figure 32.12a). In monocots, the vascular bundles are scattered throughout the stem (figure 32.12b). This difference in vascular tissue organization, in addition to other characteristics discussed in chapter 17, illustrate the differences between these two major groups of angiosperms. The vascular bundles contain both primary xylem and primary phloem. At the stage when only primary growth has occurred, the inner portion of the ground tissue of a dicot stem is called the **pith** (the center pink-stained cells in figure 32.12a), and the outer portion is the **cortex** (the light green-stained cells located toward the outside).

Secondary Growth

In stems, secondary growth (the thickening of the stem) is initiated by the differentiation of a lateral meristem called the **vascular cambium,** a thin cylinder of actively dividing cells located between the bark and the inte-

Figure 32.12 A comparison of dicot and monocot stems.

(a) Transection of a young stem of a dicot, the common sunflower, *Helianthus annuus*, in which the vascular bundles are arranged around the outside of the stem. (b) Transection of a monocot stem, corn, *Zea mays*, with the scattered vascular bundles characteristic of the group.

rior region of the stem in woody plants. The vascular cambium develops from cells within the vascular bundles of the stem, between the xylem (the purple-colored areas in figure 32.13) and the phloem (the light green area). The cylindrical form of the vascular cambium is completed by the differentiation of some of the parenchyma cells that lie between the bundles. Once established, the vascular cambium consists of elongated and somewhat flattened cells with large vacuoles. The cells that divide from the vascular cambium outwardly, toward the bark, become secondary phloem; those that divide from it inwardly become secondary xylem.

While the vascular cambium is becoming established, a second kind of lateral meristem, the **cork cambium,** develops in the stem's outer layers. The cork cambium usually consists of plates of dividing cells that move deeper and deeper into the stem as they divide. Outwardly, the cork cambium splits off densely packed cork cells; they contain a fatty substance and are nearly impermeable to water. Cork cells are dead at maturity. Inwardly, the cork cambium divides to produce a layer of parenchyma cells. The cork, the cork cambium that produces it, and this layer of parenchyma cells make up a layer called the **periderm** (see figure 32.13), which is the plant's outer protective covering.

Cork covers the surfaces of mature stems or roots. The term **bark** refers to all of the tissues of a mature stem or root outside of the vascular cambium. Because the vascular cambium has the thinnest-walled cells that occur anywhere in a secondary plant body, it is the layer at which bark breaks away from the accumulated secondary xylem.

Wood is one of the most useful, economically important, and beautiful products obtained from plants. Anatomically, wood is accumulated secondary xylem (the light purple pie-shaped areas in figure 32.13). As the secondary xylem ages, its cells become infiltrated with gums and resins, and the wood may become darker. For this reason, the wood located nearer the central regions of a given trunk, called heartwood, can be darker and denser than the wood nearer the vascular cambium, called sapwood, which is still actively involved in water transport within the plant.

Because of the way it is accumulated, wood often displays rings. The rings that you see in the section of pine in figure 32.14 reflect the fact that the vascular cambium of trees divides more actively in the spring and summer, when water is plentiful and temperatures are suitable for growth, than in the fall and winter, when water is scarce and the weather is cold. As a result, layers of larger, thinner-walled cells formed during the growing season (the lighter rings) alternate with the smaller, darker layers of thick-walled cells formed during the rest of the year. New rings are laid down each year toward the outer edge of the stem. A count of such annual rings in a tree trunk can be used to calculate the tree's age, and the width of rings can reveal information about environmental factors. For example, the region of thinner rings could indicate a period of prolonged drought conditions that was followed by wetter years. Can you estimate the age of the tree shown in figure 32.14?

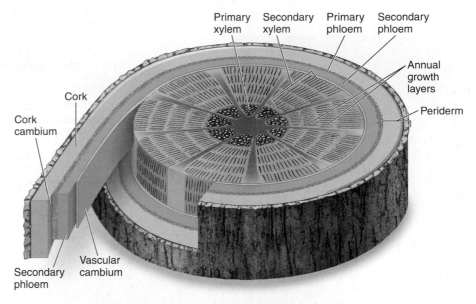

Figure 32.13   **Vascular cambium and secondary growth.**

The vascular cambium and cork cambium (lateral meristems) produce secondary tissues, causing the stem's girth to increase. Each year, a new layer of secondary tissue is laid down, forming rings in the wood.

IMPLICATION   Musical instruments like violins, guitars, clarinets, and recorders are made mostly or entirely of wood. The choice of wood makes a significant difference to the tone of the instrument. The bodies of clarinets are made from hard, dense African blackwood, while light spruce and sycamore are used in violins. How do you think the nature of the wood affects the sound of the instrument?

Figure 32.14   **Annual rings in a section of pine.**

To test your understanding of how they form, answer this question: Are the inner rings older or younger than the outer rings?

32.5 Leaves

CONCEPT PREVIEW: Leaves, the photosynthetic organs of the plant body, are varied in shape and arrangement.

Leaves are usually the most prominent shoot organs and are structurally diverse (figure 32.15). As outgrowths of the stem apex, leaves are the major light-capturing organs of most plants. Most of the chloroplast-containing cells of a plant are within its leaves, and it is there where the bulk of photosynthesis occurs (see chapter 6). Exceptions to this are found in some plants, such as cacti, whose green stems have largely taken over the function of photosynthesis for the plant. Photosynthesis is conducted mainly by the "greener" parts of plants because they contain more chlorophyll, the major photosynthetic pigment.

The apical meristems of stems and roots are capable of growing indefinitely under appropriate conditions. Leaves, in contrast, grow by means of **marginal meristems,** which flank their thick central portions. These marginal meristems grow outward and ultimately form the **blade** (flattened portion) of the leaf, while the central portion becomes the midrib. Once a leaf is fully expanded, its marginal meristems cease to grow.

In addition to the flattened blade, most leaves have a slender stalk, the **petiole.** Two leaflike organs, the **stipules,** may flank the base of the petiole where it joins the stem (see figure 32.11). Veins, consisting of both xylem and phloem, run through the leaves. In most dicots the pattern is net, or reticulate, venation—as you can see in figure 32.16a. In many monocots, the veins are parallel, like the parallel veins that pass vertically up through the monocot leaf in figure 32.16b.

Leaf blades come in a variety of forms from oval to deeply lobed to having separate leaflets (the blade being divided but attached to a single petiole like the black walnut leaf in figure 32.15c). In *simple leaves* (see figure 32.15a, b), such as those of birch or maple trees, there is a single blade, undivided, but some simple leaves may have teeth, indentations, or lobes, such as the leaves of maples and oaks. In *compound leaves,* such as

(a)

(b)

(c)

(d)

(e)

(f)

Figure 32.15 Leaves.

Leaves are stunningly variable. (a) *Simple leaves* from a gray birch, in which there is a single blade. (b) A simple leaf, its margin lobed, from the vine maple. (c) A *pinnately compound* leaf of a black walnut tree, where leaflets occur in pairs along the central axis of the main vein. (d) *Palmately compound* leaves of a horse chestnut tree, in which the leaflets radiate out from a single point. (e) The leaves of pine trees are tough and needlelike. (f) Many unusual types of modified leaves occur in different kinds of plants. For example, some plants produce floral leaves or bracts; the most conspicuous parts of this poinsettia flower are the red bracts, which are not petals, but rather are modified leaves that surround the small yellowish true flowers in the center.

those of ashes, box elders, and walnuts, the blade is divided into leaflets. If the leaflets are arranged in pairs along a common axis—the equivalent of the main central vein, or *midrib,* in simple leaves—the leaf is *pinnately compound,* such as in the black walnut (see figure 32.15*c*). If, however, the leaflets radiate out from a common point at the blade end of the petiole, the leaf is *palmately compound,* such as in buckeyes, horse chestnuts (figure 32.15*d*), and Virginia creepers. Leaves may be alternately arranged (alternate leaves usually spiral around a stem), or they may be in opposite pairs. Less often, three or more leaves may be in a whorl, a circle of leaves at the same level at a node (figure 32.17).

A typical leaf contains masses of parenchyma, called **mesophyll** ("middle leaf"), through which the vascular bundles, or veins, run. Beneath the upper epidermis of a leaf are one or more layers of closely packed, columnlike parenchyma cells called **palisade mesophyll** (the red-stained cells in the photo in figure 32.18). These cells contain more chloroplasts than other cells in the leaf and so are more capable of carrying out photosynthesis. This makes sense when you consider that the cells on the surface receive more sun. The rest of the leaf interior, except for the veins, consists of a tissue called **spongy mesophyll.** Between the spongy mesophyll cells are large intercellular spaces that function in gas exchange and particularly in the passage of carbon dioxide from the atmosphere to the mesophyll cells. You can see the spongy mesophyll in the photo of figure 32.18, but the air spaces that are the basis of this tissue's function might be easier to see in the drawing. These intercellular spaces are connected, directly or indirectly, with the stomata in the lower epidermis.

> During photorespiration, discussed on page 109, the stomata close and carbon dioxide cannot enter the cell. Likewise, oxygen gas, produced as a by-product of photosynthesis, builds up inside these air spaces and interferes with photosynthesis.

(a) (b)

Figure 32.16 **Dicot and monocot leaves.**

(a) The leaves of dicots have netted, or reticulate, veins; (b) those of monocots have parallel veins.

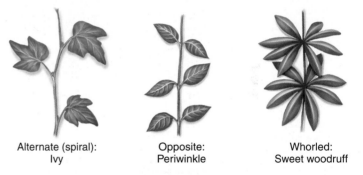

| Alternate (spiral): Ivy | Opposite: Periwinkle | Whorled: Sweet woodruff |

Figure 32.17 **Types of leaf arrangements.**

Concept Check

1. How do root and stem branching differ?
2. Of what tissue is wood composed?
3. Where on a plant would you expect to find marginal meristems?

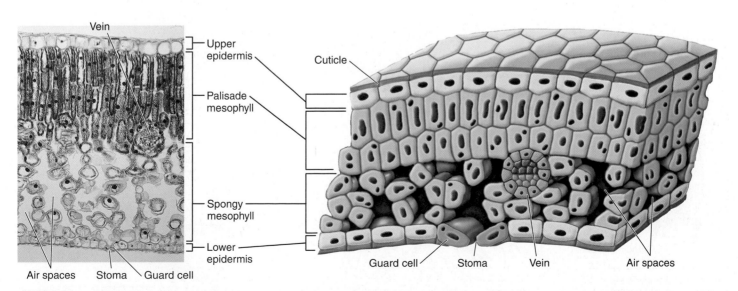

Figure 32.18 **A leaf in cross section.**

Cross section of a leaf, showing the arrangement of palisade and spongy mesophyll, a vascular bundle or vein, and the epidermis, with paired guard cells flanking the stoma.

Plant Transport and Nutrition

How Tall Can a Tree Grow? Owner of the title "The World's Tallest Tree" goes to a California redwood called Hyperion that stands at a height of 379.1 feet (115.5 meters). Laid on its side, this tree would span just over the full length of a football field including the endzones. And Hyperion is still growing taller. How tall can we realistically expect it to grow? The pull of gravity and friction that builds up between the water molecules and the walls of the xylem vessels limits how high a column of water can be supported. The high end for redwoods appears to be about 427 feet (130 meters). So, how tall will Hyperion get? According to this research, maybe only another 50 feet, but that is still taller than a 30-story building. That is a long way up indeed!

32.6 Water Movement

CONCEPT PREVIEW: Water is drawn up the plant stem from the roots by transpiration from the leaves.

Vascular plants have a conducting system, as humans do, for transporting fluids and nutrients from one part to another. Functionally, a plant is essentially a bundle of tubes with its base embedded in the ground. At the base of the tubes are roots, and at their tops are leaves. For a plant to function, two kinds of transport processes must occur: First, the carbohydrate molecules produced in the leaves by photosynthesis must be carried to all of the other living cells in the plant. To accomplish this, liquid with dissolved carbohydrate molecules, must move both up and down the tubes. Second, minerals and water in the soil must be taken up by the roots and ferried to the leaves and other plant cells. In this process, liquid moves up the tubes. Plants accomplish these two processes by using chains of specialized cells. Cells of the phloem transport photosynthetically produced carbohydrates up and down the plant (red arrows in figure 32.19), and those of the xylem carry water and minerals upward (blue arrows in figure 32.19).

Cohesion-Adhesion-Tension Theory

Many of the leaves of a large tree may be more than 10 stories off the ground. How does a tree manage to raise water so high? Several factors are at work to move water up the height of a plant. The initial movement of water into the roots of a plant involves osmosis. Water moves into the cells of the root because the fluid in the xylem contains more solutes than the surroundings—recall from chapter 4 that water will move across a membrane from an area of lower solute concentration to an area of higher solute concentration. However, this force, called *root pressure,* is not by itself strong enough to "push" water up a plant's stem.

Figure 32.19 **The flow of materials into, out of, and within a plant.**

Water and minerals enter through the roots of a plant and are transported through the xylem to all parts of the plant body (*blue arrows*). Water leaves the plant through the stomata in the leaves. Carbohydrates synthesized in the leaves are circulated throughout the plant by the phloem (*red arrows*).

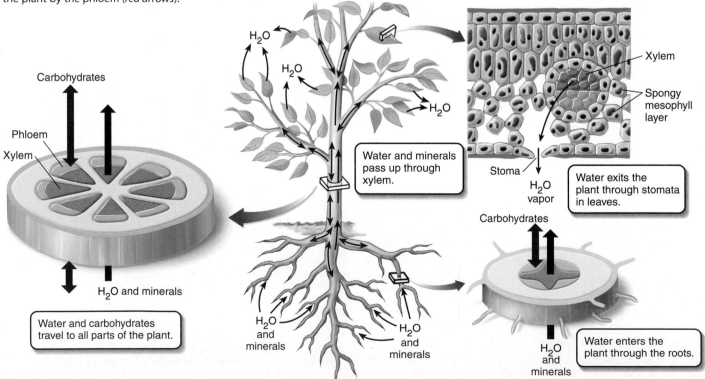

Capillary action adds a "pull" to the process. *Capillary action* results from the tiny electrical attractions of polar water molecules to surfaces that carry an electrical charge, a process called *adhesion.* In the laboratory, a column of water rises up a tube of glass because the attraction of the water molecules to the charged molecules on the interior surface of the glass tube "pull" the water up in the tube. In figure 32.20, which illustrates this process, why does the water travel higher up in the narrower tube? The water molecules are attracted to the glass molecules, and the water travels up farther in the narrower tube because the amount of surface area available for adhesion is greater than in the larger-diameter tube.

However, although capillary action can produce enough force to raise water a meter or two, it cannot account for the movement of water to the tops of tall trees. A second very strong "pull" accomplishes this, provided by transpiration. Blowing air across the upper end of the tube in figure 32.20 demonstrates how transpiration draws water up a plant stem. The stream of relatively dry air causes water molecules at the water column's exposed top surface to evaporate from the tube. The water level in the tube does not fall, because as water molecules are drawn from the top through evaporation, they are replenished by new water molecules pulled up from the bottom. This, in essence, is what happens in plants. The passage of air across leaf surfaces results in the loss of water by evaporation, creating a "pull" at the open upper end of the plant. New water molecules that enter the roots are pulled up the plant. Adhesion of water molecules to the walls of the narrow vessels in plants also helps to maintain water flow to the tops of plants.

A column of water in a tall tree does not collapse under its weight because water molecules have an inherent strength that arises from their tendency to form hydrogen bonds with one another. These hydrogen bonds cause *cohesion* of the water molecules; in other words, a column of water resists separation. The beading of water droplets illustrates the property of cohesion. This resistance, called *tensile strength,* varies inversely with the diameter of the column; that is, the smaller the diameter of the column, the greater the tensile strength. Therefore, plants must have very narrow transporting vessels to take advantage of tensile strength.

How the combination of gravity, adhesion, and tensile strength due to cohesion affects water movement in plants is called the **cohesion-adhesion-tension theory.** It is important to note that the movement of water up through a plant is a passive process and requires no expenditure of energy on the part of the plant.

Transpiration

The process by which water leaves a plant is called **transpiration.** More than 90% of the water taken in by plant roots is ultimately lost to the atmosphere, almost all of it from the leaves. It passes out primarily through the stomata in the evaporation of water vapor, as you can see in panel 1 of *Essential Biological Process 32A.* On its journey from the plant's interior to the outside, a molecule of water first diffuses from the xylem into the spongy mesophyll cells of the leaf (see figure 32.18). Then, water passes into the pockets of air within the leaf by evaporating from the walls of the spongy mesophyll that line the

> Adhesion and cohesion are properties of water, as discussed on page 39. Water molecules form hydrogen bonds with other polar surfaces (adhesion) and with each other (cohesion). These properties allow water to move up the stem of a plant.

> The water cycle, as discussed on page 391, involves both environmental and organismal components. The organismal component is the cycling of water through plants in transpiration, as described here. Transpiration is key to the functioning of the water cycle.

Figure 32.20   Capillary action.
The attraction of water molecules to the glass surface of narrow tubes draws water upwards, while the force of gravity tends to draw it down. The narrower the tube, the greater the surface area available for adhesion for a given volume of water, and the higher the water rises in the tube.

Essential Biological Process 32A
Transpiration

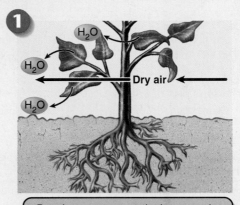

Dry air passes across the leaves and causes water vapor to evaporate out of the stomata.

The loss of water from the leaves creates a type of "suction" that draws water up the stem through the xylem.

New water enters the plant through the roots to replace the water moving up the stem.

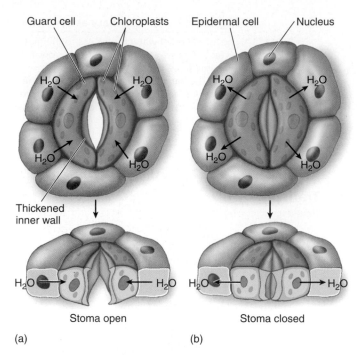

Guard cell   Chloroplasts   Epidermal cell   Nucleus

Thickened
inner wall

Stoma open   Stoma closed

(a)   (b)

Figure 32.21   How guard cells regulate the opening and closing of stomata.

(a) When guard cells contain a high level of solutes, water enters the guard cells by osmosis, causing them to swell and bow outward. This bowing opens the stoma. (b) When guard cells contain a low level of solutes, water leaves the guard cells, causing them to become flaccid. This flaccidity closes the stoma.

Figure 32.22   Root hairs.

Abundant fine root hairs can be seen in the back of the root apex of this germinating seedling of radish, *Raphanus sativus*.

intercellular spaces. These intercellular spaces open to the outside of the leaf by way of the stomata. The water that evaporates from these surfaces of the spongy mesophyll cells is continuously replenished from the xylem in the leaves. Because the strands of xylem conduct water within the plant in an unbroken stream all the way from the roots to the leaves, when a portion of the water vapor in the intercellular spaces passes out through the stomata, the supply of water vapor in these spaces is continually renewed from lower down in the column (panel 2) and ultimately from the roots (panel 3). Because the process of transpiration is dependent upon evaporation, factors that affect evaporation also affect transpiration. In addition to the movement of air across the stomata, mentioned earlier, humidity levels in the air will affect the rate of evaporation—high humidity reducing it and low humidity increasing it. Temperature will also affect the rate of evaporation—high temperatures increasing it and lower temperatures reducing it. This temperature effect is especially important because evaporation also acts to cool plant tissues.

Regulation of Transpiration: Open and Closed Stomata

The only way plants can control water loss on a short-term basis is to close their stomata. Many plants can do this when subjected to water stress. But the stomata must be open at least part of the time so that carbon dioxide, which is necessary for photosynthesis, can enter the plant. In its pattern of opening or closing its stomata, a plant must respond to both the need to conserve water and the need to admit carbon dioxide. A number of environmental factors affect the opening and closing of stomata. The most important is water loss.

The stomata open and close because of changes in the water pressure of their guard cells. Stomatal guard cells are long, sausage-shaped cells attached at their ends. These are the green cells in figure 32.21. The cellulose microfibrils of their cell wall wrap around the cell such that when the guard cells are **turgid** (plump and swollen with water), they expand in length. This causes the cells to bow, opening the stomata, shown on the left side of the figure. Turgor in guard cells results from the active uptake of ions, causing water to enter osmotically as a consequence (discussed in chapter 33). Loss of water, as in a wilted plant, causes the guard cells to become flaccid and the stomata closes.

Water Absorption by Roots

Most of the water absorbed by plants comes in through the root hairs, extensions of epidermal cells. These give a root the feathery appearance shown in figure 32.22. These root hairs greatly increase the surface area and therefore the absorptive powers of the roots. Root hairs are turgid—plump and swollen with water—because they contain a higher concentration of dissolved minerals and other solutes than does the water in the soil solution; water, therefore, tends to move into them steadily. Once inside the roots, water passes inward to the conducting elements of the xylem.

Water is not the only substance that enters the roots by passing into the cells of root hairs. Minerals also enter the root. Membranes of root hair cells contain a variety of ion transport channels that actively pump specific ions into the plant, even against large concentration gradients. These ions, many of which are plant nutrients, are then transported throughout the plant as a component of the water flowing through the xylem.

32.7 Carbohydrate Transport

CONCEPT PREVIEW: Carbohydrates move through the plant by the passive osmotic process of translocation.

Most of the carbohydrates manufactured in plant leaves and other green parts move through the phloem to other parts of the plant. This process, known as **translocation,** makes carbohydrate building blocks available at the plant's actively growing regions. The carbohydrates are concentrated in storage organs such as underground stems (potatoes), roots (carrots), and leaves (onions and cabbage), often in the form of starch. The starch is converted into transportable molecules such as sucrose, and moves through the plant.

> Many plants store carbohydrates as starch, but starch is not produced during photosynthesis. The Calvin cycle, as discussed on page 108, produces a three-carbon sugar molecule, glyceraldehyde 3-phosphate, and it is this molecule that is used to build starch, or other sugars.

The pathway by which sugars and other substances travel within the plant has been demonstrated precisely by using radioactive isotopes and aphids, a group of insects that suck the sap of plants. Aphids thrust their piercing mouthparts into the phloem cells of leaves and stems to obtain the abundant sugars there. When the aphids are cut off of the leaf, the liquid continues to flow from the detached mouthparts protruding from the plant tissue and is thus available in pure form for analysis. The liquid in the phloem contains 10% to 25% sucrose.

> Sucrose is a disaccharide formed from the linking together of two monomers, glucose and fructose, as discussed page 54 and in table 3.1. Sucrose is what makes maple syrup sweet and is the substance harvested from sugar cane.

The harvesting of sap from maple trees uses a similar process. A hole is drilled in the tree and the sugar-rich fluid is drained from the tree into buckets. The sap is then processed into maple syrup.

Using aphids to obtain the critical samples and radioactive tracers to mark them, researchers have learned that movement of substances in the phloem can be remarkably fast—rates of 50 to 100 centimeters per hour have been measured. This translocation movement is a passive process that does not require the expenditure of energy by the plant. The **mass flow** of materials transported in the phloem occurs because of water pressure, which develops as a result of osmosis. *Essential Biological Process 32B* walks you through the process of translocation. Sugar produced as a result of photosynthesis is actively "loaded" into the sieve tubes (or sieve cells) of the vascular bundles (panel 1). This loading increases the solute concentration of the sieve tubes, so water passes into them by osmosis (panel 2). The area where the sugar is made is called a *source* and the area where sugar is delivered from the sieve tubes is called a *sink*. Sinks include the roots and other regions of the plant that are not photosynthetic, such as young leaves and fruits. Water flowing into the phloem forces the sugary substance in the phloem to flow down the plant (panel 3). The sugar is unloaded and stored in sink areas (panel 4). There the solute concentration of the sieve tubes is decreased as the sugar is removed. As a result of these processes, water moves through the sieve tubes from the areas where sucrose is being added into those areas where it is being withdrawn, and the sugar moves passively with the water. This is called the *pressure-flow hypothesis.*

Concept Check

1. What is the difference between cohesion and adhesion?
2. Describe how stomata open and close.
3. Describe how sugar made in a leaf reaches the cells of a root.

Essential Biological Process 32B

Translocation

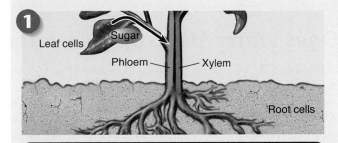

Sugar created in the leaves by photosynthesis ("source") enters the phloem by active transport.

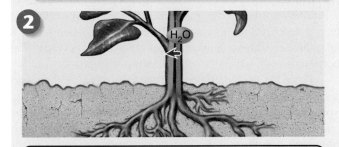

When the sugar concentration in the phloem increases, water is drawn into phloem cells from the xylem by osmosis.

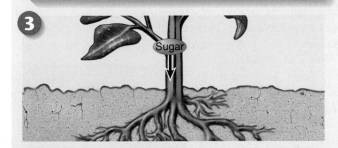

The addition of water from the xylem causes pressure to build up inside the phloem and pushes the sugar down.

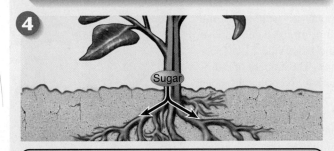

Sugar from the phloem enters the root cells ("sink") by active transport.

Does Water Move Up a Tree Through Phloem or Xylem?

Before reading this chapter, you may have wondered how water gets to the top of a tree, 10 stories above its roots. Earlier scientists also wondered about this. A column of water that tall weighs an awful lot. If you were to make a tube of drinking straws that tall and fill it with water, you would not be able to lift it. The answer to this puzzle was first proposed by biologist Otto Renner in Germany in 1911. He suggested that dry air moving across the tree's leaves captured water molecules by evaporation, and that this water was replaced with other water molecules coming in from the roots. Renner's idea, which was essentially correct, forms the core of the cohension-adhesion-tension theory described in this chapter. Essential to the theory is that there is an unbroken water column from leaves to roots, a "pipe" from top to bottom through which the water can move freely.

There are two candidates for the role of water pipe, each a long series of narrow vessels that runs the length of the stem of a tree. As you have learned earlier, these two vessel systems are called xylem and phloem. In principle, either xylem or phloem could provide the plumbing through which water moves up a tree trunk or other stem. Which is it?

An elegant experiment demonstrates which of these vessel systems carries water up a tree stem. A section of a stem was placed in water containing the radioactive potassium isotope ^{42}K. A piece of wax paper was carefully inserted between the xylem and the phloem in a 23-cm section of the stem to prevent any lateral transport of water between xylem and phloem.

After enough time had elapsed to allow water movement up the stem, the 23-cm section of the stem was removed, cut into six segments, and the amounts of ^{42}K measured both in the xylem and in the phloem of each segment, as well as in the stem immediately above and below the 23-cm section. The amount of radioactivity recorded provides a direct measure of the amount of water that has moved up from the bottom of the stem through either the xylem or phloem.

The results are presented in the graph.

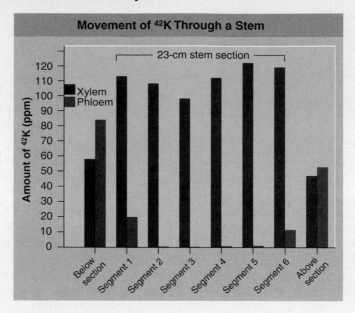

Movement of ^{42}K Through a Stem

Analysis

1. **Interpreting Data**
   a. In the portion of the stem below where the 23-cm section was removed, do xylem and phloem both contain radioactivity? How about in the portion above where the 23-cm section was removed?
   b. In the central portion of the 23-cm segment of the stem (segments 2, 3, 4, 5), do both xylem and phloem contain radioactivity?
2. **Making Inferences**
   a. In the 23-cm section, is more ^{42}K found in xylem or phloem? What might you conclude from this?
   b. Above and below the 23-cm section, is more ^{42}K found in xylem or phloem? How would you account for this? [Hint: These sections did not contain the wax paper barrier that prevents lateral transport between xylem and phloem.]
   c. Within the 23-cm section, the phloem in segments 1 and 6 contains more ^{42}K than interior segments. What best accounts for this?
   d. Is it fair to infer that water could move through either xylem or phloem vessel systems?
3. **Drawing Conclusions** Does water move up a stem through phloem or xylem? Explain.

Concept Summary

Structure and Function of Plant Tissues

32.1 Organization of a Vascular Plant

- Most plants possess roots, stems, and leaves, although they may not always look the same. Vascular tissue extends throughout the plant, connecting roots, stems, and leaves (**figure 32.1**).
- Growth occurs in regions called meristems. The tips of the roots and shoots contain apical meristems, which are the sites of primary growth. Primary growth extends the plant body lengthwise. Extending the thickness or girth of the plant, called secondary growth, occurs at the lateral meristems, which are cylinders of meristematic tissue.

32.2 Plant Tissue Types

- Ground tissue makes up the main body of the plant and contains several different cell types. Parenchyma cells are the most common type of cell in plants. They have thin cell walls and are alive at maturity. They carry out functions such as photosynthesis, food and water storage. The edible parts of fruits and vegetables are primarily parenchyma cells (**figure 32.2**).
- Collenchyma cells form strands that provide support, especially for plants that do not have secondary growth. They are usually elongated cells with unevenly thickened primary cell walls (**figure 32.3**).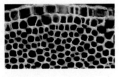
- Sclerenchyma cells have thick secondary cell walls that provide strength and rigidity. The secondary cell wall is laid down after the cell has stopped growing. They form long fibers or branched structures called sclereids or "stone cells" (**figure 32.4**).
- Dermal tissue makes up the outer layer of the plant body, which consists of epidermal cells covered with a waxy layer called the cuticle (root cells lack a cuticle). This outer layer protects the plant and the cuticle provides a barrier to water loss. Paired guard cells are specialized cells in the epidermis (**figure 32.5a**). The space between the two guard cells, called the stoma, opens (allowing for gas exchange) and closes in response to external factors.
- Trichomes and root hairs are extensions of epidermal cells (**figure 32.5b**). Trichomes help regulate heat and water balance in the leaves, and some secrete toxic substances as defense mechanism. Root hairs increase the surface area of the roots, allowing them to take up more water and minerals from the soil.
- Vascular tissue is composed of xylem and phloem. Xylem contains water-conducting cells, like the tracheids and vessel elements shown here from **figure 32.6.** They have thick secondary cell walls that provide structural support. They form long strands that are connected by pits and pores through which water passes.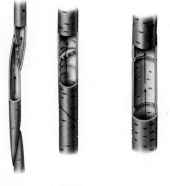
- Phloem contains food-conduction cells, sieve cells and sieve-tube members. The cells fit end-to-end, and food is conducted through pores between the cells. Companion cells associated with the phloem cells carry out metabolic functions needed to maintain the sieve-tube members (**figure 32.7**).

The Plant Body

32.3 Roots

- Roots are organs adapted to absorb water and minerals from the soil (**figure 32.8**). Vascular tissue forms the core of the root.
- A single layer of cells called the endodermis surrounds the vascular tissue. A waxy Casparian strip encircles the endodermis cells and blocks the passage of water between the cells. Water is forced through the cells into the xylem (**figure 32.9**).
- The root grows at the tip in the area called the zone of elongation. New cells are added by the apical meristem. Symbiotic relationships with fungi and bacteria can increase the absorption of water and minerals. Sometimes roots will emerge from the stems of a plant and are called adventitious roots.

32.4 Stems

- The stem serves as a framework for positioning the leaves. Primary growth occurs at the apical meristem. Leaves grow out of the stems at node areas, shown here from **figure 32.11.** Secondary growth occurs at the lateral meristems with the differentiation of vascular cambium into xylem and phloem, and the cork cambium into the layers of cork inside bark (**figures 32.12** and **32.13**). Wood forms from the accumulation of secondary xylem, which is thicker in spring and summer resulting in rings in the wood (**figure 32.14**).

32.5 Leaves

- Leaves are the primary site for photosynthesis. They grow out from the stem by means of marginal meristems that form the blade. They vary in size, shape, and arrangement (**figures 32.15–32.17**). Photosynthetic palisade mesophyll cells lie toward the surface. An underlying spongy mesophyll cell layer has large intercellular spaces that function in gas exchange (**figure 32.18**).

Plant Transport and Nutrition

32.6 Water Movement

- Carbohydrates and water are transported by phloem and xylem respectively (**figure 32.19**).
- Water enters the roots by osmosis. Root pressure and capillary action (**figure 32.20**) cause the water to pass up into the tissues. However, for water to travel up the length of the stem, it requires a stronger force, the combination of cohesion and adhesion, known as the cohesion-adhesion-tension theory. Transpiration, the evaporation of water vapor from the leaves, creates the "pull" that raises water through the xylem (***Essential Biological Process 32A***).
- The guard cells flanking stomata will swell up when water is plentiful (**figure 32.21**). This turgid pressure opens stomata, letting water vapor out. Under conditions of water stress, water leaves the guard cells and stomata close, reducing water loss.
- Minerals enter the plant along with water through ion channels in the cells of root hairs (**figure 32.22**).

32.7 Carbohydrate Transport

- Carbohydrates produced in the leaves travel throughout the plant in phloem tissue. Translocation, involves osmotic movement of water into the phloem cells, forcing the sugars to "sinks" for carbohydrate storage until needed (***Essential Biological Process 32B***).

Self-Test

1. Growth in vascular plants originates in
   a. photosynthetic tissue.
   b. root tissue.
   c. meristematic tissue.
   d. leaf epidermal tissue.

2. In vascular plants, phloem tissue primarily
   a. transports water.
   b. transports carbohydrates.
   c. transports minerals.
   d. supports the plant.

3. The ground tissue that carries out most of the metabolic and storage functions is
   a. parenchyma cells.
   b. collenchyma cells.
   c. sclerenchyma cells.
   d. sclereid cells.

4. In roots, growth of lateral branches begins
   a. on the root epidermis.
   b. on the root hairs.
   c. at the ground meristem.
   d. at the pericycle.

5. In stems, the tissue responsible for secondary growth is the
   a. collenchyma.
   b. pith.
   c. cambium.
   d. cortex.

6. One difference between monocot and dicot plant stems is the
   a. absence of buds in monocots.
   b. organization of vascular tissue.
   c. presence of guard cells.
   d. absence of stomata.

7. In vascular plant leaves, gases enter and leave the plant through pores called
   a. cuticle.
   b. trichomes.
   c. stomata.
   d. guard cells.

8. Which of the following is *not* a process that directly assists in water movement from the roots to the leaves?
   a. photosynthesis
   b. root pressure
   c. capillary action
   d. transpiration

9. The passive process of moving carbohydrates throughout a plant is called
   a. transpiration.
   b. translocation.
   c. translation.
   d. evaporation.

10. The movement of carbohydrates through the plant is driven by
    a. facilitated diffusion.
    b. active transport.
    c. evaporation.
    d. osmosis.

Visual Understanding

1. **Figure 32.8** What is the purpose of the root cap covering the apical meristem of the root?

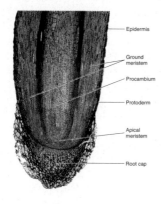

Epidermis
Ground meristem
Procambium
Protoderm
Apical meristem
Root cap

2. **Figure 32.14** Besides wet and dry years, and the age of the tree, what else might the study of tree rings tell us?

3. **Figure 32.21** If you went on vacation for several days and left your house plants in a warm, stuffy apartment, would the stomata look like the one on the left or the one on the right? Explain.

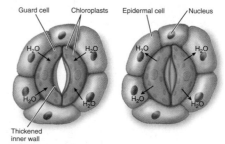

Guard cell Chloroplasts Epidermal cell Nucleus
H_2O H_2O H_2O H_2O
H_2O H_2O H_2O H_2O
Thickened inner wall

Challenge Questions

1. Why do land plants need sclerenchyma cells?

2. In desert climates some plants, such as palo verde and ocotillo, may lose some or all of their leaves in the hot, dry summer to minimize water loss. These plants have green stems. Why?

3. A friend just returned from a family trip to northern Michigan, where he visited a maple tree farm where they made maple syrup. On the maple trees, they made just one small hole (or two small holes on larger trees) and hung a bucket beneath to catch the sap. Why, he asks you, wouldn't they just make a cut completely around the tree and collect much more sap, much faster? Why isn't this feasible?

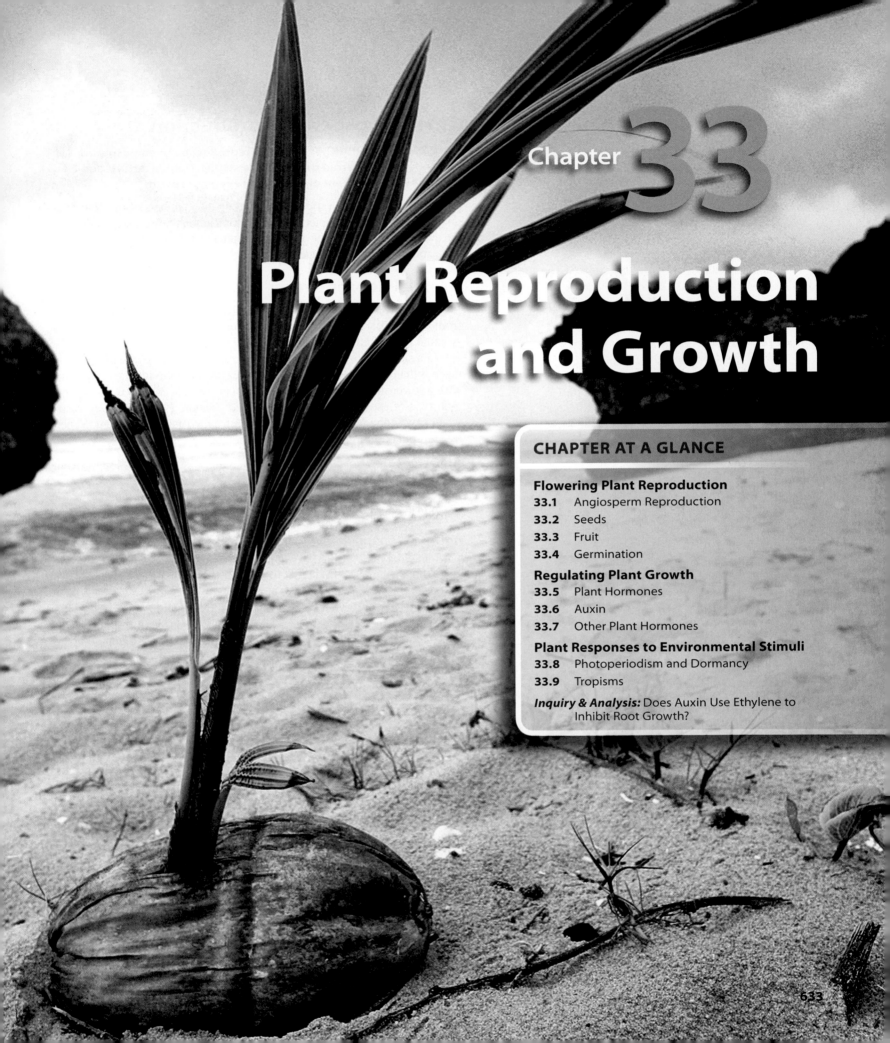

Plant Reproduction and Growth

Flowering Plant Reproduction

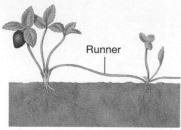

Runner

(a)

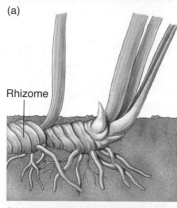

Rhizome

(b)

(c)

Figure 33.1   Vegetative reproduction.

(a) Runners are slender stems that grow along the ground, sending out roots and shoots at the nodes. (b) Rhizomes are underground horizontal stems that give rise to new shoots. (c) Small plants arise from notches along the leaves of the house plant *Kalanchoë daigremontiana.*

IMPLICATION   Bamboo, a kind of grass, is the fastest-growing and most prolific woody plant in the world. If you plant an individual bamboo plant in a corner of your yard, within a few years you will have bamboo plants growing all over your yard. How would you determine if bamboo individuals growing far from the one you planted were the result of reproduction by seeds or rhizomes?

33.1 Angiosperm Reproduction

CONCEPT PREVIEW: In sexual reproduction of angiosperms, pollen is transferred to the female stigma. Double fertilization leads to the development of an embryo and endosperm.

Although reproduction varies greatly among the members of the plant kingdom, we focus in this chapter on reproduction among flowering plants. While the evolution of their unique sexual reproductive features, flowers and fruits, have contributed to their success, angiosperms also reproduce asexually.

Asexual Reproduction

In **asexual reproduction,** an individual inherits all of its chromosomes from a single parent and is, therefore, genetically identical to that parent. Asexual reproduction produces a "clone" of the parent.

In a stable environment, asexual reproduction may prove more advantageous than sexual reproduction because it allows individuals to reproduce with a lower investment of energy and maintain successful traits. A common type of asexual reproduction, called *vegetative reproduction,* results when new individuals are simply cloned from parts of the parent. Vegetative reproduction in plants varies and includes:

Runners.   Some plants reproduce by means of runners—long, slender stems that grow along the surface of the soil. The strawberry plant shown in figure 33.1*a* reproduces by runners. Notice that at node regions on the stem, adventitious roots form, extending into the soil. Leaves and flowers form, and a new stem is sent out, continuing the runner.

Rhizomes.   Rhizomes are underground horizontal stems that create a network underground. As in runners, nodes give rise to new flowering shoots (figure 33.1*b*). The noxious character of many weeds results from this type of growth pattern, but so do grasses and many garden plants such as irises. Other specialized stems, called tubers, function in food storage and reproduction. White potatoes are specialized underground stems that store food, and the "eyes" give rise to new plants.

Suckers.   The roots of some plants produce "suckers," or sprouts, which give rise to new plants, such as found in cherry, apple, raspberry, and blackberry plants. When the root of a dandelion is broken, which may occur if one tries to pull it from the ground, each root fragment may give rise to a new plant.

Adventitious Plantlets.   In a few species, even the leaves are reproductive. One example is the house plant *Kalanchoë daigremontiana,* familiar to many people as the "maternity plant," or "mother of thousands." The common names of this plant are based on the fact that numerous plantlets arise from meristematic tissue located in notches along the leaves. You can see the numerous little plantlets in figure 33.1*c.*

Sexual Reproduction

Plant sexual life cycles are characterized by an alternation of generations, in which a diploid *sporophyte generation* gives rise to a haploid *gametophyte generation,* as described in chapter 17. In angiosperms, the developing gametophyte generation is completely enclosed within the tissues of the parent sporophyte (see figure 17.17). The male gametophytes are **pollen grains,** and they develop from *microspores.* The female gametophyte is the **embryo sac,** which develops from a *megaspore.* Pollen grains and

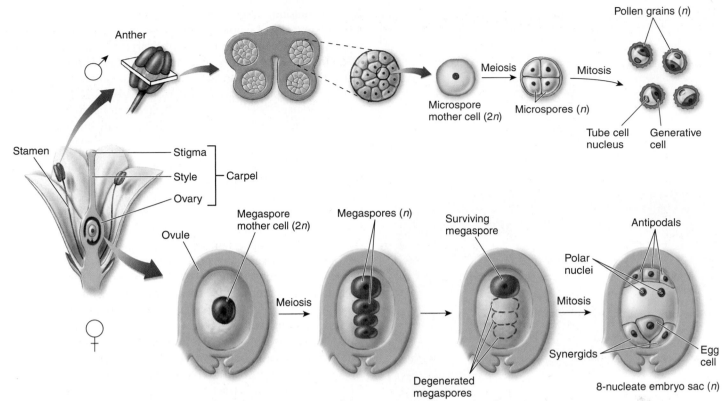

Figure 33.2 **Formation of pollen and egg.**

Diploid microspore mother cells are housed in the anther and divide by meiosis to form four haploid (n) microspores. Each microspore develops by mitosis into a pollen grain with a generative cell and a tube cell nucleus. The generative cell will later divide to form two sperm cells. Within the ovule, one diploid megaspore mother cell divides by meiosis to produce four haploid megaspores. Usually only one megaspore survives, and the other three degenerate. The surviving megaspore divides by mitosis to produce an embryo sac with eight nuclei. Upon fertilization, the egg cell becomes the embryo and the polar nuclei become the endosperm.

the embryo sac both are produced in separate, specialized structures of the angiosperm flower.

Flowers contain the organs for sexual reproduction in angiosperms. Like animals, angiosperms have separate structures for producing male and female gametes (sperm and eggs), but the reproductive organs of angiosperms are different from those of most animals in two ways. First, in angiosperms, both male and female structures usually occur together in the same individual flower. Second, angiosperm reproductive structures are not permanent parts of the adult individual. Angiosperm flowers and reproductive organs develop seasonally; these flowering seasons correspond to times of the year most favorable for pollination.

Structure of the Flower. Most flowers contain male and female parts. The male parts, called *stamens,* are the long filament structures you see in the cutaway flower in figure 33.2. At the tip of each filament is a swollen portion, called the *anther,* that contains pollen. The female part, called the *carpel,* is the vase-shaped structure in figure 33.2. The carpel consists of a lower bulging portion called the *ovary,* a slender stalk called the *style,* and a sticky tip called the *stigma* that receives pollen. In some plants, there are separate male and female flowers, but they occur on the same plant. These plants are called *monoecious,* meaning "one house." In monoecious plants, the male and female flowers may mature at different times, which keeps the plant from pollinating itself.

Pollen Formation. If you were to cut an anther in half, you would see four pollen sacs, each sac containing a collection of microspores. Pollen grains develop from these microspores. Each pollen sac contains specialized chambers in which the microspore mother cells are enclosed and protected. Each microspore mother cell undergoes meiosis to form four haploid microspores. Subsequently, mitotic divisions form pollen grains that contain a generative cell (the purple cell in the pollen grain) and a tube cell nucleus. The tube cell nucleus forms the pollen tube; the generative cell will later divide to form two sperm cells.

IN THE NEWS

Petal Color. It comes as no surprise that petal color, a key trait in flowers, is controlled by which pigments a plant is able to produce. It now turns out petal color is also controlled by the pH at which the pigments are stored. Anthocyanins—the main kind of flower pigment—reside in the vacuoles of a petal's cells. In 2008, researchers discovered that a proton pump is located in the vacuolar membrane. Experimentally disabling this proton pump (an ATPase) in petunias raises the pH within the plant's vacuoles, leading to a blue flower, a color never seen under natural conditions! Only by strictly controlling pH within its petal vacuoles does the petunia plant succeed in producing flowers with the proper pink–purple color.

Egg Formation. Eggs develop in the **ovule** of the angiosperm flower, which is contained within the ovary at the base of the carpel. Within each ovule is a megaspore mother cell (the large brown cell in the lower portion of figure 33.2). Each megaspore mother cell undergoes meiosis to produce four haploid megaspores (the column of four cells). In most plants, only one of these megaspores survives; the rest are absorbed by the ovule. The lone remaining megaspore undergoes repeated mitotic divisions to produce eight haploid nuclei, which are enclosed within a structure called an *embryo sac.*

Pollination. The process by which pollen is transferred from the anther to the stigma (the top of the carpel) is called **pollination.** The pollen may be carried to the flower by wind or by animals, or it may originate within the individual flower itself. When pollen from a flower's anther pollinates the same flower's stigma, the process is called *self-pollination,* which can lead to *self-fertilization.*

In many angiosperms, the pollen grains are carried from flower to flower by insects and other animals that visit the flowers for food or other rewards or are deceived into doing so because the flower's characteristics suggest such rewards. A liquid called **nectar,** which is rich in sugar as well as amino acids and other substances, is often the reward sought by animals.

For pollination by animals to be effective, a particular insect or other animal must visit plant individuals of the same species. A flower's color and form have been shaped by evolution to promote such specialization. Yellow and blue flowers are particularly attractive to bees (figure 33.3*a*), whereas red flowers attract birds but are not particularly noticed by most insects. Some flowers have very long floral tubes with the nectar produced deep within them; only the long, slender beaks of hummingbirds or the long, coiled proboscis of moths or butterflies (figure 33.3*b*) can reach such nectar supplies.

In certain angiosperms and all gymnosperms, pollen is blown about by the wind and reaches the stigmas passively. For such a system to operate efficiently, the individuals of a given plant species must grow relatively close together because wind does not carry pollen very far compared to transport by insects or other animals. Because gymnosperms, such as spruces or pines, grow in dense stands, wind pollination is very effective. Wind-pollinated angiosperms, such as birches, grasses, and ragweed, also tend to grow in dense stands. The flowers of wind-pollinated angiosperms are usually small, greenish, and odorless.

Fertilization. Once a pollen grain has been spread by wind, an animal, or self-pollination, it adheres to the sticky, sugary substance that covers the stigma and begins to grow a **pollen tube,** which pierces the style. The pollen tube, nourished by the sugary substance, grows until it reaches the ovule in the ovary.

> A detailed description of fertilization was presented in chapter 17, page 327 and in figure 17.17. Refer back to that discussion to review the specific steps in the process of fertilization.

Figure 33.4 traces the steps from fertilization through seed formation. When the pollen tube reaches the entry to the embryo sac in the ovule ❶, the tip of the tube bursts and releases the two sperm cells that form from the generative cell. In ❷ you can see that one of the sperm cells fertilizes the egg cell, forming a zygote. The other sperm cell fuses with the two polar nuclei located at the center of the embryo sac, forming the triploid (3*n*) primary endosperm nucleus. This process of fertilization in angiosperms in which two sperm cells are used is called **double fertilization.** Once fertilization is complete, the cells of the zygote divide numerous times. The primary endosperm nucleus eventually develops into the endosperm, shown in ❹, which nourishes the embryo.

(a)

(b)

Figure 33.3 Insect pollination.

(a) Bees are usually attracted to yellow flowers. (b) This alfalfa butterfly (*Colias eurytheme*) has a long proboscis that allows it to feed on nectar deep in the flower.

33.2 Seeds

CONCEPT PREVIEW: A seed contains a dormant embryo and substantial food reserves, encased within a tough drought-resistant coat.

After fertilization, active cell division forms an organized mass of cells, the embryo shown in figure 33.4 ❻. The differentiation of cell types within the embryo begins almost immediately after fertilization. By the fifth day, the principal tissue systems can be detected within the embryo mass, and within another day, the root and shoot apical meristems can be detected, as shown in ❽. The developing embryo is first nourished by the endosperm, and then later in some plants by the seed leaves, thick leaflike food storage structures called **cotyledons.**

Early in the development of an angiosperm embryo, a profoundly significant event occurs: The embryo simply stops developing and becomes dormant as a result of drying. In many plants, embryo development is arrested at the point shown in ❽, soon after apical meristems and the cotyledons are differentiated. The ovule of the plant has now matured into a **seed,** which includes the dormant embryo and a source of stored food, both surrounded by a protective and relatively impermeable seed coat that develops from the outer cover of the ovule.

Once the seed coat fully develops around the embryo, most of the embryo's metabolic activities cease; a mature seed contains only about 10% water. Under these conditions, the seed and the young plant within it are very stable.

Germination, or the resumption of metabolic activities that leads to the growth of a mature plant, cannot take place until water and oxygen reach the embryo, a process that sometimes involves cracking the seed. Seeds of some plants have been known to remain viable for hundreds, and in some cases thousands, of years.

BIOLOGY & YOU

Deadly Seeds. While many seeds are edible, providing humans with the bulk of their nutrition, some seeds are poisonous, containing deadly chemicals that discourage herbivores and seed predators. While in some cases these compounds simply taste bad to insects (such as in mustard plants), other compounds are quite toxic to humans. One of the most famous is ricin, a chemical found in the seeds of the castor bean—two to eight seeds provide enough ricin for a lethal dose. Another deadly seed chemical is cyanide, released by the breakdown of a defensive chemical called amygdalin that is found in the seeds of bitter almond, apricot, peach, and others. The seeds of many legumes, including the common bean and soybeans, contain proteins called lectins which can cause gastric distress if the beans are eaten without cooking, which degrades lectins to a harmless form.

Figure 33.4 Development in an angiosperm embryo.

After the zygote forms, the first cell division is asymmetric ❸. After another division, the basal cell, the one nearest the opening through which the pollen tube entered, undergoes a series of divisions and forms a narrow column of cells called the suspensor ❹. The other three cells continue to divide and form a mass of cells arranged in layers ❺. By about the fifth day of cell division, the principal tissue systems of the developing plant can be detected within this mass ❼.

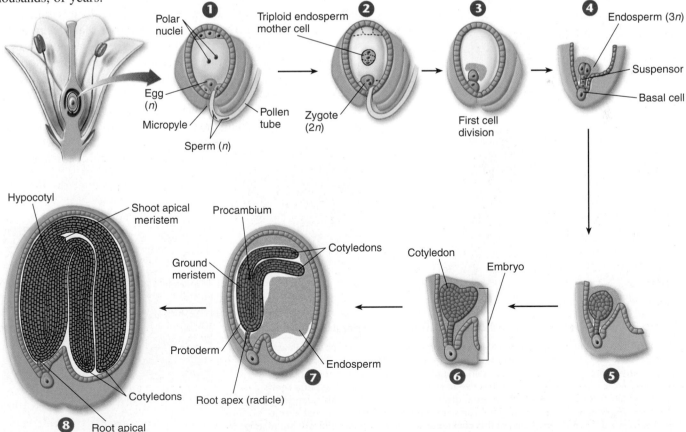

(a) Berries

(b) Drupes

(c) Pomes

(d) Eaten by animals

33.3 Fruit

CONCEPT PREVIEW: Fruits are specialized to achieve widespread dispersal by wind, by water, by attachment to animals, or, in the case of fleshy fruits, by being eaten.

During seed formation, the flower ovary begins to develop into fruit. The evolution of flowers was key to the success and diversification of the angiosperms. But of equal importance to angiosperm success has been the evolution of these fruits in response to their modes of dispersal.

There are three main kinds of fleshy fruits: berries, drupes, and pomes. In *berries*—such as grapes, tomatoes (figure 33.5*a*), and peppers—which are typically many-seeded, the inner layers of the ovary wall are fleshy. In *drupes*—such as peaches (figure 33.5*b*), olives, plums, and cherries—the inner layer of the fruit is stony and adheres tightly to a single seed. In *pomes*—such as apples (figure 33.5*c*) and pears—the fleshy portion of the fruit forms from the portion of the flower that is embedded in the receptacle (the swollen end of the flower stem that holds the petals and sepals). The inner layer of the ovary is a leathery membrane that encloses the seeds.

Fruits that have fleshy coverings, often black, bright blue, or red (as in figure 33.5*d*), are normally dispersed by birds and other vertebrates. By feeding on these fruits, the animals carry seeds from place to place before excreting the seeds as solid waste. The seeds, not harmed by the animal's digestive system, are carried to another habitat.

> Seeds withstand harsh conditions going through the digestive systems of birds and mammals. Recall on page 508 that food enters the gizzards of birds and is churned up with pebbles. In mammalian stomachs, the seeds are exposed to strong acids, as discussed on page 510.

Fruits that are dispersed by wind, or by attaching themselves to the fur of mammals or the feathers of birds, are called dry fruits because they lack the fleshy tissue of edible fruits, with their ovaries forming hard layers rather than fleshy tissue. Dry fruits can have structures that aid in their dispersion, as seen by the fluffy "parachute" structure of the wind-dispersed dandelion seed in figure 33.5*e*, or the spiny cocklebur in figure 33.5*f* which catches onto fur (or socks or pants!) and is carried to new habitats. Still other fruits, such as those of mangroves, coconuts, and certain other plants that characteristically occur on or near beaches or swamps, are spread from place to place by water, as illustrated in the photo at the beginning of this chapter.

(e) Dispersed by wind

(f) Dispersed by attaching to animals

Figure 33.5   **Types of fruits and common modes of dispersion.**

(a) Tomatoes are a type of fleshy fruit called berries that have multiple seeds. (b) Peaches are a type of fleshy fruit called drupes that contain a single large seed. (c) Apples are a type of fleshy fruit called pomes that contain multiple seeds. (d) The bright red berries of this honeysuckle, *Lonicera,* are highly attractive to birds. Birds may carry the berry seeds either internally or stuck to their feet for great distances. (e) The seeds of this dandelion, *Taraxacum officinale,* are enclosed in a dry fruit with a "parachute" structure that aids its dispersal by wind. (f) The spiny fruits of this cocklebur, *Xanthium strumarium,* adhere readily to any passing animal.

33.4 Germination

CONCEPT PREVIEW: Germination is the resumption of a seed's growth and reproduction, triggered by water.

What happens to a seed when it encounters conditions suitable for its germination? First, it absorbs water. Seed tissues are so dry at the start of germination that the seed takes up water with great force, after which metabolism resumes. Initially, the metabolism may be anaerobic, but when the seed coat ruptures, aerobic metabolism takes over. At this point, oxygen must be available to the developing embryo because plants require oxygen for active growth (see chapter 7) and plants can "drown" for the same reason people do if submersed in water. Few plants produce seeds that germinate successfully underwater, although some, such as rice, have evolved a tolerance of anaerobic conditions and can initially respire anaerobically. Figure 33.6 shows the development of a dicot (on the left) and monocot (on the right) from germination through early stages. The first stage in both cases is the emergence of the roots. Following that, in dicots, the cotyledons emerge from underground along with the stem. The cotyledons eventually wither and the first leaves begin the process of photosynthesis. In monocots, the cotyledon doesn't emerge from underground; instead a structure called the *coleoptile* (a sheath wrapped around the emerging shoot) pushes through to the surface where the first leaves emerge and begin photosynthesis.

> Although plants produce their own food through photosynthesis, they still undergo cellular respiration to break down food molecules to obtain energy for other metabolic activities. As discussed on pages 120 and 121, oxygen is required for cellular respiration.

Concept Check

1. How does a runner differ from a rhizome?
2. What are the male parts of a flower?
3. What two structures found in seeds are produced through double fertilization?

Figure 33.6   Development of angiosperms.

Dicot development in a soybean. The first structure to emerge is the embryonic root followed by the two cotyledons of the dicot. The cotyledons are pulled up through the soil along with the hypocotyl (the stem below the cotyledons). The cotyledons are the seed leaves that provide nutrients to the growing plant. As other leaves develop, they provide nutrients through photosynthesis, and the cotyledons shrivel and fall off the stem. Flowers develop in buds at the nodes. **Monocot development in corn.** The first structure to emerge is the radicle or primary root. Monocots have one cotyledon, which does not emerge from underground. The coleoptile is a tubular sheath; it encloses and protects the shoot and leaves as they push their way up through the soil.

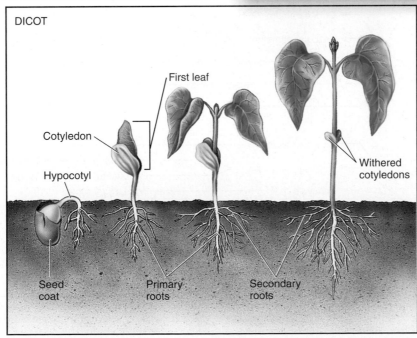

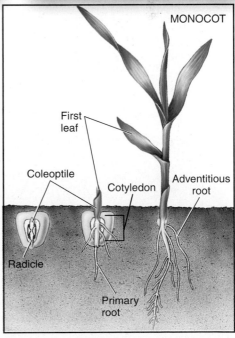

Regulating Plant Growth

1. Small bits of phloem tissue were cut from the root of a mature carrot.

2. Phloem tissue was placed in a flask with a liquid growth medium.

3. The flask was rocked on a motorized table, dislodging cells from the clumps of tissue.

4. Isolated cells grew rapidly; placed in agar, they developed roots and shoots.

5. Eventually, plants grew and were able to flower and reproduce normally.

Figure 33.7   **How Steward regenerated a plant from differentiated tissue.**

33.5 Plant Hormones

CONCEPT PREVIEW: The development of plant tissues is controlled by the actions of hormones. They act on the plant by regulating the expression of key genes.

After a seed germinates, the pattern of growth and differentiation that was established in the embryo is repeated indefinitely until the plant dies. But differentiation in plants, unlike that in animals, is largely reversible. Botanists first demonstrated in the 1950s that individual differentiated cells isolated from mature individuals could give rise to entire individuals. F. C. Steward's experiment is shown in figure 33.7 where he was able to induce isolated bits of phloem tissue taken from carrots to form new plants, plants that were normal in appearance and fully fertile. Regeneration of entire plants from differentiated tissue has since been carried out in many plants, including cotton, tomatoes, and cherries. These experiments clearly demonstrate that the original differentiated phloem tissue still contains cells that retain all of the genetic potential needed for the differentiation of entire plants. No information is lost during plant tissue differentiation in these cells, and no irreversible steps are taken.

Once a seed has germinated, the plant's further development depends on the activities of the meristematic tissues, which interact with the environment through hormones (discussed later). The shoot and root apical meristems give rise to all of the other cells of the adult plant. Differentiation, or the formation of specialized tissues, occurs in five stages in plants and is shown in figure 33.8. The establishment of the shoot and root apical meristems occurs at stage 2; after that point, the tissues become more and more differentiated.

The tissue regeneration experiments of Steward and many others have led to the general conclusion that some nucleated cells in differentiated plant tissue are capable of expressing their hidden genetic information when provided with suitable environmental signals. What halts the expres-

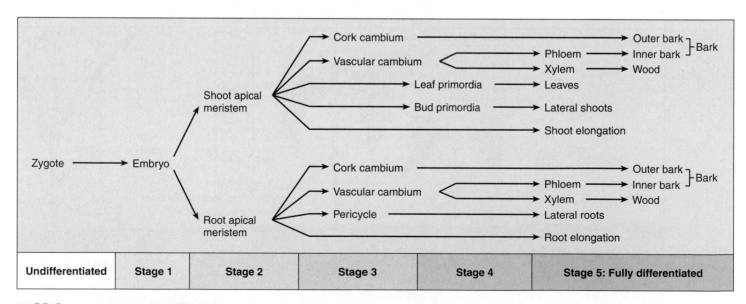

Figure 33.8   **Stages of plant differentiation.**

As this diagram shows, the different cells and tissues in a plant originate from the shoot and root apical meristems. It is important to remember, however, that this is showing the origin of the tissue, not the location of the tissue in the plant. For example, the vascular tissues of xylem and phloem arise from the vascular cambium, but these tissues are present throughout the plant, in the leaves, stems, and roots.

sion of genetic potential when the same kinds of cells are incorporated into normal, growing plants? As we will see, the expression of some of these genes is controlled by plant hormones.

In animals, there are several organs, called endocrine glands, whose sole function is hormone production (although hormones are produced in other organs as well). In plants, on the other hand, all hormones are produced in tissues that are not specialized for that purpose and carry out many other functions.

At least five major kinds of hormones are found in plants: auxin, gibberellins, cytokinins, ethylene, and abscisic acid. Their chemical structures and descriptions are provided in table 33.1. Other kinds of plant hormones certainly exist but are less well understood. Hormones have multiple functions in the plant; the same hormone may work differently in different parts of the plant, at different times, and interact with other hormones in different ways. The study of plant hormones, especially how hormones produce their effects, is today an active and important field of research.

TABLE 33.1	Functions of the Major Plant Hormones		
Hormone	**Major Functions**	**Where Produced or Found in Plant**	**Practical Applications**
Auxin (IAA)	Promotes stem elongation and growth; forms adventitious roots; inhibits leaf abscission; promotes cell division (with cytokinins); induces ethylene production; promotes lateral bud dormancy	Apical meristems; other immature parts of plants	Seedless fruit production; synthetic auxins act as herbicides
Gibberellins (GA$_1$, GA$_2$, GA$_3$, etc.)	Promotes stem elongation; stimulates enzyme production in germinating seeds	Root and shoot tips; young leaves; seeds	Uniform seed germination for production of barley malt used in brewing; early seed production of biennial plants; increasing size of grapes by allowing more space for growth
Cytokinins	Stimulates cell division, but only in the presence of auxin; promotes chloroplast development; delays leaf aging; promotes bud formation	Root apical meristems; immature fruits	Tissue culture and biotechnology; pruning trees and shrubs, which cause them to "fill out"
Ethylene	Controls leaf, flower, and fruit abscission; promotes fruit ripening	Roots, shoot apical meristems; leaf nodes; aging flowers; ripening fruits	Fruit ripening of agricultural products that are picked early to retain freshness
Abscisic acid (ABA)	Controls stomatal closure; some control of seed dormancy; inhibits effects of other hormones	Leaves, fruits, root caps, seeds	Research on stress tolerance in plants, specifically drought tolerance

33.6 Auxin

CONCEPT PREVIEW: The primary growth-promoting hormone of plants is auxin, which increases the plasticity of plant cell walls, allowing growth in specific directions.

In his later years, the great evolutionist Charles Darwin became increasingly devoted to the study of plants. In 1881, he and his son Francis published a book called *The Power of Movement in Plants,* in which they reported their systematic experiments concerning the way in which growing plants bend toward light, a phenomenon known as **phototropism.**

After conducting a series of experiments shown in figure 33.9, they observed that plants grew toward light ❶. If the tip of the seedling was covered, the plant didn't bend toward the light ❷. A control experiment showed that the cap was not interfering with the directional growth pattern ❸. Another control showed that covering the lower portions of the plant did not block the directional growth ❹. The Darwins hypothesized that when plant shoots were illuminated from one side, an "influence" that arose in the uppermost part of the shoot was then transmitted downward, causing the shoot to bend. Later, several botanists conducted experiments that demonstrated that the substance causing the shoots to bend was a chemical we call **auxin.**

How auxin controls plant growth was discovered in 1926 by Frits Went, a Dutch plant physiologist, in the course of studies for his doctoral dissertation. From his experiments, described in figure 33.10, Went was able to show that the substance that flowed into agar from the tips of the light-grown grass seedlings (steps ❶ and ❷) enhanced cell elongation (shown in step ❸). This chemical messenger caused the tissues on the side of the seedling into which it flowed to grow more than those on the opposite side (step ❹). Control experiments indicated that the effects were not due to properties of the agar (steps ❶ₐ and ❷ₐ). He named the substance that he had discovered auxin, from the Greek word *auxin,* meaning "to increase."

Figure 33.9 **The Darwins' experiment with phototropism.**

From these experiments, the Darwins concluded that, in response to light, an "influence" that causes bending was transmitted from the tip of the seedling to the area below the tip, where bending usually occurs.

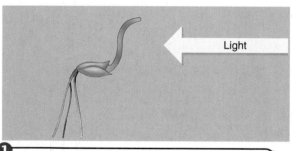

❶ Charles and Francis Darwin found that a young grass seedling normally bent toward the light.

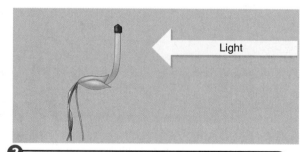

❷ If the tip of the seedling was covered with a lightproof cap, the seedling did not bend toward the light.

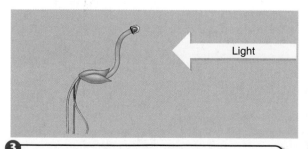

❸ When the tip of the seedling was covered with a transparent cap, the bending did occur.

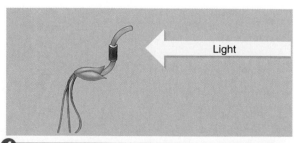

❹ When the Darwins placed a lightproof collar below the tip, the seedling bent toward the light.

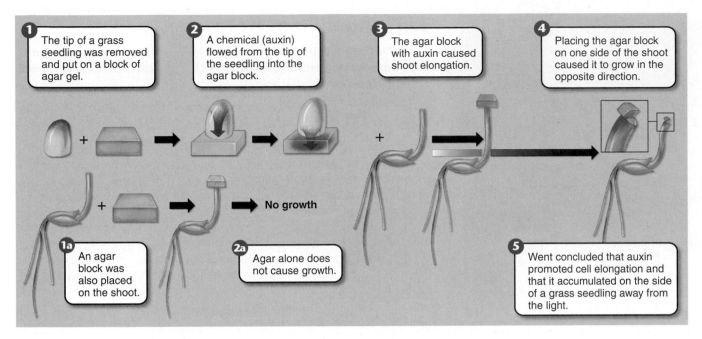

1 The tip of a grass seedling was removed and put on a block of agar gel.

2 A chemical (auxin) flowed from the tip of the seedling into the agar block.

3 The agar block with auxin caused shoot elongation.

4 Placing the agar block on one side of the shoot caused it to grow in the opposite direction.

No growth

1a An agar block was also placed on the shoot.

2a Agar alone does not cause growth.

5 Went concluded that auxin promoted cell elongation and that it accumulated on the side of a grass seedling away from the light.

Figure 33.10 **How Went demonstrated the effects of auxin on plant growth.**

The experiment showed how a chemical at the tip of the seedling caused the shoot to elongate and to bend. Steps 1a and 2a show the control experiment.

Went's experiments provided a basis for understanding the responses the Darwins had obtained some 45 years earlier: Grass seedlings bend toward the light because the side of the shoot that is in the shade has more auxin; therefore, its cells elongate more than those on the lighted side, bending the plant toward the light as in the enlargement in figure 33.11. Later experiments showed that auxin in normal plants migrates away from the illuminated side toward the dark side in response to light and thus causes the plant to bend toward the light.

> Plant cell walls are made up of cellulose, long strands of glucose molecules linked together, as shown in figure 3.15 and table 3.1 on pages 54 and 55. Auxin may exert its effects by altering the covalent bonds within the cellulose molecule, allowing the cells to lengthen.

Auxin appears to act by increasing the "stretchability" of the plant cell wall within minutes of its application. Researchers speculate that the covalent bonds linking the polysaccharides of the cell wall to one another change extensively in response to auxin, allowing the cells to take up water and thus enlarge.

Synthetic auxins are routinely used to control weeds. When applied as herbicides, they are used in higher concentrations than those at which auxin normally occurs in plants. One of the most important of the synthetic auxins used in this way is 2,4-dichlorophenoxyacetic acid, usually known as 2,4-D. It kills weeds in lawns without harming the grass because 2,4-D affects only broad-leaved dicots. When treated, the weeds literally "grow to death," rapidly reducing ATP production so that no energy remains for transport or other essential functions.

Closely related to 2,4-D is the herbicide 2,4,5-trichlorophenoxyacetic acid (2,4,5-T), which is widely used to kill woody seedlings and weeds. Notorious as the Agent Orange of the Vietnam War, 2,4,5-T is easily contaminated with a by-product of its manufacture, dioxin. Dioxin is harmful to people because it is an *endocrine disrupter*, a chemical that interferes with the course of human development. The growing release of endocrine disrupters as by-products of modern chemical manufacturing is a subject of great environmental concern.

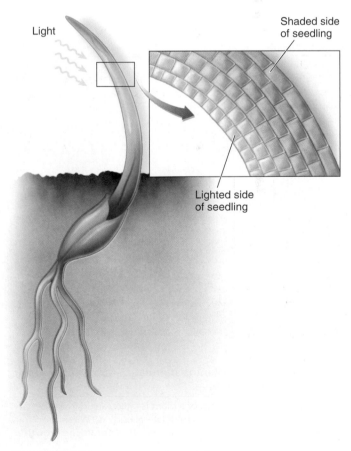

Light

Shaded side of seedling

Lighted side of seedling

Figure 33.11 **Auxin causes cells to elongate.**

Plant cells that are on the shaded side have more auxin and grow faster, elongating more, than cells on the lighted side, causing the plant to bend toward light.

Figure 33.12   **The effect of a gibberellin.**

This rosette mutant (*left*) of the mustard family plant (*Brassica rapa*) is defective in producing gibberellins, but it can be rescued by applying gibberellin (*right*).

(a)   (b)

Figure 33.13   **Cytokinins stimulate lateral bud growth in the absence of auxin.**

(a) When the apical meristem of a plant is intact, auxin from the apical bud will inhibit the growth of lateral buds. (b) When the apical bud is removed, cytokinins are able to promote the growth of lateral buds.

33.7 Other Plant Hormones

CONCEPT PREVIEW: Other plant hormones work with auxin and each other to control growth. These include gibberellins, cytokinins, ethylene, and abscisic acid.

Gibberellins

Gibberellins are a large class of over 100 naturally occurring hormones. Synthesized in the apical portions of both shoots and roots, gibberellins characteristically promote elongation within the spaces between nodes. This is very apparent if you compare the plant on the left in figure 33.12, a mutant that is defective in producing gibberellin, and a similarly mutant plant on the right, which was exposed to gibberellin. Gibberellins also affect a number of other aspects of plant growth and development. The application of gibberellins can hasten seed germination, apparently because they can substitute for the effects of cold or light requirements in this process. Gibberellins are used commercially to space out the flowers of grape vines by extending the internode lengths so the fruits have more room to grow and become larger.

Cytokinins

A **cytokinin** is a plant hormone that, in combination with auxin, stimulates cell division in plants and determines the course of differentiation. In vascular plants, most cytokinins seem to be produced in the roots, from which they are then transported throughout the rest of the plant. Cytokinins apparently stimulate cell division by influencing the synthesis or activation of proteins specifically required for mitosis.

Cytokinins promote growth of lateral buds into branches but only when the influence of auxin is removed. Auxin released from the apical meristem inhibits the formation of lateral branches as in figure 33.13*a*, but when the apical meristem is removed as in figure 33.13*b*, cytokinins are able to promote lateral branch growth. This is why pruning bushes and trees makes the plants fuller. Thus, along with auxin and ethylene (discussed next), cytokinins play a role in the control of apical dominance and lateral bud growth. Cytokinins inhibit formation of lateral roots, while auxins promote their formation. The balance between cytokinins and auxin determines the appearance of a mature plant.

Ethylene

Ethylene is a gas that is produced in relatively large quantities during fruit ripening, when respiration is proceeding at its most rapid rate. At this phase, complex carbohydrates are broken down into simple sugars, cell walls become soft, and the volatile compounds associated with flavor and scent in the ripe fruits are produced. When applied to fruits, ethylene hastens their ripening.

One of the first lines of evidence that led to the recognition of ethylene as a plant hormone was the observation that gases from oranges caused premature ripening in bananas. Such relationships have led to major commercial uses. Tomatoes are often picked green and then artificially ripened as desired by the application of ethylene. Ethylene is widely used to speed the color formation of lemons and oranges as well.

Genetic engineers, using techniques described in chapter 13, have placed genes that interfere with the synthesis of ethylene into tomatoes to slow the ripening process. Until now, commercial tomatoes had to be picked very early in order to get them to market before they become over-ripe. The ripening of genetically engineered tomatoes is delayed, so the tomatoes can be left on the vine longer, improving their taste.

Ethylene also plays an important ecological role. Ethylene production increases rapidly when a plant is exposed to ozone and other toxic chemicals, temperature extremes, drought, attack by pathogens or herbivores, and other stresses. The increased production of ethylene that occurs can accelerate the abscission (the dropping) of leaves or fruits that have been damaged by these stresses. The experiment in figure 33.14 demonstrates this property of ethylene. The holly twig on the left hasn't lost its leaves, even after a week. The holly twig on the right, exposed to ethylene from the ripe apple, lost its leaves. It now appears that some of the damage associated with exposure to ozone is due to the ethylene produced by the plants. Some studies suggest that the production of ethylene by plants subjected to attack by herbivores or infected with diseases may be a signal to activate the defense mechanisms of the plants. Such mechanisms may include the production of molecules toxic to the animals or pests attacking them. A full understanding of these relationships is obviously important for agriculture and forestry.

Figure 33.14 **The effects of ethylene.**

Ethylene released from the ripe apple on the *right* causes the holly twig to drop its leaves, while the twig on the *left* keeps its leaves.

Abscisic Acid

Abscisic acid (ABA) is a naturally occurring plant hormone that is synthesized mainly in mature green leaves, fruits, and root caps. The hormone was given its name because it was thought that it stimulated leaves to age rapidly and fall off (the process of abscission), but evidence that abscisic acid plays an important natural role in this process is scant. In fact, it is believed that abscisic acid may cause ethylene synthesis, and that it is actually the ethylene that promotes senescence and abscission. When abscisic acid is applied to a green leaf, the areas of contact turn yellow. Thus, abscisic acid has the exact opposite effect on a leaf from that of the cytokinins; a yellowing leaf remains green in an area where cytokinins are applied.

Abscisic acid was also initially thought to induce the formation of winter buds—dormant buds that remain through the winter—by suppressing growth, but recent evidence does not support this. Abscisic acid levels increase during seed development and decrease during germination, and so it is likely that ABA plays a role in causing the dormancy of many seeds.

ABA may also function in transpiration. During drought conditions, leaves produce large amounts of ABA, which induces the closing of stomata. ABA's effects on stomata occur on the order of minutes, suggesting that this action does not involve turning genes on and off, but

> Transpiration, as discussed on page 627, is the process whereby water will leave a plant through evaporation of water vapor from the stomata in the leaves. Abscisic acid acts to close the stomata thereby reducing water loss in drought conditions.

rather occurs by influencing the membrane permeability of guard cells. Stomata open when water flows into guard cells, causing them to swell. Water flows into the guard cells osmotically, driven by the influx of potassium ions into the guard cells (figure 33.15). The lower panel shows how ABA most likely stimulates the transport of potassium ions out of the guard cells, causing water to pass out of the guard cells by osmosis. This loss of water causes the stomata to close.

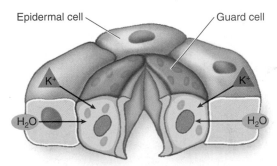

In the absence of ABA—stoma is open

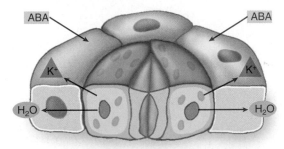

In the presence of ABA—stoma is closed

Figure 33.15 **Effects of abscisic acid.**

Abscisic acid (ABA) affects the closing of stomata by influencing the movement of potassium ions out of guard cells. As potassium ions are transported out of the guard cells, water also passes out of the cells due to osmosis, causing the stoma to close.

Concept Check

1. What five major kinds of hormones are found in plants?
2. How does auxin cause a plant to bend toward light?
3. How can oranges cause nearby bananas to ripen early?

Plant Responses to Environmental Stimuli

33.8 Photoperiodism and Dormancy

CONCEPT PREVIEW: Plant growth and reproduction are sensitive to photoperiod, using chemicals to link flowering to seasons.

Photoperiodism

Essentially all eukaryotic organisms are affected by the cycle of night and day, and many features of plant growth and development are keyed to changes in the proportions of light and dark in the daily 24-hour cycle. Such responses constitute **photoperiodism,** a mechanism by which organisms measure seasonal changes in relative day and night length. One of the most obvious of these photoperiodic reactions concerns angiosperm flower production.

Day length changes with the seasons; the farther from the equator you are, the greater the variation. Plants' flowering responses fall into three basic categories in relation to day length: long-day plants, short-day plants, and day-neutral plants. Long-day plants like the iris in figure 33.16 ❶ initiate flowers in the summer, when nights become shorter than a certain length (and days become longer). Short-day plants, on the other hand, begin to form flowers when nights become longer than a critical length (and days become shorter); the goldenrod in ❷ doesn't flower in summer, but instead flowers in fall. Thus, many spring and early summer flowers are long-day plants, and many fall flowers are short-day plants. An "interrupted night" experiment ❸ makes it clear that it is actually the length of uninterrupted dark that is the flowering trigger. The flash of light during a long night triggers flowering in the iris and inhibits flowering in the goldenrod, even though the day is shorter.

Figure 33.16   How photoperiodism works in plants.

❶ *Early summer.* Short periods of darkness induce flowering in long-day plants, such as iris, but not in short-day plants, such as goldenrod.

❷ *Late fall.* Long periods of darkness induce flowering in short-day plants, such as goldenrod, but not in long-day plants, such as iris.

❸ *Interrupted night.* If the long night of winter is artificially interrupted by a flash of light, the goldenrod will not bloom and the iris will.

The Chemical Basis of Photoperiodism

Flowering responses to daylight and darkness are controlled by several chemicals that interact in complex ways. Plants contain a pigment, *phytochrome,* that exists in two interconvertible forms. In short-day plants, the presence of the active form leads to a biological reaction that suppresses flowering. The amount of the active form steadily declines in darkness, the molecules converting to the inactive form. When the period of darkness is long enough, the suppression reaction ceases and the flowering response is triggered. However, a single flash of red light (wavelength of 660 nanometers) converts most of the phytochrome molecules from inactive to active, and the flowering reaction is blocked.

Dormancy

Plants respond to their external environment largely by changes in growth rate. A plant's ability to stop growing altogether when conditions are not favorable—to become dormant—is critical to its survival. In temperate regions, dormancy is generally associated with winter, when low temperatures and the unavailability of water because of freezing make it impossible for plants to grow. During this season, the buds of deciduous trees and shrubs remain dormant, and the apical meristems remain well protected inside enfolding scales. Perennial herbs spend the winter underground as stout stems or roots packed with stored food. Many other kinds of plants, including most annuals, pass the winter as seeds.

33.9 Tropisms

CONCEPT PREVIEW: Growth of the plant body is often sensitive to light, gravity, or touch.

Tropisms are directional and irreversible growth responses of plants to external stimuli. They control patterns of plant growth and thus plant appearance. Three major classes of plant tropisms include phototropism (figure 33.17*a,* and discussed earlier), gravitropism, and thigmotropism.

Gravitropism

Gravitropism causes stems to grow upward and roots downward, in response to gravity. Both of these responses clearly have adaptive significance. Stems, like the one growing from a tipped over flower pot in figure 33.17*b,* grow upward and are apt to receive more light than those that do not; roots that grow downward are more apt to encounter a more favorable environment than those that do not.

Thigmotropism

Still another commonly observed response of plants is **thigmotropism,** a name derived from the Greek root *thigma,* meaning "touch." Thigmotropism is defined as the response of plants to touch. Examples include plant tendrils, which rapidly curl around and cling to stems or other objects, and twining plants, such as bindweed, which also coil around objects (figure 33.17*c*). These behaviors result from rapid growth responses to touch.

Concept Check

1. What controls flowering in plants, the length of day or night?
2. What changes in a plant are seen during dormancy?
3. Are growth patterns called tropisms reversible? Explain.

(a)

(b)

(c)

Figure 33.17   Tropism guides plant growth.

(a) Phototropism is exhibited by this *Coriandrum* plant growing toward light. (b) Gravitropism is exhibited by this plant, *Phaseolus vulgaris.* Note the negative gravitational response of the shoot. (c) The thigmotropic response of these twining stems causes them to coil around the object with which they have come in contact.

Does Auxin Use Ethylene to Inhibit Root Growth?

The plant hormone auxin (the name is from the Greek word *auxein,* "to increase") promotes plant growth. Released at low concentrations from cells at the growing tip of a plant, auxin diffuses downward, causing the stem to elongate. At similarly low concentrations, auxin also promotes elongation of roots, and formation of new roots at cut surfaces. In the photo, the stalk of the leaf on the left sits in a solution of auxin, the one on the right in water. You can see auxin has promoted root growth. However, at high concentrations auxin has quite the opposite effect, inhibiting root growth (**inhibition** is the stopping or restraining of a process). It has long been assumed that auxin achieves this inhibition by triggering the production of the plant hormone ethylene, a well-known inhibitor of growth. Supporting this hypothesis is the observation that auxin does stimulate many kinds of plant cells to produce ethylene. But just because the "suspect is present at the scene of a crime" does not prove the suspect is guilty of the crime. Perhaps auxin is directly inhibiting root growth by some unknown mechanism.

The graph to the right displays the results of an experiment designed to test the hypothesis that high concentrations of auxin inhibit root growth by stimulating ethylene production. Seedlings with roots were grown in varying concentrations of auxin, with ethylene assays taken at the end of the experiment. The rate of pea root elongation (blue points) and the production of ethylene in the pea roots (red points) are determined for a range of auxin concentrations. Auxin concentrations are plotted on a log scale (a **log** or logarithmic **scale** is a series of numbers plotted as powers of 10; because the scale is exponential [1, 10, 100, 1000...] rather than linear [1, 2, 3, 4...], a broad range of values can be visualized on a single graph).

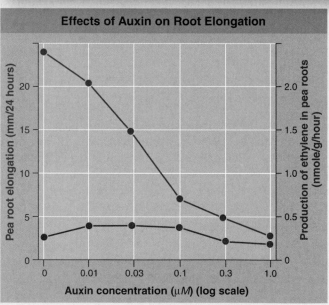

Effects of Auxin on Root Elongation

Analysis

1. **Applying Concepts**
   a. **Inhibition.** What is the rate of pea root elongation when there is no auxin present? at the highest auxin concentration tested? Is there any concentration of auxin in this study that promotes root growth? Why?
   b. **Log scale.** What is the effect of auxin on the rate of pea root elongation over the range of concentrations studied? What would the elongation rate curve look like plotted on a linear scale? What does this say about the sensitivity of the inhibition to auxin concentration?
2. **Interpreting Data** Assess the effect of these high auxin concentrations on the production of ethylene by pea root cells. Do the cells produce more ethylene when exposed to higher concentrations of auxin?
3. **Making Inferences**
   a. Construct a plot of ethylene concentration (*y* axis) versus elongation rate (*x* axis). Does this plot provide any evidence that these two variables are being influenced in the same way by auxin? that they are *not* being influenced in the same way?
   b. Is the inhibition of root elongation by high concentrations of auxin accompanied by corresponding increases in the concentration of ethylene?
4. **Drawing Conclusions** Does auxin exert its inhibition of root elongation by stimulating ethylene production?

Concept Summary

Flowering Plant Reproduction

33.1 Angiosperm Reproduction

- Angiosperms reproduce sexually and asexually. In asexual reproduction, offspring are genetically identical to the parent; this often involves vegetative reproduction. Plants use many forms of vegetative reproduction including runners, rhizomes, suckers, and adventitious plantlets (**figure 33.1**).

- Sexual reproduction in plants involves an alternation of generations. A diploid sporophyte gives rise to a haploid gametophyte that produces gametes—egg and pollen. Pollination and fertilization bring the gametes together.

- Flowers have male and female parts. The male parts include the stamen and the pollen-producing anthers. Pollen grains, the male gametophyte, are produced in the anthers. The female parts include the stigma, style, and ovary, which make up the carpel. The embryo sac is the female gametophyte, and it forms in the ovule, within the ovary (**figure 33.2**).

- Pollen grains are carried to flowers by wind or animals (**figure 33.3**). A pollen grain lands on the stigma and extends a pollen tube through the style to the base of the ovule. Two sperm cells travel down the pollen tube. One sperm fertilizes the egg and the other fuses with the two polar nuclei (**figure 33.4**). This process is called double fertilization.

33.2 Seeds

- The fertilized egg begins dividing, forming the embryo. After the shoot and root apical meristems form, the embryo stops growing and becomes dormant in a structure called a seed.

- The endosperm provides a source of food for the plant embryo. In some cases, the endosperm is consumed in the seed and stored in structures called cotyledons (seed leaves). The outer layer of the ovule becomes the seed coat (**figure 33.4**). Seed development resumes when conditions are favorable.

33.3 Fruit

- During seed formation, the flower's ovary begins to develop into fruit that surrounds the seed. The walls of the ovary can develop differently which accounts for the variety of fruits. Fruits are dispersed in different ways. Fleshy fruits are eaten by animals and then dispersed through their feces. Dry fruits are usually dispersed by wind, water, or animals. Dry fruits have structures that aid in their dispersal (**figure 33.5**).

33.4 Germination

- A seed germinates under favorable conditions. The seed absorbs water and uses the endosperm or cotyledons as a food source. The seed coat cracks open and the plant begins to grow. The overall process is similar in monocots and dicots, but there are differences in the structures involved (**figure 33.6**).

Regulating Plant Growth

33.5 Plant Hormones

- Differentiation in plants is largely reversible. New plants can be grown from parts of adult plants (**figure 33.7**). The shoot and root meristems differentiate early in development and give rise to all of the other cell types (**figure 33.8**).

- Hormones are chemicals produced in very small quantities that effect growth and development. There are at least five major kinds of plant hormones: auxin, gibberellins, cytokinins, ethylene, and abscisic acid (**table 33.1**).

33.6 Auxin

- Early researchers, including Darwin and his son, described a process, now called phototropism, where plants grow toward light. In a series of experiments, they showed that the tip of a plant will grow toward light (as shown here from **figure 33.9**). Went identified a chemical he called auxin as the hormone involved in phototropism (**figure 33.10**). When a plant is exposed to light on one side, auxin is released from the tip and causes cells on the shady side of the plant to elongate. This causes the plant to grow toward the light (**figure 33.11**).

33.7 Other Plant Hormones

- Gibberellins affect stem elongation, promoting elongation between the node regions (**figure 33.12**). The effects of gibberellins are enhanced in the presence of auxin.

- Cytokinins stimulate cell division and act in combination with auxin. They also stimulate lateral bud growth in the absence of auxin (**figure 33.13**).

- Ethylene is released in relatively large quantities as a gas during fruit ripening. When applied to fruit, ethylene can hasten ripening and so has commercial applications. Ethylene also accelerates abscission of leaves and fruit (**figure 33.14**).

- Abscisic acid (ABA) works in combination with other hormones. ABA stimulates the release of ethylene, which triggers the dropping of leaves. Abscisic acid seems to play a role in seed dormancy. It is also produced in large amounts by the leaves during drought conditions, which causes the closing of stomata (**figure 33.15**).

Plant Responses to Environmental Stimuli

33.8 Photoperiodism and Dormancy

- The length of daylight affects flowering, a process called photoperiodism and dormancy. Some plants flower in response to short days, some to long days, like the iris shown here from **figure 33.16,** and some are day-neutral. A plant pigment, phytochrome, exists in two forms converted by darkness. The active form of phytochrome inhibits flowering. Darkness converts the active form to the inactive form which allows flowering to occur. Plants survive unfavorable conditions by entering a phase of dormancy, where they stop growing.

33.9 Tropisms

- Tropisms are irreversible growth patterns in response to external stimuli. Phototropism is a growth response toward light. Gravitropism is growth in response to the pull of gravity; this causes stems to grow upward and roots to grow downward. Thigmotropism is growth in response to touch (**figure 33.17**).

Self-Test

1. Sexual reproduction in angiosperms requires
   a. pollen.   **c.** rhizomes.
   b. runners.   **d.** clones.
2. The male parts of a plant include all of the follow except
   a. anthers.   **c.** stamen.
   b. stigma.   **d.** microspores.
3. The flower shape, scent, color, and presence of nectar in the flowers of some angiosperms are to
   a. deter predators.   **c.** deter insect pests.
   b. attract animal pollinators.   **d.** enhance symbiosis.
4. For a seed to germinate, the dormant plant embryo must get
   a. carbon dioxide and water. **c.** oxygen and nitrogen.
   b. nitrogen and water.   **d.** oxygen and water.
5. Fruit forms from a flower's
   a. ovary.   **c.** carpels.
   b. sepals.   **d.** stigma.
6. Meristematic tissue regulates plant growth and development through the use of
   a. carbohydrates.   **c.** hormones.
   b. amount of available water. **d.** phototropism.

7. The hormone auxin causes
   a. cells in plant stems to shorten by releasing water.
   b. fruit to ripen.
   c. cells in plant stems to elongate.
   d. growth of more lateral branches.
8. The hormone ethylene causes
   a. cells in plant stems to shorten by releasing water.
   b. fruit to ripen.
   c. increase in fruit size.
   d. growth of more lateral branches.
9. Angiosperm flowering is controlled by
   a. temperature.   **c.** photoperiod.
   b. gibberellins.   **d.** magnesium.
10. Sensitivity of plants to touch is known as
    a. thigmotropism.   **c.** phototropism.
    b. photoperiodism.   **d.** gravitropism.

Visual Understanding

1. **Figure 33.3** Bees tend to pollinate yellow flowers and so would not be attracted to the pink flowers that the butterfly is visiting, but that is not the only reason. Give another reason why a bee probably wouldn't pollinate these pink flowers.

2. **Figure 33.5** From a culinary standpoint, plant products that are sweet and juicy are considered fruits, while other edible plant parts are often considered vegetables. For the following plant products, indicate if you eat the fruit, the seed, or other parts of the plant.

   (1) corn   (2) tomato   (3) cucumber   (4) peanut   (5) carrot
   (6) potato   (7) peas   (8) green peppers (9) lettuce

(a) Berries   (b) Drupes   (c) Pomes   (d) Eaten by animals   (e) Dispersed by wind   (f) Dispersed by attaching to animals

Challenge Questions

1. A friend says her teacher made the comment that "nectar is to pollination as fruit is to dispersal." She asks you to explain.

2. An old saying is "one bad apple spoils the whole barrel." We even refer to troublemakers as "a bad apple." Can you explain the biological basis for these statements?

3. If it is out in the open, trailing ivy grows along the ground and can form dense mats of ground cover. If planted in a hanging pot, the stems dangle down, forming a graceful flow. If ivy is planted next to a building, however, its stems attach to the side of the building and, in time, it will cover the building. What causes this?

Appendix A

Answers to Self-Test Questions

Chapter 1 *The Science of Biology*

1. a. kingdoms 2. c. cellular organization 3. b. atom, molecule, organelle, cell, tissue, organ, organ system, organism, population, species, community, ecosystem 4. d. emergent properties 5. b. evolution, energy flow, cooperation, structure determines function, and homeostasis 6. b. test each hypothesis, using appropriate controls, to rule out as many as possible 7. c. After sufficient testing, you can accept it as probable, being aware that it may be revised or rejected in the future 8. d. all living organisms consist of cells, and all cells come from other cells 9. c. is contained in a long molecule called DNA 10. c. Darwin

Chapter 2 *The Chemistry of Life*

1. b. an atom 2. c. an ion 3. d. ionic, covalent, and hydrogen 4. b. it can form four single covalent bonds 5. d. All of these are correct 6. a. hydrogen bonds between the individual water molecules 7. c. heat storage and heat of vaporization 8. a. cohesion 9. a. (1) acids and (2) bases 10. c. A buffer stops water from ionizing.

Chapter 3 *Molecules of Life*

1. b. proteins, carbohydrates, lipids, and nucleic acids 2. c. polypeptides 3. c. structure, function 4. d. All of the above 5. d. are information storage devices found in body cells 6. a. Adenine forms hydrogen bonds with thymine 7. a. structure and energy 8. a. glycogen 9. d. All of these are characteristics of fat molecules. 10. c. energy storage and for some hormones

Chapter 4 *Cells*

1. a. cells are the smallest living things. Nothing smaller than a cell is considered alive 2. d. prokaryotes, eukaryotes 3. b. a double lipid layer with proteins inserted in it, which surrounds every cell individually 4. a. a nucleolus 5. c. endoplasmic reticulum and the Golgi complex 6. d. mitochondria and the chloroplasts 7. b. cristae 8. d. slowly disperse throughout the water; this is because of diffusion 9. b. endocytosis and phagocytosis 10. c. energy and specialized pumps or channels

Chapter 5 *Energy and Life*

1. c. energy 2. d. says that energy can change forms, but cannot be made or destroyed 3. b. says that entropy, or disorder, continually increases in a closed system 4. a. exergonic and release energy

5. b. enzymes 6. c. temperature and pH 7. d. All of the above 8. c. an inhibitor molecule competes with the substrate for the same binding site on the enzyme 9. a. active site 10. a. the breaking of phosphate bonds in ATP

Chapter 6 *Photosynthesis: Acquiring Energy from the Sun*

1. c. photosynthesis 2. b. with molecules called pigments that absorb photons and use their energy 3. c. a small portion in the middle of the spectrum 4. a. red and blue 5. d. All of these are true 6. c. only go through the system once; they are obtained by splitting a water molecule 7. b. chemiosmosis 8. a. electron transport system of photosystem I, Calvin cycle 9. c. build sugar molecules 10. b. use C_4 photosynthesis or CAM

Chapter 7 *How Cells Harvest Energy from Food*

1. a. breaking down the organic molecules that were consumed 2. c. substrate-level phosphorylation 3. b. glycolysis 4. b. makes ATP by splitting a molecule of glucose in half and capturing the energy 5. b. NAD^+, electron transport chain 6. c. mitochondria of the cell and are broken down in the presence of O_2 to make more ATP 7. d. during the electron transport chain 8. b. fermentation 9. a. pyruvate 10. c. each type of macromolecule is broken down into its subunits, which enter the oxidative respiration pathway

Chapter 8 *Mitosis*

1. a. copying DNA then undergoing binary fission 2. c. in the production of genetically identical daughter cells 3. c. and most eukaryotes have between 10 and 50 pairs of chromosomes 4. c. carry information about the same traits located in the same places on the chromosome 5. d. All of the above 6. b. metaphase 7. c. cytokinesis 8. a. Interphase/DNA replication 9. d. All of the above 10. b. cancer

Chapter 9 *Meiosis*

1. d. the egg and sperm have only half the number of chromosomes found in the parents because of meiosis 2. d. 23 3. a. 1*n* gametes (haploid), followed by 2*n* zygotes (diploid) 4. a. homologous chromosomes exchange sections of chromosomes 5. b. Homologous chromosomes randomly orient themselves on the metaphase plate, called

independent assortment. 6. c. The duplicated sister chromatids separate. 7. a. prophase I 8. c. homologous chromosomes become closely associated 9. a. cells that are genetically identical to the parent cell/haploid cells 10. d. All of the above

Chapter 10 *Foundations of Genetics*

1. d. All of these 2. a. all purple flowers 3. c. $^3/_4$ purple and $^1/_4$ white flowers 4. b. some factor, or information, about traits to their offspring and it may or may not be expressed 5. a. dihybrid cross 6. d. multiple genes 7. c. codominant traits 8. b. sex-linked eye color in fruit flies 9. d. All of the above. 10. b. chromosome karyotype

Chapter 11 *DNA: The Genetic Material*

1. b. hereditary information can be added to cells from other cells 2. d. DNA 3. a. structure of DNA 4. c. the type of nitrogen base 5. d. adenine and guanine 6. b. TAACGTA 7. b. splits down the middle into two single strands, and each one then acts as a template to build its complement 8. a. a primer 9. c. by mutation or by transformation 10. a. germ-line tissues and are passed on to future generations

Chapter 12 *How Genes Work*

1. a. nRNA (nuclear RNA) 2. d. a codon 3. c. a codon consists of three nucleotides 4. c. transcription 5. a. promoter 6. d. translation 7. c. AUG 8. d. Some genes remain off as long as a repressor is bound. 9. d. All of these statements are correct. 10. c. translating a gene as it is being transcribed

Chapter 13 *The New Biology*

1. c. genome 2. b. the exons used to make a specific mRNA can be rearranged to form different proteins. 3. c. restriction enzyme 4. b. exposing the mRNA of the desired eukaryotic gene to reverse transcriptase 5. c. DNA fingerprinting becomes more and more reliable as more probes are used 6. a. the drug to be produced in far larger amounts than in the past 7. a. increased yield 8. d. harm to the crop itself from mutations 9. a. immunological rejection of the tissue by the patient 10. c. viruses

Chapter 14 *Evolution and Natural Selection*

1. c. populations are capable of geometric increase, yet remain at constant levels 2. a. natural selection 3. b. seems to agree with Darwin's original ideas

4. a. homologous structures 5. b. microevolution
6. c. no mutation within the population 7. a. 0.20
8. d. genetic drift 9. c. directional selection 10. d.
reproductive isolation

Chapter 15 *Exploring Biological Diversity*

1. a. the red fox is in the same family, but different
genus than dogs and wolves 2. b. phylogeny 3. a.
physical, behavioral, and molecular characteristics
4. c. share a more recent common ancestor than
those organisms that are farther apart 5. d. cell
structure and DNA sequence 6. b. Archaea 7. d. are
prokaryotes 8. c. Protista 9. b. multicellular 10. d.
endosymbiosis of bacteria

Chapter 16 *Evolution of Microbial Life*

1. c. RNA 2. d. All answers are correct 3. b.
peptidoglycan 4. a. protein coats that contain DNA
or RNA 5. c. birds 6. c. cysts 7. d. mushrooms 8. a.
hyphae 9. a. both sexually and asexually 10. b. algae
and fungi

Chapter 17 *Evolution of Plants*

1. c. dehydration 2. a. chloroplasts 3. b. they do not
have specialized vascular tissue to transport water
very high 4. d. seeds 5. c. sporophyte 6. c.
nourishment 7. c. fertilization occurs in seed cones
8. c. fruits and flowers 9. a. pollination 10. b. the
process of double fertilization

Chapter 18 *Evolution of Animals*

1. b. choanocytes 2. c. specialization of digestive
tract 3. b. endoderm; mesoderm 4. c. deuterostome
development 5. b. an animal with a sessile lifestyle,
rather than one that moves through the environment
6. b. jaws 7. c. an internal skeleton made of cartilage
8. c. middle ear bones 9. d. thin hollow bones in the
skeleton 10. c. hair

Chapter 19 *Populations and Communities*

1. a. population 2. c. carrying capacity 3. b.
increased competition for food 4. a. short life span
5. d. community 6. c. resource partitioning 7. b.
commensalism 8. b. decreasing competitive
exclusion between prey species 9. a. aposematic
coloration 10. c. secondary succession

Chapter 20 *Ecosystems*

1. c. producers 2. d. All answers are correct 3. d.
amount of energy transferred to the top carnivores
4. a. from plants 5. b. combustion 6. c. ATP 7. d.
desert conditions on the downwind side of a
mountain due to increased moisture-holding
capacity of the winds as the air heats up 8. a.
decreases, temperature increases 9. d. thermocline
10. c. tundra

Chapter 21 *Behavior and the Environment*

1. b. cannot be modified, as these behaviors seem
built into the brain and nervous system 2. a. there is

a clear link between presence or absence of a
specific gene, a specific metabolic pathway, and a
specific behavior 3. b. operant conditioning 4. c.
reproductive fitness 5. d. foraging behavior 6. c.
intersexual selection 7. b. polyandry 8. c. reciprocity
9. b. kin selection 10. d. helpers at the nest

Chapter 22 *Planet Under Stress*

1. d. sulfur oxides as a major air pollutant 2. b.
coal-powered industry 3. c. ozone levels 4. a. habitat
loss 5. c. chlorofluorocarbons 6. a. environmental
costs are hardly ever recognized as part of the
economy 7. a. needed to preserve possible direct value
from species, such as new medicines 8. c. increasing
amounts of open space as countries develop 9. d.
captive propagation 10. a. keystone species

Chapter 23 *The Animal Body and How It Moves*

1. c. the thoracic cavity 2. a. cells, tissues, organs,
organ systems, organism 3. c. move the body 4. d.
red blood cells 5. c. osteoblasts; osteoclasts 6. b.
neurotransmitters 7. a. skeletal 8. b. axial skeleton
9. a. a single muscle can only pull and not push
10. c. to reset the myosin head so that it can reattach
to the actin

Chapter 24 *Circulation*

1. c. A series of tubular hearts pump hemolymph
into cavities of the body 2. d. All of the above are
functions of the circulatory system 3. c. by passing
warm blood near cold blood in the extremities to
warm the blood 4. a. capillaries 5. d. carry fluids
6. d. erythrocyte 7. d. Blood would flow back into
the right ventricle. 8. b. lungs->veins->heart->
arteries-> arteriole->capillaries->venules->veins->
heart->lungs 9. a. Only arteries carry oxygenated
blood 10. b. a stethoscope

Chapter 25 *Respiration*

1. a. gills 2. b. fish gills 3. c. the animal's blood to be
continually exposed to water of higher oxygen
concentration 4. c. nostrils->nasal cavity->trachea->
bronchus->alveoli 5. b. active transport of carbon
dioxide 6. c. contracting the muscles around the
thoracic cavity pulls your chest out 7. a. hemoglobin
in red blood cells 8. b. as bicarbonate in the blood
plasma 9. a. bicarbonate 10. d. All of the above

Chapter 26 *The Path of Food Through the Animal Body*

1. b. specialization of different regions of the
digestive system 2. a. herbivores 3. d. begin the
physical digestion of food 4. c. soft palate 5. c.
stomach 6. b. The surface area of the large intestine
is greater than that in the small intestine. 7. c. small
intestine 8. d. increase the surface area of the small
intestine for absorption of nutrients 9. c. the
concentration of solid wastes 10. a. pancreas

Chapter 27 *Maintaining the Internal Environment*

1. b. homeostasis 2. d. negative feedback loop 3. b.
glycogen to break down 4. a. pancreas 5. b. ants 6. b.

nephron 7. a. glomerulus 8. c. loop of Henle
9. d. osmosis 10. d. urea

Chapter 28 *How the Animal Body Defends Itself*

1. a. sweat and oil glands 2. b. have cell surface
proteins that are different from the body's own cell
surface receptors 3. c. increase the number of
immune system cells in an infected area 4. b.
pathogenic bacteria do not grow well at high
temperatures 5. a. T cells 6. c. B cells 7. d. destroy
cells infected by pathogens 8. b. memory T and B
cells 9. c. an autoimmune response 10. a. helper
T cells

Chapter 29 *The Nervous System*

1. c. association neurons 2. a. influx of sodium ions
3. a. sodium ion gates in the postsynaptic cell 4. b.
into the synaptic cleft 5. c. hypothalamus 6. b. relay
messages to skeletal muscles 7. b. interoceptors 8. c.
semicircular canals 9. c. taste 10. c. Rod cells detect
different colors and cone cells detect different
shades of gray, working to see in dim light.

Chapter 30 *Chemical Signaling Within the Animal Body*

1. c. chemical signals stick around longer than
electrical signals and can be used for slow processes
2. a. hypothalamus 3. a. all fit into receptors
specifically shaped for them 4. b. steroid hormones
must enter the cell to begin action, whereas peptide
hormones must begin action on the external surface
of the cell membrane 5. d. posterior pituitary gland
6. a. ACTH 7. a. pancreas 8. c. too much calcium in
the blood 9. d. sympathetic nervous system
10. a. cortisol

Chapter 31 *Reproduction and Development*

1. b. sexual reproduction 2. c. hermaphroditism 3. b.
if both sex chromosomes are X 4. d. scrotum 5. b.
FSH and LH 6. a. embryo releasing hCG 7. d.
gastrulation is complete 8. c. neural 9. c. oxytocin
10. a. destruction of the egg

Chapter 32 *Plant Form and Function*

1. c. meristematic tissue 2. b. transports
carbohydrates 3. a. parenchyma cells 4. d. at the
pericycle 5. c. cambium 6. b. organization of
vascular tissue 7. c. stomata 8. a. photosynthesis
9. b. translocation 10. d. osmosis

Chapter 33 *Plant Reproduction and Growth*

1. a. pollen 2. b. stigma 3. b. attract animal
pollinators 4. d. oxygen and water 5. a. ovary 6. c.
hormones 7. c. cells in plant stems to elongate 8. b.
fruit to ripen 9. c. photoperiod 10. a. thigmotropism

Glossary

Terms and Concepts

A

absorption (L. *absorbere,* to swallow down) The movement of water and substances dissolved in water into a cell, tissue, or organism.

acid Any substance that dissociates to form H^+ ions when dissolved in water. Having a pH value less than 7.

acoelomate (Gr. *a,* not + *koiloma,* cavity) A bilaterally symmetrical animal not possessing a body cavity, such as a flatworm.

actin (Gr. *actis,* ray) One of the two major proteins that make up myofilaments (the other is myosin). It provides the cell with mechanical support and plays major roles in determining cell shape and cell movement.

action potential A single nerve impulse. A transient all-or-none reversal of the electrical potential across a neuron membrane. Because it can activate nearby voltage-sensitive channels, an action potential propagates along a nerve cell.

activation energy The energy a molecule must acquire to undergo a specific chemical reaction.

activator A regulatory protein that binds to the DNA and makes it more accessible for transcription.

active transport The transport of a solute across a membrane by protein carrier molecules to a region of higher concentration by the expenditure of chemical energy. One of the most important functions of any cell.

adaptation (L. *adaptare,* to fit) Any peculiarity of structure, physiology, or behavior that promotes the likelihood of an organism's survival and reproduction in a particular environment.

adenosine triphosphate (ATP) A molecule composed of ribose, adenine, and a triphosphate group. ATP is the chief energy currency of all cells. Cells focus all of their energy resources on the manufacture of ATP from ADP and phosphate, which requires the cell to supply 7 kilocalories of energy obtained from photosynthesis or from electrons stripped from foodstuffs to form 1 mole of ATP. Cells then use this ATP to drive endergonic reactions.

adhesion (L. *adhaerere,* to stick to) The molecular attraction exerted between the surfaces of unlike bodies in contact, as water molecules to the walls of the narrow tubes that occur in plants.

aerobic (Gr. *aer,* air + *bios,* life) Oxygen-requiring.

allele (Gr. *allelon,* of one another) One of two or more alternative forms of a gene.

allele frequency The relative proportion of a particular allele among individuals of a population. Not equivalent to gene frequency, although the two terms are sometimes confused.

allosteric interaction (Gr. *allos,* other + *stereos,* shape) The change in shape that occurs when an activator or repressor binds to an enzyme. These changes result when specific, small molecules bind to the enzyme, molecules that are not substrates of that enzyme.

alternation of generations A reproductive life cycle in which the multicellular diploid phase produces spores that give rise to the multicellular haploid phase and the multicellular haploid phase produces gametes that fuse to give rise to the zygote. The zygote is the first cell of the multicellular diploid phase.

alveolus, *pl.* alveoli (L. *alveus,* a small cavity) One of the many small, thin-walled air sacs within the lungs in which the bronchioles terminate.

amniotic egg An egg that is isolated and protected from the environment by a more or less impervious shell. The shell protects the embryo from drying out, nourishes it, and enables it to develop outside of water.

anaerobic (Gr. *an,* without + *aer,* air + *bios,* life) Any process that can occur without oxygen. Includes glycolysis and fermentation. Anaerobic organisms can live without free oxygen.

anaphase In mitosis and meiosis II, the stage initiated by the separation of sister chromatids, during which the daughter chromosomes move to opposite poles of the cell; in meiosis I, marked by separation of replicated homologous chromosomes.

angiosperms The flowering plants, one of five phyla of seed plants. In angiosperms, the ovules at the time of pollination are completely enclosed by tissues.

anterior (L. *ante,* before) Located before or toward the front. In animals, the head end of an organism.

anther (Gr. *anthos,* flower) The part of the stamen of a flower that bears the pollen.

antibody (Gr. *anti,* against) A protein substance produced by a B cell lymphocyte in response to a foreign substance (antigen) and released into the bloodstream. Binding to the antigen, antibodies mark them for destruction by other elements of the immune system.

anticodon The three-nucleotide sequence of a tRNA molecule that is complementary to, and base pairs with, an amino acid-specifying codon in mRNA.

antigen (Gr. *anti,* against + *genos,* origin) A foreign substance, usually a protein, that stimulates lymphocytes to proliferate and secrete specific antibodies that bind to the foreign substance, labeling it as foreign and destined for destruction.

apical meristem (L. *apex,* top + Gr. *meristos,* divided) In vascular plants, the growing point at the tip of the root or stem.

aposematic coloration An ecological strategy of some organisms that "advertise" their poisonous nature by the use of bright colors.

appendicular skeleton (L. *appendicula,* a small appendage) The skeleton of the limbs of the human body containing 126 bones.

archaea A group of prokaryotes that are among the most primitive still in existence, characterized by the absence of peptidoglycan in their cell walls, a feature that distinguishes them from bacteria.

asexual Reproducing without forming gametes. Asexual reproduction does not involve sex. Its outstanding characteristic is that an individual offspring is genetically identical to its parent.

association neuron A nerve cell found only in the CNS that acts as a functional link between sensory neurons and motor neurons. Also called interneuron.

atom (Gr. *atomos,* indivisible) A core (nucleus) of protons and neutrons surrounded by an orbiting cloud of electrons. The chemical behavior of an atom is largely determined by the distribution of its electrons, particularly the number of electrons in its outermost level.

atomic number The number of protons in the nucleus of an atom. In an atom that does not bear an electric charge (that is, one that is not an ion), the atomic number is also equal to the number of electrons.

autonomic nervous system (Gr. *autos,* self + *nomos,* law) The motor pathways that carry commands from the central nervous system to regulate the glands and nonskeletal muscles of the body. Also called the involuntary nervous system.

autosome (Gr. *autos,* self + *soma,* body) Any of the 22 pairs of human chromosomes that are similar in size and morphology in both males and females.

autotroph (Gr. *autos,* self + *trophos,* feeder) An organism that can harvest light energy from the sun or from the oxidation of inorganic compounds to make organic molecules.

axial skeleton The skeleton of the head and trunk of the human body containing 80 bones.

axon (Gr., axle) A process extending out from a neuron that conducts impulses away from the cell body.

B

bacterium, *pl.* bacteria (Gr. *bakterion*, dim. of *baktron*, a staff) The simplest cellular organism. Its cell is smaller and prokaryotic in structure, and it lacks internal organization.

basal body In eukaryotic cells that contain flagella or cilia, a form of centriole that anchors each flagellum.

base Any substance that combines with H^+ ions thereby reducing the H^+ ion concentration of a solution. Having a pH value above 7.

Batesian mimicry After Henry W. Bates, English naturalist. A situation in which a palatable or nontoxic organism resembles another kind of organism that is distasteful or toxic. Both species exhibit warning coloration.

B cell A lymphocyte that recognizes invading pathogens much as T cells do, but instead of attacking the pathogens directly, it marks them with antibodies for destruction by the nonspecific body defenses.

bilateral symmetry (L. *bi*, two + *lateris*, side; Gr. *symmetria*, symmetry) A body form in which the right and left halves of an organism are approximate mirror images of each other.

binary fission (L. *binarius*, consisting of two things or parts + *fissus*, split) Asexual reproduction of a cell by division into two equal, or nearly equal, parts. Bacteria divide by binary fission.

binomial system (L. *bi*, twice, two + Gr. *nomos*, usage, law) A system of nomenclature that uses two words. The first names the genus, and the second designates the species.

biomass (Gr. *bios*, life + *maza*, lump or mass) The total weight of all of the organisms living in an ecosystem.

biome (Gr. *bios*, life + *-oma*, mass, group) A major terrestrial assemblage of plants, animals, and microorganisms that occur over wide geographical areas and have distinct characteristics. The largest ecological unit.

buffer A substance that takes up or releases hydrogen ions (H^+) to maintain the pH within a certain range.

C

calorie (L. *calor*, heat) The amount of energy in the form of heat required to raise the temperature of 1 gram of water 1 degree Celsius.

calyx (Gr. *kalyx*, a husk, cup) The sepals collectively. The outermost flower whorl.

cancer Unrestrained invasive cell growth. A tumor or cell mass resulting from uncontrollable cell division.

capillary (L. *capillaris*, hairlike) A blood vessel with a very small diameter. Blood exchanges gases and metabolites across capillary walls. Capillaries join the end of an arteriole to the beginning of a venule.

carbohydrate (L. *carbo*, charcoal + *hydro*, water) An organic compound consisting of a chain or ring of carbon atoms to which hydrogen and oxygen atoms are attached in a ratio of approximately 1:2:1. A compound of carbon, hydrogen, and oxygen having the generalized formula $(CH_2O)_n$, where n is the number of carbon atoms.

carcinogen (Gr. *karkinos*, cancer + -gen) Any cancer-causing agent.

cardiovascular system (Gr. *kardia*, heart + L. *vasculum*, vessel) The blood circulatory system and the heart that pumps it. Collectively, the blood, heart, and blood vessels.

carpel (Gr. *karpos*, fruit) A leaflike organ in angiosperms that encloses one or more ovules.

carrying capacity The maximum population size that a habitat can support.

catabolism (Gr. *katabole*, throwing down) A process in which complex molecules are broken down into simpler ones.

catalysis (Gr. *katalysis*, dissolution + *lyein*, to loosen) The enzyme-mediated process in which the subunits of polymers are positioned so that their bonds undergo chemical reactions.

catalyst (Gr. *kata*, down + *lysis*, a loosening) A general term for a substance that speeds up a specific chemical reaction by lowering the energy required to activate or start the reaction. An enzyme is a biological catalyst.

cell (L. *cella*, a chamber or small room) The smallest unit of life. The basic organizational unit of all organisms. Composed of a nuclear region containing the hereditary apparatus within a larger volume called the cytoplasm bounded by a lipid membrane.

cell cycle The repeating sequence of growth and division through which cells pass each generation.

cellular respiration The process in which the energy stored in a glucose molecule is released by oxidation. Hydrogen atoms are lost by glucose and gained by oxygen.

central nervous system The brain and spinal cord, the site of information processing and control within the nervous system.

centromere (Gr. *kentron*, center + *meros*, a part) A constricted region of the chromosome joining two sister chromatids, to which the kinetochore is attached.

chemical bond The force holding two atoms together. The force can result from the attraction of opposite charges (ionic bond) or from the sharing of one or more pairs of electrons (a covalent bond).

chemiosmosis The cellular process responsible for almost all of the adenosine triphosphate (ATP) harvested from food and for all the ATP produced by photosynthesis.

chemoautotroph An autotrophic bacterium that uses chemical energy released by specific inorganic reactions to power its life processes, including the synthesis of organic molecules.

chiasma, *pl.* chiasmata (Gr. a cross) In meiosis, the points of crossing over where portions of chromosomes have been exchanged during synapsis. A chiasma appears as an X-shaped structure under a light microscope.

chloroplast (Gr. *chloros*, green + *plastos*, molded) A cell-like organelle present in algae and plants that contains chlorophyll (and usually other pigments) and is the site of photosynthesis.

chromatid (Gr. *chroma*, color + L. *-id*, daughters of) One of two daughter strands of a duplicated chromosome that is joined by a single centromere.

chromatin (Gr. *chroma*, color) The complex of DNA and proteins of which eukaryotic chromosomes are composed.

chromosome (Gr. *chroma*, color + *soma*, body) The vehicle by which hereditary information is physically transmitted from one generation to the next. In a eukaryotic cell, long threads of DNA that are associated with protein and that contain hereditary information.

cilium, *pl.* cilia (L. eyelash) Refers to flagella, which are numerous and organized in dense rows. Cilia propel cells through water. In human tissue, they move water or mucus over the tissue surface.

cladistics A taxonomic technique used for creating hierarchies of organisms based on derived characters that represent true phylogenetic relationship and descent.

class A taxonomic category ranking below a phylum (division) and above an order.

clone (Gr. *klon*, twig) A line of cells, all of which have arisen from the same single cell by mitotic division. One of a population of individuals derived by asexual reproduction from a single ancestor. One of a population of genetically identical individuals.

codominance In genetics, a situation in which the effects of both alleles at a particular locus are apparent in the phenotype of the heterozygote.

codon (L. code) The basic unit of the genetic code. A sequence of three adjacent nucleotides in DNA or mRNA that codes for one amino acid or for polypeptide termination.

coelom (Gr. *koilos*, a hollow) A body cavity formed between layers of mesoderm and in which the digestive tract and other internal organs are suspended.

coenzyme A cofactor of an enzyme that is a nonprotein organic molecule.

coevolution (L. *co-*, together + *e-*, out + *volvere*, to fill) A term that describes the long-term evolutionary adjustment of one group of organisms to another.

commensalism (L. *cum*, together with + *mensa*, table) A symbiotic relationship in which one species benefits while the other neither benefits nor is harmed.

community (L. *communitas*, community, fellowship) The populations of different species that live together and interact in a particular place.

competition Interaction between individuals for the same scarce resources. Intraspecific competition is competition between individuals of a single species. Interspecific competition is competition between individuals of different species.

competitive exclusion The hypothesis that if two species are competing with one another for the same limited resource in the same place, one will be able to use that resource more efficiently than the other and eventually will drive that second species to extinction locally.

complement system The chemical defense of a vertebrate body that consists of a battery of proteins that insert in bacterial and fungal cells, causing holes that destroy the cells.

concentration gradient The concentration difference of a substance as a function of distance. In a cell, a greater concentration of its molecules in one region than in another.

condensation The coiling of the chromosomes into more and more tightly compacted bodies begun during the G_2 phase of the cell cycle.

conjugation (L. *conjugare*, to yoke together) An unusual mode of reproduction in unicellular organisms in which genetic material is exchanged between individuals through tubes connecting them during conjugation.

consumer In ecology, a heterotroph that derives its energy from living or freshly killed organisms or parts thereof. Primary consumers are herbivores; secondary consumers are carnivores or parasites.

cortex (L. bark) In vascular plants, the primary ground tissue of a stem or root, bounded externally by the epidermis and internally by the central cylinder of vascular tissue. In animals, the outer, as opposed to the inner, part of an organ, as in the adrenal, kidney, and cerebral cortexes.

cotyledon (Gr. *kotyledon*, a cup-shaped hollow) Seed leaf. Monocot embryos have one cotyledon, and dicots have two.

countercurrent flow In organisms, the passage of heat or of molecules (such as oxygen, water, or sodium ions) from one circulation path to another moving in the opposite direction. Because the flow of the two paths is in opposite directions, a concentration difference always exists between the two channels, facilitating transfer.

covalent bond (L. *co-*, together + *valare*, to be strong) A chemical bond formed by the sharing of one or more pairs of electrons.

crossing over An essential element of meiosis occurring during prophase when nonsister chromatids exchange portions of DNA strands.

cuticle (L. *cutis*, skin) A very thin film covering the outer skin of many plants.

cytokinesis (Gr. *kytos*, hollow vessel + *kinesis*, movement) The C phase of cell division in which the cell itself divides, creating two daughter cells.

cytoplasm (Gr. *kytos*, hollow vessel + *plasma*, anything molded) A semifluid matrix that occupies the volume between the nuclear region and the cell membrane. It contains the sugars, amino acids, proteins, and organelles (in eukaryotes) with which the cell carries out its everyday activities of growth and reproduction.

cytoskeleton (Gr. *kytos*, hollow vessel + *skeleton*, a dried body) In the cytoplasm of all eukaryotic cells, a network of protein fibers that supports the shape of the cell and anchors organelles, such as the nucleus, to fixed locations.

D

deciduous (L. *decidere*, to fall off) In vascular plants, shedding all the leaves at a certain season.

dehydration reaction Water-losing. The process in which a hydroxyl (OH) group is removed from one subunit of a polymer and a hydrogen (H) group is removed from the other subunit, linking the subunits together and forming a water molecule as a by-product.

demography (Gr. *demos*, people + *graphein*, to draw) The statistical study of population. The measurement of people or, by extension, of the characteristics of people.

density The number of individuals in a population in a given area.

deoxyribonucleic acid (DNA) The basic storage vehicle or central plan of heredity information. It is stored as a sequence of nucleotides in a linear nucleotide polymer. Two of the polymers wind around each other like the outside and inside rails of a circular staircase.

depolarization The movement of ions across a cell membrane that wipes out locally an electrical potential difference.

deuterostome (Gr. *deuteros*, second + *stoma*, mouth) An animal in whose embryonic development the anus forms from or near the blastopore, and the mouth forms later on another part of the blastula. Also characterized by radial cleavage.

dicot Short for dicotyledon; a class of flowering plants generally characterized by having two cotyledons, netlike veins, and flower parts in fours or fives.

diffusion (L. *diffundere*, to pour out) The net movement of molecules to regions of lower concentration as a result of random, spontaneous molecular motions. The process tends to distribute molecules uniformly.

dihybrid (Gr. *dis*, twice + L. *hibrida*, mixed offspring) An individual heterozygous for two genes.

dioecious (Gr. *di*, two + *eikos*, house) Having male and female flowers on separate plants of the same species.

diploid (Gr. *diploos*, double + *eidos*, form) A cell, tissue, or individual with a double set of chromosomes.

directional selection A form of selection in which selection acts to eliminate one extreme from an array of phenotypes. Thus, the genes promoting this extreme become less frequent in the population.

disaccharide (Gr. *dis*, twice + *sakcharon*, sugar) A sugar formed by linking two monosaccharide molecules together. Sucrose (table sugar) is a disaccharide formed by linking a molecule of glucose to a molecule of fructose.

disruptive selection A form of selection in which selection acts to eliminate rather than favor the intermediate type.

diurnal (L. *diurnalis*, day) Active during the day.

division Traditionally, a major taxonomic group of the plant kingdom comparable to a phylum of the animal kingdom. Today divisions are called phyla.

dominant allele An allele that dictates the appearance of heterozygotes. One allele is said to be dominant over another if an individual heterozygous for that allele has the same appearance as an individual homozygous for it.

dorsal (L. *dorsum*, the back) Toward the back, or upper surface. Opposite of ventral.

double fertilization A process unique to the angiosperms, in which one sperm nucleus fertilizes the egg and the second one fuses with the polar nuclei. These two events result in the formation of the zygote and the primary endosperm nucleus, respectively.

E

ecdysis (Gr. *ekdysis*, stripping off) The shedding of the outer covering or skin of certain animals. Especially the shedding of the exoskeleton by arthropods.

ecology (Gr. *oikos*, house + *logos*, word) The study of the relationships of organisms with one another and with their environment.

ecosystem (Gr. *oikos*, house + *systema*, that which is put together) A community, together with the nonliving factors with which it interacts.

ectoderm (Gr. *ecto*, outside + *derma*, skin) One of three embryonic germ layers that forms in the gastrula; giving rise to the outer epithelium and to nerve tissue.

ectothermic Referring to animals whose body temperature is regulated by their behavior or their surroundings.

electron A subatomic particle with a negative electrical charge. The negative charge of one electron exactly balances the positive charge of one proton. Electrons orbit the atom's positively charged nucleus and determine its chemical properties.

electron transport chain A collective term describing the series of membrane-associated electron carriers embedded in the inner mitochondrial membrane. It puts the electrons harvested from the oxidation of glucose to work driving proton-pumping channels.

electron transport system A collective term describing the series of membrane-associated electron carriers embedded in the thylakoid membrane of the chloroplast. It puts the electrons harvested from water molecules and energized by photons of light to work driving proton-pumping channels.

element A substance that cannot be separated into different substances by ordinary chemical methods.

emergent properties Novel properties in the hierarchy of life that were not present at the simpler levels of organization.

endergonic (Gr. *endon*, within + *ergon*, work) Reactions in which the products contain more energy than the reactants and require an input of usable energy from an outside source before they can proceed. These reactions are not spontaneous.

endocrine gland (Gr. *endon*, within + *krinein*, to separate) A ductless gland producing hormonal secretions that pass directly into the bloodstream or lymph.

endocrine system The dozen or so major endocrine glands of a vertebrate.

endocytosis (Gr. *endon*, within + *kytos*, cell) The process by which the edges of plasma membranes fuse together and form an enclosed chamber called a vesicle. It involves the incorporation of a portion of an exterior medium into the cytoplasm of the cell by capturing it within the vesicle.

endoderm (Gr. *endon*, outside + *derma*, skin) One of three embryonic germ layers that forms in the gastrula; giving rise to the epithelium that lines internal organs and most of the digestive and respiratory tracts.

endoplasmic reticulum (ER) (L. *endoplasmic*, within the cytoplasm + *reticulum*, little net) An extensive network of membrane compartments within a eukaryotic cell; attached ribosomes synthesize proteins to be exported.

endoskeleton (Gr. *endon*, within + *skeletos*, hard) In vertebrates, an internal scaffold of bone or cartilage to which muscles are attached.

endosperm (Gr. *endon*, within + *sperma*, seed) A nutritive tissue characteristic of the seeds of angiosperms that develops from the union of a male nucleus and the polar nuclei of the embryo sac. The endosperm is either digested by the growing embryo or retained in the mature seed to nourish the germinating seedling.

endosymbiotic (Gr. *endon,* **within +** *bios,* **life) theory** A theory that proposes how eukaryotic cells arose from large prokaryotic cells that engulfed smaller ones of a different species. The smaller cells were not consumed but continued to live and function within the larger host cell. Organelles that are believed to have entered larger cells in this way are mitochondria and chloroplasts.

endothermic The ability of animals to maintain a constant body temperature.

energy The capacity to bring about change, to do work.

enhancer A site of regulatory protein binding on the DNA molecule distant from the promoter and start site for a gene's transcription.

entropy (Gr. *en,* **in +** *tropos,* **change in manner)** A measure of the disorder of a system. A measure of energy that has become so randomized and uniform in a system that the energy is no longer available to do work.

enzyme (Gr. *enzymos,* **leavened; from** *en,* **in +** *zyme,* **leaven)** A protein capable of speeding up specific chemical reactions by lowering the energy required to activate or start the reaction but that remains unaltered in the process.

epidermis (Gr. *epi,* **on or over +** *derma,* **skin)** The outermost layer of cells. In vertebrates, the non-vascular external layer of skin of ectodermal origin; in invertebrates, a single layer of ectodermal epithelium; in plants, the flattened, skinlike outer layer of cells.

epistasis (Gr. *epistasis,* **a standing still)** An interaction between the products of two genes in which one modifies the phenotypic expression produced by the other.

epithelium (Gr. *epi,* **on +** *thele,* **nipple)** A thin layer of cells forming a tissue that covers the internal and external surfaces of the body. Simple epithelium consists of the membranes that line the lungs and major body cavities and that are a single cell layer thick. Stratified epithelium (the skin or epidermis) is composed of more complex epithelial cells that are several cell layers thick.

erythrocyte (Gr. *erythros,* **red +** *kytos,* **hollow vessel)** A red blood cell, the carrier of hemoglobin. Erythrocytes act as the transporters of oxygen in the vertebrate body. During the process of their maturation in mammals, they lose their nuclei and mitochondria, and their endoplasmic reticulum is reabsorbed.

estrus (L. *oestrus,* **frenzy)** The period of maximum female sexual receptivity. Associated with ovulation of the egg. Being "in heat."

estuary (L. *aestus,* **tide)** A partly enclosed body of water, such as those that often form at river mouths and in coastal bays, where the salinity is intermediate between that of saltwater and freshwater.

ethology (Gr. *ethos,* **habit or custom +** *logos,* **discourse)** The study of patterns of animal behavior in nature.

euchromatin (Gr. *eu,* **true +** *chroma,* **color)** Chromatin that is extended except during cell division, from which RNA is transcribed.

eukaryote (Gr. *eu,* **true +** *karyon,* **kernel)** A cell that possesses membrane-bounded organelles, most notably a cell nucleus, and chromosomes whose DNA is associated with proteins; an organism composed of such cells. The appearance of eukaryotes marks a major event in the evolution of life, as all organisms on earth other than bacteria and archaea are eukaryotes.

eumetazoan (Gr. *eu,* **true +** *meta,* **with +** *zoion,* **animal)** A "true animal." An animal with a definite shape and symmetry and nearly always distinct tissues.

eutrophic (Gr. *eutrophos,* **thriving)** Refers to a lake in which an abundant supply of minerals and organic matter exists.

evaporation The escape of water molecules from the liquid to the gas phase at the surface of a body of water.

evolution (L. *evolvere,* **to unfold)** Genetic change in a population of organisms over time (generations). Darwin proposed that natural selection was the mechanism of evolution.

exergonic (L. *ex,* **out +** **Gr.** *ergon,* **work)** Any reaction that produces products that contain less free energy than that possessed by the original reactants and that tends to proceed spontaneously.

exocytosis (Gr. *ex,* **out of +** *kytos,* **cell)** The extrusion of material from a cell by discharging it from vesicles at the cell surface. The reverse of endocytosis.

exoskeleton (Gr. *exo,* **outside +** *skeletos,* **hard)** An external hard shell that encases a body. In arthropods, comprised mainly of chitin.

experiment The test of a hypothesis. An experiment that tests one or more alternative hypotheses and those that are demonstrated to be inconsistent with experimental observation and are rejected.

F

facilitated diffusion The transport of molecules across a membrane by a carrier protein in the direction of lowest concentration.

family A taxonomic group ranking below an order and above a genus.

feedback inhibition A regulatory mechanism in which a biochemical pathway is regulated by the amount of the product that the pathway produces.

fermentation (L. *fermentum,* **ferment)** A catabolic process in which the final electron acceptor is an organic molecule.

fertilization (L. *ferre,* **to bear)** The union of male and female gametes to form a zygote.

fitness The genetic contribution of an individual to succeeding generations, relative to the contributions of other individuals in the population.

flagellum, *pl.* **flagella (L.** *flagellum,* **whip)** A fine, long, threadlike organelle protruding from the surface of a cell. In bacteria, a single protein fiber capable of rotary motion that propels the cell through the water. In eukaryotes, an array of microtubules with a characteristic internal 9 + 2 microtubule structure that is capable of vibratory but not rotary motion. Used in locomotion and feeding. Common in protists and motile gametes. A cilium is a short flagellum.

food web The food relationships within a community. A diagram of who eats whom.

founder effect The effect by which rare alleles and combinations of alleles may be enhanced in new populations.

frequency In statistics, defined as the proportion of individuals in a certain category, relative to the total number of individuals being considered.

fruit In angiosperms, a mature, ripened ovary (or group of ovaries) containing the seeds.

G

gamete (Gr. **wife)** A haploid reproductive cell. Upon fertilization, its nucleus fuses with that of another gamete of the opposite sex. The resulting diploid cell (zygote) may develop into a new diploid individual, or in some protists and fungi, may undergo meiosis to form haploid somatic cells.

gametophyte (Gr. *gamete,* **wife +** *phyton,* **plant)** In plants, the haploid (n), gamete-producing generation, which alternates with the diploid ($2n$) sporophyte.

ganglion, *pl.* **ganglia (Gr.** **a swelling)** A group of nerve cells forming a nerve center in the peripheral nervous system.

gastrulation The inward movement of certain cell groups from the surface of the blastula.

gene (Gr. *genos,* **birth, race)** The basic unit of heredity. A sequence of DNA nucleotides on a chromosome that encodes a polypeptide or RNA molecule and so determines the nature of an individual's inherited traits.

gene expression The process in which an RNA copy of each active gene is made, and the RNA copy directs the sequential assembly of a chain of amino acids at a ribosome.

gene frequency The frequency with which individuals in a population possess a particular gene. Often confused with allele frequency.

genetic code The "language" of the genes. The mRNA codons specific for the 20 common amino acids constitute the genetic code.

genetic drift Random fluctuations in allele frequencies in a small population over time.

genetic map A diagram showing the relative positions of genes.

genetics (Gr. *genos,* **birth, race)** The study of the way in which an individual's traits are transmitted from one generation to the next.

genome (Gr. *genos,* **offspring +** **L.** *oma,* **abstract group)** The genetic information of an organism.

genomics The study of genomes as opposed to individual genes.

genotype (Gr. *genos,* **offspring +** *typos,* **form)** The total set of genes present in the cells of an organism. Also used to refer to the set of alleles at a single gene locus.

genus, *pl.* **genera (L.** **race)** A taxonomic group that ranks below a family and above a species.

germination (L. *germinare,* **to sprout)** The resumption of growth and development by a spore or seed.

gland (L. *glandis,* **acorn)** Any of several organs in the body, such as exocrine or endocrine, that secrete substances for use in the body. Glands are composed of epithelial tissue.

glomerulus (L. **a little ball)** A network of capillaries in a vertebrate kidney, whose walls act as a filtration device.

glycolysis (Gr. *glykys,* **sweet +** *lyein,* **to loosen)** The anaerobic breakdown of glucose; this enzyme-catalyzed process yields two molecules of pyruvate with a net of two molecules of ATP.

golgi complex Flattened stacks of membrane compartments that collect, package, and distribute molecules made in the endoplasmic reticulum.

gravitropism (L. *gravis*, heavy + *tropes*, turning) The response of a plant to gravity, which generally causes shoots to grow up and roots to grow down.

greenhouse effect The process in which carbon dioxide and certain other gases, such as methane, that occur in the earth's atmosphere transmit radiant energy from the sun but trap the longer wavelengths of infrared light, or heat, and prevent them from radiating into space.

guard cells Pairs of specialized epidermal cells that surround a stoma. When the guard cells are turgid, the stoma is open; when they are flaccid, it is closed.

gymnosperm (Gr. *gymnos*, naked + *sperma*, seed) A seed plant with seeds not enclosed in an ovary. The conifers are the most familiar group.

H

habitat (L. *habitare*, to inhabit) The place where individuals of a species live.

half-life The length of time it takes for half of a radioactive substance to decay.

haploid (Gr. *haploos*, single + *eidos*, form) The gametes of a cell or an individual with only one set of chromosomes.

Hardy-Weinberg equilibrium After G. H. Hardy, English mathematician, and G. Weinberg, German physician. A mathematical description of the fact that the relative frequencies of two or more alleles in a population do not change because of Mendelian segregation. Allele and genotype frequencies remain constant in a random-mating population in the absence of inbreeding, selection, or other evolutionary forces. Usually stated as: If the frequency of allele A is p and the frequency of allele a is q, then the genotype frequencies after one generation of random mating will always be $(p + q)^2 = p^2 + 2pq + q^2$.

Haversian canal After Clopton Havers, English anatomist. Narrow channels that run parallel to the length of a bone and contain blood vessels and nerve cells.

helper T cell A class of white blood cells that initiates both the cell-mediated immune response and the humoral immune response; helper T cells are the targets of the AIDS virus (HIV).

hemoglobin (Gr. *haima*, blood + L. *globus*, a ball) A globular protein in vertebrate red blood cells and in the plasma of many invertebrates that carries oxygen and carbon dioxide.

herbivore (L. *herba*, grass + *vorare*, to devour) Any organism that eats only plants.

heredity (L. *heredis*, heir) The transmission of characteristics from parent to offspring.

heterochromatin (Gr. *heteros*, different + *chroma*, color) That portion of a eukaryotic chromosome that remains permanently condensed and therefore is not transcribed into RNA. Most centromere regions are heterochromatic.

heterokaryon (Gr. *heteros*, other + *karyon*, kernel) A fungal hypha that has two or more genetically distinct types of nuclei.

heterotroph (Gr. *heteros*, other + *trophos*, feeder) An organism that does not have the ability to produce its own food. *See also* autotroph.

heterozygote (Gr. *heteros*, other + *zygotos*, a pair) A diploid individual carrying two different alleles of a gene on its two homologous chromosomes.

hierarchical (Gr. *hieros*, sacred + *archos*, leader) Refers to a system of classification in which successively smaller units of classification are included within one another.

histone (Gr. *histos*, tissue) A complex of small, very basic polypeptides rich in the amino acids arginine and lysine. A basic part of chromosomes, histones form the core around which DNA is wrapped.

homeostasis (Gr. *homeos*, similar + *stasis*, standing) The maintaining of a relatively stable internal physiological environment in an organism or steady-state equilibrium in a population or ecosystem.

homeotherm (Gr. *homeo*, similar + *therme*, heat) An organism, such as a bird or mammal, capable of maintaining a stable body temperature independent of the environmental temperature. "Warm-blooded."

hominid (L. *homo*, man) Human beings and their direct ancestors. A member of the family Hominidae. *Homo sapiens* is the only living member.

homologous chromosome (Gr. *homologia*, agreement) One of the two nearly identical versions of each chromosome. Chromosomes that associate in pairs in the first stage of meiosis. In diploid cells, one chromosome of a pair that carries equivalent genes.

homology (Gr. *homologia*, agreement) A condition in which the similarity between two structures or functions is indicative of a common evolutionary origin.

homozygote (Gr. *homos*, same or similar + *zygotos*, a pair) A diploid individual whose two copies of a gene are the same. An individual carrying identical alleles on both homologous chromosomes is said to be homozygous for that gene.

hormone (Gr. *hormaein*, to excite) A chemical messenger, often a steroid or peptide, produced in a small quantity in one part of an organism and then transported to another part of the organism, where it brings about a physiological response.

hybrid (L. *hybrida*, the offspring of a tame sow and a wild boar) A plant or animal that results from the crossing of dissimilar parents.

hybridization The mating of unlike parents of different taxa.

hydrogen bond A molecular force formed by the attraction of the partial positive charge of one hydrogen atom of a water molecule with the partial negative charge of the oxygen atom of another.

hydrolysis reaction (Gr. *hydro*, water + *lyse*, break) The process of tearing down a polymer by adding a molecule of water. A hydrogen is attached to one subunit and a hydroxyl to the other, which breaks the covalent bond. Essentially the reverse of a dehydration reaction.

hydrophilic (Gr. *hydro*, water + *philic*, loving) Describes polar molecules, which form hydrogen bonds with water and therefore are soluble in water.

hydrophobic (Gr. *hydro*, water + *phobos*, hating) Describes nonpolar molecules, which do not form hydrogen bonds with water and therefore are not soluble in water.

hydroskeleton (Gr. *hydro*, water + *skeletos*, hard) The skeleton of most soft-bodied invertebrates that have neither an internal nor an external skeleton. They use the relative incompressibility of the water within their bodies as a kind of skeleton.

hypertonic (Gr. *hyper*, above + *tonos*, tension) A cell that contains a higher concentration of solutes than its surrounding solution.

hypha, *pl.* hyphae (Gr. *hyphe*, web) A filament of a fungus. A mass of hyphae comprises a mycelium.

hypothalamus (Gr. *hypo*, under + *thalamos*, inner room) The region of the brain under the thalamus that controls temperature, hunger, and thirst and that produces hormones that influence the pituitary gland.

hypothesis (Gr. *hypo*, under + *tithenai*, to put) A proposal that might be true. No hypothesis is ever proven correct. All hypotheses are provisional—proposals that are retained for the time being as useful but that may be rejected in the future if found to be inconsistent with new information. A hypothesis that stands the test of time—often tested and never rejected—is called a theory.

hypotonic (Gr. *hypo*, under + *tonos*, tension) A solution surrounding a cell that has a lower concentration of solutes than does the cell.

I

inbreeding The breeding of genetically related plants or animals. In plants, inbreeding results from self-pollination. In animals, inbreeding results from matings between relatives. Inbreeding tends to increase homozygosity.

incomplete dominance The ability of two alleles to produce a heterozygous phenotype that is different from either homozygous phenotype.

independent assortment Mendel's second law: The principle that segregation of alternative alleles at one locus into gametes is independent of the segregation of alleles at other loci. Only true for gene loci located on different chromosomes or those so far apart on one chromosome that crossing over is very frequent between the loci.

industrial melanism (Gr. *melas*, black) The evolutionary process in which a population of initially light-colored organisms becomes a population of dark organisms as a result of natural selection.

inflammatory response (L. *inflammare*, to flame) A generalized nonspecific response to infection that acts to clear an infected area of infecting microbes and dead tissue cells so that tissue repair can begin.

integument (L. *integumentum*, covering) The natural outer covering layers of an animal. Develops from the ectoderm.

interneuron A nerve cell found only in the CNS that acts as a functional link between sensory neurons and motor neurons. Also called association neuron.

internode The region of a plant stem between nodes where stems and leaves attach.

interoception (L. *interus*, inner + Eng. *[re]ceptive*) The sensing of information that relates to the body itself, its internal condition, and its position.

interphase That portion of the cell cycle preceding mitosis. It includes the G_1 phase, when cells grow, the S phase, when a replica of the genome is synthesized, and a G$_2$ phase, when preparations are made for genomic separation.

intron (L. *intra*, within) A segment of DNA transcribed into mRNA but removed before translation. These untranslated regions make up the bulk of most eukaryotic genes.

ion An atom in which the number of electrons does not equal the number of protons. An ion carries an electrical charge.

ionic bond A chemical bond formed between ions as a result of the attraction of opposite electrical charges.

ionizing radiation High-energy radiation, such as X rays and gamma rays.

isolating mechanisms Mechanisms that prevent genetic exchange between individuals of different populations or species.

isotonic (Gr. *isos*, equal + *tonos*, tension) A cell with the same concentration of solutes as its environment.

isotope (Gr. *isos*, equal + *topos*, place) An atom that has the same number of protons but different numbers of neutrons.

J

joint The part of a vertebrate where one bone meets and moves on another.

K

karyotype (Gr. *karyon*, kernel + *typos*, stamp or print) The particular array of chromosomes that an individual possesses.

kinetic energy The energy of motion.

kinetochore (Gr. *kinetikos*, putting in motion + *choros*, chorus) A disk of protein bound to the centromere to which microtubules attach during cell division, linking chromatids to the spindle.

kingdom The chief taxonomic category. This book recognizes six kingdoms: Archaea, Bacteria, Protista, Fungi, Animalia, and Plantae.

L

lamella, *pl.* lamellae (L. a little plate) A thin, plate-like structure. In chloroplasts, a layer of chlorophyll-containing membranes. In bivalve mollusks, one of the two plates forming a gill. In vertebrates, one of the thin layers of bone laid concentrically around the Haversian canals.

ligament (L. *ligare*, to bind) A band or sheet of connective tissue that links bone to bone.

linkage The patterns of assortment of genes that are located on the same chromosome. Important because if the genes are located relatively far apart, crossing over is more likely to occur between them than if they are close together.

lipid (Gr. *lipos*, fat) A loosely defined group of molecules that are insoluble in water but soluble in oil. Oils such as olive, corn, and coconut are lipids, as well as waxes, such as beeswax and earwax.

lipid bilayer The basic foundation of all biological membranes. In such a layer, the nonpolar tails of phospholipid molecules point inward, forming a nonpolar zone in the interior of the bilayers. Lipid bilayers are selectively permeable and do not

permit the diffusion of water-soluble molecules into the cell.

littoral (L. *litus*, shore) Referring to the shoreline zone of a lake or pond or the ocean that is exposed to the air whenever water recedes.

locus, *pl.* loci (L. place) The position on a chromosome where a gene is located.

loop of Henle After F. G. J. Henle, German anatomist. A hairpin loop formed by a urine-conveying tubule when it enters the inner layer of the kidney and then turns around to pass up again into the outer layer of the kidney.

lymph (L. *lympha*, clear water) In animals, a colorless fluid derived from blood by filtration through capillary walls in the tissues.

lymphatic system An open circulatory system composed of a network of vessels that function to collect the water within blood plasma forced out during passage through the capillaries and to return it to the bloodstream. The lymphatic system also returns proteins to the circulation, transports fats absorbed from the intestine, and carries bacteria and dead blood cells to the lymph nodes and spleen for destruction.

lymphocyte (Gr. *lympha*, water + Gr. *kytos*, hollow vessel) A white blood cell. A cell of the immune system that either synthesizes antibodies (B cells) or attacks virus-infected cells (T cells).

lyse (Gr. *lysis*, loosening) To disintegrate a cell by rupturing its plasma membrane.

M

macromolecule (Gr. *makros*, large + L. *moliculus*, a little mass) An extremely large molecule. Refers specifically to carbohydrates, lipids, proteins, and nucleic acids.

macrophage (Gr. *makros*, large + *-phage*, eat) A phagocytic cell of the immune system able to engulf and digest invading bacteria, fungi, and other microorganisms, as well as cellular debris.

marrow The soft tissue that fills the cavities of most bones and is the source of red blood cells.

mass flow The overall process by which materials move in the phloem of plants.

mass number The mass number of an atom consists of the combined mass of all of its protons and neutrons.

meiosis (Gr. *meioun*, to make smaller) A special form of nuclear division that precedes gamete formation in sexually reproducing eukaryotes. It results in four haploid daughter cells.

Mendelian ratio After Gregor Mendel, Austrian monk. Refers to the characteristic 3:1 segregation ratio that Mendel observed, in which pairs of alternative traits are expressed in the F$_2$ generation in the ratio of three-fourths dominant to one-fourth recessive.

menstruation (L. *mens*, month) Periodic sloughing off of the blood-enriched lining of the uterus when pregnancy does not occur.

meristem (Gr. *merizein*, to divide) In plants, a zone of unspecialized cells whose only function is to divide.

mesoderm (Gr. *mesos*, middle + *derma*, skin) One of the three embryonic germ layers that form in the gastrula. Gives rise to muscle, bone, and other connective tissue; the peritoneum; the circulatory

system; and most of the excretory and reproductive systems.

mesophyll (Gr. *mesos*, middle + *phyllon*, leaf) The photosynthetic parenchyma of a leaf, located within the epidermis. The vascular strands (veins) run through the mesophyll.

metabolism (Gr. *metabole*, change) The process by which all living things assimilate energy and use it to grow.

metamorphosis (Gr. *meta*, after + *morphe*, form + *osis*, state of) Process in which form changes markedly during postembryonic development—for example, tadpole to frog or larval insect to adult.

metaphase (Gr. *meta*, middle + *phasis*, form) The stage of mitosis characterized by the alignment of the chromosomes on a plane in the center of the cell.

metastasis, *pl.* metastases (Gr. to place in another way) The spread of cancerous cells to other parts of the body, forming new tumors at distant sites.

microevolution (Gr. *mikros*, small + L. *evolvere*, to unfold) Refers to the evolutionary process itself. Evolution within a species. Also called adaptation.

microtubule (Gr. *mikros*, small + L. *tubulus*, little pipe) In eukaryotic cells, a long, hollow cylinder about 25 nanometers in diameter and composed of the protein tubulin. Microtubules influence cell shape, move the chromosomes in cell division, and provide the functional internal structure of cilia and flagella.

mimicry (Gr. *mimos*, mime) The resemblance in form, color, or behavior of certain organisms (mimics) to other more powerful or more protected ones (models), which results in the mimics being protected in some way.

mitochondrion, *pl.* mitochondria (Gr. *mitos*, thread + *chondrion*, small grain) A tubular or sausage-shaped organelle 1 to 3 micrometers long. Bounded by two membranes, mitochondria closely resemble the aerobic bacteria from which they were originally derived. As chemical furnaces of the cell, they carry out its oxidative metabolism.

mitosis (Gr. *mitos*, thread) The M phase of cell division in which the microtubular apparatus is assembled, binds to the chromosomes, and moves them apart. This phase is the essential step in the separation of the two daughter cell genomes.

mole (L. *moles*, mass) The atomic weight of a substance, expressed in grams. One mole is defined as the mass of 6.0222×10^{23} atoms.

molecule (L. *moliculus*, a small mass) The smallest unit of a compound that displays the properties of that compound.

monocot Short for monocotyledon; flowering plant in which the embryos have only one cotyledon, the flower parts are often in threes, and the leaves typically are parallel-veined.

monomers (Gr. *mono*, single + *meris*, part) Simple molecules that can join together to form polymers.

monosaccharide (Gr. *monos*, one + *sakcharon*, sugar) A simple sugar.

morphogenesis (Gr. *morphe*, form + *genesis*, origin) The formation of shape. The growth and differentiation of cells and tissues during development.

motor endplate The point where a neuron attaches to a muscle. A neuromuscular synapse.

multicellularity A condition in which the activities of the individual cells are coordinated and the cells themselves are in contact. A property of eukaryotes alone and one of their major characteristics.

muscle (L. *musculus*, mouse) The tissue in the body of humans and animals that can be contracted and relaxed to make the body move.

muscle cell A long, cylindrical, multinucleated cell that contains numerous myofibrils and is capable of contraction when stimulated.

muscle spindle A sensory organ that is attached to a muscle and sensitive to stretching.

mutagen (L. *mutare*, to change) A chemical capable of damaging DNA.

mutation (L. *mutare*, to change) A change in a cell's genetic message.

mutualism (L. *mutuus*, lent, borrowed) A symbiotic relationship in which both participating species benefit.

mycelium, *pl.* mycelia (Gr. *mykes*, fungus) In fungi, a mass of hyphae.

mycology (Gr. *mykes*, fungus) The study of fungi. A person who studies fungi is called a mycologist.

mycorrhiza, *pl.* mycorrhizae (Gr. *mykes*, fungus + *rhiza*, root) A symbiotic association between fungi and plant roots.

myofibril (Gr. *myos*, muscle + L. *fibrilla*, little fiber) An elongated structure in a muscle fiber, composed of myosin and actin.

myosin (Gr. *myos*, muscle + *in*, belonging to) One of two protein components of myofilaments. (The other is actin.)

N

natural selection The differential reproduction of genotypes caused by factors in the environment. Leads to evolutionary change.

nematocyst (Gr. *nema*, thread + *kystos*, bladder) A coiled, threadlike stinging structure of cnidarians that is discharged to capture prey and for defense.

nephron (Gr. *nephros*, kidney) The functional unit of the vertebrate kidney. A human kidney has more than 1 million nephrons that filter waste matter from the blood. Each nephron consists of a Bowman's capsule, glomerulus, and tubule.

nerve A bundle of axons with accompanying supportive cells, held together by connective tissue.

nerve impulse A rapid, transient, self-propagating reversal in electrical potential that travels along the membrane of a neuron.

neuromodulator A chemical transmitter that mediates effects that are slow and longer lasting and that typically involve second messengers within the cell.

neuromuscular junction The structure formed when the tips of axons contact (innervate) a muscle fiber.

neuron (Gr. nerve) A nerve cell specialized for signal transmission.

neurotransmitter (Gr. *neuron*, nerve + L. *trans*, across + *mitere*, to send) A chemical released at an axon tip that travels across the synapse and binds a specific receptor protein in the membrane on the far side.

neurulation (Gr. *neuron*, nerve) The elaboration of a notochord and a dorsal nerve cord that marks the evolution of the chordates.

neutron (L. *neuter*, neither) A subatomic particle located within the nucleus of an atom. Similar to a proton in mass, but as its name implies, a neutron is neutral and possesses no charge.

neutrophil An abundant type of white blood cell capable of engulfing microorganisms and other foreign particles.

niche (L. *nidus*, nest) The role an organism plays in the environment; realized niche is the niche that an organism occupies under natural circumstances; fundamental niche is the niche an organism would occupy if competitors were not present.

nitrogen fixation The incorporation of atmospheric nitrogen into nitrogen compounds, a process that can be carried out only by certain microorganisms.

nocturnal (L. *nocturnus*, night) Active primarily at night.

node (L. *nodus*, knot) The place on the stem where a leaf is formed.

node of Ranvier After L. A. Ranvier, French histologist. A gap formed at the point where two Schwann cells meet and where the axon is in direct contact with the surrounding intercellular fluid.

nondisjunction The failure of homologous chromosomes to separate in meiosis I. The cause of Down syndrome.

nonrandom mating A phenomenon in which individuals with certain genotypes sometimes mate with one another more commonly than would be expected on a random basis.

notochord (Gr. *noto*, back + L. *chorda*, cord) In chordates, a dorsal rod of cartilage that forms between the nerve cord and the developing gut in the early embryo.

nucleic acid A nucleotide polymer. A long chain of nucleotides. Chief types are deoxyribonucleic acid (DNA), which is double-stranded, and ribonucleic acid (RNA), which is typically single-stranded.

nucleosome (L. *nucleus*, kernel + *soma*, body) The basic packaging unit of eukaryotic chromosomes, in which the DNA molecule is wound around a ball of histone proteins. Chromatin is composed of long strings of nucleosomes, like beads on a string.

nucleotide A single unit of nucleic acid, composed of a phosphate, a five-carbon sugar (either ribose or deoxyribose), and a purine or a pyrimidine.

nucleolus A region inside the nucleus where rRNA and ribosomes are produced.

nucleus (L. *a kernel*, dim. Fr. *nux*, nut) A spherical organelle (structure) characteristic of eukaryotic cells. The repository of the genetic information that directs all activities of a living cell. In atoms, the central core, containing positively charged protons and (in all but hydrogen) electrically neutral neutrons.

O

oocyte (Gr. *oion*, egg + *kytos*, vessel) A cell in the outer layer of the ovary that gives rise to an ovum. A primary oocyte is any of the 2 million oocytes a female is born with, all of which have begun the first meiotic division.

operon (L. *operis*, work) A cluster of functionally related genes transcribed onto a single mRNA molecule. A common mode of gene regulation in prokaryotes; it is rare in eukaryotes other than fungi.

order A taxonomic category ranking below a class and above a family.

organ (L. *organon*, tool) A complex body structure composed of several different kinds of tissue grouped together in a structural and functional unit.

organelle (Gr. *organella*, little tool) A specialized compartment of a cell. Mitochondria are organelles.

organism Any individual living creature, either unicellular or multicellular.

organ system A group of organs that function together to carry out the principal activities of the body.

osmoconformer An animal that maintains the osmotic concentration of its body fluids at about the same level as that of the medium in which it is living.

osmoregulation The maintenance of a constant internal solute concentration by an organism, regardless of the environment in which it lives.

osmosis (Gr. *osmos*, act of pushing, thrust) The diffusion of water across a membrane that permits the free passage of water but not that of one or more solutes. Water moves from an area of low solute concentration to an area with higher solute concentration.

osmotic pressure The increase of hydrostatic water pressure within a cell as a result of water molecules that continue to diffuse inward toward the area of lower water concentration (the water concentration is lower inside than outside the cell because of the dissolved solutes in the cell).

osteoblast (Gr. *osteon*, bone + *blastos*, bud) A bone-forming cell.

osteocyte (Gr. *osteon*, bone + *kytos*, hollow vessel) A mature osteoblast.

outcross A term used to describe species that interbreed with individuals other than those like themselves.

oviparous (L. *ovum*, egg + *parere*, to bring forth) Refers to reproduction in which the eggs are developed after leaving the body of the mother, as in reptiles.

ovulation The successful development and release of an egg by the ovary.

ovule (L. *ovulum*, a little egg) A structure in a seed plant that becomes a seed when mature.

ovum, *pl.* ova (L. egg) A mature egg cell. A female gamete.

oxidation (Fr. *oxider*, to oxidize) The loss of an electron during a chemical reaction from one atom to another. Occurs simultaneously with reduction. Is the second stage of the 10 reactions of glycolysis.

oxidative metabolism A collective term for metabolic reactions requiring oxygen.

oxidative respiration Respiration in which the final electron acceptor is molecular oxygen.

P

parasitism (Gr. *para*, beside + *sitos*, food) A symbiotic relationship in which one organism benefits and the other is harmed.

parthenogenesis (Gr. *parthenos*, virgin + Eng. *genesis*, beginning) The development of an adult from an unfertilized egg. A common form of reproduction in insects.

partial pressures (P) The components of each individual gas—such as nitrogen, oxygen, and carbon dioxide—that together constitute the total air pressure.

pathogen (Gr. *pathos*, suffering + Eng. *genesis*, beginning) A disease-causing organism.

pedigree (L. *pes*, foot + *grus*, crane) A family tree. The patterns of inheritance observed in family histories. Used to determine the mode of inheritance of a particular trait.

peptide (Gr. *peptein*, to soften, digest) Two or more amino acids linked by peptide bonds.

peptide bond A covalent bond linking two amino acids. Formed when the positive (amino, or NH_2) group at one end and a negative (carboxyl, or COOH) group at the other end undergo a chemical reaction and lose a molecule of water.

peristalsis (Gr. *peri*, around + *stellein*, to wrap) The rhythmic sequences of waves of muscular contraction in the walls of a tube.

pH Refers to the concentration of H^+ ions in a solution. The numerical value of the pH is the negative of the exponent of the molar concentration. Low pH values indicate high concentrations of H^+ ions (acids), and high pH values indicate low concentrations (bases).

phagocyte (Gr. *phagein*, to eat + *kytos*, hollow vessel) A cell that kills invading cells by engulfing them. Includes neutrophils and macrophages.

phagocytosis (Gr. *phagein*, to eat + *kytos*, hollow vessel) A form of endocytosis in which cells engulf organisms or fragments of organisms.

phenotype (Gr. *phainein*, to show + *typos*, stamp or print) The realized expression of the genotype. The observable expression of a trait (affecting an individual's structure, physiology, or behavior) that results from the biological activity of proteins or RNA molecules transcribed from the DNA.

pheromone (Gr. *pherein*, to carry + [hor]mone) A chemical signal emitted by certain animals as a means of communication.

phloem (Gr. *phloos*, bark) In vascular plants, a food-conducting tissue basically composed of sieve elements, various kinds of parenchyma cells, fibers, and sclereids.

phosphodiester bond The bond that results from the formation of a nucleic acid chain in which individual sugars are linked together in a line by the phosphate groups. The phosphate group of one sugar binds to the hydroxyl group of another, forming an—O—P—O bond.

photon (Gr. *photos*, light) The unit of light energy.

photoperiodism (Gr. *photos*, light + *periodos*, a period) A mechanism that organisms use to measure seasonal changes in relative day and night length.

photorespiration A process in which carbon dioxide is released without the production of ATP or NADPH. Because it produces neither ATP nor NADPH, photorespiration acts to undo the work of photosynthesis.

photosynthesis (Gr. *photos*, light + *-syn*, together + *tithenai*, to place) The process by which plants, algae, and some bacteria use the energy of sunlight to create from carbon dioxide (CO_2) and water (H_2O) the more complicated molecules that make up living organisms.

phototropism (Gr. *photos*, light + *trope*, turning to light) A plant's growth response to a unidirectional light source.

phylogeny (Gr. *phylon*, race, tribe) The evolutionary relationships among any group of organisms.

phylum, *pl.* phyla (Gr. *phylon*, race, tribe) A major taxonomic category, ranking above a class.

physiology (Gr. *physis*, nature + *logos*, a discourse) The study of the function of cells, tissues, and organs.

pigment (L. *pigmentum*, paint) A molecule that absorbs light.

pili (pilus) Short flagella that occur on the cell surface of some prokaryotes.

pinocytosis (Gr. *pinein*, to drink + *kytos*, cell) A form of endocytosis in which the material brought into the cell is a liquid containing dissolved molecules.

pistil (L. *pistillum*, pestle) Central organ of flowers, typically consisting of ovary, style, and stigma; a pistil may consist of one or more fused carpels and is more technically and better known as the gynoecium.

plankton (Gr. *planktos*, wandering) The small organisms that float or drift in water, especially at or near the surface.

plasma (Gr. form) The fluid of vertebrate blood. Contains dissolved salts, metabolic wastes, hormones, and a variety of proteins, including antibodies and albumin. Blood minus the blood cells.

plasma membrane A lipid bilayer with embedded proteins that control the cell's permeability to water and dissolved substances.

plasmid (Gr. *plasma*, a form or something molded) A small fragment of DNA that replicates independently of the bacterial chromosome.

platelet (Gr. dim of *plattus*, flat) In mammals, a fragment of a white blood cell that circulates in the blood and functions in the formation of blood clots at sites of injury.

pleiotropy (Gr. *pleros*, more + *trope*, a turning) A gene that produces more than one phenotypic effect.

polarization The charge difference of a neuron so that the interior of the cell is negative with respect to the exterior.

polar molecule A molecule with positively and negatively charged ends. One portion of a polar molecule attracts electrons more strongly than another portion, with the result that the molecule has electron-rich (−) and electron-poor (+) regions, giving it magnetlike positive and negative poles. Water is one of the most polar molecules known.

pollen (L. fine dust) A fine, yellowish powder consisting of grains or microspores, each of which contains a mature or immature male gametophyte. In flowering plants, pollen is released from the anthers of flowers and fertilizes the pistils.

pollen tube A tube that grows from a pollen grain. Male reproductive cells move through the pollen tube into the ovule.

pollination The transfer of pollen from the anthers to the stigmas of flowers for fertilization, as by insects or the wind.

polygyny (Gr. *poly*, many + *gyne*, woman, wife) A mating choice in which a male mates with more than one female.

polymer (Gr. *polus*, many + *meris*, part) A large molecule formed of long chains of similar molecules called subunits.

polymerase chain reaction (PCR) A process by which DNA polymerase is used to copy a sequence of DNA repeatedly, making millions of copies of the same DNA.

polymorphism (Gr. *polys*, many + *morphe*, form) The presence in a population of more than one allele of a gene at a frequency greater than that of newly arising mutations.

polynomial system (Gr. *polys*, many + [bi]nomial) Before Linnaeus, naming a genus by use of a cumbersome string of Latin words and phrases.

polyp A cylindrical, pipe-shaped cnidarian usually attached to a rock with the mouth facing away from the rock on which it is growing. Coral is made up of polyps.

polypeptide (Gr. *polys*, many + *peptein*, to digest) A general term for a long chain of amino acids linked end to end by peptide bonds. A protein is a long, complex polypeptide.

polysaccharide (Gr. *polys*, many + *sakcharon*, sugar) A sugar polymer. A carbohydrate composed of many monosaccharide sugar subunits linked together in a long chain.

population (L. *populus*, the people) Any group of individuals of a single species, occupying a given area at the same time.

posterior (L. *post*, after) Situated behind or farther back.

potential difference A difference in electrical charge on two sides of a membrane caused by an unequal distribution of ions.

potential energy Energy with the potential to do work. Stored energy.

predation (L. *praeda*, prey) The eating of other organisms. The one doing the eating is called a predator, and the one being consumed is called the prey.

primary growth In vascular plants, growth originating in the apical meristems of shoots and roots, as contrasted with secondary growth; results in an increase in length.

primary plant body The part of a plant that arises from the apical meristems.

primary producers Photosynthetic organisms, including plants, algae, and photosynthetic bacteria.

primary structure of a protein The sequence of amino acids that makes up a particular polypeptide chain.

primordium, *pl.* primordia (L. *primus*, first + *ordiri*, begin) The first cells in the earliest stages of the development of an organ or structure.

productivity The total amount of energy of an ecosystem fixed by photosynthesis per unit of time. Net productivity is productivity minus that which is expended by the metabolic activity of the organisms in the community.

prokaryote (Gr. *pro*, before + *karyon*, kernel) A simple organism that is small, single-celled, and has little evidence of internal structure.

promoter An RNA polymerase binding site. The nucleotide sequence at the end of a gene to which RNA polymerase attaches to initiate transcription of mRNA.

prophase (Gr. *pro*, before + *phasis*, form) The first stage of mitosis during which the chromosomes become more condensed, the nuclear envelope is reabsorbed, and a network of microtubules (called the spindle) forms between opposite poles of the cell.

protein (Gr. *proteios*, primary) A long chain of amino acids linked end to end by peptide bonds. Because the 20 amino acids that occur in proteins have side groups with very different chemical properties, the function and shape of a protein is critically affected by its particular sequence of amino acids.

protist (Gr. *protos*, first) A member of the kingdom Protista, which includes unicellular eukaryotic organisms and some multicellular lines derived from them.

proton A subatomic particle in the nucleus of an atom that carries a positive charge. The number of protons determines the chemical character of the atom because it dictates the number of electrons orbiting the nucleus and available for chemical activity.

protostome (Gr. *protos*, first + *stoma*, mouth) An animal in whose embryonic development the mouth forms at or near the blastopore. Also characterized by spiral cleavage.

protozoa (Gr. *protos*, first + *zoion*, animal) The traditional name given to heterotrophic protists.

pseudocoel (Gr. *pseudos*, false + *koiloma*, cavity) A body cavity similar to the coelom except that it forms between the mesoderm and endoderm.

punctuated equilibrium A hypothesis of the mechanism of evolutionary change that proposes that long periods of little or no change are punctuated by periods of rapid evolution.

Q

quaternary structure of a protein A term to describe the way multiple protein subunits are assembled into a whole.

R

radial symmetry (L. *radius*, a spoke of a wheel + Gr. *summetros*, symmetry) The regular arrangement of parts around a central axis so that any plane passing through the central axis divides the organism into halves that are approximate mirror images.

radioactivity The emission of nuclear particles and rays by unstable atoms as they decay into more stable forms. Measured in curies, with 1 curie equal to 37 billion disintegrations a second.

radula (L. scraper) A rasping, tonguelike organ characteristic of most mollusks.

recessive allele An allele whose phenotype effects are masked in heterozygotes by the presence of a dominant allele.

recombination The formation of new gene combinations. In bacteria, it is accomplished by the transfer of genes into cells, often in association with viruses. In eukaryotes, it is accomplished by reassortment of chromosomes during meiosis and by crossing over.

reducing power The use of light energy to extract hydrogen atoms from water.

reduction (L. *reductio*, a bringing back; originally, "bringing back" a metal from its oxide) The gain of an electron during a chemical reaction from one atom to another. Occurs simultaneously with oxidation.

reflex (L. *reflectere*, to bend back) An automatic consequence of a nerve stimulation. The motion that results from a nerve impulse passing through the system of neurons, eventually reaching the body muscles and causing them to contract.

refractory period The recovery period after membrane depolarization during which the membrane is unable to respond to additional stimulation.

renal (L. *renes*, kidneys) Pertaining to the kidney.

repression (L. *reprimere*, to press back, keep back) The process of blocking transcription by the placement of the regulatory protein between the polymerase and the gene, thus blocking movement of the polymerase to the gene.

repressor (L. *reprimere*, to press back, keep back) A protein that regulates transcription of mRNA from DNA by binding to the operator and so preventing RNA polymerase from attaching to the promoter.

resolving power The ability of a microscope to distinguish two points as separate.

respiration (L. *respirare*, to breathe) The utilization of oxygen. In terrestrial vertebrates, the inhalation of oxygen and the exhalation of carbon dioxide.

resting membrane potential The charge difference that exists across a neuron's membrane at rest (about 70 millivolts).

restriction endonuclease A special kind of enzyme that can recognize and cleave DNA molecules into fragments. One of the basic tools of genetic engineering.

restriction fragment-length polymorphism (RFLP) An associated genetic mutation marker detected because the mutation alters the length of DNA segments.

retrovirus (L. *retro*, turning back) A virus whose genetic material is RNA rather than DNA. When a retrovirus infects a cell, it makes a DNA copy of itself, which it can then insert into the cellular DNA as if it were a cellular gene.

ribose A five-carbon sugar.

ribosome A cell structure composed of protein and RNA that translates RNA copies of genes into protein.

RNA polymerase The enzyme that transcribes RNA from DNA.

S

saltatory conduction A very fast form of nerve impulse conduction in which the impulses leap from node to node over insulated portions.

sarcoma (Gr. *sarx*, flesh) A cancerous tumor that involves connective or hard tissue, such as muscle.

sarcomere (Gr. *sarx*, flesh + *meris*, part of) The fundamental unit of contraction in skeletal muscle. The repeating bands of actin and myosin that appear between two Z lines.

sarcoplasmic reticulum (Gr. *sarx*, flesh + *plassein*, to form, mold; L. *reticulum*, network) The endoplasmic reticulum of a muscle cell. A sleeve of membrane that wraps around each myofilament.

scientific creationism A view that the biblical account of the origin of the earth is literally true, that the earth is much younger than most scientists believe, and that all species of organisms were individually created just as they are today.

secondary growth In vascular plants, growth that results from the division of a cylinder of cells around the plant's periphery. Secondary growth causes a plant to grow in diameter.

secondary structure of a protein The folding and bending of a polypeptide chain, which is held in place by hydrogen bonds.

second messenger An intermediary compound that couples extracellular signals to intracellular processes and also amplifies a hormonal signal.

seed A structure that develops from the mature ovule of a seed plant. Contains an embryo and a food source surrounded by a protective coat.

selection The process by which some organisms leave more offspring than competing ones and their genetic traits tend to appear in greater proportions among members of succeeding generations than the traits of those individuals that leave fewer offspring.

self-fertilization The transfer of pollen from an anther to a stigma in the same flower or to another flower of the same plant.

sepal (L. *sepalum*, a covering) A member of the outermost whorl of a flowering plant. Collectively, the sepals constitute the calyx.

septum, *pl.* septa (L. *saeptum*, a fence) A partition or cross-wall, such as those that divide fungal hyphae into cells.

sex chromosomes In humans, the X and Y chromosomes, which are different in the two sexes and are involved in sex determination.

sex-linked characteristic A genetic characteristic that is determined by genes located on the sex chromosomes.

sexual reproduction Reproduction that involves the regular alternation between syngamy and meiosis. Its outstanding characteristic is that an individual offspring inherits genes from two parent individuals.

shoot In vascular plants, the aboveground parts, such as the stem and leaves.

sieve cell In the phloem (food-conducting tissue) of vascular plants, a long, slender sieve element with relatively unspecialized sieve areas and with tapering end walls that lack sieve plates. Found in all vascular plants except angiosperms, which have sieve-tube members.

soluble Refers to polar molecules that dissolve in water and are surrounded by a hydration shell.

solute The molecules dissolved in a solution. *See also* solution, solvent.

solution A mixture of molecules, such as sugars, amino acids, and ions, dissolved in water.

solvent The most common of the molecules in a solution. Usually a liquid, commonly water.

somatic cells (Gr. *soma*, body) All the diploid body cells of an animal that are not involved in gamete formation.

somite A segmented block of tissue on either side of a developing notochord.

species, *pl.* **species (L. kind, sort)** A level of taxonomic hierarchy; a species ranks next below a genus.

sperm (Gr. *sperma*, sperm, seed) A sperm cell. The male gamete.

spindle The mitotic assembly that carries out the separation of chromosomes during cell division. Composed of microtubules and assembled during prophase at the centrioles of the dividing cell.

spore (Gr. *spora*, seed) A haploid reproductive cell, usually unicellular, that is capable of developing into an adult without fusion with another cell. Spores result from meiosis, as do gametes, but gametes fuse immediately to produce a new diploid cell.

sporophyte (Gr. *spora*, seed + *phyton*, plant) The spore-producing, diploid (2*n*) phase in the life cycle of a plant having alternation of generations.

stabilizing selection A form of selection in which selection acts to eliminate both extremes from a range of phenotypes.

stamen (L. thread) The part of the flower that contains the pollen. Consists of a slender filament that supports the anther. A flower that produces only pollen is called staminate and is functionally male.

steroid (Gr. *stereos*, solid + L. *ol*, from oleum, oil) A kind of lipid. Many of the molecules that function as messengers and pass across cell membranes are steroids, such as the male and female sex hormones and cholesterol.

steroid hormone A hormone derived from cholesterol. Those that promote the development of the secondary sexual characteristics are steroids.

stigma (Gr. mark) A specialized area of the carpel of a flowering plant that receives the pollen.

stoma, *pl.* **stomata (Gr. mouth)** A specialized opening in the leaves of some plants that allows carbon dioxide to pass into the plant body and allows water vapor and oxygen to pass out of them.

stratum corneum The outer layer of the epidermis of the skin of the vertebrate body.

substrate (L. *substratus*, strewn under) A molecule on which an enzyme acts.

substrate-level phosphorylation The generation of ATP by coupling its synthesis to a strongly exergonic (energy-yielding) reaction.

succession In ecology, the slow, orderly progression of changes in community composition that takes place through time. Primary succession occurs in nature on bare substrates, over long periods of time. Secondary succession occurs when a climax community has been disturbed.

sugar Any monosaccharide or disaccharide.

surface tension A tautness of the surface of a liquid, caused by the cohesion of the liquid molecules. Water has an extremely high surface tension.

surface-to-volume ratio Describes cell size increases. Cell volume grows much more rapidly than surface area.

symbiosis (Gr. *syn*, together with + *bios*, life) The condition in which two or more dissimilar organisms live together in close association; includes parasitism, commensalism, and mutualism.

synapse (Gr. *synapsis*, a union) A junction between a neuron and another neuron or muscle cell. The two cells do not touch. Instead, neurotransmitters cross the narrow space between them.

synapsis (Gr. *synapsis*, contact, union) The close pairing of homologous chromosomes that occurs early in prophase I of meiosis. With the genes of the chromosomes thus aligned, a DNA strand of one homologue can pair with the complementary DNA strand of the other.

syngamy (Gr. *syn*, together with + *gamos*, marriage) Fertilization. The union of male and female gametes.

T

taxonomy (Gr. *taxis*, arrangement + *nomos*, law) The science of the classification of organisms.

T cell A type of lymphocyte involved in cell-mediated immune responses and interactions with B cells. Also called a T lymphocyte.

tendon (Gr. *tenon*, stretch) A strap of connective tissue that attaches muscle to bone.

tertiary structure of a protein The three-dimensional shape of a protein. Primarily the result of hydrophobic interactions of amino acid side groups and, to a lesser extent, of hydrogen bonds between them. Forms spontaneously.

test cross A cross between a heterozygote and a recessive homozygote. A procedure Mendel used to further test his hypotheses.

theory (Gr. *theorein*, to look at) A well-tested hypothesis supported by a great deal of evidence.

thigmotropism (Gr. *thigma*, touch + *trope*, a turning) The growth response of a plant to touch.

thorax (Gr. a breastplate) The part of the body between the head and the abdomen.

thylakoid (Gr. *thylakos*, sac + *-oides*, like) A flattened, saclike membrane in the chloroplast of a eukaryote. Thylakoids are stacked on top of one another in arrangements called grana and are the sites of photosystem reactions.

tissue (L. *texere*, to weave) A group of similar cells organized into a structural and functional unit.

trachea, *pl.* **tracheae (L. windpipe)** In vertebrates, the windpipe.

tracheid (Gr. *tracheia*, rough) An elongated cell with thick, perforated walls that carries water and dissolved minerals through a plant and provides support. Tracheids form an essential element of the xylem of vascular plants.

transcription (L. *trans*, across + *scribere*, to write) The first stage of gene expression in which the RNA polymerase enzyme synthesizes an mRNA molecule whose sequence is complementary to the DNA.

translation (L. *trans*, across + *latus*, that which is carried) The second stage of gene expression in which a ribosome assembles a polypeptide, using the mRNA to specify the amino acids.

translocation (L. *trans*, across + *locare*, to put or place) In plants, the process in which most of the carbohydrates manufactured in the leaves and other green parts of the plant are moved through the phloem to other parts of the plant.

transpiration (L. *trans*, across + *spirare*, to breathe) The loss of water vapor by plant parts, primarily through the stomata.

transposon (L. *transponere*, to change the position of) A DNA sequence carrying one or more genes and flanked by insertion sequences that confer the ability to move from one DNA molecule to another. An element capable of transposition (the changing of chromosomal location).

trophic level (Gr. *trophos*, feeder) A step in the movement of energy through an ecosystem.

tropism (Gr. *trop*, turning) A plant's response to external stimuli. A positive tropism is one in which the movement or reaction is in the direction of the source of the stimulus. A negative tropism is one in which the movement or growth is in the opposite direction.

turgor pressure (L. *turgor*, a swelling) The pressure within a cell that results from the movement of water into the cell. A cell with high turgor pressure is said to be turgid.

U

unicellular Composed of a single cell.

urea (Gr. *ouron*, urine) An organic molecule formed in the vertebrate liver. The principal form of disposal of nitrogenous wastes by mammals.

urine (Gr. *ouron*, urine) The liquid waste filtered from the blood by the kidneys.

V

vaccination The injection of a harmless microbe into a person or animal to confer resistance to a dangerous microbe.

vacuole (L. *vacuus*, empty) A cavity in the cytoplasm of a cell that is bound by a single membrane and contains water and waste products of cell metabolism. Typically found in plant cells.

van der Waals forces Weak chemical attractions between atoms that can occur when atoms are very close to each other.

variable Any factor that influences a process. In evaluating alternative hypotheses about one variable, all other variables are held constant so that the investigator is not misled or confused by other influences.

vascular bundle In vascular plants, a strand of tissue containing primary xylem and primary phloem. These bundles of elongated cells conduct water with dissolved minerals and carbohydrates throughout the plant body.

vascular cambium In vascular plants, the meristematic layer of cells that gives rise to secondary phloem and secondary xylem. The activity of the vascular cambium increases stem or root diameter.

ventral (L. *venter*, belly) Refers to the bottom portion of an animal. Opposite of dorsal.

vertebrate An animal having a backbone made of bony segments called vertebrae.

vesicle (L. *vesicula*, a little (ladder) Membrane-enclosed sacs within eukaryotic cells.

vessel element In vascular plants, a typically elongated cell, dead at maturity, that conducts water and solutes in the xylem.

villus, *pl.* **villi (L. a tuft of hair)** In vertebrates, fine, microscopic, fingerlike projections on epithelial cells lining the small intestine that serve to increase the absorptive surface area of the intestine.

vitamin (L. *vita*, life + *amine*, of chemical origin) An organic substance that the organism cannot synthesize, but is required in minute quantities by an organism for growth and activity.

viviparous (L. *vivus,* **alive +** *parere,* **to bring forth)** Refers to reproduction in which eggs develop within the mother's body and young are born free-living.

voltage-gated channel A transmembrane pathway for an ion that is opened or closed by a change in the voltage, or charge difference, across the cell membrane.

W

water vascular system The system of water-filled canals connecting the tube feet of echinoderms.

whorl A circle of leaves or of flower parts present at a single level along an axis.

wood Accumulated secondary xylem. Heartwood is the central, nonliving wood in the trunk of a tree. Hardwood is the wood of dicots, regardless of how hard or soft it actually is. Softwood is the wood of conifers.

X

xylem (Gr. *xylon,* **wood)** In vascular plants, a specialized tissue, composed primarily of elongate, thick-walled conducting cells, that transports water and solutes through the plant body.

Y

yolk (O.E. *geolu,* **yellow)** The stored substance in egg cells that provides the embryo's primary food supply.

Z

zygote (Gr. *zygotos,* **paired together)** The diploid (2*n*) cell resulting from the fusion of male and female gametes (fertilization).

Credits

Photographs

Chapter 0

Opener: © PIER/Getty; 0.1: © BananaStock/
JupiterImages RF; 0.3: © Corbis RF; p. 5: © Randy
Faris/Corbis RF; 0.8: The McGraw-Hill Companies,
Inc./Lars A. Niki, photographer. All other photos
credited in later chapters.

Chapter 1

Opener: © Frans Lanting; 1.1(archaea): © R. Robinson/
Visuals Unlimited; 1.1(bacteria): © Alfred Pasieka/
Science Photo Library/Photo Researchers; 1.1(protista)
(plantae): © Corbis RF; 1.1(fungi) (anamalia): © Getty
RF; 1.2: © Tom E. Adams/Visuals Unlimited; 1.3:
© Corbis RF; 1.4(Canada geese on lake): © Raymond
Gehman/Corbis; 1.4(Sandhill cranes on lake):
© Corbis RF; 1.4(Canada goose–left) (Sandhill
crane–right): © Jim Bailey; 1.4(cranes and geese
on lake): © Winfried Wisniewski/zefa/Corbis; p. 18
(rock pigeon): © Joe McDonald/Animals Animals;
p. 18(Red fantail pigeon): © Kenneth Fink/Photo
Researchers; p. 18(Fairy swallow pigeon): © Tom
McHugh/Photo Researchers; p. 18(Bald eagle)
(ants) (hippopotamus): © Corbis RF; p. 19(moth):
© George Grall/National Geographic/Getty; 1.6:
Ozone Processing Team, Goddard Space Flight
Center, NASA; 1.7a: © Stocktrek/age fotostock RF;
1.7b: © Tom Pepeira, photographer/Iconotec RF; 1.7c:
© Digital Vision/Getty RF; 1.8: © Dennis Kunkel/
Phototake; 1.11: © Leonard Lessin/Peter Arnold.

Chapter 2

Opener: © Digital Archive Japan/Alamy RF; 2.1:
© Thinkstock Images/Jupiter Images RF; 2.7:
Courtesy Todd M. Blodgett, MD, University of
Pittsburgh Medical Center; 2.8: The McGraw-Hill
Companies; p. 37: Courtesy Who Zoo (whozoo.
org); 2.12, 2.13a: © Corbis RF; 2.13b: © Hermann
Eisenbeiss/National Audubon Society Collection/
Photo Researchers; 2.15(top): © The McGraw-Hill
Companies/Bob Coyle, photographer; 2.15(middle):
© The McGraw-Hill Companies/Jacques Cornell
photographer; 2.15(bottom): © The McGraw-Hill
Companies/Jill Braaten photographer; p. 41: © Gilbert
S. Grant/National Audubon Society Collection/Photo
Researchers; I&A 2.1(both): © Photo Archives South
Tyrol Museum of Archaeology – www.iceman.it.

Chapter 3

Opener: © Michael Viard/Peter Arnold; 3.1(both):
© The McGraw-Hill Companies; 3.5a: © Corbis;
3.5b-c: © Getty RF; 3.5d: © Suza Scalora/Photodisc/
Getty RF; 3.5e: © Phototake/Alamy; 3.5f: © Dennis
Kunkel Microscopy, Inc.; 3.7: McGraw-Hill
Companies, Inc./Gary He, photographer; 3.10a:

© Dr. Gopal Murt/Visuals Unlimited; 3.15:
© J.D. Litvay/Visuals Unlimited; Table 3.1(cow)
(potatoes) (leaves): © Corbis RF; Table 3.1(sugarcane):
© H. Wiesenhofer/PhotoLink/Getty RF; Table 3.1(arm):
© Getty RF; Table 3.1(lobster): © Scott Johnson/
Animals Animals; 3.17a: © Getty RF; 3.17b:
© C Squared Studios/Getty RF; p. 57:
© Otto Greule Jr/Getty; I&A 3.2: © Dr. David
M. Phillips/Visuals Unlimited.

Chapter 4

Opener: © Manfred Kage/Peter Arnold; 4.2: © Dr. Tony
Brain & David Parker/Photo Researchers; p. 64(light
microscope): © David M. Phillips/Visuals Unlimited;
p. 64(TEM): © Microworks/Phototake; p. 64(SEM):
© Stanley Flegler/Visuals Unlimited; 4.4a: © Andrew
Syred/Photo Researchers; 4.4b: © Alfred Pasieka/
Photo Researchers; 4.4c: © Microfield Scientific
Ltd/Photo Researchers; 4.14b: Courtesy Dr. Charles
Flickinger, Medical Cellular Biology, W.B. Saunders,
1979; 4.16: © Biophoto Associates/Photo Researchers;
4.17: © Don W. Fawcett/Visuals Unlimited; 4.18:
Courtesy Dr. Kenneth Miller, Brown University; 4.22:
© Stanley Flegler/Visuals Unlimited; 4.25: Courtesy
Dr. Birgit Satir, Albert Einstein College of Medicine;
Table 4.4: © SIU/Visuals Unlimited.

Chapter 5

Opener: © Andrew Darrington/Alamy; 5.1a:
© Nice One Productions/Corbis RF; 5.2:
© Robert A. Caputo/Aurora & Quanta Productions;
5.3(both): © Keith Eng, 2008 RF.

Chapter 6

Opener: © Skip Moody/Dembinsky Photo Associates;
Table 6.1: © Roger Brooks/Beateworks/Corbis RF;
p. 103: © Corbis RF; 6.5: © Andrew Syred/Photo
Researchers; I&A 6.2: © Philip Sze/Visuals
Unlimited.

Chapter 7

Opener: © John Gerlach/Animals Animals; 7.1:
© BananaStock/PunchStock RF; p. 119(tree):
© Corbis RF; p. 119(zebra) (lion): © Getty RF;
p. 119(hyena and vultures): © Anup Shah/Photodisc/
Getty RF; p. 119(butterflies): © Edward S. Ross; p. 125:
© Corbis RF; I&A 7.2: U.S. Fish and Wildlife Service.

Chapter 8

Opener: © Andrew S. Bajer; 8.1b: © Lee D. Simon/
Photo Researchers; 8.3: © SPL/Photo Researchers;
EBP 8B1-6: © Andrew S. Bajer; 8.5a: © David M.
Phillips/Visuals Unlimited; 8.6: © Cabisco/Visuals
Unlimited; 8.7: © Moredun Animal Health LTD/SPL/
Photo Researchers; I&A 8.2: Courtesy Priv. Doz.
Dr. Roland Zell.

Chapter 9

Opener: © Adrian T Sumner/SPL/Photo Researchers;
9.2: © David Cavagnaro/Peter Arnold; 9.4a:
© MedicalRF.com/Corbis RF; 9.4b: © Creatas/
PunchStock RF; 9.4c: © Wetzel and Company;
EBP 9a(all): © C.A. Hasenkampf/Biological
Photo Service; I&A 9.2: Reprinted with
permission from Science Vol. 311,
no. 5759 20 January 2006. Image:
Khodjakov. Copyright 2006 AAAS.

Chapter 10

Opener: © Richard Gross/Biological Photography; 10.1:
© Digital Vision/PunchStock RF; Table 10.1: Image
238141 American Museum of Natural history; 10.4:
Courtesy R. W. Van Norman; Table 10.2: Look-alike
Meter graphic © MyHeritage.com, used with permission;
Table 10.2: © Dave M. Benett/Getty; Table 10.2:
© Marcel Thomas/FilmMagic/Getty; Table 10.2:
© Dave M. Benett/Getty; 10.11: From Albert &
Blakeslee "Corn and Man," Journal of Heredity V. 5,
pg. 511, 1914, Oxford University Press; 10.14a-b:
© Fred Bruemmer; p. 168: © RubberBall Productions/
Getty RF; 10.15: © Corbis; 10.17(both): © CBS/
Phototake; 10.21: © Joseph Sohm ChromoSohm/
Corbis; 10.25: © Bettmann/Corbis; 10.26(both):
© Stanley Flegler/Visuals Unlimited; 10.31: © Yoav Levy/
Phototake; I&A 10.1: Journal of Heredity 1932, v.23,
pp. 345-352, author O.L. Mohr. Used with permission.

Chapter 11

Opener: © Michael Dunning/Getty; 11.4(Franklin):
© Science Source/Photo Researchers; 11.4a:
From J.D. Watson, The Double Helix Atheneum,
New York, 1968. Cold Spring Harbor Lab; 11.4b:
© A.C. Barrington Brown/Photo Researchers; 11.9:
© David Scharf; p. 194: © Brand X Pictures/
PunchStock RF; p. 195: © Corbis RF.

Chapter 12

Opener: N. Ban, P. Nissen, J. Hansen, P.B. Moore
& T.A. Steitz, "The Complete Atomic Structure of
the Large Ribosomal Subunit at 2.4 A Resolution,"
reprinted with permission from Science v. 289 #5481,
p917, © American Association for the Advancement
of Science; 12.8: Courtesy Dr. Oscar L. Miller;
12.13: Courtesy Dr. Victoria Foe.

Chapter 13

Opener: © OSF/Bromhall, Derek/Animals Animals;
13.3: Courtesy Loris McGavran, Denver Children's
Hospital; 13.4(doctor): Courtesy Dr. Ken Culver,
Photo by John Crawford, National Institutes of Health;
13.4(fish): Robert H. Devlin/Fisheries & Oceans
Canada; 13.4(cotton): Courtesy Monsanto Corporation;
13.7: Courtesy Lifecodes Corp., Stamford, CT;

13.8: R.L. Brinster, U. of Pennsylvania Sch. of Vet. Med.; p. 224: Courtesy Affymetrix; p. 225: © Chad Baker/Getty RF; 13.11: Monsanto Co.; 13.12: Courtesy Ingo Potrykus & Peter Beyer, photo by Peter Beyer; 13.13: © Getty; 13.14: © AP Photo/Paul Clements; 13.15(left): © AP Photo/Hwang Woo-suk; 13.15(right): © Seoul National University/Handout/Reuters/Corbis; 13.16: © University of Wisconsin-Madison News & Public Affairs; 13.17: © Alex Wong/Getty; 13.19: © SUW-Madison News & Public Affairs University of Wisconsin-Madison, Photo by Jeff Miller; 13.21: C. Garon & J. Rose/NIAID/NIH; I&A 13.2: © Jack Hollingsworth/Getty RF.

Chapter 14

Opener: © Peter Arnold, Inc./Alamy; 14.1: © Huntington Library/Superstock; 14.2: From Darwin, the Life of a Tormented Evolutionist, by Adrian Desmond; 14.7: © Rene Frederick/ Digital Vision/Getty RF; 14.8: © Mary Evans Picture Library/Photo Researchers; 14.10(both): Courtesy Dr. Arkhat Abzhanov, Harvard School of Dental Medicine, Photographer: Dr. Peter Grant of Princeton University; 14.13: © Andy Crawford/ Getty; 14.15: Courtesy Michael Richardson and Ronan O'Rahilly; 14.19: © Ted Daeschler/VIREO; 14.21a: © Dr. David Phillips/Visuals Unlimited; 14.23: Courtesy Dr. Victor A. McKusick, Johns Hopkins University; 14.26: Courtesy the University of Chicago Library/Dept. of Special Collections and Todd L. Savitt; 14.29(both): © Breck P. Kent/Animals Animals; p. 267(all): The genetic basis of adaptive melanism in pocket mice. Michael W. Nachman*, Hopi E. Hoekstra, and Susan L. D'Agostino. PNAS April 29, 2003 vol. 100 no. 9 5268-5273; 14.32b: Courtesy H. Rodd; p. 270: © Peter E. Smith/ The Natural Sciences Image Library; 14.33a-b: © Corbis RF; 14.33c: © Porterfield/Chickering/ Photo Researchers; 14.34(1): © John Shaw/Tom Stack & Associates; 14.34(2): © Rob & Ann Simpson/Visuals Unlimited; 14.34(3): © Suzanne L. Collins & Joseph T. Collins/ National Audubon society Collection/Photo Researchers; 14.34(4): © Phil A. Dotson/ National Audubon Society Collection/ Photo Researchers.

Chapter 15

Opener: © Dave Watts/Tom Stack & Associates; 15.1: © Time & Life Pictures/Getty; 15.3(corn): © Dwight Kuhn; 15.3(wheat): © Henry Ausloos/ Animals Animals; 15.3(brown bear): © Heather Angel; 15.3(koala): © John Cancalosi/Peter Arnold; 15.5(horse): © Gerard Lacz/Peter Arnold; 15.5(donkey): © Ralph Reinhold/Animals Animals; 15.5(mule): © Steve Taylor/Alamy RF; 15.8(cat) (cheetah) (lynx) (tiger) (lion) (leopard): © Corbis RF; 15.8(puma) (bobcat) (ocelot) (caracal) (snow leopard) (jaguar): © Getty RF; Table 15.1: © Jim Pickerell/ Alamy; 15.11b: © OSF/Animals Animals; I&A 15.2(all): U.S. Fish and Wildlife Service.

Chapter 16

Opener: © Dwight Kuhn; p. 296(top): NASA; p. 296(bottom): NASA/JPL/Cornell; p. 300(duck): © Royalty-Free/Index Stock RF; p. 300(chimp): © Getty RF; p. 300(bat): © Eric and David Hosking/ Corbis; p. 300(crow): © Tim Zurowski/Corbis; 16.5: © Dennis Kunkel Microscopy, Inc.; 16.6a: © Brian Parker/Tom Stack & Associates; 16.6b: © Manfred Kage/Peter Arnold; 16.7-8: © John D. Cunningham/

Visuals Unlimited; 16.10a-b: © Corbis RF; 16.12: © Bill Keogh/Visuals Unlimited; 16.13: © L. West/ Photo Researchers; 16.14a: © Image Source/Getty RF; 16.14b: © Gallo Images-Nigel Dennis/Digital Vision/ Getty RF; 16.15: © D.H. Marx/Visuals Unlimited; 16.16: © Corbis RF; I&A 16.2: © Sue Ford/Photo Researchers.

Chapter 17

Opener, 17.1: © Corbis RF; 17.2: © Terry Ashley/ Tom Stack & Associates; 17.4a: © Edward S. Ross; 17.4b: © Richard Gross/Biological Photography; 17.6: © Edward S. Ross; 17.7: Courtesy Hans Steur, The Netherlands; 17.8: © E.J. Cable/Tom Stack & Associates; 17.9a, d: © Edward S. Ross; 17.9b: © Corbis RF; 17.9c: © Kingsley R. Stern; 17.9e: © Rod Planck/Tom Stack & Associates; 17.11: © Kingsley R. Stern; 17.13 and inset: © R.J. Delorit, Agronomy Publications; 17.14a: © CBS/ Visuals Unlimited; 17.14b: © Walter H. Hodge/ Peter Arnold; 17.14c: © imagebroker/ Alamy RF; 17.14d: © Runk/Schoenberger/Grant Heilman Photography; 17.16: © Ed Pembleton Photography; I&A 17.2: © blickwinkel/Alamy.

Chapter 18

Opener: © James H. Robinson/Animals Animals; Table 18.1(bear) (brittlestar) (butterfly) (jellyfish): © Corbis RF; Table 18.1(cancer cells): © David M. Phillips/ Visuals Unlimited; Table 18.1(milipede): © Edward S. Ross; Table 18.1(turtles): © Cleveland P. Hickman; Table 18.1(frog egg): © Cabisco/Phototake; Table 18.1(Neuromuscular junctions): © Ed Reschke; p. 334(radial): © Frank & Joyce Burek/Getty RF; p. 334(bilateral): © Corbis RF; p. 334(segmentation): © IT Stock Free/Alamy RF; 18.4a: © ImageState/ Alamy RF; 18.4b: © Corbis RF; 18.5a: © Gwen Fidler/ Tom Stack & Associates; 18.5b: © Corbis RF; 18.8a: © T.E. Adams/Visuals Unlimited; 18.8b: © Stan Elems/ Visuals Unlimited; 18.10: © T.E. Adams/Visuals Unlimited; 18.11: © Larry Jensen/Visuals Unlimited; 18.12(top): © Comstock Images/PictureQuest RF; 18.12(bottom): © Fred Bavendam/Peter Arnold; 18.13a: © David M. Dennis; 18.13b: © Darlyne A. Murawski/ National Geographic/Getty; 18.14a: © Photostock – Dieter Heinemann/Alamy RF; 18.14b: © Comstock Images/PictureQuest RF; 18.14c: © Alex Kerstitch/ Visuals Unlimited; 18.14d: © IT Stock Free/Alamy RF; 18.16a: © Alex Kerstitch/Visuals Unlimited; 18.16b: © Andrew J. Martinez/Photo Researchers; 18.16c: © Daniel W. Gotshall; 18.17a: © Corbis RF; 18.17b: © Eric N. Olsen/The University of Texas MD Anderson Cancer Center; 18.18a: © Laurie O'Keefe/ Photo Researchers; 18.18b: © Alex Kerstitch/Visuals Unlimited; 18.19: © Andrew Ward/Life File/Photodisc/ Getty RF; 18.20: © Michael Long/Natural History Museum, London; 18.21: © Louie Psihoyos/Corbis; 18.22: © E.R. Degginger/Animals Animals; 18.24: © Stephen Frink Collection/Alamy; 18.27: © John Shaw/Tom Stack & Associates; 18.32a: © Erwin & Peggy Bauer/Tom Stack & Associates; 18.32b: © Charles Philip/Corbis; 18.32c: © Corbis RF; I&A 18.2: © Sinclair Stammers/Photo Researchers.

Chapter 19

Opener: © ABPL/Gavin Thomson/Animals Animals; 19.1a: © Vanessa Vick/Photo Researchers; 19.1b: © Edward S. Ross; 19.1c: © Ken Lucas/Visuals Unlimited; 19.1d: © Edward S. Ross; 19.2: © Ingram Publishing/age Fotostock RF; 19.3: © Tom Pepeira, photographer/Iconotec RF; 19.4a:

© Manfred Danegger/Peter Arnold; 19.9: Courtesy National Museum of Natural History, Smithsonian Institution; 19.10: © Corbis RF; 19.12a: © Tim Davis/Photo Researchers; 19.12b: © Digital Vision RF; 19.12c: © Corbis RF; 19.15a-d: Courtesy J.B. Losos; 19.17: © Merlin D. Tuttle, Bat Conservation International; 19.18: © Fred Bavendam/Minden Pictures; 19.19: © Jeremy Woodhouse/Photodisc Blue/Getty RF; 19.20: © Graphic Science/Alamy; 19.21: Courtesy Rolf O. Peterson; 19.23: © Tom J. Ulrich/Visuals Unlimited; 19.24: © Alex L. Fradkin/Stockbyte RF; 19.25: © Edward S. Ross; 19.26: © Royalty Free/ PictureQuest RF; 19.27: © Peter Arnold/Alamy; 19.28a: © Edward S. Ross; 19.28b: © Paul A. Opler; 19.29a-d: © Edward S. Ross; 19.30: © Peter Chew; 19.31: © Tom Bean; I&A 19.2: © Jim Bailey.

Chapter 20

Opener: © Bill Ross/Corbis; 20.3a: © Dave G. Houser/ Corbis; 20.3b-c: © Corbis RF; 20.3d-e: © Edward S. Ross; 20.8: © Gerry Ellis/Minden Pictures; 20.19a: © Digital Vision/PictureQuest RF; 20.19b: © Jeff Hunter/Photographer's Choice RF/Getty RF; 20.20a: © David Fleetham/Alamy; 20.20b: Courtesy J. Frederick Grassel, Woods Hole Oceanographic Institution; 20.20c: © Kenneth L. Smith; 20.21: © Dwight Kuhn; 20.22b-c: © Corbis RF; 20.25: © Michael Graybill & Jan Hodder/Biological Photo Service; 20.26: © E.R. Degginger/Photo Researchers; 20.27: © S.J. Krasemann/Peter Arnold; 20.28: © J. Weber/Visuals Unlimited; 20.29: © IFA/Peter Arnold; 20.30: © Charlie Ott/The National Audubon Society Collection/Photo Researchers; 20.31: © John Shaw/Tom Stack & Associates; 20.32: © Tom McHugh/Photo Researchers; 20.33: © Dave Watts/Tom Stack & Associates; 20.34: © E.R. Degginger/Animals Animals; I&A 20.2: U.S. Forest Service.

Chapter 21

Opener: © Francois Gohier/Photo Researchers; 21.1: © Stone/Getty; 21.3: © William C. Dilger, Cornell University; 21.4a-b: From J.R. Brown et al, "A defect in nurturing mice lacking . . . gene for fosB" Cell v. 86, 1996 pp 297-308, © Cell Press; 21.5: © Thomas McAvoy, Life Magazine/Time/Getty; 21.7: © Jeff Foott/Tom Stack & Associates; 21.8: Courtesy Bernd Heinrich; 21.9: © Nina Leen/Time & Life Pictures/ Getty; 21.10: © David R. Frazier Photolibrary/Alamy RF; 21.11: © Bios(C.Thouvenin)/Peter Arnold; Table 21.1(bird): © Eric & David Hosling/Corbis; Table 21.1(seals): © Marc Moritsch/National Geographic Image Collection; Table 21.1(herd): © Alamy RF; Table 21.1(frog): Courtesy James Traniello; Table 21.1(lions): © K. Ammann/Bruce Coleman Inc.; Table 21.1(ants): © Mark Moffett/Minden Pictures; 21.14: © Dr. Don W. Fawcett/Visuals Unlimited; 21.15b: © Scott Camazine/Photo Researchers; 21.16a: © S. Osolinski/OSF/Animals Animals; 21.17: © Nigel Dennis/National Audubon Society Collection/ Photo Researchers; 21.19: © David Hosking/Photo Researchers; 21.20: © Jim Pickerell/Alamy; I&A 21.2: © Kevin Schafer/Alamy RF.

Chapter 22

Opener: © Steve McCurry/Magnum Photos; 22.2: © Grant Heilman/Grant Heilman Photography; 22.3: © Rob & Ann Simpson/Visuals Unlimited; p. 438: © 1999 Ed Ely/Biological Photo Service; 22.7: NASA; 22.9: © Gary Griffen/Animals Animals;

Index

Applications Index